D1087179

# New Directions in Conservation Medicine

# New Directions in Conservation Medicine

## Applied Cases of Ecological Health

Edited by A. Alonso Aguirre,

Richard S. Ostfeld,

and Peter Daszak

OXFORD

UNIVERSITY PRESS

Oxford University Press, Inc., publishes works that further
Oxford University's objective of excellence
in research, scholarship, and education.

Oxford  New York
Auckland  Cape Town  Dar es Salaam  Hong Kong  Karachi
Kuala Lumpur  Madrid  Melbourne  Mexico City  Nairobi
New Delhi  Shanghai  Taipei  Toronto

With offices in
Argentina  Austria  Brazil  Chile  Czech Republic  France  Greece
Guatemala  Hungary  Italy  Japan  Poland  Portugal  Singapore
South Korea  Switzerland  Thailand  Turkey  Ukraine  Vietnam

Published by Oxford University Press, Inc.
198 Madison Avenue, New York, New York 10016

www.oup.com

Oxford is a registered trademark of Oxford University Press

Library of Congress Cataloging-in-Publication Data
New directions in conservation medicine : applied cases of ecological health /
edited by A. Alonso Aguirre, Richard S. Ostfeld, and
Peter Daszak.
p. cm.
Includes bibliographical references and index.
ISBN 978-0-19-973147-3
1. Environmental health. 2. Medical geography. 3. Environmental toxicology.
I. Aguirre, A. Alonso. II. Ostfeld, Richard S., 1954-
III. Daszak, Peter.
RA566.N49 2012
614.4'2—dc23                                              2011031645


1 3 5 7 9 8 6 4 2
Printed in the United States of America
on acid-free paper

*This book is dedicated to the next generation of earth rescuers and planet doctors who will transform the conservation of biodiversity into a mainstream social movement.*

# CONTENTS

# FOREWORD: PLANET DOCTORS

## Thomas E. Lovejoy

In the late 1960s, I was living in Belem, the port city of the Amazon. I was doing field research in the forests on the outskirts, for a Ph.D. combining the ecology of Amazon forest birds with the epidemiology of arthropod-borne viruses. That background enabled me to grasp the importance of conservation medicine when it began to emerge as a field decades later.

For two years, I was based at the Belem Virus Laboratory, one of a series of laboratories around the world created by the Rockefeller Foundation with the purpose of ending the great scourges of humanity. Prominent among such diseases was yellow fever. The laboratory was quite successful in this endeavor in two ways. The first was by understanding the urban cycle of the vector *Aedes aegypti* and identifying ways to greatly diminish its power through control of the breeding sites where water would collect (e.g., old tires). The second involved the development of a vaccine, for which Max Theiler received a Nobel Prize in 1951.

The process of addressing yellow fever revealed a wealth of other arthropod-borne viruses (soon termed arboviruses). This led to the network of Rockefeller Foundation laboratories around the world to explore this important frontier of human health. Each laboratory included field naturalists as well as laboratory scientists in recognition of the fundamental importance of natural cycles. It is interesting that, after a long period in which laboratory work has dominated, the link with field biology is now very much back in practice—witness conservation medicine.

At Belem, I had the honor of sharing a large office with Dr. Jorge Boshell, a legendary Colombian who became director of the very first of the laboratories in Villavicencio. In that post, he followed Marston Bates, who wrote *The Forest and the Sea* (a popular book on tropical forests) and whose wife, Nancy Bell Bates, also a naturalist, penned *East of the Andes and West of Nowhere*. Both books were on my bookshelf.

At that point, one of the great mysteries involved the other cycle of yellow fever, the sylvan cycle (or more popularly, "jungle yellow fever"). It was known to move nomadically through the forest canopy, causing major mortality in howler monkeys before moving on. How it was able to jump to the occasional human on the forest floor 100 feet below seemed an intractable mystery. Intractable, that is, until Dr. Boshell watched some woodcutters bring down a tree and instantly become surrounded by a cloud of little blue (canopy) mosquitoes in the genus *Haemagogus*.

That was a brilliant example of the prepared mind and the central importance of field biology, but it is also a sterling metaphor for human self-inflicted impact through disturbance of nature. Ecological health and conservation medicine are full of such examples today. We have come to understand that the health of humans, animals (and plants!), and ecosystems are all inextricably intertwined. Indeed, given the major disturbance to the biophysical system of the planet itself through climate change and ecosystem destruction and degradation, we also must include the "health" of the biosphere.

It has long been axiomatic that disturbance tips the balance in favor of weedy species, vectors, and disease agents. But we also are seeing and learning the obverse, namely that biodiversity itself is of real value in addressing these kinds of problems. David Tilman's magnificent set of experiments at Cedar Creek has demonstrated conclusively that greater productivity, stability, and resilience of ecosystems are directly related to their biodiversity.

Rick Ostfeld and other scientists have slowly and steadily built a strong case for the role of biodiversity in cycles of Lyme disease. Certainly our management of the landscape and predators has greatly increased the populations of the adult blacklegged tick's host, the white-tailed deer. Of greater importance is that when the larval stages of the tick vector encounter less biodiversity (because both forest fragmentation and acorn yields in years of "mast" fruiting magnify populations of the ideal host, the white-footed mouse), the nymphal ticks greatly increase in abundance and become more likely to transmit the spirochete pathogen, *Borrelia burgdorferi*, to humans. On a bigger scale, it has long been clear that maintaining the Darien Gap of Panama and Colombia in tropical forest provides a *cordon sanitaire* against the spread of foot and mouth disease north from South America.

This volume shows that conservation medicine has clearly become a richer and more textured field than a decade earlier. There is not only more scholarship but also much more to study, understand, and manage. For example, the amphibian crisis is a long way from being understood and addressed with appropriate measures. However, scientists have been able to study a chytrid fungus impact as it occurs and to find some frog species resistant to it. In general, I suspect that conservation medicine as a discipline has barely scratched the surface of the complex problems we face. How, for example, will increased ocean acidity affect the health of marine organisms? What will be the disease sequelae to the immense conifer die-back in western North America? What kinds of pathology might we anticipate in the estuarine and coastal dead zones that double in number every decade? As species change their annual cycles and geographic distributions in response to climate change, will that create new opportunities for pathogens?

Some in the field of conservation medicine have started to think as "planet doctors," essentially taking conservation medicine to an even greater scale. That includes ecosystem restoration on a global scale to reduce global burdens of excess $CO_2$, therefore avoiding some of the otherwise inevitable climate change and its impacts. Almost by definition, this means boosting the resiliency of all ecosystems. Indeed, it seems increasingly clear that conservation and biodiversity restoration at a planetary scale is practicing conservation medicine in itself.

# PREFACE

## A. Alonso Aguirre and Sara E. Howard

Ten years have passed since the publication of *Conservation Medicine: Ecological Health in Practice*. In 2002, the book presented the thesis that ecosystem health, animal health, and human health are inextricably connected. It challenged practitioners and students of diverse disciplines to consider new, collaborative ways to address ecological health concerns. As a new discipline, conservation medicine provided the framework for exploring the connections between animal and human health; tracing the environmental sources of pathogens and pollutants; understanding the ecological causes of changes in human and animal health; and addressing the consequences of diseases to populations and ecological communities.

When writing the introduction for the first edition of *Conservation Medicine*, Else and Pokras were "struck by the problems presented in chapter after chapter." The common obstacle, they concluded, was a lack of definitive data for effective policy and prevention. Conservation medicine was still a relatively young field, and more transdisciplinary research was necessary to finding real solutions. Now, with the first decade of the 21st century already behind us, we continue to observe dramatic changes in the ecology of the earth, manifested by novel and potentially catastrophic health consequences.

The field of infectious disease ecology has undergone major changes since the early 1980s, when one of the authors (AAA) was a master's student at Colorado State University. Due to his veterinary background, he was perplexed by the population dynamics course or the wildlife ecology course that described disease as a part of ecosystems and animal populations. At that time, the field of disease ecology was largely a documentation of epidemics or models assuming that diseases were just another part in the puzzle of carrying capacity and compensatory mortality of wild populations. The field has evolved dramatically in the past 30 years. It now acknowledges the dynamic phenomenon involving the ecology of the infectious agents, pathogenesis in the host, reservoirs and vectors, and the complex mechanisms that lead to the spread of infection across species and ecosystem barriers.

The environmental causes of health problems are complex, global, and poorly understood. Traditional approaches to the development of health strategies and environmental protection offer limited solutions. Rapid globalization, bushmeat use, and wildlife trade have led to unprecedented interest in infectious diseases and their effect on complex human population dynamic interactions, including migration, famine, natural disasters, war, and terrorism. It is more important than ever for public health officials, veterinarians, economists, politicians, and the general public to understand the basic transdisciplinary science behind conservation medicine.

Conservation medicine offers a new "toolbox" to understand the underlying causes driving these problems. It is solution-oriented, using an integrative approach that includes microbiologists, pathologists, physicians, veterinarians, modelers, economists, toxicologists, epidemiologists, public health professionals, landscape analysts, modelers, and wildlife and marine biologists, among others.

This book, *New Directions in Conservation Medicine: Applied Cases of Ecological Health*, brings theory into practice. We suggest strategies built from diverse disciplines to analyze complex relationships, promote transdisciplinary cooperation, increase efficiency, and minimize bias when investigating large-scale ecological drivers such as climate change, wildlife trade, environmental pollution, disease emergence, and land use change. The case studies and techniques presented herein are intended to complement traditional approaches to health and ecological studies and hypothesis testing. We suggest novel ways to determine epidemiological causal criteria with strong inference, diagrams of causation, molecular epidemiology, and mathematical model selection. We present several examples from recent emerging infectious disease investigations where these techniques were successfully applied.

*New Directions in Conservation Medicine: Applied Cases of Ecological Health* not only overviews and examines the problems, but also describes solutions in detail. It is intended to be a complement to the first book, *Conservation Medicine: Ecological Health in Practice*, and a standard reference for human and veterinary medicine, public health, conservation biology, ecology, wildlife biology, epidemiology, economics, and public policy professionals and students interested in the fields of conservation medicine and ecological health.

# ACKNOWLEDGMENTS

This book is the result of a collaborative, transdisciplinary approach to health, ecology, and conservation. Many risk-taking foundations and funders provided continuing support and vision throughout the development of the discipline of conservation medicine: the Eppley Foundation for Research, Marisla Foundation, New York Community Trust, Overbrook Foundation, Panaphil Foundation, Rockefeller Foundation, and V. Kann Rasmussen Foundation. The National Institutes of Health, the National Science Foundation, the U.S. Agency for International Development, U.S. Department of Agriculture APHIS Wildlife Services, and the U.S. Environmental Protection Agency provided generous research support relevant to the materials presented in this book. We also would like to acknowledge the Cary Institute and EcoHealth Alliance.

Our thanks go to Christine Banks, who supported us in the crucial first stages of moving this project from conception to reality. We gladly recognize Felicia Keesing and Hannia M. Smith for intellectual and moral support during book preparation. Chapters of this book were peer-reviewed by at least two experts in the topic. We are deeply grateful for the many anonymous hands that edited these chapters and supported the final product. Special recognition goes to Michael Balick, Val Beasley, Christiana Grim, Parviez Hosseini, William Karesh, Hamish McCallum, Mindy Rostal, Gerardo Suzán, Gary Tabor, and B. Zimmerman. This book could never have been published without the support of Phyllis Cohen of Oxford University Press, who played the essential role of enabling authors and editors to reach completion.

# CONTRIBUTORS

**Roberto F. Aguilar, D.V.M.**
Veterinarian
Cape Wildlife Center
Humane Society of the United States
Barnstable, Massachusetts

**A. Alonso Aguirre, D.V.M., M.S., Ph.D.**
Associate Professor
Department of Environmental Science and
    Policy
George Mason University
Executive Director
Smithsonian-Mason School of
    Conservation
Front Royal, Virginia

**Anne M. Alexander, M.A., Ph.D.**
Director of International Programs
Department of Economics and
    Finance
University of Wyoming
Laramie, Wyoming

**Emily Almberg, M.S.**
Department of Biology
Pennsylvania State University
University Park, Pennsylvania

**Sonia Altizer, Ph.D.**
Associate Professor
Odum School of Ecology
University of Georgia
Athens, Georgia

**John Arnason, Ph.D.**
Professor
Department of Biology
University of Ottawa
Ottawa, Ontario, Canada

**Michael J. Balick, Ph.D.**
Vice President for Botanical Science
Director and Philecology Curator
Institute of Economic Botany
The New York Botanical Garden
Bronx, New York

**Rebecca Bartel, M.S., Ph.D.**
Terrestrial Ecologist
Inventory and Monitoring Program
National Wildlife Refuge System
U.S. Fish and Wildlife Service
Manteo, North Carolina

**Val R. Beasley, D.V.M., Ph.D., Dipl.A.B.V.T.**
Professor Emeritus of Veterinary, Wildlife,
    and Ecological Toxicology
Department of Comparative Biosciences
College of Veterinary Medicine
University of Illinois at Urbana-Champaign,
    Urbana, Illinois

**Julieta Benítez-Malvido, M.S., Ph.D.**
Centro de Investigaciones en Ecosistemas
Universidad Nacional Autónoma de
    México
Morelia, Michoacán, Mexico

**Aaron Bernstein, M.P.H., M.D.**
Associate Director, Center for Health and the
    Global Environment
Instructor in Medicine
Harvard Medical School
Physician in Medicine, Children's Hospital
    Boston
Boston, Massachusetts

**Jon Bielby, Ph.D.**
Postdoctoral Researcher
Institute of Zoology
Zoological Society of London
London, Great Britain

**Leslie Bienen, D.V.M., M.F.A.**
Center for Large Landscape Conservation
Bozeman, Montana

**Evan S Blumer, V.M.D., M.S.**
Principal
OsoMono, LTD
Gahanna, Ohio
and
Retired Executive Director
The Wilds
Cumberland, Ohio

**Tiffany L. Bogich, Ph.D.**
Postdoctoral Fellow
Fogarty International Center
National Institutes of Health
Bethesda, Maryland

**Bob H. Bokma, D.V.M., M.P.V.M.**
Export Animal Products
National Center for Import and Export
Veterinary Services
USDA Animal and Plant Health Inspection
    Service
Riverdale, Maryland

**Christopher J. Brand, M.S., Ph.D.**
Branch Chief, Field and Laboratory
    Research
U.S. Geological Survey National Wildlife
    Health Center
Madison, Wisconsin

**Edward B. Breitschwerdt, D.V.M.**
Professor of Internal Medicine
College of Veterinary Medicine

North Carolina State University
Raleigh, North Carolina

**Cheryl J. Briggs, Ph.D.**
Professor
Department of Ecology, Evolution and
    Marine Biology
University of California, Santa Barbara
Santa Barbara, California

**Ilana L. Brito, Ph.D.**
Postdoctoral Research Fellow
Civil and Environmental Engineering
    Department
Massachusetts Institute of Technology
Cambridge, Massachusetts

**John S. Brownstein, Ph.D.**
Director and Assistant Professor
Children's Hospital Boston
Harvard Medical School
Harvard-MIT Division of Health Sciences
    and Technology
Boston, Massachusetts

**Victor Cal**
Belize Indigenous Training Institute
Punta Gorda, Belize

**Roberto Carrasco-Hernández, M.S.**
Facultad de Medicina Veterinaria y Zootecnia
Universidad Nacional Autónoma de México
Mexico City, Mexico

**Colin A. Chapman, Ph.D.**
Professor
Canada Research Chair
Fellow of the Royal Society
Department of Anthropology
McGill School of Environment
McGill University
Montreal, Québec, Canada

**Edward E. Clark, Jr., B.A.**
President
Wildlife Center of Virginia
Waynesboro, Virginia

**Axel Cloeckaert, Ph.D.**
Institut National de la Recherche Agronomique
Infectiologie Animale et Santé Publique
Nouzilly, France

**Paul C. Cross, Ph.D.**
U.S. Geological Survey Northern Rocky
  Mountain Science Center
Bozeman, Montana

**Patricia Curry, D.V.M.**
Ph.D. Candidate, Medical Sciences
Department of Ecosystem and Public Health
Faculty of Veterinary Medicine, University of
  Calgary
Calgary, Alberta, Canada

**Peter Daszak, Ph.D.**
President
EcoHealth Alliance
New York, New York

**Joseph Le Doux Diffo, M.Sc.**
Ecology Team Leader
Global Viral Forecasting
Yaounde, Cameroon

**Cyrille F. Djoko, M.S.**
Regional Laboratory Director for Central Africa
Global Viral Forecasting
Yaounde, Cameroon

**Marion L. East, Ph.D.**
Research Scientist
Evolutionary Ecology Research Group
Leibniz Institute for Zoo and Wildlife Research
Berlin, Germany

**Jonathan H. Epstein, D.V.M., M.P.H.**
Associate Vice President
EcoHealth Alliance
New York, New York

**Fernando Esponda, Ph.D.**
Computer Science Department
Instituto Tecnológico Autónomo de Mexico
Mexico City, Mexico

**Joseph N. Fair, M.P.H., Ph.D.**
Vice President, Executive Director –
  Government Services
Global Viral Forecasting
San Francisco, California

**Anne Fairbrother, D.V.M., M.S., Ph.D.**
Senior Managing Scientist
EcoSciences
Exponent
Seattle, Washington

**Hume E. Field, B.V.Sc., M.S., Ph.D.,
  M.A.C.V.S.**
Principal Scientist
Queensland Centre for Emerging Infectious
  Diseases
Department of Employment, Economic
  Development & Innovation
Brisbane, Queensland, Australia

**Claudia Filoni, M.S., D.V.M., Ph.D.**
Professor
Infectious Diseases and Microbiology
Universidade Paulista
Researcher and Member
Instituto Brasileiro para Medicina da
  Conservação (TRÍADE)
São Paulo, Brazil

**David C. Finnoff, Ph.D.**
Associate Professor
Department of Economics and Finance
University of Wyoming
Laramie, Wyoming

**Spencer E. Fire, M.S., Ph.D.**
Biologist
Marine Biotoxins Program
NOAA National Ocean Service
Center for Coastal Environmental Health and
  Biomolecular Research
Charleston, South Carolina

**Matthew C. Fisher, Ph.D.**
Reader in Molecular Epidemiology
Department of Infectious Disease
  Epidemiology
Imperial College London
St. Mary's Campus
London, Great Britain

**Patrick Foley, M.S., Ph.D.**
Department of Biological Sciences
California State University
Sacramento, California

**Janet E. Foley, D.V.M., Ph.D.**
Professor, Research Epidemiologist
  Department of Medicine and
  Epidemiology
School of Veterinary Medicine
University of California
Davis, California

**Geoffrey Foster, C.Sci., F.I.B.M.S., Ph.D.**
Veterinary Centre Manager
SAC C Veterinary Services
Scottish Agricultural College
Inverness, United Kingdom

**Sebastian Funk, Ph.D., Dipl. Phys.**
Postdoctoral Researcher
Institute of Zoology
Zoological Society of London
London, Great Britain

**Trenton W. J. Garner, M.S., Ph.D.**
Senior Research Fellow
Institute of Zoology
Zoological Society of London
London, Great Britain

**Claire Geoghegan, M.Res.**
Ph.D. Candidate and Researcher
Mammal Research Institute
Department of Zoology and
    Entomology
University of Pretoria
Pretoria, Guateng, South Africa

**E. Paul Gibbs, B.V.Sc., Ph.D., F.R.C.V.S.**
Associate Dean for Students and
    Instruction
Professor of Virology
College of Veterinary Medicine
University of Florida
Gainesville, Florida

**Jacques Godfroid, D.V.M., M.S., Ph.D.**
Professor, Head of the Section of Arctic
    Veterinary Medicine
Norwegian School of Veterinary Science
Tromsø, Norway

**Tony L. Goldberg, M.S., D.V.M., Ph.D.**
Professor of Epidemiology
Department of Pathobiological Sciences
School of Veterinary Medicine
University of Wisconsin-Madison
Madison, Wisconsin

**Andrés Gómez, Ph.D.**
Biodiversity Scientist
Center for Biodiversity and Conservation
American Museum of Natural History
New York, New York

**Larry J. Gorenflo, Ph.D.**
Associate Professor
Department of Landscape Architecture
Stuckeman School of Architecture and
    Landscape Architecture
University Park, Pennsylvania

**Christian Gortázar, D.V.M., Ph.D.**
Head, Wildlife Disease Department
National Wildlife Research Institute IREC
(CSIC-UCLM-JCCM)
Ciudad Real, Spain

**K. Christiana Grim, D.V.M.**
Research Associate
Center for Species Survival
Smithsonian Conservation Biology Institute
Front Royal, Virginia

**Timm Harder, Ph.D.**
Senior Scientist
Friedrich Loeffler Institute
Institute of Diagnostic Virology
OIE and National Reference Library for Avian
    Influenza
Insel Riems, Greifswalk, Germany

**Craig A. Harms, D.V.M., Ph.D., Dipl. A.C.Z.M.**
Associate Professor of Aquatic, Wildlife, and
    Zoo Medicine
College of Veterinary Medicine
Center for Marine Sciences and Technology
North Carolina State University
Morehead City, North Carolina

**Katherine Herrera, B.S.**
Research Assistant
Institute of Economic Botany
New York Botanical Garden
Bronx, New York

**Heribert Hofer, D.Phil.**
Evolutionary Ecology Research Group
Director
Leibniz Institute for Zoo and Wildlife Research
Berlin, Germany

**Pierre Horwitz, Ph.D.**
Associate Professor
School of Natural Sciences
Edith Cowan University
Joondalup, Western Australia, Australia

**Parviez R. Hosseini, Ph.D.**
Senior Research Fellow
EcoHealth Alliance
New York, New York

**Sara E. Howard, B.A.**
Program Assistant
EcoHealth Alliance
New York, New York

**Peter J. Hudson, F.R.S.**
Willaman Professor of Biology
Director of the Huck Institutes of Life
    Sciences
Pennsylvania State University
University Park, Pennsylvania

**Thierry Jauniaux, D.V.M., Ph.D.**
Assistant Professor
Veterinary Pathology
University of Liege
Liege, Belgium

**David A. Jessup, D.V.M., M.P.V.M.,
    Dipl. A.C.Z.M.**
Senior Wildlife Veterinarian, Retired
Marine Wildlife Veterinary Care and Research
    Center
California Department of Fish and Game
Executive Manager
Wildlife Disease Association
Santa Cruz, California

**Damien O. Joly, Ph.D.**
Associate Director
Wildlife Health Monitoring and
    Epidemiology
Wildlife Conservation Society
Nanaimo, British Columbia,
    Canada

**Menna Jones, Ph.D.**
Australian Research Council Future Fellow
School of Zoology
University of Tasmania
Hobart, Tasmania, Australia

**Kate E. Jones, Ph.D.**
Senior Research Fellow
Institute of Zoology
Zoological Society of London
London, Great Britain

**Rodrigo Silva Pinto Jorge, D.V.M., Ph.D.**
Environmental Analyst
Instituto Chico Mendes de Conservação da
    Biodiversidade (ICMBio/MMA)
Researcher
Instituto Brasileiro para Medicina da
    Conservação (TRÍADE)
Brasilia, D.F., Brazil

**Laura H. Kahn, M.D., M.P.H., M.P.P.**
Research Scholar
Program on Science and Global Security
Woodrow Wilson School of Public and
    International Affairs
Princeton University
Princeton, New Jersey

**William B. Karesh, D.V.M.**
Executive Vice President for Health
    and Policy
EcoHealth Alliance
New York, New York

**Thomas J. Keefe, Ph.D.**
Professor, Biostatistics
Department of Environmental and
    Radiological Health Sciences
Colorado State University
Fort Collins, Colorado

**Felicia Keesing, Ph.D.**
Associate Professor of Biology
Bard College
Annandale, New York

**Kevin Knight**
Department of Biology
University of Ottawa
Ottawa, Ontario, Canada

**Thijs Kuiken, D.V.M., Ph.D., Dipl. A.C.V.P.**
Professor of Comparative Pathology
Department of Virology
Erasmus University Medical Center
Rotterdam, The Netherlands

**Susan Kutz, D.V.M., Ph.D.**
Associate Professor
Department of Ecosystem and Public Health
Faculty of Veterinary Medicine, University
    of Calgary
Calgary, Alberta, Canada

**Wayne Law, Ph.D.**
Postdoctoral Research Associate
Institute of Economic Botany
New York Botanical Garden
Bronx, New York

**Matthew LeBreton, B.Sc.**
Ecology Director
Global Viral Forecasting
Yaounde, Cameroon

**Roberta A. Lee, M.D.**
Vice Chair
Department of Integrative Medicine
Co-Director Fellowship in Integrative Medicine
Continuum Center for Health and Healing
Beth Israel Medical Center
New York, New York

**Jeffrey M. Levengood, Ph.D.**
Wildlife Toxicologist
Department of Comparative Biosciences and
Illinois Natural History Survey
University of Illinois at Urbana-Champaign
Urbana, Illinois

**Marc A. Levy, M.A.**
Deputy Director
Center for International Earth Science
  Information Network
Earth Institute
Palisades, New York

**Elizabeth Loh, M.S.**
Research Scientist
EcoHealth Alliance
New York, New York

**Guillermo López, D.V.M., Ph.D.**
EGMASA-Consejería de Medio Ambiente de
  la Junta de Andalucia
Córdoba, Spain

**Thomas E. Lovejoy, Ph.D.**
Professor
Environmental Science and Policy
George Mason University
Fairfax, Virginia
Biodiversity Chair
The H. John Heinz III Center for Science,
  Economics and the Environment
Washington, D.C.

**Ricardo G. Maggi, M.S., Ph.D.**
Research Assistant Professor
College of Veterinary Medicine
North Carolina State University
Raleigh, North Carolina

**Paulo Rogerio Mangini, D.V.M.,
  M.S., Ph.D.**
Associate Researcher
Instituto de Pesquisas Ecológicas (IPÊ)
Researcher and Council of Directors
Instituto Brasileiro para Medicina da
  Conservação (TRÍADE)
Curitiba, Paraná, Brazil

**Daniel Martineau, D.V.M., M.S., Ph.D.**
Associate Professor
Département de Pathologie et Microbiologie
Faculté de Médecine Vétérinaire
Université de Montréal
St. Hyacinthe, Québec, Canada

**Fernando Martínez, D.V.M., M.Sc.**
Programa de Conservación Ex Situ del Lince
  Ibérico
Centro de Cría en Cautividad El Acebuche
Matalascañas, Huelva, Spain

**Maria Fernanda Vianna Marvulo,
  M.S., D.V.M., Ph.D.**
Professor of Veterinary Medicine and
  Public Health
Universidade Federal Rural de Pernambuco
  (UFRPE)
Researcher and Council of Directors
Instituto Brasileiro para Medicina da
  Conservação (TRÍADE)
Recife, Pernambuco, Brazil

**Jonna A. K. Mazet, D.V.M., M.P.V.M., Ph.D.**
Professor and Director
Wildlife Health Center
One Health Institute
University of California
Davis, California

**Hamish McCallum, Ph.D., D.I.C.**
Head
Griffith School of Environment
Griffith University
Nathan, Queensland, Australia

**Dave L. McRuer, M.S., D.V.M., Dipl. A.C.V.P.M.**
Director of Veterinary Services
Wildlife Center of Virginia
Waynesboro, Virginia

**Melissa A. Miller, D.V.M., M.S., Ph.D.**
Senior Wildlife Veterinarian
Marine Wildlife Veterinary Care and Research Center
California Department of Fish and Game
Santa Cruz, California

**Thomas P. Monath, M.D.**
Partner
Pandemic and Biodefense Fund
Kleiner Perkins Caufield & Byers
Harvary, Massachusetts

**Edgardo Moreno, Ph.D.**
Professor of Cellular Biology
Programa de Investigación en Enfermedades Tropicales
Escuela de Medicina Veterinaria
Universidad Nacional
Heredia, Costa Rica

**Stephen S. Morse, Ph.D.**
Professor
Department of Epidemiology
Columbia University Mailman School of Public Health
New York, New York

**Alessandra Nava, D.V.M., Ph.D.**
Senior Scientist
EcoHealth Alliance
Associate Researcher
Biotropicos Institute
São Paulo, Brazil

**Elizabeth S. Nichols, M.A., PhD.**
Visiting Scientist
Center for Biodiversity and Conservation
American Museum of Natural History
New York, New York

**Ingebjørg Helena Nymo, D.V.M.**
Ph.D. Candidate
Section of Arctic Veterinary Medicine
Norwegian School of Veterinary Science
Tromsø, Norway

**Kevin J. Olival, Ph.D.**
Senior Research Scientist
EcoHealth Alliance
New York, New York

**Glenn H. Olsen, D.V.M., Ph.D.**
Veterinary Medical Officer
U.S. Geological Survey Patuxent Wildlife Research Center
Laurel, Maryland

**Nancy Ortiz, M.P.H.**
Project Coordinator
Global Viral Forecasting
San Francisco, California

**Richard S. Ostfeld, Ph.D.**
Senior Scientist
Cary Institute of Ecosystem Studies
Millbrook, New York

**Sarah B. Paige, M.P.H., Ph.D.**
Research Associate/Program Manager
Center for Community Health and Evaluation
Group Health Research Institute
Seattle, Washington

**Susan L. Perkins, Ph.D.**
Associate Curator and Professor
Department of Invertebrate Zoology
Sackler Institute for Comparative Genomics
Richard Gilder Graduate School
American Museum of Natural History
New York, New York

**Todd J. Pesek, M.D.**
Associate Professor
School of Health Sciences
Director
Center for Healing Across Cultures
College of Sciences and Health Professions
Cleveland State University
Cleveland, Ohio

**Brian L. Pike, M.P.H., Ph.D.**
Director of Laboratory Sciences
Global Viral Forecasting
San Francisco, California

**Raina K. Plowright, B.V.Sc., M.S., Ph.D.**
David H. Smith Conservation
    Research Fellow
Center for Infectious Disease Dynamics
Pennsylvania State University
University Park, Pennsylvania

**Julia B. Ponder, D.V.M.**
Executive Director
The Raptor Center
Assistant Clinical Specialist
College of Veterinary Medicine
University of Minnesota
St. Paul, Minnesota

**Wolfgang Preiser Prof. Dr. med. Dr. med.
    habil, D.T.M.&H., M.R.C.Path.**
Professor and Head: Division of Medical
    Virology
Department of Pathology
Faculty of Health Sciences
Stellenbosch University and National
    Health Laboratory Service
    Tygerberg
Tygerberg, South Africa

**Margo J. Pybus, Ph.D.**
Provincial Wildlife Disease Specialist
Fish and Wildlife Division
Edmonton, Alberta, Canada

**Barnett A. Rattner, Ph.D.**
Ecotoxicologist
U.S. Geological Survey Patuxent Wildlife
    Research Center
Beltsville Laboratory
Beltsville, Maryland

**William C. Raynor, M.S.**
Director
Micronesia Program
The Nature Conservancy
Kolonia, Pohnpei, Federated State of
    Micronesia

**Anne W. Rimoin, Ph.D.**
Associate Professor
Department of Epidemiology
UCLA School of Public Health
Los Angeles, California

**Gail E. Rosen, M.S.**
Center for Infection and Immunity
Mailman School of Public Health
Columbia University
New York, New York

**Melinda K. Rostal, M.P.H., D.V.M.**
Field Veterinarian
EcoHealth Alliance
New York, New York

**Karen E. Saylors, Ph.D.**
Director, Behavioral Sciences
Global Viral Forecasting
San Francisco, California

**Lisa M. Schloegel, B.A.**
Consulting Research Scientist
EcoHealth Alliance
New York, New York

**Jason F. Shogren, Ph.D.**
Department Chair
Stroock Professor of Natural
    Resource Conservation and
    Management
Department of Economics and
    Finance
University of Wyoming
Laramie, Wyoming

**Todd K. Shury, D.V.M.**
Wildlife Health Specialist
Parks Canada
Adjunct Professor
Department of Veterinary Pathology
Western College of Veterinary
    Medicine
Saskatoon, Saskatchewan, Canada

**Jean Carlos Ramos Silva, M.S.,
    D.V.M., Ph.D.**
Professor Veterinary Medicine and Public
    Health
Universidade Federal Rural de Pernambuco
    (UFRPE)
Researcher
Instituto Brasileiro para Medicina da
    Conservação (TRÍADE)
Recife, Pernambuco, Brazil

Marcelo Renan de Deus Santos,
D.V.M., M.S.
President
Instituto de Ensino Pesquisa e Preservação
Ambiental Marcos Daniel
Professor
Vila Velha University
Conservation Medicine
Espíritu Santo, Brazil

Jonathan M. Sleeman, M.A., Vet. M.B.,
Dipl. A.C.Z.M., Dipl. E.C.Z.M.,
M.R.C.V.S.
Director
U.S. Geological Survey National Wildlife
Health Center
Madison, Wisconsin

Katherine F. Smith, Ph.D.
Assistant Professor
Department of Ecology and Evolutionary
Biology
Brown University
Providence, Rhode Island

Francisca Sohl Obispo
The Conservation Society of Pohnpei
Kolonia, Pohnpei, Micronesia

Gerardo Suzán, Ph.D.
Facultad de Medicina Veterinaria y Zootecnia
Universidad Nacional Autónoma
de México
Mexico City, Mexico

Gary M. Tabor, V.M.D., M.E.S.
Director
Center for Large Landscape
Conservation
Bozeman, Montana

Ubald Tamoufe, M.S., M.P.H.
Central Africa Regional Director
Global Viral Forecasting
Yaounde, Cameroon

Morten Tryland, D.V.M., Ph.D.
Professor
Section of Arctic Veterinary Medicine
Norwegian School of Veterinary Science
Tromsø, Norway

Frances M. Van Dolah, Ph.D.
Research Biochemist
Marine Biotoxins Program
NOAA National Ocean Service
Center for Coastal Environmental Health and
Biomolecular Research
Charleston, South Carolina

Carlos Eduardo da Silva Verona, D.V.M.,
M.S., Ph.D.
Professor and Postgraduation Coordinator
Estácio de Sá University
Vertebrate Zoology and Wildlife Management
Researcher
Instituto Brasileiro para Medicina da
Conservação (TRÍADE)
Rio de Janeiro, Brazil

Lin-Fa Wang, Ph.D.
CEO Science Leader
CSIRO Livestock Industries
Australian Animal Health Laboratory
Geelong, Australia

E. Scott Weber III, V.M.D., M.S.
Associate Professor of Clinical Aquatic Animal
Medicine
UC Davis School of Veterinary Medicine
VM: Medicine and Epidemiology
Davis, California

Adrian M. Whatmore, M.S., Ph.D.
FAO/WHO Collaborating Centre for Brucellosis
OIE Brucellosis Reference Centre
Department of Bacteriology
Veterinary Laboratories Agency
Addlestone, Surrey, United Kingdom

Bruce A. Wilcox, M.S., Ph.D.
Director, Global Health Asia Program
Adjunct Professor
Faculty of Public Health
Mahidol University
Bangkok, Thailand

Michelle M. Willette, D.V.M.
Staff Veterinarian, Assistant Clinical Specialist
The Raptor Center
College of Veterinary Medicine
University of Minnesota
St. Paul, Minnesota

**Nathan D. Wolfe, M.A., D.Sc.**
Founder & CEO
Global Viral Forecasting
Lorry Lokey Visiting Professor
Program in Human Biology, Stanford
    University
San Francisco, California

**Barbara A. Wolfe, D.V.M., Ph.D., Dipl.
    A.C.Z.M.**
Director
Wildlife and Conservation Medicine
The Wilds
Cumberland, Ohio

**Scott D. Wright, Ph.D.**
Branch Chief, Disease Investigations
U.S. Geological Survey National Wildlife
    Health Center
Madison, Wisconsin

**Carlos Zambrana-Torrelio, M.S.**
Research Scientist
EcoHealth Alliance
New York, New York

**B. Zimmerman, M.S.**
President
Wildlife Health International
Fort Collins, Colorado

# PART ONE

## CONSERVATION MEDICINE

### Ecological Health in Practice

# 1

# CONSERVATION MEDICINE

## Ontogeny of an Emerging Discipline

## A. Alonso Aguirre, Gary M. Tabor, and Richard S. Ostfeld

In recent years, the term *conservation medicine* has been used in several contexts within different scientific communities, national/international organizations, and research groups (Meffe 1999; Pokras et al. 1999; Deem et al. 2000; Osofsky et al. 2000; Speare 2000; Dierauf et al. 2001; Norris 2001; Pullin and Night 2001; Lafferty and Gerber 2002). The book *Conservation Medicine: Ecological Health in Practice* (Aguirre et al. 2002) was published in an attempt to define a new discipline that links human and animal health with ecosystem health and global environmental change. This novel approach challenged scientists and practitioners in the health, natural, and social sciences to think about new, collaborative, transdisciplinary ways to address ecological health concerns in a world affected by complex, large-scale environmental threats. The objectives of this chapter are to review some of the history, lay out the formal definition of conservation medicine, and describe some key advances of the past 10 years of this emerging discipline.

## GLOBAL ENVIRONMENTAL CHANGE AND EMERGING INFECTIOUS DISEASES

The global loss of biological diversity and the disruption of ecological processes affect the well-being of all species, including humans. Human impact on ecosystems and ecological processes is well documented (Myers et al. 2000; Pimm and Raven 2000; Millenium Ecosystem Assessment 2007). Habitat destruction and species loss have led to ecological disruptions that include, among other impacts, the alteration of disease transmission patterns (i.e., emerging diseases), the accumulation of toxic pollutants, and the invasion by alien species and pathogens (Pongsiri et al. 2009, Keesing et al. 2010). Ecological perturbations are creating novel conditions for new disease patterns and health manifestations. For example, in the marine environment, new variants of *Vibrio cholerae* have been identified within red tide algal blooms (Epstein 1993). These toxic blooms are occurring with greater frequency and size throughout the temperate coastal zones of the world (Epstein et al. 1993). In various parts of the southwestern United States, Panama, Brazil, and Argentina, Hantavirus epidemics have emerged in ecosystems that exhibit habitat degradation and climatic disturbances (Monroe et al. 1999; Padula et al. 2000; Clay et al. 2009; Suzán et al. 2009). Severe acute respiratory syndrome (SARS) emerged in the early 2000s, reminding us that globalization of disease can have enormous effects beyond direct human causalities. The SARS outbreak, caused by a novel coronavirus of bat origin (Li et al. 2005), with many wild and domestic mammal hosts, cost over

$400 billion and spread within three months to over 40 countries and more than 4,000 cases before it could be contained (see Chapter 15 in this volume). More recently, H5N1 highly pathogenic avian influenza (HPAI; see Chapter 16 in this volume) and H1N1 influenza virus—both having emerged from human-aided and natural movements of viruses geographically and among zoonotic hosts—have cost the world billions of dollars and resulted in the culling of millions of animals (http://www.wpro.who.int/health_topics/avian_influenza/).

These examples illustrate our growing awareness of the complex interrelationship between species (including human) health and the environment. We are now able to observe these interactions more adeptly through a combined epidemiological and ecological lens. When the resilience of ecosystems is stressed by natural and anthropogenic impacts, conditions for disease transmission, emergence, and spread can be facilitated. Over 300 emerging infectious disease (EID) events of humans have been described since the 1940s (Taylor et al. 2001; Jones et al. 2008). Diseases like tuberculosis, temperate-zone malaria, dengue hemorrhagic fever, and diphtheria are also re-emerging as threats that were once relatively well controlled. Although the causes of emergence are complex and in many cases poorly understood, the role of environmental change is critical (McMichael 1997; McMichael et al. 1999; Keesing et al. 2010).

In recent years, species and ecosystems have been threatened by many anthropogenic factors manifested in disease emergence events causing local and global declines of populations and species. For example, chytridiomycosis threatens amphibian populations worldwide (see Chapter 36 in this volume). The disease has been linked to dramatic population declines or even extinctions of amphibian species in western North America, Central America, South America, eastern Australia, and Dominica and Montserrat in the Caribbean. *Batrachochytrium dendrobatidis* is capable of causing sporadic deaths in some amphibian populations and 100% mortality in others, contributing to a global decline in amphibian populations that apparently has affected 30% of the species (Stuart et al. 2004). Wild Tasmanian devil (*Sarcophilus harrisii*) populations in eastern Tasmania have been decimated recently by devil facial tumor disease (DFTD). DFTD is a non-viral, transmissible parasitic cancer, which likely originated in Schwann cells spread by biting, and causes large tumors to form around the mouth, interfering with feeding and eventually causing death. DFTD was first reported in 1996 (see Chapter 19 in this volume). This disease provides a good example of how low genetic diversity in a geographically restricted species can increase vulnerability to disease.

Although we currently lack data to quantify risk factors of wildlife disease emergence, diseases caused by multi-host pathogens are increasing, including several that have threatened endangered species. Several wild carnivore populations have been devastated by disease transmitted by their domestic counterparts. For example, rabies decimated populations of the Ethiopian wolf (*Canis simensis*); canine distemper caused major declines in Channel Island foxes (*Urocyon littoralis*) and lions (*Panthera leo*) in the Serengeti. Increasing human populations have brought domestic dogs, many of which carry canine distemper, parvovirus, and rabies, into frequent contact with African wild dogs (*Lycaon pictus*). These diseases are wiping out the wild packs (Cleveland 2003).

Beyond more visible terrestrial systems, growing evidence indicates that the ecological integrity of marine ecosystems is also under increasing threat. Oceanographers have been documenting changes in sea surface and deep core ocean temperatures, ocean acidification, and increased hypoxia caused by anthropogenic global environmental change. The world coastal zones face enormous human developmental and urbanization pressures as most people on the planet live at water's edge. Some ecological health symptoms of collective human impacts on the marine environment include increased frequency and intensity of harmful algal blooms e.g., brevetoxicosis in manatees and sea turtles, domoic acid poisoning in sea lions, brown tide poisoning in humans; (see Chapter 26 in this volume), increased environmental stress on species at higher trophic levels as a result of overfishing, loss of breeding/nursery habitats (see Chapter 22 in this volume), and the spread of persistent chemical pollutants such as dioxins and PCBs that bio-accumulate in the food chain (Epstein et al. 1993; Estes and Duggins 1995; Epstein 1999; Bejarano et al. 2008; see Chapters 24 and 25 in this volume). Many populations of marine mammals, marine birds, and sea turtles are exposed to pollutants from agricultural runoff, human sewage, and pathogens with a terrestrial origin (see Chapter 23 in this volume). Intensive agricultural practices resulting in increased nutrient loading and decreased water quality give rise to concerns of *Cryptosporidium* contaminations and massive fish kills linked to *Pfiesteria* toxin poisoning (Burkholder et al. 1999). An unprecedented number of emerging and re-emerging diseases such as brucellosis in dolphins, aspergillosis in coral reefs, fibropapillomatosis in sea turtles, toxoplasmosis in sea otters, and morbillivirus infections linked to large-scale marine mammal die-offs have been documented in recent times in the marine environment (Gulland and Hall 2007). The trends of diseases in the oceans have been increasing since 1970 for other taxa, including corals, urchins, and mollusks (Ward and Lafferty 2004). The human impacts on the world's oceans have devastated populations, species, and ecosystems at a rapid scale; however, methodologies to assess marine ecosystem health are very poorly developed and the scale of monitoring required is well beyond present surveillance capacity. The health consequences of these events require innovative monitoring strategies in order to promote disease prevention, health management, and conservation (see Chapters 37 and 38 in this volume). Society is ill equipped to deal with these health impacts at present, lacking professionals with the transdisciplinary skills to link ecosystem, animal, and human health issues and lacking large-scale marine health surveillance methodologies (Meffe 1999; Aguirre et al. 2002). But these skills are necessary to devise policy and management for improved health.

## DEFINING CONSERVATION MEDICINE

Although we consider that conservation medicine recently has emerged as a scientific field, the concept is the result of the long evolution of interdisciplinary thinking within the health and ecological sciences and the better understanding of the complexity within these various fields of knowledge. Aguirre et al. (1995) attempted to identify the role of biomedical sciences in the conservation of biodiversity looking at health of ecosystems. Kock (1996) described the broad ecological context of health with his wildlife health efforts in Africa. In 1996, EcoHealth Alliance (formerly known as Wildlife Trust), Tufts University Cummings School of Veterinary Medicine, and Harvard Medical School together established a novel institutional collaboration to align their complementary missions in health and the environment. Conservation medicine was born from the cross-fertilization of ideas generated by this new transdisciplinary design. By bringing the disciplines of health and ecology together, conservation medicine represents an attempt to examine the world in an inclusive way as health impacts ripple throughout the interdependent web of life (Tabor 2002).

Conservation medicine overtly recognizes the health component of conserving biodiversity. In doing so, conservation medicine addresses the effects of disease on rare or endangered species as well as the functioning of ecosystems. From a disease ecology standpoint, conservation medicine is also concerned with the impacts of changes in species diversity or rarity on disease maintenance and transmission. For example, research has shown that by conserving biodiversity we are able to dilute the transmission of a pathogen such as Lyme disease (Schmidt and Ostfeld 2001; see Chapter 5 in this volume). The dynamic balance that we term "health" is viewed on a series of widely varying spatial scales by many disciplines, including human and public health, epidemiology, veterinary medicine, toxicology, ecology, and conservation biology. Conservation medicine represents an approach that bridges these disciplines to examine the health of individuals, groups of individuals, populations, communities, ecosystems, and the landscapes in which they live as a continuum. Innovative approaches to problem-solving are born at the nexus of disciplines, and in the instance of conservation medicine, this nexus is the intersection of ecology and health. There is much work needed to bridge these two fields and much opportunity for innovation. This includes bringing biomedical research and diagnostic resources to address conservation problems, such as development of new noninvasive health monitoring techniques; training veterinarians, physicians, public health personnel, and conservation biologists in the promotion and practice of ecological health; and establishing transdisciplinary teams of health and ecological professionals to assess and address ecological health problems (Meffe 1999; Pokras et al. 1999; Tabor et al. 2001; Aguirre et al. 2002).

---

**Box 1.  Conservation Medicine: The Definition**

---

Conservation medicine examines the interactions between pathogens and disease and their linkages with the synergies that occur between species and ecosystems. Thus, it focuses on the study of the ecological context of health and the remediation of ecological health problems. In response to the growing health implications of environmental degradation, conservation medicine includes examining the relationship among (a) changes in climate, habitat quality, and land use; (b) emergence and re-emergence of infectious agents, parasites, and environmental contaminants; and (c) maintenance of biodiversity and ecosystem functions as they sustain the health of plant and animal communities, including humans (Tabor et al. 2001; Aguirre et al. 2002).

---

In comparison to human and veterinary medicine, conservation medicine is the examination of ecological health concerns beyond just a species-specific approach. Human medicine is primarily focused on the health of one species, whereas ecology and veterinary medicine, in general, examine disease and health through a multi-species comparative perspective. Both human and veterinary health have generally been attentive to the downstream effects of environmental impacts (e.g., the health consequences of natural disasters or pollution emissions) rather than encompassing a preventive approach in looking at upstream events (e.g., alteration of ecological systems that influence health outcomes or climate change) (Tabor et al. 2001; Aguirre et al. 2002; see Chapters 8 and 9 in this volume).

## MOVING BEYOND ANOTHER CRISIS DISCIPLINE

Conservation biology emerged during the early 1980s as a response to species extinctions, habitat fragmentation, pollution, and other forms of global environmental change. Soulé and Wilcox (1980), pioneers in the field, called it a "crisis discipline" to reflect the catastrophic deterioration of many ecosystems and taxa therein. Today, nearly 30 years after conservation biology solidified as a scientific field, we are still watching countless species go extinct, witnessing the continual degradation of habitats at ever-increasing rates, and detecting pollution impacts throughout the entire food chain in all parts of the biosphere. As stated by Soulé, conservation medicine was born as

"an essential response to the emergence of new diseases and the physiological threats to human beings and millions of other species caused by industry, agriculture, and commerce" (Aguirre et al. 2002, p. vi).

Conservation medicine inherently adheres to the precautionary principle (Gardiner 2006). This new paradigm contends with health and environmental risk in the context of emerging technologies. The principle implies that there is a social responsibility to protect the environment and ecological integrity when scientific investigation has found a plausible risk. Although some legal systems, like the European Union, applied the precautionary principle as a statutory requirement, this has not been widely recognized by governments in national or regional decision-making (Brown et al. 2000). Current ecological degradation requires the application of the precautionary principle. For the past 30 years, human impacts on the health of the planet can be classified within four areas of environmental concern (Tabor et al. 2001):

1. Increasing biological impoverishment, including loss of biodiversity, habitat destruction, and degradation and modification of ecological processes
2. Increasing global "toxification," including the spread of hazardous wastes and toxic substances and the impact of pervasive low-level pollutants such as endocrine-disrupting chemicals
3. Global climate change
4. Global transport of species, including but not limited to pathogens and parasites, into novel environments

These discrete and cumulative human-induced global impacts have not only diminished the environmental capital of the planet but have also yielded an array of health concerns, including the increasing spread of infectious and non-infectious diseases and the growing physiological impacts on species' reproductive health, developmental biology, and immune systems (Tabor et al. 2001; Jones et al. 2008; Milligan et al. 2009; Gore 2010; see Chapter 7 in this volume).

## EMERGING INFECTIOUS DISEASES AND CLIMATE CHANGE

Emerging and re-emerging infectious diseases are one manifestation of diminishing ecological health. The common theme in conservation medicine, regardless of whether the focus is human, zoological, or botanical, is the ecological context of disease events. Deforestation, climate change, land-use and coastal zone changes, invasive species, and chemical pollution increase the probability of many disease outbreaks or novel disease manifestations. Another troubling aspect of infectious diseases is the growing recognition of cross-species disease transmission, such as the North American epidemic of West Nile virus in animals and humans that has spread southward, infecting humans, horses, and wild birds in Mexico, the Caribbean, and Central and South America. Two modalities of cross-species contact are on the rise. Physical means of contact through transportation, trade, and travel (e.g., airplane travel) are well documented. What is less apparent but perhaps more pervasive is the notion of increasing direct interactions of humans and animals that promote transmission. Habitat fragmentation, human settlement encroachment, ecotourism, and intensive agricultural systems are creating a new milieu for the spread of disease (Pongsiri and Roman 2007; Wolfe et al. 2007; Rwego et al. 2009; Goldberg et al. 2008, 2010; Keesing et al. 2010; Altizer et al. 2011)

In essence, with increasing human numbers and the expanding human footprint, the likelihood of human-to-animal and animal-to-human transmission events increases (Chapter 42 in this volume). We are dealing with an issue of scale. For example, habitat fragmentation, bushmeat hunting, and slash-and-burn

agriculture can be considered local and regional phenomena that create new opportunities for pathogens to cross species barriers (see Chapters 10 and 12 in this volume). On the other hand, at continental and global scales human travel and wildlife trade products of globalization increase many opportunities for species from opposite sides of the planet to come in contact overnight, facilitating the exchange of pathogens that evolutionarily never co-occurred (see Chapter 11 in this volume). For example, mountain gorillas (*Gorilla gorilla*) have acquired or are at risk of acquiring human diseases like influenza, measles, and tuberculosis due to ecotourism and habitat encroachment (Woodford et al. 2002). A human metapneumovirus was found as a cause of mortality in two gorillas that died during an outbreak of respiratory disease in Rwanda in 2009 (Palacios et al. 2011).

Furthermore, the specter of climate change looms larger than ever and exacerbates the impacts of other environmental threats. As carbon emissions and associated global temperature increases now appear to exceed the worst-case scenario described in IPCC (2007), climate change may well be the most significant driver of the diminishment of ecological health and public health. The effects of climate change on health are already apparent, such as promoting the spread of infectious diseases like malaria and dengue fever from more tropical ranges to temperate areas. The impact of climate change on ecological processes in many biomes is affecting terrestrial and marine biota (Burek et al. 2008; Dobson 2009; Harvell et al. 2009; Lafferty 2009; Ostfeld 2009; Pascual and Bouma 2009; Randolph 2009). Should global temperature increases exceed 4°C, predictions indicate that up to 70% of all species are likely to go extinct, disassembling ecological communities and creating an ecological future that has no analog to the past (Patz et al. 2004; IPCC 2007; Pachauri and Reisinger 2007; Lafferty 2009; Shuman 2010; see Chapters 8 and 9 in this volume).

## GLOBAL TOXIFICATION

In addition to infectious diseases, anthropogenic environmental change involves increased levels of persistent toxic pollutants and endocrine-disrupting chemicals. The issues that Rachel Carson's *Silent*

*Spring* first brought to light in the 1960s are all the more pervasive and persistent 50 years later. While there has been documented success with the regulation of such chemicals as DDT and dieldrin, the tide of widespread chemical use and accumulation has not been stemmed and is often a legacy of rapidly increasing industrialized economies. There is much to be learned from the synergistic and cumulative effects of pervasive chemicals on ecological systems and on species health, including humans (Suter et al. 2005). Over 100,000 chemical substances have been released in the biosphere today and more than 15,000 new chemicals are being produced every year. However, we have the laboratory capacity to detect less than 10% of these substances (Colborn 1993; Colborn et al. 1996; Aguirre et al. 2002). The main classes of chemicals that pose special risks to wildlife are pesticides, pharmaceutical and personal care products, nanoparticles/nanomaterials, endocrine disruptors, ionizing radiation, industrial compounds, heavy metals, household products, fossil fuels, and energetic compounds (see Chapter 25 in this volume). So pervasive are these substances that Woodruff et al. (2011) found that pregnant women across the United States are widely exposed to multiple chemicals at different levels. Hayes (2005) detected nine pesticides used on cornfields in the midwestern United States and found that larval growth and development, sex differentiation, and immune function in leopard frogs (*Rana pipiens*) were adversely affected. Mixtures of pesticides had much greater effects on larval growth and development and reversed the typically positive correlation between time to and size at metamorphosis. The mixture also induced damage to the thymus, resulting in immunosuppression and contraction of flavobacterial meningitis. The role of pesticides in amphibian declines has been grossly underestimated (Hayes 2005; Hayes et al. 2005). Many of these classes of compounds have been of interest to wildlife toxicologists and subject to extensive research; however, some are presenting new challenges and obstacles, and some have yet to be evaluated for hazards and risks to terrestrial wildlife. Major steps have to be taken to advance the field of wildlife toxicology and ecotoxicology to develop policies that will improve our ability to maintain healthy wildlife and human populations (see Chapter 24 in this volume). We need to address from the science to the policy levels new ecotoxicological challenges, including the

Deep Water Horizon BP oil spill in the Gulf of Mexico (estimated rates of 8,400 cubic meters a day, lasting over 90 days, and releasing 205.8 million gallons of crude oil), which caused extensive damage to marine and wildlife habitats and fishing and tourism industries. Today, we do not have the technologies to deal with the aftermath of these increasingly frequent environmental disasters. Many global institutions and international agencies are obligated to help avert growing public health crises.

## CONSERVATION MEDICINE: THE PRACTICE OF ECOLOGICAL HEALTH

Simple solutions are rarely evident in addressing global environmental problems. Often a multi-pronged, transdisciplinary strategy is required. By bringing disciplines together and improving problem definition, conservation medicine can contribute to solving environmental problems. In the past ten years, new tools and institutional initiatives for assessing and monitoring ecological health concerns have emerged. Landscape epidemiology, disease ecological modeling, and Web-based Google analytics have emerged. New types of integrated ecological health assessment are being deployed; these efforts incorporate environmental indicator studies with specific biomedical diagnostic tools (Suter et al. 2005; Cox-Foster et al. 2007; Epstein et al. 2010; Lipkin 2010). Other innovations include the development of non-invasive physiological and behavioral monitoring techniques; the adaptation of modern molecular biological and biomedical techniques; the design of population-level disease monitoring strategies; the creation of ecosystem-based health and sentinel species surveillance approaches; and the adaptation of health monitoring systems for appropriate developing-country situations. The complexity of ecological health monitoring requires the development of a dashboard system of measures that are a combination of health and ecological indicators. Ultimately, some type of data-driven decision support tool must be created to help practitioners and managers devise choices for action and intervention.

Beyond monitoring and assessment, numerous training and policy actions are required. Integrating medical and ecological education is a fundamental goal of conservation medicine. There is a need for public policy development that assists in promoting ecological health. The current view of the U.S. Department of State and other agencies focused on the long-term health and prosperity of nations is that environmental instability presents real and serious global security problems (http://www.gmu.edu/programs/icar/pcs/zebich.htm). With the increasing incidence of extreme weather events, long-term droughts, and urban heat island episodes as a result of climate change and large-scale land-use change, ecological disturbances now extend over continents with profound geopolitical impacts. Sub-Saharan Africa continues to face political and cultural destabilization as a result of HIV infections in the most economically vibrant age class (16- to 45-year-olds). Massive outbreaks of Ebola hemorrhagic fever in the Congo Basin have severely decimated Western Gorilla populations as well as neighboring human communities, with extremely high case fatality rates (Leroy et al. 2004; Köndgen 2008). In North America, West Nile Virus (WNV) is affecting humans and animals in several ways: through rampant infections, through the impacts of massive pesticide spraying, and through evidence that governmental health agencies are ineffective in stopping the spread of disease. Now WNV is entrenched throughout the Americas. In reality, these issues are related to scale and spread of anthropogenic change at the global level. SARS, WNV, and H5N1 HPAI spread are signals that rapid environmental change and globalization can facilitate rapid disease consequences over large temporal and spatial scales and are obviously disruptive to human societies, creating fear and security concerns.

## THE BARRIERS TO A NEW DISCIPLINE

The current impediments to the strong development of a transdisciplinary conservation medicine include those that are typical when specialists in different disciplines try to join forces (Ostfeld et al. 2002). For example, technical languages among disciplines need to be united by clarity in the use of terms. Often different specialists use the same term with different meanings. When consistent definitions across disciplines are not possible, we urge specialists to define their terms and tolerate some diversity in usage by others. For example, even the term "health" has various subjective meanings across disciplines and needs refinement, especially if standard and comparable measures of health are to be developed. Similarly, specialists in different subdisciplines of conservation medicine need to recognize that the whole discipline is hierarchical, spanning levels of organization and a continuum from reductionist to holistic approaches. Higher levels of organization can provide the context for lower ones, and lower levels of organization can provide mechanisms underlying higher ones. It is important to embrace both individual and population-based perspectives in the practice of conservation medicine by scientists of different disciplines. It is often difficult for practitioners of individual organism medicine (human and animal health professionals) to hold a population-based understanding of disease in practice.

The science and practice of conservation medicine involves a focus on one or more of the following entities: humans; global climate; habitat destruction and alteration; biodiversity, including wildlife populations; domestic animals; and pathogens, parasites, and pollutants. Furthermore, conservation medicine focuses on explicit linkages between these entities. As a solution-oriented discipline, the usefulness of conservation medicine ultimately will depend on its applicability to addressing problems. The perspectives and scientific findings of conservation medicine provide input into education, particularly curriculum in veterinary and medical schools, and policy and management of ecosystems, habitats, and imperiled species (Kaufman et al. 2004, 2008; Aguirre and Gomez 2009). Conservation medicine receives contextual input from both the social sciences and bioinformatics. Fields such as sociology, economics, and anthropology inform the science and the practice of conservation medicine by revealing potential causes of human behavior relevant to human-induced environmental change. Top-down approaches to wildlife management are being replaced by adaptive management strategies that address uncertainty and complexity. These new management techniques are necessary to further develop the field. Bioinformatics and the creation, management, and dissemination of databases relevant to wildlife, their habitats, and their diseases may be crucial to this evolving transdisciplinary science.

## RESEARCH PRIORITIES IN CONSERVATION MEDICINE: LEARNING FROM THE PAST TEN YEARS

Four broad areas were identified by Tabor et al. (2001) for enhancing the links between conservation biology and diseases of wildlife and humans: (a) expansion of transdisciplinary interaction; (b) integrated health and ecological assessment and monitoring, and expanded diagnostic capability; (c) resolution of human and wildlife conflict; and (d) implementing conservation medicine practice in conservation reserve design and management. We have seen many advances during the past decade in disease ecology, conservation medicine and transdisciplinarity, and the links to the human ecological footprint, climate change, globalization, and the legal and illegal trade of plants and animals. Emerging zoonotic diseases remain a major threat to public health globally due to the many anthropogenic changes previously discussed (Harvell et al. 1999, 2002; Tabor et al. 2001; Aguirre et al. 2002; Jones et al. 2008; Lafferty 2009; Pongsiri et al. 2009; Keesing et al. 2010). Zoonotic diseases emerge when environmental changes and/or changes in human activities alter the relationship between humans and animals and provide new opportunities for pathogens to cause illness in people. Rather than respond to the disastrous effects after they have emerged, new approaches and collaborations have been developed in the past decade to attempt to prevent these diseases from "spilling over" from animals to humans or to halt them rapidly after that spillover by understanding what factors induce emergence and rapidly identifying ways of prevention, control, and mitigation. Conservation medicine has stimulated the rebirth of the One Health initiative (see Chapter 3 in this volume) and supported the establishment of a new scientific journal, *EcoHealth*, and the International Association of Ecology and Health to bridge ecology and health and to bring transdisciplinary teams to address and solve the challenging global environmental issues faced today (see Chapter 1 in this volume). Both One Health and EcoHealth embrace conservation medicine as a transdisciplinary science (Rapport et al. 1999; Somerville and Rapport 2000; Aguirre et al. 2002). The key factors that drive the emergence of new zoonotic diseases are related to a combination of human changes to the environment, agriculture, healthcare, and changes in demography, all against a background of a large pool of potential new and adaptable pathogens (see Chapter 42 in this volume).

Recently, the U.S. Agency for International Development (USAID) became a major investor in the global response to the emergence and spread of HPAI. Since mid-2005, it supported building local capacity in more than 50 countries for monitoring the spread of H5N1 HPAI among wild bird populations, domestic poultry, and humans, and to mount a rapid and effective containment of the virus when it is found. Recent analyses indicate that these efforts have contributed to significant downturns in reported poultry outbreaks and human infections and a dramatic reduction in the number of countries affected. The Global Avian Influenza Network for Surveillance (GAINS) conducted active surveillance of HPAI in wild bird populations. Sponsored by USAID and the CDC, GAINS was begun in 2006 and administered by the Wildlife Conservation Society. Partner institutions, such as non-governmental organizations, universities, and foreign governments, collaborated in the GAINS network to collect and analyze laboratory samples from wild birds, which were captured and released. This early warning system was intended for health officials to track viral spread in the natural hosts of the disease. The USAID Bureau for Global Health, Office of Health, Infectious Disease and Nutrition (GH/HIDN) granted in 2009 two cooperative agreements, PREDICT and RESPOND among others, under its Avian and Pandemic Influenza and Zoonotic Disease Program to continue and expand this work. The goal of PREDICT is to establish a global early warning system for zoonotic disease emergence that is capable of detecting, tracking, and predicting the emergence of new infectious diseases in high-risk wildlife (e.g., bats, rodents, and non-human primates) that could pose a major threat to human health. The goal of RESPOND is to improve the capacity of countries in high-risk areas to respond to outbreaks of emergent zoonotic diseases that pose a serious threat to human health. The intent is to respond to outbreaks while they are still within the animal community or to rapidly identify spillover to humans in the early stages of emergence. The geographic scope of this expanded effort is directed to those zoonotic "hotspots" of wildlife and domestic animal origins (Jones et al. 2008). GAINS, now known as the Global Animal Information System (https://www.gains.org/default.aspx), was transferred as the framework to develop this global database of emerging zoonotic diseases for PREDICT. These are examples of conservation medicine in action.

The scale of disease surveillance required for prediction and mitigation is massive and is well beyond one or two ambitious bilateral donor initiatives. These large governmental investments should be viewed as a the ground level foundation for building a responsive and nimble 21st Century global health surveillance system. Much more has to be done to monitor disease dynamics in wildlife and domestic animals. Comprehensive ecological surveillance and monitoring of diseases in animals requires an investment of major public health proportions. There has been a global laissez-faire attitude to this problem. There is a massive legal and illegal trade in live wildlife and wild animal products that is global in scale and poorly monitored. The global trade in wildlife has introduced species to new regions where they compete with native species for resources, alter ecosystem services, damage infrastructure, and destroy crops. It has also led to the introduction of pathogens that threaten public health, agricultural production, and biodiversity (Smith et al. 2009; Rosen and Smith 2010). A One Health/EcoHealth/conservation medicine approach involving many parties, including human and animal health professionals, economists, social scientists, ecologists, geographers, and others, can help fill this glaring health need and provide comprehensive, coordinated, and cohesive strategies in addressing this immense threat.

Beyond capacity, conservation medicine needs to lead the way in developing a dashboard for integrated health and ecological monitoring. Policymakers, health managers, and conservationists need a data-driven decision support system that allows practitioners to use adaptive management approaches to dynamic disease situations. One aspect of a dashboard may be found in the concept of sentinel species. For example, marine vertebrates have been documented as sentinels of marine ecological health (Aguirre and Lutz 2004; Aguirre and Tabor 2004; Tabor and Aguirre 2004). Such sentinels are barometers for current or potential negative impacts on individual- and population-level animal health. In turn, using sentinels permits better characterization and management of impacts that ultimately affect animal,

human, and ecosystem health associated with terrestrial and marine habitats. Today most species may be exposed to environmental stressors such as chemical pollutants, harmful algal blooms, and emerging diseases. Since many species share the same environment with humans and domestic animals and many consume the same resources, they may serve as effective sentinels for public health problems (Aguirre 2009; Bossart 2010).

There are many challenges of studying linkages between environmental change and disease emergence and transmission. Both of these are difficult to study on their own using traditional reductionist methods, so bringing them together with current techniques and collaborations is extremely hard to achieve. Investigations of disease outbreaks are impeded by time and funding constraints. An imbalance exists in the immediacy of the health implications of an emerging disease and the application of the best methodological assessments of causation. Realistically, we need to collect sufficient evidence that implicates ecological, behavioral, sociological, or economic causes of disease emergence in order to develop prevention and management actions. The only way to address the anthropogenic drivers of EIDs (i.e., climate change, environmental pollution, habitat destruction, the bushmeat trade) is through systematic, transdisciplinary collaborations (Aguirre et al. 2002; Plowright et al. 2008).

Epidemiologists, modelers, public health officials, veterinarians, and social scientists need to employ strong inference techniques, including model selection, and triangulation, to apply a rigorous approach to establishing causation in disease ecology (Plowright et al. 2008). Have we overcome these challenges in the past ten years? The answer is probably no, but we are moving forward for more realistic scenarios of prediction and prevention of EIDs. For example, we are doing a better job of collecting baseline data on systems that we anticipate will experience an EID in the future. Several creative epidemiologic methods, including SMART (Systematic, Measurable, Adaptive, Responsive and Targeted) surveillance, are being used by teams of scientists to study linkages between anthropogenic factors and disease emergence (see Chapter 42 in this volume). There are new and potentially exciting advances in "eco-immunology" which examine host-pathogen dynamics in the context of environmental stressors on host biology. We have far to go, but we are moving in the right direction. Mathematical modeling, predictive tools, and novel prevention strategies for EIDs have evolved enormously in the past decade. These exciting tools now allow for improved characterization and prediction of disease dynamics and disease behavior. For example, heterogeneities within disease hosts suggest that not all individuals have the same probability of transmitting disease or becoming infected. This heterogeneity is thought to be due to dissimilarity in susceptibility and exposure among hosts. As such, it has been proposed that many host–pathogen systems follow the general pattern that a small fraction of the population (the "super-spreaders") accounts for a large fraction of the pathogen transmission (Woolhouse et al. 1997). Furthermore, recent research shows that heterogeneities in transmission rates among individuals need to be accounted for when planning eradication and control strategies (Lloyd-Smith et al. 2005; Bolzoni et al. 2007).

At the institutional and educational level, conservation medicine work has transformed health and ecological training. Over the past decade, the growth of conservation medicine training initiatives within graduate medical and ecological programs across the globe has been remarkable. Dozens of undergraduate, graduate, and professional courses have been established, along with master's and doctorate programs, and these are expanding rapidly in all continents. There are currently several postgraduate opportunities in conservation medicine. For example, Murdoch University in western Australia has developed a Postgraduate Certificate in Veterinary Conservation Medicine and a Master of Veterinary Sciences (Conservation Medicine) with a flexible program structure and distance education. The establishment of such a program required the removal of disciplinary, institutional, cultural, experiential, and professional development boundaries. The Faculty of Agronomy and Forestry of the Pontificia Universidad Católica de Chile created a Master of Science Program in Wildlife Management. This program emphasizes conservation medicine projects in collaboration with a local wildlife conservation research group. Also, Tufts University has established the first Master of Science in Conservation Medicine program in the United States. Another Chilean veterinary school created a doctoral degree in conservation medicine (Kaufman et al. 2004, 2008; Aguirre and Gomez 2009). More recently, the National University in Costa Rica created a conservation medicine master's degree emphasizing ecosystem health and internal medicine of wildlife (www.medvet.una.ac.cr/posgrado).

Conservation medicine is the catalyst of the EcoHealth/One Health explosion. Whether the term itself has traction is less important than whether the approach is adopted, refined, and improved. Conservation medicine challenges the notion that human health is an isolated concern removed from the bounds of ecology and species interactions. Human health, animal health, and ecosystem health are moving closer together, and at some point it will be inconceivable that there was ever a clear division. May this volume facilitate the merging of these disciplines to the benefit of us all.

# REFERENCES

Aguirre, A.A., and A. Gomez. 2009. Essential veterinary education in conservation medicine and ecosystem health: a global perspective. Rev Sci Tech Off Int Epiz 28:597–603.

Aguirre, A.A., and P. Lutz. 2004. Sea turtles as sentinels of marine ecosystem health: is fibropapillomatosis an indicator? EcoHealth 1:275–283.

Aguirre, A.A., and G.M. Tabor. 2004. Marine vertebrates as sentinels of marine ecosystem health. EcoHealth 1:236–238.

Aguirre, A.A., F. Dallmeier, and J. Comiskey. 1995. Determining ecosystem health: SI/MAB techniques for long term monitoring of biological diversity. AAZV, WDA and AAWV Joint Conf, p. 536.

Aguirre, A.A., R.S. Ostfeld, G.M. Tabor, C.A. House and M.C. Pearl, eds. 2002. Conservation medicine: ecological health in practice. Oxford University Press, New York.

Altizer, S., R. Bartel and B.A. Han. 2011. Animal migration and infectious disease risk. Science 6015:296–302.

Bejarano, A.C., F.M. Van Dolah, F.M.D. Gulland, T.K. Rowles, and L.H. Schwacke. 2008. Production and toxicity of the marine biotoxin domoic acid and its effects on wildlife: a review. Human Ecol Risk Assess 14:544–567.

Bolzoni, L., L. Real, and G. De Leo. 2007. Transmission heterogeneity and control strategies for infectious disease emergence. PLoS One 2:e747 doi:10.1371/journal.pone.0000747.

Bossart, G.D. 2010. Marine mammals as sentinel species of ocean and human health. Vet Pathol doi: 10.1177/0300985810388525.

Brown, D., J. Manno, L. Westra, D. Pimentel, and P. Crabbé. 2000. Inplementing global ecological integrity: a synthesis. *In* Pimentel, D., L. Westra and R.F. Noss. Ecological integrity: integrating environment, conservation, and health, pp. 385–405. Island Press, Washington, D.C.

Burek, K.A., F.M.D. Gulland, and T. O'Hara. 2008. Effects of climate change on arctic marine mammal health. Ecological Applications 18:S126–134.

Burkholder, J.M., M.A. Mallin, and H.B. Glasgow Jr. 1999. Fish kills, bottom-water hypoxia and the toxic *Pfiesteria* complex in the Neuse River and Estuary. Mar Ecol Prog Ser 179:301–310.

Clay, C.A., E.M. Lehmer, A. Previtali, S. St. Jeor, and M.D. Dearing. 2009. Contact heterogeneity in deer mice: implications for Sin Nombre virus transmission. Proc R Soc B doi:10.1098/rspb.(2008).1693.

Cleveland, S. 2003. Emerging diseases of wildlife. Microbiology Today 30:155–156.

Colborn, T., R.S. vom Saal, and A.M. Soto. 1993. Developmental effects of endocrine-disrupting chemicals in wildlife and humans. Environ Health Persp 101:378–384.

Colborn, T., D. Dumanoski, and J.P. Meyers. 1996. Our stolen future. Dutton, New York.

Cox-Foster D.L., S. Conlan, E.C. Holmes, et al. 2007. A metagenomic survey of microbes in honeybee colony collapse disorder. Science 318:283–287.

Deem, S.L., A.M. Kilbourn, N.D. Wolfe, R.A. Cook, and W.B. Karesh. 2000. Conservation medicine. Ann NY Acad Sci 916:370–377.

Dierauf, L.A., G. Griffith, V. Beasley, and T.Y. Mashima. 2001. Conservation medicine: building bridges. J Am Vet Med Assoc 219:596–597.

Dobson, A. 2009. Climate variability, global change, immunity, and the dynamics of infectious diseases. Ecology 90:920–927.

Epstein, J., P.L. Quan, T. Briese, C. Street, O. Jabado, S. Conlan, S. Ali Khan, D. Verdugo, J. Hossein, S.P. Luby, P. Daszak, and W.I. Lipkin. 2010. Identification of GBV-D, a novel GB-like Flavivirus from old world frugivorous bats (*Pteropus giganteus*) in Bangladesh. PLoS Pathogens 6: e100972.

Epstein, P.R. 1999. Climate and health. Science 285: 347–348.

Epstein, P.R. 1993. Algal blooms in the spread and persistence of cholera. Biosystems 31:209–221.

Epstein, P.R., T.E. Ford, and R.R. Colwell. 1993. Health and climate change: marine ecosystems. Lancet 342:1216–1219.

Estes, J.A., and D.O. Duggins. 1995. Sea otters and kelp forests in Alaska: Generality and variation in a community ecological paradigm. Ecol Monog 65:75–100.

Gardiner, S.M. 2006. A core precautionary principle. J Polit Philos 14:33–60.

Goldberg, T.L., T.R. Gillespie, I.B. Rwego, E.E. Estoff, and C.A. Chapman. 2008. Forest fragmentation as cause of bacterial transmission among primates, humans, and livestock, Uganda. Emerg Infect Dis 14:1375–1382.

Goldberg, T.L., T.K. Anderson, and G.L. Hamer. 2010. West Nile virus may have hitched a ride across the Western United States on *Culex tarsalis* mosquitoes. Mol Ecol 19:1519–1519.

Gore, A. 2010. Disrupting chemicals from basic research to clinical practice. Humana Press, Totowa, New Jersey.

Gulland, F.M.D., and A. Hall. 2007. Is marine mammal health deteriorating? Trends in the global reporting of marine mammal disease. EcoHealth 4:135–150.

Harvell, C.D., K. Kim, J.M. Burkholder, R.R. Colwell, P.R. Epstein, D.J. Grimes, E.E. Hofmann, E. K. Lipp, A.D. Osterhaus, R.M. Overstreet, J.W. Porter, G.W. Smith, and G. Vasta. 1999. Emerging marine diseases–climate links and anthropogenic factors. Science 285:1505–1510.

Harvell, C.D., C.E. Mitchell, J.R. Ward, S. Altizer, A.P. Dobson, R.S. Ostfeld, and M.D. Samuel. 2002. Climate warming and disease risks for terrestrial and marine biota. Science 296:2158–2162.

Harvell, C.D., S. Altizer, I.M. Cattadori, L. Harrington, and E. Weil. 2009. Climate change and wildlife diseases: when does the host matter the most? Ecology 90:912–920.

Hayes, T.B. 2005. Welcome to the revolution: Integrative biology and assessing the impact of endocrine disruptors on environmental and public health. J Integr Comp Biol 45:321–329.

Hayes, T.B., P. Case, S. Chui, D. Chung, C. Haefele, K. Haston, M. Lee, V.P. Mai, Y. Marjoua, J. Parker, and M. Tsui. 2005. Pesticide mixtures, endocrine disruption, and amphibian declines: are we underestimating the impact? Environ Health Persp 114:40–50.

IPCC. 2007. Climate Change 2007: Synthesis Report. Contribution of Working Groups I, II and III to the Fourth Assessment Report of the Intergovernmental Panel on Climate Change [Core Writing Team, Pachauri, R.K and Reisinger, A. (eds.)]. IPCC, Geneva, Switzerland, 104 pp.

Jones, K.E., N. Patel, M. Levy, A. Storeygard, D. Balk, J.L. Gittleman, and P. Daszak. 2008. Global trends in emerging infectious diseases. Nature 451:990–994.

Kaufman, G.E., J. Else, K. Bowen, M. Anderson, and J. Epstein. 2004. Bringing conservation medicine into the veterinary curriculum: the Tufts example. EcoHealth 1:S43–S49.

Kaufman, G.E., J.H. Epstein, J. Paul-Murphy, and
J.D. Modrall. 2008. Designing graduate training
programs in conservation medicine–producing the
right professionals with the right tools. EcoHealth
5:519–527.

Keesing, F., L.K. Belden, P. Daszak, A. Dobson,
C.D. Harvell, R.D. Holt, P. Hudson, A. Joles, K.E. Jones,
C.E. Mitchell, S.S. Myers, T. Bogich, and R.S. Ostfeld.
2010. Impacts of biodiversity on the emergence
and transmission of infectious diseases. Nature
468:647–652.

Kock, M. 1996. Wildlife, people, and development. Trop
Anim Health Pro 28:68–80.

Köndgen, S., H. Kühl, P.K. N'Goran, P.D. Walsh, S. Schenk,
N. Ernst, R. Biek, P. Formenty, K. Mätz-Rensing,
B. Schweiger, S. Junglen, H. Ellerbrok, A. Nitsche,
T. Briese, W.I. Lipkin, G. Pauli, C. Boesch, F.H., and
Leendertz. 2008. Pandemic human viruses cause
decline of endangered great apes. Curr Biol 18:
260–264.

Lafferty, K.D., and L.R. Gerber. 2002. Good medicine
for conservation biology: the intersection of
epidemiology and conservation theory. Conserv Biol
16:593–604.

Lafferty, K.D. 2009. The ecology of climate change and
infectious diseases. Ecology 90:888–900.

Leroy, E.M., P. Rouquet, P. Formenty, S. Souquiere,
A. Kilbourn, J.M. Froment, M. Bermejo, S. Smit,
W.B. Karesh, R. Swanepoel, S.R. Zaki, and P.E. Rollin.
2004. Multiple Ebola virus transmission events and
rapid decline of central African wildlife. Science 303:
387–390.

Li, W., Z. Shi, M. Yu, W. Ren, C. Smith, J. H. Epstein,
H. Wang, G. Crameri, Z. Hu, H. Zhang, J. Zhang,
J. McEachern, H. Field, P. Daszak, B.T. Eaton, S. Zhang,
and L.F. Wang. 2005. Bats are natural reservoirs of
SARS-like coronaviruses. Science 310:676–679.

Lipkin, W.I. 2010. Microbe hunting. Microbiol Mol Biol Rev
74:363–377.

Lloyd-Smith, J.O., S. J. Schreiber, P.E. Kopp, and
W. M. Getz. 2005. Superspreading and the effect of
individual variation on disease emergence. Nature
438:355–359.

McMichael, A.J. 1997. Global environmental change and
human health: impact assessment, population
vulnerability, and research priorities. Ecosyst Health
3:200–210.

McMichael, A.J., B. Bolin, R. Costanza, G.C. Daily,
C. Folke, K. Lindahl-Kiessling, B. Lindgren, and
E. Niklasson. 1999. Globalization and the sustainability
of human health: an ecological perspective. Bioscience
49:205–210.

Meffe, G. 1999. Conservation medicine. Conserv Biol 13:
953–954.

Millennium Ecosystem Assessment. 2007. Ecosystems and
human wellbeing: biodiversity synthesis. World
Resources Institute, Washington, D.C.

Milligan, S.R., W.V. Holt, and R. Lloyd. 2009. Impacts of
climate change and environmental factors on
reproduction and development in wildlife. Phil Trans R.
Soc B 364:3313–3319.

Monroe, M.C., S.P. Morzunov, A.M. Johnson,
M.D. Bowen, H. Artsob, T. Yates, C.J. Peters,
P.M. Rollin, T.G. Ksiazek, and S.T. Nichol. 1999. Genetic
diversity and distribution of Peromyscus-borne
hantaviruses in North America. Emerg Infect Dis
5:75–86.

Myers, N., R.A. Mittermeier, C.G. Mittermeier, G.A.B. da
Fonseca, and J. Kent. 2000. Biodiversity hotspots for
conservation priorities. Nature 403:853–858.

Norris, S. 2001. A new voice in conservation medicine.
Bioscience 51:7–12.

Osofsky, S.A., W.B. Karesh, and S.L. Deem. 2000.
Conservation medicine: a veterinary perspective.
Conserv Biol 14:336–337.

Ostfeld, R.S. 2009. Climate change and the distribution
and intensity of infectious diseases. Ecology 90:
903–905.

Ostfeld, R.S., M. Pearl, and G. Meffe. 2002. Conservation
medicine: the birth of another crisis discipline. In
A. Aguirre, R.S. Ostfeld, C.A. House, G. Tabor, and
M. Pearl, eds. Conservation medicine: ecological
health in practice, pp. 17–26. Oxford University Press,
New York.

Pachauri, R.K., and A. Reisinger, eds. 2007. Climate Change
2007: Synthesis report—contribution of working
groups I, II and III to the 4th Assess Rep IPCC.
Geneva, Italy.

Padula, P.J., S.B. Colavecchia, V.P. Martínez, M.O. Gonzalez
Della Valle, A. Edelstein, S.D. Miguel, J. Russi,
J.M. Riquelme, N. Colucci, M. Almirón, R.D.
Rabinovich, and E.L. Segura. 2000. Genetic diversity,
distribution, and serological features of hantavirus
infection in five countries in South America. J Clin
Microbiol 38:3029–3035.

Pascual, M., and M. Bouma. 2009. Do rising temperatures
matter? Ecology 90:906–912.

Palacios, G., L.J. Lowenstine, M.R. Cranfield,
K.V.K. Gilardi, L. Spelman, M. Lukasik-Braun,
J.-F. Kinani, A. Mudakikwa, E. Nyirakaragire,
A.V. Bussetti, N. Savji, S. Hutchinson, M. Egholm, and
W.I. Lipkin. 2011. Human metapneumovirus infection in
wild mountain gorillas, Rwanda. Emerg Infect Dis
17:DOI 10.3201/eid1704100883.

Patz, J.A., P. Daszak, G.M. Tabor, A.A. Aguirre, M. Pearl, J. Epstein, N.D. Wolfe, A.M. Kilpatrick, J. Foufopoulos, D. Molyneux, and D.J. Bradley, Members of the Working Group on Land Use Change and Disease Emergence. 2004. Unhealthy Landscapes: Policy recommendations pertaining to land use change and disease emergence. Environ Health Persp 112:1092–1098.

Pimm, S.L., and P. Raven. 2000. Biodiversity—extinction by numbers. Nature 403:843–845.

Pokras, M., G.M. Tabor, M. Pearl, D. Sherman, and P. Epstein. 1999. Conservation medicine: an emerging field. *In* P. Raven and T. Williams, eds. Nature and human society: the quest for a sustainable world, pp. 551–556. National Academy Press, Washington D.C.

Pongsiri, M.J., and J. Roman. 2007. Examining the links between biodiversity and human health: an interdisciplinary research initiative at the U.S. Environmental Protection Agency. EcoHealth 4: 82–85.

Pongsiri, M.J., J. Roman, V.O. Ezenwa, T.L. Golberg, H.S. Koren, S.C. Newbold, R.S. Ostfeld, S.K. Pattanayak, and D.J. Salkeld. 2009. Biodiversity loss affects global disease ecology. Bioscience 59: 945–954.

Plowright, R.K., S.H. Sokolow, M.E. Gorman, P. Daszak, and J.E. Foley. 2008. Causal inference in disease ecology: investigating ecological drivers of disease emergence. Front Ecol Environ 6, doi:10.1890/070086

Pullin, A.S., and T.M. Knight. 2001. Effectiveness in conservation practice: pointers from medicine and public health. Conserv Biol 15:50–54.

Randolph, S. 2009. Perspectives on climate change impacts on infectious diseases. Ecology 90: 927–931.

Rapport, D.J., G. Böhm, D. Buckingham, J. Cairns Jr., R. Costanza, J.R. Karr, H.A.M. De Kruijf, R. Levins, A.J. McMichael, N.O. Nielsen, and W.G. Whitford. 1999. Ecosystem health: the concept, the ISEH, and the important tasks ahead. Ecosystem Health 5:82–90.

Rosen, G.E., and Smith, K.F. 2010. Summarizing the evidence on the international trade in illegal wildlife. EcoHealth 7:24–32

Rwego, I.B., G. Isabirye-Basuta, T.R. Gillespie, and T.L. Goldberg. 2009. Bacterial exchange between gorillas, humans, and livestock in Bwindi. Gorilla J 38:16–18.

Schmidt, K.A., and R.S. Ostfeld. 2001. Biodiversity and the dilution effect in disease ecology. Ecology 82:609–619.

Shuman, E.K. 2010. Global climate change and infectious diseases. N Engl J Med 362:1061–1063.

Smith, K.F., M. Behrens, L.M. Schloegel, N. Marano, S. Burgiel, and P. Daszak. 2009. Reducing the risks of the wildlife trade. Science 324:594–595.

Somerville, M.A., and D.J. Rapport. 2000. Transdisciplinarity: recreating integrated knowledge. EOLSS Publishers Co. Ltd., Oxford, United Kingdom.

Soulé, M.E., and B. Wilcox. 1980. Conservation biology: an evolutionary-ecological perspective. Sinauer Associates, Sunderland, Massachusetts.

Speare, J.R. 2000. Conservation medicine: the changing view of biodiversity. Conserv Biol 14: 1913–1917.

Stuart, S.N., J.S. Chanson, N.A. Cox, B.E. Young, A.S.L. Rodrigues, D.L. Fischman, and R.W. Waller. 2004. Status and trends of amphibian declines and extinctions worldwide. Science 306:1783–1786.

Suter, G.W., T. Vermeire, W.R. Munns Jr., and J. Sekisawa. 2005. An integrated framework for healthy and ecological risk assessment. Toxicol Appl Pharmacol 207:S611–616.

Suzán, G., E. Marcé, J.T. Giermakowski, J.N. Mills, G. Ceballos, R.S. Ostfeld, B. Armién, J.M. Pascale, and T.L. Yates. 2009. Experimental evidence for reduced rodent diversity causing increased Hantavirus prevalence. PLoS ONE 4(5):e5461.doi:10.1371/journal. pone.0005461

Tabor, G.M. 2002. Defining conservation medicine. *In* A.A. Aguirre, R.S. Ostfeld, G.M. Tabor, C.A. House, and M.C. Pearl, eds. Conservation medicine: ecological health in practice, pp. 8–16. Oxford University Press, New York.

Tabor, G.M., and A.A. Aguirre. 2004. Ecosystem health and sentinel species: adding an ecological element to the proverbial "canary in the mineshaft." EcoHealth 1:226–228.

Tabor, G.M., R.S. Ostfeld, M. Poss, A.P. Dobson, and A.A. Aguirre. 2001. Conservation biology and the health sciences: defining the research priorities of conservation medicine. *In* M.E. Soulé and G.H. Orians, eds. Research priorities in conservation biology. 2nd edition, pp. 165–173. Island Press, Washington D.C.

Taylor, L.H., S.M. Lathan, and M.E.J. Woolhouse. 2001. Risk factors for human disease emergence. Phil Trans R Soc Lond B 356: 983–989.

Ward, J.C., and K.D. Lafferty. 2004. The elusive baseline of marine disease: are diseases in ocean ecosystems increasing? PLoS Biology 2:542–547.

Wolfe, N.D., C.P. Dunavan, and J. Diamond. 2007. Origins of major human infectious diseases. Nature 447:279–283.

Woodford, M.H., T.M. Butynski, and W.B. Karesh. 2002. Habituating the great apes: the disease risks. Oryx 36:153–160

Woodruff, T.J., A.R. Zota, and J.M. Schwartz. 2011. Environmental chemicals in pregnant women in the United States: NHANES 2003–2004. Environ Health Perspect 119:878–885. doi:10.1289/ehp.100272

Woolhouse, M.E.J., C. Dye, J.-F. Etard, T. Smith, J.D. Charlwood, G.P. Garnett, P. Hagan, J.L.K. Hii, P.D. Ndhlove, R.J. Quinnell, C.H. Watts, S.K. Chandiwana, and R.M. Anderson. 1997. Heterogeneities in the transmission of infectious agents: implications for the design of control programs. Proc Natl Acad Sci USA 94:338–342.

# 2

# ECOHEALTH

## Connecting Ecology, Health, and Sustainability

## Bruce A. Wilcox, A. Alonso Aguirre, and Pierre Horwitz

The idea of ecohealth, a term mainly used as a contraction for either "ecology and health" or "ecosystem approach to human health," was first popularized in international development circles by the Canadian government's International Development Research Centre (IDRC). As described in Lebel's (2003) *Health: an Ecosystem Approach*, IDRC proposed ecohealth as a new transdisciplinary framework that "implies an inclusive vision of ecosystem-related health problems." In 2002 a group assembled (including the authors) at a conference titled "Healthy Ecosystems, Healthy People" held in Washington, DC, initiated the process that led to the establishment of the journal *EcoHealth*, and the eventual formation of the International Association for Ecology & Health (IAEH).

The first issue of the journal was published in March 2004 and the IAEH was formed with a Charter Board in 2006. The community of researchers and practitioners involved were motivated by a belief in the inherent interdependence of the health of humans and ecosystems (including the biodiversity within them)—in particular that the qualities of ecosystems and the health of humans, domestic animals, and wildlife are reciprocal and that disciplinary integration at the interface of ecological and health sciences was, and still is, required (Wilcox et al. 2004).

Ecohealth activities seeking this integration "inherently involve three groups of participants: researchers and other specialists; community members…; and decision-makers" (Lebel 2003).

The philosophical roots and thus epistemological basis for the idea of ecohealth can be traced to Aldo Leopold's concept of "land health" centered on the productive use of nature by society, and the maintenance (and when necessary the restoration) of the capacity for "internal self-renewal known as health," an inherent property of an organism applied metaphorically to ecosystems by Leopold (Leopold 1949; Callicott 1987; Callicott and Freyfogle 1999). Leopold's conception spanned thought across the natural sciences with an implicit attention to social sciences and systems thinking. His focus on a Hippocratic-like ethical responsibility and need for professional practitioners who can address the health of humans and all other organisms as a single "community" integrated ideas from medicine and public health with applied ecology and conservation. Leopold's approach foreshadowed latter-day "participatory research" in which scientific experts engage in public policy discourse, with and even for the citizenry. Leopold's "science in action" aimed at a real-world problem (ecological degradation), in contrast with one conceived to advance a particular discipline "in the lab," is what today is called transdisciplinary research.

The aim of this chapter is to review the theoretical and practical foundations of the concept of ecohealth, including the history of the ecohealth movement according to its strands of thought and action, and to describe ecohealth approaches in research and practice.

## FOUNDATIONS

Leopold's concept of land health emerged at the very time modern wildlife biology, particularly ecology, was coalescing with systems theory. Leopold's *Game Management* (1933), the first book on the science of wildlife conservation, was strongly influenced by the emerging field of ecology, the first textbook, *Animal Ecology*, for which had only just been published in the previous decade (Elton 1927). Subsequently, Leopold's (1949) famous series of essays in *Sand County Almanac*, published posthumously, defined the cultural value of wildlife and wilderness and humans' ethical responsibility to keep intact the "land pyramid" of agricultural landscapes as well as inherent in natural ecosystems. Leopold used "land" as shorthand for the complex of features, especially biotic components, the pyramidal structure (today called trophic structure), and processes driven by solar energy, later encompassed by "ecosystem." That term was introduced around the same time by British botanist Arthur Tansley (Golley 1993) and not widely adopted until after the publication of Eugene Odum's *Fundamentals of Ecology*, with five editions published between 1953 and 2004 in several languages (Odum and Barrett 2005). Considered the most influential organismic and environmental textbook (Wilson 2004), later editions were increasingly organized around the ecosystem concept as the field of ecosystem ecology developed (Golley 1993). Following Leopold's view, Odum's ecosystem ecology deals not only with forests, fields, and lakes, but also with humans included as ecological components of ecosystems (McDonnell and Pickett 1993).

Aldo Leopold defined "land health" as a condition under which "the land could be humanly occupied without rendering it dysfunctional" and "the capacity of the land for self-renewal"; indeed his use of the health metaphor extended to recognition of "land sickness" (Leopold 1941, 1949). This metaphorical exploration of the concept of organismal health, as we shall see, has had profound consequences for what might now be regarded as the ecohealth movement. For Leopold, the main consideration was whether or not modifications in the landscape produced by humans affected or somehow compromised essential ecological functions necessary to sustain landscape components and processes. As humans are "part of" and not "apart from" the landscape, Leopold argued that the degree to which the landscape satisfies human needs necessarily links into an assessment of landscape health (Rapport et al. 1998b). In short, Leopold's ethical and scientific philosophy was based on a holistic perspective (Hargrove 1979).

On the heels of Leopold's career, American marine biologist Rachel Carson's *The Sea Around Us* was published (Carson 1951), highlighting how people depend on the sea, and predicting the impoverishment and crisis of the world's oceans (Leisher 2008). Carson had an ability to engage wide audiences in topics of scientific discovery and the natural world (Karvonen 2008). The last of her series of best sellers, *Silent Spring* (Carson 1962), is credited with catalyzing the environmental movement in North America. In it she took aim at the growing threat to humans and wildlife of synthetic pesticides in the environment. *Silent Spring* not only spawned the environmental movement and directly led the creation of the U.S. Environmental Protection Agency, but it also foreshadowed the now-confirmed wide-ranging effects of synthetic endocrine-disruptor compounds in ecosystems (Colborn et al. 1997).

A parallel movement had been developing in Europe. Famed biologist Sir Julian Huxley, while the first Secretary General of UNESCO, initiated the establishment of the first international scientific organization devoted to conservation, the International Union for the Conservation of Nature and Natural Resources (IUCN) in 1948 (Boardman 1981). British wildlife conservationists Sir Peter Scott and Gerald Durrell, both prolific authors and counterparts of American's Leopold and Carson in the sense of publicizing the plight of nature, were central figures in IUCN's development and the establishment of World Wildlife Fund International. IUCN, now the largest conservation network and a neutral forum for governments, non-governmental organizations, scientists, business, and local communities, seeks pragmatic solutions to conservation and development challenges (http://www.iucn.org/about/). One of its most important contributions has been the creation and management of the "Red List," which provides regularly updated global assessments of the conservation status of species, subspecies, varieties, and even selected subpopulations (http://www.iucnredlist.org/). These lists have proved to be powerful advocacy tools, and they have been mirrored by intergovernmental agreements (multilateral environmental agreements like the UN Convention on Biological Diversity) and by formal listing processes adopted by many national governments. Furthermore, a working group established by the IUCN has begun to develop and implement comparable global standards for threatened ecosystems by formulating a system of quantitative categories and criteria, analogous to those used for species, for assigning levels of threat to ecosystems at local, regional, and global levels (Rodriguez et al. 2010).

In the meantime, a group of European scientists associated with the International Union of Biological Science meeting in 1961 under the theme "The Biological Basis of Human Welfare" launched what was to become the International Biological Program (IBP; Worthington 1975). It ultimately engaged a large network of biologists from across the developed world in an effort to characterize the world's biomes from an ecosystem science perspective. An analysis of ecosystems effort, with Eugene Odum as chair, evolved to become a dominant program. Odum presciently urged that IBP should develop "new thinking at the ecosystem level," including a new science of landscape ecology as a basis for landscape planning in the future (Golley 1993).

Building on the successes and failures of IBP, which was more academic than practical, in 1971 UNESCO launched the Man and the Biosphere Program (MAB). MAB introduced an ecosystem approach of conceptually integrating people, human landscapes, wildlife, wilderness, and genetic diversity, echoing Leopold's ideas and Odum's ecosystem ecology, as UNESCO policy documents referred to it as "ecology in action . . . providing a basis for ecosystem conservation . . . while linking conservation and development" (UNESCO 1981). MAB also was responsible for spawning the idea of biological diversity, which conceptually unified the relatively fragmented conservation issues of genetic resources,

endangered wildlife, and ecosystems. Raymond Dasmann, senior scientist at IUCN and a key member of the MAB subcommittee that conceived the biosphere reserve concept, coined the term "biological diversity" in a lay book (Dasmann 1968) about the time MAB was conceived. As a graduate student of University of Berkeley professor and wildlife biologist Starker Leopold, Aldo's son, Dasmann was steeped in the Leopolds' perspectives on conservation and science. By the 1980s the term "biological diversity" (and its contracted form "biodiversity") was becoming accepted in the scientific literature. "Biodiversity conservation" had begun to gain a scientific imprimatur never enjoyed by "nature preservation," "endangered species," or "wildlife conservation."

Conservation biology emerged in name simultaneously with the first use of "biological diversity" in a scientific publication (Lovejoy 1980): a collaborative effort involving ecologists, geneticists, veterinarians, and conservationists who contributed to the seminal work in the field (Soulé and Wilcox 1980). Biological diversity served as a useful "catch" term to express the idea that variety of species and ecosystems and their ultimate source and survival requirement, genetic variability, are irrevocably interconnected, functionally and in terms of conservation strategy. An explicit definition followed two years later in a paper commissioned by the IUCN (Wilcox 1984b) for 1982 Bali World National Parks Congress, based on the author's discussions with Dasmann, and at a meeting organized by Eugene Odum about the same time in Athens, Georgia, USA (Cooley and Cooley 1984; Wilcox 1984a). It centered on Odum's hierarchy of biological organization—from "genes to ecosystems" (Odum 1971, p. 5): "Biological diversity is the variety of life forms, the ecological roles they perform and the genetic diversity they contain . . . the ultimate source of biological diversity at all levels of biological systems (i.e., molecular, organismal, population, species, and ecosystem)." We now know that these roles include such functions as the regulation of pathogen transmission (Keesing et al. 2010).

By the 1990s conservation medicine, climate change, ecosystem health, and sustainable development had emerged as science and policy issues. This included the recognition that new, integrative approaches were needed to face global challenges. By the new millennium the convergence of these and other issues intensified the demand for integrative approaches and, more importantly, the breaching of disciplinary boundaries. Thus the development of thought leading to ecohealth, as conservation medicine, consisted of the multiple threads. The issue of emerging or re-emerging infectious diseases, the vast majority of which are zoonotic or otherwise primarily driven by demographic, social, technological, and environmental change, has decidedly dominated conservation medicine in particular and ecohealth in general. However, the idea of reciprocal human–nature interdependence, drawing on Odum's ecosystem perspective and Leopold's notion of the capacity for self-renewal, can be largely credited with laying the foundation for ecohealth.

A parallel stream of thought consistent with these two ideas emerged in the 1970s, with practical application to the problems of sustainable development, natural resources management, and health. Drawing on advances in complexity theory, a radically new understanding of systems developed as a collaborative effort of natural and social sciences (Jantsch and Waddington 1976). This interdisciplinary convergence included Holling's (1973) pioneering work on ecosystem resilience, along with Prigogine's discoveries in thermodynamics, for which he was awarded a Nobel Prize in Physics. Prigogine's work linked thermodynamics and General Systems Theory, providing a rigorous basis for concepts like non-linearity and self-organization, and demonstrating the "far-from-equilibrium" character of living systems (Nicolis and Prigogine 1977). These findings ultimately led ecologists to rethink the meaning of the popular idea of "the balance of nature," and how ecosystems are anything but deterministic and stable, equilibrium systems.

Recent work has been critical in the evolution of these ideas as a paradigm shift from "balance of nature" to "flux of nature" (Pickett and Ostfeld 1995). Disturbance has been found to play a major role in ecosystem structure and function. So "chance" and "meta-stable" are now thought to be more accurate characterizations (Levin 1999; O'Neill 2001). Yet, as these authors describe and Holling and colleagues have elaborated, as complex adaptive systems, ecosystems exhibit universal properties and patterns of behavior. We argue that these qualities, and how their understanding can inform environmental management and public health threats, represent an important nascent area of ecohealth and conservation medicine alike.

The development of "adaptive management" of ecosystems (Holling 1978; Walters 1986), a core element of the ecosystem approach, evolved from the above-described marriage of ecology and complex systems theory into today's "social ecological system and resilience theory." Holling and colleagues' elaboration based on regional ecosystem case studies laid a foundation for practically approaching the problem of the sustainability of human–natural systems (Gunderson et al. 1995). The resulting synthesis, centered largely on Holling's "adaptive cycle" (Holling 1986; Gunderson and Holling 2002), links adaptive management and ecological resilience (Gunderson 2000), ecosystem functioning and biodiversity (Folke et al. 2004), and formal science and local or traditional knowledge (Berkes and Folke 1998; Berkes et al. 2003). Their "resilience" is an elaboration and affirmation of Leopold's "capacity for self-renewal." The first significant attempt to integrate this stream of research with public health produced an *EcoHealth* (2005, vol. 2 no. 4) special issue on "Social–Ecological Systems and Emerging Infectious Diseases."

## THE BIOPHYSICAL QUESTION AND ITS CO-OPTION

The notion of "ecosystem health," the progenitor of the ecohealth movement, engendered a vigorous and important debate in the 1990s. The concept captured the attention of natural resource managers because it provided them with a narrative "hook" that embodied a more difficult series of questions concerning whether an ecosystem *could* be healthy and, if so, how it could be measured. The significant influence of David Rapport and his colleagues is obvious in this regard. Rapport et al. (1985) seminal paper challenged ecologists to consider ecosystem degradation in the same way that illness in humans was evaluated by a general practitioner in medicine—that is, according to a suite of indicators that together they referred to as the "ecosystem distress syndrome." The "ecosystem health" project did not deviate much from this theme and emphasis; the problematique provided momentum for the establishment of the International Society for Ecosystem Health that founded the journal *Ecosystem Health*, published with Blackwell Scientific.

Ecosystem health was described in terms of measurable properties: healthy ecosystems retain their natural vigor (productivity), their resilience (capacity to recover from disturbance, indeed self-renewal), and their organization (e.g., biodiversity and symbiotic relations between species: Rapport et al. 1998b; Table 2.1). Health of a system was more than just the absence of disease or illness, and for ecosystem health, the presence of a pathogen explained something of the nature of that system (Cook et al. 2004).

This conception of ecosystem health as advanced by Rapport and colleagues, as a contribution to understanding the functioning of natural or semi-natural ecosystems, and especially ecosystem dysfunction under different forms of stress, has been invaluable. However, these ideas did not sufficiently challenge the impermeability of the health sciences and the health sector at large. Perhaps more significantly, natural resource management continues to use the phrase "healthy ecosystems," where in latter-day use it persists as a biophysical concept only, frequently with little of the richness provided by the debate in issues of *Ecosystem Health*. Guarding its disciplinary integrity and generally viewing health as outside its scholarly jurisdiction, ecological science has yet to significantly embrace this concept of ecosystem health as a research pursuit (Costanza et al. 1992).

Paradoxically, biomedicine and public health have embraced ecological perspectives, in theory if not practice. Engle's (1977) biopsychosocial model-based systems theory has inspired "patient-centered medicine"

that is implicitly ecological and explicitly draws on general systems theory (Stewart et al. 1995). A similar evolution from a reductionist, disease focus to a holistic one developed in public health (Last 1998). This has been extended to defining and measuring health based on factors that support human health and well being rather than those causing disease, nearly identical to Holling's ecological resilience (Antonovsky 1979, 1987). Other researchers in biomedicine and public health employ an "ecological perspective" applying biomedical, behavioral, and social epidemiological models and interventions at the family, community, and population levels (McLeroy et al. 1988; IOM 2001).

These perspectives often linking concepts of stress and resilience (terms "coping" and "hardiness" with similar meanings) are being applied increasingly to a wide range of mainly global health issues (indigenous peoples, health disparities, globalization and health, migration conflict and health, and others) relevant to ecohealth. Krieger's (2001, 2008) "ecosocial" theory explicitly links ecosystem and societal levels to explain population health. This is consistent with the socio-ecological approach of the Ottawa Charter (WHO 1986; Parkes and Horwitz 2009). Power and disempowerment are critical dimensions of ecosocial theory. An ecohealth perspective considers disempowerment and the disintegration of biopsychosocial mechanisms essential for health maintenance (e.g., individual, family, and community resilience) irrevocably tied to the biophysical components of an ecosystem. This dependence of population health on ecosystems was

Table 2.1  **Examples of Measurable Indicators of Healthy Ecosystems**

| Vigor | Organization | Resilience |
|---|---|---|
| The activity, metabolism, or productivity of a system (Mageau et al. 1995); the energy or activity of a system, and identification of factors that may risk the activity of the system. Measures like net primary productivity, gross domestic product, gross ecosystem product, and metabolism of organisms. | The vital signs of a system; the diversity and number of interactions between system components; degree of specialization. Measures like diversity and multispecies indices, system-level information indices, or network analyses. | A measure of robustness, buffering capacity, magnitude of disturbance that can be absorbed before the system changes its structure by changing the components and processes that control its behavior. Measures must include identification of buffers and thresholds of change. |

After Rapport et al. 1998b.

elaborated as part of the Millennium Ecosystem Assessment (MEA 2005), discussed in more detail below.

Of course, not just human health but also human survival ultimately depends on a stable, functioning biosphere, and thus predominantly natural ecosystems (i.e., besides "healthy" agroecosystems and urban ecosystems) of sufficient quality and extent. This is the important message of MEA, despite any shortcomings. The approach and methods offered by ecohealth arguably hold significant promise toward shoring up the MEA health evaluation by linking the largely accepted body mostly dealing with the social determinants of health cited above with a growing body of literature empirically documenting the health consequences of displacement or exclusion of populations or co-option or degradation of ancestral lands and/or their natural resources. The severing of relationships with the landscape, and the associated loss of social capital and cultural attachments, as well as resources essential for economic livelihoods, can have severe psychosocial consequences. Understanding these linkages, and those between psychosocial and somatic disease, represents an ecohealth frontier that is only beginning to be explored. Albrecht et al. (1998) weave nearly all these strands together, including emerging infectious and chronic diseases, to describe a generic transdisciplinary framework consistent with ecohealth.

## IMPERATIVES OF ECOHEALTH: FRAMING THE PROBLEM STATEMENT

Upon these foundations, the development of ecohealth has occurred through the intertwining of three largely inseparable strands of thought and action: an ecology of health, the global environmental change agenda, and transdisciplinarity.

### An Ecology of Health

The time has come to more deliberately address matters of human development and well being that occur at the intersection of the ecological and health sciences in academic and practical ways. The critical point of deliberation here is that ecohealth seeks to include all matters of public health and human and veterinary medicine as being individual and population expressions of interacting systems. The systems include those at smaller scales (subsystems of the individual), and those at larger scales (ecosystems, and other levels and forms of organization). Levels, scales, and systems are the topics of consideration of a number of disciplines outside the health sciences, ecology being the main one. As they typically involve a dynamic with multiple interactions between humans or animals, or both, and the ecosystems in which they live, an understanding necessarily requires taking an ecosystem approach. The problems ecohealth seeks to address can often be seen as systemic and requiring a contextual consideration to isolate their cause(s) as a basis for designing effective intervention and prevention. In this sense, ecohealth is more concerned with "upstream" causes as espoused in public health.

Thus, as can be clearly seen in the case of infectious disease emergence, ecohealth problems often necessitate a conceptual framework (Fig. 2.1), integrative research and training models, and methods that transcend disciplines. A systems approach (Checkland 1986) is typically found necessary to build a sufficiently comprehensive theory (theoretical framework) and account for the inherent complexity of the "study system." The system, whether represented by an urban community, rural community, agricultural production system, or forest ecosystem, is inevitably a "coupled-human natural system" in which the boundary between society and nature is only an artifact of science (Berkes et al. 2003).

What exactly is it that qualifies as ecohealth research or practice? Our assumption is that our readers are likely to be more familiar with health science than ecological science, and we have already cited health models incorporating ecological thinking, so we shall focus on what is meant by ecological science. Not all activities aligned with ecohealth as a paradigm necessarily involve theory and method integration *per se*; our argument is that it should be increasingly so. Other endeavors aligned with ecohealth, for which conservation *per se* is not an operational objective, necessarily draw on ecological sciences either in the sense of employing "ecological thinking" or "ecosystem" as a paradigm (Golley 1993; Keller and Golley 2000). Thus, they draw on specific theories, principles, and methods from ecology. Besides ecosystem ecology, these often include those from population ecology, community ecology, landscape ecology, and various other subdisciplines or lines of research,

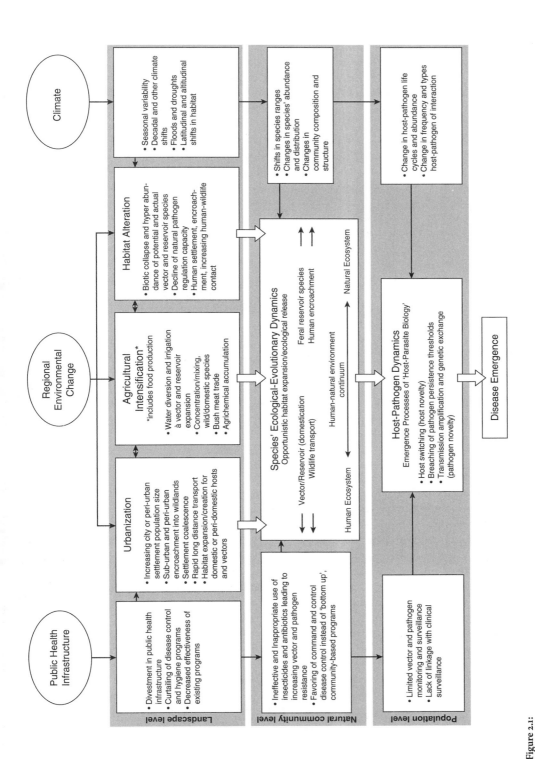

**Figure 2.1:**
The "causal schema of disease emergence" (Wilcox and Gubler 2005), serves as the theoretical framework for transdisciplinary projects as related to the project Emerging Infectious Disease and Social-Ecological Systems. (With kind permission from Springer Science+Business Media: EcoHealth 2:244–257.)

as surveyed by, among others, Odum and Barrett (2005) and Begon et al. (2006). For example, Ellis and Wilcox (2008) describe how different subdisciplines of ecology apply to the problem of vector control.

The term "disease ecology" is often used to characterize studies integrating ecology with infectious disease biology. Disease ecology can be traced at least to Sir Ronald Ross's combined field and laboratory investigations in the late 19th century that led him to the discovery of the etiological role of mosquitoes in malaria transmission. This, in turn, led to Ross's later work that laid the foundation for mathematical epidemiology (Anderson and May 1991), representing what are now considered the basic principles of "disease ecology" (Begon et al. 2006, Chapter 12). Scheiner (2009) surveys the applicability of a range of ecological theory and models in the context of biogeography (a field that significantly overlaps with ecology) to disease emergence. Numerous articles can be found in several journals reporting on studies involving the application of ecology to health issues, including those of indigenous people, dealt with in a cultural context in a special issue of *EcoHealth* (2007, vol. 4 no. 4).

Ecosystem approach in general means applying an understanding of the properties of a whole entity of relevance to the health problem of concern, an infectious disease or otherwise. This includes contextualizing a problem by situating it geographically and identifying the biophysical as well as sociocultural and economic conditions and forces contributing to human or veterinary public health issues. The point is to attempt to identify the proximal as well as the distal causative factors and how they interact mechanistically to bring about the problem. For example, to understand a zoonotic disease outbreak, in addition to conducting a routine epidemiological investigation, a study would consider all the potentially relevant underlying factors as well as the source or origin of the agent(s) responsible. The aim is to isolate the "variables" primarily responsible and develop an intervention accordingly that targets the critical variables with the goal of minimizing the likelihood of future outbreaks, or to stem transmission of the pathogen or parasite in the case of an ongoing outbreak or enzootic situation. Neudoerffer et al. (2005) provide an illustrative case example of an ecosystem approach to zoonotic disease control (Fig. 2.2).

It should be noted that the "ecosystem approach" in ecohealth is derived from "ecosystem-based management," also called "ecosystem management," developed by Holling (1978, 1986) in the context of natural resource management centered on the idea of "adaptive management" already discussed. Ecosystem ecology and failed attempts involving the development, use, or control of renewable natural resources (Gunderson et al. 1995) have led to the so-called ecosystem management approach (Slocombe 1993). It has been widely adopted by many agencies, including IUCN, as the preferred method of land and natural resources management. Social ecological systems and resilience theory can be viewed as an extension of ecosystem management. The adaptive cycle model provides a useful approach to understanding infectious disease emergence as a complex adaptive system phenomenon (Wilcox and Colwell 2005). Participatory process, described further below, is considered a key element of the ecosystem approach.

## The Global Environmental Change Agenda

The impetus for, and urgency of, developing ecohealth is the historically unprecedented ecological, social, and demographic changes taking place locally, regionally, and globally. Global challenges of climate change, overpopulation, and land use change, as well as other human-driven impacts, have severely affected the biosphere. In general terms, these impacts range from altering the physical and chemical environment of life (climate change, soil alteration and depletion) to degradation of non-human living systems (overharvesting of wildlife, habitat destruction, extinction of species) (Vitousek et al. 1997; Chapin et al. 2000). They can be said to represent dysfunction in human systems, including the very institutions established to prevent ecological degradation and control disease described as a "pathology" by Holling et al. (2002), often associated with poor governance, political corruption, and environmental as well as social injustice.

Changes have been driven by an increasing, and increasingly consumptive, human population and expansions in rural and urban environments in both developed and developing countries, which have created a significant demand for the provision of food and water; these positively feed back to each other. The changes are causally connected with degradation

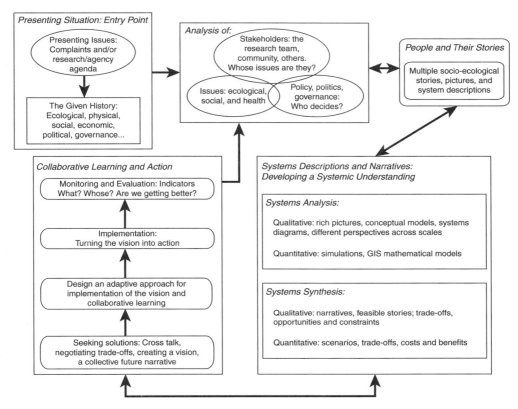

**Figure 2.2:**
Flow diagram describing the elements and steps of an adaptive approach to develop a transdisciplinary, integrative, and participatory action research approach to improve health as one outcome of that system. (Figure credit: David Walter-Toews, Network for Ecosystem Sustainability and Health; Neudoerffer et al. 2005.)

of ecosystems and ecosystem services, with declines in wildlife health and public health, and these "healths" are connected via multidirectional interactions and feedbacks in ways that science is only beginning to grasp. The association with declining renewable natural resources and environmental quality is often more clearly evident than a detailed understanding of the mechanisms involved.

These causal connections were the focus of the MEA, culminating in the final set of volumes and findings in 2005. Its principal contribution in this regard was to demonstrate the relationships between ecosystem services and the constituents of human well being. People derive benefits from ecosystems both individually and collectively, and directly and indirectly. Such benefits from ecosystem include the services of: *provisioning* (drinking water, food, genetic materials, structural products, medicinal products,

irrigation and industrial water), *regulating* (groundwater recharge, water purification, flood control, soil, sediment and nutrient retention, local climate regulation, carbon storage), *cultural* (recreational hunting and fishing, sports, tourism and education, heritage, other livelihoods, aesthetic and spiritual values), and *supporting* (those ecosystem components and processes that underpin all of these). It has been demonstrated how these ecosystem services could be related to human well being as follows: security (personal safety, secure access to resources, security for disasters), basic material for good life (adequate livelihood, sufficient nutritious food, shelter and access to goods), health (strength, feeling well, access to clean air and water), and good social relations (social cohesion, mutual respect and being able to help others), all underpinned by freedom of choice and action (an opportunity to be able to achieve what an individual values doing and being). The connections between ecosystem services and constituents of well being are mediated by socioeconomic status (Corvalan et al. 2005).

The global health challenge of newly emerging and re-emerging infectious diseases, the upsurge of which parallels the above accelerating changes, is a graphic example of these human–wildlife–ecosystem linkages and their ecological and evolutionary dynamics (Horwitz and Wilcox 2005). For example, a causal connection has been demonstrated linking infectious disease emergence to disruption of ecological communities and local loss of biodiversity and to the decline of water quality in terrestrial aquatic and coastal marine ecosystems (Keesing et al. 2010). In general, global and regional changes in the form of urbanization, agricultural intensification, and natural habitat degradation are seen as drivers of increased epidemic activity, virulence, and/or geographic range of zoonotic pathogen origin (Woolhouse et al. 2002; Wilcox and Colwell 2005; Wilcox and Gubler 2005).

The relationships expounded by the MEA (2005) provide a rational and objective account of ecohealth and targets for action in a sea of environmental change. However, this is insufficient to characterize the ecohealth project *writ large*, as stated by Val Brown (in Parkes 2006, p. 136):

"As it stands, the synthesis perpetuates the myth that objectivity and rationality alone can define and solve the problems that it describes, ignoring ethical and political change. It extends this bias in taking an anthropocentric approach to environmental resources, as if the main purpose of global ecological integrity was to provide a service to humankind. It also remains focused on the reduction of disease in a static political climate, bypassing the considerable work that confirms that social change and community development programs make essential contributions to human well-being."

The characteristic that perhaps sets ecohealth apart from other endeavors (like ecological epidemiology, environmental health, and one health) is an explicit focus on the social, cultural and political facets (Parkes et al. 2010).

## Transdisciplinarity

The third strand is an imperative that derives from the first two. The need to loosen the health sciences' dedication to their objectivist stance, as well as the need to face complex and uncertain problems of environmental change and human well being, demands a transcendence beyond disciplinary norms and their theoretical frameworks and methodologies (Albrecht et al. 1998).

"Real-world" problems require proper characterization and not only integration of theory and methods from multiple disciplines, but also perspectives and knowledge from citizens themselves, where problems will be formulated and with whom solutions will be found (post-normal science, *sensu* Funtowicz and Ravetz 2003). This participatory aspect, considered elemental to ecosystem approaches, was primarily developed in public health as "community-centered research," including intervention (Israel et al. 1998), and was adopted as a central component in the development of ecosystem management (Slocombe 1993). Cooperative community participatory interventions (Parkes and Panelli 2001) shift the locus of attention to place-based activities (Horwitz et al. 2001). They demand an exploration of case studies, each with a unique set of sociocultural circumstances. Ecohealth involves tacitly if not explicitly employing such transdisciplinary models as frameworks that combine understandings from not only different academic disciplines but also other knowledge systems (e.g., traditional knowledge held by local or indigenous groups). Such research and intervention goes beyond interdisciplinary research by employing participatory approaches that involve engaging with groups as well as drawing on expert knowledge outside academia or western science.

In such contexts, knowledge is legitimately derived from different value systems, different ethical bases, and different philosophical traditions, and this is where Max-Neef (2005) locates a transdisciplinary approach. He defines "transdisciplinary action" as any multiple vertical relations that include all four levels in a hierarchy, where the lower level refers to *what exists* (the descriptive disciplines, like biology), the second level refers to *what we are capable of doing* (the applications, like agriculture), the third to *what we want to do* (the predictive, like planning), and the top level to *what we must do, or rather how to do what we want to do* (actions stemming from our ethical concerns):

> "In other words, we travel from an empirical level, towards a purposive or pragmatic level, continuing to a normative level, and finishing at a value level" (Max-Neef 2005, p. 9)

Thus, the development of new, transdisciplinary approaches to research and practice lead to an ecohealth framework, beyond interdisciplinary, which is a less comprehensive vertical treatment across this hierarchy. *Transdisciplinary* projects combine interdisciplinary research with a participatory approach that is essential to establish values, ethics, and philosophies. Thus, they integrate both academic disciplines and non-academic knowledge systems, such as those represented by public sector agencies or communities, with a common goal and create new knowledge and theory.

Apparently few projects called interdisciplinary or transdisciplinary are carried out deliberately with the intention of being integrative. We contend that such projects typically are embarked upon without a plan including specific procedures aimed at achieving integration, or without a way of gauging the degree of the project's success in this regard. In fact, neither interdisciplinary nor transdisciplinary is defined in publications or reports purporting to represent such research. There is, however, a small literature outside biomedicine and public health that defines integrative research and describes the steps or procedures required.

We use *integrative* to refer to projects that are either interdisciplinary or transdisciplinary. New knowledge and theory emerges from the *integration* of disciplinary knowledge, where a project or activity seeks to bridge different knowledge cultures toward solving a real-world problem (as described above). Lele and Norgaard (2005) point out that the structure of scientific knowledge and differences in epistemologies, theories, and methods have little to do with what historically have been called disciplines. Yet the development and implementation of holistic, including interdisciplinary or transdisciplinary, approaches to research (and intervention), for the control of emerging infectious diseases, for example, have lagged behind the increasing interest.

# ECOHEALTH: INTEGRATIVE RESEARCH AND TRANSCENDING DISCIPLINES

The rationale for ecohealth articulated in the first issue of the journal *EcoHealth* centered on the increasing need for applying ecological concepts to understand and nurture sustainable human health and, conversely, applying health concepts to understand ecosystems; collectively, this was referred to as "integrative studies in ecology, health, and sustainability." The implicit goal was "to better understand the connections between nature, society and health, and how drivers of social and ecosystem change ultimately will also influence human health and well-being" (Wilcox et al. 2004). Specific issues identified as within its scope were (a) global loss of biological diversity, the well being of both animals and humans, the underlying drivers of human-induced changes in climate, habitat, and the use of terrestrial and marine ecosystems; (b) health of human populations influenced by large-scale environmental changes, the increasing gap between the rich and poor, degradation and pollution of natural resources, urbanization, and mass migration of people due to war or natural disasters; and (c) interactions between environment, development, and health critical to understanding sustainable development and the health of human populations on local, regional, and global scales.

Arguably, operationalizing research and practice that matches this scope and transdisciplinary approach represents the most significant challenge for ecohealth. Table 2.2 collates a series of observations of projects that have sought integration and might be regarded as having been successful.

Ecohealth is transdisciplinary in nature, as it employs perspectives and methods that transcend traditional disciplinary boundaries and engage both researchers and practitioners in addressing real-world problems. As shown in Table 2.2, transdisciplinarity requires team members to share roles and systematically cross, and even subvert, discipline boundaries. The primary purpose of this approach is to pool and integrate the team's expertise so that more efficient and comprehensive assessment and intervention may be provided in a determined field. The communication style in transdisciplinarity involves continuous give-and-take among all members on a regular,

Table 2.2 Identifiable Components of an Integrative Research Project and Rationale and Actions Required for Each, Representing the Operational Challenge for Ecohealth

| Component | Description | Rationale | Action |
|---|---|---|---|
| **Making integration part of the project** | Recognizing integration as part of the problem statement to be addressed by the project | Not having explicit integrative approach will lead to a set of isolated disciplinary efforts. | Formulate an integrative problem statement from a plan, preferably without identifying a disciplinary lead. |
| **A clear research question and project goal** | A transdisciplinary project develops a clear research question and project goal. | Identifying a goal enables team members to effectively integrate their different knowledge cultures. | Use a dialogical approach to determine a goal that represents meaningful and attractive aspects of each discipline. |
| **Inclusion of disciplines** | Involvement of disciplines is open and determined by the problem statement and the research goals. | Open but purposeful and relevant dialog facilitates an internally cohesive and efficient team. | Create a space and a dynamic in which relevant disciplines can become involved. |
| **An integrative theory, model, or approach** | Identification of an overarching theory or model that links disciplines | A conceptual model will highlight the epistemological challenge. | Plan and establish a process to develop a conceptual model that includes relevant theory. |
| **An operational efficacy** | Requirement for leadership, teamwork, and management | Team will have its personalities, organizations, and sectors, and its operational structure should match. | Develop an iterative approach to establish operational structure. |
| **An institutional environment conducive to collective learning** | The project's organizational base needs to learn and adapt to transdisciplinary collaboration. | Solutions will require a flexible, adequately resourced, and collective readiness and a willingness to relax institutional barriers. | Project management works from the outset and throughout the project to provide sufficient time and flexibility and to seek institutional and financial support and cooperation. |
| **A project plan** | An operational plan for integrative research | An agreed-on plan allows participants and supporters to recognize stages, products, and outcomes and the nature of success. | Develop and follow timelines, milestones, quality assurance, and criteria for assessing progress. |

planned basis. Assessment, intervention, and evaluation are carried out jointly. Transdisciplinarity brings together academic experts, field practitioners, community members, research scientists, political leaders, and business owners, among others, to solve some of the pressing problems facing the world, from the local to the global and the natural and social sciences, to address the ecology and health of species and ecosystems (Aguirre and Wilcox 2008).

The problem of theory development and what makes a "good" theory, as discussed among experimental researchers, is fundamental (Wacker 1998). The same is true of how we organize and manage research projects requiring the disciplinary integration. A more deliberative approach that draws on the lessons learned and criteria described here will be required to reverse the trend of increasing environmental problems we face today (Table 2.2).

As its orientation is ecosystemic, and with the aim of understanding the driving of immediate forces producing particular health states, ecohealth goes beyond inherently disease prevention or control, and beyond restoration of health. Controlling, preventing, or otherwise limiting zoonotic disease incidence, conventionally the domain of biomedicine and public health (human and animal), obviously requires spanning, at a minimum human, animal, and ecological health sciences. Ecohealth is broader: among other things, it encompasses health generally, not only infectious diseases; it is, in fact, where health is an emergent feature of every, and any, ecosystem.

# REFERENCES

Aguirre, A.A., and B.W. Wilcox. 2008. EcoHealth: envisioning and creating a truly global transdiscipline. EcoHealth 5:238–239.

Aguirre, A.A., R.S. Ostfeld, G.M. Tabor, C. House, and M.C. Pearl, eds. 2002. Conservation medicine: ecological health in practice. Oxford University Press, New York.

Albrecht, G., S. Freeman, and N. Higginbotham. 1998. Complexity and human health: the case for a new paradigm. Cult Med Psychiat 22:55–92.

Anderson, R.M., and R.M. May. 1991. Infectious diseases of humans: dynamics and control. Oxford University Press, Oxford, U.K.

Antonovsky, A. 1979. Health, stress, and coping. Jossey-Bass. San Francisco, California.

Antonovsky, A. 1987. Unraveling the mystery of health: how people manage stress and stay well. Jossey-Bass. San Francisco, California.

Begon, M., C.R. Townsend, and J.L. Harper. 2006. Ecology: From Individuals to Ecosystems (4th ed.). Blackwell Publishing, UK.

Berkes, F., J. Colding, and C. Folke, eds. 2003. Navigating social ecological systems: building resilience for complexity and change. Cambridge University Press, U.K.

Berkes, F., M. Kislalioglu, C. Folke, and M. Gadgil. 1998. Minireviews: exploring the basic ecological unit: ecosystem-like concepts in traditional societies. Ecosystems 1:409–415.

Boardman, R. 1981. International Organization and the Conservation of Nature. Indiana University Press, Bloomington, USA.

Callicott J.B., ed. 1987. Companion to a Sand County almanac: interpretive and critical essays. University of Wisconsin Press, Madison.

Callicott, J.B., and E.T. Freyfogle, eds. 1999. Aldo Leopold. For the health of the land: previously unpublished essays and other writings. Island Press. Washington, D. C.

Carpenter, S.R., J.J. Cole, M.L. Pace, R. Batt, W.R. Brock, T. Cline, J. Coloso, J.R. Hodgson, J.F. Kitchell, D.A. Seekell, L. Smith, and B. Weidel. 2011. Early warnings of regime shifts: a whole-ecosystem experiment. Science DOI: 10.1126/science.1203672.

Carson, R. 1962. Silent spring. Houghton Mifflin, Boston.

Carson, R. 1951. The sea around us. Oxford University Press, New York.

Chapin III, F.S., E.S. Zavaleta, V.T. Eviner, R.L. Naylor, P.M. Vitousek, H.L. Reynolds, D.U. Hooper, S. Lavorel, O.E. SalaI, S.E. Hobbie, M.C. Mack, and S. Díaz. 2000. Consequences of changing biodiversity. Nature 405:234–242.

Checkland, P.B. 1986. Systems thinking, systems practice. Wiley, Chichester, U.K.

Colborn, T., D. Dumanoski, and J.P. Myers. 1997. Our stolen future: are we threatening our fertility, intelligence and survival? A scientific detective story. Penguin Books, USA.

Cook, A., A. Jardine, and P. Weinstein. 2004. Using human disease outbreaks as a guide to multilevel ecosystem interventions. Environ Health Persp 112:1143–1146.

Cooley, J.L., and J.H. Cooley, eds. 1984. Natural diversity in forest ecosystems: proceedings of the workshop, 29 November–1 December 1982, Institute of Ecology, University of Georgia, Athens, Georgia.

Corvalan, C., S. Hales, and A. McMichael. 2005. Ecosystems and human well-being: health synthesis. World Health Organization, Geneva, Switzerland.

Costanza, R., B.G. Norton, and B.D. Haskell. 1992. Ecosystem health: new goals for environmental management. Island Press, Washington, D.C.

Dasmann, R.F. 1968. A different kind of country. Macmillan, New York.

Ellis, B., and B.A. Wilcox. 2008. The ecological dimensions of vector-borne disease research and control. Cad Saúde Pública 25:S155–S167.

Elton, C.S. 1927. Animal ecology. University of Chicago Press, Illinois.

Engel, G.E. 1977. The need for a new medical model: a challenge for biomedicine. Science 196:129–136.

Folke, C., S. Carpenter, B. Walker, M. Scheffer, T. Elmqvist, L. Gunderson, and C.S. Holling. 2004. Regime shifts, resilience, and biodiversity in ecosystem management. Annu Rev Ecol Evol S 35:557–581.

Folke, C., S. Carpenter, T. Elmqvist, L. Gunderson, C.S. Holling, and B. Walker. 2002. Resilience and sustainable development: building adaptive capacity in a world of transformations. AMBIO 31:437–440.

Funtowicz, S., and Ravetz J. 2003. Post-normal science. Internet Encyclopedia of Ecological Economics. International Society for Ecological Economics. http://www.ecoeco.org/education_encyclopedia.php (Accessed April 22, 2011)

Golley, F.B. 1993. A History of the ecosystem concept in ecology: more than the sum of the parts. Yale University Press, Connecticut.

Gunderson, L.H. 2000. Ecological resilience—in theory and application. Annu Rev Ecol Syst 31:425–439.

Gunderson, L.H., C.S. Holling, and S.S. Light. 1995. Barriers and bridges to the renewal of ecosystems and institutions. Columbia University Press, New York.

Gunderson, L.H., and C.S. Holling (eds.). 2002. Panarchy: understanding transformations in human and natural systems. Island Press, Washington, DC.

Hargrove, B. 1979. The sociology of religion: classical and contemporary approaches. AHM Pub. Corp. Arlington Heights, Illinois.

Holling, C.S. 1973. Resilience and stability of ecological systems. Annu Rev Ecol Syst 4:1–23.

Holling, C.S. 1978. Adaptive environmental assessment and management. John Wiley and Sons, New York.

Holling, C.S. 1986. The resilience of terrestrial ecosystems: local surprise and global change. In Clark, W.C., and R.E. Munn, eds. Sustainable development of the biosphere, pp. 292–397. Cambridge Press, Syndicate of the University of Cambridge, U.K.

Holling, C.S., L.H. Gunderson, and D.L. Ludwig. 2002. In quest of a theory of adaptive change. In L.H. Gunderson and C.S. Holling (eds.), Panarchy: understanding transformations in human and natural systems, pp. 3–22. Island Press, Washington, DC.

Horwitz, P., and Wilcox, B. 2005. Parasites, ecosystems and sustainability: an ecological and complex systems perspective. Int J Parasitol 35:725–732.

Horwitz, P., M., Lindsay, and O'Connor, M. 2001. Biodiversity, endemism, sense of place and public health: inter-relationships for Australian inland aquatic ecosystems. Ecosyst Health 7:253–265.

Institute of Medicine (IOM). 2001. Health and behavior: the interplay of biological, behavioral, and societal influences. National Academy Press, Washington, D.C.

Israel, B.A., A.J. Schulz, E.A. Parker, and A.B. Becker. 1998. Review of community-based research: assessing partnership approaches to improve public health. Annu Rev Publ Health 19:173–202.

Jantsch, E., and C.H. Waddington, eds. 1976. Evolution and consciousness: human systems in transition. Addison-Wesley, Arlington Heights, Illinois.

Karvonen, A. 2008. The gentle subversive: Rachel Carson, silent spring, and the rise of the environmental movement. Technol Cult 49:242–244.

Keesing, F., L.K. Belden, P. Daszak, A. Dobson, C.D. Harvell, R.D. Holt, P. Hudson, A. Joles, K.E. Jones, C.E. Mitchell, S.S. Myers, T. Bogich, and R.S. Ostfeld. 2010. Impacts of biodiversity on the emergence and transmission of infectious diseases. Nature 468:647–652.

Keller, D.R., and F.B. Golley. 2000. The philosophy of ecology: from science to synthesis. University of Georgia Press, Atlanta.

Krieger, N. 2001. Theories for social epidemiology in the 21st century: an ecosocial perspective. Int J Epidemiol 30:668–677.

Krieger, N. 2008. Ladders, pyramids and champagne: the iconography of health inequities. J Epidemiol Commun H 62:1098–1104.

Last, J.M. 1998. Public health and human ecology. McGraw-Hill Professional, New York.

Lebel, J. 2003. Health: an ecosystem approach. International Development Research Centre. Ottawa, Canada

Leisher, C. 2008. What Rachel Carson knew about marine protected areas. Bioscience 58:478–479.

Lélé, S., and R. B. Norgaard. 2005. Practicing interdisciplinarity. BioScience 55:967–975.

Leopold, A. 1933. Game Management. Charles Scribner's Sons, New York.

Leopold, A. 1941. Wilderness as a land laboratory. Living Wilderness 6:3.

Leopold, A. 1949. A Sand County almanac. Oxford University Press, New York.

Levin, S.A. 1999. Fragile dominion: complexity and the commons. Perseus Books Group, Reading, Massachusetts.

Lovejoy, T.E. 1980. Foreword. *In* Soulé, M E., and B.A. Wilcox, eds. Conservation biology: an evolutionary-ecological perspective, pp. ix–x. Sinauer Associates, Sunderban, Massachusetts.

Mageau, M.T., R. Costanza, and R.E. Ulanowicz. 1995. The development and initial testing of a quantitative assessment of ecosystem health. Ecosyst Health 1:201–213.

Max-Neef, M.A. 2005. Foundations of transdisciplinarity. Ecol Econ 53:5–16.

McDonnell, M.J., and S.T.A. Pickett, eds. 1993. Humans as components of ecosystems: the ecology of subtle human effects and populated areas. Springer-Verlag, New York.

McLeroy, K.R, D. Bibeau, A. Steckler, and K. Glanz. 1988. An ecological perspective on health promotion programs. Health Educ Behav 15:351.

Millennium Ecosystem Assessment. 2005. Ecosystems and human well-being: synthesis. Island Press, Washington, DC.

Neudoerffer, R. Cynthia, D. Waltner-Toews, J. Kay, D.D. Joshi, and M.S. Tamang. 2005. A diagrammatic approach to understanding complex eco-social interactions in Kathmandu, Nepal. Ecol Soc 10:12. http://www.ecologyandsociety.org/vol10/iss2/art12/

Nicolis, G., and I. Prigogine. 1977. Self-organization in nonequilibrium systems: from dissipative structures to order through fluctuations. Wiley and Sons, New York.

Odum, E.P. 1971. Fundamentals of ecology. Saunders, Philadelphia.

Odum, E.P., and G.W. Barrett. 2005. Fundamentals of ecology. Thomson Brooks/Cole, Belmont, California.

Olsson, P., C. Folke, and F. Berkes. 2004. Adaptive comanagement for building resilience in social-ecological systems. Environ Manage 34:6.

O'Neill, R.V. 2001. Is it is time to bury the ecosystem concept? (with full military honors). Ecology 82:3275–3284.

Ostfeld, R.S., F. Keesing, and V.T. Eviner. 2010. Infectious disease ecology: effects of ecosystems on disease and of disease on ecosystems. Princeton University Press, New Jersey.

Parkes, M.W. 2006. Personal commentaries on "Ecosystems and human well-being: health synthesis—a report of the Millennium Ecosystem Assessment." EcoHealth 3:136–140

Parkes, M.W., and P. Horwitz. 2009. Water, ecology and health: exploring ecosystems as a setting for promoting health and sustainability. Health Promot Int 24:94–102.

Parkes, M.W., and R. Panelli. 2001. Integrating catchment ecosystems and community health: the value of participatory research. Ecosyst Health 7:85–106.

Parkes, M., K. Morrison, M. Bunch, L. Hallstrom, C. Neudoerffer, H. Venema, and D. Waltner-Toews. 2010. Towards integrated governance for water, health and social–ecological systems: the watershed governance prism. Global Environ Chang 20: 693–704

Pickett, S.T.A., and R.S. Ostfeld. 1995. The shifting paradigm in ecology. *In* Knight, R.L., and S.F. Bates eds. A new century for natural resources management, pp. 261–278. Island Press. Washington, D.C.

Rapport, D.J., C. Gaudet, J.R. Karr, J.S. Baron, C. Bohlen, W. Jackson, B. Jones, R.J. Nalman, B. Norton, and M.M. Pollock. 1998a. Evaluating landscape health: integrating societal goals and biophysical process. J Environ Manage 53:1–15.

Rapport, D. J., R. Costanza., and A. McMichael. 1998b. Assessing ecosystem health: challenges at the interface of social, natural, and health sciences. Trends Ecol Evol 13:397–402.

Rapport, D.J., H.A. Regier, and T.C. Hutchinson. 1985. Ecosystem behavior under stress. Am Nat 125:617–640.

Rodriguez, J.P., K.M. Rodriguez-Clark, J.E.M. Baillie, N. Ash, J. Benson, T. Boucher, C. Brown, N.D. Burguess, B. Collen, M. Jennings, D.A. Keith, E. Nicholson, C. Revenga, B. Reyers, M. Rouget, T. Smith, M. Spalding, A. Taber, M. Walpole, I. Zager, and T. Zamin. 2010. Establishing IUCN red list criteria for threatened ecosystems. Conserv Biol 25:21–29.

Scheiner, S.M. 2009. The intersection of the sciences of biogeography and infectious disease ecology. EcoHealth 6:483–488

Slocombe, D.S. 1993. Implementing ecosystem-based management. BioScience 43:612–622.

Soulé, M.E., and B. A. Wilcox. 1980. Conservation biology: an evolutionary-ecological perspective. Sinauer Associates, Sunderland, Massachusetts.

Stewart M, W.W. Weston, J.B. Brown, I.R. McWhinney, C.L. McWilliam, and T.R. Freeman. 1995. Patient-centered medicine: transforming the clinical method. Sage Publications, Thousand Oaks, California.

Tress, B., G. Tress, and G. Fry. 2006. Ten steps in successful knowledge integration. *In* Tress, B., G. Tres, G. Fry, and P. Opdam, eds. Vol. 12 From landscape research to landscape planning: aspects of integration, education and application. Wageningen UR Frontis Series, Netherlands, pp. 792–807.

UNESCO. 1981. MAB activities on conservation with particular reference to biosphere reserves. United Nations Educational Scientific and Cultural Organization, International Coordinating Council of the Program on Man and the Biosphere. Seventh Session, 30 September–2 October. Washington, D.C.

Vitousek, P.M., H.A. Mooney, J. Lubchenco, and J.M. Melillo, J. M. 1997. Human domination of Earth's ecosystems. Science 277:494–499.

Wacker, J.G. 1998. A definition of theory: research guidelines for different theory-building research methods in operations management. J Oper Manag 16:361–385.

Walters, C.J. 1986. Adaptive management of renewable resources. McMillan, New York.

World Health Organisation. 1986. Ottawa Charter for Health Promotion. Geneva, Switzerland.

Wilcox, B.A. 1984a. Concepts in conservation biology: applications to the management of biological diversity. *In* Cooley, J.L., and J.H. Cooley. Natural diversity in forest ecosystems: proceedings of the workshop, 29 November–1 December 1982, Athens, Georgia. Institute of Ecology, University of Georgia.

Wilcox, B.A. 1984b. In situ conservation of genetic resources: determinants of minimum area requirements. *In* McNeely J.A. and K.R. Miller, eds. National parks, conservation and development, pp. 18–30. Proceedings of the World Congress on National Parks, Smithsonian Institution Press, Washington, D.C.

Wilcox, B.A., and R.R. Colwell. 2005. Emerging and re-emerging infectious diseases: biocomplexity as an interdisciplinary paradigm. EcoHealth 2: 244–257.

Wilcox, B.A., and D.J. Gubler. 2005. Disease ecology and the global emergence of zoonotic pathogens. Environ Health Prevent Med 10:263–272.

Wilcox, B., and C. Kueffer. 2008. Transdisciplinarity in ecohealth: status and future prospects. EcoHealth 5:111.

Wilcox, B.A., A.A. Aguirre, P. Daszak, P. Horwitz, P. Martens, M. Parkes, J.A. Patz, and D. Waltner-Toews. 2004. EcoHealth: A transdisciplinary imperative for a sustainable future. EcoHealth 1:3–5.

Wilson, E.O. 2005. Foreword. *In* Odum, E.P., and G.W. Barrett. Fundamentals of ecology, pp xv–xvi. Thomson Brooks/Cole, Belmont, California.

Woolhouse, M.E.J. 2002. Population biology of emerging and re-emerging pathogens. Trends Microbiol 10:S3–S7.

Worthington, E.B. 1975. The evolution of IBP. Cambridge University Press, New York.

# 3

## ONE HEALTH, ONE MEDICINE

## Laura H. Kahn, Thomas P. Monath, Bob H. Bokma, E. Paul Gibbs, and A. Alonso Aguirre

In recognition that the health of humans, animals, and the environment is linked, One Health seeks to increase communication and collaboration across the disciplines in order to promote, improve, and defend the health of all species on the planet. This strategy may seem simple, but unfortunately it will not be easy to implement. The explosion of medical knowledge in the 20th century led to academic, governmental, and industrial silos of specialization; these silos fostered a compartmentalized approach to health and disease. Building bridges across these silos will require leadership, joint educational programs, financial support, and other strategies that promote transdisciplinary efforts.

Before the 20th century, physicians typically worked with veterinary medical colleagues and others to improve the health of humans and animals. This chapter will describe the historical developments in medicine and veterinary medicine leading to the current status quo. It will provide examples of why the status quo is problematic and will highlight the challenges in changing the present paradigm. It will conclude with recommendations on how to implement a One Health approach in the future.

### HISTORICAL BACKGROUND

Humans have been domesticating wild animals beginning with dogs since 14,000 years BC (Trut 1999), developing agriculture, and altering the environment. In contrast to the harsh nomadic hunter-gatherer lifestyle, most humans preferred the secure and productive lifestyle that agriculture allowed. However, this novel lifestyle introduced unanticipated health risks since aggregated crops and concentrated livestock altered the interactions of humans, domestic animals, wildlife, and ecosystems. Moreover, humans lived in close proximity to animals and sometimes shared living quarters (McNeill 1977).

Small farming communities eventually grew into villages, towns, and cities, which concentrated humans into dense living conditions that facilitated the spread of microorganisms from individual to individual, allowing infectious disease epidemics to develop and propagate. As a result, infectious diseases, such as sylvatic plague, smallpox, cholera, and malaria, began to afflict humans, leading to epidemic morbidity and mortality (McNeill 1977).

Some of the diseases affecting agricultural and urbanized societies came from humans or livestock after domestication, such as bovine tuberculosis, rabies, and a wide array of food-borne bacterial and protozoan infections as transmissible zoonoses (Diamond 1999). In addition, wildlife served as reservoirs of innumerable diseases that could be transmitted back to humans and domestic animals. For example, nearly one quarter to one third of the population of Europe was decimated by plague, also called "the Black Death" during the mid-14th century (Wheelis 2002).

To complicate matters, people thought epidemics were caused by divine retribution for lapsed moral behavior, bad air "miasmas," and demons and other spirit beings, among other etiologies (Conrad 1992; De Paolo 2006). These beliefs lasted for centuries, hindering effective preventive and control efforts. However, despite a lack of understanding of infectious diseases, some individuals developed effective control measures.

For example, during the 18th century, rinderpest, a deadly viral disease of cattle, was devastating the human food supply. Pope Clement XI asked Dr. Giovanni Maria Lancisi, his personal physician, to combat the problem. Lancisi recommended that all of the ill and suspect animals be killed and buried in lime, since he suspected that the disease was communicable. His concept proved effective, and in 1762, the first school of veterinary (from the Latin "beast of burden") medicine was established in Lyon, France, to educate the next generation about the management of diseases in livestock (Palmarini 2007).

## BEGINNINGS OF SCIENTIFIC BREAKTHROUGHS AND ONE HEALTH, ONE MEDICINE

In 1827, Charles Darwin decided to leave medical school at the University of Edinburgh to pursue studies in religion and natural history at Cambridge University. Health practitioners in Darwin's time were routinely trained in natural history and zoology since these disciplines were closely aligned and were considered integral subjects in medical training. Darwin never completed medical school, but his experience aboard the HMS Beagle, and most likely his exposure to multiple disciplines, led to his publishing his

monumental book in 1859, *On the Origin of Species* (Leff 2000).

Rudolf Virchow (1821–1902), the German physician and pathologist, coined the term "zoonosis" and said, "between animal and human medicine there are no dividing lines—nor should there be." He strongly supported veterinary medicine and advocated for public health meat inspections throughout Europe. The United States eventually adopted meat inspections as well. This novel practice served as the basis for modern-day public health meat and poultry inspections by veterinarians (Kahn et al. 2007).

Sir William Osler (1849–1919), first Professor of Medicine at Johns Hopkins Hospital and considered the "father of modern medicine," had traveled from Canada to Germany to study with Virchow. Virchow impressed upon his student the importance of autopsies, pathology, and scientific methodologies. Osler returned to Canada to teach parasitology, physiology, and pathology at the Montreal Veterinary College, which eventually became affiliated with the medical school at McGill University. At the veterinary college, Osler researched hog cholera (classical swine fever), Pictou cattle disease caused by tansy ragwort (*Senecio jacobaea*) intoxication, which was believed to be a microbial infection at that time, and verminous bronchitis of dogs, among others. He worked closely with veterinarians such as Albert W. Clement, who became the President of the United States Veterinary Medical Association (USVMA; Kahn et al. 2007).

Louis Pasteur (1822–1895) and Robert Koch (1843–1910) changed the course of history by discovering that microscopic organisms caused disease. This knowledge allowed the development of effective preventive and control measures against pathogens. Pasteur developed a vaccine against rabies, and Koch discovered that *Clostridium tetani* caused tetanus, *Streptococcus pneumoniae* caused pneumonia, and *Vibrio cholerae* caused cholera (Munch 2003).

The advances in scientific knowledge spurred efforts to improve medical education. The American Medical Association (AMA) invited the Carnegie Foundation for the Advancement of Teaching to conduct a study on the status of medical education in the United States. In 1910, Abraham Flexner published the report that recommended that medical education be modeled after that at Johns Hopkins University, which emphasized a scientific approach to medical education and patient care. The ultimate effect of incorporating medical schools into universities was the emphasis on training medical specialists rather than general practitioners (Starr 1982).

The idea of an American veterinary profession was supported by agricultural societies and by physicians such as Benjamin Rush and Andrew Stone. Before the 1880s, most school-trained veterinarians were trained in Europe. The development of veterinary schools in the United States arose from concerns over animal disease epidemics following the Civil War and the interest in scientific agriculture signaled by the Morrill Land Grant Act of 1862, which provided federal funding to establish the first college-affiliated veterinary school at Iowa State University in 1879. The curriculum derived from agriculture and veterinary medicine. The University of Pennsylvania's School of Veterinary Medicine opened in 1884 and was the first accredited veterinary medical college in the United States whose origin was in medicine rather than agriculture.

By the late 19th century, a web of veterinary institutions, organizations, and periodicals were established, including the USVMA, founded in 1863 and renamed the American Veterinary Medical Association (AVMA) in 1898; the Bureau of Animal Industry (BAI), created in 1884 in the U.S. Department of Agriculture (USDA) and headed until 1905 by veterinarian Daniel E. Salmon; and the *American Veterinary Review*, begun in 1877 and renamed the *Journal of the American Veterinary Medical Association* in 1914–1915.

BAI veterinarians Fred L. Kilbourne and Cooper Curtice and physician Theobald Smith first demonstrated the role of vectors in the transmission of animal diseases. The BAI also certified and employed veterinarians in food inspection and influenced veterinary medical school curricula. Between the 1880s and 1925, graduate veterinarians sponsored state laws creating examining boards and setting graduation and licensing requirements. In late 19th and early 20th centuries, Leonard Pearson, a bovine tuberculosis expert, directed attention to the relationship between animal and human health (Palmer and Waters 2011).

As the 20th century progressed, physicians became increasingly specialized and collaborative efforts with veterinarians waned. Human and animal diseases were largely treated as separate entities. However, a few veterinarians, such as Calvin W. Schwabe (1927–2006), the renowned veterinary epidemiologist and parasitologist, continued to promote a unified human

and veterinary approach to zoonotic diseases by publishing his book *Veterinary Medicine and Human Health* (Schwabe 1984).

This need to work together has not diminished despite the professions drifting apart. Since 1940, over 330 infectious diseases have emerged from animals into human populations (Taylor et al. 2001). The threat to global health is increasing since human population density, the most significant independent predictor of disease emergence, continues to increase (Jones et al. 2008). Indeed, it is estimated that by 2050, the human population will reach 9 billion (United Nations 2007).

Human activities such as deforestation, intensive agriculture, bushmeat consumption, waste production, and greenhouse gas emissions will only intensify as growing populations demand more food, water, clothing, shelter, and energy. For example, surveillance of fruit bat health and behavior in Malaysia might have helped prevent the disaster that developed in 1998–99 after extensive deforestation destroyed the fruit bat's habitat. Millions of hectares of tropical rain forest were slashed and burned to make way for pig farms. Fruit bats (the natural reservoir of the virus), whose habitat was largely destroyed by deforestation, sought nourishment from fruit trees near the pig farms and subsequently spread the virus to livestock. The subsequent Nipah virus outbreak demonstrates that destruction of wildlife habitats has an adverse impact on livestock and human health. In this case, the flowering and fruiting trees that the fruit bats relied on for their survival were destroyed to make room for pig farms. The bats resorted to consuming fruit located next to the farms. Pigs ate the partially eaten fruit that had been contaminated by bat saliva and urine. The bats harbored the Nipah virus, a previously unknown pathogen. The economic and human health impact of the outbreak was severe: the pig farmers lost millions, and pig farming in the country largely collapsed and is now allowed in only approved areas. This set off a chain reaction that ultimately led to the development of encephalitis in hundreds of humans and over 100 fatalities (Kahn 2011).

The magnitude of the problem illustrates why human medicine, veterinary medicine, and ecology need to rejoin forces. Taylor et al. (2001) identified 1,415 infectious agents and determined that 868 (61%) were zoonotic. They found that zoonotic diseases were twice as likely to be newly emerged infections compared to other diseases. RNA viruses, in particular, are highly likely to emerge from animals and cross species barriers because they are subject to rapid mutagenesis and can readily adapt to new hosts and vectors. Examples include West Nile virus (WNV), avian influenza virus, SARS coronavirus, arenaviruses, and hantaviruses (Cleaveland et al. 2001).

## CHALLENGES AND OPPORTUNITIES OF IMPLEMENTING A NEW PARADIGM

A new paradigm requires that human, animal, and ecosystem health be addressed equally, equitably, and expeditiously. Ironically, to address future threats, we need look no further than what the medical and scientific luminaries of the 19th century developed: the One Health concept. The One Health concept seeks to integrate human, animal, and ecological health in clinical practice, public health, scientific research, and policy. Some professional organizations have recognized the importance of this paradigm. In September 2004, experts at the Wildlife Conservation Society held a "One World, One Health" conference in New York City that led to the "Manhattan Principles" calling for an international, interdisciplinary approach to protect life on the planet (Cook et al. 2004). In June 2007, the AMA House of Delegates unanimously approved a "One Health" resolution following AVMA input endorsing interdisciplinary collaboration with AVMA (Kahn et al. 2008). Then AVMA approved a similar "One Health" resolution (*JAVMA*, 2009). Other organizations that have endorsed the One Health concept include the American Society of Tropical Medicine and Hygiene, the Society for Tropical Veterinary Medicine, the American Society for Microbiology, and the Council of State and Territorial Epidemiologists.

The mission statement for the "One Health" initiative states: "Recognizing that human and animal health are inextricably linked, 'One Health' seeks to promote, improve, and protect the health and well-being of all species by enhancing cooperation and collaboration between physicians, veterinarians, epidemiologists, public health professionals and allied health scientists by promoting strengths in leadership

and management to achieve these goals." Three overarching goals are enhancing public health effectiveness, understanding anthropogenic changes and the emergence of new pathogens of animal and human origin, and accelerating biomedical research discoveries, including advances in clinical medical and surgical approaches. In June 2009, the AVMA's One Health Commission was incorporated with the mission of developing strategies to put the One Health concept into practice. The challenges are many, but the rewards would be a healthier future for humans, animals, and the Earth's ecosystems.

A similar approach was expounded by Aguirre et al. (2002), who emphasized the need to bridge disciplines, thereby linking human health, animal health, and ecosystem health under the paradigm that "health connects all species on the planet." Conservation medicine embraces the One Health concept by applying a transdisciplinary approach to the study of the health relationships between humans, animals, and ecosystems. Conservation medicine is closely allied with and primarily concentrates on the values of conservation biology by recognizing that health and disease are fundamentally related to the integrity of ecosystems. Therefore, it draws on the principles of both ecology and applied medicine in its approach to health and disease. The international peer-reviewed journal *EcoHealth* was launched in 2004 and focuses on the integration of knowledge at the interface between ecological, human, and veterinary health sciences and ecosystem sustainability. This publication, among others, links disciplines and focuses attention toward "One Health, One Medicine" (Bokma et al. 2008; Mackenzie and Jeggo 2011).

## CHALLENGES AND OPPORTUNITIES IN MEDICINE AND VETERINARY MEDICINE

There are a number of challenges in implementing the One Health concept in human and veterinary medical education and practice. First, worldwide there are a disproportionate number of accredited medical schools compared to veterinary medical schools. There are 125 accredited medical schools compared to only 29 veterinary medical schools in the United States, and only a handful of them share campuses.

Globally, there are approximately 2,161 medical schools operating in 172 countries as of 2009. These international medical schools, recognized by their respective governments, might not necessarily meet each other's standards (Bokma et al. 2008; Foundation for Advancement of International Medical Education and Research 2009).

There are five colleges of veterinary medicine in Canada (four fully accredited and one with limited accreditation) and 29 in the United States (25 fully accredited and four with limited accreditation) fulfilling AVMA standards. In addition, the AVMA Educational Commission for Foreign Veterinary Graduates (ECFVG) Veterinary Schools of the World lists 471 colleges of veterinary medicine and animal sciences in 109 countries. The majority have either not been evaluated by the AVMA or do not have comparable standards to meet AVMA accreditation, and only nine (Australia [three], Scotland [two], and England, Ireland, Netherlands, and New Zealand) fulfill AVMA standards.

The ECFVG does not represent this as a comprehensive list of all veterinary schools in the world. For example, Brazil has 46 veterinary colleges listed, but as of September 2009 there are more than 108 schools, and this may be the case for other countries. The AVMA list includes all schools listed by the World Health Organization in its 1991 *World Veterinary Directory* and in the 1983 Pan American Health Organization publication *Diagnosis of Animal Health in the Americas*. The list includes additional schools that have come to the attention of the ECFVG for reasons related to certification.

Why would foreign medical and veterinary medical colleges want to comply with AMA or AVMA standards? Global needs differ. For example, cattle production and intensification have been major concerns in developing countries. In contrast, in the developed world, canine medicine and exotic medicine have been of primary interest. Unless international educational standards are developed, it might be hard to convince many countries to accept U.S. standards as a baseline.

From a purely logistical standpoint, increasing communication and collaboration between students of these professions would be difficult since there are not as many schools of veterinary medicine, and of those that exist, relatively few are close enough to medical schools to facilitate meaningful educational

and collaborative efforts. During 2009, the World Animal Health Organization (OIE) released *Veterinary Education for Global Animal and Public Health* (Walsh 2009), which is devoted to the improvement of student education in global animal and public health. The main concern expressed by this and other publications is to determine how this education can be achieved within an already packed curriculum.

One solution might be to establish One Health Institutes in various geographic locations globally that would bring together medical and veterinary medical students for cross-species disease teaching, information-sharing, and problem-solving. For example, the Centers for Disease Control and Prevention (CDC) established a One Health program, and two veterinary colleges (UC-Davis and UM-Minneapolis) have established One Health programs within their curriculum. The trend continues to grow, and these partnerships may encourage medical and veterinary medical schools to establish "sister" institutional ties and allow their students to spend elective time at the designated sister school for courses not available at their home institution.

This arrangement could facilitate building bridges and filling gaps in areas that medical and veterinary medical schools might not emphasize. For example, medical schools do not emphasize public and environmental health, exotic pathogens, or the ecology of zoonotic diseases. In contrast, veterinary medical teaching is much more concerned with exotic pathogens (which threaten livestock if introduced), diseases affecting multiple species, and the effects of environmental health on livestock production. The lack of teaching of zoonoses in medical schools might explain why physicians are generally not comfortable discussing zoonotic disease risks with their patients (Grant and Olsen 1999).

Evidence suggests that infectious agents can jump from animals to humans and vice versa (Childs et al. 2007; CDC 2008). One bacterium of particular concern is methicillin-resistant *Staphylococcus aureus* (MRSA), which causes serious community-acquired soft-tissue and skin infections (Fridkin et al. 2005), as well as hospital-acquired infections and deaths (Klein et al. 2007). Scott et al. (2009) found that households with cats were almost eight times more likely to have MRSA on one or more household surfaces than those without cats. Members of the households in the study did not have a history of infections or antibiotic use.

The authors recommended that further study was needed to determine if MRSA cross-contamination was occurring between humans, pets, and household surfaces. Studies assessing pathogen transmission in home settings are critical for furthering our understanding of microbial dynamics and would help in developing strategies to reduce disease. Since millions of families own pets or share their homes with animals, research to prevent the spread of pathogens in homes should be given priority, especially since many pathogens are developing antibiotic resistance.

## CHALLENGES AND OPPORTUNITIES IN PUBLIC HEALTH

The WNV outbreak in New York City highlights why disease surveillance of animals is as important as disease surveillance in humans in protecting public health. This outbreak illustrates that government agencies must seamlessly integrate human and animal disease surveillance efforts. In late May 1999, residents in Queens, New York, noticed dead and dying birds, and some were brought to the local veterinary clinic. The veterinarians noted that the birds had unusual neurological signs; unfortunately, no local or state agency took responsibility for the large wildlife die-off, so nothing was done to determine why these animals were dying (U.S. General Accounting Office 2000). A month later, an infectious disease specialist at Flushing Hospital admitted eight patients with encephalitis. Three patients died and CDC found that their brain tissue contained flavivirus antigen. These were later confirmed as the first human cases of WNV in the Western Hemisphere (Asnis et al. 2000).

Before and concurrent with the human disease outbreak, exotic birds at the Bronx Zoo were noted to have died. The veterinary pathologist noted that the birds exhibited tremors, loss of coordination, and convulsions. Upon necropsy most birds had brain hemorrhages and/or meningitis similar to the human cases. Tissues from these birds were sent to the USAMRIID laboratories, where isolated viruses were sent to CDC, and WNV was diagnosed by PCR and DNA sequencing (CDC 1999). Concurrently, a group of investigators at the University of California at Irvine also used molecular techniques to show that the offending agent was WNV (Briese et al. 1999). This was the first time that the virus had appeared in the Western Hemisphere (Mahon 2003).

In response to WNV emergence, CDC established ArboNET, a cooperative surveillance system that monitors the geographic spread of WNV in mosquitoes, birds, other animals, and humans (Marfin et al. 2001). ArboNET has provided an invaluable system for tracking the spread of WNV across the United States and identifying early activity in mosquitoes and birds (CDC 2008). This surveillance system demonstrates that monitoring disease activity in arthropod vectors, animals, and humans is invaluable in tracking zoonotic disease spread and in developing successful containment and preventive strategies.

Unfortunately, surveillance of zoonotic diseases on a wider scale might be more difficult to implement. In the United States, reporting of animal diseases varies from state to state. Some states have one agency, typically departments of agriculture, responsible for domestic animal disease surveillance, while others split reporting of animal diseases between different agencies. Wildlife on non-federal lands in the United States is generally owned by the states. In some states, local public health agencies are supposed to receive reports of zoonotic diseases, primarily rabies, from veterinarians (Kahn 2006).

At the national level, surveillance of animal health is hindered because responsibility is split between many different government agencies: USDA, U.S. Department of Health and Senior Services, U.S. Department of Interior, U.S. Department of Homeland Security (USDHS), and U.S. Department of Commerce (National Academy of Sciences 2005). The USDA's Animal and Plant Health Inspection Service (APHIS) is the lead agency for livestock health and compiles disease surveillance data that are reportable to Food and Agriculture Organization (FAO) and OIE. However, there is no comparable CDC for all animals, including pets, wildlife, and zoo animals, so there are no comprehensive data available like in human disease surveillance.

At the federal level, one agency is primarily responsible for human health: the U.S. Department of Health and Human Services (USDHS). The USDHS has a subsidiary role in human health, and the U.S.

Department of Defense provides support in times of crisis, such as USAMRIID laboratory expertise during the WNV crisis. State and local governments have primary responsibility for disease surveillance in humans, and they vary in infrastructures and capabilities (Institute of Medicine 2003). They provide data to the CDC, which compiles the information on a regular basis. The CDC serves primarily as a resource for state and local health departments. The USDA is in charge of domestic animal and captive wildlife health; however, several agencies are responsible for wildlife, depending on the animal's status as a migratory or non-migratory species.

Animal health and disease surveillance are also fragmented at the international level. WHO has primary responsibility for human health and has a significant presence in UN member countries. The mission of FAO is to promote agriculture and alleviate hunger and offers limited animal health expertise to member countries. The OIE has animal health expertise, but has only a 40-person staff and no specific country presence (Institute of Medicine 2009). The OIE's primary role is in the coordination of information, and it has an early warning system for member countries. It does not have the mandate to be physically present in countries or supportive in terms of funding. These three entities are the primary players in global domestic animal health. Although they work together, their different missions, functions, and levels of support limit collaborative efforts. For example, since the OIE is not part of the UN and has a small staff and budget, it does not have the capacity to assume a role analogous to WHO's role for human health. Furthermore, none of the three has significant staff or resources focused on wildlife or ecosystem health.

The Institute of Medicine (2009) recognized that a lack of comprehensive, integrated human and animal disease surveillance systems, both in the United States and internationally, impedes an early warning system of emerging zoonotic diseases. International systems need surveillance programs and diagnostic laboratory capacities, but these are limited in developing countries, where most of the zoonotic diseases have emerged. A centralized coordinating body would be important in developing, harmonizing, and implementing integrated international human and animal health surveillance activities.

## CHALLENGES AND OPPORTUNITIES IN ECOLOGICAL HEALTH

The importance of ecological health was illustrated by the highly pathogenic avian influenza (HPAI) H5N1 outbreak in Hong Kong in 1997. Surveillance of wild waterfowl and domestic poultry in southern China during the preceding decades facilitated the early recognition of the virus in humans (Shortridge et al. 2003). In May 1997, H5N1 was isolated in a three-year-old boy who died of acute pneumonia respiratory distress syndrome (ARDS) and Reye syndrome. The isolation of this distinct avian virus subtype from a human signaled the beginning of a potentially deadly pandemic (deJong et al. 1997). By December 1997, the outbreak prompted slaughtering of all poultry in Hong Kong and introducing import control of poultry from mainland China, supervised cleaning of poultry farms, and increased surveillance of disease spread in humans and birds (Tam 2002).

These actions halted the outbreak. Unfortunately, six years later, the virus reappeared in humans in the Fujian province of China (Writing Committee of the WHO Consultation on Human Influenza 2005). In Southeast Asia, H5N1 outbreaks began in December 2003, devastating the poultry industries in the affected countries (Kuiken et al. 2005; see Chapter 16 in this volume). From 2003 to September 2009, a total of 442 laboratory-confirmed human cases were reported from 15 countries, with 262 (60%) fatalities (WHO 2009). Pathogen surveillance in wildlife was minimal to non-existent. Kuiken et al. (2005) recommended a joint expert working group to design and implement a global animal surveillance system for zoonotic pathogens. In November 2005, FAO, OIE, WHO, and World Bank officials met to discuss the worsening H5N1 HPAI crisis and agreed that surveillance systems for human and animal influenza were critical for effective responses. Veterinary infrastructures in many countries needed to be assessed and strengthened to meet OIE standards, countries needed to improve their laboratory and rapid response capabilities, and funding and investments in these efforts were urgently needed (Jong-Wook 2005).

In 2006, two animal surveillance systems were launched: the Global Early Warning and Response

System for Major Animal Diseases including Zoonoses (GLEWS) and Global Avian Influenza Network for Surveillance (GAINS). The revised 2005 International Health Regulations (IHR) require nations to notify WHO, within 48 hours, of all events that might constitute a public health emergency of international concern. WHO also has a Global Outbreak Alert and Response Network (GOARN) that shares technical expertise, supplies, and support to help coordinate outbreak response investigations. Similar to the IHR legal framework supporting WHO's central role in collecting global public health information, the OIE's Terrestrial Animal Health Code requires that member countries notify OIE within 24 hours of an animal disease event of international concern. FAO has an early warning system, Emergency Prevention System for Transboundary Animal Diseases (EMPRES), established in 1994, that collects data from a variety of sources, including from OIE, to monitor for events of concern. The goal of GLEWS is to combine the WHO, OIE, and FAO data collection systems into a joint effort to facilitate communication and collaboration between human and animal health.

Unlike GLEWS, GAINS conducts active surveillance of all strains of avian influenza in wild bird populations. Sponsored by the U.S. Agency for International Development (USAID) and the CDC, GAINS started in 2006 and is administered by the Wildlife Conservation Society. Dozens of partner institutions collaborate in the GAINS network to survey wild bird populations and collect and analyze samples from wild birds either non-invasively or from capture and release. All data, including denominator data, species and sample ownership, are publicly available via a shared, open database. This early warning system allows health officials to understand the distribution of influenza viruses as well as wild birds in country and in neighboring countries.

Much more should be done to monitor diseases in wildlife and domestic animals. There is no one international governmental agency that conducts comprehensive ecological surveillance and monitoring of diseases in animals (Karesh and Cook 2005). Even worse, many wild animals are exported from countries that conduct little or no surveillance of the pathogens they might harbor (Marano et al. 2007).

In response to a monkey pox outbreak introduced in the United States by importation of Giant Gambian rats (*Cricetomys* sp.), the CDC and the U.S. Food and Drug Administration (FDA) jointly issued an order prohibiting the importation of African rodents and banned the sale, transport, or release of prairie dogs or six specific genera of African rodents in the United States. The joint order was subsequently replaced by an interim final rule, which maintains the restrictions on African rodents, prairie dogs, and other animals. Unfortunately, the global trade in wildlife continues and poses serious threats to infectious disease ecology (GLEWS 2006; Smith et al. 2009; see Chapter 11 in this volume). There are many challenges of improving ecological health through disease surveillance of wildlife. A One Health approach involving many parties, including human and animal health professionals, modelers, ecologists, sociologists, anthropologists, and others, would help provide comprehensive, coordinated, and cohesive strategies in addressing this immense problem.

## CHALLENGES AND OPPORTUNITIES IN BIOMEDICAL RESEARCH

Society would benefit if more biomedical research was done in comparative medicine. Comparative medicine is not a new academic discipline: the first chair was established in 1862 in France (Wilkinson 1992). Comparative medicine is the study of the anatomical, physiological, pharmacological, microbiological, and pathological processes across species. A long history of collaborations between veterinarians and physicians has been documented. For example, in the 20th century, Dr. Rolf Zinkernagel, a physician, and Dr. Peter Doherty, a veterinarian, won the 1996 Nobel Prize in physiology or medicine for their discovery of how normal cells are distinguished from virus-infected cells by a body's immune system (Zinkernagel and Doherty 1974). These discoveries illustrate that cross-disciplinary collaborations help generate new scientific insights in disease.

Unfortunately, evidence suggests that the next generations of physicians and veterinarians are not collaborating with each other, and they are losing interest in pursuing careers in research. From 1970 to 1997, the number of physician-scientists receiving National Institutes of Health (NIH) grants diminished in proportion to doctoral recipients who seek and obtain funding (Rosenberg 1999). Compared to the 1980s, there are now 25% fewer physician-scientists in medical school faculties (Varki and Rosenberg 2002). To counter these trends, the NIH in 2002 established a series of competitive loan repayment programs that provide at least two years of tax-free debt relief for young physician-scientists committed to clinically oriented research training. Private foundations, such as Burroughs-Wellcome and the Howard Hughes Medical Institute, have created awards for new physician-scientists engaged in patient-oriented research. Some hospitals and medical schools are creating programs to encourage medical students to pursue research before and after receiving their medical degrees (Ley and Rosenberg 2005).

The situation is dire for veterinarian-scientists. A 2004 National Academy of Sciences (NAS) report found that the total number of veterinarians who received NIH grant support is small. In 2001, veterinarian principal investigators received only 4.7% of all NIH grants for animal research, since the NIH does not fund veterinary research, only research that is of benefit to humans. An apparent consequence of the lack of research funding available to veterinarians is that less than 1% of AVMA members are board-certified in laboratory animal medicine and less than 2% are board-certified in pathology (National Research Council 2004). Much could be done to reverse these trends. First, NIH and private foundation support for young physicians and veterinarians interested in pursuing research careers must be strengthened. Nowhere in the NIH's plans to improve biomedical research in the 21st century are comparative medicine and the importance of veterinarians mentioned, even though one of its primary goals is to foster interdisciplinary research, encouraging new pathways to discovery (Zerhouni 2003). The NIH must recognize that animal health influences human health and must be supported accordingly. Jointly sponsored comparative medicine research grants from the National Center for Research Resources (NCRR) and other institutes, such as the National Institute of Allergy and Infectious Diseases (NIAID) and the National Cancer Institute, should be offered to medical and veterinary medical research teams to promote collaborative efforts (National Research Council 2005a,b). Further, some veterinary education reimbursement funding has recently been made available by the U.S. government in the National Veterinary Medical Service Act for veterinarians who decide to

go into government positions (http://www.avma.org/advocacy/avma_advocate/jan09/aa_jan09b.asp and http://www.avma.org/fsvm/AnimalHealthcare%20(2).pdf). Also some states have begun offering veterinary student loan repayment programs (notably Ohio; http://ovmlb.ohio.gov/sl.stm). A new National Veterinary Medical Service Act will improve loan repayment options for graduating veterinarians who choose to work in certain areas that affect animal or public health (http://www.avma.org/press/releases/100420_VMLRP.asp)

## DEVELOPMENT OF NEW DRUGS AND VACCINES BY INDUSTRY

The pharmaceutical industry provides many examples of unnecessary separation of human and veterinary medicine that provide impediments to progress. Typically the animal and human health divisions of pharmaceutical companies are physically and operationally divided. The regulatory requirements and review of products for human and veterinary health also lie in separate divisions of the FDA and USDA. Since physiological and pathological underpinnings of product development are generally shared across species, there would be much to gain from a close interaction between those engaged in research and development of animal and human health products.

On the positive side, a few enlightened programs have reached in this direction. For example, when Akso Nobel created a new division devoted to development of human vaccines, it integrated scientists from its veterinary health division (Intervet). Intervet and a human vaccines biotechnology company (Acambis) collaborated on the development of vaccines against WNV. The veterinary vaccine is now commercially available (Prevenile®) and the human vaccine is in late stages of clinical testing. The development of these products required a close working relationship between scientists at both companies.

## THE FUTURE

The One Health concept has languished too long in the 20th and 21st centuries in clinical care, public and ecological health, and biomedical research. Civilization is facing many threats, including human overpopulation, the destruction of ecosystems, climate change, and emerging zoonotic pathogens. The combined, synergistic creativity and insights of transdisciplinary teams comprising physicians, veterinarians, ecologists, public health professionals, and others are needed to address these challenges.

The organizational, institutional, and financial obstacles to implementing a global One Health approach to disease threats must not be ignored. It is incumbent upon the leaders in medicine, veterinary medicine, science, ecology, and public health to alert and educate political leaders, policymakers, the media, and the public about this critical approach in global health. Implementing a One Health approach globally would significantly mitigate or possibly avert future health crises.

# REFERENCES

Aguirre, A.A., R.S. Ostfeld, G.M. Taber, C. House, and M.C. Pearl. 2002. Conservation medicine: ecological health in practice. Oxford University Press, New York.

Asnis, D.S., R. Conetta, A.A. Teixera, G. Waldman, and B.A. Sampson 2000. The West Nile virus outbreak of 1999 in New York: the Flushing Hospital experience. Clin Infect Dis 30:413–418.

Bokma, B. H., E. P. J. Gibbs, A. A. Aguirre, and B. Kaplan. 2008. A resolution by the society of tropical veterinary medicine in support of "One Health." Ann NY Acad Sci 1149:4–8.

Briese, T., X-Y Jia, C. Huang, L.J. Grady, and W.I. Lipkin. 1999. Identification of a Kunjin/West Nile-like flavivirus in brains of patients with New York encephalitis. Lancet 354:1261–1262.

Centers for Disease Control and Prevention (CDC). 1999. Outbreak of West Nile-like viral encephalitis—New York, 1999. MMWR 48:845–849.

Centers for Disease Control and Prevention (CDC). 2008. West Nile virus activity: United States, 2007. MMWR 57:720–723.

Centers for Disease Control and Prevention (CDC). 2009. Methicillin-resistant *Staphylococcus aureus* skin infections from an elephant calf: San Diego, California, 2008. MMWR 58:194–198.

Childs, J.E., J.S. Mackenzie, and J.A. Richt, eds. 2007. Wildlife and emerging zoonotic diseases: the biology, circumstances and consequences of cross-species transmission. Springer, New York.

Cleaveland, S., M.K. Laurenson, and L.H. Taylor. 2001. Diseases of humans and their domestic animals: pathogen characteristics, host range and the risk of emergence. Philos Trans Roy Soc B 356: 991–999.

Cook, R.A., W.B. Karesh, and S.A. Osofsky. 2004. One World, One Health http://www.oneworldonehealth.org/

Conrad, L.I. 1992. Epidemic disease in formal and popular thought in early Islamic society. *In* Ranger, T. and P. Slack, eds. Epidemics and ideas, pp. 77–99. Cambridge University Press, UK.

DeBoer, D.J., and R. Marsella. 2001. The ACVD task force of canine atopic dermatitis (XII): the relationship of cutaneous infections to the pathogenesis and clinical course of canine atopic dermatitis. Vet Immunol Immunopathol 81:239–249.

deJong, J.C., E.C.J. Claas, A.D.M.E Osterhaus, R.G. Webster, and W.L. Lim. 1997. A pandemic warning? Nature 389:554.

De Paolo, C. 2006. Epidemic disease and human understanding: a historical analysis of scientific and other writings. McFarland and Company, Jefferson.

Diamond, J. 1999. Guns, germs, and steel. W. W. Norton & Co., New York.

Foundation for Advancement of International Medical Education and Research. 2009. Mapping the world's medical schools. http://www.faimer.org/resources/mapping.html

Fridkin, S.K., J.C. Hageman, M. Morrison, L. Thomson Sanza, K. Como-Sabetti, J.A. Jernigan, K. Harriman, L.H. Harrison, R. Lynfield, and M.M. Farley. 2005. Methicillin-resistant *Staphylococcus aureus* disease in three communities. N Engl J Med 352:1436–1444.

Global Early Warning and Response System for Major Animal Diseases, including Zoonoses (GLEWS). FAO, OIE, WHO publication, February 2006. (www.glews.net)

Grant, S., and C.W. Olsen. 1999. Preventing zoonotic diseases in immunocompromised persons: the role of physicians and veterinarians. Emerg Infect Dis 5:159–163.

Institute of Medicine. 2003. The future of the public's health in the 21st century. National Academies Press, Washington, DC.

Institute of Medicine. 2009. Sustaining global surveillance and response to emerging zoonotic diseases. National Academies Press, Washington, DC.

Jones, K.E., N.G. Patel, M.A. Levy, A. Storeygard, D. Balk, J.L. Gittleman, and P. Daszak. 2008. Global trends in emerging infectious diseases. Nature 451:990–994.

Journal of the American Veterinary Medical Association News: July 15. 2008. Projects, policies approved http://www.avma.org/onlnews/javma/jul08/080715d.asp

Journal of the American Veterinary Medical Association News: August 1. 2009. One Health Timeline http://www.avma.org/onlnews/javma/aug09/090801a.asp

Jong-Wook, L. 2005, November 9. WHO Director General's closing remarks. FAO/OIE/WB/WHO Meeting on Avian Influenza and Human Pandemic Influenza.

Kahn, L.H. 2006. Confronting zoonoses, linking human and veterinary medicine. Emerg Infect Dis 12:556–561.

Kahn, L.H. 2011. Deforestation and emerging diseases. Bull Atomic Sci 15 February 2011 http://www.thebulletin.org/web-edition/columnists/laura-h-kahn/deforestation-and-emerging-diseases

Kahn, L.H., B. Kaplan, and J.H. Steele. 2007. Confronting zoonoses through closer collaboration between medicine and veterinary medicine (as 'one medicine'). Vet Italiana 43:5–19.

Kahn, L.H., B. Kaplan, T.P. Monath, and J.H. Steele. 2008. Teaching One Medicine, One Health. Am J Med 121:169–170.

Karesh, W.B., and R.A. Cook. 2005. The human-animal link. Foreign Affairs July/August:38–50.

Klein, E., D.La. Smith, and R. Laxminarayan. 2007. Hospitalizations and deaths caused by methicillin-resistant Staphylococcus aureus, United States, 1999–2005. Emerg Infect Dis 13:1840–1846.

Kuiken, T., F.A. Leighton, R.A.M. Fouchier, J.W. LeDuc, et al. 2005. Pathogen surveillance in animals. Science 309:1680–1681.

Leff, D. 2000. About Darwin. http://www.aboutdarwin.com/index.html

Ley, T.J., and L.E. Rosenberg. 2005. The physician-scientist pipeline in 2005: Build it, and they will come. J Am Med Assoc 294:1343–1351.

Mackenzie, J.S., and M.H. Jeggo. 2011. 1st International One Health Congress: human health, animal health, the environment and global survival. EcoHealth 7:S1–S2.

Mahon, N.M. 2003. West Nile virus: an emerging virus in North America. Clin Lab Sci. 16:43–9.

Marano, N., P.M. Arguin, and M. Pappaioanou. 2007. Impact of globalization and animal trade on infectious disease ecology. Emerg Infect Dis 13:1807–1809.

Marfin, A.A., L.R. Peterson, M. Eidson, J. Miller, J. Hadler, C. Farello, et al. 2001. Widespread West Nile virus activity, eastern United States, 2000. Emerg Infect Dis. 7:730–735.

McNeill, W.H. 1977. Plagues and peoples. Random House, New York.

Munch R. 2003. Review: On the shoulders of giants. Microbes and Infection 5:69–74

National Research Council. 2004. National need and priorities for veterinarians in biomedical research. National Academy Press, Washington, DC.

National Research Council. 2005a. Animal health at the crossroads: preventing, detecting, and diagnosing animal diseases. National Academy of Sciences, Washington, DC.

National Research Council. 2005b. Critical needs for research in veterinary science. National Academy Press, Washington, DC.

Olivry, T., R.S. Mueller, and the International Task Force on Canine Atopic Dermatitis. 2003. Evidence-based veterinary dermatology: a systematic review of the pharmacotherapy of canine atopic dermatitis. Vet Dermatol 14:121–146.

Palmarini, M. 2007. A veterinary twist on pathogen biology. PloS Pathogens 3:e12.

Palmer, M.V. and W.R. Waters. 2011. Bovine tuberculosis and establishment of an eradication program in the United States: role of veterinarians. Vet Med Int doi:10.4061/2011/816345

Rosenberg, L.E. 1999. Physician-scientists: endangered and essential. Science 283:331–2.

Schwabe, C.W. 1984. Veterinary medicine and human health. 3rd edition. Williams and Wilkins, Baltimore.

Scott, E., S. Duty, and K. McCue. 2009. A critical evaluation of methicillin-resistant Staphylococcus aureus and other bacteria of medical interest on commonly touched household surfaces in relation to household demographics. Am J Infect Cont 37: 447–453.

Shortridge, K.F., J.S.M. Peiris, and Y. Guan. 2003. The next influenza pandemic: lessons from Hong Kong. J Appl Microbiol 94:70S–79S.

Smith, K.F., M. Behrens, L.M. Schloegel, N. Marano, S. Burgiel, and P. Daszak. 2009. Reducing the risks of the wildlife trade. Science 324:594–595.

Starr, P. 1982. The social transformation of American medicine. Basic Books, New York.

Tam, J.S. 2002. Influenza A (H5N1) in Hong Kong: an overview. Vaccine 20:S77–S81.

Taylor, L.H., S.M. Latham, and M.E.J. Woolhouse. 2001. Risk factors for human disease emergence. Philos Trans R Soc Lond B Biol Sci 356:983–989.

Trut, L.N. 1999. Early canid domestication: the farm-fox experiment. Am Sci 87:160–169.

United Nations Department of Public Information. 2007 March 13. World population will increase by 2.5 billion by 2050. New York. http://www.un.org/News/Press/docs//2007/pop952.doc.htm

United States General Accounting Office. 2000. West Nile virus outbreak: lessons for public health preparedness. GAO/HEHS-00-180. http://www.gao.gov/archive/2000/he00180.pdf.

Varki, A., and L.E. Rosenberg. 2002. Emerging opportunities and career paths for the young physician-scientist. Nat Med 8:437–439.

Walsh, D. A. (coordinator). 2009. Veterinary education for global animal and public health. Rev Sci Tech Off Int Epiz 28.

Wheelis, M. 2002. Biological warfare at the 1346 siege of Caffa. Emerging Infectious Diseases 8:971–975.

Wilkinson, L. 1992. Animals and disease: an introduction to the history of comparative medicine. Cambridge University Press, Cambridge.

World Health Organization. 2009 September 24. Cumulative number of confirmed human cases of avian influenza A/(H5N1) reported to WHO. Available at: http://www.who.int/csr/disease/avian_influenza/country/cases_table_2009_09_24/en/index.html

Writing Committee of the World Health Organization (WHO) Consultation on Human Influenza A/H5. 2005. Avian influenza A (H5N1) infection in humans. N Engl J Med 353:1374–1385.

Zerhouni, E. 2003. The NIH roadmap. Science 302: 63–72.

Zinkernagel, R.M., and P.C. Doherty. 1974. Immunological surveillance against altered self components by sensitized T lymphocytes in lymphocytic choriomeningitis. Nature 251:547–548.

# 4

## BIODIVERSITY AND HUMAN HEALTH

### Aaron Bernstein

Because humans have evolved as part of the web of life, the premise that we remain embedded in it seems obvious. Yet, with our technological success, which has been driven by our needs and wants and formed with natural capital, we have made it possible to conceal this fundamental truth from ourselves. For urbanites, the sources of what keeps us healthy—food, medicine, clean water, and clean air—are hidden, and the sinks for our wastes are likewise mostly out of sight.

Scientists have steadily ratcheted up their level of concern regarding the state of the planet, citing, among other dismaying statistics, that we have cleared half of the Earth's forests (WRI 2009), degraded 40% of agricultural lands (WRI 2000), pushed a quarter of fisheries to precariously low levels (FAO 2009), and set the Earth's climate on a worrisome trajectory. Such dire statements baffle many of us as we look out our windows to view a landscape filled with greenery or visit a park where life abounds, and wonder what could possibly be wrong with nature given what we see around us. Nature's invisibility to too many of us, combined with a lapse in appreciation for our dependence on it, has brought on a human-induced, and nearly unprecedented, period of simplification in the biosphere.

Biodiversity, namely the variety of life, including genes, species, and ecosystems, is dwindling at a pace never before matched in human history, even considering the last great extinction pulse more than 60 million years ago. Fifty percent of species went extinct then, and by the best estimates that many species may be lost by century's end if we do not act to more judiciously manage the living world. This loss of species, the genes they hold and the ecosystems they form, is far more than a loss of unique creatures: with the loss of biodiversity, human health is put at risk.

Much of this book presents compelling evidence that alterations to ecosystems have unleashed infectious epidemics upon us. This chapter outlines how biodiversity underpins health more broadly and proceeds to identify a few obstacles that stand in the way of preventing the further loss of biodiversity.

### MEDICINES FROM NATURE

In most medicine cabinets can be found a drug with a natural source. Plants, for instance, have given us aspirin and the opiates, such as morphine or oxycodone. A perusal of medication prescriptions in the United States written in 2008 reveals that fully half of the 100 most prescribed drugs contain molecules either copied from or patterned after ones evolved in nature (Lamb 2009). Natural products have also been essential as a source for new drugs. From 1981 to 2006, of the 1,184 new drugs that the U.S. Food and Drug Administration (FDA) licensed, about two thirds had natural product origins (Newman and Cragg 2007). This statistic contradicts the prevailing wisdom that nature has been rendered outmoded in the era of rational drug design.

Natural products drug leads have a head start against their rationally designed counterparts. Many have already passed through a most rigorous clinical trial of sorts: natural selection. Take ziconotide, a painkiller made from a peptide secreted in the venom of the cone snail *Conus magus*. This slow-moving snail hunts fish and would surely go hungry if not for its venom's lightning-like and lethal effect, a product of 50 million years of evolutionary trial and error. Ziconotide is specific for N-type calcium channels, one of the four high-threshold voltage-gated channel subtypes that transmit pain sensation through the spinal cord. In humans, ziconotide has been shown to improve the lives of patients with advanced cancer and HIV with pain so severe that opiates, even at maximum doses, afford little if any relief (Staats et al. 2004). In comparison to opiates, which despite their ancient origins remain the mainstay of long-term pain treatment, ziconotide is tenfold more potent than morphine and also does not appear to require increasing doses over time to achieve the same degree of pain relief as the opiates do (Miljanich 2004).

A broader inspection of the genus *Conus* reveals that each of the 500 to 700 species makes its own unique assortment of 100 to 200 peptides for its venom, so the genus as a whole manufactures at least 50,000 distinct, biologically potent, peptides. Despite three decades of research, only a few hundred *Conus* peptides, or conopeptides, representing less than 1% of all the peptides in the genus, have been well studied. From these few have come ziconotide as well as five other conopeptides in clinical trials for use as painkillers and antiseizure medications. Still other conopeptides hold promise as medicines to prevent damage to the heart from a heart attack or the brain from a stroke (Han et al. 2008).

A mother lode of potential drugs may be hidden among the conopeptides. But even if many more new drugs such as ziconotide do not issue forth, the conopeptides have and will continue to make substantive contributions to our pharmacopoeia. Conopeptides are alluring to scientists because of the breathtaking specificity for, and variety of, the molecules they bind to on the cell surface. Calcium channels, sodium channels, potassium channels, nicotinic acetylcholine receptors (nAChRs), vasopressin receptors, norepinephrine receptors, and NMDA receptors, among others, are all targets of conopeptides. Prior to the discovery of conopeptides specific for nAChRs, the understanding of these receptors, which are present in muscle and brain and which have been implicated in diseases ranging from inflammatory bowel disease to neurological diseases, including Alzheimer's, epilepsy and Parkinson's, was limited by the lack of reagents that could distinguish among the receptor's myriad configurations (the neuronal subtype nAChR has 5 subunits drawn from 12 possible components) (Azam and McIntosh 2009).

Conopeptides, because of their specificity, have transformed the landscape of neuroscientific research, enabling more precise investigation into basic questions of cell biology. In so doing, they pave the way for breakthroughs in therapeutics. In Parkinson's disease research, α-conotoxinMII unlocked a new approach to the treatment of this debilitating disease by enabling the identification and characterization of a distinct nAChR subtype that contributes to the symptoms of the disease (Quik and McIntosh 2006). This nAChR subtype has become a new therapeutic target for Parkinson's disease.

α-contoxinMII, like the conopeptide that became ziconotide, comes from *Conus magus,* the magician's cone. *Conus magus,* like most cone snail species, lives on or near coral reefs, one of the most endangered habitats on Earth. Corals face a bleak future due to greenhouse gas emissions and the warmer and more acidic oceans they cause. Middle-of-the-road projections for climate change indicate that most corals will be lost by 2100 (Vernon 2008). As a result, mass die-offs of other species that rely on coral habitat are anticipated, and with them will come the loss of many coral reef habitat-dependent species, including some cone snails. If we were to lose these beautiful creatures, whose shells have been collected for centuries for their exquisite designs, the magnitude of the tragedy to aesthetics would pale in comparison to that for science and medicine.

## WHY NATURAL PRODUCTS?

Conopeptides such as α-conotoxinMII and ziconotide illustrate how natural selection may catalyze the search for drug leads. In addition, molecules in nature mutate in random ways and some of these may provide natural products with unforeseen and perhaps unforeseeable biological actions and, in turn, medicines with novel mechanisms of action. Paclitaxel, profiled in the first edition of this book (Chivian and Sullivan 2002), has become a classic example. At the time of its discovery, no cancer treatment worked by preventing the breakdown of the mitotic spindle (the cellular device that pulls chromosomes apart as a cell divides) as paclitaxel does. The same can be said for another drug, rapamycin, which prevents the rejection of transplanted organs by keeping overly aggressive immune cells at bay. Rapamycin derives from a bacteria native to Easter Island (Rapa Nui) and disables cell division through what had been a completely unrecognized signaling pathway involved in cell division. As the drug led to the identification of the previously unknown pathway, a central protein in it has been named "the molecular target of rapamycin" or mTOR. Rapamycin is now widely used in heart and kidney transplant patients. The drug also coats some coronary artery stents to prevent the cells that line the arterial wall from proliferating into the arterial lumen and causing re-occlusion. Not only do these two drugs, paclitaxel and rapamycin, demonstrate how nature may surprise us, showing us ways we might never conceive of to treat disease, but they have, because of their therapeutic novelty, established new frontiers in the treatment of cancer and organ transplantation, respectively. Multiple new drugs have already been patterned off them and they, like many other natural product drugs, have greatly advanced our understanding of how cancer cells work and how transplanted organs may be rejected, leading the way to further groundbreaking discoveries.

Aside from taking nature's lead in the design of individual compounds, medicine has also learned from how organisms use the bioactive molecules they produce. When cone snails paralyze their prey, they employ a multifaceted neurochemical assault to ensure success. The conopeptides in a snail's venom, while each potent in its own right, have far greater potency when acting in concert as they simultaneously interfere with different parts of the prey's nervous system.

Such a combinatorial approach has been adopted in human medicine, most notably in cancer chemotherapy in which patients receive drugs that subvert cancer cell propagation via different strategies, thereby maximizing the odds that cancer cells will have no means of escape. Many species also use compounds in combination for defense against microbial infection. Antimicrobial peptides secreted on the skins of certain anurans are able to prevent infections and do not seem to cultivate resistant bacterial strains (Nascimento et al. 2003). This is at odds with antimicrobial administration in human medicine, which in most cases treats infections with a single drug. While giving multiple antibiotics simultaneously may have untoward side effects and may be ill advised for many reasons, single-agent therapy for bacterial (or viral) infections selects for strains resistant to the given antibiotic, and this selective pressure may be largely responsible for the antibiotic-resistant microbes that beset medicine today.

## BIODIVERSITY AND BIOMEDICAL RESEARCH

Biomedical research has accelerated to a breakneck pace. The amount of data generated in labs around the world today is an order of magnitude greater than in 1950, and nearly twice as many papers are published in scientific journals each year as in 1970. As with new drugs and natural products, many mistakenly tag this wellspring of research as a triumph solely of high-tech science, and while it is in part, it would be impossible without biodiversity. Biodiversity contributes to discovery in essentially every domain of biomedical research, from genetics, to cell and molecular biology, physiology, and beyond.

To better understand the fundamental aspects of how our own bodies work in health and disease, scientists have always turned to other species. That much of what we know about human cell division, nerve function, and genetics comes from the study of yeast, squid, and plants, respectively, may seem improbable—and yet all are true. Furthermore, most diagnostic tests, surgical procedures, medications, and vaccines were tested at some stage in their development in an animal. Take H1N1 influenza as an example. The primary research on how the virus is acquired, how it is spread, and the efficacy of the vaccine was done largely with ferrets, which are susceptible to infection with the human influenza virus and manifest many of the symptoms humans do (Maher 2004; Munster 2009). So-called model organisms, such as the ferret, are chosen because of attributes that make them uniquely well suited to investigate a question of interest.

One question that has plagued scientists for decades is how organs harvested for transplant may be better preserved. Fortunately, several species of frogs, including the wood frog *Rana sylvatica*, because they freeze solid during the winter and thaw in the spring, make ideal subjects for research on how donated human organs may be better preserved. In an astounding metabolic feat, these frogs thaw in the spring from the *inside* out (Rubinsky et al. 1994). And despite having been frozen for months, their organs survive with no apparent injury (Devireddy 1999).

The shortage of organs available for donation in the United States is acute. As of July 2011, more than 100,000 people were awaiting organ transplants in the United States, and yet only about 20,000 organs had been donated in the entire preceding year (USDHHS 2011). Solid organs must be transplanted within hours of harvest, making time a limiting factor in organ availability; thus, the insights from frogs as to how to protect organs when frozen are potentially life-saving.

Limb regeneration in humans, once the subject of science fiction alone, is becoming ever more probable thanks in part to several amphibians, particularly salamanders. The axolotl *Ambystoma mexicanum* has fascinated scientists studying limb regeneration as it can regrow severed limbs anew in a month's time. The latest research on axolotls has shown that contrary to expectation, the so-called blastema cells that appear after limb loss are not pluripotent stem cells. Rather, they maintain some specialization to specific cell types, a finding that brings science one tantalizing step closer to unleashing the potential for limb regeneration latent in human limbs (Kragl et al. 2009). Possessing these extraordinary capabilities to regenerate limbs, as well as organs and muscle and nervous tissue, axolotls have tremendous value to science—and yet fewer than 1,500 remain alive in the wild today, as their habitat in central Mexico has all but disappeared.

Although many organisms used in research are abundant (though in some cases they have become so only because of their utility to science), some species used in research are not, and many more face uncertain futures in the wild. These include amphibians such as wood frogs and axolotls, as well as several cone snail species, all of which face multiple threats to their survival. Amphibian populations are declining around the world due to habitat loss, pollution, climate change, and a pandemic infection known as chytridiomycosis (see Chapter 36 in this volume), among others. The IUCN (2009) lists a third of frogs and toads, and half of salamanders and newts, as threatened with extinction. As amphibians have long served as experimental models for science, having had a part in major discoveries in reproductive biology, neuroscience, and developmental biology, and continue to inform cutting-edge research in these areas, a tremendous amount is at stake for human medicine with loss of amphibian species. Should axolotls go extinct in the wild we may be able to keep them alive in captivity, but the same cannot be said for countless other species.

Indeed, many species are disappearing before we even know they exist. As Gro Brundtland so aptly put it, "the library of life is burning, and we don't even know the titles of the books." Among the untold losses of species extinctions may be models for many diseases and basic questions of physiology for which we still have no adequate models to study.

## MOLECULAR DIVERSITY

Biodiversity's value to research extends beyond individual species and includes unique proteins or even genes identified within species. Consider the heat-stable DNA polymerase from *Thermus aquaticus*, the molecular engine of the polymerase chain reaction (PCR). *T. aquaticus* thrives in hot water, making use of its polymerase to replicate itself despite the heat. In PCR, paired strands of DNA must be melted apart with heat and then copied. Human DNA polymerase would cook at the nearly 94°C (200°F) required to do this, but taq polymerase from *T. aquaticus*, adapted to life at high temperatures, readily handles the heat.

For scientists in fields as disparate as forensics, evolution, and infectious diseases, PCR, made possible by a single and extraordinary gene, serves as a cornerstone of their research. Simply put, few molecules have influenced the course of modern science more than has taq polymerase. Other examples of noteworthy genes, themselves equally a reflection of adaptations to unique environments and of the scope of genetic diversity on Earth, are readily available. These include the gene for green fluorescent protein (GFP) from the jellyfish *Aequorea victoria*, which can illuminate specific proteins within a cell and has been a powerful tool in developmental and structural biology (GFP was key to the studies in the axolotl that uncovered residual specialization in the blastema cells), and restriction enzymes, which are bacterial proteins that cut DNA at specific sites, making possible refined analysis of DNA segments as well as the insertion of experimental genes, such as that for GFP, into a DNA region of interest.

In some cases, the molecular workings of several species are needed to fully describe biological mechanisms. In 1990, two teams of researchers, one in California and one in the Netherlands, were performing research on petunias to see if they might be able to increase the amount of pigment produced in their flower's petals. They inserted additional copies of the gene that controls pigmentation into the plant's genome and discovered, to their amazement, that many of these plants had *less* colorful petals. The explanation for this unexpected finding came only after many additional years of research on a different species, the worm *Caenorhabditis elegans*. Research in *C. elegans* uncovered an entirely novel mechanism controlling expression of genes that has been labeled RNA interference (RNAi), in which small pieces of double-stranded RNA in tandem with some cellular machinery silence specific messenger RNAs (mRNAs). RNAi has become a tool in the study of gene expression and has greatly advanced our understanding of the multilayered interactions that have made less tidy the once-immutable central dogma in biology that DNA becomes RNA which becomes protein (Shankar et al. 2005).

In addition to *C. elegans*, basic studies on RNAi took place in the mustard plant (*Arabidopsis thaliana*) and the fruit fly (*Drosophila melanogaster*). In the groundbreaking research on RNAi, and in other realms of biological exploration, these organisms have proven invaluable resources. While part of their utility comes from their familiarity (e.g., the complete genome is known for all three), they each possess peculiar qualities that have led to their continued use in experimentation and make them desirable choices

for the study of RNAi. *C. elegans*, for example, only need eat double-stranded RNA to realize RNA interference (Timmons et al. 2001).

## BIODIVERSITY, ECOSYSTEMS, AND HEALTH

The biodiversity of individual genes and species have readily apparent links to human well-being, as the preceding examples have shown. When it comes to human health, however, ecosystems may hold the greatest relevance. An ecosystem, such as a wetland, forest, or farm, can be defined as a group of organisms living in a defined space and their interactions between themselves and their physical environment. How ecosystems contribute to health has been covered in detail elsewhere (Corvalan et al. 2005; Sala 2009), so just a few examples will be presented here to demonstrate the links between healthy ecosystems and healthy humans (Box 4.1).

Few things are more important to health than food, and despite the many improvements in the abundance and quality of food made possible by human ingenuity, biodiversity remains an essential ingredient in our food supply. While it may appear that growing an apple, for example, requires just one species (*Malus domestica*, the apple tree) and some sun, water, and soil, producing an apple, or for that matter any crop, requires the involvement of many species. The soil itself teems with life from moles, worms, and insects we can see to microbial life we cannot. Apple trees may benefit from mycorrhizal fungi, symbiotic organisms that form expansive subterranean networks that in exchange for carbohydrates shepherd nutrients and water to the plants they colonize, and may, in some cases, divert toxic heavy metals away from plant roots (Khan et al. 2000). In experimental studies, apple trees inoculated with mycorrhizae are more robust than their non-inoculated counterparts, being taller and heavier and having larger leaves and higher leaf phosphorus concentrations (Morin et al. 1994). About 80% of all plant species have mycorrhizal counterparts, making the plant–mycorrhizal relationship one of the most widespread symbioses plants (Wang and Qiu 2006). The mycorrhizae, along with plant roots, provide the living infrastructure of the rhizosphere, the underground metropolis of life with an abundant diversity of life

---

**Box 4.1  Utilitarian Value in Ecosystem Services and Biodiversity**

Although many claims may insinuate that the value of biodiversity is tantamount to the value of ecosystem services, this is not the case. Humans obtain value from biodiversity from the uniqueness and, in some cases, the scarceness of life at the molecular, species, or ecological level. Ecosystem service value derives from the benefits that an ecosystem provides to humans.

To make this distinction more concrete, contrast 300,000 square kilometers of coral reefs with 17 million square kilometers of boreal forests. Both ecosystems provide extensive goods and services to humans. Corals support commercial fisheries, producing 10 to 20 tons of seafood per square kilometer per year (Cesar 1996), and boreal forests sequester more carbon per acre than any other forest (IPCC 2000). Yet coral reefs are home to millions of species and boreal forests to only thousands. While there may be much more biodiversity (at least of species we know of) on a per-area basis in coral reefs than in boreal forests, the value of these two ecosystems is not nearly so disparate when the goods and services they provide are examined.

Of course, ecosystem services are not independent of the biodiversity present in those ecosystems (a point that will be taken up later in this chapter). Yet distinct value to human welfare can be found in biological diversity itself, as this chapter attests.

---

forms, particularly microorganisms. One gram of fertile soil, for example, may house millions if not billions of bacteria from thousands of species. Some rhizosphere species may have little relevance to the plant roots they surround, yet others are essential to a given crop's success. By and large, though, the biodiversity of the rhizosphere remains poorly understood (Fitter et al. 2005).

Above ground, still more species contribute to agroecosystems. Roughly 100,000 species, including insects, birds, bats, and other animals, pollinate plants, including apple trees. The yearly economic value of pollinators to food production has been estimated at more than 150 billion Euros (US$225 billion), representing about 9.5% of world agricultural production (Gallai et al. 2008), and research suggests that the percentage of food crops dependent on pollination is rising (Aizen et al. 2008). Populations of pollinators, particularly honeybees and bats, have been found to be in decline in various corners of the world, including Europe and Southeast Asia. Winter losses in commercial bee colonies have been roughly double historical norms for the winters of 2006 through 2009 in the United States (vanEgelsdorp et al. 2010). As bee habitat dwindles and their populations become more concentrated, the spread of infection becomes easier. This fact, combined with pesticide exposures and long-haul transport via truck, which may foster pathogen spread and impair bee immunity, may explain multiple

plagues in bee colonies, including the varroa mite and the Israeli acute paralytic virus. These infections may cause bees to starve as they increase metabolic rates (Naug 2009). Most recently, co-infection with the fungus *Nosema cerena* and an iridovirus has been implicated as a proximal cause of honeybee declines (Bromenshenk et al. 2010).

Since the winter of 2006, 6 million or more bats have died in nine states in the United States. In a sample from two dozen hibernacula in Massachusetts, New York, and Vermont, populations have plummeted by nearly 95% since the die-off began (Chase 2009). The dead bats are almost all infected with a cold-loving fungus named *Geomyces destructans* that infects the animal's skin and coats their noses white, which is the source of the malady's name, white nose syndrome. Infected bats are emaciated, and one hypothesis to explain the emergence of this disease has been that warmer winters, the result of climate change, may increase winter arousals that deplete fat stores at a time of year when little food is available (Racey 2009). In bats as in bees and humans, poor nutritional status may compromise immune defenses. Bats control several important insect crop pests (Cleveland 2006) and bat population declines or species extinctions—events more likely to occur should the white nose syndrome continue to spread (Frick et al. 2010)—may adversely affect crop production.

How alterations to biodiversity may perturb infectious disease incidence in humans and animals has become an ever-more-pressing question and is the subject of much of this book and so will be mentioned only briefly here. Three quarters of emerging infectious diseases can affect humans and animals (Taylor et al. 2001), and several new and severe infections, including the severe acute respiratory syndrome (SARS), Nipah virus, and hantavirus, all gained access to humans through changes in the numbers and distribution of domestic or wild animals. Prevention of future plagues may be possible, but only with a deeper understanding of the pathogens themselves, their natural reservoirs, and the dynamics that govern the spread of pathogens among all species.

At the microscopic level, biodiversity may function to check the spread of certain infectious diseases. Changes to the ecology of the human microbiome, the vast and relatively unknown assemblage of microorganisms that are normal cohabitants of the human body, can predispose people to infectious disease. Altered ecology of the 700 or so microbes that live in the human mouth can cause periodontal disease (Socransky et al. 1998). In the intestine, where as many as 1,000 different microbial species may reside, overuse of antibiotics may predispose to *Clostridium difficile* colitis, which has doubled in incidence in the past 20 years in the United States, affecting 3 million people a year and causing death in 1% to 2% of those infected (Schroeder 2005).

The nascent exploration into the human microbiome gives a glimpse into how alterations to human microbial ecology affect our health, but much more research will be needed before a useful understanding of its workings is available. The workings of ecosystems at larger scales, while in some ways better understood than their microbial counterparts, are likewise still not fully mapped out. This comes as little surprise as we know precious little about most of the species that compose them. Only about 1 in 10 species have been given scientific names, many of these based upon just one or a few encounters. Our knowledge about microbes is so limited that no one knows how few of them have been identified.

Yet even with such a limited view of biodiversity, it is apparent that biodiversity is disappearing at a rapid rate. For at least the past 40 years, preventing the loss of species and degradation of ecosystems has been a focus of international attention. While many successes have been had, the tide has yet to be turned against further losses of biodiversity.

## BIODIVERSITY LOSS AND BARRIERS TO ITS PREVENTION

Species extinction rates, the most widely cited metric of biodiversity loss, are 100 to 1,000 times higher today than so-called background extinction rates (i.e., the rates observed in the fossil record when no mass extinction was occurring and before humans evolved). While pollution, invasive species, and overharvesting have all contributed to biodiversity loss, destruction of habitat has been the largest culprit, with climate change running a close second and likely to take this dubious distinction by mid-century (Thomas et al. 2004). All of these drivers result from human activities, which grow apace with the human population. Given all that ties human well-being to biodiversity, as this chapter has just begun to enumerate, how quickly we are losing that biodiversity, how little we know about life on Earth, and that fact the losses in biodiversity are mainly from human actions, why has it been so hard to prevent biodiversity loss?

The shortfall is not from lack of effort. Conservationists have made compelling appeals to preserve habitat, mitigate greenhouse gas emissions, and act broadly to curb the other forces behind biodiversity loss for decades. They have made these appeals on aesthetic, moral, and religious grounds. In sum, these appeals have met with mixed success. More recently, a shift has been made towards more quantitative justifications, and especially economics, to bolster arguments for protecting biodiversity in the belief that assigning a number, often in dollars, to life will enable a clearer, and possibly favorable, input into conservation decisions.

In the early 1980s, the field of ecological economics was founded to redress shortcomings identified with traditional economic approaches to the valuation of ecosystems (Faber 2008). Foremost among these was that most economic models gave little attention to, or outright ignored, natural capital's irreplaceability. The belief that human capital may substitute for natural capital, or that one form of natural capital may substitute for another, underlies much of traditional cost–benefit and other forms of comparative analysis. For instance, in deliberations over whether to invest in watershed protection or water filtration plants, established forms of economic analysis suppose that the two options are at least to some extent substitutable. They do not, however, generally recognize that once biodiversity is lost, it is gone forever; it cannot be purchased back into existence at some future date.

Many ecosystems appear to have functional redundancy such that others can compensate for the absence of one or a few species in that system (Naeem et al. 1998). Some have used this premise to suggest that some amount of species loss within an ecosystem is acceptable or even necessary to maintain balance with development (Arrow et al. 1995). Investigation into the interplay between biodiversity and ecosystem function has broadened in the past decade and the complexities of the relationship have begun to be clarified (Reiss et al. 2009). One of the recent and still preliminary findings in this area has been that when multiple functional endpoints are considered, apparent redundancy in ecosystem species composition declines. A study of the Great Barrier Reef, for example, revealed that despite the reef's vast diversity, a single species was needed for reef recovery from simulated overfishing stress (Bellwood et al. 2006). Such a finding, if corroborated in other ecosystems, would heighten the need to remedy the absence of consideration for biodiversity loss in ecosystem valuation (Hector and Bagchi 2007). While significant progress has been made in identifying principles that apply to loss of ecosystem function when species disappear, predicting such decrements remains an inexact science (Sala et al. 2000; Cardinale 2006; Naeem 2009).

Other challenges that arise with ecosystem valuation derive from the vagaries in models associated with long time horizons, on the order of years to decades, relevant to natural systems, as well as the measurement and inclusion of externalities. Natural systems may have non-linear responses to incremental changes (e.g., the recent separation of massive chunks from the Antarctic ice sheet due to climate change) and predicting when and to what extent these non-linear responses occur is often difficult. Forecasting externalities likewise pose problems to viable models, with the buildup of greenhouse gases in the atmosphere that causes climate change as a prime example. Attempts have been made to quantify the adverse effects for humans and wildlife from dwindling freshwater supplies, food scarcity, sea level rise, and extreme weather events, all of which are predicted to occur with climate change. Yet precisely when and where these may occur is essentially unknowable, which makes forecasting the costs (which depend heavily upon where the events occur) a problematic task. Such uncertainty limits the ability for these outcomes to be included in economic models.

Despite these and other critiques of economic valuation of ecosystems (e.g., Norton 2007), such valuations have become ever more visible in policy deliberations and have had some notable successes, including in amending the Clean Air Act (United States Environmental Protection Agency 1999) and in protecting the New York City watershed (Chichilnisky and Heal 1998). Room for greater visibility remains as tremendous sums of money are spent each year on subsidies that are both economically and ecologically harmful, perhaps as much as $1 trillion or more (Balmford et al. 2002). That being said, sound economics rarely gets in the way of good politics: economic analyses, when available, often do not dictate political decisions regardless of what position they favor (e.g., Pearce 2006).

The challenges faced in preventing biodiversity loss, of course, represent more than market failure or shortcomings in economic analysis. Why we have a hard time taking biodiversity loss seriously may have much to do with interplay between the nature of the causes of biodiversity loss and how they interface with foibles in human decision-making. Again, take climate change as an example. Mitigating climate change entails a trade-off between relatively well-defined upfront costs to move away from fossil fuel energy sources and longer-term, less well-defined, benefits to the well-being of humans and other species. Well-described cognitive biases feed into such a deliberation. Loss aversion is one: people would rather avoid losses than acquire similarly valued gains (Kahneman et al. 1991; Rachlinski 2000). On top of this, people tend to prefer immediate over future payoffs (Laibson 1997). Of course, how the issue is framed is paramount. Economists have argued that the longer-term benefits of greenhouse mitigation dwarf the short-term costs and that the costs of inaction themselves at present far exceed the costs needed to move economies off fossil fuels (Stern 2007). Framed in this way, loss aversion and future discounting may prove less of an obstacle.

However, what may motivate people most to improve the ways in which we do business with the biosphere is to lay bare the connections that bind the fate of humans and all species together. For most of humanity, these connections have become too abstract and too distant. Our relationship to nature is more tangible than the common understanding that "all life is connected" or that we are a part of a food web, and the biodiversity that sustains us is not just in a faraway and tropical place but near where we live. The challenge for conservation in the 21st century, then, is to transform the concept of biodiversity from the abstract into the concrete and the distant into the personal. Only with such a shift in culture will conservation move from an adversarial act of morality and altruism to one of custom and habit.

# REFERENCES

Aizen, M.A., L.A. Garibaldi, S.A. Cunningham, and A.M. Klein. 2008. Long-term global trends in crop yield and production reveal no current pollination shortage but increasing pollinator dependency. Curr Biol 18:1572–1575.

Arrow, K., B. Bolin, R. Costanza, P. Dasgupta, C. Folke, C.S. Holling, B.O. Jansson, S. Levin, K.-G. Mäler, C. Perrings, and D. Pimentel. 1995. Economic growth, carrying capacity, and the environment. Science 268:520–521.

Azam, L., and J.M. McIntosh. 2009. Alpha-conotoxins as pharmacological probes of nicotinic acetylcholine receptors. Acta Pharmacol Sinic 30:771–783.

Balmford, A., A. Bruner, P. Cooper, R. Costanza, S. Farber, R.E. Green, M. Jenkins, P. Jefferiss, V. Jessamy, J. Madden, K. Munro, N. Myers, S. Naeem, J. Paavola, M. Rayment, S. Trumper, and R.K. Turner. 2002. Economic reasons for conserving wild nature. Science 297:950–953

Bellwood, D.R., T.P. Hughes, and A.S. Hoey. 2006. Sleeping functional group drives coral-reef recovery. Curr Biol 16:2434–2439.

Bromenshenk, J.J., C.B. Henderson, C.H. Wick, M.F. Stanford, A.W. Zulich, R.E. Jabbour, S.V. Deshpande, P.E. McCubbin, R.A. Seccomb, P.M. Welch, T. Williams, D.R. Firth, E. Skowronski, M.M. Lehmann, S.L. Bilimoria, J. Gress, K.W. Wanner, and R.A. Cramer Jr. 2010. Iridovirus and microsporidian linked to honey bee colony decline. PLoS One 5:e13181.

Cardinale, B.J., D.S. Srivastava, J.E. Duffy, J.P. Wright, A.L. Downing, M. Sankaran, and C. Jouseau. 2006.

Effects of biodiversity on the functioning of trophic groups and ecosystems. Nature 443:989–992.

Cesar, H. 1996. Economic analyses of Indonesian coral reefs. World Bank. http://siteresources.worldbank.org/INTEEI/214574-1153316226850/20486385/EconomicAnalysisofIndonesianCoralReefs1996.pdf

Chase, S. 2009. What's killing the bats? Boston Globe. November 15. http://www.boston.com/news/science/articles/2009/11/15/whats_killing_the_bats/

Chichilnisky, G., and G. Heal. 1998. Economic returns from the biosphere. Nature 391:629–630.

Chivian, E.S., and A.S. Bernstein, eds. 2008. Sustaining life: how human health depends on biodiversity. Oxford University Press, New York.

Chivian E.S., and S. Sullivan. 2002. Biodiversity and human health. In Aguirre A.A., R.S. Ostfeld, G.M. Tabor, C. House, and M.C. Pearl, eds. Conservation medicine: ecological health in practice, pp. 182–193. Oxford University Press, New York.

Cleveland, C.J., M. Betke, P. Federico, J.D. Frank, T.G. Hallam, J. Horn, J.D. Lopez Jr., Jr., G.F. McCracken, R.A. Medellin, A. Moreno-Valdez, C.G. Sansone, J.K. Westbrook, and T.H. Kunz. 2006. Economic value of the pest control service provided by Brazilian free-tailed bats in south-central Texas. Front Ecol Environ 4:238–243.

Corvalan, C., S. Hales, and A.J. McMichael. 2005. Ecosystems and human well-being: health synthesis a report of the millennium ecosystem assessment. World Health Organization, Geneva.

Devireddy, R. 1999. Liver freezing response of the freeze-tolerant wood frog, Rana sylvatica, in the presence and absence of glucose. I. Experimental measurements. Cryobiology 38:310–326.

Faber, M. 2008. How to be an ecological economist. Ecol Econ 66:1–7.

Food and Agricultural Organization of the United Nations (FAO). 2009. The state of the world fisheries and aquaculture 2008. Figure 21, p. 33. FAO Fisheries and Aquaculture Department, Rome, Italy.

Fitter, A.H., C.A. Gilligan, K. Hollingworth, A. Kleczkowski, R.M. Twyman, J.W. Pitchford, and Members of the NERC Soil Biodiversity Programme. 2005. Biodiversity and ecosystem function in soil. Funct Ecol 19:369–377.

Frick, W.F., J.F. Pollock, A.C. Hicks, K.E. Langwig, D.S. Reynolds, G.G. Turner, C.M. Butchkoski, and T.H. Kunz. 2010. An emerging disease causes regional population collapse of a common north american bat species. Science 329:679–682.

Gallai N., J.M. Salles, J. Settele, and B.E. Vaissiere. 2008. Economic valuation of the vulnerability of world agriculture confronted with pollinator decline. Ecol Econ 68:810–821.

Han, T.S., R.W. Teichert, B.M. Olivera, and G. Bulaj. 2008. Conus venoms: a rich source of peptide-based therapeutics. Curr Pharm Design 14:2462–2479.

Hector, A., and R. Bagchi. 2007. Biodiversity and ecosystem multifunctionality. Nature 448:188–190.

Intergovernmental Panel on Climate Change (IPCC). 2000. Summary for policymakers: Land use, land-use change and forestry. Cambridge: Cambridge University Press. http://www.ipcc.ch/ipccreports/sres/land_use/index.php?idp=0

International Union for the Conservation of Nature (IUCN). 2009. Amphibians on the IUCN Red List: Red List Status. http://www.iucnredlist.org/initiatives/amphibians/analysis/red-list-status.

Kahneman, D., J.L. Knetsch, and R.H. Thaler. 1991. Anomalies: the endowment effect, loss aversion, and status quo bias. J EconPerspect 5:193–206.

Khan, A.G., C. Kuek, T.M. Chaudhry, C.S. Khoo, and W.J. Hayes. 2000. Role of plants, mycorrhizae and phytochelators in heavy metal contaminated land remediation. Chemosphere 41:197–207.

Kragl, M., D. Knapp, E. Nacu, S. Khattak, M. Maden, H.H. Epperlein, and E.M. Tanaka. 2009. Cells keep a memory of their tissue origin during axolotl limb regeneration. Nature 460:60–65.

Lamb, E. 2009. Top 200 drugs of 2008. May 15. http://www.pharmacytimes.com/issue/pharmacy/2009/2009-05/RxFocusTop200Drugs-0509.

Laibson, D. 1997. Golden eggs and hyperbolic discounting. Q J Econ 112:443–477.

Maher, J.A., and J. DeStefano. 2004. The ferret: an animal model to study influenza virus. Lab Anim 33:50–53.

Miljanich, G P. 2004. Ziconotide: neuronal calcium channel blocker for treating severe chronic pain. Curr Med Chem 11:3029–3040.

Morin, R., C. Hamel, J.A. Fortin, R.L. Granger, and D.L. Smith. 1994. Apple rootstock response to vesicular-arbuscular mycorrhizal fungi in a high phosphorous soil. J Am Soc Hort Sci 119:578–783.

Munster, V. J., E. de Wit, J.M.A. van den Brand, S. Herfst, E.J.A. Schrauwen, T.M. Bestebroer, D. van de Vijver, C.A. Boucher, M. Koopmans, G.F. Rimmelzwaan, T. Kuiken, A.D.M.E. Osterhaus, and R.A.M. Fouchier. 2009. Pathogenesis and transmission of swine-origin 2009 A(H1N1) influenza virus in ferrets. Science 325:481–483

Naeem, S. 1998. Species redundancy and ecosystem reliability. Conserv Biol 12:39–45.

Naeem, S., D.E. Bunker, A. Hector, M. Loreau, C. Perrings. 2009. Biodiversity, ecosystem functioning, and human wellbeing : an ecological and economic perspective. Oxford University Press, New York.

Nascimento, A.C.C., W. Fontes, A. Sebben, and M.S. Castro. 2003. Antimicrobial peptides from anurans skin secretions. Protein Peptide Lett 10: 227–238.

Naug, D. 2009. Nutritional stress due to habitat loss may explain recent honeybee colony collapses. Biol Conserv 142:2369–2372.

Newman, D.J., and G.M. Cragg. 2007. Natural products as sources of new drugs over the last 25 years. J Nat Prod 70:461–477.

Norton, B., and D. Noonan. 2007. Ecology and valuation: big changes needed. Ecol Econ 63:664–675.

Pearce, D. 2006. Chapter 19: The political economy of cost-benefit analysis. In Cost-benefit analysis and the environment: recent developments, pp. 279–287. Organization for Economic Cooperation and Development, Paris, France.

Quik, M., and J.M. McIntosh. 2006. Striatal alpha6* nicotinic acetylcholine receptors: potential targets for Parkinson's disease therapy. J Pharmacol Exp Ther 316:481–9.

Racey, P.A. 2009. Bats: status, threats and conservation successes. Endanger Species Res 8:1–3.

Rachlinski, J.J. 2000. Climate change psychology. Univ Illinois Law Rev 307:299–319.

Reiss, J., J.R. Bridle, J.M. Montoya, and G. Woodward. 2009. Emerging horizons in biodiversity and ecosystem functioning research. Trends Ecol Evol 24:505–514.

Rubinsky, B., S.T. Wong, J.S. Hong, J. Gilbert, M. Roos, and K.B. Storey. 1994. 1H magnetic resonance imaging of freezing and thawing in freeze-tolerant frogs. Am J Physiol Regul Integr Comp Physiol 266: R1771–1777.

Sala, O.E., F.S. Chapin, J.J. Armesto, E. Berlow, J. Bloomfield, R. Dirzo, E. Huber-Sanwald, L.F. Huenneke, R.B. Jackson, A. Kinzig, R. Leemans, D. Lodge, H.A. Mooney, M. Oesterheld, N.L. Poff, M.T. Skyes, B.H. Walker, and D.H. Wall. 2000. Global biodiversity scenarios for the year 2100. Science 287:1770–1774.

Sala, O. 2009. Biodiversity change and human health: from ecosystem services to spread of disease. Island Press, Washington, D.C.

Schroeder, M.S. 2005. Clostridium difficile-associated diarrhea. Am Fam Physician 71:921–928.

Shankar, P.N. Manjunath, and J. Lieberman. 2005. The prospect of silencing disease using RNA interference. J Am Med Assoc 293:1367–1373.

Socransky, S.S., A.D. Haffajee, M.A. Cugini, C. Smith, and R.L. Kent. 1998. Microbial complexes in subgingival plaque. J Clin Periodontol 25:134–144.

Staats, S., T. Yearwood, S.G. Charapata, R.W. Presley, M.S. Wallace, M. Byas-Smith, R. Fisher, D.A. Bryce, E.A. Mangieri, R.R. Luther, M. Mayo, D. McGuire, and D. Ellis. 2004. Intrathecal ziconotide in the treatment of refractory pain in patients with cancer or AIDS. J Am Med Assoc 291:63–70.

Stern, N. 2007. The economics of climate change: the Stern review. Cambridge University Press, New York.

Taylor, L.H., S.M. Latham, and M.E.J. Woolhouse. 2001. Risk factors for human disease emergence. Philos T Roy Soc B 356:983–989.

Thomas, C.D., A. Cameron, R.E. Green, M. Bakkenes, L.J. Beaumont, Y.C. Collingham, B.F.N. Erasmus, M. Ferreira de Siqueira, A. Grainger, L. Hannah, L. Hughes, B. Huntley, A.S. Van Jaarsveld, G.E. Midgely, L. Miles, M.A. Ortega-Huerta, A.T. Peterson, O.L. Phillips, and S.E. Williams. 2004. Extinction risk from climate change. Nature 427:145–148.

Timmons, L., D.L. Court, and A. Fire. 2001. Ingestion of bacterially expressed dsRNAs can produce specific and potent genetic interference in *Caenorhabditis elegans*. Gene 263:103–112.

United States Department of Health and Human Services (USDHHS). 2011. OrganDonor.gov. http://www.organdonor.gov/.

United States Environmental Protection Agency (EPA). 1999. The benefits and costs of the Clean Air Act 1990-2010. EPA-410-R-99-001, Washington, D.C.

vanEngelsdorp, D., J. Hayes, R.M. Underwood, and J. Pettis. 2010. A survey of honeybee colony losses in the U.S., Fall 2008 to Spring 2009. J Apicult Res 49:7–14.

Vernon, J. 2008. The ocean's canary in a reef in time: the Great Barrier Reef from beginning to end. Belknap Press of Harvard University Press, Cambridge, Massachusetts.

Wang, B., and Y.L. Qiu. 2006. Phylogenetic distribution and evolution of mycorrhizas in land plants. Mycorrhiza 16:299–363.

World Resources Institute (WRI). 2009. The world's forests from a restoration perspective. http://www.wri.org/map/worlds-forests-restoration-perspective.

World Resources Institute (WRI). 2000. A guide to world resources 2000–2001: people and ecosystems: the fraying web of life. World Resources Institute, United Nations Development Programme, United Nations Environment Programme, World Bank. Washington, D.C.

# 5

# AN ECOSYSTEM SERVICE OF BIODIVERSITY

## The Protection of Human Health Against Infectious Disease

## Felicia Keesing and Richard S. Ostfeld

In the face of unprecedented, rapid declines in global biodiversity, scientists have been investigating whether ecosystems with low biodiversity function differently than ecosystems with high biodiversity. If they do, the discovery of beneficial "ecosystem services" rendered by biodiversity has the potential to motivate conservation. Unfortunately, the kinds of ecosystem services typically investigated by scientists—ecosystem stability, resistance to invasion, productivity—often fail to capture the attention of most policymakers and the public (Biodiversity Project 2002; Ostfeld et al. 2009). Similarly, the types of biodiversity typically included in studies (e.g., wild plants or single-celled eukaryotes) are often not the types of biodiversity whose vulnerability motivates action. In this chapter, we describe an ecosystem service that has the potential to motivate conservation of biodiversity worldwide—the protection of human health by charismatic wild animals.

We will describe the role of biodiversity in the transmission of Lyme disease, a potentially debilitating condition that affects many tens of thousands of people every year in North America, Europe, and Asia. We will also describe how and why the patterns we see in Lyme disease are common to many disease systems, including not only other human diseases but also diseases of wildlife, domestic animals, and wild and cultivated plants.

## EPIDEMIOLOGY OF LYME DISEASE

About 30,000 people report confirmed cases of Lyme disease in the United States each year (CDC 2010), and an unknown number of others are also afflicted. In Europe and Asia, tens of thousands are infected each year, although reporting standards are less stringent (Lane et al. 1991; Barbour and Fish 1993; Lindgren and Jaenson 2006). In its early stages, Lyme disease is often recognized by a characteristic rash called *erythema migrans* (EM rash). Other characteristic early symptoms include headaches, fever, and pain in muscles and joints. At this stage, it can generally be treated effectively with antibiotics. However, if Lyme disease goes undetected or untreated at this early stage, it can become more severe, resulting in neurological symptoms including partial paralysis, endocarditis, and severe joint pain (Feder et al. 2007).

Lyme disease is caused by a bacterium, *Borrelia burgdorferi*, which is passed to humans through the bite of an infected ixodid tick. In the midwestern and eastern United States, the vector is the blacklegged tick, *Ixodes scapularis*. Other ixodid ticks serve as vectors in other areas, including *I. pacificus* in western North America and *I. ricinus* in Europe. Ixodid ticks have three post-egg life stages: larva, nymph, and adult. Each of these three stages takes a single blood meal from any of a wide variety of vertebrate hosts, including humans, before dropping off and molting into the next stage, in the case of larvae and nymphs, or dropping off and reproducing, in the case of adults.

There are two points about the life cycle of the tick that heavily influence the ecology of the disease. The first is that the larvae hatch out of the eggs uninfected with the bacterium that causes Lyme disease, so the larvae are not dangerous to people (Patrican 1997). Ticks can become dangerous to people only if they pick up the infection during one of their blood meals. If they do, then they will retain this infection through their nymphal and adult stages. The second key point is that most cases of Lyme disease are caused by infected nymphal ticks (Barbour and Fish 1993). These infected nymphs picked up the bacterium during their larval meal. Therefore, if we want to understand how ticks become dangerous to people, we need to focus on the sequence of events in nature during which uninfected larvae become infected nymphs.

At our study sites, larvae seek hosts in late summer through early autumn. When a host closely approaches a host-seeking ("questing") larva, the tick can get on and attempt to find a location where it can feed on the host while avoiding being groomed off. If it avoids host grooming, it will take a blood meal, during which it may acquire the bacterium. When the meal is over, the larva disengages from the host and drops off onto the ground, molts into the nymphal stage about a month later, and then overwinters before seeking another blood meal (Fig. 5.1). Blacklegged ticks take their blood meals on a wide variety of vertebrate hosts, including mammals, birds, and reptiles. Does the kind of animal a larva feeds on affect its chances of making the transition from an uninfected larva to an infected nymph? In the past few years, we have been investigating that question through a series of experiments.

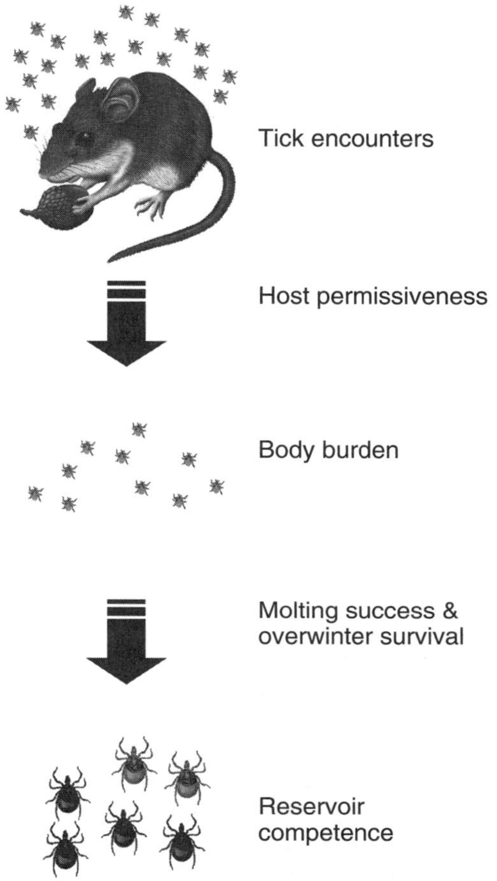


**Figure 5.1:**
Stages in the life cycle of blacklegged ticks (*Ixodes scapularis*) as they encounter and attempt to feed on a host (in this case a white-footed mouse), molt into the nymphal stage and overwinter, and either acquire the bacterium that causes Lyme disease or do not (reservoir competence). The fate of a tick through each transition is dependent on the identity of the host it encounters (see text).

## THE IMPORTANCE OF SPECIES IDENTITY

To ask whether larvae are more likely to feed successfully on some kinds of animals than others, we captured six common species of hosts for larvae at our study site in New York: white-footed mice (*Peromyscus leucopus*), eastern chipmunks (*Tamias striatus*), gray squirrels (*Sciurus carolinensis*), veeries (*Catharus fuscescens*), catbirds (*Dumatella carolinensis*), and opossums (*Didelphis virginiana*). These species were selected to provide a representative sample of hosts across a range of taxa and body sizes. We held individual hosts in cages until we were sure that they no longer carried any ticks they had acquired in the field, and then we carefully placed 100 larval ticks on each animal and let the ticks attempt to feed. We held the animals in special cages so that we could count any ticks that fell off the hosts, whether they had fed successfully or not. Any ticks that we did not recover had been destroyed during host grooming; most were apparently swallowed. Our results clearly showed that larval ticks were better able to feed on some hosts than others (Keesing et al. 2009). At one extreme, about half of the ticks that were placed on mice were able to feed successfully, while at the other extreme, almost none of the ones placed on opossums were (Fig. 5.2).

In addition to knowing the *percentage* of ticks that feed successfully on each host species and the percentage that are killed in the attempt to feed, we also know how many larvae actually feed to repletion on each species of host when they range freely in natural habitats. This is because we capture free-ranging hosts during the season of larval activity and hold each animal sufficiently long for all naturally acquired ticks to fall off after a successful meal. Using these two pieces of information—the number of ticks that feed successfully from a given host species and the percentage of ticks attempting to feed that actually do so—we can estimate how many larvae *attempt* to feed on each species; by subtraction, we can determine how many are killed while trying. For example, the average mouse has about 25 larval ticks that feed successfully on it, but we know that only about 50% of the ones that try to feed manage to feed successfully. This allows us to estimate that about 50 larvae try to feed on each mouse, while only about half of these (25) manage to do so. In contrast, only about 3% of larvae that attempt to feed on an opossum feed successfully, and yet the average opossum has about 200 ticks that have fed to repletion. This allows us to estimate that the average opossum encounters about 5,000 ticks, of which the opossum consumes 97% during grooming before the ticks can feed (Keesing et al. 2009).

After feeding, larval ticks need to molt into nymphs and survive the winter before they can take their next meal. To find out whether the different kinds of hosts affect how well larvae molt and survive over the winter, we put fed larvae from different hosts into special tubes that we had filled with intact cores of forest soil. With the ticks inside, we covered the tubes with fine mesh cloth and then returned each tube to the forest floor for the winter. When we removed the tubes in the spring, the results were very clear. Fed larvae from mice, birds, and chipmunks had molted and survived the winter much better than fed larvae

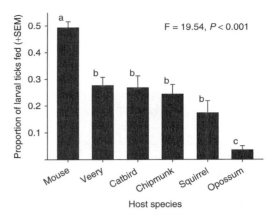

**Figure 5.2:**
The proportion of larval ticks that fed successfully (+s.e.m.) on six species that are common hosts for larval blacklegged ticks (*Ixodes scapularis*) in upstate New York. Hosts were captured in the field and held in the laboratory until ticks naturally feeding on them had fed to repletion and dropped off. Hosts were then reinfested with 100 larval ticks and monitored to determine the proportion of those ticks that fed successfully. Lowercase letters indicate results that were significantly different (one-way ANOVA; $p = 0.05$). Modified from Keesing et al. 2009.

from opossums and squirrels had (Brunner et al. 2011).

Finally, we can figure out how many larvae pick up the Lyme bacterium from each host species because we test nymphs from each host species to determine what proportion are carrying the infection. Over 90% of larval ticks that feed on white-footed mice pick up the bacterial infection from the mice, while only 50% of the ticks that feed on chipmunks do, and only 3% of the ticks that feed on opossums do (Fig. 5.3).

From these experiments, we can conclude that the most important factor determining the probability that an uninfected larval tick will become an infected nymph is what host species it feeds on. If a larva attempts to feed on a mouse, it is likely to have a successful meal, molt successfully and survive the winter, and pick up an infection that makes it potentially dangerous to humans. In contrast, if a larva tries to feed on an opossum, it is likely to get groomed off and killed. If it does manage to feed successfully, it is unlikely to molt and survive the winter; even if it does, it probably will not have picked up the Lyme bacterium anyway.

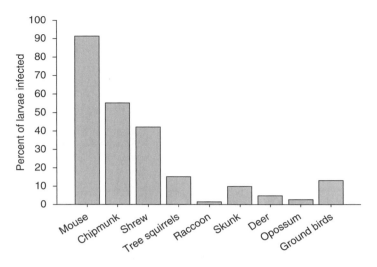

**Figure 5.3:**
The percentage of uninfected larval ticks that become infected after feeding on a particular host species. This is called the reservoir competence of the host species. Data from Keesing et al. 2009 and LoGiudice et al. 2003.

## THE ROLE OF DEER

We have discussed the importance of individual species in feeding and infecting ticks, but we have not yet discussed one host species that is frequently associated with discussions of Lyme disease: the white-tailed deer (*Odocoileus virginianus*). Many people assign deer the central role in determining Lyme disease risk because it has been considered the most important host for adult female ticks. Some researchers contend that adult females feed almost exclusively on deer during the fall. In the spring, these females lay eggs, producing the next generation of ticks that can become infected and thereby dangerous. As a consequence, the reasoning goes, if there are more deer, there will be more ticks. The origins of this dogma and its consequences have recently been thoroughly reviewed (Ostfeld 2011), so here we will only summarize the evidence that the story is not this simple.

Investigations of the role of deer in determining Lyme disease risk have typically taken one of two approaches. In the first approach, investigators reduce deer abundance by hunting or they exclude them with fencing. In the second approach, investigators determine if the abundance of ticks is correlated with the abundance of deer. Collectively, these experiments

have yielded strikingly inconsistent results (Ostfeld 2011), in part because species other than deer also serve as hosts for adult ticks, particularly when deer are at low abundance. Keirans et al. (1996) documented that adult blacklegged ticks feed on at least 17 species of mammals, and therefore clearly are not specialists on white-tailed deer. Despite their broad range of hosts, no studies to date have compared the distribution of adult blacklegged ticks among various hosts within a community. The only study in which blacklegged tick populations were extirpated in response to the complete elimination of deer occurred on Monhegan Island, Maine, where no other hosts for adult ticks were present (Rand et al. 2003). In addition, as deer populations are reduced by hunting, the abundance of adult ticks on the remaining deer increases such that virtually equivalent numbers of adult ticks feed on the reduced deer population (Deblinger et al. 1993). We suspect that the ticks also aggregate on non-deer hosts when deer decline, although no studies to date have assessed this possibility.

As we develop a more sophisticated and subtle view of the role of deer in determining tick abundance, we also have to incorporate another crucial piece of evidence. In two decades of work by our group, we have found that the abundance of larval ticks does not predict the abundance of nymphal ticks (Ostfeld et al. 2006; Fig. 5.4). In other words, even if the abundance of deer *were* related in a simple way to the abundance of larval ticks that hatch out of eggs, this relationship would not predict the abundance of the nymphal ticks that are actually dangerous to humans. Indeed, in some ways, deer are protective because they feed many larval ticks but infect almost none of them (LoGiudice et al. 2003). In this way, they are ensuring that a smaller proportion of ticks are infected than would be otherwise. We will discuss this issue in greater depth in the next section.

## THE IMPORTANCE OF THE COMMUNITY OF HOSTS

The results we have described so far demonstrate that the identity of the host species determines the probability that a larval tick will survive and become an infected nymph capable of transmitting infection to another host, including a human. The experiments described above tell us how many infected nymphs an individual of each of our representative species of hosts will produce. We have sought to use this information to predict how many infected nymphs will be

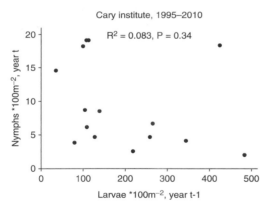

Cary institute, 1995–2010

$R^2 = 0.083, P = 0.34$

Nymphs *100m$^{-2}$, year t

Larvae *100m$^{-2}$, year t-1

**Figure 5.4:**
The relationship between the number of larvae per hectare in one year and the number of nymphs the following year over 15 years of study at the Cary Institute in Millbrook, NY. The lack of a relationship demonstrates that something other than the number of larval ticks determines how many ticks molt and survive to become nymphs the following year. This contradicts the prevailing view that deer would be critical to Lyme disease risk if they determine the number of larvae that hatch the next year (see text).

produced by a particular community of vertebrate hosts for ticks. To make quantitative predictions, we need to estimate the population densities of each of these hosts. Years of extensive live-trapping efforts in the forests at our study site, combined with data from other studies, allow us to estimate the abundance of each species in the forest. In a one-hectare patch of intact forest, for example, there are about 30 mice, 15 chipmunks, one opossum, four catbirds, and a suite of other species (LoGiudice et al. 2008; Keesing et al. 2009). We can use this information, combined with the data from our laboratory experiments described above, to do some simple calculations. To figure out how many infected nymphal ticks there should be in a typical patch of woods, we simply add up the number that should be infected from each host species—in other words, the number infected by mice plus the number infected by chipmunks plus the number infected by raccoons, and so on for all host species in the community (Keesing et al. 2009). Determining the number infected by each host is also simple; at least once we have all the numbers. Each mouse feeds about 25 larvae successfully. Of these, about half molt and survive overwinter to become nymphs, and 92% of these have the Lyme bacterium from their mouse host. So for each mouse, there should be about 11 infected nymphs. In total, then, if there are 30 mice, then there will be over 300 nymphs infected by all of those mice. We can make the same calculation for each of the other hosts in the forest. When we do, we find that most of the infected ticks in an intact forest took their larval meals from mice (Fig. 5.5).

One way to interpret this result is that mice are a critical host in the ecology of Lyme disease because most infected vectors got infected from them. Indeed, scientists are trying to figure out which hosts infect the most vectors for many diseases. One prominent example is the emerging disease West Nile virus (WNV) encephalitis, which is caused by a virus transmitted among hosts by mosquito vectors. In several studies of WNV, investigators have determined, through painstaking experimentation, which of the infected mosquitoes they capture got infected by which host species (Loss et al. 2009), as we have done for mice with Lyme disease. For WNV, American robins (*Turdus migratorius*) emerge as the host species that most infected mosquitoes have fed on, which has been interpreted as indicating that robins might be the

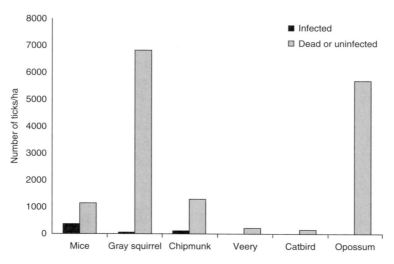

**Figure 5.5:**
The number of ticks per hectare predicted to be infected (black bars) or either killed or not infected (gray bars) after encountering one of six common hosts for ticks in northeastern U.S. forests. White-footed mice are responsible for infecting the greatest number of ticks, but other species play a much bigger role in either killing or feeding but not infecting ticks. Data from Keesing et al. 2009.

most important host species for this disease. If so, management to reduce WNV transmission could focus on reducing robin abundance. But our studies reveal that something more complicated might be going on: while the role of hosts in infecting other hosts is clearly important, the role of hosts in *preventing* infection could be just as important. Opossums and squirrels are responsible for killing, or for feeding but *not* infecting, many hundreds of ticks in the Lyme disease system (Fig. 5.5). Effective management of forest habitat to reduce Lyme disease risk, then, should focus just as much on maintaining opossum and squirrel populations as on reducing mouse populations.

Having quantified the role of each host in an intact forest community, we can now ask what would happen if each host species were to be lost from the forest. For example, how would Lyme disease risk change if there were no mice, or if there were no squirrels, in a patch of forest? We have begun to address these questions using computer simulations to remove hosts from a virtual community that initially consists of all the hosts for which we have the full complement of data described above. A fundamental issue in this computer exercise, however, is deciding whether the ticks that would have fed on the host species we remove end up feeding on the hosts that remain. If they do not—if the ticks that would have attempted to feed on a squirrel simply die in the absence of squirrels—then we would simply subtract the number of ticks feeding on squirrels from Figure 5.5. If, however, some or all of the ticks that would have fed on squirrels feed on the hosts that remain, the situation is somewhat more complicated. We have explored with our colleagues the consequences of these different scenarios for Lyme disease risk using a series of simulations.

In the simulation models, we remove each host species one at a time, and then we allow the ticks that would have fed on the removed host species to feed on the remaining hosts to varying degrees. If we remove hosts and their associated ticks simply die, then removing hosts always reduces Lyme disease risk (Fig. 5.6). However, if we allow their associated ticks to redistribute onto the hosts that remain, the picture changes considerably. When opossums are removed from our virtual forest community, for example, Lyme disease risk increases by approximately 35% if we allow just half of the ticks that would have fed on the opossums to find other hosts. Those other hosts are all more likely than opossums to allow the ticks to feed successfully, molt and overwinter, and sustain an infection with the Lyme bacterium. All of these factors contribute to the increase in risk with opossum removal.

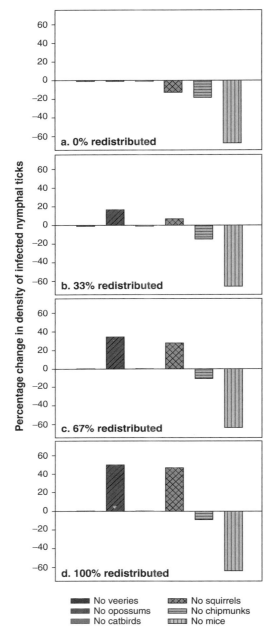

**Figure 5.6:**
The percentage change in the density of infected nymphal ticks as individual host species (see legend) were removed individually from the model. Ticks that would have fed on the removed host species were redistributed on the remaining hosts: (a) 0%; (b) 33%; (c) 67%; and (d) 100%. Figure modified from Keesing et al. 2009.

Given the different outcomes this model predicts from the removal of host species, it is critical to determine whether ticks redistribute onto other hosts and if so to what degree. This is a difficult question to test empirically, but field evidence does suggest that redistribution of ticks onto other hosts happens, and that it can be quite substantial. The best evidence comes from recent work by Brunner and Ostfeld (2008), who asked what factors affect the number of larval ticks on white-footed mice. After looking at many different factors across many different sites in many different years, they found that the best predictor of changes in the number of larval ticks on mice was the number of chipmunks in the forest at the same time. If there were lots of chipmunks around, there were few larval ticks on mice; if there were few chipmunks, there were many ticks on mice. This evidence strongly suggests that ticks redistribute from chipmunks to mice. A similar effect may occur for other hosts as well, but only further analyses and experiments can determine this with certainty.

In our previous use of the model, we removed one species at a time to observe how the loss of each individual species affected Lyme disease risk. However, conservation biologists have long known that, as forests are fragmented or degraded, some species are more likely to disappear than others. One prominent pattern is that medium-sized to large mammals often disappear from forest patches before small mammals do (Ostfeld and LoGiudice 2003). Others and we have also shown that as species disappear from forest patches, white-footed mice are always the last to go. Indeed, at our forested sites, mice are present at all sites, regardless of the number of other species present (LoGiudice et al. 2008).

We can use this information to ask what would happen to Lyme disease risk if species were lost in sequence—that is, with first one, then two, then three species disappearing, and so on. Using our data and that from other studies, we removed species from our model in a sequence that would be likely under common scenarios of biodiversity loss. The first species we removed was the veery, which is a forest-interior specialist and is sensitive to habitat disturbance. Then we removed opossums as well, followed one after another by squirrels, chipmunks, and finally catbirds. Mice were present in all runs of the model. The results of the model are clear. Lyme disease risk increases with every host removed if more than about 10% of the ticks that would have fed on removed hosts are allowed to redistribute onto the hosts that remain (Keesing et al. 2009; Fig. 5.7). The identity of the hosts removed also matters. When the two bird species are removed, Lyme disease risk increases, but not substantially, whereas the removal of chipmunks increases risk a great deal. This occurs because some species have higher tick burdens, are more abundant, or are more likely to transmit the pathogen to feeding ticks than other species are, as we have described above.

This model, like all models, simplifies the interactions in this system. One potentially crucial simplification is that it assumes that the removal of one host species does not change the abundance of the other host species. This might not in fact be realistic. For example, if chipmunks disappear from a forest fragment, the density of the mice that remain might increase because chipmunks compete with mice for food and other resources (Ostfeld et al. 2006). Similarly, LoGiudice et al. (2008) found that northeastern U.S. forest fragments with more species of vertebrate predators had lower average densities of white-footed mice. Therefore, the loss of predatory

species would be expected to release mice from regulation and thereby increase Lyme disease risk even further. Until more data are available on the direct and indirect effects of forest vertebrates on one another, we will not be able to add this important level of complexity to our models.

These modeling outputs help explain the results we obtained in a previous study (Allan et al. 2003). We compared the density of infected nymphal ticks in forest fragments that ranged in size from less than one hectare to about seven hectares. We predicted that in the smallest fragments, mice would be present and abundant; as a consequence, we expected the density of infected nymphal ticks would be high. In contrast, we predicted that mice would be less abundant and that many other species would be present in the larger fragments. As a consequence, we predicted that Lyme disease risk would be lower. The data from this comparative study confirmed our predictions. We are currently conducting a large-scale manipulation of host communities in a suite of forest fragments in upstate New York to see if fragments with fewer species have higher numbers of infected ticks.

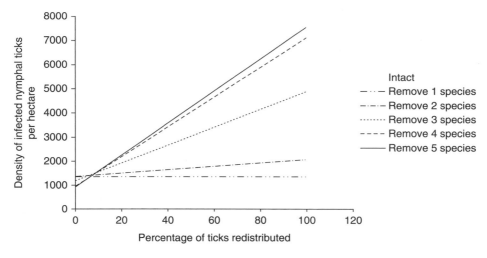

**Figure 5.7:**
The density of infected nymphal ticks (DIN) per hectare as host species were removed sequentially from our model, when the ticks that would have fed on missing hosts were redistributed among remaining hosts from 0% to 100%. Species were removed from the model in an order determined by empirical observations of the sequence of species loss in fragmented forest habitats. Veeries were removed first, and their removal did not result in differences from the density of infected ticks in "intact" forest (the two lines overlap). After veeries, opossums, squirrels, chipmunks, and then catbirds were removed, in that order; white-footed mice were present in all communities (see text). In habitats with two or more species lost, DIN was higher if ticks were redistributed on the remaining hosts, and greater rates of redistribution resulted in higher DIN values. Figure modified from Keesing et al. 2009.

## SEEKING GENERALITY

In the Lyme disease system, there are several distinct mechanisms by which the diversity of host species reduces Lyme disease risk (Keesing et al. 2006). First, in a diverse community, there are many species other than mice for ticks to feed on; in a low-diversity community, there will be mice and there might *only* be mice. Mice are important, as we have discussed, because they are abundant, they feed lots of ticks while killing few, and they are the host most likely to transmit the Lyme bacterium on to ticks. In contrast, larger species such as opossums are less likely to occur in low-diversity communities. The absence of opossums affects Lyme disease risk almost as much as the presence of mice does. This is because opossums are responsible for killing thousands of ticks that attempt to feed on them (see Fig. 5.5), which reduces risk substantially. Of the ones that do manage to feed, almost none become infected (Keesing et al. 2009). In low-diversity communities, those ticks that would have fed on opossums appear likely to feed on other hosts, including mice.

The pattern in the Lyme disease system is striking, but is it unique to Lyme disease? A growing number of studies suggests that it is not. In the past few years, researchers have found a similar pattern for a number of infectious diseases, including human diseases like hantavirus pulmonary syndrome (Suzán et al. 2009; Clay et al. 2009a; Dizney and Ruedas 2009) and West Nile virus (Ezenwa et al. 2006; Allan et al. 2009; Swaddle and Calos 2008), parasitic infections of amphibians (Johnson et al. 2008), and fungal infections of grasses (Mitchell et al. 2002; Roscher et al. 2007).

The role of diversity in affecting risk for hantavirus pulmonary syndrome provides a particularly interesting counterpoint to Lyme disease because hantaviruses are transmitted directly among their hosts, with no vector such as a tick involved. Hantaviruses are a group of negative-stranded RNA viruses that circulate primarily among wild rodents. The viruses are shed through excreta (e.g., urine, feces) and saliva; humans can be exposed if they breathe aerosolized excreta or get bitten by an infected rodent host. For people who get infected, the outcome is not good: case fatality rates are near 40% (CDC 2009).

The probability that a human will get exposed is a function of the density of infected rodent hosts

(Yates et al. 2002), so recent studies have focused on what determines this key risk factor. In Oregon, for example, Dizney and Ruedas (2009) compared the proportion of infected rodents at a series of sites. Infection prevalence varied from 2% to 14% at their sites, and the only factor that explained this variation was the diversity of mammalian species. Clay et al. (2009a) found a similar pattern in Utah: sites with a high diversity of mammals had a low infection prevalence with the virus. Based on experimental work, it appears that this effect is because the deer mice (*Peromyscus maniculatus*) that serve as the primary hosts for the virus are less likely to encounter each other—and more likely to encounter individuals of other species—in more diverse communities (Clay et al. 2009b). As a consequence, transmission of the virus occurs at a lower rate in more diverse communities.

Suzán et al. (2009) provide a third example of the effects of host diversity on hantavirus risk. In this study, rodent communities in Panama were experimentally manipulated—species that are not hosts for the virus were removed from replicated experimental plots, while communities remained intact on control plots. The density of seropositive rodents was higher on the manipulated plots (Suzán et al. 2009) where diversity had been experimentally reduced. In hantavirus pulmonary syndrome and all of the other examples we mentioned above, host communities with higher diversity are correlated with lower rates of disease transmission (reviewed in Keesing et al. 2006, Johnson and Thieltges 2010; Keesing et al. 2010). There are counter-examples (Loss et al. 2009), but the pattern appears to occur more often than not. Why should diversity so frequently reduce disease transmission? The answer is perhaps best understood from another look at the Lyme disease system.

In the Lyme disease system, the mouse is an excellent host species for both ticks and spirochetes *and* a species reliably present in low-diversity communities. It is this correlation that drives the effects of diversity loss on disease in this system. One key question, then, is whether a similar correlation occurs in other disease systems and, if it does, what is the underlying biology. One hypothesis is that small-bodied, short-lived species that are particularly resilient to disturbance might allocate their immunological resources differently than longer-lived, larger-bodied species that are more susceptible to disturbance. Another hypothesis is that ticks and spirochetes adapt to the hosts that they encounter most frequently. We have begun testing both hypotheses together. The apparent link between a host's ability to amplify pathogens and its resilience in the face of biodiversity loss (Ostfeld and Keesing 2000) suggests that low-diversity communities will often pose a strong risk of exposure to disease, which will be reduced in higher-diversity communities. If the pattern is indeed general, then protection against infectious diseases can legitimately be considered an ecosystem function that is enhanced by the protection of biodiversity.

## ACKNOWLEDGMENTS

The authors gratefully acknowledge financial support from the National Institutes of Health, the National Science Foundation, the U.S. Environmental Protection Agency, and the National Science Foundation–National Institutes of Health joint program in the ecology of infectious diseases. The work described here would not have been possible without the help of Shannon Duerr, Kelly Oggenfuss, Jesse Brunner, Kathleen LoGiudice, Mary Killilea, Ken Schmidt, Andrea Previtali, Anna Jolles, Marty Martin, Rhea Hanselmann, Laura Cheney, Charles Canham, Deanna Sloniker, and Holly Vuong.

## REFERENCES

Allan, B.F., F. Keesing, and R.S. Ostfeld. 2003. Effect of forest fragmentation on Lyme disease risk. Conserv Biol 17:267–272.

Allan, B.F., B. Langerhans, W.A. Ryberg, W.J. Landesman, N.W. Griffin, R.S. Katz, B.J. Oberle, M.R. Schutzenhofer, K.N. Smyth, A. de St. Maurice, L. Clark, K.R. Crooks, D.E. Hernandez, R.G. McLean, R.S. Ostfeld, and J.M. Chase. 2008. Ecological correlates of risk and incidence of West Nile virus in the United States. Oecologia 158:699–708.

Barbour, A.G., and D. Fish. 1993. The biological and social phenomenon of Lyme disease. Science 260:1610–1616.

Biodiversity Project. 2002. Americans and biodiversity: new perspectives for 2002. Belden Russonello and Stewart Research and Communications, Washington, DC. Available: http://www.biodiversityproject.org/docs/publicopinionresearch/americansandbiodiversitynewperspectivesfor2002.PDF

Brunner, J.L., and R.S. Ostfeld. 2008. Multiple causes of variable tick burdens on small-mammal hosts. Ecology 89:2259–2272.

Brunner, J.L., L. Cheney, F. Keesing, M. Killilea, K. Logiudice, A. Previtali, and R.S. Ostfeld. 2011. Molting success of Ixodes scapularis varies among individual blood meal hosts and species. J Med Entomol. 48:860–866.

Centers for Disease Control (CDC). 2009. Hantavirus pulmonary syndrome in five pediatric patients: four states, 2009. MMWR 58:1409–1412.

Centers for Disease Control (CDC). 2010. Reported cases of Lyme disease by year, United States, 1994-2008. URL: http://www.cdc.gov/ncidod/dvbid/lyme/ld_UpClimbLymeDis.htm accessed May 6, 2010.

Clay, C., Lehmer, E.M., St. Jeor, S., and Dearing, M.D. 2009a. Sin nombre virus and rodent species diversity: a test of the dilution and amplification hypotheses. PLoS One 4:e6467.

Clay, C.A., E.M. Lehmer, S. St. Jeor, and M.D. Dearing. 2009b. Testing mechanisms of the dilution effect: deer mice encounter rates, Sin nombre virus prevalence and species diversity. Ecohealth 6:250–259.

Deblinger, R.D., M.L. Wilson, D.W. Rimmer, and A. Spielman. 1993. Reduced abundance of immature Ixodes dammini (Acari: Ixodidae) following incremental removal of deer. J Med Entomol 30: 144–150.

Dizney, L.J., and L.A. Ruedas. 2009. Increased host species diversity and decreased prevalence of Sin nombre virus. Emerg Infect Dis 15:1012–1018.

Ezenwa, V.O., M.S. Godsey, R.J. King, and S.C. Guptill. 2006. Avian diversity and West Nile virus: testing associations between biodiversity and infectious disease risk. Proc R Soc B 273:109–117.

Feder, H.M., Jr., B.J.B. Johnson, S. O'Connell, E.D. Shapiro, A.C. Steere, G.P. Wormser, and the Ad Hoc International Lyme Disease Group. 2007. A critical appraisal of "chronic Lyme disease." N Engl J Med 357:1422–1431.

Johnson, P.T.J., and D.W. Thieltges. 2010. Diversity, decoys and the dilution effect: how ecological communities affect disease risk. J Exp Biol 213:961–970.

Johnson, P.T.J., P.J. Lund, R.B. Hartson, and T.P. Yoshino. 2009. Community diversity reduces Schistosoma mansoni transmission and human infection risk. Proc R Soc B 276:1657–1663.

Johnson, P.T.J., R.B. Hartson, D.J. Larson, and D.R. Sutherland. 2008. Linking biodiversity loss and disease emergence: amphibian community structure determines parasite transmission and pathology. Ecol Lett 11:1017–1026.

Keesing, F., L. Belden, P. Daszak, A. Dobson, D. Harvell, R.D. Holt, P. Hudson, A. Jolles, K. Jones, C. Mitchell, S. Myers, T. Bogich, and R. Ostfeld. 2010. Impacts of biodiversity on the emergence and transmission of infectious diseases. Nature 468:647–652.

Keesing, F., J. Brunner, S. Duerr, M. Killilea, K. LoGiudice, K. Schmidt, H. Vuong, and R.S. Ostfeld. 2009. Hosts as ecological traps for the vector of Lyme disease. Proc R Soc B 276:3911–3919.

Keesing, F., R.D. Holt, and R.S. Ostfeld. 2006. Effects of species diversity on disease risk. Ecol Lett 9:485–498.

Keirans, J. E., H. J. Hutcheson, L. A. Durden, and J. S. H. Klompen. 1996. *Ixodes (Ixodes) scapularis* (Acari: Ixodidae): Redescription of all active stages, distribution, hosts, geographical variation, and medical and veterinary importance. J Med Entomol 33:297–318.

Lane, R.S., J. Piesman, and W. Burgdorfer. 1991. Lyme borreliosis: relation of its causative agent to its vectors and hosts in North America and Europe. Ann Rev Entomol 36:587–609.

Lindgren, E., and T.G.T. Jaenson. 2006. Lyme borreliosis in Europe: influences of climate and climate change, epidemiology, ecology and adaptation measures. World Health Organization, Geneva, Switzerland.

LoGiudice, K., R.S. Ostfeld, K. Schmidt, and F. Keesing. 2003. The ecology of infectious disease: Effects of host diversity and community composition on Lyme disease risk. Proc Natl Acad Sci USA 100:567–571.

LoGiudice, K., S. Duerr, M. Newhouse, K.A. Schmidt, M.E. Killilea, and R.S. Ostfeld. 2008. Impact of community composition on Lyme disease risk. Ecology 89:2841–2849.

Loss, S.R., G.L. Hamer, E.D. Walker, M.O. Ruiz, T.L. Goldberg, U.D. Kitron, and J.D. Brawn. 2009. Avian host community structure and prevalence of West Nile virus in Chicago, Illinois. Oecologia 159:415–424.

Mitchell, C.E., D. Tilman, and J.V. Groth. 2002. Effects of grassland plant species diversity, abundance, and composition on foliar fungal disease. Ecology 83:1713–1726.

Ostfeld, R.S. 2011. Lyme disease: the ecology of a complex system. Oxford University Press, New York.

Ostfeld, R.S., M. Thomas, and F. Keesing. 2009. Biodiversity and ecosystem function: perspectives on disease. *In* Naeem, S., D. Bunker, A. Hector, M. Loreau, and C. Perrings, eds. Biodiversity and human impacts, pp. 209–216. Oxford University Press, New York.

Ostfeld, R.S., F. Keesing, and K. LoGiudice. 2006. Community ecology meets epidemiology: the case of Lyme disease. *In* Collinge, S., and C. Ray eds, Disease ecology: community structure and pathogen dynamics, pp. 28–40. Oxford Press, United Kingdom.

Ostfeld, R.S., and K. LoGiudice. 2003. Community disassembly, biodiversity loss, and the erosion of an ecosystem service. Ecology 84:1421–27.

Ostfeld, R.S., and F. Keesing. 2000. The function of biodiversity in the ecology of vector-borne zoonotic diseases. Can J Zool 78:2061–2078.

Patrican, L.A. 1997. Absence of Lyme disease spirochetes in larval progeny of naturally infected *Ixodes scapularis* (Acari: Ixodidae) fed on dogs. J Med Entomol 34:52–55.

Rand, P.W., C. Lubelczyk, G.R. Lavigne, S. Elias, M.S. Holman, E.H. Lacombe, and R.P. Smith. 2003. Deer density and the abundance of *Ixodes scapularis* (Acari: Ixodidae). J Med Entomol 40: 179–184.

Roscher, C., J. Schumacher, O. Foitzik, and E.D. Schulze. 2007. Resistance to rust fungi in *Lolium perenne* depends on within-species variation and performance of the host species in grasslands of different plant diversity. Oecologia 153:173–183.

Suzán, G., E. Marcé1, J.T. Giermakowski, J.N. Mills, G. Ceballos, R.S. Ostfeld, B. Armién, J.M. Pascale, and T.L. Yates. 2009. Experimental evidence for reduced rodent diversity causing increased Hantavirus prevalence. PLoS One 4:e5461.

Swaddle, J., and P. Calos. 2008. Increased avian diversity is associated with lower incidence of human West Nile infection: Observation of the dilution effect. PLoS One 3:e2488.

Yates, T.L., J.N. Mills, C.A. Parmenter, T.G. Ksiazek, R.R. Parmenter, J.R. Vande Castle, C.H. Calisher, S.T. Nichol, K.D. Abbott, J.C. Young, M.L. Morrison, B.J. Beaty, J.L. Dunnum, R.J. Baker, J. Salazar-Bravo, and C.J. Peters. 2002. The ecology and evolutionary history of an emergent disease: Hantavirus pulmonary syndrome. Bioscience, 52:989–998.

# 6

# PARASITE CONSERVATION, CONSERVATION MEDICINE, AND ECOSYSTEM HEALTH

Andrés Gómez, Elizabeth S. Nichols, and Susan L. Perkins

Conservation medicine links disparate fields of study to understand the underlying causes of ecological health problems and the ecological context of health (Tabor 2002). Often this requires analyzing complex and highly interlinked processes that are seldom well understood. The causes and consequences of changes in parasite biodiversity represent an important but overlooked aspect of conservation medicine. In this chapter we explore the quandary presented by parasites: this diverse suite of organisms threatens the health of individuals and populations but is nonetheless critical for maintaining healthy ecosystems (Hudson et al. 2006).

Parasites[1] appropriate host resources to feed and/ or to reproduce, lowering host fitness (albeit at widely varying degrees) through effects on physiology, morphology, and/or behavior. Parasites can cause mortality and severe morbidity, including disability and nutritional, growth, and cognitive impairments, and affect food production, economic trade, and biodiversity conservation. As a consequence, and for good reason, the stance of the medical sciences towards parasites is best characterized by direct antagonism. Indeed, some of the greatest achievements

in medical history have been the extirpation or eradication of pathogenic organisms. However, parasites are ubiquitous, and their effects on the overall functioning of ecosystems are complex and often positive.

Until recently, within the field of ecology, parasites have been overlooked as unimportant to the maintenance of viable ecosystems, perhaps due to their cryptic nature, intractably high diversity, small size, patchy distribution, and a general lack of observable morbidity in nature. Similarly, within conservation biology, practitioners and academics are more prone to evaluate the role of parasites as threats to host populations than to discuss the need for conserving host–parasite relationships (Gompper and Williams 1998). This may come as little surprise in a field long plagued by taxonomic biases, in which invertebrates, fungi, and microscopic organisms have historically received the least attention.

It is within this mixture of open antipathy, disregard, and lack of knowledge and awareness that parasite biodiversity is most often viewed today. However, over the past two decades, several lines of research have highlighted the critical importance of parasite biodiversity to the central goals of both

---

[1] Here we use the term "parasite" to refer to macro and microparasite taxa. We do not consider nest parasites or parasitoids in this chapter.

conservation biology (the long-term maintenance of biodiversity) and conservation medicine (maintaining ecological health). The numerical dominance of parasites in ecosystems and their roles in ecological and evolutionary processes suggest a critical need for explicit attention to parasite conservation. This task is both important and challenging. As parasites are deleterious for their host, they have the potential to negatively affect the conservation of biodiversity (Cunningham and Daszak 1998; Leendertz et al. 2006). To conserve parasites is to explicitly conserve morbidity and mortality in host populations. This is not a charge to take lightly, but it is likely a necessary step towards the conservation of overall ecosystem integrity.

This chapter outlines the current state of parasite conservation. We make a general case for *why* parasite conservation should be considered by briefly covering the contributions of parasites to ecosystems in both ecological and evolutionary time. To discuss *how* parasite conservation is currently achieved, we focus on the few existing parasite-focused conservation strategies, and offer some suggestions on how such strategies could be implemented. Finally, we highlight some of the challenges and conflicts that will likely arise in the implementation of such strategies.

## WHY CONSERVE PARASITES?

As with all biodiversity, the overarching drive to conserve parasite species or host–parasite interactions can be outlined in terms of utilitarian or intrinsic value arguments. Utilitarian arguments are invoked to engender support for the conservation of functions that result in benefits for human health and well-being, and as a means to quantify the potential losses (e.g., in economic or in quality of life metrics) that could result from biotic impoverishment. The scientific literature has shed light on the functional importance of parasites at different spatial and temporal scales. Much has been written about the importance of parasite biodiversity to ecosystem structure and function (Marcogliese 2004; Thompson et al. 2005; Christe et al. 2006; Lefevre et al. 2009); their role in the evolution of complex structures, mechanisms, and behaviors through evolutionary time (Loehle 1995; Wegner et al. 2003; Blanchet et al. 2009); and their ability to

inform our understanding of ecology and evolution (Whiteman and Parker 2005; Nieberding and Olivieri 2007). In the following sections we highlight some of these roles, as well as ecosystem services mediated by parasites.

Any dialogue about biodiversity conservation is incomplete without mention of intrinsic value. Ethical, moral, and aesthetic arguments for conserving parasites are equally strong as those invoked for the conservation of their hosts (Daszak and Cunningham 2002). Eloquent discussions of this issue are provided by Gompper and Williams (1998) and Windsor (1997, 1998a, 1998b).

## PARASITE DIVERSITY

Of the approximately 42 recognized phyla, nine are entirely parasitic, 22 are predominantly parasitic, and most of the remaining have multiple parasitic clades (Poulin and Morand 2000; DeMeeûs and Renaud 2002). Recent studies estimate that parasitic helminths alone are about twice as speciose as their vertebrate hosts (Poulin and Morand 2000; Dobson et al. 2008) (Table 6.1). Estimating global parasite diversity remains challenging; current estimates do not yet incorporate most host taxa and focus only on macroscopic parasites, most vertebrates are inadequately sampled for parasites, and many of the parasite species upon which these estimates are based are likely to represent clusters of cryptic species (Poulin and Morand 2004; Dobson et al. 2008). Notably, microscopic parasitic diversity is also vast (Angly et al. 2006; Cotterill et al. 2008).

Overall, while we know that parasites outnumber free-living species, much of parasite biodiversity, which includes an astonishing variety of taxa, morphologies, life histories, and transmission modes, remains largely unknown. This non-monophyletic group also spans broad variation in the magnitude and direction of effects on individuals, host populations, and ecosystems, such that the overall effect of parasitism in one ecosystem or in a particular function can be quite different in another (Lafferty 2008). Therefore, while generalizations about the functional relevance or threat levels of parasite biodiversity are difficult, we suggest that common perceptions of parasites severely overlook relevance and underestimate threat.

**Table 6.1  Estimated Number of Vertebrate Host Species Endangered or Extinct Based on the 2009 IUCN Red List, and Corresponding Estimates of Parasitic Helminth Species at Risk of Vertebrate Definitive Host-Dependent Extinction**

|  | Chondrichthyes | Osteichthyes | Amphibia | Reptilia | Aves | Mammalia | Total |
|---|---|---|---|---|---|---|---|
| Critically endangered and endangered hosts | 67 | 535 | 1,238 | 243 | 554 | 637 | 3,274 |
| Associated estimate of critically endangered and endangered parasitic helminths | | | | | | | |
| Trematoda | 2 | 27 | 54 | 82 | 203 | 254 | 623 |
| Cestoda | 64 | 21 | 15 | 19 | 365 | 337 | 821 |
| Acanthocephala | – | 2 | 5 | 1 | 6 | 10 | 24 |
| Nematoda | 5 | 8 | 126 | 116 | 171 | 67 | 492 |
|  |  |  |  |  |  | Total | 1,959 |

|  | Chondrichthyes | Osteichthyes | Amphibia | Reptilia | Aves | Mammalia | Total |
|---|---|---|---|---|---|---|---|
| Extinct and extinct in the wild | 0 | 103 | 39 | 22 | 137 | 78 | 379 |
| Associated estimate of extinct parasitic helminths | | | | | | | |
| Trematoda | 0 | 5 | 2 | 7 | 50 | 31 | 96 |
| Cestoda | 0 | 7 | 1 | 5 | 60 | 25 | 97 |
| Acanthocephala | – | 1 | 0 | 0 | 1 | 1 | 3 |
| Nematoda | 0 | 3 | 5 | 30 | 28 | 5 | 71 |
|  |  |  |  |  |  | Total | 266 |

Based on the methods and estimated vertebrate host and global parasitic helminth diversity data from Dobson et al. (2008).

## PARASITES AND ECOSYSTEMS

Beyond their preponderance in global biodiversity estimates, within individual ecosystems, parasite biomass can surpass the biomass of other much larger-bodied groups and dominate food webs (Suttle 2005; Lafferty et al. 2006). Recent research has shown that food webs contain on average more parasite–host links than predator–prey links (Lafferty et al. 2008). This sheer abundance confers a remarkable role in maintaining a "cohesive matrix" of interactions within food webs (Lafferty et al. 2006). While our theoretical understanding of these relationships remains broadly based on free-living organisms, empirical evidence demonstrates a positive influence of parasites on food web connectance: the percent of a food web's realized resource use links (Lafferty et al. 2006). Connectance has empirical and theoretical positive relationships with food web stability (Dunne et al. 2002; Lafferty et al. 2006) and is positively associated with food web robustness in the face of secondary extinctions

(Bascompte and Jordano 2007; Dunne and Williams 2009).

We now understand parasites as major drivers of ecosystem organization, capable of shaping community and ecosystem ecology (Hudson et al. 2006; Holdo et al. 2009). Parasites can influence host distribution across all spatial scales. For example, within a single water column, trematodes (*Microphallus papillorobustus*) effectively split populations of their gammarid shrimp (*Gammarus insensibilis*) hosts by causing infected individuals to live closer to the surface than their uninfected counterparts (Ponton et al. 2005); at a continental scale, trypanosomiasis has long had a role in determining the distribution of humans and animals in Africa (Knight 1971; Rogers and Randolph 1988). Parasites can also affect host abundance in a variety of ways. Direct mortality from infection can modulate host abundance, but sub-lethal infection can also result in changes in host population size, for instance by altering host susceptibility to predation (Mouritsen and Poulin 2003). Parasites can lower host and/or vector populations by causing altered sex ratios, lowered reproductive output, infertility, or abortion. In the most extreme case, parasitic castrators will prevent the host from reproducing altogether. The host then becomes the parasite's "extended phenotype": the parasitized host is removed from the gene pool and its metabolic output will only contribute to the parasite's reproduction (Lafferty and Kuris 2009).

Overall, parasites play significant roles in mediating species coexistence, community composition, and species diversity across space and time (Mouritsen and Poulin 2005; Freckleton and Lewis 2006; Wood et al. 2007). Many parasites have differential effects on sympatric host species, becoming the modulators of direct and apparent competition (situations in which the presence of one species lowers another's fitness through the presence of a shared enemy) outcomes. Models suggest that the replacement of the native red squirrel (*Sciurus vulgaris*) by the introduced gray squirrel (*S. carolinensis*) in the United Kingdom has been driven by apparent competition mediated by a parapoxvirus (Tompkins et al. 2003). These effects are not limited to free-living species: parasites can affect the host's morphology and immunological landscape, thereby altering its suitability for other parasite species, both promoting and inhibiting parasite coexistence (Thomas et al. 2005; Jolles et al. 2008; Lefevre et al. 2009)

Often these modulating roles of parasite activity translate to large effects on energy transfer across ecosystems. In some cases, parasites manipulate host behavior to increase their (the parasites) predation risk, thereby promoting increased transfers of matter and energy between different compartments within an ecosystem (Kuris 2005). For example, infection with a nematomorph worm (*Paragordius tricuspidatus*) induces crickets (*Nemobius sylvestris*) to commit "suicide" by jumping into water (Thomas et al. 2002); in the absence of the parasite, the trophic interactions that complete the transmission cycle in the water would not take place. Also, recent research has shown that viruses are critical players in global geochemical, nutrient, and energy cycles. Marine viruses kill an estimated 20% of the total oceanic bacterial biomass each day, and the resultant movement of matter from living organisms to dissolved nutrient pools ultimately channels as much as 25% of the ocean's primary production (Suttle 2007), and influence respiratory rates, productivity, and the physical-chemical properties of sea water (Suttle 2005).

Future research will undoubtedly uncover further additional parasite effects on ecosystem processes.

## PARASITES AND EVOLUTION

Parasitic organisms have also played a large role in the evolutionary processes of their hosts. Some of the classic examples of this are the "arms races" that occur between parasites and hosts, with genetic changes in one species resulting in a selective pressure on the other species, which then evolves as well. This general phenomenon has been termed the "Red Queen" hypothesis, after Lewis Carroll's character that must always keep running "just to keep in the same place." For example, *Microphallus* trematode parasites in New Zealand ponds track the genotypes of their *Potamopyrgus* snail hosts (Lively 1989). Parasites also have contributed to the maintenance of sexual reproduction in their hosts (as opposed to clonal reproduction, which would leave twice as many daughters as those produced in a biparental population), via the advantage that rare genetic variants, produced from recombination, have in environments where there are parasites. This has been observed in nature in both the same snail species described above (Lively 1987; Jokela et al. 2009) and in topminnow fish (*Poeciliopsis monacha*) (Lively et al. 1990).

The presence of parasites has also clearly shaped other components of the evolutionary pathways of their hosts. Guralnick et al. (2004) showed that large species of *Cyclocalyx* clams were more than 12 times more likely to be parasitized by allocreadiid trematodes than small species and postulated that tradeoffs between reproductive output and risk of parasitism may have resulted in numerous transitions in body size in the evolutionary diversification in this group of hosts. A similar trend has been observed in fish with parasites imposing selection pressures on their hosts to mature at a smaller body size (Morand 2003). Diverse parasites may have also shaped fundamental physiological properties of organisms. Morand and Harvey (2000) showed a significant positive correlation between parasite richness and the basal metabolic rate in mammals and a significant negative correlation between parasite diversity and host longevity. They hypothesized that parasites cause increased metabolic demands due to maintenance of the immune system.

The powerful selective force of parasites has also shaped human evolution, and the most famous pathogen for this (perhaps because it exerts the strongest pressure) is the malaria parasite, *Plasmodium falciparum*, and the maintenance of the sickle-cell gene. Although people homozygous for the allele suffer from a severe reduction in life expectancy (Platt et al. 1994), heterozygotes enjoy a resistance to malaria.

## PARASITE-MEDIATED ECOSYSTEM SERVICES

### Parasites and Health

In today's world, parasites are significant contributors to the global burden of human disease, have important demographic consequences, and can be obstacles to socioeconomic development (Sachs and Malaney 2002; Lopez et al. 2006). Parasite control and extirpation efforts have indeed increased health and well-being for millions. Yet might the loss of certain host–parasite interactions translate into impaired ecosystem services and diminished human well-being (Table 6.2)?

In developed regions, extreme reductions in the abundance of some parasites may contribute to a new suite of negative human health effects. Recent evidence suggests that contact with parasites (and non-pathogenic microorganisms) modulates the immune system's response to pathogens and allergens. Studies suggest that contact with saprophytic bacteria and parasitic helminths can reduce the risk of immune-mediated disorders such as inflammatory bowel disease and asthma (Falcone and Pritchard 2005; Rook 2009). This idea, termed the "hygiene hypothesis," might contribute to observed patterns of incidence of these disorders in developed countries. Consequently, there is currently empirical support and ongoing clinical trials for the therapeutic use of helminth infestations (Falcone and Pritchard 2005).

Certain parasites provide other kinds of direct benefits to humans. Leeches are used in reconstructive surgery and pain and wound management (Michalsen et al. 2003; Frodel et al. 2004). Parasites can also be potential sources of novel drugs and immune-modulating compounds (Fallon and Alcami 2006; McKay 2006). Parasite molecular processes are used in the research, development, and delivery of medically important molecules: using a virus to bind to specific receptors, therapeutic agents can be

**Table 6.2  Summary of Parasite-Mediated Functions Associated with Actual or Potential Benefits for Hosts**

| Category | Function |
| --- | --- |
| Ecosystem organization | |
| | Regulation of host distribution and abundance |
| | Maintenance of food web structure |
| | Mediation of competitive interactions and apparent competition |
| | Regulation of parasite community |
| | Ecosystem engineering |
| | Nutrient and energy cycling |
| Benefits to hosts | |
| | Modulation of host immune response (hygiene hypothesis) |
| | Provision of molecules and biochemical processes |
| | Sentinels species for environmental stress |
| | Accumulation of heavy metals |
| Information sources | |
| | Biodiversity and environmental indicators/sentinels |
| | Host ecology, ontogeny, and phylogeny |
| | Ecological models |

delivered to target tissues or organs (Douglas and Young 2006). Some parasites (predominantly intestinal helminths) bioaccumulate circulating heavy metals, which often concentrate in parasite tissue at levels orders of magnitude higher than in host tissue, potentially providing a direct service for hosts living in polluted environments (Sures 2003).

Finally, relatively little is known about the competitive interactions across the entire parasite community within a host individual. Loss of highly virulent species from this community is clearly beneficial for the host. However, parasite diversity loss can also have negative ecological and epidemiological consequences for the host (Gompper and Williams 1998). For example, recent research shows that hosts infected with multiple strains of a pathogen may experience reduced mortality (Balmer et al. 2009), and that infection with vertically transmitted parasites can offer protection against viral infection (Hedges et al. 2008). Theoretically, the loss of specialist parasites could increase the risk of infection with generalist species, which often produce more severe morbidity (Dobson and Foufopoulos 2001; Dunn et al. 2009). While much more research is needed to understand the nature, strength, and local ecological context of these complex interactions, these examples suggest that blanket parasite eradication may lead to unintended costs for host populations.

## Environmental and Biodiversity Indicators

Parasite species are also important indicators of environmental stress and can be incorporated into environmental monitoring programs (Marcogliese 2004; Sures 2004). The extent to which toxins accumulate in parasites at rates proportional to their environmental abundance determines the parasite's function as an accumulation indicator. As environmental fluctuations in pollutant levels are more rapidly reflected in parasite tissue (Sures 2003), bioaccumulating parasites can be used as sentinels, capable of signaling health-threatening environmental conditions.

Parasites, particularly those with transmission cycles involving several species, can also function as biodiversity indicators. Parasites may be excellent indicators of food web structure and reflect the presence of diverse intermediate and definitive hosts in

the ecosystem participating in parasite life cycles (Marcogliese 2004; Hechinger and Lafferty 2005; Lafferty et al. 2006). Hechinger and Lafferty (2005) found positive associations between bird diversity and abundance and trematode richness in their intermediate snail hosts. These trematodes can thus be used as useful biodiversity monitors providing information about bird communities through longer time periods and with a lower sampling effort than with host surveys (Hechinger and Lafferty 2005).

## Information Sources

An increased understanding of parasite biodiversity offers us a range of new tools with which to understand the natural world. For example, the inclusion of parasites in food webs alters our understanding of the consequences of biodiversity loss. Traditionally, ecological models have suggested that the highest vulnerability to cascade effects is faced by those species occupying top trophic levels, while alternative models including parasites suggest that mid-trophic-level species are in fact most sensitive, because they are subject to both parasitism and predation (Lafferty et al. 2006). A range of examples of parasites as information models are found in the study of evolution, where the study of parasite diversity can aid in our understanding of host population dynamics, ontogeny, and phylogeny and strengthen hypotheses about niche specialization, adaptation, and speciation (Marcogliese 2005; Whiteman and Parker 2005).

## PARASITE CONSERVATION

### Threats to Parasite Persistence

The modern-day biodiversity crisis is most often portrayed in vertebrate terms, yet it is overwhelmingly a loss of invertebrate life, in which affiliate species such as parasites face the risk of co-extinction as their hosts decline (Dunn 2005; Dunn et al. 2009). Although the extensive diversity of this multiphyletic group means that generalizations about threat levels are difficult, parasites face synergistic threats to their persistence that are cause for conservation concern.

Medical and/or veterinary parasite interventions focused on active control and extirpation represent obvious direct threats to parasite survival. Parasite extirpation is a common goal in public health strategies and captive breeding and wildlife management programs. The subsequent attrition of non-target parasites can be expected to be high when and where control programs use broad-spectrum techniques. For example, broad-spectrum molluscicides are used in campaigns to control several parasites of public health and veterinary importance. However, snails are also hosts to a variety of other parasite species that become unintended and unseen targets of these strategies (Kristensen and Brown 1999; McClymont et al. 2005). Parasite eradication in *ex situ* situations represents a challenge for both the conservation status of the parasite and the evolutionary potential of the host (Gompper and Williams 1998). The deliberate removal or destruction of a given parasite species removes an entire lineage of co-evolutionary past and future co-evolutionary potential.

Beyond deliberate attempts to reduce parasite abundance, parasite conservation is intimately tied to host status. Factors threatening the host (e.g., habitat loss and fragmentation, overexploitation, emerging diseases, climate change, species invasions, and interactions among all of the above) imperil parasite persistence. This risk level increases with host specialization (Table 6.2). Compounding the issue, host populations must remain above parasite-specific thresholds of abundance to maintain viable parasite populations, which in fact might be greater than that required to cue conservation intervention (Altizer et al. 2007). Depending on the epidemiological factors that modulate this threshold (e.g., host specificity and transmission efficiency), it can be expected that parasite species will disappear long before their hosts do (Dunne and Williams 2009). These effects will be greater in multi-species transmission cycles as extinctions or significant decreases of just one link of the transmission chain can cause parasite extinction.

Only one parasite species is currently included in the 2009 IUCN Red List, the pygmy hog-sucking louse (*Haematopinus oliveri*), listed as critically endangered because of the rarity of its host, the pygmy hog (*Porcula salvania*) (IUCN 2009). Actual documented cases of parasite extinction are also scarce (see Koh et al. 2004 and Dunn et al. 2009). Estimates of parasite endangerment suggest that anywhere from around 2,000 (Table 6.3; Dobson et al. 2008) to over 5,700 (Poulin and Morand 2004) parasitic helminth species alone are likely be to threatened with extinction and at least 266 are already likely extinct (Table 6.3). The vast disconnect between one endangered host species and even the most conservative estimates of endangerment suggest that single-species listing efforts for

---

**Table 6.3  Three Potential Prioritization Categories for Parasite Species Conservation**

Risk-based

      Affiliation with endangered, rare, or geographically restricted hosts
      Host specialists
      Transmitted inefficiently
      Complex transmission involving several hosts/vectors

Function-based

      Species with higher link diversity in food webs
      Specialists of primary producers
      Ecosystem engineers
      Unique, non-redundant functions
      Causing significant morbidity/mortality in top-down controllers

Uniqueness-based

      Phylogenetically unique
      Endemic

parasites are both a poor reflection of risk and unlikely to successfully chronicle their conservation plight (Dunn et al. 2009).

## Present and Potential Approaches

Existing measures to conserve parasites are remarkably scarce. Academic calls for increased inclusion into endangered species listing (Dunn et al. 2009; Dunne and Williams 2009), pleas for attention to the high extinction risk associated with control interventions in *ex situ* conservation situations (Gompper and Williams 1998), and calls for inclusion in broader conservation strategies (Windsor 1995; Perez et al. 2006; Pizzi 2009) have grown louder in recent years. Yet practical solutions and explicit targets are rare. The intentional conservation management of host–parasite interactions within and outside protected area boundaries requires preserving a background level of transmission between host and parasite. In most cases, setting such transmission targets will be extremely complex.

In this section, we briefly discuss the extent to which parasites are currently incorporated in conservation approaches, and discuss ways to strengthen their representation, as well as potential challenges in this endeavor. As with all biodiversity targets, limited funding and resources need to be strategically allocated. Below we suggest a list of broad guidelines to highlight categories of parasite species that might require the most immediate attention from conservation managers. These include priorities based on extinction risk, key ecosystem roles, and functional or phylogenetic uniqueness (see Table 6.2). These general guidelines should be refined in the context of local conditions and updated with expanding information about the ecological relevance of parasite species.

### Single-Species Conservation

Single-species conservation measures, ranging from population assessments to species management or recovery plans, are often focused on concerns for species that are formally listed as threatened or endangered. This almost by definition implies they are in short supply for parasites. Poor representation on endangered species lists is a product of several factors.

Survey efforts for invertebrate and microscopic taxa are often insufficiently replicated to complete dependable species lists for a given area (Samways and Grant 2007; Cotterill et al. 2008). A low and often geographically biased degree of taxonomic expertise may compound these limited survey efforts. If one adds to this the basic bioinformatics challenges faced by hyperdiverse groups (Clark and May 2002; Samways and Grant 2007), considerable ground must be made up before the distribution and therefore the extinction risk of parasites are properly assessed (Dunn 2005). Judging from the mismatch between conservation threat and the extent of inclusion in the IUCN Red List, parasites exemplify problematic single-species conservation targets. The benefits of listing a given species or group are readily apparent only when it has been adequately assessed (Regnier et al. 2009). This demands data that are unlikely to exist for parasites (Maudsley and Stork 1995). Creating high-quality population data for small-bodied, elusive, or cryptic taxa demands a costly and often difficult blend of geographic coverage, thorough host surveys, and molecular analyses. Yet without these data, the conservation potential of listing efforts and other conservation strategies is greatly diminished.

Save for one parasitic plant (Ecroyd 1995) we know of no published parasite recovery or management plan, even for those parasites that are obligates upon hosts whose long-term conservation is clearly challenged, such as geographically restricted or threatened host species (Durben and Keirans 1996) or those extinct in the wild (Gompper and Williams 1998). How might this situation change? We advocate that a minimal starting point is increased parasite surveys for host taxa currently globally assessed on the IUCN Red List. Parasite surveys could minimally be conducted for three easily targetable classes of hosts: those that are (1) geographically restricted, (2) endangered, or (3) in active decline. Their small or declining population sizes make them less likely to support the host–parasite interaction rates needed to sustain parasite persistence. These surveys can improve our understanding of the identity, degree of specialization, and threats faced by the subset of parasite species at a known high risk of co-extinction or co-decline.

*In situ* conservation planning and management for maintaining specific host–parasite interactions is related to those designed for their hosts: conservation targets and potential management actions should be identified, cost–benefit analyses carried out, feasible actions implemented, and effectiveness in reaching the goals continually monitored (Margules and Pressey 2000; Bottrill et al. 2008). Monitoring changes in a parasite spatial and host distribution, as well as changes in prevalence, is always necessary but especially important following environmental alteration or changes in the conservation management plans for ecosystems or host species (Lebarbenchon et al. 2006). *Ex situ* conservation strategies are also applicable for parasite biodiversity. In other cases, parasites of rare or threatened hosts could be isolated and maintained under controlled conditions in closely related abundant or non-threatened species until recovery efforts allow the original host populations to rebound and the host–parasite interaction can be re-established in the wild (Gompper and Williams 1998).

## Conservation by Proxy

Conservation by proxy or systems-level conservation refers to decision-making for entire ecosystems or landscapes, and includes a variety of interventions, from large-scale conservation planning to natural resource management. Conservation by proxy is in practice the most common type of conservation strategy for parasites, as conservation plans for hosts are considered umbrellas for parasites and other associates, typically without clear objectives, monitoring, or evaluation for these dependent taxa. However, to assume that host conservation will directly result in parasite conservation is dangerously complacent. Due to their threshold requirements for host population size, this assumption breaks down for small, captive, or fragmented populations of any of the species involved in a parasite transmission cycle (host, vector, or reservoir). Therefore, maintenance of the parasite fundamental niche requires a complex, systems-level view of conservation. For these reasons, parasite conservation requires its own assessments, targets, and monitoring.

The kinds of ecological changes that accompany host conservation strategies (e.g., protected-area design and management) affect parasite diversity and infection patterns (Ezenwa 2003; Lebarbenchon et al. 2006). *In situ* and *ex situ* health programs, food supplementation, vaccination, and translocations will also have effects on infection prevalence and incidence.

This highlights the obvious fact that changes in host ecology will affect the ecology of its parasites, and that the net result can be detrimental for the host–parasite interaction, even if beneficial for the host in the short term. While we know of no clear conservation-by-proxy endeavor purposefully targeting the conservation of a parasite–host interactions, we suggest that the combination of comprehensive parasite surveys and long-term adaptive monitoring should be added to host conservation strategies whenever possible (Lindenmayer and Likens 2009). Important increases in our understanding of parasite species identity, the size and variation of effective parasite populations at the individual and population levels, and geographic variation among allopatric host populations are prerequisites to successful conservation management.

Finally, recent research showing that parasite ecology reflects the ecology of vertebrate hosts (Hechinger and Lafferty 2005; Hechinger et al. 2008) suggests that the current hosts-as-umbrellas-for-parasites approach might deserve to be turned around: maintaining endemic parasite transmission can, in some cases, provide a conservation target that, when met, will signify that other biodiversity components of an ecosystem remain viable. In these cases, rather than being an additional monitoring task, parasite biodiversity can be an inclusive conservation target, and parasite ecology an effective indicator of overall ecosystem health.

## Conservation by Inspiration

Conservation-by-inspiration approaches rely upon capturing the attention and imagination of the public, funding bodies, and decision-makers to inspire investment in the protection of a particular species group or ecosystem. These approaches aim directly at highlighting the intrinsic or aesthetic value of single species, and espousing the utilitarian value of multiple species assemblages at larger spatial scales. Both approaches serve to motivate engagement with a given conservation issue yet focus on different biological targets (Caro and O'Doherty 1999; Lindenmayer et al. 2007).

Popular science books and newspaper articles that emphasize the positive roles of parasites in ecosystems and dangers of parasite biodiversity loss are a recent and welcome contrast to the usual publications unlikely to improve public perception of parasites (LaFee 2006; Zuk 2007). The scarcity of pro-parasite messages extends to the scientific literature: from journal inception through 2009, a single published article with "ecosystem services" in the title from the journals *Conservation Biology, Biological Conservation,* and *Biodiversity and Conservation* makes any reference to parasite biodiversity in the abstract. Within the textbooks that serve as the academic foundation in conservation biology education, cogent arguments of parasites as organisms deserving conservation attention are a rarity (Moritz and Kikkawa 1994; Groom et al. 2006; Riordan et al. 2007). In a survey of 76 English-language conservation biology textbooks published between 1970 and 2008, 72% either do not mention the terms "parasite" or "pathogens" in the index, or limit their discussions to descriptions of parasites as a threat to species of conservation interest (Nichols and Gómez 2011). This lack of inclusion in basic conservation education helps perpetuate the scarcity of knowledge about the practice of parasite conservation.

Overall, parasite species (as much of free-living biodiversity) still fail to capture the public's attention. The academic consilience among the fields of ecology, conservation, and epidemiology embodied by conservation medicine can offer knowledge and tools to discover and effectively communicate the roles of parasites as dominant, important, and vital members of every ecological system on Earth. Bridging this gap within the scientific community is still necessary, and a fundamental reframing of epidemiological issues in an ecological context may help demystify the study and conservation of parasites. Further engagement from conservation biologists and conservation medicine practitioners is needed to temper messages focused solely on disease causation with mentions of positive attributes of parasite biodiversity.

## CHALLENGES AND CONFLICTS

Human aversion to most invertebrates is well documented (Kellert 1993). These aversions extend to microorganisms and are compounded through the funding, training, and academic networks that support conservation research and practice (Clark and May 2002; Samways 2005). The active inclusion of parasites in conservation strategies may open a Pandora's box of financial constraints, basic scientific hurdles, and public relations concerns. Anticipating and navigating these challenges will be critical to the future of parasite biodiversity and to the practice of conservation medicine.

Recent focus on emerging diseases may create a public relations issue for proponents of parasite conservation (Daszak and Cunningham 2002). Recent studies have shown that wildlife species are the most common reservoir of human emerging infectious diseases, and that emergence events have increased through time (Jones et al. 2008). For the lay public, it is hard to escape news reports concerning the H1N1 influenza pandemic, West Nile virus fever, or Ebola virus outbreaks. Conserving parasite biodiversity ultimately conserves the pool of pathogens that may spill over to populations of humans, domestic animals, and crops. This is a difficult idea to sell to donors and stakeholders, let alone impose upon local communities. Conceptually, however, this situation is similar to human–wildlife conflict, and much can be learned from the large body of literature dealing with strategies to prevent and mitigate it (e.g., buffer zones).

Another challenge for conserving parasites is simply the global lack of taxonomic expertise for most of the relevant groups (Brooks and Hoberg 2000). Given current funding opportunities and the enormous estimates of parasite biodiversity, there are simply too few researchers who can describe new taxa within the timeframe that is needed to know what species are out there, let alone identify those in need of conservation. As an example, Hugot et al. (2001) estimate that at the current rate of taxonomic effort, an additional 1,300 years of study will be necessary to produce a detailed record of nematode diversity.

Conflicts can also arise within the conservation community. Since parasites are still often overlooked by ecologists and conservation biologists, there may be conflicts related to allocating limited resources toward the conservation of parasites and away from other more conventional targets. Successful parasite conservation might also pose significant challenges for managers. Similar to poorly targeted top predator conservation measures, inappropriately estimated population targets for parasites may have strong, cascading effects on their host organisms. Managers of parasite biodiversity would be required to maintain an often-delicate balance between endemic and epidemic levels of transmission, and errors on either side could threaten either parasite or host. Further, given the potentially insurmountable public relations challenges of parasite-oriented conservation (e.g., a park for parasites), managers might need to set parasite conservation targets and strategies that are nested within those for other species.

Increasing the awareness of parasites as intrinsically valuable and critical elements in functioning ecosystems would seem an unchallenged good. However, calls for parasite conservation may create further concerns for conservation biologists and advocates to resolve: issue fatigue and negative public perception. We define conservation "issue fatigue" as disengagement from a conservation issue on the basis of the perception that effective action has not occurred or cannot occur; therefore, the issue in question deserves reduced attention or funding. When coupled with the daily stream of environmental admonitions, it is possible that the audiences reached by environmental advocacy may meet persistent calls for parasite awareness, consideration, and inclusion with a sense of weariness.

Beyond fatigue, the full implications of parasite conservation may be challenging for many to understand and accept. Advocating parasite conservation implies supporting death and sickness in wild species. While parasitism is as much a natural process as photosynthesis or predation, public perceptions are generally biased against the effects of parasites on their hosts (Daszak and Cunningham 2002). Suggesting that people accept various degrees of morbidity and mortality can be difficult and might be heard by some ears as a challenge to the credibility of conservationists. Conservation medicine is largely predicated on influencing human attitudes, knowledge, and behaviors. Promoting change in these values may be critical to gaining public acceptance of the full implications of maintaining ecosystem health.

## CONCLUSIONS

Inclusion of parasite biodiversity in ecological thinking has highlighted its numerical and functional importance in ecosystems and through evolutionary time. Practitioners of conservation medicine are ideally poised to tackle the dual demands of predicting, preventing, and controlling the negative consequences of infectious diseases on all species on the one hand,

while on the other preserving the contributions of parasites to healthy ecosystems. Further research is needed to guide the development of strategies to sustain parasite diversity. Some of this research will fall within discrete disciplinary lines: parasite surveys, descriptions, and taxonomic work within and across ecosystems are still a dire necessity. However, the transdisciplinary research and training promoted by conservation medicine is a critical tool with which to analyze complex ecosystem health issues such as parasite conservation. Beyond research, parasite conservation requires effective communication with a variety of audiences, emphasizing parasites and disease as natural components of the web of life.

Incorporating parasite biodiversity conservation into the science and practice of conservation medicine does not mean that all parasite species, at all times and in all ecosystems, deserve conservation efforts. It does, however, imply that there are important tradeoffs in conserving individual, population, and ecosystem health; that preserving healthy ecosystems requires maintaining infection; and that the costs and benefits of control and eradication strategies need to be considered in a systematic and inclusive manner. Parasite control, extirpation, and eradication will still be desirable outcomes. We only ask that these tradeoffs are considered and that parasite conservation (or elimination) is given appropriate weight in conservation decision-making and in the practice of ecological health.

## REFERENCES

Altizer, S., C.L. Nunn, and P. Lindenfors. 2007.
    Do threatened hosts have fewer parasites? A comparative study in primates. J Anim Ecol 76:304–314.
Angly, F.E., B. Felts, M. Breitbart, P. Salamon,
    R.A. Edwards, C. Carlson, A.M. Chan, M. Haynes,
    S. Kelley, H. Liu, J.M. Mahaffy, J.E. Mueller, J. Nulton,
    R. Olson, R. Parsons, S. Rayhawk, C.A. Suttle, and
    F. Rohwer. 2006. The marine viromes of four oceanic
    regions. PLoS Biol 4:e368.
Balmer, O., S.C. Stearns, A. Schotzau, and R. Brun. 2009.
    Intraspecific competition between co-infecting parasite
    strains enhances host survival in African trypanosomes.
    Ecology 90:3367–3378.
Bascompte, J., and P. Jordano. 2007. Plant-animal
    mutualistic networks: the architecture of biodiversity.
    Annu Rev Ecol Evol Syst 38:567–593.

Blanchet, S., O. Rey, P. Berthier, S. Lek, and G. Loot. 2009. Evidence of parasite-mediated disruptive selection on genetic diversity in a wild fish population. Mol Ecol 18:1112–1123.

Bottrill, M.C., L.N. Joseph, J. Carwardine, M. Bode, C. Cook, E.T. Game, H. Grantham, S. Kark, S. Linke, E. Mcdonald-Madden, R.L. Pressey, S. Walker, K.A. Wilson, and H.P. Possingham. 2008. Is conservation triage just smart decision making? Trends Ecol Evol 23:649–654.

Brooks, D.R., and E.P. Hoberg. 2000. Triage for the biosphere: the need and rationale for taxonomic inventories and phylogenetic studies of parasites. Comp Parasitol 67:1–25.

Caro, T.M., and G. O'Doherty. 1999. On the use of surrogate species in conservation biology. Conserv Biol 13:805–814.

Christe, P., S. Morand, and J. Michaux. 2006. Biological conservation and parasitism. In S. Morand, B. Krasnov, and R. Pouin, eds. Micromammals and macroparasites, pp. 593–613. Springer, New York.

Clark, J.A., and R.M. May. 2002. Taxonomic bias in conservation research. Science 297:191–192.

Cotterill, F.P.D., K.S. Al-Rasheid, and W. Foissner. 2008. Conservation of protists: is it needed at all? Biodivers Conserv 17:427–443.

Cunningham, A.A., and P. Daszak. 1998. Extinction of a species of land snail due to infection with a microsporidian parasite. Conserv Biol 12:1139–1141.

Daszak, P., and A.A. Cunningham. 2002. Emerging infectious diseases: a key role for conservation medicine. In A.A. Aguirre, R.S. Ostfeld, G.M. Tabor, C. House, and M.C. Pearl, eds. Conservation medicine: ecological health in practice, pp. 40–61. Oxford University Press, New York.

Demeeûs, T., and F. Renaud. 2002. Parasites within the new phylogeny of eukaryotes. Trends Parasitol 18:247–251.

Dobson, A., and J. Foufopoulos. 2001. Emerging infectious pathogens of wildlife. Phil Trans R Soc B 356:1001–1012.

Dobson, A., K.D. Lafferty, A.M. Kuris, R.F. Hechinger, and W. Jetz. 2008. Homage to Linnaeus: How many parasites? How many hosts? Proc Natl Acad Sci USA 105:11482–11489.

Douglas, T., and M. Young. 2006. Viruses: making friends with old foes. Science 312:873–875.

Dunn, R.R. 2005. Modern insect extinctions, the neglected majority. Conserv Biol 19:1030–1036.

Dunn, R.R., N.C. Harris, R.K. Colwell, L.P. Koh, and N.S. Sodhi. 2009. The sixth mass coextinction: are most endangered species parasites and mutualists? Phil Trans Roy Soc B Biol Sci 276:3037–3045.

Dunne, J.A., R.J. Williams, and N.D. Martinez. 2002. Network structure and biodiversity loss in food webs: robustness increases with connectance. Ecol Lett 5:558–567.

Dunne, J.A., and R.J. Williams. 2009. Cascading extinctions and community collapse in model food webs. Phil Trans R Soc B 364:1711–1723.

Durben, L., and J. Keirans. 1996. Host-parasite coextinction and the plight of tick conservation. Am Entomol 42:87–92.

Ecroyd, C.E. 1995. Dactylanthus taylorii recovery plan, Threatened species recovery plan. Series No. 16. Department of Conservation, Threatened Species Unit, Forest Research Institute Ltd. Wellington, New Zealand.

Ezenwa, V.O. 2003. Parasite infection rates of impala (Aepyceros melampus) in fenced game reserves in relation to reserve characteristics. Biol Conserv 118:397–401.

Falcone, F.H., and D.I. Pritchard. 2005. Parasite role reversal: worms on trial. Trends Parasitol 21:157–160.

Fallon, P.G., and A. Alcami. 2006. Pathogen-derived immunomodulatory molecules: future immunotherapeutics? Trends Immunol 27:470–476.

Freckleton, R.P., and O.T. Lewis. 2006. Pathogens, density dependence and the coexistence of tropical trees. Proc R Soc B 273:2909–2916.

Frodel, J.L., P. Barth, and J. Wagner. 2004. Salvage of partial facial soft tissue avulsions with medicinal leeches. Otolaryng Head Neck Surg 131:934–939.

Gompper, M.E., and E.S. Williams. 1998. Parasite conservation and the black-footed ferret recovery program. Conserv Biol 12:730–732.

Groom, M.J., G. Meffe, and C.R. Caroll. 2006. Principles of conservation biology. 3rd ed. Sinauer Associates, Sunderland, Mass., USA.

Guralnick, R., E. Hall, and S. Perkins. 2004. A comparative approach to understanding causes and consequences of mollusc-digenean size relationships: a case study with allocreadiid trematodes and Cyclocalyx clams. J Parasitol 90:1253–1262.

Hechinger, R.F., and K.D. Lafferty. 2005. Host diversity begets parasite diversity: bird final hosts and trematodes in snail intermediate hosts. Proc R Soc B 272:1059–1066.

Hechinger, R.F., K.D. Lafferty, and A.M. Kuris. 2008. Trematodes indicate animal biodiversity in the Chilean intertidal and Lake Tanganyika. J Parasitol 94:966–968.

Hedges, L.M., J.C. Brownlie, S.L. O'Neill, and K.N. Johnson. 2008. Wolbachia and virus protection in insects. Science 322:702.

Holdo, R.M., A.R.E. Sinclair, A.P. Dobson, K.L. Metzger, B.M. Bolker, M.E. Ritchie, and R.D. Holt. 2009.

A disease-mediated trophic cascade in the Serengeti and its implications for ecosystem C. PLoS Biol 7:e1000210.

Hudson, P.J., A.P. Dobson, and K.D. Lafferty. 2006. Is a healthy ecosystem one that is rich in parasites? Trends Ecol Evol 21:381–385.

Hugot, J.P., P. Baujard, and S. Morand. 2001. Biodiversity in helminths and nematodes as a field of study: an overview. Nematology 3:199–208.

IUCN. 2009. Red list of threatened species. Version (2009).2.

Jokela, J., M.F. Dybdahl, and C.M. Lively. 2009. The maintenance of sex, clonal dynamics, and host-parasite coevolution in a mixed population of sexual and asexual snails. Am Nat 174:S43–S53.

Jolles, A.E., V.O. Ezenwa, R.S. Etienne, W.C. Turner, and H. Olff. 2008. Interactions between macroparasites and microparasites drive infection patterns in free-ranging African buffalo. Ecology 89:2239–2250.

Jones, K.E., N.G. Patel, M.A. Levy, A. Storeygard, D. Balk, J.L. Gittleman, and P. Daszak. 2008. Global trends in emerging infectious diseases. Nature 451:990–993.

Kellert, S.R. 1993. Values and perceptions of invertebrates. Conserv Biol 7:845–855.

Knight, C.G. 1971. Ecology of African sleeping sickness. Ann Assoc Am Geogr 61:23–44.

Koh, L.P., R.R. Dunn, N.S. Sodhi, R.K. Colwell, H.C. Proctor, and V.S. Smith. 2004. Species coextinctions and the biodiversity crisis. Science 305:1632–34.

Kristensen, T.K., and D.S. Brown. 1999. Control of intermediate host snails for parasitic diseases: a threat to biodiversity in African freshwaters? Malacologia 41:379–391.

Kuris, A.M. 2005. Trophic transmission of parasites and host behavior modification. Behav Process 68:215–217.

Lafee, S. 2006. Parasites lost: of lice and men and the value of small, disgusting things. Nov. 2, San Diego Union-Tribune, California.

Lafferty, K.D., A.P. Dobson, and A.M. Kuris. 2006. Parasites dominate food web links. Proc Natl Acad Sci USA 103:11211–11216.

Lafferty, K.D. 2008. Effects of disease on community interactions and food web structure. In R.S. Ostfeld, F. Keesing, and V.T. Eviner, eds. Infectious disease ecology: effects of ecosystems on disease and of disease on ecosystems, pp. 205–222. Princeton University Press, New Jersey.

Lafferty, K.D., S. Allesina, M. Arim, C.J. Briggs, G. De Leo, A.P. Dobson, J.A. Dunne, P.T.J. Johnson, A.M. Kuris, D.J. Marcogliese, N.D. Martinez, J. Memmott, P.A. Marquet, J.P. Mclaughlin, E.A. Mordecai, M. Pascual, R. Poulin, and D.W. Thieltges. 2008. Parasites in food webs: the ultimate missing links. Ecol Lett 11:533–546.

Lafferty, K.D., and A.M. Kuris. 2009. Parasitic castration: the evolution and ecology of body snatchers. Trends Parasitol 25:564–572.

Lebarbenchon, C., R. Poulin, M. Gauthier-Clerc, and F. Thomas. 2006. Parasitological consequences of overcrowding in protected areas. EcoHealth 3:303–307.

Leendertz, F.H., G. Pauli, K. Maetz-Rensing, W. Boardman, C. Nunn, H. Ellerbrok, S.A. Jensen, S. Junglen, and C. Boesch. 2006. Pathogens as drivers of population declines: the importance of systematic monitoring in great apes and other threatened mammals. Biol Conserv 131:325–337.

Lefevre, T., C. Lebarbenchon, M. Gauthier-Clerc, D. Misse, R. Poulin, and F. Thomas. 2009. The ecological significance of manipulative parasites. Trends Ecol Evol 24:41–48.

Lindenmayer, D.B., J. Fischer, A. Felton, R. Montague-Drake, A.D. Manning, D. Simberloff, K. Youngentob, D. Saunders, D. Wilson, A.M. Felton, C. Blackmore, A. Lowe, S. Bond, N. Munro, and C.P. Elliott. 2007. The complementarity of single-species and ecosystem-oriented research in conservation research. Oikos 116:1220–1226.

Lindenmayer, D.B., and G.E. Likens. 2009. Adaptive monitoring: a new paradigm for long-term research and monitoring. Trends Ecol Evol 24:482–486.

Lively, C.M. 1987. Evidence from a New Zealand snail for the maintenance of sex by parasitism. Nature 328:519–521.

Lively, C.M. 1989. Adaptation by a parasitic trematode to local populations of its snail host. Evolution 43:1663–1671.

Lively, C.M., C. Craddock, and R.C. Vrijenhoek. 1990. Red Queen hypothesis supported by parasitism in sexual and clonal fish. Nature 344:864–866.

Loehle, C. 1995. Social barriers to pathogen transmission in wild animal populations. Ecology 76:326–335.

Lopez, A.D., C.D. Mathers, M. Ezzati, D.T. Jamison, and C.J.L. Murray. 2006. Global and regional burden of disease and risk factors, 2001: systematic analysis of population health data. Lancet 367:1747–1757.

Marcogliese, D.J. 2004. Parasites: small players with crucial roles in the ecological theater. EcoHealth 1:151–164.

Marcogliese, D.J. 2005. Parasites of the superorganism: are they indicators of ecosystem health? Int J Parasitol 35:705–716.

Margules, C.R., and R.L. Pressey. 2000. Systematic conservation planning. Nature 405:243–253.

Maudsley, N.A., and N.E. Stork. 1995. Species extinctions in insects: ecological and biogeographical considerations. In R. Harrington and N.E. Stork, eds. Insects in a changing environment, pp. 321–69. Academic Press, London, England.

Mcclymont, H.E., A.M. Dunn, R.S. Terry, D. Rollinson, D. Timothy, J. Littlewood, and J.E. Smith. 2005. Molecular data suggest that microsporidian parasites in freshwater snails are diverse. Int J Parasitol 35:1071–1078.

Mckay, D.M. 2006. The beneficial helminth parasite? Parasitology 132:1–12.

Michalsen, A., Stefanie Klotz, R. Ludtke, S. Moebus, G. Spahn, and G.J. Dobos. 2003. Effectiveness of leech therapy in osteoarthritis of the knee. Ann Intern Med 139:724–730.

Morand, S., and P.H. Harvey. 2000. Mammalian metabolism, longevity and parasite species richness. Proc R Soc B 267:1999–2003.

Morand, S. 2003. Parasites and the evolution of host life history traits. Int Congr Zool 18:213–218.

Moritz, C., and J. Kikkawa. 1994. Conservation biology in Australia and Oceania. Surrey Beatty & Sons Pty Ltd., Chipping Norton, Australia.

Mouritsen, K.N., and R. Poulin. 2003. Parasite-induced trophic facilitation exploited by a non-host predator: a manipulator's nightmare. Int J Parasitol 33:1043–1050.

Mouritsen, K.N., and R. Poulin. 2005. Parasites boosts biodiversity and changes animal community structure by trait-mediated indirect effects. Oikos 108:344–350.

Nichols, E., and A. Gómez. 2011. Conservation education needs more parasites. Biol Conserv 144:937–941.

Nieberding, C.M., and I. Olivieri. 2007. Parasites: proxies for host genealogy and ecology? Trends Ecol Evol 22:156–165.

Perez, J.M., P.G. Meneguz, A. Dematteis, L. Rossi, and E. Serrano. 2006. Parasites and conservation biology: the `ibex-ecosystem.' Biodivers Conserv 15:2033–2047.

Pizzi, R. 2009. Veterinarians and taxonomic chauvinism: the dilemma of parasite conservation. J Exot Pet Med 18:279–282.

Platt, O.S., D.J. Brambilla, W.F. Rosse, P.F. Milner, O. Castro, M.H. Steinberg, and P.P. Klug. 1994. Mortality in sickle-cell disease: Life expectancy and risk factors for early death. N Engl J Med 330:1639–1644.

Ponton, F., D.G. Biron, C. Joly, S. Helluy, D. Duneau, and F. Thomas. 2005. Ecology of parasitically modified populations: a case study from a gammarid-trematode system. Mar Ecol Prog Ser 299:205–215.

Poulin, R., and S. Morand. 2000. The diversity of parasites. Q Rev Biol 75:277–293.

Poulin, R., and S. Morand. 2004. Parasite biodiversity. Smithsonian Institute Scholarly Press, Washington D.C.

Regnier, C., B. Fontaine, and P. Bouchet. 2009. Not knowing, not recording, not listing: numerous unnoticed mollusk extinctions. Conserv Biol 23:1214–1221.

Riordan, P., P. Hudson, and S. Albon. 2007. Do parasites matter? Infectious disease and the conservation of host populations. In D. Macdonald and K. Service, eds. Key topics in conservation biology, pp. 156–172. Blackwell Publishing, Oxford, U.K.

Rogers, D.J., and S.E. Randolph. 1988. Tsetse flies in Africa: bane or boon? Conserv Biol 2:57–65.

Rook, G.A. 2009. Review series on helminths, immune modulation and the hygiene hypothesis: the broader implications of the hygiene hypothesis. Immunology 126:3–11.

Sachs, J., and P. Malaney. 2002. The economic and social burden of malaria. Nature 415:680–685.

Samways, M.J. 2005. Insect diversity conservation. Cambridge University Press, Cambridge, U.K.

Samways, M.J., and P.B. Grant. 2007. Honing Red List assessments of lesser-known taxa in biodiversity hotspots. Biodivers Conserv 16:2575–2586.

Sures, B. 2003. Accumulation of heavy metals by intestinal helminths in fish: an overview and perspective. Parasitology 126:S53–S60.

Sures, B. 2004. Environmental parasitology: relevancy of parasites in monitoring environmental pollution. Trends Parasitol 20:170–177.

Suttle, C.A. 2005. Viruses in the sea. Nature 437:356–361.

Suttle, C.A. 2007. Marine viruses: major players in the global ecosystem. Nat Rev Micro 5:801–812.

Tabor, G.M. 2002. Defining conservation medicine. In A.A. Aguirre, R. S. Ostfeld, G.M. Tabor, C. House, and M.C. Pearl, eds. Conservation medicine: ecological health in practice, pp. 3–7. Oxford University Press, New York.

Thomas, F., A. Schmidt-Rhaesa, G. Martin, C. Manu, P. Durand, and F. Renaud. 2002. Do hairworms (Nematomorpha) manipulate the water seeking behaviour of their terrestrial hosts? J Evol Biol 15:356–361.

Thomas, F., S. Adamo, and J. Moore. 2005. Parasitic manipulation: where are we and where should we go? Behav Processes 68:185–199.

Thompson, R.M., K.N. Mouritsen, and R. Poulin. 2005. Importance of parasites and their life cycle characteristics in determining the structure of a large marine food web. J Anim Ecol 74:77–85.

Tompkins, D.M., A.R. White, and M. Boots. 2003. Ecological replacement of native red squirrels by invasive greys driven by disease. Ecol Lett 6:189–196.

Wegner, K.M., T.B.H. Reusch, and M. Kalbe. 2003. Multiple parasites are driving major histocompatibility complex polymorphism in the wild. J Evol Biol 16:224–232.

Whiteman, N.K., and P.G. Parker. 2005. Using parasites to infer host population history: a new rationale for parasite conservation. Anim Conserv 8:175–181.

Windsor, D.A. 1995. Equal rights for parasites. Conserv Biol 9:1–2.

Windsor, D.A. 1997. Equal rights for parasites. Perspect Biol Med 40: 222–229.

Windsor, D.A. 1998a. Most of the species on earth are parasites. Int J Parasitol 28:1939–1941.

Windsor, D.A. 1998b. Equal rights for parasites. Bioscience 48:244.

Wood, C.L., J.E. Byers, K.L. Cottingham, I. Altman, M.J. Donahue, and A.M.H. Blakeslee. 2007. Parasites alter community structure. Proc Natl Acad Sci USA 104:9335–9339.

Zuk, M. 2007. Riddled with life: friendly worms, ladybug sex, and the parasites that make us who we are. Harcourt Books, Orlando, Florida.

# 7

# STRESS AND IMMUNOSUPPRESSION AS FACTORS IN THE DECLINE AND EXTINCTION OF WILDLIFE POPULATIONS

## Concepts, Evidence, and Challenges

## Heribert Hofer and Marion L. East

The decline or extinction of wildlife populations is the consequence of processes that substantially reduce the survival and reproductive success (Darwinian fitness, hereafter termed fitness) of individual members (Soulé 1986; Caughley 1994). While ecology investigates how natural and anthropogenic environmental factors affect the fate and fitness of individuals (Soulé 1986; Begon et al. 2006), conservation medicine places greater emphasis on health status, factors that affect health of species, and remedies against negative impacts (Aguirre et al. 2002).

Highly virulent pathogens and toxic pollutants often cause an immediate population decline. More subtle challenges may have little short-term impact but in the long term can significantly reduce an individual's fitness because they cause "stress" or decrease immunocompetence. Here we consider stress and immunosuppression as prominent examples of multifactorial etiologies of the impairment of health (Fig. 7.1, paths a and b) and review the evidence for environmental stimuli, particularly anthropogenic ones, to act as stressors or suppress immunocompetence in wildlife and thereby influence each individual's contribution to population change. We also assess the validity of common methods to determine the consequences of

stressors because their interpretation and differences in their conceptual foundation are important.

## CONCEPTUAL ISSUES

Environmental stimuli that significantly alter an individual's fitness belong to three classes: (1) abiotic factors such as temperature, precipitation, solar radiation, pH, salinity, photoperiod, and catastrophes; (2) interspecific factors, from top-down processes by predators or pathogens (Box 7.1), via processes caused by competitors for resources, to bottom-up processes such as resource availability (Begon et al. 2006); and (3) intra-specific factors such as competition and social processes that act as a force of natural, social, or sexual selection (Andersson 1994; Tanaka 1996; Hofer and East 2003; Silk 2007).

Environmental stimuli cause stress if they are unpredictable or uncontrollable and disturb the homeostasis of an organism (Levine and Ursin 1991) (Figs. 7.2 and 7.3). They may directly, or through immunosuppression (Fig. 7.3), lead to a decline in individual survival or reproduction. If this repeatedly occurred in the evolutionary history of a species,

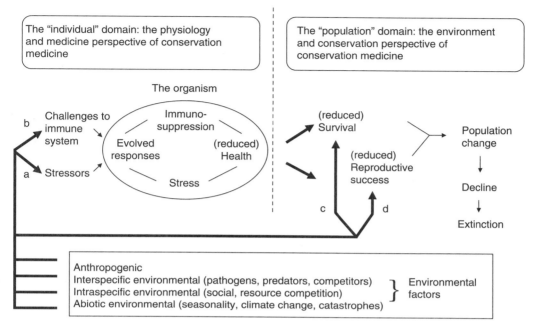

**Figure 7.1:**
The dual perspective of conservation challenges. Conventionally, ecology and conservation biology focus on the outcome of direct effects of environmental drivers in the "population domain" in terms of reducing (c) the survival, or (d) the reproductive success of organisms. Conservation medicine expands this perspective by also examining how the health of an individual modifies or results in such effects. Key mechanisms within this context are environmental drivers as (a) stressors, or as (b) challenging the immune system. Stress or immunosuppression may reduce individual health and hence survival or reproductive success.

then organisms would be expected to evolve a stress response (Fig. 7.2) molded by natural selection (McNamara and Buchanan 2005). Research on response mechanisms should therefore determine their adaptive value as an evolutionary trait and the ecological and social contexts in which they evolved (Wingfield et al. 1998; Lochmiller and Deerenberg 2000; Abbott et al. 2003).

Experimental research on stress has mostly been conducted in a biomedical context, with little attention to its evolutionary foundation or its ecological context (Martin 2009). Biomedical research mainly aims to detail the mechanisms by which humans or laboratory animals respond to stress. As a conceptual tool, this approach has not greatly contributed to the development of a theory of stress (Romero et al. 2009). Research that includes the ecological context of stress investigates the physiological mechanisms that organize the stress response in wildlife, which

may underlie the observed variation in life histories or reproductive careers within or between species (see McEwen and Wingfield 2003; Martin 2009; Romero et al. 2009).

Stress is therefore a term with different meanings in different scientific disciplines (von Holst 1998) and appears to have no rigorous conceptual framework. Hence, researchers in animal welfare and behavioral ecology suggested using an evolutionary concept of stress. In its rigorous version, it considers stressors to be environmental stimuli likely to reduce an individual's fitness without being immediately lethal or causing immediate sterility (Fraser and Broom 1990; Hofer and East 1998). This concept draws attention to traits where natural selection may have acted in the past (McNamara and Buchanan 2005), encompasses stress responses (see Fig. 7.2) that deflect homeostasis briefly or for an extended period, or cause pre-pathological or pathological states when responses fail to cope with

**Box 7.1 Glossary**

Terms commonly used in studies that discuss stress, immunosuppression, or pathogen–host dynamics are used by different authors to mean different things. We therefore define below how we use terms and place in *italics* all terms used in these definitions that are explained elsewhere in the glossary.

**Allostasis:** The physiological processes that maintain *homeostasis* through change in both environmental stimuli and physiological mechanisms. Allostasis essentially describes adjustments through *physiological mediators* that maintain physiological parameters within narrow ranges, including the anticipation of predictable *environmental stimuli*.

**Disease:** A disorder of body functions, systems, or organs. In severe cases such disorder results in pathologies. Infection with viruses, bacteria, fungi, or *parasites* or exposure to toxins may lead to disease but need not if the host is immune to the infectious agent or can tolerate the toxin.

**Environmental stimulus:** Any environmental factor that impinges on the organism and may affect its *homeostasis* or change its *allostasis*.

**Environmental stressor:** The subset of *environmental stimuli* that causes *stress* in an organism.

**(Darwinian) fitness:** The success of an organism in passing its genes on to the next generation. Survival and reproductive success are essential components of Darwinian fitness.

**Homeostasis:** The state of mental and physical stability and health of an organism. In a strict sense in the literature, often applied only to a limited number of systems, such as pH, body temperature, glucose levels, and oxygen tension.

**Host:** An individual, population, or species infected with a *pathogen*.

**Parasites:** Protozoa, helminths, or arthropods that live a parasitic lifestyle (i.e., require *hosts* for survival or reproduction).

**Pathogen:** Any organism that lives inside or on a *host* and causes pathologies (*disease*). Pathogens may be viruses, bacteria, fungi, or *parasites* (protozoa, helminths, or arthropods). Transmissible agents for transmissible spongiform encephalopathies (TSEs), which cause the abnormal folding of prion proteins in the brain cells of infected animals, do not strictly fit this definition because they are not an organism. Even so, the impact of TSEs on infected individuals may be similar to those caused by pathogens.

**Physiological mediators:** Physiological parameters (such as glucocorticoid concentration) that effect a change in *homeostasis* in response to both predictable and unpredictable environmental stimuli.

**Reactive scope:** The range of physiological (rather than pathological) concentrations of *physiological mediators* available to an organism to return to homeostasis in response to both its own current state and *environmental stimuli*. Concentrations of *physiological mediators* outside the reactive scope lead to pathological effects such as homeostatic overload (excess concentrations) or failure (absence) and are detrimental to the short-term or long-term health of an organism. The reactive scope is affected by the *wear and tear* of an organism.

**Stress:** (1) Any effect caused by an unpredictable and/or uncontrollable *environmental stressor* to disturb *homeostasis* (a traditional, widely used definition). (2) Any effect caused by an *environmental stressor* that is likely to reduce an individual's *(Darwinian) fitness* without being immediately lethal or leading to immediate sterility (a rigorous definition consistent with evolutionary theory that directs the focus on the most important issue relevant to biological conservation, the contribution of each individual to population dynamics).

Box 7.1  Glossary *Cont.*

**Virulence:** *Pathogens* differ in their level of virulence, from relatively benign *pathogens* that have a limited impact on *host homeostasis* or *(Darwinian) fitness*, to highly virulent *pathogens* that induce a strong negative effect on *host* survival. Highly virulent *pathogens* typically have a high replication rate coupled to an efficient mechanism to move from infected *host* individuals to susceptible non-infected "naïve" *host* individuals and cause severe pathological effects (*disease*) in their *host*.

**Wear and tear:** The gradual breakdown in responsiveness of an organism's physiological system because its maintenance for the purpose of enabling *physiological mediators* is costly. With age, these systems lose their ability to effectively respond to threatening and unpredictable stimuli.

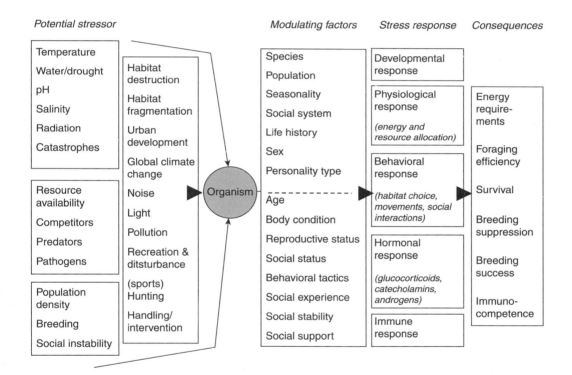

**Figure 7.2:**
A framework for the study of stress in conservation medicine that emphasizes the population consequences of stress. Stressors are environmental stimuli that impinge on the organism and include natural environmental (the three boxes in the first column) and anthropogenic environmental stimuli (box in the second column). Organisms possess a stress response that consists of several components (column 4) and has evolved in response to natural selection exerted by natural environmental stressors in the past. Empirical evidence shows that the stress response may be modulated by a wide variety of species-specific factors (column 3) that do not (upper part of column 3) or may potentially change during an individual's lifetime (lower part of column 3). If the stimulus and the (modulated) response increase the organism's energy requirements and decrease foraging efficiency, survival, and reproductive activity and success, then population development will be adversely affected.

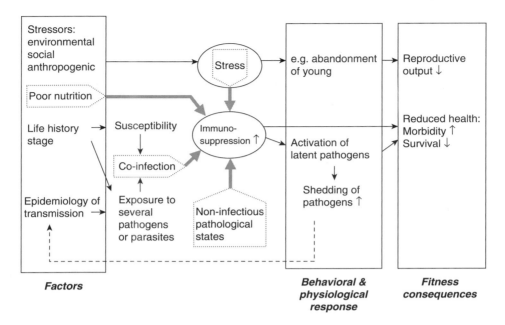

**Figure 7.3:**
The causes and population consequences of stress and immunosuppression. Immunosuppression may be a consequence of the interaction of an infection with several pathogens or parasites (co-infection) within the organism, of stress, of poor nutrition, or of the synergistic effect of an infection with a non-infectious pathological state. The likelihood of infection with one or several pathogens or parasites depends on the organism's susceptibility and the suite of factors that determine the epidemiology of pathogen or parasite transmission. Immunosuppression may trigger the activation of latent pathogens or directly impair health by increasing morbidity or decreasing survival, thereby creating detrimental fitness consequences for the organism.

the stressor (stress *sensu* Toates 1995). It also includes single, unpredictable events (e.g., an attempt to capture an animal, stress *sensu* Hoffmann & Parsons 1991) and continuous stimuli (e.g., environmental pollution, stress *sensu* Rollo 1995). Definitions of stress commonly used in psychology, ethology, and welfare ignore fitness consequences and restrict stress to situations when the stress response fails to return the organism to homeostasis (Broom & Johnson 1991).

We think an evolutionary concept of stress is useful because it connects the biological effects of stressors to fitness and thus links stress to the decline of populations of conservation concern (see Fig. 7.1). It also mirrors conceptual advances in physiology, where the recently developed reactive scope model (Romero et al. 2009) fused biomedical and ecological approaches to stress into a unified concept. Here, pathological effects occur when physiological mediators move outside their usual range (reactive scope) to create homeostatic overload or failure, with detrimental health effects. The model recognizes that

physiological systems evolved divergent characteristics to respond to different sets of environmental stimuli and provides a mechanism for why and how modulating factors (see Fig. 7.2) influence the response to environmental stimuli.

Here, we apply a relaxed version of the evolutionary concept of stress. We consider environmental stimuli as putative stressors if they are unpredictable or uncontrollable and likely to disturb homeostasis, and consider it stronger evidence if detrimental fitness consequences were documented. This permits the inclusion of many studies and accommodates the difficult issue that absence of evidence through lack of measurement is not evidence of absence of an effect.

## THE INDIVIDUAL DOMAIN: STRESS

A key issue is the modification of an organism's stress response by modulating factors (see Fig. 7.2) that may change during an individual's lifetime (Hofer and

East 1998; Wingfield et al. 1998; Romero 2004; Korte et al. 2005). Because few studies simultaneously measured the response of several physiological mediators to stimuli, we focus on the commonly studied hypothalamic-pituitary-adrenocortical (HPA) axis and glucocorticoids as associated mediators.

## Factors Modulating the Stress Response

Important factors modulating the stress response include reproductive status, social status, age, and body condition (see Fig. 7.2). Noninvasively measured fecal glucocorticoid concentrations were significantly higher in lactating than pregnant or reproductively inactive females in spotted hyenas (*Crocuta crocuta*, Goymann et al. 2001) and other species (Hunt et al. 2006; Starling et al. 2010). In baboons (*Papio cynocephalus*), maternal social status had long-lasting effects on fecal glucocorticoid concentrations of sons even after the period of infant dependence (Onyango et al. 2008). Glucocorticoid concentrations may also vary as a function of age or body condition (e.g., Sockman and Schwabl 2001).

Social status is important because in many social species it is the key to fitness (East and Hofer 2010). Social status can determine how an individual copes with stressful challenges (Sapolsky 2005), yet evidence from social mammals suggests that there is no simple relationship between social status and stress (Goymann and Wingfield 2004; Sapolsky 2005; Goymann and Hofer 2010). If dominance is obtained by assertive behavior and subordinates are frequently threatened by dominants, a rank-related pattern of glucocorticoid levels may emerge, with subordinates showing higher concentrations (Goymann et al. 2001). In contrast, when dominance hierarchies are nonlinear or dominants frequently challenged by subordinates, dominants may have chronically elevated glucocorticoid concentrations (Creel 2001). Where dominance is obtained by queuing conventions, as in many cooperatively breeding species, glucocorticoid concentrations are not rank-related and reproductive suppression is not an indicator of stress (Hofer and East 1998). Hence, the processes by which rank is achieved and maintained determine the physiological and psychosocial impact of social status (Sapolsky 1992; Abbott et al. 2003; Goymann and Wingfield 2004). Social instability also increased glucocorticoid

concentrations (Goymann et al. 2001), an important result since anthropogenic factors may facilitate social instability and decrease fitness by changing group composition or increasing turnover of high-ranking individuals (Johanos et al. 2010).

## Pathogens as Stressors

Pathogens may affect the energy budget or other physiological parameters of their host (Delahay et al. 1995; Lochmiller and Deerenberg 2000; Ots et al. 2001; Mougeot et al. 2007; Poulin 2007) and thus might be stressors. Consider a parasite that usurps 2% of host energy budget. If this is taken from energy allocated to host production (growth or reproduction), in mammals and birds often in the order of only 10% of total energy budget, a loss of 2% effectively leads to a 20% drop in energy allocated to production. Such a loss may decrease fitness (Munger and Karasov 1994). Accumulating evidence suggests that the reduction in host fitness caused by stressful effects of pathogens has been underestimated (Merino et al. 2002; Martinez-Padilla et al. 2004; Careau et al. 2010).

## Anthropogenic Stressors

Putative anthropogenic stressors include habitat destruction, modification, and fragmentation, urban development, pollution (chemical, noise, and light), disturbance, hunting, global climate change, and intervention for research and conservation (see Fig. 7.2). The study of their effects has rapidly expanded since noninvasive measurements of fecal glucocorticoid concentrations became available; these will be the focus of discussion.

### Habitat Change, Pollution, and Global Climate Change

Habitat destruction and fragmentation, urbanization, and pollution often destroy the basis of existence for wildlife populations of conservation concern. Subtler changes may act as potential stressors. Black howler monkeys (*Alouatta pigra*) in Mexican forest fragments had significantly higher fecal glucocorticoid concentrations than those in continuous forest, after possible differences in confounding factors were considered (Martinez-Mota et al. 2007). Although fitness consequences were not measured, this study and other studies indicate that habitat modifications may reduce habitat quality and thereby raise a difficult problem. Because low-quality habitats are expected to be inhabited by "low-quality" individuals (Fretwell and Lucas 1970) (e.g., late arrivals at breeding grounds, Newton 2008), any observed fitness difference between individuals in high- and low-quality habitats may then either be a consequence of differences in individual quality but not stress or a mixture of differences in individual quality and stress (Hofer and East 1998).

Sometimes anthropogenic changes increase aspects of habitat quality. Brazilian free-tailed bats (*Tadarida brasiliensis*) gained fitness benefits by roosting under bridges compared to in natural caves (Allen et al. 2010). African forest elephants (*Loxodonta cyclotis*) in an "industrial corridor" of active oil fields showed significantly lower glucocorticoid concentrations than those in adjacent national parks, after accounting for age and gender effects (Munshi-South et al. 2008), although fitness consequences were not measured.

Changes in the physical environment may be more important than currently realized. Artificial light as an ecological factor has only recently received attention (Longcore and Rich 2004; Rich and Longcore 2006; Stone et al. 2009; Kempenaers et al. 2010). Observational and experimental studies typically show detrimental effects of noise or demonstrate positive effects of noise reduction (Richardson et al. 1995; Wysocki et al. 2006; Schaub et al. 2008; Francis et al. 2009; Purser and Radford 2011). For instance, low-flying jets reduce caribou (*Rangifer tarandus*) calf survival (Harrington and Veitch 1992), and anthropogenic noise alters the behavior of many marine animals (see Myrberg 1990 and bibliography at www.nmfs.noaa.gov/pr/bibliography.htm). For example, sonar signals can induce a panic response in beaked whales during deep dives, disrupting the end of dives and causing "gas and fat embolic syndrome," mass stranding, and death (Fernández et al. 2005). The establishment of wind turbines prompted considerable research on the noise and mortality caused by turbine blades, models of long-term consequences of wind farms on wildlife populations, modifications of behavior, and measures to reduce their impact on wildlife (e.g., Carrete et al. 2009).

The effects of pollution and global climate change are difficult to assess because they are rarely amenable to experimental assessment (Hofer and East 2010) and because the response of tagged individuals may differ from that of untagged ones (Saraux et al. 2011). Direct lethal impacts of catastrophic toxic spills are well known; more difficult to assess are subtle long-term effects on the polluted individuals that survived. A case where global climate change, pollution, and immunosuppression operate in synergy is described in Box 7.2.

## Recreation, Ecotourism, and Disturbance

Outdoor recreational activities and ecotourism may cause the decline of threatened species in the United States (Czech 2000). It is therefore important to know whether and how recreation and tourism cause stress and how they should be managed to minimize stressful effects. Research on these topics has rapidly increased and parameters to assess disturbance have proliferated (Gill et al. 2001; Mallord et al. 2007; Tarlow and Blumstein 2007; Thiel et al. 2008).

Several experimental studies quantified the fitness consequences of disturbance and tourism. Experimentally disturbed grazing pink-footed geese (*Anser brachyrhynchus*) bred less successfully than nondisturbed geese (Madsen 1995). Adult Eastern bluebirds (*Sialia sialis*) buffered their developing brood from human activity by sacrificing their own self-maintenance before compromising the growth and survival of their young (Knight and Swaddle 2007). Recreational disturbance of denning sites of American black bears (*Ursus americanus*) resulted in cub abandonment and a delay in hibernation (Goodrich and Berger 1994). Experimental harassment of mule deer (*Odocoileus hemionus*) by all-terrain vehicles induced reproductive pauses during subsequent breeding seasons (Yarmoloy et al. 1988). Such experimental verification of stressful effects can be challenging, as research on the impact of tourism on Antarctic penguins (Box 7.3) and on threatened species in alpine ecosystems (Arlettaz et al. 2007) showed.

Traditionally, habituation to anthropogenic stimuli is considered evidence of coping. Habituation does not, however, prevent detrimental fitness consequences of tourism or recreation. Take the guidelines for low-impact whale watching advocated by the United Nations Environment Programme (UNEP) and others. Most include "no approach zones" and prohibition of the direct pursuit of animals. They assume that if guidelines are followed, habitat use or fitness will not be affected. A long-term comparison of

---

**Box 7.2  Global Climate Change, Pollution, Immunosuppression, and Seals**

---

Direct evidence of the effect of global climate change on the dynamics of pathogen–wildlife host interactions is difficult to assemble because pathogens are part of the natural ecology of their hosts and the effect of global climate change cannot be easily separated from confounding factors (Lafferty 2009; Hofer and East 2010; and Chapter 8 in this volume). Some of the best examples of direct effects of global climate change come from Arctic ecosystems where the accelerating effect of climate change on pathogen development and its consequences for host population dynamics have been well documented (see Kutz et al. 2005 and Chapter 9 in this volume). Other examples, such as the emergence of phocine distemper virus (PDV) infection in European seal populations, which we discuss below, suggest that global climate change may indirectly operate by causing stress and immunosuppression. Here, available knowledge is often suggestive and instructive but illustrates how much information is missing.

Traditionally, it was thought that harbor seals (*Phoca vitulina*) in European waters were free of PDV. Harp seals (*Phoca groenlandica*) are known to be infected with PDV, suffer little in terms of clinical disease, and are thought to be viral carriers. Harp seals were observed in Danish waters first in 1987–1988, having extended their range unusually far to the south, and were thought to have come into contact with harbor seals (Dietz et al. 1989a), although the appropriate molecular genetic evidence documenting possible PDV transmission is not (yet) available (Müller et al. 2008). Why harp seals moved south is unclear—possibly as a consequence of climate-related movements of prey or human overexploitation of arctic fish stocks. From April 1988 onwards, hundreds of harbor seals died of PDV infection in Danish waters; all main colonies in the North Sea were infected within 7 months. Overall mortality rates in Wadden Sea and Danish populations were 50% to 60% and in UK colonies rates were 10% to 20%; approximately 18,000 seals died (Dietz et al. 1989b). In May 2002, another outbreak occurred, with high mortality throughout all harbor seal populations and a total loss of approximately 22,000 animals. Apparently, no more than 7% of the population exposed in 1988 were still alive in 2002 (Müller et al. 2004).

The ultimate cause(s) of the PDV epidemics remain unclear. Several hypotheses invoke global climate change, and available evidence is consistent with this idea: (1) A positive correlation exists between temperature and several seal mass die-offs in the 20th century; (2) unusually high mean monthly temperatures cause seals to leave the seawater and aggregate on land in high densities outside the breeding season, thereby facilitating efficient transmission of PDV, with a basic reproductive rate of pathogen transmission $R_0$ (Keeling and Rohani 2008) of around 2.0 to 2.5; (3) in early 1988, a temperature-related exceptional bloom of the algae *Chrysochromulina polylepis*, known to secrete toxic substances, may have contributed to substantial stress and immunosuppression in harbor seals; and (4) documented stress-related skull bone lesions indicated a state of immunosuppression in affected harbor seals in 1988, possibly caused by high levels of environmental pollutants (particularly PCBs) in North Sea waters (Lavigne and Schmitz 1990; Simmonds 1991; Klepac et al. 2009). Global climate change may therefore have amplified the stressful effects of environmental pollutants, causing immunosuppression and increased susceptibility to PDV infection plus increased clinical responsiveness, morbidity, and subsequent mortality. Global climate change would then have operated in both a direct and indirect manner, with a potentially multiplicative impact.

---

habituated bottlenosed dolphins (*Tursiops truncatus*) in an area with unhabituated dolphins in an adjacent area without dolphin-watching revealed a significant decline in dolphin densities in the dolphin-watching area as the number of tour operators increased (Bejder et al. 2006). Incubating great-crested grebes (*Podiceps cristatus*) exposed to high-intensity boat traffic permitted closer approaches by boats before leaving their

---

**Box 7.3  Ecotourism and Antarctic Penguins**

---

Increasing numbers of tourists visit breeding colonies of Antarctic penguins since ship-borne tourism to the Antarctic began in 1958 (Enzenbacher 1993). Concern about the impact of disturbance on breeding performance increased when colonies near Antarctic research stations declined but recovered after termination of operations (Culik and Wilson 1991). Early studies (Wilson et al. 1991, Nimon et al. 1994, 1995; Culik and Wilson 1995) produced contradictory results on penguin responses to experimental disturbance because the studies differed in terms of species, breeding stage, pattern of experimental disturbance, and physiological mediators.

Recent studies suggest that non-disruptive, managed behavior of visitors is needed to protect penguins from disturbance (Burger and Gochfeld 2007); that the behavior of penguins does not reflect the true costs of disturbance; and that the impact of visitor disturbance varies between species, breeding stages, and sites. Habituation of Magellanic penguins (*Spheniscus magellanicus*) to visitors in areas frequented by tourist groups and in control areas varied between both locations; was reflected in behavioral responses and glucocorticoid concentrations; and varied between chicks and adults (Walker et al. 2005, 2006). The reproductive performance of gentoo penguins (*Pygoscelis papua*) with daily visits by tourist groups and in a control area showed no differences (Cobley and Shears 1999; Holmes et al. 2006). Other environmental factors exerted a stronger effect on breeding success of Adélie penguins (*Pygoscelis adeliae*) than disturbance by people, with little evidence of any impact by the latter (Carlini et al. 2006). On Petermann Island, gentoo penguin reproductive success increased as visitor disturbance declined, whereas the opposite trend occurred in Adèlie penguins (Lynch et al. 2010). Breeding Humboldt penguins (*Spheniscus humboldti*) showed no obvious behavioral response but significantly raised their heart rate and maintained it at elevated levels long after experimental disturbance by people approaching as close as 150 meters. Breeding success was significantly reduced at colonies frequently visited by visitors (Ellenberg et al. 2006).

---

nests than those less often disturbed. Grebes with shorter flight distances had a higher reproductive success than those with longer ones, yet even the most successful grebes at sites with high boat disturbance had a lower reproductive success than those on sites without (Keller 1989a,b). Incubating hoatzins (*Opisthocomus hoazin*) in Amazonian rainforest lakes habituated to the presence of tourist boats had 50% shorter flight distances than those at undisturbed sites. Even though hatching success was similar between sites, chick survival was significantly lower in disturbed nests owing to increased juvenile mortality prior to fledging, lower body mass, and elevated glucocorticoid concentrations (Müllner et al. 2004). Clearly, assessing disturbance will be improved by measuring fitness consequences and potentially confounding factors, using experiments, and assessing the impact of activities designed to reduce disturbance (e.g., Finney et al. 2005).

## Hunting

Here, hunting covers killing for meat, killing as recreational activity, and culling to reduce population size or control diseases, and as a potential stressor may resemble predation. Few studies have investigated the consequences of hunting on stress responses or reproductive success, or its effect on the fitness of individuals crippled (Hofer et al. 1993; Quiatt et al. 1994, 2002) by shooting or snaring. Experimentally pursued cougars (*Felis concolor*) significantly altered their adrenal physiological response (Harlow et al. 1992). An African savanna elephant (*Loxodonta africana*) population subjected to poaching in the 1980s showed persistent effects of significantly increased fecal glucocorticoid concentrations, disrupted group structure, and reduced reproduction in groups affected by poaching (Gobush et al. 2008). Red deer (*Cervus elaphus*) killed after pursuit by sports staghunts showed more detrimental physiological effects than those killed in traffic accidents or stalked and shot (Bradshaw and Bateson 2000), suggesting that deer escaping from pursuit hunts were likely to die from capture myopathy or other debilitating consequences (Bateson and Bradshaw 1997). Other species also showed capture myopathy after pursuit, and in dugongs (*Dugong dugon*) even short pursuits led to severe physiological problems and mortality (Anderson 1981).

Culling of species viewed as pests or disease vectors (Lovegrove 2007) may change social structure, create social instability, increase dispersal, and alter the frequency of conspecific encounters, thereby changing pathogen transmission. For example, "randomized culling trials" in Eurasian badgers (*Meles meles*) to reduce the incidence of bovine tuberculosis (*Mycobacterium bovis*) in cattle decreased badger group size, changed their group composition, and left vacant group territories. Subsequently, survivors and neighbors extended their ranging behavior, thereby potentially increasing pathogen transmission to susceptible conspecifics and cattle (McDonald et al. 2008).

## Interventions for Research and Conservation

Interventions ("handling") for wildlife research and conservation may affect research results or have a negative impact on conservation aims (Hofer and East 1998). For instance, a comparison of capture methods for grizzly (*Ursus arctos*) and black bears revealed that leghold snares significantly increased biochemical indicators of capture myopathy compared to helicopter darting or barrel traps. Capture reduced mean rates of movement for three weeks; repeatedly captured bears had poorer body condition than those captured only once (Cattet et al. 2008).

Chemical immobilization can reduce fertility in black rhino (*Diceros bicornis*) females (Alibhai et al. 2001) and moose (*Alces alces*) females (Ballard and Tobey 1981) and decrease social status and reproduction in bighorn (*Ovis canadensis*) rams (Pelletier et al. 2004). It may also increase mortality: Arnemo et al. (2006) observed mortality of between 0.7% and 3.9% in several large mammals, and DelGiudice et al. (2005) observed capture-related mortality in 5.4%, capture accidents in 2.9%, and mortality within 14 days of release in 2.5% of handled white-tailed deer (*Odocoileus virginianus*). Immobilized and radio-collared recent

immigrants to African wild dog packs (*Lycaon pictus*) had reduced survival (Burrows et al. 1994), as did radio-collared female watervoles (*Arvicola terrestris*, Moorhouse and Macdonald 2005). Technical advances that cut the size of radio-transmitters will help reduce side effects of this useful tool (Cooke et al. 2004), provided that methods of attachment are also improved.

Translocation of animals involves capture and immobilization and extensive "handling" periods, yet its effects are rarely studied. For instance, wild rabbits (*Oryctolagus cuniculus*) were translocated to bolster declining prey populations of the endangered Iberian lynx (*Lynx pardinus*). After translocation, rabbits with high levels of serum cortisol had poorer body condition and lower survival after release (Cabezas et al. 2007).

Without a theory to predict circumstances in which handling may be harmful, its impact should be evaluated on a case-by-case basis. Whether handling has an effect is likely to remain unknown without a good study design, including the long-term monitoring of unhandled "control" individuals (Saraux et al. 2011). Many studies that report no handling effects have insufficient statistical power to demonstrate effects even if they existed (White and Garrott 1990). These aspects equally apply to planned interventions for conservation management (Hofer and East 1998; Stearns and Stearns 1999), which should be designed as scientific experiments with "untreated" controls and properly monitored target populations (Heinsohn 1992).

## Methodological Issues

The choice of physiological mediator and stress response system determines the quality of stress measurements (Hofer and East 1998; von Holst 1998; Schwarzenberger 2007).

## Measuring Stress

Measurements include glucocorticoids (Schwarzenberger 2007) or catecholamines (Dehnhard 2007) in blood, urine, feces, or hair (Pereg et al. 2011); pheromones (Dehnhard 2011); heat-shock proteins and other "stress proteins" in various tissues (Moreno et al. 2002; Herring and Gawlik 2007); hyperthermia (Meyer et al. 2008a,b); the ratio of heterophils to lymphocytes in blood (Vleck et al. 2000); leukocyte coping capacity in blood (McLaren et al. 2003); heart rate and energy turnover (Wilson et al. 1991; Bisson et al. 2009); activity rhythms (Berger 2011); and the fractal structure of behavior sequences (Alados et al. 1996). These mediators vary in their response speed to stressors, are sensitive to different environmental stimuli, are subjected to endogenous rhythms at different time scales, and may be affected by handling.

Because of modulating factors and potential interference from handling, true stress-less baselines are known for few species. Either very rapid sampling (Romero and Reed 2005) or the application of ingenious methods for blood sampling that do not involve handling, such as the application of blood-sucking bugs (Voigt et al. 2004, 2006; Becker et al. 2006; Arnold et al. 2008), will deliver such baselines. As a consequence, noninvasive techniques to measure glucocorticoid metabolites in feces and urine (Goymann et al. 1999; Möstl et al. 2002) are of great interest to conservation physiology, leading to their widespread and sometimes uncritical adoption. Fecal samples provide an integrated measure of glucocorticoids over a longer time period, are supposedly free from handling effects, and dampen endogenous rhythms, provided (Goymann and Trappschuh 2011) that feces are not excreted frequently. These benefits are tempered by the issues that diet composition may alter hormone metabolite concentrations without affecting true hormone excretion (Wasser et al. 1993; Goymann 2005) and may have to be corrected for feces excretion rate (Goymann et al. 2006), that assays are needed to measure several metabolites, and that noninvasive assays need to be validated for every species. For many study designs, knowledge of individual identity or physiological status will strengthen the interpretation of the data, and can be achieved even for whales or other species in challenging habitats such as oceans (Hunt et al. 2006).

## Fitness Consequences of Stress

Glucocorticoids are an important component of the HPA axis and may reflect individual fitness (Pride 2005, Pereg et al. 2011). Intervention by scientists in free-ranging populations caused a failure to breed in baboons (Sapolsky 1985), abandonment of offspring by shorebirds (Colwell et al. 1988), reduced survival and reproduction in king penguins (*Aptenodytes* *patagonicus*, Saraux et al. 2011), declines in the number of surviving young in Atlantic puffins (*Fratercula arctica*, Rodway et al. 1996), increased mortality in rock ptarmigan (*Lagopus mutus*, Cotter and Gratto 1995), and a decline in longevity in African wild dogs (Burrows et al. 1994).

The relationship between short-term and long-term measures of fitness consequences of stress is complex. Short-term monitoring sometimes detects significant effects of intervention. Long-term monitoring of individuals subjected to intervention is useful because interventions can result in delayed mortality, as with capture myopathy (Beringer et al. 1996) or other effects (Thorne and Williams 1988; Putman 1995; Meyer et al. 2008a,b), and the development of pathological states may require weeks (Lee and Cockburn 1985; Suleman et al. 2000) or longer (Kitaysky et al. 2010; Saraux et al. 2011). Short-term measures would then conclude that there are no detrimental effects when in fact they do occur.

## THE INDIVIDUAL DOMAIN: IMMUNOCOMPETENCE AND IMMUNOSUPPRESSION

Life-history theory predicts tradeoffs between the allocation of resources to self-maintenance, growth, and reproduction (Kirkwood 1977, 1981, 1997; Kirkwood and Rose 1991; Roff 1992; Stearns 1992; van der Most et al. 2011). Important components of self-maintenance are mechanisms that prevent infections and clear pathogens following infection. Much is known about the cellular and molecular mechanisms of host responses to pathogen invasion in humans and laboratory species. Little is known about rules of resource allocation to defense mechanisms (Ardia et al. 2011) and how they change between life-history stages, physiological states, or species (Martin 2009; Michalakis 2009; Sorci et al. 2009; Graham et al. 2011; Hawley and Altizer 2011). Few studies have considered the effect of nutrition, including specific dietary components, on immunity in most wildlife species. For instance, nutritional status is recognized as an important driver of hendra virus infections in Australian red-flying foxes (*Pteropus scapulatus*), suggesting that epidemics depend on environmental and seasonal factors (Plowright et al. 2008). Clearly, this hampers the development of a conceptual framework that links individual health to nutrition or other factors. Here we discuss ecological and evolutionary concepts of immune defense and consider how stress, nutrition, and concurrent infections with several pathogens may influence the consequences of infection (Fig. 7.3).

## Life-History Approaches to Immunocompetence

Organisms with high reproductive rates should invest less in immune defense than those with low rates (Sheldon and Verhulst 1996; Rigby and Moret 2000); life history tradeoffs probably determine which part of the immune system receives preferential allocation of resources (Ricklefs and Wikelski 2002). Lee et al. (2008) tested these ideas by measuring two components of the immune response in 70 neotropical bird species with different reproductive rates: cheaply and rapidly produced complement proteins and costly antibodies produced by B cells that take time to differentiate. There was a significant positive association between complement protein activity and clutch size, and a positive association between natural antibody levels and incubation period. This suggests that species with high reproductive rates use inexpensive immune defenses to minimize juvenile resource demands, whereas species with low reproductive rates afford longer developmental periods to produce a costly adaptive immune system.

A second perspective links immunity, aging, and longevity (Kirkwood 1977, 1997; Kirkwood and Rose 1991). When immune defense against infections usurp resources otherwise allocated to anti-aging mechanisms such as telomerase production to repair telomere attrition (Blackburn 2000), aging processes can accelerate. Thus, serious infectious and inflammatory diseases can accelerate aging and curtail lifespan and lifetime reproductive success (Aviv 2008; Ilmonen et al. 2008), thereby affecting population growth. In long-lived seabirds such as the European shag (*Phalacrocorax aristotelis*) and the wandering albatross (*Diomedea exulans*), unfavorable environmental conditions during chick growth caused rapid telomere attrition and poor survival (Hall et al. 2004). In tree swallows (*Tachycineta bicolor*), individuals with relatively short telomeres at 1 year of age had lower survival than same-aged individuals with longer telomeres (Haussmann et al. 2005). Minimum lifespan in adult sand martins (*Riparia riparia*) was positively correlated with residual telomere length, and lifetime reproductive success in male dunlin (*Calidris alpina*) was higher in males with longer-than-average telomeres (Pauliny et al. 2006).

The social environment also influences telomere attrition. Female mice (*Mus musculus*) exposed to males and experiencing reproductive stress had greater lymphocyte telomere attrition than control females, and male mice exposed to crowding had greater telomere attrition than male controls (Kotrschal et al. 2007). In humans, substantial psychological stress is associated with lymphocyte telomere shortening, equivalent to a lifespan reduction by 9 to 17 years (Epel et al. 2004).

## Immunosuppression as a Consequence of Stress

The relationship between stress and immunocompetence is complex. In the past two decades, studies on humans and laboratory rodents revealed mechanisms by which putative stressors cause physiological changes; how they affect the immune system; how immunocompetence interacts with the central nervous system, reproduction, and the endocrine system; and how such interactions affect health. For example, circulating glucocorticoids can alter immune function by inducing a shift from cellular to humoral immunity (Dhabhar and McEwen 1996, 1999). A key conclusion is that physiological stress can induce immunosuppression, exacerbate the effects of infection and inflammation, and significantly contribute to morbidity and mortality (Glaser and Kiecolt-Glaser 2005).

The impact of stressors on immunocompetence in wildlife has received little attention (Martin 2009). For instance, olfactory detection of conspecifics infected with *Babesia microti* by pregnant female mice increased their glucocorticoid levels and kidney growth and led to an accelerated response to *B. microti* infection in their offspring (Curno et al. 2009). Boxes 7.2 and 7.4 further consider the complex links between immunocompetence, stress, and reproduction in detail.

Social processes in group-living mammals can be sufficiently stressful to reduce immunocompetence and thus health (Abbott et al. 2003; Sapolsky 2005). Both positive and negative social relationships can affect well-being, and physiological alterations

**Box 7.4  Reproductive Performance, Immunocompetence, Pathogen Exposure, Reintroductions, and the Endangered Cheetah**

Captive breeding and reintroduction programs are considered important conservation activities for vulnerable and endangered species (Ebenhard 1995; Seddon et al. 2005), yet captive populations of some large mammals are characterized by poor reproductive performance. Recent studies of cheetahs (*Acinonyx jubatus*) monitored cub survival, measured exposure to pathogens and its consequences, and assessed female reproductive status to test predictions from all four hypotheses that explain this phenomenon: stress, genetic monomorphism, innate rhythms, or schedules of reproduction management in relation to individual age.

In captivity, low cub survival (Marker-Kraus and Grisham 1993) and irregular estrus cycles and anestrus periods of female cheetahs (Wildt et al. 1993; Brown et al. 1996) were thought to be a consequence of a lack of genetic diversity (the genetic monomorphism hypothesis, O'Brien et al. 1983; 1985), or unfavorable husbandry conditions (the stress hypothesis, see below, Wildt et al. 1993; Caro and Laurenson 1994; Caughley 1994; Merola 1994; Brown et al. 1996; Wielebnowski 1996; Terio et al. 2003). Free-ranging cheetah cubs had low post-emergence survival in the Serengeti, Tanzania, because of predation by lions (*Panthera leo*) and spotted hyenas (Laurenson 1994), but high survival in a predator-free habitat in Namibia (Wachter et al. 2011) where the population was exposed to a wide variety of viral pathogens (Munson et al. 2004, Thalwitzer et al. 2010), showed no manifestation of clinical disease (Munson et al. 2005; Thalwitzer et al. 2010) and had a higher genetic diversity than previously posited (Castro-Prieto et al. 2011). These results suggest that the relatively low genetic diversity of cheetah is currently of little practical concern.

The mechanism responsible for impairment of female fertility in captivity remained elusive at first. The captive stress hypothesis blamed husbandry conditions to suppress ovarian activity (the stress hypothesis, Brown et al. 1996; Jurke et al. 1997). Terio et al. (2003) posited that reproductive cycling was triggered by an endogenous circannual rhythm not apparent in the conditions of zoos abroad (the innate rhythm hypothesis, Terio et al. 2003). Alternatively, frequent fluctuations of estrogen concentrations and maturation of follicles in older, nulliparous females are likely to cause faster aging of reproductive organs than in breeding females, the development of hormone-dependent and age-related pathological lesions of the reproductive tract, and a premature post-reproductive status (the asymmetric reproductive aging hypothesis, Hermes et al. 2004, 2006). Wachter et al. (2011) simultaneously tested these hypotheses and showed that reproductive activity and health of cheetah females were determined by reproductive history and age rather than innate rhythms or captive or chronic stress.

In species where asymmetric reproductive aging is conceivable (Wildt et al. 1993; Hildebrandt et al. 2000; Lu et al. 2000), management of captive breeding and reintroduction programs might therefore encourage early reproduction in females to induce healthy reproductive performance. This implies that management actions to prevent genetic adaptations to captivity (Kraaijeveld-Smit et al. 2006) should not delay female reproduction for the purpose of minimizing the number of generations in captivity (Frankham 2008; Williams & Hoffman 2009).

associated with social interactions can modulate immune and neuroendocrine functions and disease susceptibility (Bartolomucci 2007). In male social mammals, immune and reproductive functions can be influenced by social status, as in Brandt's vole (*Lasiopodomys brandtii*), where social status was negatively correlated with spleen mass and serum antibody levels and positively correlated with serum cortisol levels (Li et al. 2007).

The timing and duration of a stressor's impact on an organism can have very different effects on the immune system (Dhabhar 2009a,b; Martin 2009).

The short-term fight-or-flight stress response not only activates the neuroendocrine, cardiovascular, and musculoskeletal systems but may also enhance immunocompetence, presumably because the immune system needs to respond quickly if wounding or infections follow an attack. Korte et al. (2005) describe how behavioral strategies of two extreme phenotypes ("hawks" and "doves") for coping may underlie the physiological stress response, and how they are associated with different costs and benefits and tradeoffs between health and disease.

There is some evidence that anthropogenic disturbance of wildlife can induce sufficient physiological stress to reduce host immunity and thereby increase susceptibility to infection (Hofer and East 1998; Wikelski and Cooke 2006; Martin 2009). The colony collapse disorder of honey bee (*Apis mellifera*) populations may be an example of the synergistic effects of anthropogenically induced nutritional stress and viral infection that led to fatal immunosuppression (see Chapter 20 in this volume).

## Immunity, Tradeoffs, and Concurrent Infections

Infection can influence life-history tradeoffs in hosts (Lee 2006; Barrett et al. 2008; Jones et al. 2008) because body resources channeled into fighting infections will not be available for other functions, such as body maintenance, growth, costly sexual displays, gestation, or lactation (Kirkwood 1981; Hamilton and Zuk 1982; McNamara and Buchanan 2005), and in males for reproductive functions, particularly testosterone concentrations (Naguib et al. 2004; Muehlenbein and Bribiescas 2005; Boonekamp et al. 2008; Nunn et al. 2009; Greiner et al. 2010).

Because lactation is energetically expensive, higher levels of parasite infection in lactating than non-lactating bighorn ewes were probably a result of insufficient investment of body reserves in immune function (Festa-Bianchet 1989). Many factors, such as age, sex, social status, competition for food resources, and seasonal changes, may influence investment in immune function (Lee 2006; Martin et al. 2008; Martin 2009). When nutritionally demanding immune responses decrease body reserves, future investment in immune function may also be curtailed, making individuals more vulnerable to future infections (Beldomenico et al. 2008).

Immune-mediated interactions between co-infecting pathogens occur through cross-reactivity when exposure to one pathogen enhances or reduces immunity to a second pathogen. The nature of these interactions can modulate host susceptibility and influence patterns of co-infection, rates of transmission, and distribution patterns of pathogens across hosts (Behnke et al. 2001; Graham et al. 2007; Poulin 2007; Lello and Hussell 2008). For example, co-infection of European rabbits with the myxoma virus disrupted their ability to clear helminths (*Trychostrongylus retortaeformis*) from the gastrointestinal system (Cattadori et al. 2007). In Box 7.5 we discuss how co-infection leads to apparent immunosuppression and facilitates widespread population declines (Augustine 1998) in koalas (*Phascolarctos cinereus*). Co-infections have been little studied but are probably common (e.g., Goller et al. 2010); there is therefore an urgent need to assess the impact of co-infections on immunity and host survival (Fig. 7.3).

## THE POPULATION DOMAIN: THE MULTIPLICATIVE EFFECTS OF INDIVIDUAL FITNESS IMPAIRED BY STRESS AND IMMUNOSUPPRESSION

Although it is widely accepted that emerging infectious diseases can cause population decline and extinction, it is difficult to find evidence that directly links these to stress or immunosuppression. Here, it is helpful to distinguish between small and declining populations (Caughley 1994).

### The Small Population Paradigm

In free-ranging, long-lived mammals, juvenile survival is more important for population growth than fecundity (Heppell et al. 2000) because fecundity is usually high. However, if populations are small or suffer from anthropogenic stress, fecundity may become an important issue, as demonstrated by the Allee effect (Courchamp et al. 2008). If pathogens or stress cause immunosuppression and thereby reduce fecundity or survival, then the theory of host–parasite population dynamics (Anderson and May 1982, 1991) may also be relevant, although it is usually not applied for this purpose.

---

**Box 7.5  Retroviruses, *Chlamydia*, Immunosuppression, and Koalas**

---

The Australian koala (*Phascolarctos cinereus*) was hunted extensively until the 1920s, and its habitat was destroyed in many places, leading to fragmented and isolated populations (Phillips 1987; Lee and Martin 1988; Martin and Handasyde 1999). Koalas are an example for the synergistic effects of co-infection and the co-evolution of pathogens and their hosts because they are widely infected with a gammaretrovirus, the koala retrovirus (KoRV, Hanger et al. 2000), and two bacteria, *Chlamydia pecorum* and *C. pneumoniae*.

KoRV is an endogenous retrovirus transmitted via the germline that is still actively transcribed within the host and has only been recently introduced into its host population, as some koala populations are free of KoRV (Tarlinton et al. 2006). Infection with KoRV induces lymphomas and leukemias in 3% to 5% of cases; in captivity, mortality of infected individuals may reach 80%. KoRV induces immune deficiencies associated with opportunistic infections such as chlamydiosis (Fiebig et al. 2006), and there is an association of neoplasia with high KoRV titers (Tarlinton et al. 2005). Infections with gammaretroviruses are known to be immunosuppressive, and KoRV is no exception (Fiebig et al. 2006). Of the two *Chlamydia*, infection with *C. pecorum* is more severe, with clinical manifestations and impaired reproduction, infertility, or blindness (Jackson et al. 1999; McLean and Handasyde 2006).

The temporal coincidence of the occurrence and spread of KoRV and *Chlamydia* within koala populations is intriguing. It seems conceivable that individual susceptibility to infection with *Chlamydia* is enhanced by the presence of KoRV in an individual host. If so, then the incidence of infection with *Chlamydia* should be associated with the presence of KoRV titers, and major outbreaks of *Chlamydia* should be associated with KoRV in historical as well as recent times. Experimental studies are under way to establish whether KoRV has indeed immunosuppressive effects and to use museum specimens as well as recent samples to establish whether the spread of infection with *Chlamydia* and the KoRV are associated events (A.D. Greenwood, personal communication).

---

## The Declining Population Paradigm

The recent decline of passerine populations in agricultural habitats in the Western world is well documented (Fuller et al. 1995). This decline is a consequence of habitat destruction and associated reductions in breeding and feeding opportunities owing to the intensification of agriculture. Little is known about the effect of such changes on the immunocompetence of individuals or populations, or possible synergistic effects of reduced foraging opportunities and increased predation owing to impaired health. We clearly need more studies that consider both the state of individuals and the resulting population dynamics in order to move away from the current retrospective, descriptive perspective.

## Population Extinction

In the case of mammals, there are few well-corroborated instances of pathogens and diseases having caused or significantly contributed to the extinction of populations or species through stress or immunosuppression. Here, a plethora of theoretical models face few empirical tests of their predictions, whereas the multiplicity of factors invoked in observed crashes are rarely reflected in the construction of theoretical models (de Castro and Bolker 2005; Nunn and Altizer 2006). Possible examples include the extinction of the free-ranging populations of the black-footed ferret (*Mustela nigripes*, Thorne and Williams 1988) or the Serengeti population of the African wild dog (Burrows 1992; Burrows et al. 1994); the possible extinction of the Mednyi Arctic fox (*Alopex lagopus semenovi*, Goltsman et al. 1996); the population crash of the Spanish ibex (*Capra pyrenaica hispanica*, Léon-Vizcaino et al. 1999) and the Iberian lynx (*Lynx pardus*, Meli et al. 2009); and the population crashes and local extinctions of the Tasmanian devil (*Sarcophilus harrisii*, Hawkins et al. 2006; McCallum et al. 2007; Jones et al. 2008; McCallum 2008).

Retrospective studies using museum material may help identify such processes. One example may be the extinction of the endemic Christmas Island rat (*Rattus macleari*) around 1900, for which a pathogenic trypanosome in fleas carried by recently introduced black rats (*Rattus rattus*) may have been responsible. Endemic rats collected prior to the introduction of black rats showed no evidence of infection with the trypanosome but were infected after the introduction (Wyatt et al. 2008). To what extent competition for food between native and introduced rats exacerbated the effect of trypanosome infection remains to be determined.

## Studying the Population Consequences of Stress and Immunosuppression Mediated by Pathogens of Wildlife Hosts

Several problems hamper quick advances in understanding the stressful or immunosuppressive impact of pathogens on their wildlife hosts. First, field ecologists often overlook pathogens and thus do not design studies to consider their impact (East et al. 2011). Second, pathogen-mediated host mortality may be misassigned. Mortality of an ungulate killed by a predator may be recorded as predation unless samples from the "kill" are screened for pathogens that may have weakened the prey and thus increased its chances of being caught by the predator (Kutz et al. 2004).

Third, the prevalence of infection and its clinical relevance is often assessed in adults, where it might be benign, but not in the most vulnerable early life stages, when infections may be fatal. For instance, African carnivores display a high prevalence (>90%) of infection with the protozoan *Hepatozoon* in adults of several species, including spotted hyenas, but no clinical signs (Averbeck et al. 1990; Van Heerden et al. 1995; East et al. 2008). In 18% of infected juvenile spotted hyenas, *Hepatozoon* infection was identified as cause of morbidity and subsequent mortality (East et al. 2008). Abortions caused by pathogens may also be difficult to detect (Marco et al. 2007).

Fourth, the impact of many pathogens on species of conservation concern will not be clarified without long-term monitoring of the fitness consequences of infection. For example, canine distemper virus (CDV) is considered a threat to endangered African

wild dogs, although only three possible incidences of diseased packs have been described. Even if CDV infection may threaten small, isolated populations, there is currently little evidence that the virus threatens large, healthy populations (Goller et al. 2010).

## CONCLUSIONS

Practitioners of conservation medicine should carefully design protocols for handling so that the effects of their actions can be assessed. They need to ensure adequate controls and post-intervention monitoring, check in advance whether target species are sensitive to the type of planned intervention, use noninvasive sampling methods if possible, and minimize disturbance. For instance, when vaccinating social species, it may be unwise to pursue group members all day as this is likely to incite capture myopathy, decrease the chance of the vaccine working, and increase handling-associated stress.

The study of the population consequences of stress and immunosuppression is only beginning. It is an area of research at the interface of behavioral ecology, population biology, evolutionary epidemiology, physiology, infection biology, (eco)immunology, endocrinology, conservation biology, and conservation medicine—fields with little contact, rarely linked during academic training (Hofer 2007). Conservation medicine has an interest to strengthen these links because it needs to understand the mechanisms and fitness consequences by which populations of conservation concern meet anthropogenic challenges. Only then will a theory be developed that can predict the population consequences of anthropogenic challenges. We therefore need to encourage the transfer of conceptual, methodical, and technological advances between disciplines in order to make progress soon.

## REFERENCES

Abbott, D.H., E.B. Keverne, F.B. Bercovitch, C.A. Shively, S.P. Mendoza, W. Saltzman, C.T. Snowdon, T.E. Ziegler, M. Banjevic, T. Garland, and R.M. Sapolsky. 2003. Are subordinates always stressed? A comparative analysis of rank differences in cortisol levels among primates. Horm Behav 43:67–82.

Aguirre, A.A., R.S. Ostfeld, G.M. Tabor, C. House, and M.C. Pearl. 2002. Conservation medicine: ecological health in practice. Oxford University Press, New York.

Alados, C.L., J.M. Escos, and J. M. Emlen. 1996. Fractal structure of sequential behaviour patterns: an indicator of stress. Anim Behav 51:437–443.

Allen, L.C., C.S. Richardson, G.F. McCracken, and T.H. Kunz. 2010. Birth size and postnatal growth in cave- and bridge-roosting Brazilian free-tailed bats. J Zool Lond 280:8–16.

Alibhai, S.K., Z.C. Jewell, and S.S. Towindo. 2001. Effect of immobilization on fertility in female black rhino (*Diceros bicornis*). J Zool Lond 253:333–345.

Anderson, P.K. 1981. The behavior of the dugong (*Dugong dugon*) in relation to conservation and management. Bull Mar Sci 31:640–647.

Anderson, R.M., and R.M. May. 1982. Coevolution of hosts and parasites. Parasitol 85:411–426.

Anderson, R.M., and R.M. May. 1991. Infectious diseases of humans: dynamics and control. Oxford University Press, Oxford.

Andersson, M. 1994. Sexual selection. Princeton University Press, Princeton.

Ardia, D.R., H.K. Parmentier, and L.A. Vogel. 2011. The role of constraints and limitation in driving individual variation in immune response. Funct Ecol 25:61–73.

Arlettaz, R., P. Patthey, M. Baltic, T. Leu, M. Schaub, R. Palme, and S. Jenni-Eiermann. 2007. Spreading free-riding snow sports represent a novel serious threat for wildlife. Proc R Soc B 274:1219–1224.

Arnemo, J.M., P. Ahlqvist, R. Andersen, F. Berntsen, G. Ericsson, J. Odden, S. Brunberg, P. Segerström, and J.E. Swenson. 2006. Risk of capture-related mortality in large free-ranging mammals: experiences from Scandinavia. Wildl Biol 12:109–113.

Arnold, J.M., S.A. Oswald, C.C. Voigt, R. Palme, A. Braasch, C. Bauch, and P.H. Becker. 2008. Taking the stress out of blood collection: comparison of field blood-sampling techniques for analysis of baseline corticosterone. J Avian Biol 39:588–592.

Augustine, D. 1998. Modelling *Chlamydia*–koala interactions: coexistence, population dynamics and conservation implications. J Appl Ecol 35: 261–272.

Averbeck, G.A., K.E. Bjork, C. Packer, and L. Herbst. 1990. Prevalence of hematozoans in lions (*Panthera leo*) and cheetah (*Acinonyx jubatus*) in Serengeti National Park and Ngorongoro Crater, Tanzania. J Wildl Dis 26:392–394.

Aviv, A. 2008. The epidemiology of human telomeres: faults and promises. J Gerontol 63A:979–983.

Ballard, W.B., and R.W. Tobey. 1981. Decreased calf production of moose immobilized with acetine administered from helicopters. Wildl Soc Bull 9:207–209.

Barrett, L.G., P.H. Thrall, J.J. Burdon, and C.C. Linde. 2008. Life history determines genetic structure and evolutionary potential of host–parasite interactions. Trends Ecol Evol 23:678–685.

Bartolomucci, A. 2007. Social stress, immune functions and disease in rodents. Front Neuroendocrinol 28:28–49.

Bateson, P.P.G., and E.L. Bradshaw. 1997. Physiological effects of hunting red deer (*Cervus elaphus*). Proc R Soc B 264:1707–1714.

Becker, P.H., C.C. Voigt, J. M. Arnold, and R. Nagel. 2006. A technique to bleed incubating birds without trapping: A blood-sucking bug in a hollow egg. J Ornithol 147:115–118.

Begon, M., J.L. Harper, and C.R. Townsend. 2006. Ecology: from individuals to ecosystems. Blackwell Publishing, Oxford.

Behnke, J.M., A. Bajer, E. Sinski, and D. Wakelin. 2001. Interactions involving intestinal nematodes of rodents: experimental and field studies. Parasitol 122:S39–S49.

Bejder, L., A. Samuels, H. Whitehead, N. Gales, J. Mann, R. Connor, M. Heithaus, J. Watson-Capps, C. Flaherty, and M. Krützen. 2006. Decline in relative abundance of bottlenose dolphins exposed to long-term disturbance. Conserv Biol 20:1791–1798.

Beldomenico, P.M., S. Telfer, S. Gebert, L. Lukomski, M. Bennett, and M. Begon. 2008. Poor condition and infection: a vicious cycle in natural populations. Proc R Soc B 275:1753–1759.

Berger, A. 2011. Activity patterns, chronobiology and the assessment of stress and welfare in zoo and wild animals. Int Zoo Yearb 45: 80–90.

Beringer, J., L.P. Hansen, W. Wilding, J. Fischer, and S.L. Sheriff. 1996. Factors affecting capture myopathy in white-tailed deer. J Wildl Manage 60:373–380.

Bisson, I.-A., L.K. Butler, T.J. Hayden, L.M. Romero, and M.C. Wikelski. 2009. No energetic cost of anthropogenic disturbance in a songbird. Proc R Soc B 276:961–969.

Blackburn, E.H. 2000. Telomere states and cell fates. Nature 408:53–56.

Boonekamp, J.J., A.H.F. Ros, and S. Verhulst. 2008. Immune activation suppresses plasma testosterone level: a meta-analysis. Biol Lett 4:741–744.

Bradshaw, E.L., and P.P.G. Bateson. 2000. Welfare implications of culling red deer (*Cervus elaphus*). Anim Welfare 9:3–24.

Broom, D.M., and K.G. Johnson. 1991. Animal welfare. Chapman & Hall, London.

Brown, J.L., D.E. Wildt, N. Wielebnowski, K.L. Godrowe, L.H. Graham, S. Wells, and J.G. Howard. 1996. Reproductive activity in captive female cheetahs

(*Acinonyx jubatus*) assessed by faecal steroids. J Reprod Fertil 106:337–346.

Burger, J., and M. Gochfeld. 2007. Responses of Emperor Penguins (*Aptenodytes forsteri*) to encounters with ecotourists while commuting to and from their breeding colony. Polar Biol 30:1303–1313.

Burrows, R. 1992. Rabies in wild dogs. Nature 359:277.

Burrows, R., H. Hofer, and M.L. East. 1994. Demography, extinction and intervention in a small population: the case of the Serengeti wild dogs. Proc R Soc B 256:281–292.

Cabezas, S., J. Blas, T.A. Marchant, and S. Moreno. 2007. Physiological stress levels predict survival probabilities in wild rabbits. Horm Behav 51:313–320.

Careau, V., D.W. Thomas, and M.M. Humphries. 2010. Energy cost of bot fly parasitism in free-ranging eastern chipmunks. Oecologia 162:303–312.

Carlini, A.R., N.R. Coria, M.M. Sanots, M.M. Libertelli, and G. Donini. 2006. Breeding success and population trends in Adélie penguins in areas with low and high levels of human disturbance. Polar Biol 30:917–924.

Caro, T.M., and K.M. Laurenson. 1994. Ecological and genetic factors in conservation: a cautionary tale. Science 263:485–486.

Carrete, M., J.A. Sánchez-Zapata, J.R. Benítez, M. Lobón, and J.A. Donázar. 2009. Large scale risk-assessment of wind-farms on population viability of a globally endangered and long-lived raptor. Biol Conserv 142:2954–2961.

Castro-Prieto, A., B. Wachter, and S. Sommer. 2011. Cheetah paradigm revisited: MHC diversity in the world's largest free-ranging population. Mol Biol Evol 28:1455–1468.

Cattadori, I.M., R. Albert, and B. Boag. 2007. Variation in host susceptibility and infectiousness generated by co-infection: the myxoma–*Trichostrongylus retortaeformis* case in rabbits. J R Soc Interface 4:831–840.

Cattet, M., J. Boulanger, G. Stenhouse, R.A. Powell, and M.J. Reynolds-Hogland. 2008. An evaluation of long-term capture effects in ursids: implications for wildlife welfare and research. J Mammal 89: 973–990.

Caughley, G. 1994. Directions in conservation biology. J Anim Ecol 63:215–244.

Cobley, N.D., and J.R. Shears. 1999. Breeding performance of gentoo penguins (*Pygoscelis papua*) at a colony exposed to high levels of human disturbance. Polar Biol 21:355–360.

Colwell, M.A., C.L. Gratto, L.W. Oring, and A.J. Fivizzani. 1988. Effects of blood sampling on shorebirds: injuries, return rates and clutch desertions. Condor 90:942–945.

Cooke, S.J., S.G. Hinch, M. Wikelski, R.D. Andrews, L.J. Kuchel, T.G. Wolcott, and P.J. Butler. 2004. Biotelemetry: a mechanistic approach to ecology. Trends Ecol Evol 19:334–343.

Cotter, R.C., and C.J. Gratto. 1995. Effects of nest and brood visits and radio transmitters on rock ptarmigan. J Wildl Manage 59:93–98.

Courchamp, F., L. Berec, and J. Gascoigne. 2008. Allee effects in ecology and conservation. Oxford University Press, Oxford.

Creel, S. 2001. Social dominance and stress hormones. Trends Ecol Evol 16:491–497.

Culik, B.M., and R.P. Wilson. 1991. Penguins crowded out? Nature 351:340.

Culik, B.M., and R.P. Wilson. 1995. Penguins disturbed by tourists. Nature 376:301–302.

Curno, O., J.M. Behnke, A.G. McElligott, T. Reader, and C.J. Barnard. 2009. Mothers produce less aggressive sons with altered immunity when there is a threat of disease during pregnancy. Proc R Soc B 276:1047–1054.

Czech, B. 2000. Economic associations among causes of species endangerment in the United States. Bioscience 50:593–601.

de Castro, F., and B. Bolker. 2005. Mechanisms of disease-induced extinction. Ecol Lett 8:117–126.

Dehnhard, M. 2007. Characterisation of the sympathetic nervous system of Asian (*Elephas maximus*) and African (*Loxodonta africana*) elephants based on urinary catecholamine analyses. Gen Comp Endocrinol 151:274–284.

Dehnhard, M. 2011. Mammal semiochemicals: understanding pheromones and signature mixtures for better zoo-animal husbandry and conservation. Int Zoo Yearb 45: 55–79.

Delahay, R.J., J.R. Speakman, and R. Moss. 1995. The energetic consequences of parasitism – effects of a developing infection of *Trichostrongylus tenuis* (Nematoda) on red grouse (*Lagopus lagopus scoticus*): energy balance, body weight and condition. Parasitol 110:473–482.

DelGiudice, G.D., B.A. Sampson, D.W. Kuehn, M.C. Powell, and J. Fieberg. 2005. Understanding margins of safe capture, chemical immobilization and handling of free-ranging white-tailed deer. Wildl Soc Bull 33:677–687.

Dhabhar, F.S., and B.S. McEwen. 1996. Stress induced enhancement of antigen specific cell-mediated immunity. J Immunol 156:2608–2615.

Dhabhar, F.S., and B.S. McEwen. 1999. Enhancing versus suppressive effects of stress hormones on skin immune function. Proc Natl Acad Sci USA 96:1059–1064.

Dhabhar, F.S. 2009a. A hassle a day may keep the pathogens away: The fight-or-flight stress response and the augmentation of immune function. Integr Comp Biol 49:215–236.

Dhabhar, F.S. 2009b. Enhancing versus suppressive effects of stress on immune function: implications for immunoprotection and immunopathology. Neuroimmunomodul 16:300–317.

Dietz, R., C.T. Hansen, P. Have,and M.P. Heide-Jørgensen. 1989a. Clue to seal epizootic? Nature 338:627.

Dietz, R., C.T. Hansen, P. Have, and M.P. Heide-Jørgensen. 1989b. Mass deaths of harbor seals (Phoca vitulina) in Europe. Ambio 18:258–264.

East, M.L., B. Bassano, and B. Ytrehus. 2011. The role of pathogens in the population dynamics of European ungulates. In R.J. Putman, M. Apollonio, and R. Andersen, eds. European ungulates and their management in the 21st century, pp. 319–348. Cambridge University Press, Cambridge.

East, M.L., and H. Hofer. 2010. Social environments, social tactics and their fitness consequences in complex mammalian societies. In T. Szekely, A.J. Moore, and J. Komdeur, eds. Social behaviour: genes, ecology and evolution, pp. 360–390. Cambridge University Press, Cambridge.

East, M.L., G. Wibbelt, D. Lieckfeldt, A. Ludwig, K.V. Goller, K. Wilhelm, G. Schares, D. Thierer, and H. Hofer. 2008. A Hepatozoon species genetically distinct from H. canis infecting spotted hyena in the Serengeti ecosystem, Tanzania. J Wildl Dis 44:45–52.

Ebenhard, T. 1995. Conservation breeding as a tool for saving animal species from extinction. Trends Ecol Evol 10:438–443.

Ellenberg, U., T. Mattern, P.J. Seddon, and G. Luna Jorquera. 2006. Physiological and reproductive consequences of human disturbance in Humboldt penguins: the need for species-specific visitor management. Biol Conserv 133:95–106.

Enzenbacher, D.J. 1993. Tourists in Antarctica: numbers and trends. Tourism Manage 14:142–146.

Epel, S.E., E.H. Blackburn, J. Lin, F.S. Dhabhar, N.E. Adler, J.D. Morrow, and R.M. Cawthon. 2004. Accelerated telomere shortening in response to life stress. Proc Natl Acad Sci USA 101:17312–17315.

Fernández, A., J.F. Edwards, F. Rodrígues, A. Espinosa de los Monteros, P. Herráez, P. Castro, J.R. Jaber, V. Martin, and M. Arbelo. 2005. "Gas and fat embolic syndrome" involving a mass stranding of beaked whales (family Ziphiidae) exposed to anthropogenic sonar signals. Vet Pathol 42:446–451.

Festa-Bianchet, M. 1989. Individual differences, parasites and the cost of reproduction for bighorn ewes (Ovis canadensis). J Anim Ecol 58:785–795.

Fiebig, U., M.G. Hartmann, N. Bannert, R. Kurth, and J. Denner. 2006. Transspecies transmission of the endogenous koala retrovirus. J Virol 80:5651–5654.

Finney, S.K., J.W. Pearce-Higgins, and D.W. Yalden. 2005. The effect of recreational disturbance on an upland breeding bird, the golden plover Pluvialis apricaria. Conserv Biol 121:53–63.

Francis, C.D., C.P. Ortega, and A. Cruz. 2009. Noise pollution changes avian communities and species interactions. Curr Biol 19:1415–1419.

Frankham, R. 2008. Genetic adaptation to captivity in species conservation programs. Mol Ecol 17:325–333.

Fraser, A.F., and D.M. Broom. 1990. Farm animal behaviour and welfare. Baillière Tindall, London.

Fretwell, S.D., and H.L. Lucas. 1970. On territorial behaviour and other factors influencing habitat distribution in birds. Acta Biotheor 19:16–36.

Fuller, R.J., R.D. Gregory, D.W. Gibbons, J.H. Marchant, J.D. Wilson, S.R. Baillie, and N. Carter. 1995. Population declines and range contractions among lowland farmland birds in Britain. Conserv Biol 9:1425–1441.

Gill, J.A., K. Norris, and W.J. Sutherland. 2001. Why behavioural responses may not reflect the population consequences of human disturbance. Biol Conserv 97:265–268.

Glaser, R., and J.K. Kiecolt-Glaser. 2005. Stress-induced immune dysfunction: implications for health. Nature Rev Immunol 5:243–251.

Gobush, K.S., B.M. Mutayoba, and S.K. Wasser. 2008. Long-term impacts of poaching on relatedness, stress physiology, and reproductive output of adult female African elephants. Conserv Biol 22:1590–1599.

Goller, K.V., R.D. Fyumagwa, V. Nikolin, M.L. East, M. Kilewo, S. Speck, T. Müller, M. Matzke, and G. Wibbelt. 2010. Fatal canine distemper infection in a pack of African wild dogs in the Serengeti ecosystem, Tanzania. Vet Microbiol 146:245–252.

Goltsman, M., E.P. Kruchenkova, and D.W. Macdonald. 1996. The Mednyi arctic foxes: treating a population imperilled by disease. Oryx 30:251–258.

Goodrich, J.M., and J. Berger. 1994. Winter recreation and hibernating black bears Ursus americanus. Biol Conserv 67:105–110.

Goymann, W. 2005. Non-invasive monitoring of hormones in bird droppings: biological validations, sampling, extraction, sex differences, and the influence of diet on hormone metabolite levels. Ann New York Acad Sci 1046:35–53.

Goymann, W., M.L. East, B. Wachter, O. Höner, E. Möstl, T.J. Van't Hof, and H. Hofer. 2001. Social, state-dependent and environmental modulation of faecal corticosteroid levels in free-ranging female spotted hyaenas. Proc R Soc B 268:2453–2459.

Goymann, W., and H. Hofer. 2010. Mating systems, social behaviour and hormones. In P. Kappeler, ed. Animal behaviour: evolution and mechanisms, pp. 465–501. Springer Verlag, Heidelberg, New York.

Goymann, W., E. Möstl, T. Van't Hof, M.L. East, and H. Hofer. 1999. Non-invasive fecal monitoring of glucocorticoids in spotted hyenas, Crocuta crocuta. Gen Comp Endocrinol 114:340–348.

Goymann, W., and M. Trappschuh. 2011. Life-history and diel variation of hormone metabolites in European stonechats—on the importance of high signal-to-noise-ratios in non-invasive hormone studies. J Biol Rhythms 26:44–54.

Goymann, W. M. Trappschuh, W. Jensen, and I. Schwabl. 2006. Low ambient temperature increases food intake and dropping production, leading to incorrect estimates of hormone metabolite concentrations in European stonechats. Horm Behav 49:644–653.

Goymann, W., and J.C. Wingfield. 2004. Allostatic load, social status and stress hormones: the costs of social status matter. Anim Behav 67:591–602.

Graham, A., I.M. Cattadori, J. Lloyd-Smith, M. Ferrari, and O.N. Bjørnstad. 2007. Transmission consequences of co-infection: cytokines writ large? Trends Parasitol 23:281–291.

Graham, A.L., D.M. Shuker, L.C. Pollitt, S.K.J.R. Auld, A.J. Wilson, and T.J. Little. 2011. Ecological immunology: fitness consequences of immune responses: strengthening the empirical framework for ecoimmunology. Funct Ecol 25:5–17.

Greiner, S., V. Stefanski, M. Dehnhard, and C.C. Voigt. 2010. Plasma testosterone levels decrease after activation of skin immune system in a free-ranging mammal. Gen Comp Endocrinol 168:466–473.

Hall, M.E., L. Nasir, F. Daunt, E.A. Gault, J.P. Croxall, S. Wanless, and P. Monaghan. 2004. Telomere loss in relation to age and early environment in long-lived birds. Proc R Soc B 271:1571–1576.

Hamilton, W.D., and M. Zuk. 1982. Heritable true fitness and bright birds: a role for parasites? Science 218:384–387.

Hanger, J.J., L.D. Bromham, J.J. McKee, T.M. O'Brien, and W.F. Robinson. 2000. The nucleotide sequence of koala (Phascolarctos cinereus) retrovirus: a novel type C endogenous virus related to gibbon ape leukemia virus. J Virol 74:4264–4272.

Harlow, H.J., F.G. Lindzey, W.D. Van Sickle, and W.A. Gern. 1992. Stress response of cougars to nonlethal pursuit by hunters. Can J Zool 70:136–139.

Harrington, F.H., and A.M. Veitch. 1992. Calving success of woodland caribou exposed to low-level jet fighter overflights. Arctic 45:213–218.

Haussmann, M.F., D.W. Winkler, and C.M. Vleck. 2005. Longer telomeres associated with higher survival in birds. Biol Lett 1:212–214.

Hawkins, C.E., C. Baars, H. Hesterman, G.J. Hocking, M.E. Jones, B. Lazenby, D. Mann, N. Mooney, D. Pemberton, M. Pyecroft, M. Restani, and J. Wiersma. 2006. Emerging disease and population decline of an island endemic, the Tasmanian devil Sarcophilus harrisii. Biol Conserv 131:307–324.

Hawley, D.M., and S.M. Altizer. 2011. Disease ecology meets ecological immunology: understanding the links between organismal immunity and infection dynamics in natural populations. Funct Ecol 25:48–60.

Heinsohn, R. 1992. When conservation goes to the dogs. Trends Ecol Evol 7:214–215.

Heppell, S.S., H. Caswell, and L.B. Crowder. 2000. Life histories and elasticity patterns: perturbation analysis for species with minimal demographic data. Ecology 81:654–665.

Hermes, R., T.B. Hildebrandt, and F. Göritz. 2004. Reproductive problems directly attributable to long-term captivity—asymmetric reproductive aging. Anim Reprod Sci 82–83:49–60.

Hermes, R., T.B. Hildebrandt, C. Walzer, F. Göritz, M.L. Patton, S. Silinski, M.J. Anderson, C.E. Reid, G. Wibbelt, K. Tomasova, and F. Schwarzenberger. 2006. The effect of long non-reproductive periods on the genital health in captive female white rhinoceroses (Ceratotherium simum simum, C.s.cottoni). Theriogenology 65:1492–1515.

Herring, G., and D.E. Gawlik. 2007. The role of stress proteins in the study of allostatic overload in birds: use and applicability to current studies in avian ecology. Sci World J 7:1596–1602.

Hildebrandt, T.B., F. Göritz, N.C. Pratt, J.L. Brown, R.J. Montali, D.L. Schmitt, G. Fritsch, and R. Hermes. 2000. Ultrasonography of the urogenital tract in elephants (Loxodonta africana and Elephas maximus): an important tool for assessing female reproductive function. Zoo Biol 19:321–332.

Hofer, H. 2007. Wildlife conservation as a veterinary and interdisciplinary task. Nova Acta Leopold NF 95, Nr. 353:199–210.

Hofer, H., and M.L. East. 1998. Biological conservation and stress. Adv Study Behav 27:405–525.

Hofer, H., and M.L. East. 2003. Behavioral processes and costs of co-existence in female spotted hyenas: a life history perspective. Evol Ecol 17:315–331.

Hofer, H., and M.L. East. 2010. Impact of global climate change on wildlife hosts and their pathogens. Nova Acta Leopold NF 111, Nr. 381:103–110.

Hofer, H., M.L. East, and K.L.I. Campbell. 1993. Snares, commuting hyaenas, and migratory herbivores: humans as predators in the Serengeti. Symp Zool Soc Lond 65:347–366.

Hoffmann, A.A., and P.A. Parsons. 1991. Evolutionary genetics and environmental stress. Oxford University Press, Oxford.

Holmes, N.D., M. Giese, H. Achurch, S. Robinson, and L.K. Kriwoken. 2006. Behaviour and breeding success of gentoo penguins *Pygoscelis papua* in areas of low and high human activity. Polar Biol 29:399–412.

Hunt, K.E., R.A. Rolland, S.D. Kraus, and S.K. Wasser. 2006. Analysis of fecal glucocorticoids in the North Atlantic right whale (*Eubalaena glacialis*). Gen Comp Endocrinol 148:260–272.

Ilmonen, P., A. Kotrschal, and D.J. Penn. 2008. Telomere attrition due to infection. PloS One 3:e2143.

Jackson, M., N. White, P. Giffard, and P. Timms. 1999. Epizootiology of *Chlamydia* infections in two free-range koala populations. Vet Microbiol 65:255–264.

Johanos, T.C., B.L. Becker, J.D. Baker, T.J. Ragen, W.G. Gilmartin, and T. Gerrodette. 2010. Impacts of sex ratio reduction on male aggression in the critically endangered Hawaiian monk seal *Monachus schauinslandi*. Endang Species Res 11:123–132.

Jones, M.E., A. Cockburn, R. Hamede, C. Hawkins, H. Hesterman, S. Lachish, D. Mann, H. McCallum, and D. Pemberton. 2008. Life-history change in disease-ravaged Tasmanian devil populations. Proc Natl Acad Sci USA 105:10023–10027.

Jurke, M.H., N.M. Czekala, D.G. Lindburg, and S.E. Millard. 1997. Fecal corticoid metabolite measurement in the cheetah (*Acinonyx jubatus*). Zoo Biol 16:133–147.

Keeling, M.J., and P. Rohani. 2008. Modeling infectious diseases in humans and animals. Princeton University Press, Princeton.

Keller, V. 1989a. Variations in the response of great crested grebes *Podiceps cristatus* to human disturbance: a sign of adaptation? Biol Conserv 49:31–45.

Keller, V. 1989b. Egg-covering behaviour by great crested grebes *Podiceps cristatus*. Ornis Scand 20:129–131.

Kempenaers, B., P. Borgström, P. Loe, E. Schlicht, and M. Valcu. 2010. Artificial night lighting affects dawn song, extra-pair siring success, and lay date in songbirds. Curr Biol 20:1735–1739.

Kirkwood, T.B.L. 1977. Evolution of aging. Nature 270:301–304.

Kirkwood, T.B.L. 1981. Repair and its evolution: survival versus reproduction. *In* C. R. Townsend and P. Calow eds. Physiological ecology: an evolutionary approach to resource use, pp. 165–189. Blackwell Scientific Publications, Oxford.

Kirkwood, T.B.L. 1997. The origin of human aging. Phil Trans R Soc Lond B352:1765–1772.

Kirkwood, T.B.L., and M.R. Rose. 1991. Evolution of senescence—late survival sacrificed for reproduction. Phil Trans R Soc B 332:15–24.

Kitaysky, A.S., J.F. Piatt, S.A. Hatch, E.V. Kitaiskaia, M.Z. Benowitz-Fredericks, M.T. Shultz, and J.C. Wingfield. 2010. Food availability and population processes: severity of nutritional stress during reproduction predicts survival of long-lived seabirds. Funct Ecol 24:625–637.

Klepac, P., L.W. Pomeroy, O.N. Bjørnstad, T. Kuiken, A.D.M.E. Osterhaus, and J. Rijks. 2009. Stage-structured transmission of phocine distemper virus in the Dutch (2002) outbreak. Proc R Soc B 272:2469–2476.

Knight, C.R., and J.P. Swaddle. 2007. Association of anthropogenic activities and disturbance with fitness metrics of eastern bluebirds (*Sialia sialis*). Biol Conserv 138:189–197.

Korte, S.M., J.M. Koolhaas, J.C. Wingfield, and B.S. McEwen. 2005. The Darwinian concept of stress: benefits of allostasis and costs of allostatic load and the trade-offs in health and disease. Neurosci Biobehav 29:3–38.

Kotrschal, A., P. Ilmonen, and D.J. Penn. 2007. Stress impacts telomere dynamics. Biol Lett 3:128–130.

Kraaijeveld-Smit, F.J.L., R.A. Griffiths, R.D. Moore, and T.J.C. Beebee. 2006. Captive breeding and the fitness of reintroduction species: a test of the response to predators in a threatened amphibian. J Appl Ecol 43:360–365.

Kutz, S.J., E.P. Hoberg, J. Nagy, L. Polley, and B. Elkin. 2004. "Emerging" parasitic infections in Arctic ungulates. Integr Comp Biol 44:109–118.

Kutz, S.J., E.P. Hoberg, L. Polley, and E.J. Jenkins. 2005. Global warming is changing the dynamics of Arctic host-parasite systems. Proc R Soc B 272:2571–2576.

Lafferty, K.D. 2009. The ecology of climate change and infectious diseases. Ecology 90:888–900.

Laurenson, K.M. 1994. High juvenile mortality in cheetahs (*Acinonyx jubatus*) and its consequences for maternal care. J Zool Lond 234:387–408.

Lavigne, D.M., and O.J. Schmitz. 1990. Global warming and increasing population densities: a prescription for seal plagues. Mar Poll Bull 21:280–284.

Lee, A.K., and A. Cockburn. 1985. Evolutionary ecology of marsupials. Cambridge University Press, Cambridge.

Lee, A.K., and R. Martin. 1988. The koala: a natural history. University of New South Wales Press, Sydney.

Lee, K.A. 2006. Linking immune defenses and life history at the levels of the individual and the species. Integr Comp Biol 46:1000–1015.

Lee, K.A., M. Wikelski, W.D. Robinson, T.R. Robinson, and K.C. Klasing. 2008. Constitutive immune defences correlate with life-history variables in tropical birds. J Anim Ecol 77:356–363.

Lello, J., and T. Hussell. 2008. Functional group/guild modelling of inter-specific pathogen interactions: a potential tool for predicting the consequences of co-infections. Parasitology 135:825–839.

Léon-Vizcaino, L., M.R. Ruiz de Ybanez, M.J. Cubero, J.M. Ortiz, J. Espinosa, L. Perez, M.A. Simon, and F. Alonso. 1999. Sarcoptic mange in Spanish ibex from Spain. J Wildl Dis 35:647–659.

Levine, S., and H. Ursin. 1991. What is stress? In M.R. Brown, G.F. Koob, and C. Rivier, eds. Stress: neurobiology and neuroendocrinology, pp. 3–21. Marcel Dekker, New York.

Li, F.-H., W.-Q. Zhong, Z. Wang, and D.-H. Wang. 2007. Rank in a food competition test and humoral immune functions in male Brandt's voles (*Lasiopodomys brandtii*). Physiol Behav 90:490–495.

Lochmiller, R.L., and C. Deerenberg. 2000. Trade-offs in evolutionary immunology: just what is the cost of immunity? Oikos 88:87–98.

Longcore, T., and C. Rich. 2004. Ecological light pollution. Front Ecol Env 2:191–198.

Lovegrove, R. 2007. Silent fields: the long decline of a nation's wildlife. Oxford University Press, Oxford.

Lu, Z., W. Pan, X. Zhu, D. Wang, and H. Wang. 2000. What has the panda taught us? *In* A. Entwistle and N. Dunstone, eds. Priorities for the conservation of mammalian diversity: has the panda had its day?, pp. 325–334. Cambridge University Press, Cambridge.

Lynch, H.J., W.F. Fagan, and R. Naveen. 2010. Population trends and reproductive success at a frequently visited penguin colony on the western Antarctic Peninsula. Polar Biol 33:493–503.

Madsen, J. 1995. Impacts of disturbance on migratory waterfowl. Ibis 137:S67–S74.

Mallord, J.W, P.M. Dolman, A.F. Brown, and W.J. Sutherland. 2007. Linking recreational disturbance to population size in a ground-nesting passerine. J Appl Ecol 44:185–195.

Marco, I., J.R. López-Olvera, R. Rosell, E. Vidal, A. Hurtado, R. Juste, M. Pumarola, and S. Lavín. 2007. Severe outbreak of disease in the southern chamois (*Rupicapra pyrenaica*) associated with border disease virus infection. Vet Microbiol 120:33–41.

Marker-Kraus, L., and J. Grisham. 1993. Captive breeding of cheetahs in North American zoos: 1987–1991. Zoo Biol 12:5–18.

Martin, L.B. 2009. Stress and immunity in wild vertebrates: Timing is everything. Gen Comp Endocrinol 163:70–76.

Martin, L.B., Z.M. Weil, and R.J. Nelson. 2008. Seasonal changes in vertebrate immune activity: mediation by physiological trade-offs. Phil Trans R Soc B 363: 321–339.

Martin, R.W., and K.A. Handasyde. 1999. The koala: natural history, conservation and management. University of New South Wales Press, Sydney.

Martinez-Mota, R., C. Valdespino, M.A. Sánchez-Ramos, and J.C. Serio-Silva. 2007. Effects of forest fragmentation on the physiological stress response of black howler monkeys. Anim Conserv 10:374–379.

Martínez-Padilla, J., J. Martínez, J.A. Dávilla, S. Merino, J. Moreno, and J. Millán. 2004. Within-brood size differences, sex and parasites determine blood stress protein levels in Eurasian kestrel nestlings. Funct Ecol 18:426–434.

McCallum, H. 2008. Tasmanian devil facial tumour disease: lessons for conservation biology. Trends Ecol Evol 23:631–637.

McCallum, H., D.M. Tompkins, M.E. Jones, S. Lachish, S. Marvanek, B. Lazenby, G. Hocking, J. Wiersma, and C. E. Hawkins. 2007. Distribution and impacts of Tasmanian devil facial tumour disease. EcoHealth 4:318–325.

McDonald, R.A., R.J. Delahay, S.P. Carter, G.C. Smith, and C.L. Cheeseman. 2008. Perturbing implications of wildlife ecology for disease control. Trends Ecol Evol 23:53–56.

McEwen, B.S., and J.C. Wingfield. 2003. The concept of allostasis in biology and biomedicine. Horm Behav 43:2–15.

McLaren, G.W., D.W. Macdonald, C. Georgiou, F. Mathews, C. Newman, and R. Mian. 2003. Leukocyte coping capacity: a novel technique for measuring the stress response in vertebrates. Exp Physiol 88.4:541–546.

McLean, N., and K.A. Handasyde. 2006. Sexual maturity, factors affecting the breeding season and breeding in consecutive seasons in populations of overabundant Victorian koalas (*Phascolarctos cinereus*). Austral J Zool 54:385–392.

McNamara, J.M., and K. L. Buchanan. 2005. Stress, resource allocation, and mortality. Behav Ecol 16:1008–1017.

Meli, M.L., V. Cattori, F. Martínez, G. López, A. Vargas, M.A. Simón, I. Zorrilla, A. Muñoz, F. Palomares, J.V. López-Bao, J. Pastor, R. Tandon, B. Willi, R. Hofmann-Lehmann, and H. Lutz. 2009. Feline leukemia virus and other pathogens as important threats to the survival of the critically endangered Iberian lynx (*Lynx pardinus*). PLoS One 4(3):e4744.

Merino, S., J. Martínez, A.P. Møller, A. Barbosa, F. de Lope, and F. Rodríguez-Caabeiro. 2002. Blood stress protein levels in relation to sex and parasitism of barn swallows (*Hirundo rustica*). Ecoscience 9:300–305.

Merola, M. 1994. A reassessment of homozygosity and the case for inbreeding depression in the cheetah, *Acinonyx jubatus*: implications for conservation. Conserv Biol 8:961–971.

Meyer, L.C.R., L. Fick, A. Matthee, D. Mitchell, and A. Fuller. 2008a. Hyperthermia in captured impala (*Aepyceros melampus*): a fright not flight response. J Wildl Dis 44:404–416.

Meyer, L.C.R., R.S. Hetem, L.G. Fick, A. Matthee, D. Mitchell, and A. Fuller. 2008b. Thermal, cardiorespiratory and cortisol responses of impala (*Aepyceros melampus*) to chemical immobilisation with 4 different drug combinations. J S Afr Vet Assoc 79:121–129.

Michalakis, Y. 2009. Parasitism and the evolution of life-history traits. *In* F. Thomas, J.-F. Guégan, and F. Renaud, eds. Ecology and evolution of parasitism, pp. 19–30. Oxford University Press, Oxford.

Moorhouse, T.P., and D.W. Macdonald. 2005. Indirect negative impacts of radio-collaring: sex-ratio variation in water voles. J Appl Ecol 42:91–98.

Moreno, J., S. Merino, J. Martínez, J.J. Sanz, and E. Arriero. 2002. Heterophil/lymphocyte ratios and heat shock protein levels are related to growth of nestling birds. Ecoscience 9:434–439.

Möstl, E., J.L. Maggs, G. Schrötter, U. Besenfelder, and R. Palme. 2002. Measurements of cortisol metabolites in faeces of ruminants. Vet Res Commun 26: 127–139.

Mougeot, F., L. Perez-Rodriguez, J. Martinez-Padilla, S. Redpath, and F. Leckie. 2007. Parasites, testosterone and honest carotenoid-based signalling of health. Funct Ecol 21:886–898.

Muehlenbein, M.P., and R.G. Bribiescas. 2005. Testosterone-mediated immune functions and male life histories. Am J Human Biol 17:527–558.

Müller, G., P. Wohlsein, A. Beineke, L. Haas, I. Greiser-Wilke, U. Siebert, S. Fonfara, T. Harder, M. Stede, A. D. Gruber, and W. Baumgärtner. 2004. Phocine distemper in German seals, 2002. Emerg Infect Dis 10:723–725.

Müller, G., U. Kaim, L. Haas, I. Greiser-Wilke, P. Wohlsein, U. Siebert, and W. Baumgärtner. 2008. Phocine distemper virus: characterization of the morbillivirus causing the seal epizootic in northwestern Europe in 2002. Archiv Virol 153:951–956.

Müllner, A., K.E. Linsenmair, and M. Wikelski. 2004. Exposure to ecotourism reduces survival and affects stress response in hoatzin chicks (*Opisthocomus hoazin*). Biol Conserv 118:549–558.

Munger J.C., and W.H. Karasov. 1994. Costs of bot fly infection in white-footed mice: energy and mass-flow. Can J Zool 72:166–173.

Munshi-South, J., L. Tchignoumba, J.L. Brown, N. Abbondanza, J.E. Maldonado, A. Henderson, and A. Alonso. 2008. Physiological indicators of stress in African forest elephants (*Loxodonta africana cyclotis*) in relation to petroleum operations in Gabon, Central Africa. Divers Distrib 14:995–1003.

Munson, L., L. Marker, E. Dubovi, J.A. Spencer, J.F. Evermann, and S.J. O'Brien. 2004. Serosurvey of viral infections in free-ranging Namibian cheetahs (*Acinonyx jubatus*). J Wildl Dis 40:23–31.

Munson, L., K.A. Terio, M. Worley, M. Jago, A. Bagot-Smith, and L. Marker. 2005. Extrinsic factors significantly affect patterns of disease in free-ranging and captive cheetah (*Acinonyx jubatus*) populations. J Wildl Dis 41:542–548.

Myrberg, A.A. 1990. The effects of man-made noise on the behavior of marine animals. Env Int 16:575–586.

Naguib, M., K. Riebel, A. Marzal, and D. Gil. 2004. Nestling immunocompetence and testosterone covary with brood size in a song bird. Proc R Soc B 271:833–838.

Newton, I. 2008. The migration ecology of birds. Academic Press, London.

Nimon, A.J., R.K.C. Oxenham, R.C. Schroter, and B. Stonehouse. 1994. Measurement of resting heart rate and respiration in undisturbed and unrestrained, incubating gentoo penguins (*Pygoscelis papua*). J Physiol (Cambr) 481:57P–58P.

Nimon, A.J., R.C. Schroter, and B. Stonehouse. 1995. Heart rate of disturbed penguins. Nature 374:415–415.

Nunn, C.L., P. Lindenfors, E. Pursall, and J. Rolff. 2009. On sexual dimorphism in immune function. Phil Trans R Soc B 364:61–69.

Nunn, C.L., and S. Altizer. 2006. Infectious disease in primates: behavior, ecology and evolution. Princeton University Press, Princeton.

O'Brien, S.J., M.E. Roelke, L. Marker, A. Newman, C.A. Winkler, D. Meltzer, L. Colly, J.F. Evermann, M. Bush, and D.E. Wildt. 1985. Genetic basis for species vulnerability in the cheetah. Science 227: 1428–1434.

O'Brien, S.J., D.E. Wildt, D. Goldman, C.R. Merril, and M. Bush. 1983. The cheetah is depauperate in genetic variation. Science 221:459–462.

Onyango, P.O., L.R. Gesquiere, E.O. Wango, S.C. Alberts, and J. Altmann. 2008. Persistence of maternal effects in baboons: Mother's dominance rank at son's conception predicts stress hormone levels in subadult males. Horm Behav 54:319–324.

Ots, I., A.B. Kerimov, and E.V. Ivankina. 2001. Immune challenge affects basal metabolic rates in wintering great tits. Proc R Soc B 268:1175–1181.

Pauliny, A., R.H. Wagner, J. Augustin, T. Szép, and D. Blomqvist. 2006. Age-independent telomere length predicts fitness in two bird species. Mol Ecol 15: 1681–1687.

Pelletier, F., J.T. Hogg, and M. Festa-Bianchet. 2004. Effect of chemical immobilization on the social status of bighorn rams. Anim Behav 67:1163–1165.

Pereg, D., R. Gow, M. Mosseri, M. Lishner, M. Rieder, S. Van Uum, and G. Koren. 2011. Hair cortisol and the risk for acute myocardial infarction in adult men. Stress 14:73–81.

Phillips, S. 1987. The koala and mankind. In L. Cronin, ed. Koala—Australia's endearing marsupial, pp. 112–127. Reed Books, Frenchs Forest, Australia.

Plowright, R.K., H.E. Field, C. Smith, A. Divljan, C. Palmer, G. Tabor, P. Daszak, and J.E. Foley. 2008. Reproduction and nutritional stress are risk factors for hendra virus infection in little red flying foxes (Pteropus scapulatus). Proc R Soc B 275:861–869.

Poulin, R. 2007. Evolutionary ecology of parasites. Princeton University Press, Princeton.

Pride R.E. 2005. High faecal glucocorticoid levels predict mortality in ring-tailed lemurs (Lemur catta). Biol Lett 1:60–63.

Purser, J., and A.N. Radford. 2011. Acoustic noise induces attention shifts and reduces foraging performance in three-spined sticklebacks (Gasterosteus aculeatus). PLoS One 6(2):e17478.

Putman, R.J. 1995. Ethical considerations and animal welfare in ecological field studies. Biodiv Conserv 4:903–915.

Quiatt, D., B. Rutan, T. Stone, and V. Reynolds. 1994. Budongo forest chimpanzees: composition of feeding groups during the rainy season, with attention to the spatial integration of disabled individuals. Budongo Forest Project Report 21. Institute of Biological Anthropology, Oxford, UK.

Quiatt, D., V. Reynolds, and E.J. Stokes. 2002. Snare injuries to chimpanzees (Pan troglodytes) at 10 study sites in east and west Africa. Afr J Ecol 40:303–305.

Rich, C., and T. Longcore. 2006. Ecological consequences of artificial night lighting. Island Press, Washington, DC.

Richardson, W.J., C.R. Greene, C.I. Malme, and D.H. Thomson. 1995. Marine mammals and noise. Academic Press, San Diego.

Ricklefs, R.E., and M. Wikelski. 2002. The physiology/ life-history nexus. Trends Ecol Evol 17:462–468.

Rigby, M.C., and Y. Moret. 2000. Life-history trade-offs with immune defenses. Dev Anim Vet Sci 32: 129–142.

Rodway, M.S., W.A. Montevecchi, and J.W. Chardine. 1996. Effects of investigator disturbance on breeding success of Atlantic puffins Fratercula arctica. Biol Conserv 76:311–319.

Roff, D.A. 1992. The evolution of life histories. Chapman & Hall, London.

Rollo, C.D. 1995. Phenotypes: their epigenetics, ecology and evolution. Chapman & Hall, London.

Romero, L.M. 2004. Physiological stress in ecology: lessons from biomedical reearch. Trends Ecol Evol 19:249–255.

Romero, L.M., M.J. Dickens, and N.E. Cyr. 2009. The reactive scope model – a new model integrating homeostasis, allostasis, and stress. Horm Behav 55:375–389.

Romero, L.M., and J.M. Reed. 2005. Collecting baseline corticosterone samples in the field: is under 3 min good enough? Comp Biochem Physiol A 140:73–79.

Sapolsky, R.M. 1985. Stress-induced suppression of testicular function in the wild baboon: role of glucocorticoids. Endocrinology 116: 2273–2278.

Sapolsky, R.M. 1992. Neuroendocrinology of the stress response. In Becker, J.B., S.M. Breedlove, and D. Crews eds. Behavioral endocrinology, pp. 287–324. MIT Press, Cambridge, Massachusetts.

Sapolsky, R.M. 2005. The influence of social hierarchy on primate health. Science 308:648–652.

Saraux C., C. Le Bohec, J.M. Durant, V.A. Viblanc, M. Gauthier-Clerc, D. Beaune, Y-H. Park, N.G. Yoccoz, N.C. Stenseth, and Y. Le Maho. 2011. Reliability of flipper-banded penguins as indicators of climate change. Nature 469:203–206.

Schaub, A., J. Ostwald, and B.M. Siemers. 2008. Foraging bats avoid noise. J Exp Biol 211:3174–3180.

Schwarzenberger, F. 2007. The many uses of non-invasive faecal steroid monitoring in zoo and wildlife species. Int Zoo Yearb 41:52–74.

Seddon, P.J., P.S. Soorae, and F. Launay. 2005. Taxonomic bias in reintroduction projects. Anim Conserv 8: 51–58.

Sheldon B.C., and S. Verhulst. 1996. Ecological immunology: Costly parasite defences and trade-offs in evolutionary ecology. Trends Ecol Evol 11: 317–321.

Silk, J.B. 2007. Social components of fitness in primate groups. Science 317:1347–1351.

Simmonds, M. 1991. The involvement of environmental factors in the 1988 European seal epidemic. Rév int d'oceanogr médic 101–104:109–114.

Sockman, K.W., and H. Schwabl. 2001. Plasma corticosterone in nestling American kestrels: effects of age, handling stress, yolk androgens, and body condition. Gen Comp Endocrinol 122:205–212.

Sorci, G., T. Boulinier, M. Gauthier-Clerc, and B. Faivre. 2009. The evolutionary ecology of the immune response. In F. Thomas, J.-F. Guégan, F. Renaud, eds. Ecology and evolution of parasitism, pp. 5–18. Oxford University Press, Oxford.

Soulé, M.E. 1986. Conservation biology: the science of scarcity and diversity. Sinauer Associates, Sunderland, Mass., USA.

Starling, A.P., M.J.E. Charpentier, C. Fitzpatrick, E.S. Scordato, and C.M. Drea. 2010. Seasonality, sociality, and reproduction: Long-term stressors of ring-tailed lemurs (*Lemur catta*). Horm Behav 57 (Spec Issue):76–85.

Stearns, B.T., and S.C. Stearns. 1999. Watching from the edge of extinction. Yale University Press, New Haven, Conn.

Stearns, S.C. 1992. The evolution of life histories. Oxford University Press, Oxford.

Stone, E.L., G. Jones, and S. Harris. 2009. Street lighting disturbs commuting bats. Curr Biol 19:1123–1127.

Suleman, M.A., E. Wango, I.O. Farah, and J. Hau. 2000. Adrenal cortex and stomach lesions associated with stress in wild male African green monkeys (*Cercopithecus aethiops*) in the post-capture period. J Med Primatol 29:338–342.

Tanaka, Y. 1996. Social selection and the evolution of animal signals. Evolution 50:512–523.

Tarlinton, R., J. Meers, J. Hanger, and P. Young. 2005. Real-time reverse transcriptase PCR for the endogenous koala retrovirus reveals an association between plasma viral load and neoplastic disease in koalas. J Gen Virol 86:783–787.

Tarlinton, R.E., J. Meers, and P.R. Young. 2006. Retroviral invasion of the koala genome. Nature 442:79–81.

Tarlow, E.M., and D.T. Blumstein. 2007. Evaluating methods to quantify anthropogenic stressors on wild animals. Appl Anim Behav Sci 102:429–451.

Terio, K.A., L. Marker, E.W. Overstrom, and J.L. Brown. 2003. Analysis of ovarian and adrenal activity in Namibian cheetahs. S Afr J Wildl Res 33:71–78.

Thalwitzer, S., B. Wachter, N. Robert, G. Wibbelt, T. Müller, J. Lonzer, M.L. Meli, G. Bay, H. Hofer, and H. Lutz. 2010. Seroprevalences to viral pathogens in free-ranging and captive cheetahs (*Acinonyx jubatus*) on Namibian farmland. Clin Vacc Immunol 17:232–238.

Thiel, D., S. Jenni-Eiermann, V. Braunisch, R. Palme, and L. Jenni. 2008. Ski tourism affects habitat use and evokes a physiological stress response in capercaillie *Tetrao urogallus*: a new methodological approach. J Appl Ecol 45:845–853.

Thorne, E.T., and E.S. Williams. 1988. Disease and endangered species: the black-footed ferret as a recent example. Conserv Biol 2:66–74.

Toates, F. 1995. Stress: conceptual and biological aspects. John Wiley, Chichester.

van der Most, P.J., B. de Jong, H.K. Parmentier, and S. Verhulst. 2011. Trade-off between growth and immune function: a meta-analysis of selection experiments. Funct Ecol 25:74–80.

Van Heerden, J., M.G.L. Mills, M.J. Van Vuuren, P.J. Kelly, and M.J. Dreyer. 1995. An investigation into the health status and diseases of wild dogs (*Lycaon pictus*) in the Kruger National Park. J S Afr Vet Assoc 66:18–27.

Vleck, C.M., N. Vertalino, D. Vleck, and T.L. Bucher. 2000. Stress, corticosterone, and heterophil to lymphocyte ratios in free-living Adelie penguins. Condor 102:392–400.

Voigt, C.C., M. Faßbender, M. Dehnhard, K. Jewgenow, G. Wibbelt, H. Hofer, and G.A. Schaub. 2004. Validation of a minimally invasive blood sampling technique for hormonal analysis in domestic rabbits. Gen Comp Endocrinol 135:100–107.

Voigt, C.C., U. Peschel, G. Wibbelt, and K. Frölich. 2006. An alternative, less invasive blood sample collection technique for serologic studies utilizing Triatomine bugs (Heteroptera; Insecta). J Wildl Dis 42:466–469.

von Holst, D. 1998. The concept of stress and its relevance for animal behavior. Adv Study Behav 27:1–109.

Wachter, B., S. Thalwitzer, H. Hofer, J. Lonzer, T.B. Hildebrandt, and R. Hermes. 2011. Reproductive history and absence of predators are important determinants of reproductive fitness: the cheetah controversy revisited. Conserv Lett 4:47–54.

Walker, B.G., P.D. Boersma, and J.C. Wingfield. 2005. Physiological and behavioural differences in Magellanic penguin chicks in undisturbed and tourist-visited locations of a colony. Conserv Biol 19:1571–1577.

Walker, B.G., P.D. Boersma, and J.C. Wingfield. 2006. Habituation of adult Magellanic penguins to human visitation as expressed through behavior and corticosterone secretion. Conserv Biol 20:146–154.

Wasser, S.K., R. Thomas, P.P. Nair, C. Guidry, J. Southers, J. Lucas, D.E. Wildt, and S.L. Monfort. 1993. Effects of dietary fibre on faecal steroid measurements in baboons (*Papio cynocephalus cynocephalus*). J Reprod Fertil 97:569–574.

White, G.C., and R.A. Garrott. 1990. Analysis of wildlife radio-tracking data. Academic Press, New York.

Wielebnowski, N. 1996. Reassessing the relationship between juvenile mortality and genetic monomorphism in captive cheetah. Zoo Biol 15:353–369.

Wikelski, M., and S.J. Cooke. 2006. Conservation physiology. Trends Ecol Evol 21:38–46.

Wildt, D.E., J.L. Brown, M. Bush, M.A. Barone, K.A. Cooper, J. Grisham, and J.G. Howard. 1993. Reproductive status of cheetahs (*Acinonyx jubatus*) in North American Zoos: the benefits of physiological surveys for strategic planning. Zoo Biol 12:45–80.

Williams, S.E., and E.A. Hoffman. 2009. Minimizing genetic adaptation in captive breeding programs: a review. Biol Conserv 142:2388–2400.

Wilson, R.P., B.M. Culik, R. Danfeld, and D. Adelung. 1991. People in Antarctica: how much do Adélie penguins (*Pygoscelis adeliae*) care? Polar Biol 11:363–370.

Wingfield, J.C., D.L. Maney, C.W. Breuner, J.D. Jacobs, S. Lynn, M. Ramenofsky, and R.D. Richardson. 1998. Ecological bases of hormone–behavior interactions: the "emergency life history stage." Am Zool 38:191–206.

Wyatt, K.B., P.F. Campos, M.T.P. Gilbert, S.-O. Kolokotronis, W.H. Hynes, R. DeSalle, P. Daszak, R.D.E. MacPhee, and A.D. Greenwood. 2008. Historical mammal extinction on Christmas Island (Indian Ocean) correlates with introduced infectious disease. PLoS One 3:e3602.

Wysocki, L.E., J.P. Dittami, and F. Ladich. 2006. Ship noise and cortisol secretion in European freshwater fishes. Biol Conserv 128:501–508.

Yarmoloy, C., M. Bayer, and V. Geist. 1988. Behavior responses and reproduction of mule deer, *Odocoileus hemionus*, does following experimental harassment with an all-terrain vehicle. Can Field Natural 102: 425–429.

# PART TWO

## ANTHROPOGENIC CHANGE AND CONSERVATION MEDICINE

# 8

# CLIMATE CHANGE AND INFECTIOUS DISEASE DYNAMICS

Raina K. Plowright, Paul C. Cross, Gary M. Tabor,
Emily Almberg, Leslie Bienen, and Peter J. Hudson

The International Panel on Climate Change has made an unequivocal case that the earth's climate is changing in profound ways, and that human activities are contributing significantly to climate disruption (IPCC 2007). The weight of evidence demonstrates warming global temperatures, changing patterns of precipitation, and increasing climate variability, with more extreme events. Thus, the physical underpinnings of ecology are changing, with pervasive effects on disease dynamics. Interactions among environment, hosts, and pathogens drive disease processes, and climate change will influence every interaction in this triad, directly and indirectly.

Direct effects of climate change on hosts include the health consequences of diminishing water availability and increasing thermal stress. Some host species will succumb to the indirect effects of climate disruption through an array of mechanisms, including increased competition from invasive species, decreasing food availability, changes in parasite ranges and/or virulence, and many others. Some species will flourish under changed climate scenarios; ultimately, survival of each species will depend on its ability to adapt to climate change. Species adaptation will take on various forms, some unpredictable, such as through behavioral changes that may lead to different patterns of aggregation; changes in timing of life-history events to follow changing phenology or other climate signals;

changes to physiology to allow for survival in different climatic regimes; and development of coping mechanisms based on phenotypic plasticity. Under worst-case scenarios of potential future climate change, many species may survive at their ecological limits and face narrowing climatic and ecological constraints, including the rise of novel ecological conditions.

Of course, species adaptation to climate change will not happen in a vacuum, isolated from other environmental stressors; existing environmental threats create constraints for species and ecosystems that will affect their ability to adapt to climate change. For example, large-scale habitat fragmentation hinders species adaptation capacity by impeding animal movement necessary for reproduction, dispersal, seasonal migration, and tracking of new environmental conditions. Climate change will likely exacerbate the impacts of existing environmental threats, and this may include disease.

Extracting the effects of climate change on disease from the effects of other environmental changes is challenging, in part because we have only recently begun to understand the complexity of the interactions between climate change and disease dynamics. To date, some of the strongest system-specific evidence elucidating the effects of climate change on disease dynamics comes from vector-borne disease systems and from diseases that are environmentally transmitted and have obviously climate-dependent processes. For example, many insect vectors are highly sensitive to both temperature and precipitation, with climate change predicted to influence life-cycle completion times, biting rates, and overwintering survival, among other factors (Harvell et al. 2002; Ostfeld 2009). For parasites with free-living stages, temperature affects their development times and transmission windows (see Chapter 9 in this volume). Nevertheless, very little work on climate change and disease has focused on how changes in host population structure, dynamics, and phenology will influence disease dynamics under global warming scenarios—a key element of research as we begin considering how to integrate climate-change adaptation into conservation efforts. The links between climate change and host-population response—and how those links are manifested in disease ecology—are highlighted in this chapter.

## DENSITY, CONNECTIVITY, AND COMPOSITION: EFFECTS OF CLIMATE CHANGE ON HOST-POPULATION ECOLOGY AND SUBSEQUENT IMPACT ON DISEASE DYNAMICS

### Population Density

As climate warms, vegetation communities shift in composition and distribution, and the availability of surface water changes. Wildlife population structures and aggregation patterns will be sensitive to these changes as animals adapt to new conditions or seek alternative habitats, especially if temperature and precipitation change very quickly. In many regions water and/or snow accumulation are major drivers of host aggregation, but the direction and strength of the climatic effects will vary between species, region, and time. For example, diminishing water availability may concentrate wildlife around remaining water sources in the short term, while reducing population densities over the long term and on broader spatial scales. Elevational migration is another climate-driven process that may concentrate wildlife populations locally on high-elevation "islands" while wildlife populations decline globally (Parmesan and Yohe 2003). Some species, or populations, may experience

increased densities due to increased population size. For example, elk (*Cervus elaphus*) from Montana are predicted to increase in density as snow accumulation decreases (Creel and Creel 2009). Finally, rapid ecological changes brought about by climate change, such as the massive outbreak of mountain pine beetle (*Dendroctonus ponderosa*) across Canada and the United States, will have varying effects on different species; moose (*Alces alces*) may gain increased foraging resources and hence increase in density, while caribou (*Rangifer tarandus caribou*), fisher (*Martes pennanti*), and marten (*Martes americana*) will be adversely affected by the loss of coniferous trees critical for forage, cover, and denning (Ritchie 2008).

These changes in host density and spatial structure will have cascading impacts on pathogen transmission rates and alter thresholds for pathogen establishment and persistence. Models predict that pathogens with density-dependent transmission are unable to persist when host density is reduced below some threshold (Kermack and McKendrick 1927; Getz and Pickering 1983), a result well supported by classic studies of measles (Bartlett 1957; Bjornstad et al. 2002; Grenfell et al. 2002). These studies provide an understanding to influence policy, for example school closures to control human pandemics (Glass and Barnes 2007; Cauchemez et al. 2008; Halloran et al. 2008), as well as for using hunting or culling as a disease-control measure in wildlife populations (Caley et al. 1999; Lloyd-Smith et al. 2005; Conner et al. 2007).

For systems in which transmission is correlated with host density, if climate change aggregates hosts, transmission rates may increase and contribute to larger disease outbreaks locally, but the relationship between host density and disease transmission is complex and not always possible to predict. Several studies suggest that host density is unassociated with transmission or parasitism, even for directly transmitted parasites (Bouma et al. 1995; Rogers et al. 1999; Delahay et al. 2000; Joly and Messier 2004), or even that decreasing density can increase transmission. For example, culling badgers in the United Kingdom counter-intuitively increased the incidence of tuberculosis on adjacent cattle properties, perhaps due to culling resulting in increased movement rates of badgers (Donnelly et al. 2006). Such contradictory results may also be explained by the fact that the density–transmission relationship is complicated by parasite life history (Côté and Poulin 1995), by seasonal fluctuations in both host density and transmission (Altizer et al. 2006; Cross et al. 2007), by movement among groups (Ball et al. 1997; Cross et al. 2005), by the effects of alternative hosts (Dobson 2004; Craft et al. 2008), and by interactions with other pathogens (Jolles et al. 2008).

Human diseases are also influenced by changing patterns of aggregation; Ferrari et al. (2008) hypothesized that the large seasonal variation in measles transmission in Niger was driven by human migrations from agricultural to urban areas during the dry season. Drought in combination with habitat loss may have driven a similar phenomenon of urban aggregations of Australian flying foxes (*Pteropus* spp.) with consequences for disease emergence into humans. Though flying foxes were historically nomadic, they increasingly prefer well-watered urban landscapes where food is abundant and reliable year-round (Parry-Jones and Augee 1991; Birt 2004; Markus and Hall 2004; McDonald-Madden et al. 2005). Concomitant with an increase in the number and size of urban flying fox camps was the emergence of Hendra virus, a fatal zoonotic disease that has repeatedly emerged from flying foxes into horses and humans since 1994. Many hypotheses exist to explain the cause of emergence (Plowright et al. 2008a), including higher rates of intra-specific transmission within urban flying fox colonies—consistent with density-dependent transmission, and closer and more frequent contact between flying foxes and high-density human and horse populations (Fig. 8.1) (Plowright et al. 2008a). The above examples highlight the complexity of the relationship between host density and disease dynamics. Although there are special cases where increasing host density does not increase disease transmission, the default hypothesis is for a positive association. More data are needed to explore these relationships, and generalize predictions across systems.

## Population Connectivity and Metapopulation Structure

Climate change will affect connectivity among populations and species by affecting the spatial distribution of habitat; dispersal and migration rates; and species distributions. One mechanism predicted to have profound effects on connectivity is range contraction, particularly affecting species in arid or montane regions, which are predicted to shift to areas

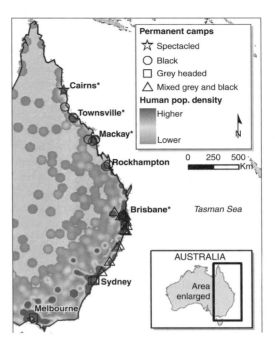

**Figure 8.1:**
Permanently occupied flying fox colonies (known as camps) and their relation to urban areas in Australia (depicted by high human population density). Spectacled = spectacled flying fox (*Pteropus conspicillatus*); Black = black flying fox (*P. alecto*); Grey Headed = grey headed flying fox (*P. poliocephalus*). Modified from Plowright et al. (2011).

of higher elevation and greater precipitation. For example, elevation shifts in distribution have already been observed for the Edith checkerspot butterfly in California (Parmesan 1996), for montane woodlands plant assemblages (Fisher 1997), for tropical species in Costa Rica (Pounds et al. 1999), and for desert bighorn sheep (*Ovis canadensis nelsoni*) (Epps et al. 2004). In the short term, reduced resource availability may concentrate animals on remaining habitat, thus increasing local densities and intensifying disease transmission. However, at the same time, densities and connectivity at a broader scale may decrease, effectively isolating local populations and decreasing metapopulation function. Thus, climate change may lead to the loss of parasites and pathogens, and the potential release of hosts from this source of regulation.

Epps et al. (2004) use desert bighorn sheep as the basis for a convincing case study to show how climate-induced changes to population structure may impose conditions that both decrease population viability while, theoretically, also decreasing the threat of infectious diseases. The study demonstrates that climate change has altered the metapopulation structure of desert bighorn sheep, which occupy isolated mountain ranges in the Sonoran, Mojave, and Great Basin deserts of the southwestern United States, where recent trends in warming have been particularly severe, annual precipitation has declined by approximately 20% over the past century, and mean annual temperatures have simultaneously increased 0.12°C per decade. Epps et al. (2004) demonstrated that populations inhabiting lower, drier mountain ranges were more likely to be extirpated, essentially contracting the metapopulation into isolated high-elevation islands. Long-term persistence of desert bighorn sheep metapopulations depends on some connectivity for gene flow and demographic rescue, but immigration also promotes the transmission and persistence of infectious diseases (Hess 1996; McCallum and Dobson 2002). Thus, desert bighorn sheep are an excellent example of the conundrum facing wildlife-disease managers: as climate change decreases population sizes and connectivity, isolation and decreased connectivity diminish long-term population viability; however, these conditions may also reduce disease threats. Many pathogens, particularly those with short infectious periods, no environmental reservoir, and no alternative hosts have a difficult time persisting in isolated, spatially structured systems (Cross et al. 2005; Lloyd-Smith et al. 2005; Cross et al. 2007). Climate change may drive such pathogens to extinction.

Geographic isolation is not the only mechanism that may reduce animal migrations as climate changes. Climate change may reduce migration in areas where seasonal ranges become hospitable year-round (Harvell et al. 2009). For example, increasingly urbanized Australian flying foxes (*Pteropus* spp.) are choosing a resident lifestyle because food supplies in well-watered urban gardens reduce their need for energy-expensive long-distance migrations (Parry-Jones and Augee 1991; Birt 2004; Markus and Hall 2004; McDonald-Madden et al. 2005). It has been hypothesized that decreasing flying fox migration, and subsequent declining population connectivity, may lead to a reduction in herd immunity. When virus is reintroduced into populations where herd immunity has declined, more intense outbreaks may result—a

mechanism that potentially explains the recent emergence of Hendra virus from flying foxes into humans and domestic animals (Plowright et al. 2011).

Migration rates may also decline when some seasonal ranges become inhospitable, rather than more hospitable. For example, a complex ecological scenario may lead to increasing year-round resident populations of elk, while migratory populations decline. Some populations of elk on the eastern side of Yellowstone National Park migrate to higher elevations during the summer and return to lower-elevation areas in winter, while non-migratory elk remain at lower-elevation sites year-round. Over the past 20 years, pregnancy and recruitment rates in migratory elk have been declining relative to non-migratory populations (Middleton et al., in review). These declines appear to be the result of increasing grizzly bear (*Ursus arctos*) and wolf (*Canis lupus*) populations at higher elevations, as well as of shorter growing seasons at those elevations. This pattern of increased resident behavior may be a recurring one in northern latitudes, where higher-elevation areas are more likely to still have intact predator communities, with cascading impacts upon herbivore migrations, population dynamics, and diseases. Harvell et al. (2002) describe increasing resident behavior as a growing phenomenon, where higher population densities increase host contact, parasite transmission, and the concentration of parasites in the environment. Additionally, increasing resident behavior may reduce the opportunity for migration to "weed out" weak or infected animals, leading to a higher prevalence of parasites in resident populations compared to migratory populations (Bradley and Altizer 2005).

## Changing Community Composition

Climate change has been projected to change the composition of host communities, not only through new species assemblages, but also via the exchange of parasite communities (Dobson and Carper 1992), thus providing a mechanism for increased inter-species connectivity. Many species are responding to climate change by shifting geographic ranges towards the poles or higher altitudes (Parmesan and Yohe 2003), which is a pattern reinforced by habitat conversion. Species do not move as communities, but rather at individual directions and rates (Lovejoy 2008). Therefore, we anticipate that differential rates in range shifts and survival will lead to new species assemblages (Williams et al. 2007). As new species come together, we anticipate greater opportunities for pathogen spillover, which may in some cases result in severe decline of novel susceptible hosts. For example, elk and white-tailed deer have expanded their ranges northward, in part due to warming temperatures, and consequently have introduced novel pathogens to caribou and musk ox (Dobson et al. 2003; Kutz et al. 2005). By rearranging communities we are also potentially breaking coevolved host–pathogen life cycles, a disruption that could potentially outpace spillover and reduce the disease burden in some species. Parasites that play an import2ant role in mediating competitive interactions among hosts may also experience shifts in abundance and distribution associated with changes to the larger parasite community, potentially triggering cascading changes in host communities (Dobson and Carper 1992). Feedback loops between changing host assemblages and parasite communities may be most pronounced in temperate regions, where the buffering effects of biodiversity are absent (Harvell et al. 2002).

## TIMING AND SYNCHRONY: CHANGING THE PHENOLOGY OF HOST BEHAVIORS IMPORTANT FOR PATHOGEN TRANSMISSION

### Host Behavior

Climate change also has potential to change the timing of host behaviors that are fundamental to pathogen transmission. For example, climate-mediated shifts in phenology have been documented across a wide range of taxa and may act to decouple the spatial and temporal overlap between hosts and their parasites (Hoberg et al. 2008). We may see changes in the timing, intensity, and location of seasonal aggregations, and in migrations of hosts in response to changes in temperature or to the timing of food availability on the landscape (Buskirk et al. 2009). These shifts may increase, or decrease, the likelihood of pathogen transmission if they alter the timing of host contacts with infective stages of parasites in the environment.

Research on brucellosis in elk highlights connections between climate, timing of host aggregation, and disease. In this system, both elk density and

disease transmission have strong seasonal patterns. The largest elk aggregations occur in late winter prior to snowmelt and spring migration (Cross et al. 2007), and can be particularly large (several hundred to several thousand individuals per group) in areas of Wyoming where elk receive supplementary feed from managers. Brucellosis is typically transmitted by abortion events, which tend to occur in winter and spring (Cheville et al. 1998). The overlap between the timing of high densities of animals and the timing of disease-transmission windows is critical (Fig. 8.2).

The duration of the supplemental feeding season explained almost 60% of variation in elk seroprevalence for brucellosis across numerous feeding areas. Meanwhile, there was no relationship between elk seroprevalence and peak winter elk densities. Brucellosis exposure was determined by the length of time animals were aggregated during the transmission window; low-density elk aggregations that occurred during the transmission window may have similar levels of pathogen exposure to sites with short aggregations but high densities (Cross et al. 2007) (Fig. 8.2). Further, the duration that elk were aggregated each winter was highly correlated with snowpack, because elk were concentrated for longer during years with more snow. Climate change is likely to alter the timing and duration of winter elk aggregations. Over the past 50 years, snowpack in areas of the southern Greater Yellowstone Ecosystem has fallen as much as 50%, though the declines are stronger in early winter (Fig. 8.3). As a result of earlier snowmelt we might expect brucellosis transmission to decline. However, the timing of birth and abortion (and thus brucellosis transmission) may also shift to earlier in the year.

### Synchrony of Host Population Dynamics and Parasite Life Cycles

An important component of climate disruption is an increase in the frequency of extreme climatic events, which may act to synchronize disease outbreaks over large spatial areas. More than 50 years ago, Moran (1953) proposed that independent populations that have the same underlying density-dependent structure will fluctuate in synchrony when their climatic conditions are correlated. The extent of this synchrony between populations will depend on the correlation of these climatic events between localities,

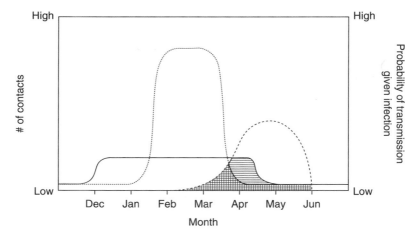

**Figure 8.2:**
Conceptual illustration of the interaction between contact and transmission rates when both are varying over time in the elk brucellosis system. Some elk populations are aggregated in smaller groups for long periods during winter (*solid line*) while others are in large groups for shorter periods (*dotted line*). Brucellosis transmission is associated with abortion events, which may be more likely in late spring (*dashed line*). The cumulative annual transmission should be more correlated with the amount overlap between high contact rates and brucellosis transmission than with maximum density.

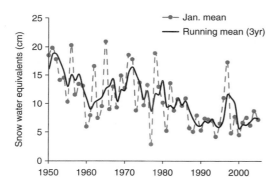

**Figure 8.3:**
The average snow water equivalent (cm) measured on January 1 at 16 SNOwpack TELemetry sites in the Jackson and Pinedale regions of Wyoming. Data source: USDA Natural Resources Conservation Service.

able to identify the years when populations were unpredictably brought into synchrony. Analyses of climatic conditions together with levels of parasitism by the nematode *Trichostrongylus tenuis* in these years identified the weather variables that were associated with these changes. Subsequent model fitting provided evidence to support that climatic conditions acted on parasite transmission, either to (1) accelerate transmission, leading to a common, large-scale disease outbreak and a decline in the host population or (2) reduce transmission, and this would result in low parasite intensity and an increase in the host population. In effect, extreme climatic events may bring populations into synchrony by acting on the transmission stages. This is important since it provides good evidence to support that an increase in the frequency of extreme events could lead to spatially synchronized disease outbreaks.

The spatial scale of these disturbances in relation to species distributions will obviously influence the impact on infectious diseases. For example, small-scale disturbances such as tornadoes will not have the same effect as widespread drought. Large-scale impacts are more likely to reduce the host population across large areas, rendering these populations

and this correlation would increase with increased frequency of large-scale extreme events, as expected with climate change. Cattadori et al. (2005) examined the spatial synchrony of grouse in northern England and found that spatially independent grouse populations exhibit unstable cyclic fluctuations, and they were vulnerable to all effects. Large-scale extreme climatic events could also synchronize resource availability and so influence disease transmission by changing host susceptibility over large scales. For example, nutritional stress increases Hendra virus seroprevalence in flying foxes (Plowright et al. 2008b), suggesting that processes that alter flying fox food sources—such as climate-change–associated drought—could both increase and synchronize the risk of Hendra virus spillover. Temporal correlation of some Hendra virus outbreaks supports this hypothesis of "stress synchrony" and is worth further exploration (Plowright et al. 2011).

## Environmental Variation and Parasite Development Rate

Climate may influence transmission by killing the infective stage faster with directly transmitted pathogens; however, with helminths and vector-borne infections, the free-living stages are subject to temperature variations that may influence development. Generally, there is a minimum temperature below which free-living stages of parasites do not develop, and as temperature increases, the parasite development rate increases linearly. This is a pattern well documented in a wide range of nematode species where the free-living stages were incubated at set temperatures and development rates were tracked (Anderson 2002). As such, development rate can be modeled using a degree-day approach, based on the assumption that a fixed amount of thermal energy is required by a parasite to develop, and this accumulates daily according to the local temperature (Grenfell and Smith 1983). Implicit in this assumption is that variation in temperature is not important but development is achieved after the thermal energy is accumulated; this assumption has indeed been confirmed in some systems. Interestingly, there is mounting evidence that variation in temperature may accelerate development faster than the degree-day models predict. In a study of the development rate of eggs of the caecal nematode Heterakis gallinarum, Saunders et al. (2000) found that an increase in temperature conditions resulted in a linear increase in the development rate, which could be described by a simple degree-day model. However, when the parasite was placed in a daily temperature cycle, development started significantly earlier than that expected from the linear model. Furthermore, when eggs were placed in stochastic fluctuations (with temperatures giving the same thermal energy as the temperature cycle) they developed significantly earlier, indicating that fluctuations in temperature, and particularly increased variation, could accelerate parasite development rates (Saunders et al. 2000). An aspect of climate disruption may well be increased variability in temperature, which could have a profound effect on the development of free-living stages of parasites. It is important to note that increasing temperatures also increase parasite mortality rates for non-feeding organisms, and therefore the relationship between temperature and an organism's performance may follow a convex function (Lafferty 2009).

## STRESS: CLIMATE-CHANGE IMPACTS ON HOST SUSCEPTIBILITY TO DISEASE

A substantial body of evidence suggests that climate change has the potential to mediate stress, and hence host susceptibility to infectious disease. Thermal stress has been documented among numerous ectotherms, ranging from amphibians to corals and fish (Marcogliese 2008; Harvell et al. 2009). Increasing temperatures have raised the metabolic rates of common toads in the UK during hibernation, resulting in their spring emergence in a reduced physiological state (Reading 2007). Corals exhibit stress responses to increased water temperatures, and recent increases in pathogen loads and distributions have been, in part, attributed to increased coral susceptibility due to thermal stress (Bruno et al. 2007; Mydlarz et al. 2010). Changes in behaviors due to increased environmental temperatures may also result in suboptimal foraging, and thus in compromised body condition and immune function. Endotherms with narrow temperature envelopes are also expected to be susceptible to thermal stress—moose are known to have extremely inflexible temperature thresholds, above which their metabolism rapidly speeds up (Renecker and Hudson 1986). Thermal stress in moose is thought to contribute, in part, to reduced immune function and ability to handle parasite loads (Murray et al. 2006). However, increasing temperatures may not always equate to negative effects on immune function. In some cases, increased temperatures may increase the productivity of primary producers (although decreased primary productivity may be just as likely), providing additional nutritional resources to hosts, and thereby increasing body condition and immune function. For example, on St. Kilda, the site of the long-term study on Soay sheep (*Ovis aries*), primary productivity is related to sheep body condition and winter survival in the face of parasites (Crawley et al. 2004; Wilson et al. 2004). If the trends of increasing temperature and less severe winters documented on St. Kilda result in increased primary productivity (Ozgul et al. 2009), there may be a suite of climate-mediated effects on host–parasite dynamics.

The flying fox scenario may also illustrate how climate can serve as an ecological vise, with water and temperature stress influencing disease dynamics. While water stress may have altered flying fox colonization ecology, the associated temperature increases in urban environments due to climate-related urban heat island effects have pushed some flying foxes to their maximum physiological thresholds of heat tolerance (Welbergen et al. 2008). Temperature-related stress is, therefore, one more potential causal factor to consider when understanding Hendra virus emergence (Plowright et al. 2008a,b).

Other forms of climate-mediated stress may also have immunosuppressive effects. Host crowding, as individuals respond to shifting habitats, may increase intra-specific aggression and stress, resulting in increased pathogen and parasite susceptibility. Climate change may also affect the accumulation and persistence of pollutants in the Arctic, for example, via changes to air and ocean currents (Burek et al. 2008). If these changes result in increased exposure of marine life to toxins, we may see increased susceptibility to disease among these organisms (Bustnes et al. 2004).

## CONCLUSIONS

In this chapter we have discussed several modalities of host population response to climate change and the disease implications of these responses. Density, connectivity, metapopulation structure, phenology, and stress responses are presented as examples of host population parameters involved in species adaptation and reaction to climate change. A key challenge now for disease ecologists is to develop tools and

methods necessary to accurately detect the potential effects of climate change within already-complex disease dynamics.

A second challenge will be to anticipate and understand the unforeseen consequences of host responses to climate change on disease dynamics, such as through species shifting to new ranges and transporting their diseases into novel situations and hosts. The consequences range from amplification of disease impacts, to changes in geographic disease distribution, to potential pathogen extinction (Lafferty 2009; Randolph 2009).

Because it is difficult to predict precisely what will happen in any individual ecological scenario, conservationists must be as prepared as possible to make rapid assessments and potentially intervene in worst-case scenarios as they arise. However, a more preemptive role for conservation managers and researchers—and one that conservation medicine practitioners can prepare for by following the precepts of this discipline—would be to actively facilitate ecological resilience, an emerging concept often defined as the ability of a system to buffer disruption (Holling 1996). Examples of strategies that promote ecological resilience include maintaining biodiversity and preserving large landscapes for conservation purposes. To date, strategies linking the concept of ecological resilience with ecological health are lacking, and ecological health benchmarks to guide preventive actions do not exist. However, we can presume that strategies that enhance ecological resilience will also buffer the effects of changing infectious disease dynamics. Furthermore, a sound management concept is that once disease is introduced into a susceptible system it is very difficult to control, and almost always impossible to eradicate. Thus, a crucial strategy is to take measures to limit new disease introductions. The most important tool in this endeavor is surveillance and monitoring and, fortunately, climate-change monitoring efforts are increasing exponentially worldwide, offering opportunities to incorporate disease monitoring within these efforts. The growing body of data from these surveillance efforts should form the basis of adaptive management responses for disease interventions (Plowright et al. 2008a). Understanding the interactions between climate change, species responses, and disease is an emerging area within ecological health investigation, and a vital one if we are to succeed in limiting the damaging effects of climate change on species and ecosystems already under stress.

## ACKNOWLEDGMENTS

Thanks to Kevin Lafferty and Greg Pederson for their insightful reviews, which significantly improved this manuscript.

## REFERENCES

Altizer, S., A. Dobson, P. Hosseini, P. Hudson, M. Pascual and P. Rohani. 2006. Seasonality and the dynamics of infectious diseases. Ecol Lett 9:467–484.

Anderson, R.C. 2002. Nematode parasites of vertebrates: Their development and transmission. CABI Publishing, London.

Ball, F., D. Mollison and G. Scalia-Tomba. 1997. Epidemics with two levels of mixing. Ann App Probab 7:46–89.

Bartlett, M.S. 1957. Measles periodicity and community size. J R Stat Soc 120:48–71.

Birt, P. (2004). Mutualistic interactions between the nectar-feeding little red flying fox Pteropus scapulatus (Chiroptera: Pteropodidae) and flowering eucalypts (Myrtaceae): habitat utilisation and pollination. University of Queensland PhD Thesis, Brisbane.

Bjornstad, O.N., B.F. Finkenstadt and B.T. Grenfell. 2002. Dynamics of measles epidemics: Estimating scaling of transmission rates using a Time series SIR model. Ecol Monogr 72:169–184.

Bouma, A., M.C.M. de Jong and T.G. Kimman. 1995. Transmission of pseudorabies virus within pig-populations is independent of the size of the population. Prev Vet Med 23:163–172.

Bradley, C., and S. Altizer. 2005. Parasites hinder monarch butterfly flight ability: implications for disease spread in migratory hosts. Ecol Lett 8:290–300.

Bruno, J.F., E.R. Selig, K.S. Casey, C.A. Page, B.L. Willis, C.D. Harvell, H. Sweatman and A.M. Melendy. 2007. Thermal stress and coral cover as drivers of coral disease outbreaks. PLoS Biol 5:e124.

Burek, K.A., F.M.D. Gulland and T.M. O'Hara. 2008. Effects of climate change on Arctic marine mammal health. Ecol Appl 18:S126–S134.

Buskirk, J.V., R.S. Mulvihill and R.C. Leberman. 2009. Variable shifts in spring and autumn migration phenology in North American songbirds associated with climate change. Glob Change Biol 15:760–771.

Bustnes, J.O., S.A. Hanssen, I. Folstad, K.E. Erikstad, D. Hasselquist and J. U. Skaare. 2004. Immune function and organochlorine pollutants in arctic breeding glaucous gulls. Arch Environ Con Tox 47:530–541.

Caley, P., G.J. Hickling, P.E. Cowan and D.U. Pfeiffer. 1999. Effects of sustained control of brushtail possums on levels of Mycobacterium bovis infection in cattle and brushtail possum populations from Hohotaka, New Zealand. N Z Vet J 47:133–142.

Cattadori, I.M., D.T. Haydon and P.J. Hudson. 2005. Parasites and climate synchronize red grouse populations. Nature 433:737–741.

Cauchemez, S., A. Valleron, P. Boëlle, A. Flahault and N.M. Ferguson. 2008. Estimating the impact of school closure on influenza transmission from sentinel data. Nature 452:750–754.

Cheville, N.F., D.R. McCullough and L.R. Paulson. 1998. Brucellosis in the Greater Yellowstone Area. National Academy Press, Washington, D.C.

Conner, M.M., M.W. Miller, M.R. Ebinger and K.P. Burnham. 2007. A meta-BACI approach for evaluating management intervention on chronic wasting disease in mule deer. Ecol Appl 17:140–153.

Côté, I.M., and R. Poulin. 1995. Parasitism and group-size in social animals—a metaanalysis. Behav Ecol 6:159–165.

Craft, M.E., P.L. Hawthorne, C. Packer and A.P. Dobson. 2008. Dynamics of a multihost pathogen in a carnivore community. J Anim Ecol 77:1257–1264.

Crawley, M.J., S.D. Albon, D.R. Bazely, J.M. Milner, J.G. Pilkington and A.L. Tuke. 2004. Vegetation and sheep population dynamics. In Clutton-Brock, T.H. and J.M. Pemberton, eds. Soay Sheep: Dynamics and Selection in an Island Population, pp. 89–112. Cambridge University Press, Cambridge.

Creel, S., and M. Creel. 2009. Density dependence and climate effects in Rocky Mountain elk: an application of regression with instrumental variables for population time series with sampling error. J Anim Ecol 78: 1291–1297.

Cross, P.C., W.H. Edwards, B.M. Scurlock, E.J. Maichak and J.D. Rogerson. 2007. Effects of management and climate on elk brucellosis in the Greater Yellowstone Ecosystem. Ecol Appl 17:957–964.

Cross, P.C., P.L.F. Johnson, J.O. Lloyd-Smith and W.M. Getz. 2007. Utility of R0 as a predictor of disease invasion in structured populations. J R Soc Interface 4:315–324.

Cross, P.C., J.O. Lloyd-Smith, P.L.F. Johnson and W.M. Getz. 2005. Duelling timescales of host mixing and disease recovery determine disease invasion in structured populations. Ecol Lett 8:587–595.

Delahay, R.J., S. Langton, G.C. Smith, R.S. Clifton-Hadley and C.I. Cheeseman. 2000. The spatio-temporal distribution of Mycobacterium bovis (bovine tuberculosis) infection in a high-density badger population. J Anim Ecol 69:428–441.

Dobson, A. 2004. Population dynamics of pathogens with multiple host species. Am Nat 164:S64–S78.

Dobson, A., S. Kutz, M. Pascual and R. Winfree. 2003. Pathogens and parasites in a changing climate. In Hannah, L., & T. Lovejoy, eds. Climate change and biodiversity: synergistic impacts, pp. 33–38. Yale University Press, New Haven.

Dobson, A.P., and R.J. Carper. 1992. Global warming and potential changes in host–parasite and disease vector relationships. In Peters, R., & T. Lovejoy, eds. Global warming and biodiversity, pp. 201–217. Yale University Press, New Haven.

Donnelly, C.A., R. Woodroffe, D.R. Cox, F.J. Bourne, C.L. Cheeseman, R.S. Clifton-Hadley, G. Wei, G. Gettinby, P. Gilks, H. Jenkins, W.T. Johnston, A.M. Le Fevre, J.P. McInerney and W.I. Morrison. 2006. Positive and negative effects of widespread badger culling on tuberculosis in cattle. Nature 439:843–846.

Epps, C.W., D.R. McCullough, J.D. Wehausen, V.C. Bleich and J.L. Rechel. 2004. Effects of climate change on population persistence of desert-dwelling mountain sheep in California. Conserv Biol 18:102–113.

Ferrari, M.J., R.F. Grais, N. Bharti, A.J.K. Conlan, O.N. Bjornstad, L.J. Wolfson, P.J. Guerin, A. Djibo and B.T. Grenfell. 2008. The dynamics of measles in sub-Saharan Africa. Nature 451:679–684.

Fisher, M. 1997. Decline in the juniper woodlands of Raydah Reserve in southwestern Saudi Arabia: a response to climate changes? Global Ecol Biogeogr 6:379–386.

Getz, W.M. and J. Pickering. 1983. Epidemic models: thresholds and population regulation. Am Nat 121:892–898.

Glass, K. and B. Barnes. 2007. How much would closing schools reduce transmission during an influenza pandemic? Epidemiology 18:623–628.

Grenfell, B.T., O.N. Bjornstad and B.F. Finkenstadt. 2002. Dynamics of measles epidemics: Scaling noise, determinism, and predictability with the TSIR model. Ecol Monogr 72:185–202.

Grenfell, B.T. and G. Smith. 1983. Population biology and control of ostertagiasis in first-year grazing calves. P Vet Epid Preventative Med: 54–61.

Halloran, M.E., N.M. Ferguson, S. Eubank, I.M. Longini, D.A. Cummings, B. Lewis, S. Xu, C. Fraser, A. Vullikanti, T.C. Germann, D. Wagener, R. Beckman, K. Kadau, C. Barrett, C.A. Macken, D.S. Burke and P. Cooley. 2008. Modeling targeted layered containment of an influenza pandemic in the United States. P Acad Nat Sci Phila 105:4639–4644.

Harvell, C.D., C.E. Mitchell, J.R. Ward, S. Altizer, A.P. Dobson, R.S. Ostfeld and M.D. Samuel. 2002. Ecology—Climate warming and disease risks for terrestrial and marine biota. Science 296:2158–2162.

Harvell, D., S. Altizer, I.M. Cattadori, L. Harrington and E. Weil. 2009. Climate change and wildlife diseases: when does the host matter the most? Ecology 90:912–920.

Hess, G. 1996. Disease in metapopulation models: Implications for conservation. Ecology 77:1617–1632.

Hoberg, E.P., L. Polley, E.J. Jenkins and S.J. Kutz. 2008. Pathogens of domestic and free-ranging ungulates: global climate change in temperate to boreal latitudes across North America. Rev Sci Tech 27:511–528.

Holling, C.S. 1996. Engineering resilience versus ecological resilience. In Schulze, P., ed. Engineering within ecological constraints, pp. 31–44. National Academy, Washington, D.C.

IPCC. 2007. Fourth Assessment Report of the Intergovernmental Panel on Climate Change.

Jolles, A.E., V.O. Ezenwa, R.S. Etienne, W.C. Turner and H. Olff. 2008. Interactions between macroparasites and microparasites drive infection patterns in free-rangeing African buffalo. Ecology 89:2239–2250.

Joly, D.O. and F. Messier. 2004. Factors affecting apparent prevalence of tuberculosis and brucellosis in wood bison. J Anim Ecol 73:623–631.

Kermack, W.O. and A.G. McKendrick. 1927. Contributions to the mathematical theory of epidemics. P Royal Soc Edin 115:700–721.

Kutz, S.J., E.P. Hoberg, L. Polley and E.J. Jenkins. 2005. Global warming is changing the dynamics of Arctic host-parasite systems. P Biol Sci 272:2571–2576.

Lafferty, K.D. 2009. The ecology of climate change and infectious diseases. Ecology 90:888–900.

Lloyd-Smith, J.O., P.C. Cross, C.J. Briggs, M. Daugherty, W.M. Getz, J. Latto, M.S. Sanchez, A.B. Smith and A. Swei. 2005. Should we expect population thresholds for wildlife disease? Trends Ecol Evol 20:511–519.

Lovejoy, T. 2008. Climate change and biodiversity. Rev Sci Tech 27:331–338.

Marcogliese, D.J. 2008. The impact of climate change on the parasites and infectious diseases of aquatic animals. Rev Sci Tech 27:467–484.

Markus, N. and L. Hall. 2004. Foraging behaviour of the black flying-fox (Pteropus alecto) in the urban landscape of Brisbane, Queensland. Wildlife Res 31:345–355.

McCallum, H. and A. Dobson. 2002. Disease, habitat fragmentation and conservation. P Roy Soc Lond B Biol 269:2041–2049.

McDonald-Madden, E., E.S.G. Schreiber, D.M. Forsyth, D. Choquenot and T.F. Clancy. 2005. Factors affecting grey-headed flying-fox (Pteropus poliocephalus:

Pteropodidae) foraging in the Melbourne metropolitan area, Australia. Austral Ecol 30:600–608.

Middleton A.D., M.J. Kauffman, D.E. McWhirter, J.G. Cook R.C. Cook, A.A. Nelson, M.D. Jimenez, and R.W. Klaver (in review) Animal migration amid shifting patterns of phenology and predation: lessons from a Yellowstone elk herd.

Moran, P.A.P. 1953. The statistical analysis of the Canadian Lynx cycle II. Synchronization and meteorology. Aust J Zool 1:291–298.

Murray, D.L., E.W. Cox, W.B. Ballard, H.A. Whitlaw, M.S. Lenarz, T.W. Custer, T. Barnett and T.K. Fuller. 2006. Pathogens, nutritional deficiency, and climate influences on a declining moose population. Wildlife Monog 1–29.

Mydlarz, L.D., E.S. McGinty and C.D. Harvell. 2010. What are the physiological and immunological responses of coral to climate warming and disease? J Exp Biol 213:934–945.

Ostfeld, R.S. 2009. Climate change and the distribution and intensity of infectious diseases. Ecology 90:903–905.

Ozgul, A., S. Tuljapurkar, T.G. Benton, J.M. Pemberton, T.H. Clutton-Brock and T. Coulson. 2009. The dynamics of phenotypic change and the shrinking sheep of St. Kilda. Science 325:464–467.

Parmesan, C. 1996. Climate and species' range. Nature 382:765–766.

Parmesan, C. and G. Yohe. 2003. A globally coherent fingerprint of climate change impacts across natural systems. Nature 421:37–42.

Parry-Jones, K.A. and M.L. Augee. 1991. Food selection by gray-headed flying foxes (Pteropus poliocephalus) occupying a summer colony site near Gosford, New South Wales. Wildl Res 18:111–124.

Plowright, R.K., H.E. Field, C. Smith, A. Divljan, C. Palmer, G.M. Tabor, P. Daszak and J.E. Foley. 2008b. Reproduction and nutritional stress are risk factors for Hendra virus infection in little red flying foxes (Pteropus scapulatus). Proc R Soc B 275:861–869.

Plowright, R.K., P. Foley, H.E. Field, A.P. Dobson, J.E. Foley, P. Eby and P. Daszak. 2011. Urban habituation, ecological connectivity and epidemic dampening: the emergence of Hendra virus from flying foxes (Pteropus spp.). P R Soc B doi:10.1098/rspb.(2011).0522.

Plowright, R.K., S.H. Sokolow, M.E. Gorman, P. Daszak and J.E. Foley. 2008a. Causal inference in disease ecology: investigating ecological drivers of disease emergence. Front Ecol Environ DOI: 101890/070086.

Pounds, J.A., M.P.L. Fogden and J.H. Campbell. 1999. Biological responses to climate change on a tropical mountain. Nature 398:611–615.

Randolph, S.E. 2009. Perspectives on climate change impacts on infectious diseases. Ecology 90:927–931.

Reading, C. 2007. Linking global warming to amphibian declines through its effects on female body condition and survivorship. Oecologia 151:125–131.

Renecker, L.A. and R.J. Hudson. 1986. Seasonal energy expenditures and thermoregulatory responses of moose. Can J Zool 64:322–327.

Ritchie, C. 2008. Management and challenges of the mountain pine beetle infestation in British Columbia. Alces 44:127–135.

Rogers, L.M., R.J. Delahay, C.L. Cheeseman, G.C. Smith and R.S. Clifton-Hadley. 1999. The increase in badger (Meles meles) density at Woodchester Park, south-west England: a review of the implications for disease (Mycobacterium bovis) prevalence. Mammalia 63:183–192.

Saunders, L.M., D.M. Tompkins and P.J. Hudson. 2000. The role of oxygen availability in the embryonation of Heterakis gallinarum eggs. Int J Parasitol 30:1481–1485.

Welbergen, J.A., S.M. Klose, N. Markus and P. Eby. 2008. Climate change and the effects of temperature extremes on Australian flying-foxes. Proc Biol Sci 275:419–425.

Williams, J.W., S.T. Jackson and J.E. Kutzbach. 2007. Projected distributions of novel and disappearing climates by 2100 AD. Proc Natl Acad Sci USA 104:5738–5742.

Wilson, K., B.T. Grenfell, J.G. Pilkington, H.E.G. Boyd and F.M.D. Gulland. 2004. Parasites and their impact. In Clutton-Brock, T.H., & J.M. Pemberton, eds. Soay Sheep: Dynamics and selection in an island population, pp. 113–165. Cambridge University Press, Cambridge.

# 9

# WILDLIFE HEALTH IN A CHANGING NORTH

## A Model for Global Environmental Change

## Morten Tryland, Susan Kutz, and Patricia Curry

The "North," defined in this chapter as arctic and sub-arctic regions of the circumpolar world, is a vast expanse that is generally sparsely populated. It is inhabited by numerous distinct groups of indigenous peoples, many of whom retain close ties to the land and depend to varying extents on wildlife and semi-domesticated wildlife for food, income, and culture. The North is also a land tremendously rich in nonrenewable resources, including oil, gas, and minerals, and under current climate-warming scenarios there is growing interest in augmenting the exploitation of these resources.

The climate is changing much more rapidly in the North than elsewhere in the world, as reported by the Intergovernmental Panel on Climate Change (IPCC) and the Arctic Climate Impact Assessment (ACIA). Some recently observed changes in the physical environment include melting of ice caps, sea ice, and permafrost, as well as warming of oceans and a rise in sea level. Biological changes, although more difficult to detect and monitor, have also occurred. These include shifts in phenology of animals and plants, including earlier greening-up vegetation; advanced migration and reproduction, including egg laying; as well as shifts in species ranges to higher altitudes and latitudes. Current climate-change scenarios predict ongoing warming (3.5° to 7.5°C by 2099) over most

high-northern latitudes, reduced snow cover and sea ice, greater precipitation (5% to 20% increase), and more frequent severe weather events, such as heat and precipitation extremes (IPCC 2007).

People in the circumpolar North maintain a close relationship with the land and animals and, much like farmers, are intimately aware of and affected by changes in climate and the landscape. Primary food sources, such as caribou (*Rangifer tarandus*), muskox (*Ovibos moschatus*), migratory waterfowl, fish, seals, and berries, remain important parts of the diets of northern residents. For example, in 2001, such resources represented about half of the meat or fish eaten in the majority of Inuit and Inuvialuit households in Canada (Anonymous 2009). In Russia and Fennoscandia (Norway, Sweden, and Finland), herding of reindeer (*Rangifer tarandus tarandus*) is an important livelihood, and trapping and outfitting are important wildlife-based economic activities in communities around the Arctic. These traditional activities not only serve as a source of income and food but also provide opportunities to maintain various aspects of culture. Climate change is rapidly and dramatically influencing the relationship that Northerners have with the land and wildlife to the extent that traditional knowledge about the land, the animals, the water, and the ice may no longer be valid. It is a considerable

**Figure 9.1:**
Semi-domesticated reindeer in Finnmark, Norway. Food restriction can be a serious problem on winter pasture. If climate change introduces more freeze–thaw and rain–frost weather cycles, supplementary feeding may become more common, also increasing the contact between animals and the potential for a disease outbreak.

challenge for people in the North to adjust to these changes and continue to maintain cultural traditions and derive livelihoods from nature (Fig. 9.1).

## WHY STUDY THE IMPACTS OF CLIMATE CHANGE ON WILDLIFE DISEASE IN THE NORTH?

Infectious wildlife diseases can limit population growth and, in the worst case, precipitate extinction events. Climate change, through its effects on disease ecology and host physiology, can have significant impacts on the health and sustainability of animals of ecological and socioeconomic importance across the North. Understanding the impacts of climate change on disease provides critical insights that inform wildlife managers and promote the safety and availability of traditional foods, as well as the stability of animal-based income, such as trapping, reindeer herding, and tourism, for northern peoples.

Transmission rates for most pathogens are closely linked to environmental conditions. Many agents have free-living stages with survival and development rates that are sensitive to temperature and humidity. Some free-living stages are transmitted through invertebrate hosts whose activities and abundance are also closely linked to environmental conditions. For endemic pathogens, life cycles may be accelerated and transmission rates amplified, and for pathogens at the northern extent of their range, a milder climate may allow invasion of new habitats and host species. Changes in the hydrological cycle, such as flooding or droughts that cause crowding around water sources, may increase transmission rates for water-borne pathogens. Climate change can also alter habitats, host physiology (i.e., heat stress), behavior, migration patterns, range use, and community structure, including human–animal interactions. Together, these climate-induced changes in pathogen diversity and transmission rates, as well as host exposure and susceptibility, may lead to enhanced pathogen abundance and shifts in pathogen biodiversity and distribution, and ultimately to animal diseases with impacts at the population level.

For wildlife, the consequences of new pathogens and changed transmission rates, together with other climate-related cumulative stressors, can be substantial. Impacts may include altered abundance, geographic range, and meat quality for important species of ecological, subsistence, and economic values, and may lead to shifts in food web dynamics (Tryland et al. 2009). In the Canadian North, reduced abundance of wildlife species may have a negative impact on the way of life of many Northerners, may result in increased hunting efforts for the same amount of food (Anonymous 2009), and may also affect traditional cultural activities and knowledge across generations. This issue was raised as a significant concern when eradication of diseased wood bison (*Bison bison athabascae*) was recommended for Wood Buffalo National Park in western Canada (see Chapter 28 in this volume). Local hunters were concerned that if bison were removed from the ecosystem for an extended period, their children would grow up without learning how to be bison hunters (Environmental Assessment Panel 1990).

Another significant consequence of climate change in the North is the potential emergence of zoonotic diseases in wildlife that could affect human health (Jones et al. 2008) (Box 9.1). A higher prevalence of such diseases can lead to increased health risks and reduced confidence in traditional food sources, which may lead to greater consumption of more expensive and often less nutritious store-bought foods.

## LINKS BETWEEN CLIMATE CHANGE AND INFECTIOUS DISEASE IN NORTHERN WILDLIFE

Climate change works in multiple ways to alter diversity and abundance of hosts and pathogens and the interactions among them. It is anticipated that climate

---

### Box 9.1  Animal and Human Health in the Canadian North

In 2002, representative leaders, elders, and hunters from all five communities in the Sahtu Settlement Region, Northwest Territories, Canada, met to discuss concerns with respect to wildlife management and research in the region. Participants identified climate change and its effects on wildlife and human health, as well as the need for youth to be better prepared to deal with future wildlife management issues, as major concerns (Brook et al. 2009). Subsequent to this workshop, a series of focus-group interviews were held in Inuit and Dené aboriginal communities in the western Canadian Arctic to investigate whether hunters had observed changes in wildlife health over their lifetimes (Kutz 2007).

These interviews, together with interactions with hunters, identified several concerns related to animal health. One consistent finding was the apparent increase in appearance of "greenish yellow tea-colored slimy stuff" under the skin on the legs and bodies of caribou (Kutz et al. 2009). Community members also expressed concerns about diseases that they had heard of in the news, such as "bird flu" (avian influenza) and "mad cow disease" (bovine spongiform encephalitis). The concern over bird flu was such that people were considering not going on the traditional spring duck and goose hunts for fear of contracting this disease. It became evident that people were observing changes in animal health and hearing about other emerging diseases that were outside of their traditional knowledge framework, yet there was no other readily available source of knowledge to inform them about these new issues. Since meat inspection of hunted northern game is done by the hunters themselves, people are left on their own, with few resources to distinguish "normal" from abnormal and possibly hazardous conditions. Elders also expressed considerable concern about a perceived increase in meat wastage by younger hunters. It was felt that the reason for this might be lack of familiarity with "regular" abnormalities, suggesting a breakdown in transfer of traditional knowledge, but also the possible emergence of syndromes previously unknown to the communities (Kutz 2007).

change will affect the endemic pathogen fauna as well as lead to invasion of new host and pathogen species. In both cases, novel host–pathogen interactions may result.

## Endemic Pathogens: Warmer and Wetter Climate

A warmer and wetter climate, manifested as shorter and milder winters and longer warmer summers, may release many climate-sensitive pathogens from their temperature constraints. In general, this may lead to increased development and survival rates of parasites in the environment. Such changes can result in non-linear shifts in transmission patterns, parasite range expansion, and disease outbreaks. A protostrongylid nematode lungworm (*Umingmakstrongylus pallikuukensis*) of muskoxen (*Ovibos moschatus*) was discovered in 1988 in the Northwest Territories and Nunavut, Canada. The adults of *U. pallikuukensis* live in large cysts in the lung, where they produce first-stage larvae (L1) that are passed in feces. L1 develop into third-stage larvae (L3) in suitable gastropod (slug or snail) intermediate hosts, and these L3 are infective to muskoxen. Development of L1 to L3 in the gastropods is temperature-dependent and occurs more rapidly at warmer temperatures. The effect of *U. pallikuukensis* in muskoxen is unknown, but anecdotal evidence suggests that infected individuals have exercise intolerance and epistaxis, and it is thought that such effects could affect muskoxen at the population level (Kutz et al. 2001). Predictive models based on field and laboratory studies have demonstrated that, before 1990, temperature conditions and season length in the core of the parasite range were insufficient to permit development of L1 to L3 in a single season. However, since 1990, temperatures in this region have been warm enough to permit development in a single season, thus accelerating the life cycle from a 2-year period to 1 year only. Range expansion of the parasite and increased prevalence and intensity of infection are anticipated (Kutz et al. 2005).

Victoria Island, with a population of approximately 45,000 muskoxen, lies immediately to the north of the mainland distribution of *U. pallikuukensis*. Recent modeling indicates that, currently, this parasite could establish and develop to L3 in a single year there (Peacock and Kutz, unpublished data 2007). This is of considerable concern for the residents of Ikaluktutiak,

an Inuit community of approximately 1,300 people on Victoria Island. An annual commercial muskox harvest is an important source of income for this community. A collaborative effort among hunters, biologists, and researchers is underway to determine what pathogens are present and how these may affect the health and sustainability of the muskox population, as well as implications for zoonotic transmission.

Insect-transmitted filaroid nematodes, *Setaria* spp., are commonly found in the abdominal cavity of ungulates and are usually considered harmless (Urquhart et al. 1996). Recently, however, *S. yehi* has been associated with chronic peritonitis in Alaskan reindeer (*Rangifer tarandus*) and *S. tundra* has emerged as a sporadic cause of peritonitis, poor body condition, and death in semi-domesticated reindeer in Fennoscandia. In 1973, tens of thousands of semi-domesticated reindeer died of *Setaria* infection in the northern reindeer-herding areas of Finland alone (Rehbinder et al. 1975; Kummeneje 1980). Wild ungulates were also affected (Laaksonen 2010). In 2003, an outbreak of *S. tundra* caused severe disease in semi-domesticated reindeer. The outbreak started in reindeer husbandry areas in southern Finland and moved northwards by approximately 100 km/year (Laaksonen 2007). The development of the parasite in mosquitoes (*Aedes* spp. but also *Anopheles* spp.) is temperature-dependent, and both of the two major disease outbreaks in reindeer (1973 and 2003) followed summers with unusually warm weather and high mosquito abundance (Laaksonen et al. 2009). It is thought that mean summer temperatures exceeding 14°C result in an increased prevalence of disease in reindeer the following year (Laaksonen et al. 2009). Outbreaks have resulted in extensive carcass and organ condemnation and significant economic losses for the reindeer industry, and although treatment is possible, this is a significant economic and labor cost for herders.

## Temperature and Humidity Extremes

More frequent freeze–thaw cycles or rain-on-snow events in the winter, and increased amplitude of climate extremes, all of which are predicted under current climate-change scenarios, may have variable influences on pathogen development, survival, and transmission rates. Excessive heat and desiccation may be detrimental to the survival of the free-living stages of pathogens such as gastrointestinal tri-chostrongylid nematodes of ungulates (Lettini and Sukhdeo 2006; O'Connor 2006). Conversely, precipitation extremes are likely to enhance dispersal of water-borne pathogens.

The protozoan parasite *Giardia* sp. is commonly known as the cause of "beaver fever" in people and animals around the world. In the Canadian Arctic, *Giardia* sp. has been reported in domestic dogs, wildlife, and people. *Giardia duodenalis* Assemblage A, a zoonotic type, was identified in 21% of muskox feces sampled near the community of Sachs Harbour, on Banks Island in Northwest Territories, Canada (Kutz et al. 2008). *Giardia* sp. has since been diagnosed in people with clinical symptoms in the region, but the parasite has not been typed. Until recently, some reports suggested that this parasite would not be present at this high latitude because of the absence of beavers. This misunderstanding of the ecology of *Giardia* may have resulted in under-diagnosis in people in the past. Current water and sewage treatment procedures, and hunting, camping, and traveling practices on Banks Island permit many opportunities for transmission of *Giardia* sp. between people and muskoxen. Climate change in the form of severe storms and flooding may increase dissemination of *Giardia* sp. on the landscape and contamination of other water bodies. Broad-based surveys and molecular epidemiology are required to determine the transmission pathways for *Giardia* sp. among muskoxen and other wildlife, domestic dogs, and people, and to evaluate the risks to people, in this northern environment (Kutz et al. 2008; Salb et al. 2008).

Climate extremes in the form of increased and/ or prolonged heat may exceed the physiologic tolerances of north-adapted species and lead to direct effects on host susceptibility to infection. During August and September 2006 a severe pneumonia die-off occurred in a muskox population in Dovrefjell, Norway (Ytrehus et al. 2008). The summer had been particularly warm and wet, with mean air temperature and relative humidity 3.2°C and 26% higher, respectively, than the long-term averages. The pneumonia was thought to have developed secondary to multiple simultaneous stressors (i.e., unusually warm and humid weather, increased human disturbance coinciding with the hunting season for wild reindeer). These muskoxen are an introduced population living considerably further south than any natural

populations, and they are occasionally sympatric with domestic sheep and cattle. The particularly warm and humid conditions likely led to heat stress, increased respiration rates, and resultant enhanced opportunities for pulmonary colonization by opportunistic bacteria such as *Pasteurella* spp. Heat stress has also been considered a limiting factor for moose (*Alces alces*) at the southern extent of their range in North America (Dussault et al. 2004).

## Natural Animal Movements and Invasions

Migratory birds may be an ideal form of transport for pathogens, acting as a seasonal link between ecosystems that are separated by huge distances. The coccidian parasite *Toxoplasma gondii* (protozoan of the phylum Apicomplexa) has a global distribution. The domestic cat (*Felis catus*) and other felids are the definitive hosts, harboring the sexual stages of the parasite in the intestines and shedding infective oocysts in feces. A wide range of mammals and birds, including humans, can serve as intermediate hosts in which asexual reproduction and tissue cyst formation occur (Dubey and Beattie 1988). Recently, *T. gondii* was found in arctic fox and polar bears on the high arctic archipelago of Svalbard, where there are no definitive felid hosts (Prestrud et al. 2007; Oksanen et al. 2008). Further investigation revealed *Toxoplasma* antibodies in migratory barnacle geese (*Branta leucopsis*; 7%) but not in kittiwakes (*Rissa tridactyla*) or glaucous gulls (*Larus hyperboreus*) (Prestrud et al. 2007). The authors suggested the possibility that migratory birds sustain the *T. gondii* cycle in this remote archipelago. The birds likely ingest oocysts on their wintering grounds and maintain them as tissue cysts, which then enter the Svalbard food chain through predation by foxes. Other possible means of maintaining the parasite in arctic foxes and polar bears may include transplacental transmission, as described for sheep (Innes et al. 2009). Climate-linked changes in bird migration patterns and changes in the number and diversity of migratory bird species may influence the abundance and types of *Toxoplasma* as well as other pathogens in such ecosystems.

The wild boar (*Sus scrofa*) has recently re-colonized Denmark, Sweden, and Finland from Germany and Russia. In Sweden, the wild boar population counted around 80,000 animals in 2006, and approximately 25,000 were hunted. Wild boars have also started to invade southern regions of Norway. For some, the wild boar is a welcomed game species. However, the wild boar also represents a new pathogen reservoir. The boars may breed with and share infectious diseases with free-ranging domestic pigs, and they are also a possible source of zoonotic pathogens, including *Trichinella* sp., *Leptospira* sp., *Brucella* sp., mycobacteria, and hepatitis E virus (Meng et al. 2009). Expanded wild boar populations, increased contact between wild boars and free-ranging domestic pigs, and increased hunting and human consumption of wild boar meat could mean more human contact with these pathogens in future.

## Anthropogenic Introduction of a New Host

The raccoon dog (*Nystereutes procyonides*), a member of the family *Canidae*, is indigenous to East Asia but was introduced to western and central Europe during the period 1930 to 1955 as a fur and game animal. It is estimated that approximately 9,100 of these animals escaped or were released, and the species is now abundant in many European countries, including Finland, where more than 100,000 animals are shot each year. This species has not yet established in Sweden, and in Norway there have been only a few observations to date. It is not known which pathogens raccoon dogs carry, but it is clear that introduction of this species has complicated the control of rabies in eastern Europe (Zienius et al. 2007). These animals adapt easily to new conditions, fitting the same ecological niche as the red fox (*Vulpes vulpes*), and may be able to maintain rabies virus over winter, when the red fox population often decreases (Finnegan et al. 2002). This situation is an example of a new species shifting the epidemiology of an existing pathogen, rabies virus.

## Insect Vectors and Climate Change: Introduction of Bluetongue to the Arctic?

Changes in the distribution of vectors (mechanical and biological) can have a dramatic impact on the distribution of infectious agents, and such changes may be the most important and immediate effects of a changing climate. Expanded distribution of a vector may not only introduce viruses to hosts in new regions, but also introduce such viruses into the range of other potential vector species. Such expansion may even enhance the competence of a vector and increase the survival of viruses from year to year (Wittmann and Baylis 2000). Bluetongue virus (BTV; genus *Orbivirus*, family *Reoviridae*) causes bluetongue disease in ruminants (Wilson and Mellor 2009). This virus (serotype 8; BTV-8) was discovered in northern Europe in June 2006 at a location approximately 800 km further north than previously recorded. After transmission was interrupted during winter (due to lack of vectors), the virus spread to additional countries the following summer, and the same pattern was repeated in the summer of 2008. The disease caused substantial losses in farm animals during these years and led to restrictions on animal movements (Carpenter et al. 2009). The change in distribution of the virus was probably due to a combination of mechanisms, including movement of infected livestock but also passive movement of *Culicoides* sp. vectors via winds (Purse et al. 2008; Wilson and Mellor 2009). Screening for the presence of possible BTV insect vectors on Swedish farms revealed 33 *Culicoides* species, of which two (*C. obsoletus* and *C. scoticus*) are considered vectors for BTV and represented the largest fraction (Nielsen et al. 2010). These two species were also found in northern regions of Sweden, indicating that, based on presence of vectors, BTV has the potential to establish in arctic and sub-arctic fauna (Nielsen et al. 2010). Since BTV has been detected in Swedish and Norwegian livestock and has appeared in a range of wildlife species in Europe (Ruiz-Fons et al. 2008; Linden et al. 2010), it is possible that reindeer populations may be affected in the future.

## Pathogens Introduced Through Globalization and Other Socioeconomic Drivers

Global movement of people and global trade of animals, animal products, flowers, fruits, and vegetables are transporting animal pathogens and vectors to new environments where they may be maintained and amplify (Harrus et al. 2005; Karesh et al. 2005). For example, the establishment and persistence of the zoonotic tapeworm *Echinococcus multilocularis* near human settlements in the high arctic Svalbard archipelago (Norway) is linked to accidental intro- duction, probably by boat transport, of the sibling vole (*Microtus levis*) from eastern Europe between 1920 and 1960. *Echinococcus multilocularis* was probably introduced to Svalbard through migration of arctic foxes from Siberia (Russia) to these islands (Henttonen et al. 2001); however, it would not have established in the absence of the vole intermediate host. The arctic fox is the definitive host for *E. multilocularis*, and people can be dead-end hosts in which invasive larval cysts develop and potentially cause serious disease. Almost 100% of the overwintering sibling voles on Svalbard are infected with *E. multilocularis* (Henttonen et al. 2001), and this high prevalence is directly corre- lated to the local abundance of voles (Stien et al. 2010). Based on a recent survey of arctic fox feces, the risk of human exposure is limited to the region occupied by the sibling vole (Fuglei et al. 2008). Anthropogenic introduction of the sibling vole to Svalbard is an inter- esting example of how a parasite can invade and estab- lish in a new ecosystem provided that suitable hosts are available (Box 9.2).

## WILDLIFE DISEASES AND CONSERVATION MEDICINE IN THE NORTH

Investigation of the ecology of wildlife disease in the circumpolar world is often hampered by challenges of huge distances, harsh climatic conditions, generally sparse human settlements, and logistical and financial constraints. However, there is now a sense of urgency to better understand how climate change may affect animal and human health in the North. This requires a transdisciplinary approach that addresses the immense geographic area and recognizes the critical human dimensions that are involved, including social, cul- tural, and economic issues. Recent disease ecology research in the North has focused on establishment of baseline data, ongoing surveillance, experimental hypothesis-based research, and generating predictive models (Hoberg et al. 2008). The backdrop for all of these scientific endeavors is the human component, which must be understood and incorporated for any wildlife conservation program to be successful.

---

### Box 9.2  Seals and Phocine Distemper

In 1988 and 2002, phocine distemper virus (PDV) (genus *Morbillivirus*, family *Paramyxoviridae*) caused two major die-offs in the harbor seal (*Phoca vitulina*) colonies of northwestern Europe, killing more than 20,000 individuals in each outbreak (Visser et al. 1993; Härkönen et al. 2006). After the first epizootic, the seal populations returned to immunologically naïve status and were thus susceptible to the virus once again. Several theories were put forth to explain the outbreaks, one of which was transmission of canine distemper virus from infected sled dog carcasses dumped into the Greenland Sea, but evidence suggested that the seals were infected by a different virus. It has also been hypothesized that the seal virus was endemic in pelagic seal species, such as harp seals (*Phoca groenlandica*) and hooded seals (*Cystophora cristata*), which were shown to have antibodies against PDV (Stuen et al. 1994). This virus was thought to have been introduced to the Norwegian Sea and the Kattegat and Wadden Seas via southern seal migrations.

It has also been considered that persistent organic pollutants may have induced immunosuppression, thus making the harbor seals more susceptible to infectious agents (Van Loveren et al. 2000; Troisi et al., 2001). Although an immunomodulating effect of persistent organic pollutants may play a role in the infection biology of phocine distemper, a more recent theory for the outbreaks is that grey seals (*Halichoerus grypus*), which also were hit by the epizootic but sustained much lower mortality, were the natural reservoir for the virus and reintroduced it to the harbor seal colonies in adjacent waters; however, this was not supported by recent PCR screening that included grey seals (Kreutzer 2008). It is clear that PDV is not enzootic in the harbor seal populations, since it causes high-mortality epizootics, but the epidemiology of these disease outbreaks is not fully understood.

## Disease Ecology

The general lack of comprehensive, quantitative baseline data for pathogens in the Arctic and sub-Arctic is a frequent barrier to detecting changes in the ecology or distribution of disease agents. One example of an existing pathogen baseline is the serological surveys that were done for multiple wildlife species in Alaska from the 1980s through 2006 (Zarnke 1983, 2000, 2002; Zarnke and Ballard 1987; Zarnke and Evans 1989; Zarnke and Rosendal 1989; Zarnke et al. 2000, 2004, 2006). Importantly, this information is readily available in the published literature. The Beringian Coevolution Project (BCP), initiated in 1999 and focused on documenting biodiversity of small mammals and their pathogens in Beringia (the Russian and Alaskan land masses on either side of the Bering Strait), is also a primary model for integrated survey and inventory of northern fauna (Hoberg et al. 2003; Cook et al. 2005). Some important outcomes from the BCP include the archiving of mammal and pathogen specimens in museum collections, and ensuring the availability of these specimens for future reference and research. During International Polar Year (2007–2009), the Circum Arctic Rangifer Monitoring and Assessment Network (CARMA) conducted systematic synchronized sampling of multiple circumpolar caribou and reindeer herds with the goal of establishing baselines for *Rangifer* health (http://www.carmanetwork.com/display/public/Projects). Other datasets also exist, but many reflect the interests of individual researchers and tend to be focused on a single host, pathogen, or geographic region. Unfortunately, even in cases where baseline data have been collected, follow-up sampling is rare and thus trends in disease ecology and diversity are not tracked.

## Surveillance and Monitoring

Disease surveillance facilitates detection of and response to new disease patterns. Globally, there is a need for stronger, more widespread surveillance of wildlife disease from both public-health and climate-change perspectives (Kuiken et al. 2005; Parkinson and Butler 2005). Wildlife disease surveillance can be opportunistic or targeted and may involve hunters, scientists, and the general public. In the North, there are several examples of different types of surveillance

and monitoring systems operated by regional or federal government agencies, or less frequently by nongovernmental organizations.

On a smaller scale are northern hunter-based surveillance programs, such as Community-Based Monitoring of Wildlife Health in the Sahtu Settlement Region of the Northwest Territories, Canada (Brook et al. 2009) (Box 9.3). Hunters serve as the "eyes on the land," provide a standardized set of samples and data for scientific analyses, and typically receive compensation for their efforts. These programs benefit from the highly integrated species knowledge and experience of hunters, and from hunters' observations and insights that cover extensive geographic and temporal scales. Further, such collaborative systems can build community capacity for participation in wildlife management (Lyver and Gunn 2004; Lyver and Lutsel Ke Dené First Nation 2005; Kutz et al. 2009).

## Hypotheses, Experiments, and Models

Baseline data and tools for surveillance provide a foundation for developing hypotheses on host–pathogen interactions and formulating predictive models. For example, hunter-based surveillance programs have raised concerns about a new disease syndrome in caribou that is currently under investigation (Kutz et al. 2009). Similarly, researcher-driven targeted surveillance demonstrated that the proto-strongylid nematode *Parelaphostrongylus odocoilei* was present in Dall's sheep (*Ovis dalli dalli*) across this species range south of the Arctic Circle (Jenkins et al. 2005a). Subsequent laboratory and field research established the life cycle, clinical signs, and pathology. Empirically based predictive models derived from the experimental and surveillance results indicated that if *P. odocoilei* were to be translocated to regions further north, climatic conditions would be suitable for it to establish (Jenkins et al. 2005b). This provided important insights and recommendations with respect to Dall's sheep management and translocations.

## The Human Dimension

Scientific data and new knowledge do not make or change policy, or translate policy into action. It is people who implement animal conservation and management strategies, and novel findings and

---

**Box 9.3  Community-Based Wildlife Health Monitoring in Northern Canada**

---

This strategy was introduced in the Sahtu Settlement Region of the Northwest Territories in 2003 and focuses on hunter-based monitoring of caribou and moose health (Brook et al. 2009). Both species are very important to Sahtu residents as food that has high nutritional quality and is traditional and afford-able. The program was spearheaded by a team of government and university collaborators in response to concerns voiced by the five small remote communities of the Sahtu (populations 125 to 800). Teams of hunters and scientists work together to exchange observations and knowledge. Community visits are arranged that gather hunters (Wildlife Health Monitors [WHMs]) for discussion and for training in sampling protocols. WHMs collect a small set of samples and data from animals that are hunted for subsistence only. The scientists then analyze these for health indicators and evidence of disease.

An incentive, such as fuel, is provided for each sample set submitted. One goal is to report results to communities as soon as possible. A video has been distributed to aid in hunter training and raise awareness and interest in the program, and also to engage young minds on the subject of wildlife health. The monitoring program has been constantly adapted in response to feedback on which methods and sample types will work best. The success of initiatives such as the Sahtu WHM program relies on two main elements: the issue (wildlife health) must be viewed as *relevant to the community*, and the *methods and logistics must "work"* for all those involved.

---

"education" do not necessarily trump entrenched attitudes, perceptions, or emotions. Bath and others have highlighted the importance of such human factors relative to wildlife management, and have pointed to the need for robust social-science inquiry to gain understanding of the viewpoints, motivations, conflicts, and rationale surrounding any issue (Bath and Buchanan 1989; Bath 1998; Bath and Enck 2003). However, this area of investigation is relatively new, and most examples have involved issues and locations south of the Arctic and sub-Arctic.

For conservation medicine to succeed in the North, practitioners must recognize and respect the human issues involved and embrace transdisciplinary methods. However, the realities of the region pose unique challenges. The many circumpolar indigenous peoples have differing traditional and cultural values, different languages, varied beliefs and ways of living, differing views on animals and nature, and traditional hunting and fishing rights that may set them apart from other Northerners or other residents of a given country. These can all create barriers and tensions when attempting to implement and assess a program. Also, sensitivities surrounding historical and current governmental policies related to aboriginal people may charge an already challenging situation. In some political jurisdictions of the North, wildlife co-management boards, a partnership between indigenous groups and the government, have been established to facilitate wildlife management. Small community populations severely restrict the number of potential leaders and participants in wildlife management and hunter sampling programs. Also, the relatively few northern communities tend to receive intensive research pressure, and this may exhaust the enthusiasm of residents for collaborative wildlife monitoring.

## Norwegian Health Surveillance Program for Deer Species and Muskoxen (HOP)

This program was established in 1998 to document the health status of Norwegian red deer (*Cervus elaphus*), moose, roe deer (*Capreolus capreolus*), and wild reindeer (*R. t. tarandus*), as well as free-ranging muskoxen, and to monitor zoonoses and diseases transferable to domestic livestock. The program is based on routine diagnostic work on submitted material, sample collections from selected populations through cooperation with hunters, and samples (serum) obtained through management and research projects. An important feature of this program is that samples are banked to enable surveys and future investigations that can be linked to ecosystem changes. The HOP is funded by the Directorate for Nature Management and is run by the National Veterinary Institute in Oslo, Norway.

## Surveillance of Reindeer and Other Deer Species in Finland

The Finnish Food Safety Authority (EVIRA), in cooperation with hunters, developed a cervid health program in 2007 to monitor infectious diseases (i.e., chronic wasting disease, bluetongue, brucellosis) as well as persistent organic pollutants in cervid species. It maintains a serum bank for future needs and also provides hunter education on biology, diseases, and meat hygiene in accordance with European Union legislation. Since 2004, EVIRA has also run a reindeer health program that features special education and training of veterinarians and reindeer herders, and collection of data on health indicators and disease records from slaughtered animals.

## Surveillance of Carnivores in Sweden and Norway

Sweden and Norway share populations of grey wolves (*Canis lupus*), Eurasian lynx (*Lynx lynx*), and brown bears (*Ursus arctos*). These carnivores are generally protected, but they are also in conflict with human activities, especially sheep farming. Surveillance programs have been launched to help manage these carnivore populations and inform the general public about population dynamics, genetics, and their impacts as predators. The wolf program (SKANDULV) started in 2000 and includes investigation of predation impact on deer populations. The Scandinavian lynx program (SCANDLYNX) started in 1993 with collection of ecological data on the Eurasian lynx and roe deer, the most important prey for lynx in different regions of Scandinavia. The research project on the Scandinavian brown bear started in Sweden in 1984 and has included Hedmark County (Norway) since 1987. A blood and tissue bank has been set up

and data related to life history and mother–offspring relationships are collected.

## CONCLUSIONS

The North is changing at a rapid rate with respect to both climate change and anthropogenic activities. In the circumpolar world, climate change is happening in real time with documented physical and biological effects. In the North, there are relatively few confounding factors that complicate the understanding of climate-change impacts on infectious disease (Kutz et al. 2009). The North is generally a system of fairly low biological diversity, and most of the indigenous wildlife species are naïve to many pathogens that occur elsewhere. As such, the circumpolar region may be particularly vulnerable to invasions of new pathogens caused by climate-mediated changes. Finally, indigenous peoples of the North maintain a close relationship with wildlife (or semi-domesticated wildlife in the case of reindeer) and are quick to detect changes in animal health. Health of animals is a priority for Northerners relative to ensuring the sustainability of this important resource and relative to food safety. Thus, by examining the host–pathogen interactions and impacts of climate change at northern latitudes we can gain new and insightful perspectives on the implications of climate change for infectious and zoonotic diseases of wildlife in general, and use these to develop and test a framework for understanding more complex human and animal systems globally (Bradley et al. 2005; Burek et al. 2008; Kutz et al. 2009). The diversity and abundance of infectious diseases are anticipated to increase under existing climate-change scenarios, and some endemic pathogens may disappear due to altered life conditions. Climate change and exposure to persistent organic pollutants and various socioeconomic drivers are having important effects in the North. To track changes over time, it is crucial to conduct surveillance and research on wildlife populations, and to monitor exposure to infectious agents, disease dynamics, and population-level impacts. However, knowledge of these biological processes alone is not sufficient to bring action; such knowledge must be translated into political motivations that change policy in order to mitigate the negative effects that climate change and other socioeconomic factors are having on wildlife health in northern ecosystems and societies.

# REFERENCES

ACIA (http://www.acia.uaf.edu/pages/overview.html).

Anonymous. 2009. Food Safety Network. Safe preparation and storage of aboriginal traditional/country foods: a review. Vancouver, BC, Canada: National Collaborating Centre for Environmental Health.

Bath, A. 1998. The role of human dimensions in wildlife resource research in wildlife management. Ursus 10:349–355.

Bath, A.J., and T. Buchanan. 1989. Attitudes of interest groups in Wyoming toward wolf restoration in Yellowstone National Park. Wildl Soc Bull 17:519–525.

Bath, A., and J. Enck. 2003. Wildlife-human interactions in National Parks in Canada and the USA. NPS Soc Sci Rev 4:1–29.

Bradley, M.J., S.J. Kutz, E. Jenkins, and T.M. O`Hara. 2005. The potential impact of climate change on infectious diseases of Arctic fauna. Int J Circumpol Heal 64:468–477.

Brook, R.K., S.J. Kutz, A.M. Veitch, R. Popko, B.T. Elkin, and G. Guthrie. 2009. Generating synergy in communicating and implementing community based wildlife health monitoring and education. EcoHealth. 6:266–278.

Burek, K.A., F.M. Gulland, and T.M. O'Hara. 2008. Effects of climate change on Arctic marine mammal health. Ecol Appl 18:126–134.

Carpenter, S., A. Wilson, and P.S. Mellor. 2009. Culicoides and the emergence of bluetongue virus in northern Europe. Trends Microbiol 17:172–178.

Cook, J.A., E.P. Hoberg, A. Koehler, H. Henttonen, L. Wickström, V. Haukisalmi, K. Galbreath, F. Chernyavski, N. Dokuchaev, A. Lahzuhtkin, S.O. MacDonald, A. Hope, E. Waltari, A. Runck, A. Veitch, R. Popko, E. Jenkins, S. Kutz, and R. Eckerlin. 2005. Beringia: Intercontinental exchange and diversification of high latitude mammals and their parasites during the Pliocene and Quaternary. Mammal Study 30:S33–S44.

Dubey, J.P., and C.P. Beattie. 1988. Toxoplasmosis of animals and man. CRC Press, Boca Raton, Florida, USA.

Dussault, C., C. Dussault, J.P. Ouellet, R. Courtois, J. Huot, L. Breton, and J. Larochelle. 2004. Behavioural responses of moose to thermal conditions in the boreal forest. Le Naturaliste Canadien 11:321–328.

Environmental Assessment Panel. 1990. Northern diseased bison report of the Environmental Assessment Panel.

Environmental Assessment Review Office, Report 35. Hull, Quebec. 47 pp.

Finnegan, C.J., S.M. Brookes N. Johnson, J. Smith, K.L. Mansfield, V.L. Keene L. McElhinney, and A.R. Fooks. 2002. Rabies in North America and Europe. J R Soc Med 95: 9–13.

Fuglei, E., A. Stien, N.G. Yoccoz, R.A. Ims, N.E. Eide, P. Prestrud, P. Deplazes, and A. Oksanen. 2008. Spatial distribution of Echinococcus multilocularis, Svalbard, Norway. Emerg Infect Dis 14:73–75.

Härkönen, T., R. Dietz, P. Reijnders, J. Teilmann, K. Harding, A. Hall, S. Brasseur, U. Siebert, S.J. Goodman, P.D. Jepson, T.D. Rasmussen, and P. Thompson. 2006. A review of the 1988 and 2002 phocine distemper virus epidemic in European harbour seals. Dis Aquat Organ 68:115–130.

Harrus, S., and G. Baneth. 2005. Drivers for the emergence and re-emergence of vector-borne protozoal and bacterial diseases. Int J Parasitol 35:1309–1318.

Henttonen, H., E. Fuglei, C.N. Gower, V. Haukisalmi, R.A. Ims J. Niemimaa, and N.G. Yoccoz. 2001. Echinococcus multilocularis on Svalbard: introduction of an intermediate host has enabled the local life cycle. Parasitology 123:547–552.

Hoberg, E.P., S.J. Kutz, K.E. Galbreath, and J. Cook. 2003. Arctic biodiversity: From discovery to faunal baselines—revealing the history of a dynamic system. J Parasitol 89:S84–S95.

Hoberg, E.P., Polley, L., Jenkins, E.J., Kutz, S.J., Veitch, A.M., and Elkin, B.T. 2008. Integrated approaches and empirical models for investigation of parasitic diseases in northern wildlife. Emerg Inf Dis 14:10–17.

Innes, E.A., P.M. Bartley, D. Buxto, and F. Katzer. 2009. Ovine toxoplasmosis. Parasitology 136:1887–1894.

IPCC. 2007. Fourth Assessment Report, AR4; 17 Nov. 2007, http://www.ipcc.ch/pdf/assessment-report/ar4/syr/ar4_syr.pdf.

Jenkins, E.J., G.D. Appleyard, E.P. Hoberg, B.M. Rosenthal S.J. Kutz, A.M. Veitch, H.M. Schwantje, B.T. Elkin, and L. Polley. 2005a. Geographic distribution of the muscle-dwelling nematode Parelaphostrongylus odocoilei in North America, using molecular identification of first-stage larvae. J Parasitol 91: 574–584.

Jenkins, E.J., A.M.Veitch, S.J. Kutz E.P. Hoberg, and L. Polley. 2005b. Climate change and the epidemiology of protostrongylid nematodes in northern ecosystems: Parelaphostrongylus odocoilei and Protostrongylus stilesi in Dall's sheep (Ovis d. dalli). Parasitology 7:1–15.

Jones, K.E., N.G. Patel, M.A. Levy, A. Storeygard, D. Balk, J.L. Gittleman, and P. Daszak. 2008. Global trends in emerging infectious diseases. Nature 451: 990–993.

Karesh, W.B., R.A. Cook, E.L. Bennett, and J. Newcomb. 2005. Wildlife trade and global disease emergence. Emerg Infect Dis 11:1000–1002.

Kreutzer, M., R. Kreutzer, U. Siebert, G. Müller, P. Reijnders, S. Brasseur, T. Härkönen, R. Dietz, C. Sonne, E.W. Born, and W. Baumgärtner. 2008. In search of virus carriers of the 1988 and 2002 phocine distemper virus outbreaks in European harbour seals. Arch Virol 153:187–92.

Kuiken, T., F.A. Leighton, R.A. Fouchier, J.W. Leduc, J.S. Peiris, A. Schudel, K. Stohr, and A.D. Osterhaus. 2005. Pathogen surveillance in animals. Science 309:1680–1681.

Kummeneje, K. 1980. Diseases in reindeer in Northern Norway. Proceedings of the Second International Reindeer-Caribou Symposium, pp. 456–458. Røros, Norway.

Kutz, S.J., E.P. Hoberg, and I. Polley. 2001. A new lungworm in muskoxen: an exploration in arctic parasitology. Trends Parasitol 17:276–280.

Kutz, S.J., E.P. Hoberg, L. Polley, and E.J. Jenkins. 2005. Global warming is changing the dynamics of Arctic host–parasite systems. Proc Royal Soc B 271:2571–2576.

Kutz, S.J. 2007. An evaluation of the role of climate change in the emergence of pathogens and diseases in Arctic and sub-Arctic caribou populations. Climate Change Action Fund, Project A760. Report to Government of Canada.

Kutz, S.J., R.C.A.Thompson, L. Polley, K. Kandola J. Nagy, C. Wielinga, and B.T. Elkin. 2008. Giardia Assemblage A: human genotype in muskoxen in the Canadian Arctic. Parasite Vector 1:32doi:10.1186/1756-3305-1-32.

Kutz, S.J., E.J. Jenkins, A.M. Veitch, J. Ducrocq, L. Polley, B.T. Elkin, and S. Lair. 2009. The Arctic as a model for anticipating, preventing, and mitigating climate change impacts on host-parasite interactions. Vet Parasitol 163:217–228.

Laaksonen, S., J. Kuusela, S. Nikander, M. Nylund, and A. Oksanen. 2007. Outbreak of parasitic peritonitis in reindeer in Finland. Vet Rec 160:835–841.

Laaksonen, S., M. Solismaa, R. Kortet, J. Kuusela, and A. Oksanen. 2009. Vectors and transmission dynamics for Setaria tundra (Filaroidea; Onchocercidae), a parasite of reindeer in Finland. Parasite Vector 2:3 doi: 10.1186/1756-3305-2-3.

Laaksonen S., J. Pusenius, J. Kumpula, A. Venäläinen, R. Kortet, A. Oksanen, and E. Hoberg. 2010. Climate change promotes the emergence of serious disease outbreaks for filaroid nematodes. EcoHealth 7:7–13.

Lettini, S.E., and M.V.K. Sukhdeo. 2006. Anydrobiosis increases survival of trichostrongyle nematodes. J Parasitol 92:1002–1009.

Linden, A., F. Gregoire, A. Nahayo, D. Hanrez, B. Mousset, A.L. Massart, I. De Leeuw, E. Vandemeulebroucke, F. Vandenbussche, and K. De Clercq . 2010. Bluetongue virus in wild deer, Belgium, 2005-2008. Emerg Infect Dis 16:833–836.

Lyver, P.O'B., and A. Gunn. 2004. Calibration of hunters' impressions of female caribou body condition indices to predict probability of pregnancy. Arctic 57: 233–241.

Lyver, P.O'B., and Lutsël K'é Dene First Nation. 2005. Monitoring barren-ground caribou body condition with Denésoliné traditional knowledge. Arctic 58:44–54.

Meng, X.J., D.S. Lindsay, and N. Sriranganathan. 2009. Wild boars as sources for infectious diseases in livestock and humans. Philos Trans R Soc Lond B 364:2697–707.

Nielsen, S.A., B.O. Nielsen, and J. Chirico. 2010. Monitoring of biting midges (Diptera: Ceratopogonidae: Culicoides Latreille) on farms in Sweden during the emergence of the 2008 epidemic of bluetongue. Parasitol Res 106:1197–1203.

O'Connor, L.J., S.W. Walkden-Brown, and L.P. Kahn. 2006. Ecology of the free-living stages of major trichostrongylid parasites of sheep. Vet Parasitol 142:1–15.

Oksanen, A., K. Åsbakk, K.W. Prestrud, J. Aars, A. Derocher, M. Tryland , Ø. Wiig, J.P. Dubey, C. Sonne, R. Dietz, M. Andersen, and E.W. Born. 2008. Prevalence of antibodies against Toxoplasma gondii in polar bears (Ursus maritimus) from Svalbard and East Greenland. J Parasitol 95:89–94.

Parkinson, A.J., and J.C. Butler. 2005. Potantial impact of climate change on infectious diseases in the Arctic. Int J Circumpol Heal 64:478–486.

Prestrud, K.W., K. Åsbakk, E. Fuglei, T. Mørk, A. Stien, E. Ropstad, M. Tryland, G.W. Gabrielsen, C. Lydersen, K. Kovacs, M. Loonen, K. Sagerup, and A. Oksanen. 2007. Serosurvey for Toxoplasma gondii in arctic foxes and possible sources of infection in the high Arctic of Svalbard. Vet Parasitol 150:6–12.

Purse, B.V., H.E. Brown, L. Harrup, P.P.C. Mertens, and D.J. Rogers. 2008. Invasion of bluetongue and other orbivirus infections into Europe: the role of biological and climate processes. Rev Sci Tech Off Int Epiz 27:427–442.

Rehbinder, C., D. Christensson, and V. Glatthard. 1975. Parasitic granulomas in reindeer. A histopathological, parasitological and bacteriological study. Nord Vet Med 27:499–507.

Ruiz-Fons, F., A.R. Reyes-Garcia, V. Alcaide, and C. Gortázar. 2008. Spatial and temporal evolution of bluetongue virus in wild ruminants, Spain. Emerg Infect Dis 14:951–953.

Salb, A.L., H.W. Barkema, B.T. Elkin, R.C.A. Thompson, D. Whiteside, S.R. Black, J.P. Dubey, and S.J. Kutz. 2008. Domestic dogs as sources and sentinels of parasites in northern people and wildlife. Emerg Inf Dis 14:60–63.

Stien, A., L. Voutilainen, V. Haukisalmi, E. Fuglei, T. Mørk, N.G. Yoccoz, R.A. Ims, and H. Henttonen. 2010. Intestinal parasites of the Arctic fox in relation to the abundance and distribution of intermediate hosts. Parasitology 137:149–157.

Stuen S., P. Have, A.D. Osterhaus, J.M. Arnemo, and A. Moustgård. 1994. Serological investigation of virus infections in harp seals (*Phoca groenlandica*) and hooded seals (*Cystophora cristata*). Vet Rec 134:502–3.

Troisi, G.M., K. Haraguchi, D.S. Kaydoo, M. Nyman, A. Aguilar, A. Borrell, U. Siebert, and C.F. Mason. 2001. Bioaccumulation of polychlorinated biphenyls (PCBs) and dichlorodiphenylethane (DDE) methyl sulfones in tissues of seal and dolphin morbillivirus epizootic victims. J Toxicol Env Heal A 62:1–8.

Tryland, M., J. Godfroid, and P. Arneberg, eds. 2009. Impact of climate change on infectious diseases of animals in the Norwegian Arctic. Brief Report Series, Norwegian Polar Institute 2009, Norway.

Urquhart, G.M., J. Armour, J.L. Duncan, A.M. Dunn, and F.W. Jennings. 1996. Veterinary parasitology. 2nd ed. Blackwell Science Ltd., Oxford.

Van Loveren, H., P.S. Ross, A.D. Osterhaus, and J.G. Vos. 2000. Contaminant-induced immunosuppression and mass mortalities among harbor seals. Toxicol Lett 15:112–113, 319–24.

Visser, I.K.G., M.F. van Bressem, T. Barrett, and A.D.M.E. Osterhaus. 1993. Morbillivirus infections in aquatic mammals. Vet Res 24:169–178.

Wilson, A.J., and P.S. Mellor. 2009. Bluetongue in Europe: past, present and future. Phil Trans Royal Soc B 364:2669–2681.

Wittmann, E.J., and M. Baylis. 2000. Climate change: effects on culicoides-transmitted viruses and implications for the UK. Vet J 160:107–117.

Ytrehus, B., T. Bretten, B. Bergsjo, and K. Isaksen. 2008. Fatal pneumonia epizootic in muskox (*Ovibos moschatus*) in a period of extraordinary weather conditions. EcoHealth 5:213–223.

Zarnke, R.L. 1983. Serologic survey for selected microbial pathogens in Alaskan wildlife. J Wildl Dis 19:324–329.

Zarnke, R.L. 2000. Alaska Wildlife Serologic Survey, 1975–2000. Alaska Department of Fish and Game. Federal Aid in Wildlife Restoration, Juneau, Alaska.

Zarnke, R.L. 2002. Serum antibody prevalence of malignant catarrhal fever viruses in seven wildlife species from Alaska. J Wildl Dis 38:500–504.

Zarnke, R.L., and W.B. Ballard. 1987. Serologic survey for selected microbial pathogens of wolves in Alaska, 1975-1982. J Wildl Dis 23:77–85.

Zarnke, R.L., and M.B. Evans. 1989. Serologic survey for infectious canine hepatitis virus in grizzly bears (*Ursus arctos*) from Alaska, 1973 to 1987. J Wildl Dis 25:568–573.

Zarnke, R.L., and S. Rosendal. 1989. Serologic survey for *Mycoplasma ovipneumoniae* in free-ranging dall sheep (*Ovis dalli*) in Alaska. J Wildl Dis 25:612–613.

Zarnke, R.L., J.P. Dubey, O.C.H. Kwok, and J.M. Ver Hoef. 2000. Serologic survey for *Toxoplasma gondii* in selected wildlife species from Alaska. J Wildl Dis 36:219–224.

Zarnke R.L., J.M. Ver Hoef, and R.A. DeLong. 2004. Serologic survey for selected disease agents in wolves (*Canis lupus*) from Alaska and the Yukon Territory, 1984-2000. J Wildl Dis 40:632–8.

Zarnke, R.L., J.T. Saliki, A.P. Macmillan, S.D. Brew, C.E. Dawson, J.M. Ver Hoef, K.J. Frost, and R.J. Small. 2006. Serologic survey for Brucella spp., phocid herpesvirus-1, phocid herpesvirus-2, and phocine distemper virus in harbor seals from Alaska, 1976–1999. J Wildl Dis 42:290–300.

Zienius, D., V. Sereika, and R. Lelešius. 2007. Rabies occurrence in red fox and raccoon dog population in Lithuania. Ekologija 53:59–64.

# 10

# HABITAT FRAGMENTATION AND INFECTIOUS DISEASE ECOLOGY

Gerardo Suzán, Fernando Esponda,
Roberto Carrasco-Hernández, and A. Alonso Aguirre

In this chapter we address the effects of habitat loss and fragmentation on infectious disease ecology, first on a theoretical basis and second by reviewing published empirical studies that examine this relationship. We report the mechanisms proposed in the empirical studies to explain the observed results. Finally, we suggest avenues for future research describing other drivers that should be considered for disease emergence in fragmented habitats and the development of alternative theoretical models and empirical studies on multispatial scales that will help to understand patterns and processes.

The integration of both theoretical and empirical findings on habitat changes and infectious diseases ecology will help scientists and policymakers to build transdisciplinary thinking to predict and understand disease dynamics, and simultaneously develop conservation strategies including the design of natural reserve systems and landscape management.

## THEORETICAL BACKGROUND

A fragmented habitat is defined as "a large expanse of habitat transformed into a number of smaller patches of smaller total area, isolated from each other by a matrix of habitats unlike the original" (Wilcove 1986). Consequences of habitat fragmentation have been analyzed from the perspective of landscape ecology, which establishes that some characteristics of any ecological system will be directly or indirectly affected by the spatial context that surrounds it. Landscapes are conceptualized as units of higher organization level formed by a collection of ecosystems in a geographic area. A landscape is a spatially heterogeneous area that can be mathematically described based on patterns of shapes, sizes, clustering, and diversity of different surface units. Many of the present ideas regarding ecological landscape studies and disease ecology are based on the theory of island biogeography, developed by MacArthur and Wilson (1967). This theory identifies size and isolation of islands as important factors influencing diversity and suggests that smaller and more isolated islands are likely to show lower species diversity than larger islands. Landscape ecology extends this theory to different levels, including populations, metapopulations, communities, and ecosystems. In relation to disease ecology, it is suggested that smaller and more isolated patches are prone to invasions of both invasive species and infectious diseases, affecting native hosts of plants and animals in low-diversity areas (Allan et al. 2003; Keesing et al. 2006; Suzán et al. 2008).

In terrestrial landscapes, a large patch from which species migrate to colonize smaller patches is analogous to the continents on the theory of MacArthur and Wilson (1967). However, in the absence of large patches, community structure will only depend on the connectivity between small patches. A simple rule regarding patch size and connectivity is generally addressed in the literature of landscape ecology (Fig. 10.1).

In terrestrial systems the form, size, and connectivity of patches are dynamic through time and space and can also be affected by human activities. With this in mind, we can regard "habitat fragmentation" as a particular case in the dynamics of landscape patch structure, where matrix area between patches increases, consequently reducing patch area and enhancing isolation. Habitat fragmentation also increases the size of the patch in close proximity to an edge. The underlying mechanisms of this phenomenon are diverse; however, most are anthropogenic in nature, including deforestation, invasion of natural vegetated areas by human communities, and climate change. Some definitions consider that habitat fragmentation is different from habitat loss, the latter being defined as the complete elimination of a localized or regional ecosystem, leading to the total loss of its former biological function (Dodd and Smith 2003).

Under such definitions, habitat fragmentation should be restricted to changes in patch configuration that result from the breaking apart of habitat, independent of habitat loss. However, some other descriptions of habitat fragmentation often impede separating the effects of habitat loss from those of fragmentation.

Although the effects of habitat loss alone on biodiversity and infectious disease dynamics may be stronger than those of fragmentation, studies in which the effects of habitat loss are not controlled may obscure the effects of fragmentation *per se*. If the particular consequences of fragmentation are not evaluated independently from those of habitat loss, researchers will be unable to understand the independent contribution of each to the observed changes in pathogen prevalence and occurrence (Langlois et al. 2001; Fahrig 2003).

## ANALYTICAL MODELS AND SIMULATIONS IN FRAGMENTATION THEORY AND DISEASE ECOLOGY

A variety of models have been developed to understand how reservoirs, infectious agents, and vectors respond to habitat fragmentation. Models are simplifications of reality that permit researchers to focus on specific outcomes. The art and science of modeling is in abstracting the right traits from reality such that the observations made on the model apply to reality as well. Different models capture distinct features of the world and offer different insights on the underlying phenomena. The two most used techniques are analytical and simulation techniques. Analytical models, from which differential equation techniques have been the most widely used, produce a set of equations that characterize the phenomenon of interest. In the particular case of infectious diseases and habitat fragmentation, the equations describe the dynamics of the disease or the population. A key example of the former is the SIR model (Anderson and May 1979; Chapter 41 in this volume), where individuals in a community are categorized as being in one of three states: as being susceptible (S) to the disease, as being infected (I) by the disease, or as having recovered (R) from it. The dynamics are governed by the rate at which individuals are born and die, by the virulence of the disease, and by the population ability to recover from it. The resulting model is thus a set of differential equations that describe how the individuals move between the different states.

These models provide valuable insights into the dynamics of a system and are a useful method for identifying critical parameters (e.g., migration rates) for which the model and presumably the real system

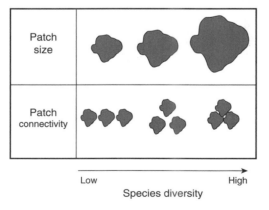

**Figure 10.1:**
Relations of species diversity to patch size and connectivity, according to the theory of island biogeography (McArthur and Wilson 1967).

exhibit noteworthy phenomena. Also, this basic SIR model provides a useful starting point for modeling more complex phenomena; for instance, Hess (1996) and McCallum (2008) included the notion of patches and migration from a metapopulation perspective, and McCallum and Dobson (2002) extended it to model recovered patches and a reservoir host. Analytical models of this kind have also been useful in aiding the study of habitat fragmentation and its effects, as in Zanette (2000), where population growth is modeled as a function of the fragments' connectivity, and Murrell and Law (2000), who modeled the dynamics of beetles in heterogeneous landscapes. Hanski and Ovaskainen (2003) provide an excellent review of some equation-based models for fragmented landscapes. On occasion it is of interest to include further detail into models such as geographic accidents, specific community arrays, as well as heterogeneous traits in individuals. It is often the case that equations that describe systems in such detail and with so many exceptions are difficult to write down and infeasible to solve. The advances in computer power now make it possible to simulate systems of heterogeneous agents in considerable detail. Following this approach the agents and the environment are modeled as computer programs and the researcher studies the results of running these programs with different parameter settings. For example, to study the impact of the spread of disease between populations, Hess (1996) first used an analytical model of the type described above and then resorted to a simulation model to examine the impact of geographic connectivity between populations. Brooker and Brooker (1994) used computer simulations to model population variability under the effects of fire and landscape fragmentation. Tracey et al. (2008) use an agent-based model to examine the relationships between some human activities and fragmentation of animal habitats. Parker and Meretsky (2004) used these same techniques to explore the economic and agricultural impact of landscape fragmentation.

## CHANGES IN POPULATION DENSITY

Habitat fragmentation can change the density, relative abundance, and geographic distribution of species involved in pathogen transmission cycles both within fragments and across landscapes. These changes can alter the dynamics of density-dependent transmission and result in different spatial patterns of disease prevalence. It has been shown that high densities in the host population often drive an increase in transmission rates and the success of parasite reproduction (Anderson and May 1979; Mitchell 2002). Host population size has a profound effect on the dynamics and virulence of the parasite. In theory, a parasite will not be established in the host if the host population is not sufficiently large. Initially, after any fragmentation event it is common that population density and thus transmission rates increase. In the very short term the hosts can retreat to the remaining habitat, and in the somewhat longer term those species that can manage to survive in the fragmented habitat may increase in population density because many of their predators and competitors cannot survive. This may explain the following examples.

White-footed mice (*Peromyscus leucopus*) reach high densities in forest fragments, most likely due to the lower abundance of predators and competitors (Nupp and Swihart 1998). Page et al. (2001) suggested that the increased prevalence of the raccoon roundworm *Baylisascaris procyonis* in habitat fragments can be due to the higher density of both raccoons and white-footed mice (which are intermediate hosts) and the distribution of raccoon denning trees in a fragmented landscape. In the northeastern United States, white-footed mice are the most competent hosts of Lyme disease. Allan et al. (2003) found that the risk of exposure to Lyme disease in this ecosystem is higher in small forest remnants, and higher density of mice and reduced species diversity in these forest fragments lead to a higher fraction of tick meals taken from the most competent host, thereby increasing the proportion of infected vectors. High density of deer mice (*Peromyscus maniculatus*) in fragmented landscapes may also contribute to higher prevalence of hantavirus infection (Langlois et al. 2001).

## COMMUNITIES AND PATHOGEN DYNAMICS IN FRAGMENTED LANDSCAPES

One of the most important consequences of habitat loss and fragmentation is the loss of species diversity and ecosystem function. Reduced species diversity in small isolated patches will have an influence on disease occurrence. According to the theory of the dilution effect on vector-borne diseases (Ostfeld and Keesing 2000; Schmidt and Ostfeld 2001), a higher number of alternative host species would reduce the prevalence of diseases like Lyme disease. This has been supported by empirical studies (LoGiudice et al. 2003) including those on directly transmitted diseases like hantavirus demonstrating that the prevalence is higher in communities with poor biodiversity associated with fragmented landscapes (Suzán et al. 2009). Mechanisms clarifying these buffering effects on community approach, including encounter reduction, transmission reduction, and susceptible host regulation, among others, are explained thoroughly (Keesing et al. 2006).

However, not only reductions in species richness and diversity affect pathogen dynamics; other important community attributes, like composition and dominance, also have a great influence on diseases occurrence (LoGiudice et al. 2008). A widespread pattern in altered landscapes is that the resulting species assemblage is dominated by generalist species. These species take advantage of habitats like agricultural plots, small remnants of native vegetation, and even human settlements (Daily et al. 2003). Increased densities of generalist species have amplified contact rates with humans, and emerging and re-emerging outbreaks are more often reported. Also, Keesing et al. (2006) argue that these "generalist" species are also the most competent reservoirs for some pathogens and hosts for some vectors. This is the case for rabies and hantavirus reservoirs worldwide (Daszak et al. 2000). Similarly, it is estimated that abrupt changes in landscape structure have a substantial effect on wild carnivores and rabies (Real et al. 2005). Rabies is transmitted by invasive reservoirs that have become habituated to urban settlements and are able to maintain and introduce the disease in areas with different habitat fragmentation levels (Suzán and Ceballos 2005).

Habitat fragmentation results in decreased area and increased patch isolation, along with microclimatic and biogeographic changes (Saunders et al. 1991). For some species, habitat fragmentation is detrimental, with some populations going locally extinct in fragmented landscapes (Reed 2004). Other species will experience increases in density, either temporarily, as habitat loss forces individuals to congregate in the last remaining patches, or more permanently

because they are favored by the ecological conditions in fragmented landscapes. When these effects alter the richness and/or the relative abundance of species involved in disease transmission cycles, they can lead to changes in pathogen persistence, prevalence, and distribution in the landscape. For example, Hulbert and Boag (2001) found a higher prevalence of parasitic helminths in hares from continuous forested areas. In contrast, studies conducted in Costa Rica (Daily et al. 2003), Venezuela (Utrera et al. 2000), the United States (Diffendorfer et al. 1995), Chile (Torres-Perez et al. 2004), Panama (Suzán 2005), and Paraguay (Yahnke et al. 2001) have demonstrated that human-induced habitat loss is favoring those rodent species that tolerate human activities and human presence, and that these generalist rodents are often involved in the transmission of hantavirus to humans.

## THE EDGE EFFECT AND MATRIX COMPOSITION

Patches in terrestrial landscapes will not always operate as actual islands, like the ones in the island biogeography theory. Environmental contrast between patches and the matrix may not be as evident as the contrast between islands and the surrounding ocean, resulting in the so-called "edge effect," which refers to the interactions between patches and the adjacent matrix. In many landscapes an active exchange of energy, matter, and information takes place among patches and the matrix, especially when environmental conditions do not differ greatly between them; in this case, animal species can move in and out of patches, colonizing them or carrying other species (i.e., microorganisms, plant seeds, parasites). The edge effect has a complex assembly of benefits and costs on diversity, given that many species living inside the matrix obtain advantages from the edges, and edge area would theoretically increase beta-diversity (spatial heterogeneity). However, in most of the cases, a few opportunists constitute matrix biota that lower diversity numbers when expanding their populations. Furthermore, edges facilitate the introduction of exotic and invasive species (predators, competitors, or parasites) into patches, which may endanger local populations.

Edges can modify the abiotic conditions, distribution and abundance of species, and species interactions in fragmented landscapes. As continuous habitat becomes fragmented, the remnants are increasingly exposed to the interaction with the surrounding matrix, producing altered reservoir contact rates between ecosystems and transforming disease dynamics (Loye and Carroll 1995; Murcia 1995; Ries et al. 2004; Siitonen et al. 2005).

Current landscape-management practices are trying to reduce the amount of edge perimeter around patches. This perimeter will be a consequence of patch size and shape: geometrically smaller forms have a higher perimeter versus area ratio, tending to have more perimeter linear units for each area unit than larger forms. From a landscape point of view this means that in a large patch a large proportion of area units (e.g., hectares) are inside the patch and a low proportion are under the edge effects. In addition, any geometric form that moves away from a circular shape will also show a higher perimeter versus area ratio; thus, it is expected that the circular shapes of patches would be a better way to reduce the edge effect (Fig. 10.2).

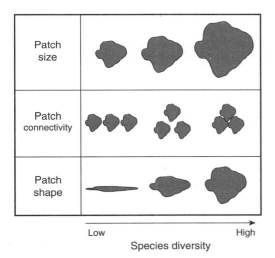

**Figure 10.2:**
Relations of species diversity to patch size, connectivity. and shape according to the theory of island biogeography (McArthur and Wilson 1967) and considering the edge effect in terrestrial landscapes. Notice that species diversity is a function of patch structure. When the drawings on Figures 10.1 and 10.2 are examined from right to left, it symbolizes the effect of habitat loss and fragmentation, with the consequential decrease in diversity or genetic variability numbers.

Several micro- and macroparasites have been associated with the increase of edges, including botflies (Wolf and Batzli 2001); leaf fungus (Benitez-Malvido and Lemus-Albor 2005); gastrointestinal parasites (Chapman et al. 2006); mistletoe, *Psittacantus schiedeanus* (de Buen et al. 2002); and hantavirus (Suzán et al. 2008). Carrasco-Hernández et al. (2009) found that the total amount of edge area surrounding human settlements (counties) was one of the most important landscape variables for modeling the risk of paralytic rabies in cattle.

On the other hand, matrix composition influences the dispersal and abundance of organisms and represents an important topic in disease ecology. Matrix composition may support competent reservoirs to dominate matrix communities and allow persistence and transmission for infectious diseases. For example, at local spatial scales in Costa Rica and Panama, hantavirus reservoirs' distribution and abundance depend on the matrix composition (Daily et al. 2003; Armien et al. 2009). This may suggest that matrix, rather than habitat patches, affects disease transmission, and it should be considered a research priority in disease ecology.

## CURRENT RESEARCH ON HABITAT FRAGMENTATION AND DISEASE

To assess the magnitude of habitat loss and fragmentation on disease ecology, we reviewed published empirical studies that examine this relationship. We performed a literature search using the Science Citation Index (January 2011) to identify published empirical studies reporting pathogen presence or prevalence in fragmented habitats, using the following search string: [(landscape* OR forest* OR habitat*) AND (fragmented OR fragmentation OR fragments) AND (disease* OR antibod* OR patho* OR parasit* OR virus* OR protoz* OR helminth* OR infect* OR zoono* OR infest* OR serolog* OR viral OR epid* OR epizoo*)]. We included all papers in which pathogen presence or prevalence was reported specifically in the context of habitat fragmentation. We considered macro- and microparasites but excluded hemoparasites, structural and nest parasites, and parasitoids; also, topic reviews were excluded. A total of 953 papers were found, but only 33 matched our search criteria. These findings suggest that the consequences of fragmentation on disease dynamics are poorly understood. Several papers suggested that habitat fragmentation is an important factor for their findings; however, no specific data on fragment size, isolation, landscape structure, or disease prevalence were reported.

The results of the literature search identified that the majority of studies focused on specific host–parasite relationships; however, in all cases in which the prevalence of more than one parasite was reported, the different parasite species were closely related. The studies included in this survey varied greatly in habitat type, matrix composition, fragment size, duration of the study, and landscape history. The studies revealed pronounced differences and contradictory responses of parasite prevalence to habitat fragmentation, connectivity, and isolation (Table 10.1).

Some empirical studies demonstrated that infectious or parasitic disease increased in small and isolated patches—for example *Escherichia coli* infection (Goldberg et al. 2008); hemoparasites (Cottontail et al. 2009); pine beetle, *Dendroctonus ponderosae* (Coops et al. 2010); and *Trypanosoma* infection (Vaz et al. 2007). In contrast, other studies demonstrated a decrease of infectious or parasitic disease in small and isolated patches—for example, with nematodes (Vandergast and Roderick 2003); ticks (Wilder and Meinkle 2004); fungal infections in plants (Lienert and Fischer 2003; Colling and Matthies 2004); and Acari infestation (Rubio and Simonetti 2009).

The specific nature of different host–parasite systems and the ecological context in each study are partly responsible for this lack of consistency. General patterns are also obscured by difficulties in comparing empirical studies due to differences in definitions, scales, and study designs. To properly recognize the effect of habitat fragmentation on infection dynamics, future studies should carefully define habitat fragmentation, control for the effects of habitat loss and spatial configuration, and determine the appropriate spatial and temporal scales for the host–pathogen system under study. Under habitat loss and fragmentation, synergistic processes are occurring that make isolated populations prone to both non-infectious (genetic isolation and inbreeding) and infectious diseases. We found few examples in which the effects of habitat fragmentation *per se* are clearly discernible, and further research is needed to understand the relationship between habitat fragmentation and infectious disease dynamics.

Definitions of habitat fragmentation and how to measure it abound in the ecological literature (Fahrig 2003; Ferraz 2007). Although all the studies included in this survey mention habitat fragmentation, their definitions and scale differ widely and range from small experimental plots to large natural forest remnants. Some studies focused on fragment size (Wilder and Meikle 2004) or distance from the edge (Schlaepfer and Gavin 2001), while others include attributes such as landscape composition and configuration (Langlois et al. 2001; Page et al. 2001) in their definitions. Therefore, differing ways of describing habitat fragmentation make meaningful comparisons difficult. The studies included in this survey varied greatly in habitat type, matrix composition, fragment size, duration of the study, and landscape history. In general, empirical studies are clustered around three general topics: (1) changes in community structure and composition in fragments; (2) influence of edge effects; and (3) pathogen ecology in small and isolated populations.

## ABIOTIC CHANGES IN FRAGMENTED LANDSCAPES AND INFECTIOUS DISEASES

Habitat fragmentation alters the microclimatic conditions inside patches (Saunders et al. 1991; Murcia 1995), as habitat fragments may experience higher solar radiation and lower moisture near their margins (Murcia 1995). The effects of such changes may be beneficial or detrimental to different pathogens. For example, according to Roland and Kaupp (1995), increased light incidence may inactivate the polyhedrosis virus of tent caterpillars (Family Lasiocampidae), leading to reduced virus transmission along habitat edges; also, parasitic helminths of hares (Hulbert and Boag 2001) and ectoparasites of *Norops* lizards (Schlaepfer and Gavin 2001) may prefer the cooler, moister, and more shaded areas of forest interiors. Vandergast and Roderick (2003) suggest that wet soils are more suitable for mermithid nematode persistence. In contrast, increased light penetration may be responsible for the higher prevalence of botflies near fragment edges (Wolf and Batzli 2001). Finally, abiotic conditions in and around remnants may lead to

**Table 10.1   Published Reports of Pathogen Presence or Prevalence in Fragmented Habitats**

| Pathogen–Host | Primary Findings | Fragment Sizes | Mechanisms | Reference |
|---|---|---|---|---|
| *Ixodes scapularis* | Higher Lyme disease risk in small forest fragments | 0.7–7.6 ha | Changes in community structure and/or distribution | Allan et al. 2003 |
| Leaf fungus, three plant species | Contrasting effects on different host species | 1–100 ha, continuous forest | Changes in connectivity and changes in community structure | Benitez-Malvido et al. 1999 |
| Leaf fungus: woody seedling species | Pathogen damage was three times greater in edge plots (1.85%) than in interior plots (0.57%). | — | Changes in community structure | Benitez-Malvido and Lemus-Albor 2005 |
| *Borrelia burgdorferi*, *Ixodes scapularis* | Higher entomologic risk in fragmented habitats | 3.1–76.1 ha | Changes in community structure and/or distribution | Brownstein et al. 2005 |
| *Mycrobotryum violaceum*, *Lychnis alpina* | Lower incidence but higher prevalence in isolated populations | ~48 to ~520 m$^2$ | Changes in genetic diversity and connectivity | Carlsson-Graner and Thrall 2002 |
| Gastrointestinal parasites: *Piliocolobus tephrosceles*, *Colobus guereza* | The parasite communities from both the edge and forest interior were similar in terms of the species. Percentage of infected individuals with more than one species of parasite was greater in edge groups (29.6%) than in forest interior groups (13.8%). | 3.7–4.1 ha | Abiotic changes, landscape context and habitat suitability | Chapman et al. 2006 |
| *Ustilago scorzonerae*, *Scorzonera humilis* | Lower prevalence in small and isolated populations | NA | Changes in genetic diversity and stochastic changes | Colling and Matthies 2004 |
| Pine beetle (*Dendroctonus ponderosae*)—trees | Parasite prevalence increased in forest fragments. | — | Landscape context | Coops et al. 2010 |
| Hemoparasites: *Artibeus jamaicensis* | Prevalence was significantly higher in bats from forest fragments than in bats captured in continuous forest. | 2.5–50 ha | Changes in community structure and/or distribution. Abiotic changes. | Cottontail et al. 2009 |

*Continued*

**Table 10.1  Continued**

| Pathogen–Host | Primary Findings | Fragment Sizes | Mechanisms | Reference |
|---|---|---|---|---|
| *Psittacanthus schiedeanus, Liquidambar styraciflua* | Fragments with more edge length had more infected *L. styraciflua* trees than those with less edge length. | <5 ha | Changes in community structure and/or distribution. Abiotic changes. | de Buen et al. 2002 |
| *Trypanosoma cruzy, Glossina morsitans morsitans* | Heavily fragmented areas have lower numbers of tsetse flies, but when the fragmentation of natural vegetation decreases, the number of tsetse flies increases. | — | Changes in community structure and/or distribution | Ducheyne et al. 2009 |
| Crimean-Congo hemorrhagic fever virus— *Hyalomma m. marginatum* | Highest prevalence occurs at areas of high habitat suitability and high habitat fragmentation. Areas with a low suitability for *Hyalomma* ticks or with low fragmentation showed zero risk rates. | — | Landscape context and habitat suitability | Estrada-Peña, Zatansever et al. 2007 |
| Parasites: *Piliocolobus tephrosceles* | Infection prevalence and the magnitude of multiple infections were greater in fragmented than in unfragmented forest. | 0.8–130 ha | Changes in community structure and/or distribution | Gillespie and Chapman 2006 |
| *Escherichia coli: Procolobus rufomitratus, Colobus guereza, Cercopithecus ascanius* | Forest fragmentation increases bacterial transmission between primates and humans. | 1.2–8.7 ha | Landscape context and habitat suitability | Goldberg, Gillespie et al. 2008 |
| *Epichloe bromicola, Bromus erectus* | Higher number of diseased plants in fragments | 0.25–20.25 m² | Abiotic changes | Groppe et al. 2001 |
| Intestinal helminths, *Lepus timidus* | Lower parasite loads in natural habitats | NA | Abiotic changes | Hulbert and Boag 2001 |
| *Podosphaera plantaginis, Plantago lanceolata* | The rate of spread is low in the highly fragmented landscape. | <1 ha | Landscape context and stochastic changes | Laine and Hanski 2006 |

**Table 10.1  Continued**

| Pathogen–Host | Primary Findings | Fragment Sizes | Mechanisms | Reference |
|---|---|---|---|---|
| Hantavirus, *Peromyscus maniculatus* | Higher viral prevalence in fragmented habitats | NA | Changes in connectivity and changes in community structure | Langlois et al. 2001 |
| *Urocystis primulicola, Primula farinosa* | Lower prevalence in small and isolated populations | 40–46,300 m² | Changes in genetic diversity and stochastic changes | Lienert and Fischer 2003 |
| *Baylisascaris procyonis, P. maniculatus* | Higher parasite prevalence in fragmented habitats | 0.6–1,500 ha | Changes in community structure and landscape context | Page et al. 2001 |
| Parasites— rodents and marsupials | Low or no correlation between parasite burden and fragmentation was detected, suggesting little effect of fragmentation on population health. | 14–30 ha | Changes in community structure and/or distribution | Puttker et al. 2008 |
| *Plasmodium* sp., birds | Higher prevalence in continuous forests | <30 ha | Changes in community structure and/or distribution | Ribeiro et al. 2005 |
| Polyhedrosis virus, *Malacosoma disstria* | Lower viral prevalence in fragment edges | — | Changes in genetic diversity and abiotic changes | Roland and Kraupp 1995 |
| *Eutrombicula alfreddugesi* larvae, *Liolaemus tenuis* | Lower intensity of infestation at forest fragment edges compared with either large forest interiors or forest edges | 1.5–20 ha | Abiotic changes | Rubio and Simonetti 2009 |
| *Giardia* sp. and *Cryptosporidium* sp.—primates | The presence of *Cryptosporidium* and *Giardia* species in primates living in forest fragments, but not in primates in undisturbed forest | 1.2–8.7 ha | Landscape context | Salzer et al., 2007 |
| Nematodes, *Microcebus murinus* | Differences in parasite load are unrelated to fragment size. | 42–188 ha | Landscape context and habitat suitability | Schad et al. 2005 |
| Ectoparasites, *Norops* lizards | Lower parasite prevalence in edges | <1 to <25 ha | Abiotic changes | Schlaepfer and Gavin 2001 |

*Continued*

**Table 10.1  Continued**

| Pathogen–Host | Primary Findings | Fragment Sizes | Mechanisms | Reference |
|---|---|---|---|---|
| Hantavirus—small mammals | Two hantavirus reservoirs were more commonly found in both disturbed habitats and edge habitats than in forest. | <5 to >5 ha | Changes in community structure. Landscape context. | Suzán et al. 2008 |
| Midgut gregarines, *Calopteryx maculata* | Higher prevalence in continuous forests | NA | Changes in connectivity or movement rates | Taylor and Merriam 1996 |
| Helminth parasites: *Alouatta palliata mexicana* and *A. pigra* | Populations in conserved/protected forests and in fragmented landscapes had similar parasitic specific richness, but some parasites occurred only in fragmented habitat. Parasitic prevalence in populations of both howler species was higher in the fragmented habitat than in continuous and/or protected forests. | — | Changes in connectivity and changes in community structure | Trejo-Macias et al. 2007 |
| *Trypanoxyuris minutus*—primates | Lower prevalence associated with larger fragments | 6.57–57.18 ha | Landscape context | Valdespino et al. 2010 |
| Mermithid nematodes, *Tetragnatha* sp. | Parasite prevalence decreases with decreasing fragment size. | 0.6–43.6 ha | Changes in community structure and/or distribution | Vandergast and Roderick 2003 |
| *Trypanosoma cruzi*—wild mammals | The fragmented habitat showed the higher seroprevalence. | <10 to >40 ha | Changes in community structure and/or distribution. Stochastic changes. | Vaz et al. 2007 |
| *I. scapularis, P. maniculatus* | Parasite prevalence decreases with decreasing fragment size. | 2–100 ha | Changes in community structure and/or distribution | Wilder and Meinkle 2004 |
| *Cuterebra fontinella, P. maniculatus* | Higher parasite prevalence in edges | 6–610 ha | Abiotic changes | Wolf and Batzli 2001 |

Science Citation Index, January 2011.

increased disease expression in infected plants or increased fungal transmission rates (Groppe et al. 2001).

## ISOLATION, GENETIC DIVERSITY LOSS, AND DISEASE ECOLOGY

Small and isolated populations are more susceptible to demographic and genetic stochasticity, which results in lower genetic diversity and higher extinction probability of local populations. Lienert and Fischer (2003) suggested that small and isolated populations may be more susceptible to smut fungi due to higher levels of inbreeding. In vertebrate communities, disease epizootics are rare unless they result from introduced pathogens against which the native fauna have no resistance or have suffered loss of genetic diversity, or where one dominant species becomes susceptible to a pathogen (Acevedo-Whitehouse et al. 2003). Often introduced pathogens have ravaged wild populations where one species dominates. The black-footed ferret (*Mustela nigripes*) is a highly endangered North American mammal whose population decline was attributed to habitat loss and land conversion during the 20th century (IUCN 1988). The last wild population (40 adults) experienced epizootics of canine distemper in 1985 and a few years later of sylvatic plague (*Yersinia pestis*), leading to a decline in genetic variability (Leberg 1992) and lowering individual fitness and adaptability (Frankel and Soulé 1981). On the other hand, there is a possibility of reducing the risk of pathogen spread from infected subpopulations into small and isolated patches that do not reach density thresholds for pathogen persistence. Maintaining genetic diversity in natural populations is increasingly becoming an important issue when dealing with diseases, since it is often reported that those populations with reduced genetic variability exhibit increased susceptibility to infectious diseases (Acevedo-Whitehouse et al. 2003).

## SPATIAL AND TEMPORAL SCALES IN LANDSCAPE ECOLOGY

The identification of significant patterns in landscape ecology may depend on the scale of observation. For example, it has been demonstrated that the number, sizes, and shapes of patches in a landscape depend on the spatial scale of the map (Gardner 1987). The problem of scale in landscape ecology can be intuitively understood if we think that scales are comparable to sample size in any statistical survey. On one hand, the information obtained from small samples will not allow us to see the whole picture, given that important variables may stay out of our survey if the sample is very small. On the other hand, very large scales may fail to recognize important information contained in small areas that go unnoticed in comparison with the entire region; as a result, observed patterns and their significance are a function of scale. To assess which spatial scale describes better a landscape pattern or process, a researcher may compare the resulting patterns after sampling at different scales. For example, at coarse regional scales, anthropogenic habitat fragmentation has been associated with increased abundance of hantavirus hosts; however, at finer scales other landscape attributes favor reservoir abundance, including slope and crop type (Suzán 2005). The same considerations should be applied to temporal scales. For example, many studies record observations during a single year or season but make inferences that extend beyond the time scale used to perform analyses. For example, most studies do not consider the amount of time that has occurred since habitats became fragmented. Conducting studies on fragmented landscapes of different ages may produce inconsistent patterns of disease transmission, making comparisons between landscapes with widely different fragmentation histories unwarranted. The effects of habitat fragmentation should be understood as dynamic; species composition and abundance patterns, along with pathogen prevalence, will change over time, and it may take several decades for stabilization to take place (Gascon et al. 2000; Laurance et al. 2001).

In some cases the effects of habitat fragmentation may also depend on the scale of biological aggregation at which one performs the observation (Andren 1994). Future studies must therefore carefully choose the appropriate spatial, temporal, and biological scale for the host–pathogen system under consideration. Explaining landscape patterns by analyzing only one hypothetically ideal scale may not be sufficient, since underlying mechanisms may operate at different scales than those on which the patterns are observed: "in some cases, the patterns must be understood as emerging from the collective behaviors of large ensembles of smaller scale units. In other cases, the pattern is imposed by larger scale constraints" (Levin 1992). This means that both bottom-up and top-down forces build landscape patterns. Designing any study in multiple scales allows an analysis of these complex interactions of phenomena observed at different scales of space, time, and biological organization. The use of integrated approaches such as the geographic analysis coupled with information on spatial dynamics and community ecology will become indispensable tools in understanding patterns of disease emergence, especially in those infectious diseases in which ecology depends on the community dynamics of host species, like rabies, hantavirus pulmonary syndrome, and Lyme disease. Planning future studies at multiple scales will allow both the finding of an ideal scale of study and also an analysis of complex interactions of events occurring at different scales in space, time, and biological organization.

## FROM THEORY TO PRACTICE

Theoretical studies are for the most part analytical, whereby a set of differential equations is used to describe and predict the dynamics of a population under different conditions. Specific characteristics of a system that are simulation-based—such as individual traits, complex migration patterns, species assemblages, landscape configuration—are programmed into a computer and then run to observe their impact. The use of theoretical models can help increase our understanding of phenomena and even suggest interesting research lines. Particular computer simulations now allow us to represent individuals in considerable detail so that we can study how individual traits lead to population, metapopulation, and even community dynamics (Reynolds 1987; Mitchell 2009). Empirical studies are developed on several landscape and community ecology approaches, including island biogeography, matrix composition, and edge effect concepts. Empirical studies are focused on changes in species assemblages and changes on reservoirs, and infectious agent distributions and abundances resulting from landscape changes. Empirical studies have shown the relationship of habitat loss and fragmentation to disease dynamics with contrasting results.

Our literature survey shows that several studies are focused on single parasites or infectious agents and few studies consider the patch context (i.e., the

characteristics of the matrix surrounding habitat remnants; Fahrig 2001); this may be a critical issue in understanding the generality of the observations (Benitez-Malvido and Lemus-Albor 2005). We suggest that such an approach will be better poised to predict the magnitude, direction, spatial context, and timing of the potential health consequences of habitat fragmentation. Furthermore, combining empirical studies with theoretical modeling will lead to a deeper understanding of the problem and the consequences of proposed actions. While there are examples in which habitat fragmentation is associated with higher disease risks, the opposite can also be true.

It is not our intention to imply that the consequences of habitat fragmentation on disease transmission are inherently unpredictable, but rather that the immediate anthropogenic drivers of changes in pathogen transmission processes must be carefully evaluated. Habitat fragmentation, like other forms of anthropogenic disturbance, has the potential to have a negative impact on species, habitats, and the ecosystem services on which human health ultimately depends. The development of alternative theoretical models and empirical studies at multiple spatial and temporal scales will help to understand patterns and processes associated with this relationship and will help conservation biologists, public health professionals, modelers, veterinary epidemiologists, and policymakers to prevent infectious disease outbreaks in human, domestic animal, and wildlife populations and simultaneously protect ecological health.

## ACKNOWLEDGMENTS

We thank A. Gómez, T. Giermakoski, E. Marcé, P. Martínez, A. Vigueras, L. de la Rosa, H. Zarza, and R. Ávila, who performed literature searches and reviewed earlier versions of this manuscript. We are grateful to H. McCallum and R. S. Ostfeld for their thoughtful comments on this chapter, which considerably improved our perspectives on fragmentation and disease ecology.

## REFERENCES

Acevedo-Whitehouse, K., F. Gulland, D. Greig, and W. Amos. 2003. Disease susceptibility in California sea lions. Nature 422:35–35.

Allan, B.F., F. Keesing, and R.S. Ostfeld. 2003. Effect of forest fragmentation on Lyme disease risk. Conserv Biol 17:267–272.

Anderson R.M., and R.M. May. 1979. Population biology of infectious diseases. Nature 280:361–367.

Andren, H. 1994. Effects of habitat fragmentation on birds and mammals in landscapes with different proportions of suitable habitat—a review. Oikos 71:355–366.

Armien, A.G., B. Armien, F. Koster, J.M. Pascale, M. Avila, P. Gonzalez, M. de la Cruz, Y. Zaldivar, Y. Mendoza, F. Gracia, B. Hjelle, S. J. Lee, T.L. Yates, and J. Salazar-Bravo. 2009. Hantavirus infection and habitat associations among rodent populations in agroecosystems of Panama: implications for human disease risk. Am J Trop Med Hyg 81:59–66.

Benitez-Malvido, J., G. Garcia-Guzman, and I.D. Kossmann-Ferraz. 1999. Leaf-fungal incidence and herbivory on tree seedlings in tropical rainforest fragments: an experimental study. Biol Conserv 91:143–150.

Benitez-Malvido, J., and A. Lemus-Albor. 2005. The seedling community of tropical rain forest edges and its interaction with herbivores and pathogens. Biotropica 37:301–313.

Brooker, L., and M. Brooker. 1994. Dispersal and population dynamics of the blue-breasted fairywren *Malurus pulcherrimus* in fragmented habitat in the western Australian wheatbelt. Wildl Res 29:225–233.

Brownstein, J.S., D.K. Skelly, T.R. Holford, and D. Fish. 2005. Forest fragmentation predicts local scale heterogeneity of Lyme disease risk. Oecologia 146:469–475.

Carlsson-Graner, U., and P.H. Thrall. 2002. The spatial distribution of plant populations, disease dynamics and evolution of resistance. Oikos 97:97–110.

Carrasco-Hernández, R., M.D. Manzano-Martínez, C. Bautista, A. deVega-García, A. Flisser, R.A. Medellín, and G. Suzán. 2009. Ecogeographic model of bovine paralytic rabies risk in Puebla, México. XX Conf RITA. Quebec, Canada.

Chapman, C.A., M.L. Speirs, T.R. Gillespie, T. Holland, and K. Austad. 2006. Life on the edge: gastrointestinal parasites from the forest edge and interior primate groups. Am J Primatol 68:397–409.

Colling, G., and D. Matthies. 2004. The effects of plant population size on the interactions between the endangered plant *Scorzonera humilis*, a specialised herbivore, and a phytopathogenic fungus. Oikos 105:71–78.

Coops, N.C., S.N. Gillanders, M.A. Wulder, S.E. Gergel, T. Nelson, and N.R. Goodwin. 2010. Assessing changes in forest fragmentation following infestation using time series Landsat imagery. Forest Ecol Manag 259: 2355–2365.

Cottontail, V.M., N. Wellinghausen, and E.K.V. Kalko. 2009. Habitat fragmentation and haemoparasites in the common fruit bat, *Artibeus jamaicensis* (Phyllostomidae) in a tropical lowland forest in Panama. Parasitology 136:1133–1145.

Daily, G.C., G. Ceballos, J. Pacheco, G. Suzán, and A. Sanchez-Azofeifa. 2003. Countryside biogeography of neotropical mammals: conservation opportunities in agricultural landscapes of Costa Rica. Conserv Biol 17:1814–1826.

Daszak, P., A.A. Cunningham, and A.D. Hyatt. 2000. Emerging infectious diseases of wildlife -threats to biodiversity and human health. Science 287:443–449.

Daszak, P., G.M. Tabor, A.M. Kilpatrick, J. Epstein, and R. Plowright. 2004. Conservation medicine and a new agenda for emerging diseases. Ann NY Acad Sci 1026:1–11.

de Buen, L.L., Ornelas, J.F., and J.G. Garcia-Franco. 2002. Mistletoe infection of trees located at fragmented forest edges in the cloud forests of central Veracruz, Mexico. Forest Ecol Manag 164:293–302 .

Diffendorfer J.E., M.S. Gaines, and R.D. Holt. 1995. Habitat fragmentation and movements of three small mammals (*Sigmodon, Microtus*, and *Peromyscus*). Ecology 76:827–839.

Dodd, C.K., and L.L. Smith. 2003. Habitat destruction and alteration: historical trends and future prospects for amphibians. *In* R.D. Semlitsch, ed. Amphibian conservation, pp. 94–112. Smithsonian Institution, Washington D.C.

Dobson, A. 2004. Population dynamics of pathogens with multiple host species. Am Nat 164:S64–S78.

Ducheyne, E., C. Mweempwa, C. De Pus, H. Vernieuwe, R. De Deken, G. Hendrickx, and P. Van den Bossche. 2009. The impact of habitat fragmentation on tsetse abundance on the plateau of eastern Zambia. Prev Vet Med 91:11–18.

Estrada-Peña, A., Z. Zatansever, A. Gargili, M. Aktas, R. Uzun, O. Ergonul, and F. Jongejan. 2007. Modeling the spatial distribution of Crimean-Congo hemorrhagic fever outbreaks in Turkey. Vector-Borne Zoonot 7:667–678.

Fahrig, L. 2001. How much habitat is enough? Biol Conserv 100:65–74.

Fahrig, L. 2003. Effects of habitat fragmentation on biodiversity. Annu Rev Ecol Evol S 34:487–515.

Ferraz, G., J.D. Nichols, J.E. Hines, P.C. Stouffer, R.O. Bierregaard Jr., and T. Lovejoy. 2007. A large scale deforestation experiment: effects of patch area and isolation on Amazon birds. Science 315: 238–241.

Frankel, O., and M. Soulé. 1981. Conservation and evolution. Cambridge University Press, Cambridge, England.

Gardner, R.H., B.T. Milne, M.G. Turner, and R.V. O'Neill. 1987. Models for the analysis of broad-scale landscape pattern. Landscape Ecol 1:19–28.

Gascon, C., G.B. Williamson, and G.A.B. da Fonseca. 2000. Receding forest edges and vanishing reserves. Science 288:1356–1358.

Gillespie, T.R., and C.A. Chapman. 2006. Prediction of parasite infection dynamics in primate metapopulations based on attributes of forest fragmentation. Conserv Biol 20:441–448

Goldberg, T.L., T.R. Gillespie, and I.B. Rwego. 2008. Forest fragmentation and bacterial transmission among nonhuman primates, humans and livestock, Uganda. Emerg Infect Dis 14:1375–1382.

Groppe, K., T. Steinger, B. Schmid, B. Baur, and T. Boller. 2001. Effects of habitat fragmentation on choke disease (*Epichloe bromicola*) in the grass *Bromus erectus*. J Ecol 89:247–255.

Hanski, I., and O. Ovaskainen. 2003. Metapopulation theory for fragmented landscapes. Theor Popul Biol 64:119–127.

Hess, G. 1996. Disease in metapopulation models: Implications for conservation. Ecology 77:1617–1632.

Hess, G.R. 1994. Conservation corridors and contagious disease—a cautionary note. Conserv Biol 8:256–262.

Hulbert, I.A.R., and B. Boag. 2001. The potential role of habitat on intestinal helminths of mountain hares, *Lepus timidus*. J Helminthol 75:345–349.

International Union for the Conservation of Nature. 1988. IUCN red list of threatened mammals. Conservation Monitoring Centre, Cambridge, UK.

Keesing, F., R.D. Holt, and R.S. Ostfeld. 2006. Effects of species diversity on disease risk. Ecol Lett 9:485–498.

Laine, A.L., and I. Hanski. 2006. Large-scale spatial dynamics of a specialist plant pathogen in a fragmented landscape. J Ecol 94:217–226.

Langlois, J.P., L. Fahrig, G. Merriam, and H. Artsob. 2001. Landscape structure influences continental distribution of hantavirus in deer mice. Landscape Ecol 16: 255–266.

Laurance, W.F., D. Perez-Salicrup, P. Delamonica, P.M. Fearnside, S. D'Angelo, A. Jerozolinski, L. Pohl, and T.E. Lovejoy. 2001. Rain forest fragmentation and the structure of Amazonian liana communities. Ecology 82:105–116.

Leberg, L.P. 1992. Effects of population bottlenecks on genetic diversity as measured by allozyme electrophoresis. Evolution 46:477–494.

Levin, S.A. 1992. The problem of pattern and scale in ecology. Ecology 73:1943–1967.

Lienert, J., and M. Fischer. 2003. Habitat fragmentation affects the common wetland specialist *Primula farinosa* in northeast Switzerland. J Ecol 91:587–599.

LoGiudice, K., R.S. Ostfeld, K.A. Schmidt, and F. Keesing. 2003. The ecology of infectious disease: effects of host diversity and community composition on Lyme disease risk. Proc Nat Acad Sci USA 100:567–571.

LoGiudice, K., S.T.K. Duerr, M.J. Newhouse, K.A. Schmidt, M.E. Killilea, and R.S. Ostfeld. 2008. Impact of host community composition on Lyme disease risk. Ecology 89:2841–2849.

Loye, J., and S. Carroll. 1995. Birds, bugs and blood—avian parasitism and conservation. Trends Ecol Evol 10:232–235.

MacArthur, R.H., and E.O. Wilson. 1967. The theory of island biogeography. Princeton University Press. Princeton, New Jersey.

McCallum, H., and A. Dobson. 2002. Disease, habitat fragmentation and conservation. P Roy Soc Lond B 269:2041–2049.

McCallum, H. 2008. Landscape structure, disturbance and disease dynamics. In R. S. Ostfeld, F. Keesing, and V. T. Eviner, eds. Infectious disease ecology, pp. 100–122. Princeton University Press, New Jersey.

Mitchell, C.E., D. Tilman, and J.V. Groth. 2002. Effects of grassland plant species diversity, abundance, and composition on foliar fungal disease. Ecology 83:1713–1726.

Mitchell, M. 2009. Complexity: a guided tour. Oxford University Press, New York.

Murcia, C. 1995. Edge effects in fragmented forests— implications for conservation. Trends Ecol Evol 10:58–62.

Murrell, D.J., and R. Law. 2000. Beetles in fragmented woodlands: a formal framework for dynamics of movement in ecological landscape. J Anim Ecol 69:471–483.

Nupp, T.E., and R.K. Swihart. 1998. Effects of forest fragmentation on population attributes of white-footed mice and eastern chipmunks. J Mammal 79:1234–1243

Ostfeld, R., and F. Keesing. 2000. The function of biodiversity in the ecology of vector-borne zoonotic diseases. Can J Zool 78:2061–2078

Page, L.K., R.K. Swihart, and K.R. Kazacos. 2001. Changes in transmission of *Baylisascaris procyonis* to intermediate hosts as a function of spatial scale. Oikos 93:213–220.

Parker, D.C., and V. Meretsky. 2004. Measuring pattern outcomes in an agent-based model of edge-effect externalities using spatial metrics. Agr Ecosyst Environ 101: 233–250.

Patz, J.A., and N.D. Wolfe. 2002. Global ecological change and human health. In A.A. Aguirre, R.S. Ostfeld, G.M. Tabor, C. House, and M.C. Pearl, eds. Conservation medicine: ecological health in practice, pp. 167–181. Oxford University Press, New York.

Patz, J.A., P. Daszak, G.M. Tabor, A.A. Aguirre, M. Pearl, J. Epstein, N.D. Wolfe, A.M. Kilpatrick, J. Foufopoulos, D. Molyneux, D.J. Bradley, and Working Group Land Use Change. 2004. Unhealthy landscapes: Policy recommendations on land use change and infectious disease emergence. Environ Health Persp 112: 1092–1098.

Peixoto, I.D., and G. Abramson. 2006. The effect of biodiversity on the hantavirus epizootic. Ecology 87:873–879.

Puttker, T., Y. Meyer-Lucht, and S. Sommer. 2008. Effects of fragmentation on parasite burden (nematodes) of generalist and specialist small mammal species in secondary forest fragments of the coastal Atlantic Forest, Brazil. Ecol Res 23:207–215.

Real, L.A., J.C. Henderson, R. Biek, J. Snaman, T.L. Jack, J.E. Childs, E. Stahl, L. Waller, R. Tinline, and S. Nadin-Davis. 2005. Unifying the spatial population dynamics and molecular evolution of epidemic rabies virus. Proc Nat Acad Sci USA 102: 12107–12111.

Reed, D.H. 2004. Extinction risk in fragmented habitats. Anim Conserv 7:181–191

Reynolds, C.W. 1987. Flocks, herds, and schools: A distributed behavioral model, in computer graphics. SIGGRAPH 87 Conf Proc 21:25–34.

Ribeiro, S.F., F. Sebaio, F.C.S. Branquinho, M.A. Marini, A.R. Vago, and E.M. Braga. 2005. Avian malaria in Brazilian passerine birds: parasitism detected by nested PCR using DNA from stained blood smears. Parasitology 130:261–267.

Ries, L., R.J. Fletcher, J. Battin, and T.D. Sisk. 2004. Ecological responses to habitat edges: mechanisms, models, and variability explained. Ann R Ecol Evol Syst 35:491–522.

Roland, J., and W.J. Kaupp. 1995. Reduced transmission of forest tent caterpillar (Lepidoptera, Lasiocampidae) nuclear polyhedrosis-virus at the forest edge. Environ Entomol 24:1175–1178.

Rubio, A.V., and J.A. Simonetti. 2009. Ectoparasitism by Eutrombicula alfreddugesi larvae (acari: trombiculidae) on Liolaemus tenuis lizard in a Chilean fragmented temperate forest. J Parasitol 95:244–245.

Salzer, J.S., I.B. Rwego, T.L. Goldberg, M.S. Kuhlenschmidt, and T.R. Gillespie. 2007. Giardia sp and Cryptosporidium sp infections in primates in fragmented and undisturbed forest in western Uganda. J Parasitol 93:439–440.

Saunders, D.A., R.J. Hobbs, and C.R. Margules. 1991. Biological consequences of ecosystem fragmentation—a review. Conserv Biol 5:18–32.

Schad, J., J.U. Ganzhorn, and S. Sommer. 2005. Parasite burden and constitution of major histocompatibility complex in the malagasy mouse lemur, Microcebus murinus. Evolution 59:439–450.

Schauber, E.M., and R.S. Ostfeld. 2002. Modeling the effects of reservoir competence decay and demographic turnover in Lyme disease ecology. Ecol Appl 12:1142–1162.

Schlaepfer, M.A., and T.A. Gavin. 2001. Edge effects on lizards and frogs in tropical forest fragments. Conserv Biol 15:1079–1090.

Schmidt, K.A., and R.S. Ostfeld. 2001. Biodiversity and the dilution effect in disease ecology. Ecology 82:609–619.

Sih, A., B.G. Jonsson, and G. Luikart. 2000. Habitat loss: ecological, evolutionary and genetic consequences. Trends Ecol Evol 15:132–134.

Siitonen, P., A. Lehtinen, and M. Siitonen. 2005. Effects of forest edges on the distribution, abundance, and regional persistence of wood-rotting fungi. Conserv Biol 19:250–260.

Suzán, G. 2005. The responses of hantavirus host communities to habitat fragmentation and biodiversity loss in Panama. Dissertation. Department of Biology. The University of New Mexico, Albuquerque, New Mexico, United States.

Suzán, G., E. Marcé, J. Tomasz Giermakowski, B. Armién, J. Pascale, J. Mills, G. Ceballos, A. Gómez, A. A. Aguirre, R. Ostfeld, J. Salazar-Bravo, A. Armién, R. Parmenter, and T. Yates. 2008. The effect of habitat fragmentation and species diversity loss on hantavirus prevalence in Panama. Ann NY Acad Sci 1149:80–83.

Suzán, G., E. Marcé, J. T. Giermakowski, J. N. Mills, G. Ceballos, R. S. Ostfeld, B. Armién, J. M. Pascale, and T. L. Yates. 2009. Experimental evidence for reduced rodent diversity causing increased Hantavirus prevalence. PLoS ONE 4(5):e5461.doi:10.1371/journal.pone.0005461

Valdespino C., G. Rico-Hernández, and S. Mandujano S. 2010. Gastrointestinal parasites of howler monkeys (Alouatta palliata) inhabiting the fragmented landscape of the Santa Marta Mountain Range, Veracruz, Mexico. Am J Primatol 72:539–548.

Taylor, P.D., and G. Merriam. 1996. Habitat fragmentation and parasitism of a forest damselfly. Landscape Ecol 11:181–189.

Tracey, J., S. Vandewoude, S., S. Bevins, and K. Crooks. 2008. Agent-based modeling, functional connectivity, and disease transmission for felids in fragmented landscapes. Intl Congr Conserv Biol, Chattanooga, Tennessee.

Trejo-Macias, G., A. Estrada, and M.A.M. Cabrera. 2007. Survey of helminth parasites in populations of Alouatta palliata mexicana and A. pigra in continuous and in fragmented habitat in southern Mexico. Int J Primatol 28:931–945.

Tobler, W.R. 1970. Computer movie simulating urban growth in Detroit region. Econ Geogr 46:234–240.

Torres-Perez, F., J. Navarrete-Droguett, R. Aldunate, T.L. Yates, G.J. Mertz, P.A. Vial, M. Ferres, P.A. Marquet, and R.E. Palma. 2004. Peridomestic small mammals associated with confirmed cases of human hantavirus disease in south central Chile. Am J Trop Med Hyg 70:305–309.

Utrera, A., G. Duno, B.A. Ellis, R.A. Salas, N. de Manzione, C.F. Fulhorst, R.B. Tesh, and J.N. Mills. 2000. Small mammals in agricultural areas of the western llanos of Venezuela: Community structure, habitat associations, and relative densities. J Mammal 81:536–548.

Vandergast, A.G., and G.K. Roderick. 2003. Mermithid parasitism of Hawaiian Tetragnatha spiders in a fragmented landscape. J Invert Pathol 84:128–136.

Vaz, V.C., P.S. D'Andrea, and A.M Jansen. 2007. Effects of habitat fragmentation on wild mammal infection by *Trypanosoma cruzi*. Parasitology 134:1785–1793

Wilcove, D.S., C.H. McLellan, and A.P. Dobson. 1986. Habitat fragmentation in the temperate zone. *In* M. Soulé ed. Conservation biology: science of scarcity and diversity. Sinauer Associates, Sunderland, Massachusetts.

Wilcove, D.S., D. Rothstein, J. Dubow, A. Phillips, and E. Losos. 1998. Quantifying threats to imperiled species in the United States. Bioscience 48:607–615.

Wilder, S.M., and D.B. Meikle. 2004. Prevalence of deer ticks (*Ixodes scapularis*) on white-footed mice (*Peromyscus leucopus*) in forest fragments. J Mammal 85:1015–1018.

Wolf, M., and G.O. Batzli. 2001. Increased prevalence of bot flies (*Cuterebra fontinella*) on white-footed mice (*Peromyscus leucopus*) near forest edges. Can J Zool 79:106–109.

Wolfe, N.D., W.M. Switzer, J.K. Carr, V.B. Bhullar, V. Shanmugam, U. Tamoufe, A.T. Prosser, J.N. Torimiro, A. Wright, E. Mpoudi-Ngole, F.E. McCutchan, D.L. Birx, T.M. Folks, D.S. Burke, and W. Heneine. 2004. Naturally acquired simian retrovirus infections in central African hunters. Lancet 363:932–937.

Yahnke, C.J., P.L. Meserve, T.G. Ksiazek, and J.M. Mills. 2001. Patterns of infection with Laguna Negra virus in wild populations of *Calomys laucha* in the central Paraguayan Chaco. Am J Trop Med Hyg 65:768–776.

Zanette, L. 2000. Fragment size and the demography of an area-sensitive songbird. J Anim Ecol 69:458–470.

# 11

## WILDLIFE TRADE AND THE SPREAD OF DISEASE

Katherine F. Smith, Lisa M. Schloegel, and Gail E. Rosen

Most readers of a book on conservation medicine would agree that humans are having increasingly profound effects on the planet and changing the environment in significant ways. They would also agree that emerging infectious diseases are on the rise and of growing concern to all nations. As conservation medicine attests, the two are not mutually exclusive. Indeed, evidence is mounting to support causal links between anthropogenic environmental change and disease emergence in humans, livestock, and wildlife. Of particular concern is the dramatic increase in emerging zoonotic diseases, those that originate in animals but spill over into the human population. The rise in emerging zoonoses implies increased contact between humans and animals, in all corners of the globe, for myriad reasons. Yet human–animal interactions have been a mainstay of cultures and societies for millennia, so what has changed to promote such a dramatic increase in zoonotic infectious diseases?

Two factors are at least partially responsible for the majority of emergent zoonoses over the past century: (1) increased movement of human populations into uninhabited regions, where they come into contact with a large and novel diversity of wildlife (i.e., following deforestation and increased bushmeat hunting in central African nations) and (2) increased trade and transport of wildlife over vast distances on a sizeable

scale that continues to escalate (i.e., for food, pet trade, zoos, and research). The commingling of human and wildlife populations, at both regional and global scales, creates numerous opportunities for pathogens to spread between host groups that may have had only minimal contact in the ancestral environment. In this chapter we focus on one of these factors—the global trade in wildlife (non-domesticated animals), and its role in disease introduction, spread, and emergence in susceptible populations, including humans, wildlife, and livestock.

Today, live animals are moved inter-continentally for a variety of personal, social, and economic gains. Tens of millions of live animals are traded annually for human consumption alone (Schloegel et al. 2009). The desire for "fresh" meat still predominates in many cultures and not only encourages this trade but ensures the continued existence of wet markets. Such markets provide breeding grounds for disease emergence as countless animals are brought into close proximity to each other and to humans. A 2000–2003 survey of food markets in southern China revealed the presence of 36 mammal species, 212 bird species, 84 reptile species, and 5 amphibian species (Lee et al. 2004). The severe acute respiratory syndrome (SARS) pandemic was traced back to the wet markets of southwest China, where the virus appears to have jumped from its reservoir host (bats) to masked palm civets (*Paguma larvata*) before infecting humans and eventually spreading to 51 countries (Guan et al. 2003; Li et al. 2005; Chapter 15 in this volume). Perhaps an even larger number of animals are traded annually as pets. It has been inferred that revenue from the U.S. exotic pet trade (legal and illegal) is second only to the country's illegal drug trade, as avid enthusiasts are willing to pay top dollar for rare and extraordinary animals (Ebrahim and Solomon 2006). Today, the intercontinental movement of live animals and their associated pathogens continues to grow, largely unstudied and typically unregulated.

In this chapter, we discuss the spread of diseases, both known and unknown, through the wildlife trade. We offer a history of disease emergence resulting from animal trade and transport (Box 11.1) and summarize recently compiled data on the scope and scale of present-day wildlife trade. We highlight the specific aspects of trade that facilitate pathogen introduction and emergence and use the best available scientific evidence and insights from disease ecology to consider risk. We illustrate how pathogens emerge through trade by highlighting three examples relevant to humans, native wildlife, and livestock

(Boxes 11.2–11.4), and identify areas for future research. Because of space limitations, we do not fully discuss the specific national and international systems in place to regulate wildlife trade (Table 11.1). Indeed, this topic would be a stand-alone chapter, and we refer readers to Defenders of Wildlife (2007) and Smith et al. (2009) for more detailed information on this topic. Likewise, we do not consider the illegal trade in wildlife but refer readers to two sources (Gomez and Aguirre 2008; Rosen and Smith 2010).

## PATHOGEN POLLUTION THROUGH WILDLIFE TRADE

Pathogen pollution is the human-mediated introduction of a pathogen to a new host species, population, or geographic region (Cunningham 2003). While the term was only recently coined, it is not a novel trend and appears to be an important driver of emerging infectious diseases, particularly zoonoses both historically and at present (Box 11.1). Today, in a century characterized by the rapid breakdown of old barriers between people and nature, pathogen pollution and

---

**Box 11.1  Pathogen Pollution Through Time**

Humans have a longstanding history of animal exploitation, domestication, and trafficking. The domestication of dogs and cattle has led to a global distribution in these animals. Interactions between domestic animals and native wildlife have facilitated spillover events and disease emergence. Once abundant throughout southern Africa, painted dog—also known as the African wild dog (*Lycaon pictus*)—numbers are now dwindling due to encroachment of habitat by humans and their domestic canine companions. This increased interaction with domestic dogs has decimated wild dog populations due to infection with rabies and canine distemper (van de Bildt et al. 2002). In 1887, the importation into Somalia of Indian cattle by the Italian army led to an outbreak of rinderpest that swept across the continent and led to mass mortalities in native ungulates, altering the trophic structure of the African savannas and causing starvation among the African peoples. The herding of cattle has been linked to numerous other disease-emergence events throughout history, including brucellosis, foot and mouth disease, and bovine tuberculosis (Seimenis 2008). Beginning in the 1500s, acclimation societies born in lands newly settled by Europeans began populating the landscape with the flora and fauna of their homelands. This practice led to the introduction of non-native birds and their pathogens to Hawaii, and the subsequent emergence of diseases, including avian pox and avian malaria, in the native bird populations. Perhaps the most significant emerging infectious disease so far described, HIV/AIDS, had its origins in wild-caught animals. Transmission from non-human primates to humans through the local bushmeat trade in Africa led to a viral emergence event that has now become a global human pandemic (Hahn et al. 2000).

---

**Box 11.2 Chytridiomycosis and Global Amphibian Declines**

In the late 1970s, researchers around the globe began to report mass mortality events and declines in anuran populations on an alarming scale. Habitat destruction, chemical pollution, UV-B radiation, and climate change all were suspected of contributing to the declines. Numerous authors, however, suggested that the pattern of declines in many regions fit that of a recently emerged epidemic (Laurance et al. 1996; Lips et al. 1998). It wasn't until 1998 that the culprit was discovered. *Batrachochytrium dendrobatidis* (*Bd*), a parasitic fungus, proliferates on the keratinized skin cells of amphibians, leading to clinical signs of the disease chytridiomycosis and often death.

What led to the seemingly simultaneous emergence of this deadly pathogen in such distant regions as Australia and Central and South America? The live animal trade is one theory gaining credibility. The North American bullfrog (*Lithobates catesbeiana*), for instance, is globally traded and farmed for the trade in frog's legs and has established feral populations outside its historical range spanning the European, American, and Asian continents. Captive-reared and introduced animals of this species are testing positive for *Bd* infection in all regions (Hanselmann et al. 2004; Garner et al. 2006; Schloegel et al. 2009, Schloegel et al. unpublished data 2011). The African clawed frog (*Xenopus laevis*), a widely used specimen for laboratory research, and the White's tree frog (*Litoria caerulea*), a popular pet trade species, can also carry *Bd* (Pessier et al. 1999; Weldon et al. 2004). Reintroduction efforts of the Mallorcan mid-wife toad (*Alytes muletensis*) appear to have led to the inadvertent introduction of the pathogen to Mallorca. While the origins of *Bd* are still under dispute, scientists are increasingly aware of the dangers associated with the live transport of *Bd*-infected anurans. A recent study found that *Bd* infections in farm-reared North American bullfrogs in Brazil are genetically similar (and even identical in some instances) to *Bd* infections found in native amphibians in Central and South America, suggesting a common source of infection and/or the occurrence of transmission between wild and captive populations (Schloegel et al. 2010b).

In light of recent studies, and the seriousness of the injuries suffered by amphibians due to disease, the World Organization for Animal Health made chytridiomycosis a notifiable disease in 2008, which ensures the reporting of the pathogen's occurrence in trade animals and implementation of guidelines and recommendations to minimize its spread through the trade (Schloegel et al. 2010a).

---

the continued emergence of infectious diseases are more important than ever. The introduction of non-native species around the world, both accidental and intentional, is a significant source of pathogen pollution. There are many pathways by which humans facilitate worldwide species invasions, including via ship ballast water, agricultural commerce, and international globalization and travel. But the global trade in wildlife, given its enormity alone, is arguably of greatest concern.

It is impossible to know the true magnitude of the global wildlife trade. The data simply are not available. Many nations do not track the import/export of wildlife shipments across their borders, and those that do make it difficult to obtain the records. Even if all nations did collect this information, in the absence of

a globalized database that tracked shipments, it would be an enormous undertaking to collate the data and quantify the trade. Nevertheless, attempts to estimate the trade have been made.

In 2001 the World Wildlife Fund estimated that more than 40,000 primates, 4 million birds, 640,000 reptiles and 350 million tropical fish are traded globally each year (WWF 2001). Four years later, Karesh and colleagues (2005) suggested that many billions of live animals and products are traded globally each year, generating commodities totaling in the hundred-billions of dollars. At first glance these numbers appear shockingly high. However, recent summaries of the wildlife trade of just one nation—the United States—indicate that these global estimates may be extremely low.

**Box 11.3  Monkeypox Virus: A Wake-Up Call for the United States**

The 2003 outbreak of human monkeypox virus in the United States illustrates the public health risks associated with live animal importation. Monkeypox is a zoonotic disease endemic to central and west Africa. African rodents are considered to be the natural host of the virus, which in humans causes rashes similar to smallpox, fever, chills, and headache (CDC 2003a). Human infections during the 2003 outbreak resulted from contact with pet prairie dogs that contracted monkeypox from infected African rodents imported for the commercial pet trade (DiGuilio and Eckburg 2004; Reed et al. 2004; Hutson et al. 2007). The shipment of mammals imported from Ghana contained more than six species and a total of 762 African rodents, some of which were confirmed to be infected with monkeypox (Hutson et al. 2007). The monkeypox outbreak resulted in 72 human cases, with 37 of those cases being laboratory-confirmed (CDC 2003a). Most patients had direct or close contact with the infected prairie dogs, including 28 children at a daycare center and veterinary clinic staff (Reynolds et al. 2007). In June 2003, CDC and the FDA issued a joint order prohibiting the interstate transportation, sale, or release into the environment of prairie dogs and the six implicated species of African rodents (CDC 2003b; FDA 2003). CDC also implemented an immediate embargo on the importation of all rodents (order Rodentia) from Africa. This emergency order was superseded in November 2003, when the two agencies issued an interim final rule creating two complementary regulations restricting both domestic trade and importation, intended to prevent the further introduction, establishment, and spread of the monkeypox virus in the United States This interim final rule successfully mitigated the risk of human exposure to monkeypox associated with importation of African rodents. However, nearly all government initiatives to regulate live animal imports have been reactive, focusing on detecting and preventing the spread of non-native species already established, or initiated as an urgent response to an emerging public health issue (Smith et al. 2009). There is an obvious need for new regulation of live animal importation in the United States, and such action must be precautionary.

## WILDLIFE TRADE AND DISEASE RISK

Over half a million shipments containing more than 1.68 billion live animals were traded by the U.S. between 2000 and 2006. Seventy-nine percent of shipments (506,025) were imports containing less than 1.4 billion live animals. Twenty-one percent of shipments were exports containing less than 198 million live individuals. Only 13% of shipments imported contained animals identified to the species level, indicating a huge hole in border enforcement, inspection, and regulation (Smith et al. 2009).

It is estimated that about 2,200 non-native species representing all major taxonomic classes have been imported at some point in time (Defenders of Wildlife 2007; Smith et al. 2009). A 2010 report by the U.S. Government Accountability Office (GAO 2010) revealed that more than 1 billion live animals were legally imported into the United States alone from 2005 to 2008 for various uses (e.g., agriculture, research, education). Annually, this amounts to about 200 million to 300 million individual live animals traded by the United States alone, representing one fifth of Karesh et al.'s (2005) estimate for the entire annual global trade. Few other nations have published similar accounts of live animal imports and exports, making it difficult to quantify the global trade, and it is not our intention to do so here. We can nevertheless use the U.S. data to represent the extreme minimum number and diversity of animals in the global trade (Smith et al. 2009). At a minimum, the millions of live animals traded annually provide ample justification for utilizing pathogen pollution as a platform for evaluating the risks of disease emergence.

Of the more than 2,240 non-native species believed to have been imported into the United States, about 13% (302 non-native species) were identified by a coarse literature review as potentially high risk, given the negative ecological, economical, and/or health

**Box 11.4  H5N1 Identifying the Animal Pathways that Matter**

The early spread of H5N1 highly pathogenic avian influenza through China demonstrated how wildlife migration pathways and trade in wild and domestic animals could interact to promote disease emergence. While wild birds act as reservoirs, carrying virus strains between domestic flocks, chickens and other poultry amplify H5N1 in the stressful and crowded conditions of live markets (Shinya et al. 2006). Early outbreaks across Southeast Asia, the Middle East, and several African countries appeared to be the result of legal and illegal poultry trade, as well as trade in poultry products (e.g., feces used in agricultural fertilizer). In contrast, European outbreaks came on the heels of wild bird migrations from H5N1-endemic countries. Just as the synergy of wild bird migration and the poultry trade was the principal driver behind the spread of H5N1 through China and other Asian countries, it is the interaction of these pathways that poses the greatest threat of H5N1 introduction to North America. The monitoring efforts currently in place are focused on the carrier potential of migratory birds coming from Siberia over the Bering Strait, but wild and domestic birds arriving from other countries in the Americas are a more likely source of an initial outbreak (Kilpatrick et al 2006). Poultry import bans based on the presence of infected domestic birds are not a complete protection; wild birds may still harbor the virus, which could spill over into domestic flocks and reach international markets before the damage is recognized. Risk for H5N1 spread to non-U.S. countries in the Americas is thus substantial unless all imported birds are tested for influenza or trade with the Old World is restricted. The 9 million migratory birds entering North America every year, migrating between the wintering and breeding ranges, including 2.7 million American wigeon (*Anas americana*) and 6.1 million blue-winged teal (*Anas discors*), among other species, may becoming bearers of infection of H5N1 in their countries of origin (Kilpatrick et al 2006). Avian influenza does have pandemic potential in humans, but it has yet to demonstrate efficient human-to-human transmission (Shinya et al. 2006). Despite widespread infection in tens of millions of birds, only about 510 human cases and 303 deaths have been confirmed as of Dec. 31, 2010 (WHO 2010). Although most of these are attributable to close contact with sick or dead birds, particularly with small household flocks (Vong et al. 2006), the potential for viral re-assortment or evolution remains. The species barrier appears to be high, but avian influenza should still be considered a potentially serious threat to public health as well as to livestock and wildlife.

**Table 11.1  Key Steps for Nations to Reduce Disease Emergence from Wildlife Trade**

- Require pre-import risk analysis, as defined by the Convention of Biological Diversity, using the best available data from science and industry.*
- Allocate sufficient resources to support empirical science on disease risks associated with national wildlife imports.
- Allocate sufficient manpower, training, and resources to national agencies charged with enforcing trade.
- Decline entry to improperly labeled shipments containing imported wildlife.
- Work with stakeholders to develop incentives for industry participation in risk-reduction programs, including disease screening and scientific studies on animals in trade.
- Work with other nations to implement third-party screening for high-priority animals and pathogens prior to import.

*Risk analysis, as defined by the Convention on Biological Diversity, involves assessing the consequences of introduction, the likelihood of establishment of non-native species, and the identification of measures to reduce or manage these risks, taking into account socioeconomic and cultural considerations.

implications they pose in other nations (Defenders of Wildlife 2007). More than 70 of these species, including 8 amphibian, 36 bird, 23 mammal, and 7 reptile species, were identified as potential carriers of pathogens that may threaten native wildlife, livestock, or humans (Defenders of Wildlife 2007). These included 10 species of rodents imported from South America, Asia, and Europe, places where rodents are known to carry agents of viral hemorrhagic fevers that are zoonotic pathogens foreign to the United States (Schroeder et al 2008). Fish species were noticeably absent from this list, although they also carry harmful pathogens, including multidrug-resistant *Salmonella* and numerous species of mycobacteria that threaten native wildlife and humans (Smith et al. 2008). Furthermore, analyses of U.S. mammalian imports from 2000 to 2005 identified a number of high-risk genera (*Canis, Felis, Rattus, Macaca,* and *Lepus*) with associated high-risk pathogens in the trade (*Bacillus anthracis, Mycobacterium tuberculosis, Echinococcus* spp., and *Leptospira* spp.) (Pavlin et al. 2009).

Non-native species introductions are forecast to increase up to 24% in the United States over the next 20 years (Levine and D'Antonio 2003). The volume and diversity of live animals imported since 2000 (Smith et el. 2009) bolster this forecast, intensify the need for better regulation, and beg the question: How likely is disease introduction via animal importation?

## MISSING PARASITES: INSIGHTS FROM INVASION BIOLOGY AND DISEASE ECOLOGY

There is a relatively new concept from invasion biology and disease ecology that provides a starting point for considering the risk of disease emergence from wildlife trade. Introduced species leave behind a large proportion (up to 50%) of the parasites and pathogens normally present in their native range populations, leading to a decrease in the number of infectious species and the proportion of hosts infected in the introduced range (Torchin and Mitchell 2004). This widely documented pattern has been termed the missing parasites rule (Torchin and Mitchell 2004).

Ecologists have proposed several mechanisms by which an invasion pathway, such as wildlife trade, may "filter out" native host pathogens. First, the total number of pathogens found in a host population is higher than the total number of pathogens found in any one individual (i.e., very few individuals will be infected with an extremely diverse assemblage of pathogens) (Dobson and May 1986). Animals selected for transport are often derived from a relatively small subset of individuals. As a result, it is unlikely that the "founder population" selected for export will be dominated by highly infected individuals. This suggests that most animals will enter the wildlife trade with only a small fraction of the potentially harmful pathogens available.

Second, many pathogens have complex life cycles requiring more than one host. If suitable hosts for all pathogen life-cycle stages are not present, then the pathogen will not become established in the introduced range. There may be up to an order of magnitude difference in the successful establishment of directly transmitted pathogens versus those with complex life cycles. While animals transported for the wildlife trade are removed from their natural surroundings, thus disrupting normal vector–host pathways, they may also come into close contact with various "novel" species along the trade route, any number of which could act as suitable new hosts.

Third, for pathogens with density-dependent transmission, there is a host threshold density below which a pathogen cannot persist in a host population (Kermack and McKendrick 1927). Thus, host population bottlenecks (i.e., low host density) in the initial stages of invasion may break transmission, eliminating pathogens present in the founder population. Shipments in wildlife trade are renowned for harboring very high densities of animals in single containers or crates. Indeed, it is to the benefit of the dealer to consolidate animals into as small an area as possible to reduce shipping costs. Animals in wildlife trade are likely to experience much higher population densities than those moving along other invasion pathways.

Fourth, an introduced species may accumulate new pathogens from its introduced range. Indeed, the longer an invader is established, the more non-native pathogens it should acquire (Blaustein et al 1983). However, literature surveys suggest that this accumulation is rarely sufficient to compensate for the parasites lost (Torchin et al. 2003). On average, introduced animals accumulate about 25% of the pathogen species they leave behind in their native range. This will undoubtedly hold for animals introduced via wildlife trade.

For those that establish successful populations that thrive over long periods (i.e., domestic dogs), pathogen accumulation may eventually catch up and surpass prevalence and diversity in the ancestral population.

## IDENTIFYING THE RISKY SIDE OF TRADE

Given the quantity and characteristics of wildlife trade (using the United States as a case study) and the lessons from invasion biology and disease ecology, we can identify several aspects of the trade that should facilitate disease emergence.

### Size Matters

The extraordinary number of individual animals and diversity of species shipped around the world each year raises the likelihood that a novel pathogen will emerge. It's a simple matter of probability. Undoubtedly, the missing parasites rule occurs along wildlife trade invasion pathways, but because exported species are harvested repeatedly from a single source, the effect of secondary introductions should speed up the rate at which the imported population accumulates the full suite of pathogens found in the harvested population.

### Trade from Disease Hotspots

Animal importations into the United States can originate in any one of approximately 194 countries. Current analyses indicate that more than 69% of U.S. imports originate in Southeast Asia, which has recently been identified as a hotspot for future emerging zoonotic diseases (Jones et al. 2008). Southeast Asia appears to be both a hub for local and international wildlife trade. This, coupled with high species biodiversity and increasing human population density and travel, creates the likelihood that this region will harbor some of humankind's more threatening zoonoses that have yet to emerge.

### Industry Practices that Promote Disease

At various points along wildlife trade routes, animals are often held at unnaturally high densities (e.g., in shipping containers and at captive breeding facilities) and may be aggregated in such a way as to allow for close inter-specific interactions, thus facilitating the transmission of pathogens between unnatural species pairs. Shipping containers and holding facilities at ports may also become contaminated with pathogens capable of surviving prolonged periods in the environment, and potentially infect additional animals if not properly disposed of or sanitized. Models of multi-host epidemics indicate that the probability that a new disease will emerge is directly correlated with the number of infected individuals (McCormack and Allan 2007). Through the international live animal trade, zoonotic pathogens are given access to a wide range of densely packed, highly mobile hosts and are thus able to rapidly emerge locally or in very distant regions from their original source (Karesh et al. 2005).

### Novel Pathogens, Susceptibility, and Evolution

A staggering 82% of wildlife species imported into the United States are not indigenous to the country, and the pathogens they harbor, therefore, should also be foreign. This novelty increases the risk that an introduced pathogen will harm humans, native wildlife, or domesticated animals because these hosts lack the benefit of host–pathogen co-evolution and are unlikely to have the necessary immune defenses with which to fight new infections. Even if imported animals do not harbor pathogens of concern, they still may pose health risks. Immunologically naïve, non-native animals that are released or escape to the wild may serve as new hosts for pathogens already established in a resident wild population. Finally, seemingly harmless pathogens introduced with imported animals may have the potential to evolve and subsequently infect humans, domestic animals, and/or native animals. Indeed, when faced with novel environmental conditions, pathogens are likely to undergo selection, leading to new strains (Defenders of Wildlife 2007).

### Lack of Hard Science

The U.S. wildlife trade is estimated to have imported more than 2,240 non-native species at one time or another, with those species representing every major taxonomic group (Defenders of Wildlife 2007). Only 13.6% of shipments assessed by Smith et al. (2009), however, were identified at the species level. Nearly 80% of imported shipments contained animals from wild populations, the majority of which are not tested for pathogens before or after shipment, nor held in quarantine after arrival to assess their health status. This is a particular concern as recent studies suggest that imports from wild populations are a greater ecological risk than those from captive-bred populations (Carrete and Tella 2008). The truth is that we know very little about the general biology and ecology of the multitude of animals in the trade, let alone their pathogens. Our lack of knowledge is arguably our greatest risk.

## SCIENCE AND POLICY TO REDUCE PATHOGEN POLLUTION

### Policy Needs

Efforts to minimize pathogen pollution through the live animal trade must begin with a standardized system of import/export documentation protocols. At present, it is impossible to acquire exact figures of wildlife trade in and out of China, given the absence of surveillance systems (Lau et al. 1995). While the U.S. Fish and Wildlife Service's Law Enforcement Management Information System (LEMIS) maintains an electronic database of U.S. wildlife imports and exports, animals in only 13.6% of imported shipments between 2000 and 2006 were identified to the level of species (24.7% to genus and 36.5% to family), making it nearly impossible to assess the diversity of wildlife imported (Smith et al. 2009). The United States has in place four different agencies to oversee the importation of live animals, each entity dealing with a different aspect of the trade, with no one agency wholly responsible for oversight of animal and zoonotic disease risks (GAO 2010). Barriers to collaboration among agencies exist, leaving open gaps that urgently need to be addressed (GAO 2010). Apart from New Zealand, Australia, Israel, and a small handful of other well-regulated nations, most countries are likely to be in the same position as the United States and China.

Many national policies regarding wildlife trade and pathogen introduction have so far been reactionary

(e.g., the U.S. ban on African rodent imports following the 2003 outbreak of monkeypox virus; Box 11.3). At present, only a tiny fraction of the hundreds of millions of animals traded annually are required to undergo quarantine before given clearance to enter their destination country. Furthermore, there is an increasing tendency to make unlicensed sales through the use of the Internet, newspaper advertisements, and hobbyist shows (Reaser et al. 2008). Once inside the border, importers can easily ship animals anywhere in the country without restraint, provided it is a non-CITES animal and was imported through legal means (CITES 2010).

The OIE (World Organization for Animal Health) is an international body that establishes guidelines and standards that ensure the sanitary safety of international trade in live animals and their products. OIE recommendations provide a strong foundation from which policy can be influenced in participating countries around the world. In addition to quarantine measures prior to authorized entrance into the country, it should be required that wild sourced animals be maintained in facilities that limit the interaction with, and the exchange of bodily fluids and wastes from, one species to another. Detailed data on source populations should be made readily available. Captive-reared animals should be given import priority over wild-caught animals, especially where health certification has been conducted prior to exportation.

### Science Needs

Conducting risk analyses for the large number of species crossing international borders would be an indispensable tool for the development of guidelines and regulations, but it is a massive undertaking years in the making. While some data already exist for certain taxa, the data are minimal. Targeting high-profile species for immediate assessment should be a top priority. The United States, for instance, currently has in place a pre-border surveillance protocol for human immigrants and refugees. Individuals are subjected to a medical examination to determine whether they suffer from any inadmissible health related conditions (CDC 2008). A similar system could be adopted for key species prior to import, whereby they undergo testing for high-priority pathogens—for example, *Batrachochytrium dendrobatidis* in frogs, hemorrhagic viruses in rodents, influenza viruses in birds, novel strains of *Salmonella* in reptiles and *Geomyces destructans* in bats—and port-of-entry screening to identify sick animals.

There is a great need for scientists to use the best available evidence to determine species, pathogens, and regions of concern for pathogen spread via wildlife trade. It is equally important to construct a general understanding of host–pathogen dynamics along international trade routes, as industry practices are likely to influence these relationships. Where possible, disease-free stocks should be established and construction of breeding enclosures and waste disposal should be carefully maintained so that the potential for pathogen spread between captive and wild populations is minimized. Likewise, wildlife breeders should be included in the surveillance process. Individual nations could implement incentive programs to encourage these practices.

## CONCLUSIONS

The Convention on Biological Diversity identified wildlife trade as the most glaring gap in the international legal system related to trade and invasive species (CBD 2004). With few national and international initiatives poised to reduce the volume of wildlife trade, resulting disease and species introductions should continue. We have demonstrated that this is not a new problem; indeed, humans, livestock, and wildlife species have been plagued by pathogen introductions resulting from long-distance animal transport for millennia. Globalization and myriad desires for wildlife have only increased the opportunity for new host–pathogen encounters. We may never stop pathogen pollution from the wildlife trade, but using the best available science to reduce its risk is worth striving for and well within our reach.

## REFERENCES

Allan, S.A., L.A. Simmons, and M.J. Burridge. 1998. Establishment of the tortoise tick *Amblyomma marmoreum* (Acari: Ixodidae) on a reptile-breeding facility in Florida. J Med Entomol 35:621–624.

Ashraf, H. 2003. China finally throws full weight behind efforts to contain SARS. Lancet 361:1439–1439.

Avashia, S.B., J.M. Petersen, C.M. Lindley, M.E. Schriefer, K.L. Gage, M. Cetron, T.A. DeMarcus, D.K. Kim, J. Buck, J.A. Montenieri, J.L. Lowell, M.F. Antolin, M.Y. Kosoy, L.G. Carter, M.C. Chu, K.A. Hendricks, D.T. Dennis, and J.L. Kool. 2004. First reported prairie dog-to-human tularemia transmission, Texas, 2002. Emerg Infects Dis 10:483–486.

Blaustein, A.R., A.M. Kuris, and J.J. Alio. 1983. Pest and parasite species richness problems. Am Nat 122:556–566.

Brown, C. 2004. Emerging zoonoses and pathogens of public health significance—an overview. Rev Sci Tech Off Int Epiz 23:435–442.

Brown, R. 2006. Exotic pets invade United States ecosystems: legislative failure and a proposed solution. Indiana Law J 81:713.

Burnham, B.R., D.H. Atchley, R.P. DeFusco, K.E. Ferris, J.C. Zicarelli, J.H. Lee, and F.J. Angulo. 1998. Prevalence of fecal shedding of *Salmonella* organisms among captive green iguanas and potential public health implications. J Am Vet Med Assoc 213:48–50.

Burridge, M.J. 1997. Heartwater: An increasingly serious threat to the livestock and deer populations of the United States. *In* Proceedings of 101st Annual Meeting of US Animal Health Association, pp. 582–597. Louisville, Kentucky. Spectrum Press, Richmond, Virginia.

Burridge, M.J., L.A. Simmons, and S.A. Allan. 2000. Introduction of potential heartwater vectors and other exotic ticks into Florida on imported reptiles. J Parasitol 86:700–704.

Burridge, M.J., L.A. Simmons, B.H. Simbi, T.F. Peter, and S.M. Mahan. 2009. Evidence of *Cowdria ruminantium* infection (heartwater) in *Amblyomma sparsum* ticks found on tortoises imported into Florida. J Parasitol 86:1135–1136.

Camus, E., N. Barre, D. Martinez, and G. Uilenberg. 1996. Heartwater (cowdriosis) a review. Office International des Epizooties, Paris, France.

Carrete, M., and J.L. Tella. 2008. Wild-bird trade and exotic invasions: a new link of conservation concern? Front Ecol Environ 6: 207–211.

Centers for Disease Control and Prevention (CDC). 2003a. Update: Multistate outbreak of monkeypox—Illinois, Indiana, Kansas, Missouri, Ohio, and Wisconsin, 2003. MMWR dispatch. Atlanta: Centers for Disease Control and Prevention, July 2, 2003. (Accessed March 23, 2011 at http://www.cdc.gov/mmwr/preview/mmwrhtml/mm52d702a1.htm).

Centers for Disease Control and Prevention (CDC). 2003b. Restrictions on African rodents and prairie dogs, interim final rule, November 4, 2003. (Accessed 23 March 2011, at http://www.cdc.gov/ncidod/monkeypox/animals.htm).

Center for Disease Control and Prevention (CDC). 2008. Global migration and quarantine: medical examination of aliens (refugees and immigrants), October 6, 2008. (Accessed 23 March 2011 at http://www.cdc.gov/ncidod/dq/health.htm).

Chan, P.K.S. 2002. Outbreak of avian influenza A (H5N1) virus infection in Hong Kong in 1997. Clin Infect Dis 34:S58–S64.

Chen, H., G.J.D. Smith, S.Y. Zhang, K. Qin., J. Wang, K.S. Li, R.G. Webster, J.S.M. Peiris, and Y. Guan. 2005. Avian flu H5N1 virus outbreak in migratory waterfowl. Nature 436:191–192.

Convention on Biological Diversity (CBD). 2004. Seventh Conference of the Parties: Decision VII/13: Alien species that threaten ecosystems, habitats or species (Article 8 (h)), Kuala Lumpur, Malaysia. (Available at www.biodiv.org).

Convention on International Trade in Endangered Species of Wild Fauna and Flora (CITES). 2010. (Available at http://www.cites.org/)

Cunningham, A.A., P. Daszak, and J.P. Rodríguez. 2003. Pathogen pollution: defining a parasitological threat to biodiversity conservation. J Parasitol 89:S78–S83.

Defenders of Wildlife. 2007. Broken screens: the regulation of live animal imports in the United States. Washington, DC.

De Schrijver, K. 1998. A psittacosis outbreak in customs officers in Antwerp (Belgium). Bull Inst Marit Trop Med Gdynia 49:97–99.

Di Giulio, D.B., and P.B. Eckburg. 2004. Human monkeypox: an emerging zoonosis. Lancet Infect Dis 4:15–25.

Dobson, A.P., and R.M. May. 1986. Patterns of invasions by pathogens and parasites. In H.A. Mooney and J.A. Drake, eds. Ecology and biological invasions of North America and Hawaii, pp. 58–76. Berlin: Springer-Verlag.

Donovan, D.G. 2004. Cultural underpinnings of the wildlife trade in Southeast Asia. In J. Knight, ed. Wildlife in Asia: cultural perspectives, pp. 88–111. Routledge Curzon, New York.

Drosten, C., S. Gunther, W. Preiser, S. VanderWerf, H.R. Brodt, S. Becker, H. Rabenau, M. Panning, L. Kolesnikova, R.A.M. Fouchier, A. Berger, A.-M. Burgière, J. Cinati, M. Eickmann, N. Escriou, K. Grywna, S. Kramme, J.-C. Manuguerra, S. Müller, V. Rickerts, M. Stürmer, S. Vieth, H.-D. Klenk, A.D.M.E. Osterhaus, H. Schmitz, and H.W. Doerr. 2003. Identification of a novel coronavirus in patients with severe acute respiratory syndrome. N Engl J Med 348:1967–1976.

Ebrahim, M., and J. Solomon. Exotic pet trade booming in U.S. Associated Press, Nov. 27, 2006 (Accessed March 23, 2011, at http://dsc.discovery.com/news/2006/11/27/exoticpet_ani.html).

Ellis, T.M., B.R. Bousfield, L.A. Bissett, K.C. Dyrting, G.S. Luk, S.T. Tsim, K. Sturm-Ramirez, R.G. Webster, Y. Guan, and J.S. Malik Peiris. 2004. Investigation of outbreaks of highly pathogenic H5N1 avian influenza in waterfowl and wild birds in Hong Kong in late 2002. Avian Pathol 33:492–505.

Favoretto, S.R., C.C. de Mattos, N.B. Morais, F.A. Alves Araujo, and C.A. de Mattos. 2001. Rabies in marmosets (Callithrix jacchus), Ceará, Brazil. Emerg Infect Dis 7:1062–1065.

Feng, Y., M. Meng, G. Songhui, S. Zhao, R. Chen, and H. Cui. 2006. Survey on wildlife consumption as food in China. China Wildlife 27:2–5.

Food and Drug Administration. 2003. 68 FR 36566 Joint order of the Centers for Disease Control and Prevention and the Food and Drug Administration, Department of Health and Human Services ( June 18, 2003).

Garner, T.W.J., M.W. Perkins, P. Govindarajulu, D Seglie., S. Walker, A.A. Cunningham, and M.C. Fisher. 2006. The emerging amphibian pathogen Batrachochytrium dendrobatidis globally infects introduced populations of the North American bullfrog, Rana catesbeiana. Biol Letters 2:455–459.

Gomez, A., and A.A. Aguirre. 2008. Infectious diseases in the illegal wildlife trade. Ann NY Acad Sci 1149:16–19.

Guan, Y., B.J. Zheng, Y.Q. He, X.L. Liu, Z.X. Zhuang, C.L. Cheung, S.W. Luo, P.H. Li, L.J. Zhang, Y.J. Guan, K.M. Butt, K.L. Wong, K.W. Chan, W. Lim, K.F. Shortridge, K.Y. Yuen, J.S. Peiris, and L.L. Poon. 2003. Isolation and characterization of viruses related to the SARS coronavirus from animals in southern China. Science 302:276–278.

Hahn, B.H., G.M. Shaw, K.M. De Cock, and P.M. Sharp. 2000. AIDS as a zoonosis: scientific and public health implications. Science 287:607–614.

Hanselmann, R.A., Rodriguez, M. Lampo, L. Fajardo-Ramos, A.A. Aguirre, A.M. Kilpatrick, J.P. Rodriguez, and P. Daszak. 2004. Presence of an emerging pathogen of amphibians in introduced bullfrogs (Rana catesbeiana) in Venezuela. Biol Conserv 120:115–119.

He, J.F., X.R. Yu, D.W. Peng, G.W. Liu, Y.Y. Liang, L.H. Li, R.N. Guo, Y. Fang, X.C. Zhang, H.Z. Zheng, H.M. Luo, and J.Y. Lin. 2003. Severe acute respiratory syndrome in Guangdong Province of China: epidemiology and control measures. Chin J Prev Med 37:227–232.

Heymann, D.L., and G. Rodier. 2004. Global surveillance, national surveillance, and SARS. Emerg Infect Dis 10:173–5.

Hutson, C.L., K.N. Lee, J. Abel, D.S. Carroll, J.M. Montgomery, V.A. Olson, Y. Li, W. Davidson, C. Hughes, M. Dillon, P. Spurlock, J.J. Kazmierczak, C. Austin, L. Miser, F.E. Sorhage, J. Howell, J.P. Davis,

M.G. Reynolds, Z. Braden, K.L. Karem, I.K. Damon, and R.L. Regnery. 2007. Monkeypox zoonotic associations: insights from laboratory evaluation of animals associated with the multi-state U.S. outbreak. Am J Trop Med Hyg 76:757–768.

Janies, D., F. Habib, B. Alexandrov, A. Hill, and D. Pol. 2008. Evolution of genomes, host shifts and the geographic spread of SARS-CoV and related coronaviruses. Cladistics 24:111–130.

Jezek, Z., and Fenner, F. 1988. Human monkeypox. In J.L. Melnick, ed. Monographs in virology 17. Karger, Basel.

Jones, K.E., N. Patel, M. Levy, A. Storeygard, D. Balk, J.L. Gittleman, and P. Daszak. 2008. Global trends in emerging infectious diseases. Nature 451:990–993.

Karesh, W., R. Cook, E. Bennett, and J. Newcomb. 2005. Wildlife trade and global disease emergence. Emerg Infect Dis 11:1000–1002.

Kermack, W.O., and, A.G. McKendrick. 1927. Contributions to the mathematical theory of epidemics. Proc R Soc Lon A 115:700–21.

Kilpatrick, A., A. Chmura, D. Gibbons, R. Fleischer, P. Marra, and P. Daszak. 2006. Predicting the global spread of H5N1 avian influenza. P Natl Acad Sci 103:19368.

Ksiazek, T.G., D. Erdman, C.S. Goldsmith, S.R. Zaki, T., Peret, S. Emery, S. Tong, C. Urbani, J.A. Comer, W. Lim, P.E. Rollin, S.F. Dowell, A.-E. Ling, C.D. Humphrey, W.-J. Shieh, J. Guarner, C.D. Paddock, P. Rota, B. Fields, J. DeRisi, J.-Y. Yang, N. Cox, J.M. Hughes, J.W. LeDuc, W.J. Bellini, L.J. Anderson, and the SARS Working Group. 2003. A novel coronavirus associated with severe acute respiratory syndrome. N Engl J Med 348:1953–1966.

Lau, M.W.N., G. Ades, N. Goodyer, and F.S. Zou. 1995. Wildlife trade in southern China including Hong Kong and Macao. Biodiversity Working Group of the China Council for International Cooperation on Environment and Development Project. Hong Kong.

Laurance, W.F., K.R. McDonald, and R. Speare. 1996. Epidemic disease and the catastrophic decline of Australian rain forest frogs. Conserv Biol 10: 406–413.

Lee, K.S., M.W.N. Lau, and B.P.L. Chan. 2004. Wild animal trade monitoring in selected markets in Guangzhou and Shenzhen, South China 2000–2003. Kadoorie Farm and Botanic Garden, Hong Kong, China.

Levine, J.M., and C. D'Antonio. 2003. Forecasting invasions with increasing international trade. Conserv Biol 17:322–326.

Li, K.S., Y. Guan, J. Wang, G.J.D. Smith, K.M., Xu, L. Duan, A.P. Rahardjo, P. Puthavathana, C. Buranathai, T.D. Nguyen, A.T. Estoepangestie, A. Chaisingh,

P. Auewarakul, H.T. Long, N.T. Hanh, R.J. Webby, L.L. Poon, H. Chen, K.F. Shortridge, K.Y. Yuen, R.G. Webster, and J.S. Peiris. 2004. Genesis of a highly pathogenic and potentially pandemic H5N1 influenza virus in eastern Asia. Nature 430:209–213.

Li, W., Z. Shi, M. Yu, W. Ren, C. Smith, J. Epstein, et al. 2005. Bats are natural reservoirs of SARS-like coronaviruses. Science 310:676–679.

Likos, A.M., S.A. Sammons, V.A. Olson, A.M. Frace, Y. Li, M. Olsen-Rasmussen, W. Davidson, R. Galloway, M.L. Khristova, M.G. Reynolds, H. Zhao, D.S. Carroll, A. Curns, P. Formenty, J.J. Esposito, R.L. Regnery, and I.K. Damon. 2005. A tale of two clades: monkeypox viruses. J Gen Virol 86:2661–2672.

Lin, Jinyan W., J. Liang, H. Li, Y.D. Yu. M. Liao, K. Ming, J. Wen, L. Li, R. Zouli, P. Huang, H. Tang, and H. Luo. 2006. SARS-CoV contamination in wild animal market after intervention in Guangdong province, South China. J Prev Med 32:5–8.

Lips, K.R. 1998. Decline of a tropical montane amphibian fauna. Conserv Biol 12:106–117.

Mahan. S.M., G.E. Smith, D. Kumbula, M.J. Burridge, and R.F. Barbet. 2001. Reduction in mortality from heartwater in cattle, sheep and goats exposed to field challenge using an inactivated vaccine. Vet Parasitol 97:295–308.

McCormack, R.K., and L. Allan. 2007. Disease emergence in multi-host epidemic models. Math Med and Biol. 24:17–34.

Mukhebi. A.W., T. Chamboko, C.J. O'Callaghan, T.F. Peter, R.L. Kruska, G.F. Medley, S.M. Mahan, and B.D. Perry. 1999. An assessment of the economic impact of heartwater (Cowdria ruminantium infection) and its control in Zimbabwe. Prev Vet Med 39:173–189.

Pavlin, B.I., L.M. Schloegel, and P. Daszak. 2009. Risk of importing zoonotic diseases through wildlife trade, United States. Emerg Infect Dis 15:1721–1726.

Perry, N., B. Hanson, W. Hobgood, R. Lopez, C. Okraska, K. Karem, I.K. Damon, and D.S. Carroll. 2006. New invasive species in southern Florida: Gambian rat (Cricetomys gambianus). J Mammal 87:262–264.

Pessier, A.P., D.K. Nichols, J.E. Longcore, and M.S. Fuller. 1999. Cutaneous chytridiomycosis in poison dart frogs (Dendrobates spp.) and White's tree frog (Litoria caerulea). J Vet Diagn Invest 11:194–199.

Phalen, D.N. 2004. Prairie dogs: vectors and victims. Semin Avian Exot Pet 13:105–107.

Peiris, J.S.M., M.D. de Jong, and Y. Guan. 2007. Avian influenza virus (H5N1): a threat to human health. Clin Microbiol Rev 20:243–267.

Peter, T.F., M.J. Burridge, and S.M. Mahan. 2009. Competence of the African tortoise tick (Amblyomma marmoreum Acari: Ixodidae), as a vector of the agent of heartwater (*Cowdria ruminantium*). J Parasitol 86:438–441.

Reaser, J.K., E.E. Clark, and N.M. Meyers. 2008. All creatures great and minute: a public policy primer for companion animal zoonoses. Zoo Public Health 55:385–401.

Reed, K.D., J.W. Melski, M.B. Graham, R.L. Regnery, M.J. Sotir, M.V. Wegner, J.J. Kazmierczak, E.J. Stratman, Y. Li, J.A. Fairley, G.R. Swain, V.A. Olson, E.K. Sargent, S.C. Kehl, M.A. Frace, R. Kline, S.L. Foldy, J.P. Davis, and I.K. Damon. 2004. The detection of monkeypox in humans in the Western hemisphere. New Engl J Med 350:342–350.

Reynolds, M.G., W.B. Davidson, A.T. Curns, C.S. Conover, G. Huhn, J.P. Davis, M. Wegner, D.R. Croft, A. Newman, N.N. Obiesie, G.R. Hansen, P.L. Hays, P. Pontones, B. Beard, R. Teclaw, J.F. Howell, Z. Braden, R.C. Holman, K.L. Karem, and I.K. Damon. 2007. Spectrum of infection and risk factors for human monkeypox, United States, 2003. Emerg Infect Dis 13:1332.

Riley, S., C. Fraser, C.A. Donnelly, A.C. Ghani, L.J. Abu-Raddad, A.J. Hedley, G.M. Leung, L.M. Ho, T.H. Lam, T.Q. Thach, P. Chau, K.P. Chan, S.V. Lo, P.Y. Leung, T. Tsang, W. Ho, K.H. Lee, E.M. Lau, N.M. Ferguson, and R.M. Anderson. 2003. Transmission dynamics of the etiological agent of SARS in Hong Kong: impact of public health interventions. Science 300:1961–1966.

Rosen, G.E., and K.F. Smith. 2010. Summarizing the evidence on the international trade in illegal wildlife. EcoHealth 7:24–32.

Rota, P.A., M.S. Oberste, S.S. Monroe, W.A. Nix, R. Campagnoli, J.P. Icenogle, S. Penaranda, B. Bankamp, K. Maher, M.H. Chen, S. Tong, A. Tamin, L. Lowe, M. Frace, J.L. DeRisi, Q. Chen, D. Wang, D.D. Erdman, T.C. Peret, C. Burns, T.G. Ksiazek, P.E. Rollin, A. Sanchez, S. Liffick, B. Holloway, J. Limor, K. McCaustland, M. Olsen-Rasmussen, R. Fouchier, S. Gunther, A.D. Osterhaus, C. Drosten, M.A. Pallansch, L.J. Anderson, and W.J. Bellini. 2003. Characterization of a novel coronavirus associated with severe acute respiratory syndrome. Science 300:1394–1399.

Schloegel, L.M., P. Daszak, A.A. Cunningham, R. Speare, and B. Hill. 2010a. Two amphibian diseases, chytridiomycosis and ranaviral disease, are now globally notifiable to the World Organization for Animal Health (OIE). Dis Aquat Organ 92:101–108.

Schloegel, L.M., C.M. Ferreira, T.Y. James, M. Hipolito, J.E. Longcore, A.D. Hyatt, M.Y. Yabsley, A.M.C.R.P.F. Martins, R. Mazzoni, A.J. Davies, and P. Daszak. 2010b. The North American bullfrog as a reservoir for the spread of *Batrachochytrium dendrobatidis* in Brazil. Anim Conserv 13:53–61.

Schloegel, L.M., A.M. Picco, A.M. Kilpatrick, A.J. Davies, A.D. Hyatt, and P. Daszak. 2009. Magnitude of the US trade in amphibians and presence of *Batrachochytrium dendrobatidis* and ranavirus infection in imported North American bullfrogs (*Rana catesbeiana*). Biol Conserv 142: 1420–1426.

Seimenis, A.M. 2008. The spread of zoonoses and other infectious diseases through the international trade of animals and animal products. Veter Ital 44: 591–599

Schroeder, B., J. McQuiston, R. Marquis, G. Galland, and N. Marano. 2008. Anticipating the next monkeypox: trends in rodent importation, 1999–2006. International Conference on Emerging Infectious Diseases, Atlanta, March 2008.

Shinya, K., M. Ebina, S. Yamada, M. Ono, N. Kasai, and Y. Kawaoka. 2006. Avian flu: influenza virus receptors in the human airway. Nature 440:435–436.

Shortridge, K.F., N. N. Zhou, Y. Guan, et al. 1998. Characterization of avian H5N1 influenza viruses from poultry in Hong Kong. Virology 252:331–334.

Sims, L.D., T.M. Ellis, K.K. Liu, K. Dyrting, H. Wong, M. Peiris, Y. Guan, and K.F. Shortridge. 2003. Avian influenza in Hong Kong 1997–2002. Avian Dis 47:832–838.

Smith, K.F., M.D. Behrens, L.M. Schloegel, N. Marano, S. Burgiel, and P. Daszak. 2009. Reducing the risks of the wildlife trade. Science 324:594–595.

Smith, K.F., M.D. Behrens, L.M. Max, and P. Daszak. 2008. U.S. drowning in unidentified fishes: scope, implications and regulation of live fish import. Conserv Lett 1:103–109.

Stadler, K., V. Masignani, M. Eickmann, S. Becker, S. Abrignani, H.D. Klenk, and R. Rappouli. 2003. SARS—beginning to understand a new virus. Nat Rev Microbiol 1:209–218.

Sturm-Ramirez, K.M., T. Ellis, B. Bousfield, L. Bissett, K. Dyrting, J.E. Rehg, L. Poon, Y. Guan, M. Peiris, and R.G. Webster. 2004. Re-emerging H5N1 influenza viruses in Hong Kong in 2002 are highly pathogenic to ducks. J Virol 78:4892–901.

Svoboda, T, B. Henry, L. Shulman, E. Kennedy, E. Rea, W. Ng, T. Wallington, B. Yaffe, E. Gournis, E. Vicencio, S. Basrur, and R.H. Glazier. 2004. Public health measures to control the spread of the severe acute respiratory syndrome during the outbreak in Toronto. N Engl J Med 350:2352–2361.

Tai, Z., and T. Sun. 2007. Media dependencies in a changing media environment: the case of the 2003 SARS epidemic in China. New Media Soc 9: 987–1009.

Telecky, T. 2001. United States import and export of live turtles and tortoises. Turtle and Tortoise Newsletter 4:8–13.

Torchin, M.E., and C.E. Mitchell. 2004. Parasites, pathogens, and invasions by plants and animals. Front Ecol Environ 2:183–190.

Torchin, M.E., K.D. Lafferty, A.P. Dobson, V.J. McKenzie, and A.M. Kuris. 2003. Introduced species and their missing parasites. Nature 421:628–630.

Tu, C.C., G. Crameri, X.G. Kong, J.D. Chen, Y.W. Sun, M. Yu, B.T. Eaton, H. Xuan, and L.F. Wang. 2004. Antibodies to SARS coronavirus in civets. Emerg Infect Dis 10:2244–2248.

U.S. Government Accountability Office (GAO). (2010). Live animal imports: agencies need better collaboration to reduce the risk of animal-related diseases. Report to the Committee on Homeland Security and Governmental Affairs, U.S. Senate, Washington, D.C.

U.S. Department of Agriculture (USDA). (1966). National Tick Surveillance Program, Calendar Year 1965. Animal Health Division, Agricultural Research Service, US Department of Agriculture, Hyattsville, Maryland.

Van Amstel, S., F. Reyers, A. Guthrie, P. Oberem, and H. Bertschinger. 1988. The clinical pathology of heartwater. I. Haematology and blood chemistry. Onderstepoort J Vet 55:37.

Van de Bildt, M.W.G., T. Kuiken, A.M. Visee, S. Lema, T.R. Fitzjohn, and A.D.M.E. Osterhaus. 2002. Distemper outbreak and its effect on African wild dog conservation. Emerg Infect Dis 82:211–213.

Vong, S., B. Coghlan, S. Mardy, D. Holl, H. Seng, S. Ly, M.J. Miller, P. Buchy, Y. Froehlich, J.B. Dufourcq, T.M. Uyeki, W. Lim, and T. Sok. 2006. Low frequency of poultry-to-human H5N1 virus transmission, southern Cambodia, 2005. Emerg Infect Dis 12:1542–1547.

Weldon, C.W., L.H. du Preez, A.D. Hyatt, R. Muller, and R. Speare. 2004. Origin of the amphibian chytrid fungus. Emerg Infect Dis 10:2100–2105.

WildAid/China Wildlife Conservation Association (CWCA). 2005. Report on the survey of wildlife consumption and public attitudes to wildlife consumption in China. Available: http://www.wildaid.org/PDF/reports/CWCA%20&%20WildAid%20survey%20report%20-%20English.pdf [Accessed March 24, 2010].

World Health Organization (WHO). 2004. Summary of probable SARS cases with onset of illness from Nov. 1,

2002, to July 31, 2003. Available at: http://www.who.int/csr/sars/country/table 2004 _04_21/en/ [Accessed March 23, 2011]

World Health Organization (WHO). 2005a. Confirmed human cases of avian influenza A (H5N1), Sept. 7, 2004. Available: www.who.int/csr/disease/avian_influenza/country/cases_table_2004_09_07/en/print.ht [Accessed March 23, 2011]

World Health Organization (WHO). 2005b. Inter-country consultation: influenza A/H5N1 in humans in Asia, Manila, May 6–7, 2005 Available: http://www.who.int/csr/resources/publications/influenza/WHO_CDS_CSR_GIP_2005_7/en/index.html

World Health Organization (WHO). 2010. Cumulative number of confirmed human cases of avian influenza A/(H5N1) Reported to WHO. Available: http://www.who.int/csr/disease/avian_influenza/country/cases_table_2010_12_09/en/index.html

World Wildlife Fund (WWF). 2001. Souvenir alert highlights deadly trade in endangered species. Available: http://www.wwf.org.uk/news/scotland/n_0000000409.asp

Xu, X., K. Subbarao, N.J. Cox, and Y. Guo. 1999. Genetic characterization of the pathogenic influenza a/goose/Guangdong/1/96 (H5N1) virus: similarity of its hemagglutinin gene to those of H5N1 viruses from the 1997 outbreaks in Hong Kong. Virology 261:15–19.

Xu, H.F., R.H. Xu, J.G. Xu, et al. 2006. Study on the dynamic prevalence of serum antibody against severe acute respiratory syndrome coronavirus in employees from wild animal market in Guangzhou. China J Epidemiol 27:950–953.

Xu, R.H., J.F. He, M.R. Evans, et al. 2004. Epidemiologic clues to SARS origin in China. Emerg Infect Dis 10:1030–1037.

Zhong, N.S., B.J. Zheng, Y.M. Li, Poon, Z.H. Xie, K.H. Chan, N.S. Zhong, B.J. Zheng, Y.M. Li, L.L.M. Poon, Z.H. Xie, K.H. Chan, P.H. Li, S.Y. Tan, Q. Chang, J.P. Xie, X.Q. Liu, J. Xu, D.X. Li, K.Y. Yuen, J.S.M. Peiris, and Y. Guan. 2003. Epidemiology and cause of severe acute respiratory syndrome (SARS) in Guangdong, People's Republic of China, in February 2003. Lancet 362:1353–1358.

# 12

# BUSHMEAT AND INFECTIOUS DISEASE EMERGENCE

Matthew LeBreton, Brian L. Pike, Karen E. Saylors,
Joseph Le Doux Diffo, Joseph N. Fair, Anne W. Rimoin,
Nancy Ortiz, Cyrille F. Djoko, Ubald Tamoufe,
and Nathan D. Wolfe

Pandemics of animal origin, such as those caused by the influenza virus or human immunodeficiency virus (HIV), are poignant reminders of our vulnerability to animal diseases. The majority of human infectious diseases originate through the transmission of pathogenic agents from animals to humans (Taylor et al. 2001; Wolfe et al. 2007; Jones et al. 2008); however, accurately predicting how, when, and where a pandemic of animal origin will emerge remains elusive due to the diversity of animal species that may harbor pathogens and the multiple pathways of emergence. These pathways are determined by the nature and frequency of human interactions with animals, and the behavioral, economic, and environmental factors that may influence them (Morse 1995; Brashares et al. 2004; Jones et al. 2008). Each of these factors helps to create or avert the conditions favorable for cross-species transmission that may mark the beginning of a pandemic. Monitoring these factors and the interface between humans and animals is of major importance, as preventing zoonotic transmission will likely be key to future pandemic prevention efforts. However, most animal pathogens will not easily establish themselves in human populations and result in pandemics (May et al. 2001; Antia et al. 2003), as they must possess or evolve the capacity for successful, long-term transmission within their new host to become pandemic (Wolfe et al. 2007). Surveillance to detect secondary spread of newly emergent disease may be one way of tracking this poorly understood evolutionary process and will also be important in stemming emergent pandemics.

One particularly important interface for disease emergence is the interface created between humans and animals by the hunting and butchering of wildlife. The intimate blood contact between humans and animals at this interface makes hunting and butchering a risk-prone behavior in terms of microbial transmission (Figs. 12.1, 12.2, and 12.3). As many people worldwide participate in hunting and butchering activities, a large number of people are constantly being challenged by zoonotic agents (Kalish et al. 2005; Wolfe et al. 2004b, 2005b; Calattini et al. 2007). Knowing the diversity of microbes in wild animal populations and having a better understanding of both the factors contributing to the transmission of these microorganisms from animals to humans and the capacity for the novel pathogens to spread and cause disease in humans will assist future efforts aimed at predicting and preventing pandemics

**Figure 12.1:**
Butchering an African brush-tailed porcupine (*Atherurus africanus*) in Cameroon. (Photo by Matthew LeBreton, GVF)

**Figure 12.2:**
Butchering an unidentified antelope in Cameroon. (Photo by Joseph LeDoux Diffo, GVF. Photo previously appeared on the cover of *Animal Conservation* Volume 9, iss. 4.)

(Pike et al. 2010). However, a significant modern challenge is for us to use existing knowledge to promote awareness and to reduce current levels of exposure and transmission. In this chapter, we discuss the role of hunting in disease emergence in central Africa, some of the microbes associated with hunting, and opportunities for surveillance and prevention of diseases within hunting communities.

## HUNTING IN CENTRAL AFRICA

In central Africa, hunted wildlife, often referred to as bushmeat, play important nutritional, cultural, and economic roles for communities (Bennett and Robinson 2000), especially in rural areas where the majority of people hunt, butcher, or eat wild animals at some stage of their life. Bushmeat is likely to remain an important source of animal protein for many people due to deficits in domestic animal meat production (Fa et al. 2003); the difficulties in reliably transporting and storing commercially produced domestic animals

and meats from production areas to rural areas; and rural poverty combined with the low cost and high availability of bushmeat in rural areas (Wilcox and Nambu 2007). Also, in areas where large-scale commercial activities such as logging occur, there is often an increased demand for bushmeat due to disposable income and few dietary alternatives (Auzel and Wilkie 2000; Karesh and Noble 2009). In urban areas, bushmeat is also popular, and hunting often takes place in remote forest villages to supply these urban markets (Wilcox and Nambu 2007).

Ungulates, primates, and rodents make up the majority of hunted animals in central Africa (see Wilkie and Carpenter 1999 for a review), and ungulates and/or rodents are usually the most preferred meat types (Fa et al. 2002b; Schenck et al. 2006) (Figs. 12.1–12.3). The intensity and extent of the bushmeat trade in central Africa has increased over the past two decades (Karesh et al. 2009), and it is estimated that 4.9 million metric tons of bushmeat are harvested annually from the Congo Basin (Fa et al. 2002a).

**Figure 12.3:**
Butchering an unidentified antelope in Cameroon. (Photo by Pien Huang)

Among other things, the modernization of hunting has permitted increased levels of bushmeat harvesting, primarily through increased access to previously remote forested areas (Wilkie et al. 1992; Milner-Gulland et al. 2003; Wolfe et al. 2005a; Laurance et al. 2006), and changes in hunting techniques (Milner-Gulland et al. 2003) that have resulted in increased efficiency (Noss 1998b). Snares, while previously made from forest products, today are mostly made from cheap and widely available cable or wire, and these allow fewer escapes and the capture of larger animals. Firearms have largely replaced traditional arms, which historically included bows, crossbows, spears, and poisons. For example, in Cameroon, 81% of hunters report using modern wire snares, versus only 4% using traditional non-wire snares; 31% report using firearms compared to just 4% who reported using bows and arrows (LeBreton et al. 2006). With these transformations in techniques over the past century, hunting of large game, which previously would have required the efforts of dozens of individuals using nets and spears (Merfield 1956; Noss 1997), can now be undertaken by a single individual and a firearm.

## TRANSMISSION ROUTES AND RISKY BEHAVIOR

Direct contact with animal blood and body fluids, through both hunting and butchering, is the fundamental risk factor for acquisition of zoonotic pathogens from hunted wildlife. Accidents involving bites or scratches while handling live and dying animals, and cuts resulting from the use of knives and machetes during butchering, pose a considerable risk for transmission. Infections can also occur during transport of animals when blood and other fluids drain from the carcass onto the torso of the person carrying the hunted animal. Any cuts or abrasions on the body could allow for the transmission of pathogens. Also, while blood-borne pathogens pose a high level of risk, and the exposures mentioned above are cause for concern, it is worth noting that a number of other pathogens can be acquired via skin or transdermal infections (e.g., anthrax, Ebola).

The hunting of non-human primates, which can make up over 10% of the captured animals in some areas (Wilkie and Carpenter 1999; Fa et al. 2006;

Wilcox and Nambu 2007), poses a particular risk to human health. Non-human primates are our closest relatives and as such represent the weakest barrier to cross-species transmission (Wolfe et al. 2007) and share many microbes with humans.

## HUNTING-RELATED ZOONOTIC MICROBES

A number of zoonotic agents are likely to have originated from the hunting and butchering of wild animals. Some, such as HIV and human T-cell lymphotropic virus (HTLV), are now of international significance and have caused large numbers of infections and deaths over a wide geographic area. Others, such as simian foamy virus (SFV), have only recently been detected among a small number of people, though such infections may be more common and widespread than we realize. The threat that such newly discovered zoonotic viruses pose to human health is unknown; however, in retrospect, the same could have been said of HIV in the earliest stages of the current pandemic. Therefore, from a public health perspective, we should be taking notice of such microbes and their potential threat to human health before they expand into local, regional, or even global problems.

### Simian Immunodeficiency Virus (SIV)

Hunting and butchering of wild non-human primates is now widely recognized as the factor that led to the introduction of SIV into the human population and that gave rise to the HIV pandemic (Worobey et al. 2008). The high diversity of SIV among more than 40 species of non-human primates (Peeters et al. 2002; VandeWoude and Apetrei 2006) and known HIV sequence data suggest that at least 12 separate transmission events from non-human primates to humans must have occurred over the past century: four to account for the diversity of HIV-1 (Plantier et al. 2009) and eight to account for the diversity of HIV-2 (Van Heuverswyn and Peeters 2007). The oldest known HIV infections date to 1959 (Zhu et al. 1998) and the 1960s in Kinshasa, Democratic Republic of Congo (Worobey et al. 2008), and molecular clocks indicate that HIV-1 originated sometime in the late 1800s or early 1900s (Worobey et al. 2008). This timeframe corresponds with a period of rapid growth of

a number of large cities in central Africa and their increasing regional and intercontinental connectivity, factors that would have facilitated the spread of a previously rare virus.

Interestingly, SIV infections have never been documented in people hunting or butchering wild non-human primates despite the high diversity of SIV among hunted animals (Peeters et al. 2002), the fact that other closely related viruses are transmitted through the same pathway (Wolfe et al. 2004b, 2005b; Calattini et al. 2007), and the fact that SIV infections have been reported among laboratory workers exposed to non-human primates (Khabbaz et al. 1994). However, indirect evidence does point to human SIV infections in people exposed to non-human primates through hunting and butchering (Kalish et al. 2005). Human SIV infections may simply occur at much lower frequencies than other recently discovered zoonotic retroviruses (Wolfe et al. 2004b), or alternatively, infections may be transient, although that would be exceptional among the retroviruses, which typically cause chronic infections.

As hunting and butchering of non-human primates is ongoing, there is a possibility that continued cross-species transmission of viruses from the diverse existing pool of SIVs may give rise to novel HIVs or HIV diversity in the human population (Van Heuverswyn and Peeters 2007). Such emergence could potentially generate new forms of treatment-resistant HIV and may also compromise the efficacy of any future HIV vaccines; hence, there is an acute need for education programs and surveillance to reduce the possibility of ongoing transmission.

### Human T-Cell Lymphotropic Virus

HTLV was discovered in the 1980s (Poiesz et al. 1980), and an estimated 15 million to 20 million people worldwide live with HTLV-1 and HTLV-2 (de Thé and Kazanji 1996). These viruses are spread from person to person primarily by infected cells in body fluids such as milk, blood, and semen (Proietti et al. 2005). Of those infected with HTLV-1, 5% develop adult T-cell lymphoma or one of several inflammatory disorders, including HTLV-1-associated myelopathy or tropical spastic paraparesis (Proietti et al. 2005). HTLV-2 appears to be less pathogenic but has been associated with the same disorders (Feuer and Green 2005).

As with SIV, a high diversity of simian T-cell lymphotropic viruses (STLVs) has been detected among wild non-human primates (Courgnaud et al. 2004; Liégeois et al. 2008; Sintasath et al. 2009a), and there is a growing consensus that HTLVs also originated from multiple cross-species transmissions of STLV (Ina and Gojobori, 1990; Suzuki and Gojobori 1998; Van Dooren et al. 2001; Calattini et al. 2005; Wolfe et al. 2005b; Reitz & Gallo 2010). If this is the case, hunting and butchering may well have played an important role. Our recent work in central Africa demonstrates that individuals with high levels of contact with non-human primate blood through hunting and butchering have a wide array of these viruses (Wolfe et al. 2005b), including two novel viruses, HTLV-3 (Calattini et al. 2005; Wolfe et al. 2005b) and HTLV-4 (Wolfe et al. 2005b), which likely originated from non-human primates. Importantly, examination of the genomes of the recently discovered HTLV-3s (Switzer et al. 2006), HTLV-4 (Switzer et al. 2009), and STLV-3 (Sintasath et al. 2009a) indicates that pathogenic components of the genomes remain intact in the novel viruses, suggesting that like HTLV-1 and -2, these newly discovered viruses could cause disease in humans. As such, these viruses should also be considered a high priority for prevention and surveillance, particularly because no curative treatment for HTLV-related disease exists at present (Proietti et al. 2005).

## Simian Foamy Virus (SFV)

Simian foamy viruses are ubiquitous among the non-human primates with which they have co-evolved (Switzer et al. 2004b). While they are absent among people who do not have contact with non-human primates, a number of cases have been reported among people hunting and butchering non-human primates in central Africa (Wolfe et al. 2004b; Calattini et al. 2007); among veterinarians, zoo workers, and research scientists working with non-human primates (Switzer et al. 2004a); and among people exposed to macaques in villages and temples in Asia (Jones-Engel et al. 2005a, 2005b). Among people exposed to non-human primates, this infection appears to be somewhat common. Around 1% of people surveyed in rural Cameroon had antibodies to SFV (Wolfe et al. 2004b), and 24% of people reporting contact with apes and 4% of people reporting contact with monkeys were infected with SFV (Calattini et al. 2007). The 16 known

reported SFV infections in Cameroon originated from a number of non-human primates, including gorilla, chimpanzee, drill, and De Brazza's monkey; the infected individuals generally reported hunting-related contacts and injuries with animals that were consistent with their infections (Wolfe et al. 2004b; Calattini et al. 2007).

No disease has been definitively associated with SFV (Schweizer et al. 1997; Switzer et al. 2004a), and secondary transmission of this virus between humans has not been recorded (Schweizer et al. 1997; Switzer et al. 2004a; Calattini et al. 2007). However, the number of known infections remains low and most of the contemporary infections are probably occurring in areas with limited access to facilities capable of detecting and monitoring this virus. Clinical monitoring of infected individuals and their contacts is of particular importance to determine what threat, if any, these viruses might pose to human health. The frequent presence of these viruses in exposed populations also suggests that other viruses, such as SIV, are probably being transmitted via the same pathway but at much lower rates (Wolfe et al. 2004b), which we are not able to detect at the current levels of surveillance.

## Human Monkeypox

Sporadic cases of human monkeypox have been reported from central and west Africa over the past 40 years (Heymann et al. 1998), and the Democratic Republic of Congo is a particular focus for this disease (Rimoin et al. 2007). Monkeypox causes fever and a severe rash, somewhat similar to the eradicated smallpox virus, but it can also be confused with chickenpox (Rimoin et al. 2007). Fatality rates can be as high as 17%, but vaccinia immunization is somewhat effective for this virus (Di Giulio and Eckburg 2004).

Up to 80% of human monkeypox cases are thought to result from wild animal contact (Arita et al. 1985), although a definitive animal reservoir in Africa remains unknown (Hutson et al. 2007). Monkeypox virus has been isolated from a rope squirrel (*Funisciurus anerythrus*) in the Democratic Republic of Congo (Khodakevich et al. 1986), and monkeypox DNA has been detected in pouched rats (*Cricetomys*), dormice (*Graphiurus*), and ground squirrels (*Xerus*) in Ghana (Reynolds et al. 2010) and in imported and domestic rodents during the 2003 U.S. monkeypox outbreak

(Hutson et al. 2007). Also a variety of animals from areas where human monkeypox is endemic, including primates, squirrels, rats, and shrews, have demonstrated antibody evidence of exposure to viruses from the same family as monkeypox (Hutson et al. 2007; Reynolds et al. 2010).

The exact mode of transmission from animals to humans is still not well known (Jezek et al. 1986), but it likely occurs by direct contact with infected animals, either through bites, scratches, or contact with the blood or body fluids of infected animals. While human-to-human transmission has been reported (Jezek et al. 1986), it is currently believed that the virus cannot be maintained in the human population without continuous reintroduction from the wild reservoir (Jezek et al. 1987; Hutin et al. 2001). This means that present-day cases of human monkeypox will not be likely to cause large-scale epidemics. However, human monkeypox cases will continue to be an issue for populations hunting and butchering rodents and primates in endemic areas, and the current inability of monkeypox to maintain sustained human-to-human transmission does not preclude the possibility that a more transmissible human monkeypox virus may one day arise. This possibility necessitates a better understanding of the zoonotic acquisition and prevention of this virus.

## Ebola Hemorrhagic Fever Virus

Ebola is intermittently responsible for localized but high-mortality epidemics in humans (Pourrut et al. 2005) and wild animals in central Africa (Huijbregts et al. 2003; Walsh et al. 2003; Leroy et al. 2004; Pourrut et al. 2005; Bermejo et al. 2006; Lahm et al. 2007; Rizkalla et al. 2007) and elsewhere. A number of recent Ebola hemorrhagic fever outbreaks in central Africa were associated with the collection, handling, and butchering of Ebola virus-infected carcasses of gorillas, chimpanzees, or ungulates found in the forest (Georges et al. 1999; Huijbregts et al. 2003; Leroy et al. 2004; Rouquet et al. 2005; Ascenzi et al. 2008; Leroy et al. 2009), indicating that once Ebola virus is in susceptible wild animals, humans are at high risk of infection from contact with infected carcasses. While the natural reservoirs of this virus remain unknown, a number of species of fruit bats (*Epomops franqueti*, *Hypsignathus monstrosus, Myonycteris torquata*) appear to maintain asymptomatic infections (Leroy et al.

2005; Pourrut et al. 2007). An exceptional Ebola outbreak in 2007 in the Democratic Republic of Congo, which was responsible for 186 human deaths, was not associated with the usual wild animal mortality or contact with infected animal carcasses found in the forest. However, the assumed first victim bought freshly killed bats, of the same species already known to be natural reservoirs of the Ebola virus, to prepare as food (Leroy et al. 2009). The transmission of Ebola virus through body fluids (Bausch et al. 2007) and the detection of Ebola virus DNA in bat liver and spleen (Leroy et al. 2005) indicate that both handling of hunted carcasses (Leroy et al. 2009) and butchering of bats may pose a risk due to contact with potentially infected body fluids, blood, and organs.

## Anthrax

Anthrax is an acute infectious disease caused by the spore-forming bacterium *Bacillus anthracis*. It is one of the oldest known zoonotic diseases and has been associated with die-offs of wild ungulates in east Africa (Prins and Weyerhaeuser 1987; Wafula et al. 2007) and other fauna, including primates, in central and west Africa (Leendertz et al. 2004, 2006; Klee et al. 2006). Anthrax may even be a significant contributor to the decline of great ape populations in central Africa (Leendertz et al. 2006).

In humans, anthrax presents in three main clinical forms—cutaneous, inhalation, or gastrointestinal—depending on the route of entry. However, regardless of the form, septicemia rapidly ensues (Dixon et al. 1999). At least one human anthrax case has been associated with contact with wild animal carcasses (Van den Enden 2006), and the skinning, butchering, and eating of meat suspected of being infected with anthrax are significant risk factors for human infections (Beatty et al. 2003; Woods et al. 2004). As carcasses of dead animals found in the forest are sometimes collected for consumption in central Africa, this bacterium has the potential to cause infections in the region. While there is an absence of human case reports in the region, the general lack of secondary transmission of anthrax (Beatty et al. 2003; Holty et al. 2006) and the limited access to medical and laboratory diagnostic facilities in the remote forested areas of central Africa where this bacteria has been recorded, mean that isolated cases may have gone undiagnosed or unreported.

## Herpes Viruses

Although in general herpes viruses are highly host specific (Tischer and Osterrieder 2010), Cercopithecine herpesvirus-1 (CeHV-1) has caused human infections in a number of individuals working with captive macaques or macaque tissues (Elmore and Eberle, 2008). In addition to CeHV-1, diverse herpes viruses closely related to pathogenic human counterparts have been described in non-human primates (Lacoste et al. 2001), indicating that monkeys and apes represent a potential reservoir of new herpes viruses that could cross over into human populations (Lacoste et al. 2000a, b, 2001). Nothing is known of zoonotic transmission routes or risk factors for infections with such viruses, and a better understanding of herpes virus diversity in wild primates and exposed individuals is urgently needed to enable us to better address the threats posed by this group of viruses.

## Hepatitis Viruses

Worldwide an estimated 350 million people are infected with hepatitis B virus (HBV) (Lavanchy 2004) and 170 million with hepatitis C virus (HCV) (Markov et al. 2009). HBV is among the leading causes of human mortality worldwide (Lavanchy, 2004). Transmission is through contact with infected body fluids (Lavanchy, 2004). Among chimpanzees and other non-human primates there is a range of hepatitis viruses, including some closely related to HBV and HCV (Simmonds 2001), each of which may pose a risk of zoonosis. Among people exposed to HBV-infected non-human primates there are currently no known cases of zoonotic hepatitis (Noppornpanth et al. 2003), yet these viruses present a particular risk for hunters and butchers given their contact with blood and other body fluids. Surveillance of both animals and exposed individuals for a range of these viruses is essential to determine the risk these viruses pose and to better understand the origins of the widespread human forms of these viruses.

## Paramyxoviruses

Hendra and Nipah and other paramyxoviruses occur in bats (*Pteropus* spp.) on a number of continents, and human infections have resulted from contact with bat-exposed swine and horses (Field et al. 2001) or indirect contact with bats through contaminated palm sap (Luby et al. 2006). In humans these viruses can cause encephalitis, respiratory disease, and death (Field et al. 2001). While there are currently no known cases of human infections resulting from the hunting of bats, serological and molecular evidence of Henipavirus infections in a commonly hunted species, the straw-coloured fruit bat (*Eidolon helvum*) (Mickleburgh et al. 2009), has been reported (Hayman et al. 2008; Drexler et al. 2009); hence, surveillance of hunted bat populations and hunters will be key to ensure that hunting communities are not at risk of infections with these viruses.

## Intestinal Parasites

A number of helminths and protozoans reported in wild non-human primates have the potential to be transmitted to humans (Muller-Graf et al. 1996; Muriuki et al. 1998; Munene et al. 1998; Bezjian et al. 2008; Krief et al. 2010), and in Kibale National Park, Uganda, researchers, tourists, and chimpanzees were found to be infected with similar strains of *Escherichia coli* (Goldberg et al. 2007; Chapter 31 in this book). Zoonotic transmission during hunting and butchering may occur during transport or butchering via contamination of hands with infected material from the intestinal tract or feces. If hands are not washed prior to eating or preparing food, the food may be contaminated and transmission may occur on consumption. Also, certain hookworms (*Ancylostoma* spp.) can be transmitted transcutaneously by infectious larvae, which can be frequent and abundant in the gastrointestinal tracts of hunted monkeys (Pourrut et al. 2011).

## BEHAVIORAL CHANGE

The hunting, butchering, and eating of wild animals has great cultural and economic importance in many parts of the world. Risk-reduction education programs can help to ensure people are aware of the hazards and the means by which they can reduce exposure and potential infections from contact with wild animals. Such educational programs may be an important strategy in reducing the risk of future epidemics and pandemics.

Since 2003, Global Viral Forecasting (GVF) has been undertaking a "Healthy Hunting" program in rural central Africa to provide vulnerable populations with information on the risk of zoonotic transmission from hunted wild animals (Fig. 12.4). Sessions are designed to encourage hunters to reduce their contact with wild animal blood and body fluids by informing them of the pathogens that can be found in wild animals, which species are believed to pose the greatest risk with regards to the transmission of zoonotic agents, and what steps can be taken to avoid possible infections. While it is important to explain that the best way to prevent infections is to avoid handling animals and to reduce exposure to animal blood and body fluids, for many people hunting and butchering is an important daily activity that is unlikely to be discontinued. Thus, to help people minimize their risk, we identified a number of simple and mostly cost-free behavioral changes that may offer a protective benefit if adopted. For example, individuals are encouraged not to butcher when there are injuries on their hands or limbs and to ask someone else to do the work instead; to immediately wash cuts, bites, and scratches with soap and water; to prevent children from handling animal carcasses and uncooked meat; to reduce unnecessary contact with blood and organs; to reduce their exposure to carcasses by using leaves or plastic sheeting to wrap meat, to prevent blood from draining onto the person transporting the animals; and to avoid all contact with animals found dead in the forest. Such actions may decrease the risk to both individuals and communities. Furthermore, as they are interventions that do not increase cost, reduce availability, or fundamentally change current practices, they can be easily adopted. We also encourage people to redirect hunting efforts away from riskier species such as apes and monkeys towards species that are less risky, such as antelopes or rodents. Because of the phylogenetic relatedness of non-human primates to our own species, reducing the opportunities for cross-species transmission of microbes that are already adapted to apes and other non-human primates will likely be of major importance in reducing future human disease emergence. A shift away from the hunting of non-human primates would also benefit the conservation of these species in central Africa, in particular gorillas (*Gorilla gorilla*), chimpanzees (*Pan troglodytes*), bonobos (*Pan paniscus*), mandrills (*Mandrillus sphinx*),

**Figure 12.4:**
Healthy Hunting education sessions, this one being undertaken in the Ngoïla area in Cameroon by Joseph Le Doux Diffo, are an important part of GVF's strategy for preventing disease and engaging communities. (Photo by Matthew LeBreton, GVF)

drills (*M. leucophaeus*), and the localized colobus (e.g., *Piliocolobus preussi*) and guenon species (e.g., *Cercopithecus preussi*).

However, even with proactive measures to facilitate change, the cultural, nutritional, and economic importance of bushmeat may limit people's ability or desire to embrace such change. Providing training, opportunities for alternative activities, or incentives for hunters to reduce hunting will play an important role, in particular in areas where it is imperative to reduce hunting of endangered or localized species such as chimpanzees (*P. troglodytes*), gorillas (*G. gorilla*), mandrills (*M. sphinx*), and Preuss's colobus (*P. preussi*). Also, enhanced commercial meat production will be required to provide viable market alternatives to bushmeat; improved road accessibility is required to allow such alternatives to easily arrive in rural areas; reliable rural electricity is required to permit conservation of butchered meats; and economic development is required to allow people to be able to afford such purchases. However, it is also unlikely that poverty alleviation and economic development alone will drive a decline in bushmeat harvesting (Robinson and Bennett 2002), as in some areas bushmeat consumption increases with household wealth (Wilkie et al. 2005; Fa et al. 2009) and expanded road construction will likely facilitate trade in bushmeat. Importantly, fish rather than domestic animal meat is a substitute for bushmeat (Wilkie et al. 2005; Brashares et al. 2004) and, as such, enhancing fish production and fisheries management programs should be undertaken concurrently with any other activities designed to reduce reliance on wild animals. Taxation has also been suggested (Wilkie et al. 2005); however, price-based interventions should prudently consider potential unintended outcomes. For instance, increasing bushmeat prices through taxes may increase the economic viability of hunting of certain species and put additional pressure on species such as non-human primates (Damania et al. 2005).

## RESEARCH OPPORTUNITIES, SURVEILLANCE, AND EMERGING INFECTIOUS DISEASE PREVENTION

While current global disease control focuses almost exclusively on responding to pandemics after they have already spread globally, we should not neglect the challenging goal of predicting and preventing diseases prior to their emergence (Wolfe et al. 2005a; Jones et al. 2008). As more than half of all human diseases have resulted from zoonotic transmission (Taylor et al. 2001; Wolfe et al. 2007; Jones et al. 2008) and as the human–animal interface is pivotal to the process of disease emergence, it stands to reason that one of the most effective strategies in terms of early detection of an emergent pathogenic threat would be to focus surveillance among people who are highly exposed to animals and on the animal populations to which they are exposed. Since 2003, GVF has been establishing a worldwide network of surveillance sites, consisting of laboratory scientists, veterinarians, wildlife ecologists, behavioral scientists, hunters, and butchers to help predict, detect, and prevent zoonotic disease emergence. One component of this surveillance network has included the collection of dried blood spots on filter paper from over 20,000 hunted animals. This was undertaken in collaboration with people who hunt and butcher wild animals and through the "Healthy Hunting" program. The success of this approach stems from the fact that the technique requires little training, no cold chain, and few special shipping requirements. Using this approach and molecular analytical techniques has led to the discovery of a number of new retroviruses (Sintasath et al. 2009a), including a full genome sequence from one of these viruses (Sintasath et al. 2009b). Such collections require close engagement with communities to develop rapport and acceptance, especially as it is unethical to make payments for specimens collected from hunted animals to avoid further increases in hunting that already represents a conservation issue. Time dedicated to developing community acceptance for such collaborative surveillance also provides opportunities to communicate the risks involved with hunting and butchering, presents opportunities for long-term surveillance of people at risk, and allows for long-term follow-up of people who are infected with novel zoonotic viruses. Such long-term surveillance facilitates the assessment of the pathogenicity and secondary transmission of novel pathogens, an important component of surveillance, particularly because more than just zoonotic transfer is required to start an epidemic (Van Heuverswyn and Peeters 2007; Wolfe et al. 2007).

## CONCLUSIONS

The monitoring of communities and individuals involved in the hunting and butchering of wild animals will likely allow the discovery of a number of pathogens that may have the potential to cause epidemics or even pandemics. Where such pathogens are known or discovered, minimizing hunting/butchering-associated infections will be an essential but complex challenge. Such interventions must be carefully tailored to ensure that cultural, economic, and conservation issues are considered, and that outcomes do not compromise food security or conservation while accomplishing the ultimate goal of reducing the risk of disease emergence.

## ACKNOWLEDGMENTS

N.D.W. is supported by an award from the National Institutes of Health Director's Pioneer Award (Grant DP1-OD000370). Global Viral Forecasting is supported by google.org, the Skoll Foundation, the Henry M. Jackson Foundation for the Advancement of Military Medicine, the United States Armed Forces Health Surveillance Center Division of GEIS Operations, and the United States Agency for International Development (USAID) Emerging Pandemic Threats Program, PREDICT project, under the terms of Cooperative Agreement Number GHN-A-OO-09-00010-00. The contents are the responsibility of the authors and do not necessarily reflect the views of USAID or the United States Government. The authors would also like to thank the Government of Cameroon and the United States Embassy in Cameroon; the entire staff of GVFI-Cameroon for their support; and Dr. Bruno Sainz for assistance in preparing this manuscript.

## REFERENCES

Antia, R., R.R. Regoes, J.C. Koella, and C.T. Bergstrom. 2003. The role of evolution in the emergence of infectious diseases. Nature 426:658–661.

Arita, I., Z. Jezek, L. Khodakevich, and K. Ruti. 1985. Human monkeypox: a newly emerged orthopoxvirus zoonosis in the tropical rain forests of Africa. Am J Trop Med Hyg 34:781–9.

Ascenzi, P., A. Bocedi, J. Heptonstall, M.R. Capobianchi, A. Di Caro, E. Mastrangelo, M. Bolognesi, and G. Ippolito. 2008. Ebolavirus and Marburgvirus: insight the Filoviridae family. Mol Aspects Med 29:151–85.

Auzel, P., and D.S. Wilkie. 2000. Wildlife use in northern Congo: hunting in a commercial logging concession. In J.G. Robinson and E.L. Bennett, eds. Hunting for sustainability in tropical forests, pp. 413–426. Columbia University Press, New York.

Bausch, D.G., J.S. Towner, S.F. Dowell, F. Kaducu, M. Lukwiya, A. Sanchez, S.T. Nichol, T.G. Ksaizek, and P.E. Rollin. 2007. Assessment of the risk of Ebola virus transmission from bodily fluids and fomites. J Infect Dis 196:S142–147.

Beatty, M.E., D.A. Ashford, P.M. Griffin, R.V. Tauxe, and J. Sobel. 2003. Gastrointestinal anthrax: review of the literature. Arch Intern Med 163: 2527–2531.

Bennett, E.L., and J.G. Robinson. 2000. Hunting of wildlife in tropical forests: implications for biodiversity and forest peoples. Biodiversity series—impact studies paper no 76. The World Bank, Washington D.C.

Bermejo, M., J.D. Rodríguez-Teijeiro, G. Illera, A. Barroso, C. Vilà, and P. Walsh. 2006. Ebola outbreak killed 5000 gorillas. Science 314:1564.

Bezjian, M., T.R. Gillespie, C.A. Chapman, and E.C. Greiner. 2008. Coprologic evidence of gastrointestinal helminths of forest baboons, *Papio anubis*, in Kibale National Park, Uganda. J Wildl Dis 44:878–87.

Brashares, J.S., P. Arcese, M.K. Sam, P.B. Coppolillo, A.R. Sinclair, and A. Balmford. 2004. Bushmeat hunting, wildlife declines, and fish supply in West Africa. Science 306:1180–1183.

Calattini, S., S.A. Chevalier, R. Duprez, S. Bassot, A. Froment, R. Mahieux, and A. Gessain. 2005. Discovery of a new human T-cell lymphotropic virus (HTLV-3) in Central Africa. Retrovirology 9:30.

Calattini, S., E.B. Betsem, A. Froment, P. Mauclère, P. Tortevoye, C. Schmitt, R. Njouom, A. Saib, and A. Gessain. 2007. Simian foamy virus transmission from apes to humans, rural Cameroon. Emerg Infect Dis 13:1314–20.

Chapman, C.A., T.R. Gillespie, and T.L. Goldberg. 2005. Primates and ecology of their infectious diseases: how will anthropogenic change affect host-parasite interaction? Evol Anthropol 14:134–144.

Courgnaud, V., S. Van Dooren, F. Liegeois, X. Pourrut, B. Abela, S. Loul, E. Mpoudi-Ngole, A. Vandamme, E. Delaporte, and M. Peeters. 2004. Simian T-cell leukemia virus (STLV) infection in wild primate populations in Cameroon: evidence for dual STLV type 1 and type 3 infection in agile mangabeys (*Cercocebus agilis*). J Virol 78:4700–4709.

Damania, R., E.J. Milner-Gulland, and D.J. Crookes. 2005. A bioeconomic analysis of bushmeat hunting. Proc Biol Sci 272:259–66.

de Thé, G., and M. Kazanji. 1996. An HTLV-I/II vaccine: from animal models to clinical trials? J Acquir Immune Defic Syndr Hum Retrovirol 1:S191–198.

Di Giulio, D.B., and P.B. Eckburg. 2004. Human monkeypox: an emerging zoonosis. Lancet Infect Dis 4:15–25.

Dixon, T.C., M. Meselson, J. Guillemin, and P.C. Hanna. 1999. Anthrax. N Engl J Med 341:815–26.

Drexler, J.F., V.M. Corman, F. Gloza-Rausch, A. Seebens, A. Annan, A. Ipsen, T. Kruppa, M.A. Müller, E.K. Kalko, Y. Adu-Sarkodie, S. Oppong, and C. Drosten. 2009. Henipavirus RNA in African bats. PLoS One 4:e6367.

Engel, G.A., L. Jones-Engel, M.A. Schillaci, K.G. Suaryana, A. Putra, A. Fuentes, and R. Henkel. 2002. Human exposure to herpesvirus B-seropositive macaques, Bali, Indonesia. Emerg Infect Dis 8:789–795.

Elmore, D., and R. Eberle. 2008. Monkey B virus (Cercopithecine herpesvirus 1). Comp Med 58:11–21.

Fa, J.E., C.A. Peres, and J. Meeuwig. 2002a. Bushmeat exploitation in tropical forests: an intercontinental comparison. Conserv Biol 16:232–237.

Fa, J.E., J. Juste, R.W. Burn, and G. Broad. 2002b. Bushmeat consumption and preferences of two ethnic groups in Bioko Island, West Africa. Human Ecol 30:397–416.

Fa, J.E., D. Currie, and J. Meeuwig. 2003. Bushmeat and food security in the Congo Basin: linkages between wildlife and people's future. Environ Conserv 31: 71–78.

Fa, J.E., S.F. Ryan, and D.J. Bell. 2005. Hunting vulnerability, ecological characteristics and harvest rates of bushmeat species in afrotropical forests. Biol Conserv 121:167–176.

Fa, J.E., Albrechtsen, L., Johnson, P.J., Macdonald, D.W. 2009. Linkages between household wealth, bushmeat and other animal protein consumption are not invariant: evidence from Rio Muni, Equatorial Guinea. Anim Conserv 12:599–610.

Fa, J.E., S. Seymour, J. Dupain, R. Amin, L. Albrechtsen, and D. Macdonald. 2006. Getting to grips with the magnitude of exploitation: Bushmeat in the Cross-Sanaga rivers region, Nigeria and Cameroon. Biol Conserv 129:497–510.

Feuer, G., and P.L. Green. 2005. Comparative biology of human T-cell lymphotropic virus type 1 (HTLV-1) and HTLV-2. Oncogene 24:5996–6004.

Field, H., P. Young, J.M. Yob, J. Mills, L. Hall, and J. Mackenzie. 2001. The natural history of Hendra and Nipah viruses. Microbes Infect 3: 307–14.

Georges, A.J., E.M. Leroy, A.A. Renaut, C.T. Benissan, R.J. Nabias, M.T. Ngoc, P.I. Obiang, J.P. Lepage, E.J. Bertherat, D.D. Bénoni, E.J. Wickings, J.P. Amblard, J.M. Lansoud-Soukate, J.M. Milleliri, S. Baize, and M.C. Georges-Courbot. 1999. Ebola hemorrhagic fever outbreaks in Gabon, 1994–1997: epidemiologic and health control issues. J Infect Dis 179:S65–75.

Goldberg, T.L., T.R. Gillespie, I.B. Rwego, E. Wheeler, E.L. Estoff, and C.A. Chapman. 2007. Patterns of gastrointestinal bacterial exchange between chimpanzees and humans involved in research and tourism in western Uganda. Biol Conserv 135:511–517.

Hahn, B.H., G.M. Shaw, K.M. De Cock, and P.M. Sharp. 2000. AIDS as a zoonosis: scientific and public health implications. Science 287:607–614.

Hayman, D.T., R. Suu-Ire, A.C. Breed, J.A. McEachern, L. Wang, J.L., and A.A. Cunningham. 2008. Evidence of henipavirus infection in West African fruit bats. PLoS One 3:e2739.

Heymann, D.L., M. Szczeniowski, and K. Esteves. 1998. Re-emergence of monkeypox in Africa: a review of the past six years. Br Med Bull 54:693–702.

Huijbregts, B., P. DeWachter, S. Ndong Obiang, and M. Akou Ella. 2003. Ebola and the decline of gorilla Gorilla gorilla and chimpanzee Pan troglodytes populations in Minkebe forest, north-eastern Gabon. Oryx 37:437–443.

Hutin, Y.J., R.J. Williams, P. Malfait, R. Pebody, V.N. Loparev, S.L. Ropp, M. Rodriguez, J.C. Knight, F.K. Tshioko, A.S. Khan, M.V. Szczeniowski, and J.J. Esposito. 2001. Outbreak of human monkeypox, Democratic Republic of Congo, 1996–1997. Emerg Infect Dis 7:434–438.

Hutson, C.L., K.N. Lee, J. Abel, D.S. Carroll, J.M. Montgomery, V.A. Olson, Y. Li, W. Davidson, C. Hughes, M. Dillon, P. Spurlock, J.J. Kazmierczak, C. Austin, L. Miser, F.E. Sorhage, J. Howell, J.P. Davis, M. G. Reynolds, Z. Braden, K.L. Karem, I.K. Damon, and R.L. Regnery. 2007. Monkeypox zoonotic associations: insights from laboratory evaluation of animals associated with the multi-state US outbreak. Am J Trop Med Hyg 76:757–768.

Ina, Y., and T. Gojobori. 1990. Molecular evolution of human T-cell leukemia virus. J Mol Evol 31:493–499.

Jezek, Z., B. Grab, and H. Dixon. 1987. Stochastic model for interhuman spread of monkeypox. Am J Epidemiol 26:1082–1092.

Jezek, Z., I. Arita, M. Mutombo, C. Dunn, J.H. Nakano, and M. Szczeniowski . 1986. Four generations of probable person-to-person transmission of human monkeypox. Am J Epidemiol 123:1004–1012.

Holty, J.C., D.M. Bravata, H. Liu, R.A. Olshen, K.M. McDonald, and D.K. Owens. 2006. Systematic Review: A Century of Inhalational Anthrax Cases from 1900 to 2005. Ann Intern Med 144:270–280.

Jones, K.E., N.G. Patel, M.A. Levy, A. Storeygard, D. Balk, J.L. Gittleman, and P. Daszak. 2008. Global trends in emerging infectious diseases. Nature 451:990–993.

Jones-Engel, L., C.C. May, G.A. Engel, K.A. Steinkraus, M.A. Schillaci, A. Fuentes, A. Rompis, M.K. Chalise, N. Aggimarangsee, M.M. Feeroz, R. Grant, J.S. Allan, A. Putra, N. Wandia, R.Watanabe, L.R. Kuller, S. Thongsawat, R. Chaiwarith, R.C. Kyes, and M.L. Linial. 2005a. Diverse contexts of zoonotic transmission of simian foamy viruses in Asia. Emerg Infect Dis 14:1200–1208.

Jones-Engel, L., G.A. Engel, M.A. Schillaci, A. Rompis, A. Putra, K.G. Suaryana, A. Fuentes, B. Beer, S. Hicks, R. White, B. Wilson, and J.S. Allan. 2005b. Primate-to-human retroviral transmission in Asia. Emerg Infect Dis 14:1028–1035.

Kalish, M.L., N.D. Wolfe, C.B. Ndongmo, J. McNicholl, K.E. Robbins, M. Aidoo, P.N. Fonjungo, G. Alemnji, C. Zeh, C.F. Djoko, E. Mpoudi-Ngole, D.S. Burke, and T.M. Folks. 2005. Central African hunters exposed to simian immunodeficiency virus. Emerg Infect Dis 11:1928–30.

Karesh, W.B., and E. Noble. 2009. The bushmeat trade: increased opportunities for transmission of zoonotic disease. Mt Sinai J Med 76:429–34.

Khabbaz, R.F., W. Heneine, J.R. George, B. Parekh, T. Rowe, T. Woods, W.M. Switzer, H.M. McClure, M. Murphey-Corb, and T.M. Folks. 1994. Infection of a laboratory worker with simian immunodeficiency virus. N Engl J Med 330:172–177.

Klee, S.R., M. Ozel, B. Appel, C. Boesch, H. Ellerbrok, D. Jacob, G. Holland, F.H. Leendertz, G. Pauli, R. Grunow, and H. Nattermann. 2006. Characterization of Bacillus anthracis-like bacteria isolated from wild great apes from Cote d'Ivoire and Cameroon. J Bacteriol 188:5333–5344.

Khodakevich, L., Z. Jezek, and K. Kinzanzka. 1986. Isolation of monkeypox virus from wild squirrel infected in nature. Lancet 1:98–99.

Krief, S., B. Vermeulen, S. Lafosse, J.M. Kasenene, A. Nieguitsila, M. Berthelemy, M. L'hostis, O. Bain, and J. Guillot. 2010. Nodular worm infection in wild chimpanzees in Western Uganda: a risk for human health? PLoS Negl Trop Dis 4:e630.

Lacoste, V., P. Mauclere, G. Dubreuil, J. Lewis, M.-C. Georges-Courbot, J. Rigoulet, T. Petit, and A. Gessain. 2000a. Simian homologues of human gamma-2 and betaherpesviruses in mandrill and drill monkeys. J Virol 74:11993–11999.

Lacoste, V., P. Mauclère, G. Dubreuil, J. Lewis, M.-C. Georges-Courbot, and A. Gessain. 2000b. KSHV-like herpesviruses in chimps and gorillas. Nature 407:151–152.

Lacoste, V., P. Mauclere, G. Dubreuil, J. Lewis, M.-C. Georges-Courbot, and A. Gessain. 2001. A novel gamma-2-herpesvirus of the Rhadinovirus 2 lineage in chimpanzees. Genome Res 11:1511–1519.

Lahm, S.A., M. Kombila, R. Swanepoel, and R.F. Barnes. 2007. Morbidity and mortality of wild animals in relation to outbreaks of Ebola haemorrhagic fever in Gabon, 1994-2003. Trans R Soc Trop Med Hyg 101:64–78.

Laurance, W.E., B.M. Croes, L. Tchignoumba, S.A. Lahm, A. Alonso, M.E. Lee, P. Campbell, and C. Ondzeano. 2006. Impacts of roads and hunting on central African rainforest mammals. Conserv Biol 20:1251–1261.

Lavanchy, D. 2004. Hepatitis B virus epidemiology, disease burden, treatment, and current and emerging prevention and control measures. J Viral Hepat 11:97–107.

LeBreton, M., A.T. Prosser, U. Tamoufe, W. Sateren, E. Mpoudi-Ngole, J.L.D. Diffo, D.S. Burke, and N.D. Wolfe. 2006. Patterns of bushmeat hunting and perceptions of disease risk among central African communities. Anim Conserv 9:357–363.

Leendertz, F.H., H. Ellerbrok, C. Boesch, E. Couacy-Hymann, K. Matz-Rensing, R. Hakenbeck, C. Bergmann, P. Abaza, S. Junglen, Y. Moebius, L. Vigilant, P. Formenty, and G. Pauli. 2004. Anthrax kills wild chimpanzees in a tropical rainforest. Nature 430:451–452.

Leendertz, F.H., F. Lankester, P. Guislain, C. Neel, O. Drori, J. Dupain, S. Speede, P. Reed, N. Wolfe, S. Loul, E. Mpoudi-Ngole, M. Peeters, C. Boesch, G. Pauli, H. Ellerbrok, and E.M. Leroy. 2006. Anthrax in Western and Central African great apes. Am J Primatol 68:928–33.

Leroy, E.M., P. Rouquet, P. Formenty, S. Souquière, A. Kilbourne, J.M. Froment, M. Bermejo, S. Smit, W. Karesh, R. Swanepoel, S.R. Zaki, and P.E. Rollin. 2004. Multiple Ebola virus transmission events and rapid decline of central African wildlife. Science 303:387–390.

Leroy, E.M., B. Kumulungui, X. Pourrut, P. Rouquet, A. Hassanin, P. Yaba, A. Délicat, J.T. Paweska, J.P. Gonzalez, and R. Swanepoel. 2005. Fruit bats as reservoirs of Ebola virus. Nature 438:575–576.

Leroy, E.M., A. Epelboin, V. Mondonge, X. Pourrut, J.P. Gonzalez, J.J. Muyembe-Tamfum, and P. Formenty. 2009. Human Ebola outbreak resulting from direct exposure to fruit bats in Luebo, Democratic Republic of Congo, 2007. Vector Borne Zoonotic Dis 9:723–728.

Liégeois, F., B. Lafay, W.M. Switzer, S. Locatelli, E. Mpoudi-Ngolé, S. Loul, W. Heneine, E. Delaporte, and M. Peeters. 2008. Identification and molecular characterization of new STLV-1 and STLV-3 strains in wild-caught nonhuman primates in Cameroon. Virology 371:405–17.

Luby, S.P., M. Rahman, M.J. Hossain, L.S. Blum, M.M. Husain, E. Gurley, R. Khan, B.N. Ahmed, S. Rahman, N. Nahar, E. Kenah, J.A. Comer, and T.G. Ksiazek. 2006. Foodborne transmission of Nipah virus, Bangladesh. Emerg Infect Dis 12:1888–1894.

Markov, P.V., J. Pepin, E. Frost, S. Deslandes, A.C. Labbé, and O.G. Pybus. 2009. Phylogeography and molecular epidemiology of hepatitis C virus genotype 2 in Africa. J Gen Virol 90:2086–2096.

May, R.M., S. Gupta, and A.R. McLean. 2001. Infectious disease dynamics: What characterizes a successful invader? Philos Trans R Soc Lond B Biol Sci 356: 901–910.

Merfield, F. 1956. Gorilla hunter: The African Adventures of a hunter extraordinary. Farrar, Straus, and Cudahy, New York.

Mickleburgh, S., K. Waylen, and P. Racey. 2009. Bats as bushmeat: a global review. Oryx 43:217–234.

Milner-Gulland, E.J., E.L. Bennett, and the SCB 2002 Annual Meeting Wild Meat Group. 2003. Wild meat: the bigger picture. Trend Ecol Evol 18:351–357.

Morse, S.S. 1995. Factors in the emergence of infectious diseases. Emerg Infect Dis 1:7–15.

Muller-Graf, C.D., D.A. Collins, and M.E. Woolhouse. 1996. Intestinal parasite burden in five troops of olive baboons (Papio cynocephalus anubis) in Gombe Stream National Park, Tanzania. Parasitology 112:489–97.

Munene, E., M. Otsyula, D.A. Mbaabu, W.T. Mutahi, S.M. Muriuki, and G.M. Muchemi. 1998. Helminth and protozoan gastrointestinal tract parasites in captive and wild-trapped African non-human primates. Vet Parasitol 78:195–201.

Muriuki, S.M., R.K. Murugu, E. Munene, G.M. Karere, and D.C. Chai. 1998. Some gastro-intestinal parasites of zoonotic (public health) importance commonly observed in old world non-human primates in Kenya. Acta Tropica 71:73–82.

Nopppornpanth, S., B.L. Haagmans, P. Bhattarakosol, P. Ratanakorn, H.G. Niesters, A.D. Osterhaus, and Y. Poovorawan. 2003. Molecular epidemiology of gibbon hepatitis B virus transmission. J Gen Virol 84:147–155.

Noss, A.J. 1997. The economic importance of communal net hunting among the BaAka of the Central African Republic. Human Ecol 25:71–89.

Noss, A.J. 1998a. Cable snares and bushmeat markets in a central African forest. Environ Conserv 25: 228–233.

Noss, A.J. 1998b. The impacts of cable snare hunting on wildlife populations in the forests of the Central African Republic. Conserv Biol 12:390–398.

Peeters, M., V. Courgnaud, B. Abela, P. Auzel P, X. Pourrut, F. Bibollet-Ruche, S. Loul, F. Liegeois, C. Butel, D. Koulagna, E. Mpoudi-Ngole, G.M. Shaw, B.H. Hahn, and E. Delaporte. 2002. Risk to human health from a plethora of simian immunodeficiency viruses in primate bushmeat. Emerg Infect Dis 8:451–457.

Pike, B.L., K.E. Saylors, J.N. Fair, M. LeBreton, U. Tamoufe, C.F. Djoko, A.W. Rimoin, and N.D. Wolfe. 2010. The origin and prevention of pandemics. Clin Infect Dis 50:1636–1640.

Plantier, J.C., M. Leoz, J.E. Dickerson, F. De Oliveira, F. Cordonnier, V. Lemee, F. Damond, D.L. Roberton, and F. Simon. 2009. A new human immunodeficiency virus derived from gorillas. Nat Med 15:871–872.

Poiesz, B.J., F.W. Ruscetti, A.F. Gazdar, P.A. Bunn, J.D. Minna, and R.C. Gallo. 1980. Detection and isolation of type C retrovirus particles from fresh and cultured lymphocytes of a patient with cutaneous T-cell lymphoma. Proc Natl Acad Sci USA 77: 7415–7419.

Pourrut, X., B. Kumulungui, T. Wittmann, G. et al. 2005. The natural history of Ebola virus in Africa. Microbes Infect (7-8):1005–1014.

Pourrut, X., A. Délicat, P.E. Rollin, T.G. Ksiazek, J.P. Gonzalez, and E.M. Leroy. 2007. Spatial and temporal patterns of Zaire ebolavirus antibody prevalence in the possible reservoir bat species. J Infect Dis 196:S176–83.

Pourrut, X., J.L. Diffo, R.M. Somo, C.F. Bilong Bilong, E. Delaporte, M. LeBreton, and J.P. Gonzalez. 2011. Prevalence of gastrointestinal parasites in primate bushmeat and pets in Cameroon. Vet Parasitol 175:187–191.

Prins, H.H.T., and F.J. Weyerhaeuser. 1987. Epidemics in populations of wild ruminants: anthrax and impala, rinderpest and buffalo in Lake Manyara National Park, Tanzania. Oikos 49:28–38.

Proietti, F.A., A.B. Carneiro-Proietti, B.C. Catalan-Soares, and E.L. Murphy. 2005. Global epidemiology of HTLV-I infection and associated diseases. Oncogene 24:6058–6068.

Reitz Jr., M.S., and R.C. Gallo. 2010. HTLV and HIV. In R. Kurth and N. Bannert, eds. Molecular biology, genomics and pathogenesis, Chapter 16. Caister Academic Press, Norfolk, UK.

Reynolds, M.G., D.S. Carroll, V.A. Olson, C. Hughes, J. Galley, A. Likos, J.M. Montgomery, R. Suu-Ire, M.O. Kwasi, J. Jeffrey Root, Z. Braden, J. Abel, C. Clemmons, R. Regnery, K. Karem, and I.K. Damon. 2010. A silent enzootic of an Orthopoxvirus in Ghana, West Africa: evidence for multi-species involvement in the absence of widespread human disease. Am J Trop Med Hyg. 82:746–754.

Rimoin, A.W., N. Kisalu, B. Kebela-Ilunga, T. Mukaba, L.L. Wright, P. Formenty, N.D. Wolfe, R.L. Shongo, F. Tshioko, E. Okitolonda, J.J. Muyembe, R.W. Ryder, and H. Meyer. 2007. Endemic human monkeypox, Democratic Republic of Congo, 2001–2004. Emerg Infect Dis 13:934–937.

Rizkalla, C., F. Blanco-Silva, and S. Gruver. 2007. Modeling the impact of Ebola and bushmeat hunting on western lowland gorillas. EcoHealth 4:151–155.

Robinson, J.G., and E.L. Bennett. 2002. Will alleviating poverty solve the bushmeat crisis? Oryx 36:332.

Rouquet, P., J.M. Froment, M. Bermejo, A. Kilbourn, B. Karesh, P. Reed, B. Kumulungui, P. Yaba, A. Délicat, P.E. Rollin, and E.M. Leroy. 2005. Wild animal mortality monitoring and human Ebola outbreaks, Gabon and Republic of Congo, 2001–2003. Emerg Infect Dis 11:283–290.

Schenck, M., E.N. Effa, M. Starkey, D. Wilkie, K. Abernethy, P. Telfer, R. Godoy, and A. Treves. 2006. Why people eat bushmeat: results from two-choice, taste tests in Gabon, Central Africa. Human Ecol 34:433–445.

Schweizer, M., V. Falcone, J. Gange, R. Turek, and D. Neumann-Haefelin. 1997. Simian foamy virus isolated from an accidentally infected human individual. J Virol 71:4821–4824.

Sellin, B., E. Simonkovich, and L. Ovazza. 1980. *Schistosoma mansoni*: an experimental schistosomiasis in *Erythrocebus patas* monkeys. Médecine Tropicale 40(3):237–41.

Simmonds, P. 2001. The origin and evolution of hepatitis viruses in humans. J Gen Virol 82:693–712.

Sintasath, D.M., N.D. Wolfe, M. LeBreton, H. Jia, A.D. Garcia, J. Le Doux-Diffo, U. Tamoufe, J.K. Carr, T.M. Folks, E. Mpoudi-Ngole, DS. Burke, W. Heneine, and W.M. Swizer. 2009a. Simian T-lymphotropic virus diversity among nonhuman primates, Cameroon. Emerg Infect Dis 2015:175–184.

Sintasath, D.M., N.D. Wolfe, H.Q. Zheng, M. LeBreton, M. Peeters, U. Tamoufe, C.F. Djoko, J.L. Diffo, E. Mpoudi-Ngole, W. Heneine, and W.M. Switzer. 2009b. Genetic characterization of the complete genome of a highly divergent simian T-lymphotropic virus (STLV) type 3 from a wild *Cercopithecus mona* monkey. Retrovirology 27:97.

Suzuki, Y., and T. Gojobori. 1998. The origin and evolution of human T-cell lymphotropic virus types I and II. Virus Genes 16:69–84.

Switzer, W.M., V, Bhullar, V. Shanmugam, M. Cong, B. Parekh, N.W. Lerche, J.L. Yee, J.J. Ely, R. Boneva, L.E. Chapman, T.M. Folks, and W. Heneine. 2004a. Frequent simian foamy virus infection in persons occupationally exposed to nonhuman primates. J Virol 78:2780–2789.

Switzer, W.M., M. Salemi, V. Shanmugam, F. Gao, M. Cong, C. Kuiken, V. Bhullar, B.E. Beer, D. Vallet, A. Gautier-Hion, Z. Tooze, F. Villinger, E.C. Holmes, and W. Heneine. 2004b. Ancient co-speciation of simian foamy viruses and primates. Nature 434: 376–380.

Switzer, W.M., S.H. Qari, N.D. Wolfe, D.S. Burke, T.M. Folks, and W. Heneine. 2006. Ancient origin and molecular features of the novel human T-lymphotropic virus type 3 revealed by complete genome analysis. J Virol 80:7427–38.

Switzer, W.M., M. Salemi, S.H. Qari, H. Jia, R.R. Gray, A. Katzourakis, S.J. Marriott, K.N. Pryor, N.D. Wolfe, D.S. Burke, T.M, Folks, and W. Heneine. 2009. Ancient, independent evolution and distinct molecular features of the novel human T-lymphotropic virus type 4. Retrovirology 2:6–9.

Taylor, L.H., S.M. Latham, and M.E. Woolhouse. 2001. Risk factors for human disease emergence. Philos Trans R Soc Lond B Biol Sci 356:983–989.

Ter Meulen, J., I. Lukashevich, K. Sidibe, A. Inapogui, M. Marx, A. Dorlemann, M.L. Yansane, K. Koulemou, J. Chang-Claude, and H. Schmitz. 1996. Hunting of peridomestic rodents and consumption of their meat as possible risk factors for rodent-to-human transmission of Lassa virus in the Republic of Guinea. Am J Trop Med Hyg 55:661–666.

Tischer, B.K., and N. Osterrieder. 2010. Herpesviruses—a zoonotic threat? Vet Microbiol 140:266–270.

VandeWoude, S., and C. Apetrei. 2006. Going wild: lessons from naturally occurring T-lymphotropic lentiviruses. Clin Microbiol Rev 19:728–762.

Van den Enden, E., A. Van Gompel, and M. Van Esbroeck. 2006. Cutaneous anthrax, Belgian traveler. Emerg Infect Dis 12:523–525.

Van Dooren, S., M. Salemi, and A.M. Vandamme. 2001. Dating the origin of the African human T-cell lymphotropic virus type-i (HTLV-I) subtypes. Mol Biol Evol 18:661–671.

Van Heuverswyn, F., and M. Peeters. 2007. The origins of HIV and implications for the global epidemic. Curr Infect Dis Rep 9:338–346.

Van Lieshout, L., M. Johanna, J.M. De Gruijter, M. Adu-Nsiah, M. Haizel, J.J. Verweij, E.A. Brienen, R.B. Gasser, and A.M. Polderman. 2005. *Oesophagostomun bifurcum* in non-human primates is not a potential reservoir for human infection in Ghana. Trop Med Int 10:1315–1320.

Wafula, M.M., A. Patrick, and T. Charles. 2007. Managing the 2004/05 anthrax outbreak in Queen Elizabeth and Lake Mburo National Parks, Uganda. Afr J Ecol 46:24–31.

Walsh, P.D., K.A. Abernethy, M. Bermejo, R. Beyers, P. De Wachter, M.E. Akou, B. Huijbregts,

D.I. Mambounga, A.K. Toham, A.M. Kilbourn, S.A. Lahm, S. Latour, F. Maisels, C. Mbina, Y. Mihindou, S.N. Obiang, E.N. Effa, M.P. Starkey, P. Telfer, M. Thibault, C.E. Tutin, L.J. White, and D.S. Wilkie. 2003. Catastrophic ape decline in western equatorial Africa. Nature 422:611–614.

Wilkie, D.S., J.G. Sidle, and G.C. Boundzanga. 1992. Mechanized logging, market hunting and a bank loan in Congo. Conserv Biol 6:570–580.

Wilkie, D.S., M. Starkey, K. Abernethy, E. Nstame Effa, P. Telfier, and R. Godoy. 2005. Role of prices and wealth in consumer demand. Conserv Biol 19: 268–274.

Wilkie, D.S., and J.F. Carpenter. 1999. Bushmeat hunting in the Congo Basin: an assessment of impacts and options for mitigation. Biodiv Conserv 8:927–955.

Wilcox, A.S., and D.M. Nambu. 2007. Wildlife hunting practices and bushmeat dynamics of the Banyangi and Mbo people of southwestern Cameroon. Biol Conserv 134: 251–261.

Wolfe, N.D., T.A. Prosser, J.K. Carr, U. Tamoufe, E. Mpoudi-Ngole, J.N. Torimiro, M. LeBreton, F.E. McCutchan, D.L. Birx, and D.S. Burke. 2004. Exposure to nonhuman primates in rural Cameroon. Emerg Infect Dis 10:2094–2099.

Wolfe, N.D., W.M. Switzer, J.K. Carr, V.B. Bhullar, V. Shanmugam, U. Tamoufe, A.T. Prosser, J.N. Torimiro, A. Wright, E. Mpoudi-Ngole, F.E. McCutchan, D.L. Birx, T.M. Folks, D.S. Burke, and W. Heneine. 2004b. Naturally acquired simian retrovirus infections in central African hunters. Lancet 363:932–937.

Wolfe, N.D., P. Daszak, A.M. Kilpatrick, and D.S. Burke. 2005a. Bushmeat hunting, deforestation, and prediction of zoonotic disease emergence. Emerg Infect Dis 11:1822–1827.

Wolfe, N.D., W. Heneine, J.K. Carr, A.D. Garcia, V. Shanmugam, U. Tamoufe, J.N. Torimiro, A.T. Prosser, M. LeBreton, E. Mpoudi-Ngole, F.E. McCutchan, D.L. Birx, T.M. Folks, D.S. Burke, and W.M. Switzer. 2005b. Emergence of unique primate T-lymphotropic viruses among central African bushmeat hunters. Proc Natl Acad Sci 102:7994–7999.

Wolfe, N.D., C.P. Dunavan, and J. Diamond. 2007. Origins of major human infectious diseases. Nature 447:279–283.

Woods, C.W., K. Ospanov, A. Myrzabekov, M. Favorov, B. Plikaytis, and D.A. Ashford. 2004. Risk factors for human anthrax among contacts of anthrax-infected livestock in Kazakhstan. Am J Trop Med Hyg 71: 48–52.

Woolhouse, M.E.J., R. Howey, E. Gaunt, E. Reilly, M. Chase-Topping, and N. Savill. 2008. Temporal tends in the discovery of human viruses. Proc R Soc 275:2111–2115.

Worobey, M., M. Gemmel, D.E. Teuwen, T. Haselkorn, K. Kunstman, M. Bunce, J.J. Muyembe, J.M. Kabongo, R.M. Kalengayi, E. Van Marck, M.T. Gilbert, and S.M. Wolinsky. 2008. Direct evidence of extensive diversity of HIV-1 in Kinshasa by 1960. Nature 455:661–664.

Zhu, T., B.T. Korber, A.J. Nahmias, E. Hooper, P.M. Sharp, and D.D. Ho. 1998. An African HIV-1 sequence from 1959 and implications for the origin of the epidemic. Nature 391:594–597.

# 13

# HUMAN MIGRATION, BORDER CONTROLS, AND INFECTIOUS DISEASE EMERGENCE

## Anne M. Alexander, David C. Finnoff, and Jason F. Shogren

Outbreaks of infectious diseases pose significant threats to human health across the globe. Economists have studied such natural hazards for decades (Kunreuther and Rose 2004). They help formulate cost-effective policies by examining (a) the resources at risk from an outbreak, (b) the driving forces of the outbreak, (c) how policies affect (a) and (b), and (d) the costs to implement alternative policy options. While some effort has been expended to estimate the consequences of disease outbreaks (Meltzer et al. 1999; Haacker 2004) and the driving forces (Jones et al. 2008; Smith et al. 2009), little work exists on how policies influence the consequences or driving forces given the costs.

These evaluations are complicated by the intricacies of the disease process. Infectious diseases spread from one locality to another through human contact or through other vectors, depending on the disease. For example, zoonotic infectious diseases spread by natural processes like species movements, and through human-mediated processes such as travel

and the trade of infected animals, and directly from non-human vertebrates to humans (zoonoses), waterborne transmission, and environmental contaminants. Economic incentives underlie the human-mediated processes. Humans travel and trade for a reason. Private firms and people also respond and react to the risk of an outbreak, as do collective decision makers in governments. The risk[1] of disease outbreaks is endogenous (Shogren and Crocker 1991), as human behaviors can exacerbate or mitigate risks. Risk of an outbreak depends not only on exogenous forces, such as ecological conditions, but also on the human behaviors that govern human migration and trade, and on how people, firms, and policymakers respond to the risk. Since most people can protect themselves (take actions to reduce the chances they are affected) or adapt (take actions to lower the harmful effects of the outbreak), they alter the risk they face. People can remove themselves from or limit interactions with society and the environment around them. By undertaking such actions, regardless of the exogenous

---

1 The economic definition of risk combines two elements: the likelihood the event occurs and the severity of the outcome if it is realized.

components of the risk they face, they are able to influence the risk they face. The risk is endogenous.[2]

The intertwined actions of humans and nature make defining optimal policy responses to disease outbreaks difficult. If social policies are made that do not address individual responses of people and firms, scarce resources will be wasted. In addition, the timing of policies complicates matters. Policies may be enacted before the cause or transmission mechanism of the disease outbreak has been understood. This can lead to broad-based "scattershot" policies, wherein a large number of varied measures are implemented regardless of efficacy or broader cost–benefit considerations. Such policies enacted on social or collective levels include restrictions on human movement and trade into an unaffected area to deter the risk of further transmission. Other policies are enacted following an outbreak. These may be private or collective and include vaccinations, quarantines, and any measures employed to avoid further infections or to alleviate the suffering of infections. As the benefits of *ex ante* policies are hard to enumerate and those of *ex post* easier, most policies tend to be *ex post* and violate the well-known rule of thumb "an ounce of prevention is worth a pound of cure" (Leung et al. 2002).

In the context of cross-border trade and migration, policy decisions take on additional layers of complexity. For example, the gene New Delhi metallo-beta-lactamase-1 (NDM-1) makes bacteria produce an enzyme that makes people resistant to antibiotics. The gene originates in Pakistan and India where infected patients have received medical care and has recently been detected in the Western Hemisphere, including Canada, the United Kingdom, Sweden, Australia, Japan, and the United States. These "superbugs" raise policy questions about appropriate levels of response: Should all travelers from Pakistan and India be isolated and tested for the bacteria? Should only those who have received medical care there be isolated? Is it possible that a returnee has already affected his or her community by passing the bacteria to others, and if so will and should the policy response be more stringent than if the status quo continues? In turn it is also

necessary to understand the impact this response would have on trade flows between India and the United States, Canada, and the European Union.

Compounding the difficulty to find and implement an economically optimal policy response is that reactions to outbreaks may be motivated by political interests, not by public health concerns or economic efficiency. These non-optimal levels of restrictions on trade or human migration are not intended to deter disease movement, but rather they serve as a non-tariff trade or migration barrier. Xenophobia, retaliatory trade intent, and appellations to public hysteria are examples of such non-public health, non-economic policies. Models of the political economy (how political forces affect policy choice) of such behavior provide insights into how the added dimension of this behavior can have long-term negative implications for economic growth. In this chapter we present a basic economic model of the problem of infectious disease outbreaks that arise through human migration and demonstrate the implications of policies directed at the cause of disease outbreaks and their evaluation, including the implications of political motivations for policies. We illustrate the issues by appealing to an intriguing example detailing the spread of typhus across the U.S.–Mexican border in the early 20th century. We find that while the policies directed at the spread of typhus reduced the rate of incidence, they also likely reduced the long-term rate of economic growth of the region. The implication is that there is a need to pay attention to the appropriate level of trade-restricting policies such that the goal of protecting public health remains the objective, rather than political objectives that are detrimental to economic development.

## MIGRATION, DISEASE, AND BORDER CONTROLS: AN ECONOMIC MODEL

We illustrate the consequences and policy responses for a society facing the threat of a disease outbreak

---

2 The term "endogenous risk" reflects the idea that changes in behavior change the risk faced by a person, all else constant. While the exogenous components to the spread of a disease matter too, a change in behavior changes the risk faced by individuals. Examples abound. People reduce personal risk by wearing facemasks while traveling, applying sunscreen, filtering drinking water, and so on.

that is driven by cross-border human travel and migration. We use an economic model of resource allocation at a national level (i.e., for the United States), as shown in Figure 13.1. The model is a derivative from Finnoff et al. (2010). The figure represents the economy of the United States, containing individual consumers, firms, and government. To focus on disease emergence through migration, all cross-border trade in goods is omitted, although it could be represented in a basic extension of the model. Firms in the economy use resources such as labor and raw materials to produce goods that are bought and consumed by consumers. Cross-border migration expands the resources an economy has to produce goods, although the skill characteristics of those migrating may have differential impacts on different industries in the economy.

The economy is assumed to have two goods, $A$ and $X$. How much of the two goods can be produced from the available resources is given by any point inside the bowed line. This line traces out efficient production combinations of the two goods and is called the production possibilities frontier ($PPF$). An economy uses its resources efficiently when it is on the $PPF$.

Any point inside the $PPF$ is feasible but leaves resources under- or unemployed. The slope of the $PPF$ is a key technological description of the economy as it describes how the economy can substitute from the production of one good to the other. This relationship is called the marginal rate of transformation and describes what must be given up of one good to produce more of the other. In the model, increasing an economy's available resources (such as an inflow of undifferentiated migratory labor) shifts the $PPF$ out and to the right as in panel (i), allowing more of both goods to be produced. If migratory labor were to be differentiated and productive in only a single industry (for example, industry A) the $PPF$ would be pivoted upwards in the direction of the input-employing industry, as in panel (ii).

Determining the best allocation of the economy's resources requires a balancing of the available resources with the optimal decisions by firms and optimal decisions by consumers. Firms in the model in both industries make their optimal choices by maximizing their profits given the resource availability, costs, and output prices. These optimization problems are jointly reflected by a balance between the marginal

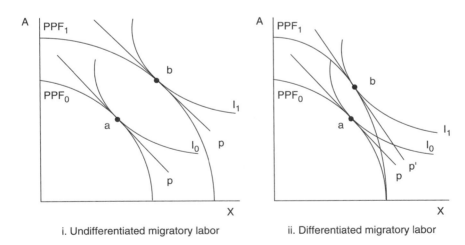

i. Undifferentiated migratory labor            ii. Differentiated migratory labor

**Figure 13.1:**
Temporary migration in an economic framework. With no migration the domestic equilibrium is point *a*. Inflows of migratory labor can be undifferentiated (i) or differentiated (ii) and productive only in industry A. In either case the migration expands the resource base and allows the equilibrium to move to a higher level of indifference from point *a* to point *b* (the gains from migration). Note that undifferentiated migratory labor leaves the price ratio *p* unchanged but that differentiated migratory labor lowers the relative price of good *A* (as it can be produced cheaper), seeing a larger expansion in the production of *A* than *X*.

rate of transformation (the rate at which one good can be substituted for the other in production) and the market rate of exchange in the two goods as given by the price ratio $p=p_X/p_A$. The profit-maximizing point requires a balance between a technological rate of change (how much resources would have to be taken away from the production of one good to produce another unit of the other) and the rate of market exchange between the goods (how much of one good has to be given up to buy a unit of the other good).

In contrast, consumers own resources and stock in firms that provide their income, and they buy the goods firms produce. What and how much they optimally buy depend on their income, output prices, and their preferences. In the figure a single representative consumer represents all consumers. Consumer incomes come from resource sales and firm profits and are reflected by a "budget line" with slope equal to the price ratio $p=p_X/p_A$. Consumer preferences are given by bowed upwards lines known as indifference curves $(I)$, which trace out alternative combinations of both goods that provide the consumer a certain level of satisfaction (or utility). Higher (to the right) indifference curves indicate higher levels of satisfaction to the consumer. The slope of the indifference curve also conveys key information. It is known as the marginal rate of substitution and describes how consumers are willing to exchange units of one good for the other. The consumer problem is to find the highest indifference curve that is affordable (the consumer's utility maximum), which occurs when the marginal rate of substitution (consumers' rate of exchange between the goods) just equals the price ratio (the market rate of exchange between the goods).

Equilibrium in the economy is when profits and utility are being maximized, when both the marginal rate of transformation and the marginal rate of substitution are equal to the price ratio $p=(p_X/p_A)$. If the PPF and indifference curves have the shapes given in Figure 13.1, and all agents in the economy pay and receive the same prices, this equilibrium is at a single point (production and consumption points overlay one another at point $a$ in the figure, where $PPF_o$ is tangent to $I_o$).

To introduce migration to the model in a straightforward way, and to be relevant to our application, we focus on temporary labor migration from "abroad" to our "home" country, where the migrant labor "works" in the home country. If the influx of labor is

un-differentially productive in the economy, this influx is seen as a shift out of the PPF, as in panel (i) in Figure 13.1. If the labor is differentially productive, for example productive only in industry $A$ but not $X$, the PPF pivots, as in panel (ii). Migration in either of these cases expands the resource base of the economy and the level of equilibrium satisfaction to rise from point $a$ to $b$. These are the gains to the economy from migration.

But with migration comes the risk of some types of disease outbreaks. The economic definition of risk combines both the likelihood the event occurs and the outcome—risk is determined in part by human behavior and endogenous depending on conditions inducing or deterring migration, and on how firms, consumers, and policymakers respond to reduce the risk privately. The consequences of disease outbreaks are represented in the model as an inward shift of the production possibilities frontier (with migration) due to a real loss in resources, as shown for differentiated migratory labor in Figure 13.2. As we focus on diseases that affect humans through mortality or morbidity impacts, the impact of a disease outbreak in the model is loss in labor or labor productivity to the economy. This reduces what the economy can produce and reduces the incomes of consumers. But since the

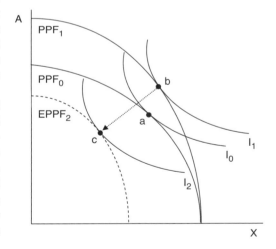

**Figure 13.2:**
Disease outbreak associated with differentiated migratory labor. With the migration comes the threat of an outbreak, with expected impacts being the shift from $PPF_1$ to the expected $EPPF_1$ and expected reductions in welfare from point $b$ to point $c$ (a major outbreak illustrated here).

outbreak is uncertain, it leads to a lower *expected* production possibilities frontier with disease *EPPF*, where the magnitude of the outbreak drives the magnitude of the expected shift in the *EPPF*. In the case illustrated in Figure 13.2, for a major disease outbreak the consequences more than eliminate the gains from migration. There is a continuum of cases in which the expected outcomes can range from no impact to a diminishment of the gains from migration to the case shown. The point is that these migratory flows are beneficial to an economy, and can remain so even if they result in some unintended consequences as long as the gains from migration are positive.

If the consequences are detrimental in net to the economy (where gains from migration are more than exhausted, as in Fig. 13.2) public health and migration/trade-deterring policy responses may be justified. But policies come at a cost. For a policy to be worthwhile, it needs to result in a net improvement to the economy (or in the terms of the model, a higher indifference curve). Policies then introduce further layers of tradeoffs: gains of migration versus consequences of disease outbreak versus consequences of policy versus costs of policy. For example, there are many current migration policies intended to deter disease outbreaks

that have been thought to be effective and beneficial. For example, China requires submission of chest X-rays with all long-term visa applications in order to deter tuberculosis introduction. The United States refuses visa status to any applicant without proof of current MMR and DPT vaccinations. These policies do not assume that the source of such diseases is migrants; rather, they are coupled with domestic public health measures intended to minimize the chance of emergence of those diseases.

The tradeoffs can be illustrated in the model. Collective policies (policies enacted at the national level such as border controls) are typically aimed at reducing the chance of an outbreak or reducing the consequences if an outbreak occurs. The policies may be strict (Fig. 13.3), such as a closing of the border and complete elimination of migration, moving the economy back to point *a* (in the absence of an outbreak). If the policy is assumed to be perfectly effective and a major outbreak is avoided, there could be an expected gain to society from the policy, the gain being the difference between point *c* and *a* in panel (i). But if only a minor outbreak is avoided, then the strict policy would result in an expected welfare loss, as the difference between point *c* and *a* in panel (ii).

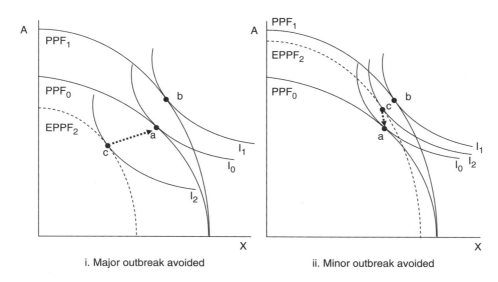

i. Major outbreak avoided                    ii. Minor outbreak avoided

**Figure 13.3:**
A strict policy (migration ban) that avoids an outbreak. The strict policy eliminates the gains from migration yet avoids the outbreak. If the outbreak avoided is major, the policy provides an expected gain in welfare from point *c* to point *a* (panel (i)). But if the outbreak avoided is minor, the policy provides an expected loss in welfare from point *c* to point *a* (panel (ii)).

The policies may not be as complete as a border closing, rather a policy that makes migration safer, more expensive to the migrant, or just more difficult, with the end result being a reduction in the risk of an outbreak. In these cases there is then a distinction between the PPF with and without the policy (the policy pivots $PPF_1$ down to $PPF_2$) and EPPF with no policy with policy (a shift out from $EPPF_2$ to $EPPF_3^p$), as illustrated in Figure 13.4. Here the case of a major outbreak is illustrated for two cases, a small reduction in risk and a large reduction in risk. The figure demonstrates the potential for a continuum of outcomes ranging from where the partial policy is inferior to a strict policy to where it is superior, in this case depending on the magnitude of the reduction in risk. In such cases, evaluating the policies is likely to be tricky but necessary if a policy is to be used. Such a partial policy and its ambiguous impacts are illustrated in the example in the next section.

An additional complication to policymaking and evaluation is the potential for an outbreak to not occur.

In the absence of an outbreak, both policies reduce welfare in comparison to doing nothing ($b$ is always the highest and best possible outcome if there are no adverse consequences). This can cut both ways: it may lead to less use of policies, as if the events are avoided and there are no consequences, there is nothing to show for the cost of the policy. Alternatively, it can lead to political gaming, in which the risks are overstated for means other than public health.

Political gaming where policy is employed largely for political ends (those not intended in the policy's initial design) can be demonstrated in policies over trade with a political economy model, for example, using the example of invasive species (Margolis et al. 2005). The point made in this work is that there are damages due to invasive species that are delivered to new locales through international trade. In formulating policies in response to these invasions the authors show that driving the invader's population to zero is not practical (it is far too costly), but a tariff introduced on trade with the source nation(s) may be

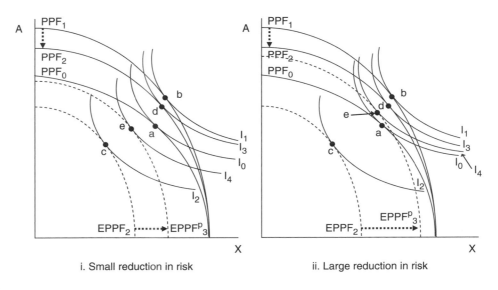

**Figure 13.4:**
A partial policy that reduces the chance or consequences of a major outbreak. The partial policy pivots down $PPF_1$ to $PPF_2$ and reduces the risk of an outbreak by shifting out $EPPF_2$ to $EPPF_3^p$. Panel (i) depicts a small reduction in risk where the shift in $EPPF_3^p$ is only large enough to provide an expected welfare gain over the major outbreak (difference between points $c$ and $e$) but is a worse outcome than the strict policy that avoided an outbreak (difference between points $e$ and $a$). Panel (ii) depicts a large reduction in risk where the shift in $EPPF_3^p$ is large enough to provide an expected welfare gain over both the major outbreak (difference between points $c$ and $e$) and the strict policy that avoided an outbreak (difference between points $e$ and $a$).

optimally designed to reduce the rate of invasion of the species. This tariff's optimal level will internalize the damage of the invasive species costs, deterring but not stopping trade and reducing trade levels to those commensurate with reducing the rate of invasive species damages, balancing social costs and benefits. However, the political process used to design the tariff may result in a tariff that is higher than optimal. The difference between the optimal tariff and the actual tariff represent disguised protectionism (policies that protect domestic industry at the expense of domestic consumers and foreign welfare) and deters trade more than the optimal amount. In this case, certain domestic interests are favored in policymaking, and the social benefits of the resulting policy will be outweighed by the social costs of the non-optimal tariff level.

In sum, deciding on optimal policies in these situations is a challenge and requires the policymaker to understand the epidemiology of the disease, the workings of the disease vector, and the direct and indirect economic consequences. In addition, the incentive to deter trade for protectionist reasons rather than to protect health can add a layer of complexity to the policymaker's decision. We next illustrate the application of the model and some of these issues for the spread of typhus across the U.S.–Mexican border in the early 20th century.

## DISEASE AND BORDER CONTROLS: UNITED STATES–MEXICO AND TYPHUS

The border area between the United States and Mexico was, from the late 1840s until about 1910, a fluid and porous region characterized by high degree of trade and large amounts of temporary and permanent human migration. Cultures and lives mixed closely because of the historical evolution of the border that delineated the United States and Mexico, and both the U.S. and Mexican governments encouraged close integration with the creation of several "free trade zones." Both these policies and the growing transport network pushed the growth of northern Mexico's industries, such as silver and copper, between the Southwest and the East and West Coasts of the United States. During the stable era of President Porfirio Diaz's rule, between 1884 and 1910, average annual economic growth in Mexico was 8%, which has yet to be matched in the 20th or 21st centuries. Workers from Mexico were an integral part of the economic boom of the Southwest during this period as well, providing a reliable labor pool for agriculture, the railroads, and mining. In terms of Figure 13.1, applied to the United States in this example, let good A be a composite of agriculture, railroads, and mining where migrant labor was productive and let good X be "everything else" in the differentiated case (panel ii).

The border around El Paso del Norte, Texas, became a focal point of national security for the United States in the 1880s not because of this flow of labor from Mexico to the United States, but because other ethnic groups began using the southern border (and especially El Paso) as a way of bypassing stricter measures at other points of entry. Increasing security measures along the U.S.–Mexican border were targeted at the Chinese, Mediterranean, and Eastern European immigrants using the southern border as a "back door." The Immigration Service set up headquarters in El Paso not because of Mexican migration, but because of the nature of other entrants at the border.

Political instability and growing tensions along the border began boiling in the early 20th century, when the El Paso del Norte/Juarez twin cities became hotbeds of revolutionary activity. The U.S. government, at the Mexican government's request, stationed increasing numbers of troops under General John J. Pershing along the border to deter northern revolutionaries, such as Pancho Villa, from destabilizing Mexico. This unstable political period heralded the beginning of heightened U.S. xenophobia about Mexico. Among the most vocal xenophobes were U.S. public health officials; as a result, the first real border restrictions between Mexico and the United States came about because of a public health scare, the appearance of typhus in border cities.

### The Typhus Scare and the U.S.–Mexican Border

Epidemic typhus (more commonly called typhus) is a lice-borne disease often associated in history with breakouts in military camps, jails, and post-famine and post-disaster regions. It is noted historically because of its high mortality rate, between 10% and 60% (in terms of our model developed above, a major outbreak as in Figure 13.2). The disease is caused by the *Rickettsia prowazekii* bacterium and spread through lice feces.

In 1916, the first reported case of typhus was investigated in El Paso, leading to concern that the disease was about to become epidemic in both El Paso and Juarez. This led to the initiation of numerous policies. During the time Gen. Pershing was stationed in El Paso, he assigned his troops to assist the residents of south El Paso to "clean up" their neighborhoods. These residents were predominantly of Mexican descent, and though the cleanup exercises were done in the name of general hygiene, the undertone of the actions was that Mexican neighborhoods were subject to inferior hygiene and the source of potential public health threats to the United States.

Later in 1916, a facility designed for "soft soap and warm water" baths for border-crossers at the El Paso port of entry was re-tasked as a disinfection and quarantine station by the U.S. Public Health Service. The disinfection station was publicly announced as a measure to stop the spread of typhus from Mexico to the United States via human migration or cross-border travel. In our model (as illustrated by Fig. 13.4) this would be classified a "partial policy," as cross-border migration was allowed, yet restricted, by the policy.

Under the surface, however, there was limited evidence that typhus was spreading from Mexico via the El Paso border crossing. Literature examining typhus outbreaks at the time show that public health threats were found tied to migrants arriving in eastern U.S. ports from Europe. Analyses performed by the authors on the public health reports of the time show that around 30% of cases presented outside of the border region (U.S. Public Health Service *Public Health Reports*). Many of these reports cite incidence on the East and West Coast immigration ports. By comparison, around 50% of cases presented in the entire state of Texas, with around 10% of total U.S. cases presenting in El Paso itself. El Paso presented less of a threat

from a public health point of view than immigrants entering along the coasts.

Underlying this measure were the instability and political turmoil of the Mexican Revolution, and a growing eugenics movement, nativism, and racial tensions. In relation to the model developed above, the implication is that the policies were convoluted with political pressures on top of public health threats, such that in Figure 13.4 the policy pivots of $PPF_2$ were not being made to reduce the risk of a typhus outbreak (i.e., to shift $EPPF^P_3$) but for other reasons. This is akin to the use of trade tariffs to in excess of optimal levels for protectionist reasons. The disinfection process that border-crossers endured for the first few months after the opening of the disinfection facility were unpleasant but innocuous—migrants arriving at the border had their clothes steamed and then took a bath in kerosene. Use of this method spread from the disinfection station to El Paso.[3]

In January 1917 a stricter disinfection and quarantine procedure was implemented for all Mexican citizens. Upon entering the disinfection station building, all migrants would strip, have their clothing "chemically scoured," have their scalp inspected, shower in kerosene, be vaccinated for smallpox, and finally be examined for any "mental or physical defects." The public health literature makes it clear these procedures far surpassed anything immigrants at other ports of entry on the East and West Coasts were subjected to. In our model these increasingly severe measures can be reflected in Figure 13.4 with increasingly large pivots downward of $PPF_2$, which serve to lower point *d* and reduce any gains seen by migrant labor.

In the 1920s, agricultural interests in the Southwest provided a strong counterweight to the nativists and eugenicists, reducing the pressure to place restrictions on Mexican migration. The shrill tone seen in the early public health measures faded, but some change ensuing from this period was permanent: as already mentioned, the health inspections and disinfections

---

3  Interestingly, among the early kerosene-bath recipients was a group of about a dozen Mexican citizens suspected of being part of Francisco Villa's brigade. Unfortunately, their bath ended in tragedy. Someone—no one seems to know who—struck a match, and the jail exploded, killing almost all the incarcerated persons and guards. Francisco Villa, believing the tragedy to be an intentional act of violence against the men, responded by raiding Columbus, N.M., less than a week later, killing many civilians and burning the presidio. Pershing spent the next several months trying to find Villa and exact retribution for his incursion onto U.S. soil.

continued for another 40 years. In 1924, the U.S. Border Patrol was established and tasked with the regulation of migration at the U.S.–Mexican border. Finally, the border between the countries was treated as a demarked area of difference, a border rather than a frontier between the two countries, and the demarcation was more strictly enforced. In the span of 20 years, the fluid and unified region of northern Mexico and the U.S. Southwest were effectively separated and cut off from each other.

To put into perspective the number of typhus-related examinations carried out in U.S.–Mexican border sites, it is useful to look at the numbers from 1916, the first year such measures were instituted. It is reported that there were "871,639 bodies inspected; 69,674 disinfected, 30,970 vaccinated for smallpox, 420 excluded on account of illness, 7 denied entry for refusing disinfection, and 8 retained for observation" (Stern 2005). Along the Texas border alone, it is reported "officials inspected 39,620 bodies per week, or 5,660 a day" (Stern 2005). Stern estimates that in El Paso, this translated into 236 inspections per hour, surpassing inspection rates at East and West Coast ports substantially.

## Can Economics Be Used to Evaluate the Typhus Measures?

Did the extraordinary measures instituted at the El Paso port of entry have an impact on U.S.–Mexican economic activity in a fashion as illustrated by our model? As shown by Figure 13.4, what is necessary is both the policy costs (reduction in labor migration or shift down in the PPF to $PPF_2$) and the reduction in risk, or shift in EPPF to $EPPF^P_3$.

Assessing these tradeoffs directly is impossible given the paucity of data available, but we are able to gain some indirect insight and make the point that these evaluations are necessary to determine whether the policy responses are socially advantageous. Researchers have documented that making definitive evaluations of the material social consequences of the mortality and morbidity of disease outbreaks is neither straightforward nor easy to generalize (Hirshleifer 1966; Brainerd and Siegler 2002). We can gain insight by looking at aggregate U.S. data in the light of Adam Smith. In *The Wealth of Nations*, published in 1776, Smith stressed that the appropriate measure of the well-being of society is measured by per capita income or product (in terms of available data, per capita gross domestic product [GDP]). The higher the income per capita of a society, the better off are individuals in the society. But problems arise with this viewpoint because it neglects non-material losses. For example, if there is a disease outbreak that has a pulse of high mortality and no loss of material assets, social welfare measured in terms of per capita GDP could increase. The loss of human life and livelihoods needs to be included in the balance, but considering the effects on per capita GDP provides one (restive) view of the problem.[4]

In this light, Figure 13.5 presents the annual percent change from one year to the next of population and GDP for the United States and (for comparison) Mexico from 1910 to 1929. Either variable is growing if the percentage change is positive and vice versa. The figure demonstrates that the U.S. population consistently grew throughout the years (although it did decline in its growth rate during the years of the typhus scare, World War I, and the Spanish flu outbreak). GDP for the United States fluctuated significantly but still contracted in only 4 years of the interval. Per capita GDP (shown in Fig. 13.6) shrank in 6 years of the interval and declined following the end of hostilities in Europe and over the course of the Spanish flu pandemic.

However, understanding the changes in welfare in relation to the typhus outbreak and its measures is tricky. Trade measures were not reliably reported between the countries before the 1940s. At any rate, trade between the most affected areas in terms of labor mobility, the city pairs across the border such as El Paso/Juarez, would be the more reliable measure, and those numbers are not recorded until the mid-20th century. A proxy for labor mobility is examined here, mindful of the fact that extreme turmoil existed along the border at the same time as the quarantine and disinfection measures, and that the numbers will be volatile.

Figure 13.6 also shows changes in crossings at the El Paso port of entry for temporary migrants (i.e., classified as non-immigrants, those who cross for

---

4 The point is that these evaluations are not obvious and require serious scientific analysis.

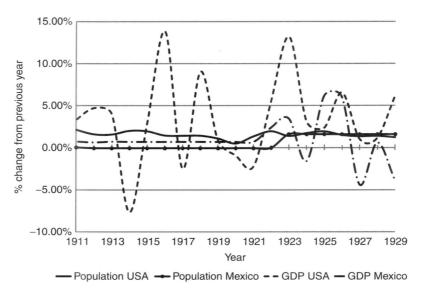

**Figure 13.5:**
Changes in population and GDP for the United States and Mexico. (Accessed from The World Economy: Historical Statistics, OECD Development Centre, Paris 2003.)

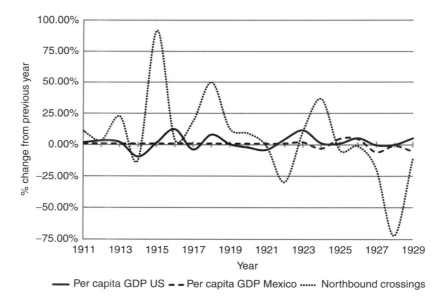

**Figure 13.6:**
Changes in GDP per capita and crossings at the El Paso port of entry for temporary migrants. (Percent changes from the previous year for GDP per capita for United States and Mexico accessed from The World Economy: Historical Statistics, OECD Development Centre, Paris 2003. Percent change in non-immigrant northbound migration at El Paso port of entry from Annual Reports of the Commissioner General of Immigration.)

temporary reasons—working on one side and living on the other, tourism). The numbers take into account multiple-crossers; that is, those who cross the border daily, weekly, or monthly are counted only once. As predicted, the numbers reveal extreme volatility from year to year. In addition, it is obvious that non-immigrant crossings increased significantly during the year the quarantine and disinfection measures were instituted. This could be an artifact of the better recordkeeping that ensued when the public health measures were put in place.

During the period including World War I, analyses showed that public health attention shifted substantially to the outbreak of the so-called "Spanish flu" pandemic (U.S. Public Health Service *Public Health Reports*). This is evidenced by increased inspection and quarantine measures being instituted at the East and West Coast ports of entry to the United States, and likely softening the previous focus on Mexican migrants. The number of cross-border migrants may reflect this, and something definitely affected population, GDP, and per capita GDP during this period. However, teasing out the effects of the flu pandemic itself, the end of World War I, and those associated with the typhus measures is impossible with the available data.

One can also see that there were declining non-immigrant crossings in 1920–21, 1921–22, and again from 1924 through the end of the decade. These may be due to the inspection measures but are also likely due to the increasing nativist sentiments in the United States. The declining numbers cannot be put down to any particular U.S. measure put in place to reduce the flow of Mexican workers into the United States; rather, it is probably due to the sum total of these measures. In addition, as the 1920s unfolded, non-immigrant (and immigrant) crossings moved to other ports as Mexican workers moved to fill jobs in agriculture (not a predominant industry near El Paso). Finally, the world economy began to slow down in the later years of the 1920s, reducing demand for all labor, including Mexican workers. Whether the border controls caused the U.S. *PPF* to contract is unclear (i.e., whether $PPF_2$ shifted down). The correlation coefficient between the annual growth of per capita GDP and northbound border crossings is 0.139, demonstrating a weak yet positive link between the two. While this says nothing of causality or magnitude, the implications are that less restrictive border crossing

policies are associated with higher levels of per capita GDP and that the restrictive border control measures are associated with lower levels of per capita GDP.

Any impacts could be more evident and concentrated in Mexico, especially in Juarez—not north of the border. At the national level in terms of per capita GDP there is little evidence of this for Mexico. If data could be found to approximate unemployment, poverty, or income in areas like Juarez, it is likely that a reduction in Mexican citizens' quality of life would be evident. In contrast, these policies reduced typhus incidence on the border. A total of 70% of all cases and 74% of all deaths from typhus occurred along the border, with 51% of cases and 63% of deaths in Texas alone. Once the measures were instituted, morbidity and mortality dropped to nearly zero on the southern border. This decline might also been because people stopped calling the doctor once the measures went into effect, but regardless the data show a distinct drop in cases. In our model, it suggests the $EPPF_3$ did shift to the right.

## CONCLUSIONS

In the typhus story, we saw how policy measures like severe quarantine impede mobility and reduce productivity and output. The challenge is to determine to what degree these policy restrictions improve health at the expense of lower economic activity. Whether the costs of the policy exceed the benefits from reduced health risk is unclear given the information that currently exists. As a comparison, the flu pandemic of 1918 is thought to have reduced economic productivity in the United States by a relatively small level. Flows of labor and mobility of people were not deterred, however, and increased significantly during this period, and so productivity effects of public health measures are conflated with the impacts of World War I.

Likewise, confounding factors also existed when the typhus measures were instituted along the United States–Mexico border. Confounding factors include (1) nativist and racial tensions in the United States, (2) concerns about the destabilizing impact of the Mexican revolutionaries operating along the U.S. border, and (3) the internal impact within Mexico of the revolution itself (which likely pushed a large number of Mexicans towards the north). These three factors are not controlled for in the data for non-immigrant border crossings in El Paso del Norte. Nonetheless, the evidence from the existing literature suggests that the typhus measures along the border were not purely a public health measure. They were a shrill response both to the potential spread of a deadly disease, and to other underlying fears about Mexicans. Our model predicts, as shown in Figure 13.4, that the industries contributing substantially to the economic robustness of both the U.S. Southwest and northern Mexico would have seen drops in production. The measures likely contributed to reductions in economic growth and may well have pushed up poverty rates on both sides of the border. It is plausible the measures helped put Mexico and some regions in the United States on a downward economic path.

The border measures appear to have lowered the incidence of typhus in the region at the cost of reduced inflows of migratory labor and human suffering. Our analysis reveals the significant economic tradeoffs in policies directed at reducing the risk of outbreaks of infectious disease. Neglecting any aspect of these tradeoffs is likely to result in less health protection provided at greater costs. A more complete accounting is required: one that measures the direct and indirect effects of the disease outbreaks and policies directed at the disease. While challenging, this is what the federal government requires of major policy directed at U.S. citizens and for public initiatives such as healthcare reform. Such evaluations are required by Executive Order 12291, which was signed by President Reagan and continued on by all presidents since: all regulations having over a $100 million impact on the economy have to be evaluated in terms of costs and benefits. Assessing the economic tradeoffs using cost-benefit analysis should also be applied to the control of infectious diseases. Behavioral responses to emerging diseases should be measured and accounted for in such a cost-benefit analysis, with the recognition that the political atmosphere will affect the final outcome of the policy options. Should emerging infectious diseases originate in the developing world, politically motivated and reactionary policy measures could affect the economies of these developing nations. If such a situation emerged in China or India, the domestic reaction could be vociferous, likely leading to needed public health measures, and to other measures veiled in the guise of public health protection that are in actuality economic protectionism. We should be mindful that some may use a pandemic as an excuse to reduce mobility of labor, capital, and net exports quite separately from the need to protect public health. Public health authorities should take care not let their measures be used for political purposes and protectionism.

# REFERENCES

Brainerd, E., and M.V. Siegler. 2002. The economic effects of the 1918 influenza epidemic. Unpublished manuscript, accessed from http://birdflubook.com/resources/brainerd1.pdf.

Finnoff, D., C.R. McIntosh, J.F. Shogren, C. Sims, and T. Warziniack. 2010. Invasive species and endogenous risk. Ann Rev Res Econ 2:77–100.

Haacker, M. 2004. The macroeconomics of HIV/AIDS. International Monetary Fund, Washington D.C.

Hirshleifer, J. 1966. Disaster and recovery: The black death in Western Europe. Rand Corporation Memorandum RM-4700-TAB, accessed from http://www.rand.org/pubs/research_memoranda/RM4700/RM4700.pdf.

Kunreuther, H., and A.Z. Rose. 2004. The economics of natural hazards. Edward Elgar Publishing Company, Cheltenham, UK.

Leung, B., D.M. Lodge, D. Finnoff, J.F. Shogren, M.A. Lewis, & G. Lamberti. 2002. An ounce of prevention or a pound of cure: bioeconomic risk analysis of invasive species. P Roy Soc B Biol Sci 269:2407–2413.

Jones, K., N. Patel, M. Levy, A. Storeygard, D. Balk, J. Gittleman, and P. Daszak. 2008. Global trends in emerging infectious diseases. Nature 451:990–993.

Lorey, D.E. 1999. US–Mexican border in the 20th century. Roman and Littlefield, Lanham Maryland.

Margolis, M., J.F. Shogren, and C. Fischer. 2005. How trade politics affect invasive species control. Ecol Econ 52:305–313.

Meltzer, M.I., N.J. Cox, and K. Fukuda. 1999. The economic impact of pandemic influenza in the United States: priorities for intervention. Emerg Infect Dis 5:659–71.

Romo, D.D. 2005. Ringside seat to a revolution: an underground cultural history of El Paso and Juarez: 1893–1923. Cinco Puntos Press, El Paso, Texas.

Shogren, J.F., and T. Crocker. 1991. Risk, self-protection, and ex ante economic value. J Environ Econ Manag 20:1–15.

Smith, K., M. Behrens, L. Schloegel, N. Marano, S. Burgiel, and P. Daszak. 2009. Reducing the risks of the wildlife trade. Science 324: 594–595.

Stern, A. M. 1999. Buildings, boundaries, and blood: Medicalization and nation-building on the U.S. Mexico border, 1910–1930. Hispanic Am Hist Rev 79:41–81.

Stern, A. M. 2005. Eugenic nation: Faults and frontiers of "better breeding" in America. University of California Press, Berkeley California.

Tyler, P.S. 1931. critique of the official statistics of Mexican migration to and from the United States. In W.S. Wilcox, ed. International migrations, vol. II: Interpretations, pp. 581–590. NBER, Cambridge, Massachusetts.

U.S. Public Health Service, Public Heath Reports, vols. 30, 31, 32, and 33, 1915–18 (published monthly), accessed online at http://www.ncbi.nlm.nih.gov/pmc/journals/333/

# PART THREE

## EMERGING INFECTIOUS DISEASES AND CONSERVATION MEDICINE

# 14

# ARE BATS EXCEPTIONAL VIRAL RESERVOIRS?

Kevin J. Olival, Jonathan H. Epstein, Lin-Fa Wang, Hume E. Field, and Peter Daszak

In recent years bats (Order Chiroptera) have received growing attention as reservoirs for emerging infectious diseases (EIDs). Particularly, a number of high-profile zoonotic viruses with significant human and animal morbidity and mortality have been linked to bat reservoirs, including severe acute respiratory syndrome (SARS) and Ebola, Marburg, Hendra, and Nipah virus (Chua et al. 2000; Halpin et al. 2000; Leroy et al. 2005; Li et al. 2005; Swanepoel et al. 2007; Rahman et al. 2010). Ebola and Marburg viruses have caused outbreaks in Africa with associated mortality rates as high as 89%, and Zaire ebolavirus (ZEBOV) has been linked to mass gorilla die-offs, making it both a public health and conservation concern (Rouquet et al. 2005; Bermejo et al. 2006; Groseth et al. 2007). Nipah virus (NiV), which emerged in Malaysia in 1997, initially as a porcine respiratory and neurological disease, affected over 260 people and had a case fatality rate of 40% (Chua et al. 2000). In Bangladesh, near-annual localized outbreaks or spillover events of NiV have occurred since 2001 with mortality rates as high as 90% (Luby et al. 2009). The outbreak of respiratory illness caused by SARS coronavirus spread from China to 26 other countries and cost an estimated $50 billion in lost tourism and trade (Ksiazek et al. 2003; Lipsitch et al. 2003). This series of high-impact viruses that have

emerged from bats has led to an increase in media interest, funding, and scientific research on these and other bat-borne pathogens. In turn, the new viral species or strains isolated or sequenced from bats have garnered continued attention from both the scientific community and the public. There have also been a growing number of international conferences focused on pathogens and microbes from bats (Wibbelt et al. 2010). The interest in bat microbes has led to debate on the propensity of bats to be reservoirs for zoonotic pathogens. In particular, authors have proposed that bats have unique biological traits (e.g., life in dense colonies, or immunological differences from other mammals) that make them unlike other mammals in their ability to act as viral reservoirs (Box 14.1) (Dobson 2005, 2006; Fenton et al. 2006). Recently published reviews have exhaustively listed currently known viruses from bats (Calisher et al. 2006; Wong et al. 2007), and others have focused on the ecology of emerging bat viruses, including henipaviruses and coronaviruses (Field et al. 2001; Breed et al. 2006; Daszak et al. 2006; Halpin et al. 2007; Field 2009). However, whether bats truly represent a unique mammalian reservoir for viruses is still an open and little-examined question.

The primary goal of this chapter is to address the questions: "Are bats exceptional among vertebrates in

---

**Box 14.1  What Is a "Natural Reservoir" for a Pathogen?**

---

Unfortunately, there are no hard-and-fast rules about what criteria are minimally necessary to call a wildlife species a "reservoir" for a particular disease. In general, a disease reservoir is defined as a species that is essential for the maintenance and transmission of an infectious agent (Haydon et al. 2002; Ashford 2003). There are several types of reservoirs, characterized by their role in transmission cycles. Natural reservoirs are the species that maintain the infectious agent in nature. Incidental or accidental reservoir hosts are species that may get infected by the pathogen, and even transmit it, but are not part of the normal maintenance cycle of the pathogen (i.e., involved in a very small number of transmission incidents) (Ashford 2003). Other language is used to describe specific involvement of different host species in a life-cycle or transmission event (e.g., dead-end, amplifier, final, intermediate hosts), but this is beyond the scope of this chapter.

It is important to note that when investigating the ecology of a zoonotic pathogen, more than one species may act as natural reservoirs in the same ecosystem. For example, West Nile virus is maintained in the wild by hundreds of species of birds, each contributing, to a greater or lesser degree, to the overall transmission of the virus (Kilpatrick et al. 2006). Different hosts can act as reservoirs in geographically distinct ecosystems, such as arenaviruses like Guanarito virus carried by the short-tailed cane mouse (*Zygodontomys brevicauda*) in the New World, and Lhasa virus carried by the natal multimammate mouse (*Mastomys natalensis*) in the Old World (Emonet et al. 2006).

Several criteria can be used to support the identification of a natural reservoir of a pathogen in an epidemiological investigation—in descending order of importance: (1) isolation of the agent from individuals of the target species; (2) detection of nucleic acid from a significant proportion of individuals from a population; (3) serological evidence that a significant proportion of the population has been previously exposed to the pathogen; (4) nucleic acid detected in a biological sample. While the detection of DNA or RNA does not necessarily indicate active infection, it does support the presence of the agent in the tissue sampled; (5) detection of antibodies in a significant proportion of the target population, which would indicate both exposure to the agent and survival of infection—often with viral pathogens, there is minimal disease in the natural host, while significant disease may occur in a naïve or novel host species; (6) experimental inoculation of the target species with the agent, followed by infection (determined by histopathology) and shedding (determined by isolation or culture of the agent from a biological sample); and (7) experimental evidence for transmission of the agent between conspecifics.

---

their ability to harbor and transmit viruses to other vertebrates?" and "Is there a higher likelihood of future emerging diseases originating in bats?" We will examine and discuss four hypotheses: (1) Bats are reservoirs of a disproportionate number of previously emerging diseases given their diversity relative to that of other mammalian orders; (2) Bats are subject to a disproportionate amount of disease-related research effort, and the bat–virus relationship is not unique; (3) Bat ecology and behavior make the bat–human–virus relationship unique compared with other mammals and the cause for recent spillover events to humans and domestic animals; and (4) Bats have a unique immune system and evolutionary history with viruses, making them better reservoirs compared to other mammals. We test these four hypotheses using data from, and trends in, the recently published literature, data from our compiled database on mammal viruses, and our collective experience from field and laboratory studies in Chiropteran virology. In doing so, we hope to take a provocative look at whether or not bats truly represent exceptional hosts for emerging viruses and/or a taxonomic group of special concern for human and animal health.

We finish this chapter with a discussion on how to reconcile public health issues with conservation when wildlife are reservoirs of potentially deadly diseases.

## KNOWN VIRAL DIVERSITY IN BATS: AN UPDATE

Calisher et al. (2006) listed 66 different viruses that have been isolated from or detected in bat tissue. Since this review, other novel viruses have been discovered in bats, including Broome virus, an Orthoreovirus from *Pteropus* spp. in Australia (Thalmann et al. 2010); Tuhoko virus 1, 2, and 3, paramyxoviruses from fruit bats from China (Lau et al. 2010); Xi River virus, a reovirus related to Nelson Bay virus (Du et al. 2010); and GBV-D virus, a flavivirus from *Pteropus* spp. in Bangladesh (Epstein et al. 2010). Similarly, in the past year, expanded host and/or geographic ranges for several previously known viruses have been recorded for Lagos bat virus (Dzikwi et al. 2010), Henipaviruses (Li et al. 2008), and SARS-like coronaviruses (Rihtaric et al. 2010).

We have compiled our own database on published bat-virus records using ISI Web of Science literature searches over the past 50 years incorporating various combinations of bat and virus keywords, and from published reviews (Calisher et al. 2006; Wong et al. 2007; Turmelle and Olival 2009). The currently known tally of viral diversity in bats at a viral family and bat genus level is summarized in Table 14.1. Data are shown for species with confirmed detection of a given virus (viral isolation or molecular detection) and for those studies only showing evidence of previous exposure (serological assays). While we recognized that cross-reactivity in serological studies might cause potentially misleading results at the viral species level, we believe these summarized data accurately reflect infection status and host range within bat genera at a viral family level. In total, 15 viral families— 10 RNA virus families (shown in italic font) and 5 DNA virus families (normal font)—are known to infect 75 bat genera. We also show the number of viral species, according to International Committee on Taxonom of Viruses (ICTV) taxonomy, that are shared between each bat genus and humans (Olival, Bogich, et al. unpublished data).

## HYPOTHESIS 1: BATS ARE RESERVOIRS IN A DISPROPORTIONATE NUMBER OF EIDs IN HUMANS

Previous authors have suggested that bats are responsible for a surprisingly large number of EIDs (Wong et al. 2007; Li et al. 2010). However, bats are a very diverse group of mammals, with nearly 1,200 known species (Simmons 2005), representing over 20% of the known mammal diversity (species richness), and likely have a similarly high viral diversity. Despite some high-profile EIDs from bats, are bats the source of more zoonotic EIDs compared to other mammals? Using a database of known emerging diseases, Woolhouse and Gowtage-Sequeria (2005) analyzed the proportion of known zoonotic emerging infections that have different non-human reservoirs. Their analysis shows that bats are not responsible for the largest proportion of EIDs originating in non-human animals, and that ungulates, carnivores, rodents, and non-human primates are all responsible for a greater proportion of viruses, bacteria, and other pathogens that have emerged in humans.

To further examine this, we used the Jones et al. (2008) database, with host range information for 335 known human EIDs (including viruses, rickettsia, bacteria, protozoa, and fungal pathogens), and found that bats are known to harbor only 4.5% of the known zoonotic EID pathogens. Rodents, ungulates, carnivores, rabbits, primates, and other mammals are known to harbor a greater percentage of these zoonotic EIDs (Fig. 14.1). Note that these are not strict taxonomic categories (e.g., squirrels should be included in rodents and horses with other ungulates), but rather are presented as they are recorded in the database (Jones et al. 2008). However, a number of issues need further analysis. First, ungulates include a number of domesticated species, which have close contact with humans, and therefore may be more likely to act as reservoirs or amplifiers of zoonoses. Secondly, rodents are relatively easy to catch compared to bats, and because of this, and the peri-urban habits of some known rodent reservoirs, they have likely been studied more intensively by infectious disease researchers and linked to more EIDs. To that point, taxonomic sampling bias overall needs to be better accounted for than simple observation of the raw data.

Table 14.1 Current Known Tally of Viruses that Have Been Detected in Bat Genera (at the virus family level). Also Showing Number of Viral Species Shared with Humans for Each Bat Genus (from Olival, Bogich, et al., unpublished).

| Bat Genus | Adenoviridae | Arenaviridae | Bunyaviridae | Coronaviridae | Filoviridae | Flaviviridae | Herpesviridae | Orthomyxoviridae | Papillomaviridae | Paramyxoviridae | Parvoviridae | Polyomaviridae | Reoviridae | Rhabdoviridae | Togaviridae | Number of Viral Species Shared with Humans |
|---|---|---|---|---|---|---|---|---|---|---|---|---|---|---|---|---|
| Anoura | | | X | | | X | | | | | | | | X | X | 1 |
| Antrozous | | | | | | | | | | | | | | X | | 0 |
| Ariteus | | X | | | | | | | | | | | | | | 0 |
| Artibeus | | X | X | | | X | | | | | | | | X | X | 4 |
| Barbastella | | | | | | X | | | | | | | | | | 0 |
| Cardioderma | | | | X | | | | | | | | | | | | 0 |
| Carollia | | | X | X | | X | X | | | | | | | X | X | 2 |
| Chaerephon | | | | X | | X | | | | | | | | | X | 0 |
| Chrotopterus | | | | | | | | | | | | | | X | | 0 |
| Corynorhinus | | | | | | | | | | | | | | X | | 0 |
| Cynomops | | | | | | | | | | | | | | X | | 0 |
| Cynopterus | | | X | X | | X | | X | | X | | | | X | | 2 |
| Desmodus | | X | X | X | | X | | | | | | | | X | X | 2 |
| Diaemus | | | | | | | | | | | | | | X | | 1 |
| Diclidurus | | | | | | | | | | | | | | X | | 0 |
| Diphylla | | | | | | | | | | | | | | X | | 1 |
| Eidolon | | | X | X | X | X | | | | X | | | X | X | | 0 |
| Eonycteris | | | X | X | | X | | | | X | | | | X | | 1 |
| Epomophorus | | | | | | | | | | X | | | | X | | 0 |
| Epomops | | | X | | X | | | | | | | | | X | | 2 |
| Eptesicus | | | X | X | | X | X | | | | | | | X | X | 3 |
| Euderma | | | | | | | | | | | | | | X | | 0 |

| Genus | | | | | | | | Count |
|---|---|---|---|---|---|---|---|---|
| *Eumops* | | | | | | X | X | 0 |
| *Glossophaga* | | | X | | X | X | X | 2 |
| *Hipposideros* | X | X | X | X | X | X | X | 3 |
| *Histiotus* | | X | | | | X | | 0 |
| *Hypsignathus* | | X | X | | X | | | 1 |
| *Lasionycteris* | | | | | | X | | 0 |
| *Lasiurus* | | | | | | X | | 0 |
| *Lissonycteris* | | X | | X | | | | 0 |
| *Lonchophylla* | | | | | X | X | | 0 |
| *Lonchorhina* | | | | | | X | | 0 |
| *Lophostoma* | | | | | | X | | 0 |
| *Macroglossus* | | | X | | X | X | | 0 |
| *Megaderma* | | | | X | | X | | 0 |
| *Micronycteris* | | | | | | X | | 0 |
| *Micropteropus* | X | X | X | | | X | | 1 |
| *Miniopterus* | X | X | X | X | X | X | | 2 |
| *Molossus* | X | | X | | X | X | X | 1 |
| *Mormoops* | | | | | | X | | 0 |
| *Murina* | | X | X | | | X | | 0 |
| *Myonycteris* | X | | | | | X | | 1 |
| *Myotis* | X | X | X | X | X | X | X | 2 |
| *Natalus* | X | X | X | | | X | | 0 |
| *Nyctalus* | X | X | X | X | | X | | 1 |
| *Nycteris* | | | | | X | X | | 0 |
| *Nycticeius* | | | | | | X | | 0 |
| *Nyctinomops* | | X | X | | | X | | 0 |
| *Otomops* | | | | | | | | 0 |
| *Philetor* | | X | X | | | X | | 0 |
| *Phyllostomus* | X | X | X | X | X | X | X | 2 |
| *Pipistrellus* | X | X | X | | X | X | | 0 |
| *Platyrrhinus* | X | X | X | | X | X | X | 2 |
| *Plecotus* | | X | X | | X | | | 0 |

(Continued)

**Table 14.1 (Continued)**

| Bat Genus | Adenoviridae | Arenaviridae | Bunyaviridae | Coronaviridae | Filoviridae | Flaviviridae | Herpesviridae | Orthomyxoviridae | Papillomaviridae | Paramyxoviridae | Parvoviridae | Polyomaviridae | Reoviridae | Rhabdoviridae | Togaviridae | Number of Viral Species Shared with Humans |
|---|---|---|---|---|---|---|---|---|---|---|---|---|---|---|---|---|
| Promops | | | | | | | | | | | | | | X | | 0 |
| Ptenochirus | | | | X | | | | | | | | | | X | | 0 |
| Pteronotus | | | X | | | X | | | | | | | | X | | 1 |
| Pteropus | | | X | X | X | X | | | | X | | | X | X | | 4 |
| Rhinolophus | | | | | | X | | | | X | | | X | X | | 3 |
| Rhinophylla | | | | | X | X | | | | | X | | | | | 0 |
| Rhynchonycteris | | | | | | | | | | | | | | | X | 0 |
| Rousettus | | | X | X | X | X | | X | X | X | | | X | X | X | 2 |
| Saccolaimus | | X | | | | | | | | | | | | X | | 1 |
| Scotophilus | X | | X | X | | X | | | | X | X | | | X | X | 2 |
| Sturnira | | | | | | X | | | | X | | | | X | X | 1 |
| Syconycteris | | | | | | | | | | | | | X | | | 0 |
| Tadarida | | | X | X | X | X | | | | | | | | X | X | 2 |
| Taphozous | | | X | | | X | | | | | | | | X | | 0 |
| Tonatia | | X | | | | | | | | | | | | | | 0 |
| Trachops | | | | | | | | | | | | | | X | | 0 |
| Tylonycteris | | X | | X | | | | | | | | | | | | 0 |
| Uroderma | | | | | | | | | | | | | | X | X | 1 |
| Vampyrodes | | | | | | X | | | | | | | | X | | 0 |
| Vespadelus | | | X | | | | | | | | | | | | | 0 |
| Vespertilio | | | | | | | | | | | | | | X | | 0 |
| **# of Genera positive** | 4 | 8 | 25 | 22 | 10 | 31 | 9 | 5 | 1 | 10 | 5 | 1 | 5 | 59 | 17 | |
| **% Genera positive** | 5 | 10 | 32 | 29 | 13 | 40 | 12 | 6 | 1 | 13 | 6 | 1 | 6 | 77 | 22 | |

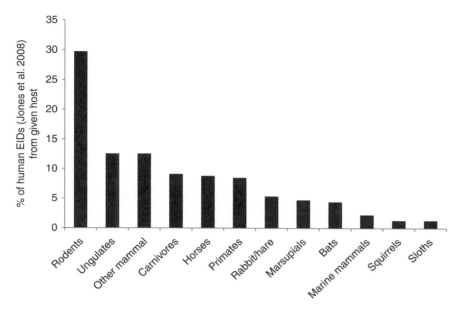

**Figure 14.1:**
Proportion of known human emerging infectious diseases from each host group. (Data taken from Jones et al. 2008)

## HYPOTHESIS 2: BATS ARE SUBJECT TO A DISPROPORTIONATE DISEASE-RELATED RESEARCH EFFORT, AND BAT–VIRUS RELATIONSHIP IS NOT UNIQUE

It is possible that the perceived role of bats as important reservoirs for emerging viruses is a function of sampling bias and research effort, and not representative of a truly greater number of viruses in bats relative to other groups. In other words, has more research been focused on bats than other taxonomic groups, such as rodents or birds? Can we compare the number of known viruses with a correction for research intensity for each group? To assess this, we looked at trends in the number of publications (research effort) from ISI Zoological Record by using the keyword search string: "TS=(disease* OR pathogen* OR parasite*) AND [Taxonomic terms for Bats, Rodents, or Birds]." We also compiled known numbers of viruses for different mammalian groups from extensive literature searches (Olival, Bogich, et al. unpublished data) and compared these data using a simple, but novel, metric of relative viral diversity.

There has been a steady growth in the cumulative number of disease-related papers published for three groups of vertebrates—birds, rodents, and bats—over the past 40 years (Fig. 14.2). At the end of 2009, the cumulative number of disease publications from 1864 to present was 33,503 for birds, 2,972 for bats, 10,125 for carnivores, and 14,754 for rodents. These numbers may be misleading for several reasons: (1) This measure of disease-related research uses data on papers focused on non-viral investigations and, given the special techniques required to isolate or characterize viruses, it is possible that the conclusions are skewed. We are here assuming that research on all mammalian orders is uniformly allotted to research on viruses and other parasites or microbes; (2) This analysis is not focused on emerging viruses; rather, it covers all known pathogens, and diversity of viruses may not equate to their propensity to emerge in the future; (3) Not all disease papers will be captured by searching for indexed studies in Zoological Record, although results were comparable in ISI Web of Science searches using the same keyword terms; and (4) Rodents may be over-represented in the literature due to their use as experimental models.

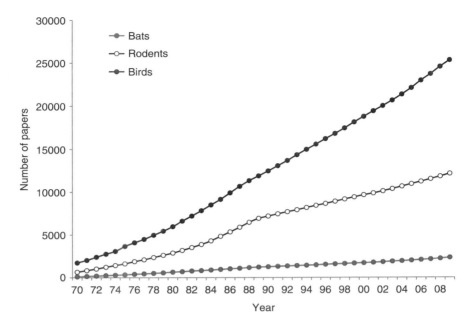

**Figure 14.2:**
Cumulative number of disease-related publications for birds, rodents, and bats over the past 40 years.

These measures of disease-related research bias are misleading without considering the number of species within each taxonomic group. If we divide the number of papers by the number of described species for each group (birds 10,050, rodents 2,227, carnivores 286, bats 1,116) (Wilson and Reeder 2005; Gill 2006), this gives the average number of disease-related publications per species: birds 3.33, rodents 6.48, carnivores 35.40, and bats 2.66. For mammals we can use this rough correction of sampling bias and number of species in each Order to obtain a simple corrected metric for the number of viruses, per the following equation:

# known viruses in Order/(average # of disease-related publications per species)

We can apply this corrected metric for the three mammalian Orders (bats, rodents, and carnivores) using the raw number of known viruses from the literature, compiled using an updated list of Calisher et al. (2006) for bats and from a recently compiled database of host species range for all known mammals viruses (Olival, Bogich, et al. unpublished data) for all other mammal genera (Fig. 14.3). Excluding humans, Artiodactyla have the most viruses with 99, followed by bats with 88 viral species, and then non-human primates (Fig. 14.3). Unfortunately we do not yet have a reliable source for the total number of unique known bird viruses.

Using these data to calculate the corrected measure of viral diversity (per the equation above) gives an interesting result. The corrected index of viral diversity is 9.54 for rodents, 1.21 for carnivores, and 33.08 for bats. Thus, on a per-research effort and per-species basis, bats have 3.5 times as many known viruses than rodents and more than 25 times more than carnivores. We also examined trends in research effort over the past 40 years for bats separated by pathogen type: viruses, bacteria, protozoa, fungi, and helminths (Fig. 14.4). Overall, there has been a marked rise in the number of papers investigating bat viruses annually over the past 7 years. These data also show that while the number of studies on fungi in bats was nearly flat for 35 years, there has been a recent spike in the past 2 years, likely spurred by the white-nose syndrome (WNS) epidemic in North America.

To examine the viral discovery curve for bat viruses, we used a linear regression for the number of known viral species for each bat genus against the amount of disease research (number of disease-related publications) for that genus. This includes only species

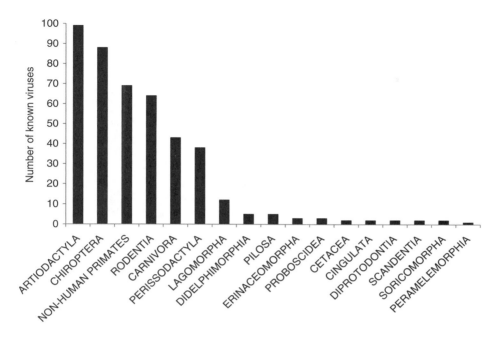

**Figure 14.3:**
Number of unique viral species (from ICTV taxonomy) for each mammalian order from extensive literature reviews. (From Olival, Bogich et al., unpublished)

described by the ICTV in 2009 and excludes any strain or subspecies designations and currently unclassified viruses. There was a significant positive linear relationship ($r^2$ = 0.528, $p$ < 0.001). Although non-linear relationships may also fit the data, the significant positive slope suggests that further study of bats for new pathogens is likely to be fruitful, and may act as a "self-fulfilling" prophecy for those proponents of bats as key reservoirs for EIDs.

An ideal set of comparative data (beyond literature reviews) would include information from field surveillance across different mammalian orders. Unfortunately, few studies have done this, as most studies tend to be taxonomically specific. Chu et al. quantified astrovirus prevalence in both rodents and bats in two separate studies (Chu et al. 2008, 2010). Overall, they found that bats had a greater number of astroviruses than rodents, but this was for only one viral family and two mammal orders, and only for viruses that could be identified with primers derived from known viral sequence. Truly assessing differences in viral diversity across mammalian taxa will require using field data with globally standardized sampling efforts and standardized laboratory methodologies for

viral discovery (e.g., consensus PCRs at the viral family level using multiplex assays such as mass-tag PCR and 454 sequencing technology; Lipkin 2008). One such global effort is underway with the USAID-funded Emerging Pandemic Threats: PREDICT project (http://www.usaid.gov/press/releases/2009/pr091021_1.html), which aims to quantify known and unknown viral diversity across a broad range of wildlife species in 20 countries.

Another factor that may have increased the rate of detection of novel bat viruses is the availability of new diagnostic technologies. In other words, are bats being studied more at a time when our abilities to detect viruses—through PCR techniques and next-generation sequencing—are more sensitive and efficient? There is little doubt that methods such as next-generation sequencing have contributed to an acceleration of microbial discovery in bats, but these techniques are also being applied to other vertebrate groups. This technology has been used to characterize entire bat viromes—the diversity of all viruses found in a given clinical or environmental sample—including the virome of bat guano (Li et al. 2010) and viromes contained in other excreta for three sympatric North

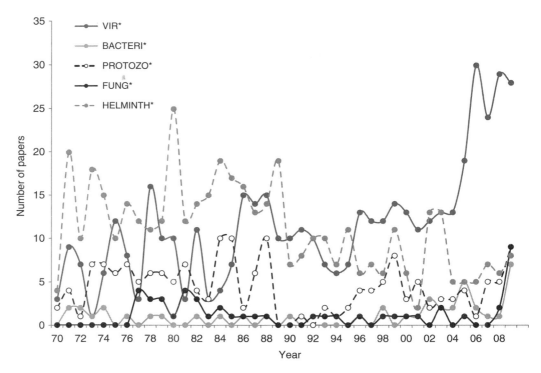

**Figure 14.4:**
Number of disease-related publications per year (not cumulative) for bats over the past 40 years, separated by pathogen type.

American bat species (Donaldson et al. 2010). When techniques such as these are applied to organisms for the first time, the potential for discovering microbial agents is large and one expects "bursts" of information during the early phases of application; this rate may decrease once effort shifts to other vertebrate hosts.

## HYPOTHESIS 3: BAT ECOLOGY AND BEHAVIOR MAKE THE BAT–HUMAN–VIRUS RELATIONSHIP UNIQUE AND THE CAUSE FOR RECENT SPILLOVER EVENTS TO HUMANS AND DOMESTIC ANIMALS

### Higher Rates of Contact with Humans and Other Animals

Do bats have a greater degree of contact with humans than other taxonomic groups that makes them exceptional reservoirs in terms of their ability to pass on

pathogens to other hosts? This has been suggested by many researchers, and can be examined by looking broadly at bat ecology and what is known from a few well-documented cases of bat zoonoses spillover. Bats are ubiquitous, being present on every continent except Antarctica, and occupying almost every habitat that people do. They are also the second most diverse group of mammals in the world, after rodents (Simmons 2005). The potential for interactions with people, either direct or indirect, is high. Bats rarely seek direct contact with humans; however, bats often roost in or near human dwellings, which can lead to accidental contact or human exposure to bat excreta.

People also engage in recreational or work activities inside caves, which commonly house large bat colonies. Histoplasmosis has been transmitted to people exposed to bat feces while caving. Marburg virus, a member of the Filoviridae family that is carried by Egyptian rousette bats (*Rousettus egyptiacus*) in Africa, has infected people who were exposed while entering a mine in which thousands of bats were roosting. While the exact route of transmission remains unknown, infection may have occurred via inhalation of aerosolized bat excreta. Indirect exposure to frugivorous bats in the tropics may occur via food-borne routes such as the consumption of fruits or plant products that have been contaminated with bat excreta. For example, NiV has been transmitted to people who have eaten bat-contaminated date palm sap (Luby et al. 2006). Rabies, which is transmitted via biting, is one of the rare examples of a disease that requires direct bat–human contact. Rabies transmission from vampire bats (only 3 of the 1,200 species of bats are sanguivorous) has been documented in humans living in Latin America, where these species are common (Carnieli et al. 2009; Castilho et al. 2010).

Zoonotic transmission of bat-borne pathogens has also occurred via domestic animal or wildlife amplifier hosts. Nipah and Hendra viruses, members of a recently recognized genus of paramyxovirus (*Henipavirus*), have been transmitted to pigs and horses respectively, and then to humans, causing a fatal encephalitis in humans. ZEBOV is carried by bat species in Central Africa and has caused fatal outbreaks in gorillas. Human outbreaks have occurred when people have been around these infected domestic animals or, in the case of ZEBOV, harvested infected gorilla carcasses for bushmeat.

## Bats: Volant Mammals with Unusual Life History Traits

It has been argued that the unique life history and ecological characteristics of bats make them better positioned to harbor and transmit viruses to humans than other mammals. Bats have a wide variety of life-history traits, morphologies, and ecologies (Nowak 1999; Simmons 2005). Other groups of mammals (i.e., Order Dermoptera) rely on gliding for short-distance dispersal, but bats are the only order of mammals that are truly capable of self-powered flight. Most bats, with the exception of the family Pteropodidae, have also evolved the ability to echolocate, a sensory trait shared with marine mammals that makes bats unique among terrestrial mammals. The ability to be volant and disperse over long distances may have consequences for the degree of pathogen dispersal, but it is unclear if this would lead to greater or lesser diversity of pathogens. For example, integration of bat population

genetic analyses with disease ecology has shown that some "panmictic" species (showing no genetic structuring characteristic of high levels of dispersal) share similar viruses across their range, as with European bat lyssavirus-2 and *Myotis daubentonii* (Atterby et al. 2010) and NiV and *Pteropus* spp. in Southeast Asia (Olival 2008; Rahman et al. 2010). Likewise, an ability to fly does not necessarily mean that the pathogen dynamics within bat populations would be significantly different from other mammalian groups. For example, many rodents have a high capacity for intrapopulation pathogen transmission via the rapid dispersal of a large number of juveniles with each successive generation.

Bat species found in northern and southern latitudes have evolved the ability to hibernate and enter states of torpor that allow them to conserve energy and survive long winters. There is some reason to suspect that the ability to hibernate may confer a unique immunological advantage in some bat species in their ability to live with viral infections (see Hypothesis 4). In contrast to rodents that have large litters at a young age, bats have a very low fecundity (generally giving birth to only one young per year) and a late age of maturity (often individuals will not reproduce until 3 to 4 years old) (Barclay and Harder 2003). Bats can also be extremely long-lived for mammals of such small size, living up to 20 to 30 years (Brunet-Rossinni and Wilkinson 2009). This longevity may have implications for viral transmission and viral maintenance within and between bat populations. For example, Turmelle et al. (2010) demonstrated reduced mortality in big brown bats (*Eptesicus fuscus*) following multiple experimental challenges with rabies virus, suggesting that long-lived adults have immunological memory that allows them to cope with enzootic viruses. However, this is only one virus from one species of bat, and it is unknown how longevity affects bats' ability to act as reservoirs.

Bats also exhibit several important behaviors that may affect their role as reservoirs for viruses. Bats live in some of the densest aggregations of any mammal, and this has been cited as a reason why bats might be important disease reservoirs (Dobson 2005; Calisher et al. 2006). For example, straw-coloured fruit bats (*Eidolon helvum*) are known to form seasonal aggregations of 5 million to 10 million individuals (Richter and Cumming 2006), and the Mexican free-tailed bat (*Tadarida brasiliensis*) can roost in caves with tens of millions of individuals (McCracken 2003). These very large populations can facilitate the persistence of viruses because of the sheer number of individuals in a population, even considering low fecundity. Also the extremely dense and vertically stratified conditions that bats live in, combined with very social behaviors such as grooming, fighting, territorial displays, and simple exposure to aerosolized excreta (Markus 2002; Markus and Blackshaw 2002), can allow viruses to pass from one individual to the next through saliva, urine, or feces (Calisher et al. 2006). We should note two caveats. First, that there are many other groups of wildlife (e.g. rodents, birds) that live in large social groups, and a recent study found that colony size and migratory capacity were not correlated with viral species richness in bats (Turmelle and Olival 2009). Secondly, many bat species do not live in large social gatherings.

To identify which life-history traits and ecologies might be most important in determining viral diversity in bats, a meta-analysis approach can be useful. Turmelle and Olival (2009) conducted a phylogenetically controlled, multivariate analysis with the number of known viruses for 33 bat species used as the dependent variable. In summary, they found that sampling effort, endangered status, and a measure of population genetic structure ($F_{ST}$) significantly correlated with the observed viral richness across bat species, and this model explained 37% of the variance in viral richness data. As noted, colony size and migratory ability were not found to be significant and were excluded from the model. The significance of $F_{ST}$ is of interest as this suggested that population subdivision may promote viral divergence in isolated bat populations. The approach of Turmelle and Olival (2009) warrants further investigation once more data from bat species and their viruses become available.

## HYPOTHESIS 4: BATS HAVE A UNIQUE IMMUNE SYSTEM AND EVOLUTIONARY HISTORY WITH VIRUSES AND THEREFORE ARE BETTER RESERVOIRS COMPARED TO OTHER MAMMALS

This hypothesis is driven mainly by the following observations: (1) Many of the bat viruses identified in the past few decades are apparently harmless in bats

of different species; (2) Under experimental infection conditions, there appears to be elimination of viruses or effective control of virus infection; and (3) Both of these phenomena also appear to occur in species other than known natural reservoirs. For example, infection of Australian fruit bats with SARS-CoV failed to cause any clinical signs seen in most other experimentally infected mammals (G. Crameri and L.-F. Wang, unpublished data 2010); (4) Many bat-borne viruses seem to be present at low level and the infection is persistent; and (5) Isolation of viruses from primary bat cell cultures derived from healthy bats seems to be more common than expected (Maeda et al. 2008, G. Crameri and L.-F. Wang, unpublished data 2010).

These seemingly unusual findings have led some to suggest that bats have an unusual innate immune system that deals with viruses in a different way to other mammalian groups. Some have argued that their capacity to fly has led to changes in physiology that have bearing on their immune system (Dobson 2005, 2006). In this context, it is important to point out that our understanding of the mammalian innate immune system itself is still evolving. Recent studies have suggested that exposure of immune systems to one pathogen may alter the effectiveness of the innate immunity against other unrelated pathogens (Barton et al. 2007). If this type of symbiotic enhancement of innate immunity is common among different animals and it plays an important role in determining the host antiviral ability, it would be worth further testing as a mechanism driving bats' viral diversity and their apparent resistance to virus-induced diseases.

Unfortunately, our current knowledge on bat biology in general and on bat immunology in particular is very poor. Before we can address these challenging questions in any depth, it is essential that we first conduct basic research. We identify the following research priorities to better understand the bat–virus relationship from an immunological perspective.

## Genomics/Transcriptomics

It is essential to have the genome sequence as a starting blueprint to facilitate the design, analysis, and interpretation of experiments. The little brown bat (*Myotis lucifugus*) genome was recently made available (http://www.ncbi.nlm.nih.gov/nuccore/306753024) and partial genome for the large flying fox (*Pteropus vampyrus*) is being processed. There is still an urgent need for more complete and annotated bat genome sequences of different species.

## Bat Immunology

One area of great interest is the change of immune status during hibernation or torpor. Studies from other mammals indicate that hibernation affects both the innate and adaptive immune systems (Bouma et al. 2010). It is believed that the bat's immune functionality also decreases during hibernation. It is hypothesized that the recent mass death of bats in North America as a result of the WNS, caused by the fungus *Geomyces destructans*, may be directly related to immune system changes during hibernation and the infection of a fungus that thrives at low temperatures (Bouma et al. 2010; Frank et al. 2010). Also, the role of hibernation in facilitating persistent infection at low body temperatures has been raised during experimental infection of bats with Japanese encephalitis virus (Sulkin and Allen 1974). The apparent immunosuppression during hibernation is probably beneficial to the host in terms of energy conservation. Since most viruses do not replicate well at low temperature, the chance of establishing productive infection is small.

Much of the current research focus in bat immunology is on orthologs of human or mouse genes with known importance in innate immunity. This includes genes of interferons (Omatsu et al. 2008; He et al. 2010; Kepler et al. 2010), cytokines (Omatsu et al. 2006), Toll-like receptors (Cowled et al. 2010), and STAT1 molecule (Fujii et al. 2010). There is also a study conducted on the basic structure and function of antibody genes (Baker et al. 2010). Although subtle differences are found among some of these genes in bats in comparison to other mammals, their biological significance is yet to be elucidated. The focus of such studies on orthologs of known immune-relevant genes from other mammals is driven by the ease of the study design, but limits us from uncovering anything completely new or unique in bat immunology. It is therefore essential to conduct research on a discovery basis.

## Establishment of Bat Cell Lines

One of the difficulties in bat biology research is the lack of appropriate cell lines. As an extremely diverse group of outbreed animals, conducting live virus infection studies in bats is challenging due to the difficulty in obtaining genetically homogenous animals and in sourcing appropriate animals under current stringent animal ethics codes. It is envisaged that the majority of the basic bat immunology research will be conducted at the cell level. In recent years, multiple groups have invested significant effort in this area, and it is pleasing to see that several stable bat cell lines have been established (Crameri et al. 2009; Jordan et al. 2009).

## Discovery-Driven Investigation

Several groups have already begun bat immunology studies focusing on the innate immunity genes to see whether there is any significant functional difference between bats and other mammals. It is almost certain that the research activities in this field will experience great growth in the years to come. However, if we believe that bats may have taken a different evolutionary path at their branching point from other mammals, it is possible that we will miss bat-specific immunity genes or pathways if we focus only on the innate immunity gene orthologs based on previous human or mouse studies. Two new technological platforms are being adopted to conduct discovery-driven research: mass parallel sequencing (Shendure and Ji 2008) to identify genes specifically up- or down-regulated during virus infection in bats, and the genome-wide siRNA library screening (Horn et al. 2010) to identify specific genes or pathways that play a key and common role in controlling virus infections.

## RECONCILING CONSERVATION AND PUBLIC HEALTH

A major concern expressed by conservationists and ecologists is that the risk of human deaths linked to bat pathogens would induce efforts to exterminate bat populations. There are examples in the media of wildlife being culled and natural habitats being destroyed after having been implicated as the source of human disease—for instance, China's destruction of over 10,000 civets in response to the SARS outbreak; Queensland's call to displace or destroy flying foxes following Hendra virus cases; wetland drainage in Russia after highly pathogenic influenza H5N1; and the closing of an economically important cave in Angola following a Marburg infection and death of a tourist. However, there are several important reasons why culling bats would be ineffective for disease control. First of all, overall we have shown that bats are responsible for only a small proportion of human EIDs compared to other mammal groups. Second, culling efforts would directly increase the contact between people and bats and thus increase, not decrease, the risk interface. Third, a recent analysis of the underlying causes of zoonotic disease emergence clearly demonstrates that emergence is correlated with human activity that facilitates contact with other animals, often through large-scale environmental change (Jones et al. 2008). In other words, the bats are not to blame, but rather disease spillover happens through human disruption of natural ecosystems. For example, the emergence of NiV in Malaysia has been fairly well reconstructed through the combined efforts of many studies (Daszak et al. 2006), and it has been shown that anthropogenic activity—specifically pig farm expansion and intensification—caused the NiV outbreak in Malaysia (Pulliam et al. 2007).

Further, bats are important to our daily lives and critical to ecosystem health for a number of reasons. Insectivorous bats, which represent the majority of bat species globally, are essential for regulating invertebrate pest populations that threaten agricultural crops (Cleveland et al. 2006; Federico et al. 2008). Frugivorous bats, particularly the Old World family *Pteropodidae*, are critical pollinators and seed dispersers, responsible for the propagation of nearly 300 tropical rainforests species in the Old World (Fujita and Tuttle 1991), including a number of highly valuable crops like durian (Bumrungsri et al. 2009).

One solution to preventing spillover and emergence of bat viruses in human populations is to increase public education about the way that land use change affects human–wildlife interactions in general, and how bat-borne diseases emerge because of human activity. The challenge will be to identify ways in which human activities can be changed in a way that decreases the risk of pathogen spillover from wildlife to domestic animals or people. It is neither practical nor responsible to try to stop cultural practices that have persisted for generations. Changes at the bat-human interface must be simple, cost-effective, and practical. For example, following the discovery of NiV, the Malaysian government ordered that fruit trees be removed from pig farms and that in the future a buffer zone between fruit trees and animal enclosures be observed. Also, decreasing activities such as habitat destruction will protect natural resources that bats and other wildlife use for foraging, and will decrease the risk of spillover to domestic animal or people via food-borne routes. Thus, the protection of bat habitat, coupled with a greater awareness of anthropogenic changes that bring bats into closer contact with people or domestic animals, will serve to protect both bats and human health. These types of systemic interventions are critical. We cannot easily predict which virus carried by bats has the potential to cause disease in humans, and so by limiting contact, this serves the dual purpose of universal precaution and creates a win–win situation for both conservation and public health. This approach requires strong and deliberate communication between the scientific community, rural people, farmers, hunters, and policymakers so that bats are not vilified and people are made aware of the importance of bats as part of our ecosystem, without which we would be far worse off.

In reconciling public health studies with conservation, we also encourage investigators to "piggyback" ecological and conservation studies onto disease investigations. For example, Epstein et al. (2009), through their research on NiV ecology in Malaysia, conducted regular roost surveys and satellite telemetry and quantified hunting pressure on flying foxes. All of these related studies, partially funded through disease research, raised awareness of unsustainable levels of flying fox hunting and led to conservation action by the government.

## CONCLUSIONS

Our preliminary review has not given a definitive answer on whether bats are disproportionately reservoirs of emerging pathogens, although our preliminary analysis suggests that on a per-species and per-research-unit basis, bats may overall harbor more viruses than rodents or carnivores. Nonetheless, we have presented some undeniable facts. First, bats harbor some microbes that are extremely lethal to people, and for which there are no useful therapies or vaccines. Second, there is a great deal of interest in identifying microbes in bats and seeking the likely next potential zoonosis. Trends in virus research effort in bats show an increase in the past decade, but overall disease research in bats has historically been much lower than in birds or rodents. We have presented a series of hypotheses as to why bats are reservoirs of some zoonoses, and there are tantalizing glimpses of data that could support many of them.

Ultimately, it is not unreasonable to expect all of these hypotheses to have some elements of fact. For example, it is possible, given the shifting focus of research over the past few decades, that there are inherent biases in the way we have conducted research on microbial diversity in bats. It is also possible that there are aspects of bat immunology, and changes to the dynamics of their contact with people, that make them more likely biologically and in recent years to be the wildlife origin of new EIDs. Given this, and the growing interest in working on bat microbes, we conclude this section with a call for a concerted, trans-disciplinary, international focus on research on this issue. Pathogen discovery will be a central part of this effort, but rather than groups in different countries competing to generate the most novel viral sequences from bats, these efforts could be fused by collaboration among national academies and national research funding agencies. Similar efforts have begun in research on disease ecology and social science (http://www.nsf.gov/funding/pgm_summ.jsp?pims_id=5269) and have been extremely successful in climate research, research on nuclear physics, and genomic research. These efforts could be focused specifically on the issue of understanding the importance of bats as reservoirs. A research effort that combines infectious disease pathology and microbiology with ecology and a deeper understanding of bat immunology, behavior, and conservation would be particularly fruitful, building on the depth of knowledge of each of these fields to test these and other key hypotheses. Indeed, we propose that it is the only plausible way to deal with the biological complexity of these fascinating hosts and the diversity of pathogens that they harbor.

## ACKNOWLEDGMENTS

This study was supported by a NIAID Non-biodefense emerging infectious disease research opportunities award 1 R01 AI079231-01, an NIH/NSF "Ecology of Infectious Diseases" award from the John E. Fogarty International Center (2R01-TW005869), an NIH Fogarty ARRA award 3R01TW005869-06S1 to KJO, an NIH NIAID award (AI067549) to JHE; the Rockefeller Foundation, New York Community Trust, and USAID Emerging Pandemic Threats Program, PREDICT project under Cooperative Agreement Number GHN-A-00-09-00010-00. We thank Liam Brierley for assistance with searching the literature and compiling data for Figures 14.1 and 14.2, and Tiffany Bogich for her contribution to the database of mammal viruses used for comparative analysis.

## REFERENCES

Ashford, R.W. 2003. When is a reservoir not a reservoir? Emerg Infect Dis 9:1495–1496.

Atterby, H., J.N. Aegerter, G.C. Smith, C.M. Conyers, T.R. Allnutt, M. Ruedi, and A.D. Macnicoll. 2010. Population genetic structure of the Daubenton's bat (*Myotis daubentonii*) in western Europe and the associated occurrence of rabies. Eur J Wildl Res 56:67–81.

Baker, M.L., M. Tachedjian, and L.F. Wang. 2010. Immunoglobulin heavy chain diversity in Pteropid bats: evidence for a diverse and highly specific antigen binding repertoire. Immunogenetics 62:173–184.

Barclay, R.M.R., and L.D. Harder. 2003. Life histories of bats: life in the slow lane. *In* T.H. Kunz and B. Fenton, eds. Bat ecology, pp. 209–253. The University of Chicago Press, Illinois.

Barton, E.S., D.W. White, J.S. Cathelyn, K.A. Brett-Mcclellan, M. Engle, M.S. Diamond, V.L. Miller, and H.W.T. Virgin. 2007. Herpesvirus latency confers symbiotic protection from bacterial infection. Nature 447:326–9.

Bermejo, M., J.D. Rodriguez-Teijeiro, G. Illera, A. Barroso, C. Vila, and P.D. Walsh. 2006. Ebola outbreak killed 5000 gorillas. Science 314:1564.

Bouma, H.R., H.V. Carey, and F.G. Kroese. 2010. Hibernation: the immune system at rest? J Leukoc Biol 88:619–624.

Breed, A.C., H.E. Field, J.H. Epstein, and P. Daszak. 2006. Emerging henipaviruses and flying foxes - Conservation and management perspectives. Biol Conserv 131: 211–220.

Brunet-Rossinni, A., and G. Wilkinson. 2009. Methods for age estimation and the study of senescence in bats. *In* T.H. Kunz and S. Parsons, eds. Ecological and behavioral methods for the study of bats, 2nd ed. The Johns Hopkins University Press, Baltimore.

Bumrungsri, S., E. Sripaoraya, T. Chongsiri, K. Sridith, and P.A. Racey. 2009. The pollination ecology of durian

(*Durio zibethinus*, Bombacaceae) in southern Thailand. J Trop Ecol 25:85–92.

Calisher, C.H., J.E. Childs, H.E. Field, K.V. Holmes, and T. Schountz. 2006. Bats: Important reservoir hosts of emerging viruses. Clin Microbiol Rev 19:531–545.

Carnieli, P., J.G. Castilho, W.D. Fahl, N.M.C. Veras, and M. Timenetsky. 2009. Genetic characterization of Rabies virus isolated from cattle between (1997) and (2002) in an epizootic area in the state of Sao Paulo, Brazil. Virus Res 144:215–224.

Castilho, J.G., P. Carnieli, E.A. Durymanova, W.D. Fahl, R.D. Oliveira, C.I. Macedo, E.S.T. Da Rosa, A. Mantilla, M.L. Carrieri, and I. Kotait. 2010. Human rabies transmitted by vampire bats: Antigenic and genetic characterization of rabies virus isolates from the Amazon region (Brazil and Ecuador). Virus Res 153:100–105.

Chu, D.K.W., L.L.M. Poon, Y. Guan, and J.S.M. Peiris. 2008. Novel astroviruses in insecitvorous bats. J Virol 82:9107–9114.

Chu, D.K.W., A.W.H. Chin, G.J. Smith, K.H. Chan, Y. Guan, J.S.M. Peiris, and L.L.M. Poon. 2010. Detection of novel astroviruses in urban brown rats and previously known astroviruses in humans. J Gen Virol 91: 2457–2462.

Chua, K.B., W.J. Bellini, P.A. Rota, B.H. Harcourt, A. Tamin, S.K. Lam, T.G. Ksiazek, P.E. Rollin, S.R. Zaki, W.J. Sheih, C.S. Gouldsmith, D.J. Gubler, J.T. Roehrig, B. Eaton, A.R. Gould, J. Olson, H. Field, P. Daniels, A.E. Ling, C.J. Peters, L.J. Anderson, and B.W.J. Mahy. 2000. Nipah virus: A recently emergent deadly Paramyxovirus. Science 288:1432–1435.

Cleveland, C.J., M. Betke, P. Federico, J.D. Frank, T.G. Hallam, J. Horn, J.D.J. Lopez, G.F. Mccracken, R.A. Medellin, A. Moreno-Valdez, C.G. Sansone, J.K. Westbrook, and T.H. Kunz. 2006. Estimation of the economic value of the pest control services provided by the Brazilian free-tailed bat in the winter garden region of south-central Texas. Front Ecol Environ 4:238–243.

Cowled, C., M. Baker, M. Tachedjian, P. Zhou, D. Bulach, and L.F. Wang. 2010. Molecular characterisation of Toll-like receptors in the black flying fox *Pteropus alecto*. Dev Comp Immunol 35:7–18.

Crameri, G., S. Todd, S. Grimley, J.A. Mceachern, G.A. Marsh, C. Smith, M. Tachedjian, C. De Jong, E.R. Virtue, M. Yu, D. Bulach, J.P. Liu, W.P. Michalski, D. Middleton, H.E. Field, and L.F. Wang. 2009. Establishment, immortalisation and characterisation of pteropid bat cell lines. PLoS One 4:e8266.

Daszak, P., R. Plowright, J.H. Epstein, J.H. Pulliam, S. Abdul Rahman, H.E. Field, C.S. Smith, K.J. Olival, S. Luby, K. Halpin, A.D. Hyatt, and H.E.R.G. 2006.

The emergence of Nipah and Hendra virus: pathogen dynamics across a wildlife-livestock-human continuum. *In* S. Collinge and C. Ray, eds. Disease ecology: community structure and pathogen dynamics, pp. 188–203. Oxford University Press, Oxford.

Dobson, A.P. 2005. What links bats to emerging infectious diseases? Science 310:628–629.

Dobson, A.P. 2006. Linking bats to emerging diseases— reply. Science 311:1099.

Donaldson, E., A. Haskew, J. Gates, J. Huynh, C. Moore, and M. Frieman. 2010. Metagenomic analysis of the viromes of three North American bat species: viral diversity among different bat species that share a common habitat. J Virol 84:13004–13018.

Du, L.F., Z.J. Lu, Y. Fan, K.Y. Meng, Y. Jiang, Y. Zhu, S.M. Wang, W.J. Gu, X.H. Zou, and C.C. Tu. 2010. Xi River virus, a new bat reovirus isolated in southern China. Arch Virol 155:1295–1299.

Dzikwi, A.A., Kuzmin, Ii, J.U. Umoh, J.K.P. Kwaga, A.A. Ahmad, and C.E. Rupprecht. 2010. Evidence of Lagos bat virus circulation among Nigerian fruit bats. J Wildl Dis 46:267–271.

Emonet, S., J.J. Lemasson, J.P. Gonzalez, X. De Lamballerie, and R.N. Charrel. 2006. Phylogeny and evolution of Old World arenaviruses. Virology 350:251–257.

Epstein, J.H., K.J. Olival, J.R.C. Pulliam, C. Smith, J. Westrum, T. Hughes, A.P. Dobson, A. Zubaid, S.A. Rahman, M.M. Basir, H.E. Field, and P. Daszak. 2009. *Pteropus vampyrus*, a hunted migratory species with a multinational home-range and a need for regional management. J Appl Ecol 46:991–1002.

Epstein, J.H., P.L. Quan, T. Briese, C. Street, O. Jabado, S. Conlan, S.A. Khan, D. Verdugo, M.J. Hossain, S.K. Hutchison, M. Egholm, S.P. Luby, P. Daszak, and W.I. Lipkin. 2010. Identification of GBV-D, a novel GB-like flavivirus from Old World frugivorous bats (*Pteropus giganteus*) in Bangladesh. Plos Path 6:e1000972.

Federico, P., T. Hallam, G. Mccracken, S. Purucker, W. Grant, A. Correa Sandoval, J. Westbrook, R. Medellin, C. Cleveland, C. Sansone, J.J. Lopez, M. Betke, A. Moreno-Valdez, and T. Kunz. 2008. Brazilian free-tailed bats (*Tadarida brasiliensis*) as insect pest regulators in transgenic and conventional cotton crops. Ecol Appl 18:826–837.

Fenton, M.B., M. Davison, T.H. Kunz, G.F. Mccracken, P.A. Racey, and M.D. Tuttle. 2006. Linking bats to emerging diseases. Science 311:1098–1099.

Field, H., P. Young, J.M. Yob, J. Mills, L. Hall, and J. Mackenzie. 2001. The natural history of Hendra and Nipah viruses. Microb Infect 3:307–314.

Field, H.E. 2009. Bats and emerging zoonoses: henipaviruses and SARS. Zoonoses Public Health 56:278–284.

Frank, C.L., D. Reeder, A. Hicks, and R. Rudd. 2010. The effects of white nose syndrome (WNS) on bat hibernation. Integr Comp Biol 50:E56.

Fujii, H., S. Watanabe, D. Yamane, N. Ueda, K. Iha, S. Taniguchi, K. Kato, Y. Tohya, S. Kyuwa, Y. Yoshikawa, and H. Akashi. 2010. Functional analysis of *Rousettus aegyptiacus* "signal transducer and activator of transcription 1" (STAT1). Develop Comp Immunol 34:598–602.

Fujita, M.S., and M.D. Tuttle. 1991. Flying foxes (Chiroptera, Pteropodidae)—Threatened animals of key ecological and economic importance. Conserv Biol 4:455–463.

Gill, F. 2006. Birds of the world: recommended English names. Princeton University Press, New Jersey.

Groseth, A., H. Feldmann, and J.E. Strong. 2007. The ecology of Ebola virus. Trends Microbiol 15: 408–416.

Halpin, K., P.L. Young, H.E. Field, and J.S. Mackenzie. 2000. Isolation of Hendra virus from pteropid bats: a natural reservoir of Hendra virus. J Gen Virol 81:1927–1932.

Halpin, K., A.D. Hyatt, R.K. Plowright, J.H. Epstein, P. Daszak, H.E. Field, L.F. Wang, and P.W. Daniels. 2007. Emerging viruses: coming in on a wrinkled wing and a prayer. Clin Infect Dis 44:711–717.

Haydon, D.T., S. Cleaveland, L.H. Taylor, and M.K. Laurenson. 2002. Identifying reservoirs of infection: a conceptual and practical challenge. Emerg Infect Dis 8:1468–73.

He, G.M., B.B. He, P.A. Racey, and J. Cui. 2010. Positive selection of the bat interferon alpha gene family. Biochem Genet 48:840–846.

Horn, T., T. Sandmann, and M. Boutros. 2010. Design and evaluation of genome-wide libraries for RNA interference screens. Genome Biol 11:R61.

Jones, K.E., N.G. Patel, M.A. Levy, A. Storeygard, D. Balk, J.L. Gittleman, and P. Daszak. 2008. Global trends in emerging infectious diseases. Nature 451: 990–993.

Jordan, I., D. Horn, S. Oehmke, F.H. Leendertz, and V. Sandig. 2009. Cell lines from the Egyptian fruit bat are permissive for modified vaccinia Ankara. Virus Res 145:54–62.

Kepler, T.B., C. Sample, K. Hudak, J. Roach, A. Haines, A. Walsh, and E.A. Ramsburg. 2010. Chiropteran types I and II interferon genes inferred from genome sequencing traces by a statistical gene-family assembler. BMC Genomics 11:444.

Kilpatrick, A.M., P. Daszak, M.J. Jones, P.P. Marra, and L.D. Kramer. 2006. Host heterogeneity dominates West Nile virus transmission. Proc Royal Soc B 273: 2327–2333.

Ksiazek, T.G., D. Erdman, C.S. Goldsmith, S.R. Zaki, T. Peret, S. Emery, S.X. Tong, C. Urbani, J.A. Comer, W. Lim, P.E. Rollin, S.F. Dowell, A.E. Ling, C.D. Humphrey, W.J. Shieh, J. Guarner, C.D. Paddock, P. Rota, B. Fields, J. Derisi, J.Y. Yang, N. Cox, J.M. Hughes, J.W. Leduc, W.J. Bellini, L.J. Anderson, and S.W. Grp. 2003. A novel coronavirus associated with severe acute respiratory syndrome. N Engl J Med 348:1953–1966.

Lau, S.K.P., P.C.Y. Woo, B.H.L. Wong, A.Y.P. Wong, H.W. Tsoi, M. Wang, P. Lee, H.F. Xu, R.W.S. Poon, R.T. Guo, K.S.M. Li, K.H. Chan, B.J. Zheng, and K.Y. Yuen. 2010. Identification and complete genome analysis of three novel paramyxoviruses, Tuhoko virus 1, 2 and 3, in fruit bats from China. Virology 404:106–116.

Leroy, E.M., B. Kumulungui, X. Pourrut, P. Rouquet, A. Hassanin, P. Yaba, A. Delicat, J.T. Paweska, J.P. Gonzalez, and R. Swanepoel. 2005. Fruit bats as reservoirs of Ebola virus. Nature 438:575–576.

Li, L.L., J.G. Victoria, C.L. Wang, M. Jones, G.M. Fellers, T.H. Kunz, and E. Delwart. 2010. Bat guano virome: predominance of dietary viruses from insects and plants plus novel mammalian viruses. J Virol 84:6955–6965.

Li, W., Z. Shi, M. Yu, W. Ren, C. Smith, J.H. Epstein, H. Wang, G. Crameri, Z. Hu, H. Zhang, J. Zhang, J. Mceachern, H. Field, P. Daszak, B.T. Eaton, S. Zhang, and L.-F. Wang. 2005. Bats are natural reservoirs of SARS-like coronaviruses. Science 310:676–679.

Li, Y., J.M. Wang, A.C. Hickey, Y.Z. Zhang, Y.C. Li, Y. Wu, H.J. Zhang, J.F. Yuan, Z.G. Han, J. Mceachern, C.C. Broder, L.F. Wang, and Z.L. Shi. 2008. Antibodies to Nipah or Nipah-like viruses in bats, China. Emerg Infect Dis 14:1974–1976.

Lipkin, W.I. 2008. Pathogen discovery. PLoS Pathog 4:e1000002.

Lipsitch, M., T. Cohen, B. Cooper, J.M. Robins, S. Ma, L. James, G. Gopalakrishna, S.K. Chew, C.C. Tan, M.H. Samore, D. Fisman, and M. Murray. 2003. Transmission dynamics and control of severe acute respiratory syndrome. Science 300:1966–1970.

Luby, S.P., M. Rahman, M.J. Hossain, L.S. Blum, M.M. Husain, E. Gurley, R. Khan, B.N. Ahmed, S. Rahman, N. Nahar, E. Kenah, J.A. Comer, and T.G. Ksiazek. 2006. Foodborne transmission of Nipah virus, Bangladesh. Emerg Infect Dis 12: 1888–1894.

Luby, S.P., M.J. Hossain, E.S. Gurley, B.N. Ahmed, S. Banu, S.U. Khan, N. Homaira, P.A. Rota, P.E. Rollin, J.A. Comer, E. Kenah, T.G. Ksiazek, and M. Rahman. 2009. Recurrent zoonotic transmission of Nipah virus into humans, Bangladesh, 2001–2007. Emerg Infect Dis 15:1229–1235.

Maeda, K., E. Hondo, J. Terakawa, Y. Kiso, N. Nakaichi, D. Endoh, K. Sakai, S. Morikawa, and T. Mizutani. 2008. Isolation of novel adenovirus from fruit bat (*Pteropus dasymallus yayeyamae*). Emerg Infect Dis 14:347–349.

Markus, N. 2002. Behaviour of the black flying fox *Pteropus alecto*: 2. Territoriality and courtship. Acta Chiropt 4:153–166.

Markus, N., and J.K. Blackshaw. 2002. Behaviour of the black flying fox *Pteropus alecto*: 1. An ethogram of behaviour, and preliminary characterisation of mother–infant interactions. Acta Chiropt 4:137–152.

McCracken, G.F. 2003. Estimates of population sizes in summer colonies of Brazilian free-tailed bats (*Tadarida brasiliensis*). In T.J. O'Shea and M.A. Bogan, eds. Monitoring trends in bat populations of the United States and territories: problems and prospects, pp. 21–30. U.S. Geological Survey, Washington, D.C.

Nowak, R.M. 1999. Walker's mammals of the world, 2 vols. 6th ed. Johns Hopkins University Press, Baltimore.

Olival, K.J. 2008. Population genetic structure and phylogeography of Southeast Asian flying foxes: implications for conservation and disease ecology. Ph.D. Dissertation, Columbia University, New York.

Omatsu, T., Y. Nishimura, E.J. Bak, Y. Ishii, Y. Tohya, S. Kyuwa, H. Akashi, and Y. Yoshikawa. 2006. Molecular cloning and sequencing of the cDNA encoding the bat CD4. Vet Immunol Immunopathol 111:309–313.

Omatsu, T., E.J. Bak, Y. Ishii, S. Kyuwa, Y. Tohya, H. Akashi, and Y. Yoshikawa. 2008. Induction and sequencing of Rousette bat interferon alpha and beta genes. Vet Immunol Immunopathol 124:169–176.

Pulliam, J.R., J. Dushoff, H.E. Field, J.H. Epstein, A.P. Dobson, and P. Daszak. 2007. Understanding Nipah virus emergence in peninsular Malaysia: The role of epidemic enhancement in domestic pig populations. Am Soc Trop Med Hyg 56th Ann Meet Abstr Book 956.

Rahman, S.A., S.S. Hassan, K.J. Olival, M. Mohamed, L.-Y. Chang, L. Hassan, A.S. Suri, N.M. Saad, S.A. Shohaimi, Z.C. Mamat, M.S. Naim, J.H. Epstein, H.E. Field, P. Daszak, and Herg. 2010. Characterization of Nipah virus from naturally infected *Pteropus vampyrus* bats, Malaysia. Emerg Inf Dis 16:1990–1993.

Richter, H.V., and G.S. Cumming. 2006. Food availability and annual migration of the straw-colored fruit bat (*Eidolon helvum*). J Zool 268:35–44.

Rihtaric, D., P. Hostnik, A. Steyer, J. Grom, and I. Toplak. 2010. Identification of SARS-like coronaviruses in horseshoe bats (*Rhinolophus hipposideros*) in Slovenia. Arch Virol 155:507–514.

Rouquet, P., J.M. Froment, M. Bermejo, A. Kilbourn, W. Karesh, P. Reed, B. Kumulungui, P. Yaba, A. Delicat, P.E. Rollin, and E.M. Leroy. 2005. Wild animal mortality monitoring and human Ebola outbreaks, Gabon and Republic of Congo, 2001–2003. Emerg Infect Dis 11:283–290.

Shendure, J., and H. Ji. 2008. Next-generation DNA sequencing. Nat Biotechnol 26:1135–1145.

Simmons, N.B. 2005. Order Chiroptera. In D. Wilson and D. Reeder, eds. Mammal species of the world: a taxonomic and geographic reference, 3rd ed. Smithsonian Institution Press, Washington D.C.

Sulkin, S.E., and R. Allen. 1974. Virus infections in bats. Monogr Virol 8:1–103.

Swanepoel, R., S. Smit, P. Rollin, P. Formenty, P. Leman, A. Kemp, and E. Al. 2007. Studies of reservoir hosts for Marburg virus. Emerg Infect Dis 13:1847–1851.

Thalmann, C.M., D.M. Cummins, M. Yu, R. Lunt, L.I. Pritchard, E. Hansson, S. Crameri, A. Hyatt, and L.F. Wang. 2010. Broome virus, a new fusogenic Orthoreovirus species isolated from an Australian fruit bat. Virology 402:26–40.

Turmelle, A.S., and K.J. Olival. 2009. Correlates of viral richness in bats (Order Chiroptera). EcoHealth 6:522–539.

Turmelle, A.S., F.R. Jackson, D. Green, G.F. Mccracken, and C.E. Rupprecht. 2010. Host immunity to repeated rabies virus infection in big brown bats. J Gen Virol 91:2360–2366.

Wibbelt, G., M.S. Moore, T. Schountz, and C.C. Voigt. 2010. Emerging diseases in Chiroptera: why bats? Biol Lett 6:438–440.

Wilson, D., and D. Reeder. 2005. Mammal species of the world: a taxonomic and geographic reference. Smithsonian Institution Press, Washington D.C.

Wong, S., S. Lau, P. Woo, and K.Y. Yuen. 2007. Bats as a continuing source of emerging infections in humans. Rev Med Virol 17:67–91.

Woolhouse, M.E.J., and S. Gowtage-Sequeria. 2005. Host range and emerging and reemerging pathogens. Emerg Infect Dis 11:1842–1847.

# 15

## SARS

### A Case Study for Factors Driving Disease Emergence

Wolfgang Preiser

## THE HISTORY OF SARS

During the first half of 2003, the world experienced a widespread outbreak caused by a novel, severe infectious disease that was termed severe acute respiratory syndrome (SARS) (Kamps and Hoffmann 2003). The emergence of an apparently new, transmissible pneumonic disease in southeastern China's Guangdong province was first noted outside mainland China in February 2003 (Whaley and Mansoor 2006). Media reports painted a picture of panic among the local population, obviously worse than what official reports purported: on February 11, the World Health Organization (WHO) said it had received reports from the Chinese Ministry of Health of an outbreak of acute respiratory syndrome with 300 cases and 5 deaths in Guangdong Province. No further details were given at that time.

Shortly thereafter, WHO reported the detection of avian influenza virus strain H5N1 in a child in Hong Kong and in the child's father, who had died from a severe influenza-like illness; two more family members, one of whom also died, had experienced similar illnesses but no virological test results were available. This family outbreak had begun during a recent visit to Fujian province in mainland China. Hong Kong had previously experienced an outbreak of H5N1 avian

influenza in 1997, during which 6 out of 18 known infected individuals had died and which had been brought under control after stringent measures had been taken to regulate the trade in imported live poultry. Thus it was obvious the zoonosis had reappeared in mainland China, and it could be speculated that it might possibly also be responsible for the mysterious outbreak in Guangdong. Confusingly, however, one week after their first report the Chinese authorities stated that investigations had revealed *Chlamydia pneumoniae* as the probable cause of the Guangdong outbreak and that anthrax, pulmonary plague, leptospirosis, and hemorrhagic fever had been ruled out; influenza was not mentioned. The earliest cases were said to have occurred from mid-November 2002 in the province. It was now claimed that the outbreak was "coming under control."

In the meantime, however, the novel disease was spreading. The index case for most of the cases that occurred outside China was a medical doctor from Guangdong. Having himself treated patients suffering from the novel type of atypical pneumonia there, he traveled to Hong Kong, where he briefly stayed in a hotel before falling ill and later succumbing to the illness. While in Hong Kong, he transmitted the infection to a number of contacts, including some who subsequently traveled further. Some of their destinations experienced massive hospital-associated outbreaks: Vietnam (from where an infected medical doctor, Carlo Urbani, later traveled to Thailand), Singapore (from where an infected medical doctor traveled with his family to the United States and then to Germany), and Canada (from where an infected healthcare worker traveled to the Philippines). Thus, the early global spread of SARS was due to one individual "super-spreader" who was very contagious and gave rise to local and international chains of transmission.

On March 12, 2003, WHO issued a global alert about cases of atypical pneumonia, as SARS was still officially referred to, warning that cases of severe respiratory illness may spread to hospital staff. The world owes this early alert to Dr. Carlo Urbani, working for WHO in Vietnam, who realized that he was facing a previously unknown infectious disease. Three days later, an "emergency travel advisory" was published that included first definitions for suspect and probable cases (Table 15.1). The following day, WHO reported a multi-country outbreak and gave guidance for infection control in healthcare settings. WHO also initiated and organized multicenter collaborative networks dealing with laboratory and clinical aspects of SARS, to bring together those confronted with the new disease and internationally available expertise, with the aim of expediting research and ensuring fast progress.

In the network set up to investigate the cause of SARS, laboratories with access to samples from SARS patients and centers with longstanding experience in emerging or respiratory viruses cooperated rather than competed. Patient samples were shared, and daily telephone conferences and a password-protected website ensured real-time exchange of latest findings and ideas. Influenza and other potentially outbreak-causing respiratory pathogens were quickly ruled out. Several network laboratories diagnosed active infections with *Chlamydophila spp.* or paramyxoviruses in some SARS patients. However, these agents were not found in all patients investigated, nor were they plausible etiological agents for what appeared for all intents and purposes to be a genuinely new infectious disease. The paramyxovirus found in some SARS patients was human metapneumovirus (van den Hoogen et al. 2001); although paramyxoviruses have characteristics that have made them agents of emerging viral diseases (Field et al. 2007) and they need to be considered potential culprits, human

metapneumovirus causes a different clinical presentation and is a very widespread pathogen unable to cause major outbreaks due to the high proportion of immune individuals in most populations (van der Hoek 2007).

Within weeks, different laboratories had identified a previously unknown coronavirus in SARS patients (Drosten et al. 2003a; Ksiazek et al. 2003; Peiris et al. 2003). The isolates or genomic sequences obtained by scientists in different countries were all those of the same new coronavirus (Drosten et al. 2003b). Furthermore, SARS patients were shown to seroconvert against this virus during their illness, while healthy controls did not have specific antibodies. These facts combined made it unlikely that a mere "innocent bystander" (an agent that is present but not causative for the disease in question) had been discovered in the course of unprecedentedly thorough laboratory studies, and experimental studies in macaque monkey formally proved that the new coronavirus fulfilled Koch's postulates as the etiological agent for SARS. WHO announced the discovery of SARS-associated coronavirus (SARS-CoV) as the causative agent of SARS on April 16, 2003 (WHO 2003a). This allowed the rapid development and introduction of specific laboratory tests (Drosten et al. 2004); however, these assays would play only a minor role in managing cases or trying to control the outbreak. The discovery triggered a boom in coronavirus research, leading to the identification of several new species of the genus (see below), and efforts to develop vaccines, antiviral compounds, and antibodies for prevention or therapy (ter Meulen et al. 2004; Cinatl et al. 2005; Roper and Rehm 2009).

SARS is typically a rather nonspecific clinical illness, necessitating the exclusion of a range of alternative causes. The differential diagnosis must take into account season and epidemiologic data and should include common respiratory pathogens such as influenza. Laboratory testing for SARS-CoV most commonly uses viral genome detection from respiratory samples. A diagnostic reverse transcriptase-polymerase chain reaction (RT-PCR) was developed and reagents were made available free of charge to affected countries only weeks into the international outbreak. Due to the confusion that may be caused by false-positive results, WHO has put into place guidelines to ensure a reliable diagnosis of suspect SARS cases. Other available methods are the isolation of SARS-CoV on

**Table 15.1 WHO Case Definitions for SARS Surveillance, revised May 1, 2003**

**Suspect case**

1. A person presenting after 1 November 2002 with history of:
   - high fever (>38°C)
   *AND*
   - cough or breathing difficulty
   *AND*
   one or more of the following exposures during the 10 days prior to onset of symptoms:
   - close contact (defined as having cared for, lived with, or had direct contact with respiratory secretions or body fluids of a suspect or probable case of SARS) with a person who is a suspect or probable case of SARS;
   - history of travel, to an area with recent local transmission of SARS
   - residing in an area with recent local transmission of SARS

2. A person with an unexplained acute respiratory illness resulting in death after 1 November 2002 (The surveillance period begins on 1 November 2002 to capture cases of atypical pneumonia in China now recognized as SARS. International transmission of SARS was first reported in March 2003 for cases with onset in February 2003) but on whom no autopsy has been performed
*AND*
one or more of the following exposures during to 10 days prior to onset of symptoms:
   - close contact with a person who is a suspect or probable case of SARS
   - history of travel to an area with recent local transmission of SARS
   - residing in an area with recent local transmission of SARS

**Probable case**

1. A suspect case with radiographic evidence of infiltrates consistent with pneumonia or respiratory distress syndrome on chest X-ray
2. A suspect case of SARS that is positive for SARS coronavirus by one or more assays (cf. WHO guidelines "Use of laboratory methods for SARS diagnosis")
3. A suspect case with autopsy findings consistent with the pathology of respiratory distress syndrome without an identifiable cause

**Exclusion criteria**

A case should be excluded if an alternative diagnosis can fully explain their illness.

**Reclassification of cases**

As SARS is currently a diagnosis of exclusion, the status of a reported case may change over time.
   A patient should always be managed as clinically appropriate, regardless of their case status.
   - A case initially classified as suspect or probable, for whom an alternative diagnosis can fully explain the illness, should be discarded after carefully considering the possibility of co-infection.
   - A suspect case who, after investigation, fulfils the probable case definition should be reclassified as "probable."
   - A suspect case with a normal chest X-ray should be treated, as deemed appropriate, and monitored for 7 days. Those cases in whom recovery is inadequate should be re-evaluated by chest X-ray.
   - Those suspect cases in whom recovery is adequate but whose illness cannot be fully explained by an alternative diagnosis should remain as "suspect."
   - A suspect case who dies, on whom no autopsy is conducted, should remain classified as "suspect." However, if this case is identified as being part of a chain transmission of SARS, the case should be reclassified as "probable."
   - If an autopsy is conducted and no pathological evidence of respiratory distress syndrome is found, the case should be "discarded."

*Source*: WHO, http://www.who.int/csr/sars/casedefinition/en/, accessed May 31, 2011.

Vero and other types of cells and antibody testing for the indirect diagnosis of SARS-CoV infection by demonstration of seroconversion (Drosten and Leitmeyer 2006).

The number of SARS cases as well as the number of affected areas continued to increase during March 2003. This led WHO to use unprecedented measures. On March 27, 2003, exit screening of passengers leaving affected areas by airplane was advised to prevent febrile people from traveling, and on April 2, 2003, for the first time ever, a "travel advice" was issued. This "advice" was in fact a warning, recommending that those planning to travel to Hong Kong or Guangdong consider postponing their travel unless it was essential. Over the following weeks, such travel warnings were issued for all areas where there was evidence for transmission taking place outside the healthcare setting (i.e., in the community), and they were lifted once the situation had improved. This unprecedented step reflects the great concern over the rapid spread of SARS across widely spread parts of the globe and the steep rise in the number of suspect or proven patients. For the affected countries, these official warnings meant serious economic losses and social disruption, and a sharp decline in tourism and business-related travel caused a major downturn in certain sectors of the economy.

Over the following two months, the pandemic was gradually brought under control. Adequate case management and stringent infection prevention and control policies for healthcare facilities were based on the experiences of the Guangdong outbreaks in February and March 2003 and evidence from subsequent outbreaks. Measures included source isolation of symptomatic cases (fortunately there does not seem to be transmissibility before onset of symptoms), strict infection control precautions in all healthcare facilities, as well as the quarantining of exposed contacts. The extent to which increased social distancing, whether brought about by travel warnings and screening of travelers or resulting from widespread panic, often fueled by media reports, contributed towards effective control is difficult to ascertain. Once these measures were introduced and well implemented, the reproductive rate $R_0$ was reduced to less than 1 so that chains of human-to-human transmission were interrupted (Merianos and Plant 2006) (Fig. 15.1). However, there were a number of events where either poor foresight, poor implementation of necessary

measures, or insufficiencies in health systems had temporarily catastrophic consequences. Such failures affected areas such as the Chinese capital, Beijing, Toronto in Canada, a number of Chinese provinces, and Taiwan. Once there, too, tried-and-tested "old-fashioned" measures such as isolation and quarantine had been put into place, this eventually brought an end to the SARS outbreak (Anderson et al. 2004; Balasegaram and Schnur 2006). On July 5, 2003, WHO officially declared that the last chain of person-to-person transmission of SARS had been interrupted. In total, 8,096 cases of probable SARS were recorded between Nov. 1, 2002, and July 31, 2003. The worst-affected country was mainland China, with 5,327 cases, followed by Hong Kong, with 1,755, and Taiwan, with 346 cases. Canada had recorded 251, Singapore 238, and Vietnam 63 cases. The overall case fatality ratio was 9.6%. Of the 774 deaths in total, 349 occurred in China, 299 in Hong Kong, 43 in Canada, 37 in Taiwan, and 33 in Singapore (Table 15.2). Since the end of the SARS outbreak in mid-2003, there have been four known reappearances of SARS. One could be traced back to exposure to infected animals, but three were due to infection acquired in the course of laboratory work. These cases stress the need to adhere to strict biosafety procedures and practices at all times, and one of them also illustrates the risk of laboratory escapes giving rise to community transmission. To reduce the risk of SARS being re-established, ongoing surveillance should be in place based on WHO recommendations.

## WHERE DID SARS ORIGINATE?

The discovery of a novel coronavirus as the etiological agent of SARS took most of the scientific and public health community by surprise. In contrast to veterinary medicine, where different coronaviruses had long played important roles causing a variety of disease manifestations in domestic (including pigs, turkeys, and chickens) and companion animals (including cats, dogs, and rabbits), human virology had rather neglected this group of viruses (Cavanagh 2004). The coronaviruses form the subfamily Coronavirinae of the family Coronaviridae in the order Nidovirales, enveloped viruses characterized by non-segmented, single-stranded positive-sense RNA genomes from which a set of subgenomic mRNAs are produced.

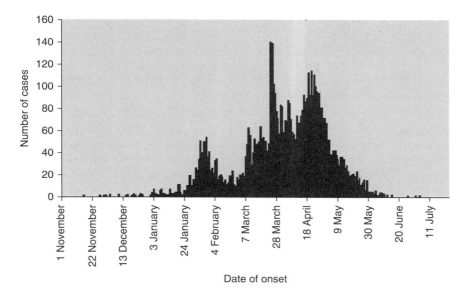

**Figure 15.1:**
Epidemic curve of probable SARS cases worldwide (n = 5.910; this excludes 2.527 cases for whom date of onset is not available), by week of onset, from Nov. 1, 2002, until July 11, 2003. (*Source*: WHO, http://www.who.int/csr/sars/epicurve/epiindex/en/index1.html, accessed May 31, 2011.)

The members of the subfamily Coronavirinae have been classified into the three genera Alpha-, Beta- and Gammacoronavirus on the basis of antigenic and genomic relatedness (Table 15.3) (Carstens and Ball 2009; Carstens 2010; International Committee on Taxonomy of Viruses 2011).

Two coronaviruses, HCoV-229E and HCoV-OC43, had been known to cause common cold-like illnesses in human beings that usually warrant neither laboratory diagnostics nor specific therapy. In addition, these viruses are not easy to isolate and propagate *in vitro*. While therefore only a few human virology institutions were performing research on coronaviruses, there was much more interest and activity in the veterinary sector. Several different coronaviruses were known as important pathogens affecting economically relevant species such as pigs, cattle, and poultry or companion animals such as cats and dogs.

The vast majority of SARS cases during the 2002–03 outbreak originated from human-to-human transmission. In Guangdong, about 50% of cases were known to have been in close contact with another case, often in a hospital or family setting. For the remaining patients, no such contact could be elucidated, initially giving rise to fears about transmission from infected but not yet clinically ill individuals during the incubation period. Fortunately, detailed case investigations ruled out that individuals were contagious during the incubation period. This was crucial for the successful control of the SARS outbreak, which could rely on "traditional" methods without the need for laboratory studies.

This left unanswered the question of where and how this new outbreak had originated. The possibility that this was a preexisting disease entity that had not been recognized previously seemed unlikely. Examples for preexisting diseases being recognized as distinct entities with an infectious origin exist from the recent past (e.g., hantavirus pulmonary syndrome in the Americas). Nevertheless, the ability of SARS to cause large outbreaks in hospital settings is unlikely to have been overlooked for long (Berger et al. 2004). More likely was the possibility of an environmental source. Clues for finding this had to be derived from the peculiarities of the "first" cases at the beginning of chains of human-to-human transmission. As the outbreak had originated in Guangdong in southeastern China, this was the obvious place to investigate. As mentioned before, a thorough retrospective analysis of hospital records had revealed that clusters of cases

Table 15.2 Countries Reporting Probable SARS cases, until July 31, 2003: Summary of Cases and Deaths and Percentage of Healthcare Workers (HCW) Affected

| Country/territory (listed in chronological order of first cases) | Cumulative number of probable SARS cases | Number of deaths | Percentage of HCW among cases (%)* |
|---|---|---|---|
| China | 5,327 | 349 | 19% |
| Hong Kong SAR | 1,755 | 299 | 22% |
| Vietnam | 63 | 5 | 57% |
| Canada | 251 | 43 | 43% |
| United States | 27 | 0 | 0 |
| Singapore | 238 | 33 | 41% |
| Taiwan | 346 | 37 | 20% |
| Philippines | 14 | 2 | 29% |
| Australia | 6 | 0 | 0 |
| Ireland | 1 | 0 | 0 |
| United Kingdom | 4 | 0 | 0 |
| Germany | 9 | 0 | 11% |
| Switzerland | 1 | 0 | 0 |
| Thailand | 9 | 2 | 11% |
| Italy | 4 | 0 | 0 |
| Malaysia | 5 | 2 | 0 |
| France | 7 | 1 | 29% |
| Romania | 1 | 0 | 0 |
| Spain | 1 | 0 | 0 |
| Sweden | 5 | 0 | 0 |
| Mongolia | 9 | 0 | 0 |
| South Africa | 1 | 1 | 0 |
| Indonesia | 2 | 0 | 0 |
| Kuwait | 1 | 0 | 0 |
| New Zealand | 1 | 0 | 0 |
| India | 3 | 0 | 0 |
| Korea | 3 | 0 | 0 |
| Russian Federation | 1 | 0 | 0 |
| Macao SAR | 1 | 0 | 0 |
| Total | 8096 | 774 (overall case fatality ratio 9.6%) | 21.1% (1706 HCW affected) |

*This includes HCW who acquired infection in other areas.
Countries in bold suffered local transmission outside hospitals, leading to community outbreaks.
*Source:* WHO, http://www.who.int/csr/sars/country/table2004_04_21/en/, accessed May 31, 2011.

fulfilling the SARS definition had occurred in different locations around the province since mid-November 2002, with the first such event, in Foshan city, dating back to November 16 and affecting a total of 19 patients (Zhong et al. 2003). Thorough studies of these early index cases to identify any features that these patients may have had in common found that a substantial proportion of them were food handlers or

lived close to places where animals were kept or traded (Breiman et al. 2003; Xu et al. 2004). Guangdong is notorious for its culinary habits. Numerous markets supply an enormous variety of domestic and wild animals, including many threatened species, which is also of great conservation concern. So-called "wet markets" trade in live animals. These may be wild-caught, captivity-bred, or farmed and are usually slaughtered

**Table 15.3 The Revised Taxonomy of Coronaviruses**

| Genus | Species | New species comprising previously recognized coronaviruses |
|---|---|---|
| Alphacoronavirus type species: Alphacoronavirus 1 | Alphacoronavirus 1<br>Human coronavirus 229E<br>Human coronavirus NL63<br>Porcine epidemic diarrhea virus<br>Miniopterus bat coronavirus 1<br>Miniopterus bat coronavirus HKU8<br>Rhinolophus bat coronavirus HKU2<br>Scotophilus bat coronavirus 512 | New Alphacoronavirus 1: porcine transmissible gastroenteritis virus (TGEV, swine), feline coronavirus (FCoV, cats), canine coronavirus (CCoV, dogs) |
| Betacoronavirus type species: Murine coronavirus | Betacoronavirus 1<br>Human coronavirus HKU1<br>Murine coronavirus<br>Severe acute respiratory syndrome (SARS)-related coronavirus<br>Pipistrellus bat coronavirus HKU5<br>Rousettus bat coronavirus HKU9<br>Tylonycteris bat coronavirus HKU4 | New Betacoronavirus 1: human enteric coronavirus, human coronavirus OC43 (human beings), bovine coronavirus, porcine hemagglutinating encephalomyelitis virus, equine coronavirus<br>New Murine coronavirus: murine hepatitis virus, rat coronavirus, puffinosis coronavirus (shearwaters)<br>New SARS-related coronavirus: SARS coronavirus (human beings), SARS-related bat coronavirus (bats) |
| Gammacoronavirus type species: Avian coronavirus | Avian coronavirus<br>Beluga whale coronavirus SW1 | New Avian coronavirus: Infectious bronchitis virus, duck coronavirus, goose coronavirus, pheasant coronavirus, pigeon coronavirus, turkey coronavirus |

Data from Carstens and Ball 2009; Carstens 2010; International Committee on Taxonomy of Viruses 2011.

and prepared at the site of consumption, often in or near restaurant kitchens, creating opportunities for direct and massive exposure of people to animal excreta and fresh blood and tissues potentially containing viable infectious agents. While the (by now clearly human-to-human) outbreak was still ongoing, researchers started studying animals traded at markets in Guangdong. Infection with novel coronaviruses closely related to SARS-CoV was found in all six *Paguma larvata* (masked palm civet or Himalayan palm civet) (Duckworth et al. 2008) tested (two by isolation, two by PCR, and two by both), one of one

*Nyctereutes procyonoides* (raccoon dog; by isolation and PCR), and one of two *Melogale moschata* (Chinese ferret badger; positive neutralizing antibody titer). Furthermore, 40% of animal traders and 20% of animal slaughterers had detectable serum antibodies, compared to only 5% of vegetable traders (Guan et al. 2003). Interestingly, animal isolates contained a 29-nucleotide sequence absent from most human isolates, making "reverse" transmission from human beings to animals unlikely (Ellis et al. 2006).

While these findings suggest that "wet markets" were the source of the SARS outbreak, they do not prove that the species found to be infected are the reservoir hosts of the virus. Subsequent studies found *Paguma larvata* to be highly susceptible to infection with SARS-CoV, which in them causes multiorgan pathology (Xiao et al. 2008); the authors suggest their use as an animal model for testing candidate vaccines and antiviral drugs. However, the pathogenicity of the infection almost certainly ruled out this species as the natural host. In addition, another study found there were few antibody-positive *Paguma larvata* on farms, while four fifths of these animals at one animal market in Guangdong were seropositive. The authors concluded that infections were probably taking place at or near the market, possibly linked to conditions of overcrowding and close mixing of various animal species (Tu et al. 2004).

This was confirmed by another SARS outbreak that occurred, again in Guangdong, in the winter of 2003–04. Two of four affected SARS patients had been directly exposed to caged *Paguma larvata* at a restaurant (the ban introduced earlier in 2003 did not last). All six animals were infected with SARS-CoV; sequence analysis demonstrated these to be most closely related to previous animal isolates. The virus in the two patients was phylogenetically identical to animal isolates and distinct from previous human SARS-CoV isolates, demonstrating that these cases resulted immediately from inter-species transmission from *Paguma larvata* at the venue (Wang et al. 2005). The natural host of SARS-CoV was likely to be found in a different species to which all the traded animals had been exposed, perhaps as a food item or through close contact over prolonged periods of time under conditions of poor hygiene. The species found to be infected may, however, play an important role in providing an epidemiologically important link between an as-yet-unknown reservoir host and the human population and possibly also in amplifying SARS-CoV infection.

Studies of different wildlife species later identified various different previously undescribed coronaviruses in different species of bats. One of these bat coronaviruses, found in different species of Chinese horseshoe bat, genus *Rhinolophus*, is closely related to SARS-CoV and probably the progenitor of the virus that, via *Paguma larvata* and other animals traded in Guangdong, caused the 2002–03 SARS outbreak (Lau et al. 2005; Li et al. 2005). It is thought that while being kept in captivity prior to being sold for consumption,

*Paguma larvata* and the other species implicated were exposed to infected bats, either through close proximity resulting in environmental contamination or perhaps ingestion of dead bats, became infected with their coronavirus, and later passed it on to human beings.

This spillover from a wildlife reservoir via an intermediate host to the human population was possible due to coronaviruses being viral "generalists," *a priori* able to infect a wide range of hosts—"pre-adapted" (Childs et al. 2007; Wolfe et al. 2007; Parrish et al. 2008). Following the "species jump," rapid viral evolution took place and helped SARS-CoV to rapidly adapt to replication in a new species (CSMEC 2004; Cheng et al. 2007; Wang and Eaton 2007).

## WHY DID SARS HAPPEN?

The rapid containment of the SARS outbreak in 2003 was not a consequence of preparedness. Sadly, as Cleaveland et al. (2007) illustrate, there would have been every reason to expect such an event. The emergence of infectious diseases, and in particular viruses, among whom viruses with RNA genomes are the most likely candidates, is anything but uncommon (Ludwig et al. 2003; Morse et al. 2004). Among the factors in emergence listed by the latest Institute of Medicine report on microbial threats to health (Smolinski et al. 2003), only the first, microbial adaptation and change, concerns an intrinsic property of infectious agents. It certainly applies to SARS. To identify virus families posing a high risk for causing pandemics, three criteria have been proposed (Burke 1998). Coronaviruses fulfilled two of the three even before the SARS outbreak (now they obviously fulfill all three) and should for that reason alone have received more attention also in the fields of human medicine and public health (Table 15.4) than was the case prior to 2003 (Saif 2004).

Not all of the remaining 12 factors listed by the 2003 report apply to SARS; however, the subset of five of these additional factors that had already been listed by the predecessor report (Lederberg et al. 1992) all apply to what happened in 2003. Firstly, rapid economic development had led to a surge in demand for "exotic" meat in China, fueling a massively increased trade in wild animals originating from East and Southeast Asia (the Far Eastern equivalent of the

Table 15.4 Criteria for Virus Families with a High Pandemic Risk and Examples

| Criterion | Applicable to coronaviridae? |
|---|---|
| 1. Has caused pandemics in human populations in the recent past | Prior to the SARS outbreak of 2002/2003: **No** |
| 2. Has the proven ability to cause larger epizootics | **Yes:** Porcine respiratory coronavirus (PRCoV) has caused epizootics since the 1980s |
| 3. Has the intrinsic propensity to rapidly undergo evolution ("intrinsic evolvability") on the basis of a high mutation rate or a genome organization favoring recombination | **Yes:** Two animal coronaviruses with altered tissue tropism and different virulence are known to have arisen from progenitor viruses: <br> 1. The less virulent feline enteric CoV (FeCoV) evolved into the virulent and systemic feline infectious peritonitis virus (FIPV) <br> 2. The less virulent porcine respiratory CoV (PRCoV) evolved from transmissible gastroenteritis virus (TGEV) through deletion mutations in the S gene on two occasions during the 1980s |

After Burke 1998; Saif 2004.

African "bushmeat" trade), which is of concern not only in the context of emerging infectious diseases (Bell et al. 2004), but also from biodiversity conservation and animal welfare points of view (Jiao and Thomas 2008; Nijman 2010). The other four factors—human demographics and behavior, international travel and commerce, technology and industry, as well as breakdown of public health measures—all played roles in the further spread of SARS once it had jumped across the species barrier into the human population. High population density and poor hygiene behavior, high travel activity nationally and internationally, modern healthcare equipment (such as respirators), and often, at least initially, poor public health systems all contributed to allowing the rapid dissemination of the new disease (Hufnagel et al. 2004). One could argue that technology and industry also contributed to SARS before it reached the human population, if referring to the farming and trading of wildlife species. Of the seven factors listed in the 2003 report but not in the previous one, human susceptibility to infection (individuals with preexisting medical conditions were more prone to severe and often fatal disease and also more likely to become "super-spreaders") and lack of political will contributed in the case of SARS.

Therefore, if coronaviruses should have been among the "usual suspects," and the conditions for the subsequent pandemic emergence in the human population were rife, what about the source of the virus?

Zoonotic pathogens are particularly prone to becoming emerging (Taylor et al. 2001), and a zoonotic source must be considered for all apparently new infectious diseases (Bengis et al. 2004; Kruse et al. 2004). With approximately 1,000 known species, the order Chiroptera (bats) accounts for a quarter of all known mammal species. Bats are now recognized as one of the most important sources of newly emerging viruses of humans (Calisher et al. 2006; Wong et al. 2007; Field 2009; Turmelle and Olival 2009). Consequently, substantial research efforts are now being directed to viruses carried by bats and have within a few years led to the discovery of a number of hitherto unknown viruses, including a number of different coronaviruses, some closely related to SARS-CoV and others not (Tang et al. 2006; Wang et al. 2006; Müller et al. 2007; Gloza-Rausch et al. 2008; Pfefferle et al. 2009; Tong et al. 2009; Reusken et al. 2010; Drexler et al. 2010; Drexler et al. 2011).

The link between potentially zoonotic coronaviruses, bats, and the human population was, in the case of SARS, provided by the capture or farming and subsequent trading of wildlife species in "wet markets." The display and sale of live animals means that animal excreta and, if butchered at point of sale, animal blood and tissues will contaminate stalls. In addition, a mixture of different species may be kept in close vicinity that might under natural circumstances never have encountered one another, let alone have had the chance of inadvertently exchanging excreta and body fluids; this provides for potentially intensive inter-species contact, which may lead to spillover of pathogens. These conditions make "wet markets" ideal breeding grounds for infectious agents waiting for an opportunity to emerge (Webster et al. 2004; Woo et al. 2006). In summary, a multitude of factors and conditions favoring the emergence of novel infectious agents came together in the case of SARS. One might be tempted to ask, why did SARS not happen before? Figure 15.2 illustrates and annotates the episode of emergence as it took place in 2003.

## LESSONS FOR THE FUTURE

In his chapter "What did we learn from SARS?" Doberstyn (2006) lists 13 lessons that should have been learned. I will here concentrate on the two that I consider most pertinent in the context of this book and neglect those referring to the later stages of the outbreak.

One could indeed say that the world was "lucky this time." A number of factors were in our favor, including certain viral characteristics; the "human factor" may have taken some time, but eventually all necessary steps were taken. However, it would be more comforting had the remarkable and remarkably rapid success in containing and even terminating the SARS outbreak been the result of preparedness. Alas, it was not, despite the fact that the writing had been on the wall, waiting to be read. Interestingly, the hugely increased interest in coronaviruses has led to the discovery of novel ones not only in various species of animals but also in humans (Kahn 2006; Wevers and van der Hoek 2009). HCoV-NL63 and HCoV-HKU1 are associated with respiratory tract infections in children and have probably a worldwide distribution (Gaunt et al. 2010). Other types of coronaviruses were found in bats (see above) and some other animal species (Dong et al. 2007; Shi and Hu 2008). SARS did demonstrate very vividly how animal husbandry and marketing practices seriously affect human health. The ongoing epizootic caused by highly pathogenic influenza A (H5N1) virus started shortly after SARS had been controlled (Kamps et al. 2006). The agent belongs to a virus group notorious for causing pandemics. So far, almost all cases of transmission to human beings have occurred by direct exposure to

infected, sick poultry. Nevertheless, there is a possibility that this avian influenza virus might still become fully "humanized," which could cause an influenza pandemic far worse than the one recently seen (Maritz et al. 2010). While considerable efforts are being undertaken to stem the epizootic, some actions (or inactions) by governments and individuals do not contribute towards that goal and are reminiscent of the misjudgments and short-sighted, non-constructive behaviors seen during some stages of the SARS outbreak in 2003. As with SARS, avian influenza is a natural phenomenon; its emergence, however, is driven by factors and determinants that are related to human activities.

In summary, SARS was a novel, severe infectious disease that originated in Guangdong, China. It arose from a natural reservoir in bats through an intermediary host, linked to the commerce in and consumption of wild animal species. It then established itself as a human pathogen, being transmitted from human to human predominantly via the respiratory route. It showed considerable nosocomial potential. SARS spread rapidly around the globe through air travel after a single super-spreading event in Hong Kong. Knowledge about the new disease was gained rapidly through global networking and international collaboration, led by the WHO (2003b). Interestingly, the SARS outbreak was controlled rather quickly through "traditional" sanitary measures, including the consequent diagnosis of suspect cases and isolation of the diseased. Whenever and wherever this was not done adequately, larger outbreaks ensued that required more rigorous measures to be contained. Also, industrialized countries with modern healthcare systems were not "immune." SARS has not been able to maintain itself in the human population permanently. The risk of re-emergence is unclear; the post-outbreak experience has shown that monitoring of the animal trade and of laboratories is necessary to prevent it. More importantly even, measures should be developed to discourage risky practices relating to animal trade and consumption. This would also benefit the conservation of biodiversity (Zhang et al. 2008). While it would be overly optimistic to claim that humankind has now gained the ability to successfully contain infectious disease threats, it is encouraging to see what open international cooperation, modern biomedical science, and modern information technology were able to achieve, even though there

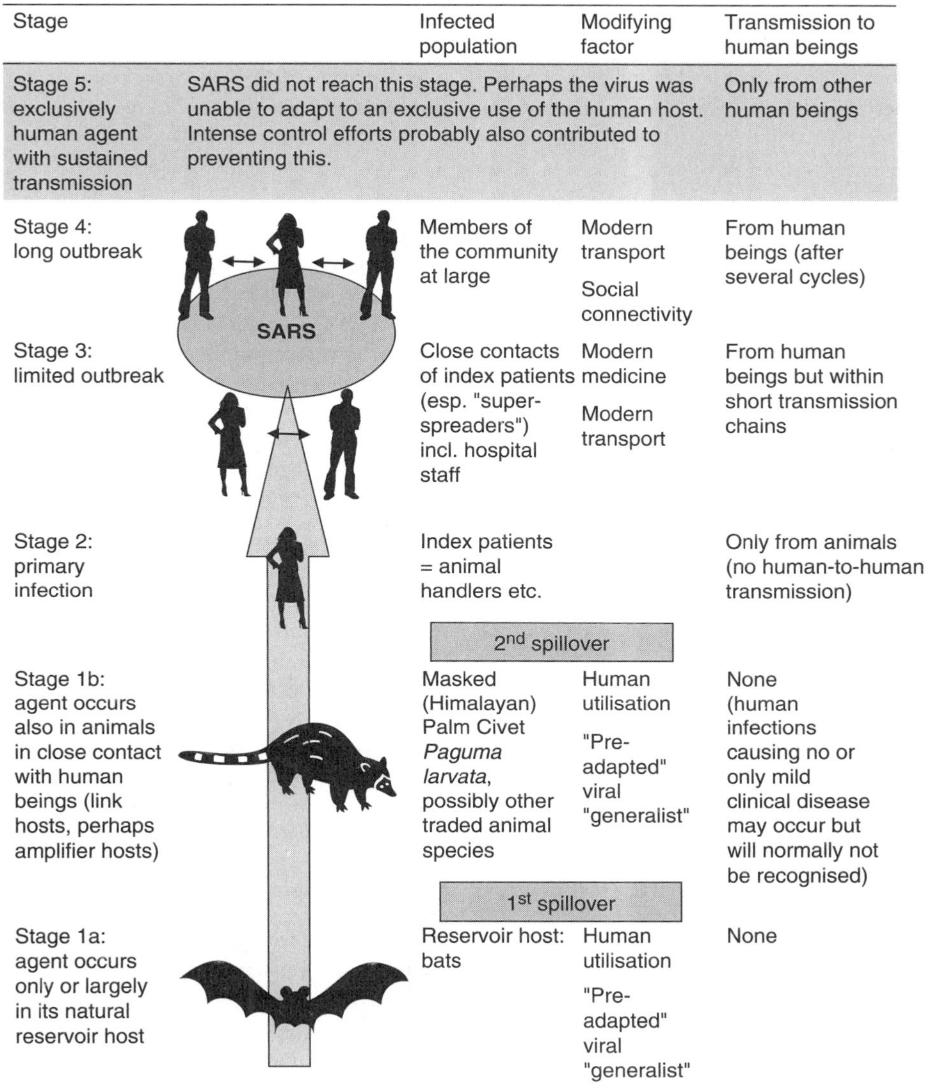

| Stage | Infected population | Modifying factor | Transmission to human beings |
|---|---|---|---|
| Stage 5: exclusively human agent with sustained transmission | SARS did not reach this stage. Perhaps the virus was unable to adapt to an exclusive use of the human host. Intense control efforts probably also contributed to preventing this. | | Only from other human beings |
| Stage 4: long outbreak | Members of the community at large | Modern transport<br><br>Social connectivity | From human beings (after several cycles) |
| Stage 3: limited outbreak | Close contacts of index patients (esp. "super-spreaders") incl. hospital staff | Modern medicine<br><br>Modern transport | From human beings but within short transmission chains |
| Stage 2: primary infection | Index patients = animal handlers etc. | | Only from animals (no human-to-human transmission) |
| Stage 1b: agent occurs also in animals in close contact with human beings (link hosts, perhaps amplifier hosts) | Masked (Himalayan) Palm Civet *Paguma larvata*, possibly other traded animal species | 2nd spillover<br><br>Human utilisation<br><br>"Pre-adapted" viral "generalist" | None (human infections causing no or only mild clinical disease may occur but will normally not be recognised) |
| Stage 1a: agent occurs only or largely in its natural reservoir host | 1st spillover<br><br>Reservoir host: bats | Human utilisation<br><br>"Pre-adapted" viral "generalist" | None |

**Figure 15.2:**
By modifying a previously proposed schema (Wolfe et al. 2007) and combining it with some elements of the one developed by Childs et al. (2007), the emergence of SARS can best be illustrated by splitting stage 1 into 1a and 1b (to allow for the reservoir host and the secondary host from whom infection spilled over into the human population) and extending as far as in between stages 3 and 4 but not reaching stage 5.

was heavy reliance on "traditional" approaches of disease control.

# REFERENCES

Anderson, R.M., C. Fraser, A.C. Ghani, C.A. Donnelly, S. Riley, N.M. Ferguson, G.M. Leung, T.H. Lam, and A.J. Hedley. 2004. Epidemiology, transmission dynamics and control of SARS: the 2002–2003 epidemic. Philos Trans R Soc B 359: 1091–1105.

Balasegaram, M., and A. Schnur. 2006. China: from denial to mass mobilization. In World Health Organization Regional Office for the Western Pacific Region. SARS: how a global epidemic was stopped, pp. 73–85. World Health Organization, Geneva.

Bell, D., S. Roberton, and P.R. Hunter. 2004. Animal origins of SARS coronavirus: possible links with the international trade in small carnivores. Philos Trans R Soc B 359:1107–1114.

Bengis, R.G., F.A. Leighton, J.R. Fischer, M. Artois, T. Mörner, and C.M. Tate. 2004. The role of wildlife in emerging and re-emerging zoonoses. Rev Sci Tech 23:497–511.

Berger, A., C. Drosten, H.W. Doerr, M. Stürmer, and W. Preiser. 2004. Severe acute respiratory syndrome (SARS)—paradigm of an emerging viral infection. J Clin Virol 29:13–22.

Breiman, R.F., M.R. Evans, W. Preiser, J. Maguire, A. Schnur, A. Li, H. Bekedam, and J.S. MacKenzie. 2003. Role of China in the quest to define and control severe acute respiratory syndrome. Emerg Infect Dis 9:1037–1041.

Burke, D.S. 1998. The evolvability of emerging viruses. In C.R. Horsburgh, ed. Pathology of emerging infections, pp. 1–12. American Society for Microbiology, Washington, D.C.

Calisher, C.H., J.E. Childs, H.E. Field, K.V. Holmes, and T. Schountz. 2006. Bats: important reservoir hosts of emerging viruses. Clin Microbiol Rev 19:531–545.

Carstens, E.B., and L.A. Ball. 2009. Ratification vote on taxonomic proposals to the International Committee on Taxonomy of Viruses 2008. Arch Virol 154: 1181–1188.

Carstens, E.B. 2010. Ratification vote on taxonomic proposals to the International Committee on Taxonomy of Viruses 2009. Arch Virol 155:133–146.

Cavanagh, D. 2004. Coronaviruses and Toroviruses. In A.J. Zuckerman, J.E. Banatvala, J.R. Pattison, P.D. Griffiths, and B.D. Schoub eds. Principles and practice of clinical virology, 5th ed., pp. 379–397. John Wiley & Sons, Chichester, U.K.

Cheng, V.C., S.K. Lau, P.C. Woo, and K.Y. Yuen. 2007. Severe acute respiratory syndrome coronavirus as an agent of emerging and reemerging infection. Clin Microbiol Rev 20:660–694.

Childs, J.E., J.A. Richt, and J.S. Mackenzie. 2007. Introduction: conceptualizing and partitioning the emergence process of zoonotic viruses from wildlife to humans. Curr Top Microbiol Immunol 315:1–31.

Cinatl, J. Jr., M. Michaelis, G. Hoever, W. Preiser, and H.W. Doerr. 2005. Development of antiviral therapy for severe acute respiratory syndrome. Antiviral Res 66:81–97.

Cleaveland, S., D.T. Haydon, and L. Taylor. 2007. Overviews of pathogen emergence: which pathogens emerge, when and why? Curr Top Microbiol Immunol 315:85–111.

CSMEC (Chinese SARS Molecular Epidemiology Consortium). 2004. Molecular evolution of the SARS coronavirus during the course of the SARS epidemic in China. Science 303:1666–1669.

Doberstyn, B. 2006. What did we learn from SARS? In World Health Organization Regional Office for the Western Pacific Region. SARS: how a global epidemic was stopped, pp. 243–254. World Health Organization, Geneva.

Dong, B.Q., W. Liu, X.H. Fan, D. Vijaykrishna, X.C. Tang, F. Gao, L.F. Li, G.J. Li, J.X. Zhang, L.Q. Yang, L.L. Poon, S.Y. Zhang, J.S. Peiris, G.J. Smith, H. Chen, and Y. Guan. 2007. Detection of a novel and highly divergent coronavirus from Asian leopard cats and Chinese ferret badgers in Southern China. J Virol 81:6920–6926.

Drexler, J.F., F. Gloza-Rausch, J. Glende, V.M. Corman, D. Muth, M. Goettsche, A. Seebens, M. Niedrig, S. Pfefferle, S. Yordanov, L. Zhelyazkov, U. Hermanns, P. Vallo, A. Lukashev, M.A. Müller, H. Deng, G. Herrler, and C. Drosten. 2010. Genomic characterization of severe acute respiratory syndrome-related coronavirus in European bats and classification of coronaviruses based on partial RNA-dependent RNA polymerase gene sequences. J Virol 84:11336–11349.

Drexler, J.F., V.M. Corman, T. Wegner, A.F. Tateno, R.M. Zerbinati, F. Gloza-Rausch, A. Seebens, M.A. Müller, and C. Drosten. 2011. Amplification of emerging viruses in a bat colony. Emerg Infect Dis 17:449–456.

Drosten, C., S. Günther, W. Preiser, S. van der Werf, H.R. Brodt, S. Becker, H. Rabenau, M. Panning, L. Kolesnikova, R.A. Fouchier, A. Berger, A.M. Burguière, J. Cinatl, M. Eickmann, N. Escriou, K. Grywna, S. Kramme, J.C. Manuguerra, S. Müller, V. Rickerts, M. Stürmer, S. Vieth, H.D. Klenk, A.D. Osterhaus, H. Schmitz, and H.W. Doerr. 2003a. Identification of a novel coronavirus in patients with severe acute respiratory syndrome. N Engl J Med 348:1967–1976.

Drosten, C., W. Preiser, S. Günther, H. Schmitz, and H.W. Doerr. 2003b. Severe acute respiratory syndrome: identification of the etiological agent. Trends Mol Med 9:325–327.

Drosten, C., L.L. Chiu, M. Panning, H.N. Leong, W. Preiser, J.S. Tam, S. Günther, S. Kramme, P. Emmerich, W.L. Ng, H. Schmitz, and E.S. Koay. 2004. Evaluation of advanced reverse transcription-PCR assays and an alternative PCR target region for detection of severe acute respiratory syndrome-associated coronavirus. J Clin Microbiol 42:2043–2047.

Drosten, C., and K. Leitmeyer. 2006. Laboratory diagnostics. In World Health Organization

Regional Office for the Western Pacific Region. SARS: how a global epidemic was stopped, pp. 218–224. World Health Organization, Geneva.

Duckworth, J.W., C. Wozencraft, and B. Kanchanasaka. 2008. Paguma larvata. In IUCN 2010. IUCN Red List of Threatened Species. Version 2010.4 www.iucnredlist.org.

Ellis, A., Y. Guan, and E. Miranda. 2006. The animal connection. In World Health Organization Regional Office for the Western Pacific Region. SARS: how a global epidemic was stopped, pp. 225–229. World Health Organization, Geneva.

Field, H.E., J.S. Mackenzie, and P. Daszak. 2007. Henipaviruses: emerging paramyxoviruses associated with fruit bats. Curr Top Microbiol Immunol 315:133–159.

Field, H.E. 2009. Bats and emerging zoonoses: henipaviruses and SARS. Zoonoses Public Health 56:278–284.

Gaunt, E.R., A. Hardie, E.C. Claas, P. Simmonds, and K.E. Templeton. 2010. Epidemiology and clinical presentations of the four human coronaviruses 229E, HKU1, NL63, and OC43 detected over 3 years using a novel multiplex real-time PCR method. J Clin Microbiol 48:2940–2947.

Gloza-Rausch, F., A. Ipsen, A. Seebens, M. Göttsche, M. Panning, J.F. Drexler, N. Petersen, A. Annan, K. Grywna, M. Müller, S. Pfefferle, and C. Drosten. 2008. Detection and prevalence patterns of group I coronaviruses in bats, northern Germany. Emerg Infect Dis 14:626–631.

Guan, Y., B.J. Zheng, Y.Q. He, X.L. Liu, Z.X. Zhuang, C.L. Cheung, S.W. Luo, P.H. Li, L.J. Zhang, Y.J. Guan, K.M. Butt, K.L. Wong, K.W. Chan, W. Lim, K.F. Shortridge, K.Y. Yuen, J.S. Peiris, and Poon, L.L. 2003. Isolation and characterization of viruses related to the SARS coronavirus from animals in southern China. Science 302:276–278.

Hufnagel, L., Brockmann, D., Geisel, T. 2004. Forecast and control of epidemics in a globalized world. Proc Natl Acad Sci USA 101:15124–15129.

International Committee on Taxonomy of Viruses (ICTV). 2011. ICTV Master Species List 2009—Version 9. Accessed March 26, 2011. Available at http://www.ictvonline.org.

Jiao, P., and R. Thomas. 2008. The state of wildlife trade in China. TRAFFIC East Asia—China Programme, Beijing.

Kahn, J.S. 2006. The widening scope of coronaviruses. Curr Opin Pediatr 18:42–47.

Kamps, B.S., and C. Hoffmann, eds. 2003. SARS reference–10/2003, 3rd ed. Flying Publisher, Paris–Cagliari–Wuppertal–Sevilla. Downloaded from SARSReference.com

Kamps, B.S., C. Hoffmann, and W. Preiser eds. 2006. Influenza report 2006. Flying Publisher, Paris–Cagliari–Wuppertal–Sevilla. Available at http://www.InfluenzaReport.com

Kruse, H., A.M. Kirkemo, and K. Handeland. 2004. Wildlife as source of zoonotic infections. Emerg Infect Dis 10:2067–2072.

Ksiazek, T.G., D. Erdman, C.S. Goldsmith, S.R. Zaki, T. Peret, S. Emery, S. Tong, C. Urbani, J.A. Comer, W. Lim, P.E. Rollin, S.F. Dowell, A.E. Ling, C.D. Humphrey, W.J. Shieh, J. Guarner, C.D. Paddock, P. Rota, B. Fields, J. DeRisi, J.Y. Yang, N. Cox, J.M. Hughes, J.W. LeDuc, W.J. Bellini, L.J. Anderson; SARS Working Group. 2003. A novel coronavirus associated with severe acute respiratory syndrome. N Engl J Med. 348:1953–1966.

Lau, S.K., P.C. Woo, K.S. Li, Y. Huang, H.W. Tsoi, B.H. Wong, S.S. Wong, S.Y. Leung, K.H. Chan, and K.Y. Yuen. 2005. Severe acute respiratory syndrome coronavirus-like virus in Chinese horseshoe bats. Proc Natl Acad Sci USA 102:14040–14045.

Lederberg, J., R.E. Shope, and S.C. Oaks Jr. eds. 1992. Emerging infections: microbial threats to health in the United States. National Academies Press, Washington D.C.

Li, W., Z. Shi, M. Yu, W. Ren, C. Smith, J.H. Epstein, H. Wang, G. Crameri, Z. Hu, H. Zhang, J. Zhang, J. McEachern, H. Field, P. Daszak, B.T. Eaton, S. Zhang, and L.F. Wang. 2005. Bats are natural reservoirs of SARS-like coronaviruses. Science 310:676–679.

Ludwig, B., F.B. Kraus, R. Allwinn, H.W. Doerr, and W. Preiser. 2003. Viral zoonoses—a threat under control? Intervirology 46:71–78.

Maritz, J., L. Maree, and W. Preiser. 2010. Pandemic influenza A (H1N1) 2009: the experience of the first six months. Clin Chem Lab Med 48:11–21.

Merianos, A., and A. Plant. 2006. Epidemiology. In World Health Organization Regional Office for the Western Pacific Region. SARS: how a global epidemic was stopped, pp.185–198. World Health Organization, Geneva.

Morse, S.S. 2004. Factors and determinants of disease emergence. Rev Sci Tech 23:443–451.

Müller, M.A., J.T. Paweska, P.A. Leman, C. Drosten, K. Grywna, A. Kemp, L. Braack, K. Sonnenberg, M. Niedrig, and R. Swanepoel. 2007. Coronavirus antibodies in African bat species. Emerg Infect Dis 13:1367–1370.

Nijman, V. 2010. An overview of international wildlife trade from Southeast Asia. Biodivers Conserv 19:1101–1114.

Parrish, C.R., E.C. Holmes, D.M. Morens, E.C. Park, D.S. Burke, C.H. Calisher, C.A. Laughlin, L.J. Saif, and P. Daszak. 2008. Cross-species virus transmission and the emergence of new epidemic diseases. Microbiol Mol Biol Rev 72:457–470.

Peiris, J.S., S.T. Lai, L.L. Poon, Y. Guan, L.Y. Yam, W. Lim, J. Nicholls, W.K. Yee, W.W. Yan, M.T. Cheung, V.C. Cheng, K.H. Chan, D.N. Tsang, R.W. Yung, T.K. Ng, K.Y. Yuen; SARS study group. 2003. Coronavirus as a possible cause of severe acute respiratory syndrome. Lancet 361:1319–1325.

Pfefferle, S., S. Oppong, J.F. Drexler, F. Gloza-Rausch, A. Ipsen, A. Seebens, M.A. Müller, A. Annan, P. Vallo, Y. Adu-Sarkodie, T.F. Kruppa, and C. Drosten. 2009. Distant relatives of severe acute respiratory syndrome coronavirus and close relatives of human coronavirus 229E in bats, Ghana. Emerg Infect Dis 15:1377–1384.

Reusken, C.B., P.H. Lina, A. Pielaat, A. de Vries, C. Dam-Deisz, J. Adema, J.F. Drexler, C. Drosten, and E.A. Kooi. 2010. Circulation of group 2 coronaviruses in a bat species common to urban areas in Western Europe. Vector Borne Zoonotic Dis 10:785–791.

Roper, R.L., and K.E. Rehm. 2009. SARS vaccines: where are we? Expert Rev Vaccines 8:887–898.

Saif, L.J. 2004. Animal coronaviruses: what can they teach us about the severe acute respiratory syndrome? Rev Sci Tech 23:643–660.

Shi, Z., and Z. Hu. 2008. A review of studies on animal reservoirs of the SARS coronavirus. Virus Res 133:74–87.

Smolinski, M.S., M.A. Hamburg, and J. Lederberg eds. 2003. Microbial threats to health: emergence, detection, and response. National Academies Press, Washington D.C.

Tang, X.C., J.X. Zhang, S.Y. Zhang, P. Wang, X.H. Fan, L.F. Li, G. Li, B.Q. Dong, W. Liu, C.L. Cheung, K.M. Xu, W.J. Song, D. Vijaykrishna, L.L. Poon, J.S. Peiris, G.J. Smith, H. Chen, and Y. Guan. 2006. Prevalence and genetic diversity of coronaviruses in bats from China. J Virol 80:7481–7490.

Taylor, L.H., S.M. Latham, and M.E. Woolhouse. 2001. Risk factors for human disease emergence. Philos Trans R Soc B 356:983–989.

ter Meulen, J., A.B. Bakker, E.N. van den Brink, G.J. Weverling, B.E. Martina, B.L. Haagmans, T. Kuiken, J. de Kruif, W. Preiser, W. Spaan, H.R. Gelderblom, J. Goudsmit, and A.D. Osterhaus. 2004. Human monoclonal antibody as prophylaxis for SARS coronavirus infection in ferrets. Lancet 363:2139–2141.

Tong, S., C. Conrardy, S. Ruone, I.V. Kuzmin, X. Guo, Y. Tao, M. Niezgoda, L. Haynes, B. Agwanda, R.F. Breiman, L.J. Anderson, and C.E. Rupprecht. 2009. Detection of novel SARS-like and other coronaviruses in bats from Kenya. Emerg Infect Dis 15:482–485.

Tu, C., G. Crameri, X. Kong, J. Chen, Y. Sun, M. Yu, H. Xiang, X. Xia, S. Liu, T. Ren, Y. Yu, B.T. Eaton, H. Xuan, and L.F. Wang. 2004. Antibodies to SARS coronavirus in civets. Emerg Infect Dis 10:2244–2248.

Turmelle, A.S., and K.J. Olival. 2009. Correlates of viral richness in bats (order Chiroptera). Ecohealth 6:522–539.

van den Hoogen, B.G., J.C. de Jong, J. Groen, T. Kuiken, R. de Groot, R.A. Fouchier, and A.D. Osterhaus. 2001. A newly discovered human pneumovirus isolated from young children with respiratory tract disease. Nat Med 7:719–724.

van der Hoek, L. 2007. Human coronaviruses: what do they cause? Antivir Ther 12:651–658.

Wang, M., M. Yan, H. Xu, W. Liang, B. Kan, B. Zheng, H. Chen, H. Zheng, Y. Xu, E. Zhang, H. Wang, J. Ye, G. Li, M. Li, Z. Cui, Y.F. Liu, R.T. Guo, X.N. Liu, L.H. Zhan, D.H. Zhou, A. Zhao, R. Hai, D. Yu, Y. Guan, and J. Xu. 2005. SARS-CoV infection in a restaurant from palm civet. Emerg Infect Dis 11:1860–1865.

Wang, L.F., Z. Shi, S. Zhang, H. Field, P. Daszak, and B.T. Eaton. 2006. Review of bats and SARS. Emerg Infect Dis 12:1834–1840.

Wang, L.F., and B.T. Eaton. 2007. Bats, civets and the emergence of SARS. Curr Top Microbiol Immunol 315:325–344.

Webster, R.G. 2004. Wet markets—a continuing source of severe acute respiratory syndrome and influenza? Lancet 363:234–236.

Wevers, B.A., and L. van der Hoek. 2009. Recently discovered human coronaviruses. Clin Lab Med 29:715–724.

Whaley, F., and O.D. Mansoor. 2006. SARS chronology. In World Health Organization Regional Office for the Western Pacific Region. SARS: how a global epidemic was stopped, pp 3–48. World Health Organization, Geneva.

WHO (World Health Organization). 2003a. Multicentre Collaborative Network for Severe Acute Respiratory Syndrome (SARS) Diagnosis: A multicentre collaboration to investigate the cause of severe acute respiratory syndrome. Lancet 361:1730–1733.

WHO (World Health Organization). 2003b. SARS: lessons from a new disease. Chapter 5, The World Health Report 2003: shaping the future. World Health Organization, Geneva.

Wolfe, N.D., C.P. Dunavan, and J. Diamond. 2007. Origins of major human infectious diseases. Nature 447:279–283.

Wong, S., S. Lau, P. Woo, and K.Y. Yuen. 2007. Bats as a continuing source of emerging infections in humans. Rev Med Virol 17:67–91.

Woo, P.C., S.K. Lau, and K.Y. Yuen. 2006. Infectious diseases emerging from Chinese wet-markets: zoonotic origins of severe respiratory viral infections. Curr Opin Infect Dis 19:401–407.

Xiao, Y., Q. Meng, X. Yin, Y. Guan, Y. Liu, C. Li, M. Wang, G. Liu, T. Tong, L.F. Wang, X. Kong, and D. Wu. 2008. Pathological changes in masked palm civets experimentally infected by severe acute respiratory syndrome (SARS) coronavirus. J Comp Pathol 138:171–179.

Xu, R.H., J.F. He, M.R. Evans, G.W. Peng, H.E. Field, D.W. Yu, C.K. Lee, H.M. Luo, W.S. Lin, P. Lin, L.H. Li, W.J. Liang, J.Y. Lin, and A. Schnur. 2004. Epidemiologic clues to SARS origin in China. Emerg Infect Dis 10:1030–1037.

Zhang, L., N. Hua, and S. Sun. 2008. Wildlife trade, consumption and conservation awareness in southwest China. Biodivers Conserv 17:1493–1516.

Zhong, N.S., B.J. Zheng, Y.M. Li, L.L.M. Poon, Z.H. Xie, K.H. Chan, P.H. Li, S.Y. Tan, Q. Chang, J.P. Xie, X.Q. Liu, J. Xu, D.X. Li, K.Y. Yuen, J.S.M. Peiris, and Y. Guan. 2003. Epidemiology and cause of severe acute respiratory syndrome (SARS) in Guangdong, People's Republic of China, in February, 2003. Lancet 362:1353–1358.

# 16

## H5N1 HIGHLY PATHOGENIC AVIAN INFLUENZA

### Breaking the Rules in Disease Emergence

## Thijs Kuiken and Timm Harder

It took several decades after the isolation of influenza viruses from domestic animals and humans before wild waterbirds were identified as the original reservoir of influenza A viruses. In 1931 a virus was identified that, in concert with *Haemophilus influenzae*, caused influenza in experimentally infected swine (Shope 1931). Two years later the virus causing influenza in humans was identified by replicating the disease through intranasal inoculation of domestic ferrets (Smith et al. 1933). In 1955, the virus causing so-called fowl plague in chickens was found to be closely related to influenza virus (Schäfer 1955). More than 10 years later, the first serological (Easterday et al. 1968) and virological (Slemons et al. 1974) evidence was obtained that led to the understanding that healthy wild birds, primarily in the orders Anseriformes and Charadriiformes, are the ultimate reservoir of all subtypes of influenza A virus, as reviewed recently (Olsen et al. 2006).

The tissue tropism of influenza in wild waterbirds differs from that in mammals. While influenza in mammals is primarily a respiratory tract infection, the first researchers collecting samples for virus culture from wild ducks were surprised that influenza virus was isolated at a higher frequency from cloacal than from tracheal swabs (Slemons and Easterday 1978). Experimental infection of domestic ducks with these influenza viruses showed that virus replication in the digestive tract rather than in the respiratory tract was the most likely source of the virus found in cloacal swabs (Slemons and Easterday 1978; Webster et al. 1978).

The pathogenicity of the infection in wild ducks appears to be attenuated compared to that in humans and other mammals. Even uncomplicated influenza in humans typically causes nasal obstruction, cough, sore throat, headache, fever, chills, anorexia, and myalgia (Wright et al. 2007), and corresponds histopathologically with a diffuse, superficial, necrotizing tracheobronchitis (Walsh et al. 1961). In contrast, clinical signs are absent or very rare for natural influenza virus infection in wild ducks (Olsen et al. 2006), and histopathological changes associated with virus replication in the digestive tract are absent (Daoust et al. 2011).

What are the selective factors in the wild waterbird metapopulation for avian influenza virus (AIV) to develop digestive tract tropism and low pathogenicity, assuming that AIV has the plasticity (based on mutation and reassortment) to change both tissue tropism and level of pathogenicity? We speculate that selection for digestive tract tropism may be driven by the feeding behavior of wild waterbirds, many of which feed in large groups at the water surface. Therefore, infected hosts excrete infective feces into the water, which is subsequently ingested by new hosts, resulting in fecal–oral transmission. The superior efficiency of this mode of AIV transmission may explain why these viruses are found primarily in wild waterbirds and far less in other bird groups with a different feeding behavior.

Hypothetically, selection for low pathogenicity may be driven by the mobility of wild waterbirds, both during and in between migration periods. Even during wintering and breeding seasons, wild waterbirds rarely remain in the water at one specific location for any length of time. This high mobility may select for low pathogenicity of AIV, because high pathogenicity would render the infected host less mobile, as well as inducing it to separate from the rest of its group. This would reduce the contact rate and therefore the transmission rate. In addition, selection for low pathogenicity may be driven by the circulation of multiple AIV subtypes in the metapopulation of wild waterbirds. Because these subtypes give cross-protection, even if temporary and limited, the proportion of naïve hosts in the metapopulation is decreased. This would necessitate a higher contact rate among hosts for effective transmission, and thus also favor low pathogenicity.

## HPAIV Infection in Poultry and Wildlife

The evolution of AIV in wild waterbirds towards an innocuous digestive tract infection contrasts with that of some AIV subtypes after introduction into commercial poultry flocks. The AIV subtypes H5 and H7 are able to evolve a biotype that causes high mortality in chickens and domestic turkeys. This high mortality is associated with a change from replication restricted to digestive (and respiratory) tracts to replication in multiple organs, with an associated expansion of tissue damage. Based on their pathogenicity for chickens, the AIVs normally occurring in wild waterbirds are classified as low pathogenic AIV (LPAIV), while the H5 and H7 subtypes that cause high mortality in poultry are classified as highly pathogenic AIV (HPAIV) (Swayne and Halvorson 2003).

The mechanism of this change from LPAIV to HPAIV is based at least partially on the cleavability of the precursor hemagglutinin (HA), the surface protein by which virus particles attach to and fuse with host cells. This precursor HA needs to be cleaved by host proteases to become fully infective. The precursor HA of LPAIV is cleaved by extracellular proteases,

present only in the digestive and respiratory tracts. In contrast, the precursor HA of HPAIV can be cleaved by ubiquitous intracellular proteases (Swayne and Halvorson 2003). Apart from mutations at the HA cleavage site, other genome segments are likely to contribute to pathogenicity (Bogs et al. 2010).

The selective factors for the evolution from LPAIV to HPAIV in commercial poultry are poorly understood (Bin Muzaffar et al. 2006). There are at least three contrasting aspects in the ecology of commercial poultry with that of wild waterbirds that may play a role. First, poultry are highly sedentary, so that infected hosts do not need to remain mobile to infect new hosts. Second, there is no endemic circulation of AIV in poultry populations, and the turnover rate of poultry flocks is high, resulting in frequent introduction of immunologically naïve hosts into the population. Third, spread of AIV among poultry farms is not dependent on mobility of infected hosts, but instead may occur through contaminated equipment or personnel. Together, these factors may select for higher pathogenicity of AIV.

An exception to the rule that HPAIV develops in gallinaceous poultry may come from the detection of HPAIV H5N2 from wild waterbirds (Gaidet et al. 2008). One white-faced whistling duck (*Dendrocygna viduata*) and three spur-winged geese (*Plectropterus gambensis*) sampled in Nigeria in 2007 had an oropharyngeal swab or fresh fecal samples positive for two HPAI H5N2 viruses, one in each host species. There was no evidence of related HPAI H5N2 viruses circulating in poultry in Nigeria at the time; the HA sequences showed little or no homology with recent HPAI H5N2 or H5N1 viruses isolated from gallinaceous poultry in Europe and Africa, but clustered with sequences from contemporary LPAIV strains isolated from ducks in south and central Europe; and no mutations—such as potential additional glycosylation sites in the HA and stalk deletion in the neuraminidase—commonly observed in gallinaceous poultry were detected. Together, these findings suggest that the acquisition of a highly pathogenic viral genotype could have taken place in ducks, either wild or domestic, or in other non-gallinaceous species.

Once HPAIV has developed in gallinaceous poultry, wild birds are traditionally not considered to play an important role in its spread. The spread of HPAIV among poultry flocks appears to be mainly by infected poultry, their excreta, and equipment and personnel contaminated by these excreta (Swayne and Halvorson 2003). Wild birds like European starlings (*Sturnus vulgaris*) (Lipkind et al. 1982) and house sparrows (*Passer domesticus*) (Morgan and Kelly 1990) that live in the proximity of poultry farms occasionally are infected by HPAIV but have not been implicated in its further spread. A historical exception in the involvement of wild birds in HPAIV infection is the HPAIV H5N3 outbreak in 1961, when 1,300 common terns (*Sterna hirundo*) in South Africa died of the infection (Becker 1966). This outbreak was not connected to any known occurrence of this virus in poultry, and the source of infection for the common terns is unknown.

Wild mammals, too, were generally considered to be resistant to disease from influenza virus infection prior to the recent HPAIV H5N1 epidemic. The known exception is the harbor seal (*Phoca vitulina*), in which five outbreaks of influenza were reported on the New England coast of the United States between 1979 and 1992. These outbreaks involved AIV of the subtypes H3N3, H4N5, H4N6, and H7N7 (Reperant et al. 2009). The outbreaks caused variable mortality, with up to 600 deaths, representing 25% of the local population. Clinical signs included prostration, respiratory distress, nasal discharge, and subcutaneous emphysema. The main lesion was a necrotizing or hemorrhagic pneumonia. Interestingly, influenza B virus—until then considered to be restricted to humans—also has been isolated from harbor seals held captive at a rehabilitation center (Osterhaus et al. 2000). What renders harbor seals so sensitive to disease from influenza virus infection is not understood. Cross-species transmission of AIV from birds to seals may occur because of close contact with waterbirds at haul-out sites or at sea. Transmission may occur through direct physical contact, including predation and ingestion of infected bird carcasses; indirect contact with bird feces or contaminated environment; or inhalation of virus particles excreted by birds as aerosols (Reperant et al. 2009).

## THE HPAIV H5N1 OUTBREAK, 1997 TO 2004

In 1997, a HPAIV H5N1 outbreak in poultry in Hong Kong startled the world by spreading directly to humans and causing the deaths of 6 out of 18 people with confirmed infection (de Jong et al. 1997; Subbarao et al. 1998). Besides its unusual pathogenicity for humans, the emergence of this HPAIV H5N1 challenged several of the generally held concepts about influenza in both wild birds and wild mammals. Until then, influenza virus infections—including those by HPAIV—had rarely been reported to cause clinical disease in wild birds. However, in 2002, there were two spillover events of HPAIV H5N1 from poultry to free-living waterbirds in two parks in Hong Kong. These events resulted in mortality of multiple species of ducks, geese, swans, and other waterbirds. At necropsy, they had evidence of systemic disease similar to that caused by HPAIV infection in chickens (Ellis et al. 2004). These outbreaks were isolated and did not appear to lead to further spread of the virus. More or less at the same time, in 2001, increased pathogenicity of HPAIV H5N1 isolates from poultry was demonstrated: domestic ducks that previously had showed no clinical signs from this or other HPAIV showed high mortality associated with viral encephalitis (Hulse-Post et al. 2005).

HPAIV H5N1 also caused unusual mortality in wild mammals. It caused fatal disease in captive tigers (*Panthera tigris*), leopards (*P. pardus*), and Owston's palm civets (*Chrotogale owstoni*), and a free-living stone marten (*Mustela foina*) and American mink (*M. vison*) (Reperant et al. 2009). Experimentally, red foxes (*Vulpes vulpes*) also were found to be susceptible to disease (Reperant et al. 2008). These infections usually resulted from contact with or feeding upon chickens or wild birds infected with the virus. An unusual aspect of HPAIV H5N1 infection in these mammals was that infection and associated tissue damage were not limited to the respiratory tract, which is the normal target for influenza in mammals, but that the virus spread systemically. Systemic infection of carnivores by HPAIV H5N1 was confirmed by experimental inoculation of domestic cats, which showed virus-associated necrosis and inflammation in tissues of the nervous, cardiovascular, urinary, digestive, lymphoid, and endocrine systems (Rimmelzwaan et al. 2006).

## THE RAPID WESTWARD SPREAD OF HPAIV H5N1

In 2005, the range of HPAIV H5N1 expanded dramatically and again involved wild birds. Until that time,

HPAIV H5N1 had spread gradually among countries in Southeast Asia. However, in April 2005, about 1,500 wild birds died from HPAIV H5N1 infection on Qinghai Lake, in western China, in the absence of disease in poultry in the same area (Chen et al. 2005; Liu et al. 2005). The main species involved were bar-headed geese (*Anser indicus*), brown-headed gulls (*Larus brunnicephalus*), and great black-headed gulls (*Larus ichthyaetus*). Gene sequence analysis suggested that the virus causing the outbreak at Qinghai Lake was a single introduction, most probably from poultry in southern China (Chen et al. 2005). The Qinghai Lake outbreak heralded the rapid westward spread of the infection. In the next four months, between June and October 2005, H5N1 spread several thousand kilometers westward, being isolated from both wild birds and poultry, and reached Croatia on Oct. 16, 2005 (www.oie.int).

There was an extensive debate on how the virus spread so rapidly from Southeast Asia to Europe, Africa, and the Middle East in the second half of 2005. This debate revolved around the question whether spread had occurred by transport of infected poultry and associated equipment, or by long-distance flights of infected wild birds (Gilbert et al. 2006; Kilpatrick et al. 2006). Favoring the former was the correlation between the spread of infection and the route of the Trans-Siberian Express. Favoring the latter was the correlation between the spread of infection and the migration route of some waterbird species from Asia to western Europe, and the detection of HPAIV H5N1 in wild birds in absence of its presence in poultry despite intensive surveillance.

An unknown factor was the pathogenicity of HPAIV H5N1 for wild waterbirds. Although the term "highly pathogenic" suggests that HPAIV H5N1 causes severe disease in all bird species, this term only reflects its pathogenicity for chickens. Several research groups performed experimental infections in ducks, geese, and swans to determine whether wild waterbirds could excrete HPAIV H5N1 in the absence of clinical signs of disease and thus potentially carry the virus over a long distance (Kalthoff et al. 2008; Keawcharoen et al. 2008). These studies showed a marked variability in the pathogenicity of HPAIV H5N1 for wild waterbirds. For example, dabbling duck species—mallard (*Anas platyrhynchos*), common teal (*A. crecca*), Eurasian wigeon (*A. penelope*), and gadwall (*A. strepera*)—showed no clinical signs, while

diving ducks—tufted ducks (*Aythya fuligula*) and Eurasian pochards (*A. ferina*)—exposed to the virus according to an identical protocol showed high morbidity and mortality associated with central nervous system disease (Keawcharoen et al. 2008). These experimental studies demonstrated that some species of wild waterbirds, such as mallards, can shed HPAIV H5N1 for several days in the absence of clinical signs and thus potentially play a role in its long-distance spread. Other species, such as tufted ducks and mute swans (*Cygnus olor*), are more susceptible to severe disease from HPAIV H5N1 infection, suggesting that they would be less likely candidates to spread H5N1 (Keawcharoen et al. 2008). However, it is a mistake to rule out a species as a pathogen carrier just because a proportion of the population suffers severe disease or mortality from infection with that pathogen. Examples in birds where the host species is the reservoir for the pathogen despite fatal infection in a proportion of the population are Newcastle disease in double-crested cormorants (*Phalacrocorax auritus*) (Kuiken 1999) and avian cholera in snow geese (Samuel et al. 2007).

Besides poultry and wild bird migration, legal and illegal transport of wild birds is another mode of spread for HPAIV H5N1. HPAIV H5N1 was detected in aviary birds imported legally into the United Kingdom in 2005 (van den Berg 2009). Also, HPAIV H5N1 was detected in illegally imported crested hawk-eagles (*Spizaetus nipalensis*) in Brussels, Belgium, in 2004, and the virus was transmitted to a customs official (Van Borm et al. 2005). These examples highlight the need to be aware of unexpected routes of spread of HPAIV H5N1.

## SPREAD OF THE QINGHAI LINEAGE (CLADE 2.2) OF HPAIV H5N1 TO EUROPE AND ITS OCCURRENCE IN WILDLIFE

The infection of more than 6,000 wild birds by a clade 2.2 HPAIV H5N1 at Lake Qinghai in Northern China in the spring of 2005 (Chen et al. 2005) heralded the start of a stepping stone-like westward spread of clade 2.2 viruses via central Asia into Europe and Africa, where the viruses surfaced in single wild birds, mainly of aquatic species, in fall 2005 in southeastern Europe.

In February 2006, an outbreak started on the Isle of Rügen in Northern Germany (see below). This was part of a broad-scale virus incursion affecting wild birds, especially aquatic birds, in the wider region of the southern Baltic Sea, including Denmark and Sweden. In Danish and Swedish waters, diving ducks, especially tufted ducks and Eurasian pochards, were more prominently affected (Bragstad et al. 2007). From mid-February 2006 onwards, increased mortality in wild birds due to HPAIV H5N1 infection also was detected in southern Germany. The majority of cases originated around Lake Constance and in Franconia; these outbreaks showed less of the centrifugal spreading pattern but a similar spectrum of affected species compared to the Rügen outbreak. Again, the cases in southern Germany were part of a broad-scale incursion of the virus from southeastern Europe into Slovakia, Czech Republic, Austria, Switzerland, Italy, and France. With the dispersal of wild birds from April 2006 onwards, the number of cases receded until May, when the last infected birds, five goosanders (*Mergus merganser*), in the course of this outbreak were detected in southern Germany. Still, single and geographically widely isolated cases of HPAIV H5N1 infections were observed in Europe in 2006: a great crested grebe (*Podiceps cristatus*) in northern Spain (July 2006) and a black swan (*Chenopis atrata*) in a zoo in Germany (August 2006). A total of 21 European countries became affected in 2006, mainly with wild bird cases.

The broad-scale westward spread of the virus apparently came to a halt along a line drawn from the southern tip of Sweden via Denmark and the northeast of Germany southward following the Rhine River in Germany and the Rhone River in France. Because no specific separating landmarks eventually blocking migratory movements are present along this line, other factors must have been acting. Recently it has been shown that migratory movements of aquatic wild birds during the winter season seem to float around the 0°C isotherm (Reperant et al. 2010). This isotherm was swiftly moving westwards since late December 2005 until January 2006 (Ottaviani et al. 2010). It is interesting to note that the appearance of HPAIV H5N1 in northeastern Germany was preceded by the arrival of whooper swans (*Cygnus cygnus*) from the eastern coasts of the Baltic Sea in mid-January, as was evident from the analysis of bird ringing data (Globig et al. 2009).

A recurrence of HPAIV H5N1 of clade 2.2 in wild birds of Europe was observed in Russia in January 2007, and in June in southeastern Germany, southeastern France, and the Czech Republic. Compared to the 2006 outbreaks, a different geographic region was affected, and also the spectrum of affected species changed slightly. Although mute swans still dominated the list of infected dead birds, this time grebes were in the forefront of affected species. An epizootic focus among black-necked grebes (*Podiceps nigricola*) was reported from a small reservoir lake at Kelbra, Thuringia, Germany. More than 200 grebes of an estimated population of approximately 500 birds gathering here succumbed to the infection. The status of this population (resident-breeding vs. migratory-molting) has not been fully resolved. Although further species such as Eurasian pochards and mute swans were present in substantial numbers at the Kelbra reservoir when the outbreak among grebes started, these birds apparently remained largely unaffected. This outbreak ceased in August 2007. Finally, cases were reported in November and December from Romania and Poland. Poultry was also affected in these and other European countries at that time.

From 2008 onwards, only sporadic cases of HPAIV H5N1 infections have been reported from wild birds in Europe. These comprise semi-domesticated mute swans in England (Alexander et al. 2010) and clinically healthy wild birds in Switzerland in 2008 (one Eurasian pochard) (Baumer et al. 2010), and in Germany in 2009 (one mallard). The most recent detection of HPAIV H5N1 in Europe was in March 2010 in poultry in the Romanian Danube delta and in a common buzzard (*Buteo buteo*) at the Black Sea coast of Bulgaria; the latter viruses proved to belong to clade 2.3.2, and their closest phylogenetic relatives are found to circulate in Nepal and Bangladesh (Guan et al. 2009). Apparently, Europe remains an open field to new HPAIV H5N1 incursions from the east. In addition, occasional HPAIV H5N1 infections in wild birds continue to be reported from other regions, especially Southeast Asia, to date.

## HPAIV H5N1 OUTBREAK AROUND THE ISLE OF RÜGEN, GERMANY

In early February 2006 the first two wild birds infected with HPAIV H5N1, whooper swans, were detected in

Germany at the coast of the Isle of Rügen in the Baltic Sea. During the following weeks this island and the surrounding regions appeared to be a focal point of spread of the virus. Due to unusually low temperatures in December 2005 and January 2006 the majority of shallow inlets of the Baltic Sea around the Isle of Rügen, which are used as a staging and wintering site by tens of thousands of migrating aquatic wild birds, had frozen up. Only a few passages remained free of sea ice due to stronger tidal currents, and the majority of aquatic wild birds crowded in these small zones, thereby grossly enhancing intra- and interspecies contacts. Mainly whooper and mute swans as well as several species of geese—greylag (*Anser anser*), greater white-fronted (*A. albifrons*) and Canada geese (*Branta canadensis*)—were affected by the HPAIV H5N1 infection. Centrifugally, the infection spread from the Isle of Rügen along the German coastline of the Baltic Sea westward, reaching the estuary of the river Elbe, and southeastward along the Oder River. In addition to the geographic expansion, an increased spectrum of affected species was found, comprising birds of prey, gulls, diving ducks, and, in April after their return from African wintering sites, even white storks (*Ciconia ciconia*) (Globig et al. 2009; Müller et al. 2009). In addition, three stray cats and a stone marten died due to an HPAIV H5N1 infection, which they obviously acquired while feeding on infected fallen wild birds. Due to the generally tenuous feeding situation that winter, the observed high mortality among aquatic wild birds was at least partially due also to emaciation, especially of juveniles. However, numerous experimental infections in several of the affected wild bird species clearly demonstrated the causal connection between an HPAIV H5N1 infection and increased mortality (Kalthoff et al. 2008; Keawcharoen et al. 2008).

## SUSCEPTIBILITY OF BIRDS OF PREY: INFECTION BY INGESTION

A considerable proportion of wild bird species infected by HPAIV H5N1 in Europe as well as in the epicenter of the H5N1 epizootic in Southeast Asia consisted of birds of prey. In Europe, mainly common buzzards but also peregrine falcons (*Falco peregrinus*) and eagle owls (*Bubo bubo*) were found virus-positive

and died in the course of the infection in 2006 (Globig et al. 2009). It is highly likely that these birds became naturally infected after contact with and ingestion of diseased or fallen infected prey. Susceptibility of birds of prey had been observed earlier when smuggled crested hawk-eagles, which had obviously been fed on infected chickens, were found affected (Van Borm et al. 2005). Experimentally infected cross-bred peregrine falcons developed fatal disease dominated by neurological symptoms within one week after infection; vaccination with an inactivated adjuvanted whole virus vaccine based on an Eurasian H5N2 isolate of low pathogenicity prevented disease development (Lierz et al. 2007). Enteral infection is also at the basis of cases in mammalian carnivores (Kuiken et al. 2004; Rimmelzwaan et al. 2006; Vahlenkamp et al. 2010).

## GEOGRAPHIC RESTRICTION VERSUS HEMIGLOBAL SPREAD OF HPAIV H5N1 PHYLOGENETIC LINEAGES

Phylogenetic analyses of the HA sequences of HPAIVs H5N1 of Asian origin revealed a complex and deeply furcated phylogenetic tree with at least ten different clades, some with deeper fragmentations into several subclades and lineages (Guan et al. 2009). The highest phylogenetic diversity within the smallest geographic region is seen in Southeast Asia, emphasizing that this region not only is the origin of the current HPAIV H5N1 epizootic but also continues to serve as a motor of development of new lineages, especially in poultry. Currently, no species-specific lineages can be discerned. Viruses from all clades have affected poultry, with spillover infections into wild birds as well as into mammals, including humans. Some lineages show a clear geographic restriction while others apparently had a limited temporal span of existence and are no longer detected. Within these lineages predominant genotypes can be defined that reflect the composition of origin of the eight genome segments of influenza A viruses. Several of these lineages are restricted to geographically defined regions (Zhao et al. 2008). For example, Indonesian outbreaks were caused exclusively by clade 2.1 viruses (Smith et al. 2006). These viruses have developed endemic status in the local poultry population but have, so far, not appeared outside this country. Similarly, outbreaks in Vietnam were mainly associated with clade 1 viruses (Wang et al. 2008). Wild birds found infected in these countries were affected by the local H5N1 subtypes, suggestive of their infection in the respective country.

So far, only one phylogenetic lineage of H5N1, clade 2.2, has escaped from its region of origin in Southeast Asia on a larger scale and hemiglobally spread across central and southern Asia into Europe and Africa. In Africa, particularly Egypt, further diversifying evolution of this clade occurred in endemically infected poultry populations, giving rise to new sublineages. There is very recent evidence of viruses of yet another Southeast Asian lineage, clade 2.3.2, which surfaced in spring 2010 in poultry in the European Danube delta and in a common buzzard in the Bulgarian Black Sea region, setting off again the alarm bells in Europe (Reid et al. 2010). However, so far no further evidence for the presence of these viruses outside Asia has been obtained.

It is entirely unknown whether certain biological features predispose viruses of specific phylogenetic lineages to become spread on a wider geographic scale. Alternatively, epidemiological circumstances governed by accident may be responsible, as was seen during the Lake Qinghai incidence with an incursion of the clade 2.2 lineage, later referred to as the "Qinghai lineage" on a broad front, into a highly mobile, mixed species population of migratory wild birds.

Numerous attempts to back-trace the spreading pathways failed to identify clear patterns and responsible routes. Circumstantial evidence was obtained that apparently (illegal) trans-boundary trade of infected poultry and contaminated poultry products, as well as possibly also relay-like transmission by migratory birds, contributed (Feare 2010; Kilpatrick et al. 2006; Wallace et al. 2007).

Phylogenetic analysis of complete HA sequences of the HPAIV H5N1 from European wild birds revealed that at least two separate introductions of HPAIV H5N1 clade 2.2 lineages into northeast and southeast/central Europe, respectively, had occurred in 2006 (Gall-Reculé et al. 2008; Starick et al. 2008); Fink et al. 2010). Yet another phylogenetically distinguishable 2.2 lineage was represented by viruses isolated in 2007 (Nagy et al. 2007; Starick et al. 2008; Haase et al. 2010). Tracing the routes of spread and the origin of the HPAIV lineage detected in 2007 was also attempted by phylogeographic methods, indicating that an outbreak in Poland in December 2007 appeared

to be linked to viruses last found in wild birds in the Czech Republic in June of that year (Haase et al. 2010). Since neither wild bird nor poultry AIV monitoring programs revealed a hint of the presence of HPAIVs of subtype H5N1 in these regions between June and November 2007, the question is raised as to where this virus had persisted. The same question extends to a very closely related virus detected in mute swans in the United Kingdom in January 2008, one month after the last cases in Poland were reported, and further to the isolated detection of HPAIV H5N1 in a Eurasian pochard in Switzerland in 2008 and in a mallard in Bavaria, Germany, in early 2009. Unfortunately, no sequence information is available from the latter two cases.

There are several possible explanations for the sporadic detection of HPAIV H5N1 in wild birds from Europe. First, there may be an undetected low-level endemic infection in a small subpopulation of certain wild bird species. Recent studies (Globig et al. 2009; Wilking et al. 2009) indicated that the AIV monitoring programs in place in the European Union are insufficient to statistically exclude a continuous presence at low prevalence of HPAIV H5N1 infection in Europe. Second, there may be a persistent and clinically silent HPAIV H5N1 infection of poultry somewhere in Europe that fuels occasional spillover transmissions into aquatic wild birds. A study from Germany showed that silent endemic infections can easily be established in fattening Pekin ducks (Harder et al. 2009). Again, the results of the continuous poultry AIV monitoring programs in Europe seem to exclude this possibility. Finally, an environmental source of HPAIV H5N1 may initiate from time to time new transmission chains. Environmentally deposited AIVs of low pathogenicity have been detected in lake water, benthal sediments, and Arctic ice (Lang et al. 2008), and such viral deposits have been identified as an important factor in a model explaining the annual fluctuations of the prevalence of AIV (Rohani et al. 2009), although phylogenetic evidence appears to argue against a role of long-time natural abiotic reservoirs of influenza viruses (Worobey 2008).

## CONCLUSIONS

The currently ongoing HPAIV H5N1 outbreak in poultry is the largest in history in terms of duration, geographic range, and number of animals involved

(Alexander and Brown 2009). Although the extent of the outbreak has decreased since 2005–2006, the virus persists in poultry in several countries. Current control strategies of stamping out and vaccination have not succeeded in eradicating HPAIV H5N1 from poultry populations so far (Eagles et al. 2009). Novel strategies, such as the use of genetically modified poultry that are resistant to infection (Lyall et al. 2011), are being developed but still are far from being ready for use. As long as HPAIV H5N1 persists in the world's poultry population, the risk for it to spill over to other species remains.

Special vigilance is required for HPAIV H5N1 because of the unusual properties of this virus. First, its potential to spread over long distance by infected wild birds means that it can appear unexpectedly in geographic regions that were previously free of the virus (Keawcharoen et al. 2008). Second, its high pathogenicity for a wide range of host species, including carnivores and birds of prey, means that HPAIV H5N1 needs to be considered in the differential diagnosis in species and in clinical or pathological presentations with which influenza virus infection normally is not associated (de Jong et al. 2005; Rimmelzwaan et al. 2006). Finally, we need to remain alert for new properties of HPAIV H5N1. This virus has shown a remarkable genetic diversification since the first record of its appearance in 1996 (Guan et al. 2009). Therefore, the possibility remains that the virus will develop additional properties by which it "breaks the rules" of disease emergence.

Further research is required to determine the underlying factors that allowed HPAIV H5N1 to develop its unusual properties, and the mechanisms by which these properties operate. One important question is whether HPAIV H5N1 is able to persist in the free-living wild bird population in the absence of a reservoir in poultry. To answer this question will require collaboration between scientists from multiple disciplines, including virologists, ornithologists, mathematical modelers, and wildlife disease specialists. Another important question is whether HPAIV H5N1 will be able to acquire the property to transmit efficiently among humans, and so cause a pandemic in humans.

Continued surveillance is required for cases of HPAIV H5N1 infection in humans, wildlife, and domestic animals, including poultry. Regarding wildlife, especially those people who routinely handle and examine sick or dead wild birds need to be alert.

Such people include those at wildlife health centers and other diagnostic centers where wild animals are submitted, wildlife rehabilitation centers, and research institutes investigating wild animals and their diseases. Experience from active and passive surveillance of wild birds for HPAIV H5N1 in past years has shown that examination of sick or dead wild birds is the most sensitive method to detect the incursion of HPAIV H5N1 in a geographic area (Hesterberg et al. 2009). Unless we make a concerted effort both to understand the properties of HPAIV H5N1 better and to monitor the geographic spread of this virus, the risk remains that the virus will surprise us by appearing in an unexpected location or developing in an unexpected way—and the surprise that it causes may not be a pleasant one.

## REFERENCES

Alexander, D.J., and I.H. Brown. 2009. History of highly pathogenic avian influenza. Rev Sci Tech 28:19–38.

Alexander, D.J., R.J. Manvell, R. Irvine, B.Z. Londt, B. Cox, V. Ceeraz, J. Banks, and I.H. Brown. 2010. Overview of incursions of Asian H5N1 subtype highly pathogenic avian influenza virus into Great Britain, 2005–2008. Avian Dis 54:194–200.

Baumer, A., J. Feldmann, S. Renzullo, M. Müller, B. Thür, and M.A. Hofmann. 2010. Epidemiology of avian influenza virus in wild birds in Switzerland between 2006 and 2009. Avian Dis 54:875–884.

Becker, W.B. 1966. The isolation and classification of Tern virus: influenza A-Tern South Africa—1961. J Hyg (Lond) 64:309–320.

Bin Muzaffar, S., R.C. Ydenberg, and I.L. Jones. 2006. Avian influenza: An ecological and evolutionary perspective for waterbird scientists. Waterbirds 29:243–257.

Bogs, J., J. Veits, S. Gohrbandt, J. Hundt, O. Stech, A. Breithaupt, J.P. Teifke, T.C. Mettenleiter, and J. Stech. 2010. Highly pathogenic H5N1 influenza viruses carry virulence determinants beyond the polybasic hemagglutinin cleavage site. PLoS One 5:e11826-

Bragstad, K., P.H. Jørgensen, K. Handberg, A.S. Hammer, S. Kabell, and A. Fomsgaard. 2007. First introduction of highly pathogenic H5N1 avian influenza A viruses in wild and domestic birds in Denmark, Northern Europe. Virol J 4:43.

Chen, H., G.J. Smith, S.Y. Zhang, K. Qin, J. Wang, K.S. Li, R.G. Webster, J.S. Peiris, and Y. Guan. 2005. Avian flu: H5N1 virus outbreak in migratory waterfowl. Nature 436:191–192.

Daoust, P.-Y., F.S.B. Kibenge, R.A.M. Fouchier, M.W.G. van de Bildt, D. van Riel, and T. Kuiken. 2011. Replication of low-pathogenicity avian influenza virus in naturally infected mallard ducks (*Anas platyrhynchos*) causes no morphological lesions. J Wildl Dis 47:401–409.

de Jong, J.C., E.C.J. Claas, A.D.M.E. Osterhaus, R.G. Webster, and W.L. Lim. 1997. A pandemic warning? Nature 389:554.

de Jong, M.D., V.C. Bach, T.Q. Phan, M.H. Vo, T.T. Tran, B.H. Nguyen, M. Beld, T.P. Le, H.K. Truong, V.V. Nguyen, T.H. Tran, Q.H. Do, and J. Farrar. 2005. Fatal avian influenza A (H5N1) in a child presenting with diarrhea followed by coma. N Engl J Med 352:686–691.

Eagles, D., E.S. Siregar, D.H. Dung, J. Weaver, F. Wong, and P. Daniels. 2009. H5N1 highly pathogenic avian influenza in Southeast Asia. Rev Sci Tech 28:341–348.

Easterday, B.C., D.O. Trainer, B. Tůmová, and H.G. Pereira. 1968. Evidence of infection with influenza viruses in migratory waterfowl. Nature 219:523–524.

Ellis, T.M., R.B. Bousfield, L.A. Bissett, K.C. Dyrting, G.S. Luk, S.T. Tsim, K. Sturm-Ramirez, R.G. Webster, Y. Guan, and J.S.M. Peiris. 2004. Investigation of outbreaks of highly pathogenic H5N1 avian influenza in waterfowl and wild birds in Hong Kong in late 2002. Avian Pathol 33:492–505.

Feare, C.J. 2010. Role of wild birds in the spread of highly pathogenic avian influenza virus H5N1 and implications for global surveillance. Avian Dis 54:201–212.

Fink, M., S.R. Fernández, H. Schobesberger, and J. Koefer. 2010. Geographical spread of highly pathogenic avian influenza virus H5N1 during the 2006 outbreak in Austria. J Virol 84:5815–5823.

Gaidet, N., G. Cattoli, S. Hammoumi, S.H. Newman, W. Hagemeijer, J.Y. Takekawa, J. Cappelle, T. Dodman, T. Joannis, P. Gil, I. Monne, A. Fusaro, I. Capua, S. Manu, P. Micheloni, U. Ottosson, J.H. Mshelbwala, J. Lubroth, J. Domenech, and F. Monicat. 2008. Evidence of infection by H5N2 highly pathogenic avian influenza viruses in healthy wild waterfowl. Plos Pathog 4:e1000127.

Gall-Reculé, G.L., F.X. Briand, A. Schmitz, O. Guionie, P. Massin, and V. Jestin. 2008. Double introduction of highly pathogenic H5N1 avian influenza virus into France in early 2006. Avian Pathol 37:15–23.

Gilbert, M., X. Xiao, J. Domenech, J. Lubroth, V. Martin, and J. Slingenbergh. 2006. Anatidae migration in the western Palearctic and spread of highly pathogenic avian influenza H5NI virus. Emerg Infect Dis 12:1650–1656.

Globig, A., C. Staubach, M. Beer, U. Köppen, W. Fiedler, M. Nieburg, H. Wilking, E. Starick, J.P. Teifke, O. Werner, F. Unger, C. Grund, C. Wolf, H. Roost,

F. Feldhusen, F.J. Conraths, T.C. Mettenleiter, and T.C. Harder. 2009. Epidemiological and ornithological aspects of outbreaks of highly pathogenic avian influenza virus H5N1 of Asian lineage in wild birds in Germany, 2006 and 2007. Transbound Emerg Dis 56:57–72.

Guan, Y., G.J. Smith, R. Webby, and R.G. Webster. 2009. Molecular epidemiology of H5N1 avian influenza. Rev Sci Tech 28:39–47.

Haase, M., E. Starick, S. Fereidouni, G. Strebelow, C. Grund, A. Seeland, C. Scheuner, D. Cieslik, K. Smietanka, Z. Minta, O. Zorman-Rojs, M. Mojzis, T. Goletic, V. Jestin, B. Schulenburg, O. Pybus, T. Mettenleiter, M. Beer, and T. Harder. 2010. Possible sources and spreading routes of highly pathogenic avian influenza virus subtype H5N1 infections in poultry and wild birds in Central Europe in 2007 inferred through likelihood analyses. Infect Genet Evol 10:1075–1084.

Harder, T.C., J. Teuffert, E. Starick, J. Gethmann, C. Grund, S. Fereidouni, M. Durban, K.H. Bogner, A. Neubauer-Juric, R. Repper, A. Hlinak, A. Engelhardt, A. Nöckler, K. Smietanka, Z. Minta, M. Kramer, A. Globig, T.C. Mettenleiter, F.J. Conraths, and M. Beer. 2009. Highly pathogenic avian influenza virus (H5N1) in frozen duck carcasses, Germany, 2007. Emerg Infect Dis 15:272–279.

Hesterberg, U., K. Harris, D. Stroud, V. Guberti, L. Busani, M. Pittman, V. Piazza, A. Cook, and I. Brown. 2009. Avian influenza surveillance in wild birds in the European Union in 2006. Influenza Other Respi Viruses 3:1–14.

Hulse-Post, D.J., K.M. Sturm-Ramirez, J. Humberd, P. Seiler, E.A. Govorkova, S. Krauss, C. Scholtissek, P. Puthavathana, C. Buranathai, T.D. Nguyen, H.T. Long, T.S. Naipospos, H. Chen, T.M. Ellis, Y. Guan, J.S. Peiris, and R.G. Webster. 2005. Role of domestic ducks in the propagation and biological evolution of highly pathogenic H5N1 influenza viruses in Asia. Proc Natl Acad Sci USA 102:10682–10687.

Kalthoff, D., A. Breithaupt, J.P. Teifke, A. Globig, T. Harder, T.C. Mettenleiter, and M. Beer. 2008. Highly pathogenic avian influenza virus (H5N1) in experimentally infected adult mute swans. Emerg Infect Dis 14:1267–1270.

Keawcharoen, J., D. van Riel, G. van Amerongen, T. Bestebroer, W.E. Beyer, R. van Lavieren, A.D. Osterhaus, R.A. Fouchier, and T. Kuiken. 2008. Wild ducks as long-distance vectors of highly pathogenic avian influenza virus (H5N1). Emerg Infect Dis 14:600–607.

Kilpatrick, A.M., A.A. Chmura, D.W. Gibbons, R.C. Fleischer, P.P. Marra, and P. Daszak. 2006.

Predicting the global spread of H5N1 avian influenza.
Proc Natl Acad Sci USA 103:19368–19373.

Kuiken, T. 1999. Review of Newcastle disease in
cormorants. Waterbirds 22:333–347.

Kuiken, T., G. Rimmelzwaan, D. van Riel, G. van
Amerongen, M. Baars, R. Fouchier, and A. Osterhaus.
2004. Avian H5N1 influenza in cats. Science 306:241.

Lang, A.S., A. Kelly, and J.A. Runstadler. 2008. Prevalence
and diversity of avian influenza viruses in
environmental reservoirs. J Gen Virol 89:509–519.

Lierz, M., H.M. Hafez, R. Klopfleisch, D. Lüschow,
C. Prusas, J.P. Teifke, M. Rudolf, C. Grund, D. Kalthoff,
T. Mettenleiter, M. Beer, and T. Harder. 2007.
Protection and virus shedding of falcons vaccinated
against highly pathogenic avian influenza A virus
(H5N1). Emerg Infect Dis 13:1667–1674.

Lipkind, M., E. Shihmanter, and D. Shoham. 1982. Further
characterization of H7N7 avian influenza virus isolated
from migrating starlings wintering in Israel. Zentralbl
Veterinarmed B 29:566–572.

Liu, J., H. Xiao, F. Lei, Q. Zhu, K. Qin, X.W. Zhang,
X.L. Zhang, D. Zhao, G. Wang, Y. Feng, J. Ma, W. Liu,
J. Wang, and G.F. Gao. 2005. Highly pathogenic H5N1
influenza virus infection in migratory birds. Science
309:1206.

Lyall, J., R.M. Irvine, A. Sherman, T.J. McKinley, A. Núñez,
A. Purdie, L. Outtrim, I.H. Brown, G. Rolleston-Smith,
H. Sang, and L. Tiley. 2011. Suppression of avian
influenza transmission in genetically modified chickens.
Science 331:223–226.

Morgan, I.R., and A.P. Kelly. 1990. Epidemiology of an
avian influenza outbreak in Victoria in 1985. Aust Vet J
67:125–128.

Müller, T., A. Hlinak, C. Freuling, R.U. Mühle,
A. Engelhardt, A. Globig, C. Schulze, E. Starick,
U. Eggers, B. Sass, D. Wallschläger, J. Teifke, T. Harder,
and F.J. Conraths. 2009. Virological monitoring of white
storks (Ciconia ciconia) for avian influenza. Avian Dis
53:578–584.

Nagy, A., J. Machova, J. Hornickova, M. Tomci, I. Nagl,
B. Horyna, and I. Holko. 2007. Highly pathogenic avian
influenza virus subtype H5N1 in mute swans in the
Czech Republic. Vet Microbiol 120:9–16.

Olsen, B., V.J. Munster, A. Wallensten, J. Waldenström,
A.D. Osterhaus, and R.A. Fouchier. 2006. Global
patterns of influenza A virus in wild birds. Science
312:384–388.

Osterhaus, A.D.M.E., G.F. Rimmelzwaan, B.E.E. Martina,
T.M. Bestebroer, and R.A.M. Fouchier. 2000. Influenza
B virus in seals. Science 288:1051–1053.

Ottaviani, D., S. de la Rocque, S. Khomenko, M. Gilbert,
S.H. Newman, B. Roche, K. Schwabenbauer, J. Pinto,
T.P. Robinson, and J. Slingenbergh. 2010. The cold

European winter of 2005–2006 assisted the spread and
persistence of H5N1 influenza virus in wild birds.
Ecohealth 7:226–236.

Reid, S.M., W.M. Shell, G. Barboi, I. Onita, M. Turcitu,
R. Cioranu, A. Marinova-Petkova, G. Goujgoulova,
R.J. Webby, R.G. Webster, C. Russell, M.J. Slomka,
A. Hanna, J. Banks, B. Alton, L. Barrass, R.M. Irvine,
and I.H. Brown. 2010. First reported incursion of highly
pathogenic notifiable avian influenza A H5N1 viruses
from clade 2.3.2 into European poultry. Transbound
Emerg Dis 58:76–78.

Reperant, L.A., N.S. Fučkar, A.D. Osterhaus, A.P. Dobson,
and T. Kuiken. 2010. Spatial and temporal association of
outbreaks of H5N1 influenza virus infection in wild
birds with the 0 degrees C isotherm. PLoS Pathog
6:e1000854.

Reperant, L.A., G.F. Rimmelzwaan, and T. Kuiken. 2009.
Avian influenza viruses in mammals. Rev Sci Tech
28:137–159.

Reperant, L.A., G. van Amerongen, M.W. van de Bildt,
G.F. Rimmelzwaan, A.P. Dobson, A.D. Osterhaus, and
T. Kuiken. 2008. Highly pathogenic avian influenza
virus (H5N1) infection in red foxes fed infected bird
carcasses. Emerg Infect Dis 14:1835–1841.

Rimmelzwaan, G., D. van Riel, M. Baars, T.M. Bestebroer,
G. van Amerongen, R.A.M. Fouchier,
A.D.M.E. Osterhaus, and T. Kuiken. 2006.
Influenza A virus (H5N1) infection in cats causes
systemic disease with potential novel routes of virus
spread within and between hosts. Am J Pathol
168:176–183.

Rohani, P., R. Breban, D.E. Stallknecht, and J.M. Drake.
2009. Environmental transmission of low pathogenicity
avian influenza viruses and its implications for pathogen
invasion. Proc Natl Acad Sci USA 106:10365–10369.

Samuel, M.D., R.G. Botzler, and G.A. Wobeser. 2007. Avian
cholera. In N.J. Thomas, D.B. Hunter, and C.T. Atkinson
eds. Infectious diseases of wild birds, pp. 239–269.
Blackwell, Oxford.

Schäfer, W. 1955. Vergleichende sero-immunologische
Untersuchungen über die Viren der Influenza und der
klassischen Geflügelpest. Z Naturforsch 10b:81–91.

Shope, R.E. 1931. Swine influenza : iii. filtration experiments
and etiology. J Exp Med 54:373–385.

Slemons, R.D., and B.C. Easterday. 1978. Virus replication in
the digestive tract of ducks exposed by aerosol to
type-A influenza. Avian Dis 22:367–377.

Slemons, R.D., D.C. Johnson, J.S. Osborn, and F. Hayes.
1974. Type-A influenza viruses isolated from wild
free-flying ducks in California. Avian Dis 18:119–124.

Smith, G.J., T.S. Naipospos, T.D. Nguyen, M.D. de Jong,
D. Vijaykrishna, T.B. Usman, S.S. Hassan, T.V. Nguyen,
T.V. Dao, N.A. Bui, Y.H. Leung, C.L. Cheung,

J.M. Rayner, J.X. Zhang, L.J. Zhang, L.L. Poon, K.S. Li, V.C. Nguyen, T.T. Hien, J. Farrar, R.G. Webster, H. Chen, J.S. Peiris, and Y. Guan. 2006. Evolution and adaptation of H5N1 influenza virus in avian and human hosts in Indonesia and Vietnam. Virology 350:258–268.

Smith, W., C.H. Andrewes, and P.P. Laidlaw. 1933. A virus obtained from influenza patients. Lancet 222:66–68.

Starick, E., M. Beer, B. Hoffmann, C. Staubach, O. Werner, A. Globig, G. Strebelow, C. Grund, M. Durban, F.J. Conraths, T. Mettenleiter, and T. Harder. 2008. Phylogenetic analyses of highly pathogenic avian influenza virus isolates from Germany in 2006 and 2007 suggest at least three separate introductions of H5N1 virus. Vet Microbiol 128:243–252.

Subbarao, K., A. Klimov, J. Katz, H. Regnery, W. Lim, H. Hall, M. Perdue, D. Swayne, C. Bender, J. Huang, M. Hemphill, T. Rowe, M. Shaw, X. Xu, K. Fukuda, and N. Cox. 1998. Characterization of an avian influenza A (H5N1) virus isolated from a child with a fatal respiratory illness. Science 279:393–396.

Swayne, D.E., and D.A. Halvorson. 2003. Influenza. In Y.M. Saif, H.J. Barnes, J.R. Glisson, A.M. Fadly, L.R. McDougald, and D.E. Swayne, eds. Diseases of poultry, pp. 135–160. Iowa State University Press, Ames, Iowa.

Vahlenkamp, T.W., J.P. Teifke, T.C. Harder, M. Beer, and T.C. Mettenleiter. 2010. Systemic influenza virus H5N1 infection in cats after gastrointestinal exposure. Influenza Other Resp Viruses 4:379–386.

Van Borm, S., I. Thomas, G. Hanquet, B. Lambrecht, M. Boschmans, G. Dupont, M. Decaestecker, R. Snacken, and T. Van Den Berg. 2005. Highly pathogenic H5N1 influenza virus in smuggled Thai eagles, Belgium. Emerg Infect Dis 11:702–705.

van den Berg, T. 2009. The role of the legal and illegal trade of live birds and avian products in the spread of avian influenza. Rev Sci Tech 28:93–111.

Wallace, R.G., H. Hodac, R.H. Lathrop, and W.M. Fitch. 2007. A statistical phylogeography of influenza A H5N1. Proc Natl Acad Sci USA 104:4473–4478.

Walsh, J.J., L.F. Dietlein, F.N. Low, G.E. Burch, and W.J. Mogabgab. 1961. Bronchotracheal response in human influenza: Type A, Asian strain, as studied by light and electron microscopic examination of bronchoscopic biopsies. Arch Intern Med 108: 376–388.

Wang, G., D. Zhan, L. Li, F. Lei, B. Liu, D. Liu, H. Xiao, Y. Feng, J. Li, B. Yang, Z. Yin, X. Song, X. Zhu, Y. Cong, J. Pu, J. Wang, J. Liu, G.F. Gao, and Q. Zhu. 2008. H5N1 avian influenza re-emergence of Lake Qinghai: phylogenetic and antigenic analyses of the newly isolated viruses and roles of migratory birds in virus circulation. J Gen Virol 89:697–702.

Webster, R.G., M. Yakhno, V.S. Hinshaw, W.J. Bean, and K.G. Murti. 1978. Intestinal influenza: replication and characterization of influenza viruses in ducks. Virology 84:268–278.

Wilking, H., M. Ziller, C. Staubach, A. Globig, T.C. Harder, and F.J. Conraths. 2009. Chances and limitations of wild bird monitoring for the avian influenza virus H5N1— detection of pathogens highly mobile in time and space. PLoS One 4:e6639.

Worobey, M. 2008. Phylogenetic evidence against evolutionary stasis and natural abiotic reservoirs of influenza A virus. J Virol 82:3769–3774.

Wright, P.F., G. Neumann, and Y. Kawaoka. 2007. Orthomyxoviruses. In D.M. Knipe and P.M. Howley, eds. Fields virology, pp. 1691–1740. Wolters Kluwer Health/Lippincott, Williams and Wilkins, Philadelphia.

Zhao, Z.M., K.F. Shortridge, M. Garcia, Y. Guan, and X.F. Wan. 2008. Genotypic diversity of H5N1 highly pathogenic avian influenza viruses. J Gen Virol 89:2182–2193.

# 17

## BARTONELLOSIS

### An Emerging Disease of Humans, Domestic Animals, and Wildlife

## Ricardo G. Maggi, Craig A. Harms, and Edward B. Breitschwerdt

Zoonotic diseases represent a constant threat to humans as a direct consequence of social, economic, and cultural evolution. These interrelated factors contribute to ongoing changes in the number and diversity of vector-borne and zoonotic infections, which continue to emerge and re-emerge. Companion animals and wildlife species, due to their potentially high exposure to arthropod vectors, may become reservoirs for human infection or serve as sentinels for the detection of zoonotic pathogens. In conjunction with more rapid changes in social, economic, and cultural activities, the advent of molecular techniques, in particular polymerase chain reaction (PCR) amplification of bacterial, protozoal, or viral DNA or RNA (reverse transcriptase PCR) from patient blood or tissue samples, has begun to revolutionize the current understanding and medical practices related to the management of many infectious diseases. In particular, molecular techniques have facilitated rapid change in our understanding of emerging pathogens that are medically important and previously unrecognized in animal and human populations.

Clearly, a "One Health/EcoHealth" approach is critical for the prevention of zoonotic infectious diseases. Bubonic plague, which killed over half of the European population within a three-year period, monkeypox, brucellosis, tick-borne encephalitis, Rocky Mountain spotted fever, West Nile virus, and Lyme disease are just few examples of regional or geographically widespread zoonotic diseases that emerged as a consequence of changes in social, economic, or cultural practices (Telford et al. 1991; Murphy 1994; Dennis 1995; Walker et al. 1996; Antonijevic et al. 2007; Jones et al. 2008, Murphy 1998; Murphy 1999; Irwin 2002; Harrus and Baneth 2005; Chapman et al. 2006; Vorou et al. 2007; Forman et al. 2008; Murphy 2008; Beugnet and Marie 2009; Gubler 2009; Johnson et al. 2010). Many zoonotic infectious diseases are vector-transmitted among various wildlife reservoir hosts, domestic animals, and people. With more than 30 zoonotic infectious diseases newly emerging and/or re-emerging in the past 25 years, vector-borne organisms represent a constant threat to humankind and in many instances undermine social and economic advances, particularly in developing nations (White and Hospedales 1994; Mara and Alabaster 1995; Meslin 1995; Meslin 1997; Chomel 1998; Meslin et al. 2000; Morgan 2000; Anan'ina Iu 2002; Feldmann et al. 2002; Kock et al. 2002; Woolhouse 2002; Chomel 2003; Bengis et al. 2004; Russell 2004; Higgins 2004; Blancou et al. 2005; Polley 2005; Boulware 2006; Butler 2006; Chomel and Osburn 2006; Hawkes and Ruel 2006; Heeney 2006; Tapper 2006; Jost et al. 2007; Kelly and Marshak 2007; Seki and Kohno 2007;

Zinsstag et al. 2007; Chugh 2008; Dalton et al. 2008; Forman et al. 2008; Mavroidi 2008; Seimenis 2008; Dorny et al. 2009; Leibler et al. 2009; Mathews 2009; Maudlin et al. 2009; Bruckner 2009). These examples illustrate the need for a more holistic "One Health/EcoHealth" approach when attempting to control and prevent infectious diseases in animal and human populations.

## HISTORY AND DESCRIPTION OF *BARTONELLA* SPP.

The genus *Bartonella* comprises facultative intracellular gram-negative bacteria that infect a large number of vertebrate species. An increasingly long list of *Bartonella* spp. are being identified as important opportunistic pathogens that induce acute and chronic infections in immunocompetent and immunocompromised people. The genus (formerly known as *Rochalimaea*) comprises more than 26 different species and subspecies that are transmitted among different animal species by a large number of vectors, including fleas, sand flies, lice, and potentially biting flies, wingless flies, mosquitoes, and ticks (Table 17.1). On an evolutionary basis, each *Bartonella* species has become adapted to one or more mammalian reservoir hosts, within which these bacteria cause a long-lasting (months to years), intra-erythrocytic and endotheliotropic infection. The discovery of *B. quintana* and *B. henselae* bacteremia in HIV patients during the 1990s initiated environmental, entomological, and medical research related to the role of *Bartonella* spp. as a cause of disease in animals and human patients. The first described *Bartonella* species, *Bartonella bacilliformis*, is transmitted to people by sand flies. The organism invades human red blood cells, causing a severe, acute, life-threatening hemolytic anemia known as Carrion's disease. Following the acute illness, chronic infection can cause vasoproliferative skin lesions known as *verruga peruana*, which are anatomically similar to bacillary angiomatosis, lesions described in immunocompromised patients infected with *B. quintana* or *B. henselae* (Noguchi 1928a,b; Gray et al. 1990; Alexander 1995; Caceres-Rios et al. 1995; Ihler 1996; Ellis et al. 1999; Piemont and Heller 1999; Kosek et al. 2000; Chamberlin et al. 2002; Hambuch et al. 2004; Maco et al. 2004; Eremeeva et al. 2007; Lydy et al. 2008).

Although *B. bacilliformis* was originally identified in Peru in 1905 by Alberto Barton Thompson, Carrion's disease was described in pre-Inca times. *B. quintana*, a human pathogen discovered during World War I as the cause of trench fever or Quintan fever, was the second described *Bartonella* species. Although humans are considered the primary reservoir host for both *B. bacilliformis* and *B. quintana*, gerbils and potentially cats may also serve as reservoir hosts for *B. quintana* (Baneth et al. 1996; Chomel et al. 2003; La et al. 2005; Marie et al. 2006; Breitschwerdt et al. 2007). Trench fever, transmitted from human to human by the body louse, was first thought to be viral in origin. However, the bacterial etiology was subsequently confirmed, and more recent findings suggest that *B. quintana* may be transmitted from animals to humans by fleas or cat bites (Breitschwerdt et al. 2007). In immunocompromised individuals, *B. quintana* may cause endocarditis, bacillary angiomatosis, and infection of the liver, spleen, bone marrow, and lymph nodes. From a comparative medical perspective, *B. quintana* has also been associated with endocarditis in dogs (Kelly et al. 1996, 2006). As in the case of *B. bacilliformis*, human infection with *B. quintana* predates modern times, as demonstrated by the detection of bacterial DNA in the dental pulp of a 4,000-year-old human tooth (Drancourt et al. 2005). *B. henselae* became the third *Bartonella* species to be characterized, which was a direct result of the HIV epidemic and the association of this bacteria with febrile illness and bacillary angiomatosis in immunocompromised patients. *B. henselae* is perhaps the most prevalent *Bartonella* species that infects cats (the primary reservoir host), dogs, and humans; however, this bacterium has been found in several other animal species, including cattle, horses, feral swine, and several species of marine mammals (Maggi et al. 2005; Harms et al. 2008; Jones et al. 2008; Maggi et al. 2008; Cherry et al. 2009; Morick et al. 2009). *B. henselae* is one of the most intensively studied species of this genus, historically due to its association with cat scratch disease (CSD), which is typically a self-limiting disease, reported to affect more than 24,000 people in the United States alone (Chen and Gilbert 1994; Drancourt and Raoult 1995; Maurin et al. 1997; Maurin and Raoult 1998; Chomel 2000; Rolain et al. 2003b; Chomel et al. 2004; Gouriet et al. 2007). However, recent advances in *Bartonella* molecular diagnostic tests, in conjunction with the use of a specialized insect cell culture-based

**Table 17.1** *Bartonella* **Species or Subspecies Currently Described, Their Main Reservoir, Confirmed or Potential (\*) Vector**

| *Bartonella* Species | Main Reservoir | Vector or potential vector |
|---|---|---|
| *B. alsatica* | Rabbits (*Oryctolagus cuniculus*) | Fleas\*, ticks\* |
| *B. bacilliformis* | Humans, cats\*, dogs\* | Fleas, sandflies (*Phlebotomus* spp.) |
| *B. birtlesii* | Wood mice (*Apodemus* spp.) | Fleas\*, ticks\* |
| *B. bovis* (*weissii*) | Domestic cattle (*Bos taurus*) | Biting flies\*, ticks |
| *B. capreoli* | Roe deer (*Capreolus capreolus*) | Biting flies\*, ticks\* |
| *B. chomelii* | Domestic cattle (*Bos taurus*) | Biting flies\*, ticks\* |
| *B. clarridgeiae* | Cats (*Felis catus*) | Fleas, ticks\* |
| *B. doshiae* | Meadow voles (*Microtus agrestis*), rats (*Rattus* spp.) | Fleas |
| *B. elizabethae* | Rats (*Rattus norvegicus*) | Fleas |
| *B. grahamii* | Voles (*Clethrionomys spp.*), mice (*Apodermus spp*) | Fleas |
| *B. henselae* | Cats (*Felis catus*), dogs (*Canis familiaris*) | Fleas, ticks\* |
| *B. koehlerae* | Cats (*Felis catus*), gerbils (*Meriones lybicus*)? | Fleas |
| *B. melophagi*\* | Sheep (*Ovis* spp.) | Sheep keds (*Melophagus ovinus*) |
| *B. peromysci*\* | Field mice (*Peromyscus* spp.) | Fleas\*, ticks\* |
| *B. quintana* | Humans, gerbils (*Meriones lybicus*)\* | Human body lice, fleas |
| *B. rattimassiliensis* | Rats (*Rattus* spp.) | Fleas\*, ticks\* |
| *B. rochlimaea* | Dogs (*Canis familiaris*) | Sandflies\* |
| *B. schoenbuchensis* | Roe deer (*Capreolus capreolus*) | Deer keds, biting flies, ticks\* |
| *B. talpae* | Moles (*Talpa europaea*) | Fleas\*, ticks\* |
| *B. tamiae* | Rat (*Rattus* spp.) | Fleas\*, ticks\* |
| *B. taylorii* | Mice (*Apodemus* spp.), gerbils (*Meriones lybicus*), voles (*Clethrionomys spp.*) | Fleas |
| *B. tribocorum* | Rats (*Rattus* spp.), mice (*Apodermus* spp.) | Fleas\*, ticks\* |
| *B. vinsonii* subsp. *arupensis* | White-footed mice (*Peromyscus leucopus*) | Fleas\*, ticks\* |
| *B. vinsonii* subsp. *berkhoffii* | Coyotes (*C. latrans*), dogs (*Canis familiaris*), foxes (*Urocyon* spp.) | Ticks\* |
| *B. vinsonii* subsp. *vinsonii* | Meadow voles (*Microtus pennsylvanicus*) | Ear mites (*Trombicula microti*)\* |
| *B. volans* | Southern flying squirrels (*Glaucomys volans*) | Fleas\*, ticks\* |
| *B. washoensis* | California ground squirrel (*Spermophilus beecheyi*), rabbits (*Oryctolagus cuniculus*) | Fleas\*, ticks |

enrichment culture medium, has generated microbiological evidence of persistent infection in immunocompetent individuals with cardiac, neurological, and rheumatological symptoms that span years to decades (Breitschwerdt et al. 2009, 2010b,c). In the context of evolving data, a spectrum of newly discovered, highly fastidious *Bartonella* species, found in a diversity of domestic animal and wildlife reservoir hosts and transmitted by an increasing number of transmission-competent insect vectors, appear to be contributing to chronic disease manifestations in immunocompetent as well as immunocompromised patients.

## NATURAL RESERVOIR HOSTS OF *BARTONELLA*

The long-term co-evolution and/or co-speciation of these intravascular bacteria with numerous and specific mammal species appears to be the major factor that defines the current geographic distribution, the prevalence of bacteremia among reservoir hosts, and the evolutionary diversity of *Bartonella* species that exist in nature. This co-evolutionary relationship has translated into a narrow bacteria–host specificity, in which the host acts as the natural bacteremic reservoir: felines as the main reservoir for *B. henselae*, canids for *B. vinsonii* subsp. *berkhoffii*, rabbits for *B. alsatica*, and ruminants for *B. bovis*, *B. shoenbuensis*, and *B. chomelii* (Rolain et al. 2003g; Maillard et al. 2004a,b, 2006, 2007; Raoult et al. 2005; Martini et al. 2008). Nevertheless, rodents are perhaps the most abundant natural reservoir host group and also support the largest number of different *Bartonella* species in nature (Table 17.1). New *Bartonella* species have been recently characterized in other rodent species: *Candidatus B. washoensis* in ground squirrels (*Spermophilus* spp.), *Candidatus B. durdenii* in grey squirrels (*Sciurus carolinensis*), *Candidatus B. volans* in flying squirrels (*Glaucomys volans*), and *Candidatus B. monaxi* in ground hogs (*Marmota monax*) (Breitschwerdt et al. 2010a).

Chronic bacteremia with a *Bartonella* spp. can frequently be detected in outwardly healthy, asymptomatic reservoirs using conventional microbiological approaches, whereas detection of *Bartonella* species in a non-reservoir (accidental) host can be extremely difficult, even when symptoms and evidence of ongoing pathology are present. Bacterial load in natural reservoirs seems to be very high (i.e., the infectious load of *B. henselae* in cats can be as high as 1 million bacteria per microliter of blood) as compared with the level of bacteremia typically found in accidental hosts (i.e., a single *B. henselae* bacterium per microliter of dog or human blood). In addition to the large number of documented reservoir hosts, there are an increasing number of arthropod vectors, including biting flies, fleas, keds, lice, sandflies, and ticks, that have been confirmed or suspected to be associated with the transmission of *Bartonella* spp. among animal populations (Billeter et al. 2008). Considering the diversity of *Bartonella* species and subspecies, the large number of reservoir hosts, and the spectrum of arthropod vectors, the clinical and diagnostic challenges posed by

*Bartonella* transmission in nature appear to be much more complex than is currently appreciated in either human or veterinary medicine.

## *BARTONELLA* SPP. IN CANIDS

As both cats and dogs can maintain a longstanding infection with one or more *Bartonella* spp., the role of pets in the epidemiology of human bartonellosis requires additional study. In addition, experimentally cats and dogs have been infected with *Bartonella* spp. by blood transfusion, needle inoculation, or flea infestation to better understand disease pathogenesis, immunopathology, modes of transmission, and effective treatment and preventive strategies. Domestic and wild canids are the main reservoir for two *Bartonella* species (*B. vinsonii* subsp. *berkhoffii*, and *B. rochalimae*). However, at least eight *Bartonella* species have been detected (PCR amplification and DNA sequencing) or isolated from domestic canids, including *B. henselae*, *B. clarridgeiae*, *B. washoensis*, *B. quintana*, *B. rochalimae*, *B. vinsonii* subsp. *berkhoffii*, *B. koehlerae*, and *B. elizabethae* (Breitschwerdt et al. 1995; Breitschwerdt and Kordick 2000; Duncan et al. 2007a,b, 2008; Breitschwerdt et al. 2010a). All of these species have been recognized as pathogenic in dogs and are known or believed to induce subclinical to chronic infections that can be accompanied with intermittent lethargy and severe weight loss. Other clinical and pathological manifestations of canine bartonellosis include epistaxis, endocarditis, myocarditis, arthritis, and granulomatous inflammatory lesions in various tissues (Chomel et al. 2001, 2006a, 2009; Diniz et al. 2009). Due to the intracellular nature of *Bartonella* species, which can invade endothelial cells (Brouqui and Raoult 1996; Dehio et al. 1997; Dehio 1999, 2001, 2003, 2004, 2005; Dehio et al. 2005; McCord et al. 2005; Schmid et al. 2006; Schulte et al. 2006; McCord et al. 2007; Pitassi et al. 2007), erythrocytes (Kordick and Breitschwerdt 1995; Dehio 2001; Rolain et al. 2001; Schulein et al. 2001; Greub and Raoult 2002; Rolain et al. 2002; Schulein and Dehio 2002; Seubert et al. 2002; Rolain et al. 2003a,b,c,e,f; Dehio 2004; Foucault et al. 2004; Dehio 2005; Schroder and Dehio 2005; Pitassi et al. 2007), CD34 progenitor cells (Mandle et al. 2005; Marsilia et al. 2006), microglia (Munana et al. 2001), and dendritic cells (Arrese Estrada and Pierard 1992; Vermi et al.

2006), the clinical spectrum of disease manifestation induced by these bacteria appears to be very diverse and is the focus of several recent reviews (Hennemann 1963; Agan and Dolan 2002; Chomel et al. 2003, 2004; Fenollar and Raoult 2004; Birtles 2005; Blanco and Raoult 2005; Chomel and Boulouis 2005; Enia et al. 2005; Maruyama 2005; Brouqui and Raoult 2006; Chomel et al. 2006a; Foucault et al. 2006; Hajjaji et al. 2007).

## Bartonella vinsonii berkhoffii

Since the isolation of B. vinsonii subsp. berkhoffii from the blood of a dog with intermittent epistaxis and endocarditis in 1993 (Breitschwerdt et al. 1995; Kordick et al. 1996), this organism has emerged as an important pathogen in dogs and humans. Current evidence indicates that canids, including coyotes (Canis latrans), dogs (Canis lupus familiaris), and grey foxes (Urocyon cinereoargenteus), potentially serve as reservoir hosts. In dogs, B. vinsonii subsp. berkhoffii has been identified as an important cause of endocarditis and has been associated with cardiac arrhythmias, myocarditis, granulomatous rhinitis, and anterior uveitis (Breitschwerdt and Kordick 1995; Breitschwerdt et al. 1995, 1999; Pappalardo et al. 2000, 2001 Michau et al. 2003; Breitschwerdt et al. 2004, 2005; Maggi et al. 2006; Cockwill et al. 2007). More recently this species has been detected in asymptomatic dogs (Duncan et al. 2007b), in dogs with systemic granulomatous disease involving the many organs (Tsukahara et al. 1998; Saunders and Monroe 2006), and in the blood or lymph nodes of golden retrievers with lymphoma (Duncan et al. 2008). In addition, co-infection with two of the four previously described B. v. berkhoffii genotypes has been identified in the same dog. Furthermore, B. v. berkhoffii appears to be an important and previously unrecognized human pathogen. Recently, we described a veterinarian who most likely became infected with this bacteria by needle stick transmission, which represents an alternative mode of transmission among people with occupational canine contact (Breitschwerdt et al. 2007, 2008).

## Bartonella henselae

Based upon serological or molecular evidence, B. henselae has been implicated in association with several pathological conditions in dogs, including peliosis hepatis (Kitchell et al. 2000), granulomatous hepatitis (Gillespie et al. 2003), and granulomatous sialoadenitis (Saunders and Monroe 2006). B. henselae has been implicated as a possible cause or a cofactor in the development of granulomatous hepatitis (Gillespie et al. 2003), pyogranulomatous lymphadenitis (Morales et al. 2007), and endocarditis (De Paiva Diniz et al. 2007, Diniz et al. 2007). It is currently unclear as to whether dogs serve as accidental hosts or as reservoirs for B. henselae. More recently, B. henselae was also detected in the blood and lymph nodes of golden retrievers with lymphoma and in an equal number of healthy age- and sex-matched control dogs (Duncan et al. 2007, 2008). The molecular prevalence of Bartonella spp. infection was 18% in both study populations. In addition, co-infection with B. henselae and B. v. berkhoffii has been reported in both sick dog and human patients (Breitschwerdt et al. 2007; Diniz et al. 2007; Breitschwerdt et al. 2008).

## Bartonella quintana

To date, documentation of B. quintana infection in dogs has been based upon molecular evidence only. DNA of B. quintana was amplified and sequenced from two dogs with endocarditis (Kelly et al. 2006) and has been detected in dog saliva (Duncan et al. 2007a) and in blood and lymph node samples from healthy dogs and dogs with lymphoma (Duncan et al. 2007b, 2008).

## Other Bartonella Species

Other Bartonella species have also been associated with different pathological conditions in dogs. B. clarridgeiae, B. washoensis, and B. rochalimae (Chomel et al. 2001; Gillespie et al. 2003) have been associated with endocarditis, whereas B. elizabethae (Mexas et al. 2002) has been associated with sudden death. In addition, a species closely related with B. volans has been detected in both sick dog and human blood samples from the southeastern United States (Maggi, Breitschwerdt, unpublished data).

## BARTONELLA SPECIES IN FELIDS

At least six Bartonella species have been detected or isolated from cats. B. henselae and B. clarridgeiae are the most prevalent species found in cats and *B. henselae* is the primary, if not the sole, etiological agent for CSD. *Bartonella*-infected cats are normally asymptomatic, although myocarditis, kidney disease and urinary tract infections, stomatitis, and lymphadenopathy may appear as clinical signs in consequence of *Bartonella* infection (Chomel et al. 2006b). In experimentally infected cats, fever, lymphadenopathy, mild neurological signs, and reproductive disorders have been reported (Chomel et al. 2006b).

## *Bartonella henselae*

The worldwide prevalence of *B. henselae* bacteremia varies considerably among stray and domestic cat populations, ranging from 0% in Norway to as high as 68% in the Philippines (Boulouis et al. 2005; Chomel et al. 2006b). Of the two most distinctive genotypes reported in cats, *B. henselae* Marseille (16S rDNA type II) is the predominant strain in Europe, the western United States, and Australia, whereas *B. henselae* Houston I (16S rDNA type I) is the predominant strain in Asia and the eastern United States. Nevertheless, the prevalence of each genotype varies among cat populations in a given country (Boulouis et al. 2005; Chomel et al. 2006b; Breitschwerdt et al. 2010a; Kelly et al. 2010a).

## *Bartonella clarridgeiae*

*Bartonella clarridgeiae* was first isolated in the United States from a pet cat belonging to an HIV-positive patient. In most regions, this species is less frequently isolated from cats than is *B. henselae*, and the prevalence of bacteremia has ranged from 10% to 36% among the *Bartonella* isolates obtained from cats from France, the Netherlands, the Philippines, Thailand, the United States, Japan, and Taiwan (Chomel et al. 2004; Boulouis et al. 2005; Chomel et al. 2006a,b). In *B. henselae* type II and *B. clarridgeiae* experimentally co-infected cats, histopathologic examination identified peripheral lymph node hyperplasia, splenic follicular hyperplasia, lymphocytic cholangitis/pericholangitis, lymphocytic hepatitis, lymphoplasmacytic myocarditis, and interstitial lymphocytic nephritis. However, clinical signs were minimal and gross necropsy results were unremarkable (Kordick et al. 1999).

## *Bartonella koehlerae*

*Bartonella koehlerae* has been rarely isolated from domestic cats worldwide. This highly fastidious species had been isolated from only two cats in California and one cat in France (Rolain et al. 2003c,d; Avidor et al. 2004; Boulouis et al. 2005) and from stray cats from Israel (Boulouis et al. 2005), where the first human case of *B. koehlerae* endocarditis was reported in 2004. More recently, this species was isolated from three cats from the southeastern United States (R. Maggi, unpublished data, 2011).

## *Bartonella quintana*, *B. bovis*, and *B. v. berkhoffii*

Bacillary angiomatosis, endocarditis, and a few suspect cases of CSD have been associated with *B. quintana* infection, for which the only risk factor identified was contact with cats or cat fleas (La et al. 2005). The identification of *B. quintana* DNA in cat fleas (Rolain et al. 2003d), dental pulp of a cat (La et al. 2005), and blood (Breitschwerdt et al. 2007) has raised the question as to whether cats might be a possible source of human infection with this *Bartonella* sp.; historically humans were considered the sole reservoir host. The putative transmission of *B. quintana* to a woman following the bite of a feral cat provides further support for the hypothesis that cats can maintain and transmit this bacterium (Breitschwerdt et al. 2007). *Bartonella bovis* (formerly *B. weissii*) infections have been reported in cats from Illinois and Utah in the United States (Chomel et al. 2004, 2006b) and more recently have been detected in cats from North Carolina (Breitschwerdt/Maggi, unpublished data). *B. v. berkhoffii*, a canid-adapted species, has been recently described in a cat with recurrent osteomyelitis, hypercalcemia, hyperglobulinemia, thrombocytopenia, and lesional plasma cell infiltrates (Varanat et al. 2009).

## *Bartonella* Species in Exotic Animals

The frequent introduction of exotic animals as pets represents an ongoing public health challenge. Rodents, amphibians, reptiles, and small exotic mammals are traded worldwide and an increasing number of exotic pets are maintained in households throughout the world, particularly over the past 50 years

(Chapter 11 in this book; http://www.animallaw.info/ articles/ovusexoticpets.htm; http://www.aemv.org/ Documents/2006_AEMV_proceedings_1.pdf). Because many of these outwardly healthy exotic animals, which are sold as pets, maintain chronic intravascular infections with a spectrum of *Bartonella* spp., the human risk of acquiring infection with a rodent-adapted *Bartonella* from these animals warrants further research. A study from Japan found that over 26% of 546 small mammals, including 28 different animal species that had been imported for pet sales from Asia, North America, Europe, and the Middle and Near East, were infected with at least 10 different *Bartonella* species (Inoue et al. 2009). This study also found that 17.6% of the rodents tested were co-infected with more than one *Bartonella* spp. From a public health perspective, a recent study involving febrile patients from Thailand found bacteremia associated with numerous rodent *Bartonella* spp. (Kosoy et al. 2010).

## *BARTONELLA* IN TERRESTRIAL WILDLIFE

Although evolutionary adaptation is important, the currently recognized range of animal species that can be infected with a specific *Bartonella* species appears to be limited primarily by the time and resources necessary to investigate additional potential host species. This has been particularly true of *B. henselae,* which has been detected in cats, cattle, dogs, horses, marine mammals, and sea turtles. In most instances, *Bartonella* species in wildlife have been investigated primarily in the context of their role as reservoir hosts. Rodents and lagomorphs in particular have been extensively investigated as *Bartonella* reservoirs. *Bartonella* species have been detected by culture, PCR, and/or serology in a wide range of rodent species (or their ectoparasites), including the black rat (*Rattus rattus*), Norway rat (*R. norvegicus*) (Ellis et al. 1999), cotton rat (*Sigmodon hispidus*), white-footed mouse (*Peromyscus leucopus*) (Kosoy et al. 1997, 1998, 1999, 2000, 2004a,b), grey squirrels (*Sciurus carolinensis*), red squirrels (*Sciurus vulgaris*) (Bown et al. 2002), Richardson's ground squirrel (*Spermophilus richardsonii*), Franklin's ground squirrel (*S. franklinii*), thirteen-lined ground squirrels (*S. tridecimlineatus*), least chipmunk (*Tamias minimus*),

meadow vole (*Microtus pennsylvanicus*), deer mouse (*Peromyscus maniculatus*), southern red-backed vole (*Myodes gapperi*) (Jardine et al. 2005), bank vole (*Myodes glareolus*), wood mouse (*Apodemus sylvaticus*) (Telfer et al. 2007), Southern flying squirrel (*Glaucomys volans*), groundhog (*Marmota monax*) (Breitschwerdt et al. 2009), gerbil (*Meriones lybicus*) (Marie et al. 2006), black-tailed prairie dog (*Cynomys ludovicianus*) (Bai et al. 2009), and European wild rabbits (*Oryctolagus cuniculus*) (Heller et al. 1999), among others.

As domestic cats and dogs have been identified as both natural reservoirs and accidental hosts for several *Bartonella* species (see above), it is not surprising that infection with *Bartonella* species has been detected in wild canids and felids, including coyotes (*Canis latrans*), grey foxes (*Urocyon cinereoargenteus*) (Maggi et al. 2006; Henn et al. 2007), African lions (*Panthera leo*), pumas (*Felis concolor*), cheetahs (*Acinonyx jubatus*) (Kelly et al. 1996), Florida panthers (*Felis concolor coryi*), and bobcats (*Lynx rufus*) (Rotstein et al. 2000; Chomel et al. 2004). Other wildlife species reported to harbor *Bartonella* species include ruminants: roe deer (*Capreolus capreolus*) (Dehio et al. 2004), mule deer (*Odocoileus hemionus*), elk/red deer (*Cervus elaphus*) (Chang et al. 2000), the gray kangaroo (*Macropus giganteus*) (Fournier et al. 2007), shrew (*Sorex araneus*) (Bray et al. 2007), and bats (*Pipistrellus* sp., *Myotis mystacinus*, *M. daubentonii*, *Nyctalus noctula*) (Concannon et al. 2005). Thus, of epidemiological and medical relevance, a wide range of established and potential reservoir hosts exists in nature for a large number of *Bartonella* species. Competent and suspected vectors that facilitate efficient transmission among reservoir and at times accidental hosts have likewise received considerable attention, with sand flies, lice, and fleas serving as confirmed vectors, and ticks, mites, keds, and biting flies included as suspected vectors (Billeter et al. 2008). Considerably less is known regarding the impact of *Bartonella* species as a cause of pathology in wildlife, although some level of bacterial-induced debility in prey species could theoretically increase the transmission potential to a predator species through bites, scratches, or vectors.

Laboratory strains of mice (*Mus musculus*, BALB/c) have been infected with *B. henselae*, leading to regional lymphadenopathy (Karem et al. 1999), and with *B. birtlesii* isolated from wild *Apodemus* spp., which resulted in reproductive disorders in the recipient mice, including increased fetal death rates and placental vasculitis (Boulouis et al. 2001). The relevance of these studies to wild rodent populations is uncertain; however, in those situations in which nutrition is marginal and stress excessive, it is possible that a reservoir-adapted *Bartonella* can become pathogenic. In homologous mouse models, experimental infection of laboratory-bred cotton rats (*Sigmodon hispidus*) produced persistent bacteremia but no detectable illness (Kosoy et al. 1999), and no obvious clinical signs were observed in *Bartonella*-infected grey squirrels or red squirrels (Bown et al. 2002), European rabbits (Heller et al. 1999), and grey foxes (Henn et al. 2007). None of these studies were designed to detect subtle effects of infection, however. An etiological role for *Bartonella* species has been suggested for a clinical condition in roe deer and red deer (Dehio et al. 2004). Deer ked dermatitis, which has a similar clinical presentation as CSD in humans, is associated with bites of the deer ked, *Lipotena cervi*. *B. schoebuchensis* has been isolated from deer keds and localized to the ked midgut. The impact of infection with various *Bartonella* species on the individual and population health of terrestrial wildlife remains under-investigated.

## BARTONELLA IN MARINE WILDLIFE

Among marine mammals, *Bartonella* species have recently been detected from cetaceans, pinnipeds, and sea otters. *B. henselae* has been detected (PCR and sequencing) or isolated from the harbor porpoise (*Phocoena phocoena*) (Maggi et al. 2005), bottlenose dolphin (*Tursiops truncatus*), striped dolphin (*Stenella coeruleoalba*), Risso's dolphin (*Grampus griseus*), pygmy sperm whale (*Kogia breviceps*) (Harms et al. 2008), and beluga whale (*Delphinapterus leucas*) (Maggi et al. 2008). A *B. henselae*-like organism was also found in a harbor seal (Morick et al. 2009), and a *B. volans*-like organism in Northern sea otters (*Enhydra lutris kenyoni*) (Chomel et al. 2009). *B. henselae* and an organism genetically similar to *B. vinsonii* subsp. *berkhoffii* have been detected from blood of apparently healthy rehabilitated and free-ranging loggerhead turtles (*Caretta caretta*), the only evidence to date of *Bartonella* infection in a non-mammalian vertebrate (Valentine et al. 2007). *Bartonella* prevalence was

found to be higher in captive bottlenose dolphins cohabiting with a *B. henselae*-positive dolphin that was experiencing recurrent leukocytosis and anemia, and in stranded cetaceans of multiple species, than in free-ranging, presumably healthier bottlenose dolphins (Harms et al. 2008). Similarly, four of five captive beluga whales cohabiting with a *Bartonella*-positive whale that became anemic rapidly and expired were *B. henselae* PCR-positive, but a high prevalence of *B. henselae* DNA was also found in tissue samples from a small number of hunter-harvested wild belugas from the Mackenzie River Delta, Canada (Maggi et al. 2008). The harbor seal from which a *Bartonella* sp. was detected died in rehabilitation with interstitial pneumonia, but the role played by *Bartonella* is unclear, and there was no comparison of prevalence with other populations (Morick et al. 2009). Investigation of an unusual mortality event of Northern sea otters implicated vegetative valvular endocarditis as a predominant cause of death. Among other potential pathogens identified from the lesions was a *B. volans*-like species (Chomel et al. 2009). Thus, while no morbidity or mortality has been definitively ascribed to *Bartonella* species in marine mammals, some circumstantial evidence suggests these bacteria could play pathogenic roles similar to the conditions that are currently better described in humans and terrestrial mammals.

Finding *Bartonella* species in the aquatic environment raises interesting questions regarding transmission. Seal lice (*Echinophthirius horridus*) from harbor seals were PCR-positive for a *Bartonella* identical to that detected from the spleen of a harbor seal (Morick et al. 2009). Based on the known role of lice as vectors of *Bartonella* species in terrestrial mammals (Billeter et al. 2008), seal lice are likely candidate vectors for seals. A cyamid amphipod ectoparasite, *Isocyamus delphinii*, colloquially known as a "whale louse" but actually a parasitic crustacean, carried the same *B. henselae*-like strain as was amplified and sequenced from its Risso's dolphin host (Harms et al. 2008). However, unlike sand flies, lice, fleas, ticks, and mites, cyamids lack piercing mouthparts and feed on superficial skin rather than blood, making them somewhat less likely vectors, although they could still ingest *Bartonella* species from their hosts. Sea otters harbor nasal mites (*Halarachne miroungae*) and sea turtles can carry leeches (*Ozobranchus* spp.), which could potentially serve as vectors. Scratches and bites ("raking" in odontocete cetaceans) could also play a role in transmission of *Bartonella* species in the marine environment, although transmission of *B. henselae* by cat bites and scratches is thought most likely to occur as a result of inoculating contaminated flea feces into the wound, for which a marine analog is not readily apparent (Harms et al. 2008). Vertical transmission, as suggested for rodent hosts (Kosoy et al. 1998), could also occur. Because domestic cats are a primary reservoir for *B. henselae*, some consideration of cats as a source of *Bartonella* species in marine animal habitats may be warranted, along the lines strongly implicated for *Toxoplasma gondii* in Southern sea otters (Miller et al. 2002). Transmission of *Bartonella* from vertebrate feces (as opposed to flea feces) has not been described, however, and the finding of *B. henselae* from Arctic beluga whales makes it difficult to implicate run-off of cat feces in marine bartonellosis. *Bartonella* species in marine mammals may be of zoonotic concern for marine mammal handlers and pathologists, albeit probably of no greater concern than for several other zoonotic pathogens of marine mammals (Harms et al. 2008).

## CONCLUSIONS

The high prevalence of *Bartonella* species in wild, peri-domestic, and domestic animals and the high levels of bacteremia normally associated with these intravascular bacteria among natural reservoir hosts highlight the ongoing need for enhanced diagnostic detection modalities, effective treatment strategies, and comprehensive control programs for vector-borne infectious diseases. As diagnostic techniques have advanced, the epidemiology of *Bartonella* infection has been shown to be more complex than was historically appreciated. The bartonellosis paradigm may be even more important for tropical and subtropical areas, where environmental conditions allow year-round propagation of pathogenic organisms and the vectors that transmit them. As our peri-urban and urban areas expand, and with increasing use of exotic animals as pets, public health professionals, physicians, and veterinarians should proactively evaluate the risks for emerging and re-emerging animal and human infections caused by zoonotic *Bartonella* species. In many parts of the world, the lack of a strong and well-organized public health infrastructure makes the assessment of the impact of these diseases nearly impossible. In addition, the indiscriminate use of antibiotics makes these diseases less likely to be reported in humans in most tropical and subtropical regions, where several thousand undiagnosed cases of bartonellosis may occur every year.

Domestic animals, despite sharing the same environment with their owners, frequently have more extensive vector exposure through contact with environmental niches that favor the vector, but are less often frequented by a pet human counterpart. Consequently, it may be possible that the detection and identification of known or novel vector-borne pathogens can be accomplished more readily using blood samples and arthropod vectors obtained from cats and dogs. Molecular epidemiological examination of canine and feline blood samples can greatly enhance our knowledge of those pathogens most likely to affect these companion animals and their human counterparts. Much remains unknown regarding the epidemiology and pathogenesis of bartonellosis in wildlife, and epidemiology at the interfaces of wildlife with humans and with domestic animals. Exacting culture and diagnostic techniques limit much of the needed research to highly specialized laboratories. The rapidly expanding list of *Bartonella* spp., host species, vectors, and potential vectors defies easy synthesis. Even the name "cat scratch disease" for infection with *B. henselae* in humans has served as an impediment to the improved understanding of bartonellosis, by incorrectly implying that the transmission, host range, and disease pathogenesis in immunocompetent humans are already well understood. The detection of *B. henselae* in cetaceans and in the blood of animal health professionals clearly belies those assumptions.

Verifying competent vectors and reservoir hosts is essential for predictive models of potential range expansion or contraction, or disease outbreaks resulting from climate cycles and climate change, habitat fragmentation, and dispersal of invasive species. Because infection with *Bartonella* spp. can occur in apparently healthy animals, establishing links between pathogen and disease is a complex deductive process, requiring detailed epidemiological, diagnostic, pathological, and therapeutic response information. These data are often not obtainable for wildlife cases and are complex and difficult to obtain for domestic animal cases. *In situ* hybridization will be a valuable tool to link the pathogen with pathology in the context of what appears to be a spectrum of disease syndromes, but this approach is expensive and labor-intensive and has not been well validated. Co-factors (i.e., environmental, infectious, toxicologic) that may contribute to or exacerbate the pathogenic manifestations of latent *Bartonella* spp. infections need to be identified. The finding of *Bartonella* spp. infections in aquatic habitats suggests the possibility of alternate modes of infection and dispersal, either by previously unsuspected vectors or possibly even by other routes. Although *Bartonella* spp. have been identified in a wide range of mammals, ectoparasites of mammals, and one reptile species, other phyla remain to be investigated. We conclude that the spectrum of vector-borne pathogens, including over 26 characterized *Bartonella* spp., as well as the extent to which these microorganisms contribute to chronic debilitating disease manifestations in both animal and human patients, remains substantially under-appreciated in veterinary and human medicine.

In summary, knowledge relative to modes of transmission and the pathological potential of *Bartonella* spp. in pets and human patients is evolving rapidly. Importantly, an increased emphasis on the pathogenic potential and reservoir role of *Bartonella* spp. in wildlife is needed to better understand the ecology and epidemiology and to more accurately define risk factors for disease transmission. The ecology and competence of various insect species for the transmission of *Bartonella* spp. among reservoir hosts or to opportunistic hosts deserve critical reappraisal. The large number of newly emerging and/or re-emerging zoonotic infectious diseases associated with vector transmission, as described in the past 25 years, should be used to guide social, behavioral, and developmental policies in the future. The genus *Bartonella* provides yet another important example of the utility of a "One Health/EcoHealth" approach to animals, the environment, the organisms, and the vectors that influence animal and human health throughout the world on a daily basis.

## REFERENCES

Agan, B.K., and M.J. Dolan. 2002. Laboratory diagnosis of Bartonella infections. Clin Lab Med 22:937–962.

Alexander, B. 1995. A review of bartonellosis in Ecuador and Colombia. Am J Trop Med Hyg 52:354–359.

Anan'ina Iu, V. 2002. [Bacterial zoonoses with natural focality: current trends in epidemic manifestation]. Zh Mikrobiol Epidemiol Immunobiol 86–90.

Antonijevic, B., N. Madle-Samardzija, V. Turkulov, G. Canak, C. Gavrancic, I. Petrovic-Milosevic. 2007. [Zoonoses—a current issue in contemporary infectology]. Med Pregl 60:441–443.

Arrese Estrada, J., and G.E. Pierard. 1992. Dendrocytes in verruga peruana and bacillary angiomatosis. Dermatology 184:22–25.

Avidor, B., M. Graidy, G. Efrat, C. Leibowitz, G. Shapira, A. Schattner, O. Zimhony, and M. Giladi. 2004. *Bartonella koehlerae*, a new cat-associated agent of culture-negative human endocarditis. J Clin Microbiol 42:3462–3468.

Bai, Y., M.Y. Kosoy, K. Lerdthusnee, L.F. Peruski, and J.H. Richardson. 2009. Prevalence and genetic heterogeneity of *Bartonella* strains cultured from rodents from 17 provinces in Thailand. Am J Trop Med Hyg. 81:811–816.

Baneth, G., D.L. Kordick, B.C. Hegarty, and E.B. Breitschwerdt. 1996. Comparative seroreactivity to *Bartonella henselae* and *Bartonella quintana* among cats from Israel and North Carolina. Vet Microbiol 50:95–103.

Bengis, R.G., F.A. Leighton, J.R. Fischer, M. Artois, T. Morner,and C.M. Tate. 2004. The role of wildlife in emerging and re-emerging zoonoses. Rev Sci Tech. 23:497–511.

Beugnet, F., and J.L. Marie. 2009. Emerging arthropod-borne diseases of companion animals in Europe. Vet Parasitol 163:298–305.

Billeter, S.A., M.G. Levy, B.B. Chomel, and E.B. Breitschwerdt. 2008. Vector transmission of *Bartonella* species with emphasis on the potential for tick transmission. Med Vet Entomol 22:1–15.

Birtles, R.J. 2005. Bartonellae as elegant hemotropic parasites. Ann NY Acad Sci. 1063:270–279.

Blanco, J.R., and D. Raoult. 2005. [Diseases produced by *Bartonella*]. Enferm Infecc Microbiol Clin 23: 313–320.

Blancou, J., B.B. Chomel, A. Belotto, and F. X. Meslin. 2005. Emerging or re-emerging bacterial zoonoses: factors of emergence, surveillance and control. Vet Res 36:507–522.

Boulouis, H.J., F. Barrat, D. Bermond, F. Bernex, D. Thibault, R. Heller, J.J. Fontaine, Y. Piemont, and B.B. Chomel, 2001. Kinetics of *Bartonella birtlesii* infection in experimentally infected mice and pathogenic effect on reproductive functions. Infect Immun 69:5313–5317.

Boulouis, H.J., C.C. Chang, J.B. Henn, R.W. Kasten, and B.B. Chomel. 2005. Factors associated with the rapid emergence of zoonotic *Bartonella* infections. Vet Res 36:383–410.

Boulware, D.R. 2006. Travel medicine for the extreme traveler. Dis Mon 52:309–325.

Bown, K.J., B.A. Ellis, R.J. Birtles, L.A. Durden, J. Lello, M. Begon, and M. Bennett. 2002. New world origins for haemoparasites infecting United Kingdom grey squirrels (*Sciurus carolinensis*), as revealed by phylogenetic analysis of *Bartonella* infecting squirrel populations in England and the United States. Epidemiol Infect 129:647–653.

Bray, D.P., K.J. Bown, P. Stockley, J.L. Hurst, M. Bennett, and R.J,Birtles. 2007. Haemoparasites of common shrews (*Sorex araneus*) in Northwest England. Parasitology 134:819–826.

Breitschwerdt, E.B., C.E. Atkins, T.T. Brown, D.L. Kordick, and P.S. Snyder. 1999. *Bartonella vinsonii subsp. berkhoffii* and related members of the alpha subdivision of the Proteobacteria in dogs with cardiac arrhythmias, endocarditis, or myocarditis. J Clin Microbiol 37:3618–3626.

Breitschwerdt, E.B., K.R., Blann, M.E. Stebbins, K.R. Munana, M.G. Davidson, H.A. Jackson, and M.D. Willard. 2004. Clinicopathological abnormalities and treatment response in 24 dogs seroreactive to *Bartonella vinsonii (berkhoffii)* antigens. J Am Anim Hosp Assoc 40:92–101.

Breitschwerdt, E.B., B.C. Hegarty, R. Maggi, E. Hawkins, and P. Dyer. 2005. *Bartonella* species as a potential cause of epistaxis in dogs. J Clin Microbiol. 43:2529–2533.

Breitschwerdt, E.B., and D.L. Kordick. 1995. Bartonellosis. J Am Vet Med Assoc 206:1928–1931.

Breitschwerdt, E.B., and D.L. Kordick. 2000. *Bartonella* infection in animals: carriership, reservoir potential, pathogenicity, and zoonotic potential for human infection. Clin Microbiol Rev 13:428–438.

Breitschwerdt, E.B., D.L. Kordick, D.E. Malarkey, B. Keene, T.L. Hadfield, and K. Wilson. 1995. Endocarditis in a dog due to infection with a novel *Bartonella* subspecies. J Clin Microbiol 33:154–160.

Breitschwerdt, E.B., R.G. Maggi, M.B. Cadenas, and P.P. de Paiva Diniz. 2009. A groundhog, a novel *Bartonella* sequence, and my father's death. Emerg Infect Dis 15:2080–2086.

Breitschwerdt, E.B., R.G. Maggi, B.B. Chomel, and M.R. Lappin. 2010a. Bartonellosis: an emerging infectious disease of zoonotic importance to animals and human beings. J Vet Emerg Crit Care 20:8–30.

Breitschwerdt, E.B., R.G. Maggi, A.W. Duncan, W.L. Nicholson, B.C. Hegarty, and C.W. Woods. 2007. *Bartonella* species in blood of immunocompetent persons with animal and arthropod contact. Emerg Infect Dis 13:938–941.

Breitschwerdt, E.B., R.G. Maggi, P. Farmer, and P.E. Mascarelli. 2010b. Molecular evidence of perinatal transmission of *Bartonella vinsonii subsp. berkhoffii* and *Bartonella henselae* to a child. J Clin Microbiol 48:2289–2293.

Breitschwerdt, E.B., R.G. Maggi, P.M. Lantos, C.W. Woods, B.C. Hegarty, amd J.M. Bradley. 2010c. *Bartonella vinsonii subsp. berkhoffii* and *Bartonella henselae* bacteremia in a father and daughter with neurological disease. Parasit Vectors 3:29.

Breitschwerdt, E.B., R.G. Maggi, W.L. Nicholson, N.A. Cherry, and C.W. Woods. 2008. *Bartonella* sp. bacteremia in patients with neurological and neurocognitive dysfunction. J Clin Microbiol 46:2856–2861.

Breitschwerdt, E.B., R.G. Maggi, B. Sigmon, and W. L. Nicholson. 2007. Isolation of *Bartonella quintana* from a woman and a cat following putative bite transmission. J Clin Microbiol 45:270–272.

Brouqui, P., and D. Raoult. 1996. *Bartonella quintana* invades and multiplies within endothelial cells in vitro and in vivo and forms intracellular blebs. Res Microbiol 147:719–731.

Brouqui, P., and D. Raoult. 2006. New insight into the diagnosis of fastidious bacterial endocarditis. FEMS Immunol Med Microbiol 47:1–13.

Bruckner, G.K. 2009. The role of the World Organisation for Animal Health (OIE) to facilitate the international trade in animals and animal products. Onderstepoort J Vet Res 76:141–146.

Butler, D. 2006. Disease surveillance needs a revolution. Nature 440:6–7.

Caceres-Rios, H., J. Rodriguez-Tafur, F. Bravo-Puccio, C. Maguina-Vargas, C.S. Diaz, D.C. Ramos, and R. Patarca. 1995. Verruga peruana: an infectious endemic angiomatosis. Crit Rev Oncog 6:47–56.

Chamberlin, J., L.W. Laughlin, S. Romero, N. Solorzano, S. Gordon, R.G. Andre, P. Pachas, H. Friedman, C. Ponce, and D. Watts. 2002. Epidemiology of endemic *Bartonella bacilliformis*: a prospective cohort study in a Peruvian mountain valley community. J Infect Dis 186:983–990.

Chang, C.C., B.B. Chomel, R.W. Kasten, R.M. Heller, H. Ueno, K. Yamamoto, V.C. Bleich, B.M. Pierce, B.J. Gonzales, P.K. Swift, W.M. Boyce, S.S. Jang, H.J. Boulouis, Y. Piemont, G.M. Rossolini, M.L. Riccio, G. Cornaglia, L. Pagani, C. Lagatolla, L. Selan, and R. Fontana. 2000. *Bartonella* spp. isolated from wild and domestic ruminants in North America. Emerg Infect Dis 6:306–311.

Chapman, A.S., S.M. Murphy, L.J. Demma, R.C. Holman, A.T. Curns, J.H. McQuiston, J.W. Krebs, and D.L. Swerdlow. 2006. Rocky Mountain spotted fever in the United States, 1997–2002. Vector Borne Zoon Dis 6:170–178.

Chen, S.C., and G.L. Gilbert. 1994. Cat scratch disease: past and present. J Paediatr Child Health 30:467–469.

Cherry, N.A., R.G. Maggi, A.L. Cannedy, and E.B. Breitschwerdt. 2009. PCR detection of *Bartonella bovis* and *Bartonella henselae* in the blood of beef cattle. Vet Microbiol 135:308–312.

Chomel, B.B. 1998. New emerging zoonoses: a challenge and an opportunity for the veterinary profession. Comp Immunol Microbiol Infect Dis 21:1–14.

Chomel, B.B. 2000. Cat-scratch disease. Rev Sci Tech 19:136–150.

Chomel, B.B. 2003. Control and prevention of emerging zoonoses. J Vet Med Educ 30:145–147.

Chomel, B.B., and H.J. Boulouis. 2005. Zoonotic diseases caused by bacteria of the genus *Bartonella* genus: new reservoirs? new vectors? Bull Acad Nat Med 189:465–480.

Chomel, B.B., and B.I. Osburn. 2006. Zoological medicine and public health. J Vet Med Educ 33:346–351.

Chomel, B.B., K.A. Mac Donald, R.W. Kasten, C.C. Chang, A.C. Wey, J.E. Foley, W.P. Thomas, and M.D. Kittleson. 2001. Aortic valve endocarditis in a dog due to *Bartonella clarridgeiae*. J Clin Microbiol 39:3548–3554.

Chomel, B.B., R.W. Kasten, J.E. Sykes, H.J. Boulouis, and E.B. Breitschwerdt. 2003. Clinical impact of persistent *Bartonella* bacteremia in humans and animals. Ann NY Acad Sci 990:267–278.

Chomel, B.B., H.J. Boulouis, and E.B. Breitschwerdt. 2004. Cat scratch disease and other zoonotic *Bartonella* infections. J Am Vet Med Assoc 224:1270–1279.

Chomel, B.B., H.J. Boulouis, S. Maruyama, and E.B. Breitschwerdt. 2006a. *Bartonella* spp. in pets and effect on human health. Emerg Infect Dis 12: 389–394.

Chomel, B.B., R.W. Kasten, J.B. Henn, and S. Molia. 2006b. *Bartonella* infection in domestic cats and wild felids. Ann NY Acad Sci 1078:410–415.

Chomel, B.B., R.W. Kasten, C. Williams, A.C. Wey, J.B. Henn, R. Maggi, S. Carrasco, J. Mazet, H.J. Boulouis, R. Maillard, and E.B. Breitschwerdt. 2009. *Bartonella* endocarditis: a pathology shared by animal reservoirs and patients. Ann NY Acad Sci 1166:120–126.

Chugh, T.D. 2008. Emerging and re-emerging bacterial diseases in India. J Biosci 33:549–555.

Cockwill, K.R., S.M. Taylor, H.M. Philibert, E.B. Breitschwerdt, and R.G. Maggi. 2007. *Bartonella vinsonii subsp. berkhoffii* endocarditis in a dog from Saskatchewan. Can Vet J 48:839–844.

Concannon, R., K. Wynn-Owen, V.R. Simpson, and R.J. Birtles. 2005. Molecular characterization of

haemoparasites infecting bats (Microchiroptera) in Cornwall, UK. Parasitology 131:489–496.

Dalton, H.R., R. Bendall, S. Ijaz, and M. Banks. 2008. Hepatitis E: an emerging infection in developed countries. Lancet Infect Dis 8:698–709.

De Paiva Diniz, P.P., D.S. Schwartz, H.S. De Morais, and E.B. Breitschwerdt, 2007. Surveillance for zoonotic vector-borne infections using sick dogs from outheastern Brazil. Vector-Borne Zoonot 7: 689–697.

Dehio, C. 1999. Interactions of *Bartonella henselae* with vascular endothelial cells. Curr Opin Microbiol 2:78–82.

Dehio, C. 2001. *Bartonella* interactions with endothelial cells and erythrocytes. Trends Microbiol 9:279–285.

Dehio, C. 2003. Recent progress in understanding *Bartonella*-induced vascular proliferation. Curr Opin Microbiol 6:61–65.

Dehio, C. 2004. Molecular and cellular basis of *Bartonella* pathogenesis. Annu Rev Microbiol 58:365–390.

Dehio, C. 2005. *Bartonella*-host-cell interactions and vascular tumour formation. Nat Rev Microbiol 3:621–631.

Dehio, C., M. Meyer, J. Berger, H. Schwarz, and C. Lanz. 1997. Interaction of *Bartonella henselae* with endothelial cells results in bacterial aggregation on the cell surface and the subsequent engulfment and internalisation of the bacterial aggregate by a unique structure, the invasome. J Cell Sci 110:2141–2154.

Dehio, C., U. Sauder, and R. Hiestand. 2004. Isolation of Bartonella schoenbuchensis from *Lipoptena cervi*, a blood-sucking arthropod causing deer ked dermatitis. J Clin Microbiol 42:5320–5323.

Dehio, M., M. Quebatte, S. Foser, and U. Certa. 2005. The transcriptional response of human endothelial cells to infection with *Bartonella henselae* is dominated by genes controlling innate immune responses, cell cycle, and vascular remodelling. Thromb Haemost 94:347–361.

Dennis, D.T. 1995. Lyme disease. Dermatol Clin 13:537–551.

Diniz, P.P., R.G. Maggi, D.S. Schwartz, M.B. Cadenas, J.M. Bradley, B. Hegarty, and E.B. Breitschwerdt. 2007. Canine bartonellosis: serological and molecular prevalence in Brazil and evidence of co-infection with Bartonella henselae and Bartonella vinsonii subsp. berkhoffii. Vet Res 38:697–710.

Diniz, P.P., M. Wood, R.G. Maggi, S. Sontakke, M. Stepnik, and E.B. Breitschwerdt. 2009. Co-isolation of Bartonella henselae and *Bartonella vinsonii subsp. berkhoffii* from blood, joint and subcutaneous seroma fluids from two naturally infected dogs. Vet Microbiol 138:368–372.

Dorny, P., N. Praet, N. Deckers, and S. Gabriel. 2009. Emerging food-borne parasites. Vet Parasitol 163:196–206.

Drancourt, M., and D. Raoult. 1995. Cat-scratch disease and disease caused by Bartonella (Rochalimaea). Presse Med 24:183–188.

Drancourt, M., L. Tran-Hung, J. Courtin, H. Lumley, and D. Raoult. 2005. Bartonella quintana in a 4000-year-old human tooth. J Infect Dis 191:607–611.

Duncan, A.W., R.G. Maggi, and E.B. Breitschwerdt. 2007a. Bartonella DNA in dog saliva. Emerg Infect Dis 13:1948–1950.

Duncan, A.W., R.G. Maggi, and E.B. Breitschwerdt. 2007b. A combined approach for the enhanced detection and isolation of Bartonella species in dog blood samples: pre-enrichment liquid culture followed by PCR and subculture onto agar plates. J Microbiol Meth 69:273–281.

Duncan, A.W., H.S. Marr, A.J. Birkenheuer, R.G. Maggi, L.E. Williams, M.T. Correa, and E.B. Breitschwerdt. 2008. Bartonella DNA in the blood and lymph nodes of golden retrievers with lymphoma and in healthy controls. J Vet Intern Med 22:89–95.

Ellis, B.A., L.D. Rotz, J.A. Leake, F. Samalvides, J. Bernable, G. Ventura, C. Padilla, P. Villaseca, L. Beati, R. Regnery, J.E. Childs, J.G. Olson, and C.P. Carrillo. 1999. An outbreak of acute bartonellosis (Oroya fever) in the Urubamba region of Peru, 1998. Am J Trop Med Hyg 61:344–349.

Enia, F., G. Di Stefano, A.M. Floresta, and C. Matassa. 2005. New etiologies responsible for infective endocarditis with negative blood cultures. Ital Heart J 6:S128–134.

Eremeeva, M.E., H.L. Gerns, S.L. Lydy, J.S. Goo, E.T. Ryan, S.S. Mathew, M.J. Ferraro, J.M. Holden, W.L. Nicholson, G.A. Dasch, and J.E. Koehler. 2007. Bacteremia, fever, and splenomegaly caused by a newly recognized Bartonella species. N Engl J Med 356:2381–2387.

Feldmann, H., M. Czub, S. Jones, D. Dick, M. Garbutt, A. Grolla, and H. Artsob. 2002. Emerging and re-emerging infectious diseases. Med Microbiol Immunol 191:63–74.

Fenollar, F., and D. Raoult. 2004. Molecular genetic methods for the diagnosis of fastidious microorganisms. Apmis 112:785–807.

Forman, S., N. Hungerford, M. Yamakawa, T. Yanase, H.J. Tsai, Y.S. Joo, D.K. Yang, and J.J. Nha. 2008. Climate change impacts and risks for animal health in Asia. Rev Sci Tech 27:581–597.

Foucault, C., P. Brouqui, and D. Raoult. 2006. Bartonella quintana characteristics and clinical management. Emerg Infect Dis 12:217–223.

Foucault, C., J.M. Rolain, D. Raoult, and P. Brouqui. 2004. Detection of Bartonella quintana by direct immunofluorescence examination of blood smears of a patient with acute trench fever. J Clin Microbiol 42:4904–4906.

Fournier, P.E., C. Taylor, J.M. Rolain, L. Barrassi, G. Smith, and D. Raoult. 2007. Bartonella australis sp. nov. from kangaroos, Australia. Emerg Infect Dis 13:1961–1962.

Gillespie, T.N., R.J. Washabau, M.H. Goldschmidt, J.M. Cullen, A.R. Rogala, and E.B. Breitschwerdt. 2003. Detection of Bartonella henselae and Bartonella clarridgeiae DNA in hepatic specimens from two dogs with hepatic disease. J Am Vet Med Assoc 222:47–51.

Gouriet, F., H. Lepidi, G. Habib, F. Collart, and D. Raoult. 2007. From cat scratch disease to endocarditis, the possible natural history of Bartonella henselae infection. BMC Infect Dis 7:30.

Gray, G.C., A.A. Johnson, S.A. Thornton, W.A. Smith, J. Knobloch, P.W. Kelley, L. Obregon Escudero, M. Arones Huayda, and F.S. Wignall. 1990. An epidemic of Oroya fever in the Peruvian Andes. Am J Trop Med Hyg 42:215–221.

Greub, G., and D. Raoult. 2002. Bartonella: new explanations for old diseases. J Med Microbiol 51:915–923.

Gubler, D.J. 2009. Vector-borne diseases. Rev Sci Tech. 28:583–588.

Hajjaji, N., L. Hocqueloux, R. Kerdraon, and L. Bret. 2007. Bone infection in cat-scratch disease: a review of the literature. J Infect 54:417–421.

Hambuch, T.M., S.A. Handley, B. Ellis, J. Chamberlin, S. Romero, and R. Regnery. 2004. Population genetic analysis of Bartonella bacilliformis isolates from areas of Peru where Carrion's disease is endemic and epidemic. J Clin Microbiol 42:3675–3680.

Harms, C., R.G. Maggi, E.B. Breitschwerdt, C.L. Clemons-Chevis, M. Solangi, D.S. Rotstein, P.A. Fair, L.J. Hansen, A.A. Hohn, G.N. Lovewell, W.A. McLellan, D.A. Pabst, T.K. Rowles, L.H. Schwacke, F.I. Townsend, and R.S. Wells. 2008. Bartonella species detection in captive, stranded and free-ranging cetaceans. Vet Res 39:59.

Harrus, S., and G. Baneth. 2005. Drivers for the emergence and re-emergence of vector-borne protozoal and bacterial diseases. Int J Parasitol 35:1309–1318.

Hawkes, C., and M. Ruel. 2006. The links between agriculture and health: an intersectoral opportunity to improve the health and livelihoods of the poor. B World Health Organ 84:984–990.

Heeney, J.L. 2006. Zoonotic viral diseases and the frontier of early diagnosis, control and prevention. J Intern Med 260:399–408.

Heller, R., M. Kubina, P. Mariet, P. Riegel, G. Delacour, C. Dehio, F. Lamarque, R. Kasten, H.J. Boulouis, H. Monteil, B. Chomel, and Y. Piemont. 1999. *Bartonella alsatica sp. nov.*, a new *Bartonella* species isolated from the blood of wild rabbits. Int J Syst Bacteriol 49:283–288.

Henn, J.B., M.W. Gabriel, R.W. Kasten, R.N. Brown, J.H. Theis, J.E. Foley, and B.B. Chomel. 2007. Gray foxes (*Urocyon cinereoargenteus*) as a potential reservoir of a *Bartonella clarridgeiae*-like bacterium and domestic dogs as part of a sentinel system for surveillance of zoonotic arthropod-borne pathogens in northern California. J Clin Microbiol 45:2411–2418.

Hennemann, H.H. 1963. Oroya fever (Carrion's disease): an acute acquired hemolytic anemia. Dtsch Med Wochenschr 88:1759–1767.

Higgins, R. 2004. Emerging or re-emerging bacterial zoonotic diseases: bartonellosis, leptospirosis, Lyme borreliosis, plague. Rev Sci Tech 23:569–581.

Ihler, G.M. 1996. *Bartonella bacilliformis*: dangerous pathogen slowly emerging from deep background. FEMS Microbiol Lett 144:1–11.

Inoue, K., S. Maruyama, H. Kabeya, K. Hagiya, Y. Izumi, Y. Une, and Y. Yoshikawa. 2009. Exotic small mammals as potential reservoirs of zoonotic *Bartonella* spp. Emerg Infect Dis 15:526–532.

Irwin, P.J. 2002. Companion animal parasitology: a clinical perspective. Int J Parasitol. 32:581–593.

Jardine, C., Appleyard, G., Kosoy, M.Y., McColl, D., Chirino-Trejo, M., Wobeser, G., Leighton, F.A. 2005. Rodent-associated *Bartonella* in Saskatchewan, Canada. Vector Borne Zoon 5:402–409.

Johnson, P.T., A.R. Townsend, C.C. Cleveland, P.M. Glibert, R.W. Howarth, V.J. McKenzie, E. Rejmankova, and M.H. Ward. 2010. Linking environmental nutrient enrichment and disease emergence in humans and wildlife. Ecol Appl 20:16–29.

Jones, K.E., N.G. Patel, M.A. Levy, A. Storeygard, D. Balk, J.L. Gittleman, and P. Daszak. 2008. Global trends in emerging infectious diseases. Nature 451:990–993.

Jones, S.L., R. Maggi, J. Shuler, A. Alward, and E.B. Breitschwerdt. 2008. Detection of *Bartonella henselae* in the blood of 2 adult horses. J Vet Intern Med 22:495–498.

Jost, C.C., J.C. Mariner, P.L. Roeder, E. Sawitri, and G.J. Macgregor-Skinner. 2007. Participatory epidemiology in disease surveillance and research. Rev Sci Tech. 26:537–549.

Karem, K.L., K.A. Dubois, S.L. McGill, and R.L. Regnery. 1999. Characterization of *Bartonella henselae*-specific immunity in BALB/c mice. Immunology 97:352–358.

Kelly, A.M., and R.R. Marshak. 2007. Veterinary medicine, global health. J Am Vet Med Assoc. 231:1806–1808.

Kelly, P.J., L.A. Matthewman, D. Hayter, S. Downey, K. Wray, N.R. Bryson, and D. Raoult. 1996. *Bartonella (Rochalimaea) henselae* in southern Africa—evidence for infections in domestic cats and implications for veterinarians. J S Afr Vet Assoc 67:182–187.

Kelly, P., J.L. Rolain, R. Maggi, S. Sontakke, B. Keene, S. Hunter, H. Lepidi, K.T. Breitschwerdt, and E.B. Breitschwerdt. 2006. *Bartonella quintana* endocarditis in dogs. Emerg Infect Dis 12:1869–1872.

Kelly, P.J., L. Moura, T. Miller, J. Thurk, N. Perreault, A. Weil, R. Maggi, R., H. Lucas, and E. Breitschwerdt. 2010. Feline immunodeficiency virus, feline leukemia virus and *Bartonella* species in stray cats on St Kitts, West Indies. J Feline Med Surg 12:447–450.

Kitchell, B.E., T.M. Fan, D. Kordick, E.B. Breitschwerdt, G. Wollenberg, C.A. Lichtensteiger. 2000. Peliosis hepatis in a dog infected with *Bartonella henselae*. J Am Vet Med Assoc 216:517, 519–523.

Kock, R., B. Kebkiba, B., R. Heinonen, and B. Bedane. 2002. Wildlife and pastoral society—shifting paradigms in disease control. Ann NY Acad Sci 969:24–33.

Kordick, D.L., and E.B. Breitschwerdt. 1995. Intraerythrocytic presence of *Bartonella henselae*. J Clin Microbiol 33:1655–1656.

Kordick, D.L., T.T. Brown, K. Shin, K., and E.B. Breitschwerdt. 1999. Clinical and pathologic evaluation of chronic *Bartonella henselae* or *Bartonella clarridgeiae* infection in cats. J Clin Microbiol 37:1536–1547.

Kordick, D.L., B. Swaminathan, C.E. Greene, K.H. Wilson, A.M. Whitney, S. O'Connor, D.G. Hollis, G.M. Matar, A.G. Steigerwalt, G.B. Malcolm, P.S. Hayes, T.L. Hadfield, E.B. Breitschwerdt, and D.J. Brenner. 1996. *Bartonella vinsonii subsp. berkhoffii subsp. nov.*, isolated from dogs; *Bartonella vinsonii subsp. vinsonii*; and emended description of *Bartonella vinsonii*. Int J Syst Bacteriol 46:704–709.

Kosek, M., R. Lavarello, R.H. Gilman, J. Delgado, C. Maguina, M. Verastegui, A.G. Lescano, V. Mallqui, J.C. Kosek, S. Recavarren, and L. Cabrera. 2000. Natural history of infection with *Bartonella bacilliformis* in a nonendemic population. J Infect Dis 182:865–872.

Kosoy, M.Y., R.L. Regnery, T. Tzianabos, E.L. Marston, D.C. Jones, D. Green, G.O. Maupin, J.G. Olson, and J.E. Childs. 1997. Distribution, diversity, and host specificity of *Bartonella* in rodents from the southeastern United States. Am J Trop Med Hyg 57:578–588.

Kosoy, M.Y., Regnery, R.L., Kosaya, O.I., Jones, D.C., Marston, E.L., Childs, J.E., 1998. Isolation of

*Bartonella* spp. from embryos and neonates of naturally infected rodents. J Wildl Dis 34:305–309.

Kosoy, M.Y., R.L. Regnery, O.L. Kosaya, and J.E. Childs. 1999. Experimental infection of cotton rats with three naturally occurring *Bartonella* species. J Wildl Dis 35:275–284.

Kosoy, M.Y., E.K. Saito, D. Green, E.L. Marston, D.C. Jones, and J.E. Childs. 2000. Experimental evidence of host specificity of *Bartonella* infection in rodents. Comp Immunol Microbiol Infect Dis 23:221–238.

Kosoy, M., E. Mandel, D. Green, E. Marston, and J. Childs. 2004a. Prospective studies of *Bartonella* of rodents. Part I. Demographic and temporal patterns in population dynamics. Vector-Borne Zoon 4:285–295.

Kosoy, M., E.Mandel, D. Green, E. Marston, D. Jones, and J. Childs. 2004b. Prospective studies of *Bartonella* of rodents. Part II. Diverse infections in a single rodent community. Vector-Borne Zoon 4:296–305.

Kosoy, M., Y. Bai, K. Sheff, C. Morway, H. Baggett, S.A. Maloney, S. Boonmar, S. Bhengsri, S.F. Dowell, A. Sitdhirasdr, K. Lerdthusnee, J. Richardson, and L.F. Peruski. 2010. Identification of *Bartonella* infections in febrile human patients from Thailand and their potential animal reservoirs. Am J Trop Med Hyg 82:1140–1145.

La, V.D., L. Tran-Hung, G. Aboudharam, D. Raoult, and M. Drancourt. 2005. *Bartonella quintana* in domestic cat. Emerg Infect Dis 11:1287–1289.

Leibler, J.H., J. Otte, D. Roland-Holst, D.U. Pfeiffer, R. Soares Magalhaes, J. Rushton, J.P. Graham, and E.K. Silbergeld. 2009. Industrial food animal production and global health risks: exploring the ecosystems and economics of avian influenza. EcoHealth 6:58–70.

Lydy, S.L., M.E. Eremeeva, D. Asnis, C.D. Paddock, W.L. Nicholson, D.J. Silverman, and G.A. Dasch. 2008. Isolation and characterization of *Bartonella bacilliformis* from an expatriate Ecuadorian. J Clin Microbiol 46:627–637.

Maco, V., C. Maguina, A. Tirado, V. Maco, and J.E. Vidal. 2004. Carrion's disease (Bartonellosis bacilliformis) confirmed by histopathology in the High Forest of Peru. Rev Inst Med Trop Sao Paulo 46:171–174.

Maggi, R.G., B. Chomel, B.C. Hegarty, J. Henn, and E.B. Breitschwerdt, E.B. 2006. A *Bartonella vinsonii berkhoffii* typing scheme based upon 16S-23S ITS and Pap31 sequences from dog, coyote, gray fox, and human isolates. Mol Cell Probes 20:128–134.

Maggi, R.G., C.A. Harms, A.A. Hohn, D.A. Pabst, W.A. McLellan, J.W. Walton, D.S. Rotstein, and E.B. Breitschwerdt. 2005. *Bartonella henselae* in porpoise blood. Emerg Infect Dis 11:1894–1898.

Maggi, R.G., S.A. Raverty, S.J. Lester, D.G. Huff, M. Haulena, S.L. Ford, O. Nielsen, J.H. Robinson, and E.B. Breitschwerdt. 2008. *Bartonella henselae* in captive and hunter-harvested beluga (*Delphinapterus leucas*). J Wildl Dis 44:871–877.

Maillard, R., B. Grimard, S. Chastant-Maillard, B. Chomel, T. Delcroix, C. Gandoin, C. Bouillin, L. Halos, M. Vayssier-Taussat, and H.J. Boulouis. 2006. Effects of cow age and pregnancy on *Bartonella* infection in a herd of dairy cattle. J Clin Microbiol 44:42–46.

Maillard, R., E. Petit, B. Chomel, C. Lacroux, F. Schelcher, M. Vayssier-Taussat, N. Haddad, and H.J. Boulouis. 2007. Endocarditis in cattle caused by *Bartonella bovis*. Emerg Infect Dis 13:1383–1385.

Maillard, R., P. Riegel, F. Barrat, C.Bouillin, D. Thibault, C. Gandoin, L. Halos, C. Demanche, A. Alliot, J. Guillot, Y. Piemont, H.J. Boulouis, and M. Vayssier-Taussat. 2004a. *Bartonella chomelii sp. nov.*, isolated from French domestic cattle (*Bos taurus*). Int J Syst Evol Microbiol 54:215–220.

Maillard, R., M. Vayssier-Taussat, C. Bouillin, C. Gandoin, L. Halos, L. B. Chomel, Y. Piemont, and H.J. Boulouis. 2004b. Identification of *Bartonella* strains isolated from wild and domestic ruminants by a single-step PCR analysis of the 16S-23S intergenic spacer region. Vet Microbiol 98:63–69.

Mandle, T., H. Einsele, M. Schaller, D. Neumann, W. Vogel, I.B. Autenrieth, and V.A. Kempf. 2005. Infection of human CD34+ progenitor cells with *Bartonella henselae* results in intraerythrocytic presence of *B. henselae*. Blood 106:1215–1222.

Mara, D.D., and G.P. Alabaster. 1995. An environmental classification of housing-related diseases in developing countries. J Trop Med Hyg 98:41–51.

Marie, J.L., P.E. Fournier, J.M. Rolain, S. Briolant, B. Davoust, and D. Raoult. 2006. Molecular detection of *Bartonella quintana*, *B. elizabethae*, *B. koehlerae*, *B. doshiae*, *B. taylorii*, and *Rickettsia felis* in rodent fleas collected in Kabul, Afghanistan. Am J Trop Med Hyg 74:436–439.

Marsilia, G.M., A. La Mura, R. Galdiero, E. Galdiero, G. Aloj, and A. Ragozzino. 2006. Isolated hepatic involvement of cat scratch disease in immunocompetent adults: Enhanced magnetic resonance imaging, pathological findings, and molecular analysis—two cases. Int J Surg Pathol 14:349–354.

Martini, M., M.L. Menandro, A. Mondin, D. Pasotto, S. Mazzariol, S. Lauzi, and C. Stelletta. 2008. Detection of *Bartonella bovis* in a cattle herd in Italy. Vet Rec 162:58–59.

Maruyama, S. 2005. Diagnostic tests: Cat-scratch disease. Nippon Rinsho 63:S237–240.

Mathews, F. 2009. Zoonoses in wildlife integrating ecology into management. Adv Parasitol 68:185–209.

Maudlin, I., M.C. Eisler, and S.C. Welburn. 2009. Neglected and endemic zoonoses. Philos Trans R Soc Lond B 364:2777–2787.

Maurin, M., R. Birtles, and D. Raoult. 1997. Current knowledge of *Bartonella* species. Eur J Clin Microbiol Infect Dis 16:487–506.

Maurin, M., and D. Raoult. 1998. *Bartonella* infections: diagnostic and management issues. Curr Opin Infect Dis 11:189–193.

Mavroidi, N. 2008. Transmission of zoonoses through immigration and tourism. Vet Ital 44:651–656.

McCord, A.M., A.W. Burgess, M.J. Whaley, and B.E. Anderson. 2005. Interaction of *Bartonella henselae* with endothelial cells promotes monocyte/macrophage chemoattractant protein 1 gene expression and protein production and triggers monocyte migration. Infect Immun 73:5735–5742.

McCord, A.M., J. Cuevas, and B.E. Anderson. 2007. *Bartonella*-induced endothelial cell proliferation is mediated by release of calcium from intracellular stores. DNA Cell Biol 26:657–663.

Meslin, F.X. 1995. Zoonoses in the world: current and future trends. Schweiz Med Wochenschr 125:875–878.

Meslin, F.X. 1997. Emerging and re-emerging zoonoses: local and worldwide threats. Med Trop 57:7–9.

Meslin, F.X., K. Stohr, and D. Heymann. 2000. Public health implications of emerging zoonoses. Rev Sci Tech 19:310–317.

Mexas, A.M., S.I. Hancock, and E.B. Breitschwerdt. 2002. *Bartonella henselae* and *Bartonella elizabethae* as potential canine pathogens. J Clin Microbiol 40:4670–4674.

Michau, T.M., E.B. Breitschwerdt, B.C. Gilger, and M.G. Davidson. 2003. *Bartonella vinsonii subspecies berkhoffi* as a possible cause of anterior uveitis and choroiditis in a dog. Vet Ophthalmol 6:299–304.

Miller, M.A., I.A. Gardner, C. Kreuder, D.M. Paradies, K.R. Worcester, D.A. Jessup, E. Dodd, M.D. Harris, J.A. Ames, A.E. Packham, and P.A. Conrad. 2002. Coastal freshwater runoff is a risk factor for *Toxoplasma gondii* infection of Southern sea otters (*Enhydra lutris nereis*). Int J Parasitol 32:997–1006.

Morales, S.C., E.B. Breitschwerdt, R.J. Washabau, I. Matise, R.G. Maggi, and A.W. Duncan. 2007. Detection of *Bartonella henselae* DNA in two dogs with pyogranulomatous lymphadenitis. J Am Vet Med Assoc 230:681–685.

Morgan, U.M. 2000. Detection and characterisation of parasites causing emerging zoonoses. Int J Parasitol 30:1407–1421.

Morick, D., N. Osinga, E. Gruys, and S. Harrus. 2009. Identification of a *Bartonella* species in the harbor seal (*Phoca vitulina*) and in seal lice (*Echinophtirius horridus*). Vector-Borne Zoon 9:751–753.

Munana, K.R., S.M. Vitek, B.C. Hegarty, D.L. Kordick, and E.B. Breitschwerdt. 2001. Infection of fetal feline brain cells in culture with *Bartonella henselae*. Infect Immun 69:564–569.

Murphy, F.A. 1994. New, emerging, and reemerging infectious diseases. Adv Virus Res 43:1–52.

Murphy, F.A. 1998. Emerging zoonoses. Emerg Infect Dis 4:429–435.

Murphy, F.A. 1999. The threat posed by the global emergence of livestock, food-borne, and zoonotic pathogens. Ann NY Acad Sci 894:20–27.

Murphy, F.A. 2008. Emerging zoonoses: the challenge for public health and biodefense. Prev Vet Med 86:216–223.

Noguchi, H. 1928a. Etiology of Oroya fever: IX. *Bacterium peruvianum*, N. Sp., a secondary invader of the lesions of verruga peruana. J Exp Med 47:165–170.

Noguchi, H., 1928b. Etiology of Oroya fever: X. Comparative studies of different strains of *Bartonella bacilliformis*, with special reference to the relationship between the clinical types of Carrion's disease and the virulence of the infecting organism. J Exp Med 47:219–234.

Pappalardo, B.L., T. Brown, J.L. Gookin, C.L. Morrill, and E.B. Breitschwerdt. 2000. Granulomatous disease associated with *Bartonella* infection in 2 dogs. J Vet Intern Med 14:37–42.

Pappalardo, B.L., T.T. Brown, M. Tompkins, and E.B. Breitschwerdt, E.B. 2001. Immunopathology of *Bartonella vinsonii* (*berkhoffii*) in experimentally infected dogs. Vet Immunol Immunopathol 83:125–147.

Piemont, Y., and R. Heller. 1999. Bartonellosis. II. Other *Bartonella* responsible for human diseases. Ann Biol Clin 57:29–36.

Pitassi, L.H., R.F. Magalhaes, M.L. Barjas-Castro, E.V. de Paula, M.R. Ferreira, and P.E. Velho. 2007. *Bartonella henselae* infects human erythrocytes. Ultrastruct Pathol 31:369–372.

Polley, L. 2005. Navigating parasite webs and parasite flow: emerging and re-emerging parasitic zoonoses of wildlife origin. Int J Parasitol 35:1279–1294.

Raoult, D., B. La Scola, P.L. Kelly, B. Davoust, and J. Gomez. 2005. *Bartonella bovis* in cattle in Africa. Vet Microbiol 105:155–156.

Rolain, J.M., B. La Scola, Z. Liang, B. Davoust, and D. Raoult. 2001. Immunofluorescent detection of intraerythrocytic *Bartonella henselae* in naturally infected cats. J Clin Microbiol 39:2978–2980.

Rolain, J.M., C. Foucault, R. Guieu, B. La Scola, P. Brouqui, and D. Raoult. 2002. *Bartonella quintana* in human erythrocytes. Lancet 360:226–228.

Rolain, J.M., D. Arnoux, D. Parzy, J. Sampol, and D. Raoult. 2003a. Experimental infection of human erythrocytes from alcoholic patients with *Bartonella quintana*. Ann NY Acad Sci 990:605–611.

Rolain, J.M., V. Chanet, H. Laurichesse, H. Lepidi, J. Beytout, and D. Raoult. 2003b. Cat scratch disease with lymphadenitis, vertebral osteomyelitis, and spleen abscesses. Ann NY Acad Sci 990:397–403.

Rolain, J.M., P.E. Fournier, D. Raoult, and J.J. Bonerandi. 2003c. First isolation and detection by immunofluorescence assay of *Bartonella koehlerae* in erythrocytes from a French cat. J Clin Microbiol 41:4001–4002.

Rolain, J.M., M. Franc, B. Davoust, and D. Raoult. 2003d. Molecular detection of *Bartonella quintana, B. koehlerae, B. henselae, B. clarridgeiae, Rickettsia felis,* and *Wolbachia pipientis* in cat fleas, France. Emerg Infect Dis 9:338–342.

Rolain, J.M., M. Maurin, M.N. Mallet, D. Parzy, and D. Raoult. 2003e. Culture and antibiotic susceptibility of *Bartonella quintana* in human erythrocytes. Antimicrob Agents Chemother 47:614–619.

Rolain, J.M., S. Novelli, P. Ventosilla, C. Maguina, H. Guerra, and D. Raoult. 2003f. Immunofluorescence detection of *Bartonella bacilliformis* flagella in vitro and in vivo in human red blood cells as viewed by laser confocal microscopy. Ann NY Acad Sci 990:581–584.

Rolain, J.M. E. Rousset, B. La Scola, R. Duquesnel, and D. Raoult. 2003g. *Bartonella schoenbuchensis* isolated from the blood of a French cow. Ann NY Acad Sci 990:236–238.

Rotstein, D.S., S.K. Taylor, J. Bradley, and E.B. Breitschwerdt. 2000. Prevalence of *Bartonella henselae* antibody in Florida panthers. J Wildl Dis 36:157–160.

Russell, L.H. 2004. The needs for public health education: reflections from the 27th World Veterinary Congress. J Vet Med Educ 3:17–21.

Saunders, G.K., and W.E. Monroe. 2006. Systemic granulomatous disease and sialometaplasia in a dog with *Bartonella* infection. Vet Pathol 43:391–392.

Schmid, M.C., F. Scheidegger, M. Dehio, N. Balmelle-Devaux, R. Schulein, P. Guye, C.S. Chennakesava, B. Biedermann, and C. Dehio. 2006. A translocated bacterial protein protects vascular endothelial cells from apoptosis. PLoS Pathog 2:e115.

Schroder, G., and C. Dehio. 2005. Virulence-associated type IV secretion systems of *Bartonella*. Trends Microbiol 13:336–342.

Schulein, R., and C. Dehio. 2002. The VirB/VirD4 type IV secretion system of *Bartonella* is essential for establishing intraerythrocytic infection. Mol Microbiol 46:1053–1067.

Schulein, R., A. Seubert, C. Gille, C. Lanz, Y. Hansmann, Y. Piemont, and C. Dehio. 2001. Invasion and persistent intracellular colonization of erythrocytes. A unique parasitic strategy of the emerging pathogen *Bartonella*. J Exp Med 193:1077–1086.

Schulte, B., D. Linke, S. Klumpp, M. Schaller, T. Riess, I.B. Autenrieth, and V.A. Kempf. 2006. *Bartonella quintana* variably expressed outer membrane proteins mediate vascular endothelial growth factor secretion but not host cell adherence. Infect Immun 74:5003–5013.

Seimenis, A.M. 2008. The spread of zoonoses and other infectious diseases through the international trade of animals and animal products. Vet Ital 44:591–599.

Seki, M., and S. Kohno. 2007. Emerging, re-emerging infectious diseases. Nippon Rinsho 65:S15–20.

Seubert, A., R. Schulein, and C. Dehio. 2002. Bacterial persistence within erythrocytes: a unique pathogenic strategy of *Bartonella* spp. Int J Med Microbiol 291:555–560.

Tapper, M.L. 2006. Emerging viral diseases and infectious disease risks. Haemophilia 12:S3–7 and 26–28.

Telfer, S., M. Begon, M. Bennett, K.J. Bown, S. Burthe, X. Lambin, G. Telford, and R. Birtles. 2007. Contrasting dynamics of *Bartonella* spp. in cyclic field vole populations: the impact of vector and host dynamics. Parasitology 134:413–425.

Telford, S.R. III, R.J. Pollack, and A. Spielman. 1991. Emerging vector-borne infections. Infect Dis Clin North Am 5:7–17.

Tsukahara, M., H. Tsuneoka, H. Iino, K. Ohno, and I. Murano. 1998. *Bartonella henselae* infection from a dog. Lancet 352:1682.

Valentine, K.H., C.A. Harms, M.B. Cadenas, A.J. Birkenheuer, H.S. Marr, J. Braun-McNeill, R.G. Maggi, and E.B. Breitschwerdt. 2007. *Bartonella* DNA in loggerhead sea turtles. Emerg Infect Dis 13:949–950.

Varanat, M., A. Travis, W. Lee, R.G. Maggi, S.A. Bissett, K.E. Linder, and E.B. Breitschwerdt. 2009. Recurrent osteomyelitis in a cat due to infection with *Bartonella vinsonii subsp. berkhoffii* genotype II. J Vet Intern Med 23:1273–1277.

Vermi, W., F. Facchetti, E. Riboldi, H. Heine, S. Scutera, S. Stornello, D. Ravarino, P. Cappello, M. Giovarelli, R. Badolato, M. Zucca, F. Gentili, M. Chilosi, C. Doglioni, A.N. Ponzi, S. Sozzani, and T. Musso. 2006. Role of dendritic cell-derived CXCL13 in the pathogenesis of *Bartonella henselae* B-rich granuloma. Blood 107:454–462.

Vorou, R.M., V.G. Papavassiliou, and S. Tsiodras. 2007. Emerging zoonoses and vector-borne infections affecting humans in Europe. Epidemiol Infect 135:1231–1247.

Walker, D.H., A.G. Barbour, J.H. Oliver, R.S. Lane, J.S. Dumler, T.D. Dennis, D.H. Persing, A.F. Azad, and E. McSweegan. 1996. Emerging bacterial zoonotic and vector-borne diseases. Ecological and epidemiological factors. J Am Med Assoc 275:463–469.

White, F., and C.J. Hospedales. 1994. Communicable disease control as a Caribbean public health priority. Bull Pan Am Health Organ 28:73–76.

Woolhouse, M.E. 2002. Population biology of emerging and re-emerging pathogens. Trends Microbiol 10:S3–7.

Zinsstag, J., E. Schelling, F. Roth, B. Bonfoh, D. de Savigny, and M. Tanner. 2007. Human benefits of animal interventions for zoonosis control. Emerg Infect Dis 13:527–531.

# 18

## *BRUCELLA CETI* AND *BRUCELLA PINNIPEDIALIS* INFECTIONS IN MARINE MAMMALS

Jacques Godfroid, Ingebjørg Helena Nymo, Morten Tryland, Axel Cloeckaert, Thierry Jauniaux, Adrian M. Whatmore, Edgardo Moreno, and Geoffrey Foster

Marine mammals consist of a diverse group of roughly 120 species, which live in or depend on the ocean and the marine food chain. They include cetaceans (of which there are two suborders: *Mysticeti* [baleen whales] and *Odontoceti* [toothed whales, such as dolphins and porpoises], pinnipeds (true seals, eared seals, and walrus), sirenians (manatees and dugong), polar bear (*Ursus maritimus*), and several species of otters. The polar bear is included because this species spends large parts of the year on the ice around the coastline of the Arctic Ocean, in close association with the marine environment, feeding of its major prey, the ringed seal (*Pusa hispida*) (Born et al. 1997; Stirling 2009). The sea otter (*Enhydra lutris*), native to the coasts of the northern and eastern North Pacific Ocean, is fully aquatic with no association to the terrestrial environment, whereas the marine otter (*Lontra felina*), found in littoral areas of southwestern South America, goes to shore to eat, rest, give birth, and rear pups but also feeds exclusively from the sea, and these species are thus considered to be marine mammals (Miller et al. 2001). In addition, some populations of freshwater otters are almost exclusively marine-living and should also be considered as marine mammals, such as the southern river otter (*Lontra provocax*), the

North American river otter (*Lontra canadensis*), the European otter (*Lutra lutra*), and the African clawless otter (*Aonyx capensis*) (Estes et al. 2009). Cetaceans have great ecological and commercial value, since they are a fundamental part of the food chain and a source for protein and fat for many people around the world (Endo et al. 2005). The presence of marine mammals in the seas and littorals is a significant indicator for ocean health and can be used to gauge the magnitude at which the marine resources are protected. These mammals are also an important tourist attraction in aquariums and littorals (Lloret and Riera 2008). In addition, dolphins are used in therapies (Antonioli and Reveley 2005). One frequent phenomenon that brings people in close contact with these attractive animals is the arrival to the shorelines of disoriented dolphins and whales displaying swimming problems. During recent years, these actions and contacts between marine mammals and humans have increased worldwide (Hernandez-Mora et al. 2008; Lloret and Riera 2008), augmenting the risk of transmission of pathogens from these marine animals to people and possibly terrestrial animals. Within this context, infectious diseases, such as brucellosis, should be taken into consideration in conservation programs.

Based upon what is known about *Brucella* spp. infection of the reproductive organs of some cetaceans (e.g., *Phocoena* spp., *Turciops* spp., and *Stenella* spp.), it is likely that brucellosis can have a negative impact on reproduction and thus efforts at protecting and increasing the genetic diversity in sparse populations, captive collections, or endangered species. For instance, it is worth mentioning that the endangered species Vaquita (*Phocoena sinus*), living in the upper Gulf of Baja California, Mexico, inhabits the same area visited by striped (*Stenella coeruleoalba*) and bottlenose (*Tursiops truncatus*) dolphins (http://www.iobis.org), species that have been demonstrated to exhibit severe clinical brucellosis (Hernandez-Mora et al. 2008; Gonzalez-Barrientos et al. 2010).

The Saimaa ringed seal (*Pusa hispida saimensis*), a subspecies of ringed seal is the most endangered seal species in the world, having a total population of only about 260 individuals. The only existing population of these seals is found in Lake Saimaa, Finland. There are three documented species of monk seals. The Caribbean monk seal (*Monachus tropicalis*) was last sighted in the 1950s and was officially declared extinct in June 2008. The Mediterranean monk seal (*Monarchus monarchus*) is believed to be the world's second-rarest pinniped and one of the most endangered mammals in the world, with only 350 to 450 individuals. In 2010, it was estimated that only 1,100 Hawaiian monk seals (*Monachus schauinslandi*) remain; it is listed as critically endangered. Anti-*Brucella* antibodies have been found in the Hawaiian monk seal (Nielsen et al. 2005; Aguirre et al. 2007). However, as for other seal species, no evidence of gross pathology consistent with clinical brucellosis was noted in any of the seropositive animals tested (Nielsen et al. 2005).

## WHALING AND SEALING

The primary species hunted during modern commercial whaling are the common minke whale (*Balaneoptera acurostrata*) and Antarctic minke whale (*Balaenoptera bonaerensis*), two of the smallest species of baleen whales. The International Whaling Commission (IWC) was set up under the International Convention for the Regulation of Whaling (ICRW) to decide hunting quotas and other relevant matters based on the findings of its Scientific Committee. The IWC voted on July 23, 1982 to establish a moratorium on commercial whaling beginning in the 1985–86 season. Since 1992, the IWC's Scientific Committee has requested that it be allowed to give quotas for some whale stocks, but this has so far been refused by the Plenary Committee (http://iwcoffice.org/). Faroese whaling of long-finned pilot whales (*Globicephala melaena*, actually a species of dolphin) is regulated by Faroese authorities but not by the IWC, which does not regulate the catching of small cetaceans. Modern commercial whaling is done for human food consumption. It is worth noting that *Brucella* spp. have been isolated from minke whales in the Atlantic (Clavareau et al. 1998; Foster et al. 2002) and that *Brucella* DNA has been amplified by PCR from minke whales in the Northern Pacific (Ohishi et al. 2004).

Seal hunting, or sealing, is the personal or commercial hunting of seals. Hunting is currently practiced in five countries: Canada, where most of the world's seal hunting takes place, Namibia, Greenland (Denmark), Norway, and Russia. Some 900,000 seals are hunted each year around the globe, with the commercial hunts in Canada, Greenland, and Namibia accounting for some 60% of the seals killed each year.

Seal skins have been used by aboriginal people for millennia to make waterproof jackets and boots, and seal fur has been used to make fur coats. Pelts account for over half the processed value of a seal. The European Union banned the importation of any seal product in May 2009, with the exception of seal products resulting from hunts that are traditionally conducted by Inuit and other indigenous communities and that contribute to their subsistence. The main commercial seal species in the Northern Hemisphere are the harp seal (*Phoca groenlandica*) and the hooded seal (*Cystophora cristata*). A high prevalence of antibodies to *Brucella* spp. has been found in the hooded seal, which is traditionally consumed by people in northern Norway (Tryland et al. 2005).

## *BRUCELLA CETI* AND *BRUCELLA PINNIPEDIALIS* IN MARINE MAMMALS

*Brucella* spp. are gram-negative, facultative intracellular bacteria that can infect many mammalian species, including humans. Ten species are recognized within the genus *Brucella*: the six "classical" *Brucella* species, some of which include different biovars: *B. abortus* (biovars 1, 2, 3, 4, 5, 6, 7, 9), *B. melitensis* (biovars 1, 2, 3), *B. suis* (biovars 1, 2, 3, 4, 5), *B. ovis*, *B. canis*, and *B. neotomae* (Corbel and Brinley-Morgan 1984; Alton et al. 1988) and the recently described *B. ceti* and *B. pinnipedialis* (Foster et al. 2007), *B. microti* (Scholz et al. 2008), and *B. inopinata* (Scholz et al. 2010).

The classification for the classical species was mainly based on differences in phenotypic characteristics, host preference(s), and pathogenicity. Distinction between species and biovars is currently performed by differential laboratory tests (Corbel and Brinley-Morgan 1984; Alton et al. 1988). The overall characteristics of the marine mammal strains are different from those of any of the six "classical" *Brucella* species (Jahans et al. 1997; Clavareau et al. 1998; Bricker et al. 2000; Cloeckaert et al. 2001), and since 2007, *B. ceti* and *B. pinnipedialis* (infecting preferentially cetaceans and pinnipeds, respectively) have been recognized as new *Brucella* species (Foster et al. 2007).

Since the first description of an abortion due to *Brucella* spp. in a captive dolphin in California in 1994 (Ewalt et al. 1994) and the first isolation of *Brucella* spp. in marine mammals in their natural habitat, reported in 1994 from stranded harbour seals (*Phoca vitulina*), harbour porpoises (*Phocoena phocoena*), and common dolphins (*Delphinus delphis*) on the Scottish coast (Ross et al. 1994), several studies have described the isolation and characterization of *Brucella* spp. from a wide variety of marine mammals, raising both conservation and zoonotic concerns.

*B. ceti* and *B. pinnipedialis* have been isolated from cetaceans (*Mysticeti* and *Odontoceti*), true seals inhabiting seas and oceans of Europe and North and Central America, and from an European otter, thus in animals inhabiting almost all the seas covering the globe except Antarctic waters (Ross et al. 1994; Foster et al. 1996; Ross et al. 1996; Garner et al. 1997; Jahans et al. 1997; Clavareau et al. 1998; Miller et al. 1999; Forbes et al. 2000; Maratea et al. 2003; Watson et al. 2003; Tryland et al. 2005; Dawson et al. 2006; Muñoz et al. 2006; Dagleish et al. 2007, 2008; Prenger-Berninghoff et al. 2008; Hernandez-Mora et al. 2008; Davison et al. 2009; Gonzalez-Barrientos et al. 2010; Jauniaux et al. 2010). *Brucella* spp. DNA has also been

isolated from common minke whale in the western North Pacific (Ohishi et al. 2004).

Anti-*Brucella* antibodies have since been detected in serum samples from several species of marine mammals from the Northern and Southern Hemispheres (Nielsen et al. 1996; Jepson et al. 1997; Tryland et al. 1999; Nielsen et al. 2001; Van Bressem et al. 2001; Hanni et al. 2003; Ohishi et al. 2003; Burek et al. 2005; Dawson 2005; Nielsen et al. 2005; Rah et al. 2005; Munoz et al. 2006; Tachibana et al. 2006; Zarnke et al. 2006; Aguirre et al. 2007; Hernandez-Mora et al. 2009;,Gonzalez-Barrientos et al. 2010). Although no *Brucella* spp. strain has been isolated from marine mammals in Antarctic waters, anti-*Brucella* antibodies have been identified (Retamal et al. 2000; Blank et al. 2002). No anti-*Brucella* antibodies were detected in marine mammals in New Zealand (Mackereth et al. 2005). Recently, a high prevalence of anti-*Brucella* antibody titers was documented in Australian fur seal (*Arctocephalus pusillus doriferus*). Inflammatory lesions suggestive of infectious agents were found in 14 of 39 (36%) aborted Australian fur seal pups. Anti-*Brucella* antibodies in pregnant females had no influence on the likelihood of abortion. No *Brucella* spp. was isolated and no positive *Brucella* PCR results were obtained from fresh or frozen tissues collected from fetuses, juveniles, or adults (Lynch et al. 2011). These results suggest that *Brucella* spp. was not the cause of abortion and that cross-reactive bacteria may have induced false-positive results in brucellosis serological tests as suggested earlier (Aguirre et al. 2007).

The polar bear is the apex predator in the Arctic marine food web, and in the Svalbard area ringed seals, bearded seals (*Erignathus barbatus*), and harp seals are the main prey. Anti-*Brucella* antibodies were found in ringed seals and harp seals in Svalbard (Tryland et al. 1999). A seroprevalence of 5.4% of anti-*Brucella* antibodies was found in serum samples from 297 polar bears from Svalbard and the Barents Sea (Tryland et al. 2001). Antibodies have also been found in polar bears from Alaska (Rah et al. 2005). To date, there is no indication of disease caused by *Brucella* spp. in polar bear populations.

In terrestrial mammals, horizontal transmission usually takes place through direct or indirect contact with aborted material, most often through ingestion but also through respiratory exposure (aerosols), conjunctival inoculation, udder inoculation during milking, and contamination of damaged skin or mucosal membranes. Mating and lactation also pose a transmission risk (Corbel 2006). *Brucella* spp. generally does not multiply outside the host (apart from *B. microti*) but can persist in the environment for long periods of time, depending on the conditions.

It is not known to what extent these characteristics are also valid for marine mammal *Brucella* infections. Some species of sea mammals are social animals often found in large groups where there is ample opportunity for transmission (e.g., seal haul-out sites). Some other species are largely solitary animals, coming together only infrequently primarily for mating (venereal transmission) and giving birth, thereby creating fewer opportunities for transmission.

Ewalt et al. (1994) documented that *Brucella* spp. isolated from an aborted bottlenose dolphin fetus may indicate the cause of abortion. In 1999, it was reported that two bottlenose dolphins aborted fetuses died as a result of *Brucella* spp. infection at the Space and Naval Warfare Systems Center, San Diego, California. Placentitis occurred in both cases (Miller et al. 1999). The authors suggested that dolphin brucellosis is a naturally occurring disease that can have an adverse impact on reproduction in cetaceans and may thus play an important role in the population dynamics of these species. However, to date abortion has not been reported in cetaceans in their natural habitat, although the isolation of *Brucella* spp. from milk, fetal tissues, and secretions in a stranded striped dolphin (*Stenella coeruleoalba*) has been described in Costa Rica (Hernandez-Mora et al. 2008).

Garner et al. (1997) demonstrated *Brucella* spp. in *Parafilaroides* spp. in the lung of a Pacific harbour seal and suggested that transmission in pinnipeds may occur by infected lungworms. This hypothesis was also suggested following the description of *Brucella* spp. infection within the uterine tissue of lung nematodes *Pseudalius inflexus* collected from the lungs of a stranded juvenile male harbour porpoise in Cornwall, United Kingdom (Dawson et al. 2008). Lastly, the presence of *Brucella* spp. was demonstrated by electron microscopy in tattoo-like lesions in a stranded porpoise in Belgium. *Brucella* spp. was cultured and identified as *B. ceti* (Jauniaux et al. 2010). The interested reader is directed towards a recently published update on marine mammal brucellosis with special emphasis on *Brucella pinnipedialis* infections in hooded seals (Nymo et al. 2011).

## *BRUCELLA CETI*- AND *BRUCELLA PINNIPEDIALIS*-INDUCED PATHOLOGY

Brucellosis in terrestrial animals is clinically characterized by one or more of the following clinical signs: abortion, retained placenta, orchitis, epididymitis, and excretion of the organisms in uterine discharge and in milk (Godfroid et al. 2005). It is important to note that pathology induced by *Brucella* spp. is different in cetaceans versus seals. As a rule, no gross pathology has been associated with *B. pinnipedialis* infections in seals, whereas different acute and chronic pathological changes have been associated with *B. ceti* infection in whales both in *Odontoceti* and *Mysticeti*.

No gross pathology was documented in stranded or by-caught seals in Scotland (Foster et al. 2002), although *Brucella* spp. has been isolated from the testes of a grey seal without any associated pathology (Foster et al. 1996). *Brucella* spp. was isolated from the spleen, gastric lymph node, and colorectal lymph node of one stranded, dead, adult hooded seal from the coast of Scotland without any signs of pathology (Foster et al. 1996). In Norway, during scientific sealing operations, hooded seals were sampled and investigated for brucellosis. Despite the high seroprevalence rates (i.e., 35%, n = 48/137 [Tryland et al. 1999] and 31%, n = 9/29 [Tryland et al. 2005]) and the high number of bacteriological positive animals (i.e., 38%, n = 11/29 [Tryland et al. 2005]) recorded for the hooded seals in the Greenland Sea population, no gross pathological changes have been seen in association with the organism. These results suggest that there is a persistent *B. pinnipedialis* bacteremia and that limited pathology and immune responses are induced in hooded seals. Sampling occurred in May–June, after the pupping season, so that the potential abortifacient effect of *B. pinnipedialis* could not be assessed. Moreover, since embryonic diapause (i.e., the blastocyst does not immediately implant in the uterus, but is maintained in a state of dormancy and no development takes place as long as the embryo remains unattached to the uterine lining) occurs in seals, no fetus could be sampled in order to measure early *B. pinnipedialis* infection of the pregnant uterus. The prevalence of seropositive hooded seals in the Northwest Atlantic population (4.9%) is much lower, and no decline in this hooded seal population was observed (Nielsen et al. 2001).

The gross pathology in cetaceans is associated with skin lesions, sub-blubber abscessation, hepatic and splenic necrosis, macrophage infiltration in liver and spleen, epididymitis, spinal discospondylitis, meningitis, lymphadenitis, and mastitis. Neurological signs linked to *Brucella* infections have been associated with primary standings of cetaceans. Indeed, *B. ceti* has been isolated from the brain and cerebrospinal fluid of harbour porpoises, a white-beaked dolphin, a white-sided dolphin, and, for the most part, stranded striped dolphins. A chronic, non-suppurative meningoencephalitis was found in three young striped dolphins (Gonzalez et al. 2002). *B. ceti* was isolated from the mammary gland of sperm whales (*Physeter macrocephalus*) and dolphins, suggesting parasitism of resident macrophages in these glands (Foster et al. 2002), as in the case of terrestrial mammals. In another report, a minke whale from the western North Pacific, displaying *Brucella*-positive serology, showed several nodular granulomatous lesions in the uterine endometrium (Ohishi et al. 2003). These lesions presented significant mononuclear infiltration and had epithelioid and giant cells, suggesting *Brucella*-associated pathology. *B. ceti* was also isolated from a diseased atlanto-occipital joint of an Atlantic white-sided dolphin (*Lagenorhynchus acutus*) (Dagleish et al. 2007) and in the testes of a harbour porpoise (Dagleish et al. 2008). Suppurative granulomatous lesions in both female and male reproductive organs have been observed in minke whales and Bryde's whales (*Balaenoptera brydei*) that had anti-*Brucella* spp. antibodies (Ohishi et al. 2003). *B. ceti* has also been isolated from the uterus of a striped dolphin without any associated pathology (Muñoz et al. 2006). Notwithstanding these reports, there is currently no information on the occurrence of *B. ceti* abortion in cetaceans in their natural habitat.

In one conspicuous case of brucellosis in a pregnant striped dolphin (Hernandez-Mora et al. 2008; Gonzalez-Barrientos et al. 2010), the bacteria was isolated and directly observed by immunofluorescence in placenta, umbilical cord, milk, allantoic and amniotic fluids as well as in multiple fetal organs. In this case, a necrotizing severe placentitis with multiple necrotic foci and a dead fetus close to seven-month gestation was found. Neurobrucellosis in stranded striped dolphins has been described. The lesions presented hyperemic vessels and mononuclear cell infiltrates (Fig. 18.1) (Hernandez-Mora et al. 2008).

By transmission electron microscopy, large numbers of relatively small (diameter 380 to 450 nm) intracellular coccoid bacteria that suggested *Brucella* spp. were observed in a genital ulcer in a harbour porpoise. *B. ceti* was also found in a longitudinal ulcer between flippers of the same animal (Fig. 18.2) (Jauniaux et al. 2010).

## LABORATORY DIAGNOSTICS

Brucellosis does not present pathognomonic lesions. Diagnosis depends partly on clinical investigations but mainly on laboratory testing. Laboratory diagnosis includes indirect tests that can be applied to serum as well as direct tests (classical bacteriology, PCR-based methods). Only the isolation of *Brucella* spp. (or *Brucella* spp. DNA detection) allows definite confirmation. Several techniques are available to identify *Brucella* spp. The Stamp staining is still often used, and even if this technique is not specific—other abortive agents such as *Chlamydophila abortus* (formerly *Chlamydia psittaci*) or *Coxiella burnetii* are also stained—it provides valuable information for the analysis of abortive material (Alton et al. 1988).

Bacterial isolation is nevertheless always preferable and even required for the typing of the strain. For the definitive diagnosis of brucellosis, the choice of samples depends on the observed clinical signs. For the isolation of *Brucella* spp., the most commonly used medium is the Farrell medium (FM), which contains antibiotics able to inhibit the growth of other bacteria present in clinical samples. While the majority of cetacean isolates will appear on FM after 4 days of incubation, seal isolates will often be recovered on FM only at about 10 days. It is therefore recommended that the incubation period is extended to 14 days before cultures are discarded as negative. Most cetacean isolates will grow in the absence of an increased $CO_2$ concentration, whereas most seal isolates require $CO_2$ for growth. It is therefore recommended that all primary cultures be incubated in 10% carbon dioxide at 37°C (Foster et al. 2002). The identification and typing of *Brucella* spp. is done by analysis of morphology, staining, control of the biochemical profile (catalase, oxidase, and urease), anti-polysaccharide "O" chain (O-LPS) specific for the A or M epitopes, the lysis by phages, the dependence on $CO_2$ for growth, production of $H_2S$, growth in the presence of basal

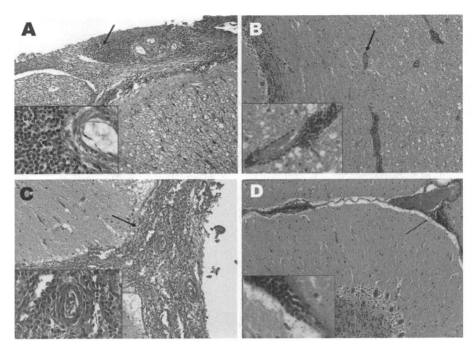

**Figure 18.1:**
Main histopathologic finding of neurobrucellosis in stranded striped dolphin (*Stenella coeruleoalba*). A. Mononuclear infiltrates in the meninges (*arrow*) surrounding the spinal cord. B. Mononuclear infiltrate around vessels (*arrow*) of the cerebellum. C. Mononuclear infiltrate (*arrow*) in the meninges around the brain. D. Hyperemic vessels and mononuclear cell infiltrate in the meninges around the cerebellum (*arrow*). The insets correspond to amplified sections of each figure demonstrating the mononuclear cell infiltrate (Hernandez-Mora et al. 2008).

fuchsine or thionin, and the crystal violet or acryflavin tests (Corbel and Brinley-Morgan 1984; Alton et al. 1988).

Several PCR-based methods have been developed. The best-validated methods are based on the detection of specific sequences of *Brucella* spp. such as the 16S-23S genes, the IS711 insertion sequence (which has so far been detected only in *Brucella* spp.), or the *bcsp31* gene encoding for a 31Kda protein (Halling et al. 1993; Baddour and Alkhalifa, 2008). New PCR techniques allowing the identification and sometimes a quick typing of *Brucella* spp. have been developed and are currently implemented in certain diagnostic laboratories (Bricker and Ewalt 2005; Le Fleche et al. 2006; Lopez-Goni et al. 2008; Maquart et al. 2009a; Whatmore 2009).

New techniques such as single nucleotide polymorphism signatures (SNPs, aiming at detecting DNA sequence variation occurring when a single nucleotide in the genome differs between members of a species),

multiple locus sequence analysis (MLSA, aiming at directly measuring the DNA sequence variations in a set of housekeeping genes and characterizing strains by their unique allelic profiles), and multiple locus variability analysis (MLVA, aiming at analyzing the variability of loci presenting repeated sequences) are currently used for the typing of marine mammal *Brucella* spp. (Le Fleche et al. 2006; Maquart et al. 2009a; Whatmore 2009).

The earliest molecular studies related to marine mammal *Brucella* strains in the late 1990s confirmed their distinction from classical species associated with terrestrial mammals (Clavareau et al. 1998). A marker specific for the marine mammal strains was identified when amplification of the gene encoding the immunodominant BP26 protein revealed a larger-than-expected PCR product reflecting the insertion of an IS711 element downstream of the gene (Cloeckaert et al. 2000). A PCR based on this IS711 element downstream of the *bp26* gene and its flanking regions

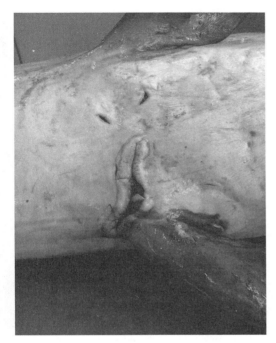

**Figure 18.2:**
Longitudinal ulcer between flippers of a harbour porpoise (*Phocoena phocoena*) with *B. ceti* infection (Jauniaux et al. 2010).

has become a well-used test for differentiation of *Brucella* spp. associated with marine mammals from classical species associated with terrestrial mammals. Following molecular characterization of the *omp2* genes, a division into two species (*B. pinnipediae* and *B. cetaceae*), compatible with the classical criteria of host preference and DNA polymorphism at the omp2 locus, was suggested (Cloeckaert et al. 2001). Eventually, two new *Brucella* species labeled (with corrected etymology) as *B. ceti* for isolates from cetaceans and *B. pinnipedialis* for isolates from pinnipeds were validly published (Foster et al. 2007). This was in line with the decision of the *Brucella* Taxonomic Subcommittee (Osterman and Moriyon 2006) and would allow the prospective inclusion of biovars within these two species. Further, MLSA studies suggested that *Brucella* strains from marine mammals corresponded to a cluster of five sequence types (STs) distinct from all previously described *Brucella* species from terrestrial mammals (Whatmore et al. 2007). The first large-scale application of both MLVA and MLSA techniques specifically to the marine mammal

*Brucella* group was published in 2007 and, examining over 70 isolates, described the clear existence of three groups with distinct host preferences (Groussaud et al. 2007). Recently the largest study to date examined 294 isolates from 173 marine mammals by MLVA (Maquart et al. 2009a). More than 100 genotypes were identified and divided into five clusters that related to previous MLSA findings. On the basis of emerging data, the taxonomic descriptions of marine mammal *Brucella* may need to be reconsidered in the future (Whatmore 2009).

Taxonomic classification of *Brucella* spp. is very often made difficult by the lack of, or high degree of similarity in, the marker genes traditionally used for this (Foster et al. 2009). Recently such methods were used to compare 32 sequenced genomes from the *Brucella* genus, representing the six classical species, *B. ceti*, and *B. pinnipedialis* (Bohlin et al. 2010). The findings were remarkably consistent with the current taxonomy, indicating that phylogenetic classification of *Brucella* spp. based on MLSA and marker genes (Whatmore et al. 2007) shows a surprising similarity with the actual whole gene content of the *Brucella* organism (Bohlin et al. 2010).

Brucellosis serology in marine mammals is usually performed using the same antigens as in domestic ruminant serology because the *Brucella* immunodominant antigens are associated with the surface "smooth" lipopolysaccharide (LPS) and are to a large extent shared by all the naturally occurring strains of *B. abortus*, *B. melitensis*, *B. suis*, *B. microti*, *B. ceti*, *B. pinnipedialis*, and *B. inopinata*. According to their reactivity, three different type of immunochemical techniques have been used: (i) direct serological assays, such as agglutination tests (Rose Bengal, RB test) and fluorescence polarization method (FPA), in which the antibodies modified the physical properties of the antigen, a phenomenon that is visually or photometrically recorded in a short period of time; (ii) displacing methods, such as competitive ELISA (cELISA), in which the antibodies have to compete with monoclonal antibodies directed against the main epitope associated with the O-chain; and finally (iii) indirect serological assays, mostly designed to detect anti-LPS antibodies, such as protein G-ELISA (gELISA), protein A-ELISA (aELISA), recombinant protein G/A-ELISA (g/aELISA), antibody-ELISA, using species-specific anti-IgG conjugates (iELISA), Western blot (WB), dot blot (DB), immunofluorescence (IF), and complement fixation (CF). Indirect ELISAs rely on species-specific reagents that are not commercially available. This limitation of the lack of polyclonal or monoclonal antibodies to many wildlife species' immunoglobulins can be partly overcome by the use of either protein A or protein G conjugates (Nielsen et al. 2004). Other techniques, like competitive ELISAs or the fluorescent polarization assay, which do not rely on species-specific reagents, have been proven useful in marine mammals (Nielsen et al. 2005). In cetaceans, a broad cross-reaction among immunoglobulins of different *Odontoceti* families has been documented, allowing the use of antiserum raised against one species of dolphin as a general reagent for detecting antibodies against different species of this suborder (Hernandez-Mora et al. 2009).

## MARINE MAMMAL *BRUCELLA* SPP. IN LIVESTOCK AND FISH

An experimental inoculation of three pregnant cattle with a *Brucella* spp. isolated from a Pacific harbour seal resulted in two of the animals aborting. This study indicated that marine mammal *Brucella* spp. is capable of producing antibodies and abortion in cattle but is less pathogenic than *B. abortus* (Rhyan et al. 2001). Another experimental investigation demonstrated colonization, limited establishment of infection, transmission, and low pathogenicity of the three marine mammal *Brucella* spp. strains in pregnant sheep (Perrett et al. 2004). Lastly, ten weaned piglets were challenged by the oral and ocular routes with a human *Brucella* spp. strain (02/611) that was isolated from a patient with spinal osteomyelitis (McDonald et al. 2006) and is closely related to a *Brucella* spp. originating from a bottlenose dolphin from the United States (Sohn et al. 2003; McDonald et al. 2006; Whatmore et al. 2008). Low and transient antibody titers were detected in only three pigs, two of which were culture-negative. Thus, *Brucella* spp. strain 02/611 does not seem to replicate readily in pigs, and it is unlikely that pigs are maintenance hosts for these *Brucella* spp. (Bingham et al. 2008).

Brucella spp. was not known to infect poikilotherms until recently. the Nile catfish (*Clarias gariepinus*) has been experimentally infected with *B. melitensis* biovar 3. The fish seroconverted and *B. melitensis* was isolated from internal organs, but the bacterium was not transferred to noninfected sentinel fishes (Salem and Mohsen 1997). Nile catfish captured from Nile canals were shown to be seropositive for *Brucella* spp. by Rose Bengal and Rivanol tests. Further, *B. melitensis* biovar 3 was cultured from skin swabs, and PCR confirmed the identity of the bacterium (El-Tras et al. 2010). These findings suggest that fish are susceptible to *Brucella* spp. infection and thus may also be susceptible to marine *B. ceti* and *B. pinnipedialis*. If infection with marine mammal *Brucella* spp. is proven to occur in fish, this would have a tremendous economic impact on the fish industry and significant veterinary public health implications, given the potential zoonotic concern of these *Brucella* species. This clearly warrants further investigation.

## ZOONOTIC CONSIDERATIONS

Today, brucellosis in humans is mainly occupational (abattoir, animal industry, hunters, and health workers). Symptoms like undulant fever, tiredness, night sweats, headaches, and chills may drag on as long as three months before the illness becomes severe and debilitating enough to require medical attention (Godfroid et al. 2005). Zoonotic concerns regarding marine mammal strains were initially raised following the recovery of a cetacean strain of *Brucella* spp. from a laboratory worker at the Central Veterinary Laboratory, Weybridge, United Kingdom, who had seroconverted after suffering from headaches, lassitude, and severe sinusitis (Brew et al. 1999). People at risk of zoonotic transfer of marine mammal brucellosis are individuals in traditional communities where products from whales and seals are still an important part of the diet. Also people with only occasional consumption of marine mammal meat, people handling stranded marine mammals, whale and seal hunters, people handling products from marine mammals, people in contact with raw products from the ocean, veterinary meat inspectors, and researchers could be exposed. Because of the unspecific and varied symptoms of human brucellosis and the very recent awareness of the existence of marine mammal brucellosis, transfer from marine mammals to humans could pass unrecognized.

In April 2003, the first report of community-acquired human infections with marine mammal-associated *Brucella* spp. was published. The authors described the identification of these strains in two patients with neurobrucellosis and intracerebral granulomas. Despite a more than 15-year separation, these cases have similarities: both patients were from Peru and denied significant exposure to marine mammals (Sohn et al. 2003). In 2006, the isolation and characterization of a marine *Brucella* from a New Zealand patient was reported (McDonald et al. 2006). It was suggested that all three reported cases of natural human infection associated with *Brucella* spp. from marine mammals were associated with ST27 (Whatmore et al. 2008). Unfortunately the natural host of ST27 (first isolated from a captive dolphin in the United States) has not been identified, although there is molecular evidence of the presence of this genotype in minke whale from the Northern Pacific Ocean (Ohishi et al. 2004).

Norwegians have a long tradition of consumption of meat from harp seals, hooded seals, and minke whales, all of which have been found to be infected with *Brucella* spp. Native subsistence hunting also accounts each year for about 4,000 to 6,000 seals in Greenland, and up to 100 seals are hunted by natives in northern Canada. In 2000, it has been estimated that Alaska natives harvested 2,229 harbour seals and 205 sea lions/fur seals. For 2008 through 2010, the annual subsistence needs were 1,645 to 2,000 seals on St. Paul and 300 to 500 seals on St. George (the Pribilof Islands). In spite of this, brucellosis has not been reported in humans at risk. Marine mammal *Brucella* spp. isolates were tested for their ability to infect human and murine macrophage cells. The study showed that some *B. ceti* and *B. pinnipedialis* isolates were found to be virulent in these models of infection, whereas other isolates were not. In fact, all the *B. pinnipedialis* isolated from hooded seals did not demonstrate ability to infect human and murine macrophage cells (Maquart et al. 2009b), which may explain the absence of records of human infection with hooded seal *B. pinnipedialis*.

## SIGNIFICANCE AND IMPLICATIONS FOR CONSERVATION

Several of the cetacean and seal species diagnosed with brucellosis are listed in the IUCN Red List of threatened species (World Conservation Union, http://www.iucnredlist.org/search). In spite of this, the level of endemism of cetacean species in the

Central American littorals generally is not estimated for their protection or epidemiological surveillance. In this sense, it would be desirable that future conservation and management efforts would focus on whales and dolphin species that occupy neritic waters, where human activities are most intense and more likely to affect their populations, and promote the spreading of infectious diseases. Indeed, practices such as littoral pollution, microorganism contamination, and fishing and hunting, which jeopardize the food resources of the cetaceans, may promote malnutrition and competition and clustering of different species in reduced areas where food is available. These phenomena could favor the number of susceptible animals and increase the transmission of *B. ceti* within and between species of cetaceans. It has been shown that the pup production of the Greenland Sea hooded seal has decreased substantially since the 1950s and has stabilized at a low level since the 1970s, despite reduced hunting. Population fertility is one important parameter that varies in response to environmental changes, but other factors, like infections, may also be contributing factors. Although it is not known if *B. pinnipedialis* induces abortion in hooded seals, its importance in inducing reproductive failure should be investigated. Some *B. ceti* and *B. pinnipedialis* strains seem to be well adapted to their preferential marine mammal hosts, which could serve as the primary reservoir hosts. This may be the case of porpoises, in which anti-*Brucella* antibodies are relatively frequent, but pathology is limited to a few cases. On the contrary, some cetaceans, such as striped dolphins, may be highly susceptible to brucellosis, as demonstrated by the number of fatal cases recorded in different latitudes of the world. Alternatively, some strains of *B. ceti* and *B. pinnipedialis* are more virulent than others, as suggested by some limited *in vitro* experiments (Maquart et al. 2009b). In any case, these conjectures remain open questions until more *Brucella*-related pathologies are documented in cetaceans and pinnipeds. Lastly, there are very few data on the transmission of *Brucella* spp. in marine mammals, and the role of fish as reservoirs has not yet been investigated.

## REFERENCES

Aguirre, A.A., T.J. Keefe, J.S. Reif, L. Kashinsky, P.K. Yochem, J.T. Saliki, J.L. Stott, T. Goldstein,

J.P. Dubey, R. Braun, and G. Antonelis. 2007. Infectious disease monitoring of the endangered Hawaiian monk seal. J Wildl Dis 43:229–241.

Alton, G.G., L.M. Jones, R.D. Angus, and R. Saint-Louis. 1988. Techniques for the brucellosis laboratory. Institut National de la Recherche Agronomique, Paris, France.

Antonioli, C., and M.A. Reveley. 2005. Randomised controlled trial of animal facilitated therapy with dolphins in the treatment of depression. Br Med J 331:1231–1234.

Baddour, M.M., and D.H. Alkhalifa. 2008. Evaluation of three polymerase chain reaction techniques for detection of Brucella DNA in peripheral human blood. Can J Microbiol 54:352–357.

Bingham, J., T.K. Taylor, J.E. Swingler, G. Meehan, D.J. Middleton, G.F. Mackereth, J.S. O'Keefe, and P.W. Daniels. 2008. Infection trials in pigs with a human isolate of Brucella (isolate 02/611 "marine mammal type"). New Zeal Vet J 56:10–14.

Blank, O., P. Retamal, P. Abalos, and D. Torres. 2002. Detection of anti-Brucella antibodies in Weddell seals (Leptonychotes weddellii) from Cape Shirreff, Antarctica. Arch Med Vet 34:117–122.

Bohlin, J., L. Snipen, A. Cloeckaert, K. Lagesen, D. Ussery, A.B. Kristoffersen, and J. Godfroid. 2010. Genomic comparisons of Brucella spp. and closely related bacteria using base compositional and proteome based methods. BMC Evol Biol 10:249.

Born, E.W., Ï. Wiig, and J. Thomassen. 1997. Seasonal and annual movements of radio-collared polar bears (Ursus maritimus) in northeast Greenland. J Mar Syst 10:67–77.

Brew, S.D., L.L. Perrett, J.A. Stack, A.P. MacMillan, and N.J. Staunton. 1999. Human exposure to Brucella recovered from a sea mammal. Vet Rec 144:483.

Bricker, B.J., D.R. Ewalt, A.P. MacMillan, G. Foster, and S. Brew. 2000. Molecular characterization of Brucella strains isolated from marine mammals. J Clin Microbiol 38:1258–1262.

Bricker, B., and D. Ewalt. 2005. Evaluation of the HOOF-Print assay for typing Brucella abortus strains isolated from cattle in the United States: results with four performance criteria. BMC Microbiol 5:37.

Burek, K.A., F.M.D. Gulland, G. Sheffield, K.B. Beckmen, E. Keyes, T.R. Spraker, A.W. Smith, D.E. Skilling, J.F. Evermann, J.L. Stott, J.T. Saliki, and A.W. Trites. 2005. Infectious disease and the decline of Steller sea lions (Eumetopias jubatus) in Alaska, USA: Insights from serologic data. J Wildl Dis 41:512–524.

Clavareau, C., V. Wellemans, K. Walravens, M. Tryland, J.M. Verger, M. Grayon, A. Cloeckaert, J. J. Letesson, and J. Godfroid. 1998. Phenotypic and molecular characterization of a Brucella strain isolated from a minke whale (Balaenoptera acutorostrata). Microbiol-Sgm 144:3267–3273.

Cloeckaert, A., M. Grayon, and O. Grepinet. 2000. An IS711 element downstream of the bp26 gene is a specific marker of Brucella spp. isolated from marine mammals. Clin Diag Lab Immunol 7:835–839.

Cloeckaert, A., J.M. Verger, M. Grayon, J.Y. Paquet, B. Garin-Bastuji, G. Foster, and J. Godfroid. 2001. Classification of Brucella spp. isolated from marine mammals by DNA polymorphism at the omp2 locus. Microbes Infect 3:729–738.

Corbel M.J., Elberg, S.S., and Cosivi, O. 2006. Brucellosis in humans and animals. Geneva, World Health Organization, Switzerland.

Corbel, M.J., and W.J. Brinley-Morgan. 1984. Genus Brucella Meyer and Shaw 1920. In N. R. Krieg and J. G. Hold, eds. Bergey manual of systematic bacteriology, Vol. 1, pp. 377–388. Williams and Wilkins, Baltimore.

Dagleish, M.P., J. Barley, J. Finlayson, R.J. Reid, and G. Foster. 2008. Brucella ceti associated pathology in the testicle of a harbour porpoise (Phocoena phocoena). J Comp Pathol 139:54–59.

Dagleish, M.P., J. Barley, F.E. Howie, R.J. Reid, J. Herman, and G. Foster. 2007. Isolation of Brucella species from a diseased atlanto-occipital joint of an Atlantic white-sided dolphin (Lagenorhynchus acutus). Vet Rec 160:876–878.

Davison, N.J., M.P. Cranwell, L.L. Perrett, C.E. Dawson, R. Deaville, E.J. Stubberfield, D.S. Jarvis, and P.D. Jepson. 2009. Meningoencephalitis associated with Brucella species in a live-stranded striped dolphin (Stenella coeruleoalba) in south-west England. Vet Rec 165:86–89.

Dawson, C.E. 2005. Anti-Brucella antibodies in pinnipeds of Australia. Microbiol Aust 26:87–89.

Dawson, C.E., L.L. Perrett, E.J. Young, N.J. Davison, and R.J. Monies. 2006. Isolation of Brucella species from a bottlenosed dolphin (Tursiops truncatus). Vet Rec 158:831–832.

Dawson, C.E., E.J. Stubberfield, L.L. Perrett, A.C. King, A.M. Whatmore, J.B. Bashiruddin, J.A. Stack, and A.P. MacMillan. 2008. Phenotypic and molecular characterisation of Brucella isolates from marine mammals. BMC Microbiol 8:224.

El-Tras, W.F., A.A. Tayel, M.M. Eltholth, and J. Guitian. 2010. Brucella infection in fresh water fish: Evidence for natural infection of Nile catfish, Clarias gariepinus, with Brucella melitensis. Vet Microbiol 141:321–325.

Endo, T., Y. Hotta, K. Haraguchi, and M. Sakata. 2005. Distribution and toxicity of mercury in rats after oral administration of mercury-contaminated whale red meat marketed for human consumption. Chemosphere 61:1069–1073.

Estes, J.A., J.L. Bodkin, and M. Ben-David. 2009. Otters, marine. *In* W. F. Perrin, B. Wursig, and J. G. M. Thewissen, eds. Encyclopedia of marine mammals, pp. 807–816. Elsevier Inc., Academic Press, New York.

Ewalt, D.R., J.B. Payeur, B.M. Martin, D.R. Cummins, and W.G. Miller. 1994. Characteristics of a *Brucella* species from a bottlenose dolphin (*Tursiops truncatus*). J Vet Diag Inv 6:448–452.

Forbes, L.B., O. Nielsen, L. Measures, and D.R. Ewalt. 2000. Brucellosis in ringed seals and harp seals from Canada. J Wildl Dis 36:595–598.

Foster, G., K.L. Jahans, R.J. Reid, and H.M. Ross. 1996. Isolation of *Brucella* species from cetaceans, seals and an otter. Vet Rec 138:583–586.

Foster, G., A. P. MacMillan, J. Godfroid, F. Howie, H. M. Ross, A. Cloeckaert, R. J. Reid, S. Brew, and I. A. P. Patterson. 2002. A review of *Brucella* sp infection of sea mammals with particular emphasis on isolates from Scotland. Veterinary Microbiology 90:563–580.

Foster, G., B.S. Osterman, J. Godfroid, I. Jacques, and A. Cloeckaert. 2007. *Brucella ceti* sp. nov and *Brucella pinnipedialis* sp. nov. for *Brucella* strains with cetaceans and seals as their preferred hosts. Int J Syst Evol Microbiol 57:2688–2693.

Foster, J.T., S.M. Beckstrom-Sternberg, T. Pearson, J. S. Beckstrom-Sternberg, P.S.G. Chain, F.F. Roberto, J. Hnath, T. Brettin, and P. Keim. 2009. Whole-genome-based phylogeny and divergence of the genus *Brucella*. J Bacteriol 191:2864–2870.

Garner, M.M., D.M. Lambourn, S.J. Jeffries, P.B. Hall, J.C. Rhyan, D.R. Ewalt, L.M. Polzin, and N.F. Cheville. 1997. Evidence of *Brucella* infection in *Parafilaroides* lungworms in a Pacific harbor seal (*Phoca vitulina richardsi*). J Vet Diag Invest 9:298–303.

Godfroid, J., A. Cloeckaert, J.P. Liautard, S. Kohler, D. Fretin, K. Walravens, B. Garin-Bastuji, and J.J. Letesson. 2005. From the discovery of the Malta fever's agent to the discovery of a marine mammal reservoir, brucellosis has continuously been a re-emerging zoonosis. Vet Res 36:313–326.

Gonzalez, L., I.A. Patterson, R.J. Reid, G. Foster, M. Barberan, J.M. Blasco, S. Kennedy, F.E. Howie, J. Godroid, A.P. MacMillan, A. Schock, and D. Buxton. 2002. Chronic meningoencephalitis associated with *Brucella* sp. infection in live-stranded striped dolphins (*Stenella coeruleoalba*). J Comp Pathol 126:147–152.

Gonzalez-Barrientos, R., J.A. Morales, G. Hernandez-Mora, E. Barquero-Calvo, C. Guzman-Verri, E. Chaves-Olarte, and E. Moreno. 2010. Pathology of striped dolphins (*Stenella coeruleoalba*) infected with *Brucella ceti*. J Comp Pathol 142:347–352.

Groussaud, P., S.J. Shankster, M.S. Koylass, and A.M. Whatmore. 2007. Molecular typing divides marine mammal strains of *Brucella* into at least three groups with distinct host preferences. J Med Microbiol 56:1512–1518.

Halling, S.M., F.M. Tatum, and B.J. Bricker. 1993. Sequence and characterization of an insertion-sequence, IS711, from *Brucella ovis*. Gene 133:123–127.

Hanni, K.D., J. A.K. Mazet, F.M.D. Gulland, J. Estes, M. Staedler, M.J. Murray, M. Miller, and D.A. Jessup. 2003. Clinical pathology and assessment of pathogen exposure in southern and Alaskan sea otters. J Wildl Dis 39:837–850.

Hernandez-Mora, G., R. Gonzalez-Barrientos, J.A. Morales, E. Chaves-Olarte, C. Guzman-Verri, E. Baquero-Calvo, M.J. De-Miguel, C.M. Marin, J.M. Blasco, and E. Moreno. 2008. Neurobrucellosis in stranded dolphins, Costa Rica. Emerg Infect Dis 14:1825.

Hernandez-Mora, G., C.A. Manire, R. Gonzalez-Barrientos, E. Barquero-Calvo, C. Guzman-Verri, L. Staggs, R. Thompson, E. Chaves-Olarte, and E. Moreno. 2009. Serological diagnosis of *Brucella* infections in odontocetes. Clin Vaccine Immunol 16:906–915.

Jahans, K.L., G. Foster, and E.S. Broughton. 1997. The characterisation of *Brucella* strains isolated from marine mammals. Vet Microbiol 57:373–382.

Jauniaux, T.P., C. Brenez, D. Fretin, J. Godfroid, J. Haelters, T. Jacques, F. Kerckhof, J. Mast, M. Sarlet, and F.L. Coignoul. 2010. *Brucella ceti* infection in harbor porpoise. Emerg Infect Dis 16:1966–1968.

Jepson, P.D., S. Brew, A.P. MacMillan, J.R. Baker, J. Barnett, J.K. Kirkwood, T. Kuiken, I. R. Robinson, and V.R. Simpson. 1997. Antibodies to *Brucella* in marine mammals around the coast of England and Wales. Vet Rec 141:513–515.

Le Fleche, P., I. Jacques, M. Grayon, S. Al Dahouk, P. Bouchon, F. Denoeud, K. Nockler, H. Neubauer, L.A. Guilloteau, and G. Vergnaud. 2006. Evaluation and selection of tandem repeat loci for a *Brucella* MLVA typing assay. BMC Microbiol 6:9.

Lloret, J. and V.R. Riera. 2008. Evolution of a Mediterranean coastal zone: human impacts on the marine environment of Cape Creus. Environ Manag 42:977–988.

Lopez-Goni, I., D. Garcia-Yoldi, C.M. Marin, M.J. De Miguel, P.M. Muñoz, J.M. Blasco, I. Jacques, M. Grayon, A. Cloeckaert, A.C. Ferreira, R. Cardoso, M.I.C. De Sa, K. Walravens, D. Albert, and B. Garin-Bastuji. 2008. Evaluation of a multiplex PCR assay (Bruce-ladder) for molecular typing of all *Brucella* species, including the vaccine strains. J Clin Microbiol 46:3484–3487.

Lynch, M., P.J. Duignan, T. Taylor, O. Nielsen, R. Kirkwood, J. Gibbens, and J.P.Y. Arnould. 2011. Epizootiology of *Brucella* infection in Australian fur seals. J Wildl Dis 47:352–363.

Mackereth, G.F., K.M. Webb, J.S. O'Keefe, P.J. Duignan, and R. Kittelberger. 2005. Serological survey of pre-weaned New Zealand fur seals (*Arctocephalus forsteri*) for brucellosis and leptospirosis. New Zeal Vet J 53:428–432.

Maquart, M., P. Le Fleche, G. Foster, M. Tryland, F. Ramisse, B. Djonne, S. Al Dahouk, I. Jacques, H. Neubauer, K. Walravens, J. Godfroid, A. Cloeckaert, and G. Vergnaud. 2009a. MLVA-16 typing of 295 marine mammal *Brucella* isolates from different animal and geographic origins identifies 7 major groups within *Brucella ceti* and *Brucella pinnipedialis*. BMC Microbiol 9:145.

Maquart, M., M.S. Zygmunt, and A. Cloeckaert. 2009b. Marine mammal *Brucella* isolates with different genomic characteristics display a differential response when infecting human macrophages in culture. Microbes Infect 11:361–366.

Maratea, J., D.R. Ewalt, S. Frasca, J.L. Dunn, S. De Guise, L. Szkudlarek, D.J. St Aubin, and R.A. French. 2003. Evidence of *Brucella* sp. infection in marine mammals stranded along the coast of southern New England. J Zoo Wildl Med 34:256–261.

McDonald, W.L., R. Jamaludin, G. Mackereth, M. Hansen, S. Humphrey, P. Short, T. Taylor, J. Swingler, C.E. Dawson, A.M. Whatmore, E. Stubberfield, L.L. Perrett, and G. Simmons. 2006. Characterization of a *Brucella* sp strain as a marine-mammal type despite isolation from a patient with spinal osteomyelitis in New Zealand. J Clin Microbiol 44:4363–4370.

Miller, D.L., R.Y. Eywing, and G.D. Bossart. 2001. Emerging and resurging diseases. *In* L.A. Dierauf and F.M.D. Gulland, eds. CRC handbook of marine mammal medicine, pp. 15–30. CRC Press, Boca Raton, Florida.

Miller, W.G., L.G. Adams, T.A. Ficht, N.F. Cheville, J.P. Payeur, D.R. Harley, C. House, and S.H. Ridgway. 1999. *Brucella*-induced abortions and infection in bottlenose dolphins (*Tursiops truncatus*). J Zoo Wildl Med 30:100–110.

Muñoz, P.M., G. Garcia-Castrillo, P. Lopez-Garcia, J.C. Gonzalez-Cueli, M.J. De Miguel, C.M. Marin, M. Barberan, and J.M. Blasco. 2006. Isolation of *Brucella* species from a live-stranded striped dolphin (*Stenella coeruleoalba*) in Spain. Vet Rec 158:450–451.

Nielsen, K., P. Smith, W. Yu, P. Nicoletti, P. Elzer, A. Vigliocco, P. Silva, R. Bermudez, T. Renteria, F. Moreno, A. Ruiz, C. Massengill, Q. Muenks,

K. Kenny, T. Tollersrud, L. Samartino, S. Conde, G. D. de Benitez, D. Gall, B. Perez, and X. Rojas. 2004. Enzyme immunoassay for the diagnosis of brucellosis: chimeric Protein A-Protein G as a common enzyme labeled detection reagent for sera for different animal species. Vet Microbiol 101:123–129.

Nielsen, O., K. Nielsen, R. Braun, and L. Kelly. 2005. A comparison of four serologic assays in screening for *Brucella* exposure in Hawaiian monk seals. J Wildl Dis 41:126–133.

Nielsen, O., K. Nielsen, and R.E.A. Stewart. 1996. Serologic evidence of *Brucella* spp. exposure in Atlantic walruses (*Odobenus rosmarus rosmarus*) and ringed seals (*Phoca hispida*) of Arctic Canada. Arctic 49:383–386.

Nielsen, O., R.E.A. Stewart, K. Nielsen, L. Measures, and P. Duignan. 2001. Serologic survey of *Brucella* spp. antibodies in some marine mammals of North America. J Wildl Dis 37:89–100.

Nymo, I., M. Tryland, and J. Godfroid. 2011. A review of *Brucella* infection in marine mammals, with special emphasis on *Brucella pinnipedialis* in the hooded seal (*Cystophora cristata*). Vet Res 42:93; doi:10.1186/1297-9716-42-93.

Ohishi, K., R. Zenitani, T. Bando, Y. Goto, K. Uchida, T. Maruyama, S. Yamamoto, N. Miyazaki, and Y. Fujise. 2003. Pathological and serological evidence of *Brucella*-infection in baleen whales (*Mysticeti*) in the western North Pacific. Comp Immunol Microb 26:125–136.

Ohishi, K., K. Takishita, M. Kawato, R. Zenitani, T. Bando, Y. Fujise, Y. Goto, S. Yamamoto, and T. Maruyama. 2004. Molecular evidence of new variant *Brucella* in North Pacific common minke whales. Microbes Infect 6:1199–1204.

Osterman, B., and I. Moriyon. 2006. International Committee on Systematics of Prokaryotes; Subcommittee on the taxonomy of *Brucella*: Minutes of the meeting, Sept. 17, 2003, Pamplona, Spain. Int J Syst Evol Mic 56:1173–1175.

Perrett, L.L., S.D. Brew, J.A. Stack, A.P. MacMillan, and J.B. Bashiruddin. 2004. Experimental assessment of the pathogenicity of *Brucella* strains from marine mammals for pregnant sheep. Small Ruminant Res 51:221–228.

Prenger-Berninghoff, E., U. Siebert, M. Stede, A. Koenig, R. Weiss, and G. Baljer 2008. Incidence of *Brucella* species in marine mammals of the German North Sea. Dis Aquat Org 81:65–71.

Rah, H., B.B. Chomel, E.H. Follmann, R.W. Kasten, C.H. Hew, T.B. Farver, G.W. Garner, and S.C. Amstrup. 2005. Serosurvey of selected zoonotic agents in polar bears (*Ursus maritimus*). Vet Rec 156:7–13.

Retamal, P., O. Blank, P. Abalos, and D. Torres. 2000. Detection of anti-*Brucella* antibodies in pinnipeds from the Antarctic territory. Vet Rec 146:166–167.

Rhyan, J.C., T. Gidlewski, D R. Ewalt, S.G. Hennager, D M. Lambourne, and S.C. Olsen. 2001. Seroconversion and abortion in cattle experimentally infected with *Brucella* sp isolated from a Pacific harbor seal (*Phoca vitulina richardsi*). J Vet Diagn Invest 13:379–382.

Ross, H.M., G. Foster, R.J. Reid, K.L. Jahans, and A.P. MacMillan. 1994. *Brucella* species infection in sea mammals. Vet Rec 134:359.

Ross, H.M., K.L. Jahans, A.P. MacMillan, R.J. Reid, P.M. Thompson, and G. Foster. 1996. *Brucella* species infection in North Sea Seal and cetacean populations. Vet Rec 138:647–648.

Salem, S.F., and A. Mohsen. 1997. Brucellosis in fish. Vet Med (Praha) 42:5–7.

Scholz, H.C., Z. Hubalek, I. Sedlacek, G. Vergnaud, H. Tomaso, S. Al Dahouk, F. Melzer, P. Kampfer, H. Neubauer, A. Cloeckaert, M. Maquart, M. . Zygmunt, A.M. Whatmore, E. Falsen, P. Bahn, C. Gollner, M. Pfeffer, B. Huber, H.J. Busse, and K. Nockler. 2008. *Brucella microti* sp. nov., isolated from the common vole Microtus arvalis. Int J Syst Evol Microbiol 58:375–382.

Scholz, H.C., K. Nockler, C. Gollner, P. Bahn, G. Vergnaud, H. Tomaso, S. Al Dahouk, P. Kampfer, A. Cloeckaert, M. Maquart, M. S. Zygmunt, A.M. Whatmore, M. Pfeffer, B. Huber, H.J. Busse, and B.K. De. 2010. *Brucella inopinata* sp nov., isolated from a breast implant infection. Int J Syst Evol Microbiol 60:801–808.

Sohn, A.H., W.S. Probert, C.A. Glaser, N. Gupta, A.W. Bollen, J.D. Wong, E.M. Grace, and W.C. McDonald. 2003. Human neurobrucellosis with intracerebral granuloma caused by a marine mammal *Brucella* spp. Emerg Infect Dis 9:485–488.

Stirling, I. 2009. Polar bear (*Ursus maritimus*). *In* W.F. Perrin, B. Wursig, and J.G.M. Thewissen, eds. Encyclopedia of marine mammals, pp. 888–890. Elsevier Inc., Academic Press, New York.

Tachibana, M., K. Watanabe, S. Kim, Y. Omata, K. Murata, T. Hammond, and M. Watarai. 2006. Antibodies to

*Brucella* spp. in Pacific bottlenose dolphins from the Solomon Islands. J Wildl Dis 42:412–414.

Tryland, M., A. E. Derocher, O. Wiig, and J. Godfroid. 2001. *Brucella* sp. antibodies in polar bears from Svalbard and the Barents Sea. J Wildl Dis 37:523–531.

Tryland, M., L. Kleivane, A. Alfredsson, M. Kjeld, A. Arnason, S. Stuen, and J. Godfroid. 1999. Evidence of *Brucella* infection in marine mammals in the North Atlantic Ocean. Vet Rec 144:588–592.

Tryland, M., K. K. Sorensen, and J. Godfroid. 2005. Prevalence of *Brucella pinnipediae* in healthy hooded seals (*Cystophora cristata*) from the North Atlantic Ocean and ringed seals (*Phoca hispida*) from Svalbard. Vet Microbiol 105:103–111.

Van Bressem, M.F., K. Van Waerebeek, J.A. Raga, J. Godfroid, S.D. Brew, and A.P. MacMillan. 2001. Serological evidence of *Brucella* species infection in odontocetes from the south Pacific and the Mediterranean. Vet Rec 148:657–661.

Watson, C.R., R. Hanna, R. Porter, W. McConnell, D.A. Graham, S. Kennedy, and S.W.J. McDowell. 2003. Isolation of *Brucella* species from common seals in Northern Ireland. Vet Rec 153:155–156.

Whatmore, A.M. 2009. Current understanding of the genetic diversity of *Brucella*, an expanding genus of zoonotic pathogens. Infect Genet Evol 9:1168–1184.

Whatmore, A.M., C.E. Dawson, P. Groussaud, M.S. Koylass, A.C. King, S.J. Shankster, A.H. Sohn, W.S. Probert, and W.L. McDonald. 2008. Marine mammal *Brucella* genotype associated with zoonotic infection. Emerg Infect Dis 14:517–518.

Whatmore, A.M., L.L. Perrett, and A.P. MacMillan. 2007. Characterisation of the genetic diversity of *Brucella* by multilocus sequencing. BMC Microbiol 7:34.

Zarnke, R.L., J.T. Saliki, A.P. MacMillan, S.D. Brew, C.E. Dawson, J.M.V. Hoef, K.J. Frost, and R.J. Small. 2006. Serologic survey for *Brucella* spp., phocid herpesvirus-1, phocid herpesvirus-2, and phocine distemper virus in harbor seals from Alaska, 1976–1999. J Wildl Dis 42:290–300.

# 19

## INFECTIOUS CANCERS IN WILDLIFE

Hamish McCallum and Menna Jones

### NATURE AND DIVERSITY OF CANCERS IN WILDLIFE

A working definition of cancer is "a set of diseases characterized by unregulated cell growth leading to invasion of surrounding tissues and spread (metastasis) to other parts of the body" (King and Robins 2006). This definition covers a wide range of diseases in humans and other animals, but at the most fundamental level, all cancers involve dynamic changes in the genome (Hanahan and Weinberg 2000). Cancer is a genetic disease of clonal evolution within the body, characterized by multiple heritable genetic and epigenetic changes in gene expression that lead to unregulated cell growth (Hanahan and Weinberg 2000; Merlo et al. 2006; Ruddon 2007). A disease of higher multicellular organisms, cancer overcomes evolved defenses against uncontrolled cell replication, such as apoptosis and cellular senescence and differentiation, that enable cellular cooperation and thus multicellular organization (Merlo et al. 2006; Kurbel et al. 2007). The term "neoplasm" refers to any type of abnormal growth, which is termed a cancer only if it has the potential to metastasize, and a tumor is a neoplasm that forms a lump (King and Robins 2006). In this chapter, we concentrate on neoplasms that are known to be malignant and thus are cancers, but because of the difficulty in reliably distinguishing (without intensive investigation) between benign and malignant neoplasms in wildlife and poorly studied species, we will also discuss neoplasms and tumors in general. Neoplasms can be considered as complex adaptive systems, comprising ecosystems of cancer cells and normal tissue cells that have been co-opted into supporting the cancer. Tumors comprise mosaics of genetically and epigenetically variable cell lines (clones) that evolve under Darwinian selection, competing for space and resources and evading attack by the immune system (Merlo et al. 2006).

The vast majority of cancers arise independently in each individual animal in response to multiple "hits" from a range of causal factors, including chemical or irradiating carcinogens, oxygen free radicals associated with aging, and interactions of these with oncogenes and tumor-suppressing genes and viruses (Ruddon 2007). As with many human cancers, some wildlife cancers are attributable to environmental contaminants. For example, Martineau (Chapter 27 in this book) describes cancers in beluga whales (*Delphinapterus leucas*) that are related to polycyclic aromatic hydrocarbons. In some cases, an infectious agent (usually a virus) predisposes towards the development of a tumor in an individual, but the tumor cells themselves do not propagate beyond a single host. These cancers evolve within the individual but are an evolutionary dead end, dying with the death of the host.

Much more rarely, the tumor cells themselves may be the infectious agent. To date, there are only two known directly transmissible cancers in nature: canine transmissible venereal tumor (CTVT; Das and Das 2000; Murgia et al. 2006; Rebbeck et al. 2009) and Tasmanian devil (*Sarcophilus harrisii*) facial tumor disease (DFTD; Hawkins et al. 2006; Pearse and Swift 2006; McCallum 2008). These tumors do not die with their host; in fact, CTVT probably originated at least 7,500 years ago and is the oldest known somatic cell line (Murchison 2009; Rebbeck et al. 2009).

Cancer has been recorded in all of the major vertebrate classes (McAloose and Newton 2009). Nevertheless, it is not generally regarded as a major source of mortality in wildlife populations. There are probably two main reasons for this. First, although cancer is a major source of mortality in humans, with childhood cancers obviously of major concern, cancer in humans is generally an illness of old age. A major hypothesis for the increased susceptibility to cancer with age is the accumulation of DNA damage through the lifespan of an individual (Maslov and Vijg 2009). Although there is increasing evidence of senescence in wild populations of animals (O.R. Jones et al. 2008; Nussey et al. 2008), relatively few individual animals in most species survive until old age, and therefore diseases of old age will not often be observed. Secondly, wild animals with chronic diseases such as tumors are likely to be selectively removed by predators, meaning that animals with advanced tumors will relatively rarely be observed.

In zoos, neoplasms are often recorded as the cause of death of wild species (Kumar 1989). Some taxa, notably the dasyurids (carnivorous marsupials) and the felids, seem to be particularly prone to developing tumors. Reviews of spontaneous tumors in Australian marsupials have reported that neoplasms of a variety of types were among the most common causes of death of dasyurids (marsupial carnivores), whereas neoplasms were not reported in any other marsupial order (Canfield et al. 1990; Canfield and Cunningham 1993). Similarly, there are numerous reports of neoplasms in felids in the zoo literature (Harrenstien et al. 1996; de Castro et al. 2003; Marker et al. 2003; Walzer et al. 2003; Sutherland-Smith et al. 2004; Owston et al. 2008; Tucker and Smith 2008).

Determining that a cancer is infectious is not entirely straightforward. The classic criteria for determining that a disease is infectious are Koch's (or the Koch-Henle) postulates. There are various versions of these in the literature. (See, for example the Harvard University Library Open Collections Program, http://ocp.hul.harvard.edu/contagion/koch.html). These criteria were established for bacterial infections and are difficult to apply to any virus because viruses cannot be grown except in living cells (and thus not strictly in "pure culture"), and even for bacterial infections, the possibility of carriers is not allowed for, a limitation Koch himself recognized (Evans 1976). Multifactorial causation inevitably complicates the simple application of Koch's postulates, and there is increasing evidence that causation of most cancers involves multiple steps and factors (Hanahan and Weinberg 2000; Haverkos 2004)

## CANCERS WITH A VIRAL ORIGIN

Rous (1911) was the first to identify that a virus (since identified as a retrovirus) might cause cancer. Subsequently, viruses have been shown to contribute to the causation of a range of cancers in many species, including at least six viruses in humans that contribute to 10% to 15% of human cancers worldwide (Martin and Gutkind 2009). Some common wildlife cancers are associated with viral infection, although multiple factors are usually implicated in oncogenesis, often including environmental cofactors such as pollution (Haverkos 2004; McAloose and Newton 2009). Oncogenic viruses of particular significance to wildlife biologists include papilloma viruses, feline immunodeficiency virus (FIV), and feline leukemia virus (FeLV).

Papilloma viruses occur in a wide variety of animals (Aguirre et al. 1994; Woodruff et al. 2005; Newman and Smith 2006) and are associated with formation of papillomas or warts, which are (usually) benign) tumors. They are also associated with the development of malignant tumors in humans, particularly cervical cancer (Haverkos 2004). The extent to which papilloma viruses might be associated with malignant disease in nonhuman animals is less clear. However, papillomas in several wildlife species can reach intensities on individuals sufficient to cause substantial issues at the population level, irrespective of whether these tumors are benign or malignant. For example, papillomas in green turtles (*Chelonia mydas*) can hamper the ability of affected individuals to feed, swim, and dive. They may also be associated with internal tumors (Aguirre and Lutz 2004). Papilloma viruses are often found within these tumors, but this does not necessarily imply that they are the causative agent. Recently, cutaneous papillomas, which apparently developed into squamous cell carcinomas, have been reported from the endangered Western barred bandicoot (*Perameles bougainville*), and there was some evidence that these were associated with a papilloma virus (Woolford et al. 2008). Otarine herpes virus has been suggested as the cause of genital carcinomas in California sea lions (*Zalophus californianus*) (King et al. 2002).

FIV causes AIDS-like symptoms in domestic cats, including an association with some cancers (Gabor et al. 2001). It occurs at high prevalence in some wild felids (e.g., up to 100% seroprevalence in some African lion populations; Roelke et al. 2009). There is no clear evidence that FIV predisposes these wild felids to tumors. However, Roelke et al. (2009) did find CD4 T-cell depletion, poorer condition, and increased numbers of papillomas in infected versus uninfected lions.

FeLV, a retrovirus affecting domestic cats, results in persistent viremia in about one third of exposed cats, in turn leading to immunosuppression, anemia, and/or neoplasia (Cunningham et al. 2008). Transmitted by direct contact, it is relatively uncommon in wild felids, although epizootics have been reported in some threatened species such as the Florida panther (*Puma concolor coryi*) (Cunningham et al. 2008) and Iberian lynx (*Lynx pardinus*) (Lopez et al. 2009). It is likely that these outbreaks result from spillover (Daszak et al. 2000) from domestic cat populations. A vaccine has been developed for domestic cats, and this was used to successfully manage the epizootic in Florida panthers (Cunningham et al. 2008) and Iberian lynx (Lopez et al. 2009).

## DIRECTLY TRANSMISSIBLE CANCERS

A transmission mode of direct transfer of live tumor cells between hosts distinguishes directly transmissible cancers from all other forms of cancer. The infectious spread of the cancer is unrelated to the original tumorigenesis, which may have arisen through any of the range of causal mechanisms of cancer.

Cancers that can be transmitted from one host to another include CTVT and devil facial DFTD. A third directly transmitted cancer, in hamsters, is known only from laboratory colonies.

## Canine Transmissible Venereal Tumor

CTVT affects the external genitalia of dogs (*Canis familiaris*) and has been known to be transmissible for well over a century (Das and Das 2000). However, only recently have molecular techniques unequivocally demonstrated that the tumor cells themselves are the infectious agent and that all tumors are essentially a single clone (Murgia et al. 2006; Rebbeck et al. 2009). Tumor cells are most commonly transferred between dogs during coitus, but the tumor can also be spread by licking or scratching affected areas (Murchison 2009). Tumors appear as small papules but can progress into multi-lobed proliferations (Das and Das 2000). Most spontaneously regress after a few months, after which the host is usually immune (Murchison 2009). In the progressive phase of CTVT, in which the tumor is growing, major histocompatibility complex (MHC, both Class I and Class II) is not expressed or only weakly expressed in most tumor cells, but the expression increases considerably during the regressive phase (Murchison 2009). CTVT appears to have originated some thousands of years ago either in a genetically inbred population of wolves or during the early domestication of dogs, which would have created a population bottleneck (Murgia et al. 2006; Rebbeck et al. 2009). It is likely that CTVT initially acquired the capability of inter-host transmission with high genetic similarity allowing immune evasion. Since its origin between 7,800 and 78,000 years ago (Rebbeck et al. 2009), as CTVT has moved into a large genetically diverse population of domestic dogs, it has evolved more sophisticated immune evasion strategies: the ability to downregulate MHC Classes I and II and secretion of a cytokine that inhibits tumor-infiltrating lymphocytes, including natural killer cells (Murchison 2009). The tumor is now found on all continents among free-roaming dogs, irrespective of breed, and there is evidence that it can be transferred experimentally to other canids, such as coyotes (*Canis latrans*) and foxes (Rebbeck et al. 2009). Although there have been suggestions that CTVT might be a substantial threat to populations of some endangered canids, such as the African wild dog (*Lycaon pictus*) (vonHoldt and Ostrander 2006), it is not known to be an important disease in current conservation biology.

## Tasmanian Devil Facial Tumor Disease

DFTD is a new contagious cancer that, since it was first detected in 1996 in northeastern Tasmania, has spread to most of the range of wild Tasmanian devils, the world's largest (6 to 14 kg) remaining marsupial carnivore (family *Dasyuridae*). The overall population decline caused by DFTD is 70% and ongoing, with local declines now up to 94%, leading to concerns of extinction in the wild in 25 to 35 years and endangered listing at international (IUCN), national, and state levels (Hawkins et al. 2006; McCallum and Jones

2006; Lachish et al. 2007; McCallum et al. 2007) (http://www.dpiw.tas.gov.au/inter.nsf/webpages/lbun-5qf86g).

DFTD manifests as large (more than 10 cm in diameter) primary tumors around the face and neck and inside the mouth of the Tasmanian devil (Fig. 19.1), which frequently metastasize and consistently cause death within about six months of the first clinical signs (Hawkins et al. 2006; Loh et al. 2006). Allograft transmission of a clonal cell line has been established from cytogenetic evidence (Pearse and Swift 2006) and molecular genetics using microsatellite and MHC markers (Siddle et al. 2007; Murchison et al. 2010). Koch's postulates have been fulfilled with albeit small sample sizes (Pyecroft et al. 2007). Transmission most likely occurs through biting. Vertical transmission has not been reported (S. Pyecroft, personal communication 2010). The high incidence of

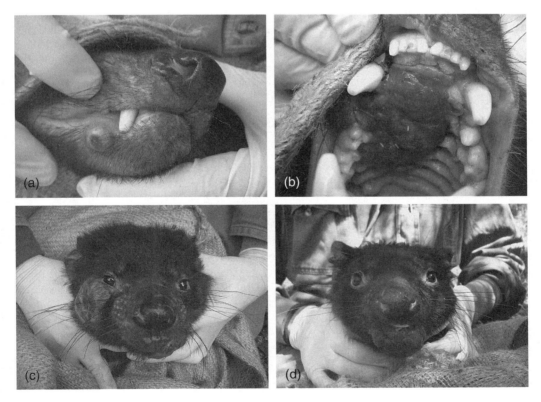

**Figure 19.1:**
Tasmanian devil facial tumor disease. (a) An early-stage tumor. (b) A tumor on the upper palate. (c) A devil with several large tumors. (d) A tumor causing complete degeneration of the lower jaw. (Photo credits: (a), (c), (d) Rodrigo Hamede, University of Tasmania. (b) Hamish McCallum.)

penetrating injuries during the annual mating season (Hamede et al. 2008), combined with the maintenance of a high force of infection even when population density declines to a low level, suggests frequency- rather than density-dependent transmission (McCallum et al. 2009).

Extremely low genetic diversity (M. Jones et al. 2004), particularly at the MHC Class I involved in tumor recognition, is thought to facilitate transmission, with the devil not recognizing the tumor as nonself (Siddle et al. 2007). Recent data suggest that there is more variation in MHC Class I loci in the currently uninfected northwest part of Tasmania than in the infected eastern and central parts (Siddle et al. 2010). Field data suggest that as the disease enters northwestern populations with increased MHC variation, there are lower infection rates and fewer effects on population size and structure, compared with impacts of disease arrival in other parts of Tasmania with less MHC variation (Hamede et al. 2012). This provides some hope for alternative epidemic outcomes other than extinction.

## Transmissible Tumors in Hamsters

There are several reports of transmissible tumors in laboratory colonies of Syrian (or golden) hamsters (*Mesocricetus auratus*). The first of these (Ashbel 1945) appeared spontaneously in an inbred population of Syrian hamsters and subsequently was experimentally passaged for up to 25 times. Some years later, a reticulum cell carcinoma also appeared spontaneously in a laboratory population of hamsters (Brindley and Banfield 1961). This was experimentally transplantable but also could be transferred simply by housing infected and uninfected animals together. Independently, transmissible tumors in the same species were also induced by exposure to a carcinogenic hydrocarbon and maintained by serial passage 288 times over 11 years (Fabrizio 1965). As with DFTD, very similar large-scale chromosomal rearrangements were observed for all occurrences of the tumor within the same cell line (Banfield et al. 1965). To our knowledge, there have been no records of such tumors in any wild hamster populations, but the observation of multiple independent origins of transmissible tumors in this species is fascinating and differs from the apparent single origin of both DFTD and CTVT.

## Why Are Directly Transmitted Tumors Apparently So Rare?

The three examples just described show that the evolution of transmissibility in cancer is biologically possible. Given that rapid within-host evolution is a characteristic of cancers (Hanahan and Weinberg 2000) and that the ability to survive the death of the host would greatly increase the fitness of a cell line, it is perhaps surprising that so few such transmissible cancers are known. However, the number of reported cases of tumor transmission between humans or in other animals, either in the medical literature or under laboratory conditions, suggests that the phenomenon may in fact not be that rare given the right conditions (Dingli and Nowak 2006; Quammen 2008; Murchison 2009). These conditions include a means of transfer of live tumor cells from the donor host, implantation of these live tumor cells in the recipient host, and failure of the recipient host's immune system to reject the recipient cells. Such conditions may rarely coincide in nature, but the first two can occur in copulation or fighting, and many populations are now inbred. Transmissible cancers in wildlife may therefore be underreported. Particularly in wildlife or poorly studied species, it may not be clear without intensive investigation whether a neoplasm has spread or has the potential to spread to other individuals.

Copulation is universal in higher vertebrates and of course involves transfer of live cells (gametes) between individuals. Transfer of live cells in some grooming activities is also conceivable. Less likely but possible transmission routes include contact with fomites such as carcasses or other food into which tumor cells have been shed, through cannibalism of an infected animal (Tasmanian devils readily scavenge dead devils), vectored transmission via blood-sucking invertebrates (e.g., mosquitoes, tabanid flies, leeches, ticks), or maternal–fetal transfer. The first two scenarios require the susceptible host to have injuries in or around the mouth, a condition that occurs frequently in devils. Transmission of cancer cells via mosquito bites has been demonstrated experimentally in the Syrian hamster (Pearse et al in press). Transmission of oncogenic viruses by marine leeches could contribute to the spread of sea turtle fibropapillomatosis (Greenblatt et al. 2004). In principle, transmission of tumor cells could follow a similar route.

A second requirement for the natural transmission of cancer cells between hosts is the capability for cells to detach from the tumor. Migration is an innate characteristic of blood-borne cancers such as leukemia that do not form neoplasia; these are capable even of crossing the maternal–fetal placental boundary to spread from mother to fetus (Tolar and Neglia 2003). With solid neoplasms, the ability for pioneer or metastasizing cells to undock from the surrounding tissues is one of the acquired capabilities that are characteristic of most cancers (Hanahan and Weinberg 2000). Metastasis within neoplasia is driven by the same evolutionary pressures that influence dispersal decisions in animals (see, for example, Hamilton and May 1977). In both naturally occurring transmissible tumors, DFTD and CTVT, transmissibility between hosts is probably also enhanced by increasing friability of the tumor with size and age. The center of the larger tumors becomes necrotic, probably as space and nutrients become limiting (Merlo et al. 2006), and they ulcerate and slough cells (Loh et al. 2006).

The third requirement for direct transmission in nature is a mechanism for non-recognition by the host immune recognition system. The MHC genes, which are involved in the recognition of self/non-self and tumor cells, are important in this context. CTVT has evolved strategies to downregulate both MHC I and MHC II gene expression (Murgia et al. 2006). DFTD appears to evade recognition by the immune system partially because the very low genetic diversity in the devil population results in very high genetic similarity in the MHC in the host and the tumor (Siddle et al. 2007, 2010), but it also appears that DFTD, like CTVT, has developed some ability to evade the host's immune system (Kreiss et al. 2011) . High genetic similarity in inbred colonies of laboratory Syrian hamsters enabled transmission of spontaneously arising sarcomas (Ashbel 1945; Brindley and Banfield 1961). More generally, Kurbel et al. (2007) suggests that the threat of transmissible tumors in early vertebrates may have contributed to the evolution of acquired immunity in an ancestral vertebrate. Murgia et al. (2006), in conceptually broad agreement with this statement, suggest that a primary function of diverse MHC is to ensure that cancer does not become an infectious disease.

Study of the histogenesis or tissue of origin of transmissible tumors may help to reveal their likely occurrence, evolution, and biology (Murchison et al. 2010). CTVT is of histiocytic origin; it is a mononuclear tissue macrophage that is part of the dog's immune system (Mozos et al. 1996). DFTD is a Schwann cell tumor, derived from the cells that create the myelin sheath in the peripheral nervous system, where they are involved in nerve repair and local immune reactions. This suggests that the plasticity and immunocompetence of Schwann cells may have significance for the evolution of DFTD as a transmissible tumor (Murchison et al. 2010).

It is possible that DFTD has an etiology in wound carcinogenesis, an uncommon origin of cancer in which repeated trauma to tissues can induce a variety of types of cancer (Trent and Kirsner 2003). Repetitive tissue damage with chronic inflammation can lead to release of toxins, which can cause mutations (Trent and Kirsner 2003). Adult male devils sustain major and repeated biting injuries to the jowls and around the vibrissae (in devils these are luxurious and well endowed with nerves and therefore with Schwann cells) in successive mating seasons, sometimes involving the removal of large pieces of skin (more than 5 cm in diameter) (Hamede et al. 2008). These wounds can take months to heal and result in thick scar tissue.

## Detecting Directly Transmitted Cancers in Wild Populations

For an emerging or novel directly transmitted tumor such as DFTD, its infectious nature may be suggested by a pattern of spatial spread and increase in prevalence similar to that observed in other emerging infectious diseases. However, if a directly transmitted tumor has been present for some time (e.g., CTVT), no such pattern of spread is likely to be detectable. Spatial spread and increasing prevalence are also likely to be observed with a novel tumor with a non-infectious cause and may, at least in its early stages, mimic epidemic spread. The tumor will initially be detected at some point location in the environment, which is likely to be close to the source of the carcinogenic agent in the environment. Subsequent detections are likely to be close to this source and decrease in prevalence with distance from the source.

The most useful indication that a tumor is directly transmitted is if there are major and consistent differences between the karyotype of the tumor cells and those of the host. Most cancer cells have karyotypes different from normal host cells. Frequently, these karyotypic changes are similar within hosts, indicating clonal origin of the cancer, but the nature of the abnormalities is highly heterogeneous between individual hosts (Nowell 1976). For both CTVT (Das and Das 2000) and the transmissible hamster tumor (Banfield et al. 1965), karyotypes of all tumor cells in all hosts are similar. It was on this basis that Pearse and Swift (2006) suggested that DFTD was transmitted as a cell line. Definitive proof that a cancer is a directly transmitted cell line requires molecular evidence that tumors have a common clonal origin and are more closely related to each other than to their hosts. Such evidence was first provided for CTVT by Murgia et al. (2006) and for DFTD by Siddle et al. (2007) and Murchison et al. (2010; online supplementary materials). Obtaining this evidence is not straightforward, as it requires multiple matched samples of tumors and their hosts and an appropriate set of molecular markers.

## Evolutionary Ecology of Directly Transmitted Tumors

The evolution in dogs and Tasmanian devils of tumors transmitted as a cell line can be regarded as the evolution of a novel type of highly degenerate mammal that is an obligate parasite. As the cell line is composed of mammalian DNA, it certainly has to be regarded as a mammal. Loss of many of the attributes of free-living relatives is typical in the evolution of parasitic forms in all taxa, but this is certainly an extreme example. Directly transmitted tumors could be regarded as a novel form of natural enemy in the spectrum described by Lafferty and Kuris (2002), but transmissible cell lines share most of the characteristics of a typical microparasite, such as the potential for rapid multiplication within the host (Anderson and May 1979).

Co-evolution typically leads to persistence of the host and optimal virulence of the pathogen in host–pathogen interactions (Dybdahl and Storfer 2003; Jensen et al. 2006; Bérénos et al. 2009). Selection operates on hosts to overcome parasitic infection and, to a good first approximation, selection acts on pathogens to maximize $R_0$ (the number of secondary cases per primary case when the pathogen is first introduced into a naïve population). This means that pathogen strains that kill the host too quickly are selected against.

Conversely, avirulent strains frequently have low transmission potential. A trade-off is therefore likely between virulence and transmissibility (Anderson and May 1982). For a tumor, transmissibility is likely to be associated with a tendency to metastasize, which is likely to be associated with virulence and immune evasion (Fassati and Mitchison 2009). Further, Darwinian selection on transmissible tumors should operate to favor immune evasion strategies that function to avoid clearance of the tumor from the host rather than increasing transmission (Schmid-Hempel 2008). Although processes of co-evolution in transmissible cancers are likely to be different from those in viral pathogens (Gupta and Hill 1995), the nature of these differences can only be speculated upon. We hypothesize that both types of naturally occurring transmissible cancers will evolve or co-evolve with the host to reduce virulence over time, thus increasing the overall fitness of the cancer (Fassati and Mitchison 2009).

DFTD is consistently lethal across eastern Tasmania, the geographic region where DFTD emerged and where host MHC genotypes are very similar to the tumor (Siddle et al. 2010), enabling the tumor to evade immune recognition. In contrast, for CTVT, co-evolution with host immune responses has resulted in a typical disease progression in immunocompetent individuals of a progressive phase with immune evasion resulting in rapid tumor growth, followed by stable and finally a regressive phase in 80% of cases, following which the individual is immune for life. In the stable and regressive phases, as numbers of tumor-infiltrating lymphocytes increase, their secretion of growth factor and interleukin-6 counteracts the suppressive effects of the tumor-derived cytokines and induces the expression of MHC I and II molecules on tumor cells, which then become visible to the immune system (Murchison 2009). The stage is set for co-evolution between DFTD and the devil in immune evasion and tumor-suppression strategies. The tumor is continuously evolving new karyotypic strains (Pearse et al. in press), and new information on variation in MHC across the infected and unaffected range of the devil suggests scope for host immune responses (Siddle et al. 2010).

Transmissible tumors are likely to be under selection for friability (Murchison 2009). Both DFTD and CTVT tumors become friable with age and size (Brown et al. 1980; Thacher and Bradley 1983; Loh et al. 2006; M. Jones et al. 2007), which will aid both detachment of cells from the tumor and transport to new hosts, via sloughing and also physical abrasion during copulation in dogs and adherence to canine teeth in devils. Although the number of tumor cells required for natural transmission of CTVT and DFTD is unknown (experimental studies of CTVT transmission use in the order of $10^8$ viable tumor cells), infection could result from a single cell (Murchison 2009).

The ability of parasites to selectively manipulate the behavior of their hosts to increase transmission is well documented (Dobson 1988; Berdoy et al. 2000) Directly transmissible cancers would apply similar selective pressures to the elements of host behavior associated with transmission. Differences between CTVT-positive and healthy dogs in estrogen receptor expression in the vaginal epithelium suggest that there may be mechanisms by which CTVT can increase sexual receptivity in dogs (de Brito et al. 2006; Rebbeck et al. 2009). In devils, DFTD potentially might induce increased aggression. There is some evidence that personality traits such as boldness and aggression in animals are heritable (Bakker 1994; van Oers et al. 2005), and candidate genes for aggressive behavior have been found in dogs (van den Berg et al. 2008; Vage et al. 2010). However, if aggressive individuals are more likely to become infected they will have lower fitness. Host behavior may thus be subject to balancing selection.

Because DFTD currently causes 100% mortality of all individuals that contract the cancer and prevalence in infected populations among animals older than 2 years exceeds 50% (McCallum et al. 2009), it causes a drastic reduction in the lifetime number of reproductive events, towards semelparity. In as little as two to three generations since disease arrival, precocial breeding has increased 16-fold (M. Jones et al. 2008; Lachish et al. 2009). Prior to DFTD, females usually bred each year from age 2 to 5 and up to 10% of females bred at age 1. In DFTD-affected populations, almost all individuals have died from DFTD by age 3, and the proportion of females breeding at age 1 has increased to 40% to 50% (M. Jones et al. 2008). Especially rapid evolution on this trait could be expected because it is already plastic and can be rapidly expressed and exposed to selection. A heritable component to rapid early maturation is suggested by variance in females' ability to reach critical size

within the seasonal limits of breeding (Lachish et al. 2009).

Genetic changes indicative of altered directional selection and a selective sweep are evident in eastern devil populations (Lachish et al. 2010b). Extreme selection pressure exists for traits that confer disease resistance, decrease the likelihood of infection, or allow infected individuals to reproduce. The question is whether sufficient genetic variation exists on which selection might operate in a species that has lost 50% of its genetic diversity (M. Jones et al. 2004) and in which inbreeding is now evident (Lachish et al. 2010b). Reduced heritability of traits is seen only in the smallest and most inbred populations (Willi et al. 2006). With around 30,000 individuals remaining in the wild, Tasmanian devils should retain significant evolutionary potential

## CAN INFECTIOUS CANCERS LEAD TO EXTINCTION?

Until recently, cancer in wildlife has not been considered of conservation concern (McAloose and Newton 2009). This situation has changed with DFTD and at least three of the virally associated cancers discussed earlier: fibropapillomatosis in green turtles, genital carcinomas in California sea lions, and malignant lesions in the Western barred bandicoot (*Perameles bougainville*). With the exception of DFTD, the cancer is just one of several factors threatening the species.

In theory, two mechanisms can lead to a pathogen alone being capable of causing host extinction: maintenance of the force of infection on a rare or reduced species by more common alternative hosts; or strongly frequency-dependent transmission (de Castro and Bolker 2005). Some virally promoted cancers have multiple hosts or other environmental reservoirs. For example, fibropapillomatosis has been described in all sea turtle species, including a number of critically endangered species, and environmental persistence has been identified as a possibility in sustaining transmission (McAloose and Newton 2009). Other viruses, such as the herpesvirus associated with genital sarcomas in sea lions, are known from only one host (Buckles et al. 2006). Although there are reports that CTVT can be transferred experimentally to other species, multiple host species are unlikely in directly transmissible cancers. Transmission depends on evasion of the host immune system, specifically the MHC genes, which are generally highly polymorphic and highly species-specific.

Most simple models of host pathogen dynamics assume that the transmission rate depends on host population density (Anderson and May 1979). This means that there is a threshold host density below which the pathogen will disappear, before host extinction occurs. Empirical evidence for a wide range of diseases suggests that transmission is often weakly related to population density (McCallum et al. 2001), with the result that the transmission rate is primarily determined by the frequency of infection in the host population. This means that there may be no threshold host density for disease maintenance and a host-specific pathogen can cause extinction. Sexually transmitted diseases typically have frequency-dependent transmission, because the rate of sexual contact is determined largely by the mating system of the species rather than population density. CTVT is a sexually transmitted disease. DFTD appears to have many of the properties of one: the majority of injurious biting occurs among adult males and females during the annual mating season and prevalence is low in sexually immature animals (Hamede et al. 2008; McCallum 2008). Furthermore, the prevalence of DFTD remains extremely high even when populations have been reduced to very low levels (more than 90% decline), the disease is spreading into geographic areas where unsuitable habitat have always limited population density to very low levels (McCallum et al. 2007), and observed population dynamics are inconsistent with density-dependent transmission but are consistent with frequency-dependent transmission (McCallum et al. 2009). It is therefore entirely possible that DFTD could cause extinction of the devil (McCallum 2008).

## MANAGING CANCER IN WILD POPULATIONS

Attempts to manage infectious cancers in wildlife are few, partly because the limited resources (human and financial) dedicated to wildlife health monitoring and surveillance and the difficulty of collecting samples and epidemiological data on wild animals mean that cancer in wildlife and resulting conservation threats largely go undetected (McAloose and Newton 2009). In human cancer management, the focus is on prevention and treatment. In contrast, for wildlife cancer, research and resources should be directed towards understanding cancer biology of spontaneous cancers and towards policy and interventions to mitigate environmental threats arising from anthropogenic activities (pollution, chemical carcinogens), particularly where these also affect livestock or human populations (McAloose and Newton 2009).

The exception is the large conservation program directed towards preventing extinction and facilitating recovery of the Tasmanian devil afflicted by DFTD (see http://www.dpipwe.tas.gov.au/). McCallum and Jones (2006) used a decision tree approach to setting priorities for how to approach researching and managing a new, poorly known wildlife disease.

Options for managing DFTD are limited (M. Jones et al. 2007). The decision process set out by McCallum and Jones (2006) is reflected in the management strategy of the Save the Tasmanian Devil Program, which is funded by the Australian and Tasmanian governments (http://www.dpipwe.tas.gov.au/) (M. Jones et al. 2007). An insurance metapopulation comprising intensive and free-range captive, and semi-wild (large island or fenced) populations was the first initiative of the program. The metapopulation is being managed for retention of a conservative or high level of genetic diversity over 25 years because devils have already lost about half of their genetic diversity (M. Jones et al. 2004). Disease suppression, implemented through a trap and cull program to detect and remove infected individuals, was instigated on a large peninsula early in the program and terminated in early 2011. Analysis of results showed that the control strategy was insufficient to reduce the force of infection to a rate that allows population recovery or pathogen decline (Lachish et al. 2010a), and modeling suggests that no feasible culling strategy could eliminate the disease (Beeton and McCallum 2011). Parallel research studies into the impacts, transmission, and epidemiology of the disease and into the development of a preclinical diagnostic test will inform an integrated approach in an adaptive management framework that could be applied at smaller landscape scales.

Research underpinning longer-term management options is in progress. Strategic research is directed towards the possible development of a vaccine (Woods et al. 2007; Kreiss et al. 2008, 2009a,b). Transdisciplinary research is investigating the genetic basis for heterogeneity in disease susceptibility (Siddle et al. 2010) and host/tumor co-evolutionary dynamics, leading to the potential for genetic management and genetic restoration of the species in the wild.

## CONCLUSIONS

Emerging infectious diseases are increasingly being recognized as significant threats to biodiversity (Daszak et al. 2000). Should we also expect increasing threats from emerging infectious cancers in wildlife? Genetic diversity is being lost at increasing rates in many wild species and is widely thought to be a factor increasing the susceptibility of wildlife to infectious disease, particularly as it affects the MHC genes involved in immune surveillance for viruses and tumor recognition (Klein 1986; O'Brien and Evermann 1988). A strong adaptive response to a pathogen can cause selective sweeps that increase future vulnerability to pathogen attack (Zinkernagel et al. 1985; O'Brien and Evermann 1988). Tasmanian devil facial tumor disease is the clearest case yet of a pathogen threatening to cause the extinction of its host because of lack of host genetic diversity (McCallum 2008).

With the combined effects of increased exposure to pollutants and chemical carcinogens, habitat fragmentation bringing previously spatially separated species and their viruses into close proximity, and loss of genetic diversity, we can expect an increase in emerging infectious cancers in wildlife, both those with viral associations and directly transmissible cancers. The ability of a cancer to disperse to other hosts is such a successful strategy that selection on cancers for this trait should be strong (Dingli and Nowak 2006). It is plausible that transmissible cancers could cross the species barrier in nature. It is not known how many transmissible cancers in wildlife have gone undetected. We predict that there are transmissible cancers yet to be discovered and that more will emerge over the next few decades.

## ACKNOWLEDGMENTS

Many of the ideas in this chapter have been developed during discussions with our collaborators Kathy Belov, Anne Maree Pearse, Greg Woods, Andrew Storfer, and Elizabeth Murchison and with our students Shelly Lachish and Rodrigo Hamede. We would like to acknowledge funding from the Australian Research Council, the Save the Tasmanian Devil Program, and the Save the Tasmanian Devil Appeal.

## REFERENCES

Aguirre, A.A., G. Balazs, B. Zimmerman, and T.R. Spraker. 1994. Evaluation of Hawaiian green turtles (Chelonia mydas) for potential pathogens associated with fibropapillomas. J Wildl Dis 30:8–15.

Aguirre, A.A., and P.L. Lutz. 2004. Marine turtles as sentinels of ecosystem health: is fibropapillomatosis an indicator? EcoHealth 1:275–283.

Anderson, R.M., and R.M. May. 1979. Population biology of infectious diseases. Part I. Nature 280:361–367.

Anderson, R.M., and R.M. May. 1982. Coevolution of hosts and parasites. Parasitology 85:411–426.

Ashbel, R. 1945. Spontaneous transmissible tumours in the Syrian hamster. Nature 155:607.

Bakker, T.C.M. 1994. Genetic correlations and the control of behavior, exemplified by aggressiveness in sticklebacks. Adv Stud Behav 23:135–171.

Banfield, W.G., P.A. Woke, C.M. Mackay, and H.L. Cooper. 1965. Mosquito transmission of reticulum cell sarcoma of hamsters. Science 148:1239–1240.

Beeton, N., and McCallum, H. 2011. Models predict that culling is not a feasible strategy to prevent extinction of Tasmanian devils from facial tumour disease. J Appl Ecol 48: 1315–1323.

Berdoy, M., J.P. Webster, and D.W. Macdonald. 2000. Fatal attraction in rats infected with Toxoplasma gondii. Proc Roy Soc Lond B 267:1591–1594.

Bérénos, C., P. Schmid-Hempel, and K. Mathias Wegner. 2009. Evolution of host resistance and trade-offs between virulence and transmission potential in an obligately killing parasite. J Evol Biol 22:2049–2056.

Brindley, D.C., and W.G. Banfield. 1961. Contagious tumor of hamster. J Nat Cancer Inst 26:949–957.

Brown, N.O., C. Calvert, and E.G. MacEwan. 1980. Chemotherapeutic management of transmissible venereal tumors in 30 dogs. J Am Vet Med Assoc 176:983–986.

Buckles, E.L., L.J. Lowenstine, C. Funke, R.K. Vittore, H.N. Wong, J.A. St Leger, D.J. Greig, R.S. Duerr, F.M.D. Gulland, and J.L. Stott. 2006. Otarine herpesvirus-1, not papillomavirus, is associated with endemic tumors in California sea lions (Zalophus californianus). J Comp Path 135:183–189.

Canfield, P.J., and A.A. Cunningham. 1993. Disease and mortality in Australasian marsupials held at London Zoo, 1872–1972. J Zoo Wildl Med 24:158–167.

Canfield, P.J., W.J. Harley, and G.L. Reddacliff. 1990. Spontaneous proliferations in Australian marsupials—a survey and review. 2. Dasyurids and bandicoots. J Comp Pathol 103:147–158.

Cunningham, M.W., M.A. Brown, D.B. Shindle, S.P. Terrell, K.A. Hayes, B.C. Ferree, R.T. McBride, E.L. Blankenship, D. Jansen, S.B. Citino, M.E. Roelke, R.A. Kiltie, J.L. Troyer, and S.J. O'Brien. 2008. Epizootiology and management of feline leukemia virus in the Florida puma. J Wildl Dis 44:537–552.

Das, U., and A.K. Das. 2000. Review of canine transmissible venereal sarcoma. Vet Res Commun 24:545–556.

Daszak, P., A.A. Cunningham, and A.D. Hyatt. 2000. Emerging infectious diseases of wildlife- threats to biodiversity and human health. Science 287:443–449.

de Brito, C.P., C.M. de Oliveira, F.A. Soares, M. Faustino, and C.A. de Oliveira. 2006. Immunohistochemical determination of estrogen receptor-alpha in vaginal and tumor tissues of healthy and TVT-affected bitches and their relation to serum concentrations of estradiol-17beta and progesterone. Theriogenology 66:1587–1592.

de Castro, F., and B. Bolker. 2005. Mechanisms of disease-induced extinction. Ecol Lett: 8117–126.

de Castro, M.B., K. Werther, G.S. Godoy, V.P. Borges, and A.C. Alessi. 2003. Visceral mast cell tumor in a captive black jaguar (Panthera onca). J Zoo Wildl Med 34:100–102.

Dingli, D., and M.A. Nowak. 2006. Cancer biology: Infectious tumour cells. Nature 443:35–36.

Dobson, A.P. 1988. The population biology of parasite-induced changes in host behaviour. Quart Rev of Biol 63:139–165.

Dybdahl, M.F., and A. Storfer. 2003. Parasite local adaptation: Red Queen versus Suicide King. Trends Ecol Evol 18:523–530.

Evans, A.S. 1976. Causation and disease—Henle-Koch postulates revisited. Yale J Biol Med 49:175–195.

Fabrizio, A.M. 1965. An induced transmissible sarcoma in hamsters—11-year observation through 288 passages. Cancer Res 25:107–117.

Fassati, A., and N.A. Mitchison. 2009. Testing the theory of immune selection in cancers that break the rules of transplantation. Cancer Immunol Immunotherapy 55:643–651.

Gabor, L.J., D.N. Love, R. Malik, and P.J. Canfield. 2001. Feline immunodeficiency virus status of Australian cats with lymphosarcoma. Aust Vet J 79:540–545.

Greenblatt, R.J., T.M. Work, G.H. Balazs, C.A. Sutton, R.N. Casey, and J.W. Casey. 2004. The Ozobranchus leech is a candidate mechanical vector for the fibropapilloma-associated turtle herpesvirus found latently infecting skin tumors on Hawaiian green turtles (Chelonia mydas). Virology 321:101–110.

Gupta, S., and A.V.S. Hill. 1995. Dynamic interactions in malaria: Host heterogeneity meets parasite polymorphism. Proc Roy Soc Lond B 261:271–277.

Hamede, R., H.I. McCallum, and M.E. Jones. 2008. Seasonal, demographic and density-related patterns of contact between Tasmanian devils: Implications for transmission of devil facial tumour disease. Austral Ecol 33:614–622.

Hamede, R.K., S. Lachish, K. Belov, G. Woods, A. Kreiss, A.-M Pearse, B. Lazenby, M. Jones, and H. McCallum. 2012. Reduced impact of Tasmanian Devil Facial Tumour Disease at the current disease front. Conserv Biol 26:124-134

Hamilton, W.D., and R.M. May. 1977. Dispersal in stable habitats. Nature 269:578–581.

Hanahan, D., and R.A. Weinberg. 2000. The hallmarks of cancer. Cell 100:57–70.

Harrenstien, L.A., L. Munson, U.S. Seal, G. Riggs, M.R. Cranfield, L. Klein, A.W. Prowten, D.D. Starnes, V. Honeyman, R.P. Gentzler, P.P. Calle, B.L. Raphael, K.J. Felix, J.L. Curtin, C.D. Page, D. Gillespie, P.J. Morris, E.C. Ramsay, C.E. Stringfield, E.M. Douglass, T.O. Miller, B.T. Baker, N. Lamberski, R.E. Junge, J.W. Carpenter, and T. Reichard. 1996. Mammary cancer in captive wild felids and risk factors for its development: A retrospective study of the clinical behavior of 31 cases. J Zoo Wildl Med 27:468–476.

Haverkos, H.W. 2004. Viruses, chemicals and co-carcinogenesis. Oncogene 23:6492–6499.

Hawkins, C.E., C. Baars, H. Hesterman, G.J. Hocking, M.E. Jones, B. Lazenby, D. Mann, N. Mooney, D. Pemberton, S. Pyecroft, M. Restani, and J. Wiersma. 2006. Emerging disease and population decline of an island endemic, the Tasmanian devil Sarcophilus harrisii. Biol Conserv 131:307–324.

Jensen, K.H., T. Little, A. Skorping, and D. Ebert. 2006. Empirical support for optimal virulence in a castrating parasite. PLoS Biol 4:1265–1269.

Jones, M.E., D. Paetkau, E.L.I. Geffen, and C. Moritz. 2004. Genetic diversity and population structure of Tasmanian devils, the largest marsupial carnivore. Mol Ecol 13:2197–2209.

Jones, M., P. Jarman, C. Lees, H. Hesterman, R. Hamede, N. Mooney, D. Mann, C. Pukk, J. Bergfeld, and H. McCallum. 2007. Conservation management of Tasmanian devils in the context of an emerging, extinction-threatening disease: devil facial tumor disease. EcoHealth 4:326–337.

Jones, M.E., A. Cockburn, C. Hawkins, H. Hesterman, S. Lachish, D. Mann, H. McCallum, and D. Pemberton. 2008. Life-history change in disease-ravaged Tasmanian devil populations. Proc Nat Acad Sci USA 205:10023–10027.

Jones, O.R., J.-M. Gaillard, S. Tuljapurkar, J.S. Alho, K B. Armitage, P.H. Becker, P. Bize, J. Brommer, A. Charmantier, M. Charpentier, T. Clutton-Brock, F.S. Dobson, M. Festa-Bianchet, L. Gustafsson, H. Jensen, C.G. Jones, B.-G. Lillandt, R. McCleery, J. Merilä, P. Neuhaus, M.A.C. Nicoll, K. Norris, M.K. Oli, J. Pemberton, H. Pietiäinen, T.H. Ringsby, A. Roulin, B.-E. Saether, J.M. Setchell, B.C. Sheldon, P.M. Thompson, H. Weimerskirch, E.J. Wickings, and T. Coulson. 2008. Senescence rates are determined by ranking on the fast-slow life-history continuum. Ecol Lett 11:664–673.

King, D.P., M.C. Hure, T. Goldstein, B.M. Aldridge, F.M.D. Gulland, J.T. Saliki, E.L. Buckles, L.J. Lowenstine, and J.L. Stott. 2002. Otarine herpesvirus-1: a novel gammaherpesvirus associated with urogenital carcinoma in California sea lions (*Zalophus californianus*). Vet Microbiol 86:131–137.

King, R.J.B., and M.W. Robins. 2006. Cancer biology. 3rd ed. Pearson Education, Harlow.

Klein, J. 1986. Natural history of the major histocompatability complex. John Wiley and Sons, New York.

Kreiss, A., N. Fox, J. Bergfeld, S.J. Quinn, S. Pyecroft, and G.M. Woods. 2008. Assessment of cellular immune responses of healthy and diseased Tasmanian devils (*Sarcophilus harrisii*). Devel Comp Immunol 32:544–553.

Kreiss, A., D.L. Obendorf, S. Hemsley, P.J. Canfield, and G.M. Woods. 2009a. A histological and immunohistochemical analysis of lymphoid tissues of the Tasmanian devil. Anat Rec 292:611–620.

Kreiss, A., B. Wells, and G.M. Woods. 2009b. The humoral immune response of the Tasmanian devil (*Sarcophilus harrisii*) against horse red blood cells. Vet Immunol Immunopath 130:135–137.

Kreiss, A., Y.Y. Cheng, F. Kimble, B. Wells, S. Donovan, K. Belov, and G.M. Woods. 2011, Allorecognition in the Tasmanian devil (*Sarcophilus harrisii*), an endangered marsupial species with limited genetic diversity. PLoS ONE 6(7): e22402. doi:10.1371/journal.pone.0022402

Kumar, S. 1989. Zoo animals not spared from cancer. J Natl Cancer Inst 81:1691–1692.

Kurbel, S., S. Plestina, and D. Vrbanec. 2007. Occurrence of the acquired immunity in early vertebrates due to danger of transmissible cancers similar to canine venereal tumors. Med Hypotheses 68:1185–1186.

Lachish, S., H. McCallum, D. Mann, C.E. Pukk, and M.E. Jones. 2010a. Evaluating selective culling of infected individuals to control Tasmanian devil facial tumor disease. Conserv Biol 24: 841–851.

Lachish, S., K.J. Miller, A. Storfer, A.W. Goldizen, and M.E. Jones. 2010b. Evidence that disease-induced population decline changes genetic structure and alters

dispersal patterns in the Tasmanian devil. Heredity 106:172–182.

Lachish, S., M.E. Jones, and H. McCallum. 2007. The impact of devil facial tumor disease on the survival and population growth rate of the Tasmanian devil. J Anim Ecol 76:926–936.

Lachish, S., H. McCallum, and M.E. Jones. 2009. Demography, disease and the devil: life-history changes in a disease affected population of Tasmanian devils (*Sarcophilus harrisii*). J Anim Ecol 78:427–436.

Lafferty, K.D., and A.M. Kuris. 2002. Trophic strategies, animal diversity and body size. Trends Ecol Evol 17:507–513.

Loh, R., J. Bergfeld, D. Hayes, A. O'Hara, S. Pyecroft, S. Raidal, and R. Sharpe. 2006. The pathology of devil facial tumor disease (DFTD) in Tasmanian devils (*Sarcophilus harrisii*). Vet Pathol 43:890–895.

Lopez, G., M. Lopez-Parra, L. Fernandez, C. Martinez-Granados, F. Martinez, M.L. Meli, J.M. Gil-Sanchez, N. Viqueira, M.A. Diaz-Portero, R. Cadenas, H. Lutz, A. Vargas, and M.A. Simon. 2009. Management measures to control a feline leukemia virus outbreak in the endangered Iberian lynx. Anim Conserv 12:173–182.

Marker, L., L. Munson, P.A. Basson, and S. Quackenbush. 2003. Multicentric T-cell lymphoma associated with feline leukemia virus infection in a captive Namibian cheetah (*Acinonyx jubatus*). J Wildl Dis 39:690–695.

Martin, D., and J.S. Gutkind. 2009. Human tumor-associated viruses and new insights into the molecular mechanisms of cancer. Oncogene 27:S31–S42.

Maslov, A.Y., and J. Vijg. 2009. Genome instability, cancer and aging. BBA Gen Subjects 1790:963–969.

McAloose, D., and A.L. Newton. 2009. Wildlife cancer: a conservation perspective. Nat Rev Cancer 9:517–526.

McCallum, H. 2008. Tasmanian devil facial tumor disease: lessons for conservation biology. Trends Ecol Evol 23:631–637.

McCallum, H., N. Barlow, and J. Hone. 2001. How should pathogen transmission be modelled? Trends Ecol Evol 16:295–300.

McCallum, H., and M. Jones. 2006. To lose both would look like carelessness. Tasmanian devil facial tumour disease. PLoS Biol 4:1671–1674.

McCallum, H., M. Jones, C. Hawkins, R. Hamede, S. Lachish, D. L. Sinn, N. Beeton, and B. Lazenby. 2009. Transmission dynamics of Tasmanian devil facial tumor disease may lead to disease-induced extinction. Ecology 90:3379–3392.

McCallum, H., D.M. Tompkins, M.E. Jones, S. Lachish, S. Marvenek, B. Lazenby, G. J. Hocking, J. Wiersma, and C. Hawkins. 2007. Distribution and impacts of

Tasmanian devil facial tumour disease. EcoHealth 4:318–325.

Merlo, L. M.F., J.W. Pepper, B. J. Reid, and C.C. Maley. 2006. Cancer as an evolutionary and ecological process. Nat Rev Cancer 6:924–935.

Mozos, E., A. Méndez, J.C. Gómez-Villamandos, J. Martin de las Mulas, and J. Pérez. 1996. Immunohistochemical characterization of canine transmissible venereal tumor. Vet Pathol 33:257–263.

Murchison, E.P. 2009. Clonally transmissible cancers in dogs and Tasmanian devils. Oncogene 27:S19–S30.

Murchison, E.P., C. Tovar, A. Hsu, H.S. Bender, P. Kheradpour, C.A. Rebbeck, D. Obendorf, C. Conlan, M. Bahlo, C.A. Blizzard, S. Pyecroft, A. Kreiss, M. Kellis, A. Stark, T.T. Harkins, J.A.M. Graves, G.M. Woods, G.J. Hannon, and A.T. Papenfuss. 2010. The Tasmanian devil transcriptome reveals Schwann cell origins of a clonally transmissible cancer. Science 327:84–87.

Murgia, C., J.K. Pritchard, S.Y. Kim, A. Fassati, and R.A. Weiss. 2006. Clonal origin and evolution of a transmissible cancer. Cell 126:477–487.

Newman, S.J., and S.A. Smith. 2006. Marine mammal neoplasia: A review. Vet Pathol 43:865–880.

Nowell, P. 1976. The clonal evolution of tumor cell populations. Science 194:23–28.

Nussey, D.H., T. Coulson, M. Festa-Bianchet, and J.M. Gaillard. 2008. Measuring senescence in wild animal populations: towards a longitudinal approach. Funct Ecol 22:393–406.

O'Brien, S.J., and J.F. Evermann. 1988. Interactive influence of infectious disease and genetic diversity in natural populations. Trends Ecol Evol 3:254–259.

Owston, M.A., E.C. Ramsay, and D.S. Rotstein. 2008. Neoplasia in felids at the Knoxville Zoological Gardens, 1979–2003. J Zoo Wildl Med 39:608–613.

Pearse, A.-M., and K. Swift. 2006. Allograft theory: Transmission of devil facial-tumour disease. Nature 439:549.

Pearse, A.-M., Swift, K.R. Hodson, P., Hua, B., McCallum, H. Pyecroft, S., Taylor, R.L., Eldridge, M. and Belov, K. (in press). Evolution in a Transmissible Cancer: a study of the chromosomal changes in Devil Facial Tumour (DFT) as it spreads through the wild Tasmanian Devil population. Cancer Genetics

Pyecroft, S.B., A.-M. Pearse, R. Loh, K. Swift, K. Belov, N. Fox, E. Noonan, D. Hayes, A.D. Hyatt, L.F. Wang, D.B. Boyle, J.S. Church, D.J. Middleton, and R.J. Moore. 2007. Towards a case definition for devil facial tumor disease: What is it? EcoHealth 4:346–351.

Quammen, D. 2008. Contagious cancer: The evolution of a killer. Harper's Mag 316:33.

Rebbeck, C.A., R. Thomas, M. Breen, A.M. Leroi, and A. Burt. 2009. Origins and evolution of a transmissible cancer. Evolution 63:2340–2349.

Roelke, M.E., M.A. Brown, J.L. Troyer, H. Winterbach, C. Winterbach, G. Hemson, D. Smith, R.C. Johnson, J. Pecon-Slattery, A. L. Roca, K.A. Alexander, L. Klein, P. Martelli, K. Krishnasamy, and S.J. O'Brien. 2009. Pathological manifestations of feline immunodeficiency virus (FIV) infection in wild African lions. Virology 390:1–12.

Rous, P. 1911. Transmission of a malignant new growth by means of a cell-free filtrate. J Am Med Assoc 56:198–201.

Ruddon, R.W. 2007. Cancer Biology. Oxford University Press, Oxford UK.

Schmid-Hempel, P. 2008. Parasite immune evasion: a momentous molecular war. Trends Ecol Evol 23:318–326.

Siddle, H.V., A. Kreiss, M.D.B. Eldridge, E. Noonan, C.J. Clarke, S. Pyecroft, G.M. Woods, and K. Belov. 2007. Transmission of a fatal clonal tumor by biting occurs due to depleted MHC diversity in a threatened carnivorous marsupial. Proc Nat Acad Sci USA 104:16221–16226.

Siddle, H.V., J. Marzec, Y. Cheng, M. Jones, and K. Belov. 2010. MHC gene copy number variation in Tasmanian devils: implications for the spread of a contagious cancer. Proc R Soc Lond B 277:2001–2006.

Sutherland-Smith, M., C. Harvey, M. Campbell, D. McAloose, B. Rideout, and P. Morris. 2004. Transitional cell carcinomas in four fishing cats (Prionailurus viverrinus). J Zoo Wildl Med 35:370–380.

Thacher, C., and R.L. Bradley. 1983. Vulvar and vaginal tumors in the dog: a retrospective study. J Am Vet Med Assoc 183:690–692.

Tolar, J., and J.P. Neglia. 2003. Transplacental and other routes of cancer transmission between individuals. J Paediat Hematol Oncol 25:430–434.

Trent, J.T., and R.S. Kirsner. 2003. Wounds and malignancy. Adv Skin Wound Care 16:31–34.

Tucker, A.R., and J.R. Smith. 2008. Prostatic squamous metaplasia in a cat with interstitial cell neoplasia in a retained testis. Vet Pathol 45:905–909.

Vage, J., C. Wade, T. Biagi, J. Fatjó, M. Amat, K. Lindblad-Toh, and F. Lingaas. 2010. Association of dopamine- and serotonin-related genes with canine aggression. Genes Brain Behav 9:372–378.

van den Berg, L., M. Vos-Loohuis, M.B. Schilder, B.A. van Oost, H.A. Hazewinkel, C.M. Wade, E.K. Karlsson, K. Lindblad-Toh, A.E. Liinamo, and P.A. Leegwater. 2008. Evaluation of the serotonergic genes htrA, htrB, htr2A, and slc6A4 in aggressive behavior of golden retriever dogs. Behav Genet 38:55–66.

van Oers, K., G. de Jong, A.J. van Noordwijk, B. Kempenaers, and P.J. Drent. 2005. Contribution of genetics to the study of animal personalities: a review of case studies. Behaviour 142:1185–1206.

von Holdt, B.M., and E.A. Ostrander. 2006. The singular
history of a canine transmissible tumor. Cell
126:445–447.

Walzer, C., A. Kubber-Heiss, and B. Bauder. 2003.
Spontaneous uterine fibroleiomyoma in a captive
cheetah. J Vet Med Assoc 50:363–365.

Willi, Y., J. Van Buskirk, and A.A. Hoffmann. 2006. Limits
to the adaptive potential of small populations. Ann Rev
Ecol Evol Syst 37:433–458.

Woodruff, R.A., R.K. Bonde, J.A. Bonilla, and
C.H. Romero. 2005. Molecular identification of a
papilloma virus from cutaneous lesions of captive and
free-ranging Florida manatees. J Wild Dis 41:437–441.

Woods, G.M., A. Kreiss, K. Belov, H.V. Siddle,
D.L. Obendorf, and H.K. Muller. 2007. The immune
response of the Tasmanian devil (*Sarcophilus harrisii*)
and devil facial tumor disease. EcoHealth 4:338–345.

Woolford, L., A.J. O'Hara, M.D. Bennett, M. Slaven,
R. Swan, J.A. Friend, A. Ducki, C. Sims, S. Hill,
P.K. Nicholls, and K.S. Warren. 2008. Cutaneous
papillomatosis and carcinomatosis in the western
barred bandicoot (*Perameles bougainville*). Vet Path
45:95–103.

Zinkernagel, R.M., H. Hengartner, and L. Stitz. 1985.
On the role of viruses in the evolution of immune
responses. Brit Med Bull 41:92–97.

# 20

## FROM PROTOZOAN INFECTION IN MONARCH BUTTERFLIES TO COLONY COLLAPSE DISORDER IN BEES

### Are Emerging Infectious Diseases Proliferating in the Insect World?

Rebecca Bartel and Sonia Altizer

GREAT FLEAS HAVE little fleas upon their backs to bite 'em, and little fleas have lesser fleas, and so ad infinitum.

*Augustus de Morgan*

In the late 1990s, scientists first noticed mysterious declines in several wild bumble bee species (*Bombus* spp.) in both eastern and western North America, with one species now possibly extinct (Colla and Packer 2008). These bumble bee losses mirrored declines in the abundance of many other native pollinators (National Research Council 2007). Interestingly, the timing of bumble bee declines in the United States coincided with reports of disease outbreaks in commercial-reared bumble bees sold for use in the production of greenhouse tomatoes and peppers (Evans et al. 2008). This observation, together with reports of a higher incidence of key pathogens, including the trypanosome *Crithidia bombi* and the microsporidian *Nosema bombi* in wild bumble bees foraging near greenhouse colonies (Colla et al. 2006), suggests that the spread of pathogens from commercial to wild bees could play a role in observed declines. Although details are still emerging on the incidence and effects of different bumble bee pathogens in North

America, this example could be one of the first cases of pathogen spillover from domestically reared to wild populations of an insect host. More generally, this example points to the potential for pathogens to cause insect declines, and underscores the need for more baseline data on pathogen prevalence in wild insect populations.

Relative to vertebrate animals, far less is known about the significance of infectious diseases for insect conservation. This crucial knowledge gap probably arises for several reasons. First, insects are both smaller in size and more diverse in numbers: in fact, the totality of described insect species outnumbers that of vertebrate species by a factor of at least 16:1. This diversity is still being described at a considerable rate; at the same time, the conservation status of the majority of insect species remains unknown (Lewis et al. 2007). Less than 1% of described insect species have been evaluated for conservation status as of 2010, compared to 44% of vertebrate species (IUCN 2010). Thus, it seems fair to infer that, aside from a small percentage of well-studied organisms, the conservation status of and major threats to most insect species remain relatively understudied and undocumented (Dunn 2005).

A common perception among scientists and the public alike is that infectious diseases pose low risks to insect hosts, in large part because insects have fast generation times, large population sizes, and high fecundity. Thus, insect hosts should be better able to rebound from disease outbreaks and have greater potential to evolve resistance to pathogens over short timescales compared to most vertebrate hosts. Yet very little is known about pathogens affecting wild insect populations in terms of their taxonomic diversity, natural host ranges, and impacts. On the one hand, many studies of insect diseases during the past 150 years focused on the use of pathogens as biological control agents to limit populations of insect pests (Lacey and Kaya 2000). On the other hand, pathogens have been known to decimate populations of cultivated insects such as honey bees and silkworm moths, with records of silkworm moth diseases in China dating back to 2700 B.C. (Tanada and Kaya 1992). Over 40 different pathogens are known to infect honey bees, and organisms such as *Varroa* mites and American foulbrood bacteria have caused local collapses and continent-wide declines in recent decades (Shimanuki et al. 1992; Martin et al. 1998). Because a high diversity of pathogens are known to infect economically important (and hence better-studied) insect species, this suggests that the majority of non-cultivated and non-pest insects probably also harbor multiple parasitic organisms that can negatively affect their survival and reproduction.

In this chapter, we discuss some of the ways that pathogens are relevant to the lives of free-living insects, and review potential concerns that emerging diseases might pose for insect conservation. We begin by considering the types of pathogens that have been most successful in targeting insect pest species, and their possible risks to non-target species. Next, we discuss how the commercial sale and long-distance transfer of captive-reared insects might cause pathogen spillover in wild populations. Insects can mount a range of behavioral and immune defenses that influence the outcome of infection, and these defenses can evolve in response to pathogen pressure and change with environmental conditions. Finally, we conclude by discussing the potential consequences of global change, including climate change, for insect–pathogen interactions, and the broader concerns of future disease threats for insect conservation.

## INSECT PATHOGENS AS BIOLOGICAL CONTROL AGENTS

Biological control involves the use of living organisms to suppress pest populations. Insects are one of the few types of organisms that can serve as natural enemies for biological control programs and that are also commonly targeted as pest species to be controlled using natural enemies. Using natural enemies as agents to control pest insects has a long history and has increased substantially in the past 30 years in response to environmental hazards and human safety issues associated with chemical pesticides. Numerous pathogens have been implemented as control agents targeting insect pests, including viruses, bacteria, fungi, and parasitic nematodes (Lacey et al. 2001). Many of these pathogens can transmit effectively and spread through an insect population following localized introductions (Bedford 1980; Zelazny et al. 1992). One of the best examples of an insect pathogen that has established and effectively controlled an insect pest in some areas is a fungal pathogen of gypsy moths (*Lymantria dispar*; Box 20.1). Other pathogens, such as the soil bacterium *Bacillus thuringiensis* (*Bt*), are mass-cultured and applied as "microbial pesticides" rather than as self-propagating agents (Lacey et al. 2001).

## Collateral Damage: Risks of Biocontrol Agents to Non-Target Insects

Irrespective of the target pest species, biological control agents require comprehensive risk assessment and evaluation of biosafety issues prior to their release (see Barratt et al. 2010 for a complete review). One of the largest concerns is the unintended impact on non-target species (Hajek 2004). Because of the immediate need to control damage during pest outbreaks, the effects of natural enemies on non-target species might not evaluated before treatment occurs. Indeed, a global review of over 5,200 classical biological control releases targeting insect pests since the late 1800s identified only 1.7% of the total cases recording the potential effects on non-target species (Lynch and Thomas 2000) and estimated that as many of 11% of past enemy releases for insect biocontrol may have had serious non-target effects (see Louda et al. 2003 for several specific cases). Sixteen percent of 313

---

**Box 20.1  When it Rains, it Spores: Fungal Control of Gypsy Moths in North America**

---

One of the most successful examples of biological control of an insect pest is the introduction of the fungal pathogen *Entomophaga maimaiga* to North American populations of the gypsy moth, *Lymantria dispar* (Fig. 20.1). Gypsy moths were brought to the United States from France in the mid-1800s with hopes of hybridization with native North American silkworms. In the late 1800s, some gypsy moths escaped in Massachusetts and spread into nearby areas, eventually causing extensive defoliation in urban and suburban forests (Hajek 2007). In 1905, state and federal agencies began to implement classical biological control programs to reduce gypsy moth populations and limit forest damage. Introduced enemies included parasitoids, a nuclear polyhedrosis virus (LdMNPV), fungal spores of *Entomophaga*, and other insects (Hajek 2007), yet none of these were effective consistently. In 1989, the fungal pathogen *E. maimaiga* was reported as causing widespread epizootics in gypsy moths across northeastern U.S. populations; scores of dead gypsy moth larvae were found clinging to the trees (Hajek et al. 1999). Whether the *E. maimaiga* strain first detected in North America in 1989 originated from *Entomophaga* spores released in 1910–11 remains debatable because there was no evidence of this fungus between 1910 and 1989 (Hajek et al. 1995; Hajek 1999). *E. maimaiga* has two spore forms: conidia, which are produced externally on cadavers, and resting spores, which are produced within cadavers. The type of spore formed after host death is determined by the fungus, host-related factors, and environmental conditions (Hajek 1999). Resting spores overwinter in the soil, and in the spring they germinate to actively eject infective conidia onto dispersing larvae and cause primary infections (Hajek et al. 2004). Infected larvae die within 7 to 10 days, after which infective conidia are actively ejected from cadavers, become airborne, and are activated following contact with lepidopteran larvae (Hajek et al. 1999, 2004).

Since 1989, multiple outbreaks caused by *E. maimaiga* subsequently occurred in gypsy moth populations (Hajek et al. 2004). Importantly, *E. maimaiga* does not appear to cause major threats to non-target hosts: infections have been documented consistently in only three species of tussock moths and laboratory-reared hawk moths (Hajek et al. 1995, 2004). Because *E. maimaiga* spores are difficult to rear in the laboratory, current biocontrol measures involve the release of field-collected resting spores into areas where gypsy moth populations have recently invaded (Hajek 2007). Field trials to date show promising results in that the fungus, when introduced into low-density gypsy moth populations, can limit population growth and slow the rate of spatial expansion of this invasive species.

---

parasitoids released in the eastern United States to control exotic pests have been documented parasitizing non-target native host species (Hawkins and Marino 1997), a conservative estimate due to lack of documentation of effects on non-target species. Perhaps more surprisingly, post-treatment impacts and population assessment of impacts on the actual target species following release of biocontrol agents are also rare (Louda et al. 2003).

Effects on non-target species can have even greater impacts in fragile ecosystems with simple food webs, particularly on islands (Hajek 2004). For example, over 675 species of natural enemies were introduced

to Hawaii for biological control in the past century, 71.6% of which were predators and parasitoids to control insects (Funasaki et al. 1988; Hajek 2004) and were not screened for host specificity (Funasaki et al. 1988). Of 84 parasitoids of lepidopteran pests released into Hawaii since 1960, 32 species have become established (Funasaki et al. 1988) and are suspected to have severe impacts on native moth populations (Zimmermann 1978; Louda et al. 2003). To examine how introduced enemies have infiltrated a native Hawaiian ecological community, Henneman and Memmott (2001) collected over 2,100 larvae from all plant species in the remote Alakai Swamp

**Figure 20.1:**
Gypsy moth larvae infected with the fungal pathogen *Entomophaga maimaiga*. Dead larvae are often found hanging vertically along tree trunks. Fungal conidia (as shown as the light-colored powder around the larva in the photo) are forcibly ejected, whereas resting spores remain within cadavers. (Photo provided by Darwin Dale and David Smitley.)

on the island of Kauai. From 216 individual reared parasitoids, 83% were of three introduced species for biocontrol and known to attack native Lepidoptera (Funasaki et al. 1988). These introduced agents probably compete with native parasitoids, as native enemy species were rarely observed (Henneman and Memmott 2001; Louda et al. 2003). Cases like this illustrate challenges in teasing apart the complex interactions between biological control agents and native natural enemies in fragile ecosystems.

With bacterial, viral, and fungal pathogens being implemented as natural enemies for pest populations, there is great potential for these pathogens to overcome invertebrate host species barriers and infect both pest and non-target species (Roy et al. 2009). Perhaps the most widely used and best-studied example of insect microbial biocontrol is the bacterial pathogen *Bt*, a widely used biopesticide. Spores of *Bt* could persist in the soil for at least a year given ideal conditions (Addison 1993). Different *Bt* strains target Lepidoptera, Diptera, Coleoptera, and a few other insect orders (Schnepf et al. 1998) and are commonly used in agricultural settings to control leaf-feeding caterpillars and beetle larvae, forest Lepidoptera, and larval mosquitoes. This pathogen

has been used successfully to control gypsy moth outbreaks in the eastern United States, but *Bt* can also have a negative impact on native non-pest populations of caterpillars. Wagner et al. (1996) found that 19 of 20 common caterpillar species showed slight population decreases in the treatment year and the following two years after application of *Bt* in forest plots in west central Virginia.

Because of the increasing number of cases of non-target effects of biocontrol agents, multiple tools are now being used to measure population-level impacts on both target and non-target species. These include molecular taxonomy to correctly identify proposed biological control agents, more comprehensive evaluation of the range of species that could be affected, and the development of food web models that incorporate native species and biocontrol agents (Barratt et al. 2010). Importantly, the fairly limited information on risks to non-target insect species from previous biocontrol efforts probably reflects a lack of knowledge and quantitative assessment rather than a lack of risk, and future studies should allow researchers to better predict direct and indirect effects of natural enemies on non-target species.

## HOW SICK IS THAT BEE IN THE WINDOW? THE COMMERCIAL SALE OF INSECTS AS A VEHICLE FOR PATHOGEN SPREAD

The growing popularity of raising and releasing insects such as ants, ladybird beetles, bees, and butterflies has resulted in the large-scale rearing and rising domestic commercial sales of many insect species. Some insects are purchased for garden biological control (e.g., ladybugs, praying mantids), others are sold for commercial use (e.g., bumble bees as greenhouse pollinators), and yet others are ordered for use in home or classroom education activities or for release at special events (e.g., butterflies). As emphasized in previous sections, the parasitic flora and fauna of most insect species and their potential for inter-specific transmission remain largely unknown. Thus, one major concern with commercial rearing, transport, and release of insects is that these activities can result in the unintentional propagation (and potential long-distance transfer) of infectious agents within and between wild insect populations.

The sale and distribution of live insects are difficult to track due to the lack of consistent legal regulations and documentation across geographic and political boundaries. Within the United States, live insects can be distributed across North America with a single permit issued under the authority of the U.S. Department of Agriculture (USDA). USDA permits are required for the domestic movement of insects that can be defined as pest species under special circumstances, including ants, beetles, butterflies, cockroaches, crickets, grasshoppers, milkweed bugs, moths, termites, and walking sticks. To import an insect species into a state, buyers or sellers must obtain a permit for each species. Permits are valid for up to four years, but the USDA does not keep track of how many insects are shipped. It is illegal to ship non-native live insects across state lines. The Endangered Species Act also prevents the collection and sale of 56 threatened and endangered insect species within the United States.

## Potential Pathogen Risks from Butterfly Releases

The commercial sale of live butterflies is a growing business fueled by the $160 million/year wedding industry and demand for butterflies for educational programs, with an estimated $40 million/year being spent in the United States on butterfly purchases. The most common species reared for festive releases are the American painted lady (Vanessa virginiensis) and monarch (Danaus plexippus). Introducing large numbers of commercially raised wild insects into potentially novel environments raises daunting implications for infectious disease risks. Rearing animals at high densities in commercial operations could increase the spread of some infectious diseases, including those transmitted by external contact with or ingestion of infectious stages. There is no mandatory testing of pathogens in these facilities, nor are sellers required by the USDA to certify that insect colonies are disease-free. Visual inspection alone may not reveal subtle infections, and sophisticated diagnostic techniques are often needed to detect bacterial, viral, protozoan, and fungal infections.

As one case study, monarchs are often infected by the vertically and horizontally transmitted protozoan Ophryocystis elektroscirrha (Fig. 20.2). Parasites infect monarchs in all populations examined to date and the prevalence varies dramatically among populations, even within North America (Leong et al. 1997; Altizer et al. 2000). O. elektroscirrha infections cause reduced adult body size, shorter adult lifespan (De Roode et al. 2007, 2008), and reduced flight performance (Bradley and Altizer 2005). However, because infections tend not cause immediate mortality in adults, individuals who harbor low-level infections may appear normal. Therefore, breeders could unknowingly release significant numbers of individuals with low levels of infection into wild populations, altering parasite prevalence in the field. This could be especially important early in the monarchs' breeding season, when natural infections are extremely rare (see Bartel et al. 2011). Because it is nearly impossible to distinguish released butterflies from wild ones, scientific investigation of infectious disease dynamics at a population level could be further complicated by the effects of captive-raised monarchs on estimated prevalence, and by the potential long-distance transfer of novel pathogen strains between populations (e.g., Brower et al. 1995).

It is difficult to estimate how many commercial operations offer butterflies for sale due to the lack of documentation and registration with the USDA. Because breeders are not required to register with the

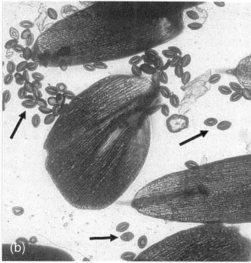

**Figure 20.2:**
(a) Newly emerged monarch infected with *O. elektroscirrha*. (b) Monarchs infected with parasites harbor dormant spores on the outsides of their bodies, as indicated by arrows pointing to smaller lemon-shaped objects next to larger abdominal scales. (Photos by Rebecca Bartel and Sonia Altizer [left to bottom].)

USDA, nor are they inspected for disease, this has ignited a debate surrounding the environmental risks of such activities. Both the North American Butterfly Association (NABA) and the Xerces Society (Pyle et al. 2010) have issued statements proposing a ban on the environmental release of commercially reared butterflies. Scientists cited several ecological threats of the releases, including the (1) spread of disease and parasites, (2) introductions of non-native species, (3) subsequent inappropriate genetic mixing, (4) creation of a commercial market for butterflies, potentially making wintering sites targets for poaching, and (5) disruption of migrations to wintering grounds if butterflies are released at the wrong time of year. Total estimates of released monarchs for all commercial breeders combined are approximately 11 million per year (Pyle et al. 2010), which amounts to a small but measurable fraction of the North American monarch population. Given the growing popularity of butterfly releases, interstate movements of sales and breeding stock, lack of required screening for infectious agents, and potential for cross-species transmission in operations where multiple butterfly species are reared together, there is little protection from the risk of pathogens released from commercial operations into wild populations.

## Disease Risks in Other Commercially Raised Insects

The risk of disease introduction into wild populations by commercially raised individuals reaches beyond Lepidoptera. Growing disease threats to bee pollinators, most notably honey bees (e.g., *Apis mellifera*), have lead to calls for stricter regulation of bee movement by the USDA. Domesticated honey bee stocks have declined by 59% since 1950 (National Research Council 2007). Previous losses were attributed to two parasitic mites, *Varroa destructor* and *Acarapis woodi*, and the bacterial pathogen *Paenibacillus larvae*, which causes American foulbrood, but more recent declines have been associated with colony collapse disorder (see below). Currently, the USDA only restricts importation of honey bees and honey bee materials into the continental United States from Australia, Canada, and New Zealand under the authority of the Honey Bee Act of 1922 (HBA) to prevent the introduction and spread of diseases and parasites.

The HBA was initially enacted by Congress primarily to prevent importing *A. woodi* into the United States. The revised regulations for the importation of honey bees were last amended in November 2004. The USDA stopped regulating interstate movement of most bee species in 1998, deferring the matter to individual state governments (Flanders et al. 2003).

Honey bees are the best-studied pollinator in terms of their infectious diseases, but disease-related declines have also caused concern for bumble bee species (*Bombus* spp.) in recent years. Since the 1990s, colonies of the native bumble bees, *B. occidentalis* and *B. impatiens*, have been mass-produced and distributed to commercial greenhouses for the pollination of tomatoes and sweet peppers in North America. The virulent strain of the microsporidian *Nosema bombi* does not naturally occur in North America, but is hypothesized to have been introduced to the United States in the early 1990s via queens of *B. terrestris* (a European species) and *B. occidentalis* that were shipped between the United States and Europe (Evans et al. 2008). Outbreaks of *N. bombi* infections in *B. occidentalis* decimated colonies in 1997 to the point that *B. occidentalis* is no longer reared commercially in the United States (Flanders et al. 2003). During the same time, wild populations of *B. occidentalis*, *B. terricola*, and *B. affinis* were declining, in part due to the spread of introduced parasites, namely *N. bombi* and *Crithidia bombi*, most likely originating from European bumble bees (National Research Council 2007; Evans et al. 2008).

Several other studies have found higher pathogen prevalence in commercially reared bumble bees than in nearby wild populations (summarized by Colla et al. 2006) including the protozoan *C. bombi* (Liu 1973; MacFarlane et al. 1995), *N. bombi* (Liu 1973; Whittington and Winston 2003), and the tracheal mite *Locustacarus buchneri* (Otterstatter and Whidden 2004). Infections by these parasites can reduce colony survival, reproduction, and worker foraging efficiency (MacFarlane et al. 1995; Otterstatter et al. 2005; Colla et al. 2006). There is evidence of *C. bombi* infecting up to 75% of wild bumble bees near industrial greenhouses that used *Bombus* species for commercial pollination, where the prevalence and intensity of *C. bombi* infections declined with increasing distance from greenhouses (Colla et al. 2006; Otterstatter and Thomson 2008).

## WILL INSECTS RESPOND EVOLUTIONARILY TO NOVEL PATHOGEN PRESSURES?

Given that many insects have a wide repertoire of behavioral and immune defenses (reviewed in Tanada and Kaya 1992; Schmid-Hempel 2005) together with high fecundity and short generation times, a common assumption is that resistance should evolve rapidly in response to lethal or debilitating pathogens—thus lowering the long-term impacts of infectious agents, including those that are newly introduced. In line with this expectation, resistance evolution has been observed in multiple case studies in recent years. One striking example of rapid evolution of behavioral defense in response to an introduced parasite was demonstrated in field crickets (*Teleogryllus oceanicus*) inhabiting the Hawaiian island of Kauai (Zuk et al. 2006). These crickets are attacked by the acoustically orienting parasitoid fly *Ormia ochracea*, common in North America and more recently introduced to Hawaii. Between 1991 and 2003, crickets on Kauai declined in numbers, and field monitoring revealed high infection rates by the flies (Zuk et al. 1993). Starting in 2003, researchers noted increasing numbers of crickets again but no calling males: in fact, nearly all males captured had female-like wings that lacked the sound-producing structures (Zuk et al. 2006). These "flatwing" males were only rarely attacked by parasitoids, yet were still accepted by females as mates. This example illustrates that strong selection pressure by introduced parasites can lead to rapid evolution of behavioral/physical resistance traits and the subsequent recovery of previously declining insect populations.

Another example of rapid evolution of host resistance in response to a potentially threatening pathogen was observed in island populations of the Polynesian butterfly *Hypolimnas bolina* infected with a male-killing *Wolbachia* (Dyson and Hurst 2004; Box 20.2). Prior to 2004, this bacterial pathogen infected nearly all butterflies on the Samoan island of Savai'i and caused an extreme sex ratio bias, with a virtual disappearance of males from the population. By destroying the male embryos from infected females, the bacteria eliminate the non-transmitting males and hence increase their own population-level spread. From 2004 to 2006, however, a dramatic turnaround was observed on some islands, where the

---

### Box 20.2 *Wolbachia*: Harmless Symbiont or Sexual Sabotage?

---

Bacteria in the genus *Wolbachia* are among the most common intracellular symbionts associated with insects. These maternally transmitted bacteria infect up to two thirds of all insect species examined to date (Weeks et al. 2002), and commonly infected orders include Lepidoptera, Hymenoptera, and Coleoptera. *Wolbachia* can interfere with insect reproduction in several ways: depending on the host species affected, these bacteria can kill male embryos, cause feminization, or cause cytoplasmic incompatibility (CI) between males and females with unrelated strains (reviewed in Islam 2007). These manipulations give the bacteria an advantage in spreading through host populations via maternal transmission by increasing the relative reproductive success of infected females or by reducing the production of non-transmitting males. However, these tactics can also cause host population declines during the process of invasion of initially healthy populations or, in the case of male-killing *Wolbachia*, due to extreme sex ratio shifts.

In some butterfly species, the population-level proportion of females and the fraction of females that produce only daughters have reached exceptionally high levels due to *Wolbachia* infections. The prevalence of *Wolbachia* across multiple populations of *Acraea encedon* in Africa ranged from 70% to 100%; at many of these sites, the proportion of males was exceedingly low, and almost none of the captured females had mated (Jiggins et al. 2000). Extreme sex ratio distortions were also reported from Polynesian populations of the butterfly *Hypolimnas bolina* infected with a male-killing *Wolbachia* (Dyson and Hurst 2004).

Given the widespread nature of these endosymbiotic bacteria and their ancient associations with insects and other invertebrates, it is tempting to assume that *Wolbachia* do not pose a significant risk for wild insect populations. However, one issue of more direct conservation concern is that these parasites can reduce host reproduction during the invasion phase, and can also lower effective population sizes (and hence genetic diversity) due to extreme sex ratio bias. These demographic changes can increase host extinction risk, especially for small or fragmented populations. A related concern is that captive breeding programs might inadvertently introduce *Wolbachia* into populations of endangered species. This risk was recently evaluated for the Karner blue (*Lycaeides melissa samuelis*, Nice et al. 2009); CI-conferring *Wolbachia* that were closely related to strains from the non-endangered Melissa (*L. m. melissa*) were found to be common in the western range of Karner blues, suggesting that cross-species transmission might have occurred following a hybridization event between the two sub-species. Importantly, model simulation suggested that *Wolbachia*-infected populations faced a two- to four-fold higher probability of extinction than uninfected populations due to the demographic impacts of CI during the invasion phase (Nice et al. 2009).

---

proportion of males increased from 1% to nearly 40% despite the fact that all adult butterflies sampled continued to show *Wolbachia* infections (Charlat et al. 2007). The authors concluded that intense selection on the host by *Wolbachia* facilitated the spread of suppressor genes to counter the male-killing properties of the bacteria, as previously described by Hornett et al. (2006). Thus, although all butterflies remained susceptible to bacterial infection, the suppressor genes minimized the deleterious fitness effects on male survival.

Although past work indicates that insect resistance to pathogens and parasites can evolve over relatively short timescales, several processes might impede the evolution of insect defenses. The most common evolutionary constraints on insect resistance to pathogens (reviewed in Schmid-Hempel 2005) are that (1) resistance traits are costly and require tradeoffs with other fitness-related traits, (2) limited resources or unfavorable environmental conditions can reduce the expression of host resistance, and (3) pathogens have evolved mechanisms to evade or counter host resistance traits. In terms of tradeoffs, insects selected to express high levels of resistance can suffer from slower development or decreased competitive ability (e.g., Sutter et al. 1968; Boots and Begon 1993; Kraaijeveld and Godfray 1997). In other cases, insects reared at high density or with poor nutrition can suffer from greater susceptibility to infection because they are in poor condition or because resources necessary for defenses are limiting (e.g., Reilly and Hajek 2008; Lindsey at al. 2009).

In considering disease risks to endangered or declining insect populations, a fourth constraint worth noting is that inbreeding and loss of genetic diversity could limit the evolution of host resistance. This concern has been noted for multiple vertebrate populations (reviewed in Acevedo-Whitehouse and Cunningham 2006; Altizer and Pedersen 2008). Although a reasonable assumption is that effective population sizes of insects will generally exceed that of many vertebrate species, heterozygosity and allelic diversity in declining or fragmented insect populations can also be quite low (e.g., Zayed and Packer 2005; Matern et al. 2009). Collectively, these issues mean that insects might not necessarily evolve effective defenses against novel and debilitating parasites over short timescales. Therefore, the assumption that insects are at lower risk of disease-induced declines than vertebrate animals due to their faster evolutionary potential might not be realized in real-world scenarios where animals face competing selection pressures, limited resources, and small, fragmented populations.

## INSECT CONSERVATION IN LIGHT OF INFECTIOUS DISEASE RISKS

Although insects on the whole remain relatively understudied in regards to extinction risk, the greatest total numbers of predicted future species extinctions are of insects, with estimates up to 57,000 losses per million species in the next 50 years (Dunn 2005). Insufficient documentation of habitat preference, geographic distributions, and abundance has resulted in the under-representation of declining insect populations in current global assessments of extinction risk, and the unfortunate lack of documentation of ongoing extinctions. The biodiversity crisis has been referred to as an undeniable insect biodiversity crisis

(Dunn 2005) with serious implications for parasites and co-extinction events (Koh et al. 2004; Dunn et al. 2009), as many specialist parasites and pathogens are expected to go extinct along with their hosts.

In terms of the effect of parasites and pathogens on insect extinctions, very little is known, especially relative to disease risks for vertebrate animals (e.g., Smith et al. 2006; Pedersen et al. 2007). The fields of insect conservation and insect pathology have made considerable progress in recent years, yet ideas from these subdisciplines seldom intersect (Roy et al. 2009). Ecologically speaking, most studies on the role of natural enemies on insect life history and non-pest population dynamics have focused on predators and parasites (rather than on microbial diseases), perhaps because predators are larger and their effects are easier to study (but see Hajek 1999, 2004; Dwyer et al. 2004). However, the diversity of insect pathogens and their importance in regulating economically important insects suggest that a high degree of insect mortality in the field probably arises from infectious diseases.

## Pollinator Declines and Loss of Pollination Services

The loss of insect species could result in a subsequent loss of ecosystem services such as waste management (dung burial), control of insect crop pests, pollination, and wildlife nutrition, the total economic value of which is at least $57 billion/year in the United States. Pollination services alone (including those by non-native species) are valued at $14.8 billion annually in the United States (Morse and Calderone 2000), and $3.1 billion annually for native pollinators (Losey and Vaughn 2006). Because insect pollinators are critical for ensuring the effective pollination of both cultivated and wild plants (Roubik 1995; Buchmann and Nabhan 1996), recent evidence of pollinator declines at local and regional scales (Biesmeijer et al 2006; National Research Council 2007) has raised concerns (Allen-Wardell et al. 1998; Winfree 2010). Bees in particular pollinate over 66% of global crop species (Roubik 1995) and are essential for an estimated 15% to 35% of food production (McGregor 1976; Klein et al. 2007). For some fruit, seed, and nut crops, production can decrease 90% without bee pollination services (Southwick and Southwick 1992).

Honey bee declines in recent decades underscore pathogen risks to pollinators and the perils of heavy reliance of crop systems on a single pollinator species (Winfree et al. 2007). Honey bees (mainly *Apis mellifera*) are one of the most economically valuable pollinators of crop monocultures (McGregor 1976; Watanabe 1994), responsible for the pollination of 100 to 150 major crops grown in the United States (Buchmann and Nabhan 1996), including alfalfa, apples, almonds, onions, broccoli, carrots, and sunflowers. Domesticated honey bee stocks have declined by 59% since 1950 (National Research Council 2007), mostly due to infestation by *Varroa destructor*, a parasitic mite thought to have originated in Asia and introduced to North America in the late 1980s (Sanford 2001). Mites feed on adults and developing workers, and if left untreated, colonies almost invariably collapse within 1 to 3 years (Wenner and Bushing 1996). The spread of *V. destructor* is thought to have wiped out nearly all feral honey bee colonies in North America (but see Seeley 2007) and forced the use of acaricide treatments and selective breeding to prevent widespread losses in commercial operations.

Between 2006 and 2008, researchers and beekeepers began to witness another decline of managed *A. mellifera* colonies in the United States, with estimated losses as high as 36% of colonies (Johnson 2010). Affected hives displayed the following symptoms: rapid loss of adult worker bees, a noticeable lack of dead worker bees within and adjacent to the hive, apparent brood abandonment, and delayed invasion of hive pests and kleptoparasitism from neighboring honey bee colonies (Cox-Foster et al. 2007; van Engelsdorp et al. 2009). The syndrome was named colony collapse disorder (CCD) and collectively resulted in the loss of 50% to 90% of colonies in beekeeping operations across 35 states (Cox-Foster et al. 2007). European beekeepers from several countries have reported similar declines in recent years. The possible causes of CCD remain widely debated and include multiple pathogens, especially single and interactive effects of *Varroa* mites, *Nosema* spp., and Israeli acute paralysis virus (e.g., Cox-Foster et al. 2007; Higes et al. 2009). Other authors have pointed to environmental stressors, including drought and pesticides, and stress caused by bee-management practices, especially a monoculture diet and "migratory beekeeping" (interstate shipping of colonies to pollinate orchards and commercial crops; Watanabe 2008; Johnson 2010). These factors could interact with

pathogens by increasing bee susceptibility to infectious agents and facilitating spatial spread.

## Pathogen Spillover in Insect Populations

Pathogens are now recognized as a significant threat to biodiversity and have been implicated in the extinction or decline of numerous wildlife populations (see Lafferty and Gerber 2002 and de Castro and Bolker 2005 for reviews). In many cases, pathogens from domesticated species or other reservoir host populations can spill over into previously unexposed wildlife populations and cause significant declines (Daszak et al. 2000; Power and Mitchell 2004). Most evidence for pathogen spillover comes from vertebrate animals, with much less known about the extent to which spillover events have caused infectious disease problems for insect populations. As noted earlier, there is growing evidence of pathogen spillover from commercially reared bumble bees causing the decline of wild bee populations (Colla et al. 2006; National Research Council 2007; Otterstatter and Thomson 2008). Because foraging bumble bees can regularly escape from greenhouses (Whittington and Winston 2004), the potential for contact between commercial and wild bumble bees near these greenhouse operations is high (Colla et al. 2006). Aside from this example, we know of no other documented cases of pathogen introductions from domesticated/reared insects leading to outbreaks or declines in native insect populations—although it is important to note that such events are extremely likely to go unnoticed due to a lack of baseline infection data and the absence of ongoing monitoring (Goulson 2003).

## How Might Climate Change Affect Insect Diseases?

Climate change can also have impacts on insect infectious disease by altering pathogen development, survival rates, disease transmission, and host susceptibility (Harvell et al. 2002, 2009). While many of these interactions have been investigated relative to disease in vertebrate hosts, effects of climate on insect–pathogen interactions could be even more notable, in large part because insects are known to be physiologically sensitive to temperature and can respond quickly in terms of life cycles and reproductive potential (Ayres and Lombardero 2000). Indeed, multiple insect species have already responded to climate change through poleward range shifts and changes in local abundance or phenology (e.g., Parmesan and Yohe 2003; Forister et al. 2010). On the one hand, warmer temperatures could increase pathogen development rates and facilitate additional transmission cycles by increasing the number of host generations per year (Harvell et al. 2002). For example, many parasitoids have generation times synchronized with host development. If changes in temperature increase the number of host generations, then parasitoids can respond through plastic changes to synchronize their development with hosts (Thomson et al. 2010). Climate warming could also increase the prevalence of insect pathogens by relaxing overwintering constraints, as cold winter temperatures are a bottleneck for many pathogens that must survive in soil or on substrates during the winter months (Andreadis and Weseloh 1990; Hajek et al. 1990). Moreover, some insect pathogens might show greater virulence in response to warmer temperatures, as demonstrated by nuclear polyhedrosis virus of a soybean-feeding caterpillar (Johnson et al. 1982).

Importantly, species range shifts (Parmesan et al. 1999) and changes in phenology (Parmesan and Yohe 2003) may alter insect movement patterns or change migration routes, increasing exposure risks to novel pathogens or facilitating cross-species transmission (Harvell et al. 2009). Many insect species migrate long distances to track seasonal changes in resources or habitats, with notable examples including monarch butterflies, multiple New World and Old World dragonfly species, milkweed bugs, convergent ladybird beetles, and other species of butterflies and moths. Migration may be beneficial to hosts in some cases where seasonal movements can reduce parasite prevalence, either by allowing hosts to periodically escape habitats where parasites build up, or because infected animals are unable to migrate long distances (Bradley and Altizer 2005; Altizer et al. 2011; Bartel et al. 2011). If the hosts' breeding season is extended through climate warming, migrations may cease, with year-round resident populations replacing migratory ones. This loss of migration could potentially increase exposure to pathogens and elevate parasite prevalence (Harvell et al. 2009; Box 20.3).

## CONCLUSIONS AND FUTURE DIRECTIONS

Our goal was to examine some of the ways that parasites and pathogens are relevant for insect conservation, and to highlight several processes that could influence pathogen risks for native or at-risk insect populations. A major theme of this chapter is that relative to knowledge of pathogens in vertebrate animals, enormous gaps remain in studies on the roles of pathogens in insect conservation. Indeed, most pathogens infecting the vast majority of non-pest insects go largely undocumented and unstudied. Thus, aside from a handful of well-studied examples, it remains difficult to evaluate the role of pathogens as regulators of insect populations, and the extent to which processes like spillover and climate change are affecting insect–pathogen interactions. With this in mind, we briefly highlight four avenues for future efforts in research and management/policy to better understand the emergence and impacts of insect pathogens in an era of rapid environmental change.

## Improved Documentation of Insects Reared for Commercial Sale and Transport

Insects are increasingly being raised and sold for commercial use (e.g., bees as pollinators) or released at special events (e.g., butterfly wedding releases). The introduction of these commercially reared insects into the environment has serious implications for infectious disease risks for wild insect populations. Given the current minimal regulations for rearing, transporting, and selling insects, there are several additional measures regulatory agencies could implement to offset some of these threats. First, commercial insect growers and breeders could be required to register in a national database to allow for better recording of commercial sales. Standardized prerequisite disease screening by trained professionals would also better ensure against the potential escape of pathogens from commercially reared individuals into wild populations. Lastly, importation and screening procedures can also be improved by requiring a quarantine period for all incoming insect species, allowing regulatory

---

**Box 20.3  Will Parasites Gain the Upper Hand if Monarch Migration Unravels?**

---

Monarch butterflies occur worldwide and inhabit a subset of the range of their larval host plants (*Asclepiadaceae*, Ackery and Vane-Wright 1984). Monarchs cannot tolerate prolonged freezing temperatures (Calvert et al. 1983) and have exploited temperate resources through the evolution of a spectacular two-way migration in parts of North America and Australia (James 1993; Brower 1995). Most research has focused on the eastern North American population that migrates up to 2,500 km each fall from as far north as Canada to wintering sites in central Mexico (Urquhart and Urquhart 1978; Brower and Malcolm 1991). In spring, the same individuals that migrated south then fly north to recolonize their breeding range in the eastern United States (Malcolm et al. 1993). Monarchs in western North America migrate shorter distances to wintering sites along the coast of California (Nagano et al. 1993). Monarchs also form non-migratory populations that breed year-round in southern Florida, coastal Texas, Hawaii, the Caribbean islands, and Central and South America (Ackery and Vane-Wright 1984).

Previous work has demonstrated how migration influences infections in monarchs by the protozoan *O. elektroscirrha* (Altizer et al. 2000; Fig. 20.3). Climate change may also play a role in disease prevalence in this system. Parasite infections are highest in monarch populations that breed year-round in warm regions as compared to the low prevalence observed in more seasonal climates where monarchs migrate long distances (Altizer et al. 2000; Fig. 20.3). Ecological niche models including forecasted climate scenarios for eastern North America have indicated the vulnerability of this impressive migration to future change due to increasing unsuitability of overwintering habitats (Oberhauser and Peterson 2003). Also, there is growing evidence in the southeastern United States of increased planting of tropical milkweed species, which do not die back seasonally in locations with mild winters, thereby providing a constant source of host plants and allowing persistent winter-breeding populations along the Gulf Coast (Howard et al. 2010). These winter-breeding monarchs that use the same habitat and host plants for an extended duration are likely to become heavily infected with *O. elektroscirrha*, based on data from between-population comparisons and modeling studies (e.g., Altizer et al. 2004), and increased transmission in non-migratory monarchs could favor the emergence of more virulent parasite strains (De Roode et al. 2008). Thus, year-round breeding associated with climate warming and planting of non-native milkweeds could ultimately increase parasite prevalence at the population level across eastern North America, with likely impacts on monarch survival and abundance.

---

agencies to test for and identify all potential pathogen risks before transport or release.

## Field and Experimental Studies of Spillover Risks to At-Risk Populations

Evidence discussed in this chapter suggests that the higher incidence of pathogens in wild bumble bee populations could represent one of the first cases of pathogen spillover from commercially raised to wild populations of insects. As a precaution, researchers and commercial producers should begin collecting baseline infection data to recognize and document pathogen spillover events. These monitoring efforts can also be applied to biological control programs. One of the largest concerns of using biological control agents is the unintended impact on non-target species, including non-pest insects (Hajek 2004). Potential effects of natural enemies are rarely documented prior to release (Lynch and Thomas 2000), and there is great potential for pathogens to overcome host species barriers (Roy et al. 2009). Therefore it is critical to experimentally test and track these complex interactions prior to the release of natural enemies or biological control agents, especially in fragile ecosystems with simple food webs that could otherwise not recover from the loss of native species.

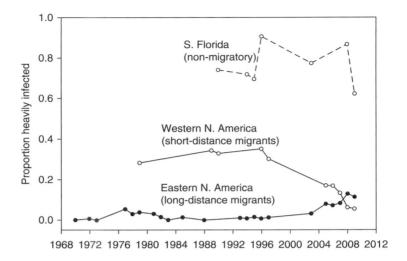

**Figure 20.3:**
Proportion of adults infected with *O. elektroscirrha* within three North American populations. Monarchs in eastern North America migrate the longest distances annually, monarchs in western North America migrate shorter distances annually, and monarchs in South Florida breed year-round and do not migrate. Sample sizes per population per year ranged from 26 to 2,730. Prevalence averages are based on heavily infected monarchs only (as described in Altizer et al. 2000; Bartel et al. 2011), averages for eastern North America are monarchs captured as adults only, and averages for western North America and South Florida include monarchs captured as both larvae and adults. (Data from eastern migrants collected by S. Altizer, L. Brower, S. Burton, A. Davis, J. De Roode, M. Maudsley, and others (see Altizer et al. 2000). Data from western migrants collected by S. Altizer, L. Brower, P. Cherubini, J. De Roode, D. Frey, K. Oberhauser, R. Rarick, and S. Stevens. Data from S. Florida collected by S. Altizer, B. Farrey, A. Knight, M. Maudsley, R. Rarick, and J. Shapiro.)

## Examine the Response of Insect Pathogens to Global Environmental Change

Global climate change may influence insect infectious diseases by affecting pathogen development, survival rates of hosts and parasites, disease transmission processes, and host susceptibility (Harvell et al. 2002). Shifts in resource distribution as a result of changes in phenology may modify insect movement patterns or disrupt migration routes, exposing novel groups of species to one another and facilitating cross-species pathogen transmission, potentially causing declines in population abundance. Moreover, habitat destruction and fragmentation could crowd some insect populations into smaller habitat patches, potentially increasing disease transmission, and exposure to pesticides or other environmental stressors could increase susceptibility to infection. At present, the role of anthropogenic change on pathogen ecology has been documented clearly for only a handful of vertebrate systems, but insects and their parasites could offer a relevant and tractable alternative for studying the impact of rapid environmental change on infectious disease dynamics.

## Baseline Monitoring of Insect Disease in Non-Pest Populations

Parasites and infectious diseases represent a major fraction of biodiversity on Earth, with half or more of all species being parasitic in nature (Price 1980). At the same time, biologists have uncovered only a miniscule percentage of the diversity of infectious organisms from natural host communities, and nowhere is this gap more extreme than for micro- and macroparasites from non-pest insects. Not only is describing insect pathogens important for inventories of biodiversity, but baseline data on the distribution and prevalence of insect pathogens are essential to detecting future changes in prevalence or novel pathogen introductions that might result from anthropogenic change. This is especially true for insect populations subject to habitat loss and exposure to environmental stressors that could make species more susceptible to disease-mediated declines or extinction.

## ACKNOWLEDGMENTS

We thank A.A. Aguirre for the invitation to contribute this chapter and for general discussion and editorial guidance. We are grateful to R. Ostfeld for suggestions and comments on the chapter draft. The section on commercial sale of insects was greatly improved by discussions with an NCEAS working group on Migration Dynamics organized by S. Altizer, L. Ries, and K. Oberhauser. L. Brower, S. Burton, P. Cherubini, A. Davis, J. De Roode, D. Frey, A. Knight, M. Maudsley, K. Oberhauser, R. Rarick, J. Shapiro, and others collected field data from North American monarch populations used in Figure 20.3. Figure photographs were contributed by D. Dale and S. Smitley. Financial support was provided to R. Bartel by a National Institutes of Health NRSA award, and to S. Altizer by a National Science Foundation grant (DEB-0643831).

## REFERENCES

Acevedo-Whitehouse, K., and A. Cunningham. 2006. Is MHC enough for understanding wildlife immunogenetics? Trends Ecol Evol 21:433–438.

Ackery, P.R., and R.I. Vane-Wright. 1984. Milkweed butterflies: their cladistics and biology. Cornell University Press, Ithaca, New York.

Addison, J.A. 1993. Persistence and nontarget effects of *Bacillus thuringiensis* in soil: a review. Can J For Res 23:2329–2342.

Allen-Wardell, G., P. Bernhardt, R. Bitner, A. Burquez, S. Buchmann, P.A. Cox, V. Dalton, P. Feinsinger, M. Ingram, D. Inouye, C.E. Jones, K. Kennedy, P. Kena,

H. Koopowitz, R. Medellin, S. Medellin-Morales, G.P. Nabhan, B. Pavlik, V. Tepedino, P. Torchio, and S. Walker. 1998. The potential consequences of pollinator declines on the conservation of biodiversity and stability of crop yields. Conserv Biol 12:8–17.

Altizer, S., and A. Pedersen. 2008. Host-pathogen evolution, biodiversity and disease risks for natural populations. *In* S. Carroll and C. Fox, eds. Conservation biology: evolution in action, pp. 259–278. Oxford University Press, New York.

Altizer, S., R. A. Bartel, and B. A. Han. 2011. Animal migration and infectious disease risk. Science 331: 296–302.

Altizer, S.M., K.S. Oberhauser, and L.P. Brower. 2000. Associations between host migration and the prevalence of a protozoan parasite in natural populations of adult monarch butterflies. Ecol Entomol 25:125–139.

Altizer, S.M., K.S. Oberhauser, and K.A. Geurts. 2004. Transmission of the protozoan parasite, *Ophryocystis elektroscirrha*, in monarch butterfly populations. *In* K. Oberhauser and M. Solensky, eds. The monarch butterfly: biology and conservation, pp. 203–218. Cornell University Press, Ithaca, New York.

Andreadis, T.G., and R.M. Weseloh. 1990. Discovery of *Entomophaga maimaiga* in North American gypsy moth, *Lymantria dispar*. Proc Natl Acad Sci USA 87:2461–2465.

Ayres, M.P., and M.J. Lombardero. 2000. Assessing the consequences of global change for forest disturbance from herbivores and pathogens. Sci Total Environ 262:263–286.

Barratt, B.I.P., F.G. Howarth, T.M. Withers, J.M. Kean, and G.S. Ridley. 2010. Progress in risk assessment for classical biological control. Biol Control 52:245–554.

Bartel, R.A., K.S. Oberhauser, J.C. De Roode, and S. Altizer. 2011. Movement patterns, seasonal habitat use and parasite transmission in eastern North American migratory monarch butterflies. Ecology 92:342–351.

Bedford, G.O. 1980. Biology, ecology, and control of palm rhinoceros beetles. Annu Rev Entomol 25:309–339.

Biesmeijer, J.C., S.P.M. Roberts, M. Reemer, R. Ohlemüller, M. Edwards, T. Peeters, A.P. Schaffers, S.G. Potts, R. Kleukers, C.D. Thomas, J. Settele, and W.E. Kunin. 2006. Parallel declines in pollinators and insect-pollinated plants in Britain and the Netherlands. Science 313:351–354.

Boots, M., and M. Begon. 1993. Trade-offs with resistance to granulosis virus in the Indian meal moth examined by laboratory evolution experiments. Funct Ecol 7:528–534.

Bradley, C., and S. Altizer. 2005. Parasites hinder monarch butterfly flight ability: implications for disease spread in migratory hosts. Ecol Lett 8:290–300.

Brower, L.P. 1995. Understanding and misunderstanding the migration of the monarch butterfly (Nymphalidae) in North America: 1857–1995. J Lepid Soc 49:304–385.

Brower, L.P., and S.B. Malcolm. 1991. Animal migrations: endangered phenomena. Am Zool 31:265–276.

Brower, L.P., L.S. Fink, A. Van Zandt, K. Oberhauser, S. Altizer, O. Taylor, D. Vickerman, T. Van Hook, A. Alonso-Meija, S.B. Malcolm, D.F. Owen, and M.P. Zalucki. 1995. On the dangers of interpopulational trade of monarch butterflies. Bioscience 45:540–544.

Buchmann, S.L., and G.P. Nabhan. 1996. The forgotten pollinators. Island Press, Washington D.C.

Calvert, W.H., W. Zuchowski, and L.P. Brower. 1983. The effect of rain, snow and freezing temperatures on overwintering monarch butterflies in Mexico. Biotropica 15:42–47.

Charlat, S., E.A. Hornett, J.H. Fullard, N. Davies, G.K. Roderick, N. Wedell, and G.D.D. Hurst. 2007. Extraordinary flux in sex ratio. Science 317:214.

Colla, S., and L. Packer. 2008. Evidence for decline in eastern North American bumblebees (Hymenoptera: Apidae), with special focus on Bombus affinis Cresson. Biodivers Conser 17:1379–1391.

Colla, S.R., M.C. Otterstatter, R.J. Gegear, and J.D. Thomson. 2006. Plight of the bumblebee: pathogen spillover from commercial to wild populations. Biol Conserv 132:461–467.

Cox-Foster, D.L., S. Conlan, E.C. Holmes, G. Palacios, J.D. Evans, N.A. Moran, P.-L. Quan, T. Briese, M. Hornig, D.M. Geiser, V. Martinson, D. vanEngelsdorp, A.L. Kalkstein, A. Drysdale, J. Hui, J. Zhai, L. Cui, S.K. Hutchison, J.F. Simons, M. Egholm, J.S. Pettis, and W.I. Lipkin. 2007. A metagenomic survey of microbes in honey bee colony collapse disorder. Science 318:283–287.

Daszak, P., A.A. Cunningham, and A.D. Hyatt. 2000. Emerging infectious diseases of wildlife- threats to biodiversity and human health. Science 287:443–449.

de Castro, F., and B. Bolker. 2005. Mechanisms of disease-induced extinction. Ecol Lett 8:117–126.

De Roode, J.C., L.R. Gold, and S. Altizer. 2007. Virulence determinants in a natural butterfly- parasite system. Parasitology 134:657–668.

De Roode, J.C., A.J. Yates, and S. Altizer. 2008. Virulence-transmission trade-offs and population divergence in virulence in a naturally occurring butterfly parasite. Proc Natl Acad Sci USA 105:7489–7494.

Dunn, R.R. 2005. Insect extinctions, the neglected majority. Conserv Biol 19:1030–1036.

Dunn, R.R., N.C. Harris, R.K. Colwell, L.P. Koh, and N.S. Sodhi. 2009. The sixth mass (co) extinction—are most endangered species parasites and mutualists. P Roy Soc B 276:3037–3045.

Dwyer, G., J. Dunshoff, and S. Harrell Yee. 2004. The combined effects of pathogens and predators on insect outbreaks. Nature 430:341–345.

Dyson, E.A., and G.D.D. Hurst. 2004. Persistence of an extreme sex-ratio bias in a natural population. Proc Natl Acad Sci USA 101:6250–6253.

Evans, E., R. Thorp, S. Jepsen, and S. Hoffman Black. 2008. Status review of three formerly common species of bumble bee in the subgenus Bombus: Bombus affinis (the rusty patched bumble bee), B. terricola (the yellowbanded bumble bee), and B. occidentalis (the western bumble bee). Available at http://www. xerces.org/wp-content/uploads/2009/03/ xerces_2008_bombus_status_review.pdf

Flanders, R.V., W.F. Wehling, and A.L. Craghead. 2003. Laws and regulations on the import, movement and release of bees in the United States. In K. Strickler and J.H. Cane, eds. For nonnative crops, whence pollinators of the future?, pp. 99–111. Proc Entomol Soc Am, Thomas Say Publications in Entomology, Lanham, Maryland.

Forister, M.L., A.C. McCall, N.J. Sanders, J.A. Fordyce, J.H. Thorne, J. O'Brien, D.P. Waetjen, and A.M. Shapiro. 2010. Compounded effects of climate change and habitat alteration shift patterns of butterfly diversity. Proc Natl Acad Sci USA 107:2088–2092.

Funasaki, G.Y., P.-Y. Lai, L.N. Nakahara, J.W. Beardsley, and A.K. Ota. 1988. A review of biological control introductions in Hawaii: 1890 to 1985. Proc Hawaii Entomol Soc 28: 105–160.

Goulson, D. 2003. Effects of introduced bees on native ecosystems. Annu Rev Ecol Syst 34:1–26.

Hajek, A.E. 1999. Pathology and epizootiology of the lepidoptera-specific mycopathogen Entomophaga maimaiga. Microbiol Molecul Biol Rev 63:814–835.

Hajek, A.E. 2004. Natural enemies: an introduction to biological control. Cambridge University Press, Cambridge.

Hajek, A.E. 2007. Introduction of a fungus into North America for control of gypsy moth. In C. Vincent, M.S. Goettel, and G. Lazarovits, eds. Biological control: a global perspective, pp. 53–62. CAB International, London.

Hajek, A.E., R.I. Carruthers, and R.S. Soper. 1990. Temperature and moisture relations of sporulation and germination by Entomophaga maimaiga (Zygomycetes: Entomophthoraceae), a fungal pathogen of Lymantria dispar (Lepidoptera: Lymantriidae). Environ Entomol 19:85–90.

Hajek, A.E., R.A. Humber, and J. Elkinton. 1995. Mysterious origin of *Entomophaga maimaiga* in North America. Am Entomol 41:31–42.

Hajek, A.E., C.H. Olsen, and J.S. Elkinton. 1999. Dynamics of airborne conidia of the gypsy moth (Lepidoptera: Lymantriidae) fungal pathogen *Entomophaga maimaiga* (Zygomycetes: Entomophthorales). Biol Control 16:111–117.

Hajek, A.E., J.S. Strzanac, M.M. Wheeler, F.M. Vermeylen, and L. Butler. 2004. Persistence of the fungal pathogen *Entomophaga maimaiga* and its impact on native Lymantriidae. Biol Control 30:466–473.

Harvell, C.D., C.E. Mitchell, J.R. Ward, S. Altizer, A.P. Dobson, R.S. Ostfeld, and M.D. Samuel. 2002. Climate warming and disease risks for terrestrial and marine biota. Science 296:2158–2162.

Harvell, D., S. Altizer, I.M. Cattadori, L. Harrington, and E. Weil. 2009. Climate change and wildlife diseases: when does the host matter the most? Ecology 90:912–920.

Hawkins, B.A., and P.C. Marino. 1997. The colonization of native phytophagus insects in North America by exotic parasitoids. Oecologia 112:566–571.

Henneman, M.L., and J. Memmott. 2001. Infiltration of a Hawaiian community by introduced biological control agents. Science 293:1314–1316.

Higes, M., R. Martín-Hernández, E. Garrido-Bailón, A.V. González-Porto, P. García-Palencia, A. Meana, M.J.d. Nozal, R. Mayo, and J.L. Bernal. 2009. Honeybee colony collapse due to *Nosema ceranae* in professional apiaries. Environ Microbiol Reports 1:110–113.

Hornett, E.A., S. Charlat, A.M.R. Duplouy, N. Davies, G K. Roderick, N. Wendall, and G.D.D. Hurst. 2006. Evolution of male-killer suppression in a natural population. PLoS Biol 4:e283.

Howard, E., H. Aschen, and A.K. Davis. 2010. Citizen science observations of monarch butterfly overwintering in the southern United States. Psyche doi:10.1155/2010/689301.

International Union for Conservation of Nature (IUCN). 2010. IUCN Red List of Threatened Species. Retrieved April 28, 2010, from http://www.iucnredlist.org.

Islam, M.S. 2007. *Wolbachia*-mediated reproductive alterations in invertebrate hosts and biocontrol implications of the bacteria: an update. Rajshahi Univ Zool J 26:1–19.

James, D.G. 1993. Migration biology of the monarch butterfly in Australia. *In* S.B. Malcolm and M.P. Zalucki, eds. Biology and conservation of the monarch biology, pp. 189–200. Natural History Museum of Los Angeles County, Los Angeles, California.

Jiggins, F.M., G.D.D. Hurst, and M.E. Majerus. 2000. Sex-ratio-distorting *Wolbachia* causes sex-role reversal in its butterfly host. Proc R Soc B 267:69–73.

Johnson, D.W., D.G. Boucias, C.S. Barfield, and G.E. Allen. 1982. A temperature–dependent development model for a nucleopolyhedrosis virus of the velvetbean caterpillar *Anticarsia gemmatalis* (Lepidoptera: Noctuidae). J Insect Pathol 40:292–298.

Johnson, R. 2010. Honey bee colony collapse disorder. Congressional Report Service for Congress, Jan. 7, 2010, 17 pages. Available online at: www.fas.org/sgp/crs/misc/RL33938.pdf

Klein, A.M., B.E. Vaissiere, J.H. Cane, I. Steffan-Dewenter, S.A. Cunningham, C. Kremen, and T. Tscharntke. 2007. Importance of pollinators in changing landscapes for world crops. Proc R Soc B 274:303–313.

Koh, L.P., R.R. Dunn, N.S. Sodhi, R.K. Colwell, H.C. Proctor, and V.S. Smith. 2004. Species co-extinctions and the biodiversity crisis. Science 305:1632–1634.

Kraaijeveld, A.R., and H.J.C. Godfray. 1997. Trade-off between parasitoid resistance and larval competitive ability in *Drosophila melanogaster*. Nature 389:278–280.

Lacey, L.A., and H.K. Kaya. 2000. Field manual of techniques in invertebrate pathology: application and evaluation of pathogens for control of insects and other invertebrate pests. Kluwer Academic, Dordrecht.

Lacey, L.A., R. Frutos, H.K. Kaya, and P. Vail. 2001. Insect pathogens as biological control agents: do they have a future? Biol Control 21:230–248.

Lafferty, K., and L. Gerber. 2002. Good medicine for conservation biology: The intersection of epidemiology and conservation theory. Conserv Biol 16:593–604.

Leong, K.L.H., M.A. Yoshimura, and H.K. Kaya. 1997. Occurrence of a neogregarine protozoan, *Ophryocystis elektroscirrha* McLaughlin and Myers, in populations of monarch and queen butterflies. Pan-Pacific Entomol 73:49–51.

Lewis, O.T., T.R. New, and A.J.A. Stewart. 2007. Insect conservation: progress and prospects. *In* A.J.A. Stewart, T.R. New, and O.T. Lewis, eds. Insect conservation biology, pp. 431–435. CABI, Cambridge, Massachusetts.

Lindsey, E., M. Mehta, V. Dhulipala, K. Oberhauser, and S. Altizer. 2009. Crowding and disease: effects of host density on parasite infection in monarch butterflies. Ecol Entomol 34:551–561.

Liu, H.J. 1973. Bombus Latr (Hymenoptera: Apidae) in Southern Ontario: its role and factors affecting it. M.Sc. thesis. University of Guelph, Guelph.

Losey. J., and M. Vaughan. 2006. The economic value of ecological services provided by insects. Bioscience 56:311–323.

Louda, S., R.W. Pemberton, M.T. Johnson, and P.A. Follett. 2003. Non-target effects—the Achilles' heel of biological control? Annu Rev Entomol 48:365–396.

Lynch, L.D., and M.B. Thomas. 2000. Nontarget effects in the biocontrol of insects with insects, nematodes, and microbial agents: the evidence. Biocontrol News Info 21:117N–130N.

MacFarlane, R.P., J.J. Lipa, and H.J. Liu. 1995. Bumble bee pathogens and internal enemies. Bee World 76: 130–148.

Malcolm, S.B., B.J. Cockrell, and L.P. Brower. 1993. Spring recolonization of eastern North America by the monarch butterfly: successive brood or single sweep? In S.B. Malcolm and M.P. Zalucki, eds. Biology and conservation of the monarch butterfly, pp. 253–267. Natural History Museum of Los Angeles County, Los Angeles, California.

Martin, S., A. Hogarth, J. van Breda, and J. Perrett. 1998. A scientific note on Varroa jacobsoni Oudemans and the collapse of Apis mellifera L. colonies in the United Kingdom. Apidologie 29:369–370.

Matern, A., K. Desender, C. Drees, E. Gaublomme, W. Paill, and T. Assmann. 2009. Genetic diversity and population structure of the endangered insect species Carabus variolosus in its western distribution range: implications for conservation. Conserv Genetics 10:391–405.

McGregor, S.E. 1976. Insect pollination of cultivated crop plants. USDA-ARS, Washington. D.C.

Morse, R., and N.W. Calderone. 2000. The value of honeybees as pollinators of U.S. crops in 2000. Bee Culture 128:1–15.

Nagano, C.D., W.H. Sakai, S.B. Malcolm, B.J. Cockrell, J.P. Donahue, and L.P. Brower. 1993. Spring migration of monarch butterflies in California. In Malcolm S.B. and M.P. Zalucki, eds. Biology and conservation of the monarch butterfly, pp. 217–232. Natural History Museum of Los Angeles County, Los Angeles, California.

National Research Council. 2007. Status of pollinators in North America. National Academies Press, Washington D.C.

2009Nice, C.C., Z. Gompert, M.L. Forister, and J.A. Fordyce. 2009. An unseen foe in arthropod conservation efforts: the case of Wolbachia in the Karner blue butterfly. Biol Conserv 142:3137–3146.

Oberhauser, K.S., and A.T. Peterson. 2003. Modeling current and future potential wintering distributions of eastern North American monarch butterflies. Proc Natl Acad Sci USA 100:14063–14068.

Otterstatter, M.C., and J.D. Thomson. 2008. Does pathogen spillover from commercially reared bumble bees threaten wild pollinators? PLoS ONE 3:e2771. doi:10.1371/journal.pone. 0002771.

Otterstatter, M.C., and T. L. Whidden. 2004. Patterns of parasitism by tracheal mites (Locustacarus buchneri) in natural bumble bee populations. Apidologie 35:351–357.

Otterstatter, M.C., R.J. Gegear, S. Colla, and J.D. Thomson. 2005. Effects of parasitic mites and protozoa on the flower constancy and foraging rate of bumble bees. Behav Ecol Sociobiol 58:383–389.

Parmesan, C., and G. Yohe. 2003. A globally coherent fingerprint of climate changes impacts across natural systems. Nature 42:37–42.

Parmesan, C., N. Ryrholm, C. Stefanescu, J.K. Hill, C.D. Thomas, H. Descimon, B. Huntley, L. Kaila, J. Kullberg, T. Tammaru, W.J. Tennent, J.A. Thomas, and M. Warrant. 1999. Poleward shifts in geographical ranges of butterfly species associated with regional warming. Nature 399:579–583.

Pedersen, A., K.E. Jones, C.L. Nunn, and S. Altizer. 2007. Infectious diseases and extinction risk in wild mammals. Conserv Biol 21:1269–1279.

Power, A.G., and C.E. Mitchell. 2004. Pathogen spillover in disease epidemics. Am Nat 64:S79–S89.

Price, P.W. 1980. Evolutionary biology of parasites. Princeton University Press, Princeton, New Jersey.

Pyle, R.M., S.J. Jepsen, S. Hoffman Black, and M. Monroe. 2010. Xerces Society policy on butterfly releases. http://www.xerces.org/wp-content/uploads/2010/08/xerces-butterfly-release-policy.pdf

Reilly, J.R., and A. E. Hajek. 2008. Density-dependent resistance of the gypsy moth, Lymantria dispar to its nucleopolyhedrovirus, and the consequences for population dynamics. Oecologia 154:691–701.

Roubik, D.W. 1995. Pollination of cultivated plants in the tropics. Food and Agriculture Organization, Rome.

Roy, H.E., R.S. Hails, H. Hesketh, D.B. Roy, and J.K. Pell. 2009. Beyond biological control: non-pest insects and their pathogens in a changing world. Insect Conserv Div 2:65–72.

Sanford, M.T. 2001. Introduction, spread, and economic impact of Varroa mites in North America. In Webster T.C. and K.S. Delaplane, eds. Mites of the honey bee, pp.149–162. Dadant and Sons, Hamilton, Illinois.

Schmid-Hempel, P. 2005. Evolutionary ecology of insect immune defenses. Annu Rev Entomol 50:529–551.

Schnepf, E., N. Crickmore, J. Van Rie, D. Lereclus, J. Baum, J. Feitelson, D. Zeigler, and D. Dean. 1998. Bacillus thuringiensis: its pesticidal crystal proteins. Microbiol Mol Biol Rev 62:775–806.

Seeley, T. 2007. Honey bees of the Arnot Forest: a population of feral colonies persisting with Varroa destructor in the northeastern United States. Apidologie 38:19–29.

Shimanuki, H., D.A. Knox, B. Furgala, D M. Caron, and
J.L. Williams. 1992. Diseases and pests of honey bees.
*In* Graham, J.M. ed. The hive and the honey bee,
pp. 1083–1151. Dadant & Sons, Inc. Hamilton,
Illinois.

Smith, K.F., D.F. Sax, and K.D. Lafferty. 2006. Evidence for
the role of infectious disease in species extinction and
endangerment. Conserv Biol 20:1349–1357.

Southwick, E.E., and L. Southwick, Jr. 1992. Estimating the
economic value of honey bees (Hymenoptera: Apidae)
as agricultural pollinators in the United States. J Econ
Entomol 85:621–633.

Sutter, G.R., W.C. Rothenbuhler, and E.S. Raun. 1968.
Resistance to American foulbrood in honey bees: VII.
Growth of resistant and susceptible larvae. J Invertebr
Pathol 12:25–28.

Tanada, Y., and H.K. Kaya. 1992. Insect pathology.
Academic Press, San Diego, California.

Thomson, L.J., S. Macfadyen, and A.A. Hoffmann. 2010.
Predicting the effects of climate change on natural
enemies of agricultural pests. Biol Control 52:296–306.

Urquhart, F.A., and N.R. Urquhart. 1978. Autumnal
migration routes of the eastern population of the
monarch butterfly (*Danaus plexippus*) (L.) (Danaidae:
Lepidoptera) in North America to the overwintering
site in the neovolcanic plateau of Mexico. Can J Zool
56:1756–1764.

van Engelsdorp, D., J.D. Evans, C. Saegerman, C. Mullin,
E. Haubruge, B.K. Nguyen, M. Frazier, J. Frazier,
D. Cox-Foster, Y. Chen, R. Underwood, D. Tarpy, and
J.S. Pettis. 2009. Colony collapse disorder: a descriptive
study. PLoS One 4:e6481.doi:10.1371/ journal.
pone.0006481.

Wagner, D.L., J.W. Peacock, J.L. Carter, and S.E. Talley. 1996.
Field assessment of *Bacillus thuringiensis*
on nontarget Lepidoptera. Environ Entomol
25:1444–1454.

Watanabe, M.E. 1994. Pollination worries arise as
honey-bees decline. Science 265:1170.

Watanabe, M.E. 2008. Colony collapse disorder: many
suspects, no smoking gun. Bioscience 58:384–388.

Weeks, A.R., K.T. Reynolds, and A.A. Hoffman. 2002.
*Wolbachia* dynamics and host effects: what has
(and has not) been demonstrated? Trends Ecol Evol
17:257–262.

Wenner, A.M., and W.W. Bushing. 1996. *Varroa* mite spread
in the United States. Bee Culture 124:342–343.

Whittington, R., and M.L. Winston. 2003. Effects of *Nosema
bombi* and its treatment fumagillin on bumble bee
(*Bombus occidentalis*) colonies. J Invertebr Pathol
84:54–58.

Whittington, R., and M.L. Winston. 2004. Comparison and
examination of *Bombus occidentalis* and *Bombus
impatiens* (Hymenoptera: Apidae) in tomato
greenhouses. J Econ Entomol 97:1384–1389.

Winfree, R. 2010. The conservation and restoration of wild
bees. Ann NY Acad Sci 1195:169–197.

Winfree, R., N.M. Williams, J. Dushoff, and C. Kremen.
2007. Native bees provide insurance against ongoing
honeybee losses. Ecol Lett 10:1105–1103.

Zayed, A., and L. Packer. 2005. Complementary sex
determination substantially increases extinction
proneness of haplodiploid populations. Proc Natl Acad
Sci USA 102:10742–10746.

Zelazny, B., A. Lolong, and B. Pattang. 1992. *Oryctes
rhinoceros* (Coleoptera: Scarabaeidae) populations
suppressed by a baculovirus. J Invertebr Pathol
59:61–68.

Zimmerman, E.C. 1978. Insects of Hawaii. University of
Hawaii Press, Honolulu.

Zuk, M., L.W. Simmons, and L. Cupp. 1993. Calling
characteristics of parasitized and unparasitized
populations of the field cricket *Teleogryllus oceanicus*.
Behav Ecol Sociobiol 33:339–343.

Zuk, M., J.T. Rotenberry, and R.M. Tinghittella. 2006. Silent
night: adaptive disappearance of a sexual signal in a
parasitized population of field crickets. Biol Lett
22:521–524.

# 21

## FUNGAL DISEASES IN NEOTROPICAL FORESTS DISTURBED BY HUMANS

Julieta Benítez-Malvido

All plant structures (roots, leaves, stems, flowers, fruits, and seeds), at all stages of the life cycle, are subject to colonization by microorganisms that may modify or interrupt their vital functions. In general terms a "disease" includes a series of microorganisms (agents) that disturb plant metabolism such as growth (hypotrophy and/or hyperplasia) in a portion of or throughout an entire plant causing even its death. An infectious agent is capable of reproducing within or on its plant host and spreading from one susceptible host to another. Infectious agents could be fungi, nematodes, viroids, viruses, bacteria. or even other flowering plants (e.g., the parasitic plant *Epifagus virginiana*). Fungi (*sensu lato*, i.e., including Oomycetes) are, however, the major causal agents of plant diseases (i.e., about 75% of all plant diseases are caused by fungi) and jointly with insects represent the major threat to wild and cultivated plant species worldwide. Fungal infection can cause local or extensive necrosis or abnormal growth in the different structures of plants. Symptoms associated with necrosis include leaf spots, blight, scab, rots, damping-off, anthracnose, dieback, and canker; whereas those that induce excessive abnormal growth include clubroot, galls, warts, and leaf curls (Anderson et al. 2004).

Fungal pathogens drive many ecological and evolutionary processes in natural ecosystems. Through their differential impacts on survival, growth, and fecundity and by reducing the competitive capability of infected individuals, pathogens influence plant recruitment and species composition (Burdon 1991; Dobson and Crawley 1994; Jarozs and Davelos 1995; Gilbert 2002). In addition, pathogens can help maintain plant species diversity (Janzen 1970; Connell 1971; Gilbert 2002), can enhance the genetic diversity and structure of host populations, and may facilitate successional processes (Gilbert 2002). In Los Tuxtlas, Mexico, the experimental elimination of soil-born pathogens (Oomycetes) by adding fungicide decreased seedling species diversity and increased seedling survival of the common species (e.g., *Brosimum alicastrum, Faramea occidentalis*) (Ayala-Orozco 2008). It is likely that pathogens help reduce competitive exclusion and enhance community diversity.

Most information concerning the study of plant–pathogen interactions in natural ecosystems comes from temperate regions, and limited attention has been given to this important biotic interaction in tropical natural systems compared to that devoted to fungal diseases in tropical agriculture (Gilbert 2002; Vurro et al. 2010). The information gathered for tropical forests illustrates the prevalence of fungal diseases in the understory and canopy vegetation affecting many plant species at all stages of their life cycles (Augspurger and Kelly 1984; Gilbert et al. 1994;

Gilbert 1995; Lodge et al. 1996; Benítez-Malvido et al. 1999; García-Guzmán and Dirzo 2001; Gilbert 2002; García-Guzmán and Morales 2007). The most common are foliar pathogens that can cause necrosis, blights, chlorosis, and abscission and deformation of leaves (Gilbert 2002; García-Guzmán and Morales 2007). Other known fungal pathogens in the tropics infect seeds and seedlings (e.g., *Phytophpora* and *Pythium* in seeds of *Miconia argentea* and *Cecropia insignis*) or the vascular system of plants (e.g., fungi of the genus *Armillaria* spp.), attack flowers and fruits (e.g., the rust *Aecidium farameae* in flowers of *Faramea occidentalis*), and cause cankers on stems and trunks (e.g., *Phytophtora* sp. in the Lauraceae) and wood decay in roots (e.g., fungi of the genus *Rhizoctonia* spp) and trunks (Augspurger 1984; Dalling et al. 1998; Travers et al. 1998; Gilbert 2002; Antonovics 2005).

In this chapter I describe several types of human disturbances in the Neotropics that have been shown to affect the interaction of plants with their fungal pathogens in natural systems. For all study cases presented, disturbance increased the levels of pathogen damage on plants. The observations are limited to foliar pathogens because of their ubiquity in tropical plant communities (Gilbert 2002; García-Guzmán and Morales 2007). First I describe leaf diseases in tropical plants, second the physical and biological factors involved in disease development and transmission, third the impact of anthropogenic disturbance in facilitating leaf fungal infection, fourth the potential consequences of disease spread, and finally the implications of such disease-induced changes in tropical rain forests' function and conservation.

## LEAF DISEASES IN THE TROPICS

Leaf diseases are very common and probably among the most important agents of plant damage in rain-forests, affecting on average 70% of the registered species and causing chlorosis, necrotic leaf spots of several shapes and colors, or leaf deformation and abscission (Agrios 1997; García-Guzmán and Dirzo 2001; Benítez-Malvido and Lemus-Albor 2005; García-Guzmán and Morales 2007). Leaf diseases in the tropics are mostly caused by Ascomycetes, many of which are facultative pathogens able to live as saprotrophs in the absence of the host plant

(García-Guzmán and Morales 2007). Apparently, all plant growth forms (i.e., trees, shrubs, herbs, vines, palms, ferns, grasses, and sedges) in the rain forests are susceptible to infection by leaf pathogens (García-Guzmán and Dirzo 2001; Benítez-Malvido and Lemus-Albor 2005; García-Guzmán and Morales 2007). Surveys of the understory plant community in two Mexican rainforests at Los Tuxtlas and Chajul indicated that 65% to 70% of the plant species registered presented symptoms of fungal infection on their leaves (García-Guzmán and Dirzo 2001; Benítez-Malvido and Lemus-Albor 2005); whereas in Manaus, Brazil, 80% of all registered plant species in the understory presented symptoms of foliar damage by fungal pathogens (Benítez-Malvido 2005 unpublished data).

Although several studies have reported that leaf area with apparent disease symptoms is very low, averaging 2% or less of leaf area (Benítez-Malvido et al. 1999; García-Guzmán and Dirzo 2001; Benítez-Malvido and Lemus-Albor 2005), even small reductions in leaf area can have large impacts on seedling survival (Clark and Clark 1985). The reduction of leaf area by pathogens affects the photosynthetic process (Burdon 1987) and, through cumulative effects, may cause reductions in growth, survivorship, reproduction, and competitive ability of the hosts (Burdon 1993; Jarosz and Davelos 1995; Esquivel and Carranza 1996). These effects may cascade from those acting on the individual plant to affect the genetic structure of host plant populations (Burdon and Jarosz 1988) and even the distribution of plant species.

Host traits have been shown to differentially affect plant susceptibility to foliar pathogen damage. There is evidence that taller and leafy seedlings present a greater proportion of infected leaves (Benítez-Malvido and Lemus-Albor 2005), as leafy plants probably have a greater surface area for fungal attack and may create a microclimate conducive for disease development, leading to greater disease incidence (G. Gilbert 2005, personal communication). Furthermore, leaf pathogen infection has been shown to be more severe in smaller and fruiting individuals of the tropical grass *Lasciacis ruscifolia* (Santos et al. 2011).

## PLANT SUSCEPTIBILITY

Fungi are the most common pathogens in the foliage of tropical rainforest plants (Gilbert 2002), and all plant species (wild or cultivated) are susceptible to fungal attack. The more susceptible the plant hosts, the quicker and more widespread will be the proliferation of a given disease (Beldomenico and Begon 2009). Each fungal species has a particular mechanism of infection, reproduction, and dispersal, and attacks its hosts with different levels of severity (Burdon 1987, 1991; Giraud et al. 2010). Fungal pathogens are spread primarily by spores, which are produced in very large numbers (Giraud et al. 2010). The spores can be carried and disseminated by wind, water (splashing, rain, and dew), soil (dust), insects, birds, and the remains of infected plants (García-Guzmán and Benítez-Malvido 2003). The experimental addition of leaf litter of previously infected plant material increased the incidence of fungal infection in *Nectandra ambigens* leaves, probably by exposing seedlings to inocula from pathogens (Gilbert 1995; García-Guzmán and Benítez-Malvido 2003). Some pathogens require wounds in order to infect plants (Giraud et al. 2010). In Los Tuxtlas, 43% of the leaves in the understory plant community were damaged by herbivores and pathogens concurrently (García-Guzmán and Dirzo 2001). Insect and other injuries provide infection sites for pathogens that, once established, can spread to neighboring plants.

Disease prevalence and infection vary in time and space depending on the physical and biological characteristics of the environment (Burdon 1987). The physical (e.g., temperature and moisture) and biotic (e.g., host density and identity) environments make plants more or less susceptible to fungal attack (Agrios 1997). However, the physical environment is the most important factor that determines the development of a disease. Environmental conditions affect the expression of disease symptoms mainly through their effect on the host (plant) prior to infection (predisposition) and on the host–fungal pathogen association once infection has occurred (Burdon 1987). Shifts in a particular physical factor (light, temperature, pH, soil fertility or moisture) can affect germination, growth, or susceptibility of the host plant; survival, germination, and growth of the pathogen (Agrios 1997); the host–pathogen interaction (Colhoun 1973); and the behavior of disease vectors (e.g., insects; García-Guzmán and Dirzo 2001; Santos and Benítez-Malvido 2011). A plant under stress (e.g., lack of nutrients and or water) is generally more susceptible to infection by a fungal pathogen; an infected plant in turn is more vulnerable to infection by secondary pathogens. Not only can poor plant condition (i.e., physiological stress) predispose individuals to infection, but also infection itself can result in a negative effect on plant condition (Beldomenico and Begon 2009; Giraud et al. 2010).

Host density and identity are also important for disease development and transmission. Cultivated plants are often more susceptible to fungal infection than their wild relatives (Vurro et al. 2010). This is because individuals of the same species with low genetic variation are grown close together in large numbers. A fungal pathogen may spread rapidly under these conditions. Similarly, under natural conditions fungal pathogens in tropical rainforests cause density- or distance-dependent mortality of the locally dominant tree species, which may prevent competitive exclusion, allowing greater survival of less common, resistant species and enhancing local tree diversity (Janzen 1970; Connell 1971). In Panama, seedlings of the tropical tree *Platypodium elegans* had a higher probability of mortality by damping-off disease (soil-born fungal pathogens, often the fungus-like Oomycetes *Phytophthora* and *Pythium*; Garrett 1970) at both high seedling densities and when close to conspecific adults (Augspurger and Kelly 1984). In saplings of *Ocotea whitei*, fungal-induced canker disease (by *Phytophthora*) increased with increasing density of conspecifics and decreasing distance to adults (Gilbert et al. 1994). With species heterogeneity, disease outbreaks are less probable.

## HUMAN DISTURBANCE AND PROLIFERATION OF DISEASES

Diseases are part of nature and one of many ecological factors that help maintain plant and animal species in balance with one another. Adverse environments, such as those originated from human activities, could make the vegetation of disturbed habitats more prone to infectious diseases (Brothers and Spingarn 1992; Gilbert and Hubbell 1996; Harvell et al. 2002; Anderson et al. 2004; Benítez-Malvido and Lemus-Albor 2006; Giraud et al. 2010; Santos and Benítez-Malvido 2011). Disturbances in tropical rainforest have major effects on vegetation structure, which in turn affects environmental conditions. Some of the most noticeable changes are those related to light intensity, soil moisture, soil and air temperature, and wind velocity and turbulence (Laurance et al. 1998; Pinto et al. 2010).

The extreme high temperatures, low atmospheric humidity, and low soil moisture of tropical disturbed habitats generally fall outside the range of conditions that occurred in the natural forest in which host plants, pathogens, and disease vectors (e.g., insects) interact. These extreme conditions stress forest plants, making them more susceptible to disease (Harvell et al. 2002; Anderson et al. 2004; Benítez-Malvido and Lemus-Albor 2006). Individuals of *Theobroma* sp. and *Herrania* sp., when exposed to full sunlight, are often killed by fungal diseases (Lodge and Cantrell 1995). Hosts in poor condition might have infections of higher intensity because pathogens would encounter less opposition to their survival and proliferation. The more "healthy" a host plant is, the more hostile an environment a pathogen will encounter (Beldomenico and Begon 2009).

Collectively, environmental changes affect plant demography, species composition (e.g., entrance of exotic and/or invasive species), and community dynamics in the remaining forests and on the vegetation that is established after forest disturbance (Peters 2001; Laurance et al. 2006a,b). Changes in faunal communities and on biotic interactions (e.g., herbivory and pathogen attack) are equally diverse (e.g., Stratford and Stouffer 1999; Benítez-Malvido and Lemus-Albor 2005, 2006). The combination of environmental changes and exotic plant and pathogen species could predispose the vegetation of disturbed habitats to infectious diseases (Anagnostakis 1987; Brothers and Spingarn 1992; Gilbert and Hubbell 1996; Peters 2001; Harvell et al. 2002; Rizzo et al. 2002; Giraud et al. 2010).

While foliar and other pathogens are generally favored by humid, cool, shady conditions (Bradley et al. 2003), some of the most important fungal pathogens, such as certain rusts (Basidiomycota) and powdery mildews (Ascomycota; G. Gilbert 2005 personal communication), are clearly adapted to drier, warmer, brighter conditions. Moreover, dew formation is generally greater on plants in open environments than in forest interiors (i.e., understory; Brewer and Smith 1997), and dew is often crucial for fungal infection of leaves (Bradley et al. 2003; G. Gilbert 2005 personal communication). Many of the tropical plant species attacked by rust (45%) and smuts (50%) are characteristically found in disturbed areas. Even certain pathogens are related to plant families characteristic of disturbed habitats such as smuts (Basidiomycetes) on Poaceae and Asteraceae, and rusts (Basidiomycetes), which are common on Fabaceae (García-Guzmán and Morales 2007). In tropical disturbed habitats flower, leaf, and stems smuts as well as rust have been described for 64 plant species (García-Guzmán and Morales 2007).

## SELECTIVE LOGGING

Machinery and various practices related to selective logging and harvesting (e.g., palm leaf collection) may seriously injure trees or plant products directly or indirectly (i.e., smash, increase canopy falling debris). Wounds are created through which pathogens may enter. Some foliar pathogens need wounds to infect plants (García-Guzmán and Dirzo 2001). Logging, road construction, and edge creation increased tree mortality and falling debris, which can physically damage the remaining vegetation (Laurance et al. 2009), providing infection sites for pathogens, which can spread to neighboring plants (Kellas et al. 1987; Gilbert and Hubbell 1996). In the Central Amazon, plants from woody species (i.e., trees, lianas, and shrubs) along abandoned logging trails were more likely to have leaves damaged by herbivores and fungal pathogens than those in the nearby conserved forest. Furthermore, early successional species (e.g., *Vismia cauliflora* and *Bellucia grossularioides*) were more common in logging trails and had a higher proportion of infected leaves than did late successional species (e.g., *Chrysophyllum amazonicum* and *Gustavia hexapetala*) (Fig. 21.1). In addition, leaves from three early successional species (i.e., *Bellucia grossularioides, Miconia* sp., and *Rinorea racemosa*) accounted for nearly 30% of all infected leaves (Benítez-Malvido 2005 unpublished data).

## ROAD CONSTRUCTION AND EDGE EFFECTS

Road creation may directly affect native populations of plant and animals through habitat loss and disturbance, including edge effects of variable magnitude (Murcia 1995). One of the main changes that roads

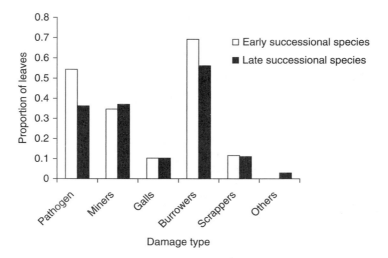

**Figure 21.1:**
Proportion of leaves within different types of biotic damage (i.e., insect herbivores and foliar pathogens). Woody plants from late and early successional species were present along trails within conserved and logged forests of the Central Amazon, north of Manaus.

bring about is an increase in the proportion of edge exposed to a different habitat, which is particularly significant where fragmentation is associated with roads (Forman and Alexander 1998). Edge effects include all physical and biotic changes caused by the interaction of two adjacent ecosystems separated by an abrupt transition and could extend well beyond the area occupied by the road surface itself (Forman and Alexander 1998; Laurance 2009). Roads, in particular,

have important ecological effects on natural ecosystems: they act as dispersal routes for some organisms and as barriers for others (Goosem 2002; Laurance 2009) and together with forest edges are recognized as the point of entry of external influences, such as the invasion of exotic flora and fauna and pathogenic organisms (Janzen 1983; Castello et al. 1995; Gelbard and Belnap 2003).

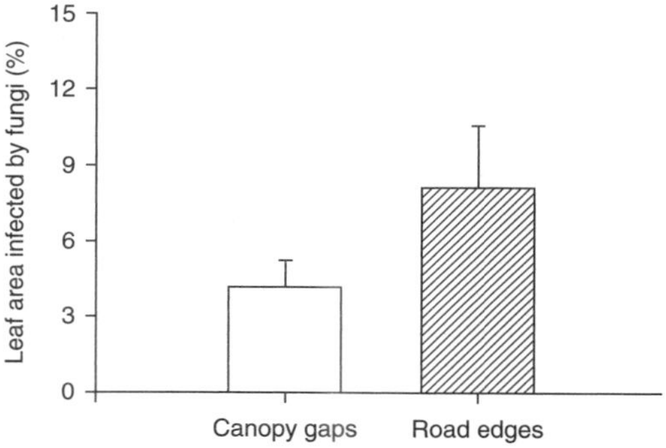

Along a 5-km track of a paved road in Chajul, southern Mexico, fungal damage on leaves of the native herb *Heliconia latispatha* was greater than within natural canopy gaps in continuous forest, where this tropical pioneer species naturally establishes (Santos and Benítez-Malvido 2011) (Fig. 21.2). In addition, despite of the fact that herbivory damage was similar in both habitats, the relative damage of different insect taxa and probable disease vectors (i.e., hispine beetles, caterpillars, leaf-cutting ants) changed dramatically, with damage by caterpillars increasing by about 10-fold in canopy gaps, while the attack by leaf-cutting ants was common along roads but absent in gaps. These findings demonstrated that in human-disturbed habitats the quality and amount of foliage damage are altered by a combination of changes in the physical and biological variables with consequences to ecosystem functioning.

For the same study region, the tree seedling communities close to forest edges showed a greater

**Figure 21.2:**
Standing levels (mean ± SE) of leaf fungal infection on leaves of the pioneer native herb *Heliconia latispatha*, present in natural canopy gaps within old-growth forest and along road-sides, in Chajul, Southeast Mexico (after Santos and Benítez-Malvido 2011).

incidence of foliar disease, represented by leaf spots and necrosis, than did those in forest interiors (Benítez-Malvido and Lemus-Albor 2005, 2006). Nearly all species (95%) present at edges and interiors showed herbivory damage, whereas 76% of the species in edge plots and 68% in interior plots showed leaf-pathogen damage. Although leaf area damaged by herbivores was similar between edges and interior habitats (average 9.2%), pathogen damage was three times greater in edges (1.85%) than forest interiors (0.57%). In contrast, damping-off disease was negligible in the lighter and dryer environments of rainforest edges of Panama (Infante 1999).

In an experimentally fragmented landscape of Amazonia the foliar-disease prevalence in three native seedling species (i.e., *Chrysophyllum pomiferum, Micropholis venulosa,* and *Pouteria caimito*) likewise increased with decreasing fragment size; smaller fragments were more affected by edge effects (Benítez-Malvido and Lemus-Albor 2006). Leaves of the understory herb *Heliconia aurantiaca* within forest fragments in Chajul showed greater rates of fungal infection than those in continuous forest (B. A. Santos and J. Benítez-Malvido 2011 unpublished data; Fig. 21.3). Forest edges and fragments, which are becoming dominant features in tropical landscapes, have hotter and drier microclimates than do forest interiors and non-fragmented forests (Williams-Linera et al. 1998; Benítez-Malvido and Martínez-Ramos 2003; González-Di Pierro et al. 2011). In tropical systems, the evident preference by smut and rust fungi for Asteraceae, Fabaceae, and Poaceae (plant families characteristic of disturbed areas) indicates that these fungal pathogens are among the most important affecting hosts that established in disturbed areas, such as roadsides, canopy gaps, abandoned crop fields, and forest edges, and therefore these diseases may be especially important in the early stages of tropical forest succession (Wennström and Ericson 1991; García-Guzmán and Morales 2007). Disease spread at forest edges may arise as a threat to tropical rainforest vegetation.

## DEFORESTATION

Deforestation has led to the replacement of rainforests by crop fields, pastures, second-growth forests of different ages, and old-growth forest remnants scattered

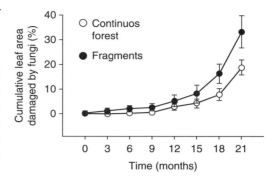

**Figure 21.3:**
Cumulative damage by foliar pathogens (mean ± SE) on the tropical herb *Heliconia aurantiaca* in forest fragments and continuous forests, Chajul, Southern Mexico.

in the landscape. Desiccation and extreme temperature stresses are greater in disturbed environments, and this can make forest plants more susceptible to disease (Burdon 1987; Harvell et al. 2002; Anderson et al. 2004; Pinto et al. 2010; Giraud et al. 2010).

Native seedling species in primary forests of Chajul and Manaus presented a low foliar-disease incidence compared to abandoned cattle pastures and young secondary forests where light and temperature were greater (Benítez-Malvido and Lemus-Albor 2006). Dense seedling carpets of the early successional tree species *Ochroma pyramidale* establishing in a young secondary forest in Chajul were devastated by an unknown fungal pathogen(s) a few weeks after establishment (R. Lombera 2009, personal communication). Furthermore, in riparian corridors of early successional vegetation at Chajul, leaves of the light-demanding herb *Heliconia collinsiana* showed greater levels of fungal infection than in canopy gaps in the nearby continuous forest. Conversely, herbivore damage in *H. collinsiana* was significantly reduced by half compared to forest gaps (3.0% vs. 6.7%), following a similar reduction in caterpillar damage (from 6.2% to 2.5%) and no leaf was recorded to be attacked by leaf-cutter ants (see above *H. latispatha*) (Santos and Benítez-Malvido 2011).

Plant species characteristic of early successional forest stages, which require high amounts of light to become established, generally lack strong resistance to fungal pathogens, whereas late-successional species that tolerate shaded conditions (*sensu* Swaine and Whitmore 1988) are more capable of escaping pathogen infection (Augspurger and Kelly 1984).

Lower resistance of early successional species to pathogen attack could result in delayed forest regeneration after forest disturbance and/or clearance.

## POTENTIAL CONSEQUENCES OF PLANT DISEASES

The limited empirical information available strongly suggests that different types of anthropogenic disturbances favor certain pathogens by altering microclimatic and other ecological conditions. In all of the study cases leaf-fungal infection was favored by tropical rainforest disturbance. Disease epidemics, in particular those caused by introduced pathogens or disturbance in an ecosystem, may cause dramatic, permanent changes in forest community structure and species composition. Local extinctions of plant populations from introduced diseases are a well-documented fact (Rizzo et al. 2002; Laidlaw and Wilson 2003; Weste 2003; Weste et al. 2003; Giraud et al. 2010). Because plants are the primary producers on which all other members of an ecosystem depend (e.g., phytophagous insects and/or nesting animals that rely on the host plant species for survival), the local extinction of a single susceptible plant species might have wide repercussions on ecosystem functioning (e.g., cascading effects) (Burdon 1993; Castello et al. 1995; Harvell et al. 2002; Anderson et al. 2004).

Plant pathogens can potentially exacerbate the deleterious effects of habitat disturbance on tropical ecosystems because the rapid proliferation of exotic plants (which can carry novel pathogens) and early successional native species (which are more susceptible to pathogen attack) in these lands could facilitate the introduction and proliferation of plant diseases (Anagnostakis 1987; Gilbert and Hubbell 1996; Peters 2001; Harvell et al. 2002; Giraud et al. 2010). The ecological costs of disease outbreaks constitute the loss of species and populations and ecosystem degradation (Rizzo et al. 2002; Laidlaw and Wilson 2003; Weste 2003; Weste et al. 2003; Anderson et al. 2004; Vurro et al. 2010). It is very likely that global environmental change may increase the loss of species from some ecosystems, especially for those species with restricted climatic and geographic ranges, low dispersal rates, and small populations (Fraser 2004). On the other hand, climate change may facilitate species with invasive characteristics (i.e., high dispersal, adaptability) to sites where climatic restrictions have previously prevented their colonization (Giraud et al. 2010).

## CONCLUSIONS AND IMPLICATIONS FOR CONSERVATION

Tropical forests are being rapidly degraded with the increasing modifications in land-use patterns and accelerated fragmentation induced by anthropogenic change. This rapid degradation might cause a large proportion of species and populations to be vulnerable to disease, and native and exotic pathogens could become a threat for both plant and animal communities (Anderson et al. 2004; Vurro et al. 2010; Giraud et al. 2010). Therefore, it is essential to understand disease dynamics, the types of potential pathogens present in different habitats, the life histories of plant hosts and of their pathogens, and the effects of the local environmental conditions on them (Gilbert and Hubbell 1996). Evidence shows that the most perturbed and homogenized environments (e.g., tropical old-fields) are more susceptible to invasion by exotic species (Qian and Ricklefs 2006), so human activities not only facilitate the movement of exotic/invasive species to new areas but also provide the conditions for their establishment (Giraud et al. 2010).

Humans cannot control the mechanisms of natural dispersal by the different species of plants and pathogens. However, it is possible to manage/control the different human activities that have major negative impacts on tropical rainforest conservation and functioning. Those activities, including deforestation, road openings, fragmentation, selective logging, mining, and land-use change, that disturb rainforest integrity represent important routes for the dispersal and dissemination of invasive and exotic species and diseases; hence, adequate preventive and monitoring measurements should be taken into consideration after each disturbance event that occurs near natural protected areas. For example, the core areas of reserves should if at all possible be free of roads, which are known to promote further human colonization, harvesting, and hunting, internally fragment natural populations of plant and animals, and facilitate invasions of exotic species (Desprez-Loustau et al. 2007; Laurance et al. 2009).

Our knowledge of the impacts of fungal pathogens on natural tropical plant populations is more than incomplete (Desprez-Loustau et al. 2007). In most cases, we do not know whether fungal diseases are endemic (a disease persistent over time, often of low incidence) or epidemic (a disease rapidly spreading, of high incidence) and the source of infection (from exotic or native pathogens). In addition, basic information on the processes involved in disease transmission (by direct contact, air or water dispersal, or insect vectors), emergence, and spread is still lacking for natural tropical systems and should become fundamental research priorities. This information is also fundamental for disease control and eradication (Desprez-Loustau et al. 2007; Giraud et al. 2010). In addition, most of the examples in this chapter come from single observations in time (i.e., standing levels of leaf fungal damage) and long-term research is crucial to understand these processes. Long-term studies are also necessary to understand the impacts of ephemeral natural events (e.g., El Niño droughts) on disease emergence and on host fitness, as such information cannot be captured in short-term studies. Fungal pathogens not only drive many ecological and evolutionary processes, but fungi also represent an essential component of biodiversity in natural ecosystems and have strong socioeconomic implications in agroecosystems (Desprez-Loustau et al. 2007). Therefore, plant diseases deserve greater attention and should be integrated into conservation biology and conservation medicine efforts (Gilbert and Hubbell 1996; Anderson et al. 2004).

## ACKNOWLEDGMENTS

Fieldwork at Los Tuxtlas, Chajul, and Manaus would have not been possible without the fieldwork support of A. Cardoso, S. Sinaca, P. Sinaca, R. Lombera, C. Ramos, A. Lemus-Albor, and A. González di Piero. This work was partly supported by the National Institute for Amazonian Research (INPA), Smithsonian Institution, the Mexican Council for Science and Technology (CONACYT grants no. 56519, 36828-V and 2007-79121), and the National Autonomous University of Mexico (UNAM, grant no. IN220008). H. Ferreira and A. Valencia provided computing technical assistance.

## REFERENCES

Anagnostakis, S. 1987. Chestnut blight: the classical problem of an introduced pathogen. Mycologia 79:23–27.

Anderson, P.K., A.A. Cunningham, N.G. Patel, F.J. Morales, P.R. Epstein, and P. Daszak. 2004. Emerging infectious diseases of plants: pathogen pollution, climate change and agrotechnology drivers. Trends Ecol Evol 19:535–544.

Agrios, G.N. 1997. Plant pathology. Academic Press, New York.

Antonovics, J. 2005. Plant venereal diseases: Insights from a messy metaphor. New Phytol 165:71–80.

Arnold, A.E., and F. Lutzoni. 2007. Diversity and host range of foliar fungal endophytes: are tropical leaves biodiversity hotspots? Ecology 88:541–549.

Augspurger, C.K. 1984. Seedling survival of tropical tree species—interactions of dispersal distance, light-gaps, and pathogens. Ecology 65:1705–1712.

Augspurger, C., and C. Kelly. 1984. Pathogen mortality of tropical tree seedlings: experimental studies of the effects of dispersal distance, seedling density, and light conditions. Oecologia 61:211–217.

Ayala-Orozco, B. 2008. Maintaining the drivers of tropical plant diversity: plant disease in conservation practice. PhD. Thesis. Environmental Studies. University of California, Santa Cruz.

Beldomenico, P.M., and M. Begon. 2009. Disease spread, susceptibility and infection intensity: vicious circles? Trends Ecol Evol 25:22–27.

Benítez-Malvido, J., G. García-Guzmán, and I.D. Kossmann-Ferraz. 1999. Leaf-fungal incidence and herbivory in tropical rainforest fragments: an experimental study. Biol Cons 91:143–150.

Benítez-Malvido, J., and M. Martínez-Ramos. 2003. Influence of edge exposure on seedling species recruitment in tropical rainforest fragments. Biotropica 35:530–541.

Benítez-Malvido, J., and A. Lemus-Albor. 2005. The seedling community of tropical rain forest edges and interactions with herbivores and leaf-pathogens. Biotropica 37:301–313.

Benítez-Malvido, J., and A. Lemus-Albor. 2006. Habitat disturbance and the proliferation of plant diseases. In W.F. Laurance and C. Peres, eds. Emerging threats to tropical forests, pp. 165–174. University of Chicago Press.

Bradley, D.J., G.S. Gilbert, and I.M. Parker. 2003. Susceptibility of clover species to fungal infection: the interaction of leaf surface traits and environment. Am J Bot 90:857–864.

Brewer, C.A., and W.K. Smith. 1997. Patterns of leaf surface wetness for montane and subalpine plants. Plant Cell Environ 20:1–11.

Brothers, T.S., and A. Spingarn. 1992. Forest fragmentation and alien plant invasion of Central Indiana old-growth forests. Cons Biol 6:91–100.

Burdon, J.J. 1987. Disease and plant population biology. Cambridge University Press, Cambridge.

Burdon, J.J. 1991. Fungal pathogens as selective forces in plant populations and communities. Aus J Ecol 16:423–432.

Burdon, J. 1993. The role of parasites in plant populations and communities. In E.D. Schulze and H.A. Mooney eds. Biodiversity and ecosystem function, pp. 165–179. Springer-Verlag, Berlin.

Burdon, J.J., and A.M. Jarosz. 1988. The ecological genetics of plant–pathogen interactions in natural communities. Phil Trans Roy Soc 321:349–363.

Castello, J.D., D.J. Leopold, and P.J. Smallidge. 1995. Pathogens, patterns, and processes in forest ecosystems. BioScience 45:16–24.

Clark, D.B., and D.A. Clark. 1985. Seedling dynamics of a tropical tree: impacts of herbivory and meristem damage. Ecology 66:1884–1892.

Connell, J.H. 1971. On the role of natural enemies in preventing competitive exclusion in some marine animals and in rain forest trees. In P.J. Boer and G. R. Graadwell, eds. Dynamics of numbers in populations: Proceedings of the Advanced Study Institute, Osterbeek 1970, pp. 298–312. Centre for Agricultural Publication and Documentation, Wageningen.

Colhoun, J. 1973. Effect of environmental factors on plant disease. Ann Rev Phytopat 11:43–364.

Dalling, J.W., M.D. Swaine, and N.C. Garwood. 1998. Dispersal patterns and seed bank dynamics of pioneer trees in moist tropical forest. Ecology 79:564–578.

Dobson, A., and M. Crawley. 1994. Pathogens and the structure of plant communities. Trends Ecol Evol 9:393–399.

Desprez-Loustau, M.R., C. Robin, M. Buée, R. Courtecuisse, J. Garbaye, F. Suffert, I. Sache and D.M. Rizzo. 2007. The fungal dimension of biological invasions. Trends Ecol Evol 22:472–480.

Esquivel, R.E., and J. Carranza. 1996. Pathogenicity of Phylloporia chrysita (Aphyllophorales: Hymenochaetaceae) on Erythrochiton gymnanthus (Rutaceae). Rev Biol Trop 44:137–145.

Fraser, J. 2004. Climate change impacts on biological systems. In T.D. Hooper, Proceedings of the Species at risk. Pathways to Recovery Conference, pp. 1–7. March 2-6, 2004. Victoria, B.C.

Forman, R.T.T., and L.E. Alexander. 1998. Roads and their major ecological effects. Ann Rev Ecol Syst 29:207–231.

García-Guzmán, G., and R. Dirzo. 2001. Patterns of leaf-pathogens infection in the understory of a Mexican rain forest: incidence, spatiotemporal variation, and mechanisms of infection. Am J Bot 88:634–645.

García-Guzmán, G., and J. Benítez-Malvido. 2003. Effect of litter on the incidence of foliar pathogens and herbivory in seedlings of the tropical tree Nectandra ambigens. J Trop Ecol 19:171–177.

García-Guzmán, G., and E. Morales. 2007. Life-history strategies of plant pathogens: distribution patterns and phylogenetic analysis. Ecology 88:589–596.

Garrett, S.D. 1970. Pathogenic root-infecting fungi. Cambridge University Press, U.K.

Gelbard, J.L., and J. Belnap. 2003. Roads as conduits for exotic plant invasions in a semiarid landscape. Cons Biol 17:420–432.

Gilbert, G.S. 1995. Rain forest plant diseases: The canopy-understory connection. Selbyana, 16:75–77.

Gilbert, G.S., S.P. Hubbell, and R.B. Foster. 1994. Density and distance-to-adult effects of a canker disease of trees in a moist tropical forest. Oecologia 98:100–108.

Gilbert, G.S., and S.P. Hubbell. 1996. Plant diseases and the conservation of tropical forests. BioScience 46:98–106.

Gilbert, G.S. 2002. Evolutionary ecology of plant diseases in natural ecosystems. Ann Rev Phytopat 40:13–43.

Giraud, T., P. Gladieux, and S. Gavrilets. 2010. Linking the emergence of fungal plant diseases with ecological speciation. Trends Ecol Evol. 25:387–395.

González-Di Pierro, J. Benítez-Malvido A. M., M. Méndez-Toribio, I. Zermeño, V. Arroyo-Rodríguez, K. E. Stoner, and A. Estrada. 2011. Effects of the physical environment and primate gut passage on the early establishment of an old-growth forest tree species (Ampelocera hottlei Standley) in tropical rainforest fragments. Biotropica 43:459–466.

Goosem, M. 2002. Effects of tropical rainforest roads on small mammals: fragmentation, edge effects and traffic disturbance. Wildlife Res 29:277–289.

Harvell, C., C.E. Mitchell, J.R. Ward, S. Atizer, A.P. Dobson, R.S. Ostfeld, and M.D. Samuel. 2002. Climate warming and disease risks for terrestrial and marine biota. Science 296:2158–2162.

Infante, L.A. 1999. Complex interactions: exploring the role of soil borne plant pathogens in tropical seedling communities. M.Sc. Thesis. Environmental Science, Policy, and Management, University of California, Berkeley.

Janzen, D.H. 1970. Herbivores and the number of tree species in tropical forests. Am Nat 104:501–527.

Janzen, D.H. 1983. No park is an island: increase in interference from outside as park size decreases. Oikos 41:402–410.

Jarosz, A.M., and A.L. Davelos. 1995. Effects of disease in wild plant populations and the evolution of pathogen aggressiveness. N Phytol 129:371–382.

Kellas, J.D., G.A. Kile, R.G. Jarrett, and J.T. Morgan. 1987. The occurrence and effects of *Armillaria luteobubalina* following partial cutting in mixed eucalypt stands in the Wombat forest, Victoria. Aus For Res 17:263–276.

Laidlaw, W.S., and B.A. Wilson. 2003. Floristic and structural characteristics of a coastal heathland exhibiting symptoms of *Phytophtora cinamomi* infestation in the eastern Otway Ranges, Victoria. Aus J Bot 51:283–293.

Laurance, W.F., L.V. Ferreira, J.M. Rankin-de Merona, S.G. Laurance, R. Hutchings, and T.E. Lovejoy. 1998. Effects of forest fragmentation on recruitment patterns in Amazonian tree communities. Cons Biol 12:460–464.

Laurance, W.F., H. Nascimento, S.G. Laurance, A. Andrade, P.M. Fearnside, and J. Ribeiro. 2006a. Rain forest fragmentation and the proliferation of successional trees. Ecology 87:469–482.

Laurance, W F., H. Nascimento, S.G. Laurance, A. Andrade, J. Ribeiro, J.P. Giraldo, T.E. Lovejoy, R. Condit, J. Chave, and S. D'Angelo. 2006b. Rapid decay of tree-community composition in Amazonian forest fragments. Proc Nat Acad Sci USA 103:19010–19014.

Laurance, W.F., M. Goosem, and S.G. Laurance. 2009. Impacts of roads and linear clearings on tropical forests. Trends Ecol Evol. 24:659–669.

Lodge, D.J., and S. Cantrell. 1995. Fungal communities in wet tropical forests: variation in time and space. Can J Bot 73:1391–1398.

Lodge, D.J., P.J. Fisher, and B.C. Sutton. 1996. Endophytic fungi of *Manilkara bidentata* leaves in Puerto Rico. Mycologia 88:733–738.

Murcia, C. 1995. Edge effects in fragmented forests: implications for conservation. Trends Ecol Evol 10:58–62.

Peters, H.A. 2001. *Clidemia hirta* invasion at the Pasoh Forest Reserve: An unexpected plant invasion in an undisturbed tropical forest. Biotropica 33:60–68.

Pinto, S.R.R., G. Mendes, A.M.M. Santos, M. Dantas, M. Tabarelli, and F.P.L. Melo. 2010. Landscape attributes drive complex spatial microclimate configuration of Brazilian Atlantic forest fragments. Trop Cons Sci 4:389–402

Qian, H., and R.E. Ricklefs. 2006. The role of exotic species in homogenizing the North American flora. Ecol Lett 9:1293–1298

Rizzo, D.M., M. Garbelotto, J.M. Davidson, G.W. Salaughter, and S.T. Koike. 2002. *Phytophtora ramorum* as the cause of extensive mortality of *Querqus* ssp. and *Lothocarpus densiflorus* in California. Plant Dis 86:205–214.

Santos, B. A., M. Quesada, F. Rosas, and J. Benítez-Malvido. 2011. Potential effects of host height and phenology on adult susceptibility to foliar attack in a tropical dry forest grass. Int Schol Res Net ISRN Ecology. Article ID 730801, 7 pages. doi:10.5402/2011/730801.

Santos, B.A. ., and J. Benítez-Malvido. 2011. Insect herbivory and leaf disease in natural and human disturbed habitats: lessons from early-successional heliconia herbs. Biotropica. Article first published online: March 9, 2011 | DOI: 10.1111/j.1744-7429.2011.00765.x

Stratford, J.A., and P.C. Stouffer. 1999. Local extinctions of terrestrial insectivorous birds in a fragmented landscape near Manaus, Brazil. Cons Biol 13:1416–1423.

Swaine, M.D., and T.C. Whitmore. 1988. On the definition of ecological species groups in tropical rain forest. Vegetatio 75:81–86.

Travers, S.E., G.S. Gilbert, and E.F. Perry. 1998. The effect of rust infection on reproduction in a tropical tree (*Faramea occidentalis*). Biotropica 30:438–443.

Vurro, M., B. Bonciani, and G. Vannacci. 2010. Emerging infectious diseases of crop plants in developing countries: impact on agriculture and socio-economic consequences. Food Security 2:113–132.

Wennström, A., and L. Ericson. 1991. Variation in disease incidence in grazed and ungrazed sites for the system *Pulsatilla pratensis–Puccinia pulsatillae*. Oikos 60:35–39.

Weste, G. 2003. The dieback cycle in Victorian forests: a 30-year study of changes caused by *Phytophtora cinnamoni* in Victorian open forests, woodlands and heathlands. Aus Plant Path 32:247–256.

Weste, G., K. Brown, J. Kennedy, and T. Walshe. 2003. *Phytophtora cinnamoni* infestation—a 24-year study of vegetation change in forests and woodlands of the Grampians, Western Victoria. Aus J Bot 50:247–274.

Williams-Linera, G., V. Domínguez-Gastelú, and M.E. García-Zurita. 1998. Microenvironment and floristics of different edges in a fragmented tropical rainforest. Cons Biol 12:1091–1102.

# 22

## EMERGING INFECTIOUS DISEASES IN FISHERIES AND AQUACULTURE

## E. Scott Philip Weber III

Fish are the most biodiverse vertebrates, numbering more than 56,000 described species and sub-species, of which over 1,100 new species have been added since 2008 (Eschmeyer et al. 2011). Fish have adapted to diverse habitats from desert to glacial lakes and from deep-sea vents to intertidal marshes. Due to their relative abundance, fish play a prominent role in the world, providing an important protein and omega-3 fatty acid source for human and animal nutrition. In addition, fish are becoming the most common laboratory animal models for toxicological, infectious disease, immunological, and genetic research. They remain the most numerous companion animals as pets in bowls, ponds, and aquaria; represent the fastest-growing taxa used for education in zoos and aquariums; serve as vital sentinel species for marine and freshwater ecosystems, habitat assessment, and ecological health; and are propagated for stocking natural fisheries for sport and natural enhancement.

An understanding of aquatic animal health and disease is crucial to incorporate a proactive approach when addressing current and future needs in conservation medicine and ecological health. This chapter will discuss the role and responsibility of veterinarians for aquaculture, summarize examples of emerging disease epidemiology from aquaculture and fisheries interactions, highlight ornamental fisheries' challenges for the hobby and aquarium trade, identify

a role for animal health professionals throughout this continuum, and conclude with environmental considerations exacerbating fish health stocks. Invertebrates are briefly discussed but have an equally significant role. Historical and current examples will be used to illustrate conservation medicine dilemmas and challenges.

## AQUACULTURE

Although aquatic organisms have been cultivated for centuries, aquaculture production is a relatively new branch of agriculture. The U.S. aquaculture in 2010 has decreased from previous peaks in tonnage (2004) and in value (2006), while world aquaculture production for inland and marine animal species has nearly doubled in both tonnage and value from 1998 to 2010 (FAO 2010). With an increase in aquaculture production, there are also increased pressures on wild fisheries to help provide broodstock, seedstock, and a nutritional source of protein and fish oils in commercial aquaculture feeds. Historically, the ecological input purported to cage-raised salmon in Norway roughly equated to five metric tons of food fish to produce each metric ton of farmed salmon (Folke 1988). Even when dietary inputs for commercial salmon feeds are not double counted for fish meal and for fish oil components, the most accurate feed conversion rates of dry weight to wet weight (1.6 to 1.8) still require fishery inputs for salmon aquaculture (Tacon 2005). In 2000 it was recognized that despite dependence on wild fisheries for fishmeal in aquaculture, production actually increased world fish supplies (Naylor et al. 2000), but in 2010 production has doubled from these 2000 numbers. Based on increases in feed conversion efficiency and scientific improvements in nutrition, some researchers are convinced that use of wild fisheries for fishmeal and fish oil in aquaculture feeds will continue to decrease (Tacon and Metian 2009). However, despite an encouraging decrease in the use ratio of wild fisheries to all farmed aquaculture of 0.63, economic and regulatory efforts are still needed to protect wild fisheries (Naylor et al. 2009).

The fastest growth in aquaculture has been the shrimp industry in Central America, South America, and Southeast Asia. FAO reports show global industry growth from 0.2 million metric tons in 1985 to 2.8 million metric tons by 2007 (FAO 2008), illustrating why shrimp are considered *oro blanco* (white gold) in countries like Ecuador. Problems with such rapid industry expansion and an increasing global shrimp demand include habitat/natural resource use, species protection, pollution, and infectious disease. As early as the 1970s shrimp was becoming the fastest-growing aquaculture industry in Thailand, and by 1995 environmental impacts included mangrove/wetland destruction, saltwater intrusion, land subsidence, water quality degradation, sediment disposal, abandoned shrimp farms, and displacement of traditional livelihoods (Dierberg and Kiattisimkul 1995). These environmental issues plus fishery depletion were associated with the shrimp industry in Honduras over a similar time period (Dewalt et al. 1996). Infectious disease problems resulted in major production declines in Taiwan (1988), China (1993), Thailand (1996–97), and Ecuador, Indonesia, and the Philippines (1999) (Primavera 1997; Kautsky et al. 2000).

Understanding basics for aquaculture health management, biosecurity, and biodiversity are vital to addressing needs of ecological health for aquatic organisms. The effects of aquaculture on fisheries are multifactorial including habitat modification, using wild seed for stocking, impacts on food web interactions, introduction of non-indigenous organisms, and effluent discharge (Naylor et al. 2000). Aquaculture is viewed as having a negative impact on biodiversity

through land use, nonrenewable energy reliance, environmental pollution, antibiotic runoff, exotic/invasive species introductions, reliance on fishmeal from wild fisheries, broodstock or seed acquisitions, and water use. However, maintaining biodiversity is essential for current and future aquaculture development (Beveridge et al. 1994; Diana 2009). Positive impacts of aquaculture on biodiversity include decreasing exploitation of wild seafood products with cultured livestock, using certain species to enhance wild stocks, and increasing natural production and species diversity (Diana 2009).

There is naturally a direct relationship between disease and production level for any type of captive management. As the intensity of production increases, the prevalence of disease also tends to increase. Intensification generally refers to increasing the number of animals raised in confined or restricted areas, requiring increasing external resource inputs. The greatest limitations to sustaining development of Asian aquaculture—which accounts for 89% of world aquaculture production (FAO 2008)—are aquatic animal health concerns, and without appropriate animal health management, major disease outbreaks and newly emerging pathogens will continue to dominate costs and socioeconomic development of aquaculture throughout Asia (Bondad-Reantaso et al. 2005).

Technological advances for disease prevention, diagnosis, and amelioration serve to optimize aquatic organism health management plans. Increased sophistication of fish health research, advances in molecular diagnostics, and improvements in fish health management can minimize infectious disease in aquaculture (e.g., vaccines and biosecurity), similar to trends noted in production of some terrestrial animals. For example, infectious disease has appeared to decline in the Atlantic salmon aquaculture industry in British Columbia today compared to 15 to 20 years ago (Gary Marty personal communication 2011).

Aquaculture poses great potential to help feed humans, improve human health through nutrition, and advance agricultural practices through more practical environmental solutions (Godfray 2010). As aquatic health practitioners, incorporating a strong foundational aquatic animal health program is essential to the management and future success of aquaculture and the maintenance or restoration of our aquatic ecosystems, including wild fisheries.

## FISHERIES INTERACTIONS AND EMERGING INFECTIOUS DISEASES

Decreasing needs for natural inputs, using less space, and raising greater biomass is a hallmark of intensive aquaculture, while extensive aquaculture uses methodologies to enhance natural food resources and often is limited by the natural resources available. Intensive aquaculture is a relatively new agricultural practice, yielding intentional and inadvertent interactions between aquaculture and wild fisheries. In the context of conservation medicine, these concerns are fisheries interactions resulting in disease outbreaks, farm-source antibiotic/chemical-resistant pathogens, exotic/invasive species introduction, and environmental contamination caused by effluent. Although aquaculture may contribute to these fisheries interactions, other industries and environmental utilization make substantially greater changes to ecosystems, posing a threat to both aquaculture and native fish populations.

These disease outbreak examples illustrate the complexity of aquaculture and fisheries interactions and the myriad of stakeholders researching, supporting, contesting, and/or debating these issues. Understanding infectious diseases in wild fish requires effectively led transdisciplinary teams with expertise in fisheries biology, medicine, public health, pathology, mathematical modeling, and ecology. By including participation of concerned citizens and multiple specialists, fish health professionals will be ultimately equipped to diagnose, treat, and/or mitigate emerging infectious disease outbreaks in fisheries and aquaculture (Georgiadis et al. 2001).

### Viral Hemorrhagic Septicemia Virus (VHSV)

VHSV is a serious disease of farmed rainbow trout (*Onchoryncus mykiss*) that appeared in Europe as early as 1931 and was first isolated by Jensen et al. (1963) in Denmark. The causative agent has the morphology of a rhabdovirus, but it differs from other salmonid-associated rhabdoviruses such as infectious hematopoietic necrosis virus (IHNV) isolated in North America (Eaton et al. 1991). It is believed that feeding offal from marine fishing boats to cultured rainbow trout transmits VHSV (Rasmussen 1965). Currently the disease has spread through most rainbow trout

farming regions in continental Europe. In North America the origins of the virus were markedly different, having been isolated from Pacific cod (*Gadus macrocephalus*) out of Alaska (Meyers et al. 1986). In marine ecosystems VHSV had potential impacts on pelagic species important for commercial fisheries (Hedrick et al. 2003a). The virus was detected in sardines (*Sardinops sagax*) and herring (*Clupea pallasi*) in the Pacific Ocean. The freshwater history of the disease is more convoluted and recent. It has involved multiple fish kills in a wide geographic area affecting many diverse freshwater species, and may have its origin in marine fish off the western Atlantic coast of Canada (Lumsden et al. 2007). Some researchers believed all marine fish were susceptible to the virus, although no records of VHSV have been found in the Southern Hemisphere (Stone et al. 1997; Snow and Smail 1999).

In 2006, VHSV type IVb was isolated from multiple species in Lake St. Clair, Lake Erie, Lake Ontario, and the St. Lawrence River (Elsayed et al. 2006). Some isolates were obtained from large mortalities. The 2006 VHSV outbreak in the Great Lakes can be considered as a multispecies epidemic. Such outbreaks are typical among naïve populations following initial exposure to an introduced pathogen, and this pathogen continues to be isolated on wild fishery surveys; it was most recently identified in Lake Superior as late as March 2010 (Gunderson 2010). The only way to control many viral pathogens in fish is through strict avoidance, proper sanitation, continued surveillance, ongoing fish health certification, and active farm biosecurity. Much more research needs to be conducted to understand changes in the host specificity of VHSV.

### Ichthyophoniasis

Ichthyophoniasis is an emerging disease in Pacific salmonids caused by a mesomycetozoean parasite of the genus *Ichthyophonus*. Prior to 1985, ichthyophoniasis was not recorded in Pacific salmonid species, although it was associated with disease outbreaks in other wild species, including many species found in the Clupidae family. In a survey by Kocan et al. (2004) of samples taken from the Yukon River, about 45% of returning adult Chinook salmon (*Oncorhynchus tshawytscha*) were infected, as confirmed by *in vitro* culture or less frequently from histopathologic evaluation.

Ichthyophoniasis in Pacific salmonids is an example of the important interaction between native fish populations and introduced species through fisheries management. American shad (*Alosa sapidissima*) had been stocked in the Pacific Northwest as early as the late 1800s (Smith 1886). Shad numbers in the Columbia River remained low until the late 1980s, when return data for adult American shad on the Columbia River were greater than the combined numbers for all returning Pacific salmonid species (Hershberger 2010). The same authors demonstrated through genetic analysis that although the fungal pathogen was most likely endemic in the Pacific Northwest, this pathogen was being amplified, causing a direct disease threat to species of Pacific salmonids sharing the Columbia River basin with shad.

## Parasitic Sea Lice

The first potential problems between sea lice (*Lepeophtheirus salmonis*) from cultured Atlantic salmon and wild fisheries occurred in 1997 in Scotland (McVicar 1997). Since then, the controversy expanded, most notably to the Broughton Archipelago region of British Columbia, Canada, where sea lice transmission from cultured Atlantic salmon to migratory wild juvenile pink salmon (*Oncorhynchus gorbuscha*) was projected to cause a 99% collapse in pink salmon populations in four salmon generations (Krkošek et al. 2007). However, these projections were based on field mortality experiments with pink salmon of unknown history (Krkošek et al. 2006; FOC 2009). The analysis did not include sea lice information from fish farms, and the paper did not cite peer-reviewed publications showing that Pacific forms of *L. salmonis* (a) were genetically different from Atlantic forms (Todd et al. 2004); (b) were more benign clinically (Saksida et al. 2007); or (c) did not cause mortality under controlled laboratory conditions (Jones et al. 2006; Jones and Hargreaves 2007; Webster et al. 2007). Later work showed that pink salmon less than 0.7 g were susceptible to very high exposures of sea lice under controlled laboratory conditions, supporting the conclusion that sea lice killed no more than 4.5% of juvenile pink salmon in any given year from 2005 to 2008 (Jones and Hargreaves 2009). To better understand these differences, sea lice counts were obtained from fish farms and compared against pink salmon data, determining that the number of adult female sea lice on farm fish in April accounted for 98% of the variability in annual sea lice prevalence on juvenile pink salmon in May (2002–09), but those same farm sea lice counts were not associated with decreased pink salmon lifetime survival (Marty et al. 2010b).

Fish health expertise is vital when developing infectious disease modeling for individual populations, and especially when these interactions are complex fishery and aquaculture interactions. Despite these dire modeling predictions (Krkošek et al. 2007) pink salmon returns in the Broughton Archipelago were up about 300% in 2009 from 2007 (Brooks and Simon 2008; Marty et al. 2010b), suggesting that more collaboration rather than divisive discourse is needed to further document the true effects sea lice may have on wild salmon populations.

## Whirling Disease: Problems and Solutions

Though speculative, the myxozoan parasite *Myxobolus cerebralis* likely reached North America in the 1950s with frozen rainbow trout or imported brown trout (*Salmo truffa*) from Europe (Hoffman 1990; Bartholomew and Reno 2002). The first recorded epizootic in North America was among brook trout at a hatchery in Pennsylvania. Subsequent episodes were reported in numerous other Eastern state trout hatcheries, frequently among rainbow trout, one of the more susceptible species of salmonids (Hoffman 1990). Under conditions of high infectivity, whirling disease may induce severe cranial and spinal deformations and death as parasite stages feed upon and destroy skeletal cartilage prior to bone formation in young fish (Halliday 1973). The myxospore stages of *M. cerebralis* in the skeletal elements of chronically infected trout may not result in the blackened tails and erratic swimming both characteristic of acute disease. If the spores are not detected, the parasite will spread to new geographic regions with the transport of hatchery fish for stocking in rivers and streams for the sport fishery. This mode of parasite movement has facilitated the spread of the parasite to over 25 of the United States, affecting most profoundly wild trout in the intermountain West (e.g., Colorado, Montana) where large-scale population declines, particularly among rainbow trout, have destroyed once highly prized sport fisheries (Nehring and Walker 1996; Vincent 1996). The two-host life cycle of *M. cerebralis*, which also includes the oligochaete *Tubifex tubifex*, has added to the complexity associated with the control and management of whirling disease. However, a combination of improved diagnostic procedures for parasite detection, in both fish and oligochaetes, and new management procedures in aquaculture are improving the ability to control whirling disease. The advent and now widespread application of PCR can detect the pathogen among fish with subclinical disease, thus preventing the inadvertent movement of infected fish (Andree et al. 1998, 2002). Improved sanitation in trout hatcheries using both treatments of the water supply with ultraviolet irradiation and several common disinfectants for ponds and equipment, including transport vehicles, has further reduced the spread of the parasite (Hedrick et al. 2007, 2008). Finally, the discovery and exploitation of rainbow trout strains resistant to whirling disease has provided a powerful management approach to significantly reducing and perhaps eliminating whirling disease from certain hatchery environments and ideally from selected wild trout sport fisheries (Hedrick et al. 2003b; Schisler et al. 2006). The most broadly exploited whirling disease-resistant rainbow trout strain is the product of 120 years of selection in a commercial hatchery in Europe. This strain is now being reared in state and commercial fish hatcheries in the United States. It is also being bred with wild trout with the aim of restoring selected wild populations of rainbow trout in Colorado.

## Iridoviruses

Iridovirus infection is an excellent example of an infectious disease that may cause havoc for fisheries and aquaculture (Palmeiro and Weber 2009). Iridoviruses are large double-stranded DNA viruses that occur in invertebrates, amphibians, and fish (Chinchar 2002; Daszak et al. 1999, 2003; Delhon et al. 2006). The three genera in fish are Ranavirus, Lymphocystivirus, and Megalocytivirus (Chinchar et al. 2005). Systemic infections in both marine and freshwater fish are commonly attributed to megalocytiviruses, while ranaviruses are more often associated with amphibians with potential for cross-infectivity between amphibians and fish (Daszak et al. 1999). Some ranaviruses have also been isolated from chelonians suffering from upper respiratory disease (Westhouse et al. 1996). In the ornamental freshwater fish trade several species

infected with iridovirus have been identified, such as the dwarf gourami (*Colisa lalia*), chromide cichlids (*Etroplus maculates*), freshwater angelfish (*Pterophyllum scalare*), and the African lampeye (*Aplocheilichthys normani*) (Armstrong et al. 1989; Anderson et al. 1993; Rodger et al. 1997; Paperna et al. 2001; Sudthongkong et al. 2002). Marine ornamentals are also susceptible to iridoviruses, with morbidity and mortality reaching 100% in imported Banggai cardinal fish (Weber et al. 2009). The danger of iridovirus is the lack of species specificity for infection with the megalocytiviruses in freshwater and marine fish species, and the ranaviruses in amphibians, chelonians, and freshwater fish. These viruses occur over great geographic areas, causing a major impact on aquaculture for both the food and ornamental fish industries (Whittington and Chong 2007), and pose a direct threat to native populations of aquatic poikilotherms.

## Herpesviruses

For temperate aquarium fish such as koi (*Cyprinus carpio*) and goldfish (*Carassius auratus auratus*), similar constraints and problems exist with monitoring, screening, and diagnosing infectious disease. The best example is the cyprinid herpesviruses, CyHV1, 2, and 3 (Palmeiro and Weber 2009). All three viruses have the potential to occur as latent infections; infected fish become asymptomatic carriers. CyHV2 in goldfish and CyHV3 in koi and common carp cause severe illness and share similar clinical signs, with high morbidity and mortality. CyHV2 or goldfish herpesvirus is a serious disease of all goldfish (*C. auratus*) varieties and has been reported with acute mortalities as high as 100% (Goodwin et al. 2006). Goldfish are seldom tested or screened for this disease despite catastrophic losses. Koi herpes virus (KHV, caused by CyHV3) infects and causes massive mortality in koi and common carp. KHV may have been around for several decades, although the first reported outbreaks were in Israel in 1998 after a large mortality of koi (Hedrick et al. 1999). Unlike the other cyprinid herpesviruses, KHV is listed as a reportable disease globally by the World Animal Health Organization (OIE) and in the United States by the USDA (Way 2008). Mortality rates are similar to goldfish infected with CyHV2, ranging from 70% to 100% (Gilad et al. 2003). The greatest risk posed by KHV is development of asymptomatic carriers from previously exposed fish. Taqman® real-time PCR is the most sensitive diagnostic tool commercially available for the diagnosis of KHV infections (Bercovier et al. 2005). An ELISA test that detects KHV antibodies in the serum of koi is available commercially to help detect fish that were previously exposed to the virus and may serve as asymptomatic carriers (Adkison et al. 2005).

There is considerable consumer misinformation circulated via the Internet through hobbyist organizations on KHV prevention, biosecurity, epidemiology, disease diagnosis, molecular testing, and treatment/management. No current regulations require screening for carrier fish, as international OIE regulations only request testing for active disease via several methods used in concert, such as PCR and virus culture. The best preventive medicine protocols include screening retailers and wholesalers using multiple diagnostic tools to both detect active KHV infections via PCR for suspected ill animals, and also screen for previous exposure to the pathogen in populations for antibodies to KHV by ELISA. An attenuated live vaccine for KHV was patented in Israel in 2003, but is not globally approved for use. This failure in international regulation is allowing fish that have been exposed to KHV naturally or at high temperatures to be shipped around the world, allowing asymptomatic carriers to spread a devastating disease.

## Mycobacteriosis

Mycobacteriosis and other diseases offer some of the greatest health challenges for managed collections or groups of fish, and have been identified in hundreds of freshwater and marine ornamentals, in laboratory animals, and also from aquaculture-reared striped bass (*M. saxatilis*) from California (Hedrick et al. 1987). In the past decade mycobacterial infections have been isolated from clinically infected wild striped bass and in juvenile Atlantic menhaden (*Brevoortia tyrannus*), with a disease prevalence ranging from 2% to 57% in tributaries of the Chesapeake Bay (Kane et al. 2007), and multiple *Mycobacteria spp.* are capable of colonizing numerous species of native fish in this region (Stine et al. 2009), threatening wild populations of fish.

## Other Pathogens

Many other disease examples exist, such as infections caused in salmonids by infectious pancreatic necrosis virus (IPNV), and the near-extinction of European crayfish (*Astacus astacus*) from crayfish plague caused by a water mold *Aphanomyces astaci* carried by American crayfish (*Pacifastacus leniusculus*). Disease outbreaks associated with wild fish, aquaculture-reared animals, and managed fisheries must be evaluated using all the diagnostic tools available to assess physical and chemical parameters, while trying to objectively understand the ecology of infectious disease through complex interactions of the host with other species in the aquatic environment (Hedrick 1998). As our understanding of fish health and aquatic animal pathogen epidemiology grows, veterinarians and aquatic animal health professionals must also be conscientious about making sound health and regulatory policy in terms of differentiating detection of pathogens from actual diagnosis of disease. This must be accomplished using strict scientific standards, ethical objectivity, and evidence-based medicine to benefit the future of robust fisheries, sustainable aquaculture, and marine ecosystem health.

## ORNAMENTAL FISHERIES

Marine and freshwater ornamental fisheries are a growing area, and they involve largely unregulated movement of livestock with global implications for potential disease outbreaks (Andrews 1990; Dentler 1993–94). The ornamental aquarium industry includes captive (more than 90%) and wild freshwater fish, with over 632 million animals produced in Malaysia alone, as recorded by the Malaysian ornamental fish industry, coupled with a far greater variety of marine ornamental species that are traded, with estimates of close to 8,000 different species of marine animals, with only about 25 species captive-bred (Helfman, 2007; FAO 2008; Smith et al. 2009). FAO (2008) suggested that 59% of all production of ornamental aquaculture-related products originated from Asian countries. Aquatic animals include fish, reptiles, amphibians, invertebrates (including corals), and live rock (dead coral that becomes encrusted with marine macro- and micro-organisms in the wild). Farms in Florida and Southeast Asia raise many freshwater ornamental varieties, with additional inputs from other countries, hobbyists, and wild-caught fisheries. Hobby specialty groups actively trade some of the animals highly prized for appearance, such as African and

South American cichlids. The freshwater temperate ornamental industry consists primarily of koi and goldfish, with major production in Israel, Arkansas, Japan, and China. There is little marine ornamental production of farm-raised stock, and the majority of marine fish for the hobbyist trade comes from captured wild stocks in the Caribbean, the Red Sea, and Indo-Pacific Ocean. Methods used for collection in the Philippines for many marine species in the past have included dynamite stunning and potassium cyanide poisoning, causing irreparable damage to stock collected as well as the environment and animals caught as by-catch (Galvez et al. 1989; Calado et al. 2003). Zoos and aquariums also contribute to these pressures on aquatic animals; as net consumers of wildlife, they are and should be held to high standards of animal care. The increased fishing pressures on these animals have made some species extremely vulnerable. In just the past decade Banggai cardinal fish (*Pterapogon kauderni*) have become one of the first marine ornamentals to be listed on the IUCN red list for threatened species (Lunn and Moreau 2004; Vagelli 2004), and institutions like the Shedd Aquarium in Chicago have sponsored conservation efforts such as Project Seahorse to help with education and research of dwindling numbers of sygnathid species. Although restoration projects are proposed to support conservation for many of these species (Ziemann 2001; Box 22.1), such endeavors can cause catastrophic effects on native populations and responsible protocols must be developed to ensure the health status of captive breeding animals and their offspring prior to release with native populations. Restoration efforts also must involve conducting health surveys on captive and wild populations to screen for the presence of common and emerging pathogens known to threaten native organisms (Smith et al. 2009).

The primary cause of mortality for many ornamental species is from stress associated with transport, resulting in death either immediately from environmental quality issues or later from disease. Mortality for wild-caught ornamentals approaches 100% of some shipments. Wholesale operations bring animals together from different geographic locations into holding systems with shared filtration. Geographic strains of common parasites and microbes can be deadly for naïve animals using a shared recirculating water source. Emerging infectious diseases are largely undetected and seldom appear in the peer-reviewed scientific literature, despite millions of animals being globally traded. This lack of knowledge is mostly the result of mortalities in the aquarium and ornamental trade being discarded without any obligation or regulation for movement of these animals and often no gross pathology and/or histopathology is performed on fish mortalities.

## CORAL REEFS

Destruction of coral reefs and coral diseases are fast outpacing new reef formation (Wilkinson 2004). Coral reefs have tremendous species diversity, and there are many new resources on conserving coral reefs. Expertise in understanding coral reef destruction through scientific research has uncovered numerous new pathogens and disease syndromes identified in corals, linking environmental quality and climate change to many problems faced by these invertebrates (Hoegh-Guldberg et al. 2007). Corals are being threatened by anthropogenic and environmental variables such as bleaching, infectious disease, climate change, predator plagues, and invasive species (Goldberg and Wilkinson 2004). It is not in the scope of this chapter to cover the plight of corals, but ecologists, marine biologists, veterinarians, and other health professionals are encouraged to become active collaborators and participants in coral reef conservation.

## AQUATIC NUISANCE/INVASIVE SPECIES

As conservation medicine professionals our role with aquatic invasives is to recognize new and emerging disease threats, including those that might be caused by human actions (Crowl et al. 2008). This role includes using our skills in education, nutrition, infectious diseases, and public health to help investigate problems associated with the current epidemic of aquatic invasive species and native animal populations. Disease risks associated with invasives have been overlooked because of poor pathology associations due to subjective data or because current emphasis focuses on the ecology of invasive organisms rather than ecology of infectious disease outbreaks or epidemiology of wildlife disease outbreaks.

---

**Box 22.1** *Proyecto Piaba*—A Conservation Medicine Example in the Rio Negro

---

Barcelos and Santa Isabel do Rio Negro (population 34,000; area 185,000 km²) are the main provinces along the river, and the fishery contributes 60% of the income revenue for this region by extracting 40 million to 70 million ornamental fish annually (Chao et al. 2001). Proyecto Piaba was established in 1989 as a community-based interdisciplinary initiative to understand the ecological and sociocultural systems of the Rio Negro basin based at Universidade do Amazonas, Manaus, Brazil. The primary objective of this program was to provide a framework for scientific research and local management of the ornamental fishery resources. The slogan for Proyecto Piaba is "buy a fish, save a tree," and this endeavor began to protect the local fishery through implementing appropriate fishery management to help preserve the Amazonian flooded-forest ecosystem. The fishery has prevented natural resources from being exploited from alternative industries in this region that include the destruction of the forest and pollution of the river through logging, mining, and ranching (Chao and Prang 1997). In 2000, the New England Aquarium signed an agreement of technical, scientific, and cultural collaboration with the Universidade do Amazonas, providing a unique opportunity to collaborate with Brazilian partners to preserve one of the largest intact regions of the Amazon. Since 2001, increased international regulatory efforts for ornamental fish in Europe and greater pressure from aquaculture-reared ornamental freshwater fish from abroad have threatened this fishery. Recent goals of Proyecto Piaba are to better evaluate fish health throughout the fishery, from capture to arrival of the fish at the importer, and in turn to decrease transportation mortality through a better understanding of shipping stress and provide health assessments of animals from local streams and tributaries (Tlusty et al. 2005).

---

## CLIMATE CHANGE AND ENVIRONMENTAL CONTAMINANTS

Climate change, whether anthropogenic or natural, has the potential to greatly alter the aquatic landscape in ways poorly appreciated or understood. Climate change may be intimately linked to many issues presented in this chapter. These include from providing sustainable aquaculture to harvesting ornamental fish, from the success or failure of invasive species (Rahel and Olden 2008) to natural changes in range for many animals, and from coral reef survival or failure (Goldberg and Wilkinson 2004) to the ecology of infectious disease organisms (Riley et al. 2008). Fish populations at risk to climate change in northern European countries include disease threats from pathogens previously occurring in warmer southern climates such as *Lactococcus garvieae* in farmed trout in the United Kingdom (Bark and McGregor 2001) and proliferative kidney disease in wild Swiss grayling (*Thymallus thymallus*) (Wahli et al. 2002).

Pollution is a common thread for many of these environmental impacts, and several examples will be highlighted to illustrate the complicated nature of these interactions. Natural populations of fish in Europe are contaminated with polychlorinated biphenyls (PCBs). Fish from the Rhone River downstream from an incineration plant had high levels of contamination, and the consequence of this exposure was a significant increase in cytochrome P-450–dependent mono-oxygenase activities (Monod et al. 1988). Pollutants can have direct environmental consequences, such as the accumulation of halogenated contaminants in farmed salmon, trout, tilapia, pangasius, and shrimp compared with wild-caught marine fisheries, showing salmon as having a greater contaminant level than all other species investigated and PCBs being found in the highest concentration of all contaminates (van Leeuwen et al. 2009). Subsequent studies investigating salmon from Europe and North America in 2004 and 2005 suggested that PCB levels were also significantly higher than levels found in wild Alaskan Chinook salmon, but that farms from Eastern Canada had decreased PCB detection in the second year (Shaw et al. 2005, 2006, 2007). Comparing over 2 metric tons of farmed and wild salmon from around the world for organochlorine contaminants, farmed salmon had significantly higher contaminant levels than wild salmon, with Europe having higher levels than both North and South America (Hites et al. 2004). However, more recent research showed that many of these differences were a result of lower lipid content in wild Pacific salmon than in farmed Atlantic salmon, rather than from increased chemical exposure risk (Ikonomou et al. 2007). Pollutant levels in wild populations of salmonids from the Great Lakes may be improving based on human studies including sport fishers that illustrated a decrease of PCBs and DDT over a 10-year period from 1995 to 2005 (Knobeloch et al. 2008).

Pollution is evident from human-induced causes and disasters from mines, factories, inadequate wastewater treatment, agricultural runoff, logging, hurricanes, tsunamis, and oil spills. After exposure to a variety of pollutants, heavy metals, acid rain, and xenobiotics were shown to cause gross and microscopic gill epithelial lesions directly linked to osmoregulatory, acid–base, or hemodynamic malfunction, suggesting that toxins are processed in the gills of fish, similar to human renal, gastrointestinal, and hepatic pathways (Evans 1987). Biomarkers of toxicity caused by crude oil have been demonstrated at very low exposure, showing the metabolic enzymes citrate synthase and lactate dehydrogenase measured in the gills of Atlantic salmon are good biomarkers of exposure to the water-accommodated fraction of Bass Strait crude oil, and to chemically dispersed crude oil (Gagnon and Holdway 1999). Fish exposed chronically to petroleum hydrocarbons at the site of the Exxon Valdez oil spill in Alaska showed a significant difference in the prevalence and intensity of trichodina parasitism between sculpin originating from an oil-free and an oil-contaminated site, with the oil-contaminated ones having the greater infection (Khan 1990). However, extensive environmental differences in both sites were not sufficiently described, as this parasite can also increase rapidly in water temperatures differing by only a couple of degrees centigrade, especially in eutrophic conditions. Among the many toxins in the aquatic environment, heavy metal toxins are of historical significance. Minamata disease in humans is characterized by severe congenital birth deformities that result from eating fish contaminated with methyl mercury waste; it was first diagnosed in Minamata Bay, Japan. Mercury concentrations in U.S. lakes from

1989 to 2005 had the greatest downward trends in fish samples when compared with data from 1969 to 1987, correlating with decreased mercury in sediment and peat cores over that period (Chalmers et al. 2010); this is positive news for mercury mitigation in this region. Effects of toxins are being newly identified, such as endocrine disruptors in fish (Mills and Chichester 2005). In U.K. rivers, a high prevalence of intersexuality and sexual disruption in the roach (*Rutilus rutilus*) was concomitant with environmental detectable levels of pollutant endocrine disruptors, documenting an initial change in vertebrate behavior related to such compounds (Jobling 1998).

## ANTIBIOTIC RESISTANCE

Researchers are concerned with the impacts of antibiotic waste on various ecosystems generated from human and agricultural uses. With improved analytical techniques, antibiotics can be identified in wastewater and post-treatment effluent, and studies suggest that of six antibiotics (ciprofloxacin, trimethoprim/sulfamethoxazole, tetracycline, ampicillin, trimethoprim, and erythromycin) tested before and after wastewater treatment, some bacteria resistance occurred to all six antibiotics before treatment and to two antibiotics after treatment (Costanzo et al. 2005). In broader surveys conducted from 16 sites in the southern North-Rhine Westphalia of Germany, antibiotics were detected in all watersheds. Both large (Rhine River) and small creeks had some substances (erythromycin and sulfamethoxazole) identified in nearly all analyzed samples, and concentrations of these pharmaceutical products ranged from just detectable to the limit quantization (Christian et al. 2003). Other regional surveys detected antibiotics in groundwater and wastewater, but groundwater in heavy livestock areas was free from antibiotic residues, suggesting that most environmental contamination in Germany is from human medical practices (Hirsch et al. 1999).

Although most environmental contamination is from human waste or terrestrial agricultural effluent, aquaculture may also contribute the problem of antimicrobial resistance. Many veterinary therapeutics are found in the aquatic environment, including insecticides, algaecides, antihelminthics, and antibiotics (Boxall et al. 2004; Kemper 2008). Inlet and outlet water samples were compared and levels of antibiotic resistance from bacterial sampling of fish, water, and sediment were assessed in four Danish trout farms. Two major fish pathogens (88 *Flavobacterium psychrophilum* isolates and 134 *Yersinia ruckeri* isolates) and 313 motile *Aeromonas* isolates had increased antibiotic resistance; the study used five MICs from antibiotics commonly used in Danish aquaculture: oxolinic acid, sulfadiazine-trimethoprim, amoxicillin, oxytetracycline, and florfenicol (Schmidt et al. 2000). In another study, oxytetracycline-resistant mesophilic aeromonads from untreated hospital effluent and from fish farm hatchery tanks in Cumbria shared tetracycline resistance-encoding plasmids identical to a similar plasmid found in *E. coli* in other distinct locations (Rhodes et al. 2000). Long-term consequences of antibiotic resistance in the environment are not fully understood, although work by Costanzo (2005) showed that certain antibiotics negatively affected environmental bionitrification. There are large research gaps for understanding the consequences of medical pollution for ecosystem health, but this should not prevent further assessment and proactive approaches to addressing the problem of environmental therapeutic accumulation.

Microbes cause environmental concerns for cultivated and harvested fisheries. Many of these pathogens can cause disease in aquatic animals and may pose a health risk for people. Some more recent findings include epidemiologic investigations of *Salmonella* outbreaks in Australia. One outbreak linked gastroenteritis outbreaks in humans with identical isolates from home aquaria for multidrug-resistant *Salmonella paratyphi* B dT+ (ApCmSmSpSuTc phenotype) (ampicillin, chloramphenicol, streptomycin, spectinomycin, sulfonamides, tetracycline) containing SGI1 (Levings et al. 2006). Subsequent investigations correlated human outbreaks of gastroenteritis with a second multidrug-resistant *S. paratyphi* B biovar Java (*S. java*) with strains resistant to ApCmSmSpSuTc, and this isolate was directly linked with home aquaria maintenance (Musto et al. 2006). Aquaculture products can be incidental vehicles to transmit pathogens, as illustrated by a 1991 outbreak of cholera in Guayaquil, Ecuador, of multiple antimicrobial-resistant *V. cholerae* recovered from a pooled sample of a bivalve mollusk and from 68% of stool samples from case patients (Weber 1994). More recently an outbreak of salmonellosis in children has occurred in the United States associated with African dwarf frogs (*Hymenochirus* spp.) sold as pets (CDC 2010).

## CONCLUSIONS

Not all changes in fisheries are simply ascribed to overfishing or environmental destruction, as non-recovery of Pacific herring populations in Alaska has recently been attributed to natural mortality events caused by several infectious pathogens and subsequent disease, which has been illustrated using an age-structured assessment model of disease and population abundance (Marty et al. 2010a). With increasing use of molecular diagnostics, infectious disease modeling, increasing globalization, and changing environmental conditions, conservation medicine specialists will need to pay attention to epidemiology, irrespective of their specialty, for the benefit of their research, clients, and patients. This chapter is far from comprehensive, but I hope it will encourage veterinarians, aquatic animal health professionals, ecologists, modelers, conservation biologists, and the general public to think outside of their routine practice, to become involved with issues that demand aquatic animal health expertise, and to work collaboratively. Given the vast number of species and current exploitation of our aquatic resources, many conservation issues have not been included but are equally relevant to conservation medicine, such as the exploitation of sharks and rays in finning fisheries, invertebrate aquaculture production, and coral reef conservation. Conservation medicine and ecosystem health are becoming increasingly complex because of the vital role that water has in conserving our terrestrial environs. Water is life. The vastness of the aquatic world should not be looked at with trepidation but rather with excitement for the vast wealth of knowledge it has yet to bestow on us. Although the challenges presented are great, the oceans, rivers, streams, and lakes have been forgiving to anthropogenic transgressions, and we need to continue striving to leave a smaller wake.

## ACKNOWLEDGMENTS

I want to thank Dr. Bryon Jacoby for all his patience during the drafting, rewriting, and editing of this chapter, and Dr. Ronald Hedrick for assistance with the whirling disease section. Dr Hedrick has led Fish Health at the University of California School of Veterinary Medicine in Davis over the past 30 years and has contributed to nearly 300 publications related to aquatic animal health. Ron is a great friend, colleague, and mentor. I also wish to acknowledge my colleague Dr. Gary Marty, who reviewed this manuscript, contributed to the sea lice section, and has made notable contributions to aquatic animal health as a veterinarian, research scientist, and pathologist.

I want to dedicate this chapter to my grandfather, Ernst Philip Weber. Despite living in the inner city of Philadelphia, my grandfather planted the seeds of conservation in my mind at a young age and fostered environmental awareness by supporting my love for animals and nature in every way possible. We traveled together to Germany, and I worked on our family-owned dairy, beef, and swine farms. He nurtured my dream of becoming a veterinarian unwaveringly, even when everyone else doubted this was a possibility. Most importantly, he gave me my first pet fish when I was five, a goldfish named Sharky who was won with the toss of a ping-pong ball into a fish bowl.

# REFERENCES

Adkison, M.A., O. Gilad, and R.P. Hedrick. 2005. An enzyme linked immunosorbent assay (ELISA) for detection of antibodies to the Koi herpesvirus in the serum of Koi, *Cyprinus carpio*. Fish Pathol 40:53–62.

Anderson I., H. Prior, B. Rodwell, and H. Go. 1993. Iridovirus-like virions in imported dwarf gourami (*Colisa lalia*) with systemic amoebiasis. Aust Vet J 70:66–67.

Andree, K.B., E. MacConnell, and R.P. Hedrick. 1998. A nested polymerase chain reaction for the detection of genomic DNA of *Myxobolus cerebralis* in rainbow trout (*Oncorhynchus mykiss*). Dis Aquat Organ 34:145–154.

Andree, K., R.P. Hedrick, and E. MacConnell. 2002. A review of approaches to detect *Myxbolus cerebralis*, the cause of salmonid whirling disease. *In* J.L. Bartholomew and J.C. Wilson, eds. Whirling disease: reviews and current topics, pp. 197–212. American Fisheries Society, Symposium 29, Bethesda, Maryland.

Andrews, C. 1990. The ornamental fish trade and fish conservation. J Fish Biol 37:S53–S59.

Armstrong, R.D., and H.W. Ferguson. 1989. Systemic viral disease of the chromide cichlid *Etroplus maculates*. Dis Aquat Organ 7:155–157.

Bark, S., and D. McGregor. 2001. The first occurrence of lactococcosis in farmed trout in England. Trout News 31:9–11.

Bartholomew, J.L., and P.W. Reno. 2002. The history and dissemination of whirling disease. *In* J.L. Bartholomew and J.C. Wilson, eds. Whirling disease: reviews and current topics, pp. 3–24. American Fisheries Society, Symposium 29, Bethesda, Maryland.

Bercovier, H., Y. Fishman, R. Nahary, S. Sinai, A. Zlotkin, M. Eyngor, O. Gilad, A. Eldar and R.P. Hedrick. 2005. Cloning of the koi herpesvirus (KHV) gene encoding thymidine kinase and its use for a highly sensitive PCR based diagnosis. BMC Microbiol 5:13.

Beveridge, M.C.M., L.G. Ross, and L.A. Kelly. 1994. Aquaculture and biodiversity . Ambio 23:497–502.

Bondad-Reantaso, M.G., R.P. Subasinghe, J.R. Arthur, K. Ogawa, S. Chinabut, R. Adlard, Z. Tan, and M. Shariff. 2005. Disease and health management in Asian aquaculture. Vet Parasitol 132:249–272.

Boxall, A.B.A., L.A. Fogg, P.A. Blackwell P. Kay, E.J. Pemberton, and A. Croxford. 2004. Veterinary medicines in the environment. Rev Environ Contam Toxicol 180:1–91.

Brooks K.M., and S.R.M. Jones. 2008. Perspectives on pink salmon and sea lice: scientific evidence fails to support the extinction hypothesis. Rev Fish Sci 16:403–412.

Calado R., J. Lin, A.L. Rhyne, R. Araújo, and L. Narciso. 2003. Marine ornamental decapods—popular, pricey, and poorly studied. J Crustac Biol 23(4):963–973.

CDC. 2010. Multistate outbreak of human *Salmonella typhimurium* infections associated with aquatic frogs. MMWR 58(51&52):1433–1436.

Chalmers A.T., D.M. Argue, D.A. Gay, M.E. Brigham, C.J. Schmitt, and D.L. Lorenz. 2010. Mercury trends in fish from rivers and lakes in the United States, 1969–2005. Environ Monit Assess DOI 10.1007/s10661-010-1504-6. Accessed April 4, 2011.

Chao, N.L., and G. Prang. 1997. Project Piaba—towards a sustainable ornamental fishery in the Amazon. Aquarium Sci Conserv 1:105–111.

Chao, N.L., P. Petry, G. Prang, L. Sonneschein, and M.F. Tlusty, eds. 2001. Conservation and management of ornamental fish resources of the Rio Negro Basin, Amazonia, Brazil—Project Piaba. Editora da Universidade do Amazonas, Manaus, Brazil.

Chinchar, V.G. 2002. Ranaviruses (family Iridoviridae): Emerging cold-blooded killers. Arch Virol 147: 447–470.

Chinchar. V.G., S. Essbauer, J.G. He, A. Hyatt, T. Miyazaki, V. Seligy, and T. Williams. 2005. Iridoviridae. *In* C.M. Fauquet, M.A. Mayo, J. Maniloff, U. Desselberger, and L.A. Ball, eds. Virus taxonomy: 8th report of the International Committee on the Taxonomy of Viruses, pp. 163–175. Elsevier, London, UK.

Costanzo, S.D., J. Murby, and J. Bates. 2005. Ecosystem response to antibiotics entering the aquatic environment. Mar Pollut Bull 51:218–223.

Crowl, T.A., T.O. Crist, R.R. Parmenter, G. Belovsky, and A.E. Lugo. 2008. The spread of invasive species and infectious disease as drivers of ecosystem change. Front Ecol Environ 6:238–246.

Christian, T., R.J. Schneider, H.A. Farber, D. Skutlarek, M.T. Meyer, and H.E. Goldbach. 2003. Determination of antibiotic residues in manure, soil, and surface waters. Acta Hydrochim Hydrobiol 31:36–44.

Daszak, P., L. Berger, A.A. Cunningham, A.D. Hyatt, D.E. Green, and R. Speare. 1999. Emerging infectious diseases and amphibian population declines. Emerg Infect Dis 5:735–748.

Daszak, P., A.A. Cunningham, and A.D. Hyatt. 2003. Infectious disease and amphibian population declines. Divers Distrib 9:141–150.

Delhon, G., E.R. Tulman, C.L. Afonso, Z. Lu, J.J. Becnel, B.A. Moser, G.F. Kutish, and D.L. Rock. 2006. Genome of invertebrate iridescent virus type 3 (mosquito iridescent virus). J Virol 80:8439–8449.

Dentler, J.L. (1993–1994). Noah's farce: the regulation and control of exotic fish and wildlife. U Puget Sound L Rev 17:192–242.

Dewalt, B.R., P. Vergne, and M. Hardin. 1996. Shrimp aquaculture development and the environment: people, mangroves and fisheries on the Gulf of Fonseca, Honduras. World Dev 24:1193–1208.

Diana, J.S. 2009. Aquaculture production and biodiversity conservation. Bioscience 59:27–38.

Dierberg, F.E., and W. Kiattisimkul. 1995. Issues, impacts, and implications of shrimp aquaculture in Thailand. Environ Manage 20:649–666.

Eaton, W.D., J. Hulett, R. Brunson, and K. True. 1991. The first isolation in North America of infectious hematopoietic necrosis virus (IHNV) and viral hemorrhagic septicemia virus (VHSV) in Coho salmon from the same watershed. J Aquat Anim Health 3:114–117.

Elsayed, E., M. Faisal, M. Thomas, G. Whelan, W. Batts, and J. Winton. 2006. Isolation of viral hemorrhagic septicemia virus from muskellunge, *Esox masquinongy* (Mitchill), in Lake St. Clair, Michigan, USA reveals a new sublineage of the North American genotype. J Fish Dis 29:611–619.

Eschmeyer, W.N., ed. Catalog of fishes: electronic version. 2011. http://research.calacademy.org/ichthyology/catalog/fishcatmain.asp. California Academy of Sciences, San Francisco. Accessed April 5, 2011.

Evans, D.H. 1987. The fish gill: site of action and model for toxic effects of environmental pollutants. Enviro Health Persp 71: 47–58.

Fisheries and Oceans Canada (FOC). 2009. Facts about sea lice. http://www.dfo-mpo.gc.ca/aquaculture/lice-pou/lice-pou04-eng.htm Accessed April 5, 2011.

Food and Agricultural Organization (FAO). 2008. Aquaculture production statistics 1998–2007. FAO, Rome. http://www.fao.org/fishery/statistics/programme/3,2,1/en Accessed April 5, 2011.

Food and Agricultural Organization (FAO). 2010. The state of world fisheries and aquaculture. FAO, Rome. http://www.fao.org/docrep/013/i1820e/i1820e00.htm. Accessed Oct. 5, 2011.

Folke, C. 1988. Energy economy of salmon aquaculture in the Baltic Sea. Environ Mgmt 12:525–537.

Gagnon, M.M., and D.A. Holdway. 1999. Metabolic enzyme activities in fish gills as biomarkers of exposure to petroleum hydrocarbons. Ecotox Environ Safe 44:92–99.

Galvez, R., G.H. Therese, C. Bautista, and M.T. Tungpalan. 1989. Sociocultural dynamics of blast fishing and sodium cyanide fishing in two fishing villages in the Lingayen Gulf area. *In* G. Silvester, E. Miclat, and T.E. Chua, eds. Towards sustainable development of the resources of Lingayen Gulf, Philippines, pp. 43–62. ICLARM Conference Proceedings Number 17.

Georgiadis, M.P., I.A. Gardner, and R.P. Hedrick. 2001. The role of epidemiology in the prevention, diagnosis, and control of infectious diseases of fish. Prev Vet Med 48:287–302.

Gilad, O., S. Yun, M.A. Adkison, K. Way, N.H. Willits, H. Becovier, and R.P. Hedrick. 2003. Molecular comparison of isolates of an emerging fish pathogen, koi herpesvirus, and the effect of water temperature on mortality of experimentally infected koi. J Gen Virol 84:2661–2668.

Godfray, H.C.J., J.R. Beddington, I.R. Crute, L. Haddad, D. Lawrence, J.F. Muir, J. Pretty, S. Robinson, S.M. Thomas, and C. Toulmin. 2010. Food security: The challenge of feeding 9 billion people. Science 327:812–818.

Goldberg, J., and C. Wilkinson. 2004. Global threats to coral reefs: coral bleaching, global climate change, disease, predator plagues, and invasive species. *In* C. Wilkinson, ed. Status of coral reefs of the world. vol. 1, p. 68–92. Australian Institute of Marine Science, Townsville, Queensland, Australia.

Goodwin, A.E., G.E. Merry, and J. Sadler. 2006. Detection of the herpesviral haematopoietic necrosis disease agent Cyprinid herpesvirus 2 in moribund and healthy goldfish: validation of a quantitative PCR diagnostic method. Dis Aquat Organ 69:137–143.

Gunderson, J. 2010. Viral hemorrhagic septicemia: are our fish doomed? University of Minnesota Sea Grant

Program http://www.seagrant.umn.edu/fisheries/
vhsv#related. Accessed April 4, 2011.

Halliday, M.M. 1973. Studies of *Myxosoma cerebralis*,
a parasite of salmonids. II. The development and
pathology of *Myxosoma cerebralis* in experimentally
infected rainbow trout (*Salmon gairdneri*) fry reared at
different water temperatures. Nordisk
Veterinaermedicin 25:349–358.

Hedrick, R.P., T. McDowell, and J. Groff. 1987.
Mycobacteriosis in cultured striped bass from
California. J Wildl Dis 23:391–395.

Hedrick, R.P. 1998. Relationships of the host, pathogen, and
environment: implications for diseases of cultured and
wild fish populations. J Aquat Anim Health 10:107–111.

Hedrick, R.P., G. Marty, R.W. Nordhausen, M. Kebus,
H. Bercovier, and A. Eldar. 1999. A herpesvirus
associated with mass mortality of juvenile and adult koi
*Cyprinus carpio*. American Fisheries Society Fish Health
Newsletter 27: 7.

Hedrick, R.P., W.N. Batts, S. Yun, G.S. Traxler, J. Kaufman,
and J.R. Winton. 2003a. Host and geographic range
extensions of the North American strain of viral
hemorrhagic septicemia virus. Dis Aquat Organ
55:211–220.

Hedrick, R.P., T.S. McDowell, G.D., Marty, G.T Fosgate,
K., Mukkatira, K. Myklebust, and M. El-Matbouli.
2003b. Susceptibility of two strains of rainbow trout
(one with a suspected resistance to whirling disease)
to *Myxobolus cerebralis* infection. Dis Aquat Organ
55:37–44.

Hedrick, R.P., B. Petri, T.S. McDowell, K. Mukkatira, and
L.J. Sealey. 2007. Evaluation of a range of doses of
ultraviolet irradiation to inactivate the waterborne
actinospore stages of *Myxobolus cerebralis*. Dis Aquat
Organ 74:113–118.

Hedrick, R.P., T.S. McDowell, K. Mukkatira,
E. MacConnell, and B. Petri. 2008. Effects of freezing,
drying, ultraviolet irradiation, chlorine and quaternary
ammonium treatments on the infectivity of myxospores
of *Myxobolus cerebralis* for *Tubifex tubifex*. J Aquat Anim
Health 20:116–125.

Helfman, G.S. 2007. Fish conservation: a guide to
understanding and restoring global aquatic biodiversity
and fishery resources. Island Press, Chicago.

Hershberger, P.K., B.K. van der Leeuw, J.L. Gregg,
C.A. Grady, K.M. Lujan, S.K. Gutenberger, M.K.Purcell,
J.C. Woodson, J.R. Winton, and M.J. Parsley.
2010. Amplification and transport of an endemic fish
disease by an introduced species. Biol Invasions
125:1387–3547.

Hirsch R., T. Ternes, K. Haberer, and K.L. Kratz. 1999.
Occurrence of antibiotics in the aquatic environment.
Sci Total Environ 225:109–118.

Hites, R.A., J.A. Foran, D.O. Carpenter, M.C. Hamilton,
B.A. Knuth, and S.J. Schwager. 2004. Global assessment
of organic contaminants in farmed salmon. Science
303:226–229.

Hoegh-Guldberg, O., P.J. Mumby, A.J. Hooten,
R.S. Steneck, P. Greenfield, E. Gomez, C.D. Harvell,
P.F. Sale, A.J. Edwards, K. Caldeira, N. Knowlton,
C.M. Eakin, R. Iglesias-Prieto, N. Muthiga,
R.H. Bradbury, A. Dubi, and M.E. Hatziolos. 2007.
Coral reefs under rapid climate change and ocean
acidification. Science 318:1737–1742.

Hoffman, G.L. 1990. *Myxobolus cerebralis*, a worldwide
cause of salmonid whirling disease. J Aquat Anim
Health 2:30–37.

Ikonomou, M.G., D. A. Higgs, M. Gibbs, J. Oakes, B. Skura,
S. McKinley, S. K. Balfry, S. Jones, R. Withler, and
C. Dubetz. 2007. Flesh quality of market-size farmed
and wild British Columbia salmon. Environ Sci Technol
41:437–443.

Jensen, M.H. 1963. Preparation of fish tissue cultures
for virus research. Bull Off Inst Epiz 59:
131–134.

Jobling, S., M. Nolan, C.R. Tyler, G. Brighty, and
J.P. Sumpter. 1998. Widespread sexual disruption in wild
fish. Environ Sci Technol 32:2498–2506.

Jones, S., E. Kim, and S. Dawe. 2006. Experimental
infections with *Lepeophtheirus salmonis* (Kroyer) on
threespine sticklebacks, *Gasterosteus aculeatus* L., and
juvenile Pacific salmon, *Oncorhynchus* spp. J Fish Dis
29:489–495.

Jones, S.R.M., and N.B. Hargreaves. 2007. The abundance
and distribution of *Lepeophtheirus salmonis* (Copepoda:
caligidae) on pink (*Oncorhynchus gorbuscha*) and chum
(*O. keta*) salmon in coastal British Columbia. J Paristol
93:1324–1331.

Jones, S.R.M., and N.B. Hargreaves. 2009. Infection
threshold to estimate *Lepeophtheirus salmonis*-associated
mortality among juvenile pink salmon. Dis Aquat Org
84:131–137.

Kane, A., C.B. Stine, L. Hungerford, M. Matsche,
C. Driscoll, and A. M. Baya. 2007. Mycobacteria as
environmental portent in Chesapeake Bay fish species.
Emerg Infect Dis 13:329–331.

Kautsky, N., P. Ronnback, M. Tedengren, and M. Troell.
2000. Ecosystem perspectives on management of
disease in shrimp pond farming. Aquaculture
191:145–161.

Kemper, N. 2008. Veterinary antibiotics in the aquatic and
terrestrial environment. Ecol Indic 8:1–13.

Khan, R.A. 1990. Parasitism in marine fish after chronic
exposure to petroleum hydrocarbons in the laboratory
and to the Exxon Valdez oil spill. Bull Environ Contam
Toxicol 44:759–763.

Kocan, R., P. Hershberger, and J. Winton. 2004. Ichthyophoniasis: An emerging disease of Chinook salmon in the Yukon River. J Aquat Anim Health 16:58–72.

Knobeloch, L., M. Turyk, P. Imm, C. Schrank, and H. Anderson. 2008. Temporal changes in PCB and DDE levels among a cohort of frequent and infrequent consumers of Great Lakes sportfish. Environ Res 109:66–72.

Krkošek, M., M.A. Lewis, A. Morton, L. N. Frazer, and J.P. Volpe. 2006. Epizootics of wild fish induced by farm fish. Proc Natl Acad Sci USA 103:15506–15510.

Krkošek, M., J.S. Ford, A. Morton, S. Lele, R.A. Myers, and M.A. Lewis. 2007. Declining wild salmon populations in relation to parasites from farm salmon. Science 318: 1772–1775.

Levings, R.S., D. Lightfoot, R.M. Hall, and S.P. Djordjevic. 2006. Aquariums as reservoirs for multidrug-resistant Salmonella paratyphi B. Emerg Infect Dis 12:507–510.

Lumsden, J.S., B. Morrison, C. Yason, S. Russell, K. Young, A. Yazdanpanah, P. Huber, L. Al-Hussinee, D. Stone, and K. Way. 2007. Mortality event in freshwater drum Aplodinotus grunniens from Lake Ontario, Canada, associated with viral haemorrhagic septicemia virus, type IV. Dis Aquat Organ 76:99–111.

Lunn, K.E., and M. A. Moreau. 2004. Unmonitored trade in marine ornamental fishes: the case of Indonesia's Banggai cardinalfish (Pterapogon kauderni). Coral Reefs 23:344–351.

Marty, G.D., P.-J.F. Hulson, S.E. Miller, T.J. Quinn II, S.D. Moffitt, and R.A. Merizon. 2010a. Failure of population recovery in relation to disease in Pacific herring. Dis Aquat Org 90:1–14.

Marty, G.D., S.M. Saksida, and T.J. Quinn II. 2010b. Relationship of farm salmon, sea lice, and wild salmon populations. Proc Natl Acad Sci USA doi: 10.1073/pnas.1009573108

McVicar, A.H. 1997. Disease and parasite implications of the coexistence of wild and cultured Atlantic salmon populations. ICES J Mar Sci 54:1093–1103.

Mills, L.J., and C. Chichester. 2005. Review of evidence: are endocrine-disrupting chemicals in the aquatic environment impacting fish populations? Sci Total Environ 343:1–34.

Meyers, T.R., A.K. Hauck, W.D. Blankenbeckler, and T. Minicucci. 1986. First report of viral erythrocytic necrosis in Alaska, USA, associated with epizootic mortality in Pacific herring, Clupea harengus pallasi (Valenciennes). J Fish Dis 9:479–491.

Monod G., A .Devaux, and J.L. Riviere. 1988. Effects of chemical pollution on the activities of hepatic

xenobiotic metabolizing enzymes in fish from the Rhone river. Sci Total Environ 73:189–201.

Musto, J., M. Kirk, D. Lightfoot, B.G. Combs, and L. Mwann. 2006. Multi-drug resistant Salmonella java infections acquired from tropical fish aquariums, Australia, 2003–04. Commun Dis Intell 30:222–227.

Naylor, R.L., R.J. Goldberg, J.H. Primavera, N. Kautsky, M.C.M. Beveridge, J. Clay, C.Folke, J. Lubchenco, H. Mooney, and M. Troell. 2000. Effects of aquaculture on world food supplies. Nature 405:1017–1024.

Naylor R. L., R.W. Hardy, D.P. Bureau, A. Chiu, M. Elliott, A.P. Farrell, I. Forster, D.M. Gatlin, R.J., Goldburg, K. Hua, and P.D. Nichols. 2009. Feeding aquaculture in an era of finite resources. Proc Natl Acad Sci USA 106:15103–15110.

Nehring, R.B., and P.G. Walker. 1996. Whirling disease in the wild: the new reality in the Intermountain West. Fisheries 21:28–30.

Palmeiro, B., and E. S. Weber III. 2009. Chapter 9: Viral pathogens of fish. In H.E. Roberts, ed. Fundamentals of ornamental fish health. Wiley-Blackwell, Ames, Iowa.

Paperna, I., M. Vilenkin, and A.P. Alves de Matos. 2001. Iridovirus infections in farm-reared tropical ornamental fish. Dis Aquat Organ 48:17–25.

Primavera, J.H. 1997. Socioeconomic impacts of shrimp culture. Aquaculture Res 28:815–827.

Rasmussen, C.J. 1965. A biological study of the Egtved disease (INUL). Ann NY Acad Sci 126:427–460.

Rahel, F.J., and J. D. Olden. 2008. Assessing the effects of climate change on aquatic invasive species. Conserv Biol 22(3):521–33.

Rhodes, G., G. Huys, J. Swings, P. Mcgann, M. Hiney, P. Smith, and R.W. Pickup. 2000. Distribution of oxytetracycline resistance plasmids between aeromonads in hospital and aquaculture environments: implications of tn1 721 in dissemination of the tetracycline resistance determinant tet A. Am Soc Microbiol 66:3883–3890.

Riley, S.C., K. R. Munkittrick, A. N. Evans, and C. C. Krueger. 2008. Understanding the ecology of disease in Great Lakes fish populations. Aquat Ecosyst Health 11:321–334.

Rodger, H., M. Kobs, A. Macartney, and G.N. Frerichs. 1997. Systemic iridovirus infection in freshwater angelfish, Pterophyllum scalare (Lichtenstein). J Fish Dis 20:69–72.

Saksida, S., J. Constantine, G.A. Karreman, and A. Donald. 2007. Evaluation of sea lice abundance levels on farmed Atlantic salmon (Salmo salar L.) located in the Broughton Archipelago of British Columbia from 2003 to 2005. Aquacult Res 38:219–231.

Schisler, G.S., K.A. Myklebust, and R.P. Hedrick. 2006. Inheritance of resistance to Myxobolus cerebralis among

F1 generation crosses of whirling disease resistant and susceptible strains of rainbow trout. J Aquat Anim Health 18:109–115.

Schmidt, A., M.S. Bruun, I. Dalsgaard, K. Pedersen, and J.L. Larsen. 2000. Occurrence of antimicrobial resistance in fish-pathogenic and environmental bacteria associated with four Danish rainbow trout farms. Appl Environ Microbiol 66:4908–4915.

Shaw, S.D., D. Brenner, A. Bourakovsky, D.O. Carpenter, K. Kannan, and C-S. Hong. 2005. PCBs, dioxin-like PCBs and organochlorine pesticides in farmed salmon (*Salmo salar*) from Maine and eastern Canada. Organohalogen Comp 67:1571–1576.

Shaw, S.D., D. Brenner, M.L. Berger, D.O. Carpenter, C-S. Hong, and K. Kannan. 2006. PCBs, PCDD/Fs, and organochlorine pesticides in farmed Atlantic salmon from Maine, eastern Canada, and Norway, and wild salmon from Alaska. Environ Sci Technol 40:5347–5354.

Shaw, S.D., D. Brenner, M.L. Berger, D.O. Carpenter, C-S. Hong, and K. Kannan. 2007. PCBs, PCDD/Fs, and organochlorine pesticides in farmed Atlantic salmon from Maine, eastern Canada, and Norway, and wild salmon from Alaska (comment/correction). Environ Sci Technol 41:4180.

Smith, H.M. 1886. A review of the history and results of the attempts to acclimatize fish and other water animals in the Pacific states. US Fish Com Bull 15:379–472.

Smith, K.F., M.D. Behrens, L.M. Schloegel, N. Marano, S. Burgiel, and P. Daszak. 2009. Reducing the risks of the wildlife trade. Science 324:594–595.

Snow, M., and D.A. Smail. 1999. Experimental susceptibility of turbot *Scophthalmus maximus* to viral haemorrhagic septicaemia virus isolated from cultivated turbot. Dis Aquat Org 38:163–168.

Stine, C.B., J.M. Jacobs, M.R. Rhodes, A. Overton, M. Fast, and A.M. Baya. 2009. Expanded range and new host species of *Mycobacterium shottsii* and *M. pseudoshottsii*. J Aquat Anim Health 21:179–183.

Stone, D.M., K. Way, and P.F. Dixon. 1997. Nucleotide sequence of the glycoprotein gene of viral haemorrhagic septicaemia (VHS) viruses from different geographical areas: a link between VHS in farmed fish species and viruses isolated from North Sea cod (*Gadus morhua* L.). J Gen Virol 78:1319–1326.

Sudthongkong, C., M. Miyata, and T. Miyazaki. 2002. Iridovirus disease in two ornamental tropical fishes: African lampeye and dwarf gourami. Dis Aquat Organ 48:163–173.

Tacon, A.G.J. 2005. Salmon aquaculture dialogue: status of information on salmon aquaculture feed and the environment. International Aquafeed 8: 22–37.

Tacon, A.G.J., and M. Metian. 2009. Fishing for aquaculture: Nonfood use of small pelagic forage fish, a global perspective. Rev Fish Sci 17:305–317.

Tlusty, M., S. Dowd, S. Weber, B. Cooper, and B. Whitaker. 2005. USA shipping cardinal tetras from the Amazon— understanding stressors to decrease shipping mortality. OFI Journal 48:21–23.

Todd, C.D., A.M. Walker, M.G. Ritchie, J.A. Graves, and A.F. Walker. 2004. Population genetic differentiation of sea lice (*Lepeophtheirus salmonis*) parasitic on Atlantic and Pacific salmonids: analyses of microsatellite DNA variation among wild and farmed hosts. Can J Fish Aquat Sci 61:1176–1190.

Vagelli, A.A. 2004. Significant increase in survival of captive-bred juvenile Banggai cardinalfish, *Pterapogon kauderni*, with an essential fatty acid enriched diet. J World Aquacult Soc 35:61–69.

van Leeuwen, S.P.J., M.J.M. van Velzen, C.P. Swart, I. van der Veen, W.A. Traag, and J. de Boer. 2009. Halogenated contaminants in farmed salmon, trout, tilapia, pangasius, and shrimp. Environ Sci Technol 43:4009–4015.

Vincent, E.R. 1996. Whirling disease and wild trout: the Montana experience. Fisheries 21:32–34.

Wahli, T., R. Knuesel, D. Bernet, H. Segner, D. Pugovkin, P. Burkhardt-Holm, M. Escher, and H. Schmidt-Posthaus. 2002. Proliferative kidney disease in Switzerland: current state of knowledge. J Fish Dis 25: 491–500.

Way, K. 2008. Koi herpesvirus and goldfish herpesvirus: an update of current knowledge and research at CEFAS. Fish Vet J 10:62–73.

Weber III E.S., T. Waltzek, D.A. Young, E.L. Twitchell, A.E. Gates, A. Vagelli, G. Risatti, R.P. Hedrick, and S. Frasca. 2009. Systemic iridovirus infection in the Banggai cardinalfish (*Pterapogon kauderni* Koumans (1933)). J Vet Diagn Invest 21:306–20.

Weber, J.T., E.D. Mintz, R. Cañizares, A. Semiglia, I. Gomez, R. Sempértegui, A. Dávila, K.D. Greene, N.D. Puhr, D.N. Cameron, F.C. Tenover, T.J. Barrett, N.H. Bean, C. Ivey, R.V. Tauxe, and P.A. Blake. 1994. Epidemic cholera in Ecuador: multi-drug resistance and transmission by water and seafood. Epidemiol Infect 112:1–11.

Webster, S.J., L.M. Dill, and K. Butterworth. 2007. The effect of sea lice infestation on the salinity preference and energetic expenditure of juvenile pink salmon (*Oncorhynchus gorbuscha*). Can J Fish Aquat Sci 64:672–680.

Westhouse, R.A., E.R. Jacobson, R.K. Harris, K.R. Winter, and B.L. Homer. 1996. Respiratory and pharyngo-esophageal iridovirus infection in a gopher tortoise (*Gopherus polyphemus*). J Wildl Dis 32:682–686.

Whittington, R.J., and R. Chong. 2007. Global trade in ornamental fish from an Australian perspective: the case for revised import risk analysis and management strategies. Prev Vet Med 81:92–116.

Wilkinson, C. 2004. Status of coral reefs of the world volumes 1 & 2. Global reef monitoring network. Australian Institute of Science. http://www.gcrmn.org/status2004.aspx. Accessed April 5, 2011.

World Organization for Animal Health (OIE). 2006. Manual of diagnostic tests for aquatic animals. Paris, France. http://www.oie.int/eng/normes/fmanual/A_00022.htm. Accessed April 5, 2011.

Ziemann, D.A. 2001. The potential for the restoration of marine ornamental fish populations through hatchery releases. Aquarium Sci Conserv 3:107–117.

# 23

# SOUTHERN SEA OTTERS AS SENTINELS FOR LAND–SEA PATHOGENS AND POLLUTANTS

## David A. Jessup and Melissa A. Miller

Southern sea otters (*Enhydra lutris nereis*) are a federally listed threatened species found only along the California coast. Despite over 70 years of state and federal legal protection, this population has failed to significantly increase their numbers or reclaim large expanses of their historical range. In addition, multiple periods of decline have been documented, and population growth has always lagged behind that of comparable otter groups in Alaska (Tinker et al. 2008). Although fecundity appears to be normal, southern sea otter mortality is extremely high, with approximately 10% of the population recovered dead each year. In addition, a high proportion (40% to 64%) of necropsied southern sea otters exhibit lesions associated with exposure to infectious agents and toxins (Thomas and Cole 1996; Kreuder et al. 2003). Deaths of prime-aged adult otters are also common, including adult females of reproductive age. These findings are unusual for a long-lived carnivore that has a low reproductive rate. Although the interactions are complex, evidence suggests that failure of population recovery and high, ongoing mortality are associated with environmental degradation, including exposure to chemical and biological pollution that appears to be spatially concentrated along the land–sea interface.

Sea otters are especially at risk for terrestrial-origin pollutant exposure due to unique aspects of their metabolism and biology. Despite swimming in relatively cold ($2°$ to $13°C$) water, sea otters are the world's smallest marine mammals. They have the densest fur of any mammal and avoid hypothermia by maintaining a high basal metabolic rate, which obligates healthy otters to consume at least 25% of their weight in invertebrate prey each day (Riedman and Estes 1990). Sea otters are nearshore feeders, often foraging within sight of land and near river mouths and embayments that concentrate and retain plumes of surface water runoff tainted with chemical and biological pollutants (Estes 2005; Jessup et al. 2004, 2007; M.A. Miller et al. 2010a). Many otters selectively feed on marine and estuarine invertebrates such as clams and mussels (Tinker et al. 2008) that are efficient pollutant bioaccumulators. These unique evolutionary strategies appear to exacerbate the risk of disease for sea otters living in environments compromised by pollution. Prior studies reveal that some bacteria, parasites, and fungi responsible for sea otter deaths have strong terrestrial connections (Nakata et al. 1998; Kannan et al. 1998, 2006, 2007; M.A. Miller et al. 2002b, 2008, 2010a,b). Similarly, high levels of anthropogenic pollutants have been detected in southern sea otter tissues (Kannan et al. 1998, 2004, 2008; Jessup et al. 2010). This chapter describes a number of terrestrial-origin pathogens and pollutants that may be interfering with sea otter recovery. Potential terrestrial sources of these pathogens and pollutants and effects on sea otter health are described.

## APICOMPLEXAN PROTOZOA

### *Toxoplasma gondii*

Sea otter infections by the protozoan parasite *Toxoplasma gondii* have stimulated great public interest, in part due to human health implications. Although the role of *T. gondii* as a primary pathogen is hotly debated, it may be an important contributor to sea otter morbidity and mortality due to its ability to reactivate in chronically infected, immunocompromised animals. First reported by Thomas et al. (1996), *T. gondii* infections rapidly became a major focus of sea otter disease research (Fayer et al. 2004; M.A. Miller et al. 2002a,b, 2004, 2008; Johnson et al. 2009; Massie et al. 2010), and there is little doubt that the sea otter/*T. gondii* connection is one of the clearest examples ever documented

of ocean pollution by terrestrial-origin pathogens. This is because a unique restriction of the *T. gondii* life cycle requires a land-based host, the cat, to serve as the only known source of the environmentally resistant oocyst stage of the parasite.

*T. gondii* is a single-celled parasite with a complex life cycle involving both definitive (oocyst-shedding) hosts and intermediate hosts that support the tissue cyst stage of the parasite (Fig. 23.1) (Conrad et al. 2005). Virtually any warm-blooded vertebrate, including humans, may serve as intermediate hosts. Only domestic and wild felids, including domestic cats (*Felis catus*), bobcats (*Lynx rufus*), and mountain lions (*Felis concolor*) are known to serve as definitive hosts. Intermediate hosts are infected through consumption of oocyst-contaminated soil, water, or paratenic hosts, or by consuming tissue cysts in muscle, brain or other organs of intermediate hosts. Transplacental infection is common in some terrestrial animals, including humans, although it does not appear to be a

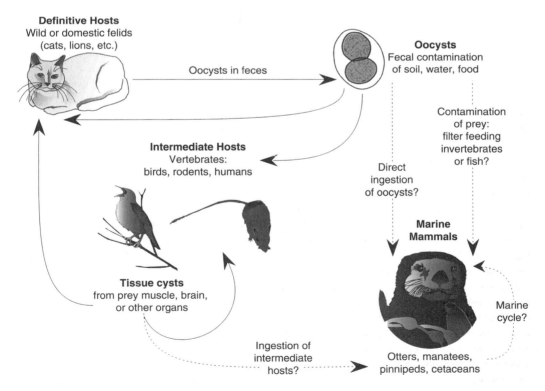

**Figure 23.1:**
Life cycle of *Toxoplasma gondii*, showing potential sources of marine mammal exposure (dashed lines).

major route for parasite propagation within sea otters (Conrad et al. 2005).

After consuming *T. gondii*-infected rodents or birds, cats may pass 100 million oocysts in their feces over 14 days. Oocysts may remain viable for months to years in the environment, and ingestion of even a single sporulated oocyst can trigger infection (Conrad et al. 2005). Once ingested by intermediate hosts, the parasites leave the intestinal tract and spread systemically. As host immunity is generated, a "resting stage" is produced that resides within tissue cysts in the cytoplasm of host cells. Tissue cysts persist for months to years, leading to chronic, perhaps lifelong infection. When host immunity wanes the parasite can reactivate, often leading to severe clinical disease (recrudescence). Parasite recrudescence in association with concurrent infection or biotoxin exposure may be important routes for development of clinical toxoplasmosis in sea otters (M.A. Miller et al. 2008, 2010b).

The majority of marine mammal infections by *T. gondii* are thought to result from oocyst ingestion; these animals rarely if ever consume recognized intermediate hosts for *T. gondii* such as rodents and birds (Thomas et al. 1996; Conrad et al. 2005). Despite this fact, as many as 70% of adult male southern sea otters may be infected with *T. gondii* (M.A. Miller et al. 2002a), suggesting that environmental contamination by oocysts is extensive. Additional factors such as high parasite pathogenicity, multiple parasite infections, and genetic homogeneity within host populations could enhance the susceptibility of marine mammals to protozoal-related infection and disease (Larson et al. 2002; Jessup et al. 2007; M.A. Miller et al. 2010b; Wendte et al. 2010).

The importance of terrestrial runoff for dissemination of *T. gondii* oocysts to humans and marine mammals is increasingly recognized (Bowie et al. 1997; Aramini et al. 1999; M.A. Miller et al. 2002b, 2008). Oocysts may contaminate water or be taken up by invertebrates (Lindsay et al. 2001; Arkush et al. 2003; M.A. Miller et al. 2008) or fish (Massie et al. 2010). *T. gondii* oocyst uptake by marine bivalves has been demonstrated experimentally (Lindsay et al. 2001; Arkush et al. 2003) and under natural conditions (M.A. Miller et al. 2008). Once ingested by invertebrates, *T. gondii* oocysts remain infectious for at least 14 days (Arkush et al. 2003).

Studies in central California have identified high-risk areas for *T. gondii* exposure and disease in sea otters (M.A. Miller et al. 2002b; Kreuder et al. 2003, 2005; Johnson et al. 2009). Characteristics shared by these "high-risk" areas include proximity to high-outflow streams and rivers or large, enclosed bays with limited tidal exchange (M.A. Miller et al. 2002a,b; Johnson et al. 2009). Given the worldwide distribution of these parasites, high-risk areas for marine mammal exposure to *T. gondii* and related pathogenic protozoa might exist elsewhere. Recent confirmation of an unusual strain of *T. gondii* (type X) in a wild marine mussel, sea otters, and terrestrial felids from Monterey Bay, California, supports the hypothesis of land–sea transfer of protozoal pathogens (M.A. Miller et al. 2004, 2008).

## *Sarcocystis neurona*

The life cycle of *Sarcocystis neurona* is similar to *T. gondii*, but with some important differences. New world opossums (*Didelphis virginiana* and *D. albiventris*) are the only known definitive hosts (Dubey et al. 2001). Importantly, Virginia opossums (*D. virginiana*) were transplanted to California from the eastern United States at the beginning of the 20th century (Grinnell 1915). The infective stage for *S. neurona* is a sporocyst, which is comparable in size to *T. gondii* oocysts (10 to 12 µm in diameter), is passed in high numbers in the feces of the opossum host, and is environmentally persistent. Intermediate hosts include birds and mammals, but not humans (Dubey et al. 2001).

Invasive parasite stages exit the sporocysts, penetrate host cells, and proliferate, forming tissue cysts in the latter stage of infection. Sea otters and other marine mammals are intermediate hosts, supporting parasite proliferation in the brain, lung, skeletal muscle, heart, and other tissues (Dubey et al. 2001). Otter mortality due to *S. neurona* is highest during the spring and summer months (Kreuder et al. 2003; M.A. Miller et al. 2010b). Similar to *T. gondii*, strain-specific variations in parasite prevalence, infectivity, and pathogenicity appear to be important in the ecology of *S. neurona* infections (M.A. Miller et al. 2010b; Wendte et al. 2010). During April 2004, 40 sea otters were found sick or dead over less than 20 km of coastline near Morro Bay, California. Postmortem examinations and molecular characterization identified *S. neurona* as the underlying cause of this epizootic (M.A. Miller et al. 2010b; Wendte et al. 2010).

Evidence to support the point-source character of this event includes the spatial and temporal clustering of cases, detection of high concentrations of anti-*S. neurona* IgM antibodies in otter serum, and confirmation of a single strain of *S. neurona* in affected animals. Sea otters were presumed to be infected through ingestion of sporocysts; no mechanism for otter-to-otter spread of *S. neurona* is known. Factors that could have contributed to the epizootic include a large rainstorm that occurred weeks prior to the event and concentration of otters along the shoreline to feed on clams capable of concentrating *S. neurona* sporocysts. Prior studies on the environmental dissemination of bacterial and parasitic pollutants along the central California coast demonstrate an increased prevalence of pathogen detection in marine bivalves after storm events (W.A. Miller et al. 2005, 2006; M.A. Miller et al. 2008).

This *S. neurona*-associated epizootic demonstrates the importance of land–sea transport as a potential source of sea otter exposure to biological pathogens. Although introduced opossums are now common in coastal California, the terrestrial-to-marine flow of fecal waste from these animals is probably patchy and episodic, entering the ocean through multiple point-source discharges (rivers and stormwater drainages) interspersed along the shoreline. Rivers that enter the ocean at sandy beaches are often intermittently blocked by sand berms deposited by ocean waves, forming large, shallow coastal lagoons that provide habitat for fish, birds, insects, and mammalian carnivores and omnivores (like opossums) that prey upon them (M.A. Miller et al. 2010b). Over time, this impounded water may become enriched with pathogens, including *S. neurona* sporocysts, which are hazardous to marine wildlife when rapidly discharged.

## ENTERIC BACTERIAL PATHOGENS

Fecal contamination of coastal marine habitat is a global problem, causing beach closures, water contact–associated illness, and shellfish harvest restrictions (Knap et al. 2002). Finding coastal waters that are not fecally impaired is difficult in some regions (Blake et al. 1983; Kaysner et al. 1987; McLaughlin et al. 2005). Bacteria that can serve as markers for fecal contamination include *Salmonella enterica*, *Campylobacter*, *Escherichia coli*, *Clostridium perfringens*, and *C. difficile*

(Lipp et al. 2002; W.A. Miller et al. 2006; Sercu et al. 2009).

Sea otters inhabit fecal-polluted coastal waters and feed on prey items that can bioaccumulate pathogens (Fayer et al. 2004; W.A. Miller et al. 2005, 2006). Although sea otters are often in contact with water containing sewage effluent and agricultural, stream, and urban runoff, potential population impacts from exposure to fecal bacteria in these terrestrial sources are poorly understood. In polluted habitats marine sediments and prey can support high concentrations of enteric bacteria (Lisle et al. 2004) and the risk of exposure is enhanced when sea otters excavate these sediments to capture and ingest prey and groom their fur. Significant bacterial exposure also occurs when sea otters move into sheltered bays to seek protection from coastal storms. These areas are exposed to high surface water influxes during storm events and dilution is inhibited by small bay openings and limited tidal exchange, thus placing sea otters directly in the path of concentrated plumes of land-based pollution (M.A. Miller et al. 2010a).

Once bacteria and protozoa enter the ocean, invertebrates can concentrate these pathogens through filter-feeding activity (Lindsay et al. 2001; Arkush et al. 2003; Fayer et al. 2004). W.A. Miller et al. (2006) tested marine and estuarine invertebrates from central California for contamination by *Salmonella*, *Campylobacter*, *Vibrio*, *C. perfringens*, and *P. shigelloides*, facilitating comparison of enteric bacterial prevalence between sea otters and their invertebrate prey. Over 80% of mussel (*Mytilus californianus*) batches were culture-positive for at least one of these bacteria. Invertebrates collected inside muddy embayments were more often contaminated than those from open, sandy areas, and freshwater clams collected near the ocean were more likely to test positive than those collected upstream (W.A. Miller et al. 2006). Mussels collected from piers and bottom-dwelling invertebrates collected at the same location and time were often culture-positive, suggesting that bacteria were passing through in the water column. Exposure to surface water runoff from precipitation events is a risk factor for detection of *Cryptosporidium* and *Giardia* in mussels (W.A. Miller et al. 2005). These findings reveal that pathogen contamination of invertebrates occurs commonly along terrestrial–marine interfaces where sea otters reside.

In central California, bacteria isolated from sea otters during the wet and dry seasons were similar to those isolated from marine invertebrates. Between 2000 and 2005, feces from 244 southern sea otters were screened for the presence of enteric bacterial pathogens (*Campylobacter, Salmonella, C. perfringens, C. difficile,* and *E. coli* O157:H7) and opportunistic pathogens that are endemic to the marine environment (*V. cholerae, V. parahaemolyticus,* and *Plesiomonas shigelloides*). Otters sampled near more urbanized coastlines were up to 5.7 times more likely to test positive for one or more of these pathogens, while animals sampled along the sparsely populated Big Sur coast were less likely to be infected. In addition, sea otters from coastal areas exposed to high freshwater runoff were up to three times more likely to test positive for enteric bacterial pathogens than otters from areas with little or no freshwater runoff. The capacity of surface runoff to contribute to marine pathogen loading is illustrated by the fact that the largest river in central California discharges 432 times more untreated water to the ocean each year than the largest municipal sewage plant (M.A. Miller et al. 2010a).

## COCCIDIOIDOMYCOSIS

The most important fungal cause of death for sea otters in California is the pathogen *Coccidioides immitis,* causal agent of the disease coccidioidomycosis or Valley fever (Higgins et al. 2000). On land, *C. immitis* detection is highest in warm, dry, alkaline environments where other fungi are less likely to dominate (Kirkland and Fierer 1996). The usual habitat for *C. immitis* is low-altitude, arid soils in areas with low rainfall and high summer temperatures. The prevalence of clinical coccidioidomycosis in humans often peaks in the dry summer period following a wet winter, when previously formed arthroconidia are transported by wind and inhaled, leading to respiratory and occasionally systemic infection (Kirkland and Fierer 1996).

Airborne hyphal structures (arthroconidia) are the primary source of infection for people and animals. In California, most human *C. immitis* infections are reported from the Central Valley. Recent molecular characterization has identified a related fungus, *C. posadasii,* that overlaps and extends the range of *Coccidioides* to include Arizona, New Mexico, and Central and South America. The pathogenicity and lesion patterns for *C. posadasii* are less well characterized. Because it

has a wider endemic range than *C. immitis,* it is possible that some marine mammals have died from *C. posadasii* infections that were misidentified as *C. immitis.* Thus, until molecular characterizations are completed, these infections will be grouped under the generic name *Coccidiodes.*

No endemic marine focus of *Coccidiodes* is recognized; at present, all marine mammal infections are presumed to have originated from exposure to airborne arthroconidia from the terrestrial environment. Most sea otters dying of coccidioidomycosis have a systemic infection, with pyogranulomas distributed throughout the lungs, pleural cavity, liver, spleen, brain, and multiple lymph nodes (Fig. 23.2a). Microscopic examination reveals the presence of spherules (Fig. 23.2b), the yeast phase of *Coccidiodes* that grows only in infected tissues (Cornell et al. 1979; Fauquier et al. 1996; Kirkland and Fierer 1996).

Lung, pleural, and hilar lymph node involvement are present in most cases, suggesting that the main route of sea otter infection is via arthroconidial inhalation, similar to humans. However, occasionally these lesions are more accentuated in abdominal tissues, while respiratory lesions are mild. Given the comparatively localized distribution of coccidioidomycosis in stranded sea otters, sometimes near regions of freshwater runoff, it is conceivable that occasional *Coccidiodes* infections arise through arthroconidial spread in water, leading to bioconcentration in marine invertebrate prey, the same route for otter exposure to other pathogens.

## ANTHROPOGENIC CHEMICAL POLLUTANTS

As origins of southern sea otter morbidity and mortality have become better understood, questions have arisen as to the role of chemical pollutants. Implicated in some studies are a broad range of anthropogenic chemicals that can enter the ocean through surface runoff and wastewater effluent. Environmental pollution by anthropogenic chemicals is a serious global problem that extends throughout the oceans. Thousands of manmade chemicals can be detected in the air and water and even in animal or human tissues collected from remote areas. The chemical composition and potential biological effects of these compounds are diverse and often poorly understood.

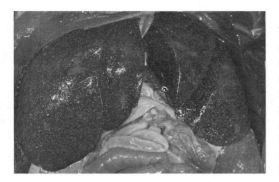

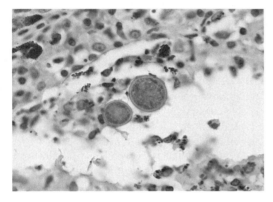

Figure 23.2:
a. Thousands of tiny pyogranulomas in the liver (and spleen, lower right) of a sea otter that died due to coccidioidomycosis. b. Microscopic view of two round, well-encapsulated, approximately 10–15-µm-diameter *Coccidiodes* spherules (yeast stages) inside the eye of an otter that died of coccidioidomycosis.

However, many compounds share properties that allow categorization into two main groups: persistent organic pollutants (POPs) and chemicals of ecological and environmental concern (COECs), which include less lipophilic pesticides and industrial chemicals.

POPs such as organochlorines (OCs), polychlorinated biphenyls (PCBs), dioxins, DDT and its breakdown products, and other legacy pesticides are often toxic and environmentally persistent. Many POPs are strongly lipophilic and have the ability to biomagnify in aquatic systems and bioaccumulate within individuals. Some studies report associations between elevated POP tissue burdens in wildlife including otters (Kannan et al. 1993) and enhanced susceptibility to infectious disease or large-scale population declines.

Female otters transfer a high proportion of POP body burdens to their pups via lactation (Jessup et al. 2010). POPs may also enter the body transplacentally or through foraging activity, transdermal absorption, and inhalation. Contaminants with lipophilic properties bioaccumulate in the blubber of some marine mammals, leading to rapid mobilization and enhanced contaminant exposure during periods of stress, fasting, starvation, or illness. For example, some striped dolphins (*Stenella coeruleoalba*) have blubber contaminant concentrations that are 10 million times higher than those in seawater (Tanabe et al. 1983), and southern resident killer whales (*Orcinus orca*) can have blubber PCB levels as high as 4,600 to 120,000 ng/g lipid weight (Krahn et al. 2009). In some marine mammal studies, high tissue POP concentrations have been linked with mass mortalities (Troisi et al. 2001; Jepson et al. 2005) and premature births (DeLong et al. 1973).

COECs include herbicides and pesticides like chlorpyrifos, dacthal, and oxadiazon that are less environmentally persistent than POPs, but are concerning due to their widespread distribution and potential for causing adverse health effects. Some COECs can impair endocrine function in laboratory animals and others accumulate in human breast milk (Colborn et al. 1997). Different risk factors and routes of exposure may govern POP and COEC accumulation and potential health effects. Despite the absence of blubber, high levels of butyltins, PCBs, organochlorines, DDTs, and other anthropogenic compounds have been detected in southern sea otter blood, organs and fat stores (Kannan et al. 1998, 2004, 2006; Nakata et al. 1998; Bacon et al. 1999). Initial efforts to characterize POP and COEC loading in sea otters focused on assessment of sea otter liver, kidney, or fat, as well as site-matched invertebrate prey (Estes et al. 1997; Norstrom et al. 1997; Nakata et al. 1998; Bacon et al. 1999; Kannan et al. 2004, 2008). Lipid-normalized concentrations of PCBs and DDTs in sea otter livers were 60- and 240-fold higher, respectively, than those in invertebrates from Monterey Bay, suggesting that otters have a significant capacity to bioaccumulate POPs (Kannan et al. 2004). These same studies showed that otters have the ability to metabolize PCBs with lower numbers of chlorine substitutions on the benzene ring structure. Other investigations have documented high tissue burdens of polychlorinated diphenyl ethers (PBDEs), perfluorinated compounds, polycyclic aromatic hydrocarbons (PAHs), and other compounds in sea otters from California and throughout the Pacific rim (Kannan et al. 1998, 2004, 2006, 2007, 2008). POP and COEC concentrations in whole blood obtained from apparently healthy sea otters in California can reach levels 20 to 50 times those seen in otters from more pristine areas of Alaska (Jessup et al. 2010).

In some cases, elevated POP and COEC burdens in sea otters reflect high levels of contamination in adjacent urban areas, coastal soils, and marine invertebrate biota. For example, high levels of DDT and DDE were detected in male otters residing in Elkhorn Slough, California, a region with substantial DDT contamination of local soils due to a century of agricultural use (Jessup et al. 2010). Some of the most intensively farmed land in the United States, the Salinas River Valley, is located immediately adjacent to southern sea otter habitat. Industrial sources also contribute to sea otter pollutant burdens; high levels of PCBs were detected in otters and invertebrates at two locations in Alaska where contamination of the nearshore marine environment was linked to previous military activity (Bacon et al. 1999; Jessup et al. 2010).

Potential relationships between elevated tissue contaminant burdens and sea otter death due to infectious disease have been reported (Kannan et al. 1998, 2006; Nakata et al. 1998; Jessup et al. 2007). Although nearly half of southern sea otter deaths are associated with infectious disease (Kreuder et al. 2003), it is unclear whether this is due to high pathogen exposure, enhanced susceptibility (including effects from POP or COEC exposure), or both. It is especially difficult to define relationships between contaminant burdens and disease because of the wide range of biological and chemical contaminants present in the nearshore marine ecosystem, the synergistic and antagonistic effects of these compounds, and the relative contributions of age, nutritional condition, reproductive status, and other biological factors.

Specific impacts of POPs and COECs on the immune system are poorly understood. Levin et al. (2007) used in vitro techniques to evaluate leukocyte function in blood collected from healthy sea otters. Blood samples were spiked with PCBs, 2,3,7,8-TCDD, and mixtures of these compounds. Results were compared with assays of non-exposed leukocytes from the same animals. Both up- and down-regulation of leukocyte function were noted in relation to POP

exposure. In addition, significant down-regulation was reported for leukocytes obtained from stressed, wild-caught otters compared to captive, minimally stressed otters.

To summarize, POP and COEC exposures are significant for southern sea otters, are mainly terrestrial in origin, and could be affecting health and population recovery. Also, because humans consume many of the same marine invertebrates, detection of high concentrations of POPs and COECs in sea otters and their prey highlights potential human health risks from consumption of locally or regionally harvested shellfish.

## DOMOIC ACID

Harmful algal blooms (HABs) are an escalating problem in coastal ecosystems, resulting in significant economic losses for commercial fisheries and illness and deaths of humans and marine fauna. The frequency and global distribution of HABs have increased worldwide in recent years, possibly as a result of large-scale ecological and climactic disturbances. *Pseudo-nitzschia*, a diatom that produces the potent neurotoxin domoic acid (DA), is frequently implicated in bloom events and is especially prevalent along the Pacific coast of North America (Work et al. 1993). During bloom events in California, DA was detected in sick sea lions (*Zalophus californianus*) and their prey, confirming that DA toxicity is sustained during trophic transfer and is a threat to marine mammal health (Scholin et al. 2000).

Acute DA poisoning is characterized by pruritus, ataxia, seizures, tremors, convulsions, lethargy, and coma (Work et al. 1993; Scholin et al. 2000; Zabka et al. 2009). Historically cases of severe neurological disease indicative of massive, acute DA intoxication have garnered the most attention. However, it is now recognized that marine mammals may be chronically exposed to DA at lower concentrations, such as during smaller events that go undetected by monitoring programs aimed at protecting human health (Zabka et al. 2009).

Preliminary data suggest that acute and chronic exposure of sea otters to DA may have significant health effects and may be affecting recovery (Kreuder et al. 2003, 2005). Sea otter prey items are known DA accumulators (Goldberg 2003). However, DA intoxication can be difficult to diagnose in sea otters due to a high prevalence of intercurrent disease, a relative paucity of "classical" brain lesions, limited access to testing, deterioration of DA in archived samples, post-mortem autolysis, and rapid clearance of DA from body fluids. Studies to clarify our understanding of the prevalence and significance of DA exposure in sea otters are needed.

DA intoxication in sea otters was first reported by Kreuder et al. (2003). A subsequent study (Kreuder et al. 2005) revealed associations between cardiac disease in sea otters and DA exposure, establishing DA as an important cause of death. DA-associated mortality in sea otters generally lags behind epizootics involving marine birds and sea lions, probably due to the time required for toxin-containing detritus to be taken up by benthic-feeding invertebrates, as opposed to epipelagic-feeding anchovies. A relationship between concurrent DA exposure and an epizootic of severe, S. neurona-associated otter disease was described, suggesting that DA exposure may exert systemic effects that enhance the pathogenicity of infectious agents (M.A. Miller et al. 2010a).

The risk of DA intoxication is not uniform throughout the sea otter range. Spatial and temporal "hot spots" for DA exposure may correspond with localized areas of recurring, high DA production due to nutrient loading of nearshore marine waters, possibly as a result of surface runoff or upwelling (Anderson et al. 2008; Kudela et al. 2008; Lane et al. 2009). Levels of urea in sea water have been shown to enhance DA production by Pseudo-nitzschia, and a temporal and spatial association between urea nitrogen detection in coastal marine waters (presumably originating from terrestrial sources) and HAB events has been documented in central California (Kudela et al. 2008). Assessment of river discharge proved to be an important component of predictive models, along with upwelling, for forecasting Pseudo-nitzschia blooms and DA production (Lane et al. 2009). Thus, while diatoms inhabiting the nearshore marine ecosystem produce this potent neurotoxin, the severity, frequency, and toxicity of bloom events may be strongly influenced by nutrients originating from terrestrial sources.

## MICROCYSTIN

During 2007, 11 southern sea otters died along the shore of Monterey Bay in central California with lesions suggestive of acute liver failure. Some animals were diffusely icteric, and their livers were enlarged, bloody, and friable. Expected causes for this condition, such as systemic bacterial infection, were excluded via postmortem examination and diagnostic testing. Livers from affected animals tested positive for a specific group of cyanotoxins, called microcystins, via liquid chromatography-tandem mass spectrophotometry (LC-MS/MS). Hepatic lesions consistent with microcystin intoxication were identified microscopically. Environmental surveillance revealed that some local lakes and rivers supported regular Microcystis blooms, triggering a broader investigation.

Cyanobacteria have a worldwide distribution and can form extensive blooms with toxin production in freshwater and estuarine habitat. Development of large-scale cyanobacterial blooms with production of potent and environmentally persistent cyanotoxins are an emerging global health issue (Guo 2007). The most common species, Microcystis aeruginosa, is capable of growth and toxin (microcystin) production when exposed to warm temperatures (Zehnder and Gorham 1960; Davis et al. 2009), nutrient loading (Jacoby et al. 2000; Davis et al. 2009), and increased light intensity (Welker and Steinburg 2000). Increasing climatic temperatures and widespread eutrophication may be contributing to the more frequent occurrence of massive cyanobacterial blooms (Paerl and Huisman 2008).

Microcystis blooms were formerly considered a public health issue limited to freshwater rivers, ponds, lakes, and reservoirs (Vasconcelos 1999). More recently, significant estuarine contamination has been reported. Large-scale Microcystis blooms are a regular occurrence in lakes and rivers throughout Washington (Johnston and Jacoby 2003), Oregon (Gilroy et al. 2000; Fetcho 2007) and California (Lehman et al. 2005; Fetcho 2007). Microcystis-contaminated outflows have been documented at the freshwater–marine interfaces of the Klamath River (Fetcho 2007), San Francisco Bay (Lehman et al. 2005), and multiple rivers flowing into Monterey Bay in California (M.A. Miller et al. 2010c). However, little monitoring of the marine environment has been performed. Microcystin accumulation has been demonstrated in marine and freshwater bivalves (Williams et al. 1997; M.A. Miller et al. 2010c), crustaceans (Vasconcelos et al. 2001), corals (Richardson et al. 2007), and fish (Malbrouck and Kestemont 2005).

Recent confirmation of deaths of southern sea otters due to microcystin intoxication extends the negative impacts of these toxins to include nearshore marine ecosystems and federally protected threatened species (M.A. Miller et al. 2010c). Land–sea flow of microcystins with trophic transfer through marine invertebrates was implicated as the most likely route of sea otter exposure. This hypothesis has been evaluated through necropsy, chemical analysis of tissues, environmental tracing of potential freshwater and marine microcystin sources, and evaluation of microcystin bioaccumulation by marine invertebrates (M.A. Miller et al. 2010c).

Microcystin concentrations up to 2,900 ppm, the second highest environmental microcystin concentration ever reported, were detected in a freshwater lake and downstream tributaries to within 1 km of Monterey Bay, California (Fig. 23.3). Environmental surveillance revealed no detectable microcystin in marine waters of Monterey Bay until the rainy season, when the marine outfalls of the three most nutrient-impaired rivers flowing into the bay tested positive. Microcystin-poisoned sea otters were often recovered near river mouths, harbors, and sloughs, including rivers with confirmed microcystin contamination.

The hypothesis that sea otters were most likely to be exposed to lethal levels of microcystins through consumption of contaminated invertebrate prey was evaluated through laboratory experiments where bioconcentration and depuration of freshwater microcystins by marine invertebrates could be assessed under defined conditions. Digestive tissues from farmed and free-living marine clams, mussels, and oysters consumed by otters and humans exhibited significant uptake of freshwater microcystins (up to 107 times ambient water levels). Marine invertebrates were also slow to depurate ingested toxins, and microcystins were relatively stable in seawater (M.A. Miller et al. 2010c). This was the first documentation of putative biotoxin transfer from the lowest trophic levels of nutrient-impaired freshwater habitat to top marine predators at the land–sea interface.

## CONCLUSIONS

There is growing scientific evidence for a land–sea connection between southern sea otter mortality and exposure to nutrient-associated biotoxins and terrestrial-origin parasites, bacteria, fungi, and anthropogenic pollutants. The cumulative impacts of these pathogens and toxicants on sea otter population recovery appear to be substantial, motivating programs aimed at enhancing public awareness and minimizing use or improper disposal of substances that could be harmful to otters and other marine wildlife.

Interconnections between coastal topography and sea otter mortality are further illustrated through examination of stranding patterns. Maps of recovery locations for live and recently deceased sea otters over the past 20 years demonstrate higher carcass recovery from river and stream mouths, embayments, harbors, and other coastal areas exposed to heavy stormwater runoff or prolonged stormwater retention. The likelihood of finding moribund and recently dead otters was at least 4.7 times higher per kilometer in high-outflow areas compared to coastal areas with lower exposure to surface runoff (Fig. 23.4). This large difference in stranding density may reflect high site fidelity and an enhanced risk of exposure to pathogens, toxins, and anthropogenic pollutants for animals residing in stormwater-impaired coastal zones.

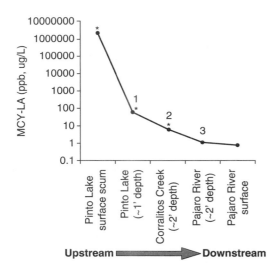

**Figure 23.3:**
Tracing freshwater contamination by microcystins from land to sea: time-matched microcystin-LA concentrations (ppb) in samples from Pinto Lake, just downstream in Corralitos Creek, and the receiving waters of the Pajaro River within 1 km of Monterey Bay. Asterisks (*) indicate sampling locations where *Microcystis* was detected microscopically.

Additional factors that could contribute to enhanced carcass recovery include variations in coastal currents, higher sea otter populations and greater recreational use of these areas by humans.

Although terrestrial sources of some chemical and biological pollutants are well recognized, associations between exposure to anthropogenic chemical pollutants and sea otter mortality are less clear. High POP and COEC tissue burdens have been associated with an increased risk of death due to infectious disease in some but not all studies. Other potential contributors to enhanced disease risk include high environmental pathogen exposure, introduction of novel pathogens by domestic and wild animals (such as cats and opossums), declines in coastal water quality, spread of more pathogenic microbe strains into the marine environment, and food limitations or nutritional deficiencies. Prey specialization, a response to food limitation, is an important risk factor for exposure to

*T. gondii* and *S. neurona* (Johnson et al. 2009), POPs (Kannan et al. 2004), DA (Goldberg 2003), microcystin (M.A. Miller et al. 2010c), and enteric bacteria (M.A. Miller et al. 2010a).

Two major sources of sea otter mortality appear to have the clearest and strongest connection to the terrestrial environment and freshwater runoff: protozoal parasite infections and microcystin intoxication. Current information on the life cycle and epidemiology of *T. gondii* and *S. neurona* suggests that terrestrial definitive hosts (cats and opossums) are the reservoir for infection of wild marine mammals, including sea otters. Potential land–sea connections are less well understood but equally statistically robust for fecal enteric bacterial pathogens. Minimizing land–sea dissemination of these pathogens will require major improvements to the ways that sewage, surface water, wastewater, and feces are handled, processed, and discharged. Elimination of all potential definitive

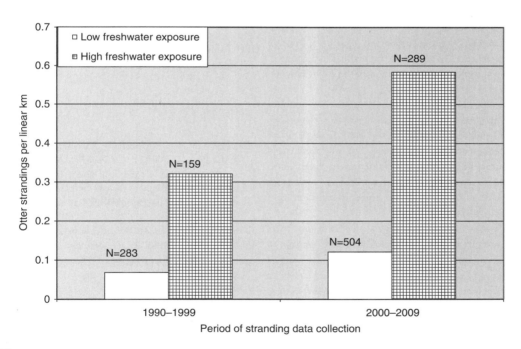

**Figure 23.4:**
Summarized recovery locations for live and recently deceased sea otters over the past 20 years. Higher carcass recoveries per kilometer are noted for river and stream mouths, embayments, harbors, and other coastal areas exposed to high stormwater runoff or prolonged retention. Although these high freshwater-outflow areas represent only about 10% of the habitat used as a primary pupping area for adult female sea otters, over 36% of live and freshly dead otters were recovered from these impaired zones over the past 20 years. Several factors may contribute to this higher stranding rate, including enhanced pathogen and toxin exposure in these high-outflow zones. (Summarized data courtesy of Brian Hatfield, USGS.)

hosts is neither achievable nor desirable, but simple steps like humane control of introduced and feral animal populations, combined with conscientious disposal of pet waste in plastic bags in approved sanitary landfill facilities, could help. A recent survey of three small coastal California communities revealed that outdoor pet and feral cats alone deposited at least 100 tonnes of feces per year adjacent to a high-risk area for *T. gondii* infection and disease for sea otters (Dabritz et al. 2007). Current efforts of institutions working for recovery of southern sea otters are focused on public outreach and education.

Given the complex interplay of host, agent, and environment in producing and sustaining disease outbreaks in natural systems, it is clear that multiple factors contribute to the longstanding mortality problems faced by southern sea otters. These factors may differ through time, by coastal location, and as a result of dietary specialization of individual animals. However, one finding that appears to be consistent across most investigations of infectious and toxic causes of southern sea otter death is a strong and statistically robust link between exposure to high freshwater runoff or more urbanized coastlines and an increased risk of death due to biotoxins and pathogens. A more complete understanding of the environmental niches occupied by these pathogens will provide clues regarding their origins and help optimize techniques for mitigation. Studies are in progress to better understand potential mechanisms of land–sea transfer of bacteria, parasites, and viruses; to explore marine mammal host ranges; and to determine methods to more effectively detect, inactivate, or remove microbes present in runoff, drinking water, wastewater, and marine foods.

Studies on causes of sea otter mortality offer important insights for guiding public policy to enhance southern sea otter population recovery and improve coastal water quality. Although this chapter was focused on sea otters, the implications of this research are far broader, because chronic land–sea pollution has the potential to simplify and destabilize marine ecosystems. Indeed, with their keystone and sentinel role in kelp forest ecosystems, the absence of sea otters results in reduced biodiversity of kelp forest-dwelling species, reduced storm surge protection provided by kelp, and reduced potential for carbon sequestration. Both sea otters and humans depend on the health of coastal ecosystems and the safety of marine foods for survival. Few better examples exist of where a conservation medicine/EcoHealth/One Health approach is sorely needed. We as a society must actively address pollution that impairs the recovery of charismatic species like sea otters, not only because of their beauty but also to retain their services as a keystone species that help maintain the kelp forest habitat, and because these same pathogens and chemicals may jeopardize our own future.

## ACKNOWLEDGMENTS

This review summarizes the products of years of collaborative research by all agency/members of the Southern Sea Otter Alliance, including the California Department of Fish and Game, the USGS Biological Resources Discipline, the U.S. Fish and Wildlife Service, the Monterey Bay Aquarium, the Marine Mammal Center, University of California-Davis, and University of California-Santa Cruz. We thank all of the members of this alliance for their efforts to protect, recover, and care for sick, stranded marine wildlife throughout California.

## REFERENCES

Anderson, D.M., J.M. Burkholder, W.P. Cochlan, P.M. Glibert, C.J. Gobler, C.A. Heil, R.M. Kudela, and M.L. Parson. 2008. Harmful algal blooms and eutrophication. Examining linkages from selected coastal regions of the United States. Harmful Algae 8:39–53.

Aramini, J.J., C. Stephen, J.P. Dubey, C. Engelstoft, H. Schwantje, and C.S. Ribble. 1999. Potential contamination of drinking water with *Toxoplasma gondii* oocysts. Epidemiol Infect 122:305–15.

Arkush, K.D., M.A. Miller, C.M. Leutenegger, I.A. Gardner, A.E. Packham, A.R. Heckeroth, A.M. Tenter, B.C. Barr, and P.A. Conrad. 2003. Molecular and bioassay-detection of *Toxoplasma gondii* oocyst uptake by mussels (*Mytilus galloprovincialis*). Int J Parasitol 33:1087–97.

Bacon, C.E., W.M. Jarman, J.A. Estes, M. Simon, and R.J. Norstrom. 1999. Comparison of organochlorine contaminants among sea otter (*Enhydra lutris*) populations in California and Alaska. Environ Toxicol Chem 18:452–458.

Blake, P.A. 1983. Vibrios on the half shell: what the walrus and carpenter didn't know. Ann Intern Med 99:558–559.

Bowie, W.R., A.S. King, D.H. Walker, J.L. Isaac-Renton, A. Bell, S.B. Eng, and S.A. Marion. 1997. Outbreak of toxoplasmosis associated with municipal drinking water. Lancet 350:173–177.

Colborn, T., D. Dumanoski, and J. P. Meyers. 1997. Our stolen future: Are we threatening our fertility, intelligence, and survival? A scientific detective story. Penguin Books. New York.

Conrad, P.A., M.A. Miller, C. Kreuder, E.R. James, J. Mazet, H. Dabritz, D.A. Jessup, F. Gulland, and M.E. Grigg. 2005. Transmission of *Toxoplasma*: Clues from the study of sea otters as sentinels of *Toxoplasma gondii* flow into the marine environment. Int J Parasitol 35:1125–1168.

Cornell, L.H., K.G. Osborne, and J.E. Antrim. 1979. Coccidiomycosis in a California sea otter. J Wildl Dis 15:373–378.

Dabritz, H.A., M.A. Miller, E.R. Atwill, I.A. Gardner, C.M. Leutenegger, A.C. Melli, and P.A. Conrad. 2007. Detection of *Toxoplasma gondii*-like oocysts in cat feces and estimates of the environmental oocyst burden. J Am Vet Med Assoc 231:1676–1684.

Davis, T.W., D.L. Berry, G.L. Boyer, and C.J. Gobler. 2009. The effects of temperature and nutrients on the growth and dynamics of toxic and non-toxic strains of *Microcystis* during cyanobacterial blooms. Harmful Algae 8:715–725.

DeLong, R.L., W.G. Gilmartin, and J.G. Simpson. 1973. Premature births in California sea lions: association with high organochlorine pollutant residue levels. Science 181:1168–1170.

Dubey, J.P., D.S. Lindsay, W.J. Saville, S.M. Reed, D.E. Granstrom, and C.A. Speer. 2001. A review of *Sarcocystis neurona* and equine protozoal myeloencephalitis (EPM). Vet Parasitol 95:89–131.

Estes, J.A. 2005. Carnivory and trophic connectivity in kelp forests. *In* J.C. Ray, K.H. Redford, R.S. Stenec, and J. Berger, eds. Large carnivores and the conservation of biodiversity, pp. 61–81. Island Press, Washington.

Estes, J.A., C.E. Bacon, W.M. Jarman, R.J. Norstrom, R.G. Anthony, and A.K. Miles. 1997. Organochlorines in sea otters and bald eagles from the Aleutian Archipelago. Mar Pollut Bull 34:486–490.

Fauquier, D.A., F.M.D. Gulland, J.G. Trupkiewlcz, T.R. Spraker, and L.J. Lowenstlne. 1996. Coccidioidomycosis in free-living California sea lions (*Zalophus californianus*) in Central California. J Wildlife Dis 32:707–710.

Fayer, R., J.P. Dubey, and D.S. Lindsay. 2004. Zoonotic protozoa: from land to sea. Trends Parasitol 20:531–6.

Fetcho, K. 2008. Final 2007 Klamath River blue-green algae summary report. Klamath: Yurok Tribe Environ Prog. http://www.klamathwaterquality.com/documents/2007YurokFINALBGAReport071708.pdf

Gilroy, D.J., K.W. Kauffman, R.A. Hall, X. Huang,and F.S. Chu. 2000. Assessing potential health risks from microcystin toxins in blue-green algae dietary supplements. Environ Health Persp 108:435–439.

Goldberg, J.D. 2003. Domoic acid in the benthic food web of Monterey Bay, California. Master of Science. Moss Landing: California State University Monterey Bay.

Grinnell, J. 1915. The Tennessee possum has arrived in California. Calif Fish Game 1:114–116.

Guo, L. 2007. Doing battle with the green monster of Taihu Lake. Science 317:1166.

Higgins, R. 2000. Bacteria and fungi of marine mammals: a review. Can Vet J 41:105–116.

Jacoby, J.M., D.C. Collier, E.B. Welch, F.J. Hardy, and M. Crayton. 2000. Environmental factors associated with a toxic bloom of *Microcystis aeruginosa*. Can J Fish Aquat Sci 57:231–240.

Jepson, P.D., P.M. Bennett, R. Deaville, C.R. Allchin, J.R. Baker, and R.J. Law. 2005. Relationships between polychlorinated biphenyls and health status in harbor porpoises (*Phocoena phocoena*) stranded in the United Kingdom. Envir Tox Chem 24:238–248.

Jessup, D.A., M. Miller, J. Ames, M. Harris, P. Conrad C. Kreuder, and J.A.K. Mazet. 2004. The Southern sea otter (*Enhydra lutris nereis*) as a sentinel of marine ecosystem health. Ecohealth 1:239–245.

Jessup, D.A., M.A. Miller, C. Kreuder-Johnson, P. Conrad, T. Tinker, J. Estes, and J. Mazet. 2007. Sea otters in a dirty ocean. J Am Vet Med Assoc 231:1648–1652.

Jessup, D.A., C.K. Johnson, J. Estes, D. Carlson-Bremer, W. Jarman, S. Reese, E. Dodd, M.T. Tinker, and M.H. Ziccardi. 2010. Persistent organic pollutants and other contaminants of concern in the blood of free ranging sea otters (*Enhydra lutris*) in Alaska and California. J Wildl Dis 46: 1214–1233.

Johnson, C.K., M.T. Tinker, J.A. Estes, P.A. Conrad., M. Staedler, M.A. Miller, D.A. Jessup, and J.K. Mazet. 2009. Prey choice and habitat use drive sea otter pathogen exposure in a resource-limited coastal system. Proc Nat Acad Sci USA 106:2242–7.

Johnston, B.R., and J.M. Jacoby. 2003. Cyanobacterial toxicity and migration in a mesotrophic lake in western Washington, USA. Hydrobiologia 495:79–91.

Kannan, K., S. Tanabe, A. Borrell, A. Aguilar, S. Focardi, and R. Tatsukawa. 1993. Isomer-specific analysis and toxic evaluation of polychlorinated biphenyls in striped dolphins affected by an epizootic in the western Mediterranean Sea. Arch Environ Contam Toxicol 25:227–233.

Kannan, K., K.S. Guruge, N.J. Thomas, S. Tanabe, and
    J.P. Giesy. 1998. Butyltin residues in southern sea otters
    (*Enhydra lutris nereis*) found dead along California
    coastal waters. Environ Sci Technol 32:1169–1175.
Kannan, K., H. Nakata, N. Kajiwara, M. Watanabe,
    N.J. Thomas, D.A. Jessup, and S. Kanabe. 2004. Profiles
    of polychlorinated biphenyl congeners, organochlorine
    pesticides and butyltins in Southern sea otters and their
    prey: implications for PCB metabolism. Environ
    Toxicol Chem 23:49–56.
Kannan, K., E. Perrotta, and N.J. Thomas. 2006. Association
    between perfluorinated compounds and pathological
    conditions in southern sea otters. Environ Sci Technol
    40:4943–4948.
Kannan, K., E. Perrotta, N. Thomas, and K. Aldous. 2007.
    A comparative analysis of polybrominated diphenyl
    ethers and polychlorinated biphenyls in southern sea
    otters that died of infectious diseases and noninfectious
    causes. Arch Environ Contam Toxicol 53:293–302.
Kannan, K., H.B. Moon, S.H. Yun, T. Agusa, N.J. Thomas,
    and S. Tanabe. 2008. Chlorinated, brominated, and
    perfluorinated compounds, polycyclic aromatic
    hydrocarbons and trace elements in livers of sea otters
    from California, Washington, and Alaska (USA), and
    Kamchatka (Russia). J Environ Monitor 10:552–558.
Kaysner, C.A., C. Abeyta Jr., M.M. Wekell, A. DePaola
    Jr., R.F. Stott, and J.M. Leitch. 1987. Incidence of *Vibrio
    cholerae* from estuaries of the United States West Coast.
    Appl Environ Microbiol 53:1344–1348.
Kirkland, T., and J. Fierer. 1996. Coccidioidomycosis:
    a re-emerging infectious disease. Emerg Infect Dis
    2:192–199.
Knap, A., É. Dewailly, C. Furgal, J. Galvin, D. Baden,
    R.E. Bowen, M. Depledge, L. Duguay, L.E. Flemming,
    T. Ford, F. Moser, R. Owen, W.A. Suk, and U. Unluata.
    2002. Indicators of ocean health and human health:
    developing a research and monitoring network. Environ
    Health Perspect 110:839–845.
Krahn, M.M., M.B. Hanson, G.S. Schorr, C.K. Emmons,
    D.G. Burrows, J.L. Bolton, R.W. Baird, and G.M. Ylitalo.
    2009. Effects of age, sex and reproduction status on
    persistent organic pollutant concentrations in "southern
    resident" killer whales from three fish-eating pods.
    Mar Pollut Bull 58:5122–1529.
Kreuder, C., M.A. Miller, D.A. Jessup, L.J. Lowenstein,
    M.D. Harris, J.A. Ames, T.E. Carpenter, P.A. Conrad,
    and J.A.K. Mazet. 2003. Patterns of mortality in
    southern sea otters (*Enhydra lutris nereis*) from
    1998-2001. J Wildlife Dis 39:495–509.
Kreuder, C., M.A. Miller, L.J. Lowenstine, P.A. Conrad,
    T.E. Carpenter, D.A. Jessup, and J.A.K. Mazet. 2005.
    Evaluation of cardiac lesions and risk factors associated
    with myocarditis and dilated cardiomyopathy in

southern sea otters (*Enhydra lutris nereis*). Am J Vet Res
    66:289–299.
Kudela, R.M., J.Q. Lane, and W.P. Cochlan. 2008. The
    potential role of anthropogenically derived nitrogen in
    the growth of harmful algae in California, USA.
    Harmful Algae 8:103–110.
Lane, J.Q., P. Raimondi, and R.M. Kudela. 2009. The
    development of a logistic regression model for the
    prediction of toxigenic *Pseudo-nitzschia* blooms
    in Monterey Bay, California. Mar Ecol-Prog Ser
    383:37–51.
Lankoff, A, W.W. Carmichael, K.A. Grasman, and M.Yuan.
    2004. The uptake kinetics and immunotoxic effects of
    microcystin-LR in human and chicken peripheral blood
    lymphocytes in vitro. Toxicology 204:23–40.
Larson, S., R. Jameson, M. Etnier, M. Fleming, and
    P. Bentzen. 2002. Loss of genetic diversity in sea otters
    (*Enhydra lutris*) associated with the fur trade of the 18th
    and 19th centuries. Mol Ecol 11:1899–1903.
Lehman, P.W., G. Boyer, C. Hall, S. Waller, and K. Gehrts.
    2005. Distribution and toxicity of a new colonial
    *Microcystis aeruginosa* bloom in the San Francisco Bay
    Estuary, California. Hydrobiologia 541:87–99.
Levin, M., H. Leibrecht, C. Mori, D. Jessup, and
    S. De Guise. 2007. Immunomodulatory effects of
    organochlorine mixtures upon in vitro exposure
    exposure of peripheral blood leukocytes differ
    between free-ranging and captive Southern sea
    otters (*Enhydra lutris*). Vet Immunol Immunopath
    119:269–277.
Lindsay, D.S., K.K. Phelps, S.A. Smith, G. Flick, and
    J.P. Dubey. 2001. Removal of *Toxoplasma gondii* oocysts
    from seawater by Eastern oysters (*Crassostrea virginica*).
    J Eukaryot Microbiol 197S–198S.
Lipp, E.K., A. Huq, and R. R. Colwell. 2002. Effects of
    global climate on infectious disease; the cholera model.
    Clin Microbiol Rev 15:757–770.
Lisle, J.T., J.J. Smith, D.D. Edwards, and G.A. McFeters.
    2004. Occurrence of microbial indicators and
    *Clostridium perfringens* in wastewater, water column
    samples, sediments, drinking water and weddell seal
    feces collected at McMurdo station, Antarctica. Appl
    Environ Microbiol 70:7269–7296.
Malbrouck, C., and P. Kestemont. 2006. Effects of
    microcystins on fish. Environ Toxicol Chem 25:72–86.
Massie, G.N., M.W. Ware, E.N. Villegas, and M.W. Black.
    2010. Uptake and transmission of *Toxoplasma gondii*
    oocysts by migratory, filter-feeding fish. Vet Parasitol
    169:296–303.
McLaughlin, J.B., A. DePaola, C.A. Bopp, K.A. Martinek,
    N.P. Napolilli, C.G. Allison, S.L. Murray,
    E.R. Thompson, M. Bird, and J.P. Middaugh. 2005.
    Outbreak of *Vibrio parahaemolyticus* gastroenteritis

associated with Alaskan oysters. N Engl J Med 353:1463–1470.

Miller, M.A., I.A. Gardner, A. Packham, J.K. Mazet, K.D. Hanni, D.A. Jessup, J. Estes, R. Jameson, E. Dodd, B.C. Barr, L.J. Lowenstine, F.M. Gulland, and P.A. Conrad. 2002a. Evaluation of an indirect fluorescent antibody test (IFAT) for demonstration of antibodies to Toxoplasma gondii in the sea otter (Enhydra lutris). J Parasitol 88:594–599.

Miller, M.A., I. Gardner, C. Kreuder, D. Paradies, K. Worcester, D. Jessup, E. Dodd, M. Harris, J. Ames, A. Packman, and P. Conrad. 2002b. Coastal freshwater runoff is a risk factor for Toxoplasma gondii infection of southern sea otters (Enhydra lutris nereis). Intl J Parasitol 32:997–1006.

Miller, M.A., M.E. Grigg, C. Kreuder, E.R. James, A.C. Melli, P.R. Crosbie, D.A. Jessup, J.C. Boothroyd, D. Brownstein, and P.A. Conrad. 2004. An unusual genotype of Toxoplasma gondii is common in California sea otters (Enhydra lutris nereis) and is a cause of mortality. Int J Parasitol 34:275–84.

Miller, M.A., W.A. Miller, P.A. Conrad, E.R. James, A.C. Melli, C.M. Leutenegger, H.A. Dabritz, A.E. Packham, D. Paradies, M. Harris, J. Ames, D.A. Jessup, K. Worcester, and M.E. Grigg. 2008. Type X Toxoplasma gondii in a wild mussel and terrestrial carnivores from coastal California: New linkages between terrestrial mammals, runoff and toxoplasmosis of sea otters. Int J Parasitol 38:1319–1328.

Miller, M.A., B.A. Byrne, S.S. Jang, E.M. Dodd, E. Dorfmeier, M.D. Harris, J. Ames, D. Paradies, K. Worcester, D.A. Jessup, and W.A. Miller. 2010a. Enteric bacterial pathogen detection in southern sea otters (Enhydra lutris nereis) is associated with coastal urbanization and freshwater runoff. Vet Res 41:01.

Miller, M.A., P.A. Conrad, M. Harris, B. Hatfield, G. Langlois, D.A. Jessup, S. Magargal, A.E. Packham, S. Toy-Choutka, A.C. Melli, M.A. Murray, F.M. Gulland, and M.E Grigg. 2010b. Localized epizootic of meningoencephalitis in southern sea otters (Enhydra lutris nereis) caused by Sarcocystis neurona. Vet Parasitol 172:183–194.

Miller, M.A., R.M. Kudela, A. Mekebri, D. Crane, S.C. Oates, M.T. Tinker, M. Staedler, W.A. Miller, S. Toy-Choutka, C. Dominik, D. Hardin, G. Langlois, M. Murray, K. Ward, and D.A. Jessup. 2010c. Evidence for a novel marine harmful algal bloom: cyanotoxin (microcystin) transfer from land to sea otters. PLoS ONE 5:e12576.

Miller, W.A., M.A. Miller, I.A. Gardner, E.R. Atwill, M.D. Harris, J.A. Ames, D.A. Jessup, A.C. Melli, D. Paradies, K. Worcester, P. Olin, N. Barnes, and P.A. Conrad. 2005. New genotypes and factors associated with Cryptosporidium detection in mussels (Mytilus spp.) along the California coast. Int J Parasitol 35:1103–1113.

Miller, W.A., M.A. Miller, I.A. Gardner, E.R. Atwill, B.A. Byrne, S. Jang, M. Harris, J. Ames, D. Jessup, D. Paradies, K. Worcester, A. Melli, and P. Conrad. 2006. Salmonella spp., Vibrio spp., Clostridium perfringens and Plesiomonas shigelloides in freshwater and marine invertebrates from coastal California ecosystems. Microb Ecol 52:198–206.

Nakata, H., K. Kannan, L. Jing, N. Thomas, S. Tanabe, and J.P. Giesy. 1998. Accumulation pattern of organochlorine pesticides and polychlorinated biphenyls in southern sea otters (Enhydra lutris nereis) found stranded along coastal California, USA. Environ Pollut 103:45–53.

Norstrom, R.J., W.M. Jarman, J.A. Estes, R.G. Anthony, C.E. Bacon, and A.K. Miles. 1997. Organochlorines in sea otters and bald eagles from the Aleutian archipelago. Mar Pollut Bull 34:486.

Paerl, H.W., and J. Huisman. 2008. Climate: Blooms like it hot. Science 320:57–58.

Richardson, L.L., R. Sekar, J.L. Myers, M. Gantar, J.D. Voss, L. Kaczmarsky, E.R. Remily, G.L. Boyer, and P.V. Zimba. 2007. The presence of the cyanobacterial toxin microcystin in black band disease of corals. FEMS Microbiol Lett 272:182–187.

Riedman, M.L., and J.A. Estes. 1990. The sea otter (Enhydra lutris): Behavior, ecology and natural history. USFWS Biol Rep 90:26.

Scholin, C.A., F. Gulland, G.J. Doucette, S. Benson, M. Busman, F.P. Chavez, J. Cordaro, R. DeLong, A. De Vogelaere, J. Harvey, M. Haulena, K. Lefebvre, T. Lipscomb, S. Loscutoff, L.J. Lowenstine, R. Marin III, P.E. Miller, W.A. McLellan, P.D.R. Moeller, C.L. Powell, T. Rowles, P. Silvagni, M. Silver, T. Spraker, V. Trainer, F.M., and Van Dolah. 2000. Mortality of sea lions along the central California coast linked to a toxic diatom bloom. Nature 403:80–84.

Sercu, B., L.C. Van De Werfhorst, J. Murray, and D. Holden. 2009. Storm drains are sources of human fecal pollution during dry weather in three urban southern California watersheds. Environ Sci Technol 43:293–298.

Tanabe, S., N. Miyazaki, T. Mori, and R. Tatsukawa. 1983. Global pollution of marine mammals by PCBs, DDTs and HCHs (BCHs). Chemosphere 12:1269–1275.

Tinker, M.T., G. Bentall, and J.A. Estes. 2008. Food limitation leads to behavioral diversification and dietary specialization in sea otters. Procs Natl Acad Sci USA 105:560–565.

Thomas, N.J., and R.A. Cole. 1996. The risk of disease and threats to the wild population. Endangered Species Update. USFWS 13:23–27.

Troisi, G. M., K. Haraguchi, D. S. Kaydoo, M. Nyman, A. Aguilar, A. Borrell, U. Siebert, and C. F. Mason. 2001. Bioaccumulation of polychlorinated biphenyls (PCBs) and dichlorodiphenylethane (DDE) methyl sulfones in tissues of seal and dolphin morbillivirus epizootic victims. J Tox Environ Health 62:1–8.

Vasconcelos, A. 1999. Dynamics of microcystin in the mussel *Mytilus galloprovincialis*. Toxicon 37:1041–1052.

Vasconcelos, V., S. Oliveira, and F.O. Teles. 2001. Impact of a toxic and a non-toxic strain of *Microcystis aeruginosa* on the crayfish *Procambarus clarkii*. Toxicon 39:1461–1470.

Welker, M., and Steinberg, C. 2000. Rates of humic substance photosensitized degradation of microcystin-LR in natural waters. Environ Sci Tech 34:3415–3419.

Wendte, J.M., M.A. Miller, D.M. Lambourn, S.L. Magargal., D.A. Jessup, and M.E. Grigg. 2010. Self-mating in the definitive host potentiates clonal outbreaks of the apicomplexan parasites *Sarcocystis neurona* and *Toxoplasma gondii*. PLoS Genet 6:e1001261.

Williams, D.E., S.C. Dawe, M.L. Kent, R.J. Andersen, M. Craig, and C.F.B. Holmes. 1997. Bioaccumulation and clearance of microcystins from salt water mussels, *Mytilus edulis*, and in vivo evidence for covalently bound microcystins in mussel tissues. Toxicon 35:1617–1625.

Work, T.M., B.C. Barr, A.M. Beale, L. Fritz, M.A. Qulliam, and J.L. Wright. 1993. Epidemiology of domoic acid poisoning in brown pelicans (*Pelecanus occidentalis*) and Brant's cormorants (*Phalacrocorax penicillatus*) in California. J Zoo Wildlife Med 24:54–62.

Zabka, T.S., T. Goldstein, C. Cross, R.W. Mueller, C. Kreuder-Johnson, S. Gill, and F.M. Gulland. 2009. Characterization of a degenerative cardiomyopathy associated with domoic acid toxicity in California sea lions (*Zalophus californianus*. Vet Pathol 46: 105–19.

Zehnder, A., and P.R. Gorham. 1960. Factors influencing the growth of *Microcystis aeruginosa*. Can J Microbiol 6:645–660.

# PART FOUR

ECOTOXICOLOGY AND CONSERVATION MEDICINE

# 24

# ECOTOXICOLOGY

## Bridging Wildlife, Humans, and Ecosystems

## Jeffrey M. Levengood and Val R. Beasley

We define *ecotoxicology* as the science of all the adverse biochemically mediated impacts of all chemicals on all living organisms and on all of their interactions with one another in the environment (Beasley 1993; Levengood and Beasley 2007). Modern ecotoxicology goes beyond measuring concentrations of toxicants in the environment or conducting straightforward laboratory toxicity testing. Determining toxic potential and potency is important, but a more holistic view of the fates and effects of contaminants is essential. Interrelationships among toxicants, soil and water chemistry, atmospheric conditions, habitat quality, food availability, infectious diseases, predator prey relationships, and other complexities of life in natural as well as human-dominated ecosystems must be considered to understand how organisms, ecosystem health, and ecosystem services can be sustained.

## DIRECT AND INDIRECT IMPACTS OF TOXICANTS

Direct impacts pertain to how toxic chemicals independently alter the health of organisms. Direct effects occur from single chemicals or multiple toxicants acting in concert after release to the environment. They can be acute, with a rapid onset and short duration, in response to a high, short-term introduction to

the environment, or chronic, with prolonged effects, often from longer, low-level exposures. An example of acute direct effects would be grasshoppers dying from cholinesterase inhibition shortly after being sprayed with an insecticide. The development of a bronchial adenocarcinoma following years of inhalation of polycyclic aromatic hydrocarbons is an example of a chronic direct health effect. Extirpation of native grass cover by herbicide use is a direct toxic impact to a component of local biodiversity. Higher mortality rates of voles and lower productivity of grassland birds due to the reduction in cover would be indirect effects of the herbicide. Both the loss of native vegetation and changes in population levels of key animal species may affect ecosystem functioning (e.g., by changing energy flow and nutrient cycling). Thus, effects of toxicants can be direct (or primary) and indirect (that is, secondary, tertiary, etc.), and both may lead to a cycle of health and ecological problems (Fig. 24.1). Toxicant-induced reductions in voles and grassland birds may lead to starvation of raptors, which may in turn undermine survival of their young. If the raptor population crashes, the highly productive rodents may rebound rapidly and then prevent recovery of the plant cover. If erosion follows depletion of plant cover due to the combination of herbicide use and overpopulated rodents, then the entire system may decline in its capacity to sustain life.

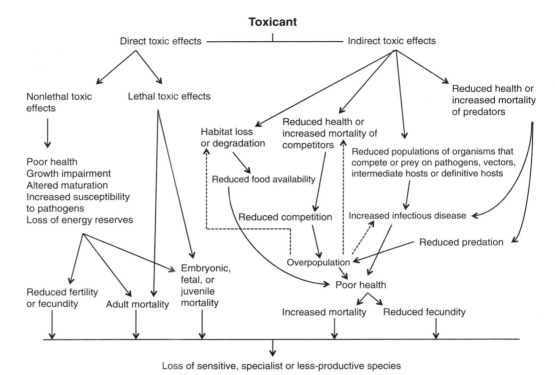

**Figure 24.1:**
Direct and indirect effects of environmental contaminants on individuals, populations, and communities.

Direct effects of exposure can be mediated by other ecosystem components. Sufficient cover, such as grass buffer strips between crop fields and riparian areas, can reduce insecticide and nutrient runoff (USDA 2000), reducing exposures of aquatic life. Mobility and bioavailability of hydrophobic chemicals are often reduced in soils and sediments rich in organic carbon. Although binding of hydrophobic toxicants to soil organic carbon may reduce runoff and passage into aquifers that feed streams and other water bodies, the local impacts on soil biota may be prolonged as compared to carbon-poor soils. The more severe toxic impacts on organisms in carbon-rich soils relate not only to localized retention of toxicants but also to delayed microbial detoxification associated with the binding.

Pesticides can harm biota directly, through toxic effects at specific or nonspecific receptors in affected organisms, or indirectly, as in reductions in food or cover (Fig. 24.1). Too often, unintended direct and indirect effects are ignored. Insecticides are intended to reduce the biomass of specific pests; however, non-target arthropods and other invertebrates, many of them "beneficial" from a human standpoint (e.g., carnivorous insects and spiders that eat pest insects; bees and moths that serve as pollinators; millipedes that shred plant debris; and earthworms that free up nutrients for microbes in soils), are negatively affected. Some studies that examined indirect effects of insecticides on wild birds in grasslands (Adams et al. 1994; Martin et al. 1998, 2000), shrub-steppe habitats (Howe et al. 1996), and forests (Rodenhouse et al. 1992) found no dramatic negative effects on productivity or nestling growth despite reductions in the target prey base, because the parents compensated for reductions in target prey by switching to alternate prey or foraging more widely. However, switching to alternate prey and foraging farther afield may have energetic or nutritional costs, especially in times of naturally occurring food shortages or adverse environmental conditions. Boatman et al. (2004) found a relationship between brood attrition and reduced abundance of arthropod prey available to yellowhammers (*Emberiza citrinella*, a bunting of European and Asian grasslands) following insecticide application. From the same study, Morris et al. (2004) reported associations between pesticide-induced reductions in invertebrate prey and increased foraging intensity of yellowhammers, as well as between insecticide use and reduced nestling body condition. In addition, the reduced arthropod abundance was associated with lower brood survival in corn buntings (*Miliaria calandra*). These studies reveal that the abundance of affected prey species, diversity of arthropod communities, and foraging plasticity of avian predators (whether they can effectively switch to an abundant, readily captured alternate prey) are important determinants of the impacts of insecticides on avian productivity. Impacts on productivity and health may be felt more strongly in concert with stress, such as unusually harsh weather.

Broad-spectrum herbicides used to control problem "weeds" in untilled field edges may reduce not only floral cover and diversity but also the abundance and diversity of associated arthropods (Moreby and Southway 1999). Such herbicide impacts can mimic the harm induced by insecticides. For example, herbicides applied directly to fields of cereal crops may reduce arthropod prey (Douglas et al. 2010), which would presumably reduce the food supply available to nestling birds. Unfortunately, few studies have examined the potential development of such indirect impacts.

# EXAMPLES OF THE EFFECTS OF CHEMICALS IN CONCERT WITH OTHER ECOSYSTEM PERTURBATIONS

## Toxicant Exposure and Infectious Disease

Phocine distemper virus (PDV) is a paramyxovirus of the genus *Morbillivirus* that is pathogenic to pinnipeds. Outbreaks of PDV occurred in 1988 and 2002 in the North Sea region, causing major mortalities, especially in harbor seals (*Phoca vitulina*; Härkönen et al. 2006). Outbreaks have occurred relatively recently in the North Pacific Ocean, with impacts on sea otters (*Enhydra lutris*), Steller sea lions (*Eumetopias jubatus*),

and harbor and fur seals (*Callorhinus ursinus*) (Goldstein et al. 2009). Some synthetic chemicals, such as polychlorinated biphenyls (PCBs) and dichlorodiphenyltrichloroethane (DDT), are immunosuppressive. Immunosuppression due to organochlorine exposure was implicated as a contributing factor in PDV infections in seals through field and laboratory studies. Seals succumbing to PDV during the 1988 outbreak had greater tissue burdens of organochlorine pesticides and PCBs than survivors (Hall et al. 1992). Researchers in the Netherlands (de Swart et al. 1994, 1996; Ross et al. 1995) demonstrated that seals fed fish from the Baltic Sea (contaminated with PCBs) had impaired immune responses compared to seals fed fish from the Atlantic Ocean. Fish from the Baltic had two to over ten times the organochlorine burden of those from the Atlantic, and organochlorine burdens of the seals fed the Baltic Sea fish were much higher than in those fed the Atlantic Ocean fish.

Immunosuppression resulting from contaminant exposure was implicated as a proximate cause of other mortality events in marine mammals. Large die-offs of bottlenose dolphins (*Tursiops truncatus*) occurred along the Atlantic coast of North America in the late 1980s and early 1990s. Dolphin and porpoise morbilliviruses were detected in carcasses of dolphins that died in these events (Taubenberger et al. 1996). The dolphin carcasses also displayed lesions consistent with infections attributable to a variety of opportunistic bacterial, viral, and fungal pathogens that can follow impairment of immune functions (Geraci 1989). Blubber from these animals contained highly elevated concentrations of PCBs and DDE (DDE is a metabolite of DDT) (Geraci 1989; Kuehl et al. 1994). Subsequently, Lahvis et al. (1995) found an inverse relationship between concanavalin A-induced lymphocyte proliferation and PCB, DDT, and DDE in the peripheral blood of free-ranging bottlenose dolphins. Of course, wildlife may be exposed to other immunotoxic chemicals. For example, concentrations of tributyltin and its breakdown products were elevated in bottlenose dolphins stranded along the Gulf of Mexico and Atlantic coasts of the United States during 1989 to 1994 (Kannan et al. 1996). Tributyltin, recently outlawed in the United States, is a known immunosuppressive agent and was used primarily as a biocide on hulls of military ships, docks, and other marine structures.

Marine mammals worldwide are beset by interactions among environmental contaminants, infectious microbes, and opportunistic pathogens. Reductions in exposures of marine mammals to immunosuppressive environmental contaminants would likely enable more effective immunological responses to highly infectious (e.g., PDV) as well as opportunistic pathogens.

## Free Nutrients and Herbicides: Facilitation of Trematode Parasitism in Frogs

Several studies have linked nutrient loadings with increased infections of developing frogs with the trematode *Ribeiroia ondatrae*, which causes limb malformations (Johnson et al. 1999, 2001; Johnson and Chase 2004). Recent studies revealed herbicide impacts that may directly and indirectly affect the health of amphibians in heavily agriculturalized regions (Fig. 24.2). Beasley et al (2005) found that in herbicide-affected farm ponds, recruitment of juvenile cricket frogs (*Acris crepitans*) was reduced and echinostome trematode infections in the kidneys of the surviving juvenile frogs were greatly increased. The reduced recruitment may have resulted from mortality due to high numbers of echinostome metacercariae, because *Rana pipiens* died when their early-stage tadpoles were exposed to the parasite (Schotthoefer et al. 2003a). Other reports showed that *R. ondatrae* are also lethal to tadpoles with high-level infections, and that lower-intensity infections cause supernumerary limbs, angular limb deformities, missing limbs, and other malformations (Johnson et al. 1999; Schotthoefer et al. 2003b). Sousa and Grosholz (1991) suggested that complex habitat structure impedes parasite transmission, and Beasley et al. (2005) hypothesized that the increased severity of trematode infections was related to the presence of fewer aquatic plants. Where herbicides reduce plant cover, tadpole predation is likely facilitated, so each survivor may receive a greater infective load of trematodes from the water (Fig. 24.2). Similarly, with less plant cover, the motile cercariae may have less difficulty finding tadpole intermediate hosts, facilitating infection. Recently, a large-scale field study was coupled with microcosm studies that examined the numbers of snails and immune responses as well as the intensity of trematode infections in developing amphibians, with and without atrazine (Rohr et al. 2008).

In the field, the authors examined over 240 variables and found that atrazine and its principal metabolite, desethylatrazine, plus phosphate accounted for most of the variation in trematode infection intensity in the amphibians. Atrazine and its metabolite seemed to have an even greater effect than the fertilizer. In the microcosm study, atrazine increased water clarity (less plankton), increased periphyton (nutrients and sunlight no longer used by the dying plankton became available to periphyton), increased numbers of snail intermediate hosts, impaired immune responses in tadpoles, and finally increased trematode infection intensities in the amphibians.

Herbicides can reduce dissolved oxygen and algal food needed by tadpoles, which may slow growth, potentially delaying metamorphosis or reducing size at metamorphosis and fitness (Diana et al. 2000). Hypoxic water and reduced food resources might also stress tadpoles so that they are less able to avoid cercarial infection and encystment. After a reduction in the algal community due to herbicide contamination, there is likely a rebound to higher-than-normal algal concentrations because competition for nutrients among surviving plant species is reduced (Diana et al. 2000). Although this increased algal food might benefit tadpoles, it might also provide food for snails that are intermediate hosts, thereby supporting massive asexual reproduction and release of infective cercariae (Fig. 24.2). When snail numbers increase relative to tadpole numbers, super-infections of tadpoles may result (Rohr et al. 2008). Nutrient enrichment and/or the effects of herbicides may account for some of the outbreaks of supernumerary limbs in frogs noted in the 1990s (Helgen et al. 1998). In earlier decades, such malformations were seen only infrequently (Hoppe 2000).

Exposure to pesticides other than atrazine can lead to immunosuppression and increased susceptibility to trematode infections in tadpoles and frogs (Kiesecker 2002; Linzey et al. 2003). Herbicides and other contaminant stressors may impair growth, development, and metamorphosis. Longer time in earlier tadpole stages may result in greater mortality due to gape-limited predators and trematodes. In addition, exposure to insecticides might influence trematode infections by affecting populations of aquatic predators that consume cercariae, as well as predators that consume snails. Schotthoefer et al. (2007) demonstrated that invertebrates, including copepods, daphnids, hydra,

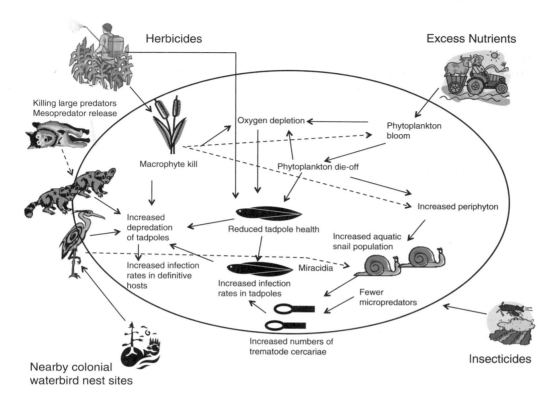

**Figure 24.2:**
Anthropogenic activities may produce a complex interaction of factors leading to increased trematode infection rates and intensity in frogs.

and especially dragonfly and damselfly larvae, consumed the motile cercariae in a controlled setting. Insecticide-mediated changes in invertebrate community structure might therefore release cercariae or other trematode life stages from this predation pressure (Fig. 24.2). Whether these "micropredators" are important in determining infective loads of cercariae in natural and human-altered aquatic environments and whether they are more or less susceptible to a wide range of pesticides than either cercariae or host snails remain to be determined.

Missing limbs are more prevalent in frogs than are supernumerary limbs. Missing limbs in metamorphic frogs were recently attributed to the impacts of predators such as dragonfly larvae that may be too small to consume the animal intact (Sessions and Ballengee 2010). It seems plausible that applications of herbicides widely marketed for control of "aquatic weeds" might deprive larval amphibians of cover and

thus set the stage for increased predation and loss of limbs. To our knowledge, investigations of such a possibility are yet to be conducted.

## Invasive Species and Contaminants: Changing Great Lakes Food Chains, Redirecting Toxicant Transmission

The food chains of the North American Great Lakes system have been dramatically altered since settlement by peoples of European descent. Notable exotic species introductions in the Great Lakes followed construction of the Welland Canal in 1829, which circumvented Niagara Falls. This allowed anadromous species, including the sea lamprey (*Petromyzon marinus*), a parasitic, jawless eel-like fish first noted in Lake Erie in 1921, and the alewife (*Alosa pseudoharengus*), a fish of the family Culpeidae first noted in Lake Erie in the early 1930s, to ascend above the falls.

Since then, approximately 140 aquatic species have invaded the Great Lakes, primarily through exchange of ballast water, as ocean-going ships take on water in a port on one side of the ocean and then dump it and take on more water on the other side. This has resulted in fresh or brackish water organisms being introduced on both sides of the Atlantic. In addition to "natural" or accidental introductions, large numbers of non-native salmon and trout have been routinely stocked for recreational fishing in the Great Lakes since the late 1800s. These introductions, coupled with declines in many native species, have dramatically changed the energy, nutrient, and contaminant flows in Great Lakes food webs. Invasion of the Great Lakes by zebra mussels (*Dreissena polymorpha*), which are native to eastern Europe and western Asia, was first noted in 1988. The aggressively colonizing zebra mussels rapidly expanded to occupy all the Great Lakes in huge abundance. The quagga mussel (*Dreissena rostriformis*), native to Ukraine, was first found in the Great Lakes in 1991. Quagga mussels are more tolerant of colder temperatures, and thus by 2005 they had expanded to the deep and cold waters of much of the Great Lakes basin. Although absent from much of Lake Michigan in 2000, by 2005 quagga mussels had replaced zebra mussels throughout much of the basin (Michigan Sea Grant 2006; Richerson 2009). Feeding by zebra and quagga mussels reduces phytoplankton availability, decreasing food for zooplankton and freeing up nutrients that may be used by other species, including toxigenic cyanobacteria.

These dressenid mussels can accumulate organic contaminants to concentrations well above those of the surrounding environment. Kannan et al. (2005) reported a bioconcentration factor of 1,000 for perfluorooctane sulfonate, a synthetic surfactant, in zebra mussels as compared to water. Willman et al. (1999) reported that zebra mussels accumulated certain PCB congeners to concentrations a million times greater than the water and their phytoplankton food.

Morrison et al. (1998) predicted that the zebra mussels would alter PCB transmission in food webs and trophic levels, such that PCB concentrations would decline in some predatory fish as the mussels removed the PCBs from phytoplankton and reintroduced them to the water. They failed, however, to account for the widespread effects of another exotic invader, the round goby (*Neogobius melanostomus*), a small, prolific benthic fish, native to central Eurasia,

first found in the upper Great Lakes in 1990. Round gobies have become an increasingly important item in the diets of Great Lakes predatory fish (Steinhart et al. 2004; Dietrich et al. 2006; Truemper et al. 2006) and waterbirds (Somers et al. 2003; Bur et al. 2008). Some positive effects of round gobies have been noted, such as increased growth rate of juvenile smallmouth bass (Steinhart et al. 2004) that fed extensively on the round gobies (Ray and Corkum 1997). However, predators of round gobies may also be exposed to elevated concentrations of contaminants. Round gobies may feed heavily on zebra mussels and quagga mussels, enabling trophic transfer of contaminants. Kwon et al. (2006) noted a three- to five-fold increase in total PCB concentrations between round gobies and smallmouth bass (*Micropterus dolomieu*) that preyed upon them. The recovery of the threatened Lake Erie water snake (*Nerodia sipedon insularum*), which now feeds extensively on round gobies, is attributed to this abundant prey (King 2007). However, incorporating round gobies into more than 90% of its diet has not changed the exposure of the snake to PCBs and DDE, body burdens of which have remained elevated in contrast to declines in other predator species of the region (Fernie et al. 2008). The full impacts of the exotic round goby on energy flow and contaminant transfer in the Great Lakes system remain to be fully realized, but it is clear that they are transferring contaminants from dreissenid mussels to larger predatory fishes.

Native species such as diving ducks (tribe Aythyini), and some benthic-feeding fish, like the freshwater drum (*Aplodinotus grunniens*), also feed on adult zebra mussels. Studies have shown that certain native fish (Bowers and de Szalay 2007) and ducks (Mitchell et al. 2000) can limit populations of zebra mussels. However, the potential ecological benefits of this predation, for example reduced populations of zebra mussels, might be offset, at least in part, by negative consequences of food chain transfer of contaminants. Nearly 100% of the diet of lesser scaup (*Aythya affinis*) and greater scaup (*Aythya marila*), closely related species of diving ducks, wintering in portions of the lower Great Lakes may consist of zebra mussels (Mazak et al. 1997; Custer and Custer 2000), and concentrations of PCBs in duck tissues were significantly higher in locales where their diets contained a higher proportion of zebra mussels than macrophytes (Mazak et al. 1997). Such exposures are of particular concern because continental scaup populations have been declining since the mid-1980s and remain below their long-term average (Austin et al. 2006). Concentrations of PCBs in scaup remain above those that trigger human consumption advisories (Smith et al. 1985; Custer and Custer 2000).

Zebra and quagga mussels may have impacts in addition to biomagnification and trophic transfer of synthetic, organic toxicants. Increased light penetration (increased water clarity) from filter feeding (Pillsbury et al. 2004) and nutrient enrichment from pseudo-feces and feces of huge populations of zebra and quagga mussels has promoted increased growth of *Cladophora*, a filamentous algae (Stankovich 2004). Dense, decomposing algal mats may provide conditions suitable for the growth of anaerobic bacteria, such as *Clostridium botulinum*. Byappanahalli and Whitman (2009) incubated *C. botulinum* in *Cladophora* collected in areas experiencing outbreaks of avian botulism. Although the pathway of transfer remains unclear, it is thought that the neurotoxic botulinum toxin from *C. botulinum* is taken up by the filter-feeding dressenid mussels, which are in turn eaten by the invasive round goby. The dead or dying gobies or mussels themselves are then eaten by predatory waterbirds, which then die from botulism. Although botulism-induced die-offs of birds occurred in the Great Lakes in the 1960s and 1970s, bans on phosphates in detergents to reduce nutrient availability coupled with improvements in municipal sewage treatment reduced algal growth, and botulism outbreaks subsided. Unfortunately, since 1998, the increased water clarity and nutrient availability due to invasive mussels have apparently combined to fuel *Cladophora* growth, resulting in large botulism-related die-offs in common loons (*Gavia immer*), mergansers, and other fish-eating birds in the Great Lakes. According to the USGS National Wildlife Health Center, more than 52,000 botulism-related avian deaths occurred on the Great Lakes during 2002 to 2006 (Zuccarino-Crowe 2007).

## Nutrient and Thermal Pollution, and Propagation of "Natural" Toxigenic Species

Hazardous blooms of toxigenic cyanobacteria are facilitated by increases in free nutrients and stagnation of water. Impoundment of water in ponds, lakes, and reservoirs, slow-flowing streams, creeks, and rivers, fertilizer or manure runoff, inadequately treated sewage effluent, and evaporation that concentrates nutrients (low water levels) favor growth of large toxigenic blooms. Seasonal warming and increases in temperature related to climate change are additional stimuli for growth of many species of cyanobacteria. Such hazards are aggravated by removal of plants that provide shade to water bodies. The ultraviolet and infrared energy, as well as the nutrients that are no longer absorbed by terrestrial and aquatic plants, are left to be used by cyanobacteria.

Cyanobacterial blooms may result in poisoning of domestic animals when their drinking water sources are contaminated. Crowding wild animals at artificial water sources (e.g., where they can be observed by tourists) can increase nutrient loading in the form of urine and feces. Recently, 52 dead wild animals were found in South Africa's Kruger National Park in relation to a bloom contaminated by microcystins in an artificial water body (Oberholster et al. 2009).

Dense cyanobacterial blooms may obstruct light from entering water, preventing oxygenation of the water by photosynthetic organisms. At night, when the cyanobacteria, algae, and/or macrophytes respire, and especially when the dominant organisms begin to die, metabolism can deplete dissolved oxygen in the water, killing aquatic organisms, including fish.

Major freshwater cyanobacterial toxins fall into two main categories, hepatotoxins (damage the liver) and neurotoxins (disrupt nerve impulses). Some cyanobacteria, such as *Anabaena flos-aquae*, can produce both neurotoxins and hepatotoxins. Commonly encountered toxigenic cyanobacteria and diagnostic procedures for confirming cyanobacterial neurotoxicoses and hepatotoxicoses have been described (Beasley et al. 1989).

The best known cyanobacterial neurotoxins paralyze respiratory muscles, terminating in suffocation. Anatoxin-A is produced by several species representing five genera of cyanobacteria and is rather commonly encountered. Anatoxin-A and its analogs stimulate nicotinic acetylcholine receptors, leading to a depolarizing blockade and thus paralysis, including paralysis of the diaphragm and intercostal muscles, which are essential for respiration. Anatoxin-A(s) (anatoxin-A sub S, which stands for salivation), produced by *A. flos-aquae* and *A. lemmermannii* (Aráoz et al. 2010), is the only known naturally occurring organophosphorus cholinesterase inhibitor. A pilot study suggests that ruminants may be tolerant of this toxin after ingestion (Cook et al. 1989). It is hypothesized that hydrolysis of the toxin in the rumen might lead to detoxification.

Both marine and freshwater harmful algal blooms produce saxitoxin and neosaxitoxin, which are sodium channel blockers. Saxitoxin from dinoflagellates is best known for its contamination of marine shellfish and its impacts on organisms that eat shellfish and fish, including people. For example, saxitoxin derivatives were implicated in mass mortalities of endangered Mediterranean monk seals (*Monachus monachus*) in the waters off Mauritania (Hernández et al. 1998; Reyero et al. 1999; van de Bildt et al. 1999). Saxitoxin and/or neosaxitoxin may also be produced by five genera of cyanobacteria (Aráoz et al. 2010). In 1991, the saxitoxin-containing cyanobacteria *A. circinalis* contaminated over 1,000 km of the Darling River in Australia, causing the deaths of sheep as well as economic losses from inability to use the water (Aráoz et al. 2010).

Recently, cyanobacteria were found to produce a toxic non-protein amino acid, L-beta-N-methylamino-L-alanine (L-BMAA) (Aráoz et al. 2010). This cyanobacterial toxin has been related to neurodegenerative diseases of people via ingestion of contaminated seeds of cycad (sago) palm (*Cycas revoluta*). Moreover, BMAA in the fleshy seed coat of the cycads may be ingested by flying foxes (*Pteropus mariannus*) (Cox et al. 2003). The Chamorro people of the Marianas Islands have at times been exposed by eating cycad flour as well as the flying foxes, which are sometimes considered a delicacy. Other sources of neurotoxic L-BMAA include strains of the cyanobacterium *Nostoc*, which can grow on moist substrates such as soils. Further studies of L-BMAA in cyanobacterial blooms are needed.

The most commonly recognized cyanobacterial hepatotoxins, microcystins and nodularins, are unusual cyclic peptides that are concentrated in the liver via bile acid carriers and act in part by inhibiting protein phosphatases. Microcystins are produced by a number of species of cyanobacteria. Nodularin, produced by *Nodularia*, is a common concern along the southern coast of Finland and has also caused problems on the coast of New Zealand. Acute impacts of microcystins and nodularin often occur within a few to several hours of ingestion and include disruption of hepatocyte structure, damage to sinusoids, severe intrahepatic hemorrhage, and lethal hypovolemic shock, hyperkalemia, and hypoglycemia. Diagnoses are established on the basis of confirming exposure, appropriate clinical manifestations and lesions, analysis of bloom material in water, and assays of digestive tract contents for the toxins. Surviving herbivores may be left with sufficient liver damage as to experience hepatogenous photosensitization or even chronic liver failure (Beasley et al. 1989).

In addition to the risks to domestic animals and wildlife, freshwater cyanobacterial hepatotoxins can also pose risks to public health. Chronic exposure in people occurs mainly via drinking water obtained from contaminated reservoirs, ponds, and stagnant rivers. Concerns include increased risks of liver cancer related in part to tumor promotion by microcystins. Groups of people drinking from surface water tend to have higher liver cancer rates than those drinking from wells. Also, in Brazil, at least 52 kidney patients died after their dialysate solution was made with contaminated water from a reservoir (Azevedo et al. 2002). In that case, cyanobacteria were lysed with an algaecide, and the toxin passed through filters that would have captured cyanobacterial cells.

Protecting water bodies from increasing risks of harmful cyanobacterial blooms requires conservation of nutrients though refined best management practices in fertilization, including careful processing and application of animal wastes; construction, maintenance, and responsible operation of human sewage treatment plants; rehabilitation of hydrologic flows; restoration of plant communities to trap nutrients and protect water from sunlight; and control of emissions of carbon dioxide, methane, and other pollutants that prompt global climate change.

## UNDERMINING ECOSYSTEM STRUCTURE AND FUNCTION

Throughout much of the world, modern agriculture and urbanization have created expanses of inhospitable habitat for many species. In addition to depriving biota of essential ecological niches, such change can alter local and regional climates, facilitating volatilization and runoff of toxic chemicals that have been introduced to the environment intentionally (e.g., pesticides, mining or industrial wastes) or accidentally (e.g., spills, leaks, fires). Wildlife populations not extirpated due to the loss of suitable habitat are vulnerable to such chemical releases, and the toxic effects may interact with stress due to other phenomena, such as weather extremes, infectious disease outbreaks, changes in predation pressure, or increased competition due to changes in community structure. Projected increases in earth surface temperatures due to global climate change (Easterling and Karl 2008) will likely mean greater volatilization of contaminants, changing rates of chemical reactions in the atmosphere, and changes in the dispersal of pollutants driven by wind and precipitation. This could accelerate the transfer of contaminants from warmer environments to polar regions, where cold climates limit both re-volatilization and microbial metabolism, and where long food chains predispose to biomagnification in top predators. Moreover, melting of ice caps may free long-trapped contaminants to cycle back into the environment. This phenomenon has been suggested as the cause of continued elevated DDT concentrations in Adelie penguins (*Pygoscelis adeliae*) in portions of the Antarctic (Geisz et al. 2008).

Drought can stress crops, exposing plant tissues, allowing insect damage, and thus enabling fungal pathogens to gain entry and produce toxins. This process may have accounted for outbreaks of fumonisin poisoning among domestic animals in the United States in 1989 (Ross et al. 1992). Also, warmer temperatures will likely result in more generations of pests per year, increasing damage to crops, which, as noted above, will often carry fungi into plant substrates, potentially resulting in more severe and widespread outbreaks of mycotoxin-induced diseases. Traditional approaches often involved broad applications of non-selective insecticides. Substantial prevention of the entire phenomenon has apparently been achieved in recent years via genetic modification of crops via introduction of genes that code for production of the insecticidal toxins of *Bacillus thurengiensis* (Munkvold and Hellmich 2000). Use of integrated pest management techniques such as pest monitoring, crop rotations, protection of native predatory insects and parasitoids, use of pest-resistant crop varieties, pest pheromones, and mechanical control, and, when necessary, very limited (targeted) spraying of selective pesticides is needed to avoid the continued introduction of large quantities of pesticides into the environment.

At sufficient concentrations, chemicals may impair ecosystem functioning by changing soil chemistry; killing soil arthropods, annelids, and microbes; or even hampering the abiotic geochemical processes involved in nutrient cycling. Soil ecology and productivity can be undermined by pesticide applications and even nutrient additions. Soils without a large complement of organic material and biota are more likely to lose their water retention capacity, become desiccated, erode, and lose natural fertility (Bot and Benites 2005). The health of soils relies upon biodiverse communities of microbes, other decomposers, and an established detritus layer (substituted by crop residue in no-tillage agricultural fields). These interact to enable nutrient cycling (Bot and Benites 2005) as well as biodegradation of plant-derived toxins and manmade chemicals, such as the fungicides azoxystrobin, tebuconazole, and chlorothalonil and the herbicide isoproturon (Bending et al. 2007; Grundmann et al. 2007). Indeed, the capacity of some soil microbes to degrade chemicals has been used to remediate contaminated soils. To have productive soils—for ecological, domestic animal, and human needs—over-fertilization and careless use of pesticides must be avoided.

In our estimation, a key change that is essential to reverse the impacts of shortsighted land-use patterns and global climate change on ecosystems—and related ecotoxic stress—is a new vision of a world in recovery. We have the option at hand to restore the biological, structural, and functional diversity of landscapes and water bodies and thereby promote improved ecosystem functions and their currency, ecosystem services. Rehabilitating highly altered landscapes might include incorporating more perennial crops into crop rotations to reduce the need for annual tillage and thus maintain a healthier soil structure; establishing wide grass buffers and riparian forests to reduce pesticide and fertilizer runoff from fields and lawns, along with reducing wind erosion and the desiccation and heating of soils; local and regional planning to protect water bodies and uplands while providing an extensive system of habitat corridors that connect parks, preserves, and conservation areas; planting native flowering vegetation in proximity to crop fields and gardens to promote a diversity of native pollinators and predatory insects; incorporating ecological functioning into the design of water retention basins and wetland complexes for stormwater retention in ways that support metapopulations of native plants and animals; promoting responsible agricultural water table management; re-meandering of channelized streams and rivers; and incorporating open space, native vegetation, and wetlands into housing developments and commercial properties. These approaches hold great promise in efforts to increase biodiversity and improve ecosystem health.

## CONCLUSIONS

The technologies and processes that have brought us advances in health care, abundant and nutritious foodstuffs, and modern transportation, communication, entertainment, and comfort have also brought contamination of our environment. Although some have suggested that the release of toxicants into the environment is an acceptable cost of "progress," ecological problems related to contaminants are now of such a scale that they are threatening the long-term survival of myriad species, including *Homo sapiens*. This chapter only touches on some of the many impacts of chemical contaminants. Herein, we have endeavored to focus on the complexity of assessment of the impacts of toxicants, rather than the usual approach of considering single toxic effects as if they occur in the absence of other stressors.

Contaminant-induced problems such as global climate change; eutrophication with associated harmful algal blooms and anoxic or "dead" zones; elimination of predatory insects, parasitoids, and pollinators; promotion of thyroid and reproductive endocrine disorders; undermining immune functions; and excessive exposures to pathogens are all entirely preventable—simply through the astute choice, use, and control of chemicals. Ecotoxicology, practiced well in concert with industry and an informed public, can improve the health of individuals, populations, species, and biotic communities. To be most effective, objectivity, humility and an appreciation of the limits of our knowledge are necessary. In many ways smarter choices are needed to enable the recovery of beleaguered native microbes, plants, and animals.

Emerging and re-emerging diseases, developmental and reproductive abnormalities, re-emergence of harmful algal blooms, global air pollution, unprecedented loss of species, and climate change are signals that we are putting undue stress on global ecosystems. As developing nations, many with very dense human populations, race to adopt agricultural, industrial, and resource extraction practices from the developed world without environmental controls, their per capita impact on the local, regional, and global environment may exceed that of many developed nations. This has already become evident in some parts of the world for contaminants such as mercury and other heavy metals, acids, particulates, and $CO_2$.

The tropical, subtropical, and temperate ecosystems, where most of the world's people live, have a tremendous capacity to cleanse themselves through natural detoxification processes. However, with repeated releases of large volumes of inorganic pollutants as well as a range of organic compounds that are highly resistant to biodegradation, we are overwhelming the capacity of resident organisms for comprehensive natural detoxification. Anthropogenic releases of toxicants will not cease in the foreseeable future, but massive reductions in toxicant releases can readily be achieved, such that environmental concentrations will decline. With steady reductions in toxic impairment, native organisms will tend to increase their capacity to survive temporary catastrophic events such as extremes of weather or habitat changes. Reduced exposure to immunotoxic chemicals will allow animals to mount a more effective immune response to highly infectious and opportunistic disease pathogens.

Reducing the release of anthropogenic toxicants will require greater public and scientific scrutiny of chemicals, including pesticides, construction and household chemicals, flame retardants, pharmaceuticals, fertilizers, and fuels—both during development and testing and after regulatory approval. This is especially true of persistent, bioaccumulative synthetic compounds and those that have neurotoxic, estrogenic, anti-estrogenic, anti-androgenic, and thyroid-antagonizing activity. Developed nations need to hold industry accountable to the same environmental controls, whether at home or abroad. Sustainable agricultural practices—those that do not require large inputs of chemicals; do not rely on burning large amounts of fossil fuels, crop residue, pasture, or forests; and allow for strategic rehabilitations of terrestrial and aquatic habitats—should be widely adopted. Efficient manufacturing processes that are more sustainable, located near sources of raw materials and allied industries, produce minimal waste streams, and incorporate cradle-to-cradle product design to ensure efficient recycling as well as upcycling of materials are sorely needed.

More ecologically friendly alternatives to many modern agricultural, industrial, and urban development practices exist. However, institutional hurdles, including the failure to design economic and regulatory incentive and disincentive systems that guarantee that productivity is coupled with good stewardship, have undermined progress around the world. With modern synthetic chemistry (i.e., "green chemistry"), creativity, and a regulatory community that hastens the adoption of responsible production and use of chemicals, humans and the many other organisms that share the Earth can synchronously benefit. Industries and citizens who dismiss ecotoxic impacts of chemical contamination as a necessary fact of life must not undermine getting about the business of ecological rehabilitation. Counteracting serious forms of environmental contamination with chemical pollutants should be the easiest challenge in ecological restoration. We can do this; and we need to get to work on it without delay.

## REFERENCES

Adams, J.S., R.L. Knight, L.C. McEwen, and T.L. George. 1994. Survival and growth of nestling vesper sparrows exposed to experimental food reductions. Condor 96:739–748.

Aráoz, R., J. Molgó, and N.T. de Marsac. 2010. Neurotoxic cyanobacterial toxins. Toxicon 56:813–828.

Austin, J.E., M.J. Anteau, J.S. Barclay, G.S. Boomer, F.C. Rohwer, and S.M. Slattery. 2006. Declining scaup populations: Reassessment of the issues, hypotheses, and research directions. Consensus Report from the Second Scaup Workshop 17–19 January, Bismarck, North Dakota.

Azevedo, S.M.F.O., W.W. Carmichael, E.M. Jochimsen, K.L. Rinehart, S. Lau, G.R. Shaw, and G.K. Eaglesham. 2002. Human intoxication by microcystins during renal dialysis treatment in Caruaru—Brazil. Toxicology 181–182:441–446.

Beasley, V.R. 1993. Ecotoxicology and ecosystem health: roles for veterinarians; goals for the Envirovet Program. J Am Vet Med Assoc 203:617–628.

Beasley, V.R., A.M. Dahlem, W.O. Cook, W.M. Valentine, R.A. Lovell, S.B. Hooser, K.I. Harada, M. Suzuki, and W.W. Carmichael. 1989. Diagnostic and clinically important aspects of cyanobacterial (blue-green algal) toxicoses. J Vet Diag Invest 1:359–365.

Beasley, V.R., S.A. Faeh, B. Wikoff, C. Staehle, J. Eisold, D. Nichols, R. Cole, A.M. Schotthoefer, M. Greenwell,

and L.E. Brown. 2005. Risk factors and declines in northern cricket frogs (Acris crepitans). In Lannoo, M. ed. Amphibian declines: The conservation status of United States species, pp. 75–86. University of California Press, Berkeley.

Bending, G.D, M.S. Rodríguez-Cruz, and S.D. Lincoln. 2007. Fungicide impacts on microbial communities in soils with contrasting management histories. Chemosphere 69:82–88.

Boatman, N.D., N.W. Brickle, J.D. Hart, T.P. Milsom, A.J. Morris, A.W. Murray, K.A. Murray, and P.A. Robertson. 2004. Evidence for the indirect effects of pesticides on farmland birds. Ibis 146:S131–S143.

Bot, A., and J. Benites. 2005. The importance of soil organic matter: Key to drought-resistant soil and sustained food production. FAO Soils Bulletin 80.

Bowers, R.W., and F.A. de Szalay. 2007. Fish predation of zebra mussels attached to Quadrula quadrula (Bivalva: Unionidae) and benthic mollusks in a Great Lakes coastal wetland. Wetland 27:203–208.

Bur, M.T., M.A. Stapanian, G. Bernhardt, and M.W. Turner. 2008. Fall diets of red-breasted merganser (Mergus serrator) and walleye (Sander vitreus) in Sandusky Bay and adjacent waters of western Lake Erie. Am Midl Nat 159:147–161.

Byappanahalli, M.N., and R.L. Whitman. 2009. Clostridium botulinum type E occurs and grows in the alga Cladophora glomerata. Can J Fish Aquat Sci 66: 879–882.

Cook, W.O., V.R. Beasley, A.M. Dahlem, R.A. Lovell, S.B. Hooser, N.B. Mahmood, and W.W. Carmichael. 1989. Consistent inhibition of peripheral cholinesterases by neurotoxins from the freshwater cyanobacterium Anabaena flos-aquae: studies of ducks, swine, mice, and a steer. Environ Toxicol Chem 8:915–922.

Cox, P.A., S.A. Banack, and S.J. Murch. 2003. Biomagnification of cyanobacterial neurotoxins and neurodegenerative disease among Chamorro people of Guam. Proc Nat Acad Sci 100:13380–13383.

Custer, C.M., and T.W. Custer. 2000. Organochlorine and trace element contamination in wintering and migrating diving dusks in the southern Great Lakes, USA, since the zebra mussel invasion. Environ Toxicol Chem 19:2821–2829.

Dietrich, J.P., B.J. Morrison, and J.A. Hoyle. 2006. Alternative ecological pathways in the eastern Lake Ontario food web—round goby in the diet of lake trout. J Great Lakes Res 32:395–400.

Douglas, D.J.T., J.A. Vickery, and T.G. Benton. 2010. Variation in arthropod abundance in barley under varying sowing regimes. Ag Ecosys Environ 135:127–131.

de Swart, R.L., P.S. Ross, L.J. Vedder, H.H. Timmerman, S. Heisterkamp, H. Van Loveren, J.G. Vos, P.J.H. Reijnders, and A.D. Osterhaus. 1994. Impairment of immune function in harbour seals (*Phoca vitulina*) feeding on fish from polluted waters. Ambio 23: 155–159.

de Swart, R.L., P.S. Ross, J.G. Vos, and A.D. Osterhaus. 1996. Impaired immunity in harbour seals (*Phoca vitulina*) exposed to bioaccumulated environmental contaminants: review of a long-term feeding study. Environ Health Perspect 104:823–828.

Diana, S.G., W.J. Resetarits, Jr., D.J. Schaeffer, K.B. Beckmen, and V.R. Beasley. 2000. Effects of atrazine on amphibian growth and survival in artificial aquatic communities. Environ Toxicol Chem 19:2961–2967.

Easterling, D., and T. Karl. 2008. Global warming: Frequently asked questions [Internet]. National Oceanic and Atmospheric Administration, National Climatic Data Center [Last updated Aug. 20, 2008]. Available at: http://www.ncdc.noaa.gov/oa/climate/globalwarming.html#q1

Fernie, K.J., R.B. King, K.G. Drouillard, and K.M. Stanford. 2008. Temporal and spatial patterns of contaminants in Lake Erie watersnakes (*Nerodia sipedon insularum*) before and after the round goby (*Apollonia melanostoma*) invasion. Sci Total Environ 406:344–351.

Geisz, H.N., R.M. Dickhut, M.A. Cochran, W.R. Fraser, and H.W. Ducklow. 2008. Melting glaciers: A probable source of DDT to the Antarctic marine ecosystem. Environ Sci Tech 42:3958–3962.

Geraci, J.R. 1989. Clinical investigation of the 1987–88 mass mortality of bottlenose dolphins along the U.S. central and south Atlantic coast. Final Report to the National Marine Fisheries Service and U.S. Navy, Office of Naval Research and Marine Mammal Commission. Ontario Veterinary College, University of Guelph.

Goldstein, T., J.A.K. Mazet, V.A. Gill, A.M. Doroff, K.A. Burek, and J.A. Hammond. 2009. Phocine distemper virus in northern sea otters in the Pacific Ocean, Alaska, USA. Emerg Infect Dis. 15:925–927.

Grundmann, S., R. Fusz, M. Schmid, M. Laschinger, B. Ruth, R. Schulin, J.C. Munch, and R. Schroll. 2007. Application of microbial hot spots enhances pesticide degradation in soils. Chemosphere 68:511–517.

Hall, A.J., R.J. Law, D.E. Wells, J. Harwood, H.M. Ross, S. Kennedy, C.R. Allchin, L.A. Campbell and P.P. Pomeroy. 1992. Organochlorine levels in common seals (*Phoca vitulina*) which were victims and survivors of the 1988 phocine distemper epizootic. Sci Total Environ 115:145–162.

Härkönen, T., R. Dietz, P. Reijnders, J. Teilmann, K. Harding, A. Hall, S. Brasseur, U. Siebert, S.J. Goodman, P.D. Jepson, T.D. Rasmussen, and P. Thompson. 2006. The 1988 and 2002 phocine distemper virus epidemics in European harbour seals. Dis Aquat Org 68:115–130.

Helgen, J., J.R. McKinnell, and M.C. Gernes. 1998. Investigations of malformed Northern leopard frogs in Minnesota. In Lannoo, M.J. ed. Status and conservation of midwestern amphibians, pp. 288–297. University of Iowa Press, Iowa City.

Hernández, M., I. Robinson, A. Aguilar, L.M. González, L. Felipe López-Jurado, M.I. Reyero, E. Cacho, J. Franco, V. López-Rodas, and E. Costas. 1998. Did algal toxins cause monk seal mortality? Nature 393:28–29.

Hoppe, D.M. 2000. History of Minnesota frog abnormalities: do recent findings represent a new phenomenon? In Kaiser H., G.S. Casper, and N. Bernstein, eds. Investigating amphibian declines: Proceedings of the 1998 Midwest Declining Amphibians Conference, pp. 86–89. Iowa Academy of Science 107, Cedar Falls.

Howe, F.P., R.L. Knight, L.C. McEwen, and T.L. George. 1996. Direct and indirect effects of insecticide applications on growth and survival of nestling passerines. Ecol Appl 6:1314– 1324.

Johnson, P.T.J., and J.M. Chase. 2004. Parasites in the food web: linking amphibian malformations and aquatic eutrophication. Ecol Lett 7:521–526.

Johnson, P.T.J., K.B. Lunde, R.W. Haight, J. Bowerman, and A.R. Blaustein. 2001. *Ribeiroia ondatrae* (Trematoda: Digenea) infection induces severe limb malformations in western toads (*Bufo boreas*). Can J Zool 79: 370–379.

Johnson, P.T.J., K.B. Lunde, E.G. Ritchie, and A.E. Launer. 1999. The effect of trematode infection on amphibian limb development and survivorship. Science 284: 802–804.

Kannan, K., K. Senthilkumar, B.G. Loganathan, S. Takahashi, D.K. Odell, and S. Tanabe. 1996. Elevated accumulation of tributyltin and its breakdown products in bottlenose dolphins (*Tursiops truncatus*) found stranded along the U.S. Atlantic and Gulf Coasts. Environ Sci Technol 311:296–301.

Kannan, K., L. Tao, E. Sinclair, S.D. Pastva, D.J. Jude, and J.P. Giesy. 2005. Perfluorinated compounds in aquatic organisms at various trophic levels in a Great Lakes food chain. Arch Env Contam Toxicol 48:559–566.

Kiesecker, J.M. 2002. Synergism between trematode infection and pesticide exposure: a link to amphibian limb deformities in nature? Proc Nat Acad Sci 99:9900–9904.

King, R.B. 2007. Invasive species, demography, and population recovery of the threatened Lake Erie watersnake. The Richard B. King Laboratory. Northern Illinois University.

Kuehl, D.W., R. Haebler, and C. Potter. 1994. Coplanar PCB and metal residues in dolphins from the U.S. Atlantic coast including Atlantic bottlenose obtained during the 1987/1988 mass mortality. Chemosphere 28:1245–1253.

Kwon, T.D., S.W. Fisher, G.W. Kim, H. Hwang, and J.E. Kim. 2006. Trophic transfer and biotransformation of polychlorinated biphenyls in zebra mussel, round goby, and smallmouth bass in Lake Erie, USA. Environ Toxicol Chem 25:1068–1078.

Lahvis, G.P., R.S. Wells, D.W. Kuehl, J.L. Stewart, H.L. Rhinehart, and C.S. Via. 1995. Decreased lymphocyte responses in free-ranging bottlenose dolphins (Tursiops truncatus) are associated with increased concentrations of PCBs and DDT in peripheral blood. Environ Health Perspect 103:S4.

Levengood, J.M., and V.R. Beasley. 2007. Principles of ecotoxicology. In Gupta, R.C., ed. Veterinary toxicology: basic and clinical principles, pp. 689–708. Academic Press, Amsterdam.

Linzey, D.W., J. Burroughs, L. Hudson, M. Marini, J. Robertson, J.P. Bacon, M. Nagarkatti, and P.S. Nagarkatti. 2003. Role of environmental pollutants on immune function, parasitic infections and limb malformations in marine toads and whistling frogs from Bermuda. Int J Environ Health Res 13:125–148.

Martin, P.A., D.L. Johnson, D.J. Forsyth, and B.D. Hill. 1998. Indirect effects of the pyrethroid insecticide deltamethrin on reproductive success of chestnut-collared longspur. Ecotoxicology 7:89–97.

Martin, P.A., D.L. Johnson, D.J. Forsyth, and B.D. Hill. 2000. Effects of two grasshopper control insecticides on food resources and reproductive success of two species of grassland songbirds. Environ Toxicol Chem 19:2987–2996.

Mazak, E.J., H.J. MacIsaac, M.R. Servos, and R. Hesslein. 1997. Influence of feeding habits on organochlorine contaminant accumulation in waterfowl on the Great Lakes. Ecol Appl 7:1133–1143.

Michigan Sea Grant. 2006. Expansion of quagga mussels in Lake Michigan adds to food web uncertainties. Upwellings Online Edition Vol 29 http://www.miseagrant.umich.edu/upwellings/issues/06june/06june-article2.html

Mitchell, J.S., R.C. Bailey, and R.W. Knapton. 2000. Effects of predation by fish and wintering ducks on dreissenid mussels at Nanticoke, Lake Erie. Ecoscience 7:398–409.

Moreby, S.J., and S.E. Southway. 1999. Influence of autumn applied herbicides on summer and autumn food available to birds in winter wheat fields in southern England. Agric Ecosys Environ 72:285–297.

Morris, A.J., J.D. Wilson, M.J. Whittington, and R.B. Bradbury. 2004. Indirect effects of pesticides on breeding yellowhammer (Emberiza citrinella). Ag Ecosys Environ 106:1–16.

Morrison, H.A., F.A.P.C. Gobas, R. Lazar, R., D.M. Whittle, and G.D. Haffner. 1998. Projected changes to the trophodynamics of PCBs in the western Lake Erie ecosystem attributed to the presence of zebra mussels (Dreissena polymorpha). Environ Sci Tech 32:3862–3867.

Munkvold, G.P., and R.L. Hellmich. 2000. Genetically modified, insect resistant maize: Implications for management of ear and stalk diseases. Plant Heal Progr http://www.plantmanagementnetwork.org/php/2000.asp

Oberholster, P.J., J.G. Myburgh, D. Govender, R. Bengis, and A.M. Botha. 2009. Identification of toxigenic Microcystis strains after incidents of wild animal mortalities in the Kruger National Park, South Africa. Ecotox Environ Saf 724:1177–82.

Pillsbury, R.D., R.L. Lowe, Y.D. Pan, and J.L. Greenwood. 2004. Why filamentous green algae dominated benthic habitats after the zebra mussel invasion in Saginaw Bay, Lake Huron. Cladophora Research and Management in the Great Lakes: Proceedings of a Workshop Held at the Great Lakes WATER Institute, pp. 17–18. University of Wisconsin-Milwaukee.

Ray, W.J., and L.D. Corkum. 1997. Predation of zebra mussels by round gobies, Neogobius melanostomus Environ Biol Fishes 50:267–273.

Reyero, M., E. Cacho, A. Martínez, J. Vázquez, A. Marina, S. Fraga, and J. M. Franco. 1999. Evidence of saxitoxin derivatives as causative agents in the 1997 mass mortality of monk seals in the Cape Blanc Peninsula. Nat Toxins 7:311–315.

Richerson, M. 2009. Dreissena species FAQs, a closer look: Dreissena rostriformis bugensis (quagga mussel) and Dreissena polymorpha (zebra mussel) [Internet]. US Department of the Interior, US Geological Survey; [last modified October 23, 2009]: http://fl.biology.usgs.gov/Nonindigenous_Species/Zebra_mussel_FAQs/

Rodenhouse, N.L., and R.T. Holmes. 1992. Results of experimental and natural food reductions for breeding black-throated blue warblers. Ecology 73:357–372.

Rohr, J.R., A.M. Schotthoefer, T.R. Raffel, H.J. Carrick, N. Halstead, J.T. Hoverman, C.M. Johnson, L.B. Johnson, C. Lieske, M.D. Piwoni, P.K. Schoff, and V.R. Beasley. 2008. Agrochemicals increase trematode infections in a declining amphibian species. Nature 455:1235–1239.

Ross, P.F., L.G. Rice, G.D. Osweiler, P.E. Nelson, J.L. Richard, and T.M. Wilson. 1992. A review and update of animal toxicoses associated with fumonisin-contaminated feeds and production of fumonisins by *Fusarium* isolates. Mycopathologia 117:109–114.

Ross, P.S., R.L. de Swart, P.J. Reijnders, H. Van Loveren, J.G. Vos, and A.D. Osterhaus. 1995. Contaminant-related suppression of delayed-type hypersensitivity and antibody responses in harbor seals fed herring from the Baltic Sea. Environ Health Perspect 103:162–167.

Sessions, S.K., and B. Ballengee. 2010. Explanations for deformed frogs: plenty of research left to do (A response to Skelly and Benard). J Exp Zool (Mol. Dev. Evol.) 314B:1–6.

Schotthoefer, A.M., R.A. Cole, and V.R. Beasley. 2003a. Relationship of tadpole stage to location of echinostome cercariae encystment and the consequences for tadpole survival. J Parasit 89:475–482.

Schotthoefer, A.M., A.V. Koehler, C.U. Meteyer, and R.A. Cole. 2003b. Influence of *Ribeiroia ondatrae* (Trematoda: Digenea) infection on limb development and survival of Northern leopard frogs (*Rana pipiens*): effects of host stage and parasite-exposure level. Can J Zoo 81:1144–1153.

Schotthoefer, A.M., K.M. Labak, and V.R. Beasley. 2007. *Ribeiroia ondatrae* cercariae are consumed by aquatic invertebrate predators. J Parasit 93:1240–1243.

Smith, V.E., J.M. Spurr, J.C. Filkins, and J.J. Jones. 1985. Organochlorine contaminants of wintering ducks foraging on Detroit River sediments. J Great Lakes Res 11:231–246.

Somers, C.M., M.N. Lozer, V.A. Kjoss, and J.S. Quinn. 2003. The invasive round goby (*Neogobius melanostomus*) in the diet of nestling double-crested cormorants (*Phalacrocorax auritus*) in Hamilton Harbour, Lake Ontario. J Great Lakes Res 29:392–399.

Sousa, W.P., and E.D. Grosholz. 1991. The influence of habitat structure on the transmission of parasites. In S.S. Bell, E.D. McCoy, and H.R. Mushinsky,

eds. Habitat structure: The physical arrangement of objects in space, pp. 300–324. Chapman and Hall, London.

Stankovich, W.S. 2004. The interaction of two nuisance species in Lake Michigan: *Cladophora glomerata* and *Dreissena polymorpha*. *Cladophora* Research and Management in the Great Lakes: Proceedings of a Workshop Held at the Great Lakes WATER Institute, pp. 31–36. University of Wisconsin-Milwaukee.

Steinhart, G.B., E.A. Marschall, and R.A. Stein. 2004. High growth rate of young-of-the-year smallmouth bass in Lake Erie: a result of the round goby invasion? J Great Lakes Res 30:381–389.

Taubenberger, J.K., M. Tsai, A.E. Krafft, J.H. Lichy, A.H. Reid, F.Y. Schulman, and T.P. Lipscomb. 1996. Two morbilliviruses implicated in bottlenose dolphin epizootics. Emerg Infect Dis 2:213–216.

Truemper, H.A., T.E. Lauer, T.S. McComish, and R.A. Edgell. 2006. Response of yellow perch diet to a changing forage base in southern Lake Michigan. J Great Lakes Res 32:806–816.

USDA-NRCS. 2000. Conservation buffers to reduce pesticide loss. 21 pp.

van de Bildt, M.W.G., E.J. Vedder, B.E.F. Martina, B.A. Sidi, A.B. Jiddou, M.E. Ould Barham, E. Androukaki, A. Komnenou, H.G.M. Niesters, and A.D.M.E. Osterhaus. 1999. Morbilliviruses in Mediterranean monk seals. Vet Microbiol 69:19–21.

Willman, E.J., J.B. Manchester-Neesvig, C. Agrell, and D.E. Armstrong. 1999. Influence of *ortho*-substitution homolog group on polychlorobiphenyl bioaccumulation factors and fugacity ratios in plankton and zebra mussels (*Dreissena polymorpha*). Environ Toxicol Chem 18:1380–1389.

Zuccarino-Crowe, C. 2007. Botulism in the Great Lakes—Frequently asked questions. U.S. EPA Great Lakes National Program Office. 14 pp.

# 25

## WILDLIFE TOXICOLOGY

### Environmental Contaminants and Their National and International Regulation

## K. Christiana Grim, Anne Fairbrother, and Barnett A. Rattner

Wildlife toxicology is the study of potentially harmful effects of toxic agents in wild animals, focusing on amphibians, reptiles, birds, and mammals. Fish and aquatic invertebrates are not usually included as part of wildlife toxicology since they fall within the field of aquatic toxicology, but collectively both disciplines often provide insight into one another and both are integral parts of ecotoxicology (Hoffman et al. 2003). It entails monitoring, hypothesis testing, forensics, and risk assessment; encompasses molecular through ecosystem responses and various research venues (laboratory, mesocosm, field); and has been shaped by chemical use and misuse, ecological mishaps, and biomedical research. While human toxicology can be traced to ancient Egypt, wildlife toxicology dates back to the late 19th century, when unintentional poisoning of birds from ingestion of lead shot and predator control agents, alkali poisoning, and die-offs from oil spills appeared in the popular and scientific literature (Rattner 2009).

By the 1930s, about 30 pesticides were commonly available (Sheail 1985), and crop-dusting aircraft facilitated their use. With the discovery of dichlorodiphenyltrichloroethane (DDT) in 1939, and related compounds shortly thereafter, use of pesticides increased exponentially. Between 1945 and 1960,

wildlife mortality events were documented following pesticide application in agricultural and forest habitats. Field and laboratory studies described effects on reproduction, survival, tissue residues, and toxicity thresholds in birds. The widespread hazards of spent lead shot, and mercurials from fungicides and industrial activities became apparent.

Scientific observations and the publication of *Silent Spring* (Carson 1962) led to contaminant monitoring programs in the United States, United Kingdom, and Canada during the 1970s. Research highlights of this era included bioaccumulation of organochlorines in food chains, exposure of diverse species (e.g., bats, marine mammals), and avian mortality due to accumulation of lethal pesticide residues. Eggshell thinning and population declines in raptorial and fish-eating birds were attributed to DDT (and its metabolite DDE), and polychlorinated biphenyls (PCB) were detected and linked to reproductive problems in mink. Screening programs were launched to examine chemical toxicity and repellency to birds and mammals.

By the 1980s, heavy metal pollution from mining and smelting activities, agrichemical practices and non-target effects, selenium toxicosis, PCB pollution, die-offs related to anticholinesterase pesticides, and environmental disasters (e.g., Chernobyl, *Exxon Valdez*) were at the forefront of ecotoxicology. Molecular biomarkers, endocrine disruption, population modeling, and studies with amphibians and reptiles dominated the 1990s. With the turn of the century, interests shifted toward pharmaceuticals, flame retardants, and surfactants; toxicogenomics; inter-specific extrapolation of toxicity data; and inter-connections between wildlife toxicology, ecological integrity, and human health (Rattner 2009).

## Mechanisms of Action and Changes Over Time

Chemicals pose risks to organisms due to a combination of factors that are intrinsic to the compound, particular species, and/or from interaction with extrinsic factors (e.g., habitat, ambient temperature), and other chemicals. The myriad physiological, behavioral, and life history characteristics of different organisms result in unique exposure pathways and unusual susceptibilities. For example, physical and chemical properties of some contaminants may cause them to preferentially accumulate in water, sediment, or soil, habitats, or ecosystems, or may cause particular toxicity to certain species. Species and individuals can be affected by contaminants directly and indirectly (see Chapter 24 in this volume). Also, individuals and populations are rarely exposed to single contaminants, and the effects of exposure to mixtures may differ from the effects of a single chemical through additive, synergistic, antagonistic, or delayed effects.

Over the past century, discoveries concerning how chemicals move through the environment and the effects they can have on wildlife and people have changed the types and properties of synthetic chemicals. Within the past decade, chemical-induced climate change and discoveries of pesticides and industrial chemicals in pristine locations have resulted in international regulation to restrict chemical manufacture and use to those with the least risk to human health and the environment. As early as the 1990s, the U.S. Environmental Protection Agency (USEPA) initiated a "Design for the Environment Program" that sought to develop cleaner product processes and reduce pollution. One goal of the program is to "minimize environmental damage and maximize efficiency during the life of the product" (www.epa.gov/dfe/pubs/comp-dic/factsheet/). The program promotes use of nontoxic manufacturing processes with minimal emissions, and recyclable materials that have few harmful effects on the environment. Characteristics of newer replacement compounds include short environmental half-lives, limited potential for bioaccumulation, and mechanisms of action that do not evoke toxic effects in non-target animals.

## MAJOR ENVIRONMENTAL CONTAMINANT CLASSES

The most widely recognized classes of contaminants that have demonstrated effects or pose risk to wildlife and human populations include pesticides, industrial chemicals, fossil and mineral fuels, pharmaceuticals (human and veterinary) and personal care products, metals, and fertilizers (Table 25.1).

## Pesticides

The vast majority of synthetic chemical substances are industrial in nature (80,000 to 100,000 registered). Even though there are only about 1,200 active pesticide compounds, their volume of use is significant. Annually, more than 5 billion pounds of pesticides are used worldwide, with the United States accounting for 20% to 30% of global usage (USEPA 2001). As of 2007, pesticides were detectable in most surface water samples, major aquifers, and fish species in the United States; they sometimes exceeded water-quality benchmarks for aquatic or fish-eating wildlife (Gilliom 2007).

Use of cyclodiene insecticides (e.g., DDT, aldrin, dieldrin, lindane, and chlordecone) became widespread after World War II. The half-lives of these chemicals are on the order of 30 years. They are very hydrophobic and accumulate in lipids within animals and humans. Consequently, these compounds biomagnify in the food chain, resulting in very high exposure to top predators and insectivorous birds. Cyclodienes cause toxicity by inhibiting ATPases, antagonizing neurotransmitters, and depolarizing nerves by interfering with calcium fluxes both along the axon and at the neuronal junctions. The neurological effects are universal across all animal species, resulting in very nonspecific insecticides. Toxic effects can occur in non-neural tissues; for example, inhibition of calcium ATPase in the shell gland of birds

**Table 25.1 Common Chemical Classes Affecting Wildlife and Human Health**

| Chemical Class | Most Common Sub-Classes or Types Found in the Environment | Unifying characteristic |
|---|---|---|
| Pesticides | Herbicides/fungicides/algaecides<br>Insecticides/nematicides/molluscicides<br>Avicides/rodenticides<br>Growth regulators (plant and insect) | Designed to kill, repel, or alter physiological mechanisms in target organisms |
| Industrial Chemicals | Volatiles (e.g., household products including paints, paint strippers, wood preservatives, aerosol sprays, cleansers and disinfectants, moth repellents, air fresheners, stored fuels, automotive products, hobby supplies, dry-cleaned clothing)<br>Semi-volatiles (e.g., industrial plasticizers (phthalates), byproducts of incomplete combustion of fossil fuels (benzo(a)pyrene), dioxins, PCBs, brominated flame retardants, lubricants)<br>Solvents (e.g., acetone, ethanol, hexane, carbon tetrachloride, ether, etc.) and surfactants<br>Nanomaterials<br>Explosives and energetic compounds | The largest class of synthetic chemicals with no definable common characteristics but used in the household, in work areas, and industrial processes |
| Fossil and Mineral Fuels | Oil/petroleum<br>Coal<br>Natural gas<br>Naturally occurring energetic compounds (e.g., perchlorate) | Natural resources that primarily consist of carbon and hydrogen, are burned to produce energy, or are used to develop consumer items (e.g., plastics) |
| Pharmaceuticals (Human and Veterinary) | Hormone agonists/antagonists (e.g., birth control pills, thyroid medications, cholesterol synthesis blockers, both synthetic and natural)<br>Antimicrobials (e.g., antibiotics, antiparasitics, antifungals, antivirals)<br>Analgesics/Neuroleptics/Anesthetics<br>Antidepressants/Antianxiety medications<br>Controlled substances (illicit)<br>Antihypertensives | Designed to be biologically active and are often introduced into the environment at steady rates through sewage treatment plants, concentrated animal feeding operations and widespread biosolid dispersal |
| Personal Care Products | Nutraceuticals<br>Food additives (e.g., caffeine)<br>Cosmetics, fragrances, soaps, and everyday use items | Individual consumer use introduced into the environment at steady rates primarily through sewage and water treatment plants |
| Metals | Heavy and/or inorganic metals<br>Metalloids (e.g., mercury, selenium, arsenic) and organotins | Non-biodegradable, cannot decompose into less harmful components, can biomagnify |
| Fertilizers | Natural (e.g., manure, water-treatment sludge)<br>Inorganic fertilizers | Chemicals used by agro-businesses to improve crop production |

causes eggshell thinning (Lundholm 1987, 1997). Cyclodienes were very effective pesticides precisely because of their persistence (a single application provided long-term effects) and their toxicity to pests of both economic and epidemiological importance. They have been, for example, instrumental in controlling mosquitoes that transmit malaria, a disease that is responsible for infecting 350 million to 500 million people annually (http://www.cdc.gov/malaria/), and DDT still is used inside homes in tropical countries for this purpose. Dieldrin was very effective in controlling cotton pests and was of considerable economic benefit in the southern United States. However, Rachel Carson's book *Silent Spring* raised the alarm about the threats that uncontrolled use of these chemicals posed to wildlife and the environment. The USEPA banned the use of DDT in the United States in 1972, confirming that continued massive use of it posed unacceptable risks to the environment and potential harm to human health (http://www.epa.gov/ history/topics/ddt/01.htm). Dieldrin was banned as an agricultural pesticide in 1974 and from other uses (e.g., termite control) in 1987 (http://www.epa.gov/pbt/ pubs/aldrin.htm). Even prior to the ban on the cyclodiene pesticides, neurotoxic chemicals in the organophosphorus and carbamate classes were available for use as pesticides. These chemicals were developed initially as nerve agents during World War II and gained widespread use as insecticides in the 1950s and 1960s. They principally act by inhibiting acetylcholinesterase at neuromuscular junctions, which also is a universal property among all animal species. However, mammals contain a-esterases that rapidly metabolize these compounds, making them much less toxic to humans than to insects (Mineau 1991). Unfortunately, birds do not have such an enzyme in blood, so they remain more vulnerable to non-target effects of the cholinesterase inhibitors. Some organophosphorus pesticides act through an alternate mechanism causing degeneration of the myelin sheath and inducing a delayed neuropathy. After the ban of DDT, dieldrin, and other cyclodiene pesticides, the cholinesterase inhibitors gained popularity due to their low toxicity to mammals and short half-life in the environment (less than 1 month to a year in soils, depending upon the formulation). They vary widely in regard to toxicity to birds and non-target species, from the highly toxic parathion and methyl parathion (no longer registered for use in the United States) to the relatively nontoxic

malathion, which is still used as a fogging agent for mosquito control. However, many of these chemicals are sufficiently toxic to people, wildlife, and the environment so that they can be used only by a licensed pesticide applicator.

The pyrethroid pesticides, developed in the late 1960s and 1970s in Britain (http://www.rothamsted. bbsrc.ac.uk/notebook/words/pyrethroids.htm), are synthetic analogs of pyrethrin, an insecticidal chemical found in flowers of plants in the genus *Chrysanthemum* (http://extoxnet.orst.edu/pips/pyrethri.htm). They are neurotoxins, opening sodium channels in nerve membranes and causing prolonged depolarization. Because the length of time the sodium channels are open is inversely related to temperature (i.e., shorter time of opening at higher temperatures), pyrethroids are much more toxic to invertebrates than to homeotherms such as birds and mammals (Narahashi et al. 1998). Pyrethroids also are rapidly metabolized in homeotherms and have a short (12 day) half-life in soils (http://npic.orst.edu/factsheets/pyrethrins. pdf). Therefore, these chemicals appear to be more environmentally benign than the organophosphorus or carbamate pesticides, although they are relatively toxic to some aquatic invertebrates, fish and tadpoles. They remain a common household pesticide used to control insects in lawns and in and around house foundations.

Other pesticide classes affect the nervous system in different ways (Brown 2006). Avermectins (abamectin) and phenyl pyrazoles (fipronil) are systemic insecticides used for flea control that block the action of the gamma-aminobutyric acid (GABA) receptor. Because GABA is an inhibitory enzyme, the nerves remain stimulated. Imidacloprid is a neonicotinoid insecticide that mimics the action of acetylcholine, also resulting in prolonged nerve stimulation. It is a more specific agonist in insects, affecting nicotinic acetylcholine receptors, than in birds or mammals, giving it greater target specificity than the older cholinesterase-inhibiting insecticides, although there is some evidence that their use may result in greater impacts on pollinator insects. Several classes of insecticides act through disruption of energy production (uncoupling of oxidative phosphorylation) and require activation within the insects. These include compounds such as hydramethylnon, sulfuryl fluoride, chlorfenapyr, and sulfluramid that may be toxic to birds, have a half-life of approximately 1 year, and can bioaccumulate in the food chain (Brown 2006).

As all animals share similar nerve physiology and energy metabolism, pesticide development has targeted insect-specific physiological mechanisms (Brown 2006). Some insect growth regulators mimic juvenile hormone, causing insects to remain in the juvenile instar and not molt into adults. These are used as mosquito control agents in some locations but are considerably more expensive than conventional pesticides (e.g., malathion, DDT). Chitin synthesis inhibitors block the production of chitin, an important component of the insect exoskeleton. Because insects need to synthesize chitin at each molt, they cannot molt or reach adult stage and eventually die without reproducing. However, some non-target organisms, such as shellfish and other crustaceans, also require chitin and can be adversely affected by these compounds.

Although most pesticides target invertebrates, there are several products directed at vertebrate pests, particularly rodents. Rodenticides include strychnine and zinc phosphide that are systemic poisons for birds and mammals. Anticoagulant baits include coumarins that kill animals after ingesting only a single dose. Indandiones (e.g., diphacinone) usually require several doses to cause death. These chemicals interfere with the synthesis of vitamin K required for post-translational processing of clotting factors and hemostasis. They increase the permeability of capillaries, allowing blood to extravasate into the body cavity. Secondary poisoning of hawks, owls, foxes, and other predators that eat poisoned rodents may be a problem in some areas. The USEPA recently restricted the use of some anticoagulant rodenticides (http://www.epa.gov/oppsrrd1/reregistration/rodenticides/finalrisk-decision.htm). It remains difficult to develop poisons for targeted wildlife pests that are sufficiently specific to reduce non-target mortality.

Herbicides are a group of pesticides that target weeds and invasive plants. They act through mechanisms specific to plants, such as disruption of photosynthesis or production of amino acids not shared by animals. Some kill on contact, whereas others are systemic and are taken up from the soil by the roots or translocated throughout the plant following a foliar spray. They may be used either pre-emergence (to inhibit seed germination and below-ground growth) or post-emergence on the above-ground plant, and can be broad-spectrum or specific to grasses or to broadleaf plants. General classes of herbicides include phenoxy acids (e.g., 2,4-D), triazine herbicides (e.g., atrazine), benzoic acids (dicamba), dinitroanilines (trifluralin), bipyrdyliums (diquat), substituted ureas (linuron), arsenicals, pyridines, and others including the widely used glyphosate and glufosinate. Historically, herbicides were considered to be minimally hazardous to people and wildlife. However, atrazine, which is widely used on corn, has been detected in groundwater throughout the Midwest, and concerns have been raised about potential exposure and toxicity to humans consuming well water. Multiple studies suggest that environmentally realistic concentrations of atrazine may affect amphibian development and reproduction (Giddings et al. 2005), although a recent critical review suggests that atrazine is not adversely affecting frogs (Solomon et al. 2008). In general, there is less concern about the potential environmental harm of herbicides than insecticides, with the possible exception of those applied directly to water for the control of vegetation (e.g., pondweed, duckweed, and *Hydrilla*). In the 1990s technological advances minimized the use of pesticides in agriculture. Some of these practices include biological pest control, crop rotation, mechanical controls and barriers that minimize pest damage, and genetically modified crops that impart disease and pest resistance. These practices can be more labor-intensive and expensive to implement. Genetically engineered crops have the potential to transfer genes to native species that may confer selective advantage and change the dynamics of natural plant communities (Ellstrand 2003). Furthermore, because of patents on insecticidal genes, farmers are prohibited from saving back seeds from one year's crop to start the next, making farming with genetically modified crops more expensive and making farmers more dependent upon agro-companies.

Mechanical means of controlling pests include silica aerogels and diatomaceous earth, which scratch through the waxy protective layer and absorb the protective oils on the insect's cuticle, causing water loss, dehydration, and death. Boric acid, which can be applied topically, also disrupts water balance in insects but has similar consequences in non-target organisms. Biopesticides include viruses and bacteria that are highly specific to certain insect species, as well as pesticidal products incorporated into plants through genetic engineering such as the insecticidal toxin from the bacterium *Bacillus thuringiensis* (http://www.epa.gov/pesticides/biopesticides/). Some biopesticides have been linked to delayed development or decreased growth in some pollinator species—for example, monarch butterfly (*Danaus plexippus*) larvae (Dively et al. 2004)—although it is not clear if there are long-term effects on populations. It is assumed that adverse effects would be less than those associated with exposure to traditional pesticides. Integrated pest management also uses a combination of strategies, with insect growth regulators and pesticides as a last resort.

## Industrial Chemicals

Compared to pesticides, much less is known about the modes of action and effects of industrial chemicals on fish, wildlife, and the environment. Industrial chemicals were not regulated until the late 1970s. At that time nearly 62,000 chemicals were put on the Toxic Substances Control Act (TSCA) Chemical Substance Inventory in the United States without requiring any information on their environmental or human health effects (http://www.epa.gov/lawsregs/laws/tsca.html). There are now over 83,000 chemicals in the TSCA Inventory, some of which have been assessed for safety, primarily through the use of structure–activity relationships (Auer et al. 1990) that compare the structures of new chemicals with those that have already been studied. Remarkably, health and environmental effects data are incomplete for over 2,000 high production volume (HPV) chemicals commonly used in the United States and elsewhere (http://www.epa.gov/HPV/pubs/general/basicinfo.htm), although considerable progress has been made to address data gaps.

The reporting of serious incidents in the media fostered public awareness of the hazards of industrial chemicals to the environment. Improper chemical disposal was highlighted at Love Canal, New York, where drums of chemical waste were dumped into ditches, covered over, and forgotten until people living in the area developed cancer and other symptoms of toxicity during the 1970s; the area was declared a state disaster area in 1978. The fire on the Cuyahoga River in Ohio in 1969 highlighted the consequences of discharging industrial wastes into rivers and other water bodies. Discovery in the 1980s of toxaphene and other organochlorine pesticides in Arctic species (Muir et al. 1988) and PCBs in breast milk

of indigenous people in the region (Dewailly et al. 1989) raised concern about the potential for long-range transport of persistent bioaccumulative chemicals.

Polychlorinated and polybrominated biphenyls (PCBs and PBBs) belong to a class of halogenated chemicals that have been used since the 1940s as flame retardants. Their mode of action is similar to that of dioxins, one of the most toxic chemicals known to humans. The USEPA banned the use of PCBs in 1979 (http://www.epa.gov/history/topics/pcbs/01.htm), and PBBs a few years later following an incident in which cattle feed was accidentally contaminated with PBBs, causing human illness by its passage through milk. Polybrominated diphenyl ethers (PBDEs) were substituted for use as flame retardants, although recently these were discovered to evoke some toxic effects in wildlife and biomagnify in the human food chain (Sjödin et al. 2004). Many are now banned from use in Europe and are being phased out elsewhere.

During the late 1990s, the potential for chemicals to cause reproductive dysfunction in people and wildlife at relatively low environmental concentrations through actions on the endocrine system was publicly highlighted (Colburn 1997). Subsequent environmental studies have demonstrated estrogenic effects in fish from streams contaminated with pulp and paper mill discharges (Mellanen et al. 1999, Tyler et al. 1998). Bisphenol A and phthalates used in plastics and nonylphenols used as surfactants in common detergents have been implicated as endocrine disruptors (Patisaul 2010), leading to their ban in Europe and certain locations in the United States. Regulatory agencies in North America and Europe are now beginning the process of screening chemicals for potential endocrine activity; for example, the USEPA issued a data call in to screen 67 pesticides in October 2009, but at the time of this writing there is no formal registry of endocrine-active chemicals.

Most plastics are high-molecular-weight synthetic polymers with carbon, hydrogen, and some with oxygen, nitrogen, chlorine, and sulfur in the backbone. Because of their strength, durability, and ease of manufacture, they are ubiquitous in modern society and have become a serious environmental issue. Although pure plastics have low toxicity, some contain additives (e.g., phthalates, bisphenol A) that may have endocrine-disruptive properties, among other toxic effects. Also, there are extensive data demonstrating that the plastic debris from finished products is harmful to wildlife through entanglement and by ingestion of litter, possibly mistaken as food items (Derraik 2002).

Nanoparticles, defined as matter that has at least one dimension in the 1- to 100-nm range (U.S. National Nanotechnology Initiative, http://www.nano.gov/), are emerging groups of industrial chemicals that have numerous applications (e.g., sunscreens, electronics, fabric coatings). Engineered nanoparticles are becoming more abundant in the environment, and the risks posed to wildlife and human health could be substantial. Toxicological studies with titanium dioxide have demonstrated effects on the respiratory and immune systems in laboratory rodents. Controlled exposure studies in fish have documented increased peroxidation of neural tissue in brain and glutathione depletion in gill tissue (Oberdörster 2004), cell cycle defects, and neurotoxicity (Smith et al. 2007). Data on potential effects in amphibians, reptiles, and higher wild vertebrates are lacking altogether (Handy et al. 2008).

Recent efforts have resulted in the replacement of commonly used hazardous chemicals with less-toxic alternatives. In the past decade, the concept of "green chemistry" has come to the forefront, entailing the use of chemicals and chemical processes designed to reduce or eliminate environmental impacts by applying contemporary processes that use nontoxic components and reduced waste production (Anastas and Warner 1998). Consumer awareness and choice are key driving elements in this change, which engages the entire supply chain for consumable products.

Some toxic compounds have been removed altogether from consumer use. Contemporary examples include some chlorofluorocarbon refrigerants linked to depletion of the ozone layer; perfluorooctane sulfonate and related perfluorinated surfactants that were detected in chemical plant workers and bioaccumulate in wildlife on a global scale; penta- and octa-brominated diphenyl ether flame retardants due to bioaccumulation in wildlife and humans; and in Europe nonylphenol, due to endocrine-disruptive characteristics. Other compounds are undergoing rigorous testing evaluation to better determine if regulatory action is in order.

## Metals

Metals are known to adversely affect wildlife and humans, especially mercury, lead, cadmium, and selenium. Their mechanisms of action, sources, and distributions in the environment are relatively well understood. Wildlife and humans continue to be exposed to and affected by metals through many sources, including mining operations, atmospheric deposition from industrial processes, and exposure through spent lead shot for hunting game.

The toxic effects of lead have been recognized for centuries, and its biocidal properties date back to ancient Egypt. Its presence in lead pipes, cosmetics, pottery, and wine preparation has been hypothesized to contribute to the fall of the Roman Empire. Lead use increased dramatically during the Industrial Revolution, with adverse effects noted in lead trade workers and miners, and it continued to be widely used in paint and as an anti-knock gasoline additive through much of the 20th century. Lead is toxic to multiple organ systems, inducing anemia, neurological impairment, nephrotoxicity, hypertension, and reproductive and endocrine system toxicity (Goyer 1986). Uses of lead that result in widespread contamination (tetraethyl lead in gasoline, water pipes, solder in food cans, paints, and ammunition for hunting) are being phased out (ATSDR 2007). There is substantial evidence that ingestion by reptiles, birds, and mammals of spent shot and bullets, lost fishing sinkers and tackle, and related lead fragments is accompanied by a range of effects (molecular to behavioral) that historically may have contributed to the population decline of some species (e.g., waterfowl, eagles, condors). Restrictions on the use of lead ammunition for hunting waterfowl (~1986), and to a lesser degree lead fishing tackle (~1987), have been instituted in many countries (Rattner et al. 2008). Safe replacements for lead used in hunting of waterfowl and for fishing tackle have been developed and approved for use, and there is evidence of reduced lead poisoning in waterfowl. Unfortunately, some alternatives to lead have also resulted in environmental problems. For example, fuel oxygenates such as methyl-*tert*-butyl ether (MTBE) were introduced in the 1980s to replace lead and enhance octane ratings in gasoline. This additive results in fuel burning more cleanly in cold weather. However, leaks and spills at pumping stations have significantly contaminated groundwater with MTBE. Human health effects have been difficult to establish unequivocally; nevertheless, reports of nausea, dizziness, and headaches and perhaps an increased potential for carcinogenesis at high exposures (e.g., from drinking contaminated groundwater) resulted in a ban on MTBE by 2004.

Mercury use by humans dates back some 2,000 years (Weiner et al. 2003), and emissions have resulted in its worldwide distribution. The environmental fate of mercury is complex in that both natural and industrial processes release various forms; it may be chemically inter-converted, metabolized by microorganisms into organic and inorganic forms, and demethylated in the liver of higher vertebrates (viz., wildlife) (Rattner et al. 2010). Today, over 50% of the waterways in the United States have fish consumption advisories related to mercury. From an ecotoxicological perspective, in the 1950s and 1960s granivorous birds and small mammals were poisoned by eating agricultural seeds coated with a mercurial fungicide (Borg et al. 1969). At about the same time, poisoning of scavenging and fish-eating birds, and humans, related to the industrial release of methyl mercury was reported in Minamata Bay, Japan (Doi et al. 1984). Pulp mills and chloralkali plants are also sources of mercury release into the environment. Affected wildlife included mink (*Mustela vison*), otter (*Lutra canadensis*), and piscivorous birds (Wobeser and Swift 1976; Wren 1985). Toxicological effects include overt neurotoxicity, reproductive failure, histopathological lesions, and outright mortality. Field and controlled dosing studies have determined tissue and dietary concentrations that cause overt toxicity or reproductive impairment. Although total mercury concentrations in liver and kidney of about 20 µg/g wet weight may be lethal, some studies have documented much greater concentrations in apparently healthy animals (e.g., ringed seal [*Pusa hispida*] and bearded seal [*Erignathus barbatus*], Smith and Armstrong 1975; wandering albatross [*Diomedea exulans*], Thompson and Furness 1989), which may be attributable to demethylation (Scheuhammer et al. 2008). Adverse effects of methylmercury on egg hatchability in birds have been noted at 1 µg/g wet weight or less (Heinz et al. 2006, 2009). Because of demethylation and the accumulation of relatively nontoxic mercury–selenium complexes in biota, toxicological field assessments cannot rely on total mercury concentration alone (Heinz et al. 2006, 2009). Concentrations of mercury in some habitats (e.g., historic mining sites and point sources, acid-affected lakes) continue to be of contemporary concern to wildlife (Wiener et al. 2003; Scheuhammer et al. 2008).

## Pharmaceuticals and Personal Care Products

As a group, pharmaceuticals and personal care products (PPCPs) have, until the past decade, been largely overlooked as active environmental contaminants. The past 100 years has been the most active in terms of drug development, beginning with the discovery of penicillin as an antibiotic in 1938. Controversy over PPCPs has primarily been generated by the release of steroids and antimicrobials into the environment from sewage treatment plants, concentrated animal feeding operations, and through widespread biosolid dispersal (Daughton and Ternes 1999). Environmentally, the most notable developments were oral contraceptives in the 1950s, several classes of antibiotics that have been associated with drug resistance, and drugs for depression and pain management. Both veterinary and human pharmaceuticals pose risks to wildlife since they are designed for their bioactive properties. Personal care products, on the other hand, such as nutraceuticals (i.e., food or food products that provide health or medical benefit), food additives, and fragrances, are not necessarily designed to be biologically active. However, like pharmaceuticals, they pose risks through the continual introduction to the environment through sewage treatment and biosolid dispersal on food crops and fields adjacent to waterways.

Few large-scale environmental catastrophes have been associated with the use of single pharmaceuticals, but the potential for such effects is demonstrated by the use of diclofenac, a nonsteroidal anti-inflammatory drug (NSAID) used extensively by veterinarians for the treatment of inflammation, fever, and pain in domestic livestock. Diclofenac appears to have been the principal cause of a population crash of Old World vultures (Genus *Gyps*) in India, Pakistan, and Nepal (Oaks et al. 2004; Green et al. 2004; Schultz et al. 2004). Vultures unintentionally ingested diclofenac when scavenging carcasses of treated livestock. It was discovered that diclofenac is extremely toxic to *Gyps* vultures ($LD_{50}$ of about 0.1 to 0.2 mg/kg), evoking visceral gout, renal necrosis, and mortality within a few days of exposure (Swan et al. 2006a). This is the best-documented instance of a pharmaceutical resulting in an adverse population-level response in non-target free-ranging wildlife. More recently another commonly used NSAID, ketoprofen, has been discovered to be causing the death of vultures in Asia (Naidoo et al. 2009b). Controlled exposure studies identified meloxicam (Swan et al. 2006b, Swarup et al. 2007) as a safe alternative, and veterinary use of diclofenac in India is being phased out. Regrettably, the sale and widespread use of diclofenac continues in Africa and elsewhere (Naidoo et al. 2009a). Recently, staggering declines of several raptor species, including vultures, have been documented in Africa. Specifically, the decline of *Gyps* spp. populations, especially during the migration season, has been linked to human activities such as poisoning of carcasses (Virani et al. 2010).

## Fossil and Mineral Fuels

The fossil and mineral fuel class includes petroleum hydrocarbons, polycyclic aromatic hydrocarbons (PAHs), coal, natural gas, and other associated chemicals. Many of these are carcinogenic compounds and also contribute to wildlife mortality from petrochemical spills in marine environments (Albers 2003). In some instances, adverse effects can be both toxicological and physical. For example, petroleum contamination reduces the insulating properties of fur and feathers, resulting in hypothermia and death. Risks from spills or burning of fossil fuels occur during exploration, drilling, refinement, transport, storage, and use, such as occurred after the explosion on April 20, 2010, of the Deepwater Horizon oil drilling platform in the Gulf of Mexico (http://www.restore-thegulf.gov/). Coal poses risks to wildlife from mining operations as well as through combustion byproducts. Some coal ash contains high levels of mercury, selenium, and arsenic that have contaminated ponds and wetlands near coal-fired power plants (Hopkins et al. 2000). Energetic compounds in this class, such as perchlorate, can affect functions of the thyroid gland and have contaminated groundwater and drinking water wells (York et al. 2001). Nitroaromatic munitions such as trinitrotoluene (TNT) and hexahyro-1,3,5-trinitro-1,3,4-triazine (RDX) are commonly found on military installations or areas where explosive devices have been detonated. These are moderately toxic to birds and amphibians when incidentally ingested (Talmage et al. 1999).

## REGULATION OF CHEMICALS

### Knowing the Pertinent Regulations and Legislation

Individuals should familiarize themselves with the general regulatory requirements of chemical use for the country in which they are located. This is particularly important for the purchase and use of pesticides, but also is applicable to chemical disposal. Chemicals should be stored properly to reduce the potential for spills or for unintentional exposure of children or animals. Sending chemicals through the mail or by courier, either within or between countries, may be illegal and requires special labeling, packaging, and handling. The Globally Harmonized System of Classification and Labeling of Chemicals (GHS) is an internationally agreed-upon approach for classifying chemicals as nonhazardous, hazardous, or extremely hazardous for shipping. It includes provisions for standardized shipping labels and safety data sheets (http://www.unece.org/trans/danger/publi/ghs/ghs_welcome_e.html). Capacity building and training for chemicals management is available through the UN Institute for Training and Research (UNITAR): http://www.unitar.org/chemicals-and-waste-management-at-unitar.

All chemicals in commerce are required to have a Materials Safety Data Sheet (MSDS) available to the public, which contains information on toxicity, flammability, solubility, vaporization, and other facts important to safe handling. These should be reviewed whenever chemicals are purchased and can be found online for most substances. Table 25.2 provides links to websites with information about occupational health and safety related to chemical use or transport.

### Who Is Responsible for What Regulations?

International treaties, negotiated and enforced through the UN, regulate international commerce of chemicals and contaminated waste. Specific regulations for individual countries can be found through a search of the Internet (suggested search words: environmental legislation [country of interest]) and is summarized for the United States in Fairbrother (2009).

Pesticides are regulated separately from industrial chemicals. Pesticides are tested for efficacy and target

Table 25.2 Worldwide Web-Based Links to Chemical Management Safety and Labeling Programs in Various Countries (Accessed August 2010)

ASEAN Occupational Safety and Health Network (ASEAN OSH-NET)
www.aseanoshnet.net
Asia-Pacific Economic Cooperation (APEC) Chemical Dialogue
www.apec.org/apec/apec_groups/committee_on_trade/chemical_dialogue.html
Chemical Hazards Communication Society
www.chcs.org.uk
Germany: Deutsche Gesellschaft für Technische Zusammenarbeit (GTZ) GmbH
www.gtz.de/en/themen/laendliche-entwicklung/7720.htm
Globally Harmonized System of Classification and Labelling of Chemicals (GHS)—official text
www.osha.europa.eu/en/news/ghs_entered_into_force_01.04022009
Globally Harmonized System of Classification and Labelling of Chemicals (GHS)—all countries' regulations
www.unece.org/trans/danger/publi/ghs/implementation_e.html
Health Canada: GHS
www.hc-sc.gc.ca/ahc-asc/intactiv/ghs-sgh/index-eng.php
ILO Safework: Programme on Safety and Health at Work and the Environment (SAFEWORK)
www.ilo.org/safework/lang—en/index.htm
International Programme on Chemical Safety (IPCS)
www.who.int/ipcs/capacity_building/ghs_statement/en/ *and* www.inchem.org/pages/about.html
Japan Ministry of Environment
www.env.go.jp/chemi/ghs/ (site in Japanese)
New Zealand, Hazardous Substances
www.ermanz.govt.nz
Pan American Health Organization (PAHO)
www.paho.org
Society for Chemical Hazard Communication
www.schc.org/home.php
U.S. Department of Labor, Occupational Health and Safety Agency (OSHA)
www.osha.gov
U.S. Environmental Protection Agency (US EPA)
www.epa.gov
WSSD Global Partnership for Capacity Building to Implement the GHS
www2.unitar.org/cwm/ghs_partnership/index.htm

specificity, and large-scale use (as for agriculture) is restricted to licensed applicators. However, home use of the same products is not regulated except through detailed labeling, although intentional poisoning of fish or wildlife is prohibited (Fairbrother 2009). In the United States, pesticides are regulated by the USEPA under the Federal Insecticide, Fungicide, and Rodenticide Act (FIFRA). The European Union (EU), Japan, Australia, and other industrialized nations have similar pesticide use laws and guidelines

for developing data required for product registration. These guidelines are harmonized through the Organization for Economic Cooperation and Development (OECD).

Other chemicals in commerce are registered prior to use in manufacturing, and are reviewed for their potential to affect the health of factory workers (under industrial hygiene laws), consumers, and the environment. In the United States, this regulation falls under TSCA. Most developed countries maintain similar lists, such as Canada's Domestic Substances List (http://www.ec.gc.ca/substances/ese/eng/DSL/DSLprog.cfm) and the Chinese Inventory of Toxic Substances (http://english.mep.gov.cn/inventory/). The UN Environment Program (UNEP) maintains a database of toxicity and environmental fate information on HPV chemicals (http://www.chem.unep.ch/irptc/sids/OECDSIDS/sidspub.html). The EU regulates chemical use through the Registration, Evaluation, Authorisation and Restriction of Chemicals (REACH) legislation that came into force in 2007. It requires manufacturers and importers of all chemicals and products into the EU to provide information about environmental transport and fate, as well as effects on people and the aquatic environment. HPV chemicals require information on toxicity to terrestrial systems to be authorized for sale. The European Chemicals Agency (EChA) handles the data submissions and hazard assessments on all registered products. The intent is to find substitutions and to phase out the most hazardous substances.

Countries also are responsible for the disposal of hazardous waste within their borders. In the United States, the Resource Conservation and Recovery Act (RCRA) defines what is meant by "hazardous waste" and regulates how it is handled and disposed. The Comprehensive Environmental Response, Compensation, and Liability Act (CERCLA, also known as the Superfund Law) provides for emergency cleanup, assessment, and compensation for improper discharge or disposal of toxic substances. Some countries, such as Germany, require manufacturers to collect and properly dispose of used products containing hazardous materials (e.g., electronics).

Most developed countries have laws against uncontrolled discharge into water bodies of chemicals or other waste (such as raw sewage) (Fairbrother 2009). These laws define what is meant by "clean water" and require the maintenance of a healthy aquatic community. The Clean Water Act in the United States and the Water Framework Directive in the EU are two examples of this type of legislation. It is recognized that clean water for drinking and bathing is essential for public health, and necessary for environmental sustainability. There is no similar legislation for clean soils, although some countries, such as the Netherlands, Canada, and Germany, have clean soil standards that define allowable limits of soil contamination. Clean air is addressed in most countries but only in relation to human health, not in terms of environmental protection, except for ozone (Fairbrother 2009).

## Responsive (Vertical) and Precautionary Regulation

Fairbrother and Fairbrother (2009) describe two schools of thought in environmental legislation. The traditional model is known as a "vertical" approach as it regulates both the manufacturing process and the end use of products. The burden of proof is on the regulators (the government) to identify potential risks. Minimal information is required and the approach assumes that market incentives will result in safe products; damage to human or environmental health will have negative feedback through consumer choices and expensive litigation. Monitoring of products in use is an important aspect of this approach to identify threats after a product reaches the market, at which time regulations are implemented. This approach looks for a balance between environmental protection and support of business and commerce, and is used by the USEPA to regulate chemicals in commerce under TSCA.

The EU takes a more precautionary approach to environmental regulation as required under the European Community Treaty, which stipulates that a high level of environmental protection be based on the precautionary principle as stated in the Rio Declaration on Environment and Development (UNCED 1992). This puts the burden of proof of safety on the manufacturer or seller of goods and assumes new materials are inherently harmful until demonstrated otherwise. The precautionary approach views science as informative rather than decisional as is the case in the vertical approach.

Most countries incorporate precaution into policy decision-making (Fairbrother and Fairbrother 2009). Pesticide legislation, for example, has historically put the burden of proof for safety on the manufacturer since pesticides are, by their very design, toxic to organisms in the environment. The Food, Drug, and Cosmetic Act in the United States that regulates PPCPs also is precautionary, and puts the burden of proof on the manufacturer of the products. Pesticide and pharmaceutical regulations both include cost–benefit analyses, as there are times when the benefits that accrue to human wellbeing may outweigh temporary environmental costs.

## International Approaches

Because pollution knows no boundaries and disputes over transboundary pollution have intensified, the UN has negotiated several international treaties to control trafficking and dumping of chemicals. The earliest recognition of the universal responsibility for marine pollution included the "London Convention and Protocol" that prohibits dumping of waste in the ocean (http://www.imo.org/OurWork/Environment/SpecialProgrammesAndInitiatives/Pages/London Convention.aspx). The Basel Convention of 1993 prohibits industrialized countries from indiscriminate waste disposal in developing countries (http://www.basel.int/). However, the international trade on e-waste (from disposal and dismantling of electronic equipment) is not covered by the Basel Convention and has been recognized as a serious international problem. The Rotterdam Convention of 2004 established codes of conduct and information exchange procedures for chemicals and pesticides in commerce, in recognition of the need to standardize datasets among countries to reduce the financial burden on chemical companies. The Stockholm Convention on Persistent Organic Pollutants (POPs) of 2004 established a process to review all persistent, bioaccumulative, and toxic substances with the intention of banning the most hazardous and finding suitable substitutes for those that are less dangerous (http://chm.pops.int/). The UN Convention on Climate Change, as embodied in the Kyoto Protocol (adopted in 1997), was the first international effort to reduce greenhouse gas emissions (http://unfccc.int/kyoto_protocol/items/2830.php). The UN Conference on Climate Change in Copenhagen in 2009 was expected to provide the framework for extension and intensification of the targets for reducing emissions that were set in Kyoto. It concluded with a nonbinding agreement with nations promising to meet future pollution reduction targets and with $30 billion over three years pledged to aid developing nations reduce pollution while still growing their economies.

## Developing World Versus Developed World

In 2006, an international conference sponsored by the UN adopted the Strategic Approach to International Chemicals Management (SAICM). The purpose of SAICM is to ensure that the goal agreed to at the 2002 Johannesburg World Summit on Sustainable Development is met, such that by the year 2020, chemicals are produced and used in ways that minimize significant adverse impacts on the environment and human health. There are five official SAICM regional focal points: Argentina, Japan, Nigeria, Romania, and the United Kingdom. SAICM identified the critical emerging issues in chemical management, particularly for developing countries, to be lead in paint, chemicals in products, commerce in e-waste, nanotechnology, and perfluorinated hydrocarbons (http://www.ipen.org/ipenweb/firstlevel/saicm.html).

## WHAT SHOULD YOU DO IN THE CASE OF ENVIRONMENTAL CONTAMINATION?

In the United States, chemical or petroleum spills are managed by the National Response Center (NRC). Observation of a chemical or petroleum spill or any other indication that a hazardous substance may be causing illness or death to plants, fish, or wildlife should be reported immediately to the NRC (http://www.nrc.uscg.mil/nrchp.html) or appropriate authorities in other countries. These can be located on the Internet using the search term "environmental protection agency [country]." Scientists performing field investigations are directed to Sheffield et al. (2005), who describe the procedures taken if animals (including fish) are found dead and poisoning is suspected. Several large databases exist that contain data on chemical characteristics, toxicity and effects in wildlife species, and fate of chemicals in the environment that are useful for contaminant investigations. The "Whole Wildlife Toxicology Catalog" (http://www.pwrc.usgs.gov/wwtc/) provides a portal to many of the databases that are Web-accessible.

## REDUCING ANTHROPOGENIC EFFECTS ON THE ENVIRONMENT

Consumer choices to buy and use less-toxic products can drive changes in the manufacturing and use of chemicals in the household and workplace.

These choices have significant impacts on the release of chemicals into the environment, potentially before governmental regulations can be set into place. Ultimately, common sense and knowledge about the resources used to develop a household and/or occupational product can be the most effective means of choosing among options—for example, taking into account the energy consumed during manufacture and shipping, the ingredients in the product and their sources, options available for disposal, and whether its use is really necessary. Participation in community decisions and thoughtful personal choices can change the impetus for the discipline of wildlife toxicology from response to environmental catastrophe to an approach based on sound science and responsible foresight.

## ACKNOWLEDGMENT

We would like to acknowledge Sarah E. Warner for her helpful suggestions and expert editorial advice.

## REFERENCES

Agency for Toxic Substance & Disease Registry (ATSDR). 2007. Toxicological profile for lead. US Department of Health and Human Services, Public Health Service, ATSDR. http://www.atsdr.cdc.gov/toxprofiles/tp13.html.

Albers, P.H. 2003. Petroleum and individual polycyclic aromatic hydrocarbons. *In* D.J. Hoffman, B.A. Rattner, G.A. Burton Jr., and J. Cairns, Jr., eds. Handbook of ecotoxicology, 2nd ed., pp. 341–371. Lewis Publishing Inc., Boca Raton, FL.

Anastas, P.T., and J.C. Warner. 1998. Green chemistry: theory and practice. Oxford University Press, New York.

Auer, C.M., J.V. Nabholz, and K.P. Baetcke. 1990. Mode of action and the assessment of chemical hazards in the presence of limited data: use of structure-activity relationships (SAR) under TSCA, Section 5. Environ Health Perspect 87:183–197.

Borg, K., H. Wanntorp, K. Erne, and E. Hanko. 1969. Alkyl mercury poisoning in terrestrial Swedish wildlife. Viltrevy 6:301–379.

Brown, A.E. 2006. Mode of action of structural pest control chemicals. Pesticide Information Leaflet No. 41, 8 pp. Department of Entomology, Maryland Cooperative Extension, College Park, MD. http://www.entmclasses.umd.edu/peap/leaflets/PIL41.pdf.

Carson, R. 1962. Silent spring. Houghton Mifflin, Boston, MA.

Colburn, T., D. Dumanoski, and J. P. Myers. 1997. Our solen future. Penguin Books, New York

Daughton, C., and T. Ternes. 1999. Pharmaceuticals and personal care products in the environment: Agents of subtle change? Environ Health Perspect 107:907–938.

Derraik, J.G.B. 2002. The pollution of the marine environment by plastic debris. a review. Mar Poll Bull 44:842–852.

Dewailly, E., A. Nantel, J.P. Weber, and F. Meyer. 1989. High levels of PCBs in breast milk of Inuit women from arctic Quebec. Bull Environ Contam Toxicol 43:641–646.

Dively, G.P., R. Rose, M.K. Dears, R.L. Hellmich, D.E. Stanley-Horn, D.D. Calvin, J.M. Russo, and P.L. Anderson. 2004. Effects of monarch butterfly larvae (Lepidoptera: Danaidae) after continuous exposure to CrylAb-expressing corn during anthesis. Environ Entomol 33:1116–1125.

Doi, R., H. Ohno, and M. Harada. 1984. Mercury in feathers of wild birds from the mercury-polluted area along the shore of the Shiranui Sea. Sci Total Environ 40:155–167.

Ellstrand, N.C. 2003. Dangerous liaisons? When cultivated plants mate with wild relatives, pp. 264. Johns Hopkins University Press, Baltimore, MD.

Fairbrother, A. and J.R. Fairbrother. 2009. Are environmental regulations keeping up with innovation? A case study of the nanotechnology industry. Ecotoxicol Environ Safe 72:1327–1330.

Fairbrother, A. 2009. Federal environmental legislation in the U.S. for protection of wildlife and regulation of environmental contamination. Ecotoxicology 18:784–790.

Giddings, J.M., T.A. Anderson, L.W. Hall, A.J. Hosmer, R.J. Kendall, R.P. Richards, K.R. Solomon, and W.M. Williams. 2005. Atrazine in North American Surface Waters: A Probabilistic Aquatic Ecological Risk Assessment, 432 pp. SETAC Press, Pensacola, Florida.

Gilliom, R.J. 2007. Pesticides in U.S. streams and groundwater. Environ Sci Technol 41:3408–3414.

Goyer, R.A. 1995. Toxic effects of metals. *In* C.D. Klaassen, M.O. Amdur, and J. Doull, eds. Cassarett and Doull's Toxicology: the basic science of poisons, 5th ed, pp. 691–736. McGraw-Hill, New York.

Green, R.E., I. Newton, S. Shultz, A.A. Cunningham, M. Gilbert, D.J. Pain, and V. Prakash. 2004. Diclofenac poisoning as a cause of vulture population declines across the Indian subcontinent. J Appl Ecol 41: 793–800.

Handy, R.D., R. Owen, and E. Valsami-Jones. 2008. The ecotoxicology of nanoparticles and nanomaterials: current status, knowledge gaps, challenges, and future needs. Ecotoxicology 17:315–325.

Heinz, G.H., D.J. Hoffman, S.L. Kondrad, and C.A. Erwin. 2006. Factors affecting the toxicity of methylmercury injected into eggs. Arch Environ Contam Toxicol 50:264–279.

Heinz, G.H., D.J. Hoffman, J.D. Klimstra, K.R. Stebbins, S.L. Kondrad, and C.A. Erwin. 2009. Species differences in the sensitivity of avian embryos to methylmercury. Arch Environ Contam Toxicol 56:129–138.

Hoffman, D.J., B.A. Rattner, G.A. Burton Jr., and J. Cairns Jr. 2003. Handbook of Ecotoxicology 2nd ed. Lewis Publishers, Boca Raton, Florida.

Hopkins, W.A., J. Congdon, and J.K. Ray. 2000. Incidence and impact of axial malformations in larval bullfrogs (Rana catebeliana) developing in sites polluted by a coal-burning power plant. Environ Toxicol Chem 19: 862–868.

Lundholm, C.E. 1987. Thinning of eggshell in birds by DDE: mode of action on the eggshell gland. Comp Biochem Physiol 88C:1–22.

Lundholm, C.E. 1997. DDE-induced eggshell thinning in birds: effects of p,p'-DDE on the calcium and prostaglandin metabolism of the eggshell gland. Comp Biochem Physiol 118C:113–128.

Mellanen, P., M. Soimasuo, B. Holmbom, A. Oikari, and R. Santti. 1999. Expression of the vitellogenin gene in the liver of juvenile whitefish (Coregonus lavaretus L. s.l.) exposed to effluents from pulp and paper mills. Ecotoxicol Environ Safe 43:133–137.

Mineau, P., ed. 1991. Cholinesterase-inhibiting insecticides: their impact on wildlife and the environment. Elsevier, Amsterdam.

Muir, D.C.G., R.J. Norstrom, and M. Simon. 1988. Organochlorine contaminants in arctic marine food chains: accumulation of specific polychlorinated biphenyls and chlordane-related compounds. Environ Sci Technol 22:1071–1079.

Naidoo, V., K. Wolter, R. Cuthbert, and N. Duncan. 2009a. Veterinary diclofenac threatens Africa's endangered vulture species. Regul Toxicol Pharm 53:205–208.

Naidoo, V., K. Wolter, D. Cromarty, M. Diekmann, N. Duncan, A. Meharg, M. Taggart, L. Venter, and R. Cuthbert. 2009b. Toxicity of non-steroidal anti-inflammatory drugs to Gyps vultures: a new threat from ketoprofen. Biol Lett DOI: 10.1098/rsbl.2009.0818.

Narahashi, T., K.S. Ginsburg, K. Nagata, J.H. Song, and H. Tatebayashi. 1998. Ion channels as targets for insecticides. Neurotoxicology 19:581–590.

Oaks, J.L., M. Gilbert, M.Z. Virani, R.T. Watson, C.U. Meteyer, B.A. Rideout, H.L. Shivaprasad, S. Ahmed, M.J.I. Chaudhry, M. Arshad, S. Mahmood, A. Ali, and A.A. Khan. 2004. Diclofenac residues as the cause of vulture population decline in Pakistan. Nature 427:630–633.

Oberdörster, E. 2004. Manufactured nanomaterials (Fullerenes, C60) induce oxidative stress in the brain of juvenile largemouth bass. Environ Health Perspect 112:1058–1062.

Patisaul, H. 2010. Assessing risks from bisphenol A. Am Sci 98:30–39.

Rattner, B.A., J.C. Franson, S.R. Sheffield, C.I. Goddard, N.J. Leonard, D. Stang, and P.J. Wingate. 2008. Sources and implications of lead ammunition and fishing tackle on natural resources. The Wildlife Society Technical Review 08-01. Bethesda, Maryland.

Rattner, B.A. 2009. History of wildlife toxicology. Ecotoxicology 18:773–783.

Rattner, B.A., A.M. Scheuhammer, and J.E. Elliott. 2010. History of wildlife toxicology and the interpretation of contaminant concentrations in tissues. In W.N. Beyer and J.P. Meador, eds. Environmental contaminants in biota: interpreting tissue concentrations, pp. 9–46. Taylor and Francis, Boca Raton, Florida.

Scheuhammer, A.M., N. Basu, N.M. Burgess, J.E. Elliott, G.D. Campbell, M. Wayland, L. Champoux, and J. Rodrigue. 2008. Relationships among mercury, selenium, and neurochemical parameters in common loons (Gavia immer) and bald eagles (Haliaeetus leucocephalus). Ecotoxicology 17:93–101.

Schultz, S., H.S. Baral, S. Charman, A.A. Cunningham, D. Das, G.R. Ghalsasi, M.S. Goudar, R.E. Green, A. Jones, P. Nighot, D.J. Pain, and V. Prakash. 2004. Diclofenac poisoning is widespread in declining vulture populations across the Indian subcontinent. P Roy Soc B 271:S458–S460.

Sheffield, S.R., J.P. Sullivan, and E.F. Hill. 2005. Identifying and handling contaminant-related wildlife mortality/morbidity. In C.E. Braun, ed. Techniques for wildlife investigations and management, pp. 213–238. The Wildlife Society. Bethesda, Maryland.

Sheail, J. 1985. Pesticides and nation conservation: the British experience, 1950–1975. Clarendon Press, Oxford, United Kingdom.

Sjödin, A., R.S. Jones, J.F. Focant, C. Lapeza, R.Y. Wang, E.E. McGahee 3rd, Y. Zhang, W.E. Turner, B. Slazyk, L.L. Needham, and D.G. Patterson Jr. 2004. Retrospective time-trend study of polybrominated diphenyl ether and polybrominated and polychlorinated biphenyl levels in human serum from the United States. Environ Health Perspect 112: 654–658.

Smith, C.J., B.J. Shaw, and R.D. Handy. 2007. Toxicity of single walled carbon nanotubes to rainbow trout (*Oncorhynchus mykiss*): Respiratory toxicity, organ pathologies, and other physiological effects. Aquatic Toxicol 82:92–109.

Smith, T.G., and F.A.J. Armstrong. 1975. Mercury in seals, terrestrial carnivores, and principal food items of the Inuit from Holman, N.W.T. J Fish Res Board Can 32:795–801.

Solomon, K.R., J.A. Carr, L.H. DuPreez, J.P. Giesy, R.J. Kendall, E.E. Smith, and G.J. Van Der Kraak. 2008. Effects of atrazine on fish, amphibians, and aquatic reptiles: A critical review. Crit Rev Toxicol 38:721–772.

Swan, G.E., R. Cuthbert, M. Quevedo, R.E. Green, D.J. Pain, P. Bartels, A.A. Cunningham, N. Duncan, A.A. Meharg, J.L. Oaks, J. Parry-Jones, S. Shultz, M.A. Taggart, G. Verdoorn, and K. Wolter. 2006a. Toxicity of diclofenac to *Gyps* vultures. Biol Lett 2:279–282.

Swan, G., V. Naidoo, R. Cuthbert, R.E. Green, D.J. Pain, D. Swarup, V. Prakash, M. Taggart, L. Bekker, D. Das, J. Diekmann, M. Diekmann, E. Killian, A. Meharg, R.C. Patra, M. Saini, and K. Wolter. 2006b. Removing the threat of diclofenac to critically endangered Asian vultures. PLoS Biol 4:e66, 1–8.

Swarup, D., R.C. Patra, V. Prakash, R. Cuthbert, D. Das, P. Avari, D.J. Pain, R.E. Green, A.K. Sharma, M. Saini, D. Das, and M. Taggart. 2007. Safety of meloxicam to critically endangered *Gyps* vultures and other scavenging birds in India. Anim Conserv 10:192–198.

Talmage, S.S., D.M. Opresko, C.J. Maxwell, C.J. Welsh, F.M. Cretella, P.H. Reno, and F.B. Daniel. 1999. Nitroaromatic munition compounds: environmental effects and screening values. Rev Environ Contam Toxicol 161:1–156.

Thompson, D.R., and R.W. Furness. 1989. The chemical form of mercury stored in South Atlantic seabirds. Environ Poll 60:305–317.

Tyler, C.R., S. Jobling, and J.P. Sumpter. 1998. Endocrine disruption in wildlife: a critical review of the evidence. Crit Rev Toxicol 28:319–361.

United Nations Conference on Environment and Development (UNCED). 1992. Rio Declaration on Environment and Development, Article 15, United Nations. No. E.73.II.A.14 and corrigendum, chap.1: http://sedac.ciesin.org/entri/texts/rio.declaration.1992.html

United States Environmental Protection Agency (USEPA). 2001. 2000-2001 Pesticide Market Estimates: Historical Data: http://www.epa.gov/oppbead1/pestsales/.

Virani, M. Z., C. Kendall, P. Njoroge, and S. Thomsett. 2010. Major declines in the abundance of vultures and other scavenging raptors in and around the Masai Mara ecosystem, Kenya. Biol Conserv DOI: 10.1016/j.biocon.2010.10.024

Wiener, J.G., D.P. Krabbenhoft, G.H. Heinz, and A.M. Scheuhammer. 2003. Ecotoxicology of mercury. *In* B.A. Rattner, G.A. Burton, and J. Cairns, eds. Handbook of ecotoxicology, 2nd ed., pp. 409–463. CRC Press, Boca Raton, Florida.

Wobeser, G., and M. Swift. 1976. Mercury poisoning in a wild mink. J Wildl Dis 12:335–340.

Wren, C.D. 1985. Probable case of mercury poisoning in a wild otter, *Lutra canadensis*, in northwestern Ontario. Can Field Nat 99:112–114.

York, G.R., W.R. Brown, M.F. Girard, and J.S. Dollardhide. 2001. Oral (drinking water) developmental toxicity study of ammonium perchlorate in New Zealand white rabbits. Internat J Toxicol 20:199–205.

# 26

## MARINE BIOTOXINS

### Emergence of Harmful Algal Blooms as Health Threats to Marine Wildlife

Spencer E. Fire and Frances M. Van Dolah

Harmful algal blooms (HABs) affect aquatic ecosystems around the world, adversely affecting marine animal and human health, coastal ecosystem integrity, and economies that depend on coastal resources. Shellfish poisoning events involving humans who had ingested bivalves contaminated with HAB toxins primarily drove early scientific and social interest in HABs. More recently, research efforts have shown that HABs are often temporally and spatially correlated with the occurrence of acute morbidity or mortality of marine animals (Landsberg et al. 2005), and to date at least four classes of algal toxins have been associated with such events. Although fish, seabirds, and many other groups of marine wildlife are affected, these mortality events frequently involve marine mammals, and as such this chapter will focus primarily on the latter. In addition, since marine mammals are important sentinel species that act as barometers of ocean health and demonstrate the link between ocean and human health, the importance placed on these species in this context is warranted (Aguirre and Tabor 2004; Tabor and Aguirre 2004; Wells et al. 2004; Bossart 2006).

The frequency of associated marine mammal mortality events and HABs appears to have increased in recent years. This may be a reflection of several factors, including (1) the increased scientific and popular attention given to marine mammal mortality events, (2) an increase in resources and observer effort dedicated to detection and study of HAB species in coastal waters, (3) an improved ability to detect algal toxins in marine mammal tissues and fluids, and (4) an apparent increase in both the frequency and the geographic distribution of HABs that has been documented over the past quarter-century (Van Dolah 2000; Hallegraeff 2003; Sellner et al. 2003). Exposure of marine mammals to algal toxins occurs via food-web transfer or directly through respiratory exposure. The susceptibility of marine mammals to algal toxins is therefore dependent not only upon the occurrence of toxin-producing algae within their habitat but, in the case of food-web transfer, on the co-occurrence of prey species at the time of a HAB to serve as vectors to higher trophic levels. Thus, management of the impacts of algal toxins on marine mammals requires an understanding of the causes and consequences of harmful algal blooms, the reasons for their apparent increase, and their dependence on large-scale oceanographic and climate changes as well as their local and regional influences.

## OVERVIEW OF ALGAL TOXINS

The harmful effects of most HABs result from production of natural toxins that disrupt normal physiological function in exposed organisms (Landsberg 2002). The origins of these toxins are single-celled microalgae that, in response to favorable conditions in their environment, proliferate and/or aggregate to form dense concentrations of cells, called "blooms." In many cases, toxic species are normally occurring members of the phytoplankton community and are present in low concentrations with no evident environmental health impacts; thus, toxic effects are generally dependent on their presence in higher-than-normal cell concentrations. Toxin-producing marine algae are primarily members of two groups, the dinoflagellates and diatoms. Many species within these groups produce natural compounds that have potent biological effects on other organisms. However, less than 5% (less than 100 species) of all known dinoflagellate species and less than 25 species of diatoms are known to produce compounds that are toxic to mammals. Many of these compounds are potent neurotoxins that target ion channels or components of the cell-signaling pathways that interact with these channels. Although the ecological and physiological reasons why such toxins are produced are not fully understood, they often provide an advantage over competing algal species or function as anti-predation mechanisms targeting zooplankton or small herbivores (Rue and Bruland 2001; Teegarden et al. 2008). Their impacts on higher trophic level species, such as marine mammals or humans, may thus be incidental.

Four major classes of algal toxins have been well studied worldwide (Fig. 26.1) because of their toxicity to wildlife or humans through seafood consumption: saxitoxins, brevetoxins, domoic acid, and ciguatoxins. These compounds are neurotoxins that are direct causes of, or have been associated with, marine mammal mortality events, generally as a result of dietary exposure from naturally contaminated items in the food web. In addition to these comparatively well-studied algal toxins, several novel algal toxins with adverse human health effects have been identified over the past decade, and these should be considered when investigating marine mammal mortality events with no obvious cause. These include diarrheic shellfish poisoning toxins, azaspiracid, yessotoxins, and spirolides. The following sections provide a brief review of each of these toxin classes and their documented impacts on marine mammals.

## MARINE MAMMAL MORBIDITY AND MORTALITY EVENTS ASSOCIATED WITH ALGAL TOXINS

### Saxitoxins

Saxitoxin (STX) and its derivatives form a suite of more than 21 water-soluble, tetrahydropurine neurotoxins. These toxin congeners are produced in varying combinations by marine dinoflagellates in three genera (*Alexandrium*, *Gymnodinium*, and *Pyrodinium*) and by several species of freshwater cyanobacteria (Landsberg 2002). In humans, STXs cause the clinical illness known as paralytic shellfish poisoning (PSP), with symptoms that include tingling and numbness of the perioral area and extremities, loss of motor control, and death by respiratory paralysis. STXs bind to site 1 of the voltage-gated sodium channel and thereby block neurotransmission (Levin 1992). However, STX is rapidly cleared from the blood (less than 24 hours in humans), so victims generally survive if they are put on life support. In the United States, STX historically posed a threat primarily on the northeast and west coasts, but it has recently been found in the Indian River Lagoon, Florida, where its occurrence is now persistent (Landsberg et al. 2002, 2006; Abbott et al. 2009).

STXs were first implicated in marine mammal deaths during a humpback whale (*Megaptera novaeangliae*) mortality event in Massachusetts between November 1987 and January 1988 (Anderson and White 1989; Geraci et al. 1989). Although baleen whales had not previously been reported to mass-strand, the unusually high frequency of strandings (14 whales within 5 weeks) in this event involved whales of robust body condition with stomachs full of undigested Atlantic mackerel (*Scomber scombrus*). In addition, since whales were observed exhibiting normal behavior 90 minutes before being found dead, death appeared to have occurred quickly, and an acutely toxic substance was suspected as the cause of death (Geraci et al. 1989). In this region, the STX-producing dinoflagellate *Alexandrium tamarense* forms blooms annually (Anderson 1997), and therefore STX was

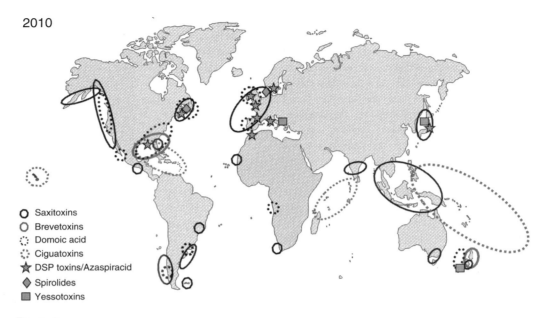

**Figure 26.1:**
Distribution of major classes of HAB toxins with known effects (open circles) or potential adverse effects (shapes) on marine wildlife. (Adapted from Van Dolah 2005.)

investigated as a potential causative agent. STX-like activity was detected by mouse bioassay in whale stomach contents, liver, and kidney, although the presence of toxin could not be confirmed by other analytical methods. Although identification of STX as the causative agent in these whale deaths remains circumstantial, detection of STX in planktivorous mackerel caught in local waters during the same timeframe suggested toxic exposure via the food web. Based on the STX concentration detected in the mackerel, it was estimated that a whale consuming 4% of its body weight in these mackerel daily would have ingested 3.2 μg STX/kg body weight (Geraci 1989). By comparison, the lethal dose of STX in an adult human is estimated at 6 to 24 μg/kg (Levin 1992). However, physiological differences that would make humpback whales more susceptible to the toxic effects of STX include (1) a large proportion of blubber (30% of body mass) into which the water-soluble STX would not partition, leaving these toxins more concentrated in target tissues; (2) the "marine mammal diving response," which shunts blood away from organs that function in detoxification, further concentrating the neurotoxin in sensitive organs such as the heart and brain; and (3) the sensitivity of the

cetacean respiratory system to anesthetic agents (Kooyman 1985; Geraci 1989).

STX exposure was also a likely factor contributing to a mass mortality of Mediterranean monk seals (*Monachus monachus*) occurring in northwest Africa during May and June 1997. In this event, over 100 animals died along the coast of Mauritania, representing more than 70% of the local population and approximately 33% of the world population of this endangered species (Osterhaus et al. 1997; Harwood 1998; Hernandez et al. 1998). Morbilliviruses, which had previously caused mass mortalities of other marine mammal species (Gulland and Hall 2007) were detected in the monk seal carcasses and therefore were identified as a likely causative agent in this event. However, unlike previous morbillivirus-associated events, the seals were in good nutritional state, appeared to experience a rapid death, and exhibited symptoms consistent with STX exposure. In addition, abundant concentrations of three toxic dinoflagellate species were identified in waters near the seal colony, and fish collected from seal feeding grounds as well as seal liver samples were positive for STX (Hernandez et al. 1998). As with the Massachusetts humpback whales, a limited understanding of STX effects on marine mammals prevented confirmation of STX as the cause of death, although both STX and morbillivirus were likely contributors (Harwood 1998). In any case, mass mortalities from PSP toxins may have far-reaching impacts on the population biology of long-lived mammals such as the Mediterranean monk seal (Forcada et al. 1999), considering that this event reduced the breeding population to fewer than 77 individuals and may therefore have reduced the population's genetic variability and ultimately compromised the survival of the species.

In addition to mortalities, STX exposure may also play a less obvious role in marine mammal health. Evidence of trophic transfer of STX from blooms of the toxic dinoflagellate *Alexandrium fundyense* to the North Atlantic right whale (*Eubalaena glacialis*) has raised speculation regarding its potential role in the unexplained decrease in reproduction rate of this endangered whale species. The North Atlantic right whale population currently consists of fewer than 350 individuals and appears to be declining (Kraus et al. 2001; Waring et al. 2009). Coastal waters of New England and the Bay of Fundy, which are major feeding grounds for these whales, experience *A. fundyense* blooms, associated STX production, and shellfish closures almost annually. The copepod *Calanus finmarchicus*, which dominates the right whale's diet, grazes on toxic *A. fundyense* and has been found to contain high concentrations of STX (Durbin et al. 2002). An estimation of STX accumulation in right whales during such blooms places their exposure levels between 4.73 and 9.65 µg STX/kg per day (Durbin et al. 2002), a value similar to the estimated lethal dose in humans (Levin 1992). STX detected in right whale feces from the area reached as high as 0.5 µg/g (Doucette et al. 2006), though the significance of this level of STX is not yet clear with respect to the level of STX circulating in the blood, or with regard to its effects on behavior and physiology. Thus it is possible that sublethal exposure to STX may affect whale behavior, leading to reduced feeding rate and fitness and ultimately reduced calving rates (Durbin et al. 2002).

A bottlenose dolphin (*Tursiops truncatus*) mortality involving 29 strandings occurring in Florida's Indian River Lagoon in June and July 2001 was also suspected to involve STX exposure. Dolphins recovered during this period were emaciated and displayed significant skin lesions, suggesting that multiple factors were associated with their poor health status (Bossart et al. 2003). However, a subsequent outbreak of seafood poisoning cases in humans in early 2002 led to the detection of STX in Indian River Lagoon puffer fish (*Sphoeroides* spp.) and the discovery of STX production by the dinoflagellate *Pyrodinium bahamense* var. *bahamense* in Indian River Lagoon waters (Landsberg 2002; Quilliam et al. 2002). Since neither of these organisms had previously been known to be associated with STX production or accumulation in the Indian River Lagoon, dolphins from the 2001 mortality event were re-examined for evidence of STX exposure. At least two dolphins examined had puffer fish in their stomachs, and stomach contents from these and other animals recovered from this event tested positive for low concentrations of STX (T. Leighfield personal communication 2010). However, puffer fish do not appear to be a normal component of the dolphin's diet (Barros and Odell 1990), and the levels of STX necessary to cause death in dolphins is unknown; therefore, it is not known if STX played a role in the compromised health and mortality of these animals.

## Domoic Acid

Domoic acid (DA) is a water-soluble amino acid produced by certain diatoms in the genus *Pseudo-nitzschia*. DA mimics the neurotransmitter glutamate and is a potent activator of certain subtypes of glutamate receptor present in the brain. Symptoms of DA poisoning in humans (known clinically as amnesic shellfish poisoning) include nausea, vomiting, diarrhea, dizziness, disorientation, lethargy, seizures, and permanent loss of short-term memory. Neurotoxicity from DA exposure results in lesions in areas of the brain where glutamate receptors are heavily concentrated, particularly in regions of the hippocampus that are responsible for learning and memory processing (Chandrasekaran et al. 2004). The first reported toxic impacts of DA poisoning resulted from a 1987 human intoxication event in eastern Canada, when over 100 people became ill after consuming contaminated mussels (Perl et al. 1990). In the United States, DA poses a threat primarily on the West Coast, although toxin-producing diatom species are present on all coasts.

The first evidence of DA poisoning in marine mammals occurred in California in 1998, when over

400 California sea lions (*Zalophus californianus*) stranded along the central California coast during a two-month period that coincided with a bloom of the diatom *Pseudo-nitzschia australis* (Scholin et al. 2000). Similar California sea lion strandings have recurred on this coast almost annually since that time, with 1,335 cases of confirmed or suspected DA poisoning cases documented between 1998 and 2006, of which nearly half have died (Bejarano et al. 2008a; Goldstein et al. 2009). Hallmarks of DA intoxication in California sea lions (scratching behavior, disorientation, ataxia, and seizures) reflect neurological dysfunction (Gulland 2000; Silvagni et al. 2005). Histopathological examination frequently reveals damage to the brain (hippocampal atrophy and ischemic neuronal necrosis) and heart (i.e., pallor of the myocardium and fibrinous pericarditis). DA intoxication also frequently causes abortion, premature births, or death due to pregnancy-related complications because females are typically in third trimester of pregnancy at the time of the spring diatom blooms (Gulland 2000; Gulland et al. 2002; Brodie et al. 2006; Goldstein et al. 2009). Sea lion pups exposed to DA during gestation have a high frequency of abnormalities and poor survival rates (Goldstein et al. 2008; Ramsdell and Zabka 2008; Goldstein et al. 2009). Animals that survive acute DA intoxication have impaired survival and reproductive potential, as these animals experience persistent neurological dysfunction and increased likelihood of restranding (Goldstein et al. 2008). Two separate clinical syndromes are now recognized: acute DA poisoning (as described above) and a second neurological syndrome characterized by epilepsy associated with chronic consequences of previous sublethal exposure. Exposure of rats to repeated subsymptomatic doses of DA confirm the ability of DA to produce delayed epileptic seizures similar to those observed in sea lions that survive acute intoxication (Ramsdell 2010).

The complex epidemiology of DA intoxication in California sea lions raises concerns regarding the population-level consequences of repeated exposures. The increased frequency of *Pseudo-nitzschia* blooms beginning in the late 1990s coincided with a shift in the North Pacific Oscillation, an approximately 25-year multidecadal climate cycle. The current regime favors cooler ocean temperatures in the eastern Pacific, and stronger upwelling supporting larger phytoplankton blooms, with anchovies dominating the herbivore community, and is thus termed the "anchovy regime" (Chavez et al. 2003). Anchovies are efficient consumers of *Pseudo-nitzschia* and are the primary vector of DA to California sea lions. Bioenergetic modeling suggests that anchovies contribute a four-fold increase in risk of toxic effects from DA as compared with sardines (Bejarano et al. 2007), the dominant grazer during the alternate phase of the North Pacific Oscillation. If the current climate regime persists for approximately 25 years, we might expect continued severe DA impacts on this species for another decade. Although DA was not observed prior to 1998, retrospective analysis reveals that clusters of stranded animals showing symptoms similar to those described in DA-related mortality events have been reported in the past along the California coast, some of which may reflect the previous "anchovy" regime (Gilmartin 1979; Beckman et al. 1995).

Other marine mammals on the California coast experience similar DA intoxications. Several weeks following the peak of sea lion strandings in 1998, an increase in southern sea otter (*Enhydra lutris nereis*) mortalities was observed in the same region (Trainer et al. 2000; Kreuder et al. 2003). The time delay likely reflects the route of exposure, as otters feed primarily upon benthic invertebrates that become toxic following the sinking-out of toxic algal material typical of a terminating bloom (Ferdin et al. 2002; Kvitek et al. 2008; Sekula-Wood et al. 2009). DA toxicosis in sea otters is associated with a high prevalence of cardiac lesions similar to those observed in California sea lions, and exposure to DA increases the probability of myocarditis and cardiomyopathy with recurring or prolonged exposure to DA (Kreuder et al. 2005). It is estimated that DA-exposed otters are 55 times more likely to die from myocarditis than unexposed otters (Kreuder et al. 2005).

Cetaceans have also suffered DA toxicosis and deaths. In 2002, the second largest marine mammal mass mortality event in history occurred in southern California and involved multiple cetacean species, including short-beaked (*Delphinus delphis*) and long-beaked (*Delphinus capensis*) common dolphins, minke (*Balaenoptera acutorostrata*) and humpback (*Megaptera novaeangliae*) whales, and harbour (*Phocoena phocoena*) and Dall's porpoises (*Phocoenoides dalli*), in addition to extensive sea lion, harbour seal (*Phoca vitulina*), and southern sea otter mortalities (Heyning 2003; Torres de la Riva et al. 2009). In 2004, a similar

DA-associated multi-species stranding event occurred in the Gulf of California, which involved common dolphins (*D. delphinus, D. capensis*) and sea lions (Sierra-Beltran et al. 2005). DA intoxication via an anchovy vector was also recently demonstrated to be a cause of death of a minke whale (*Balaenoptera acutorostrata*) stranding during an intense *Pseudo-nitzschia* bloom in southern California in 2007 (Fire et al. 2010).

The impacts of DA on wildlife are not limited to marine mammals. The first confirmed report of DA on the West Coast of the United States occurred in 1991 and involved 95 Brandt's cormorants (*Phalacrocorax penicillatus*) and 43 brown pelicans (*Pelecanus occidentalis*) that were reported dead after ingesting anchovies containing *Pseudo-nitzschia* frustules (Fritz et al. 1992; Work et al. 1993). Surviving animals exhibited typical DA-induced behavioral clinical signs, including head weaving, scratching, and vomiting (Fritz et al. 1992). Similarly, in 1996 brown pelican mortalities occurred in Baja California, causing over 150 bird deaths and decimating 50% of the colony (Ochoa et al. 1996; Sierra-Beltran et al. 1997). The 1998 *Pseudo-nitzschia* bloom in central California may likewise have contributed to the reduced 1998–99 interannual survival of the marbled murrelet (*Brachyramphus marmoratus*) (Peery et al. 2006). Invertebrates and fish are generally viewed as vectors for DA intoxication to higher trophic levels (Bejarano et al. 2008b), but reports of mass sardine mortalities associated with the 2004 Gulf of California event (Sierra-Beltran et al. 2005) and DA-associated mortalities of Humboldt squid (*Dosidicus gigas*) in 2003 and salmon sharks (*Lamna ditropis*) suggest that its toxic impacts are not limited to birds and mammals (NOAA Marine Biotoxins Program, unpublished data 2003).

## Brevetoxin

Brevetoxins (PbTx) are a suite of polyether toxins produced by the dinoflagellate *Karenia brevis*, best known as the causative organism of Florida red tides. PbTx or PbTx-like compounds are also produced by *K. brevis*-like species in New Zealand (Haywood et al. 1996) and raphidophytes of the genus *Chattonella* in Japan (Khan et al. 1995a; Khan et al. 1995b) and have been identified in a fish-killing *Chattonella* bloom in Delaware Bay (Bourdelais et al. 2002). Like STX,

PbTx target the voltage-gated sodium channel but bind to site 5, causing opening of the channel under conditions in which it is normally closed and resulting in inappropriate neuronal transmission (Ramsdell 2008). In humans, PbTx exposure causes the clinical illness known as neurotoxic shellfish poisoning (NSP), and symptoms in mammals include nausea, tingling, and numbness around the mouth, severe muscular aches, loss of motor control, and in severe cases, seizures. PbTx-containing aerosols are also a route of exposure, since fragile *K. brevis* cells are easily lysed by wind or surf action, and human exposure to aerosolized PbTx results in coughing, gagging, and a burning sensation in the upper respiratory tract (Backer et al. 2003; Pierce et al. 2003; Fleming et al. 2005).

Florida's Gulf of Mexico coast experiences PbTx-producing blooms of *K. brevis* almost annually, and these blooms may persist for several months. As evidenced by the conspicuous fish kills associated with *K. brevis* blooms, finfish are particularly sensitive to PbTx, likely because these lipophilic toxins in the water column pass across the gill epithelium and directly into the bloodstream. However, in large blooms, the effects of PbTx occur at all trophic levels, from invertebrates to birds and turtles, to marine mammals. Particularly severe *K. brevis* blooms have led to massive die-offs of polychaetes, amphipods, and gastropods (Simon and Dauer 1972; Roberts 1979; Landsberg et al. 2009) and negative sublethal effects in copepods and bivalves (Huntley et al. 1986; Summerson and Peterson 1990). Cormorants (*Phalacrocorax auritus*), ducks (*Aythya affinis*), and other seabird species are reported to sustain heavy mortalities during *K. brevis* blooms (Shumway et al. 2003; Landsberg et al. 2008). Sea turtle stranding frequencies increased four-fold during a particularly severe mortality event in 2005–06, and red tide intoxication was determined as the cause of death in over 90% of these individuals (D. Fauquier, personal communication).

Manatees (*Trichechus manatus*) and bottlenose dolphins are the marine mammal species most severely affected by *K. brevis*. The first reported association between *K. brevis* blooms and manatee deaths occurred in 1963 near Fort Myers (Layne 1965). A subsequent epizootic in 1982 involved 38 manatee deaths coinciding with a persistent *K. brevis* bloom and was associated with fish kills and cormorant deaths (O'Shea et al. 1991). Behavior of the affected manatees included disorientation, the inability to submerge or maintain a horizontal position, listlessness, and labored breathing. Most animals had stomachs full of seagrasses and associated filter-feeding tunicates (*Molgula* spp.), indicating recent feeding. However, analysis of tunicates by mouse bioassay did not detect PbTx, and the involvement of toxins as the causative agent remains circumstantial. Mass mortalities similar to the 1982 event have occurred in 1996, 2002, 2003, and 2005 (FWC 2007), all between the months of February and April, along approximately a 100-mile stretch of Florida coast centered on the mouth of the Caloosahatchee River. *K. brevis* blooms, which typically develop offshore, make landfall during fall and winter, and then dissipate (Tester and Steidinger 1997), made unusual spring appearances in these embayments during these years. These blooms coincided with large numbers of manatees wintering at warm water refuges and low-salinity areas in the Caloosahatchee River region while beginning their migration northward from the river during the early spring (Reynolds and Wilcox 1986). Thus, the co-occurrence of *K. brevis* blooms in embayments where manatees are concentrated results in a high likelihood of PbTx exposure that may precipitate the observed mass mortality events (Landsberg and Steidinger 1998; Landsberg 2002). It was not until the 1996 event, when 149 manatees died in association with a *K. brevis* bloom, that the presence of PbTx was confirmed as a factor in mortality. Stomach contents from several animals were positive for PbTx, consisting of seagrasses, tunicates, and other epifauna that were suspected in the 1982 event (Landsberg and Steidinger 1998). Evidence of PbTx exposure and severe congestion was also observed in the respiratory tract, liver, kidney, lung, and brain, as well as in lymphocytes and macrophages in these tissues (Bossart et al. 1998). Thus, PbTx exposure and possible immunosuppression in manatees may also result in part from chronic inhalation in addition to neurotoxic effects from ingestion (Baden 1996; Bossart et al. 1998). Based on extrapolation from the human symptomatic dose of 24 µg/kg body weight, a manatee dietary intake of 7% body weight per day, and PbTx concentrations detected in stomach contents, the estimated oral dose would be sufficient to cause symptoms in a 700-kg manatee (Baden 1996). In the 2002 mass mortality, manatee deaths began not during but

several weeks after termination of a February *K. brevis* bloom in the region, raising questions about the source of toxin in that event. Seagrasses collected from the area contained high levels of PbTx, both on the grass blades and in associated filter-feeding organisms, several months after termination of the bloom (Flewelling et al. 2004).

Bottlenose dolphin mortalities also have a long history of circumstantial association *K. brevis* blooms. The earliest reported association between these "red tides" and mass dolphin mortalities was a 1946–47 bloom occurring between Florida Bay and St. Petersburg that persisted for eight months (Gunter et al. 1948). At the time, the identity of *K. brevis* as the causative organism was tenuous, and the toxin was unidentified. PbTx was also proposed as a causative agent in an unprecedented die-off of over 740 bottlenose dolphins occurring between New Jersey and Florida in 1987–88 (Geraci 1989). Strandings coincided with a rare bloom of *K. brevis* originating in the Gulf of Mexico and carried via the Gulf Stream into Atlantic coastal waters, resulting in toxic shellfish and human NSP intoxications (Tester et al. 1991). Although PbTx was reported in the stomach contents and liver of several dolphins (Baden 1989), evidence of PbTx involvement remains equivocal due to inadequate confirmatory analytical methods available at the time, co-occurring morbillivirus infection, and a temporal mismatch in several PbTx-positive animals and the presence of the observed *K. brevis* bloom. A subsequent Florida bottlenose dolphin mortality event occurred in 1999–2000 coinciding with a persistent *K. brevis* bloom along the Florida panhandle near St. Joe Bay and Choctawhatchee Bay. In this event, PbTx was confirmed in stomach contents, liver, and/or kidney samples in 41% of the animals (Twiner et al. 2009). Stomach contents, consisting primarily of finfish, had the highest PbTx concentrations, and most individuals that stranded were in good body condition, indicating acute poisoning via oral PbTx exposure (Mase et al. 2000). Histopathological examination of two animals showed significant lesions in the upper respiratory tract and PbTx-specific antibody staining in lung and spleen tissue (Van Dolah 2005).

A prominent mass mortality of 107 bottlenose dolphins occurred in the spring of 2004 in the Florida panhandle, highlighting the importance of identifying food web vectors in PbTx-associated die-offs.

Although no *K. brevis* bloom was observed at the time, extremely high concentrations of PbTx were detected in dolphin stomach contents, which were full of large numbers of undigested menhaden (*Brevoortia* spp.), a planktivorous fish (Flewelling et al. 2005). Prior to this event, it was suspected that the ichthyotoxic effects of PbTx would kill finfish before they could accumulate sufficient toxin to transfer it up the food web to predators. Subsequent experimental and field work showed that PbTx could accumulate in multiple species of finfish despite a wide variety of feeding habits, indicating multiple sources of PbTx in the Florida coastal ecosystem (Naar et al. 2007; Fire et al. 2008).

In perhaps the most severe *K. brevis*-related marine mammal mortality event observed to date, large numbers of bottlenose dolphins, along with several other cetacean, finfish, and invertebrate species, died as a result of a *K. brevis* bloom occurring along the central west Florida coast during 2005–06 (Landsberg et al. 2009). At least three large peaks in dolphin strandings were observed during this time, and the presence of PbTx was confirmed in tissues and fluids from over 80 individual animals (Twiner et al. 2009). PbTx was also detected in multiple species of finfish known to be major prey items for the dolphins, showing exposure to PbTx from multiple sources in the food web. In addition, impacts of PbTx exposure on fish stocks in this region differed across trophic guilds, resulting in a shift in relative abundance of available dolphin prey species, and a dominance of clupeids, which may act as more efficient PbTx vectors to upper trophic levels (Gannon et al. 2009).

## Ciguatoxin

Ciguatoxins (CTX) are a suite of polyether toxins produced by the dinoflagellates *Gambierdiscus* spp., which cause the clinical illness in humans known as ciguatera fish poisoning (CFP). CTX is similar to PbTx in chemical structure, pharmacological target, and clinical signs; however, the potency of CTX is much greater, and neurotoxic symptoms often persist for longer periods and can include reversal of temperature sensation, tachycardia, hypertension, paralysis, and death (Lewis 2001). In the United States, ciguatera-producing *Gambierdiscus* populations primarily occupy tropical and subtropical waters of Hawaii and southern Florida. Since *Gambierdiscus* spp. are

generally benthic epiphytes that grow on filamentous algae associated with coral reefs and reef lagoons, CTX typically enters the food web via herbivorous fishes and invertebrates, and can bioaccumulate in high trophic-level reef fishes such as grouper and barracuda (Lehane and Lewis 2000; Cruz-Rivera and Villareal 2006).

Although much is known about the effects of CTX on humans, evidence of involvement of CTX as a factor in mortality and disease of marine mammals is limited and speculative. It has been proposed as one of several potential factors in the decline of Hawaiian monk seals (*Monachus schauislandi*) in the tropical Pacific. A 1978 Hawaiian monk seal mortality event occurring in Laysan Island resulted in the deaths of over 50 animals, and high levels of CTX-like activity were estimated by bioassay in the liver of two animals (Gilmartin et al. 1980). However, inconsistent additional assay results prevented unequivocal confirmation of CTX as the causative agent. A 1992–93 recovery effort that relocated a severely depleted monk seal stock from Midway Island to French Frigate Shoals resulted in only 11% of the translocated animals surviving beyond one year, and one hypothesis is that the reefs at Midway support a high incidence of CTX in the food web (Gilmartin and Antonelis 1998). Surveys of monk seal prey species in Midway lagoon in 1986 and 1992 detected CTX in over half of all fish tested (Wilson and Jokiel 1986). Again, these results were equivocal due to high false-positive rates reported by the detection method employed (Dickey et al. 1994; Wong et al. 2005).

## NOVEL AND EMERGENT CONCERNS FOR HAB TOXINS AND MARINE ANIMAL HEALTH

### DSP Toxins

Okadaic acid (OA) and its derivatives, the dinophysistoxins (DTX), are groups of polyether compounds that cause the human illness diarrhetic shellfish poisoning (DSP), and are thus collectively referred to as DSP toxins. Although DSP toxins are protein phosphatase inhibitors that cause a relatively mild intoxication resolving within a few days, OA-like toxins have been identified as potential tumor promoters in sea turtles (Landsberg et al. 1999). Produced by the dinoflagellates *Dinophysis* spp. and *Prorocentrum* spp., the first reported DSP outbreak in North America occurred in 1990 in Nova Scotia (Marr et al. 1992), but cases have also been reported in Europe, Japan, South America, South Africa, New Zealand, Australia, and Thailand. Effects of OA on marine mammals have not been reported in the literature; however, OA was detected for the first time in several bottlenose dolphins from a mortality event occurring in Texas in February through April 2008 (Fire et al. 2011). The event co-occurred with a bloom of *Dinophysis ovum* and associated shellfish closures in the region (Deeds et al. 2010; Swanson et al. 2010), but the OA concentrations detected in dolphin feces and gastric contents were very low relative to analytical detection limits, and the role of DSP toxins as a factor in the mortality of these animals remains unclear.

## Azaspiracid

Azaspiracids (AZA) are a newly identified nitrogen-containing polyether toxin, reported for the first time in 1995 in association with an outbreak of gastrointestinal illness in humans following the consumption of shellfish from Ireland (Furey et al. 2003). Toxic symptoms due to AZA poisoning are similar to DSP poisoning and include nausea, vomiting, diarrhea, and stomach cramps (Twiner et al. 2008). Production of AZA is associated with the dinoflagellate *Azadinium spinosum* and has been reported in Europe, North America, and Africa (James et al. 2003; Twiner et al. 2008; Tillmann et al. 2009). The mechanism of toxicity for AZA is still unclear, and although documented to induce teratogenic effects in fish (Colman et al. 2005), no impacts on marine mammals have been reported.

## Yessotoxins

Yessotoxin (YTX) is a sulfated polyether toxin produced by dinoflagellates belonging to the genera *Gonyaulax* (=*Protoceratium*) and *Lingulodinium*. Originally thought to be one of the DSP toxins, it has been shown to have no diarrhetic activity and little toxic potency when administered orally to mice (Tubaro et al. 2008). However, it has potent lethality to mice when injected, inducing neurological symptoms. Blooms of both *Lingulodinium polyedrum* and *Gonyaulax grindleyi* have been implicated in fish and shellfish mortality events, although the role of YTX in those events was not investigated. The impacts of YTX on marine mammals are not known.

## Spirolides

Spirolides are a group of macrocyclic amines produced by the dinoflagellate *Alexandrium ostenfeldii* and implicated in shellfish toxicity in northern Europe (Cembella et al. 2000). The mode of action of this toxin is not clear, but injection into mice causes rapid death following neurological symptoms. The distribution of *A. ostenfeldii* in North America is limited to eastern Canadian waters and the Gulf of Maine. Health impacts from spirolide exposure in marine mammals are unknown.

## NON-ACUTE EXPOSURE, MULTIPLE TOXINS, AND NOVEL IMPACTS

Acute effects of HAB toxin exposure in marine mammals are often quite conspicuous, resulting in dramatic die-offs of large numbers of animals and associated marine fauna. Dense blooms and high concentrations of toxins present in seawater, prey items, and marine mammal tissues and fluids typically precipitate these events. However, variability in the severity of HABs and the frequency in which they occur can result in repeated, sublethal exposures to one or more toxins with negative impacts on marine mammal health. Repeated nonlethal DA exposure in California sea lions has significant long-term effects such as degenerative heart disease, chronic epileptic syndrome, and *in utero* toxicity resulting in reproductive failure (Brodie et al. 2006; Goldstein et al. 2008; Ramsdell and Zabka 2008; Zabka et al. 2009). Florida bottlenose dolphins and their fish prey have been shown to contain PbTx several months to over a year following termination of a *K. brevis* bloom (Fire et al. 2007; Naar et al. 2007), suggesting year-round toxin exposure, but long-term impacts on these populations are unknown. The effects of annual STX-producing blooms or potential CTX exposure in marine mammals likely includes negative impacts to growth of already depleted populations (Gilmartin and Antonelis 1998; Durbin et al. 2002), but few data exist to support this hypothesis, in part due to insufficiently sensitive detection methods, confounding factors such as infectious disease, or a lack of longitudinal data.

As HAB observation efforts become more wide-spread, toxic species are increasingly reported in new regions, resulting in increased awareness of marine mammal exposure to multiple HAB toxins. Northern right whales in their summer feeding grounds in New England and Bay of Fundy waters experience frequent exposure to both STX and DA (Doucette et al. 2006; Leandro et al. 2010). Biomonitoring efforts have detected PbTx and DA simultaneously in Florida bottlenose dolphins for several years (Twiner et al. 2011), including live animals as well as stranded carcasses (NMFS 2004). Low levels of DA, PbTx, and OA were all detected in various bottlenose dolphins stranding during a mortality event occurring near Galveston, Texas, in 2008 (Fire et al. 2011). Currently, it is not known whether exposure to multiple HAB toxins results in additive or synergistic effects that increase the potency of one or more of the toxins present (Twiner et al. 2008).

Notwithstanding the advent of successful HAB monitoring efforts and technologies, evidence of HAB toxin exposure is seen in marine mammals even in the absence of observed toxic blooms. In a 1997–2009 survey of pygmy and dwarf sperm whales (*Kogia* spp.) stranding between North Carolina and the Atlantic coast of Florida, feces from nearly 90% of all individuals sampled were positive for DA, despite the fact that no DA-producing HABs were associated with any of the DA-positive animals (Fire et al. 2009). This may be an indication that undetected HAB activity occurring in remote regions can affect marine mammals, and highlights the importance of continued efforts to investigate such associations.

# REFERENCES

Abbott, J.P., L.J. Flewelling, and J.H. Landsberg. 2009. Saxitoxin monitoring in three species of Florida puffer fish. Harmful Algae 8:343–348.

Aguirre, A.A., and G.M. Tabor. 2004. Introduction: Marine vertebrates as sentinels of marine ecosystem health. EcoHealth 1:236–238.

Anderson, D.M., and A.W. White. 1989. Toxic dinoflagellates and marine mammal mortalities. Tech Rept WHOI-89-36, Woods Hole Oceanographic Institution, Massachusetts.

Anderson, D.M. 1997. Bloom dynamics of toxic *Alexandrium* species in the northeastern U.S. Limnol Oceanogr 42:1009–1022.

Backer, L.C., L.E. Fleming, A. Rowan, Y.S. Cheng, J. Benson, R.H. Pierce, J. Zaias, J. Bean, G.D. Bossart, D. Johnson, R. Quimbo, and D.G. Baden. 2003. Recreational exposure to aerosolized brevetoxins during Florida red tide events. Harmful Algae 2:19–28.

Baden, D.G. 1989. Toxic dinoflagellates and marine mammal mortalities. *In* D.M. Anderson and A.W. White eds. Proceedings of an Expert Consultation held at the Woods Hole Oceanographic Institution Tech Rept WHOI-89-36, Appendix 3c, pp. 47-52. Woods Hole Oceanographic Institute, Massachusetts.

Baden, D.G. 1996. Analysis of biotoxins (red tide) in manatee tissues. Report No. MR148. Marine and Freshwater Biomedical Sciences Center. National Institute of Environmental Health Sciences, Rosenstiel School of Marine and Atmospheric Sciences, University of Miami, Florida.

Barros, N.B., and D.K. Odell. 1990. Food habits of bottlenose dolphins in the southeastern United States. *In* S. Leatherwood and R.R. Reeves, eds. The bottlenose dolphin, pp. 309–328. San Diego: Academic Press, California.

Beckman, K., L.J. Lowenstein, and F. Galey. 1995. Epizootic seizures of California sea lions. Proceedings of the 11th Biennial Conference of the Marine Mammal Society, Orlando, Florida.

Bejarano, A.C., F.M.V. Dolah, F.M. Gulland, and L. Schwacke. 2007. Exposure assessment of the biotoxin domoic acid in California sea lions: application of a bioenergetic model. Mar Ecol Prog Ser 345:293–304.

Bejarano, A.C., F.M. Gulland, T. Goldstein, J.S. Leger, M. Hunter, L.H. Schwacke, F.M.V. Dolah, and T.K. Rowles. 2008a. Demographics and spatio-temporal signature of the biotoxin domoic acid in California sea lion *Zalophus californianus* stranding records. Mar Mamm Sci 24:899–912.

Bejarano, A.C., F.M. Van Dolah, F.M. Gulland, T.K. Rowles, and L.H. Schwacke. 2008b. Production and toxicity of the marine biotoxin domoic acid and its effects on wildlife: a review. Hum Ecol Risk Assess 14: 544–567.

Bossart, G.D., D.G. Baden, R.Y. Ewing, B. Roberts, and S.D. Wright. 1998. Brevetoxicosis in manatees (*Trichechus manatus latirostris*) from the 1996 epizootic: gross, histologic, and immunohistochemical features. Toxicol Pathol 26:276–282.

Bossart, G.D., R. Meisner, R. Varela, M. Mazzoil, S.D. Mcculloch, D. Kilpatrick, R. Friday, E. Murdoch, B. Mase, and R.H. Defran. 2003. Pathologic findings in stranded Atlantic bottlenose dolphins (*Tursiops truncatus*) from the Indian River Lagoon, Florida. Fla Sci 66:226–228.

Bossart, G.D. 2006. Marine mammals as sentinel species for oceans and human health. Oceanography 19:134–137.

Bourdelais, A.J., C.R. Tomas, J. Naar, J. Kubanek, and D.G. Baden. 2002. New fish-killing alga in coastal Delaware produces neurotoxins. Environ Health Persp 110:465–70.

Brodie, E.C., F.M.D. Gulland, D.J. Greig, M. Hunter, J. Jaakola, J. St. Leger, T.A. Leighfield, and F.M. Van Dolah. 2006. Domoic acid causes reproductive failure in California sea lions (Zalophus californianus). Mar Mamm Sci 22:700–707.

Cembella, A.D., N.I. Lewis, and M.A. Quilliam. 2000. The marine dinoflagellate Alexandrium ostenfeldii (Dinophyceae) as the causative organism of spirolide shellfish toxins. Phycologia 39:67–74.

Chandrasekaran, A., G. Ponnambalam, and C. Kaur. 2004. Domoic acid-induced neurotoxicity in the hippocampus of adult rats. Neurotoxicity Research 6:105–117.

Chavez, F.P., J. Ryan, S.E. Lluch-Cota, and M. Ñiquen C. 2003. From anchovies to sardines and back: Multidecadal change in the Pacific Ocean. Science 299:217–221.

Colman, J.R., M.J. Twiner, P. Hess, T. Mcmahon, M. Satake, T. Yasumoto, G.J. Doucette, and J.S. Ramsdell. 2005. Teratogenic effects of azaspiracid-1 identified by microinjection of Japanese medaka (Oryzias latipes) embryos. Toxicon 45:881–890.

Cruz-Rivera, E., and T.A. Villareal. 2006. Macroalgal palatability and the flux of ciguatera toxins through marine food webs. Harmful Algae 5:497–525.

Deeds, J.R., K. Wiles, G.B. Heideman Vi, K.D. White, and A. Abraham. 2010. First U.S. report of shellfish harvesting closures due to confirmed okadaic acid in Texas Gulf coast oysters. Toxicon 55:1138–1146.

Dickey, R.L., H.R. Granade, and F.D. McClure. 1994. Evaluation of a solid-phase immunobead assay for detection of ciguatera-related biotoxins in Caribbean finfish. Memoirs of the Queensland Museum Brisbane 34:481–488.

Doucette, G.J., A.D. Cembella, J.L. Martin, J. Michaud, T.V.N. Cole, and R.M. Rolland. 2006. Paralytic shellfish poisoning (PSP) toxins in North Atlantic right whales Eubalaena glacialis and their zooplankton prey in the Bay of Fundy, Canada. Mar Ecol Prog Ser 306:303–313.

Durbin, E., G. Teegarden, R. Campbell, A. Cembella, M.F. Baumgartner, and B.R. Mate. 2002. North Atlantic right whales, Eubalaena glacialis, exposed to paralytic shellfish poisoning (PSP) toxins via a zooplankton vector, Calanus finmarchicus. Harmful Algae 1:243–251.

Ferdin, M.E., R.G. Kvitek, C.K. Bretz, C.L. Powell, G.J. Doucette, K.A. Lefebvre, S. Coale, and M.W. Silver. 2002. Emerita analoga (Stimpson)—possible new indicator species for the phycotoxin domoic acid in California coastal waters. Toxicon 40:1259–1265.

Fire, S., D. Fauquier, L. Flewelling, M. Henry, J. Naar, R. Pierce, and R. Wells. 2007. Brevetoxin exposure in bottlenose dolphins (Tursiops truncatus) associated with Karenia brevis blooms in Sarasota Bay, Florida. Mar Biol 152:827–834.

Fire, S.E., L.J. Flewelling, J. Naar, M.J. Twiner, M.S. Henry, R.H. Pierce, D.P. Gannon, Z. Wang, L. Davidson, and R.S. Wells. 2008. Prevalence of brevetoxins in prey fish of bottlenose dolphins in Sarasota Bay, Florida. Mar Ecol Prog Ser 368:283–294.

Fire, S.E., Z. Wang, T.A. Leighfield, S.L. Morton, W.E. Mcfee, W.A. Mclellan, R.W. Litaker, P.A. Tester, A.A. Hohn, G. Lovewell, C. Harms, D.S. Rotstein, S.G. Barco, A. Costidis, B. Sheppard, G.D. Bossart, M. Stolen, W.N. Durden, and F.M. Van Dolah. 2009. Domoic acid exposure in pygmy and dwarf sperm whales (Kogia spp.) from southeastern and mid-Atlantic U.S. waters. Harmful Algae 8: 658–664.

Fire, S.E., Z. Wang, M. Berman, G.W. Langlois, S.L. Morton, E. Sekula-Wood, and C.R. Benitez-Nelson. 2010. Trophic transfer of the harmful algal toxin domoic acid as a cause of death in a minke whale (Balaenoptera acutorostrata) stranding in southern California. Aquat Mamm 36:342–350.

Fire, S.E., Z. Wang, M. Byrd, H.R. Whitehead, J. Paternoster, and S.L. Morton. 2011. Co-occurrence of multiple classes of harmful algal toxins in bottlenose dolphins (Tursiops truncatus) stranding during an unusual mortality event in Texas, USA. Harmful Algae 10:330–336.

Fleming, L.E., L.C. Backer, and D.G. Baden. 2005. Overview of aerosolized Florida red tide toxins: exposures and effects. Environ Health Persp 113:618–620.

Flewelling, L.J., J.P. Abbott, D.G. Hammond, J. Landsberg, E. Haubold, T. Pulfer, P. Speiss, and M.S. Henry. 2004. Seagrass as a route of brevetoxin exposure in the 2002 red tide-related manatee mortality event. Harmful Algae 3:205.

Flewelling, L.J., J.P. Naar, J.P. Abbott, D.G. Baden, N.B. Barros, G.D. Bossart, M.Y.D. Bottein, D.G. Hammond, E.M. Haubold, C.A. Heil, M.S. Henry, H.M. Jacocks, T.A. Leighfield, R.H. Pierce, T.D. Pitchford, S.A. Rommel, P.S. Scott, K.A. Steidinger, E.W. Truby, F.M. Van Dolah, and J.H. Landsberg. 2005. Red tides and marine mammal mortalities. Nature 435:755–756.

Forcada, J., P.S. Hammond, and A. Aguilar. 1999. Status of the Mediterranean monk seal Monachus monachus in the western Sahara and implications of a mass mortality event. Mar Ecol Prog Ser 188:249–261.

Fritz, L., M.A. Quilliam, J.L.C. Wright, A.M. Beale, and T.M. Work. 1992. An outbreak of domoic acid poisoning attributed to the pennate diatom *Pseudonitzschia australis*. J Phycol 28:439–442.

Furey, A., C. Moroney, A.B. Magdelena, M.J. Saez, F., M. Lehane, and K.J. James. 2003. Geographical, temporal, and species variation of polyether toxins, azaspiracids, in shellfish. Environ Sci Technol 37:3078–3084.

FWC. 2007. *Trichechus manatus latirostris*. *In* Florida Manatee Management Plan. Florida Fish and Wildlife Conservation Commission. Tallahassee, Florida <http://www.myfwc.com/WILDLIFEHABITATS/Manatee_index.htm>.

Gannon, D.P., E.J.B. Mccabe, S.A. Camilleri, J.G. Gannon, M.K. Brueggen, A.A. Barleycorn, V.I. Palubok, G.J. Kirkpatrick, and R.S. Wells. 2009. Effects of *Karenia brevis* harmful algal blooms on nearshore fish communities in southwest Florida. Mar Ecol Prog Ser 378:171–186.

Geraci, J.R. 1989. Clinical investigation of the 1987–88 mass mortality of bottlenose dolphins along the US central and south Atlantic coast. Final report to the National Marine Fisheries Service, US Navy Office of Naval Research, and Marine Mammal Commission, pp. 1–63. Ontario Veterinary College, University of Guelph, Canada.

Geraci, J.R., D.M. Anderson, R.J. Timperi, D.J. St. Aubin, G.A. Early, J.H. Prescott, and C.A. Mayo. 1989. Humpback whales (*Megaptera novaeangliae*) fatally poisoned by dinoflagellate toxin. Can J Fish Aquat Sci 46:1895–1898.

Gilmartin, W.G. 1979. Fetal hepatoencephalopathy in a group on California sea lions. 10th Annual International Association for Aquatic Animal Medicine Conf Proc, St Augustine, Florida.

Gilmartin, W.G., R.L. Delong, A.W. Smith, L.A. Griner, and M.D. Dailey. 1980. An investigation into an unusual mortality event in the Hawaiian Monk seal, *Monachus schauinslandi*. Proc Symp Status of Resource Investigations in the Northwestern Hawaiian Islands UNIHI-SEAGRANT Report No MR-80-04:32–41.

Gilmartin, W.G., and G.A. Antonelis. 1998. Recommended recovery actions for the Hawaiian monk seal population at Midway Island. NOAA Tech Memo NMFS NOAA-NMFS-SWFSC-253, Honolulu, Hawaii.

Goldstein, T., J.K. Mazet, T.S. Zabka, G. Langlois, K.M. Colegrove, M. Silver, S. Bargu, F. Van Dolah,
T. Leighfield, P.A. Conrad, J. Barakos, D.C. Williams, S. Dennison, M. Haulena, and F.M.D. Gulland. 2008. Novel symptomatology and changing epidemiology of domoic acid toxicosis in California sea lions (*Zalophus californianus*): an increasing risk to marine mammal health. P Roy Soc B 275:267–276.

Goldstein, T., T.S. Zabka, R.L. Delong, E.A. Wheeler, G. Ylitalo, S. Bargu, M. Silver, T. Leighfield, F. Van Dolah, G. Langlois, I. Sidor, J.L. Dunn, and F.M.D. Gulland. 2009. The role of domoic acid in abortion and premature parturition of California sea lions (*Zalophus californianus*) on San Miguel Island, California. J Wildl Dis 45:91–108.

Gulland, F.M.D. 2000. Domoic acid toxicity in California sea lions (Zalophus californianus) stranded along the central California coast, May–October 1998. NOAA Tech Memo NMFS-OPR-17.

Gulland, F.M.D., M. Haulena, D. Fauquier, G. Langlois, M.E. Lander, T. Zabka, and R. Duerr. 2002. Domoic acid toxicity in Californian sea lions (*Zalophus californianus*): clinical signs, treatment and survival. Vet Rec 150:475–480.

Gulland, F.M.D., and A.J. Hall. 2007. Is marine mammal health deteriorating? trends in the global reporting of marine mammal disease. EcoHealth 4:135–150.

Gunter, G., R.H. Williams, C.C. Davis, and F.G.W. Smith. 1948. Catastrophic mass mortality of marine animals and coincident phytoplankton bloom on the west coast of Florida, November 1946 to August 1947. Ecol Monogr 18:309–324.

Hallegraeff, G.M. 2003. Harmful algal blooms: a global overview. *In* G.M. Hallegraeff, D.M. Anderson, and A. Cembella, eds. Manual on harmful marine microalgae, pp. 25–49. UNESCO, Paris.

Harwood, J. 1998. What killed the monk seals? Nature 393:17–18.

Haywood, A., L. Mackenzie, I. Garthwaite, and N. Towers. 1996. *Gymnodinium breve* look-alikes: three *Gymnodinium* isolates from New Zealand. *In* T. Yasumoto, Y. Oshima, and Y. Fukuyo, eds. Harmful and toxic algal blooms, pp. 227–230. International Oceanographic Committee of UNESCO, Paris.

Hernandez, M., I. Robinson, A. Aguilar, L.M. Gonzalez, L.F. Lopez-Jurado, M.I. Reyero, E. Cacho, J. Franco, V. Lopez-Rodas, and E. Costas. 1998. Did algal toxins cause monk seal mortality? Nature 393:28–29.

Heyning, J.E. 2003. Final report on the multi-species marine mammal unusual mortality event along the southern California coast 2002. National Marine Fisheries Service Tech Memo.

Huntley, M.E., P. Sykes, S. Rohan, and V. Marin. 1986. Chemically mediated rejection of dinoflagellate prey by the copepods *Calanus pacificus* and *Paracalanus parvus*: mechanism, occurrence and significance. Mar Ecol Prog Ser 28:105–120.

James, K.J., C. Moroney, C. Roden, M. Satake, T. Yasumoto, M. Lehane, and A. Furey. 2003. Ubiquitous "benign" alga emerges as cause of shellfish contamination responsible for the human toxic syndrome, azaspiracid poisoning. Toxicon 41:145–151.

Khan, S., M.S. Ahmed, O. Arakawa, and Y. Onoue. 1995a. Properties of neurotoxins separated from harmful red tide organism *C hattonella marina*. Isr J Aquacult-Bamid 47:137–140.

Khan, S., M. Haque, O. Arakawa, and Y. Onoue. 1995b. Toxin profiles and ichthyotoxicity of three phytoflagellates. Bangladesh J Fish 15:73–81.

Kooyman, G.L. 1985. Physiology without restraint in diving mammals. Mar Mamm Sci 1:166–178.

Kraus, S.D., P.K. Hamilton, R.D. Kenney, A.R. Knowlton, and C.K. Slay. 2001. Reproductive parameters of the north Atlantic right whale. J Cetacean Res Manage 2:321–336.

Kreuder, C., M.A. Miller, D.A. Jessup, L.J. Lowenstine, M.D. Harris, J.A. Ames, T.E. Carpenter, P.A. Conrad, and J.A. Mazet. 2003. Patterns of mortality in southern sea otters (*Enhydra lutris nereis*) from 1998-2001. J Wildl Dis 39:495–509.

Kreuder, C., M.A. Miller, L.J. Lowenstine, P.A. Conrad, T.E. Carpenter, D.A. Jessup, and J.A.K. Mazet. 2005. Evaluation of cardiac lesions and risk factors associated with myocarditis and dilated cardiomyopathy in southern sea otters (*Enhydra lutris nereis*). Am J Vet Res 66:289–299.

Kvitek, R.G., J.D. Goldberg, G.J. Smith, G.J. Doucette, and M.W. Silver. 2008. Domoic acid contamination within eight representative species from the benthic food web of Monterey Bay, California, USA. Mar Ecol Prog Ser 367:35–47.

Landsberg, J., F. Van Dolah, and G. Doucette. 2005. Marine and estuarine harmful algal blooms: Impacts on human and animal health. *In* S. Belkin and R.R. Colwell, eds. Oceans and health: pathogens in the marine environment. pp. 165–215. Springer, New York.

Landsberg, J.H., and K.A. Steidinger. 1998. A historical review of *Gymnodinium breve* red tides implicated in mass mortalities of the manatee (*Trichechus manatus latirostris*) in Florida, USA. *In* B. Reguera, eds. Proc 8th Int Conf Harmful Algae, pp. 97–100. Xunta de Galicia and Intergovernmental Oceanographic Commission of UNESCO, Vigo, Spain.

Landsberg, J.H., G.H. Balazs, K.A. Steidinger, D.G. Baden, T.M. Work, and D.J. Russel. 1999. The potential role of natural tumor promoters in turtle fibropapillomatosis. J Aquat Anim Health 11:199–210.

Landsberg, J.H. 2002. The effects of harmful algal blooms on aquatic organisms. Rev Fish Sci 10:113–390.

Landsberg, J.H., S. Hall, J.N. Johannessen, K.D. White, S.M. Conrad, L.J. Flewelling, R.W. Dickey, F.M. Van Dolah, M.A. Quilliam, T.A. Leighfield, Z. Yinglin, C.G. Beaudry, W.R. Richardson, K. Hayes, L. Baird, R.A. Benner, P.L. Rogers, J. Abbott, D. Tremain, D. Heil, R. Hammond, D. Bodager, G. Mcrae, C.M. Stephenson, T. Cody, P.S. Scott, W.S. Arnold, H. Schurz-Rogers, A.J. Haywood, and K.A. Steidinger. 2002. Pufferfish poisoning: widespread implications of saxitoxin in Florida, p. 160. 10th Int Conf Harmful Algae, 21–25 October, St Petersburg Beach, Florida.

Landsberg, J.H., S. Hall, J.N. Johannessen, K.D. White, S.M. Conrad, J.P. Abbott, L.J. Flewelling, R.W. Richardson, R.W. Dickey, E.L.E. Jester, S.M. Etheridge, J.R. Deeds, F.M. Van Dolah, T.A. Leighfield, Y. Zou, C.G. Beaudry, R.A. Benner, P.L. Rogers, P.S. Scott, K. Kawabata, J.L. Wolny, and K.A. Steidinger. 2006. Saxitoxin puffer fish poisoning in the United States, with the first report of *Pyrodinium bahamense* as the putative toxin source. Environ Health Persp 114:1502–1507.

Landsberg, J.H., G.A. Vargo, L.J. Flewelling, and F.E. Wiley. 2008. Algal biotoxins. *In* N.J. Thomas, D.B. Hunter, and C.T. Atkinson, eds. Infectious diseases of wild birds, pp. 431–455. Blackwell Publishing, Oxford.

Landsberg, J.H., L.J. Flewelling, and J. Naar. 2009. *Karenia brevis* red tides, brevetoxins in the food web, and impacts on natural resources: Decadal advancements. Harmful Algae 8:598–607.

Layne, J.N. 1965. Observations on marine mammals in Florida waters. Bull Fl State Mus Biol Sci 9:131–181.

Leandro, L.F., R.M. Rolland, P.B. Roth, N. Lundholm, Z. Wang, and G.J. Doucette. 2010. Exposure of the North Atlantic right whale *Eubalaena glacialis* to the marine algal biotoxin, domoic acid. Mar Ecol Prog Ser 398:287–303.

Lehane, L., and R.J. Lewis. 2000. Ciguatera: recent advances but the risk remains. Int J Food Microbiol 61:91–125.

Levin, R.E. 1992. Paralytic shellfish toxins: their origins, characteristics, and methods of detection: a review. J Food Biochem 15:405–417.

Lewis, R.J. 2001. The changing face of ciguatera. Toxicon 39:97–106.

Marr, J.C., A.E. Jackson, and J.L. Mclachlan. 1992. Occurrence of *Prorocentrum lima*, a DSP-toxin producing species from the Altantic coast of Canada. J Appl Phycol 4:17–24.

Mase, B., W. Jones, R. Ewing, G. Bossart, F. Van Dolah, T. Leighfield, M. Busman, J. Litz, B. Roberts, and T. Rowles. 2000. Epizootic in bottlenose dolphins in the Florida panhandle: 1999–2000. In C.K. Baer, ed. Proc Am Assoc Zoo Vet Int Assoc Aquat An Med, pp. 522–525, Milwaukee, Wisconsin.

Naar, J.P., L.J. Flewelling, A. Lenzi, J.P. Abbott, A. Granholm, H.M. Jacocks, D. Gannon, M. Henry, R. Pierce, D.G. Baden, J. Wolny, and J.H. Landsberg. 2007. Brevetoxins, like ciguatoxins, are potent ichthyotoxic neurotoxins that accumulate in fish. Toxicon 50:707–723.

NMFS. 2004. Interim report on the bottlenose dolphin (Tursiops truncatus) unusual mortality event along the Panhandle of Florida, March-April 2004. www.nmfs.noaa.gov/pr/pdfs/health/ume_bottlenose_ 2004.pdf

O'Shea, T.J., G.B. Rathbun, R.K. Bonde, C.D. Buergelt, and D.K. Odell. 1991. An epizootic of Florida manatees associated with a dinoflagellate bloom. Mar Mamm Sci 7:165–179.

Ochoa, J.L., A. Sierra-Beltran, A. Cruz-Villacorta, and E. Nuñez. 1996. Domoic acid in Mexico. In Penney R.W., ed. Proc 5th Canadian Workshop Harmful Marine Algae, Canadian Technical Report of Fisheries and Aquatic Science No 2138:82-90.

Osterhaus, A., J. Groen, H. Niesters, M. Van De Bildt, B. Martina, L. Vedder, J. Vos, H. Van Egmond, B.A. Sidi and M.E.O. Barham. 1997. Morbillivirus in monk seal mortality. Nature 388:838–839.

Peery, M.Z., S.R. Beissinger, E. Burkett, and S.H. Newman. 2006. Local survival of marbled murrelets in central California: roles of oceanographic processes, sex, and radiotagging. J Wildlife Manage 70:78–88.

Perl, T.M., L. Bedard, T. Kosatsky, J.C. Hockin, E.C.D. Todd, and R.S. Remis. 1990. An outbreak of toxic encephalopathy caused by eating mussels contaminated with domoic acid. N Engl J Med 322:1775–1780.

Pierce, R.H., M.S. Henry, P.C. Blum, J. Lyons, Y.S. Cheng, D. Yazzie, and Y. Zhou. 2003. Brevetoxin concentrations in marine aerosol: Human exposure levels during a Karenia brevis harmful algal bloom. Bull Environ Contam Toxicol 70:161–165.

Quilliam, M., D. Wechsler, S. Marcus, B. Ruck, M. Weckell, and T. Hawryluk. 2002. Detection and identification of paralytic shellfish poisoning toxins in Florida pufferfish responsible for incidents of neurologic illness. Xth International Conference on Harmful Algae, p. 116. 21-25 October. St. Petersburg Beach, Florida.

Ramsdell, J.S. 2008. The molecular and integrative basis to brevetoxin toxicity. In L. Botana, ed. Seafood and freshwater toxins; pharmacology, physiology and detection, pp. 519–550. CRC Press, Boca Raton, Florida.

Ramsdell, J.S., and T.S. Zabka. 2008. In utero domoic acid toxicity: A fetal basis to adult disease in the California sea lion (Zalophus californianus). Mar Drugs 6:262–90.

Ramsdell, J.S. 2010. Neurological disease rises from ocean to bring model for human epilepsy to life. American Association for the Advancement of Science Annual Meeting, San Diego, California.

Reynolds, J.E., and J.R. Wilcox. 1986. Distribution and abundance of the West Indian manatee, Trichechus manatus, around selected Florida power plants following winter cold fronts: 1984–1985. Biol Conserv 38:103–13.

Roberts, B.S. 1979. Occurrence of Gymnodinium breve red tides along the west and east coasts of Florida during 1976 and 1977. In D.L. Taylor and H.H. Seliger, eds. Toxic dinoflagellate blooms, pp. 199–202. Elsevier North Holland, Inc. New York.

Rue, E., and K. Bruland. 2001. Domoic acid binds iron and copper: a possible role for the toxin produced by the marine diatom Pseudo-nitzschia. Mar Chem 76:127–134.

Scholin, C.A., F. Gulland, G.J. Doucette, S. Benson, M. Busman, F.P. Chavez, J. Cordaro, R. Delong, A. De Vogelaere, J. Harvey, M. Haulena, K. Lefebvre, T. Lipscomb, S. Loscutoff, L.J. Lowenstine, R. Marin, P.E. Miller, W.A. Mclellan, P.D.R. Moeller, C.L. Powell, T. Rowles, P. Silvagni, M. Silver, T. Spraker, V. Trainer, and F.M. Van Dolah. 2000. Mortality of sea lions along the central California coast linked to a toxic diatom bloom. Nature 403:80–84.

Sekula-Wood, E., A. Schnetzer, C.R. Benitez-Nelson, C. Anderson, W.M. Berelson, M.A. Brzezinski, J.M. Burns, D.A. Caron, I. Cetinic, J.L. Ferry, E. Fitzpatrick, B.H. Jones, P.E. Miller, S.L. Morton, R.A. Schaffner, D.A. Siegel, and R. Thunell. 2009. Rapid downward transport of the neurotoxin domoic acid in coastal waters. Nat Geosci 2:272–275.

Sellner, K.G., G.J. Doucette, and G.J. Kirkpatrick. 2003. Harmful algal blooms: causes, impacts and detection. J Ind Microbiol Biot 30:383–406.

Shumway, S.E., S.M. Allen, and P. Dee Boersma. 2003. Marine birds and harmful algal blooms: sporadic victims or under-reported events? Harmful Algae 2:1–17.

Sierra-Beltran, A., M. Palafox-Uribe, J. Grajales-Montiel, A. Cruz-Villacorta, and J.L. Ochoa. 1997. Sea bird mortality at Cabo San Lucas, Mexico: evidence that toxic diatom blooms are spreading. Toxicon 35:447–453.

Sierra-Beltran, A.P., R. Cortes-Altamirano, J.P. Gallo-Reynoso, S. Licea-Duran, and J. Egido-Villarreal. 2005. Is *Pseudo-nitzschia pseudodelicatissima* toxin the principal cause of sardines, dolphins, sea lions and pelicans mortality in 2004 in Mexico? Harmful Algae News 29:6–8.

Silvagni, P.A., L.J. Lowenstine, T. Spraker, T.P. Lipscomb, and F.M. Gulland. 2005. Pathology of domoic acid toxicity in California sea lions (*Zalophus californianus*). Vet Pathol 42:184–191.

Simon, J.L., and D.M. Dauer. 1972. A quantitative evaluation of red-tide induced mass mortalities of benthic invertebrates in Tampa Bay, Florida. Environ Lett 3:229–234.

Summerson, H.C., and C.H. Peterson. 1990. Recruitment failure of the bay scallop, *Argopecten irradians concentricus*, during the first red tide, *Ptychodiscus brevis*, outbreak recorded in North Carolina. Estuaries 13:322–331.

Swanson, K.M., L.J. Flewelling, M. Byrd, A. Nunez, and T.A. Villareal. 2010. The 2008 Texas *Dinophysis ovum* bloom: distribution and toxicity. Harmful Algae 9:190–199.

Tabor, G.M., and A.A. Aguirre. 2004. Ecosystem health and sentinel species: adding an ecological element to the proverbial "canary in the mineshaft". EcoHealth 1:226–228.

Teegarden, G.J., R.G. Campbell, D.T. Anson, A. Ouellett, B.A. Westman, and E.G. Durbin. 2008. Copepod feeding response to varying *Alexandrium* spp. cellular toxicity and cell concentration among natural plankton samples. Harmful Algae 7:33–44.

Tester, P.A., R.P. Stumpf, F.M. Vukovich, P.K. Fowler, and J.T. Turner. 1991. An expatriate red tide bloom: transport, distribution, and persistence. Limnol Oceanogr 36:1053–1061.

Tester, P.A., and K.A. Steidinger. 1997. *Gymnodinium breve* red tide blooms: Initiation, transport, and consequences of surface circulation. Limnol Oceanogr 42:1039–1051.

Tillmann, U., M. Elbrachter, B. Krock, U. John, and A. Cembella. 2009. *Azadinium spinosum* gen. et sp. nov. (Dinophyceae) identified as a primary producer of azaspiracid toxins. Eur J Phycol 44:63–79.

Torres De La Riva, G., C. Kreuder Johnson, F.M.D. Gulland, G.W. Langlois, J.E. Heyning, T.K. Rowles, and J.A.K. Mazet. 2009. Association of an unusual marine mortality event with *Pseudo-nitzschia* spp. blooms along the southern California coastline. J Wildl Dis 45:109–121.

Trainer, V.L., N.G. Adams, B.D. Bill, C.M. Stehr, J.C. Wekell, P. Moeller, M. Busman, and D. Woodruff. 2000. Domoic acid production near California coastal upwelling zones, June 1998. Limnol Oceanogr 45:1818–1833.

Tubaro, A., A. Giangaspero, M. Ardizzone, M.R. Soranzo, F. Vita, T. Yasumoto, J.M. Maucher, J.S. Ramsdell, and S. Sosa. 2008. Ultrastructural damage to heart tissue from repeated oral exposure to yessotoxin resolves in 3 months. Toxicon 51:1225–1235.

Twiner, M.J., N. Rehmann, P. Hess, and G.J. Doucette. 2008. Azaspiracid shellfish poisoning: a review on the chemistry, ecology, and toxicology with an emphasis on human health impacts. Mar Drugs 6:39–72.

Twiner, M.J., L.J. Flewelling, F. Van Dolah, L. Schwacke, Z. Wang, T. Leighfield, C.A. Heil, S. Bowen, B. Mase, and T. Rowles. 2009. Florida panhandle bottlenose dolphin mortality events between 1999 and 2006: one common toxin—three distinct events, pp. 63–64. Fifth Symposium on Harmful Algae in the US, 15–19 November, Ocean Shores, Washington.

Twiner, M.J., S.E. Fire, L. Schwacke, L. Davidson, Z. Wang, S. Morton, S. Roth, B. Balmer, T.K. Rowles, and R.S.Wells. 2011. Concurrent exposure of bottlenose dolphins (*Tursiops truncatus*) to multiple algal toxins in Sarasota Bay, Florida, USA. PLoS ONE 6(3):e17394.

Van Dolah, F.M. 2000. Marine algal toxins: origins, health effects, and their increased occurrence. Environ Health Persp 108:133–141.

Van Dolah, F.M. 2005. Effects of harmful algal blooms. In J.E. Reynolds III, W.F. Perrin, R.R. Reeves, S. Montgomery, and T.J. Ragen, eds. Marine mammal research: conservation beyond crisis, pp. 85–101. Johns Hopkins University Press, Baltimore, Maryland.

Waring, G.T., E. Josephson, C.P. Fairfield-Walsh, and K. Maze-Foley. 2009. US Atlantic and Gulf of Mexico Marine Mammal Stock Assessments 2008. NOAA Technical Memorandum NMFS-NE-210.

Wells, R.S., H.L. Rhinehart, L.J. Hansen, J.C. Sweeney, F.I. Townsend, R. Stone, D.R. Casper, M.D. Scott, A.A. Hohn, and T.K. Rowles. 2004. Bottlenose dolphins as marine ecosystem sentinels: developing a health monitoring system. EcoHealth 1:246–254.

Wilson, M.T., and P.J. Jokiel. 1986. Ciguatera at Midway: an assessment using the Hokama "stick test" for ciguatoxin. NOAA Tech Rept NOAA-SWFSC-86-1.

Wong, C.-K., P. Hung, K.L.H. Lee, and K.-M. Kam. 2005. Study of an outbreak of ciguatera fish poisoning in Hong Kong. Toxicon 46:563–571.

Work, T.M., B. Barr, A.M. Beale, L. Fritz, M.A. Quilliam, and J.L.C. Wright. 1993. Epidemiology of domoic acid poisoning in brown pelicans (*Pelecanus occidentalis*) and Brandt's cormorants (*Phalacrocorax penicillatus*) in California. J Zoo Wildlife Med 24:54–62.

Zabka, T.S., T. Goldstein, C. Cross, R.W. Mueller, C. Kreuder-Johnson, S. Gill, and F.M.D. Gulland. 2009. Characterization of a degenerative cardiomyopathy associated with domoic acid toxicity in California sea lions (*Zalophus californianus*). Vet Pathol 46:105–119.

# 27

# BELUGA FROM THE ST. LAWRENCE ESTUARY

## A Case Study of Cancer and Polycyclic Aromatic Hydrocarbons

## Daniel Martineau

The problems usually associated with defining the health status of wildlife populations, especially acute in marine animals, are partly alleviated in the small population of beluga (*Delphinapterus leucas*), which inhabits a stretch of the St. Lawrence Estuary centered on the mouth of the Saguenay Fjord, only 600 km away from Montreal. The St. Lawrence Estuary drains the Great Lakes, historically the most industrialized part of North America; hence, industrial compounds have contaminated the St. Lawrence Estuary and its beluga population for over half a century. This population, the southernmost of the species worldwide, is unique because of its accessibility to investigation and geographic isolation from the Arctic, the natural habitat of most beluga. Unfortunately, it has dwindled from an estimated 10,000 at the beginning of the 20th century to approximately 1,000 animals currently (Michaud and Béland 2001).

St. Lawrence Estuary beluga became geographically isolated from the Arctic 10,000 years ago, when the region, which had been caving under the weight of a mile-thick layer of glacial ice, bounced back because warming climates caused the melting ice to retreat northwards. The isolation of that population has been confirmed by genetic analysis and by its distinctly much higher levels of industrial contaminants such as polychlorinated biphenyls (PCBs) and polycyclic aromatic hydrocarbons (PAHs) (Martineau et al. 1987,

1988; Hobbs et al. 2003; McKinney et al. 2004, 2006a,b). Together the isolation and the severe reduction suffered by this population during the 20th century probably explain its current genetic homogeneity (Patenaude et al. 1994; Gladden et al. 1999; Murray et al. 1999).

## POSTMORTEM EXAMINATIONS

From 1982 to 2009 (28 years), 423 carcasses were observed, an average of 15 a year; slightly less than half (46%, n = 193) were examined, an average of seven carcasses a year, at the College of Veterinary Medicine (Martineau et al. 2002). The three primary causes of death of St. Lawrence Estuary beluga were cancer (15%, n = 29), metazoan and protozoan parasites (19%), and other infectious agents, mostly bacterial (15%).

### Cancer

Cancer caused the death of 21% of adults, a rate comparable only to that of people in the developed world, and higher than that in any other population of wild terrestrial or aquatic mammals, with the possible exceptions of virally induced tumors in rodents and woodchucks (*Marmota monax*), Tasmanian devils (*Sarcophilus harrisii*), and California sea lions (*Zalophus californianus*).

St. Lawrence Estuary beluga are commonly affected by infectious agents that have generally been associated with immune suppression in humans and other animals (Martineau et al. 1988; De Guise et al. 1994a, b; Mikaelian et al. 2000).

## Gastrointestinal Tract Cancer

Epithelial cancers of the gastrointestinal (GI) tract were strikingly predominant, especially intestinal cancers (n = 8), of which seven were close to the gastric compartments. Epithelial cancer of the proximal intestine is rare in all animal species and people. Epithelial gastric cancer is also rare in all animals and in people in the western world. It is frequent, however, in developing countries where it is associated with *Helicobacter pylori* infection, and in Japan where it is associated with salt- and nitrite-preserved food (Crew and Neugut 2006; Tsugane and Sasazuki 2007).

## Mammary Gland Cancer

Mammary gland cancers were the cause of death of 5 (7.8%) of the 64 adult females necropsied between 1982 and 2010. These cancers have not been reported in other marine mammals and are extremely rare in herbivores, including cattle. Only isolated cases have been reported in other free-ranging wildlife species (Gillete et al. 1978; Hruban et al. 1988; Veatch and Carpenter 1993; Raymond and Garner 2000, 2001). In zoos, mammary cancers have frequently affected captive felids treated with progestin contraceptives and occasionally affect other zoo mammals. In captivity, jaguars (*Panthera onca*) are at higher risk for mammary cancer and also have a high prevalence of ovarian papillary cystadenocarcinomas, a profile similar to women with BRCA1 mutations. In addition, 10 of 81 captive aged female black-footed ferrets (*Mustela nigripes*) had mammary adenocarcinoma. No mammary cancers have been observed in two free-ranging inbred populations of two endangered terrestrial carnivores, the cheetah (*Acinonyx juvatus*) and the island fox (*Urocyon littoralis*) (Lair et al. 2002; Munson and Moresco 2007).

In contrast, mammary cancers are common in humans, domestic carnivores, and laboratory rodents. Viruses cause mammary cancers in highly inbred laboratory mice, highlighting the important role played by genetics in virally induced cancer. In women, the increased incidence of breast cancer seen over the past decades is firmly related to increased lifelong exposures to natural estrogens and perhaps to interactions between certain cytochrome P450 (CYP) enzymes, such as CYP1A1 generic variants (see below), and estrogen-mimicking environmental contaminants (Brody et al. 2007; Keri et al. 2007).

In cattle and people in certain parts of the world, small intestinal cancers result from an interaction between alimentary carcinogens (bracken fern) and papillomaviruses (Campo 2002). It is possible that a similar interaction exists between a virus and environmental carcinogen compounds in St. Lawrence Estuary beluga (Martineau et al. 1988; De Guise et al. 1994a,b).

## Cancer in Wildlife

Cancer is rare in wildlife, both aquatic and terrestrial, with the possible exception of virus-induced cancer in woodchuck and retrovirus-induced leukemia in rodents and contaminants associated cancer in bottom-dwelling fish. Tasmanian devils (see Chapter 19 in this book) and California sea lions are also (recent) exceptions. Typically, death due to cancer accounts for less than 1% in cetaceans and less than 2% in terrestrial mammals (Martineau et al. 2002). The most frequent causes of death in terrestrial wildlife are trauma and starvation, and in cetaceans, entanglement in fishing gear, parasitic and bacterial pneumonia, and other infectious diseases (Aguirre et al. 1999; Jepson et al. 1999). St. Lawrence Estuary beluga account for more than the quarter of cancers found in cetaceans (De Guise et al. 1994a,b; Martineau et al. 2002). GI tract cancers have not been reported in other free-ranging cetaceans, except in two harbour porpoises (*Phocoena phocoena*) from European waters, which were affected by gastric cancers with multiple metastases (Breuer et al. 1989; Siebert et al. 2001; Martineau et al. 2002; Newman and Smith 2006; Siebert et al. 2010).

In bottom-dwelling fish, labial papilloma and liver cancer are strongly associated with chemical contamination of sediments (McAloose and Newton 2009). Lake whitefish (*Coregonus clupeaformis*) are the only salmonids feeding on benthic fauna. Tissue concentrations of organochlorines (OC) and of heavy metals found in lake whitefish living in the St. Lawrence Estuary were three to five times higher than those of sympatric fish species, and these high concentrations coincided with a high prevalence of liver cancer in this species (Mikaelian and Newton 1998, 2002). Thus, both beluga and lake whitefish, two aquatic vertebrates that widely diverge taxonomically, may be affected by cancer because they feed on benthic marine organisms, an unusual feature within their respective taxonomic group.

In free-ranging mammals, there are only two instances of enzootic cancer: urogenital cancers in California sea lions and facial tumor in Tasmanian devils (Lipscomb et al. 2010; McAloose and Newton 2009). A new herpesvirus, contaminants (PCBs) and genetics seem to be involved in urogenital cancers. In Tasmanian devils, the combination of genetics and aggressive reproductive behavior has allowed the transplantation of clonal tumor cells between animals. Facial tumors in Tasmanian devils and the transmissible venereal tumor of dogs are the only instances of cancer transplantation in animals, including people (Lipscomb et al. 2000; Acevedo-Whitehouse et al. 2003; Ylitalo et al. 2005; Murchison 2008; Siddle et al. 2010).

## Frequency of Cancer in Beluga and Genetic Susceptibility

In people, genetic susceptibility to cancer takes two forms: first, hereditary cancer syndromes such as familial adenomatous polyposis, and second, population susceptibility, where an ensemble of individuals has an increased risk of cancer (but not as high as in hereditary cancer syndromes). Because inbreeding has led to some degree of genetic homogeneity in St. Lawrence Estuary beluga, the possibility of a hereditary cancer syndrome within that population must be considered (Patenaude et al. 1994; Murray et al. 1995, 1998, 1999). Hereditary cancer syndromes affect multiple—and most often young—members of a same family (Fearon 1997; Perera 1997). Beluga with cancer are not younger than beluga dead of other causes (Fig. 27.1). In addition, other genetically homogeneous free-ranging wildlife populations are not affected by high cancer rates except for some small captive populations—cheetahs treated with contraceptive and highly inbred black-footed ferrets (O'Brien 1994; Munson et al. 1999). Captivity clearly plays a role in the etiology of tumors in black-footed ferrets, most likely by extending the lifespan of these animals and possibly by exposing them to carcinogenic

compounds (Lair et al. 2002). Thus, there is no evidence that cancer in St. Lawrence Estuary beluga is a hereditary cancer syndrome.

In population susceptibility, an ensemble of individuals has an increased risk of cancer because of a specific and common genetic feature caused by normal polymorphism (Fearon 1997). This feature most often influences the metabolism of carcinogenic xenobiotics. It is possible that some St. Lawrence Estuary beluga, because of normal polymorphism in the genes metabolizing xenobiotics such as CYP1A1, along with high contamination levels by PCB and PAHs, have highly induced CYP1A1 in the proximal intestinal epithelium, rendering cells susceptible to mutagenesis by DNA-damaging metabolites generated from PAHs. There is no evidence that cancer is frequent in beluga as a species. In a recent review of cancer in marine mammals, a brain carcinoma was the single cancer reported in a (captive) beluga outside the St. Lawrence Estuary beluga population (Newman and Smith 2006). We diagnosed another brain tumor in a beluga kept captive in a Canadian aquarium (Mergl et al. 2007).

## CONTAMINANTS

### Polycyclic Aromatic Hydrocarbons

PAHs are a large family of carbon-based molecules consisting of fused aromatic rings, of which many are classified as carcinogens by the International Agency for Research in Cancer (IARC) of the World Health Organization. These compounds are produced in most industrial metallurgic processes (Shopland et al. 1991; Hecht 2003) and very commonly when organic compounds are burned, for instance in forest fires, charcoal-broiled meat, coal combustion, volcanic eruptions, and smoking (80% of lung cancers are due to smoking) (Sasco et al. 2004; Vineis et al. 2004)

The carcinogenicity of PAHs was documented as early as 1761 by John Hill, who observed an association between nasal cancer and tobacco snuff, and in 1775 by Percivall Pott, who noted high rates of scrotal cancer in chimney sweeps in London. These cancers were due to exposure to soot, a PAH-rich material. Frequent showers led to an almost immediate drop in the rate of scrotal cancer (in Denmark) a few years later. In 1915, Yamagiwa reliably caused skin cancer by painting repeatedly the ears of rabbits with coal tar

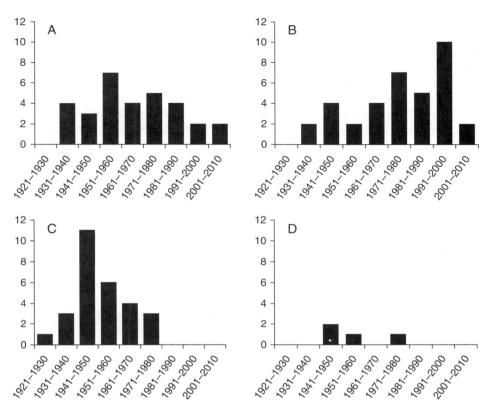

**Figure 27.1:**
Causes of mortality in beluga who inhabit the St. Lawrence Estuary and of which the age has been determined from 1982 to 2009 (n = 111). Age determined as year of birth. Some belugas are not represented because their age has not been determined yet. (A) Bacteria and viruses (n = 31). (B) Parasites (metazoans and protozoa) (n = 36). (C) All cancer types (n = 28). (D) Mammary gland cancer (n = 4, age of a single animal pending).

condensates (Weinberg 2007b). Coal tar condensates are also PAH-rich materials very similar to soot and to coal tar pitch volatiles, which are released by metallurgical industries, particularly by aluminum smelters that still use the Söderberg process (see next section).

Importantly, PAHs are found in the environment as complex mixtures. In the rare studies where complex PAH mixtures have been tested, the chronic ingestion of coal tar caused tumors in the small intestines and in other organs of laboratory mice (Goldstein 2001). Benzo(a)pyrene (B(a)P) is one of the most-studied PAHs because of its abundance and its well- and long-known carcinogenicity: it produces tumors

in all species where the effects of its exposure by any route—dermal, inhalation, intratracheal, intrabronchial, subcutaneous, intraperitoneal, intravenous—have been studied (IARC 2010).

## Saguenay, PAH Contamination, and Aluminum Production

Aluminum smelters were progressively constructed upstream and upwind of the St. Lawrence Estuary beluga habitat, on the Saguenay River shore,[1] starting as early as 1926. This site fulfilled the two conditions required for aluminum production: access to the sea

---

1 The Saguenay Fjord is a 104-km-long inland extension of the St. Lawrence Estuary, 200 to 300 m deep, with strong tidal currents at its mouth. It is the only fjord on the North American mainland.

(for shipping) and the availability of vast and cheap amounts of hydroelectricity. As a result of this rare conjunction, 4% of the world's aluminum production[2,3] now takes place in that region. In its natural state, aluminum is firmly bound to oxygen in aluminum oxide, $Al_2O_3$ (alumina), which is found in bauxite, and is shipped directly to the Saguenay smelters from Australia, Brazil, Guinea, and Jamaica. Alumina is chemically extracted from bauxite using sodium hydroxide, and aluminum is then extracted by electrolysis from alumina.

These operations are all carried out on site. In the Saguenay region, hydroelectric dams owned by the aluminum smelter company for more than 80 years provide a large proportion of the required power, and the Quebec government-run power company supplies the rest at a deep discount. Two types of anodes, both consisting of baked petroleum coke and coal tar or petroleum pitch, are employed for the electrolysis of alumina: (1) prebaked anodes, in modern smelters, and (2) baked-in-place anodes, used in the antiquated Söderberg process. This process, employed until 2004 in the Saguenay region, has been abandoned (recently) in the Western world because it releases large amounts of PAHs in ambient air owing to the anodes being burned by the strong electrical power that is run through them. This is why worldwide, the highest atmospheric B(a)P concentrations have been found in the aluminum smelters that use the Söderberg process (Straif et al. 2005; IARC 2010). Between 1937 and 1980, the Saguenay aluminum smelters released 40,000 tons of PAHs, which have accumulated in the fjord watershed (5 kg PAH/ton Al). Of these 40,000 accumulated tons, 20 tons were released every year, and 3% of these PAHs were B(a)P (Smith and Levy 1990). The tissue B(a)P concentrations in mussels (*Mytilus edulis*) transplanted in the Saguenay River increased 200 times after their transplantation (Cossa et al. 1983).

These high B(a)P concentrations probably explain the genotoxicity (the ability to cause DNA damage to living organisms) of the Saguenay sediments and particulate matter (White et al. 1998). Tissues of bivalve mollusks collected in the Saguenay were also genotoxic proportionally to their PAH tissue content. There was no gradient of genotoxicity towards the smelters; rather, genotoxicity was proportional to the drainage basin area at each collection site, consistent with airborne PAH contamination of surface waters by the smelters (Smith et al. 1990; White et al. 1997). In 1986, the sediments of the Saguenay River contained 500 to 4,500 ppb of total PAH (dry weight), a concentration significantly higher than that measured within a busy industrial harbor (Osaka harbor) (2,870 ppb) (Martel et al. 1986, 1987; Fig. 5 in Martineau et al. 2002). Both the magnitude and the persistence of this threat to public health have been clearly recognized (Smith and Levy 1990).

## Cancer in Aluminum Workers

Aluminum workers of the Saguenay–Lac Saint-Jean region are affected by a high prevalence of lung and urinary bladder cancer, for which they are now compensated financially by CSST (Quebec Workers Health and Safety Board) because these cancers have been epidemiologically related with exposure to PAH-rich coal tar pitch volatiles released by the Söderberg process (Thériault et al. 1984; Armstrong et al. 1988, 2009; Tremblay et al. 1995). This is in agreement with IARC, which states that "there is sufficient evidence in humans for the carcinogenicity of occupational exposures during aluminum production" and that "occupational exposures during aluminum production are carcinogenic to humans" (IARC Group 1, IARC 2010). Stomach cancers are also found in excess in aluminum workers in the Saguenay region and elsewhere. In the Saguenay, there was no statistically significant relationship between stomach cancer rates and exposure to coal tar pitch volatiles; in contrast, such a relationship was found in a similar plant in British Columbia (Armstrong et al. 1994; Gibbs et al. 2007; Spinelli et al. 2006). Together, these observations support that a human population, along with a population of long-lived, highly evolved mammals, may be affected by specific types of cancer because both populations share the same habitat and are exposed to the same environmental contaminants.

---

2  63% of Canada's production.
3  The production of 1 ton of aluminum requires 13 kilowatts. The aluminum company in the Saguenay region produces 1,950 megawatts annually using its own dams—that is, roughly enough electricity for three medium-sized cities (a medium-sized city of 100,000 inhabitants consumes about 600 megawatts/year).

In 2002, a single smelter in the Saguenay region was still using the Söderberg process, and the company planned to extend its operations for at least another decade. The same year, we published a paper reporting the high rate of cancer in St. Lawrence Estuary beluga and its likely relation with PAH exposure and high rates of cancer in aluminum workers and in the local general population (Martineau et al. 2002). The company replied that no such relation was possible, and that PAH emissions had been dramatically reduced and could not be blamed for cancer in the general human population. Yet the company closed the antiquated plant two years later because it was "highly polluting and no more viable economically." Thus, the lightness of aluminum, often promoted as a fuel-saver for motor vehicles, has come in fact with a high environmental cost. For more than 80 years, PAHs released by the smelters have contaminated the Saguenay river sediments, the beluga inhabiting it, the aquatic and terrestrial animals that live in that region, and the surface waters used for drinking water.

## Increased Cancer Rates in the Human Population Living in the Region

In the Saguenay region, the incidence of GI tract cancer in people not working in the aluminum industry is higher than in the rest of Quebec province (Lebel et al. 1998; Fig. 1 in Martineau et al. 2002). This may be related to the fact that the population gets 79% of its drinking water from local surface water (rivers and lakes) that have been contaminated by large quantities of PAHs released locally into the atmosphere by local aluminum smelters (Anonymous, 1995a,b; 2001; 2002).

Following our 2002 publication, it was argued that if PAHs are responsible for causing cancer in workers and in the general population, then gastric cancer and small intestinal cancer, for instance, should equally affect aluminum workers and the general population exposed to PAHs; at the time, the incidence of GI tract cancers in aluminum workers had not been examined (Martineau et al. 2002; Thériault et al. 1984, 2002). Five years later, a high rate of gastric cancer was found in these workers, but an association with B(a)P exposure was not found (Gibbs and Sevigny 2007). However, other researchers found such an association in a similar aluminum smelter in British Columbia (Spinelli et al. 2006).

The route of absorption of B(a)P often determines which organ is targeted by carcinogenesis. Aluminum workers and the general population inhabiting the area have not been exposed to B(a)P through the same route. Thériault et al. (2002) wrongly assumed that aluminum workers drink the same water as the general population. In fact, workers drank bottled water for decades (R. Lapointe, Rio Tinto, personal communication 2003), whereas at home other family members drank surface water contaminated by PAHs. Thus for decades workers have been exposed to PAH by the respiratory route at the smelter, while at home family members have been exposed orally to PAHs through drinking water, which may explain why different cancers are seen in the two populations.

The aluminum company has argued that there is no proof that PAHs cause cancer in beluga whales. This argument is reminiscent of that used by tobacco companies—the lack of experimental proof supporting that tobacco smoke causes lung cancer in people. Overwhelming evidence supports the carcinogenicity of PAHs present in tobacco smoke in laboratory animal and in vitro models (Hecht 2003; Michaels and Monforton 2005; Alberg et al. 2007). This argument ignores the historical and experimental evidence that has accumulated over half a century, obtained in all laboratory animals where B(a)P has been tested, and in a variety of in vitro cellular and molecular models. These results have been extrapolated to humans because people possess the same cellular machineries as other mammals (Fitzgerald et al. 2004; IARC 2010). As stated by R. Weinberg, the discoverer of the first human oncogene: "the embryogenesis, physiology and biochemistry of mammals is very similar, indeed, so similar that lessons learned through the study of laboratory animals are almost always transferable to an understanding of human biology" (Weinberg 2007a). It is for the same reasons that most new drugs have been and can be developed, and their toxicity can be predicted.

## Beluga, Worms, and PAHs

St. Lawrence Estuary belugas are omnivorous, as shown by various studies, such as the analysis of stomach contents from St. Lawrence Estuary beluga killed in the 1940s, and recently, stable isotope analysis of tissues from stranded St. Lawrence Estuary beluga (Vladykov 1946; Lesage et al. 2001). Along with fish,

St. Lawrence Estuary beluga eat invertebrates, particularly annelids such as *Nereis virens*. To do so, beluga dig into sediments, an unusual behavior among tooth whales. Beluga frequent the Saguenay River and Fjord extensively (Caron et al. 1988; Gosselin et al. 2007). These annelids have lived in an environment contaminated by PAH since 1926 (Cossa et al. 1983; Martel et al. 1986, 1987; Jørgensen et al. 2005; Rewitz et al. 2006; Pelletier et al. 2009). Marine polychaetes ingest large amounts of sediment: consequently, they ingest sorbed PAHs and PAHs desorbed into water through their GI tract and through their body surfaces. For instance, the common lugworm (*Arenicola marina*) ingests up to 20 times its own body weight of wet sediment per day (Jørgensen et al. 2008). Thus, it is not surprising that *Nereis* collected in the Saguenay River in 2002 were contaminated by PAHs (0.68 µg/kg of B(a)P) (Pelletier et al. 2009). Importantly, the PAH concentrations in the Saguenay sediments measured in 1982 were 10 to 30 times higher than those measured in 2002, implying that, most likely, B(a)P concentrations in worms were also 10 to 30 times higher in 1980—that is, 6.8 to 20.4 µg/kg (Martel et al. 1986, 1987; Pelletier et al. 2009). In addition, in the 1980s, the initial B(a)P levels of blue mussels transplanted from a nonpolluted area to various sites along the fjord increased 200 times within a month (Cossa et al. 1983).[4] Finally, the contamination of the Saguenay River sediments by PAHs peaked in the 1960s and started to decline in the 1980s (Pelletier et al. 2009). Thus, the figures reported in the 1980s underestimate the amount of PAHs ingested by beluga in the 1960s.

The PAH concentrations found in *Nereis* in the Saguenay are within the range found in human food in Linxian, China, where the mortality rate from esophageal and proximal stomach cancer in people is 20%, probably the highest rate in the world. In that region, these cancers are suspected to be due to high B(a)P concentrations (3.1 to 13.8 ng/g food) in food due to coal- and wood-based cooking (Roth et al. 1998;

Poirier 2004). In the 1980s, thus, the B(a)P concentration in benthic invertebrates was about twice the PAH concentration estimated to cause esophageal cancer in people.

The daily amount of fish ingested by an adult beluga in captivity is equivalent to 2% to 3% of its body mass—that is, 13 to 20 kg (Robeck et al. 2005). A series of beluga (n = 107) inhabiting the St. Lawrence Estuary were dissected in the 1930s to determine whether beluga eat significant amounts of commercial fish species. Annelids found in their stomachs were identified, weighed, and counted (Vladykov 1946). A total of 25 to nearly 1,400 annelids were present in individual stomachs; individual annelids weighed 1.3 to 9.8 g.[5] Consequently, the stomach of a single beluga could have contained up to 14 kg of worms, a number in the low range of the daily amount of fish eaten by a beluga. Thus, based on an estimate (Pelletier et al. 2009) and on the above calculation, the B(a)P dose ingested daily by an adult beluga in 14 kg of worms in the 1980s could have been between 95 and 286 µg. Comparatively, based on 500 g, the amount of staple food ingested per day by a 70-kg man, the B(a)P concentration found in the food of Linxian inhabitants implies a daily ingestion of 1.6 to 6.9 µg B(a)P per person.[6] Based on 14 kg, the potential daily input of worms of a beluga, these figures become 44.8 to 193.2 µg B(a)P. Thus, in 1980, the putative daily diet of St. Lawrence Estuary beluga that would have consisted predominantly of benthic worms contained an amount of B(a)P that was between 0.5 and 6 times higher than that found in the daily intake of Linxian people (Table 27.1).

From the maximum daily amount of B(a)P that people can safely ingest[7] (Kang et al. 1995; Rothman et al. 1990; Fitzgerald et al. 2004), it is possible to estimate roughly the "safe" daily amount of B(a)P for a 900-kg beluga: 72 µg/day, or 26 mg/year. Importantly, about 40% of beluga body weight is fat, which metabolically is relatively inert; accordingly, the latter figure is probably overestimated by 40%, and a more

---

4   The smelter company claims that, in 2001, PAH emissions have been reduced 88% from the 1983 levels.
5   Also in Vladykov: "Although beluga of both sexes eat Nereis, lactating or pregnant females show a marked preference for this type of food."
6   These figures are remarkably consistent with the maximum daily amount of B(a)P (5.6 µg) that can be ingested by a 70-kg human adult to "safeguard human health" (a value derived from studies on laboratory animals, not from studies on human subjects) (Fitzgerald et al. 2004).
7   0.08 µg/kg/day or 5.6 µg of B(a)P for a 70-kg man (a figure determined from studies carried out on mice and rats).

Table 27.1  B(a)P in the Saguenay River Sediments and Biota, Estimation of Ingestion by Beluga and Comparison to PAH Ingestion by the Inhabitants of a Region (Linxian, China) Known for a High Rate of Esophageal Cancer Associated with Ingestion of PAH-Contaminated Food

| B(a)P concentration/ Amount | B(a)P concentration (or amount) in diet | Ingested daily | Ingested annually | Ingested cumulatively | Year | Reference |
|---|---|---|---|---|---|---|
| B(a)P concentration in Saguenay sediments | NR | NR | NR | NR | 1983 | Martel et al. 1987 |
| B(a)P concentration in Saguenay sediments | NR | NR | NR | NR | 2003 | Pelletier et al. 2009 |
| B(a)P concentrations in worms | 0.68 μg/g | NR | NR | NR | 2002 | Pelletier et al. 2009 |
| Estimated B(a)P in 14 kg of worms | 9.5 μg | 9.5 μg | 3.5 mg | 87.5 mg | 2002 | Estimation (this review) |
| Estimated B(a)P in 14 kg worms | 95 to 286 μg (6.8 to 20.4 μg/kg) | 95 to 286 μg | 35 to 104 mg | 1.4 to 4.2 g (40 y) | 1980 | Estimation (this review) |
| Linxian inhabitants | 3.1 to 13.8 ng/g | 1.6 to 6.9 μg in 500 g food; 44.8 to 193.2 μg in 14 kg food | | 23 to 100 mg (40 y) | 1985 | Roth et al. 1998; Poirier 2004 |
| Cigarettes: one pack a day | NR | 500 ng | | 7.3 mg (40 y) | NR | Poirier 2004 |
| Safe dose in people | 5.6 μg (daily) | 5.6 μg | 2 mg | 80 mg (40 y) | NR | Fitzgerald et al. 2004 |
| Well-done charcoal-broiled red meat (300 g, once a week) | 2.6 to 25.2 ppb | 780 ng/wk | 40 μg | 1.6 mg (40 y) | NR | Kang et al. 1995; Rothman et al. 1990 |
| Safe daily dose for an adult beluga | NR | 72 μg | 26 mg | 650 mg | NR | Estimation (this review) |

NR, not relevant.

realistic figure would thus be 40% lower or 43 μg/day. Consequently, one can estimate the daily amount of B(a)P ingested by an adult beluga in the 1980s to be 1.3 to 4 times higher than the amount estimated here to be safe in that species, or 2.2 to 6.7 times higher if lipid tissue is considered.

In Linxian, the cumulative ingestion of 23 to 100 mg over 40 years, or 0.6 to 2.8 g adjusted for a 14-kg daily food intake, is suspected to cause esophageal cancer (Poirier 2004). Comparatively, the annual amount ingested by a beluga would have been between 35 and 104 mg; over 40 years, a beluga would have cumulatively ingested between 1.4 g and 4.2 g of B(a)P—that is, between 0.5 and 7 times the cumulative intake associated with esophageal cancer in Linxian. The amount of PAH ingested by beluga may be either over- or underestimated for various reasons. Firstly, belugas, being omnivorous, do not only ingest worms; however, individual beluga may have particular tastes. Secondly, *Nereis* density in sediments can be very high (between 35 and 3,700 individuals/m$^2$ of sediments); this species is resistant to changes in salinity and anoxia, making that invertebrate a potentially abundant food supply for beluga all year round, obtained with minimal effort (Scaps 2002). *Nereis* may have become a more important component of the beluga diet in the second half of the 20th century: in the St. Lawrence Estuary and Gulf, as elsewhere, fish abundance, especially cod, has generally plummeted owing to overfishing, anoxia, warmer temperature, loss of habitat, and possibly diseases (Lambert et al. 1997; Gilbert et al. 2005). Thirdly, PAH adducts or metabolites, not PAH, are usually detected in living tissues because CYPs, the enzymes that normally destroy exogenous and some endogenous compounds, rapidly degrade PAHs into unstable metabolites that form adducts. It is thus surprising that any PAHs at all were detected in *Nereis*. CYPs are expressed in annelids, and CYP expression is induced by PAHs, albeit at lower levels than in vertebrates (Jørgensen et al. 2005; Pelletier et al. 2009). Thus, the detection of PAH in Saguenay annelids by Pelletier's group may be explained by extensive PAH contamination overwhelming CYP degradation, and/or PAH toxicity for annelids, which can impair annelid CYP metabolic activity (Jørgensen et al. 2005). Similarly, the high PAH concentrations found in halibut in the Saguenay River may also reflect overwhelming contamination with PAH (Hellou et al. 1995).

### Carcinogenesis of PAHs Detected in Annelids in the Saguenay River

Of the 16 PAHs that were detected in the tissues of *Nereis* in the Saguenay River (Pelletier et al. 2009), nearly half (7/16 [44%]) are carcinogenic, probably carcinogenic, or possibly carcinogenic. Specifically, one PAH is classified as "carcinogenic for humans based on sufficient experimental evidence in animals and strong evidence that the mechanisms of carcinogenesis in animals also operate in exposed human beings" (IARC Group 1, IARC 2010). Two of the 16 PAHs (or 12%) are "probably carcinogenic to human beings" (IARC group 2A, IARC 2010). Four of the 16 PAHs (or 25%) are possibly carcinogenic to humans (IARC group 2B). Finally, six of the 16 PAHS (or 37%) "are not classifiable as to their carcinogenicity to humans because the evidence of carcinogenicity is inadequate in humans and inadequate or limited in experimental animals" (IARC group 3, IARC 2010).

## CYP and Metabolism of Contaminants and Drugs

Xenobiotic molecules are any absorbed molecules originating outside the body. These include drugs and environmental contaminants, of which many (PAHs and PCBs) are lipophilic. These would accumulate if enzymes were not present to metabolize them. Animals have evolved enzymes to make xenobiotics more hydrophilic so that the metabolites can be eliminated in urine or bile. These enzymes are classically grouped into two categories. Phase 1 enzymes (which include CYP) stop the biological activity of drugs and contaminants by catalyzing oxidation, reduction, or hydrolysis reactions. Phase 2 enzymes conjugate the products from phase 1 enzymes with larger endogenous hydrophilic groups to allow the elimination of these products. CYP1A1 metabolizes B(a)P to epoxide intermediates, which are further activated to highly reactive diol epoxides by the enzyme epoxide hydrolase (Xue et al. 2005). Diol epoxides bind DNA, thus forming a DNA adduct that will eventually lead to mutations in daughter cells, some of which can become cancerous (Fig. 27.2).

All mammals possess 18 CYP gene families, of which CYP 1, 2, 3, and 4 have been associated with cancers due to environmental contaminants (Shimada et al. 1989; Bartsch et al. 2000; Ko et al. 2001; Bethke et al. 2007; Singh et al. 2008; Olivieri et al. 2009). Most pro-carcinogen contaminants are lipophilic and are metabolized by the three members of the membrane-bound CYP1 enzymes (CYP1A1, CYP1A2, and CYP1B1). The latter enzymes are up-regulated by the aryl hydrocarbon receptor (AhR) when AhR ligands (molecules that bind the AhR) such as PAHs, PCBs, and dioxin (TCDD) bind the AhR. The activated complex then binds specific DNA sequences upstream of CYP1 genes, thus activating the expression of these genes. It results that AhR can be conceived of as a cellular sensor that detects lipophilic contaminants within the cell, and triggers cell defenses against these. Importantly, AhR binding by PCBs, dibenzodioxin, and polychlorinated dibenzofuran, followed by the activation of CYP1A1, triggers PCB toxicity such as thymic atrophy (which leads to immunosuppression), weight loss, and death in rodents.

PCBs are structurally similar to dioxin, explaining why PCB toxicity; like dioxin, is generally also mediated through AhR binding. Increased CYP activity in

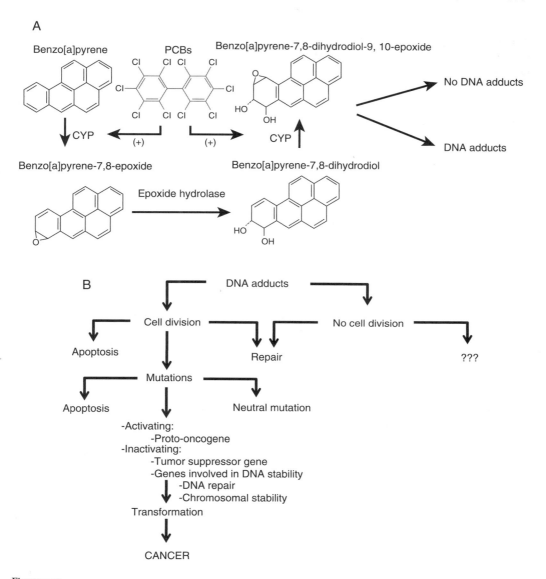

**Figure 27.2:**

Diagram showing the interactions between contaminants (PAH and PCB), CYP and DNA damage. (A) Interactions between CYPs, PCBs, and PAHs. Both PCBs and PAHs induce high levels of CYP expression. The net result is the production of larger amount of unstable metabolites (diolepoxide) by CYP. Note that PAHs are easily degraded by mammalian CYPs; PCBs, by contrast, are highly resistant to CYP-mediated metabolism. Diol epoxide also binds proteins and RNA, both very abundant in cells and located within the cytoplasm, by opposition to DNA, which is relatively more protected, highly folded in the nucleus. Thus, DNA adducts are not formed necessarily by unstable metabolites. In addition, the physical proximity of phase 2 enzymes, and the relative amounts of phase 1 and phase 2 enzymes in a given cell type, determines how much unstable metabolites will be generated and how much will be inactivated. (B) The many possible outcomes of DNA adduct formation, the crucial step in chemical carcinogenesis. Most mutations that result from DNA adduct formation and from subsequent cell division are repaired, are neutral, or lead to cell death. Rarely, a mutation will activate a proto-oncogene or inactivate a tumor cell suppressor gene or a gene involved in DNA stability. The accumulation of such mutations will eventually lead to cancer. A single mutation is unable to transform a human cell and more generally mammalian cells (rodent cells are more vulnerable to transformation). An accumulation of mutations over months and years is necessary to lead to transformation and cancer.

a given animal suggests exposure to these or to similar pollutants. The induced enzymes then degrade PCBs, slowly and partially, and mediate the toxic effects of PCB, probably by generating toxic metabolites. Consequently, PCBs may contribute indirectly to PAH carcinogenicity by inducing CYP, thus increasing the amount of mutagenic metabolites produced from PAHs. However, whereas the role of AhR in PAH-associated cancer is firmly established, that of CYP has still to be definitely demonstrated (Nebert and Dalton 2006).

The beluga AhR has been cloned and functionally assayed. It binds dioxin with an affinity at least as high as that of the AhR of strains of mice that are sensitive to PCBs, and significantly greater than that of the human AhR, suggesting that beluga might be as susceptible to PAH and PCB toxicity as susceptible mice strains and more susceptible than humans (Jensen and Hahn 2001; Jensen et al. 2010). Because contaminants such as PCBs did not exist in nature before the end of the 19th century, CYP enzymes did not encounter these compounds before that (Fig. 27.2A). In contrast to PCBs, PAHs have coexisted with mammals for millions of years because they result from the incomplete combustion of organic molecules, ubiquitous (but not continuous) in nature. This has given sufficient time to evolutionary processes to develop efficient PAH-degrading CYPs. For that reason and owing to their chemical nature, PAHs are vulnerable to and are degraded by CYPs (Fig. 27.2A). PAHs are released punctually in terms of time and space, for instance during forest fires and volcanic eruptions. Thus, evolution has come up with a compromise: PAH metabolites generated by CYPs are often unstable and carcinogenic, but overall, PAH degradation by CYP is protective in animals if exposure to PAH is limited in time and extent. During the evolution of *Homo sapiens*, however, the emergence of smoking and cooking has increased PAH exposure, and in the past two centuries, industrial, particularly metallurgical, processes that release PAHs on a massive scale have dramatically intensified PAH exposure in both space and time. Constant exposure to high amounts of PAHs can upset the evolutionary compromise between CYP beneficial and detrimental effects (Nebert and Dalton 2006; Uno et al. 2006; this review).

CYPs present in the small intestinal epithelium are among the first and major enzymes implicated in the biotransformation and subsequent detoxification or toxification of ingested xenobiotics. In rats and mice, the highest CYP concentrations occur in the duodenum, and the most abundant CYP is CYP1A1, which is also the most abundant inducible form in the duodenum, and is known to activate PAHs into carcinogenic metabolites (Zhang et al. 1997; Uno et al. 2008).

Induction of CYP1A1 in St. Lawrence Estuary beluga has been inferred from the presence of CYP-specific PCB metabolites and has been demonstrated by the detection of high CYP1A1 levels in the tissues of both Arctic and St. Lawrence Estuary beluga contaminated with PCBs (McKinney et al. 2004, 2006a,b; Wilson et al. 2005; Fossi et al. 2008). Considered together, the above observations suggest that intestinal CYP1A levels are elevated in St. Lawrence Estuary beluga, and that these high levels may trigger the development of intestinal cancer by activating ingested PAHs into carcinogenic compounds.

## CONCLUSIONS

Cancer has been observed in St. Lawrence Estuary beluga but not in other St. Lawrence Estuary marine mammals or in marine mammals elsewhere, as the health of several populations of marine mammals has been increasingly scrutinized. This observation supports that St. Lawrence Estuary beluga are uniquely exposed to chemical or biological mutagens or both, possibly through unique ecological features such as feeding on benthic invertebrates, similarly to lake whitefish. Some PAHs leave a signature on the host genome by causing mutations at specific sites in genes involved in cell proliferation or cell death. The finding of the same signature in tumors of St. Lawrence Estuary beluga, fish, and people would strongly support the etiologic role of these PAHs in carcinogenesis. The carcinogenic roles of some viruses (e.g., retroviruses and papillomaviruses) and bacteria (e.g., *Helicobacter pylori* and *H. hepaticus*) has been supported to various degrees in people and animals. Their potential role in the pathogenesis of cancer in St. Lawrence Estuary beluga remains to be clearly defined.

Biological exposure to PAHs can be measured using biomarkers such as PAH metabolites, for instance 1-hydroxypyrene (1-OHP) in urine, protein and DNA adducts in tissues[8] before cancer arises, avoiding suffering and loss of lives (Perera et al. 1992, 2000; Vineis and Perera 2007). In Quebec and elsewhere, the biological exposure to PAHs has been evaluated in people living near smelters by measuring urinary concentrations of 1-OHP. Concentrations were 2 to 2.5 times higher in persons living in the vicinity of the smelters than in persons living far from the smelter (Gilbert and Viau 1997; Jongeneelen 2001; Bouchard et al. 2009). Biological PAH exposure in the vicinity of the aluminum smelters has also been measured in woodchucks by quantifying DNA protein adducts, in benthic worms, in mussels, and in fish (Blondin and Viau 1992; Hellou et al. 1995; Pelletier et al. 2009; Picard-Bérubé et al. 1983). Sadly, in the Saguenay aluminum workers, only the rate of cancers has been used as a biomarker for PAH exposure. With sufficient resources, it would be conceivable to compare 1-OHP urinary concentrations in St. Lawrence Estuary beluga and Arctic beluga to further support the role of PAH in the cancers affecting St. Lawrence Estuary beluga.

## ACKNOWLEDGMENTS

I thank Dr. Stephane Lair for sharing his data and for his invaluable support in this work. I am indebted to Drs. Igor Mikaelian and André Dallaire for their help in this work. Many thanks also to all the veterinary students who helped throughout these years.

## REFERENCES

Acevedo-Whitehouse, K., F. Gulland, D. Greig, and W. Amos. 2003. Inbreeding: Disease susceptibility in California sea lions. Nature 422:35–35.

Aguirre, A. A., C. Brojer, and T. Morner. 1999. Descriptive epidemiology of roe deer mortality in Sweden. J Wildl Dis 35:753–762.

Alberg, A., J. Ford, and J. Samet. 2007. Epidemiology of lung cancer. Chest 132:29S.

Anonymous. 2002. Ministère de l'Environnement, Gouvernement du Québec. Effet des inondations de juillet 1996 sur les lacs et rivières de la région du Saguenay: contamination de l'eau, des sédiments et des poissons par les substances toxiques [in French]. Internal report. Québec.

Anonymous. 2001. Ministère de l'Environnement du Québec. Rapport concernant les campagnes d'échantillonnage de l'eau potable réalisées au Saguenay. Année 2000. Rapport final [in French]. Internal report. Québec.

Anonymous. 1995a. Environnement Canada. Synthèse des connaissances sur les aspects physiques et chimiques de l'eau et des sédiments du Saguenay. Rapport technique. Zones d'intervention prioritaire 22 et 23. Région du Québec. En40-216-17F. Montreal, Quebec.

Anonymous. 1995b. Environment Canada. Synthèse des connaissances sur les aspects socio-économiques du Saguenay. Rapport technique. Zones d'intervention prioritaire 22 et 23. En40–216/14F. Montreal, Quebec.

Armstrong, B., and G. Gibbs. 2009. Exposure–response relationship between lung cancer and polycyclic aromatic hydrocarbons (PAHs). Occup Environ Med 66:740–746.

Armstrong, B., C. Tremblay, D. Baris, and G. Thériault. 1994. Lung cancer mortality and polynuclear aromatic hydrocarbons: a case-cohort study of aluminum production workers in Arvida, Quebec, Canada. Am J Epidemiol 139:250–262.

Armstrong, B., C. Tremblay, and G. Thériault. 1988. Compensating bladder cancer victims employed in aluminum reduction plants. J Occup Environ Med 30:771–775.

Bartsch, H., U. Nair, A. Risch, M. Rojas, H. Wikman, and K. Alexandrov. 2000. Genetic polymorphism of CYP genes, alone or in combination, as a risk modifier of tobacco-related cancers. Cancer Epidemiol Biomark Prev 9:3–28.

Bethke, L., E. Webb, G. Sellick, M. Rudd, S. Penegar, L. Withey, M. Qureshi, and R. Houlston. 2007. Polymorphisms in the cytochrome P450 genes CYP1A2, CYP1B1, CYP3A4, CYP3A5, CYP11A1, CYP17A1, CYP19A1 and colorectal cancer risk. BMC Cancer 7:123.

Blondin, O., and C. Viau. 1992. Benzo (a) pyrene-blood protein adducts in wild woodchucks used as biological sentinels of environmental polycyclic aromatic hydrocarbons contamination. Arch Environ Contam Toxicol 23:310–315.

Bouchard, M., L. Normandin, F. Gagnon, C. Viau, P. Dumas, É. Gaudreau, and C. Tremblay. 2009. Repeated measures of validated and novel biomarkers of exposure to polycyclic aromatic hydrocarbons in

---

8  DNA adducts were measured in St. Lawrence Estuary beluga (Martineau et al. 1988).

individuals living near an aluminum plant in Quebec, Canada. J Toxicol Environ Health, Part A 72:1534–1549.

Breuer, E., B. Krebs, and R. Hofmeister. 1989. Metastasizing adenocarcinoma of the stomach in a harbor porpoise, *Phocoena phocoena*. Dis Aquat Organ 7:159–163.

Brody, J., K. Moysich, O. Humblet, K. Attfield, G. Beehler, and R. Rudel. 2007. Environmental pollutants and breast cancer. Cancer 109:2667–2711.

Campo, M. 2002. Animal models of papillomavirus pathogenesis. Virus Res 89:249–261.

Caron, L., and D. Sergeant. 1988. Yearly variation in the frequency of passage of beluga whales (*Delphinapterus leucas*) at the mouth of the Saguenay River, Quebec, over the past decade. Naturaliste canadien (Revue d'Ecologie et de Systématique) 115:111–116.

Cossa, D., M. Picard-Bérubé, and J. Gouygou. 1983. Polynuclear aromatic hydrocarbons in mussels from the estuary and northwestern gulf of St. Lawrence, Canada. Bull Environ Contam Toxicol 31:41–47.

Crew, K., and A. Neugut. 2006. Epidemiology of gastric cancer. J Gastroenterol 12:354–362.

De Guise, S., A. Lagace, and P. Beland. 1994a. Gastric papillomas in eight St. Lawrence beluga whales (*Delphinapterus leucas*). J Vet Diagn Invest 6:385–385.

De Guise, S., A. Lagacé, and P. Béland. 1994b. Tumors in St. Lawrence beluga whales (*Delphinapterus leucas*). Vet Pathol 31:444–449.

Fearon, E. 1997. Human cancer syndromes: clues to the origin and nature of cancer. Science 278:1043–1050.

Fitzgerald, D., N. Robinson, and B. Pester. 2004. Application of benzo (a) pyrene and coal tar tumor dose–response data to a modified benchmark dose method of guideline development. Environ Health Perspect 112:1341–1346.

Fossi, M., S. Casini, D. Bucalossi, and L. Marsili. 2008. First detection of CYP1A1 and CYP2B induction in Mediterranean cetacean skin biopsies and cultured fibroblasts by Western blot analysis. Mar Environ Res 66:3–6.

Gibbs, G., B. Armstrong, and M. Sevigny. 2007. Mortality and cancer experience of Quebec aluminum reduction plant workers, part 2: mortality of three cohorts hired on or before January 1, 1951. J Occup Environ Med 49:1105–1123.

Gibbs, G., and M. Sevigny. 2007. Mortality and cancer experience of Quebec aluminum reduction plant workers, part 4: cancer incidence. J Occup Environ Med 49:1351–1366.

Gilbert, D., B. Sundby, C. Gobeil, A. Mucci, and G. Tremblay. 2005. A seventy-two-year record of diminishing deep-water oxygen in the St. Lawrence estuary: The northwest Atlantic connection. Limnol Oceanogr 50:1654–1666.

Gilbert, N., and C. Viau. 1997. Biological monitoring of environmental exposure to PAHs in the vicinity of a Söderberg aluminium reduction plant. Occup Environ Med 54:619–621.

Gillete, D., H. Acland, and L. Klein. 1978. Ductular mammary carcinoma in a lioness. J Am Vet Med Assoc 173:1099–1102.

Gladden, J., M. Ferguson, M. Friesen, and J. Clayton. 1999. Population structure of North American beluga whales (*Delphinapterus leucas*) based on nuclear DNA microsatellite variation and contrasted with the population structure revealed by mitochondrial DNA variation. Mol Ecol 8:347–363.

Goldstein, L. 2001. To BaP or not to BaP? That is the question. Environ Health Perspect 109:A356.

Gosselin, J.-F., M. Hammill, and V. Lesage. 2007. Comparison of photographic and visual abundance indices of belugas in the St. Lawrence Estuary in 2003 and 2005. Mont-Joli, Quebec.

Hecht, S. 2003. Tobacco carcinogens, their biomarkers and tobacco-induced cancer. Nat Rev Cancer 3:733–744.

Hellou, J., P. Hodson, and C. Upshall. 1995. Contaminants in muscle of plaice and halibut collected from the St Lawrence Estuary and Northwest Atlantic. Chem Ecol 11:11–24.

Hobbs, K., D. Muir, R. Michaud, P. Béland, R. Letcher, and R. Norstrom. 2003. PCBs and organochlorine pesticides in blubber biopsies from free-ranging St. Lawrence River Estuary beluga whales (*Delphinapterus leucas*), 1994–1998. Environ Pollut 122:291–302.

Hruban, Z., W. Carter, T. Meehan, P. Wolff, W. Franklin, and S. Glagov. 1988. Complex mammary carcinoma in a tiger (*Panthera tigris*). J Zoo An Med 19:226–230.

IARC. 2010. IARC monographs on the evaluation of carcinogenic risks to humans. Some non-heterocyclic polycyclic aromatic hydrocarbons and some related exposures. Lyon.

Jensen, B., and M. Hahn. 2001. cDNA cloning and characterization of a high affinity aryl hydrocarbon receptor in a cetacean, the beluga, *Delphinapterus leucas*. Toxicol Sci 64:41–56.

Jensen, B.A., C.M. Reddy, R.K. Nelson, and M.E. Hahn. 2010. Developing tools for risk assessment in protected species: Relative potencies inferred from competitive binding of halogenated aromatic hydrocarbons to aryl hydrocarbon receptors from beluga (*Delphinapterus leucas*) and mouse. Aquat Toxicol 100:238–245.

Jepson, P.D., P.M. Bennett, C.R. Allchin, R.J. Law, T. Kuiken, J.R. Baker, E. Rogan, and J.K. Kirkwood. 1999. Investigating potential associations between chronic exposure to polychlorinated biphenyls and infectious disease mortality in harbour porpoises from England and Wales. Sci Total Environ 243–244:339–348.

Jongeneelen, F. 2001. Benchmark guideline for urinary 1-hydroxypyrene as biomarker of occupational exposure to polycyclic aromatic hydrocarbons. Ann Occup Hyg 45:3–13.

Jørgensen, A., A. Giessing, L. Rasmussen, and O. Andersen. 2008. Biotransformation of polycyclic aromatic hydrocarbons in marine polychaetes. Mar Environ Res 65:171–186.

Jørgensen, A., A. Giessing, L. Rasmussen, and O. Andersen. 2005. Biotransformation of the polycyclic aromatic hydrocarbon pyrene in the marine polychaete *Nereis virens*. Environ Toxicol Chem 24:2796–2805.

Kang, D., N. Rothman, M. Poirier, A. Greenberg, C. Hsu, B. Schwartz, M. Baser, J. Groopman, A. Weston, and P. Strickland. 1995. Interindividual differences in the concentration of 1-hydroxypyrene-glucuronide in urine and polycyclic aromatic hydrocarbon-DNA adducts in peripheral white blood cells after charbroiled beef consumption. Carcinogenesis 16:1079–1085.

Keri, R., S. Ho, P. Hunt, K. Knudsen, A. Soto, and G. Prins. 2007. An evaluation of evidence for the carcinogenic activity of bisphenol A. Reprod Toxicol 24:240–252.

Ko, Y., J. Abel, V. Harth, P. Bröde, C. Antony, S. Donat, H.-P. Fischer, M.E. Ortiz-Pallardo, R. Thier, A. Sachinidis, H. Vetter, H.-M. Bolt, C. Herberhold, and T. Brüning. 2001. Association of CYP1B1 codon 432 mutant allele in head and neck squamous cell cancer is reflected by somatic mutations of p53 in tumor tissue. Cancer Res 61:4398–4404.

Lair, S., I. Barker, K. Mehren, and E. Williams. 2002. Epidemiology of neoplasia in captive black-footed ferrets (*Mustela nigripes*), 1986–1996. J Zoo Wildl Med 33:204–213.

Lambert, Y., and J. Dutil. 1997. Condition and energy reserves of Atlantic cod (*Gadus morhua*) during the collapse of the northern Gulf of St. Lawrence stock. Can J Fish Aquat Sci 54:2388–2400.

Lebel, G., S. Gingras, P. Levallois, R. Gauthier, and M.-F. Gagnon. 1998. Etude descriptive de l'incidence du cancer au Québec de 1989 à 1993. Beauport, Québec, Canada.

Lesage, V., M. Hammill, and K. Kovacs. 2001. Marine mammals and the community structure of the Estuary and Gulf of St Lawrence, Canada: evidence from stable isotope analysis. Mar Ecol Prog Ser 210:203–221.

Lipscomb, T., D. Scott, R. Garber, A. Krafft, M. Tsai, J. Lichy, J. Taubenberger, F. Schulman, and F. D Gulland. 2000. Common metastatic carcinoma of California sea lions (*Zalophus californianus*): evidence of genital origin and association with novel gammaherpesvirus. Vet Pathol 37:609–619.

Lipscomb, T., D. Scott, and F.Y. Schulman. 2010. Primary site of sea lion carcinomas. Vet Pathol 47:185.

Martel, L., M. Gagnon, R. Massé, A. Leclerc, and L. Tremblay. 1986. Polycyclic aromatic hydrocarbons in sediments from the Saguenay Fjord, Canada. Bull Environ Contam Toxicol 37:133–140.

Martel, L., M. Gagnon, R. Massé, and A. Leclerc. 1987. The spatio-temporal variations and fluxes of polycyclic aromatic hydrocarbons in the sediments of the Saguenay Fjord, Québec, Canada. Water Res 21:699–707.

Martineau, D., P. Béland, C. Desjardins, and A. Lagacé. 1987. Levels of organochlorine chemicals in tissues of beluga whales (*Delphinapterus leucas*) from the St. Lawrence Estuary, Quebec, Canada. Arch Environ Contamin Toxicol 16:137–147.

Martineau, D., A. Lagace, P. Beland, R. Higgins, D. Armstrong, and L. Shugart. 1988. Pathology of stranded beluga whales (*Delphinapterus leucas*) from the St. Lawrence Estuary, Quebec, Canada. J Comp Pathol 98:287–310.

Martineau, D., K. Lemberger, A. Dallaire, P. Labelle, T. Lipscomb, P. Michel, and I. Mikaelian. 2002. Cancer in wildlife, a case study: beluga from the St. Lawrence estuary, Québec, Canada. Environ Health Perspect 110:285–292. http://www.ncbi.nlm.nih.gov/pmc/articles/PMC1240769/

McAloose, D., and A. Newton. 2009. Wildlife cancer: a conservation perspective. Nat Rev Cancer 9:517–526.

McKinney, M., A. Arukwe, S. De Guise, D. Martineau, P. Béland, A. Dallaire, S. Lair, M. Lebeuf, and R. Letcher. 2004. Characterization and profiling of hepatic cytochromes P450 and phase II xenobiotic-metabolizing enzymes in beluga whales (*Delphinapterus leucas*) from the St. Lawrence River Estuary and the Canadian Arctic. Aquat Toxicol 69:35–49.

McKinney, M., S. De Guise, D. Martineau, P. Béland, A. Arukwe, and R. Letcher. 2006a. Biotransformation of polybrominated diphenyl ethers and polychlorinated biphenyls in beluga whale (*Delphinapterus leucas*) and rat mammalian model using an in vitro hepatic microsomal assay. Aquat Toxicol 77:87–97.

McKinney, M., S. De Guise, D. Martineau, P. Béland, M. Lebeuf, and R. Letcher. 2006b. Organohalogen contaminants and metabolites in beluga whale (*Delphinapterus leucas*) liver from two Canadian populations. Environ Toxicol Chem 25:1246–1257.

Mergl, J., E. Gehring, M. Hasselblatt, and D. Martineau. 2007. Brain tumour (choroid plexus papilloma) in a captive beluga whale (*Delphinapterus leucas*). International Association for Aquatic Animal Medicine 38th Annual Conference, Orlando, FL.

Michaels, D., and C. Monforton. 2005. Manufacturing uncertainty: contested science and the protection of the public's health and environment. Am J Public Health 95:S39–S48.

Michaud, R., and P. Béland. 2001. Looking for trends in the endangered St. Lawrence beluga population: a critique of Kingsley, MCS (1998). Mar Mamm Sci 17:206–212.

Mikaelian, I., J. Boisclair, J. Dubey, S. Kennedy, and D. Martineau. 2000. Toxoplasmosis in beluga whales (Delphinapterus leucas) from the St Lawrence Estuary: two case reports and a serological survey. J Comp Pathol 122:73–76.

Mikaelian, I., Y. de Lafontaine, C. Ménard, P. Tellier, J. Harshbarger, and D. Martineau. 1998. Neoplastic and nonneoplastic hepatic changes in lake whitefish (Coregonus clupeaformis) from the St. Lawrence River, Quebec, Canada. Environ Health Perspect 106:179–183.

Mikaelian, I., Y. De Lafontaine, J. Harshbarger, L. Lee, and D. Martineau. 2002. Health of lake whitefish (Coregonus clupeaformis) with elevated tissue levels of environmental contaminants. Environ Toxicol Chem 21:532–541.

Munson, L., and A. Moresco. 2007. Comparative pathology of mammary gland cancers in domestic and wild animals. Breast Dis 28:7–21.

Munson, L., J. Nesbit, D. Meltzer, L. Colly, L. Bolton, and N. Kriek. 1999. Diseases of captive cheetahs (Acinonyx jubatus jubatus) in South Africa: a 20-year retrospective survey. J Zoo Wildl Med 30:342–347.

Murchison, E. 2008. Clonally transmissible cancers in dogs and Tasmanian devils. Oncogene 27:S19–S30.

Murray, B., S. Malik, and B. White. 1995. Sequence variation at the major histocompatibility complex locus DQ b in beluga whales (Delphinapterus leucas). Mol Biol Evol 12:582–593.

Murray, B., R. Michaud, and B. White. 1999. Allelic and haplotype variation of major histocompatibility complex class II DRB1 and DQB loci in the St Lawrence beluga (Delphinapterus leucas). Mol Ecol 8:1127–1139.

Murray, B., and B. White. 1998. Sequence variation at the major histocompatibility complex DRB loci in beluga (Delphinapterus leucas) and narwhal (Monodon monoceros). Immunogenetics 48:242–252.

Nebert, D.W., and T.P. Dalton. 2006. The role of cytochrome P450 enzymes in endogenous signalling pathways and environmental carcinogenesis. Nat Rev Cancer 6:947–960.

Newman, S.J., and S.A. Smith. 2006. Marine mammal neoplasia: a review. Vet Pathol 43:865–880.

O'Brien, S. 1994. A role for molecular genetics in biological conservation. Proc Natl Acad Sci USA 91:5748–5755.

Olivieri, E.H.R., S.D. da Silva, F.F. Mendonca, Y.N. Urata, D.O. Vidal, M.D.M. Faria, I.N. Nishimoto, C.A. Rainho, L.P. Kowalski, and S.R. Rogatto. 2009. CYP1A2*1C, CYP2E1*5B, and GSTM1 polymorphisms are predictors of risk and poor outcome in head and neck squamous cell carcinoma patients. Oral Oncol 45:E73–E79.

Patenaude, N., J. Quinn, P. Beland, M. Kingsley, and B. White. 1994. Genetic variation of the St. Lawrence beluga whale population assessed by DNA fingerprinting. Mol Ecol 3:375–381.

Pelletier, É., I. Desbiens, P. Sargian, N. Côté, A. Curtosi, and R. St-Louis. 2009. Présence des hydrocarbures aromatiques polycycliques (HAP) dans les compartiments biotiques et abiotiques de la rivière et du fjord du Saguenay. Rev Sci Eau 22:235–251.

Perera, F. 1997. Environment and cancer: who are susceptible? Science 278:1068–1073.

Perera, F., K. Hemminki, E. Gryzbowska, G. Motykiewicz, J. Michalska, R. Santella, T. Young, C. Dickey, P. Brandt-Rauf, and I. DeVivo. 1992. Molecular and genetic damage in humans from environmental pollution in Poland. Nature 360:256–258.

Perera, F., and I. Weinstein. 2000. Molecular epidemiology: recent advances and future directions. Carcinogenesis 21:517–524.

Picard-Bérubé, M., D. Cossa, and J. Piuze. 1983. Teneurs en benzo 3.4 pyrene chez Mytilus edulis L. de l'estuaire et du Golfe du Saint-Laurent [Benzo 3,4 pyrene content in Mytilus edulis of the estuary and Gulf of St. Lawrence]. Mar Environ Res 10:63–71.

Poirier, M. 2004. Chemical-induced DNA damage and human cancer risk. Nat Rev Cancer 4:630–637.

Raymond, J., and M. Garner. 2001. Spontaneous tumours in captive African hedgehogs (Atelerix albiventris): a retrospective study. J Comp Pathol 124:128–133.

Raymond, J., and M. Garner. 2000. Mammary gland tumors in captive African hedgehogs. J Wildl Dis 36:405–408.

Rewitz, K., B. Styrishave, A. Løbner-Olesen, and O. Andersen. 2006. Marine invertebrate cytochrome P450: emerging insights from vertebrate and insect analogies. Comp Biochem Physiol C Toxicol Pharmacol 143:363–381.

Robeck, T., S. Monfort, P. Calle, J. Dunn, E. Jensen, and J. Boehm. 2005. Reproduction, growth and development in captive beluga (Delphinapterus leucas). Zoo Biol 24:29–49.

Roth, M., K. Strickland, G. Wang, N. Rothman, A. Greenberg, and S. Dawsey. 1998. High levels of carcinogenic polycyclic aromatic hydrocarbons present within food from Linxian, China, may contribute to that region's high incidence of oesophageal cancer. Eur J Cancer 34:757–758.

Rothman, N., M. Poirier, M. Baser, J. Hansen, C. Gentile, E. Bowman, and P. Strickland. 1990. Formation of polycyclic aromatic hydrocarbon-DNA adducts in peripheral white blood cells during consumption of charcoal-broiled beef. Carcinogenesis 11:1241–1243.

Sasco, A.J., M.B. Secretan, and K. Straif. 2004. Tobacco smoking and cancer: a brief review of recent epidemiological evidence. Lung Cancer 45: S3–S9.

Scaps, P. 2002. A review of the biology, ecology and potential use of the common ragworm *Hediste diversicolor* (OF Müller) (Annelida: Polychaeta). Hydrobiologia 470:203–218.

Shimada, T., M.V. Martin, D. Pruess-Schwartz, L.J. Marnett, and F.P. Guengerich. 1989. Roles of individual human cytochrome P-450 enzymes in the bioactivation of benzo(a)pyrene, 7,8-dihydroxy-7,8-dihydrobenzo(a)pyrene, and other dihydrodiol derivatives of polycyclic aromatic hydrocarbons. Cancer Res 49:6304–6312.

Shopland, D.R., H.J. Eyre, and T.F. Peachacek. 1991. Smoking-attributable cancer mortality in 1991: is lung cancer now the leading cause of death among smokers in the United States? J Natl Cancer Inst 83:1142–1148.

Siddle, H.V., J. Marzec, Y. Cheng, M. Jones, and K. Belov. 2010. MHC gene copy number variation in Tasmanian devils: implications for the spread of a contagious cancer. Proc Royal Soc B 277:2001–2006.

Siebert, U., I. Hasselmeier, and P. Wohlsein. 2010. Immunohistochemical characterization of a squamous cell carcinoma in a harbour porpoise (*Phocoena phocoena*) from German waters. J Comp Pathol 143:179–184.

Siebert, U., A. Wünschmann, R. Weiss, H. Frank, H. Benke, and K. Frese. 2001. Post-mortem findings in harbour porpoises (*Phocoena phocoena*) from the German North and Baltic Seas. J Comp Pathol 124:102–114.

Singh, A.P., P.P. Shah, N. Mathur, J.T.M. Buters, M.C. Pant, and D. Parmar. 2008. Genetic polymorphisms in cytochrome P450B1 and susceptibility to head and neck cancer. Mutation Research/Fundamental and Molecular Mechanisms of Mutagenesis 639:11–19.

Smith, J., and E. Levy. 1990. Geochronology for polycyclic aromatic hydrocarbon contamination in sediments of the Saguenay Fjord. Environ Sci Technol 24:874–879.

Spinelli, J., P. Demers, N.D. Le, M. Friesen, M. Lorenzi, R. Fang, and R. Gallagher. 2006. Cancer risk in aluminum reduction plant workers (Canada). Cancer Cause Control 17:939–948.

Straif, K., R. Baan, Y. Grosse, B. Secretan, F. El Ghissassi, and V. Cogliano. 2005. Carcinogenicity of polycyclic aromatic hydrocarbons. Lancet Oncol 6:931–932.

Thériault, G., S. Cordier, C. Tremblay, and S. Gingras. 1984. Bladder cancer in the aluminium industry. Lancet 323:947–950.

Thériault, G., G. Gibbs, and C. Tremblay. 2002. Cancer in belugas from the St. Lawrence Estuary. Environ Health Perspect 110:A562.

Tremblay, C., B. Armstrong, G. Thériault, and J. Brodeur. 1995. Estimation of risk of developing bladder cancer

among workers exposed to coal tar pitch volatiles in the primary aluminum industry. Am J Ind Med 27:335–348.

Tsugane, S., and S. Sasazuki. 2007. Diet and the risk of gastric cancer: review of epidemiological evidence. Gastric Cancer 10:75–83.

Uno, S., T. Dalton, N. Dragin, C. Curran, S. Derkenne, M. Miller, H. Shertzer, F. Gonzalez, and D. Nebert. 2006. Oral benzo [a] pyrene in Cyp 1 knockout mouse lines: CYP1 A1 important in detoxication, CYP1 B1 metabolism required for immune damage independent of total-body burden and clearance rate. Mol Pharmacol 69:1103–1114.

Uno, S., N. Dragin, M.L. Miller, T.P. Dalton, F.J. Gonzalez, and D.W. Nebert. 2008. Basal and inducible CYP1 mRNA quantitation and protein localization throughout the mouse gastrointestinal tract. Free Radic Biol Med 44:570–583.

Veatch, J., and J. Carpenter. 1993. Metastatic adenocarcinoma of the mammary gland in a Père David's deer. J Vet Diagn Invest 5:639–640.

Vineis, P., M. Alavanja, P. Buffler, E. Fontham, S. Franceschi, Y.T. Gao, P.C. Gupta, A. Hackshaw, E. Matos, J. Samet, F. Sitas, J. Smith, L. Stayner, K. Straif, M.J. Thun, H.E. Wichmann, A.H. Wu, D. Zaridze, R. Peto, and R. Doll. 2004. Tobacco and cancer: recent epidemiological evidence. J Natl Cancer Inst 96:99–106.

Vineis, P., and F. Perera. 2007. Molecular epidemiology and biomarkers in etiologic cancer research: the new in light of the old. Cancer Epidemiol Biomark Prev 16:1954–1965.

Vladykov, V. 1946. Études sur les mammifères aquatiques. IV. Nourriture du marsouin blanc ou béluga (*Delphinapterus leucas*) du fleuve Saint-Laurent, Québec.

Weinberg, R.A. 2007a. The biology and genetics of cells and organisms. *In* The biology of cancer, p. 8. Garland Science, Taylor & Francis Group, New York.

Weinberg, R.A. 2007b. The nature of cancer. *In* The biology of cancer, pp. 25–56. Garland Science, Taylor & Francis Group, New York.

White, P., C. Blaise, and J. Rasmussen. 1997. Detection of genotoxic substances in bivalve molluscs from the Saguenay Fjord (Canada), using the SOS chromotest. Mutatg Res-Genet Tox En 392: 277–300.

White, P.A., J.B. Rasmussen, and C. Blaise. 1998. Genotoxic substances in the St. Lawrence system I: Industrial genotoxins sorbed to particulate matter in the St. Lawrence, St. Maurice, and Saguenay rivers, Canada. Environ Toxicol Chem 17:286–303.

Wilson, J., S. Cooke, M. Moore, D. Martineau, I. Mikaelian, D. Metner, W. Lockhart, and J. Stegeman. 2005. Systemic effects of Arctic pollutants in beluga whales indicated by CYP1A1 expression. Environ Health Perspect 113:1594–1599.

Xue, W., and D. Warshawsky. 2005. Metabolic activation of polycyclic and heterocyclic aromatic hydrocarbons and DNA damage: a review. Toxicol Appl Pharmacol 206:73–93.

Ylitalo, G.M., J.E. Stein, T. Hom, L.L. Johnson, K.L. Tilbury, A.J. Hall, T. Rowles, D. Greig, L.J. Lowenstine, and F.M.D. Gulland. 2005. The role of organochlorines in cancer-associated mortality in California sea lions (*Zalophus californianus*). Mar Pollut Bull 50:30–39.

Zhang, Q., J. Wikoff, D. Dunbar, M. Fasco, and L. Kaminsky. 1997. Regulation of cytochrome P4501A1 expression in rat small intestine. Drug Metab Dispos 25:21–26.

# PART FIVE

## PLACE-BASED CONSERVATION MEDICINE

# 28

## SENSE AND SERENDIPITY

### Conservation and Management of Bison in Canada

Margo J. Pybus and Todd K. Shury

"Stampeding hordes of bison (*Bison bison*) once thundered across the western prairies, taking flight from raging wildfires, human hunters and hunting packs of plains wolves (*Canis lupus*) and grizzlies (*Ursus arctos*). The natural ebb and flow of an intact native prairie ecosystem must have been a marvel to behold."

M.J.P.

The recent history of bison management in North America is a textbook example of conservation medicine writ large on a massive canvas with a backdrop of expansive prairie and boreal forest ecosystems. The following account attempts to portray the delicate interplay of hosts, pathogens, humans, and landscapes, an interplay driven by the effects of human decisions with lasting effects at ecological and population levels. Conservation and management of bison in Canada was a series of good and bad decisions resulting in long-lasting ramifications that rippled through various ecosystems. Although ecological health rarely entered into the decision-making, wildlife diseases remain at the core of the ecological consequences of previous decisions and the legacy they created.

In a pre-European West (pre-1700), plains bison (*Bison bison bison*) dominated grassland habitats over vast regions of North America (Hornaday 1889). Bison were a keystone species driving Prairie and

Northern Forest fringe ecosystems and the associated flora and fauna across the continental core. They also were a spiritual and cultural foundation for indigenous peoples throughout central North America (MacEwan 1995). However, in a frontier West (roughly 1750–1875), bison were viewed as a major impediment to settlement, an untapped commercial opportunity, and a challenge to modern civilization and perceived progress.

The general story of the slaughter, salvation, and subsequent recovery of plains bison in North America has been told, and told well, many times. The reader is directed to Ogilvie 1979; Lothian 1981; FEARO 1990; MacEwan 1995; Pybus 2005; and Brower 2008.

There is another side to the story of bison in North America that is largely untold. It is a fabric of sociopolitical decisions woven on a background of disease and disease management. As the story unfolds, there are a number of critical decision points at which bison and bison management ran up against social and political needs of the day, often devoid of concern for bison conservation or ecological integrity. These concepts were relatively unknown at the time, and societal mandates of progress and western settlement were the primary concerns of governments and society at large. Our aim in this work is to focus on a few specific examples that were pivotal in establishing the outcome of the interplay among bison, their diseases, and sympatric humans in North America. We also highlight the role of protected areas and national parks as key components in the conservation of bison in Canada.

Canada's national parks played a critical role in the restoration of bison in North America, encompassing over time everything from outright unpopularity to spectacular successes to bureaucratic blunders that left a legacy of complicated disease issues. Even though Rocky Mountain National Park (now Banff National Park) was Canada's first, and one of the world's first national parks, it was founded essentially as a recreational preserve to draw tourists into the new Canadian West and protect the commercial value in the local hot springs. Wildlife conservation did not become a national park priority until much later. However, the foundation of this conservation ethic is embodied in the early creation of new national parks specifically dedicated to protecting remnant bison.

Herein we provide a brief overview of the recent history of bison management in southern and northern Canada and the related diseases of concern, with a focus on two specific examples when critical decisions were made.

## GENERAL BACKGROUND

As the new century dawned in 1900, the great thundering masses of plains bison were gone from North America (Hewitt 1921). The continental population was unable to withstand the effects of modern firearms, an insatiable commercial hunt, ever-expanding railroads, and the constant westward flow of European settlers flooding across the American Great Plains and the Canadian Prairies through the late 1800s. There was no second thought regarding the effects of the loss of the primary native grazing ungulate and keystone ecological species over vast regions of western North America.

But bison did not disappear completely. In those early years of the 1900s there were a few small groups of bison in private hands (Ogilvie 1979), as well as one last remaining wild herd of plains bison on protected lands in the Yellowstone area. The privately owned bison largely went unnoticed. However, the pristine Flathead valley came to the attention of U.S. Government staff looking to expand the settled area of Montana. Grazing rights for a private herd of about

300 bison were revoked. The United States needed to grow and its citizens needed more land on which to build their nation. As in the previous century, social and political desires trumped wise stewardship of bison. The bison in Montana were sold to the Dominion of Canada, rounded up, loaded into specially built rail cars, and hauled overland to Elk Island National Park (EINP), a newly established national park in east central Alberta (Lothian 1981). Later, in 1909 once the newly created Buffalo National Park near Wainwright had been fenced, the bison were again shipped and established in both national parks. Here we see the first awakening of a social awareness towards species conservation and the directed use of protected areas as a conservation tool to preserve a species in need. The decision built on actions taken in the United States in 1872 when Yellowstone National Park, set aside for its unique geothermal landscapes, inadvertently provided protection to a small group of wild plains bison. Yet the Canadian decision too was tempered by using land deemed marginal for farmland (Brown [see Brower 2008, p. 36]).

It seems there was little thought or plan to manage the animals once they arrived at Buffalo National Park, and the focus was entirely on acquiring the bison and the parks to hold them. The bison did well, as bison do whenever they live in expansive grasslands devoid of predation pressures and commercial hunting (Fig. 28.1A). They were a popular tourist attraction and the economy of Wainwright grew on the basis of trainloads of people visiting the park to see a living history of a previous time. But the social conscience and principles of protecting an endangered species were a long way in the future. Non-native species such as yak (*Poephagus grunniens*) and cattle mingled freely with native species: some bison shared a small display enclosure with elk (*Cervus elaphus*), moose (*Alces alces*), antelope (*Antilocapra americana*), and deer (*Odocoileus* spp.). Economic decisions were the driving force in directing management actions, and visitor satisfaction was the guiding policy. The idea of conservation was still in its infancy and not well understood.

Agriculture Canada, the federal agency charged with expanding national opportunities and commercial gains in agricultural products, viewed bison as a commodity, one that could perhaps be improved to provide a bigger, faster-growing meat animal for use on lands marginal for cattle and other domestic grazers (Fig. 28.1B). Cross-breeding experiments were

**Figure 28.1:**
(A) Bison in Buffalo National Park near Wainwright Alberta, early 1920s. (B) Hybrid bison/cattle on view for visitors to the park at Wainwright, late 1920s. (C) Bison-handling corrals at Ft. McMurray for bison offloaded from trains from Wainwright and before being loaded onto river barges to float the Athabasca River to Wood Buffalo National Park, mid-1920s. (Photo credit: Provincial Archives of Alberta)

considered a viable means of improving livestock so the bison were cross-bred with various cattle breeds, as well as yaks. While the experiments failed largely due to lack of fertile male offspring, they reflect an overarching opinion that bison were not held in high regard as "wildlife" but rather were a template for commercial modification, even though the bison in Wainwright were the only substantive herd of "wild" bison in Canada. Concerns raised by a few individuals were overridden by bureaucrats intent on making the bison profitable (Ogilvie 1979).

Diseases soon became a major problem. Bovine tuberculosis, detected in 1917 (Fuller 2002), became well established in the park and the prevalence approached 75% by 1922–23 (Hadwen 1942; Lothian 1981). Introduction of potentially diseased bison, cattle, and cattalo (hybrid cattle and buffalo) was exacerbated by overcrowding, drought, and range deterioration throughout the park in the 1920s. Culls were used to reduce the bison population, limit the spread of disease, and allow overgrazed areas to regenerate. However, the plan did not succeed, and disease, in conjunction with poor forage, a series of harsh winters, and continued growth of the herd, spelled the end for Buffalo National Park in 1939.

Once again, politics and social attitudes focused on bison and this time led to an outcome that seemed favorable for bison conservation but had long-lasting hidden costs. The bison in Buffalo National Park posed a serious problem for park managers and government officials: they were diseased, they were too numerous, and they threatened an agricultural economy poised to enter the industrial revolution of the early 1900s (Isenberg 2000). Veterinarians and biologists associated with the park, and most familiar with the problems, recommended removal of the entire herd and replacement with disease-free bison from EINP (Fuller 2002). But there was widespread public outcry against this: the park was established to save bison, and the previous images of dead bison stripped of their hide and tongue and left to rot in unbelievable numbers were still very real and utterly distasteful to the society of the new century—and was thus politically unacceptable.

Officials decided to focus on young and ideally uninfected bison (originally the animals were to be tested for tuberculosis [TB], but this was discarded as "unnecessary" based on the opinions expressed by *politicians* that young animals did not get TB;

Lothian 1981), round them up, and haul them by rail and riverboat to Wood Buffalo National Park in northeastern Alberta, established in 1922 to protect a remnant population of wood bison (*Bison bison athabascae*) (Fig. 28.1C). The public was appeased by not seeing more images of dead bison, and it made good politics to put the problem in the far north, where few people would see them and they would perhaps be "out of sight, out of mind." Human settlement and agricultural operations were well protected by distance and obscurity. In addition, more bison perhaps would benefit local aboriginal populations in northern Alberta (Ogilvie 1979).

Now a new element of conservation concern arose. Despite objections, it was deemed unlikely that the two types of bison would find each other. The park was huge and the vast expanse of muskeg, jack pine ridges, and salt pans were judged to be a significant barrier to dispersal out of the Hay Camp area, where the bison from Wainwright were released. The plan was foiled by two primary factors: poor judgment in disregarding the advice from professionals about disease testing (over 6,600 bison were shipped from Wainwright to Wood Buffalo between 1925 and 1928), and a bison will go wherever a bison chooses to go. Within a relatively few decades, there was evidence that bovine TB and brucellosis were present in bison in the northern park, and that the two bison subspecies had interbred, forever obscuring the genetic divergence of the two subspecies (FEARO 1990). One positive outcome of the bison transfer from Wainwright to Wood Buffalo was the necessity to enlarge the park when bison crossed the Peace River, resulting in the largest national park in Canada at over 44,000 square kilometers (Lothian 1976; Carbyn and Watson 2001). More on this to come later.

Although government staff clearly were aware of the disease issue during the 1920s, the federal Deputy Minister of Interior publicly stated there was no bovine TB in the Wainwright herd. Modern concepts of openness and transparency apparently had not found their way into the federal government culture of the day, and the decision was so short-sighted. It is difficult for modern readers to fathom how such a decision was approved, but little was known about the conservation and management of large ungulates, and wildlife science was in its infancy.

Fortunately by the time the folly of the previous decision was recognized, there was a very different social attitude towards species conservation and also towards the management of national parks. Yet another attempt to save bison began in the mid-1960s when an isolated remnant herd, thought to be the most representative original local wood bison, was found in the north portion of Wood Buffalo National Park in the Nyarling River region. Twenty-four of these bison were relocated to a fenced isolation area within EINP (Lothian 1976). There, the animals were tested repeatedly for bovine TB and brucellosis, and although a few TB and brucellosis reactors were found and the original founder animals removed (Blyth and Hudson 1987), by the early 1980s no further infections were detected and the Elk Island wood bison herd has been free of these diseases ever since (USDA 2008). At the same time, 18 animals from the Nyarling River area were translocated to the western shores of Great Slave Lake in the Northwest Territories, where they were released into a protected area now known as the Mackenzie Bison Sanctuary. The Mackenzie herd continues to thrive and has not shown evidence of TB or brucellosis (Nishi et al. 2006).

The wood bison herd in EINP has been the primary source of wood bison translocated to various locations in Alberta, British Columbia, Yukon,

Manitoba (as well as Alaska and Russia). These populations form the basis of a highly successful national program to recover and re-establish disease-free northern wood bison populations across northern Canada (Gates et al. 2001).

As the new century opens, wood bison populations for the moment seem secure (Figs. 28.2 and 28.3). But there is an ongoing threat of eventual disease transfer from the bison in and around Wood Buffalo National Park (Nishi et al. 2006; Nishi 2010). In a seemingly incongruous situation, the Northwest Territories and Parks Canada maintain a bison-free zone between the Mackenzie herd and the Wood Buffalo herd, a *cordon sanitaire* encompassing 39,000 km$^2$. Any bison found in this area is killed, regardless of whether it is a healthy wood bison from the north or a potentially diseased wood bison from the south. Similarly, Alberta recently initiated a hunt to reduce the wood bison population in northwest Alberta in order to reduce conflicts with humans and minimize the risk that bison will disperse to the east, contact diseased animals from the Wood Buffalo area, and then return. In the meantime, cattle and bison ranching expanded in northern Alberta, and local First Nations are concerned about long-term security for the bison.

**Figure 28.2:**
Plains bison on native prairie in Grasslands National Park, southern Saskatchewan. (Photo credit: Nigel Finney, Parks Canada)

**Figure 28.3:**
Wood bison in boreal forest fringe habitat in northwestern Alberta. (Photo credit: Lyle Fullerton, Alberta Fish and Wildlife)

## APPLIED ELEMENTS OF CONSERVATION MEDICINE

Overall, it appears the story of bison management in Canada has a relatively happy ending—but not really. Moving many bison to the north and killing the remaining tuberculous animals in Buffalo National Park solved the problems at Wainwright. By saving some of the bison, it was socially and therefore politically acceptable to remove the rest. And, with removal of the primary disease reservoir, the diseases also disappeared from the Wainwright area. Livestock producers and agriculture managers breathed a sign of relief. But the transfer of diseases into northern ecosystems and the subsequent intermingling of wood and plains bison left lasting scars.

The presence of introduced livestock diseases in the bison in the Wood Buffalo area poses an ongoing threat to an increasing number of farms in northern Alberta and is a significant impediment to the full recovery of wood bison populations throughout northern regions of Canada (Gates et al. 2001). In the mid-1980s, a federal environmental panel tried to find a solution (FEARO 1990), but the issues are extremely complicated and include elements of national park management, cultural and spiritual sensitivities for First Nations people, agricultural economics, ecosystem integrity, and wise stewardship of wildlife and natural habitats. Thus, agricultural, social, cultural, environmental, and economic risks continue; stakeholders, agencies, and special interest groups are entrenched; and while dialog has been ongoing, solutions have proven evasive. All of this is overlain with

a thick layer of social attitudes and political implications that shift and change depending on a multitude of other factors. The remainder of this paper will focus on two examples of bison management in Canada and will provide a more detailed account of the lessons learned in regards to decisions made as applied elements of conservation medicine.

## ELK ISLAND NATIONAL PARK: THE "LITTLE PARK THAT COULD"

EINP is at the core of the bison story as it unfolded throughout the decades. The park repeatedly played a key role when bison conservation and management reached critical threshold points. However, it did not start out with such intentions. At the turn of the 20th century, there were few remaining elk in Alberta other than those in the southern foothills on the west side of the province (Stelfox 1964; Lothian 1981). The largest single herd outside the foothills lived in the Beaver Hills east of Edmonton (Cooper 1903, in Lothian 1976). But local residents in and around Edmonton were concerned that these elk too were at risk of extirpation, and they petitioned the federal government in Ottawa for help. Land use practices played a pivotal role in both the request and the favorable response.

The Beaver Hills are an elevated area of accumulated glacial debris in an otherwise fertile plain. Geologically, the area is designated the Cooking Lake Moraine and consists of cumulative rock debris interspersed and overlain with fine clay sediments (Godfrey1993). As the local height of land, the moraine experiences a slightly harsher microclimate as well as proportionally colder and wetter predominant weather than the adjacent flat lands. As a result, this area was viewed as relatively inhospitable for settlement as Europeans flooded into the West. The lack of interest from agriculture or mining concerns laid a foundation for setting aside lands for other purposes. The Cooking Lake Timber Reserve, established in 1892, set a precedent for protecting local areas and was a foundation on which to build. The political timing also was fortuitous: a social conscience against the wanton overuse and disregard of wildlife was slowly awakening across Canada as populations of various species declined significantly across broad landscapes. A number of national parks and national park reserves were established, primarily along the Continental Divide in the late 1800s (Lothian 1976), and there was general public support for the preservation of wildlife and wild land, particularly in areas that were not good for immediate settlement or lands that were considered too valuable to be left to private commercial enterprise. Also, the province of Alberta was established in 1905, and so maintaining federal land holdings within the newly established provincial territories was viewed favorably in the national capital (Ottawa).

In 1904, the concerned residents who pleaded for protection of the few remaining elk on the Cooking Lake Moraine were granted their request and Island Park Game reserve (41 km²) was set aside and fenced. In 1906, the federal government upgraded its status to Elk Island National Park.

Timing again proved fortuitous, as this was precisely when the negotiations regarding the fate of the bison in Montana reached a climax and there was an urgency to act quickly to secure the deal. EINP, with its fenced perimeter completed in 1907, was the only location deemed ready and appropriate to receive the first shipment of bison from the United States. The original plan was for EINP to be only a temporary stopover until a new park at Wainwright was completely fenced (Lothian 1976; Ogilvie 1979). However, bison are known for their general reluctance to do anything other than what they want, and once released, some individuals refused to return to the trains for subsequent delivery to Wainwright or were intentionally left behind by people who rounded them up (Ogilvie 1979; Blyth and Hudson 1987). These animals became the founders of the plains bison herd currently ranging within the boundaries of EINP. So, the

little park east of Edmonton spawned not one but two national foundation herds of plains bison. The driving principle behind the actions of society and the park managers was preservation of a critically diminished species. The seeds of ethical treatment of the land and respect for other species as sown by Aldo Leopold (Leopold 1948) were nurtured and expanded in the mid-20th century with the dawn of wildlife management as a guiding principle of public policy. Society took note, and this translated into political support for a shift from preservation to conservation of wild species, and the application of science to underscore management decisions. Direct application of these changing views can be seen in two key bison management examples involving EINP.

In the mid-1950s, when bovine brucellosis was identified in plains bison in the park (Lothian 1976; Blyth and Hudson 1987), a combination of rigorous testing, vaccination, and slaughter was applied. The disease was under control by 1966, and by 1972, the herd was declared free of brucellosis (Blyth and Hudson 1987). In 1967 a new era of bison conservation began with disease-free bison from EINP donated to the First Nation at Gleichen, Alberta, and others were sold to a private commercial farm in Quebec (Lothian 1976).

In a second example, unlike earlier bison transfers, salvage of wood bison from Wood Buffalo park in the mid-1960s was planned thoroughly and followed up with determination. Bison from the north were released into an isolated area of EINP separated by a major highway and two high wire fences from the area with plains bison. Public access to the new arrivals was prohibited, as was contact with livestock. Bison-appropriate handling facilities were established so that animals could be rounded up and tested for disease. A few initial tuberculous wood bison were isolated and dealt with through slaughter of all original founder animals and salvage of calves at birth, and the infection was successfully eradicated (Blyth and Hudson 1987). Park staff and programs focused specifically on the needs and issues associated with good bison management. Ongoing disease surveillance continues to this day, with no further evidence of TB or brucellosis (USDA 2008).

There was a further paradigm shift towards the end of the 20th century. As the bison herds in EINP expanded, as bison always do when left undisturbed, a framework of wise stewardship was applied to the disposition of surplus stock. Re-establishing free-roaming bison populations became the top priority. National recovery planning guided the decisions about where and when to place surplus wood bison. Security from disease also was essential to these decisions. However, as outlined elsewhere in this chapter, this latter aspect remains one of the limiting factors restricting future bison introductions across some areas of northern Canada (Gates et al. 2001).

EINP is a prime example of conservation success, and elements of wildlife disease and conservation medicine are woven into the fabric of the park history. Without this park we could have lost so much—the momentum on moving the first shipment of plains bison into Canada from Montana in 1907, the seed population of plains bison that established in EINP, the ongoing prime viewing opportunities for urban and recreational visitors to see bison and learn about

bison conservation, the opportunity to salvage animals considered remnant wood bison from the northwest corner of Wood Buffalo National Park—and ultimately we would have lost disease-free plains and wood bison to use as seed populations for subsequent bison conservation programs throughout western Canada and beyond (Fig. 28.4). We would have lost so much without the little park east of Edmonton.

## NORTHERN CANADA/WOOD BUFFALO NATIONAL PARK

The interaction of humans and bison in northern Canada provides yet another colorful example from which to view the problems and challenges of recovering large keystone mammals. The story is full of irony, intrigue, mysteries, and spectacular successes coupled

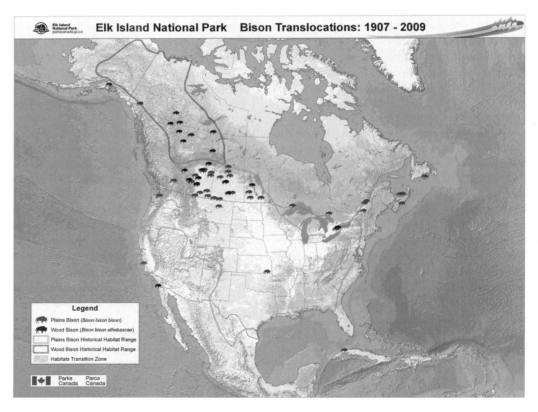

**Figure 28.4:**
Overview of bison translocations from Elk Island National Park, 1907 to 2009. (Courtesy of Elk Island National Park, Alberta)

with resounding failures. Despite almost a century of failed attempts to manage cattle diseases that spilled over into recovered bison populations, none of these attempts can be termed successful to date (Nishi 2010).

Samuel Hearne provided one of the first historical sightings of bison in the area south of Great Slave Lake in 1772 (Roe 1970; Ogilvie 1979). Other early explorers such as Alexander Mackenzie, Daniel Harmon, and John Richardson mentioned larger, darker, northern bison found in the areas near the Peace, Slave, and Athabasca rivers of northwestern Canada (Mitchell 1976; Gates et al. 1992). Up until the mid-1800s, wood bison numbers seemed relatively stable but were never well enumerated. By 1850, wood bison numbered approximately 168,000 (Ogilvie 1979). Bison populations in northern Canada were somewhat isolated from the mass slaughter of their southern ecological counterparts due to the inherent isolation in relatively inhospitable terrain that made it difficult (and unprofitable) to get hides and meat to southern markets (Mitchell 1976). However, their numbers rapidly dwindled after the mid-1850s, primarily as meat for nearby fur-trading forts, and several bad winters in the 1880s (Ogilvie 1979; Gates et al. 1992).

Roderick Macfarlane, a Hudson's Bay factor in Fort Chipewyan, first alerted authorities to the imminent decline of wood bison in 1853. Thus, alarm bells were rung and the age of wood bison preservation began; however, the response was weak. In 1877 the Northwest Territories Council enacted the Buffalo Protection Act, but the ordinance was repealed in 1878 as no enforcement personnel were present in the West (Lothian 1976; Ogilvie 1979; MacEwan 1995). Yet another Territorial Game Ordinance was amended in 1890 to save the remaining wood bison, but it too was unenforced until the Northwest Mounted Police arrived in 1897. Wood bison populations hit their lowest levels between 1896 and 1900; roughly 250 animals were the last remaining vestige of northern bison. But these bison formed the nucleus of one of the most publicized and passionate species recovery programs ever attempted and became a potent symbol of a nascent conservation movement.

A core protected area within native habitat provided a base for future efforts. Dr. Maxwell Graham of the Dominion Parks Branch was instrumental in establishing a national park to protect wood bison as early as 1912 (Fuller 2002). His efforts were bolstered

by the presence of a small band of "Bison Rangers" who helped protect wood bison in their core range south of Great Slave Lake from 1914 to 1922. The irony inherent in Maxwell's strong advocacy on behalf of the park is that he played a key role in the tragic translocation of plains bison that occurred later (Ogilvie 1979; Fuller 2002).

Wood Buffalo Dominion Park was created officially in 1922 and encompassed the primary remaining ranges of the remnant wood bison, including a northern herd of about 500 bison centered around the Little Buffalo and Nyarling rivers and a southern group of about 1,000 centered around the Slave and Peace rivers (Ogilvie 1979; McCormack 1992). Had fate not intervened, the future looked bright for wood bison. However, the decision to ship 6,673 plains bison from Buffalo National Park up the Athabasca and Slave rivers by barge into the heart of wood bison range (as outlined earlier in this chapter) has been called "one of the most tragic examples of bureaucratic stupidity in all history" (Fuller 2002) as well as "a disastrous error in judgement" and "a serious mistake." (Ogilvie 1979; Fuller 2002). But it was also a learning experience and perhaps a necessary step in the evolution of modern concepts of wildlife translocation and landscape conservation.

It is relatively clear that bovine TB was transferred to Wood Buffalo National Park with bison shipped from Wainwright (Ogilvie 1979; Gates et al. 1992; Wobeser 2009); however, the source of bovine brucellosis currently extant in northern wood bison is not as clear. Evidence of bovine brucellosis was not found in Wainwright when the plains bison herd was depopulated (Hadwen 1942), even though heavily infected bison and cattle herds typically show gross lesions at slaughter or necropsy (Tessaro 1987). There is another potential source: in 1949, 25 elk were translocated from EINP to the Lake Claire region of Wood Buffalo National Park (Law 1949). Bovine brucellosis was known to occur in EINP (Moore 1947) at the time, as 30% of the plains bison were seropositive. By 1958, 13% of elk in EINP also were seropositive and were considered a risk to cattle (Corner and Connell 1958). This seems another plausible but unexplored route for brucellosis to get to the bison in Wood Buffalo National Park.

Bison management in the Wood Buffalo National Park evolved in parallel with social and political views. Initially bison were seen primarily as an exploitable resource and were slaughtered to provide meat for local missions (Mitchell 1976). Bovine TB was found in several bison slaughtered in the late 1930s (Mitchell 1976) and routine, systematic slaughters began in 1952 (Fuller 1962). Consistent with the prevailing attitudes, game animals were valued highly in national parks and carnivores were not. Thus, wolf (*Canis lupus*) control was undertaken during the 1950s and 1960s to protect ungulates in the park. But at the same time there was an abundance of bison, two abattoirs were in use, and annual bison slaughters continued until 1976, when park programs began to shift towards ecosystem management (Sandlos 2002).

Disease management efforts associated with northern bison issues are summarized nicely in Nishi et al. (2006) and Nishi (2010). However, it is worth noting that a 10-year project near Fort Resolution, Northwest Territories, established a new baseline for

genetic salvage and disease management, but ultimately was unsuccessful due to latent TB in the herd (Nishi et al. 2002; Wilson et al. 2005; Himsworth et al. 2010). Northern bison management remains a dilemma. Bovine TB and brucellosis are enzootic in and around Wood Buffalo National Park (Tessaro 1987; Joly and Messier 2004a,b; Nishi et al. 2006) and will not resolve spontaneously (Nishi et al. 2006; Nishi 2010). Disease effects may be limited to minor changes in bison demography (Joly and Messier 2004a,b, 2005), although the herd in the Greater Wood Buffalo area is one of only three "ecologically restored" populations of wood bison in the world (Gates et al. 2010). However, the diseases are a significant ongoing risk to adjacent disease-free wood bison populations painstakingly reintroduced to northern Canada over the past 25 years (Fig. 28.5), and to commercial cattle and bison herds in northern Alberta.

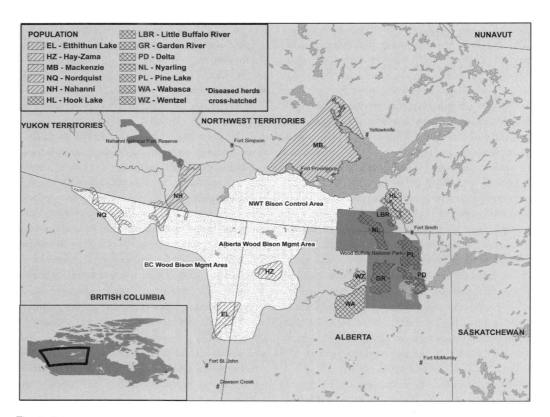

**Figure 28.5:**
Current general distribution and disease status of bison herds in northern Canada, 2010. (Courtesy of Government of Northwest Territories and Parks Canada.)

Although the actual risk to commercial herds likely is low (APFRAN 1999), transmission of either disease to the reintroduced free-ranging bison would be a major setback to wood bison recovery in Canada and would vastly complicate future efforts of disease management by the sheer increase in geographic scale of the problem.

Nishi (2010) proposed using systems thinking and adaptive management as models to approach disease management in both the Greater Yellowstone and Wood Buffalo areas. Unfortunately, such models and thinking do not come easily to government agencies, which typically operate within a very different paradigm and are easily swayed by political motivations. Incorporating systems thinking into action on the ground will be a major challenge. The type of integration and shared command necessary to achieve long-term success will not come easily despite an expanded rhetoric to the contrary from all stakeholders.

## LESSONS LEARNED

### Conservation Implications

- Wood and plains bison can rapidly expand their populations when not limited by predation or habitat, and even with substantial disease presence.
- Techniques to eradicate bovine TB and brucellosis through test/slaughter/salvage can be successful and could achieve further success if carried out under stringent protocols.
- Techniques for handling, transporting, and moving wild bison have advanced greatly in recent years, due largely to experience gained in commercial bison ranching. The high mortality rates from capture myopathy in wild bison seen in earlier programs would be much less likely with current knowledge and understanding of bison behavior.
- Information regarding persistence of bovine TB and brucellosis, most notably the 10-year experience gained during the Hook Lake wood bison recovery project (Nishi et al. 2006), is critical to inform future salvage attempts.
- Techniques being developed will allow salvage of ova and semen from infected bison and propagate

disease-free offspring using advanced reproductive technologies such as in vitro fertilization, artificial insemination, and frozen embryos and semen for long-term storage (Thundathil et al. 2007; McCorkell et al. 2008; Lessard et al. 2009). Thus, it is possible to move forward with considerable confidence in salvaging genetic material and live animals from any infected population.
- Diagnostic tests have advanced, with newer serological and cell-mediated tests providing improved differentiation of tuberculous bison (Lyashchenko et al. 2009; Himsworth et al. 2010).
- In addition to those mentioned herein, many other bison translocation initiatives were undertaken over the years (see Nishi 2010, pp. 12–13, for an excellent summary of these occurrences), setting the stage for bison conservation on a broad scale.
- In summary, the toolbox of options for managing diseased bison populations is much better stocked than it was when the last set of management decisions were initiated following the federal environmental review panel report (FEARO 1990).

### Public Policy and Approach

- Science cannot be ignored in decision-making processes involving conservation. It must form the basis of good policy and appropriate decisions, but clearly cannot be the only consideration.
- Local communities, including First Nations and all local stakeholders, must have direct input into decisions (Nishi et al. 2006; Nishi 2010). Human values, ethics, and socioeconomic concerns all influence decisions made within human societies. Previous failures in bison conservation were in part the result of neglect in one or more of these areas, resulting in poor decisions and lasting conflict.
- Time is a critical element of wise decisions. Hasty decisions in species conservation, recovery, and biology usually are not good decisions, especially when grounded exclusively in public pressure.
- Political motivation often does not serve conservation well. Governments generally avoid

---

**Box 28.1    "Sense" and "Serendipity" in Bison Management in Canada**

---

**"Sense"**

- The original decision to purchase the remnant plains bison in Montana and ship them to Canada paved the way for future bison restoration on Canadian landscapes.
- A few strongly motivated, educated individuals can make a significant difference in the conservation of a species. A small number of dedicated government and non-government individuals played key roles in the creation and salvage of both plains and wood bison.
- Bison conservation is an overall success story in North America, regardless of the pitfalls along the way.

**"Serendipity"**

- A few concerned citizens lobbied for a fence around what became Elk Island park. The fence was completed in 1907.
- A few stubborn bison in Elk Island park refused to be reloaded onto trains bound for Wainwright in 1909 (or were intentionally left behind), and thus Canada's bison "eggs" were never all in one "basket"!
- Wood Buffalo National Park was enlarged in 1926 to accommodate bison from Wainwright that wandered south across the Peace River. This provided a huge canvas on which bison could spread out and fill the ecosystem.
- The 19 bison shipped to Fort Providence to found the Mackenzie bison herd in 1969 truly were disease-free despite limited pre-movement screening.
- The remote inhospitable northern landscape isolated the remnant wood bison from complete extinction.
- The remote northern landscape continues to help isolate diseased bison from commercial farms and disease-free bison, and likely contributes to limited spread of the diseases in harsh environmental conditions.

---

controversy and divisive issues that lead to negative media attention. There is little impetus to proactively deal with diseased bison in northern Canada and the issue was/is repeatedly relegated to a "back burner." Progress on solutions to disease issues in the Greater Yellowstone Ecosystem move forward in large part because bovine brucellosis repeatedly spills over from wildlife into domestic livestock. There is direct political gain to be won if the problem can be solved. This is not the case in northern Canada—and the animals at greatest risk of disease spillover are "only" free-ranging bison. Conservationists must recognize that our primary interests—successful, self-sustaining bison populations—are not those of most publics nor many politicians.

- In the context of global economic concerns and climate change debates, we must strive to include wildlife conservation as a key component of environmental, and economic, sustainability.

## CONCLUSIONS

With some reservations, the conservation and recovery of bison in Canada has largely been a successful venture: wood bison and plains bison numbers have increased substantially since the historic, near-catastrophic, lows at the turn of the 20th century. The future of both types seems secured, although not separately as they once existed.

Wood bison were downgraded from endangered to threatened (Gates et al. 2001) and their status may be further reduced if current population trends continue. But the story does not end: a Draft Wood Bison Recovery Strategy may soon be released for public consultation, reigniting the debate over what to do with diseased bison in northern Canada. The legacy of faulty decisions continues to haunt governments, private citizens, First Nations, as well as the public in Canada, and with increasing communications, our global neighbors. Rectifying past conservation sins will be difficult, but sustained effort in similar situations has been successful in other areas of the world. Innovative solutions are required so that disease management solutions are effective and publicly acceptable.

Most stakeholders and governments agree that disease containment is an urgent first priority to protect current disease-free bison populations, but they disagree on a long-term solution. This has hampered effective dialog on containment strategies and is a major roadblock to moving forward. Similarly, public attitudes will have to adjust to living with restored free-ranging bison, which can at times damage crops, fences, and vehicles. But the values and beliefs of Canadian society change dramatically over time, as have attitudes over how public natural resources are managed and conserved for future generations. Conservationists must do what they can to convince Canadians that large populations of free-ranging bison will provide useful ecosystem services now and into the future in a sustainable way. Wildlife diseases such as bovine TB and brucellosis provide additional challenges at a landscape level to meeting all these values. By meeting the challenge, bison recovery efforts throughout North America will succeed and provide hope for endangered species recovery efforts worldwide.

# REFERENCES

Animal, Plant, and Food Risk Analysis Network (APFRAN). 1999. Risk assessment on bovine brucellosis and tuberculosis in Wood Buffalo National Park and area. Canadian Food Inspection Agency, Unpublished Report E5. Nepean, Ontario.

Blyth, C., and R.J. Hudson. 1987. A plan for the management of vegetation and ungulates: Elk Island National Park. Unpublished report. Environment Canada Parks.

Brower, J. 2008. Lost tracks: Buffalo National Park, 1909–1939. Athabasca University Press, Edmonton, Alberta.

Carbyn, L.N., and D. Watson. 2001. Translocation of plains bison to Wood Buffalo National Park: Economic and conservation implications. In S.M. Maehr, R.F. Noss, and J.L. Larkin, eds. Large mammal restoration: ecological and sociological challenges in the 21st century. Island Press, Washington, D.C.

Corner, A.H., and R. Connell. 1958. Brucellosis in bison, elk, and moose in Elk Island National Park, Alberta, Canada. Can J Comp Med Vet Sci 22:9–21.

Federal Environmental Assessment and Review Office (FEARO). 1990. Northern diseased bison: Report of the Environmental Assessment Panel. Ministry of Supply and Services Canada, Ottawa, Ontario.

Fuller, W.A. 1962. The biology and management of the bison of Wood Buffalo National Park. Can Wildl Serv Wildl Manag Bull Ser 1:1–52.

Fuller, W.A. 2002. Canada and the buffalo, Bison bison: A tale of two herds. Can Field Nat 116:141–159.

Gates, C., T. Chown, and H. Reynolds. 1992. Wood buffalo at the crossroads. In J. Foster, D. Harrison, and I. Mclaren, eds. Buffalo, pp. 139–166. University of Alberta Press, Edmonton, Alberta.

Gates, C.C., C.H. Freese, P.J.P. Gogan, and M. Kotzman, eds. 2010. American bison: Status survey and conservation guidelines 2010. IUCN, Gland, Switzerland.

Gates, C.C., R.O. Stephenson, H.W. Reynolds, C.G. van Zyll de Jong, H. Schwantje, M. Hoefs, J. Nishi, N. Cool, J. Chisholm, A. James, and B. Koonz. 2001. National recovery plan for the wood bison (Bison bison athabascae). National Recovery Plan No. 21, Recovery of National Endangered Wildlife (RENEW), Ottawa, Ontario.

Godfrey, J.D. 1993. Edmonton beneath our feet: A guide to the geology of the Edmonton region. Edmonton Geological Society, Alberta.

Hadwen, S. 1942. Tuberculosis in the buffalo. J Am Vet Med Assoc 100:19–22.

Hewitt, C.G. 1921. The conservation of the wild life of Canada. Charles Scribner's Sons, New York.

Himsworth, C.G., B.T. Elkin, J.S. Nishi, T. Epp, K.P. Lyashchenko, O. Surujballi, C. Turcotte, J. Esfandiari, R. Greenwald, and F.A. Leighton. 2010. Comparison of test performance and evaluation of novel immunoassays for tuberculosis in a captive herd of wood bison naturally infected with Mycobacterium bovis. J Wildl Dis 46:78–86.

Hornaday, W.T. 1889. The extermination of the American bison. Smithsonian Institute, Washington D.C.

Isenberg, A.C. 2000. The destruction of the bison. Cambridge University Press, New York.

Joly, D.O., and F. Messier. 2004a. Testing hypotheses of bison population decline (1970–1999) in Wood

Buffalo National Park: Synergism between exotic disease and predation. Can J Zool 82:1165–1176.

Joly, D.O., and F. Messier. 2004b. Factors affecting apparent prevalence of tuberculosis and brucellosis in wood bison. J An Ecol 73:623–631.

Joly, D.O., and F. Messier. 2005. The effect of bovine tuberculosis and brucellosis on reproduction and survival of wood bison in Wood Buffalo National Park. J An Ecol 74:543–551.

Law, C.E. 1949. Report on the introduction of elk (*Cervus canadensis*) from Elk Island National Park to Wood Buffalo Park. Unpublished government report, Wood Buffalo National Park, Fort Smith, North Western Territories, Canada.

Leopold, A. 1948. A Sand County almanac and sketches here and there. Reprinted in 1987 by Oxford University Press, New York.

Lessard, C., J. Danielson, K. Rajapaksha, G.P. Adams, and R.B. McCorkell. 2009. Banking North American buffalo semen. Theriogenology 71:1112–1119.

Lothian, W.F. 1976. A history of Canada's national parks. Volume I. Parks Canada, Ottawa.

Lothian, W.F. 1981. A history of Canada's national parks. Volume IV. Parks Canada, Ottawa.

Lyashchenko, K.P., R. Greenwald, J. Esfandiari, M.A. Chambers, J. Vicente, C. Gortazar, N. Santos, M. Correia-Neves, B.M. Buddle, R. Jackson, D.J. O'Brien, S. Schmitt, M.V. Palmer, R.J. Delahay, and W.R. Waters. 2009. Animal-side serologic assay for rapid detection of *Mycobacterium bovis* infection in multiple species of free-ranging wildlife. Vet Microbiol 132:283–292.

MacEwan, G. 1995. Buffalo—sacred and sacrificed. Alberta Sport, Recreation, Parks and Wildlife Foundation, Edmonton, Alberta.

McCorkell, R.B., M. Woodbury, and G.P. Adams. 2008. Serial ovarian ultrasonography in wild-caught wood bison (*Bison bison athabascae*). Reproduction in Domestic Animals 43:S91.

McCormack, P.A. 1992. The political economy of bison management in Wood Buffalo National Park. Arctic 454:367–380.

Mitchell, R.B. 1976. A review of bison management, Wood Buffalo National Park 1922–1976. Unpublished government report. Canadian Wildlife Service, Edmonton, Alberta.

Moore, T. 1947. A survey of buffalo and elk herds to determine the extent of *Brucella* infection. Can J Comp Med 11:131.

Nishi, J.S., B.T. Elkin, and T.R. Ellsworth. 2002. The Hook Lake wood bison recovery project: Can a disease-free captive wood bison herd be recovered from a wild population infected with bovine tuberculosis and brucellosis? Ann NY Acad Sci 969: 229–235.

Nishi, J.S., T. Shury, and B.T. Elkin. 2006. Wildlife reservoirs for bovine tuberculosis (*Mycobacterium bovis*) in Canada: Strategies for management and research. Vet Microbiol 112:325–338.

Nishi, J.S. 2010. A review of best practices and principles for bison disease issues: Greater Yellowstone and Wood Buffalo Areas. American Bison Society Working Paper No. 3, American Bison Society, Wildlife Conservation Society, New York.

Ogilvie, S.C. 1979. The park buffalo. National and Provincial Parks Association of Canada, Calgary, Alberta.

Pybus, M.J. 2005. Big game and big parks: Preservation through national parks and game preserves. *in* Fish, fur and feathers: Fish and wildlife conservation in Alberta 1905–2005, pp. 63–102. The Fish and Wildlife Historical Society and the Federation of Alberta Naturalists, Edmonton.

Roe, F.G. 1970. The North American buffalo. A critical study of the species in its wild state, 2nd ed. University of Toronto Press, Ontario, Canada.

Sandlos, J. 2002. Where the scientists roam: Ecology, management and bison in northern Canada. J Can Studies 37:93–129.

Stelfox, J.G. 1964. Elk in northwest Alberta. Land-Forest-Wildlife 6:14–23.

Tessaro, S.V. 1987. A descriptive and epizootiologic study of brucellosis and tuberculosis in bison in northern Canada. Ph.D. Dissertation. University of Saskatchewan, Canada.

Thundathil, J., D. Whiteside, B. Shea, D. Ludbrook, B. Elkin, and J. Nishi. 2007. Preliminary assessment of reproductive technologies in wood bison (*Bison bison athabascae*): Implications for preserving genetic diversity. Theriogenology 68:93–99.

U.S. Department of Agriculture (USDA). 2008. Evaluation of the brucellosis and bovine tuberculosis status of wood bison and elk in the wood bison area of Elk Island National Park, Canada. Unpublished report. Veterinary Services, Animal and Plant Health Inspection Service, Washington, D.C.

Wilson, G.A., J.S. Nishi, B.T. Elkin, and C. Strobeck. 2005. Effects of a recent founding event and intrinsic population dynamics on genetic diversity in an ungulate population. Conserv Gen 6:905–916.

Wobeser, G. 2009. Bovine tuberculosis in Canadian wildlife: an updated history. Can Vet J 50:1169–1176.

# 29

# PATHOGENS, PARKS, AND PEOPLE

The Role of Bovine Tuberculosis in South African Conservation

## Claire Geoghegan

The ecological relationships between host and pathogen species are as complicated, diverse, and important as the biological systems within which they are situated (Anderson and May 1978). However, the introduction of exotic and invasive pathogens disturbs these relationships and can result in large-scale mortality of endemic host species, as witnessed in Africa over the past 200 years (Dasak et al. 2000). Pathogens that cause chronic disease are especially problematic, as the slow progression from infection to clinical disease hinders early detection and control in naïve hosts and environments (Bengis et al. 2004). Zoonotic pathogens, which account for 73% of emerging diseases (Woolhouse et al. 2005), render traditional reactionary disease control measures ineffective, as they infect multiple hosts across human, wildlife, and livestock populations. In these circumstances, a shift towards preventive measures that span species, health services, and conservation organizations will be essential to prevent declining health from driving social, economic, and political changes, of consequence for natural resource conservation. Yet, despite a more recent appreciation of the links between ecosystem and population health, the limited capacity to implement effective disease controls in marginalized areas represents a significant challenge to long-term conservation in Africa.

In this chapter, I will explore the potential for bovine tuberculosis (BTB), a zoonotic cattle disease, to affect conservation in South Africa; examine the processes involved with detecting this exotic pathogen in wildlife, livestock, and human populations; and discuss how conservation medicine principles may be applied to aid disease control, public health, and natural resource conservation in developing areas.

## PATHOGENS: BOVINE TUBERCULOSIS IN SOUTH AFRICA

BTB is a chronic disease caused by *Mycobacterium bovis*, one of four closely related bacteria that cause tuberculosis (TB) in mammals (Brosch et al. 2002). It can be transmitted directly through inhalation of bacteria (aerosol route), via ingestion of infected tissue or body fluids (alimentary route), or less commonly through indirect exposure to contaminated water, soil, and environmental materials (Maddock 1933; O'Reilly and Daborn 1995; Keet et al. 2000b; Michel et al. 2007).

BTB was first described in South African cattle in 1880 (Hutcheon 1880), after the likely importation of infected cattle to the Western Cape Province by European settlers during the 18th and 19th centuries

(Michel et al. 2006) (Fig. 29.1). It continued to spread inland through the movement of people and their animals, and was subsequently transmitted to indigenous cattle that had no prior exposure or evolutionary resistance to the disease. By 1969, the economic impacts of widespread BTB infections for the commercial cattle industry led to the implementation of a National BTB Control and Eradication Scheme (Huchzermeyer et al. 1994), which reduced herd prevalence to less than 1% by 1995 (Michel et al. 2008). However, as these controls were poorly applied to informal and small-scale herds, BTB remained uncontrolled in community-owned livestock bordering wildlife reserves.

## PARKS: BOVINE TUBERCULOSIS IN SOUTH AFRICAN WILDLIFE

In 1929, the first evidence of BTB transmission from livestock to wildlife species was recorded in greater kudu (*Tragelaphus strepsiceros*) and common duiker (*Sylvicapra grimmia*) in the Eastern Cape Province (Paine and Martinaglia 1929). Nevertheless, the potential for BTB to infect wildlife populations was poorly recognized until 1970, when BTB was detected in a black rhinoceros (*Diceros bicornis*) in South Africa's then second largest conservation area (960 km²), Hluhluwe-iMfolozi Park (HiP) (Jolles 2004) (Fig. 29.1). Although no samples were submitted for laboratory

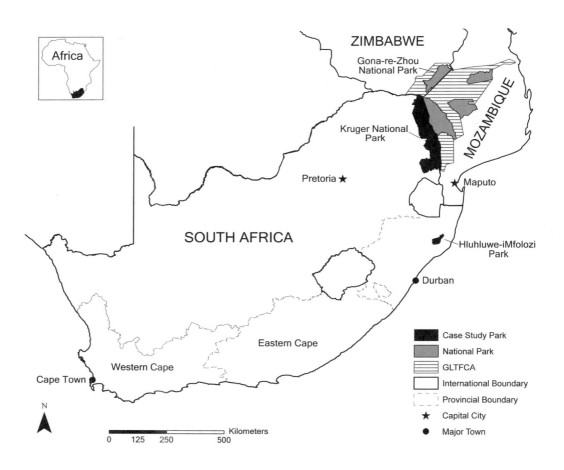

**Figure 29.1:**
Areas of current and historical importance for *Mycobacterium bovis* spread and control in southern Africa. (Illustration by Geoffrey C. Clinning)

examination, this was the first recorded case of BTB in a free-living, critically endangered species within a protected area (Jolles 2004). As BTB was primarily considered a cattle disease, the paucity of diagnostic tests for wildlife species limited further monitoring and control. Thus, by 1986, BTB had infected African buffalo (*Syncerus caffer*) throughout HiP, which was retrospectively attributed to contact with local infected cattle populations in the 1960s, before the park was fully fenced (Jolles 2004).

In 1990, BTB was opportunistically detected in a single buffalo, 300 km north in South Africa's then largest conservation area (18,989 km²), Kruger National Park (KNP) (Bengis 1999) (Fig. 29.1). A subsequent survey of 23,000 buffalo between 1991 and 1992 reported an unexpectedly high BTB prevalence of 27% in southern areas of KNP, which declined to zero in the north (Rodwell et al. 2001). Similar patterns were observed in 1998, although both the average herd prevalence (38.2%) and the number of infected herds (100%) had increased in the south, hinting at the potential importance of inter- and intra-herd transmission (Rodwell et al. 2001). Retrospective molecular analyses suggest that BTB was transmitted between domestic cattle and buffalo in the southeast of KNP between 1950 and 1960 (Kloeck 1998; Vosloo et al. 2001; Michel et al. 2009), prior to spreading north through buffalo movements (Cross et al. 2009). These lengthy delays between BTB introduction and detection in both parks illustrate the difficulty in recognizing emerging diseases in a timely manner as they infect new hosts in unexpected geographic areas.

By 1999, BTB had been confirmed in six additional wildlife species: lion (*Panthera leo*), leopard (*Panthera pardus*), cheetah (*Acinonyx jubatus*), greater kudu, bushpig (*Potamochoerus porcus*) and chacma baboon (*Papio ursinus*) (Keet et al. 2000a,b, 2001; Bengis et al. 2002; Michel et al. 2006) (Table 29.1). However, due to the complexities of detecting infections in multiple host species, little evidence could confirm the specific routes of infection, or potential for further inter- and intra-species transmission.

To address these questions and understand the long-term implications of BTB for wildlife populations, both HiP and KNP began research and monitoring programs. To understand the mechanisms of BTB transmission in the principal maintenance species, empirical data were collected on the movement of buffalo herds, and the frequency of interactions between infected and susceptible individuals (Jolles et al. 2005; Cross et al. 2009). Key determinants of herd grouping and dispersal were assessed, mapped, modeled (Cross et al. 2004), and cross-referenced with behavioral ecology data over a range of seasonal and environmental conditions (Jolles 2004). Deliberate infection trials were also established to investigate disease pathology and infection potentials across age and sex variables under natural and controlled conditions (de Klerk et al. 2006, 2009).

Large-scale testing programs were initiated in both parks to collect baseline data on the prevalence and spatial distribution of BTB in buffalo. In HiP, this was combined with the removal and analysis of positive buffalo to reduce BTB prevalence in herds, minimize

| Table 29.1   Wildlife Species with Confirmed *Mycobacterium bovis* Infections in South Africa to Date | | |
| --- | --- | --- |
| African buffalo (*Syncerus caffer*) | Spotted hyena (*Crocuta crocuta*) | Bushpig (*Potamochoerus porcus*) |
| Black rhinoceros (*Diceros bicornis*)△ | Honey badger (*Mellivora capensis*)* | Warthog (*Phacochoerus oryx*) |
| Lion (*Panthera leo*)‡ | Greater kudu (*Tragelaphus strepsiceros*) | Chacma baboon (*Papio ursinus*) |
| Leopard (*Panthera pardus*)† | Eland (*Taurotragus oryx*) | Large spotted genet (*Genetta tigrina*) |
| Cheetah (*Acinonyx jubatus*)‡ | Impala (*Aepyceros melampus*) | Meerkat (*Suricata suricatta*)* |

Species status: △ critically endangered; ‡ vulnerable; † low risk (IUCN 2010).
* Species infection documented elsewhere than Hluhluwe-iMfolozi Park and Kruger National Park.

the potential spillover to other species, and facilitate the development of new ante-mortem tests by comparison with standard pathology and laboratory analyses (Jolles 2004). In addition, the potential for non-aerosol transmission was evaluated by testing the longevity of BTB in a range of environmental conditions (Tanner and Michel 1999), and by assessing the susceptibility of predators to alimentary infection through the consumption of BTB-positive prey (Michel et al. 2006).

## PATHOGENS AND PARKS: THE IMPACT OF BOVINE TUBERCULOSIS ON AFRICAN BUFFALO POPULATIONS

In KNP, repeated surveys found an increase in the prevalence and spatial distribution of BTB, with infected buffalo herds reaching the northern park boundary (and international border with Zimbabwe) in 2006, 24 years earlier than predicted (de Vos et al. 2001). Subsequently, by 2008, epidemiologically related BTB infections were isolated from buffalo from Zimbabwe's Gona-re-Zhou National Park (Fig. 29.1) (de Garine-Wichatitsky et al. 2010). Recent studies suggest that this surprisingly rapid spread may be related to the abundance of susceptible individuals in KNP's large (27,000-strong) buffalo population (Michel et al. 2006), and the dispersal of animals across formerly separate ecosystems in the recently established Greater Limpopo Trans-Frontier Conservation Area (GLTFCA) (99,800 km$^2$) (de Garine-Wichatitsky et al. 2010).

Studies in HiP and KNP also recorded unexpected behaviors and environmental factors that increase the potential contact and transmission of BTB between infected and susceptible animals. These include the repeated fission and fusion of herds at indeterminate intervals with inconsistent alignment of individuals (Cross et al. 2005a); the frequent movement of females and juveniles between herds in splinter groups (Halley et al. 2002; Cross et al. 2004); the seasonal movement of highly infected adult males between breeding and bachelor herds during the reproductive season (Turner 2003; Hay et al. 2008); and the congregation of herds near declining vegetation and water resources during park-wide floods and droughts (Cross et al. 2004, 2005b).

Despite similar behavioral and epidemiological studies being conducted in HiP and KNP, the reported impact of BTB on buffalo populations differs markedly in each park. This highlights the inherent difficulty in attributing population changes to chronic disease rather than concomitant environmental factors (Cross et al. 2009); and reflects the potential influence of contrasting surveillance and control methodologies, buffalo population size and herd dynamics, and the relative maturity of disease in each ecosystem (Michel et al. 2006). Thus, although the impact of BTB on buffalo demographics in KNP remains statistically undetectable (Cross et al. 2005b, 2009), subtle effects were measured in HiP's smaller (3,000-strong) population. Here studies report an 11% increase in the annual buffalo mortality rate, and an estimated five-year reduction in life expectancy for infected animals (Jolles et al. 2005). This was accompanied by a reduction in the pregnancy rates of positive females, which together with impaired population fecundity lowered the number of calves available to repopulate herds (Jolles et al. 2005). When combined, these factors modify the population growth rate and age structure, but due to their counteractive influence are hard to detect through herd observations alone (Jolles et al. 2005). With potentially serious implications for population resilience and recovery rates after environmental or anthropogenic disturbances (Jolles et al. 2005), these results highlight the importance of collecting fine-scale, multi-faceted disease, population, and environmental data in order to fully determine the subtle and long-term impacts of BTB infections on buffalo populations.

BTB also impairs buffalo health by increasing the susceptibility of positive animals to other infections (de Lisle et al. 2002; Welsh et al. 2005). In HiP and KNP, high endoparasite loads in positive buffalo were observed to complicate immune system response, reduce body condition, and potentially accelerate BTB-related mortality (Caron et al. 2003; Jolles et al. 2008). However, further research will be required to fully appreciate the role of ecological immunology within this invasive pathogen context (Lee and Klassing 2004). Similarly, additional work will be required before many aspects of BTB transmission, pathology, and etiology may be effectively factored into future disease control programs. Until that time, preventing the longer-term implications of BTB for wildlife populations will be a challenge for conservation organizations.

## The Impact of Bovine Tuberculosis on Other Wildlife Species

Although few wildlife species can maintain BTB within their populations without subsequent re-infections (de Lisle et al. 2002), the high prevalence of disease in park buffalo populations increases the risk of exposure, and spillover of infection to other wildlife species (Bengis et al. 2002). To date, BTB has been confirmed in 14 South African wildlife species (Table 29.1), of which the 12 identified in HiP and KNP include rare and endangered species, predators, scavengers, omnivores and herbivores (Keet et al. 1996, 2000a, 2001; Bengis et al. 2002; Michel 2002; Michel et al. 2006; Drewe et al. 2009).

Using molecular methods to type *M. bovis* isolated from infected wildlife species enables the potential routes of infection between parks, species, and populations to be traced (Michel et al. 2009). In KNP, the identification of one primary *M. bovis* strain suggests that park-wide infections stem from an historical point source that later spread between species (Michel et al. 2006). In contrast, the identification of at least two epidemiologically unrelated *M. bovis* strains in HiP indicates that several infection events have taken place (Michel et al. 2006). Plus, as these strains show significant genetic differences to *M. bovis* in KNP, the inferred independent introduction of BTB in each ecosystem provides a baseline from which potential inter-park infections can be monitored (Michel et al. 2006).

Combining molecular techniques with information on species behavior and the clinical manifestations of BTB observed during pathology examinations, the potential routes of inter- and intra- species infections can be estimated. For many social species, the greatest risk of infection stems from aerosol transmission during periods of close contact, which poses a particular threat for ground-dwelling species that inhabit poorly ventilated areas (Drewe et al. 2009). Alimentary infections in herbivores may be related to the ingestion of contaminated browse or fecal–oral routes of transmission (Thorburn and Thomas 1940; Michel et al. 2006), although carnivores, omnivores, and scavenging species are at greatest risk through the consumption of infected prey.

Thus, in KNP, where lions are the primary cause of buffalo mortality (Radloff and du Toit 2004), the spatial distribution of lion infections since initial detection in 1995 (Keet et al. 1996) is similar to that reported in buffalo populations. These patterns are likely to continue, as although there is no evidence that lions preferentially kill BTB-infected buffalo (Cross et al. 2009), the selection of weak, old, and debilitated prey may include a disproportionate number of infected animals, especially during periods of environmental stress (Mills et al. 1995). Once infected, other characteristics of lion behavior further aid transmission through inter-pride contact (aerosol), and intra-specific competition (percutaneous) pathways (Michel et al. 2006). Consequently, these infections alter the population age structure and compromise lion immunity, survival, breeding success, and pride stability—all of which have poorly understood but potentially long-term implications for both lion conservation and broader ecosystem health (Michel et al. 2006; Cross et al. 2009; Keet et al. 2010).

Of particular interest are greater kudu, which are the only species to exhibit distinct clinical signs of BTB infection, characterized by parotid lymph node abscesses and draining fistulae (Keet et al. 2001). They are also unique in KNP for exhibiting two distinct BTB strains, and more commonly a strain that is genetically unrelated to that isolated in park buffalo (Michel et al. 2006). This suggests the potential for greater kudu to act as a maintenance host for BTB, which if confirmed will significantly complicate disease control and have unpredictable ramifications for inter-species transmission.

## The Potential for Effective Bovine Tuberculosis Control in Wildlife Populations

Controlling the spread of BTB in wildlife populations is expensive, logistically difficult, and complicated by a paucity of preventative and therapeutic options (Bengis et al. 2002). For most species, control options are limited to the removal of infected individuals, which challenges traditional conservation principles and is ethically problematic for endangered and genetically important species (Bengis et al. 2002). Park authorities therefore face pressure to find alternative control methods that balance disease prevention with longer-term species conservation.

In HiP, control efforts target the reduction of BTB prevalence in buffalo to minimize the risk of transmission to other species. Since 1999, annual test-and-slaughter programs have significantly reduced the prevalence in tested herds from over 40% to less than 10% (Jolles et al. 2005). However, as the inherent unreliability and poor capacity of ante-mortem field tests to detect newly infected and anergic animals permits positive buffalo to remain in tested populations (de Lisle et al. 2002), surveillance programs must be repeated regularly to prevent a resurgence of disease.

Unfortunately, the large financial and human resources required to test and monitor buffalo often compete with those needed for more recognizable and immediately threatening conservation priorities, including anti-poaching patrols. This jeopardizes the continuation of disease control programs, prompting both parks to incorporate additional activities to supplement funding and enhance disease control (Bengis et al. 2004). These include breeding disease-free calves to restock park buffalo populations and generate income through sales (Bengis and Grobler 2000), and targeting the management of small identifiable BTB outbreaks in spillover hosts to reduce overall park infections (Keet et al. 2000b).

Control options are limited for other vulnerable and protected species (Table 29.1), although disease monitoring may be improved by the recent development of ante-mortem tests for feline and pachyderm species (Morar 2003; Lyashchenko et al. 2008, Maas et al. 2010). Vaccination options also currently lack efficacy (Cross and Getz 2006; Perry and Scones 2007), but may provide a more appropriate method to manage the spread of BTB in the future (Blancou et al. 2005). However, as the resources required to deliver efficacious coverage potentially require similar budgets to current control methods, only through long-term financial and logistical support will BTB control become a reality in South African wildlife populations.

## PATHOGENS AND PARKS: THE DIRECT IMPACTS OF BOVINE TUBERCULOSIS ON CONSERVATION

BTB affects the ecological function of parks not only through altering population health and host–pathogen relationships, but also by increasing the complexity of park conservation strategies. As many parks in South Africa rely on capital derived through park entrance fees and the sale of excess natural resources to fund conservation initiatives, balancing the utility of parks for visitors and ecosystem conservation often requires proactive management. Accordingly, to enhance the long-term protection of wildlife species, parks often translocate animals to improve breeding opportunities, maintain genetic viability, and increase the safety and spatial distribution of threatened species (Funk et al. 2001). Translocation is also used to ensure the sustainability of sensitive habitats and ecosystem resources by periodically removing and selling surplus animals from expanding park wildlife populations. This provides additional benefits for the park, by generating revenue to support conservation activities. However, as translocating animals to new areas and distinct populations poses inherent risks for bidirectional pathogen transfer (Funk et al. 2001), these standard practices may inadvertently introduce BTB and other infectious diseases to wildlife populations and conservancies throughout Southern Africa.

To limit this possibility, international regulations restrict the trade and movement of buffalo until they can be certified free of disease. But as the prevalence and long-term ecological and economic significance of BTB increases in other species, regulations may be extended to include potential spillover hosts, of which few can currently be tested or reliably confirmed as "disease free" (Espie et al. 2009). Consequently, legal restrictions on animal movement may not only threaten the breeding, distribution, and survival of multiple endangered species, but also jeopardize park functionality and revenue.

For example, without the capacity to move animals, managers may be forced to use less desirable and lethal population control methods to maintain population health and ecosystem sustainability (Kitching et al. 2008). In turn, this may reduce tourism as well as public and political support in protest to such controversial activities (de Lisle et al. 2002; Gallagher 2000). BTB may also have wider consequences for conservation by altering the perceived value and risks associated with visiting infected parks (Marcotty et al. 2009), reducing visitor-derived revenue, especially where alternative wildlife-based tourism opportunities are locally available.

On a wider scale, the threat of BTB infections may hinder the development and success of Trans-Frontier

Conservation Areas (TFCAs) like the GLTFCA (Fig 29.1). TFCAs often integrate vast areas of park, private, and communal land to simultaneously promote regional conservation alongside the socioeconomic development of rural communities through increased access to ecotourism opportunities (Osofsky et al. 2005). However, as TFCAs also facilitate the movement of people, livestock, and wildlife across former park and international boundaries, the potential for pathogens like BTB to infect naïve populations in previously disease-free areas may have consequences for local agriculture, regional trade, and international tourism (Osofsky et al. 2005). Should these potential implications be perceived to outweigh the possible economic benefits derived from TFCA development, BTB may jeopardize the long-term public and political support required for regional conservation, regardless of the absolute levels of disease (Osofsky et al. 2008). Thus, as BTB continues to spread, park authorities will need to take a proactive approach not only to minimize the direct consequences of disease for wildlife health, but also to prevent the broader ecological and financial consequences of disease from hindering the longevity of conservation activities.

## PEOPLE AND PARKS: THE INDIRECT IMPACTS OF BOVINE TUBERCULOSIS ON CONSERVATION

### Issues of Control at the Park–Communal Land Interface

Many parks in South Africa are located in rural areas, where the close proximity of people and their livestock poses a concern for disease control. Park authorities rely on game-deterrent fences to prevent wildlife from moving across park boundaries (Bengis et al. 2002), but as these are recurrently compromised by floods, large mammal behavior, and human activities, opportunities exist for wildlife, livestock, and people to interact at the interface between park and communal lands. In doing so, a complex cycle of multidirectional disease transmission may occur and establish a reservoir of disease in community cattle that can potentially re-infect wildlife populations, and compromise the utility of park control efforts (Bengis et al. 2002).

Furthermore, as BTB can also infect people and domestic animals, limiting the spread of disease across park boundaries will be equally important to prevent detrimental effects on community health and agriculture from altering local livelihood sustainability, natural resource use, and support for conservation initiatives (Osofsky et al. 2005; Brook and McLachlan 2006).

### The Current Status of Bovine Tuberculosis Control at the Livestock–Wildlife Interface

The status of BTB in community cattle is largely unknown for many rural areas in Southern Africa. This reflects the poor surveillance capacity of remote veterinary departments, leaving subsistence farmers poorly informed, under-serviced, and vulnerable to emerging diseases. As BTB rarely exhibits clinical signs, farmers without prior knowledge or experience of the pathogen struggle to detect infected animals, permitting BTB to spread within herds and to other susceptible livestock and domestic species (Ayele et al. 2004). Inter- and intra-species transmission is further promoted by local farming practices that increase the frequency and duration of contact between animals, neighboring herds, and occasionally wildlife species. In communities neighboring HiP, such contacts commonly occur through the use of communal grazing and water resources, livestock markets, and overnight holding pens that are designed to prevent stock theft and predation (Cleaveland et al. 2007).

Since local cattle breeds have no evolutionary resistance to BTB, their susceptibility to infection and mortality results in a loss of genetic diversity, and a greater local reliance on cross-breeds that are less tolerant to endemic diseases (Ahmed et al. 2006; Steinfeld et al. 2006). With few preventive or therapeutic measures available, BTB control is limited to the testing and removal of infected animals. However, as cattle are the only appreciating assets for many households, and are culturally worth more than their commercial value, these methods are insensitive and threaten to damage the relationship between farmers and veterinary authorities (Daborn et al. 1996). Without care, these poorly balanced measures can compromise local willingness to participate in, and comply with disease testing and wider healthcare programs (Daborn et al. 1996), ultimately hindering the prospect of long-term BTB control at the wildlife and livestock interface (Cosivi et al. 1998).

## Pathogens and People: The Current Status of Bovine Tuberculosis Control at the Livestock–Human Interface

The presence of BTB in community livestock places farming households at risk of aerosol infections due to their frequent contact with animals during agricultural activities. However, in rural communities with poor access to hygienic production methods, the ingestion of Mycobacteria in contaminated animal products places a larger population at risk from alimentary infections (Grange 2001), especially via dairy products that are preferentially consumed raw or soured, and sold informally in rural communities (Walshe et al. 1991; Kazwala et al. 1998, 2005; Karimuribo et al. 2005).

Although *M. bovis* is less virulent than the causative agent of human tuberculosis (*M. tuberculosis*) (Ayele et al. 2004), in communities with high levels of malnourishment, TB, and HIV/AIDS, immunocompromised groups are susceptible to BTB infection (Macallan 1999) and rare human-to-human transmission (Cobo et al. 2001). Thus, in areas surrounding HiP where TB is a major opportunistic infection for 80% of HIV/AIDS patients (Gandhi et al. 2006), the presence of *M. bovis* in wildlife and livestock poses a potential public health threat (Daborn et al. 1996; Cosivi et al. 1998).

Quantifying the risks and role of BTB for local community health is compromised by the poor capacity of rural health services to detect, diagnose, and treat BTB infections (Cosivi et al. 1998). This is particularly relevant for alimentary infections, which often lead to clinical disease in extrapulmonary areas that are less commonly associated with TB in humans (Rodwell et al. 2008). As accurate diagnosis requires specialist knowledge and equipment that is rarely available in rural health facilities, BTB cases may remain undetected and unreported, leaving patients at greater risk of mortality (Cosivi et al. 1998). Similarly, as some BTB strains have varying resistance to locally accessible and cost-effective front-line drugs (Guerrero et al. 1997), the limited ability to identify the cause of drug resistance hinders effective treatment, and masks the extent of BTB in local TB infections (Cosivi et al. 1998). Thus, as the number of extrapulmonary and drug-resistant TB cases continue to rise in South Africa (Gandhi et al. 2006), quantifying and minimizing the contribution of BTB to these health burdens

will be essential for the well-being of high-risk rural communities.

## The Consequences of Bovine Tuberculosis for Rural Communities in South Africa

In rural communities neighboring HiP and KNP, the health and well-being of people and animals are intrinsically linked. With typically poor access to land, capital, healthcare, and public amenities, local livelihoods are based on traditional livestock and small-scale arable farming, which are easily compromised by disease. In these areas, zoonotic pathogens like BTB threaten not only the direct health of families and their animals, but also the success of agricultural systems upon which they rely for food, economic security, and social well-being.

As BTB affects the strength, immunity, fecundity, and mortality of livestock, households with infected cattle experience a steady decline in the size and economic value of their herds. Without the capacity to replace weak and deceased animals, these families subsequently struggle to complete other livestock-related activities that are essential for household health and economic security. For example, a reduction in the quantity of cattle manure and draft power needed to fertilize crops and plow fields inhibits the success of household arable activities (Goe and Mack 2005; Nara et al. 2008) reducing the volume of produce that can be sold to generate income, or that is consumed by family members. The frequency and nutritional value of household meals are also affected by dwindling milk yields, which often constitutes the most reliable, and potentially sole source of micronutrients and fats that are essential for child growth, cognitive development (Hendriks 2003; Angeles and Catelo 2006; Nara et al. 2008), and the efficacy of TB and HIV/AIDS medication (Loevinsohn and Gillespie 2003; Paton et al. 2006).

Livestock mortalities also jeopardize household access to a broader range of goods and services that customarily support household cultural status, social well-being, and long-term health. In many South African cultures, families owning fewer cattle are effectively excluded from participating in cultural and religious practices that require the slaughter or donation of animals. In turn, this prevents access to the most advantageous marriages, which traditionally help to secure improved economic opportunities, land security, and cultural rights (Hawkes and Ruel 2006). Households with fewer livestock also experience a loss of discretionary income derived from the sale of cattle and their products. This deficit hinders their ability to cope with emergencies and pay for school fees, medicines, funeral costs, transportation, communications, seeds, and food—all of which have multiple deleterious implications for family health and household survival (Goe and Mack 2005; Hawkes and Ruel 2006). In essence, for families living in marginalized areas, livestock provide the agricultural and social mechanisms to cope with unpredictable pressures arising from poor family health, which are easily exacerbated by the potential dual animal and human health burdens imposed by BTB and other zoonoses.

Such scenarios are particularly relevant near HiP, where the high prevalence of HIV/AIDS accounts for 48% of adult mortality and 75% of female deaths between 15 and 44 years of age (Hosegood et al. 2004a,b). Here families also experience extensive TB co-infections and other health burdens that increase the pressure to improve family nutrition and pay for medical and funeral expenses. In many cases, this prompts the migration of adults to find alternative sources of income (Tanser et al. 2000; Morten 2006), which, together with the morbidity and mortality of household members, drives social change and alters the size and composition of families (Hosegood et al. 2004a,b). The resulting cumulative removal of household breadwinners and decision-makers has reduced the local labor force while simultaneously inhibiting the transfer of indigenous knowledge to an increasing proportion of women- and child-headed households. The consequent reduction in household agricultural expertise often impairs arable productivity and financial resources, which, along with the forfeiture of traditional childcare, education, and fuel- and food-seeking roles (Hawkes and Ruel 2006; Hosegood et al. 2004b), hastens families towards increasing fatigue, malnutrition, and vulnerability to disease (Gillespie 2006). In these circumstances, families with fewer livestock and financial alternatives often turn to less traditional and sustainable farming methods (Goe and Mack 2005). But as these activities are rarely tenable in over-grazed, disease- and drought-prone areas (Gillespie 2006), the resulting poor yields lead to absolute poverty, loss of land, and household

dissolution (Hosegood et al. 2004b; Goe and Mack 2005; Barnighausen et al. 2007).

Studies further indicate that families with severe financial and nutritional stress engage in riskier behavior and place increasing reliance on natural resources to supplement food, traditional medicine, and materials used for shelter, clothing, and trade (Shackleton and Shackleton 2004; Dovie et al. 2007). As conservation areas usually have a plentiful supply of resources in comparison to degraded and densely populated communal lands, park authorities face increasing physical and financial pressure to protect wildlife and plant species from unsustainable harvesting. However, as many conservation areas are located on land that historically belongs to local tribal authorities, restricting access to natural resources is a politically sensitive issue that may lead to conflict, especially if local support for park activities is diminished by the perceived threat of disease stemming from park wildlife (Pfeiffer 2006).

Thus, with such great capacity for zoonoses to affect household health, agricultural productivity, and socioeconomic development in marginalized communities, preventing the spread of BTB across the wildlife, livestock, and human interfaces is important to secure the long-term future of conservation areas in South Africa.

## PATHOGENS, PARKS, AND PEOPLE: THE ROLE OF CONSERVATION MEDICINE IN DISEASE CONTROL

The potential for zoonotic diseases like BTB to compromise both human and animal health is of great concern for the longevity of wildlife conservation areas in Southern Africa. To truly limit the affects of BTB requires a greater appreciation of the links between ecosystem, animal, and population health, while developing control strategies that mirror the ability of BTB to cross multiple species and geographical boundaries. However, as this chapter has discussed, the poor capacity of health services in marginalized areas hinders the quantification, control, and prevention of disease given normal species-specific circumstances. Undertaking the control of zoonoses like BTB that infect multiple hosts is therefore beyond the traditional knowledge, financial resources, and scope of individual conservation and community health organizations. Improving the capacity for effective disease control consequently requires a change in philosophy to capitalize on local skills, knowledge, and resources, and permit the root causes of disease to be tackled holistically across species and sectors (Porter et al. 1999).

This collaborative approach proved beneficial near Hluhluwe-iMfolozi, where an integrated framework was designed to address conservation and community health problems in unison, thereby improving the potential detection and response to human, animal and ecological aspects of BTB infections across conservation, veterinary and public health authorities. Ultimately this integrative philosophy also forms the basis for a conservation medicine approach to disease control, whereby the health of people, animals, and ecosystems are combined to enhance long-term population and ecological well-being (Norris 2001; Tabor et al. 2001).

## Planning a Conservation Medicine Approach to Disease Control; The Initial Steps

To quantify the risks, prevalence, and potential consequences of BTB within parks and rural communities, an interdisciplinary project was established in Hluhluwe-iMfolozi during 2007. This project was designed to have multiple objectives, including improving the capacity for health and conservation services to plan and conduct holistic disease control across multiple species, while simultaneously benefitting local communities, supporting agricultural livelihoods, and ultimately securing greater long-term environmental sustainability.

Initial meetings were arranged between local veterinary, medical, and conservation authorities, whose experience with coordinating disease surveillance in park wildlife and human populations formed a solid basis from which to build more integrated work. To promote the inclusion of innovative ideas and increased access to technical resources, relationships were also established with private institutions and academic and government departments from a range of disciplines, including those less traditionally associated with disease control initiatives (economics, anthropology, sociology and nutritional sciences). Furthermore, collaborations were sought with local indigenous health, farming, and minority

organizations that are rarely contacted prior to the start of health programs but should be considered vital partners to increase the success, relevance and potential societal benefits of project activities (Colvin et al. 2003; Schelling et al. 2005).

Through participating in a variety of field visits and technical discussions, the unique skills and priorities of each partner were identified. Accordingly, many commonalities were identified across multiple sectors between sectors, which helped to dispel initially perceived institutional, disciplinary, and cultural barriers between different groups. Using this improved awareness, a collaborative approach was employed to design, implement, and evaluate integrated healthcare, research, and disease control activities. Developing working relationships across sectors and institutions also enhanced the multidisciplinary skills of each partner by facilitating the transfer of expertise across health and conservation sectors (Ahmed et al. 2006). Ultimately this improved the collective capacity to detect, diagnose, and quantify BTB once project activities began, while also strengthening local skills and providing incentives for the continuation of partnerships beyond the duration of the project.

Furthermore, by sharing technical, logistical, and human resources that are often prohibitively scarce in rural settings, the cost and complexity of coordinating healthcare and disease research in remote areas was significantly reduced (Zinsstag et al. 2005). In turn, this enabled the project to provide wider societal and disease control benefits than originally anticipated, by scaling up activities to include isolated communities beyond the reach of single-sector health services (Zinsstag et al. 2005).

## Project Implementation and Disease Control Activities

By combining the unique theoretical and practical resources of local health, conservation, and community groups, practical activities were initiated in 2009, which aimed to improve local health care and disease research in Hluhluwe-iMfolozi. Over a 12-month period, a field-based team comprising 12 veterinary technicians, 3 ecologists, 4 agricultural students, 4 public health professionals, and numerous local farming association members offered simultaneous veterinary and public health services to 26 farming communities within 10 km of the HiP fence line.

Due to the early involvement of local farming associations and media groups that advertised and endorsed the project, over 12,000 cattle were brought for voluntary health assessments, of which 400 were recruited for longer-term BTB testing and follow-up activities. To maximize the value of activities for local health, livelihoods, and disease control, participating livestock and domestic animals were also offered complimentary vaccinations and treatment for common debilitating, zoonotic and agriculturally important diseases. In addition, to aid local disease awareness, public health teams provided written and oral information regarding the clinical, zoonotic, and food safety aspects of BTB to 5,500 community members. And finally, to better understand and assess the vulnerability of family members and livestock to disease, 1,200 families were interviewed to establish the risks of disease during routine farming and household practices, the degree of current health burdens, household reliance on natural resources use, and their overall resilience to household disturbances that influence social, financial, and nutritional security.

## The Benefits of an Integrated Approach to Health, Conservation and Disease Control

The benefits of this intensive campaign were multifold and provided immediate advantages to local communities, as well as supporting the capacity for local health services to plan longer-term and effective disease control. Direct contact between local farmers and health services enabled vulnerable and infected groups to be identified and immediately targeted for efficient and cost-effective disease-control activities. Likewise, by increasing the trust and communication between health providers, farming associations, and individual households, the future capacity to report, detect, and respond to emerging health problems was improved, enabling faster diagnosis and containment of disease prior to the establishment of wide-scale infections.

These outcomes not only benefitted the health status, disease awareness, and livelihoods of local farming communities, but also benefitted Hluhluwe-iMfolozi Park by reducing disease in neighboring livestock populations, and buffering wildlife populations from potential contact with trans-boundary disease risks. This also helped to secure the utility of park

BTB-control strategies, while preventing any indirect consequences of poor community and livestock health from threatening the conservation of local natural resources (Osofsky et al. 2005), and vice versa.

The longer-term benefits of this collaborative projects were derived foremost through the establishment of multidisciplinary teams, willing to share the responsibility of preventing and controlling BTB, while simultaneously improving local health service delivery. As such, by combining the fine-scale clinical, social, and environmental data collected by this project with longer-term park and sociological data sets provided by project partners, local patterns and drivers of disease were able to be identified for a number of locally significant health, environmental, and social scenarios. Using this information to create a suite of "best practices" and decision-making tools for use by health and conservation services will, in time, help to optimize, simplify, and support cohesive disease-prevention and control actions across species and sectors, making health and conservation services better prepared to respond and adapt to future pathogen threats given a range of ecological and community conditions.

In addition, as these collaborative activities delivered measurable cost, societal, environmental, and health benefits that are hard for individual organizations to achieve in isolation, coordinated health programs are ideally placed to leverage greater political support, resource allocation, and policy improvements that address community health and conservation needs holistically. Plus, as effective disease control helps to secure public health, rural livelihoods, and agricultural productivity, by supporting joint health and conservation programs, government departments may have a unique way to ease rural poverty while simultaneously strengthening natural resource protection and health service capabilities.

In summary, by understanding and using the intrinsic links between human, animal, and ecosystem health, this collaborative project reduced the unpredictability of disease by establishing a unique, affordable, and locally feasible method to monitor and predict the zoonotic risks of BTB across species and sectors in Hluhluwe-iMfolozi, South Africa. By following the principles of conservation medicine, this not only promoted greater integration between human, animal, and ecosystem health authorities, but also helped to anchor the future of conservation and community development across wider disciplines, to ensure that pathogens do not kill all hope for conservation in southern Africa.

## CONCLUSIONS

The emergence of zoonotic pathogens like BTB challenge the limited capacity for disease control across conservation and health services in rural areas of South Africa. With the ability to affect agricultural productivity, human health and livelihoods, international trade, and wildlife conservation, it is clear that controlling the spread of zoonoses across park and international boundaries is essential for regional development and natural resource protection.

Analysis of a recent project reveals that conservation medicine provides a novel framework for integrating conservation and community health, to tackle the root causes of disease emergence, and provide innovative, locally-relevant, adaptable, cost-effective, and cohesive disease control options. When used sensitively, these principles can provide immediate benefits to community healthcare and service provision, and create long-term cost and societal benefits that can be applied to multiple species, pathogens, and ecosystems throughout the developing world. By simplifying the planning, processes, and priorities for disease control across multiple agencies, conservation medicine principles may provide a unique opportunity to link pathogens, parks, and people for greater conservation and community health in South Africa.

## ACKNOWLEDGEMENTS

The author wishes to acknowledge the contributions of colleagues and staff at the following institutions whom contributed to the funding and execution of the collaborative research descibed in Hluhluwe-iMfolozi Park: Ezemvelo KZN Wildlife (Hluhluwe Research Centre, Hluhluwe-iMfolozi Park); South African Department of Health (Umkanyakude); South African Department of Agricuture, Forestry and Fisheries, Directorate of Food and Veterinary Services (Hluhluwe and Mtubatuba); University of Pretoria, South Africa (Mammal Research Institute, Department of Zoology and Entomology, Department of Veterinary Tropical Diseases); University of California - Berkeley, USA (Department of Environmental Science, Policy and Management); Institute of Tropical Medicine, Antwerp, Belgium and the communities of Umkanyakude, KwaZulu-Natal.

## REFERENCES

Ahmed, J.S., H. Alf, M. Aksin, and U. Seitzer. 2006. Animal transboundary diseases: European Union and Asian Network of veterinary research cooperation for quality livestock production. J Vet Med B53:2–6.

Anderson, R.M., and R.M. May. 1978. Regulation and stability of host-parasite population interactions. I. Regulatory processes. J Anim Ecol 47:219–247.

Angeles, M. and O. Catelo. 2006. Livestock and health. In C. Hawkes and M.T. Ruel, eds. Understanding the links between agriculture and health, pp. 9–10. 2020 Focus 13, International Food Policy Research Institute.

Ayele, W.Y., S.D. Neill, J. Zinsstag, M.G. Weiss, and I. Pavlik. 2004. An old disease but a new threat to Africa. Int J Tubercul Lung Dis 8:924–937.

Barnighausen, T., V. Hosegood, I.M. Timæus, and M.-L. Newell. 2007. The socioeconomic determinants of HIV incidence: a population-based study in rural South Africa. AIDS 21:S29–S38.

Bengis, R.G. 1999. Tuberculosis in free-ranging mammals. In M.E. Fowler and R.E. Miller, eds. Zoo and wild animal medicine: current therapy 4, pp. 101–114. W.B. Saunders Company, Philadelphia, Pennsylvania.

Bengis, R.G., and D.G. Grobler. 2000. Research into the breeding of disease-free buffalo. In Proceedings of the North American Veterinary Conference, Small animal and exotic section, 15–19 January, Orlando, Florida. The Eastern States Veterinary Association, Gainesville, Florida. pp. 1032–1033.

Bengis, R.G., R.A. Kock, and J. Fischer. 2002. Infectious animal diseases: the wildlife/livestock interface. Rev Sci Tech Off Int Epizoot 21:53–65.

Bengis, R.G., F.A. Leighton, J.R. Fischer, M. Artois, T. Mörner, and C. M. Tate. 2004. The role of wildlife in emerging and re-emerging zoonoses. Rev Sci Tech Off Int Epiz 23:497–511.

Blancou, J., B.B. Chomel, A. Belotto, and F. Xavier Meslin. 2005. Emerging or re-emerging bacterial zoonoses: factors of emergence, surveillance and control. Vet Res 36:507–522.

Brook, R.K., and S.M. McLachlan. 2006. Factors influencing farmers' concerns regarding bovine tuberculosis in wildlife and livestock around Riding Mountain National Park. J Environ Manage 80:156–166.

Brosch, R., S.V. Gordon, M. Marmiesse, P. Brodin, C. Buchrieser, K. Eiglmeier, T. Garnier, C. Guiterrez,

G. Hewinson, K. Kremer, L.M. Parsons, A.S. Pym, S. Samper, D. van Soolingen, and S.T. Cole. 2002. A new evolutionary scenario for the *Mycobacterium tuberculosis* complex. Proc Natl Acad Sci USA 99:3684–3689.

Caron, A., P.C. Cross, and J.T. du Toit. 2003. Ecological implications of bovine tuberculosis in African buffalo herds. Ecol Appl 13:1338–1345.

Cobo, J., A. Asensio, S. Moreno, E. Navas, V. Pintado, J. Oliva, E. Gómez-Mampaso, and A. Guerrero. 2001. Risk factors for nosocomial transmission of multidrug-resistant tuberculosis due to *Mycobacterium bovis* among HIV-infected patients. Int J Tubercul Lung Dis 5:413–418.

Cleaveland, S., D.J. Shaw, S.G. Mfinanga, G. Shirma, R.R. Kazwala, E. Eblate, and M. Sharp. 2007. *Mycobacterium bovis* in rural Tanzania: risk factors for infection in human and cattle populations. Tuberculosis 87:30–43.

Colvin, M., L. Gumede, K. Grimwade, D. Maher, and D. Wilkinson. 2003. Contribution of traditional healers to a rural tuberculosis control programme in Hlabisa, South Africa. Int J Tubercul Lung Dis 7:S86–S91.

Cosivi, O., J.M. Grange, C.J. Daborn, M.C. Raviglione, T. Fujikura, D. Cousins, R.A. Robinson, H.F.A.K. Huchzermeyer, I. de Kantor, and F.-X. Meslin. 1998. Zoonotic tuberculosis due to *Mycobacterium bovis* in developing countries. Emerg Infect Dis 4:59–70.

Cross, P.C., J.O. Lloyd-Smith, J.A. Bowers, C.T. Hay, M. Hofmeyer, and W.M. Getz. 2004. Integrating association data and disease dynamics in a social ungulate: bovine tuberculosis in African buffalo in the Kruger National Park. Ann Zool Fenn 41:879–892.

Cross, P.C., J.O. Lloyd-Smith, and W.M. Getz. 2005a. Disentangling association patterns in fission-fusion societies using African buffalo as an example. Anim Behav 69:499–506.

Cross, P.C., J.O. Lloyd-Smith, P.L.F. Johnson, and W.M. Getz. 2005b. Duelling timescales of host movement and disease recovery determine invasion of disease in structured populations. Ecol Lett 8:587–595.

Cross, P.C., and W.M. Getz. 2006. Assessing vaccination as a control strategy in an ongoing epidemic: bovine tuberculosis in African buffalo. Ecol Modell 196: 494–504.

Cross, P.C., D.M. Heisey, J.A. Bowers, C.T. Hay, J. Wolhunter, P. Buss, M. Hofmeyer, A.L. Michel, R.G. Bengis, T.L.F. Bird, J.T. du Toit, and W.M. Getz. 2009. Disease, predation and demography: assessing the impacts of bovine tuberculosis on African buffalo by monitoring at individual and population levels. J Appl Ecol 46: 467–475.

Daborn, C.J., J.M. Grange, and R. R. Kazwala. 1996. The bovine tuberculosis cycle—an African perspective. J Appl Bacteriol 81:S27–S32.

Dasak,P., A.A. Cunningham, and A. Hyatt. 2000. Emerging infectious diseases of wildlife—threat to biodiversity and human health. Science 287:443–449.

de Garine-Wichatitsky, M., A. Caron, C. Gomo, C. Foggin, K. Dutlow, D. Pfukenyi, E. Lane, S. Le Bel, M. Hofmeyer, T. Hlokwe, and A. Michel. 2010. Bovine tuberculosis in buffaloes, Southern Africa. Emerging Infect Dis 16:884–885.

de Klerk, L., A.L. Michel, D.G. Grobler, R.G. Bengis, M. Bush, N.P.J. Kriek, M.S. Hofmeyer, J.T.F. Griffin, and C.G. Mackintosh. 2006. An experimental intratonsilar infection model for bovine tuberculosis in African buffaloes, *Syncerus caffer*. Onderstepoort J Vet Res 73:293–303.

de Klerk, L.-M., A.L. Michel, R.G. Bengis, N.P.J. Kriek, and J. Godfroid. 2009. BCG vaccination failed to protect yearling African buffaloes (*Syncerus caffer*) against experimental intratonsilar challenge with *Mycobacterium bovis*. Vet Immunol Immunopathol 137:84–92.

de Lisle, G.W., R.G. Bengis, S.M. Schmitt, and D.J. O'Brien. 2002. Tuberculosis in free-ranging wildlife: detection and management. Rev Sci Tech Off Int Epiz 21: 317–334.

de Vos, V., J.P. Raath, R.G. Bengis, N.P.J. Kriek, H. Huchzermeyer, D.F. Keet, and A. Michel. 2001. The epidemiology of tuberculosis in free-ranging African buffalo (*Syncerus caffer*) in the Kruger National Park, South Africa. Onderstepoort J Vet Res 68: 119–130.

Dovie, B.K., C.M. Shackleton, and E.T.F. Witkowski. 2007. Conceptualizing the human use of wild edible herbs for conservation in South African communal areas. J Environ Manage 84:146–156.

Drewe, J.A., A.K. Foote, R.L. Sutcliffe, and G.P. Pearce. 2009. Pathology of *Mycobacterium bovis* infection in wild meerkats (*Suricata suricatta*). J Comp Pathol 140:12–24.

Espie, I.W., T.M. Hlokwe, N.C. Gey van Pittius, E. Lane, A.S.W. Tordiffe, A.L. Michel, A. Müller, A, Kotze and P.D. van Helden. 2009. Pulmonary infection due to *Mycobacterium bovis* in a black rhinoceros (*Diceros bicornis minor*) in South Africa. J Wildl Dis 45(4):1187–1193.

Funk, S.M., C.V. Fiorello, S. Cleaveland, and M.E. Gompper. 2001. The role of disease in carnivore ecology and conservation. *In* J.L. Gittleman, S.M. Funk, D.W. MacDonald, and R.K. Wayne, eds. Carnivore conservation, pp. 433–466. Cambridge University Press.

Gallagher, J. 2000. The bovine tuberculosis dilemma—what do we do next? Vet J 160:85–86.

Gandhi, N.R., A. Moll, A.W. Sturm, R. Pawsinski, T. Govender, U. Lalloo, K. Zeller, J. Andrews, and G. Friedland. 2006. Extensively drug-resistant tuberculosis as a cause of death in patients co-infected with tuberculosis and HIV in a rural area of South Africa. Lancet 368:1575–1580.

Gillespie, S. 2006. Agriculture and HIV/AIDS. In C. Hawkes and M.T. Ruel, eds. Understanding the links between agriculture and health, pp. 9–10. 2020 Focus 13, International Food Policy Research Institute.

Goe, M.R., and S. Mack. 2005. Linkages between HIV/AIDS and the livestock sector in East and southern Africa. FAO Animal Production and Health Technical Workshop Proceedings, Addis Ababa, Ethiopia, 8-10 March.

Grange, J.M. 2001. Mycobacterium bovis infection in human beings. Tuberculosis 81:71–77.

Guerrero, A., J. Cobo, J. Fortun, E. Navas, C. Quereda, A. Asensio, J. Canon, J. Blasquez, and E. Gomez-Mampaso. 1997. Nosocomial transmission of Mycobacterium bovis resistant to 11 drugs in people with advanced HIV-1 infection. Lancet 350:1738–1742.

Halley, D.J., M.E.J. Vandewalle, and C. Taolo. 2002. Herd-switching and long-distance dispersal in female African buffalo (Syncerus caffer). Afr J Ecol 40:97–99.

Hawkes, C., and M. T. Ruel. 2006. Agriculture and nutrition linkages: old lessons and new paradigms. In C. Hawkes and M.T. Ruel eds. Understanding the links between agriculture and health 2020, pp. 9–10. Focus 13, International Food Policy Research Institute.

Hay, C.T., P.C. Cross, and P.J. Funston. 2008. Trade-offs between predation and foraging explain sexual segregation in African buffalo. J Anim Ecol 77:850–858.

Hendriks, S.L. 2003. The potential for nutritional benefits from increased agricultural production in rural KwaZulu-Natal. S Afr J Agr Ext 32:28–44.

Hosegood, V., A.-M.Vanneste, and I. A. Timæus. 2004a. Levels and causes of adult mortality in rural South Africa: the impact of AIDS. AIDS 18:663–671.

Hosegood, V., N. McGrath, K. Herbst, and I.M. Timæus. 2004b. The impact of adult mortality on household dissolution and migration in rural South Africa. AIDS 18:1585–1590.

Huchzermeyer, H.F.A.K., G.K. Brueckner, A. van Heerden, H.H. Kleeberg, I.B.J. van Rensburg, P. Koen, and R.K. Loveday. 1994. Tuberculosis. In J.A.W. Coetzer, G.R. Thompson, and R.C. Tustin, eds. Infectious diseases of livestock, pp. 1425–1445. Oxford University Press, Southern Africa, Cape Town.

Hutcheon, D. 1880. Tering: consumption, tables mesenterica. Annual Report. Colonial Veterinary Surgeon, Cape of Good Hope.

IUCN. 2010. IUCN Red List of Threatened Species. Version 2010.4 http://www.iucnredlist.org

Jolles, A.E. 2004. Disease ecology of tuberculosis in African buffalo. Ph.D. Dissertation. Princeton University, Princeton, New Jersey.

Jolles, A.E., D.V. Cooper, and S.A. Levin. 2005. Hidden effects of chronic tuberculosis in African buffalo. Ecology 86:2258–2264.

Jolles, A.E., V.O. Ezenwa, R.S. Etienne, W.C. Turner, and H. Olff. 2008. Interactions between macroparasites and microparasites drive infection patterns in free-ranging African buffalo. Ecology 89:2239–2250.

Karimuribo, E.D., L.J. Kusiluka, R.H. Mdegela, A.M. Kapaga, C. Sindato, and D.M. Kambarage. 2005. Studies on mastitis, milk quality and health risks associated with consumption of milk from pastoral herds in Dodoma and Morogoro regions, Tanzania. J Vet Sci 6:213–221.

Kazwala, R.R., C.J. Daborn, L.J. Kusiluka, S.F. Jiwa, J.M. Sharp, and D. M. Kambarage. 1998. Isolation of Mycobacterium species from raw milk or pastoral cattle of Southern Highlands of Tanzania. Trop Anim Health Prod 30:233–239.

Kazwala. R.R., C.J. Daborn, J.M. Sharp, D.M. Kambarage, S.F.H. Jiwa, and N.A. Mbembati. 2005. Isolation of Mycobacterium bovis from human cases of cervical adenitis in Tanzania: a cause for concern? Int J Tubercul Lung Dis 5:87–91.

Keet, D.F., N.P. Kriek, M.-L. Penrith, A. Michel, and H.F. Huchzermeyer. 1996. Tuberculosis in buffaloes (Syncerus caffer) in the Kruger National Park: spread of the disease to other species. Onderstepoort J Vet Res 63:239–244.

Keet, D.F., A. Michel, and D.G.A. Meltzer. 2000a. Tuberculosis in free-ranging lions (Panthera leo) in the Kruger National Park. In Proceedings of the South African Veterinary Association Biennial Congress. 20-22 September 2000. Durban, KwaZulu-Natal. pp. 232–241.

Keet, D.F., N.P.J. Kriek, R.G. Bengis, D.G., Grobler, and A.L. Michel. 2000b. The rise and fall of tuberculosis in a free-ranging chacma baboon troop in the Kruger National Park. Onderstepoort J Vet Res 67:115–122.

Keet, D.F., N.P.J. Kriek, R.G. Bengis, and A.L. Michel. 2001. Tuberculosis in kudu (Tragelaphus strepsiceros) in the Kruger National Park. Onderstepoort J Vet Res 68:225–230.

Keet, D.F., A.L. Michel, R.G. Bengis, P. Becker, D.S. Dyk, M. van Vuuren, V.P.MG. Rutten, and B.L. Penzhorn. 2010. Intradermal tuberculin testing of wild African lions (*Panthera leo*) naturally exposed to infection with *Mycobacterium bovis*. Vet Microbiol 144:384–391.

Kitching, P., M. Sabara, G. Thomson, D. Werling, and S. Albrecht. 2008. Transboundary disease management: the theory and the practice; the science and the politics. Transbound Emerg Dis 55:1–2.

Kloeck, P.E. 1998. Tuberculosis of domestic animals in areas surrounding the Kruger National Park. *In* Proceedings of the Challenges of Managing Tuberculosis in Wildlife in Southern Africa, Zunkel, July 30-31, Nelspruit, South Africa.

Lee, K.A., and K.C. Klassing. 2004. A role for immunology in invasion biology. Trends Ecol Evol 19:523–529.

Loevinsohn, M., and S. Gillespie. 2003. HIV/AIDS, Food Security and Rural Livelihoods: Understanding and Responding. Food Consumption and Nutrition Division Paper No.157, International Food Policy Research Institute, Washington D.C.

Lyashchenko, K.P., R. Greenwald, J. Esfandiari, M.A. Chambers, J. Vicente, C. Gortazar, N. Santos, M. Correia-Neves, B.M. Buddle, R. Jackson, D.J. O'Brien, S. Schmitt, M.V. Palmer, R.J. Delahay, and W.R. Waters. 2008. Animal-side serologic assay for rapid detection of *Mycobacterium bovis* infection in multiple species of free-ranging wildlife. Vet Microbiol 132:283–292.

Maas, M., I. van Rhijn, M.T.E.P. Allsopp and V.P.M.G. Rutten. 2010. Lion (*Panthera leo*) and cheetah (*Acinonyx jubatus*) IFN-γ sequences. Vet Immunol Immunopathol 134:296–298.

Macallan, D.C. 1999. Malnutrition in tuberculosis. Diagn Microbiol Infect Dis 34:153–157.

Maddock, E.C.G. 1933. Studies on the survival time of the bovine tubercle bacillus in soil, soil and dung, in dung and on grass, with experiments on the preliminary treatment of infected organic matter and the cultivation of the organism. J Hyg 33:103–117.

Marcotty, T., F. Matthys, J. Godfroid, L. Rigouts, G. Ameni, N. Gey van Pittius, R. Kazwala, J. Muma, P. van Helden, K. Walravens, L.-M. de Klerk, C. Geoghegan, D. Mbotha, M. Otte, K. Anenu, N. Abu Samra, C. Botha, M. Ekron, A. Jenkins, F. Jori, N. Kriek, C. McCrindle, A. Michel, D. Morar, F. Roger, E. Thys, and P. van den Bossche. 2009. Zoonotic tuberculosis and brucellosis in Africa: neglected zoonoses or minor public-health issue? The outcomes of a multi-disciplinary workshop. Ann Trop Med Parasit 103:401–411.

Michel, A.L. 2002. Implications of tuberculosis in African wildlife and livestock. Ann NY Acad Sci 969:251–255.

Michel, A.L., R.G. Bengis, D.F. Keet, M. Hofmeyer, L.M. de Klerk, P.C. Cross, A.E. Jolles, D. Cooper, I.J. Whyte, P. Buss, and J. Godfroid. 2006. Wildlife tuberculosis in South Africa conservation areas: Implications and challenges. Vet Microbiol 112:91–100.

Michel, A.L., L.-M. de Klerk, N. Gey van Pittius, R.M. Warren, and P.D. van Helden. 2007. Bovine tuberculosis in African buffaloes: observations regarding *Mycobacterium bovis* shedding into water and exposure to environmental mycobacteria. BMC Vet Res 3 doi:10.1186/1746-6148-3-23.

Michel, A.L., T.M. Hlokwe, M.L. Coetzee, L. Mare, L. Connoway, V.P.M.G. Rutten, and K. Kremer. 2008. High *Mycobacterium bovis* genetic diversity in a low prevalence setting. Vet Microbiol 126:151–159.

Michel, A.L., M.L. Coetzee, D.F. Keet, L. Mare, R. Warren, D. Cooper, R.G. Bengis, K. Kremer, and P. van Helden. 2009. Molecular epidemiology of *Mycobacterium bovis* isolates from free-ranging wildlife in South African game reserves. Vet Microbiol 133:335–343.

Mills, M.G.L., H.C. Biggs, and I.J. Whyte. 1995. The relationship between rainfall, lion predation and population trends in African herbivores. Proceedings of the Sixth International Theriological Congress, University of New South Wales in Sydney, Australia. Wildl Res 22:75–88.

Morar, D. 2003. The development of an interferon-gamma (IFNg) assay for the diagnosis of tuberculosis in African elephants (Loxodonta Africana) and black rhinoceros (Diceros bicornis). M.Sc. Thesis. University of Pretoria, Pretoria, South Africa.

Morten, J. 2006. Conceptualising the links between HIV/AIDS and pastoralist livelihoods. Eur J Devel Res 18:235–254.

Nara, P.L., D. Nara, R. Chaudhuri, G. Lin, and G. Tobin. 2008. Perspective on advancing preventative medicine through vaccinology at the comparative veterinary, human and conservation medicine interface: not missing the opportunities. Vaccine 26:6200–6211.

Norris, S. 2001. A new voice in conservation. BioScience 51:7–12

O'Reilly, L.M., and C.J. Daborn. 1995. The epidemiology of *Mycobacterium bovis* infections in animals and man: a review. Tuber Lung Dis 76 Supplement 1:1–46

Osofsky, S.A., R.A. Kock, M.D. Kock, G. Kalema-Zikusoka, R. Grahn, T. Leyland, and W.B. Karesh. 2005. Building support for protected areas using a "One Health" perspective. *In* J.A. McNeely, ed. Friends for Life: New Partners in Support of Protected Areas, pp. 65–79. IUCN, Gland, Switzerland.

Osofsky, S.A., D.H.M. Cumming, and M.D. Kock. 2008. Transboundary management of natural resources and the importance of a "One Health" approach: Perspectives on southern Africa. Wildlife Conservation Society. Island Press, Washington, D.C.

Paine, R., and G. Martinaglia. 1929. Tuberculosis in wild buck living under natural conditions. J Comp Pathol Ther 42:1–8.

Paton, N.I., S. Sangeetha, A. Earnest, and R. Bellamy. 2006. The impact of malnutrition on survival and the CD4 count response in HIV-infected patients starting antiretroviral therapy. HIV Med 7:323–330.

Perry, B., and K. Scones. 2007. Poverty reduction through animal health. Science 315:333–334.

Pfeiffer, D.U. 2006. Communicating risk and uncertainty in relation to development and implementation of disease control policies. Vet Microbiol 112: 259–264.

Porter, J., J. Ogden, and P. Pronyk. 1999. Infectious disease policy: towards the production of health. Health Pol Plan 144:322–328.

Radloff, F.G.T., and J.T. du Toit. 2004. Large predators and their prey in a southern African savanna: a predator's size determines its prey size range. J Anim Ecol 73:410–423.

Rodwell, T.C., N.P. Kriek, R.G. Bengis, I.J. Whyte, P.C. Viljoen, V. de Vos, and W.M. Boyce. 2001. Prevalence of bovine tuberculosis in African buffalo at Kruger National Park. J Wildl Dis 37:258–264.

Rodwell, T.C., M. Moore, K.S. Moser, S.K. Brodine, and S.A. Strathdee. 2008. Tuberculosis from *Mycobacterium bovis* in binational communities, United States. Tuberculosis 14:909–916.

Schelling, E., K. Wyss, M. Béchir, D.D. Moto, and J. Zinsstag. 2005. Synergy between public health and veterinary services to deliver human and animal health interventions in rural low-income settings. Brit Med J 331:1264–1267.

Shackleton, C.M., and S.E. Shackleton. 2004. The importance of non-timber forest products in rural livelihood security and as safety-nets: evidence from South Africa. S Afr J Sci 100:658–664.

Steinfeld, H., P. Gerber, T. Wassenaar, V. Castel, M. Rosales, and C. de Haan. 2006. Livestock's long shadow, environmental issues and options. Livestock, Environment and Development and Food and Agriculture Organisation of the United Nations, Rome.

Tabor, G.M., R.S. Ostfeld, M. Poss, A.P. Dobson, and A.A. Aguirre. 2001. Conservation biology and the health sciences: defining the research priorities of conservation medicine. *In* M.E. Soule and G.H. Orians, eds. Research priorities in conservation biology, 2nd ed., pp. 165–173. Island Press, Washington, D.C.

Tanner, M., and A.L. Michel. 1999. Investigation of the viability of *Mycobacterium bovis* under different environmental conditions in the Kruger National Park. Onderstepoort J Vet Res 66:115–122.

Thorburn, J.A., and A.D. Thomas. 1940. Tuberculosis in Cape Kudu. J S Afr Vet Med Assoc 11:3–10.

Tanser, F., D. LeSueur, G. Solarsh, and D. Wilkinson. 2000. HIV heterogeneity and proximity of homestead to roads in rural South Africa: an exploration using a geographical information system. Trop Med Int Health 5:40–46.

Turner, W.C. 2003. Activity patterns of male African buffalo. M.Sc. Dissertation. University of Witwatersrand, Johannesburg, South Africa.

Vosloo, W., A.D.S. Bastos, A. Michel, and G.R. Thomson. 2001. Tracing the movement of African buffalo in southern Africa through genetic characterisation of pathogens. Rev Sci Tech Off Int Epiz 20: 630–639.

Walshe, M.J., J. Grindle, A. Nell, and M. Bachmann. 1991. Dairy development in sub-Saharan Africa. World Bank Technical Paper No. 135. African Technical Department Series, World Bank, Washington, D.C.

Welsh, M.D., R.T. Cunningham, D.M. Corbett, R.M. Girvin, J. McNair, R.A. Skuce, D.G. Bryson, and J.M. Pollock. 2005. Influence of pathological progression on the balance between cellular and humoral immune responses in bovine tuberculosis. Immunology 114:101–111.

Woolhouse, M.E.J., and S. Gowtage-Sequeria. 2005. Host range and emerging and reemerging pathogens. Emerg Infect Dis 11:1842–1847.

Zinsstag, J., E. Schelling, K. Wyss, and M.B. Mahamat. 2005. Potential of cooperation between human and animal health to strengthen health systems. Lancet 366: 2142–2145

# 30

## DISEASE ECOLOGY AND CONSERVATION OF UNGULATES, WILD RABBITS, AND THE IBERIAN LYNX IN THE MEDITERRANEAN FOREST

Fernando Martínez, Guillermo López, and Christian Gortázar

"A MI FAMILIA, cuyos orígenes como pastores de ovejas y abejas, me han enseñado a observar, entender y respetar al monte, a sus seres y sus gentes."

*FM*

Threats to biodiversity are usually a result of anthropogenic changes in the environment, such as habitat loss and fragmentation, climate change, non-native species invasions, and resource overexploitation (Pimm et al. 1995; Wilcove et al. 1998). Although usually underestimated, infectious diseases are also recognized threats for biodiversity, since they can lead to decline and extinction (Scott 1988; Daszak and Cunningham 1999; Castro and Bolker 2005). Moreover, environmental change is likely to influence disease dynamics and emergence as a result of both direct effects on host and pathogen physiology, and indirect effects following changes in interactions with other species (Lips et al. 2008). Besides the impact of diseases on conservation, they produce severe economic losses, eventually affecting habitat conservation. In the past 500 years, 869 animal species are known to be extinct (IUCN 2009), and approximately 3.7% of those extinctions have been attributed to infectious diseases (Smith et al. 2006). Historically, the Mediterranean forest ecosystem occupied most of the Mediterranean bioclimatic part of the Iberian Peninsula (Spain and Portugal). Despite profound anthropogenic changes, it still maintains relevant habitats and species and is considered a hotspot for biodiversity (Myers et al. 2000).

We selected a group of vertebrates with an overlapping distribution and strong habitat and disease-mediated connections within the Mediterranean forest ecosystem. These species are the red deer (*Cervus elaphus*), the Eurasian wild boar (*Sus scrofa*), the European wild rabbit (*Oryctolagus cuniculus*), and the Iberian lynx (*Lynx pardinus*). Through sections devoted to rabbits, ungulates, and lynx, respectively, we demonstrate the links among changes in land use and fluctuations in wildlife populations and how often these links are disease-mediated. The wild rabbit is affected by habitat changes, ungulate overabundance, and introduced viral diseases. Ungulate populations have exploded in recent decades, on the one hand constituting an important natural resource but on the other becoming disease reservoirs and causing massive alterations in the habitat. Finally, the critically endangered Iberian lynx survives in isolated populations depending on rabbits as prey and being at risk of diseases maintained by ungulates and by other carnivores, including domestic dogs and cats. Therefore, human activities, landscape changes, and disease events interact to create a complex applied case of ecological health.

## THE MEDITERRANEAN FOREST

The Mediterranean forest is highly resistant to drought and fire, adapted to hot and dry summers and mild winters. The predominant vegetation of the Mediterranean ecosystem in the Iberian Peninsula is short-tree woods of different species of oaks (specially the evergreen oaks *Quercus ilex* and *Quercus suber*), wild olive trees (*Olea europaea sylvestris*), mastic shrubs (*Pystacea lentiscus*), strawberry trees (*Arbutus unedo*), rockroses (*Cistus* spp.), and other woody species (*Lavandula, Retama, Rubus*). For centuries, lowland woodlands have been transformed into crops or pastures. This process has been more intense in most parts of the northern Iberian Peninsula. Mediterranean woodland patches are still found throughout the southwest part of the Iberian Peninsula. In mountain ranges and pronounced slopes these woodlands are dense, while open, human-transformed savannah-like pastures and woodlands, called *dehesas* in Spain and *montados* in Portugal, predominate in the plateau. *Dehesas* are mainly used for grazing and cereal crops and represent a model of the traditional agro-sylvo-pastoral system with periodic tree pruning to increase acorn and pasture production, rotational plowing for cultivation and scrub encroachment control, as well as grazing by livestock and hunting.

The key wild vertebrate species of this ecosystem is the European wild rabbit (Delibes-Mateos et al. 2007). Rabbit abundance allows many predators to base their diet on this species. The Iberian lynx and the Spanish imperial eagle (*Aquila adalberti*) are two endemic endangered species that have specialized their diet on the wild rabbit (Ferrer and Negro 2004). Hence, viral diseases affecting rabbits have had profound effects on predator populations and land use. Game species are an important economic resource in Mediterranean habitats, often generating more profit than livestock, forest exploitation, or agriculture. The most relevant small game species other than the rabbit are the Iberian hare (*Lepus granatensis*) and the red-legged partridge (*Alectoris rufa*). Small game management, including feeding, predator control, and restocking, is still performed in areas where these species remain at high densities (Delibes-Mateos et al. 2008b).

Big game species include Eurasian wild boar, red deer, and more locally fallow deer (*Dama dama*). Populations of most Iberian wild ungulates strongly decreased in the early 20th century due to overhunting, although recovery from refuge areas occurred during the second half of the century (Acevedo et al. 2005; Delibes-Mateos et al. 2009a). Also in this period, a progressive decrease of human population and agriculture due to rural abandonment caused extensive landscape changes, mostly a marked increase in scrubland habitats and the expansion of big game species (Gortázar et al. 2000; Acevedo et al. 2005; Delibes-Mateos et al. 2009a). In addition, to increase ungulate densities for hunting purposes, fencing, food supplementation, watering, and translocations were performed (Vicente et al. 2007).

For centuries, livestock has been an important economic resource and a modulator of the Mediterranean forest. Traditionally, Spanish laws protected the rights of livestock (mainly sheep) breeders to freely access pastures rather than the rights of landowners. This changed in 1812, when a decree established the owner's right to fence his property. However, this regulation explicitly excluded the ownership of hunting rights (Gálvez 2004). In the 20th century, sheep grazing was increasingly substituted by cattle grazing, which became more profitable. In fact, the sheep population in Spain decreased by one fourth in the past 20 years, while the cattle population grew by one third (INE 2009). Diseases are a remarkable issue in this ecosystem. Rabbit viral diseases determine prey availability for several predators, and others, such as feline leukemia (FeL) or canine distemper (CD), affect them directly. In addition, diseases such as bovine tuberculosis (bTB) create conflicts between hunters and livestock producers.

## THE EUROPEAN WILD RABBIT

### Current Status

The European wild rabbit originated in the Iberian Peninsula, where it was very abundant until the 20th century. In fact, this territory was termed by the Phoenician sailors earlier than 1000 B.C. as *i-shaphan-im* or the *land of the hyraxes*, mistaking the European wild rabbit for the rock hyrax (*Procavia capensis*). The rabbit is an ecosystem engineer (Jones et al. 1994) and a keystone species in the Mediterranean forest (Delibes-Mateos et al. 2007). Rabbits are an important food resource for at least 29 predator species (Delibes and Hiraldo 1981). Furthermore rabbits, through grazing and seed dispersal, modulate plant species composition and vegetation structure in the Mediterranean landscape, their latrines improve soil fertility and provide food for many invertebrates, and their burrows provide nest sites and shelter for vertebrates and invertebrates (Delibes-Mateos et al. 2008a).

Although rabbits inhabit most Iberian habitats, they seem to prefer Mediterranean scrubland (Virgós et al. 2003; Delibes-Mateos et al. 2009b). Rabbit populations dramatically decreased during the 20th century due to viral diseases, and today most populations remain in low numbers. The main identified factors hindering the recovery of the rabbit in the Iberian Peninsula are (1) circulation of viral diseases, (2) habitat loss and fragmentation, (3) changes in the traditional scrubland management, (4) changes in agricultural uses, and (5) excessive hunting pressure (Moreno and Villafuerte 1995; Calvete et al. 2004; Williams et al. 2007; Delibes-Mateos et al. 2009a). Hence, these factors are also regulating the demography of rabbit specialist predators (Fernández et al. 2006).

Rabbit hunting is an important economic resource in Spain, with over 30,000 private hunting estates that cover 70% of the territory (Angulo and Villafuerte, 2003). Several measures, such as habitat management, artificial feeding, and restocking, are used to increase rabbit density in hunting estates and protected areas, with variable success (Angulo and Villafuerte 2003; Rouco et al. 2008).

## Viral Diseases of Wild Rabbits

Rabbit populations have declined dramatically mainly due to two introduced viral diseases: myxomatosis in the 1950s and rabbit hemorrhagic disease (RHD) in the late 1980s (Villafuerte et al. 1995; Moreno and Villafuerte 1995). Myxomatosis is caused by a poxvirus originally found in South American lagomorphs of the genus *Sylvilagus*. The virus was introduced to Australia to control the introduced rabbits that were considered a pest, and later introduced through France into Europe, causing severe changes in Mediterranean forest ecosystems and a dramatic decline of rabbit predators. This virus is mainly transmitted by vectors (fleas and mosquitoes) and tends to affect juveniles more than adults, with a higher incidence in periods of vector activity than in winter (Rosell 2000; García-Bocanegra et al. 2010a).

RHD is caused by a calicivirus originally identified in China in 1984. It was introduced into Germany through commercial rabbits, spreading like wildfire throughout Europe. In the beginning the disease was highly lethal, killing 80% to 90% of animals older than 8 weeks (Cooke and Fenner 2002). RHD arrival caused a second crash in the wild rabbit population, driving rabbit predator specialists to the verge of extinction (Delibes-Mateos et al. 2009b). Mathematical modeling suggested that RHD's impact could be highly dependent on rabbit population dynamics, with denser populations being more able to cope with the disease and recover, while small and scattered metapopulations would be more severely affected (Calvete 2006).

As a consequence of these viral diseases and habitat changes, rabbit populations declined by an estimated 70% between 1973 and 1993 (Virgós et al. 2007). The population decline of rabbits is reflected in the number shot, from 10 million in the late 1980s to 4 million during recent years (INE 2006). Despite this drastic decrease, some areas maintained good rabbit populations, probably due to habitat suitability (Blanco and Villafuerte 1993; Calvete 2006). Currently, many rabbit populations are recovering, but this tends to occur in lowland agro-ecosystems (Williams et al. 2007). Since the arrival of myxomatosis and RHD considerable efforts have been made to restore and increase rabbit populations for hunting and conservation purposes in the Iberian Peninsula. These have included scrub clearing and pasture management, construction of artificial breeding warrens, and selective fencing. Another common action is rabbit translocation from abundant populations, often without considering the existence of two different rabbit lineages present in Spain (Delibes-Mateos et al. 2008c). Vaccination of rabbits against myxomatosis and RHD is common in hunting and protected areas but has a debatable effectiveness (Calvete et al. 2004; Delibes-Mateos et al. 2008b; Ferreira et al. 2009b).

## WILD UNGULATES

### Current Status

Wild ungulates play a mayor role in the Mediterranean woodland ecosystem. The Eurasian wild boar is the most widely distributed one, covering almost the entire Iberian Peninsula. Local densities under artificial management can be as high as 90 wild boars per km², and mean densities in woodland habitats are probably higher than 10 wild boars per km² (Acevedo et al. 2007). Red deer are also abundant, with densities of up to 69 (mean 23) per km² in southern Spain (Acevedo et al. 2008). Similarly, fallow deer are only locally abundant, with mean densities of 42 per km² in Doñana National Park (Soriguer et al. 2001). Ungulate overabundance has adverse effects on the soil, vegetation, and other fauna, including rabbits. Moreover, an increased density and spatial aggregation of ungulates favors the transmission of infectious diseases (Gortázar et al. 2006).

As was explained previously, extensive changes in land use took place in Spain during the second half of the 20th century and occurred along with the development of an important commercial hunting industry. Possibly the rabbit decline also contributed to a switch towards large game production, since lower rabbit numbers made small game hunting (based mostly on rabbits and red-legged partridges) less profitable. Fencing started in the 1970s and became most popular in the 1990s, when large portions of private-owned former *dehesas* and Mediterranean woodlands were high-wire fenced for game management purposes. Today, fencing affects over 50% of Mediterranean forests in some regions (almost 100% in the current lynx range) and creates conflicts between the landowners, who defend their right to fence their properties and make profits from commercial hunting, and conservationists, who are concerned about the ecological consequences of fencing. Current Spanish law supports the owners' views but imposes conditions such as obligatory openings for protected species, limitations on fence size and shape, or limitations regarding the attachment of fences to ground level (Gálvez 2004). Unfortunately, these regulations are not applied in most instances. One consequence of ungulate overabundance is the adverse effects on wild rabbits. By competing for food and modifying the soil, high wild boar and deer densities can become a limiting factor for the rabbit (Lozano et al. 2007). Paradoxically, this contributes to the fact that rabbit availability needs to be maintained artificially in lynx habitats with overabundant wild ungulates, while the increasing rabbit densities in agro-ecosystems with no wild ungulates are causing concerns to farmers and wildlife managers (Williams et al. 2007).

## Diseases of Wild Ungulates

Diseases of wild ungulates can be shared with livestock and other wildlife species. Eurasian wild boar and red deer are important reservoir hosts of many viral diseases, including Aujezky's disease (AD) or pseudorabies (Vicente et al. 2005; Ruiz-Fons et al. 2008), swine brucellosis (Ruiz-Fons et al. 2008; Muñoz et al. 2010), and bTB caused by *Mycobacterium bovis* (Pérez et al. 2001; Gortázar et al. 2008). In both wild boar and red deer, the transmissible disease prevalence correlates with host population density and spatial aggregation (Acevedo et al. 2007; Vicente et al. 2007). The highest seroprevalence of antibodies against AD virus (ADV) in wild boar is found in Sierra Morena, Spain (Vicente et al. 2005). It has been suggested anecdotally that hunting dogs dying after consuming wild boar meat may have been infected with ADV. Since cases of AD have already been reported also among felids, the risks to lynx cannot be excluded (Vicente et al. 2005).

Infection with *M. bovis* was first noticed in Doñana National Park wild ungulates in the 1980s (León-Vizcaíno et al. 1990). In northern areas of Doñana with no cattle in recent years, wild boar are primary reservoirs of infection, in which the prevalence reached 92%. Also, the infection is a main sanitary concern in red deer, with prevalences reaching 30% in Doñana and even higher in Sierra Morena (Vicente et al. 2006; Gortázar et al. 2008). The high prevalence of TB recorded in wild boar, reed deer, and fallow deer in large areas of the Iberian Peninsula, including fenced estates without livestock, suggests that these wildlife reservoirs are able to maintain *M. bovis* in the absence of domestic cattle (Vicente et al. 2006).

## IBERIAN LYNX

### Current Status

The Iberian lynx is considered the most endangered felid in the world (IUCN 2004). It is endemic of the Iberian Peninsula (Ferrer and Negro 2004), where it was moderately abundant and widely distributed until the 18th century. During the 19th century, the Iberian lynx extinction was documented in eastern and northern Spain (Graells 1897; Cabrera 1914). The decrease continued through the 20th century, and from 1960 to

1980 the Iberian lynx population in Spain was mainly restricted to 8 to 10 isolated nuclei, distributed in central and southwestern areas (Valverde 1963; Rodríguez and Delibes 1992). During the early 21st century, Guzmán et al. (2004) found that only two isolated Iberian lynx populations survived in southern Spain: the Doñana population and the eastern Sierra Morena population, totaling no more than 160 individuals.

Most of the Iberian lynx ranges are included in protected areas but also in private properties, mostly in hunting estates. Iberian lynx conservation measures, developed mainly by governmental institutions, include creation of pastures, scrub encroachment control, and wild rabbit releases. Also, artificial supplementation of domestic rabbits in small lynx-feeding enclosures has been shown to be an effective way of maintaining lynx in low-rabbit-density zones and allowing their reproduction (Simón et al. 2009; López-Bao et al. 2010). These measures have allowed a moderate lynx population increase, and currently more than 250 individuals are known to exist (Simón et al. 2009). Like other lynx species, the Iberian lynx is eminently solitary, defending home ranges of about 600 ha with low intra-sexual and very low inter-sexual overlap (Ferreras et al. 1997). Despite being top predators, Iberian lynx are sexually mature in their second to third year, and their reproductive success ranges from zero to four cubs per year (Palomares et al. 2005). Iberian lynx diet is constituted primarily (85% to 99%) by rabbits (Delibes 1980; Palomares et al. 2001; Gil-Sánchez et al. 2006). Clearly, rabbit availability determines the presence and reproductive success of the lynx. A minimum density of 1 to 4.6 rabbits per ha is necessary to allow a female Iberian lynx to reproduce and maintain her cubs (Palomares et al. 2001). Given that rabbit availability in lynx areas is not continuous, lynx populations are structured in metapopulations around those areas where rabbits occur (Ferreras et al. 1997; Simón et al. 2009). Connectivity among metapopulations depends on both the distance and the habitat matrix between them (Rodríguez and Delibes 2003).

Current threats to the Iberian lynx include (1) rabbit depletion, (2) habitat destruction and loss of connectivity, (3) human-caused mortality (i.e., road kills, traps, poaching), and (4) diseases (Guzmán et al. 2004; Peña et al. 2006; Meli et al. 2009). A great effort is carried out by several government agencies for the conservation of the species (Simón et al. 2009),

including a captive breeding program (Vargas et al. 2009). Two important goals of the breeding program are to maintain the highest genetic variability present in the wild and to obtain optimal individuals for reintroduction. Thereby, it is necessary to avoid the effects of genetic adaptation to captivity (Gilligan et al. 2003; Frankham 2007) or the introduction of infectious agents to the free-ranging populations during reintroductions (Hess 1996).

## Health Considerations in Lynx

Health considerations have been demonstrated to be crucial in lynx conservation (Jiménez et al. 2009; López et al. 2009b; Martínez et al. 2009; Meli et al. 2009). Small host population size and genetic diversity, and pathogen generalist behavior have been described as major risk factors for disease-induced extinction (Scott 1988; Castro and Bolker 2005; Smith et al. 2009). From an epidemiological perspective the Iberian lynx is at considerable risk: there are two remaining isolated populations with a high population density in Sierra Morena, and a low genetic diversity, especially in the inbred Doñana population (Johnson et al. 2006). Also, the existence of generalist pathogens in the areas where the lynx inhabits, most of them shared with domestic species (Gortázar et al. 2008; Meli et al. 2009), make the species a candidate to face extinction by infection (Daszak and Cunningham 1999).

To obtain reliable data on the sanitary status of a population, both analyzing mortality data generated by radio-tagged individuals and testing for infectious agents are essential (Schmidt-Posthaus et al. 2002; Haines et al. 2005). Mortality data generated by radio-tagged Iberian lynxes worryingly show that the relative impact of infectious diseases as cause of mortality has increased about 10-fold in the past two decades (Ferreras et al. 1992; López et al. 2009a). Radio tracking has allowed detection of Iberian lynxes that died of TB, feline leukemia virus (FeLV), feline parvovirus (FPV), and canine distemper virus (CDV) in recent years (López et al. 2009c; Meli et al. 2009; Meli et al. 2010). Systematic screening for infectious agents performed in every lynx handled is also providing information about the exposure of the species to pathogens and their health risks. The surveys so far have found a low specific exposure to infectious agents, and consequently underlined the risk of the lynx population of suffering devastating effects due to an infectious outbreak (Roelke-Parker et al. 2008; Millán et al. 2009b; Meli et al. 2009).

Domestic cats and dogs are supposed to infect Iberian lynx with FeLV, CDV, and FPV, as domestic carnivores are reservoirs for these pathogens to wild carnivores (Williams et al. 1988; McOrist et al. 1991; Sillero-Zubiri et al. 1996; Lafferty and Gerber 2002; Woodroffe et al. 2004; Pedersen et al. 2007). CD has led to massive declines in many wild carnivores (i.e., black-footed ferrets [*Mustela nigripes*], African wild dogs [*Lycaon pictus*], African lions [*Panthera leo*]) and still constitutes a threat for these populations (Alexander et al. 1994; Roelke-Parker et al. 1996; Laurenson et al. 2004). Roelke-Parker et al. (2008) found no evidence of exposure to CDV in Iberian lynx populations, but Meli et al. (2010) detected a 14.8% seroprevalence in free-ranging animals and one case of CDV-related mortality from a radio-tagged individual in Doñana. Other non-domestic carnivores in Iberian lynx areas were also found to be PCR-positive to CDV. Likewise, CDV antibodies have been recorded in Eurasian lynx (*Lynx lynx*) and Canada lynx (*Lynx canadensis*) (Biek et al. 2002; Schmidt-Posthaus et al. 2002), and recently a CDV-associated pathology was reported in free-ranging Canada lynx (*Lynx canadensis*) and bobcats (*Lynx rufus*) (Daoust et al. 2009).

FeLV infections have been reported in lynxes from Doñana and Sierra Morena since 1994, although no related disease was described before 2007 (Luaces et al. 2008; Millán et al. 2009b). During spring and summer 2007, a FeLV outbreak affected 12 individuals of the Doñana population (López et al. 2009b; Meli et al. 2009). Seven of them died in less than a year. Sequencing revealed a common origin for FeLV. The sequences were closely related to a FeLV isolated from domestic cats (Meli et al. 2009). A control program, including removal of the infected animals from the field and vaccination, was implemented. No new cases have been detected, and a continuous surveillance of the Iberian lynx population has been in place since then (López et al. 2009b). Iberian lynx, like other felids, are prone to infection by *M. bovis* (de Lisle et al. 2001; Greene 2006), and cases have been reported from both free-ranging and captive animals. Briones et al. (2000) described the first case of an Iberian lynx killed by TB. In the following decade some other isolated cases of death due to TB were detected both in Doñana and Sierra Morena (Aranaz et al. 2004; Peña et al. 2006; Zorrilla, personal communication). Feeding on infected ungulate carcasses is considered to be the primary source of infection (Gortázar et al. 2008; Terio 2009).

*Cytauxzoon* spp. is a common protozoan in the Sierra Morena lynx population, but it has not been documented in Doñana (Luaces et al. 2005; Meli et al. 2009). Although it is considered apathogenic, Iberian lynxes from Sierra Morena that are translocated to Doñana as part of the genetic reinforcement program of the LIFE-Nature project are selected to be *Cytauxzoon*-free to avoid any risk when introducing this agent to a naïve population. Occasionally, other reported infections or contact with potential pathogens include *Toxoplasma gondii* (Sobrino et al. 2007; García-Bocanegra et al. 2010b), *Leishmania infantum* (Sobrino et al. 2008), *Leptospira interrogans* (Millán et al. 2009a), and *Encephalitozoon cuniculi* (Miró, personal communication).

## RABBITS, UNGULATES, AND LYNX IN THE MEDITERRANEAN FOREST: ECOLOGY, CONSERVATION, AND MANAGEMENT

The original Mediterranean woodland ecosystem has been largely altered by human activities that shaped for centuries a mosaic landscape of woodlands, scrublands, pastures, and small agricultural areas. Figure 30.1 shows the evolution of European wild rabbit, wild ungulates, and Iberian lynx population during the past 80 years in the Mediterranean forest. These wild species and livestock and their diseases are interconnected representatives of this ecosystem (Fig. 30.2). They need sound ecosystem approaches, wildlife management, and conservation medicine measures to ensure their conservation and sustainable use of these species. Management to conserve and restore habitat heterogeneity in hunting, agricultural, cattle-raising, forest, and protected areas is crucial (Delibes-Mateos et al. 2009a). The rabbit's role in the Mediterranean landscape makes it the target vertebrate species. Measures such as restocking or vaccinating are costly, often useless, and not sustainable. Ecologically sustainable management measures to conserve rabbits should rather rely on habitat management such as scrub clearing and promoting pastures and cereal crops (Ferreira et al. 2009a).

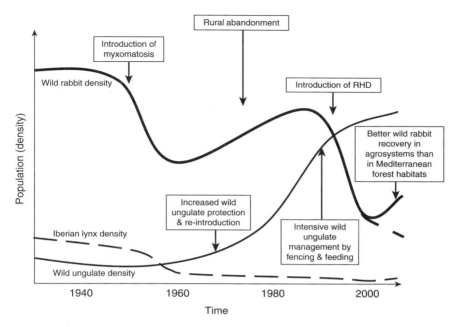

**Figure 30.1:**
Plot of European wild rabbit (*Oryctolagus cuniculus*), wild ungulate (mainly red deer [*Cervus elaphus*] and wild boar [*Sus scrofa*]), and Iberian lynx (*Lynx pardinus*) population size (or density) against time during the past 80 years. Boxed texts indicate events that significantly affected these populations.

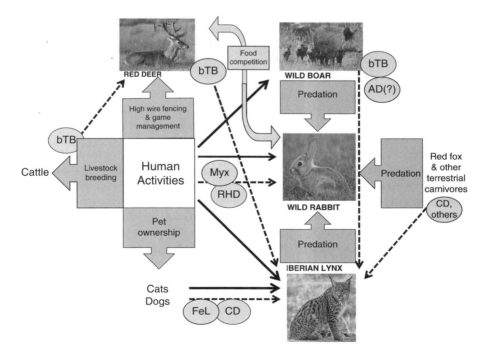

**Figure 30.2:**
Schematic representation of interactions among the group of selected species (boxes) and their main diseases in the Mediterranean forest. Arrows with dashed lines indicate disease-mediated effects. AD, Aujeszky's disease; CD, canine distemper; FeL, feline leukemia; Myx, myxomatosis; RHD, rabbit hemorrhagic disease; TB, tuberculosis.

As previously described, a relevant conservation problem is ungulate overabundance that has adverse effects on the soil, vegetation, and other fauna. Measures to reduce ungulate overabundance are difficult to implement both in hunting and protected areas for economic, political, and social reasons. A regulation of the wild ungulate harvest, along with substantially reducing supplementary feeding and watering, could reduce wild ungulate density, reducing the prevalence of infectious diseases and other impacts on the ecosystem. However, at least in Iberian Mediterranean habitats, research suggests that a supplemental feeding ban alone would have limited effect on *M. bovis* prevalence. Moreover, wildlife culling is almost never an effective means to eradicate a wildlife-related disease, and is subject to intense debate (Gortázar et al. 2008). Diseases can undermine conservation efforts if wildlife is perceived as the source of the problem affecting domestic animals and human health. This is expected to take place in some lynx conservation areas. Input from different scientific fields is needed to buffer this risk. The recent advances in wildlife vaccination against *M. bovis*, including wild boar vaccination, open new prospects in this field (Ballesteros et al. 2009). It is important to inform wildlife managers, hunters, landowners, and local authorities of the long-term benefits of these measures in terms of animal welfare, soil and vegetation conservation, and biodiversity. In parallel it is crucial to initiate and maintain extended and long-term epidemiological monitoring of the main diseases in wild ungulates and the impact of management measures (Gortázar et al. 2006).

Domestic dogs and cats are a real health risk for Iberian lynx. Lynx habitat reduction, urban spread, and extension of agricultural areas increase the numbers and density of domestic dogs and cats and consequently increase the transmission risks both within the host populations and between domestic and wild species. Infectious disease transmission risk from domestic carnivores to wild carnivores is probably more frequent in Doñana due to the larger human pressure and the associated domestic carnivore population. Moreover, domestic carnivores in lynx habitat are not usually under disease surveillance and vaccination. Implementing control measures in the domestic carnivore population to minimize the risk of entering lynx areas is difficult. An integrated health program for domestic carnivores is necessary with the participation of local authorities, protected areas managers,

and veterinarians. Microchip identification in dogs and cats (a legal requisite but not usually performed in these areas) could also be used to promote infectious disease testing, vaccination, and sterilization. Selective trapping of domestic carnivores in lynx areas and culling of non-identified animals should be maintained in a long-term basis. Epidemiological monitoring of domestic and non-domestic carnivores should become a part of the conservation measures for the Iberian lynx.

Measures to allow the conservation of the Mediterranean forest ecosystem should use a transdisciplinary approach. Often, managing one aspect (e.g., fencing of protected areas) has negative impacts on other aspects (e.g., increase ungulates' densities or decrease genetic flow). Delibes-Mateos et al. (2010) demonstrated that maintaining and favoring *dehesas* suitable for livestock and wild ungulates reduced wild rabbit habitat quality and consequently food resources for Iberian lynxes. Infectious diseases in wild rabbits, ungulates, and lynxes are interconnected and too often human-mediated. Selected diseases shared with wild ungulates, such as tick-borne diseases and TB, can directly affect humans (Gortázar et al. 2005; de la Fuente et al. 2008). A consensus management decision taking into account the Mediterranean forest as a non-separate unit, and hence considering livestock farming, hunting activities, forestry use, and ecosystem and wildlife conservation, is probably the only way to achieve its sustainable use and conservation.

## ACKNOWLEDGMENTS

Veterinarians and biologists from several institutions made possible tasks and different aspects described in this chapter—therefore, our deep gratitude to colleagues from Instituto de Investigación en Recursos Cinegéticos IREC (Ciudad Real, Spain), Estación Biológica de Doñana (Sevilla, Spain), Life-Nature Project for Iberian Lynx Conservation in Andalucia (Spain), Consejería de Medio Ambiente de la Junta de Andalucia (Spain), Ministerio de Medio Ambiente Rural y Marino (Madrid, Spain), Veterinary Clinical Laboratory (Zürich, Switzerland), Doñana National Park (Sevilla-Huelva, Spain), Centro de Análisis y Diagnóstico CAD (Málaga, Spain), Iberian Lynx *Ex Situ* Conservation Program (Spain and Portugal), Facultad de Veterinaria de Barcelona (Barcelona, Spain),

Facultad de Veterinaria de Madrid (Madrid, Spain), Facultad de Veterinaria de Córdoba (Córdoba, Spain), and many thanks to Josep Pastor, Xavier Manteca, Catarina Ferreira, and Francisco Palomares for manuscript revision.

# REFERENCES

Acevedo, P., M. Delibes-Mateos, M.A. Escudero, J. Vicente, J. Marco, and C. Gortázar. 2005. Environmental constraints in the roe deer (*Capreolus capreolus* Linnaneus, 1758) colonization sequence across the Iberian Mountains (Spain). J Biogeogr 32:1671–1680.

Acevedo, P., F. Ruiz-Fons, J. Vicente, A.R. Reyes-Garcia, V. Alzaga, and C. Gortazar. 2008. Estimating red deer abundance in a wide range of management situations in Mediterranean habitats. J Zool 276:37–47.

Acevedo, P., J. Vicente, U. Höfle, J. Cassinello, F. Ruiz-Fons, and C. Gortazar. 2007. Estimation of European wild boar relative abundance and aggregation: a novel method in epidemiological risk assessment. Epidemiol Infect 135:519–527.

Alexander, K.A., and M.J. Appel. 1994. African wild dogs (*Lycaon pictus*) endangered by a canine distemper epizootic among domestic dogs near the Masai Mara National Reserve, Kenya. J Wildl Dis 30:481–485.

Angulo, E., and R. Villafuerte. 2003. Modeling hunting strategies for the conservation of wild rabbits populations. Biol Conserv 115:291–301.

Aranaz, A., L. de Juan, N. Montero, C. Sánchez, M. Galka, C. Delso, J. Álvarez, B. Romero, J. Bezos, A.I. Vela, V. Briones, A. Mateo, and L. Domínguez. 2004. Bovine tuberculosis (*Mycobacterium bovis*) in wildlife in Spain. J Clin Microbiol 42:2602–2608.

Ballesteros, C., C. Gortázar, M. Canalesa, J. Vicente, A. Lasagna, J.A. Gamarra, R. Carrasco-García, and J. de la Fuente. 2009. Evaluation of baits for oral vaccination of European wild boar piglets. Res Vet Sci 86: 388–393.

Biek, R., R.L. Zarnke, C. Gillin, M. Wild, J.R. Squires, and M. Poss. 2002. Serologic survey for viral and bacterial infections in western populations of Canada lynx (*Lynx canadensis*). J Wild Dis 38:840–845.

Blanco, J.C., and R. Villafuerte. 1993. Factores ecológicos que influyen sobre las poblaciones de Conejos. Incidencia de la enfermedad hemorrágica. Informe Técnico. Empresa de Transformación Agraria, S.A., Madrid.

Briones, V., L. de Juan, C. Sánchez, A.I. Vela, M. Galka, N. Montero, J. Goyache, A. Aranaz, A. Mateos, and L. Domínguez. 2000. Bovine tuberculosis and the endangered Iberian lynx. Emerg Infect Dis 6:189–191.

Cabrera, A. 1914. Fauna Ibérica. Mamíferos. Museo Nacional de Ciencias Naturales, Madrid.

Calvete, C. 2006. Modeling the effects of population dynamics on the impact of rabbit hemorrhagic disease rabbit. Conserv Biol 20:1232–1241.

Calvete, C., R. Estrada, J. Lucientes, J.J. Osácar, and R. Villafuerte. 2004. Effects of vaccination against viral haemorrhagic disease and myxomatosis on long-term mortality rates of European wild rabbits. Vet Rec 155:388–392.

Castro, F., and B. Bolker. 2005. Mechanisms of disease-induced extinction. Ecol Lett 8:117–126.

Cooke, B.D., and F. Fenner. 2002. Rabbit haemorrhagic disease and the biological control of wild rabbits, *Oryctolagus cuniculis*, in Australia and New Zealand. Wild Res 29:689–706.

Daoust, P.Y., S.R. McBurney, D.L. Godson, M.W. van de Bildt, and A.D. Osterhaus. 2009. Canine distemper virus-associated encephalitis in free-living lynx (*Lynx canadensis*) and bobcats (*Lynx rufus*) of eastern Canada. J Wildl Dis 45:611–624.

Daszak, P., and A. A. Cunningham. 1999. Extinction by infection. Trends Ecol Evol 14:279.

de la Fuente, J., A. Estrada-Pena, J.M. Venzal, K.M. Kocan, and D.E. Sonenshine. 2008. Overview: Ticks as vectors of pathogens that cause disease in humans and animals. Front Biosci 13:6938–6946.

de Lisle, G.W., C.G. Mackintosh, and R.G. Bengis. 2001. *Mycobacterium bovis* in free-living and captive wildlife, including farmed deer. Rev Sci Tech 20:86–111.

Delibes, M. 1980. Feeding ecology of the Spanish lynx in the Coto Doñana. Acta Theriol 25:309–324.

Delibes, M., and F. Hiraldo. 1981. The rabbit as prey in the Iberian Mediterranean ecosystem. *In* K. Myers and C.D. MacInnes, eds. Proc I World Lagomorph Conf, pp. 614–622. University of Guelph, Ontario.

Delibes-Mateos, M., M. Delibes, P. Ferreras, and R.Villafuerte. 2008a. Key role of the European rabbits in the conservation of the western Mediterranean basin hotspot. Conserv Biol 22:1106–1117.

Delibes-Mateos, M., M.A. Farfán, J. Olivero, A.L. Márquez, and J. M. Vargas. 2009a. Long-term changes in game species over a long period of transformation in the Iberian Mediterranean landscape. Environ Manage 43:1256–1268.

Delibes-Mateos, M., M.A. Farfán, J. Olivero, and J.M. Vargas. 2010. Land-use changes as a critical factor for long-term wild rabbit conservation in the Iberian Peninsula. Environ Conser 37:169–176.

Delibes-Mateos, M., P. Ferreras, and R.Villafuerte. 2008b. Rabbit populations and game management: the situation after 15 years of rabbit haemorrhagic disease in central-southern Spain. Biodivers Conserv 17:559–574.

Delibes-Mateos, M., P. Ferreras, and R.Villafuerte. 2009b. European rabbit population trends and associated factors: a review of the situation in the Iberian Peninsula. Mammal Rev 39:124–140.

Delibes-Mateos, M., E. Ramírez, P. Ferreras, and R. Villafuerte. 2008c. Translocations as a risk for the conservation of European wild rabbit *Oryctolagus cuniculus* lineages. Oryx 42:1–6.

Delibes-Mateos, M., S.M. Redpath, E. Angulo, P. Ferreras, and R.Villafuerte. 2007. Rabbits as a keystone species in southern Europe. Biol Conserv 137:149–156.

Fernández, N., M. Delibes, and F. Palomares. 2006. Landscape evaluation in conservation: Molecular sampling and habitat modeling for the Iberian lynx. Ecol Appl 16:1037–1049.

Fernández-Alés, R., A. Martín, F. Ortega, and E.E. Ales. 1992. Recent changes in landscape structure and function in a Mediterranean region of SW of Spain (1950–1984). Landscape Ecol 7:3–18.

Ferreira, C., and P. Célio. 2009a. Influence of habitat management on the abundance and diet of wild rabbit (*Oryctolagus cuniculus algirus*) populations in Mediterranean ecosystems. Eur J Wildl Res 55: 487–496.

Ferreira, C., E. Ramirez, F. Castro, P. Ferreras, P.C. Alves, S. Redpath, and R. Villafuerte. 2009b. Field experimental vaccination campaigns against myxomatosis and their effectiveness in the wild. Vaccine 27:6998–7002.

Ferrer, M., and J.J. Negro. 2004. The near extinction of two large European predators, super specialists pay a price. Conserv Biol 18:344–349.

Ferreras, P., J.J. Aldama, J.F. Beltrán, and M. Delibes. 1992. Rates and causes of mortality in a fragmented population of Iberian lynx *Felis pardina* Temminck, 1824. Biol Conserv 61:197–202.

Ferreras, P., J.F. Beltrán, J.J. Aldama, and M. Delibes. 1997. Spatial organization and land tenure system of the endangered Iberian lynx (*Lynx pardinus*, Temminck, 1824). J Zool 243:163–189.

Frankham, R. 2007. Genetic adaptation to captivity in species conservation programs. Mol Ecol 17:325–333.

Gálvez, M.R. 2004. Régimen jurídico de la actividad cinegética en España: Análisis de las disposiciones autonómicas e intervención pública. Tesis doctoral, pp. 145–158. Universidad de Málaga, España.

García-Bocanegra, I., R.J. Astorga, S. Napp, J. Casal, B. Huerta, C. Borge, and A. Arenas. 2010a. Myxomatosis in wild rabbit: design of control programs in Mediterranean ecosystems. Prev Vet Med 93:42–50.

García-Bocanegra, I., J.P. Dubey, F. Martínez, A. Vargas, O. Cabezón, I. Zorrilla, A. Arenas, and S. Almería. 2010b. Factors affecting seroprevalence of *Toxoplasma gondii* in the endangered Iberian lynx (*Lynx pardinus*). Vet Parasitol 167:36–42.

Gil-Sánchez, J.M., E. Ballesteros-Duperón, and J.B. Bueno-Segura. 2006. Feeding ecology of the Iberian lynx *Lynx pardinus* in eastern Sierra Morena (Southern Spain). Acta Theriol 51:85–90.

Gilligan, D., and R. Frankham. 2003. Dynamics of genetic adaptation to captivity. Conserv Genet 4:189–197.

Gortázar, C., P. Acevedo, F. Ruiz-Fons, and J. Vicente. 2006. Disease risks and overabundance of game species. Eur J Wildl Res 52:81–87.

Gortázar, C., J. Herrero, R. Villafuerte, and J. Marco. 2000. Historical examination of the status of large mammals in Aragon, Spain. Mammalia 64:411–422.

Gortázar, C., M.J. Torres, J. Vicente, P. Acevedo, M. Reglero, J. de la Fuente, J.J. Negro, and J. Aznar-Martín. 2008. Bovine tuberculosis in Doñana Biosphere Reserve: The role of wild ungulates as disease reservoirs in the last Iberian lynx strongholds. PLoS One 3:e2776.

Gortázar, C., J. Vicente, S. Samper, J.M. Garrido, I.G. Fernández-de-Mera, P. Gavín, R.A. Juste, C. Martín, P. Acevedo, M. de la Puente, and U. Höfle. 2005. Molecular characterization of Mycobacterium tuberculosis complex isolates from wild ungulates in south-central Spain. Vet Res 36:43–52.

Graells, M.P. 1897. *Felis pardina* (Temminck). *In* Fauna Mastozoológica Ibérica. Memorias de la Real Academia de Ciencias, XVII, pp. 224–229, Madrid.

Greene, C.E. 2006. Infectious diseases of the dog and the cat, 3rd ed. W.B. Saunders Company, St. Louis, Missouri.

Guzmán, J.N., F.J. García, G. Garrote, R. Pérez de Ayala, and C. Iglesias. 2004. El lince ibérico (*Lynx pardinus*) en España y Portugal. Censo diagnóstico de las poblaciones. Dirección General para la Biodiversidad, Madrid.

Haines, A.M., M.E. Tewes, and L.L. Laack. 2005. Survival and sources of mortality in ocelots. J Wildl Manage 69:255–263.

Hess, H. 1996. Disease in metapopulation models: implications for conservation. Ecology 77:1617–1632.

INE. 2006. Instituto Nacional de Estadística, Madrid.

INE. 2009. Instituto Nacional de Estadística, Madrid.

IUCN. 2004. IUCN Red list of threatened species. IUCN, Gland, Switzerland.

IUCN. 2009. Wildlife in a changing world: An analysis of the 2008 IUCN Red List of Threatened Species. J-C. Vié, C. Hilton-Taylor, and S.N. Stuardt,eds. IUCN, Gland, Switzerland.

Jiménez, M.A., B. Sánchez, P. García, M.D. Pérez, M.E. Carrillo, F.J. Morena, and L. Peña. 2009. Diseases of the Iberian lynx (*Lynx pardinus*): histopathological survey, lymphoid depletion, glomerulonephritis and related clinical findings. *In* A. Vargas, C. Breitenmoser, and U. Breitenmoser, eds. Iberian lynx ex situ conservation: an interdisciplinary approach, pp. 210–219. Fundación Biodiversidad, Madrid.

Johnson, W.E., E. Eizirik, J. Pecon-Slattery, W.J. Murphy, A. Antunes, E. Teeling, and S.J. O'Brien. 2006. The late Miocene radiation of modern Felidae: a genetic assessment. Science 311:73–77.

Jones, C.G., J.H. Lawton, and M. Shachack. 1994. Organisms as ecosystem engineers. Oikos 69:373–386.

Lafferty, K.D., and L.R. Gerber. 2002. Good medicine for conservation biology: the intersection of epidemiology and conservation theory. Conserv Biol 6:593–604.

Laurenson, M.K., S. Cleaveland, M. Artois, and R. Woodroffe. 2004. Assessing and managing infectious disease threats to canids. *In* C. Sillero-Zubiri, M. Hoffman, and D.W. MacDonald, eds. Canids: foxes, wolves, jackals and dogs: status survey and conservation action plan. IUCN SSC Canid Specialist Group, pp. 246–255. Cambridge, United Kingdom and Gland, Switzerland.

León-Vizcaíno, L., A. Bernabé, A. Contreras, M.J. Cubero, S. Gómez, and R. Astorga. 1990. Outbreak of tuberculosis caused by *Mycobacterium bovis* in wild boars (*Sus scrofa*). Verh. ber. Int. Symp. Erkr. Zoo u. Wildltiere 32:185–190.

Lips, K.F., J. Diffendorfer, J.R. Mendelson III, and M.W. Sears. 2008. Riding the wave: reconciling the roles of disease and climate change in amphibian declines. Plos Biol 6:441–445.

López, G., J.M. Gil-Sánchez, I. Zorrilla, F. Martínez, M. López-Parra, L. Fernández-Pena, G. Garrote, G. Ruiz, C. Martínez-Granados, and M. A. Simón. 2009a. Análisis de las causas recientes de mortalidad en las poblaciones silvestres de lince ibérico. Resúmenes de las IX Jornadas de La SECEM, Bilbao.

López, G., M. López, L. Fernández, C. Martínez-Granados, F. Martínez, M.L. Meli, J.M. Gil-Sánchez, N. Viqueira, M.A. Díaz-Portero, R. Cadenas, H. Lutz, A. Vargas, and M.A. Simón. 2009b. Management measures to control a FeLV outbreak in the endangered Iberian lynx. Anim Conserv 12:173–182.

López, G., F. Martínez, M.L. Meli, E. Bach, C. Martínez-Granados, M. López-Parra, L. Fernández, A. Vargas, I. Molina, M.A. Díaz-Portero, J.M. Gil-Sánchez, R. Cadenas, M.A. Simón, J. Pastor, and H. Lutz. 2009c. A feline leukemia virus outbreak in the Doñana Iberian lynx population. *In* A. Vargas, C. Breitenmoser, and U. Breitenmoser eds. Iberian lynx ex situ conservation: an interdisciplinary approach, pp. 234–247. Fundación Biodiversidad, Madrid.

López-Bao, J.V., F. Palomares, A. Rodríguez, and M. Delibes. 2010. Effects of food supplementation on home-range size, reproductive success, productivity and recruitment in a small population of Iberian lynx. Anim Conserv 13:35–42.

Lozano, J., E. Virgos, S. Cabezas-Diaz, and J.G. Mangas. 2007. Increase of large game species in Mediterranean areas: Is the European wildcat (*Felis silvestris*) facing a new threat? Biol Conserv 138:321–329.

Luaces, I., E. Aguirre, M. García-Montijano, J. Velarde, M.A. Tesouro, C. Sánchez, M. Galka, P. Fernández, and A. Sainz. 2005. First report of an intraerythrocytic small piroplasm in wild Iberian lynx (*Lynx pardinus*). J Wildl Dis 41:810–815.

Luaces, I., A. Doménech, M. García-Montijano, V.M. Collado, C. Sánchez, J.G. Tejerizo, M. Galka, P. Fernández, and E. Gómez-Lucía. 2008. Detection of feline leukemia virus in the endangered Iberian lynx (*Lynx pardinus*). J Vet Diagn Invest 20:381–385.

Martínez, F., G. López, M.J. Pérez-Aspa, I. Molina, J.M. Aguilar, M.A. Quevedo, and A.Vargas. 2009. Health aspects integration in the Iberian lynx conservation. *In* A. Vargas, C. Breitenmoser, and U. Breitenmoser, eds. Iberian lynx ex situ conservation: an interdisciplinary approach, pp. 163–183. Fundación Biodiversidad, Madrid.

McOrist, S., R. Boid, T.W. Jones, N. Easterbee, A.L. Hubbard, and O. Jarrett. 1991. Some viral and protozool diseases in the European wildcat (*Felis silvestris*). J Wildl Dis 27:693–696.

Meli, M.L., V. Cattori, F. Martínez, G. López, A. Vargas, M.A. Simón, I. Zorrilla, A. Muñoz, F. Palomares, J.V. López-Bao, J. Pastor, R. Tandon, B. Willi, R. Hofmann-Lehmann, and H. Lutz. 2009. Feline leukemia virus and other pathogens as important threats to the survival of the critically endangered Iberian lynx (*Lynx pardinus*). Plos One 4:e4744.

Meli, M.L., P. Simmler, V. Cattori, F. Martínez, A. Vargas, F. Palomares, G. López-Bao, M.A. Simón, G. López, L.L. Vizcaíno, R. Hofmann-Lehmann, and H. Lutz H. 2010. Importance of canine distemper virus (CDV) infection in free-ranging Iberian lynxes (*Lynx pardinus*). Vet Microbiol 146:132–137.

Millán, J., M.G. Candela, J.V. López-Bao, M. Pereira, M.A. Jiménez, and L. León-Vizcaíno. 2009a. Leptospirosis in wild and domestic carnivores in natural areas in Andalusia, Spain. Vector-Borne Zoonot 9:549–554.

Millán, J., M.G. Candela, F. Palomares, M.J. Cubero, A. Rodríguez, M. Barral, J. de la Fuente, S. Almería, and L. León-Vizcaíno. 2009b. Disease treats to the endangered Iberian lynx (*Lynx pardinus*). Vet J 182:114–124.

Moreno, S., and R. Villafuerte. 1995. Traditional management of scrubland for the conservation of rabbits *Oryctolagus cuniculus* and their predators in Doñana National Park, Spain. Biol Conserv 73:81–85.

Muñoz, P.M., M. Boadella, M. Arnal, M.J de Miguel,
   M. Revilla, D. Martínez, J. Vicente, P. Acevedo,
   A. Oleaga, F. Ruiz-Fons, C.M. Marín, J.M. Prieto, J. de la
   Fuente, M. Barral, M. Barberán, D.F. de Luco, J.M.
   Blasco, and C. Gortázar. 2010. Spatial distribution and
   risk factors of Brucellosis in Iberian wild ungulates.
   BMC Infect Dis 10:46.
Myers, N., R.A. Mittermeier, C.G. Mittermeier, G.A.B. da
   Fonseca, and J. Kent. 2000. Biodiversity hotspots for
   conservation priorities. Nature 403:853–858.
Palomares, F., M. Delibes, E. Revilla, J. Calzada, and
   J.M. Fedriani. 2001. Spatial ecology of Iberian lynx and
   abundance of European rabbits in southwestern Spain.
   Wildl Monogr 148:1–36.
Palomares, F., E. Revilla, J. Calzada, N. Fernández, and
   M. Delibes. 2005. Reproduction and pre-dispersal
   survival of Iberian lynx in a subpopulation of the
   Doñana National Park. Biol Conserv 122:153–159.
Pedersen, A.M., K.E. Jones, C.L. Nunn, and S. Altizer. 2007.
   Infectious diseases and extinction risk in wild
   mammals. Conserv Biol 21:1269–1279.
Peña, L., P. García, M.A. Jiménez, A. Benito, M.D. Pérez
   Alenza, and B. Sánchez. 2006. Histopathological and
   immunohistochemical findings in lymphoid tissues of
   the endangered Iberian lynx (Lynx pardinus). Comp
   Immunol Microb 29:114–126.
Pérez, J., J. Calzada, L. León-Vizcaíno, M.J. Cubero,
   J. Velarde, and E. Mozos. 2001. Tuberculosis in an
   Iberian lynx (Lynx pardina). Vet Rec 148:414–415.
Pimm, S.L, G.J. Russell, J.L. Gittleman, and T.M. Brooks.
   1995. The future of biodiversity. Science 269:347–350.
Rodríguez, A., and M. Delibes. 1992. Current range and
   status of the Iberian lynx Felis pardina Temminck, 1824
   in Spain. Biol Conserv 61:189–196.
Rodríguez, A., and M. Delibes. 2003. Population
   fragmentation and extinction in the Iberian lynx. Biol
   Conserv 109:321–331.
Roelke-Parker, M.E., W. Johnson, J. Millán, F. Palomares,
   E. Revilla, A. Rodríguez, J. Calzada, P. Ferreras,
   L. León-Vizcaíno, M. Delibes, and S. O'Brien. 2008.
   Exposure to disease agents in the endangered Iberian
   lynx (Lynx pardinus). Eur J Wildl Res 54:171–178.
Roelke-Parker, M. E., L. Munson, C. Packer, R. Kock,
   S. Cleaveland, M. Carpenter, S.J. O'Brien, A. Pospischil,
   R. Hofmann-Lehmann, H. Lutz, G.L.M. Mwamengele,
   M.N. Mgasa, G.A. Machange, B.A. Summers, and
   M.J.G. Appel. 1996. A canine distemper virus epidemic
   in Serengeti lions (Panthera leo). Nature 379:441–445.
Rosell, J. 2000. Enfermedades del conejo. MundiPrensa,
   Madrid.
Rouco, C., P. Ferreras, F. Castro, and R.Villafuerte. 2008.
   The effect of exclusion of terrestrial predators on
   short-term survival of translocated European wild
   rabbits. Wildl Res 35:625–632.

Ruiz-Fons, F., J. Segalés, and C. Gortázar. 2008. A review
   of viral diseases of the European wild boar: effects of
   population dynamics and reservoir role. Vet J
   176:158–169.
Schmidt-Posthaus, H., C. Breitenmoser-Würsten,
   H. Posthaus, L. Bacciarini, and U. Breitenmoser. 2002.
   Causes of mortality in reintroduced Eurasian lynx in
   Switzerland. J Wildl Dis 38:84–92.
Scott, M.E. 1988. The impact of infection and disease on
   animal populations: Implications for Conservation
   Biology. Conserv Biol 2:40–56.
Sillero-Zubiri, C., A.A. King, and D.W. Macdonald. 1996.
   Rabies and mortality in Ethiopian wolves (Canis
   simensis). J Wildl Dis 32:80–86.
Simón, M.A., R. Cadenas, J.M. Gil-Sánchez, M. López-
   Parra, J. García, G. Ruiz, and G. López. 2009.
   Conservation of free-ranging Iberian lynx (Lynx
   pardinus) populations in Andalusia. In A. Vargas,
   C. Breitenmoser, and U. Breitenmoser, eds. Iberian lynx
   ex situ conservation: an interdisciplinary approach,
   pp. 42–55. Fundación Biodiversidad, Madrid.
Smith, K.F., K. Acevedo-Whitehouse, and A.B. Pederson.
   2009. The role of infectious diseases in biological
   conservation. Anim Conserv 12:1–12.
Smith, K.F., D.V. Sax, and K.D. Lafferty. 2006. Evidence for
   the role of infectious disease in species extinction and
   endangerment. Conserv Biol 20:1349–1357.
Sobrino, R., O. Cabezón, J. Millán, M. Pabón, M.C. Arnal,
   D.F. Luco, C. Gortázar, J.P. Dubey, and S. Almería.
   2007. Seroprevalence of Toxoplasma gondii antibodies
   in wild carnivores from Spain. Vet Parasitol
   148:187–192.
Sobrino, R., E. Ferroglio, A. Oleaga, A. Romano, J. Millán,
   M. Revilla, M.C. Arnal, A. Trisciuoglio, and C. Gortázar.
   2008. Characterization of widespread canine
   leishmaniasis among wild carnivores from Spain. Vet
   Parasitol 155:198–203.
Soriguer, R., A. Rodríguez, and L. Domínguez. 2001.
   Análisis de la incidencia de los grandes herbívoros en
   la marisma y vera del Parque Nacional de Doñana. Serie
   técnica. Organismo Autónomo de Parques Nacionales
   eds. Ministerio de Medio Ambiente, Madrid.
Terio, K. 2009. Diseases of captive and free-ranging
   non-domestic felids. In A. Vargas, C. Breitenmoser, and
   U. Breitenmoser, eds. Iberian lynx ex situ conservation:
   an interdisciplinary approach, pp. 248–263. Fundación
   Biodiversidad, Madrid.
Valverde, J.A. 1963. Información sobre el lince en España.
   Boletín Técnico, Serie Cinegética, S.N.P.F.C.,
   Ministerio de Agricultura, Madrid.
Vargas, A., I. Sánchez, F. Martínez, A. Rivas, J.A. Godoy,
   E. Roldán, M.A. Simón, R. Serra, M.J. Pérez, A. Sliwa,
   M. Delibes, M. Aymerich, and U. Breitenmoser. 2009.
   Planning and implementation of the Iberian lynx (Lynx

*pardinus*) conservation breeding programme. *In* A. Vargas, C. Breitenmoser, and U. Breitenmoser, eds. Iberian lynx ex situ conservation: an interdisciplinary approach, pp. 57–71. Fundación Biodiversidad, Madrid.

Vicente, J., U. Höfle, J.M. Garrido, I. Fernández-de-Mera, P. Acevedo, R. Juste, M. Barral, and C. Gortázar. 2007. Risk factors associated with the prevalence of tuberculosis-like lesions in fenced wild boar and red deer in south central Spain. Vet Res 38:1–15.

Vicente, J., U. Höfle, J.M. Garrido, I.G. Fernández-de-Mera, R. Juste, M. Barral, and C. Gortázar. 2006. Wild boar and red deer display high prevalences of tuberculosis-like lesions in Spain. Vet Res 37:1–11.

Vicente, J., F. Ruiz-Fons, D. Vidal, U. Höfle, P. Acevedo, D. Villanúa, I.G. Fernández-de-Mera, M.P. Martín, and C. Gortázar. 2005. Large-scale serosurvey on Aujezsky's disease virus infection in the European wild boar from Spain. Vet Rec 156:408–412.

Villafuerte, R., C. Calvete, J.C. Blanco, and J. Lucientes. 1995. Incidence of viral hemorrhagic disease in wild rabbit populations in Spain. Mammalia 59:651–659.

Virgós, E., S. Cabezas-Díaz, and J. Lozano. 2007. Is the wild rabbit (*Oryctolagus cuniculus*) a threatened species in Spain? Sociological constraints in the conservation of species. Biodivers Conserv 16:3489–3504.

Virgós, E., S. Cabezas-Díaz, A. Malo, J. Lozano, and D. López-Huertas. 2003. Factors shaping European rabbit abundance in continuous and fragmented populations of central Spain. Acta Theriol 48:113–122.

Wilcove, D.S, D. Rothstein, J. Dubow, A. Philips, and E. Losos. 1998. Quantifying threats to imperilled species in the United States. Bioscience 48:607–615.

Willi, B., C. Filoni, J.L. Catao-Dias, V. Cattori, M.L. Meli, A. Vargas, F. Martínez, M.E. Roelke, M.P. Ryser-Degiorgis, C.M. Leutenegger, H. Lutz, and R. Hofmann-Lehmann. 2007. Worldwide occurrence of feline hemoplasma infections in wild felid species. J Clin Microbiol 45:1159–1166.

Williams, D., P. Acevedo, C. Gortázar, M.A. Escudero, J.L. Labarta, J. Marco, and R. Villafuerte. 2007. Hunting for answer: rabbit (*Oryctolagus cuniculus*) populations trends in northeastern Spain. Eur J Wildl Res 53:19–28.

Williams, E.S., E.T. Thorne, M.J. Appel, and D.W. Belitsky. 1988. Canine distemper in black-footed ferrets (*Mustela nigripes*) from Wyoming. J Wildl Dis 24:385–398.

Woodroffe, R., S. Cleaveland, O. Courtenay, M.K. Laurenson, and A. Artois. 2004. Infectious disease. *In* D.W. Macdonald and C. Sillero-Zubiri, eds. The biology and conservation of wild canids, pp. 123–142. Oxford University Press, England.

# 31

## THE KIBALE ECOHEALTH PROJECT

Exploring Connections Among Human Health, Animal Health,
and Landscape Dynamics in Western Uganda

Tony L. Goldberg, Sarah B. Paige, and Colin A. Chapman

"EcoHealth" is a nebulous concept. In the broadest sense, it is a medical reiteration of the age-old philosophy that people and animals are inherently connected to each other and the physical environment. The term "EcoHealth" therefore usually needs little definition beyond the meaning it naturally evokes through the juxtaposition of its word roots. The idea of ecology segueing into health is self-evident.

In another sense, EcoHealth is a biological and ethical goal. Properly functioning ecosystems and the well-being of their constituent parts are ends to which we should all aspire. One might trace the origins of this meaning to Aldo Leopold's concept of "land health," one of the formative ideas of the modern conservation movement (Rapport 1998). The EcoHealth concept in this context exploits the health metaphor to argue that our present-day ecosystems are sick and must be healed.

Paradoxically, even its practitioners rarely treat EcoHealth as a testable scientific hypothesis. Although scientists often cite case studies showing how ecological damage has led to unhealthy people and animals, few formally acknowledge that the general trend could swing in either direction. Indeed, some of the most "successful" historical improvements to human public health have come from planned ecosystem degradations,

such as the draining of wetlands to control the mosquito vectors of malaria (Keiser et al. 2005).

The Kibale EcoHealth Project is an attempt to subject the "EcoHealth paradigm" to the scientific rigor demanded by the ideal of evidence-based medicine. Founded in 2004, the Kibale EcoHealth Project is an ecological study of animal and human health in the region of Kibale National Park, western Uganda (http://svmweb.vetmed.wisc.edu/KibaleEcoHealth/). This region's volatile ecological and political history make it a particularly useful backdrop for exploring connections among human health, animal health, and landscape dynamics (Fig. 31.1). Importantly, the Kibale EcoHealth Project treats the EcoHealth paradigm itself as a fundamentally falsifiable hypothesis. We believe that this objective approach is the right one, since acknowledging the falsifiability of any hypothesis, no matter how palatable, is the most efficient means towards scientific progress.

This chapter describes some of the insights that the Kibale EcoHealth Project has generated concerning health, disease, land use change, and their interdependency in western Uganda. We review some of our findings that support the EcoHealth paradigm, as well some that go against it. The overall picture is complex, which is not surprising. It illustrates, among other

**Figure 31.1:**
The Kibale EcoHealth Project logo. The logo was designed to convey the interdependency of human health, domestic animal health, and the health of forests, here represented by a black-and-white colobus, one of the well-studied wild non-human primates of Kibale. (Logo designed by K. Helms)

things, the importance of spatial and temporal scale to the study of health ecology, and the ways in which quantitative and qualitative approaches sometimes complement each other and other times collide. We hope that the Kibale EcoHealth Project will inspire similar efforts elsewhere in the world aimed at objectively evaluating how human health, animal health, and landscape dynamics interact.

## THE SETTING

Western Uganda is a nearly ideal place to examine the interconnections between ecology and health. Uganda is exceptionally biodiverse while at the same time having a high human and animal disease burden and a high rate of human population growth (Hartter and Southworth 2009). As a result, Uganda is an acknowledged "hotspot" for emerging zoonoses (Jones et al. 2008; Pedersen and Davies 2010). Kibale National Park is a mid-altitude, moist-evergreen forest in central-western Uganda (0 13′–0 41′ N and 30 19′–30 32′ E) near the foothills of the Ruwenzori Mountains (Struhsaker 1997). Kibale was designated a forest reserve in 1932 and became a national park in 1993. Its conservation history predates the colonial era, however; it may have served as the traditional hunting grounds for leaders of the local Batooro tribe, perhaps explaining why it persisted as a forest until the arrival of colonial powers (Naughton-Treves 1999).

Certain locations within Kibale National Park contain a remarkable diversity and biomass of non-human primates and have been the focus of primatology research for over 40 years (Chapman et al. 2005). The "core" areas of Kibale are home to 13 species of primates, from nocturnal prosimians to chimpanzees (*Pan troglodytes*), as well as an impressive variety of plants, insects, fishes, amphibians, reptiles, birds, mammals, and various other taxa that together form a diverse and "healthy" African montane forest community (Howard 1991; Struhsaker 1997).

Kibale has had a long and complex history of conservation successes and failures. On the one hand, Kibale's national park status has enhanced its protection and led to the establishment of a lucrative ecotourism industry, focused mainly on chimpanzees and birds (Archabald and Naughton-Treves 2001). However, darker episodes in Kibale's history have also occurred, including a series of forced government evictions of local people living illegally inside the park boundary in 1992 (van Orsdol 1986; Baranga 1991). Issues of park–people conflict continue to arise in Kibale, ranging from economic disparities created by tourism to the raiding of crops by now-protected wildlife (Naughton-Treves 1998; Rode et al. 2006).

Outside Kibale, and not subject to the protections afforded by the park, exist a series of community-owned forest fragments, representing what has been left after agricultural clearing (Fig. 31.2). These fragments

**Figure 31.2:**
Forest fragment outside of Kibale National Park, with Rwenzori Mountains in the background. Note the abrupt boundary between the forest fragment and the surrounding tea crop. (Photograph by Tony Goldberg)

tend to persist in areas unfavorable for agriculture, such as wet valley bottoms and steep hillsides, and they contain remnant populations of species found in the park, including primates (Onderdonk and Chapman 2000). Local people use these forest fragments for activities ranging from forest product extraction (e.g., timber, charcoal, forest plants) to slash-and-burn agriculture.

Superimposed on this spatial variation are more insidious temporal trends. Western Uganda has one of the highest rates of human population growth in sub-Saharan Africa, which has notably accelerated the rate of forest clearing and human–park conflict over the past approximately 20 years (Naughton-Treves et al. 2006; Hartter and Southworth 2009). Within the park, areas where forests had been cleared in the early 1900s and converted to grasslands are now regenerating to forest (Fig. 31.3). Long-term data document substantial local climate change, characterized by a marked increase in local yearly rainfall and maximum temperature, which are notably steeper than global averages (Fig. 31.4). Indeed, those of us fortunate

to have visited Kibale for longer than a decade have personally witnessed the receding of glaciers on the Rwenzori Mountains just to the west of the park (Taylor et al. 2006).

The Kibale EcoHealth Project has taken scientific advantage of the inherent ecological variation in this complex system. Specifically, we have treated the core protected areas of Kibale National Park as a type of "control," in which animals live more or less free of human impact (this is clearly not true, as evidenced by the presence of researchers and well-established "edge effects" that can permeate deep into forests; Murcia 1995). Comparing animals in these core areas to those in the forest fragments, which are highly degraded, thus serves as a "natural experiment," and a framework for testing our coarsest-scale hypothesis that land use change in the form of forest fragmentation affects health, defined broadly.

The Kibale EcoHealth Project strives to go beyond the level of coarse-scale statistical association and to gain more detailed insights into how disturbed landscapes alter animal and human health (or not, in

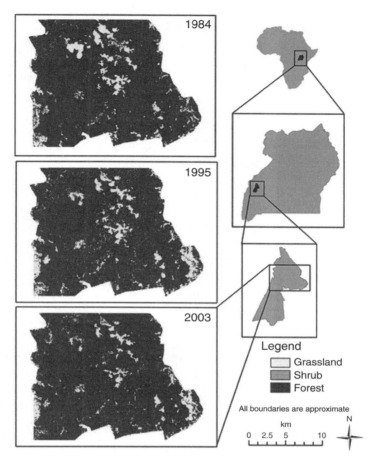

**Figure 31.3.**
Forest transition in Kibale National Park between 1984 and 2003. Note the increase in grassland from 1984 to 1995 but the reversion of grassland to forest from 1995 to 2003, when Kibale was a national park and under protection by the Uganda Wildlife Authority. The land cover assessment was based on five-class supervised classification (89% overall accuracy, kappa statistic 0.867) from three dry-season Landsat Images (May 26, 1984; January 17, 1995; January 31, 2003) at 30m spatial resolution. (Figure courtesy of J. Hartter)

the spirit of objective hypothesis testing). For this reason, we have attempted to make use of finer-scale natural ecological comparisons. The most basic of these are comparisons among forest fragments, which, for political, historical, and geographic reasons, differ widely in their nature and degree of disturbance (Chapman et al. 2006; Gillespie and Chapman 2006). Finer-scale still are contrasts among households and individuals, in the case of people and domestic animals, and social groups and individuals, in the case of primates.

The Kibale EcoHealth Project therefore operates on multiple, hierarchical spatial scales, from the forest to the social unit to the individual, framed against a background of changing human population, environment, and climate. At all levels, we make use of natural variation in the system to ask targeted questions about the relationship between environment and health. Our research methods are observational and typically non-invasive. However, these non-disruptive methodologies belie our decidedly "experimentalist" approach, in that we capitalize whenever

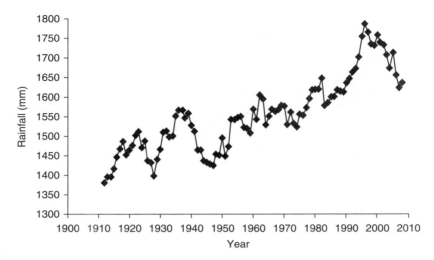

**Figure 31.4:**
Yearly rainfall (10-year running average) in Kibale National Park, Uganda, 1911–2010, showing a long-term trend of increasing precipitation. A similar trend exists for temperature, with local changes in both rainfall and temperature exceeding global averages (C. Chapman and L. Chapman, unpublished data).

we can on the experiments that politics and nature have already conducted.

## ANIMAL HEALTH AND INFECTION

The Kibale EcoHealth Project's main research emphasis has been in the area of infectious disease, focusing on how the dynamic landscapes of Kibale affect microbial transmission. Infection is clearly important, as evidenced by the striking contribution of zoonotic disease transmission to the global emergence of human pathogens (Cleaveland et al. 2001). This does not, however, mean that infection is the most important determinant of health in our system. Indeed, our broader analyses reveal that disease transmission may be a consequence of higher-level ecological changes, such as the aforementioned forest fragmentation, rather than a primary cause of health declines.

One difficulty of prospective ecological studies of multi-species infection dynamics is finding an appropriate pathogen. In primates, deadly epidemics of Ebola virus (Leroy et al. 2004; Walsh et al. 2007) and human respiratory viruses (Kaur et al. 2008; Kondgen et al. 2008) have shed considerable light on the risks of human–primate disease transmission, but these events have been thankfully infrequent.

Microscopic evaluations of gastrointestinal helminths, often the first target for primatologists venturing into the study of infection, are inadequate for understanding cross-species transmission. This is because many helminths are highly host-specific but morphologically cryptic, as illustrated by the case of the nodular worm *Oesophagostomum bifurcum* in West Africa. In this case, traditional parasitological analyses suggested a transmission link between humans and non-human primates, but subsequent molecular analyses demonstrated that the parasite in non-human primates was entirely distinct from that in humans (Gasser et al. 2006, 2009).

The ideal pathogen to serve as a "baseline" for studies of transmission ecology would be ubiquitous, biologically variable, non–species-specific, and benign. Such a pathogen would reflect transmission unbiased by non-uniform distributions, host specificity, or clinical disease. Although no pathogen meets these criteria perfectly, we initially chose to examine the common gastrointestinal bacterium *Escherichia coli*. *E. coli* inhabits the gastrointestinal tracts of all vertebrates but is highly variable genetically and clinically (Donnenberg 2002). Transmitted through food, water, and the physical environment, *E. coli* is a "generalist" microbe known for its ability to cross species barriers (Trabulsi et al. 2002). Because of the importance of certain *E. coli* serotypes to food safety (e.g., the infamous O157-H7 serotype), a suite of molecular methods is available for inferring the bacterium's movement over short time scales (i.e., "source tracking"; Foley et al. 2009).

E. coli appears to move easily between people and animals in Kibale, where ecological overlap is high. Unlike in industrialized nations, where food animals are relegated to farms, cattle, sheep, and goats (and the occasional pig) live in close proximity to households in western Uganda, often sharing an almost identical activity space with people. Not surprisingly, we found E. coli populations in people to be all but genetically indistinguishable from E. coli populations of their livestock (Rwego et al. 2008a), based on DNA fingerprinting methods optimized for this system (Goldberg et al. 2006b). Hygiene matters, since genetically inferred transmission rates between people and their livestock were twice as high when people did not regularly wash their hands before eating (Rwego et al. 2008a).

Ecological overlap between people and primates is more difficult to document, but it clearly occurs in the context of ecotourism. Uganda has developed a thriving ecotourism industry centered on apes, since the country contains the largest population of mountain gorillas (Gorilla beringei beringei) and a number of relatively accessible communities of Eastern chimpanzees (P. t. schweinfurthii). As a result, personnel employed by the Uganda Wildlife Authority as ranger-guides or as field assistants for various ape research projects spend considerable time in ape habitats as part of their daily duties. Intriguingly, this high degree of habitat overlap is reflected in genetic relationships among gastrointestinal bacterial populations: people in Kibale employed in chimpanzee-related ecotourism harbor E. coli more closely related genetically to the E. coli of the chimpanzees they track than do people from local villages (Goldberg et al. 2007). Parallel data from Bwindi Impenetrable National Park, Uganda's main site for gorilla tourism, about 150 km south of Kibale, indicate a strikingly similar pattern: gorillas interacting frequently with people and livestock at the edge of the park harbor E. coli genetically similar to those of people and livestock, whereas gorillas in the "core" areas of Bwindi harbor E. coli that are less genetically similar to those of people and their livestock (Rwego et al. 2008b).

The exact mechanisms explaining this type of microbial transmission remain elusive. We speculate a strong role of environmental contamination, perhaps of water, as a result of interacting human and ape populations living in the same watersheds and being exposed to the same water sources. However, such effects are difficult to document, partially due to small sample sizes of apes and ape-tracking personnel. Most of our recent efforts have therefore focused on more numerous primates, including the well-studied endangered red colobus monkey (Procolobus [Piliocolobus] badius tephrosceles), its common and widely distributed relative the black-and-white colobus monkey (Colobus guereza), and the red-tailed guenon (Cercopithecus ascanius), a notorious crop-raider. These three species inhabit both the protected areas of the national park and the disturbed forest fragments outside of the park, facilitating inter-population comparisons.

Our initial studies focused on three forest fragments near Kibale: Kiko-1 (most intensively used and now entirely gone), Rurama (relatively less encroached upon), and Bugembe (only moderately used). Satisfyingly, we found the "dose–response" effect predicted by this gradation of forest fragment disturbance: human–primate bacterial genetic similarity was highest in Kiko-1, followed by Rurama, followed by Bugembe (Goldberg et al. 2008c). Moreover, we were able to regress human–primate bacterial genetic similarity at the level of the individual against interview data on health and land use. These analyses indicate that people who tend livestock and experience gastrointestinal symptoms are also at elevated risk for harboring primate-like E. coli (Goldberg et al. 2008c). We note that we cannot definitively prove the direction of causality. For example, people who report gastrointestinal symptoms may have contracted primate-borne microbes, or sick people may be shedding microbes into the environment that primates subsequently encounter.

As we have come up against the limitations of our quantitative study design, we have begun to rely increasingly on observation and inference to refine our understanding of microbial transmission risk. Local human cultural practices in particular strike us as being important, one prime example being an intriguing practice that we call "maize daubing" (Goldberg et al. 2008b). In the course of our ground surveys, we have encountered ears of maize at the edges of fields abutting forest fragments that are pasted with a noxious mixture of cattle dung, sand, and hot pepper. People use this strategy to ward off wildlife intent on raiding their crops—primarily red-tailed guenons. Interestingly, red-tailed guenons harbor *E. coli* that are particularly closely related genetically to the *E. coli* of local people and their livestock (Goldberg et al. 2008c). We marvel at the likelihood that this unique human cultural practice, borne out of human–wildlife conflict, might inadvertently increase microbial transmission risk not only between people and primates, but also between people and livestock and primates and livestock.

It has been difficult to ascertain the direction of microbial transmission in our system. Although we initially thought to apply comparative phylogenetic methods to this problem (Goldberg 2003), these have been inconclusive. More informative has been our discovery of antibiotic resistance in *E. coli* in western Uganda's primates. In *E. coli* from chimpanzees, we find patterns of multiple antibiotic resistance matching the most common patterns in *E. coli* from local people (Goldberg et al. 2007). In the mountain gorillas of Bwindi Impenetrable National Park in southwest Uganda, we find carriage rates of antibiotic-resistant *E. coli* that increase with increasing contact rates between gorilla groups and people (Rwego et al. 2008b). Because wild primates only very rarely receive antibiotics (only in unusual cases where apes might be anesthetized to remove snares and treated with a single dose of antibiotics), the best explanation for these observations is transmission of bacteria or resistance-conferring bacterial genetic elements from people to primates. Indeed, our ongoing studies suggest a strong directional bias towards reverse zoonotic transmission of bacteria, perhaps indicating that, in matters of disease, non-human primates are once again disadvantaged.

Our results based on *E. coli* admittedly bias our understanding of disease transmission dynamics. *E. coli* may be a favorable "indicator system" for human–primate disease transmission, but the generality of findings based on this particular microbe should not be overstated. Although *E. coli* is benign, this is not true of other pathogens also present in Kibale's primates (Bonnell et al. 2010). For example, we have found that the pathogenic gastrointestinal protozoa *Giardia duodenalis* and *Cryptosporidium* spp. infect red colobus in forest fragments (Salzer et al. 2007). In the case of *G. duodenalis*, molecular analyses indicate that this is due to independent transmission cycles involving at least two parasite genotypes, one

moving from people to red colobus and the other moving from livestock to red colobus (Johnston et al. 2010). More troublesome still are the novel pathogens in this system. We have found evidence of a previously uncharacterized *Orthopoxvirus* in Kibale red colobus, similar but not identical to cowpox, vaccinia, and monkeypox viruses (Goldberg et al. 2008a). We have documented three novel simian retroviruses in these same red colobus (Goldberg et al. 2009), related to viruses in West Africa that are known zoonoses (Wolfe et al. 2004; Wolfe et al. 2005b), as well as two novel and highly divergent variants of simian hemorrhagic fever virus, which is an animal pathogen of biodefense concern (Lauck et al. 2011). Primates in the forest fragments where we work have frequent, antagonistic interactions with people and are unusually aggressive (Goldberg et al. 2006a). On top of this, immune compromise due to such factors as HIV/AIDS and malnutrition already burdens people in Uganda, leaving the population especially susceptible to opportunistic infections. Such a "perfect storm" of conditions may explain Uganda's status as a country of concern from the standpoint of disease emergence.

Despite our growing collection of compelling stories about how people, primates, and livestock exchange pathogens across a dynamic landscape, we nevertheless believe that enhanced microbial transmission by itself does not imperil Kibale's animals or people. Rather, disease may be the "coup de grâce" for primates forced to live in marginal habitats near Kibale and for people already burdened with the health-related challenges of poverty (Farmer 1999). Many of Kibale's forest fragments have disappeared since they were first studied almost two decades ago (Onderdonk and Chapman 2000), along with their primates (Chapman et al. 2007). Now, we are becoming aware of synergies between the nutritional and physiological stresses experienced by primates forced to live in these marginal environments and the parasites they harbor (Chapman et al. 2006). Infection does not work alone, in other words, but rather synergizes with the far more dire threats of habitat loss and population compression.

The overall outlook is unfortunately grim. In Kibale, we estimate that most primates in unprotected forest fragments will perish within the next two decades as a result of the combined stressors of habitat loss, agonistic interactions with people, nutritional stress, and disease. Based on our observations to date about the relationship between habitat overlap and cross-species microbial transmission, we predict these local primate extinctions to be accompanied by "spikes" in infectious disease transmission to other species, including humans. This fine-scale effect mirrors coarser-scale processes. Perhaps it is no coincidence that the world today appears be facing a global extinction crisis at the same time that it faces a global infection crisis.

## HUMAN DIMENSIONS OF ECOHEALTH

"Health is a state of complete physical, mental and social well-being and not merely the absence of disease or infirmary" (World Health Organization 1948). This appealing definition reflects the relativity of health and shifts our focus from disease to wellness. Unfortunately, it also makes it difficult to identify, define, and model the physical, psychological, and social factors that interact to shape health. Nevertheless, this holistic view of health can complement more traditional quantitative and "disease-centric" approaches.

Our research into the human dimensions of health in Kibale adopts such an integrated theoretical framework, focusing simultaneously on the social and biological dimensions of health and layers of interactions across scales (Mayer 1996). This approach allows us to remain tuned to the physical context of health and its connections to tropical forest ecology while also allowing us to expand beyond the limits of the traditional quantitative epidemiological approach. In this section, we describe how our holistic approach has yielded unique insights into the drivers of ecosystem health in the Kibale region. The emerging picture reveals how structural and physical forces that change landscapes also mediate human–environment relationships and health-related outcomes.

Structural forces, like poverty and inequality, are intricately linked to health and infectious disease outcomes (Marmot and Wilson 1999; Farmer 2005). In rural Uganda, the area around Kibale is experiencing significant population growth as a result of both high birth rates and immigration (Hartter and Southworth 2009). A widespread lack of livelihood options forces people to practice subsistence agriculture even as the price of arable land increases and the amount of available land decreases. This situation pushes the poorest subsistence farmers, usually immigrants, to the least expensive areas, which are plots adjacent to Kibale National Park (Naughton-Treves 1997). These economic conditions (poverty and lack of options) trigger a cascade of interactions between people and wildlife that result in negative health outcomes through two separate pathways.

One pathway to poor health is a consequence of cross-species transmission of infectious agents. As much field-based research has shown, infectious agents are shared between primates and people living next to national parks (e.g., Graczyk et al. 2001; Nizeyi et al. 2001; Graczyk et al. 2002; Nizeyi et al. 2002; Goldberg et al. 2008c; Rwego et al. 2008b; Johnston et al. 2010), particularly as wild primates transgress park boundaries to raid crops planted along the park edges (Kalema-Zikusoka 2002; Rwego et al. 2009). This type of primate behavior enhances the risk of transferring a sylvatic infectious agent to people through indirect ecological exposure, thus resulting in a higher risk of zoonotic disease for the poorest people.

Another pathway to poor health is through crop-raiding, its exacerbation of malnutrition, and the interactions between infectious disease and malnutrition. For example, one response to raiding animals has been to limit the variety of food crops planted. However, homogeneous diets make it difficult to consume the variety of nutrients necessary for good health (Krebs-Smith et al. 1987; Oldewage-Theron and Kruger 2008), especially in a location where protein deficiencies are already common (Muller and Krawinkel 2005). Ugandan law prohibits the hunting or killing of most wildlife, so a second option is to guard crops. At harvest time, local people will "camp out" in fields for days or weeks, banging cans, yelling, and burning fires day and night to keep animals out of fields. Still, findings from a study in the late 1990s found that 4% to 7% of food crops were lost due to crop-raiding (Naughton-Treves 1998), and this problem is increasing yearly, primarily due to a growing elephant population (Wanyama et al. 2010).

Malnutrition and infectious disease operate synergistically to worsen morbidity. Malnutrition weakens immunity, and infection increases nutritional needs (Scrimshaw and SanGiovanni 1997). The interaction between poverty and infectious disease is therefore manifest through conservation conflicts that represent serious barriers to good health for the poorest people.

Moreover, the strong urban bias in healthcare systems services and delivery limits access to quality care by the rural poor (Farmer 1999; Pariyo et al. 2009). If, for example, a family member becomes seriously sick with an infectious disease, the ability to seek quality care is limited further by poor health systems infrastructure. If the household is able to marshal resources to travel to the hospital to seek care, the costs incurred (transportation, hospital fees, medications, meals) frequently require liquidation of household assets, like domestic animals or even land, leading to a downward spiral.

The dynamic relationship between poverty and poor health at the border of the national park is replicated in the forest fragments that persist beyond the park edges. Forest fragments are typically surrounded by household compounds, fields, and pastures, and they are also home to primates (Onderdonk and Chapman 2000). Because of their small size and high human and primate densities, these fragment locations are magnified examples of the intense human–primate interaction and conservation conflict found in households adjacent to the park (Goldberg et al. 2008b). However, fragments are different from the park in one key way: fragments, unlike the park, are unprotected, so interaction between people and primates occurs not only in the fields, but also inside the fragment. People use fragments as sources of fuel wood, timber, medicinal plants, and materials for household use, for making charcoal, and for collecting water. The ecological role of the fragment is thus similar to that of the park boundary, in that animals leave the park to raid crops, but different in that people commonly use the fragment to access natural resources. This situation enables a sustained overlap between human activity spaces and primate activity spaces, and as a result human–primate contact, accelerating the cross-species exchange of infectious agents (Goldberg et al. 2008c).

As mentioned above, we have noted cultural adaptations to crop-raiding that likely have negative health effects, such as "maize daubing." This practice puts people and primates in direct contact with potential enteric bovine pathogens, thereby exacerbating the likelihood of disease exchange across species. Another response is the active guarding of crops throughout the night. This task is usually the responsibility of both children and adults, and the resulting sleepless nights likely have an impact on school performance and overall health. Still another practice is hunting.

Although local people do not hunt primates for food, numerous anecdotes across our study sites reference dogs hunting primates in nocturnal packs, and children hunting primates either for fun or to protect crops. Once a monkey has been killed, the head of household will usually butcher it, cook the meat, and feed it to dogs. In West Africa, the practice of butchering primates in the context of "bushmeat hunting" is known to transfer blood-borne primate pathogens to people (Wolfe et al. 2005a); this may be replicated here as a result of very different socioeconomic drivers.

The distinction between infectious disease and health emerges in a different light when people are asked to talk about their health explicitly. By asking people in communities near the park what are the most important health issues, few will respond with zoonoses as a top concern. Instead, our grounded public health/community-based health surveys suggest that the processes shaping health as physical, social, and mental well-being are profoundly rooted in the interaction between social and biological forces. Specifically, poverty and lack of access to food and healthcare are repeatedly named as the biggest barriers to good health. Moreover, local people define health in ways that more closely reflect the WHO construct and are less focused on infectious disease.

Overall, therefore, ecological interactions at the community and individual levels provide a sufficient explanation for pathogen exchange between people and animals, but they do not paint a complete picture. Rather, we believe that the knowledge, beliefs, and behaviors of individual people fundamentally shape the human–environment interactions that ultimately fuel zoonotic disease transmission as well as other health-related outcomes. The interdisciplinary approaches currently employed by the Kibale EcoHealth Project make possible the integration of inter-scalar structural and cultural factors. Going beyond traditional "disease ecology" and incorporating social science certainly complicates research questions and processes, but it also enables a more complete understanding of cross-species disease exchange and health in the broadest sense.

## ECOHEALTH INTERVENTIONS

As we have continued to identify the root causes of health declines in the Kibale region (human health, animal health, and ecosystem health), we are often asked what concrete steps can be taken to improve the situation. Interventions based on managing wildlife populations and altering agricultural practices sometimes seem like the most obvious solutions, but they tend to be difficult given the scale of implementation that would be required. Therefore, we have recently initiated an intervention that focuses on human health but that is specifically designed to have positive external benefits for wildlife health and forest conservation.

The Kibale Health and Conservation Centre was established to improve access to healthcare for rural residents around Kibale National Park. The decision to establish a primary care health clinic near the park was based on health surveys of local people conducted in 2005 indicating limited local knowledge about health, limited practice of protective health behaviors, but a strong desire for accessible healthcare. The immediate goal of the new clinic is to serve human health needs through basic primary care and public health outreach activities; however, its long-term goal is also to reduce conservation conflicts around the park. By situating the clinic within the park and framing it as a service of conservation interests, we have endeavored to put the EcoHealth paradigm into practice. Our hope is that that improving human health and local attitudes towards the park will "spill over" into improved animal health and forest conservation (Kalema-Zikusoka 2004; Kalema-Zikusoka and Gaffikin 2008). The clinic works alongside local civic organizations such as women's groups, local councils, and primary schools to improve "physical, mental, and social well-being" within catchment communities.

The clinic staff includes two nurses and a number of volunteers from the local community. Nurses provide primary care services within the clinic, make house calls, and conduct regular public health outreach activities. An Employee Health Program was initiated in August 2009 with the goal of enhancing the health of park research and conservation staff working directly with primates to limit pathogen exchange. The clinic is distinct from typical non-governmental organizations. It is integrated into the existing district public health system of Uganda, accessing the majority of its pharmaceuticals from the Ugandan national supply chain. Integration of the clinic into the existing public health infrastructure is imperative for the clinic's sustainability, and collaboration between the international and local conservation and health

professional communities exemplifies the EcoHealth spirit in action.

## CONCLUSIONS

The paradigm of EcoHealth is often represented by a Venn diagram with intersecting circles of human health, animal health, and ecosystem health. Although heuristically useful, this representation is inadequate for capturing the complexity of EcoHealth in practice. In some cases the circles fail to intersect, as exemplified by our studies showing that perceptions of health in local communities do not generally include consideration of animals or the land. In other cases, the circles are of varying sizes and the interactions are decidedly lopsided or unidirectional, as exemplified by our studies showing extensive transmission of microbes from people to wild primates but limited evidence of the reverse. In other cases still, we find incompatibility with the fundamental concept of "balance" underlying the EcoHealth paradigm, as exemplified by crop-raiding: people living in environments denuded of primates are better off economically and likely have higher nutritional status than people living in biodiverse environments containing primates. Where the optimal balance between human interests, wildlife interests, and ecosystem conservation ultimately lies is context-dependent.

Despite these complexities, we can nevertheless make some generalizations. For example, our work to date supports our coarsest-scale hypothesis that landscape changes alter infectious disease transmission dynamics, and specifically that ecological overlap enhances microbial transmission between species. In the case of E. coli, we have documented a direct relationship between the degree of human–animal overlap and the rate of microbial transmission across species (Goldberg et al. 2008c; Rwego et al. 2008a). Infection in the Kibale system appears to flow readily from people and livestock to primates, as evidenced by our findings of antibiotic-resistant bacteria in wild primates and multiple human-associated and livestock-associated G. duodenalis genotypes in endangered red colobus (Goldberg et al. 2007; Rwego et al. 2008b; Johnston et al. 2010). Infection of wild primates with human and livestock pathogens may have additional negative consequences. Primates are important seed dispersers and are central to maintaining the diversity of tropical forests and the ecosystem services they provide (Chapman and Onderdonk 1998); forests without primates, or forests with sick primates, may be poorly functioning and "unhealthy" ecosystems.

Our findings also clarify the temporal scale on which EcoHealth operates, which is relevant to the idea of EcoHealth as a biological and ethical goal. The historical forces that have shaped Kibale's present-day landscape originated decades in the past (e.g., intensive logging, forest fragmentation); however, their effects on land health are strongly evident today, even nearly two decades after Kibale's designation as a national park. It is clear that factors such as logging exert long-term influences on forest communities (Chapman et al. 2000) and that recovery is slow. It is also clear that recovery will be equally slow in the arena of health. In the Kibale system, the relevant time scale for ecological improvements to health is likely to be the time scale of forest regeneration. Health interventions based on restoring forest integrity should therefore be expected to reap benefits only decades into the future—a sobering lesson for researchers and policy planners under pressure to show tangible results quickly.

Finally, we may draw some conclusions about the knowledge, beliefs, and behaviors of the people who inhabit and shape Kibale's dynamic landscapes. Most important, perhaps, is that the paradigm of EcoHealth does not figure prominently in the minds of these people. Although people in communities near Kibale are indeed aware of the relationship between the forest and their well-being, the connection to health is tenuous at best. Because of such forces as crop-raiding, "healthy" ecosystems containing biodiverse animal communities are associated with poverty and are perceived as undesirable. Is this a failure of education, a bias in perception, or a breakdown of the utility of EcoHealth thinking? Whatever the explanation, there is a notable gap between our quantitative results showing strong ecology and health connections in the case of microbial transmission and our qualitative results showing different perceptions in the minds of local people.

As a more nuanced picture emerges of environment–health linkages in the Kibale region, it is tempting to simplify the task ahead by focusing on a small topic or a particular question. Indeed, a progressive narrowing of focus defines success in many modern fields of science. We argue, however, that this approach is antithetical to the concept of EcoHealth. If the Kibale EcoHealth Project differs from the norm in any way, it is in its dogged determination not to succumb to the temptations of reductionism. Intellectually, this is because focusing on the diverse interactions that link human health, animal health, and the environment keeps us rooted to our ecological underpinnings. Philosophically, it is because we appreciate that new insights are most likely to emerge where disciplines intersect and when multiple approaches are applied to complex problems. Pragmatically, it is because interventions that target a single relationship within a complex ecological web will surely have unforeseen consequences and are likely to fail as a result.

The Kibale EcoHealth Project ultimately provides a concrete example of how holistic science can progress, and how a "bottom-up," place-based approach can inform fundamental questions at the interface of ecology and health. As we enter the phase of intervention with such efforts as the Kibale Health and Conservation Centre, the holistic approach becomes increasingly important, since the costs of allocating resources incorrectly or sub-optimally increases commensurately. We will continue to emphasize the evidence-based approach, now studying not only the "natural experiments" that nature and politics have already conducted in western Uganda, but also the "experiments" that we are conducting ourselves in the form of targeted public health interventions. The degree to which our focused efforts to improve human health will have positive, external benefits for animal health and conservation may in the end be the ultimate test of the real-world relevance of the EcoHealth paradigm to the complex and dynamic ecosystems of western Uganda and beyond.

## REFERENCES

Archabald, K., and L. Naughton-Treves. 2001. Tourism revenue sharing around national parks in western Uganda: early efforts to identify and reward local communities. Environ Conserv 23:135–149.

Baranga, J. 1991. Kibale forest game corridor: man or wildlife? In D.A. Saunders and R.J. Hobbs, eds. Nature conservation: the role of corridors, pp. 371–375. Surrey Beatty and Sons, London.

Bonnell, T.R., R.R. Sengupta, C.A. Chapman, and T.L. Goldberg. 2010. An agent-based model of red colobus resources and disease dynamics implicates key resource sites as hot spots of disease transmission. Ecol Model 221:2491–2500.

Chapman, C.A., and D.A. Onderdonk. 1998. Forests without primates: primate/plant codependency. Am J Primatol 45:127–141.

Chapman, C.A., S.R. Balcomb, T.R. Gillespie, J. Skorupa, and T.T. Struhsaker. 2000. Long-term effects of logging on African primate communities: A 28-year comparison from Kibale National Park, Uganda. Conserv Biol 14:207–217.

Chapman, C.A., T.T. Struhsaker, and J.E. Lambert. 2005. Thirty years of research in Kibale National Park, Uganda, reveals a complex picture for conservation. Int J Primatol 26:539–555.

Chapman, C.A., M.D. Wasserman, T.R. Gillespie, M.L. Speirs, M.J. Lawes, T.L. Saj, and T.E. Ziegler. 2006. Do food availability, parasitism, and stress have synergistic effects on red colobus populations living in forest fragments? Am J Phys Anthropol 131:525–534.

Chapman, C.A., L. Naughton-Treves, M.J. Lawes, M.D. Wasserman, and T.R. Gillespie. 2007. The conservation value of forest fragments: explanations for population declines of the colobus of western Uganda. Int J Primatol 28:513–528.

Cleaveland, S., M.K. Laurenson, and L.H. Taylor. 2001. Diseases of humans and their domestic mammals: pathogen characteristics, host range and the risk of emergence. Philos Trans R Soc Lond B 356: 991–999.

Donnenberg, M., ed. 2002. Escherichia coli: Virulence mechanisms of a versatile pathogen. Academic Press, San Diego California.

Farmer, P. 1999. Infections and inequalities: the modern plagues. University of California Press, Berkeley, California,

Farmer, P. 2005. Pathologies of power: health, human rights, and the new war on the poor. University of California Press, Berkeley, California.

Foley, S.L., A.M. Lynne, and R. Nayak. 2009. Molecular typing methodologies for microbial source tracking and epidemiological investigations of Gram-negative bacterial foodborne pathogens. Infect Genet Evol 9:430–440.

Gasser, R.B., J.M. de Gruijter, and A.M. Polderman. 2006. Insights into the epidemiology and genetic make-up of Oesophagostomum bifurcum from human and non-human primates using molecular tools. Parasitology 132:453–460.

Gasser, R.B., J. M. de Gruijter, and A. M. Polderman. 2009. The utility of molecular methods for elucidating primate-pathogen relationships—the Oesophagostomum bifurcum example. In M.A. Huffman and C.A. Chapman, eds. Primate parasite ecology: the

dynamics and study of host-parasite relationships, pp. 47–62. Cambridge University Press, Cambridge, UK.

Gillespie, T.R., and C.A. Chapman. 2006. Prediction of parasite infection dynamics in primate metapopulations based on attributes of forest fragmentation. Conserv Biol 20:441–448.

Goldberg, T.L. 2003. Application of phylogeny reconstruction and character-evolution analysis to inferring patterns of directional microbial transmission. Prev Vet Med 61:59–70.

Goldberg, T.L., T.R. Gillespie, I.B. Rwego, and C. Kaganzi. 2006a. Killing of a pearl-spotted owlet (Glaucidium perlatum) by male red colobus monkeys (Procolobus tephrosceles) in a forest fragment near Kibale National Park, Uganda. Am J Primatol 68:1007–1011.

Goldberg, T.L., T.R. Gillespie, and R.S. Singer. 2006b. Optimization of analytical parameters for inferring relationships among Escherichia coli isolates from repetitive-element PCR by maximizing correspondence with multilocus sequence typing data. Appl Environ Microbiol 72:6049–6052.

Goldberg, T.L., T.R. Gillespie, I.B. Rwego, E.R. Wheeler, E.E. Estoff, and C.A. Chapman. 2007. Patterns of gastrointestinal bacterial exchange between chimpanzees and humans involved in research and tourism in western Uganda. Biol Conserv 135:511–517.

Goldberg, T.L., C.A. Chapman, K. Cameron, T. Saj, W.B. Karesh, N. Wolfe, S.W. Wong, M.E. Dubois, and M.K. Slifka. 2008a. Serologic evidence for novel poxvius in endangered red colobus monkeys, western Uganda. Emerg Infect Dis 14:801–803.

Goldberg, T.L., T.R. Gillespie, and I.B. Rwego. 2008b. Health and disease in the people, primates, and domestic animals of Kibale National Park: implications for conservation. In R. Wrangham and E. Ross, eds. Science and conservation in African forests: the benefits of long-term research, pp. 75–87. Cambridge University Press, UK.

Goldberg, T.L., T.R. Gillespie, I.B. Rwego, E.E. Estoff, and C.A. Chapman. 2008c. Forest fragmentation as cause of bacterial transmission among primates, humans, and livestock, Uganda. Emerg Infect Dis 14:1375–1382.

Goldberg, T.L., D.M. Sintasath, C.A. Chapman, K.M. Cameron, W.B. Karesh, S. Tang, N.D. Wolfe, I.B. Rwego, N. Ting, and W.M. Switzer. 2009. Coinfection of Ugandan red colobus (Procolobus [Piliocolobus] rufomitratus tephrosceles) with novel, divergent delta-, lenti-, and spumaretroviruses. J Virol 83:11318–11329.

Graczyk, T.K., A.B. Mudakikwa, M.R. Cranfield, and U. Eilenberger. 2001. Hyperkeratotic mange caused by Sarcoptes scabiei (Acariformes: Sarcoptidae) in juvenile human-habituated mountain gorillas (*Gorilla gorilla beringei*). Parasitol Res 87:1024–1028.

Graczyk, T.K., J. Bosco-Nizeyi, B. Ssebide, R.C. Thompson, C. Read, and M.R. Cranfield. 2002. Anthropozoonotic *Giardia duodenalis* genotype (assemblage) A infections in habitats of free-ranging human-habituated gorillas, Uganda. J Parasitol 88:905–909.

Hartter, J., and J. Southworth. 2009. Dwindling resources and fragmentation of landscapes around parks: wetlands and forest patches around Kibale National Park, Uganda. Landscape Ecol 24:643–656.

Howard, P.C. 1991. Nature conservation in Uganda's tropical forest reserves. IUCN Gland, Switzerland and Cambridge, UK.

Johnston, A.R., T.R. Gillespie, I.B. Rwego, T.L. Tranby McLachlan, A.D. Kent, and T.L. Goldberg. 2010. Molecular epidemiology of cross-species Giardia duodenalis transmission in western Uganda. PLoS Neglect Trop Dis 4:e683.

Jones, K.E., N.G. Patel, M.A. Levy, A. Storeygard, D. Balk, J.L. Gittleman, and P. Daszak. 2008. Global trends in emerging infectious diseases. Nature 451:990–993.

Kalema-Zikusoka, G. 2002. Scabies in free-ranging mountain gorillas (*Gorilla beringei beringei*) in Bwindi Impenetrable National Park, Uganda. Vet Rec 150:12–14.

Kalema-Zikusoka, G. 2004. Conservation through public health. Gorilla J 28:9–11.

Kalema-Zikusoka, G., and L. Gaffikin. 2008. Sharing the forest: protecting gorillas and helping families in Uganda. Smithsonian Institution, Washington, D.C.

Kaur, T., J. Singh, S. Tong, C. Humphrey, D. Clevenger, W. Tan, B. Szekely, Y. Wang, Y. Li, E. Alex Muse, M. Kiyono, S. Hanamura, E. Inoue, M. Nakamura, M.A. Huffman, B. Jiang, and T. Nishida. 2008. Descriptive epidemiology of fatal respiratory outbreaks and detection of a human-related metapneumovirus in wild chimpanzees (*Pan troglodytes*) at Mahale Mountains National Park, Western Tanzania. Am J Primatol 70:755–765.

Keiser, J., B.H. Singer, and J. Utzinger. 2005. Reducing the burden of malaria in different eco-epidemiological settings with environmental management: a systematic review. Lancet Infect Dis 5:695–708.

Kondgen, S., H. Kuhl, P.K. N'Goran, P.D. Walsh, S. Schenk, N. Ernst, R. Biek, P. Formenty, K. Matz-Rensing, B. Schweiger, S. Junglen, H. Ellerbrok, A. Nitsche, T. Briese, W.I. Lipkin, G. Pauli, C. Boesch, and F.H. Leendertz. 2008. Pandemic human viruses cause decline of endangered great apes. Curr Biol 18:260–264.

Krebs-Smith, S.M., H. Smiciklas-Wright, H.A. Guthrie, and J. Krebs-Smith. 1987. The effects of variety in food choices on dietary quality. J Am Diet Assoc 87:897–903.

Lauck, M., D. Hyeroba, A. Tumukunde, G. Weny, S.M. Lank, C.A Chapman, D.H. O'Connor, T.C. Friedrich, and T.L. Goldberg. 2011. Novel, divergent simian hemorrhagic fever viruses in a wild Ugandan red colobus monkey discovered using direct pyrosequencing. PLoS One 6:e19056.

Leroy, E.M., P. Rouquet, P. Formenty, S. Souquiere, A. Kilbourne, J.M. Froment, M. Bermejo, S. Smit, W. Karesh, R. Swanepoel, S.R. Zaki, and P.E. Rollin. 2004. Multiple Ebola virus transmission events and rapid decline of central African wildlife. Science 303:387–390.

Marmot, M.G., and R.G. Wilson. 1999. Social determinants of health. Oxford University Press, New York.

Mayer, J.D. 1996. The political ecology of disease as one new focus for medical geography. Progr Hum Geog 20:441–456.

Muller, O., and M. Krawinkel. 2005. Malnutrition and health in developing countries. CMAJ 173: 279–86.

Murcia, C. 1995. Edge effects in fragmented forests: implications for conservation. Trends Ecol Evol 10:58–62.

Naughton-Treves, L. 1997. Farming the forest edge: vulnerable places and people around Kibale National Park, Uganda. Geog Rev 87:27–46.

Naughton-Treves, L. 1998. Predicting patterns of crop damage by wildlife around Kibale National Park, Uganda. Conserv Biol 12:156–168.

Naughton-Treves, L. 1999. Whose animals? A history of property rights to wildlife in Toro, western Uganda. Land Degr Develop 10:311–328.

Naughton-Treves, L., D.M. Kammen, and C.A. Chapman. 2006. Burning biodiversity: woody biomass use by commercial and subsistence groups in western Uganda. Biodivers Conserv 34:232–241.

Nizeyi, J.B., R.B. Innocent, J. Erume, G.R. Kalema, M.R. Cranfield, and T.K. Graczyk. 2001. Campylobacteriosis, salmonellosis, and shigellosis in free-ranging human-habituated mountain gorillas of Uganda. J Wildl Dis 37:239–244.

Nizeyi, J.B., D. Sebunya, A.J. Dasilva, M.R. Cranfield, N.J. Pieniazek, and T.K. Graczyk. 2002. Cryptosporidiosis in people sharing habitats with free-ranging mountain gorillas (*Gorilla gorilla beringei*), Uganda. Am J Trop Med Hyg 66:442–444.

Oldewage-Theron, W.H., and R. Kruger. 2008. Food variety and dietary diversity as indicators of the dietary adequacy and health status of an elderly population in Sharpeville, South Africa. J Nutr Elder 27:101–133.

Onderdonk, D.A., and C.A. Chapman. 2000. Coping with forest fragmentation: The primates of Kibale National Park, Uganda. Int J Primatol 21:587–611.

Pariyo, G.W., E. Ekirapa-Kiracho, O. Okui, M.H. Rahman, S. Peterson, D.M. Bishai, H. Lucas, and D.H. Peters. 2009. Changes in utilization of health services among poor and rural residents in Uganda: are reforms benefitting the poor? Int J Equity Health 8:39.

Pedersen, A., and J. Davies. 2010. Cross species pathogen transmission and disease emergence in primates. EcoHealth 6:496–508.

Rapport, D. 1998. Defining ecosystem health. *In* D. Rapport, R. Costanza, P. Epstein, C. Gaudet, and R. Levins, eds. Ecosystem health, pp. 18–33. Blackwell Science, Inc. Malden, Massachusetts.

Rode, K.D., P.I. Chiyo, C.A. Chapman, and L.R. McDowell. 2006. Nutritional ecology of elephants in Kibale National Park, Uganda, and its relationship with crop-raiding behaviour. J Trop Ecol 22:441–449.

Rwego, I.B., T.R. Gillespie, G. Isabirye-Basuta, and T.L. Goldberg. 2008a. High rates of *Escherichia coli* transmission between livestock and humans in rural Uganda. J Clin Microbiol 46:3187–3191.

Rwego, I.B., G. Isabirye-Basuta, T.R. Gillespie, and T.L. Goldberg. 2008b. Gastrointestinal bacterial transmission among humans, mountain gorillas, and livestock in Bwindi impenetrable National Park, Uganda. Conserv Biol 22:1600–1607.

Rwego, I.B., G. Isabirye-Basuta, T.R. Gillespie, and T.L. Goldberg. 2009. Bacterial exchange between gorillas, humans, and livestock in Bwindi. Gorilla J 38:16–18.

Salzer, J.S., I.B. Rwego, T.L. Goldberg, M.S. Kuhlenschmidt, and T.R. Gillespie. 2007. *Giardia* sp. and Cryptosporidium sp. infections in primates in fragmented and undisturbed forest in western Uganda. J Parasitol 93:439–440.

Scrimshaw, N.S., and J.P. SanGiovanni. 1997. Synergism of nutrition, infection, and immunity: an overview. Am J Clin Nutr 66:S464–S477.

Struhsaker, T.T. 1997. Ecology of an African rain forest: logging in Kibale and the conflict between conservation and exploitation. University Press of Florida, Gainesville, Florida.

Taylor, R.G., L. Mileham, C. Tindimugaya, A. Majugu, A. Muwanga, and B. Nakileza. 2006. Recent glacial recession in the Rwenzori Mountains of East Africa due to rising air temperature. Geophys ResLett 33:doi:10/1029/2006GRL025962.

Trabulsi, L., R. Keller, and T. Tardelli-Gomes. 2002. Typical and atypical enteropathogenic *Escherichia coli*. Emerg Inf Dis 8:508–513.

van Orsdol, K.G. 1986. Agricultural encroachment in Uganda's Kibale Forest. Oryx 20:115–117.

Walsh, P.D., T. Breuer, C. Sanz, D. Morgan, and D. Doran-Sheehy. 2007. Potential for Ebola transmission between gorilla and chimpanzee social groups. Am Nat 169:684–689.

Wanyama, F., R. Muhabwe, A.J. Plumptre, C.A. Chapman, and J.M. Rothman. 2010. Censusing large mammals in Kibale National Park: evaluation of the intensity of sampling required to determine change. Afr J Ecol 48:953–961.

Wolfe, N.D., W.M. Switzer, J.K. Carr, V.B. Bhullar, V. Shanmugam, U. Tamoufe, A.T. Prosser, J.N. Torimiro, A. Wright, E. Mpoudi-Ngole, F.E. McCutchan, D.L. Birx, T.M. Folks, D.S. Burke, and W. Heneine. 2004. Naturally acquired simian retrovirus infections in central African hunters. Lancet 363: 932–937.

Wolfe, N.D., P. Daszak, A.M. Kilpatrick, and D.S. Burke. 2005a. Bushmeat hunting, deforestation, and predicting zoonotic emergence. Emerg Infect Dis 11:1822–1827.

Wolfe, N., W. Heneine, J.K. Carr, A.D. Garcia, V. Shanmugam, U. Tamoufe, J.N. Torimiro, A.T. Prosser, M. Lebreton, E. Mpoudi-Ngole, F.E. McCutchan, D.L. Birx, T.M. Folks, D.S. Burke, and W.M. Switzer. 2005b. Emergence of unique primate T-lymphotropic viruses among central African bushmeat hunters. Proc Natl Acad Sci USA 102:7994–7999.

World Health Organization. 1948. Preamble to the Constitution of the World Health Organization as adopted by the International Health Conference, New York, 19–June 22, 1946; signed on July 22, 1946 by the representatives of 61 States (Official Records of the World Health Organization, no. 2, p. 100) and entered into force on April 7, 1948.

# 32

## CONSERVATION MEDICINE IN BRAZIL

Case Studies of Ecological Health in Practice

Paulo Rogerio Mangini, Rodrigo Silva Pinto Jorge, Marcelo Renan de
Deus Santos, Claudia Filoni, Carlos Eduardo da Silva Verona,
Alessandra Nava, Maria Fernanda Vianna Marvulo, and
Jean Carlos Ramos Silva

Conservation medicine was introduced in Brazil in January 2001 at the First International Course on Conservation Medicine: Ecosystem Health in Practice, promoted by EcoHealth Alliance (formerly known as Wildlife Trust), and local partner Instituto de Pesquisas Ecológicas (IPÊ; Institute for Ecological Research). The conservation medicine philosophy has been incorporated into two primary areas: the impacts of disease in conservation of biodiversity and health, and surveillance of zoonotic pathogens in wildlife. Projects on conservation of biodiversity have targeted wildlife health as an important factor to assess ecosystem health. Most of these use the premise that anthropogenic change in ecosystems is the primary factor leading to diseases in wildlife, domestic animals, and humans. In Brazil, some strategies to control zoonotic pathogens applied by public health agencies tend to create conflicts with biodiversity conservation. Also, economic interests related to agribusiness expansion, animal production, and human health issues frequently emerge as serious challenges in establishing multilateral strategies to ecosystem health management. The concept that wildlife act as primary zoonotic disease reservoirs gives public health agencies the incentive to use methodologies that have a negative impact on wildlife. Fortunately, the environmental agencies and laws related to wildlife conservation and health provide a balance for the human health policies that have been developed. Emerging and re-emerging infectious diseases play a significant role in revising the Brazilian national policy for ecosystem, human, and animal health. Communication challenges among specialists and different disciplines have caused a segregation to address the environmental problems in the country. To achieve cooperation among federal and state governments, universities, and non-governmental organizations, it is important to promote integration of disciplines in the undergraduate and graduate curricula. These include health, agricultural, social, biological, and environmental sciences. Nevertheless, the evaluation of current conservation biology and public health conflicting arguments demonstrates the intersections where cooperative efforts could be applied to create a truly integrative approach to solve the seemingly divergent health problems (Fig. 32.1).

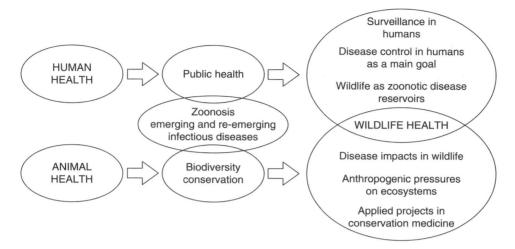

**Figure 32.1:**
Conflicting approaches of public health and conservation of biodiversity that have been presented as guidelines to cooperative initiatives on ecological health in Brazil. (Adapted from J.C.R. Silva, 2006)

## CONSERVATION MEDICINE INITIATIVES

Successful conservation medicine curricular professional activities were developed by IPÊ and EcoHealth Alliance through a 50-hour transdisciplinary continuing education course offered for several years. From that professional training, Universidade Vila Velha (UVV), in Espírito Santo state, developed and implemented a conservation medicine course comprising 60 hours of lectures for the veterinary curriculum (Aguirre and Gomez 2009). The inclusion of conservation medicine in the undergraduate program was a pioneering initiative in Brazil. Mangini and Silva (2007) produced the first conceptual reference book chapter on conservation medicine that was printed in Portuguese. The authors defined conservation medicine as "the crisis science of environmental health and the consequential biodiversity loss, which was developed by interdisciplinary efforts in research, management and public policies, aiming at the sustainability of health in all biologic communities and their ecosystems."(p. 1258) Based on a broader ecological view of interactions among biotic and non-biotic factors of ecosystems, the authors suggested an expansion of the health sphere concepts presented by Tabor (2002). Ecosystem health is the result of the balance among human, animal, and plant health;

where these three spheres overlap they mutually influence each other, contributing to ecological health (Fig. 32.2). This approach is a constructive and inclusive strategy, especially in a country where intensive deforestation and agricultural expansion are occurring at a fast rate.

The National Scientific Research Council of Brazil (CNPq) curricula database (http://www.cnpq.br, accessed March 25, 2010) demonstrates that there are 107 scientists, including 26 with doctorates, working on conservation medicine, indicating that there is a solid foundation for this emerging discipline; however, further development in the Brazilian scientific community is still needed. In July 2007, Instituto de Ensino Pesquisa e Preservação Ambiental Marcos Daniel (IMD; Institute for Teaching, Research and Environmental Conservation) and Instituto Brasileiro para Medicina da Conservação (TRÍADE; Brazilian Institute for Conservation Medicine) organized the First International Meeting in Conservation Medicine to develop a national strategy for conservation medicine by (1) promoting the writing of more scientific literature in Portuguese; (2) integrating conservation medicine themes into scientific events and undergraduate courses in the agricultural, biological, and health sciences; (3) creating programs to educate the general public; (4) involving public and private institutions; and (5) establishing an online database listing

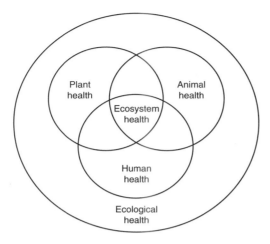

**Figure 32.2:**
Expansion of the ecological health spheres concept (formerly presented by Tabor 2002). Expanded concept is used in efforts to disseminate conservation medicine in Brazil. (Adapted from Mangini and Silva 2007)

reference laboratories that can develop the diagnostic tools for wildlife health and disease. Initiatives on conservation medicine in Brazil are disseminated across the country through professional development, field research, and public policy. In this chapter we present some selected case studies in different ecosystems.

## RABIES AND EMERGING WILDLIFE RESERVOIR HOSTS IN NORTHEASTERN BRAZIL

Human rabies control is focused on eradicating the disease in dogs and cats through vaccination in Latin America. In 2005, the sylvatic cycle, maintained by a wide variety of mammalian hosts, emerged as a new concern (Bernardi et al. 2005), and the number of human cases that originated in wildlife surpassed the number of cases caused by dogs (Schneider et al. 2005). Vampire bats (*Desmodus rotundus*) were once considered the primary wild reservoir of rabies in Brazil; in fact, recent studies have documented an increase in rabies virus exposure in humans linked to attacks by other groups of wild animals (non-vampire bats, wild canids and marmosets) in the Northeast Region (NUEND 2004). In Teresina City, Piauí state, an outbreak of rabies in wild canids led to several

human cases in 2005; a high proportion of those animals were diagnosed as rabies positive (M.S. Wada, personal communication, 2009).

The *Caatinga* biome represents 11% of the Brazil territory, and although there is a relatively low human density, more than 50% is under considerable human pressure, 30% of which is caused by agricultural use. The main threats to the region related to fragmentation and impacts on the flora and fauna include rural settlement growth; subsistence agriculture and hunting; domestic herds of sheep, goats, horses, and cattle; and deforestation to provide firewood and coal and also large-scale agriculture (i.e., sugar cane) (Silva et al. 2004). These regional changes in the landscape have created ecological changes in the epidemiology of rabies. The rabies cycle in wild canids was identified as occurring independently from the domestic dog cycle. Genetic analysis of the rabies virus isolated from free-ranging canids has been classified as variant 2, typical of domestic dogs, but sufficiently different to be classified in a different cluster, despite the fact that both cycles occur simultaneously (Bernardi et al. 2005; Carnieli Jr. et al. 2006; Favoretto et al. 2006; Carnieli Jr. et al. 2008). These results indicate that rabies spillover transmission has occurred from domestic dogs to wild canids and a sylvatic cycle has been established. Despite the newly emergent viral strain, domestic dog rabies virus continues to be isolated from wild canids (Carnieli Jr. et al. 2008), providing evidence that the domestic dog population threatens wild canids in the region. In addition, transmission occurs from wild to domestic canids, as the domestic dog virus has been grouped in the same clade of the wild canid virus (Carnieli Jr. et al. 2006). Viral genetic analysis implicates crab-eating foxes (*Cerdocyon thous*) as the primary wild reservoir (Carnieli Jr. et al. 2008). The possible expansion of rabies virus reservoirs represents a great concern for public health, since these animals are very adaptable to living near human settlements and are sometimes kept as pets (Bernardi et al. 2005).

In Ceará state, 173 rabies cases were reported in crab-eating foxes, 25 in common marmosets (*Callithrix jacchus*), and six in crab-eating raccoons (*Procyon cancrivorous*) between 1990 and 2004 (NUEND 2004). These species have become widespread in green urban areas and the outskirts of major regional cities. The adaptation for living near human settlements could result in shifting of the sylvatic and urban rabies cycles. This epidemiological situation requires active surveillance of wild carnivores and primates in northeastern Brazil. A program for oral vaccination, as implemented in Europe (Pastoret et al. 2004) and North America (Cross et al. 2007), has been recommended as an option to control rabies progress in the region for wild carnivores (Carnieli Jr. et al. 2006, 2008; Favoretto et al. 2006). For public health agencies, the primary motivation to change rabies surveillance and control strategies in northeast Brazil is based on human health, due to the increased number of cases reported. In addition, the occurrence of a sylvatic rabies cycle in wild canids represents a risk for threatened species and other terrestrial wild mammals in the *Caatinga* biome. The management of changes occurring in rabies ecology in the region demands a transdisciplinary approach including the collaboration of experts in public health, landscape management and restoration, epidemiological modeling, surveillance, and wildlife health and conservation.

## ECOLOGICAL HEALTH IN COASTAL AREAS

The primary threats to coastal ecosystems in Brazil are the growing urbanization and industrial development, improper waste disposal, expansion of harbor activities, pollution, deepwater oil extraction, and fishing. Some of these threats are significantly related and restricted to large urban centers, widely present along the coast. However, even in the relatively isolated coastal regions, small fishing communities can have a significant impact on ecosystem health (Peckham et al. 2007).

Industrial and small-scale fisheries are significant commercial activities and currently responsible for 800,000 direct jobs in Brazil (PROZEE 2006). Official reports indicate that traditional fishing communities and small-scale fisheries were responsible for catching more than 275,000 tons of fish (51.6% of the total national fish production); in the Southeast and South Regions they were responsible for 42% of production (IBAMA 2007). Small-scale fisheries predominate in Paraná, where some fishing communities are small and distant from urban centers, located inside a large continuous remnant of the Atlantic rainforest. The primary commercial fishing activity in the region is shrimp trawling, which has a major impact on other

marine natural resources, indicated by a bycatch of about 1 kg of non-commercial fish to 1 kg of shrimp captured (Krul 1999; Cattani 2010).

Mangini (2010) and Krul (2010) developed an interdisciplinary study to compare two traditional fishing villages (one with 200 and another with 900 inhabitants) located in the same estuarine ecosystem in Paraná. The authors evaluated 45 variables related to land use, fishery production, biodiversity, and environmental health. A significant correlation between forest bird community profiles (i.e., distribution in guilds, richness, and abundance) and village characteristics (i.e., size of villages, house density and distribution, demography of people and domestic animals, vegetation profile, presence of trails and hunting vestiges in surrounding forest) was identified. In the small village (VA), the forest bird community had higher richness per census (10.50 ± 1.90) and individual abundance (15.68 ± 3.75) than the larger village (VB; 7.40 ± 1.90 and 10.28 ± 4.27, respectively). Also, VA recorded 58 bird species, distributed in 23 guilds, with three guilds that were not present in VB, which had 56 bird species, distributed in 24 guilds, and five of these guilds were not present in VA. The greatest contributor to these differences was the presence of generalist species. The Sorensen's similarity coefficient for bird diversity between the two villages was 0.6; furthermore, in VB the decrease in biodiversity was associated with a significant increase in number of ectoparasites and blood parasites observed. A total of six ectoparasites and one hemoparasite were detected in birds from VA and 16 ectoparasites and three hemoparasites in those from VB. The Jaccard similarity index between the two villages was 0.16 for hemoparasites and 0.33 for ectoparasites, which are relevant differences between the sites. In addition, 52.9% unhealthy birds (i.e., poor body condition; respiratory, dermatologic, ophthalmic, or oral abnormalities) in VA had a significantly higher degree of parasitism than those without clinical abnormalities (47%), which had low or no infestation. In VB there was no significant relationship between the clinical status and the degree of parasitism, and the evaluated birds had a larger variety of clinical abnormalities. The stronger association between clinical status and degree of parasitism observed in VA could be related to the more intact diversity and ecological distribution of birds and parasites in their guilds, and the weaker correlation observed in VB was possibly determined by spillover events or the lack of a "dilution effect" (Mangini 2010). One trombiculid mite was prevalent in five guilds in VA, representing 35.2% of the ectoparasites found on the birds. In VB the same species was present in just three guilds, representing only for 5.7% of infections. The opposite situation was observed with *Trouessartia* sp., representing 25% of ectoparasites, affecting five guilds in VB. The same species represented 17.6% of ectoparasites in VA, affecting only three guilds. Data from land use, biodiversity, and health evaluations demonstrated that the size and demography of the fishing community was associated with larger impacts in the surrounding forest, which negatively influenced avian richness and abundance and produced changes in the guilds profile. At these sites a small decrease in bird richness and abundance and perturbations in guild profile in VB were followed by an expressive increase in parasite diversity, a significant increase in parasitism, and a moderate decrease in health status (Krul 2010; Mangini 2010).

The studies also evaluated fish and shrimp production in the same villages and the impact of these activities on marine bird populations. Shrimp trawlers on the Paraná coast create a yearly bycatch of approximately 5,000 tons of non-target species. Marine birds ingest a significant portion of the bycatch. Discarded fish make up 91.5% and 75.3% of the diets of magnificent frigate birds (*Fregata magnificens*) and brown boobies (*Sula leucogaster*) (Krul 1999, 2004). Evaluating the local seasonality and availability of bycatch, Mangini (2010) and Krul (2010) verified a positive influence of bycatch on reproductive activities of frigate birds and brown boobies, represented by a peak of active nesting during the periods with higher bycatch availability (between February and September). Also, brown boobies reproducing in a local colony incurred a significant decrease in their health status during periods when shrimp fishery production remained under 10 tons per month in the studied villages, when 61% of the boobies presented with health abnormalities. During periods with low shrimp and bycatch production 60.6% of brown boobies were emaciated, with a low average body mass (1,241.6 ± 170.3 g), compared with periods with elevated bycatch production, with a significantly higher body mass (1,411.5 ± 199.1 g). The authors also evaluated the seasonal registries from 1992 to 2009 for the brown booby, frigate birds, and kelp gull (*Larus dominicanus*) in a local rehabilitation center. Of

208 birds registered, more than 82% of cases were in poor body condition due to malnutrition. Also, more than 35% of the birds received each year were registered during the three months when shrimp fishery activities were very limited or prohibited. The peaks in shrimp and bycatch production during the year corresponded with peaks in marine bird reproduction, influencing the population status, and the decrease in fishery production significantly decreased the reproductive and health status of some species (Krul 2010; Mangini 2010). Extrapolation about marine bird future population size or health status in the absence of shrimp trawlers on the Paraná coast could not be assessed with the available data at present. No historical data on regional marine bird population size and fluctuation are available (Moraes and Krul 1999). Nevertheless, Krul (1999, 2004) and Carniel (2008) demonstrated that the reproductive activity of magnificent frigate birds, brown boobies, and kelp gulls is dependent on local bycatch availability. Local fishing assessment indicates that trawling shrimp fishing started during the early 1980s and became a major fishing activity on the Paraná coast (Andriguetto Filho et al. 2006, 2009), indicating three decades of unusual availability of fish resources to marine birds in the region. Bycatch availability has been demonstrated in other regions as a significant factor producing changes in feeding behavior and reproductive success, and increasing the distribution and relative abundance of marine birds (Furness 1982; Pierotti and Bellrose 1986; Oro 1996; Yorio and Caille 2004). The sudden local reduction or total unavailability of bycatch could promote artificial and intense oscillations in population size and undesirable mortality events. From a health point of view, local trawling shrimp fishing is an activity that needs urgent regulation, combining the demands of fishery communities and the protection of marine bird population health. A gradual reduction in bycatch production by applying practices that reduce the shrimp trawling impact (i.e., use of bycatch reduction devices) and better planning of a non-fishing season seems to be the ideal goal for the Paraná coast. The significant correlations among fishing village characteristics, bycatch, sea bird health, and biodiversity indicate that bird communities could be used as sentinels for ecological health.

Marine turtle fibropapillomatosis (FP) is the most important disease threatening the survival of sea turtles in Brazil. The first report of FP in Brazil was in Espírito Santo in 1986 (Baptistotte 2001). Data indicate there has been an increase from 3.2% to 10.4% of affected turtles from 1997 to 2000 (Baptistotte 2007). Locally, the disease occurs with low prevalence in loggerhead (*Caretta caretta*), olive ridley (*Lepidochelys olivacea*), and hawksbill (*Eretmochelys imbricate*) turtles but has a high incidence in green turtles (*Chelonia mydas*). Analyzing marine turtle databases (Projeto-TAMAR), Baptistotte (2007) characterized the disease status of turtles in Brazil. Of 10,170 animals sampled from 2000 to 2007, 15.4% were affected, of which 12.7% were green turtles, with a higher incidence in Ceará (36.9%), Rio Grande do Norte (31.4%), and Espírito Santo (27.4%). Turtle FP primarily affects juveniles (straight carapace length: 30 to 112 cm). In Brazil, the disease apparently occurs only along the coast, with no reports on oceanic islands. Baptistotte (2007) also determined that there was an average of 21 tumors per animal (range 1 to 121) in a cohort of green turtles in Espírito Santo state. A total of 72.5% of the tumors occurred in the anterior region, though oral tumors were not detected. Of 39 necropsied turtles, five had small internal tumors, and 8.1% of animals had tumors classified as score 3 (Work and Balazs 1999). Using the same cohort of turtles, Santos (2005) found no blood biochemical differences between healthy animals and those mildly or moderately afflicted. Torezani (2004) also analyzed the same cohort and concluded that turtles with FP had a lower growth rate than apparently healthy turtles. However, no clear link was established between the regional distribution of FP and local environmental changes that may be involved in the pathogenesis of the disease. Although the FP prevalence is low to moderate compared to that in other countries, the most interesting observation is that the disease occurs exclusively on continental coastal areas, which are the most affected by humans. The continuous monitoring of sea turtle health in Brazil will allow the investigation of the interactions between ecological, biological, and human influences on FP epidemiology, and promotion of conservation strategies. FP has a worldwide, circumtropical distribution and has been observed in all major oceans. Although reported since the late 1930s in Florida, it was not until the late 1980s that it reached epizootic proportions in several sea turtle populations. Long-term studies have demonstrated that pelagic turtles recruiting to near-shore environments are free of the disease. After exposure to these benthic ecosystems, FP manifests itself with primary tumors in epithelial tissues. One or more herpesviruses, a papillomavirus, and a retrovirus have been found associated with tumors using electron microscopy and molecular techniques; however, the primary etiological agent remains to be isolated. Field observations support that the prevalence of the disease is associated with heavily polluted coastal areas, areas of high human density, agricultural runoff, and/or biotoxin-producing algae. Marine turtles can serve as excellent sentinels of ecosystem health in these benthic environments. FP can possibly be used as an indicator, but correlations with physical and chemical characteristics of water and other factors are needed. Further research in identifying the etiologic agent and its association with other environmental variables can provide sufficient parameters to measure the ecological health of coastal marine habitats, which serve not only as ecotourism spots but also as primary feeding areas for sea turtles (Aguirre and Lutz 2004).

## HABITAT FRAGMENTATION AND HEALTH CONSEQUENCES IN THE PONTAL DO PARANAPANEMA

The highly fragmented Pontal do Paranapanema illustrates how animal health and ecology with land use characteristics can define patterns of disease and vitality in wildlife populations. During the past five decades, the region has faced several periods of instability regarding land rights and changes in landscape use, including severe deforestation in the early 1980s. At that time the primary land use was extensive sugar cane plantations and cattle ranching (Costa and Futemma 2006). However, important changes in land use dynamics occurred in the 1990s as a result of the settlement of the government-sponsored agrarian reform, which increased the presence of formerly landless families in the region, turning the landscape into a mosaic of farms and settlements surrounding the remaining forest fragments (Ditt 2002).

Morro do Diabo State Park (MDSP) (36,500 ha) and the remaining Atlantic rainforest fragments in surroundings areas (21,000 ha) represent the last refuge for local wildlife, including endangered species, such as black lion tamarins (*Leontopithecus chrysopygus*), jaguars (*Panthera onca*), and sentinel species, such as white-lipped peccaries (*Tayassu pecari*) and collared peccaries (*Pecari tajacu*). In a fragmented landscape the species profile are strongly area-dependent—for example, a larger forest fragment with higher initial biodiversity will have a lower rate of subsequent extinctions (Terborgh et al. 2001). However, the presence of white-lipped and collared peccaries in Pontal do Paranapanema forest fragments is significantly associated with the land use, landscape permeability, and domestic animal management around the forest patches (Nava and Cullen 2003; Nava 2008). Interestingly, the size of the forest fragment is not significantly related to the prevalence of infectious diseases in peccary populations. The major factor affecting peccary health is associated with the history of land use around forest patches (i.e., years of contact between wild and domestic animals, domestic animal density and vaccination) (Nava 2008). In the larger fragments, including MDSP, white-lipped peccary herds have significantly higher antibody titers to *Brucella abortus* (27% in MDSP and 7% in other fragments) and *Leptospira* spp (86% in MDSP and 28% in other fragments) than small fragments. These results are likely correlated with the history of settlements and recent deforestation and land use around smaller forest fragments, which are potentially the major factors delimiting disease prevalence in peccaries (Nava 2008). The interaction between pathogens and peccary distribution in Pontal do Paranapanema indicates the need for a larger transdisciplinary effort to understand and manage the health of fragmented landscapes.

## NON-HUMAN PRIMATES IN URBAN AREAS

The expansion of urban centers in natural areas is one of the biggest threats to biodiversity and health in the Atlantic rainforest. The majority of megacities have many green areas enclosed in an urban matrix. These islands of forest provide recreational areas for the population and ecosystem services such as protection of watershed and hillsides. All major cities in Brazil present a similar situation at different scales. The most representative example is Rio de Janeiro, where the mosaic of urban matrix and forest patches is notable. The forest wildlife diversity is composed of native and exotic species, and interaction among humans, domestic animals, and wildlife is a constant situation. Non-

human primates are important species to understand human health threats in urban green areas. One study focused on pathogens that were transmitted between humans and common marmosets (*Callithrix jacchus*), an introduced and abundant species. Verona (2008) performed clinical examinations, hematological profiles, blood biochemistry, fecal analysis, and nasopharyngeal and rectal bacterial screening in marmosets to investigate the transmission potential of pathogenic microorganisms. Three different populations were compared: marmosets in forested areas in the core of Tijuca National Park (A1), marmosets in forest/urban edges (A2), and marmosets from private owners that were kept as illegal pets (A3).

A total of 85 animals were sampled, with 34 from the A1 population, 31 from A2, and 20 from A3. All the wild marmosets from group A3 were sampled just after confiscation or donation to the Wildlife Screening Center of the IBAMA. External clinical evaluations revealed dental disease in 28.2% of the individuals, 15.3% of those from groups A1 and A2 combined, and 12.9% from group A3. Dental disease in confiscated marmosets was found only in juvenile individuals, though all animals had gingivitis. Dental disease occurred exclusively in adults in the wild, and the most frequent finding was fractures of the upper incisors. The differences described between wild and captive individuals are likely related to diet, as in captivity this is generally composed of high levels of sugars. Also, as an exotic species in the region, fractured incisors in wild individuals could be a consequence of the behavior of biting holes in the trees bark to consume sap, and may be related to the local tree characteristics (Verona 2008).

Hematological results indicated no statistical difference among the three groups, but blood films from animals in the A1 group had a high prevalence of *Trypanosoma minasense* (47%) and in two cases they were associated with *Trypanosoma devei*. In this study, *T. devei* was never found without being associated with *T. minasense*. The association of these two species of *Trypanosoma* has been described previously (Hoare 1972) as nonpathogenic in primates (Dunn et al. 1963; Hoare 1972; Stevens 1998). Also, microfilarid forms were found in 29% of the A1 group. These hemoparasites were encountered at only two sites in the core of the forested area of the A1 group. Using multivariate statistics and analyzing marmosets infected with *Trypanosoma* and other microfilarids as two separated groups, it was possible to observe significant variations in their hematological parameters. This result indicates the need to include a broader parameter in the multivariate analysis to obtain a more accurate understanding of the complexity of the association between animal metabolism and environmental conditions (Verona 2008).

Verona (2008) also performed coproparasitological analysis of 60 specimens. There was a significant difference in parasite species between groups A1 and A2 (IBAMA protocol includes treatment of marmosets with antihelmintics for the A3 group, so the results were negative). Parasite eggs found in A2 animals from the forest/urban edge areas were *Prosthenorchis elegans*, Ascaridoidea, Oxyuridae, and Physalopteroidea. All taxonomic orders found in forest/urban border areas were also found in core forested areas (group A1) as well as parasite eggs from Ancylostomatidae, Trichuridae, Gnathostomatidae, Spiruridae (*Spirura* sp.), Dylepididae (*Dypilidium* sp.), Ascaridida, Cyclophyllidea, Pseudophylidea, and trematodes. The diversity of the parasite species found in forested areas was higher than that in urban areas, and the number of parasite species found in each animal varied from one to five in urban areas and from one to 12 in forested areas. Besides the diversity of parasite species found in each animal, all parasite-infected marmosets appeared clinically healthy and had no eosinophilia. *P. elegans* is well known as parasitizing common marmosets (Bush et al. 2002; Bowman et al. 2003; Müller 2007), and although they have never been reported as causing mortality in free-ranging animals, the cause of death for the only individual that died during this research was determined on necropsy to be an intestinal embolus caused by *P. elegans*. This animal originated from the forest/urban edge, where marmosets are frequently fed by people. Oropharyngeal and rectal bacteriological screening demonstrated site-associated significant differences between individuals, but pathogenic bacteria were found only in animals living as pets. *Salmonella enteriditis* (Newlands) and *Campylobacter jejuni* biotypes I and II were three important pathogenic species found in these animals that can infect both humans and primates (Costa and Hofer 1972; Filgueiras and Hofer 1989; Verona 2008).

This study collected and analyzed information combining ecological and epidemiological data. The results showed that the population of common marmosets in the city of Rio de Janeiro carries pathogenic microorganisms for both non-human primates and humans (Verona 2008). The presence of *T. minasense*, *P. elegans*, *Campylobacter*, and *Salmonella* in common marmosets represents the potential zoonotic threat of this exotic species to humans and native wildlife.

## CANTAREIRA ECOLOGICAL HEALTH PROGRAM

Cantareira State Park (CSP) is considered the largest urban forest in the world, with a total area of 7,916 ha. It is responsible for protecting the water quality for a significant part of the city of São Paulo and also provides recreational services to local people and tourists. Among others species, CSP mammal fauna is represented by pumas (*Puma concolor*), ocelots (*Leopardus pardalis*), crab-eating foxes (*Cerdocyon thous*), coatis (*Nasua nasua*), squirrels (*Sciurus aestuans*), howler monkeys (*Alouatta clamitans*), titi monkeys (*Callicebus* sp.), and numerous species of bats and rodents. CSP faces a variety of anthropogenic pressures, such as urbanization (i.e., slums and luxury buildings), mineral extraction, and development of small farms for recreational purposes. In addition to habitat fragmentation, threats to local fauna include illegal hunting, road kills, and predation by feral dogs and cats. Illegal waste dumping and soil contamination inside the park can increase the proliferation of urban rodents and the circulation of pathogens that may affect humans and wild and domestic animals. Local veterinarians consistently report numerous cases of canine leptospirosis in the region, notably in summer, the rainiest period of the year. The impacts of leptospirosis and waste disposal within the limits of forested areas on wildlife are unknown.

Frequently feral domestic dogs that enter the park are in poor condition; they are rarely dewormed and can transmit infectious agents to wildlife and contaminate the environment. Examples of this include transmission of sarcoptic mange (*Sarcoptes scabiei*) (C. Filoni, personal communication, 2010) and fungal agents such as dermatophytes (*Microsporum* spp. and *Trichophyton* spp.) to wildlife and people. *Malassezia* sp. was isolated from healthy ear canal and skin of coatis and opossums (*Didelphis* spp.) from CSP (S.D.A. Coutinho, personal communication, 2010). In CSP, the study of *E. coli* in wild animals can serve as a measure of environmental contamination. So far, *E. coli* has been isolated from rectal swabs from the same coatis and opossum specimens, although it has not been typified (V.M. Carvalho, personal communication, 2010). Wildlife from CSP has also been implicated as source of infection to other zoonotic diseases such as leishmaniasis (Proença and Muller 1979).

To address local health challenges, TRÍADE, Universidade Paulista (UNIP), and Universidade de São Paulo (USP) established a monitoring program for ecological health in the CSP region. Infectious disease agents, including rabies, canine parvovirus (CPV), canine distemper (CDV), virulent *E. coli*, *Salmonella* sp., *Leishmania* sp., *Toxoplasma gondii*, ecto- and endoparasites, and fungi such as *Microsporum* sp., *Trichophyton* sp., and *Malassezia* sp., are under investigation in wild and domestic animals. For domestic animals, health management strategies include vaccination, antihelmintic treatment, and campaigns directed to domestic carnivore population birth control. To improve health conditions for people and domestic animals, public education through lectures and audiovisual material has been initiated. The project has allowed continuous health monitoring in CSP to be used for epidemiological investigation on infectious diseases, as well as to promote social change and contribute to public policy in local communities (C. Filoni, personal communication, 2010).

## WILD AND DOMESTIC CARNIVORES IN PANTANAL REGION

Researchers continue to emphasize the growing importance of pathogens to wild carnivore conservation (Williams et al. 1988; Randall et al. 2004). In Brazil, pathogens in wild carnivore populations have been intensively studied. Jorge (2008) analyzed the exposure of free-ranging wild carnivores and domestic animals to important pathogens for conservation of wild carnivores and for public health in a private reserve (RPPN SESC Pantanal) and surrounding areas in the Northern Pantanal. From 2002 to 2006, 76 wild carnivores were captured and blood samples were collected. Also, 103 domestic dogs and 27 horses from the same area were sampled. Serological tests were performed on wild carnivores and domestic dogs for CDV, CPV, rabies virus, and *Leptospira* spp., and

polymerase chain reaction was used for *Leishmania* spp. diagnosis. Serological tests for *Leptospira* spp. were also performed in horse samples. Among domestic dogs, 82.3% were positive for CDV, 96.1% for CPV, 26.5% for rabies virus, 17.5% for *Leptospira* spp., and 28.6% for *Leishmania* spp. Among horses, 74.1% of were positive for *Leptospira* spp. Positive samples for *Leishmania* spp. were identified as belonging to the sub-genus *Viannia*. Two of these were identified as *Leishmania (V.) braziliensis*. The results of 76 free-ranging wild carnivores captured are shown in Table 32.1.

The results indicate that wild carnivores and domestic dogs have been exposed to all five pathogens tested. This demonstrates that there is interaction between domestic dog pathogens and wild carnivore hosts, indicating that wild carnivores may be threatened by common pathogens of domestic dogs. It also demonstrates that zoonotic agents circulate among the wild carnivore population in the region (Jorge 2008). However, it has not been determined that these infectious diseases cause morbidity or mortality of wild carnivores, nor has their status as zoonotic reservoirs been demonstrated. Also, the high seroprevalence of *Leptospira* spp. in horses could indicate a complex disease cycle involving ungulates, an important group of domestic and wild mammals in the Pantanal (Jorge et al. 2011). It is clear that there is spillover of pathogens from domestic dogs to wild carnivores, and the need to prevent the pathogens' transmission to wild carnivores is urgent. Examples of such measures include domestic dog population control and vaccination in the RPPN SESC Pantanal region (Jorge 2008). In these regions, a better surveillance system for rabies, leishmaniasis, and leptospirosis, coupled with a health education program, is a fundamental step in human health and wild carnivore conservation.

## CONCLUSIONS

Since 2001, conservation medicine has been significantly developed in Brazil, making advances in field research and in institutional and educational initiatives. Many of these programs were achieved thanks to the constant work of the professionals who attended the initial conservation medicine courses at IPÊ, in partnership with EcoHealth Alliance. Those professionals, mostly local veterinarians, were able to

**Table 32.1 Serosurvey of Free-Ranging Wild Carnivores Captured in the Pantanal (Brazil) for Canine Distemper Virus, Canine Parvovirus, Rabies Virus, *Leptospira* spp., and Detection of *Leishmania* spp. by Polymerase Chain Reaction**

| Species | CDV | CPV | Rabies virus | *Leptospira* spp. | *Leishmania* spp. |
|---|---|---|---|---|---|
| *Cerdocyon thous* (crab-eating fox) | 12/43 | 42/43 | 1/43 | 17/43 | 3/9 |
| *Chrysocyon brachyurus* (maned wolf) | 3/8 | 7/8 | 0/8 | 3/8 | 1/2 |
| *Speothos venaticus* (bush dog) | * | 1/1 | 1/1 | 1/1 | * |
| *Procyon cancrivorous* (crab-eating raccoon) | 2/13 | 8/13 | 1/13 | 6/12 | 1/4 |
| *Puma concolor* (puma) | 1/7 | 7/7 | 1/7 | 2/7 | * |
| *Leopardus pardalis* (ocelot) | 3/4 | 3/4 | 0/4 | 3/4 | 1/1 |
| **Total** | 21/75 (28%) | 70/76 (92.11%) | 4/76 (5.62%) | 32/75 (40.32%) | 7/16 (43.75%) |

CDV, canine distemper virus; CPV, canine parvovirus.
* Tests not performed

integrate conservation medicine into their work; however, other specialists (i.e., ecologists, human health specialists) have not been widely integrated into the field of conservation medicine. Therefore, 10 years later, veterinarians still are the primary professionals developing local conservation medicine projects in Brazil. The synergistic strength from diverse professionals is needed to produce varied perspectives and solutions to environmental problems and their consequent effects on ecosystem health. True transdisciplinary integration will be achieved when local researchers work together to investigate and evaluate the scope and the diversity of regional environmental health problems and their relationship with Brazilian social conditions.

The cases presented here clearly illustrate that environmental changes lead to ecological and epidemiological disturbances. The primate involvement of rabies expansion in the northeast region presents an unexpected spillover situation and illustrates the need to investigate rabies epidemic cycles under a unified mission that will benefit human, domestic animal,

and wildlife health. The marine coastal example of birds and sea turtles as sentinels of ecosystem health in Brazil can be applied at different scales from local health issues (i.e., parasite spillover infections in forest birds) to global health problems (i.e., sea turtle FP). Also, the local evaluation of fishing activities and communities demonstrates that small-scale activities can cause significant changes in the health status of local bird populations. The health status of peccaries in Pontal do Paranapanema indicates that land use and occupation history are important factors of disease prevalence in fragmented habitats. The evaluation of health issues in Cantareira and Rio de Janeiro urban forests demonstrates the need to understand the role of native and exotic wildlife species and the potential threats related to public health, even in big urban centers. The evidence of spillover events from domestic animals to Pantanal carnivores demonstrates the urgent need to change domestic animal surveillance and official disease control protocols, and also reinforces the need to understand disease cycles in natural habitats to protect threatened Brazilian wildlife species. Finally, these case studies demonstrate that Brazilian researchers have integrated conservation medicine concepts into regional scientific research projects to promote regional professional training and apply field research to promote ecological health.

## ACKNOWLEDGMENTS

The authors thank all the individuals and institutions that helped local researchers to produce knowledge on conservation medicine. Special thanks to Dr. Alonso Aguirre for the incentive to produce this material and his commitment to conservation medicine in Brazil. Special thanks to these local and international organizations: Instituto de Pesquisas Ecológicas (IPÊ), Instituto Brasileiro para Medicina da Conservação (TRÍADE), Instituto de Ensino, Pesquisa e Preservação Ambiental Marcos Daniel, EcoHealth Alliance, Wildlife Conservation Society (OWOH), Brazilian Funds Program, Fundação O Boticário de Proteção À Natureza, Escola Nacional de Saúde Pública (ENSP/FIOCRUZ), Universidade Federal do Paraná, Universidade Federal Rural de Pernambuco, Universidade de São Paulo, Universidade Paulista (UNIP), Fundação de Amparo à Pesquisa do Estado de São Paulo (FAPESP), Universidade Vila Velha, Conselho Nacional de Desenvolvimento Científico e Tecnológico (CNPq), and Coordenação de Aperfeiçoamento de Pessoal de Nível Superior (CAPES).

## REFERENCES

Aguirre, A.A., and A. Gomez. 2009. Essential veterinary education in conservation medicine and ecosystem health: a global perspective. Rev Sci Tech Off Int Epiz 28:597–603.

Aguirre, A.A., and P.L. Lutz. 2004. Marine turtles as sentinels of ecosystem health: is fibropapillomatosis an indicator? EcoHealth 1:275–283.

Andriguetto Filho, J.M., P.T. Chaves, C. Santos, and S. Liberati. 2006. A pesca marinha e estuarina do Brasil no início do século XXI: Recursos, tecnologias, aspectos socioeconômicos e institucionais. *In*: V.J. Isaac, A.S. Martins, M. Haimovici, and J M. Andriguetto-Filho, orgs. Projeto RECOS: Uso e apropriação dos recursos costeiros. Grupo temático: Modelo gerencial

da pesca, pp. 117–140. Editora da Universidade Federal do Pará (UFPA), Belém, Brasil.

Andriguetto Filho, J.M., R. Krul, and S. Feitosa. 2009. Analysis of natural and social dynamics of fishery production systems in Paraná, Brazil: implications for management and sustainability. J Appl Ichthyol 25:277–286.

Baptistotte, C. 2007. Caracterização espacial e temporal da fibropapilomatose em tartarugas marinhas da costa brasileira. Ph.D. Thesis. Universidade de São Paulo, São Paulo.

Baptistotte, C., T.J. Scalfoni, B.M.G. Gallo, A.S. Santos, J.C. Castilhos, E.H.S.M. Lima, C. Bellini, and P.C.R. Barata. 2001. Prevalence of sea turtles fibropapillomatosis in Brazil. Proc 21th Ann Symp Sea Turtle Biology Conserv, NOAA-NMFS, Philadelphia.

Bernardi, F., S.A. Nadin-Davis, A.I. Wandeler, J. Armastrong, A.A.B. Gomes, F.S. Lima, F.R.B. Nogueira, and F.H. Ito. 2005. Antigenic and genetic characterization of rabies viruses isolated from domestic and wild animals of Brazil identify the hoary fox as a rabies reservoir. J Gen Virol 86:3153–3162.

Bowman, D.D., R.C. Lynn, and M.L. Eberhard. 2003. Georgi's parasitology for veterinarians. W.B. Saunders Company, Philadelphia.

Bush, A.O.F., J.C. Bush, G.W. Esch, and J.R. Seed. 2002. Acanthocephala: the thorny-headed worms. *In* A.O.F. Bush, J.C. Bush, G.W. Esch, and J.R. Seed, eds. Parasitism: the diversity and ecology of animal parasites, pp. 197–214. Cambridge University Press, England.

Carniel, V.L. 2008. Interação de aves costeiras com descartes oriundos da pesca artesanal no litoral centro-sul paranaense. Master's degree dissertation. Universidade Federal do Paraná, Curitiba, Brasil.

Carnieli Jr., P., P.E. Brandão, M.L. Carrieri, J.G. Castilho, C.I. Macedo, L.M. Machado, N. Rangel, R.C. Carvalho, V.A. Carvalho, L. Montebello, M. Wada, and I. Kotait. 2006. Molecular epidemiology of rabies virus strains isolated from wild canids in Northeastern Brazil. Virus Res 120:113–120.

Carnieli Jr., P., W.O. Fahl, J.G. Castilho, R.N. Oliveira, C.I. Macedo, E. Durymanova, R.S.P. Jorge, R.G. Morato, R.O. Spindola, L.M. Machado, J.E.U. Sá, M.L. Carrieri, and I. Kotait. 2008. Characterization of rabies virus isolated from canids and identification of the main wild canid host in northeastern Brazil. Virus Res 131:33–46.

Cattani, A.P. 2010. Avaliação de dispositivos de redução de captura incidental na pesca de arrasto do município de Pontal do Paraná - PR. Master's degree dissertation. Universidade Federal do Paraná, Curitiba, Brasil.

Costa, G.A., and E. Hofer. 1972. Isolamento e identificação de enterobactérias. Monographic. Curso do Instituto Oswaldo Cruz, FIOCRUZ, Rio de Janeiro.

Costa, R.C., and C.R.T. Futemma. 2006. Racionalidade com compromisso: os assentados do Ribeirão Bonito (Teodoro Sampaio – SP) e o projeto de conservação ambiental. Ambiente Sociedade 9:55–57.

Cross, M.L., B.M. Buddle, and F.E. Aldwell. 2007. The potential of oral vaccines for disease control in wildlife species. Vet Rec 174:472–480.

Ditt, E.H. 2002. Avaliação da nova paisagem do Pontal do Paranapanema. In E.H. Ditt, ed. Fragmentos florestais do Pontal do Paranapanema. Selo Universidade, pp.38–40, São Paulo, Brasil.

Dunn, F.L., F.L. Lambrecht, and R. Du Plessis. 1963. Trypanosomes of South American monkeys and marmosets. Am J Trop Med Hyg 12:524–534.

Favoretto, S.R., C.C. de Mattos, N.B. de Morais, M.L. Carrieri, B.N. Rolim, L.M. Silva, C.E. Rupprecht, E.L. Durigon, and C.A. de Mattos. 2006. Rabies virus maintained by dogs in humans and terrestrial wildlife, Ceará State, Brazil. Emerg Infect Dis 12:1978–1981.

Filgueiras, A.L.L., and E. Hofer. 1989. Ocorrência de Campylobacter termofílico em diferentes pontos de uma estação de tratamento de esgotos na cidade do Rio de Janeiro - RJ. Rev de Microbiol 20:303–308.

Furness, R.W. 1982. Competition between fisheries and seabird communities. Adv Mar Biol 20:225–307.

Hoare, C.A. 1972. The trypanosomes of mammals—a zoological monograph. Blackwell Scientific Publications, Oxford.

IBAMA. 2007. Estatística da pesca, Brasil. Ministério do Meio Ambiente, Coordenação Geral de Autorização de Uso e Gestão da Fauna e Recursos Pesqueiros – CGFAP. Brasília.

Jorge, R.S.P. 2008. Caracterização do estado sanitário dos carnívoros selvagens da RPPN SESC Pantanal e de animais domésticos da região. Ph.D. thesis. Universidade de São Paulo, São Paulo.

Jorge R.S.P., F. Ferreira F, J.S. Ferreira Neto, S.A. Vasconcellos, E.S. Lima, Z.M. de Morais, and G.O. de Souza. 2011. Exposure of free-ranging wild carnivores, horses and domestic dogs to Leptospira spp. in the northern Pantanal, Brazil. Mem Inst Oswaldo Cruz 106:441–444.

Krul, R. 1999. Interação de aves marinhas com a pesca do camarão no litoral paranaense. Master's degree dissertation. Universidade Federal do Paraná, Curitiba.

Krul, R. 2004. Aves marinhas costeiras do Paraná. In: J.O. Branco, ed. Aves marinhas e insulares brasileiras: bioecologia e conservação, pp. 37–56. Univali, Joinville.

Krul, R. 2010. Relações entre a pesca, a biodiversidade, a saúde e a paisagem, em duas comunidades de pescadores artesanais no litoral do Paraná. Ph.D. thesis. Universidade Federal do Paraná, Curitiba.

Mangini, P.R. 2010. A saúde e suas relações com a biodiversidade, a pesca e a paisagem, em duas comunidades de pescadores artesanais no litoral do Paraná. Ph.D. thesis. Universidade Federal do Paraná, Curitiba.

Mangini, P.R., and J.C.R. Silva. 2007. Medicina da conservação: aspectos gerais. In: Z.S. Cubas, J.C.R. Silva, and J.L. Catão-Dias, eds. Tratado de animais selvagens: medicina veterinária, pp. 1258–1268. Roca, São Paulo, Brasil.

Moraes, V.S., and R. Krul. 1995. Aves associadas a ecossistemas de influência marítima no litoral do Paraná. Arq Biol Tecnol 38:121–134.

Moraes, V.S., and R. Krul. 1999. Sugestão de um perfil descritivo da estrutura de comunidades de aves costeiras do Estado do Paraná, Brasil. Est Biol 44:55–72.

Müller, B. 2007. Determinants of the diversity of intestinal parasite communities in sympatric New World primates (Saguinus mystax, Saguinus fuscicollis, Callicebus cupreus). Ph.D. dissertation, Hannover, Germany.

Nava, A. 2008. Sentinel species for Atlantic rainforest: epidemiological consequences of forest fragmentation in Pontal do Paranapanema - São Paulo. Ph.D. thesis. Universidade de São Paulo, Brasil.

Nava, A., and L. Cullen. 2003. Peccaries as sentinel species: conservation, health and training in Atlantic forest fragments, Brazil. Suiform Soundings PPHSG Newslett 3:15–16.

NUEND. 2004. Coordenadoria de Apoio ao Desenvolvimento da Atenção à Saúde — CODAS (1998–2003). Núcleo de Controle das Endemias Transmissíveis por Vetores — NUEND. Secretaria da Saúde do Estado do Ceará, Fortaleza.

Oro, D. 1996. Effects of trawler discard availability on egg laying and breeding success in the lesser black-backed gull Larus fuscus in the western Mediterranean. Mar Ecol Prog Ser 132: 43–46.

Pastoret, P.P., A. Kappeler, and M. Aubert. 2004. European rabies control and its history. In: A.A. King. Historical perspectives of rabies in Europe and the Mediterranean Basin. Rev Sci Org Int Epizoot 337–350.

Peckham, S.H., D. Maldonado Diaz, A. Walli, G. Ruiz, L. B. Crowder, and W. J. Nichols. 2007. Small-scale fisheries bycatch jeopardizes endangered Pacific loggerhead turtles. PLoS ONE 2(10): e1041. doi:10.1371/journal.pone.0001041

Pierotti, R., and C. Bellrose. 1986. Proximate and ultimate causation of egg size and the "Third Chick Disadvantage" in the Western gull. Auk 103:401–407.

Proença, N.G., and H. Muller. 1979. Nota sobre a ocorrência de Leishmaniose tegumentar americana na Serra da

Cantareira, São Paulo - SP - Brasil. Revista de Saúde Pública 13:56–59.

PROZEE. 2006. Monitoramento da atividade pesqueira no litoral do Brasil. Fundação de Amparo a Pesquisa de Recursos Vivos na Zona Economicamente Exclusiva – Fundação PROZEE, Brasília.

Randall, D.A., S.D. Williams, I.V. Kuzmin, C.E. Rupprecht, L.A. Tallents, Z. Tefera, K. Argaw, F. Shiferaw, D.L. Knobel, C. Sillero-Zubiri, and M. K. Laurenson. 2004. Rabies in endangered Ethiopian wolves. Emerg Infect Dis 10:2214–2217.

Santos, M.R.D., 2005. Parâmetros bioquímicos, hematócrito e condição corporal no monitoramento da saúde de tartarugas marinhas Chelonia mydas (Linnaeus, 1758) juvenis selvagens. M.S. thesis. Universidade Federal do Espírito Santo, Vitória.

Schneider, M.C., A. Belotto, M. P. Ade, L.F. Leanes, E. Correa, H. Tamayo, G. Medina, and M.J. Rodrigues. 2005. Epidemiologic situation of human rabies in Latin America in 2004. Epidemiol Bull 26:2–4.

Silva, J.C.R. 2006. Oral presentation. In: Eighth Meeting of the Conference of the Parties to the Convention on Biological Diversity – COP 8, Curitiba.

Silva, J.M.C., M. Tabarelli, M.T. Fonseca, and L.V. Lins. 2004. Biodiverisdade da Caatinga: áreas e ações prioritárias para conservação. MMA – Ministério do Meio Ambiente. Brasília.

Stevens, J., H. Noyes, W. Gibson. 1998. The evolution of trypanosomes infecting humans and primates. Mem Inst Oswaldo Cruz 93:669–676.

Tabor, G.M. 2002. Defining conservation medicine. In: A.A. Aguirre, R.S. Ostfeld, G.M. Tabor, C. House, and M.C. Pearl eds. Conservation medicine: ecological medicine in practice, pp. 8–16. Oxford University Press, New York.

TAMAR. 2010. Tamar-ICMBio. Website:http//www.tamar.org.br/, Accessed April 10.

Terborgh, J., L. Lopez, P. Nuñez, M. Rao, G. Shahabuddinm, G. Orihuela, M. Riveros, R. Ascanio, G.H. Adler, T.D. Lambert, and L. Balbas. 2001. Ecological meltdown in predator-free forest fragments. Science 294:1923–1926.

Torezani, E. 2004. Abundância, tamanho e condição corporal em Chelonia mydas (Linnaeus, 1758) na área do efluente da CST (Companhia Siderúrgica de Tubarão), Espírito Santo – Brasil. M.S. dissertation. Universidade Federal do Espirito Santo, Vitória.

Verona, C.E.S. 2008. Parasitos em sagui-de-tufo-branco (Callithrix jacchus) no Rio de Janeiro. Ph.D. thesis, Fundação Oswaldo Cruz, Rio de Janeiro.

Williams, E.S., E.T. Thorne, M.J.G. Appel, and D. W. Brlitsky. 1988. Canine distemper in black-footed ferrets (*Mustela nigripes*) from Wyoming. J Wildl Dis 24:385–398.

Work, T.M., and G.H. Balazs. 1999. Relating tumor score to hematology in green turtles with fibropapillomatosis in Hawaii. J Wildl Dis 35:804–807.

Yorio, P., and G. Caille. 2004. Fish waste as an alternative resource for gulls along the Patagonian coast: availability, use and potential consequences. Mar Poll Bull 48:778–783.

# 33

# LINKING CONSERVATION OF BIODIVERSITY AND CULTURE WITH SUSTAINABLE HEALTH AND WELLNESS

## The Itzamma Model and Global Implications for Healing Across Cultures

Todd J. Pesek, Victor Cal, Kevin Knight, and John Arnason

Our environment is rapidly changing. There are a multitude of both complex anthropogenic and natural factors influencing this phenomenon. We need to creatively address a multitude of global issues in our health and well-being and in the sustainability of both people and the planet. We can learn from traditional practices and apply this wisdom to modern problems in sustaining health (Pesek et al. 2006a, b; Bell et al. 2009).

In this chapter we put forth a model project envisaged and implemented by a group of Q'eqchi' Maya traditional healers and external partners in the rainforest-covered Maya Mountains of southern Belize, Central America. The project demonstrates the utility of using a culturally appropriate development model for indigenous peoples that is rooted in their healing traditions but adapted to their changing environments.

## LOSS OF CULTURAL AND BIOLOGICAL DIVERSITY

Since the dawn of humanity, indigenous peoples the world over have lived closely with nature. These generations of shared learning in close connection to the natural world have nurtured stewardship worldviews embedded in cosmocentric perceptions of their environment. A strong utilitarian grasp of nature's resources has evolved and survives today. This traditional knowledge is recorded and propagated from generation to generation mainly via oral traditions. Traditional botanical and healing knowledge is intertwined in this knowledge network and holds strong possibilities in support of global health and wellness (Pesek et al. 2006b). This knowledge, however, is disappearing along with the loss of cultural and biological diversity. With this loss comes the loss of an opportunity to learn from millennia of shared learning in nature.

The loss of cultural diversity can be seen in the loss of languages (Maffi 2001; Buenz 2005; Sutherland 2003). There were 15,000 languages spoken 70 years ago; now roughly 6,000 are still spoken (Grimes 1996; Davis 1999; Maffi 2001; Sutherland 2003). Our cultural diversity is linked to and being lost concurrently with the loss of natural areas and the biodiversity therein. Forests are being destroyed at an unsustainable rate and biodiversity trends indicate accelerating loss. For example, the United Nations Food and Agricultural

Organization (FAO) demonstrates loss of 9 million to 12 million hectares (ha = 2.5 acres) of forests annually from 1990 to 2000. Total forest losses range from 5 million to more than 20 million ha annually (FAO 2000). Global biodiversity is usefully viewed via the Living Planet Index (LPI; WWF 2004), a marker for biodiversity that is based on hundreds of vertebrate species in global terrestrial, freshwater, and marine ecosystems. The LPI demonstrates that biodiversity is being lost at alarming rates as well. Between 1970 and 2000, the index dropped by 37% (WWF 2004). In fact, due to anthropogenic drivers, global species extinctions are 100 to 1,000 times that of the natural species extinction rate (Hanski 2005).

Jeffrey McNeely, chief scientist for the International Union for the Conservation of Nature, argues that indigenous cultures and the forests they call home are inextricably linked, and that to preserve them we must do so concurrently. It has been demonstrated that cultural and biological diversity persists to date globally in mountainous areas (Stepp et al. 2005), and that development in the context of good governance may not necessarily mean the loss of forest resources (Buenz 2005). Therefore, a viable course of action in the conservation of culture and biodiversity would focus efforts in a culturally appropriate fashion on mountainous areas of high cultural and biological diversity. Traditional botanical knowledge and traditional healing is one potentially strong way to facilitate these efforts.

With the implementation of traditional cultural healing knowledge on global levels, however, complex situations arise. These involve varied agendas; cross-cultural differences in communication, understanding, and expression; intellectual property rights; and benefit-sharing possibilities, to name a few (Nigh 2002; Berlin and Berlin 2003; Hayden 2003). Unfortunately, it has been the case that even the most well-intentioned drug-development or bioprospecting projects seldom benefit communities in substantive form (Nigh 2002). The body of literature detailing these complexities is vast. A number of studies consider these issues and develop suggestions for improvement in these regards (Alexiades 2004; Bannister and Barrett 2004). Perhaps efforts would be more productively spent on alternative paths altogether. It would seem that the drug-development or bioprospecting route is not working well with respect to community benefit and tethering of revenues to the forest, so more direct methodologies for community benefit and concurrent conservation of culture and biodiversity could be sought.

## A CULTURALLY APPROPRIATE CONSERVATION AND DEVELOPMENT MODEL FROM THE Q'EQCHI' MAYA OF BELIZE

One promising way to provide viable conservation strategies and community support is to reinforce traditional healing in local healthcare. This approach provides enhanced public health concurrently with the conservation of biodiversity and culture via support of medicinal plant cultivation practices. Specific activities include development of organizations for traditional healing, establishment of indigenous herbal gardens, and the integration of traditional healing into primary healthcare on national levels (Pesek et al. 2006a, 2007). These programs have strong community support because they are culturally relevant. Traditional healers support the health of the individual in the context of a healthful environment (Pesek et al. 2006b). Therefore, this methodology directly facilitates conservation of biodiversity and culture while improving community wellness in a traditional form. The model is based on traditional constructs in health, illness, and treatment strategies, and it is rooted in phytomedicine/herbal medicine. Therefore, it is a development model that is not focused on "developed world" priorities such as "drug development."

The Itzamma project of the Q'eqchi' Maya of southern Belize stands out as an innovative example of this approach (Fig. 33.1). The Q'eqchi' are one contemporary Maya cultural group who have inherited traditional knowledge from ancient Maya civilization. Respect for the environment and the importance of sustainability and "treading lightly" are prevalent among all the Maya groups. For example, the Itza Maya, Peten, Guatemala, have a systematic protocol for integrating agricultural crops with tropical forest stewardship. The Itza do so through synergistic species symbioses (Atran 1993). Their system facilitates cyclical forest use, which enables ongoing forest regeneration (Atran 1993). The Lacandon Maya, Chiapas, Mexico, are another group who sustainably farm the rainforest. They cultivate medicinal plants and subsistence staples in house gardens and *milpas*,

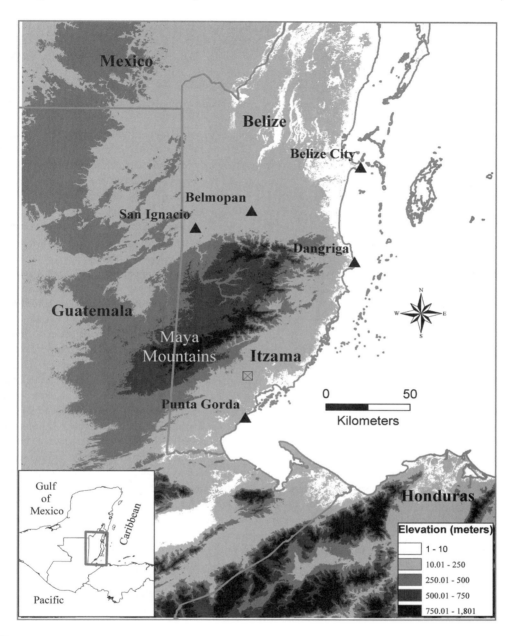

**Figure 33.1:**
Maya Mountains region: Maya Mountains Ethnobotany and Ecology Project Map: 1:1,544,427. GIS data sources: CGIAR Consortium for Spatial Information. 2008. http://srtm.csi.cgiar.org and BioGeo Berkley. 2008. [GIS and spatial data warehouses]: Pesek, L. 2010. Using: ArcView GIS [GIS Software]. Version 9.3. Redlands, CA: Environmental Systems Research Institute, Inc., 2008.

or cyclical farming plots that involve rotation of productive growth and rejuvenation phases so as to maintain the integrity of the forest. Their customs include sustainably gathering wild plants from the forest (Kashanipour and McGee 2004), a practice that supports synergistic use and continual regeneration of the forest. As a prime example of this stewardship worldview, the Yucatec Maya, Yucatan, Mexico, understand the land to be a being that needs feeding, nurturing, and other careful interactions (Barrera-Bassols and Toledo 2005). These practices are rooted in their ancestral vision of the cosmos. Indeed, it has been discovered recently that the ancient Maya of southern Belize used complex and sustainable agricultural strategies (Dunham et al. 2009). The Q'eqchi' Maya of southern Belize are proactive in the revival of their healing traditions and conservation of the Maya Mountains rainforest (Arnason et al. 2004; Pesek et al. 2006a, 2007).

## ENVIRONMENTAL AND PUBLIC HEALTH BACKDROP IN BELIZE

The remote and rugged, rainforest-covered Maya Mountains range of southern Belize (Fig. 33.1) is host to some of the most intact regions of tropical rainforest in all of Central America (Meerman and Clabaugh 2007). Rainfall is high, with roughly 431.8 cm annually (Hartshorn 1984). The complexity of the underlying geology of the mountains supports unique ecosystems (Dixon 1956; Bateson and Hall 1977; Pesek et al. 2006a), and the environmental conditions of the mountains support an abundant biodiversity (Parker et al. 1993; Iremonger and Sayre 1994; Iremonger et al. 1995; Balick et al. 2001; Pesek et al. 2006a). Surveys have found a high probability of species new to science being located in the niches of the Maya Mountains (Iremonger et al. 1995; Pesek et al. 2006a). The area is a biodiversity hotspot and as such is in critical need of conservation (Myers et al. 2000).

Most of the Q'eqchi' Maya use traditional healing for healthcare, even where modern healthcare is available (Arnason et al. 2004;Treyvaud-Amiguet et al. 2005 ; Pesek et al. 2007). This is in keeping with the fact that approximately 80% of the world's population relies on traditional healing for their primary healthcare (Farnsworth et al. 1985; WHO 2002). Indeed, it is sometimes difficult for the Maya and other rural

inhabitants to receive modern healthcare when it is sought. There is a strong need to improve access to affordable, high-quality healthcare, to ensure that the community is involved in its management, to enhance the capabilities of rural health centers, and to expand clinical care constructs, education, and awareness programming (Arnason et al. 2004). WHO has identified that an important way to address this gap is to incorporate traditional healers in the modern healthcare system as a "first line of defense" at the community level (WHO 2002). WHO has addressed this in its new programming at the district hospital in Punta Gorda, Belize.

Typical maladies in southern Belize can be categorized into several main types: tropical medicine and hygiene and maladies attributable to living conditions, infectious agents, inflammatory conditions, mental health, age-related degenerative disorders (accelerated due to lifestyle/living conditions), and emergency situations, including wild animal attacks and snake bites (Arnason et al. 2004; Pesek et al. 2007). There is unanimous concern over the increasing incidence and prevalence of cancer, HIV/AIDS, inflammatory conditions (in particular arthritis), and diabetes mellitus type 2 (Arnason et al. 2004; Pesek et al. 2007). These growing health concerns could well be addressed by preventive measures. These could include culturally relevant strategies such as the development of novel treatments from traditional medicines as well as general educational programs. The most common overall health concerns and presenting symptoms are mental health, headache, low back pain, arthritis, fever, fatigue, cough, loose stool, vomiting, skin irritations, infections, parasites, and general malaise. All of these are dealt with initially quite successfully by traditional healers, who then refer difficult cases to the district hospital in Punta Gorda (Arnason et al. 2004; Pesek et al. 2007).

Many of the plants that traditional healers and global markets rely on are taken from the wild, and this could lead to extirpation, ecosystem devastation, and biodiversity loss. Sustainable growing and harvesting programs must be adopted. Indeed, medicinal plant conservation is an important part of conservation programs in Belize and the rest of the world.

The Itzamma project, initiated by a group of elder Q'eqchi' Maya traditional healers, endeavors to support the conservation of medicinal plants for use in primary healthcare (Pesek et al. 2007). It does so both *in situ* and *ex situ* and in concurrence with goals of the Convention on Biological Diversity while promoting culturally relevant healthcare and improving public health. One main project goal is to include traditional healing in the national healthcare system by involving indigenous local communities in the inventory, conservation, sustainable growth, harvest, and use of medicinal plants.

## ITZAMMA OVERVIEW

*Itzamma* (a Q'eqchi' Maya word meaning "home of the Maya god of wisdom, Itzamna, and place of healing spiritually and with herbs") is a name given by the Belize Indigenous Training Institute (BITI) and the associated Q'eqchi' Maya Healers Association (QHA). It refers to their model community-based conservation program, which is aimed at preserving their rainforests and cultural traditions by promoting traditional healing in the national healthcare system (Arnason et al. 2004; Pesek et al. 2007) (Fig. 33.1). The Q'eqchi' Maya communities of the area lead as traditional a lifestyle as possible and maintain intact traditional healing knowledge to this day (Treyvaud-Amiguet et al. 2005). They have used nature to treat illness for millennia.

The Itzamma project addresses sustainable economic development, biodiversity and forest conservation, cultural integrity, and conservation of heritage, community, and global health and wellness via traditional healing systems and rainforest stewardship practices. The project has been implemented in southern Belize by BITI and the QHA with the support of the Government of Belize and the collaboration of external partners. These partners include the Inuit Circumpolar Council (ICC); University of Ottawa, Canada; Universidad Nacional, Costa Rica; Earth Healers and Naturaleza Foundation in the United States; and Cleveland State University, Ohio. The team is active with the physical development of the site, which includes a traditional healing center and indigenous gardens. They are also developing botanical inventories and databases in the Maya Mountains, beginning implementation of traditional sustainable plant propagation techniques, and developing culturally appropriate conservation strategies via traditional healing.

The healers developed the indigenous botanical gardens so they would have a renewable source of plant material close to their center for use in local primary healthcare. The gardens are also used as a place for healing and spiritual ceremonies, as well as a community resource and cultural center. They are used for celebration of the Maya calendar, the display of traditionally used plants, *in situ* and *ex situ* conservation of these plants, generational transmission of traditional knowledge, small-scale sustainable agricultural production for local use, and small sustainable enterprise development to generate revenue toward the conservation of biodiversity and culture.

## MILESTONES IN DEVELOPMENT OF ITZAMMA

Itzamma grew from an initiative that began in 1995 when Inuit of Canada met with four Indigenous groups of Belize (Q'eqchi' Council, Toledo Maya Cultural Council, National Garifuna Council, and the Xunantunich Organization). The purpose in meeting was to explore opportunities for collaboration and joint ventures. The Inuit had come to Belize after recognizing that they needed to consider development opportunities beyond their own circumpolar region. They considered that as newcomers to the world of international development, both their interests and those of their prospective southern partners could be best served on an indigenous-to-indigenous basis. The connected indigenous peoples could bring to the table similar development issues, strong connections to culture and traditions, similar development activities and practices, and a clear linkage to indigenous learning styles, language, and spirituality.

Over the following two years, the Inuit group, ICC, examined with its Belizean partners the needs of indigenous peoples in the southern districts of Belize. Together they determined that practical training was the principal concern of the four Belizean indigenous groups. The ICC, under a partnership agreement with its sister organization in Greenland, secured funding from the Danish International Development Agency (DANIDA) for the creation of BITI. BITI was legally incorporated in Belize in 1998 and had as its board of directors one member from each of the four Belizean indigenous groups, with an ICC representative, Kevin Knight, providing counsel, technical assistance, and project management support. While BITI was initially supported by DANIDA, the ICC continued to secure funding from the International Labor Organization, the United Nations Development Programme, the Government of Belize, ICC, Trekforce (a British non-governmental organization with a field base in Belize), and others.

The QHA, an organized group of traditional Q'eqchi' Maya healers, was established by BITI in 1999. QHA was initiated by bringing together 11 traditional healers and two understudies. The healers knew of each other but had never worked together.

In 1999, BITI/QHA secured a 50-acre site from the Government of Belize upon which they proposed developing a traditional healing garden and cultural center. They engaged Trekforce to assist with building a small house-type structure on the site. After a series of workshops, BITI/QHA had developed a set of objectives aimed at promoting respect for their traditional healing knowledge and practice, and in 1999 they invited John Arnason of the University of Ottawa to assist in these endeavors. This was a critical connection: it bridged ethnobotanical and ethnopharmacological science with traditional healing knowledge toward the furtherance of projects that augment the activities of the garden. One of the healers' objectives was respect for their knowledge via scientific findings supporting their traditional healing knowledge. The QHA and the University of Ottawa began ethnobotanical studies of the region but were set back by the regional devastation of Hurricane Iris in 2001. Iris reduced the newly built infrastructure to its foundation and felled much of the primary forest of their 50-acre site.

After recovery, the collaboration continued in 2003 with research and development, and the group invited Todd Pesek of Cleveland State University to assist in their expanding endeavors. This was another critical connection in bridging ethnobotany, ecology of medicinal plants, traditional healing, and health sciences toward the select integration of traditional healing in the national healthcare system, another objective set by the healers.

The collaboration then secured funding and support from the Naturaleza Foundation and the International Development Research Center (IDRC), which made possible "Visioning our Traditional Health Care: Workshop on Q'eqchi' Healers Center, Botanical Garden and Medicinal Plant Biodiversity Project in Southern Belize." This workshop brought together BITI, QHA, and external collaborators for the careful planning and articulation of the healers' goals for the gardens and cultural center (Arnason et al. 2004).

Following the workshop, the collaboration secured modest funding for renovation and development of the gardens and cultural center, subsequent to Hurricane Iris, from the World Bank. These funds and their deliverables, as well as the workshop report to IDRC (Arnason et al. 2004) articulating the future goals of the healers, led to a subsequent IDRC award for "Itzamma Project: Sustainable Indigenous Development Based on the Ethnobotanical Garden and Traditional Medicine Concept." This award made possible a formal policy recommendations document submitted to the Government of Belize regarding a practical proactive draft protocol for indigenous intellectual property protection in research. The framework of the protocol is the Itzamma collaborations model practice (Cal et al. 2009).

During 2005, the site was enlarged from 50 to 75 acres and the access road was improved with support from the Government of Belize and the healers continued to make improvements to the gardens, including paths, ornamental plantings, and updated facilities. In 2008, the QHA established membership criteria and a constitution for governance of their operations. They selected leadership roles and elected representative peers. They have reached out to their communities in offering healthcare and have partnered with local schools in transmission of traditional knowledge to youths.

Since 2000, the garden and cultural center has also been a site at which traditional spiritual ceremonies have been reintroduced as part of the process of Q'eqchi' Maya cultural revival in Belize. Maya spiritual ceremonies in Belize had not been practiced openly for as long as could be remembered. Given the culturally important interconnectedness of the practice of traditional healing to Q'eqchi' Maya spirituality, spiritual ceremonies were reintroduced at Itzamma.

## MILESTONES OF SCIENTIFIC AND TRADITIONAL COLLABORATION AT ITZAMMA

In 2004 and 2005, the collaboration, attuned to the healers' goals, developed the Rapid Ethnobotanical Survey (RES) methodology. This innovative ethnobotanical method was used to catalog the distribution and whereabouts of rare, disappearing, and previously unreported species, with traditional healers as researchers (Pesek et al. 2006a). In effect, the team sought to demonstrate that there are unreported medicinal plants deep within the inaccessible Maya Mountains region of southern Belize (Fig. 33.1). After two years of preparation, a team departed for an RES expedition in the spring of 2005 (Pesek et al. 2006a), and as a result 53 plant species with ethnobotanical applications were uncovered, collected, and analyzed (Pesek et al. 2006a). The species are used in the treatment of 26 distinct medical conditions recognized by the Q'eqchi' Maya (Pesek et al. 2006a). Four species were reported by the healers to be extremely powerful plants with multiple uses. The healers had not seen over half of the species in more than 20 years. Indeed, the research expedition team stopped frequently to examine and appreciate these plants, as it was unlikely that they would be seen again outside of these niches. Thirty-five accessions were identified to family and genus and eight to species, several of which are new to science (Pesek et al. 2006a).

Following the expedition, the team began developing a more systematically rigorous search. Over the years, the collaborators have gathered data from transects in targeted loci throughout the mountains and specimens for vouchers and domestication at Itzamma as a part of the studies on *in situ* and *ex situ* conservation of medicinal plant resources. These collections were made on multiple collection trips to remote areas of the Maya Mountains spanning over a decade of field research (Treyvaud-Amiguet et al. 2005; Pesek 2006a, 2009, 2010). The healers and collaborators transplanted thousands of sustainably procured plants into ecosystem microniches that they created and nurtured on site at the gardens. The healers cared for the transplants, including weeding, irrigation, provision of appropriate ratios of leaf litter, and so forth, as they consistently evolved their methodologies in developing the gardens as envisaged. The vast majority of plants on site are understory species of the primary and secondary rainforest. This represented a significant obstacle from the outset given the fact that the land granted to them for the development of Itzamma was basically now secondary forest in early succession (Bourbonnais-Spear et al. 2006) after Hurricane Iris in 2001. Establishing primary forest species therein and maintaining them

healthfully is a challenge that they overcame using various innovations supported by traditional knowledge (Pesek et al. 2007).

In a notable study of Q'eqchi' ethnobotany in the area, done as a precursor to the development of the gardens, Treyvaud-Amiguet et al. (2005) identified a collection of 169 medicinal plant species, belonging to 67 different families; these species and their uses were recorded using quantitative ethnobotanical methods. Using Trotter and Logan's (1986) informant consensus approach, this study revealed a high degree of agreement among the healers on the use of plant species and on the diseases treated, suggesting a well-defined medicinal tradition that has evolved over centuries. These data, now confirmed by further studies (Pesek et al. 2009, 2010; Otarola-Rojas et al. 2010), also show the use of a majority of species from primary or secondary rainforest of the Maya Mountains of southern Belize, rather than weedy species, and thus collectively represent a valuable new insight into the value of Central American botanical diversity. Further, in several of these studies, medicinal uses of the plants were grouped into usage categories, and the number of plants used for each category demonstrates that the healers are treating the gamut of ailments occurring in their area and that their pharmacopoeia is comprehensive and extant. These studies follow the standard for classifying symptoms and ailments as developed by Cook (1995). The category "culture-bound syndromes" (Weller et al. 2002) is used to classify folk illnesses not recognized by biomedicine as disease states.

A second quantitative analysis undertaken using Moerman's regression methods (1991) demonstrated there is a high selectivity for plant families used. The Piperaceae (pepper), Rubiaceae (coffee), and Asteraceae (sunflower) families were highly overused compared to their abundance in the flora (Treyvaud-Amiguet et al. 2006).

Follow-up ethnobotanical studies on the most-used group of medicinal plants, those used for mental health, illustrates the impressive depth of knowledge of the healers and provides evidence that some categories of traditional treatments can translate into modern concepts that are verifiable in the laboratory and then clinically. Over 70 species were recorded with high consensus use in treatments of anxiety, epilepsy, and *susto* ("fear," a culture-bound syndrome or folk illness) (Bourbonnais-Spear et al. 2005).

Bourbonnais-Spear et al. (2005) demonstrated that some of the high consensus plants used for anxiety had significant anxiolytic effects, as demonstrated in ethical animal behavioral models of anxiety such as the elevated plus maze (Bourbonnais-Spear et al. 2007). A further study has shown that the degree of healer consensus on plants used for anxiety correlated well with pharmacological activity (Awad et al. 2009).

Since the healers rely heavily on rainforest species that were found only in primary and rich secondary forests and not in areas of human influence, these species need to be protected both *in situ* and *ex situ*. Pesek et al. (2010) demonstrated that, to be most effective, a culturally relative conservation strategy should be applied. Specifically, the Q'eqchi' Maya healers have a strategy for selecting areas with the highest concentrations of medicinal plants. They note four environmental categories in which medicinal plants grow: cool areas under high forest with much humus, warm areas under low forest, hot and warm areas on rocks and cliffs, and riversides. In a telling scientific study, traditional Q'eqchi' Maya ecosystem constructs or environmental zones (Fig. 33.2) were compared with scientific ecosystems (Fig. 33.3). The study revealed that the Q'eqchi Maya environmental zones were the most salient—that is, knowledge of the Q'eqchi' Maya environmental zones improves one's ability to predict whether there will be high or low abundance of Q'eqchi' Maya medicinal plant species in a particular region, whereas knowledge of scientific ecosystems does not perform as well (Pesek et al. 2010).

The collaboration has also developed an accurate medicinal plant distribution predictive model based on traditional knowledge (Pesek et al. 2009). This work demonstrates the potential of combining ethnobotany and botanical spatial information with indigenous ecosystems concepts and Q'eqchi' Maya traditional healing knowledge via spatial evolutionary computation-based predictive modeling. Through this approach, the collaborators identify regions where species are located as a basis for prioritization and application of *in situ* and *ex situ* conservation strategies to protect them. This represents a significant step toward facilitating sustained, culturally relative health promotion and overall enhanced ecological integrity to the region and the Earth (Pesek et al. 2009).

The indigenous gardens are now growing well, but they do not have the manicured appearance of a "developed world" garden. A recent ethnobotany survey in collaboration with the healers (Otarola-Rojas et al. 2010) led to collection of 102 species from Itzamma. Of these, 40 of the previously reported 106 consensus study plants were growing in the gardens. An additional 62 plants not previously reported were also growing in the gardens. Part of the reason for this is that these garden species were transplanted from the remote, difficult-to-access niches of high ecological integrity within the Maya Mountains during years of field research.

A general comparison of these 102 garden species was made to species presented in the TRAMIL network, Caribbean Herbal Pharmacopoeia (CHP), the largest regional medicinal pharmacopoeia. The comparison shows that relatively few of the 102 garden species are found in the CHP. This demonstrates the healers' preference for primary rainforest species. Interestingly, the majority of the CHP plants are common in Belize, and many are used by the nearby Mopan and Yucatec Maya, but when specifically asked about the TRAMIL CHP plants, the Q'eqchi' Maya traditional healers are aware of them but prefer the primary rainforest species.

The gardens are also currently being used to pass along scientific and traditional knowledge to Maya teenagers from the local high school. The elders are teaching the youngsters in traditional healing and lifeways and the production of medicinal plants for small-scale commercial microenterprise. The area is further used as a community center and for educational activities with children, and is starting to be used as a visitor center to bolster the local economy via site-specific ecotourism on a prearranged basis. Plants not sacred to the Q'eqchi' Maya that have a local or international commercial value are being grown by the cooperative in adjacent fields for small-scale commercialization.

At Itzamma, traditional healing is being studied as an emerging concept in integrative healing for indigenous and worldwide communities. Indeed, the development of Itzamma is concurrent with and reciprocally supportive of the reintegration of traditional healing into the primary healthcare system of local Maya villages. Following a workshop with policymakers and government officials, local healthcare providers and traditional healers, including traditional midwives, work in a more concerted fashion now along with village elders to provide primary healthcare in the

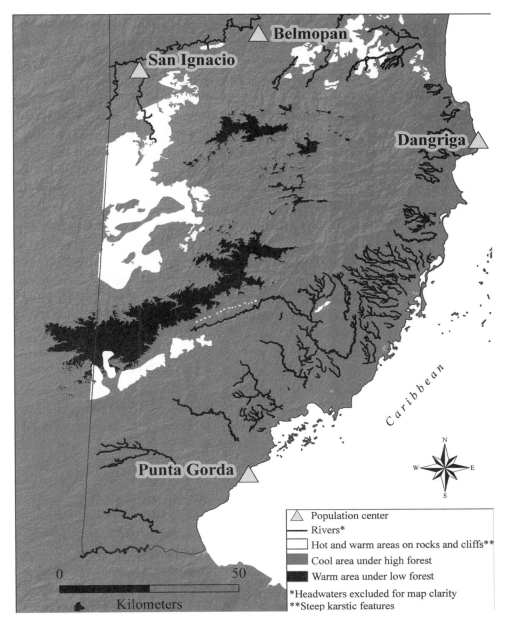

**Figure 33.2:**
Maya Mountains Q'eqchi' Maya regional environmental zones representation: Maya Mountains Ethnobotany and Ecology Project Map: 1:619,792. GIS Data Sources: BERDS. 2008. http://www.biodiversity.bz; CGIAR Consortium for Spatial Information. 2008. http://srtm.csi.cgiar.org; and BioGeo Berkley. 2008. [GIS and spatial data warehouses]: Pesek, L. 2010. Using: ArcView GIS [GIS Software]. Version 9.3. Redlands, CA: Environmental Systems Research Institute, Inc., 2008.

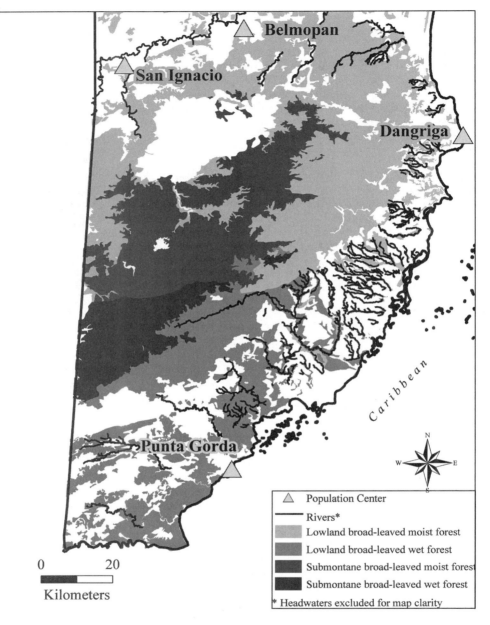

**Figure 33.3:**
Maya Mountains scientific regional ecosystems representation: Maya Mountains Ethnobotany and Ecology Project Map: 1:619,792. GIS Data Sources: BERDS. 2008. http://www.biodiversity.bz and BioGeo Berkley. 2008. [GIS and spatial data warehouses]: Pesek, L. 2010. Using: ArcView GIS [GIS Software]. Version 9.3. Redlands, CA: Environmental Systems Research Institute, Inc., 2008.

villages. Difficult medical cases are forwarded to the district hospital for medical care (Arnason et al. 2004; Cal et al. 2009).

Itzamma represents an opportunity to conserve biodiversity and culture, to study novel phytochemistry and ethnopharmacology, and to explore the diverse phenomena of traditional healing. As demonstrated by the consensus ethnobotany (Treyvaud-Amiguet et al. 2006), Q'eqchi Maya traditional healers use an ancient, intact healing tradition that differs drastically from that of the Western world. However, the tradition is similar to Eastern traditions in the sense that it promotes a type of balance (Mills and Bone 2000). In the case of the Maya, there is a focus on thermal equilibrium. Preliminary studies by our group have found that much of the Maya diagnostic and treatment paradigm focuses on symptom assessments and pulse analysis and diagnosis (consistent with findings of other workers with different Maya groups) (Balick et al. 2008). This paradigm then directs a type of thermal equilibrium promotion where "hot" and "cold" plants are used to balance this spectrum based on the symptom assessments, pulse analysis, and clinical conditions of the patient.

We recognize that the gold standard in medicine for the evaluation of a pharmaceutical intervention is the randomized double-blind placebo-controlled clinical trial. However, this tool is difficult, if not ethically impossible, to apply in traditional healing studies. This is in part due to the fact that traditional medicine, practiced by indigenous peoples since the dawn of humanity, has been based on observation and testing over long periods of time. Through this approach, effective, ineffective, or toxic plants have been discovered through closely monitored observational trials. The process of the search, discovery, use, testing by trial and error, and establishment of the particular plant in the technology of the culture is not that far removed from what modern scientists do today: discovering what already exists, reporting it, communicating it to a wider audience, and testing its efficacy. Clinical trials of well-designed observational studies, with either a cohort or a case–control design, are just as reliable as randomized, controlled trials on the same topic (Concato et al. 2000).

Therefore, given our goals of recording traditional healing practices, which are diverse phenomena, we have begun employing a qualitative methodology dovetailed with observational clinical trials. Thus far,

we have essentially used snowball sampling to home in on critical cases of traditional healers who now represent the QHA. We did so until regional representative saturation was reached, and therefore we have a robust qualitative sampling pool that can be quantified as we generate data.

We will now focus on selecting a few critical cases of patients in cohort form of these traditional healers. The cohort will be followed using a qualitative case study design, taking into consideration the healers' diagnostic and treatment scenarios in the clinical management of various conditions. It is well understood that the critical case approach will "yield the most information and have the greatest impact on the development of knowledge" (Patton 2002, p. 236).

The preservation of medicinal plants is a key objective of the biodiversity component in several conservation projects in Belize and is highlighted as a key objective in the Belize Biodiversity Action Plan as Belize embraces the goals of the Convention on Biological Diversity. Scientific output from the Itzamma project is actively being used in science and policy throughout Belize and internationally via the Biodiversity and Environmental Resource Data System of Belize (BERDS) and the Belize Biodiversity Action Strategy. Our findings are directly applicable to Belize, given the contextual nature of our work, but can also be generalized internationally in broader applications as a model. For example, the International Union for the Conservation of Nature's "Identification and Gap Analysis of Key Biodiversity Areas: Targets for Comprehensive Protected Area Systems" is enabling conservation practice to advance commensurate with scientific theory and could well add "Important Medicinal Plant Areas" supportive of community health and suggested by indigenous peoples and traditional botanical knowledge perspectives. The Itzamma model has already taken root in the Western Ghats region of southern India, another biodiversity hotspot, among the Muthuvan and Uraly peoples (Pesek et al. 2008), and it could well serve a multitude of peoples the world over. The Itzamma collaboration secured funding from Community Information and Epidemiological Technologies (CIET [http://www.ciet.org/en/about-ciet/#What]) in 2009 for "Validation of Traditional Medicine in a North–South Indigenous Collaboration," and the group is now working with Inuit and Anishinabe in Canada. The collaboration has extended invitations also to Cree people in the James Bay region in Canada; Cherokee people in the Smoky Mountains region of the United States; and the Huitoto people in the northwestern Peruvian region of Amazonia toward dissemination of the Itzamma model.

## CONCLUSIONS AND FUTURE DIRECTIONS

Itzamma serves an important purpose. It is meeting the array of outcomes already stated and is also trying to prevent the death of a knowledge base that could be a valuable source of support to the modern health system, and a cultural tradition based in a stewardship worldview.

Itzamma has its roots in an original and unique vision, one not often found in the international development community and one lacking in the fields of culturally relevant healthcare, protection of intellectual property, traditional environmental knowledge, or matters of conservation and biodiversity. Itzamma is grounded on issues of respect and rights of knowledge. It promotes recognition and regard for culture with accommodation not just of gender but also of generation. It bridges science and tradition and advances the integration of traditional healing in the national healthcare system to improve public health through prevention. And it honors and supports a long-held sense of spirituality and connection to the land, and the conservation of biodiversity.

Biodiversity provides the basis for all life on earth. It does so by providing ecosystem services that are essential to our health and well-being and also supports all aspects of our social and economic development. Animals and plants have always been used for food, fuel, medicine, clothing, building, and spirituality. Our ecosystems and ecological integrity constantly renew and provide clean water and air as well as healthful soil to all of us. Despite all these things, appreciation of biodiversity is not a high priority for those who abuse our environment and push development and globalization in unsustainable fashion.

Traditional healing practices are individualized to culture and even more so to the healers within those cultures. However, even given these diverse cultural healing traditions and traditional healers, there are cross-culturally reverberant themes in health and

wellness (Pesek et al. 2006b). Steeped in indigenous empirical observation and a cultural process akin to the scientific method in observation, hypothesis formation, testing, and then interpretation of results for application, these consistent themes have arisen in independently evolving cultural contexts and thus speak to the efficacy of these traditional healing modalities (Pesek et al. 2006b). We must learn from them (Bell et al. 2009).

Integrating traditional healing in the national healthcare system supports the conservation of culture and biodiversity while promoting culturally relevant health and wellness promotion, environmental respect, and the concept of "treading lightly" on the Earth. Traditional healers in primary healthcare enhance community health and wellness via direct care and education of their communities. They also enhance overall ecosystem health and could well provide substantive opportunities for culturally appropriate economic development in conjunction with the conservation of culture and biodiversity. A new way forward is learning from traditions and our elders. The Itzamma model demonstrates these possibilities by merging science, technology, economic development, and other modern trends with traditional lifeways and healthcare toward a promising tomorrow.

## ACKNOWLEDGMENTS

The authors thank their many supporters, without whose help this work would not have been possible. We express our gratitude to the Belize Indigenous Training Institute and the Q'eqchi' Maya Healers Association; Belizean Indigenous Councils; Government of Belize; University of Ottawa, Canada; Universidad Nacional, Costa Rica; Earth Healers and Naturaleza Foundation, United States; and Cleveland State University for their continued financial and programmatic support. The Itzamma project was funded in part by the Inuit Circumpolar Council, the Danish International Development Agency, the International Labor Organization, the United Nations Development Programme, Trekforce-Belize, the World Bank through the Grants Facility for Indigenous Peoples, International Development Research Center, and CIET.

# REFERENCES

Alexiades, M. 2004. Ethnobiology and globalization: science and ethics at the turn of the century. *In* J. Carlson and L. Maffi, eds. Ethnobotany and conservation of biocultural diversity. Advances in economic botany, 15. New York Botanical Garden Press. Bronx, New York.

Arnason, J., V. Cal, V. Assinewe, L. Poveda, J. Waldram, S. Cameron, T. Pesek, M. Cal, and N. Jones. 2004. Visioning our traditional health care: workshop on Q'eqchi' healers center, botanical garden and medicinal plant biodiversity project in southern Belize. Final report to IDRC, Ottawa, Canada.

Atran, S. 1993. Itza Maya tropical agro-forestry. Curr Anthropol 34:633–700.

Awad, R., F. Ahmed, N. Bourbonnais-Spear, M. Mullally, Ta, C. Anh, A. Tang, Z. Merali, P. Maquin, F. Caal, V. Cal, L. Poveda, P. Sanchez-Vindas, V. Trudeau, J. and Arnason. 2009. Ethnopharmacology of Q'eqchi' Maya antiepileptic and anxiolytic plants: effects on the GABAergic system. J Ethnopharmacol 125:257–264.

Balick, M., M. Nee, and D. Atha. 2001. Checklist of vascular plants of Belize, with common names and uses. New York Botanical Garden Press, New York.

Balick, M., J. De Gezelle, R. and Arvigo. 2008. Feeling the pulse in Maya medicine: an endangered traditional tool for diagnosis, therapy, and tracking patients' progress. Explore 4:113–119.

Bannister, K., and K. Barrett. 2004. Weighing the proverbial "ounce of prevention" against the "pound of cure" in a biocultural context: a role for the precautionary principle in ethnobiological research. *In* J. Carlson and L. Maffi, eds. Ethnobotany and conservation of biocultural diversity: advances in economic botany, 15. New York Botanical Garden Press, Bronx, New York.

Barrera-Bassols, N., and V. Toledo. 2005. Ethnoecology of the Yucatec Maya: symbolism, knowledge, and management of natural resources. J Lat Am Geogr 4:9–41.

Bateson, J., and I. Hall. 1977. The geology of the Maya mountains, Belize (Overseas Memoir 3). Her Majesty's Stationery Office, London, England.

Bell, P., B. Lewenstein, A. Shouse, and M. Feder, eds. 2009. Learning science in informal environments: people, places, and pursuits. Committee on Learning Science in Informal Environments. National Research Council (NRC), Washington, D.C.

Berlin, B., and E. Berlin. 2003. NGO's and the process of prior informed consent in bioprospecting research: the Maya ICBG project in Chiapas, Mexico. UNESCO, Blackwell Publishing Ltd, Oxford, UK.

Bourbonnais-Spear, N., R. Awad, P. Maquin, V. Cal, P. Sanchez-Vindas, L. Poveda, and J. Arnason. 2005. Plant use by the Q'eqchi' Maya of Belize in ethnopsychiatry and neurological pathology. Econ Bot 59:326–336.

Bourbonnais-Spear, N., R. Awad, Z. Merali, P. Maquin, V. Cal, and J. Arnason. 2007. Ethnopharmacological investigation of plants used to treat *susto*, a folk illness. J Ethnpharmacol 109:380–387.

Bourbonnais-Spear, N., J. Poissant, V. Cal, and J. Arnason. 2006. Culturally important plants from southern Belize: domestication by Q'eqchi' Maya healers and conservation. Ambio 35:138–140.

Buenz E. 2005. Country development does not presuppose the loss of forest resources for traditional medicine use. J Ethnopharmacol 100:118–123.

Cal, V., J. Arnason, B. Walshe-Rouusel, P. Audet, J. Ferrier, K. Knight, and T. Pesek. 2009. The Itzamma Project: sustainable indigenous development based on the ethnobotanical garden and traditional medicine concept. Final Report. International Development Research Center, Ottawa, Canada.

Concato, J., N. Shah, and R. Horwitz. 2000. Randomized, controlled trials, observational studies, and the hierarchy of research designs. N Engl J Med. 342: 1887–1892.

Cook, F. 1995. Economic botany data collection standard. Royal Botanic Gardens Kew, Kent, United Kingdom.

Davis, W. 1999. Clouded leopard. Douglas and MacIntyre Publishing Group, Vancouver, BC.

Dixon, C. 1956. Geology of southern British Honduras with notes on adjacent areas. Belize City: Government Printing Office.

Dunham, P., M. Abramiuk, L. Cummings, C. Yost, and T. Pesek. 2009. Ancient Maya cultivation in the southern Maya mountains of Belize: complex and sustainable strategies uncovered. Antiquity. Project Gallery. 83(319).

Farnsworth, N., O. Akerele, A. Bingel, D. Soejarto, and Z. Guo. 1985. Medicinal plants in therapy. B World Health Organ 63:965–981.

Grimes, B. 1996. Ethnologue: Languages of the world. 13th Ed. Dallas: Summer Institute of Linguistics. SIL International. Online version: http://www.ethnologue.com/.

Hanski, I. 2005. Landscape fragmentation, biodiversity loss and the societal response. EMBO Rep 6:388–392.

Hartshorn, G., L. Nicolait, L. Hartshorne, G. Bevier, R. Brightman, J. Cal, A. Cawich, W. Davidson, R. DuBois, C. Dyer, J. Gibson, W. Hawley, J. Leonard, R. Nicolait, D. Weyer, H. White, and C. Wright. 1984. Belize, country environmental profile. Trejos Hnos, San Jose, Costa Rica.

Hayden, C. 2003. When nature goes public: the making and unmaking of bioprospecting in Mexico. Princeton University Press, Princeton, New Jersey.

Iremonger, S., and R. Sayre. 1994. The Bladen nature reserve, Toledo District, Belize: a rapid ecological assessment. The Nature Conservancy, Arlington, Virginia.

Iremonger, S., R. Leisnerm, and R. Sayre. 1995. Plant records from the natural forest communities in the Bladen Nature Reserve, Maya Mountains, Belize. Caribb J Sci 31:30–48.

Kashanipour, R., and R. McGee. 2004. Northern Lacandon Maya medicinal plant use in the communities of Lacanja Chan Sayab and Naha,' Chiapas, Mexico. J Ecol Anthropol 8:47–66.

Maffi, L., ed. 2001. On biological and cultural diversity. Smithsonian Institution Press, Washington D.C.

Meerman, J., and J. Clabaugh, eds. 2007. Biodiversity and environmental resource data system (BERDS). http://www.biodiversity.bz

Mills, S., and K. Bone. 2000. Principales and practice of phytotherapy. Harcourt Publishers Ltd., London, UK.

Moerman, D. 1991. The medical flora of native North American: an analysis. J Ethnopharmacol 31:1–42.

Myers, N., R. Mittermeier, C. Mittermeier, G. da Fonseca, and J. Kent. 2000. Biodiversity hotspots for conservation priorities. Nature 403:853–858.

Nigh, R. 2002. Maya medicine in the biological gaze. Curr Anthropol 43:451–477.

Otarola-Rojas, M., S. Collins, V. Cal, F. Caal, J. Arnason, L. Poveda, P. Sanchez-Vindas, and T. Pesek. 2010. Sustaining rainforest plants, people and global health: a model for learning from traditions in holistic health promotion and community based conservation as implemented by Q'eqchi' Maya Healers, Maya Mountains, Belize. Sustainability 2:3383–3398.

Parker, T., B. Holst, L. Emmons, and J. Meyer. 1993. A biological assessment of the Columbia River Forest Reserve, Toledo District, Belize. Rapid Assessment Program Working Papers Number 3. Conservation International, Washington, D.C.

Patton, M. 2002. Qualitative research and evaluation. 3rd ed. Sage Publishing, Thousand Oaks, California.

Pesek, T., Cal, M., Cal, V., Fini, N., Minty, C., Dunham, P., and Arnason, J. 2006a. Rapid ethnobotanical survey of the Maya Mountains range in southern Belize: A pilot study. Aust S Hist 1:1–12.

ItzammaPesek, T., L. Helton, and M. Nair. 2006b. Healing across cultures: learning from traditions. EcoHealth. 3:114–118.

Pesek, T., V. Cal, N. Fini, M. Cal, M. Rojas, P. Sanchez, L. Poveda, S. Collins, K. Knight, and J. Arnason. 2007.

Itzamma: revival of traditional healing by Q'eqchi' Maya, Southern Belize, Central America. Herbal Gram 76:34–43.

Pesek, T., L. Helton, R. Reminick, D. Kannan, and M. Nair. 2008. Healing traditions of southern India and the conservation of culture and biodiversity: a preliminary study. Ethnobot Res Appl 6:471–479.

Pesek, T., M. Abramiuk, D. Garagic, N. Fini, J. Meerman, and V. Cal. 2009. Sustaining plants and people: traditional Q'eqchi' Maya botanical knowledge and interactive spatial modeling in prioritizing conservation of medicinal plants for culturally relative holistic health promotion. EcoHealth 6:79–90.

Pesek, T., M. Abramiuk, N. Fini, M. Otarola-Rojas, S. Collins, V. Cal, P. Sanchez, L. Poveda, L., and J. Arnason. 2010. Q'eqchi' Maya healers traditional knowledge in prioritizing conservation of medicinal plants: culturally relative conservation in sustaining traditional holistic health promotion. Biodivers Conserv 19:1–20.

Stepp, J., H. Castañeda, and S. Cervone. 2005. Mountains and biocultural diversity. Mountain Res Dev 25:223–227.

Sutherland, W. 2003. Parallel extinction risk and global distribution of languages and species. Nature 423:276–279.

Treyvaud-Amiguet, V., J. Arnason, P. Maquin, V. Cal, P. Sanchez-Vindaz, and L. Poveda. 2005. Consensus ethnobotany of the Q'eqchi' Maya of Southern Belize. Econ Bot 59:29–42.

Treyvaud-Amiguet, V., J. Arnason, P. Maquin, V. Cal, P. Sanchez-Vindas, and L. Poveda. 2006. A regression analysis of Q'eqchi' Maya medicinal plants from Southern Belize. Econ Bot 60:24–38.

Trotter, R., and M. Logan. 1986. Informant consensus: a new approach for identifying potentially effective medicinal plants. In N.L. Etkin, ed. Plants in indigenous medicine and diet: biobehavioural approaches, pp. 91–112. Redgrave Publishers, Bedfort Hills, New York.

United Nations Food and Agricultural Organization. 2000. Global forest resources assessment, 2000. FAO Forestry Paper 140. United Nations, http://www.fao.org/docrep/004/y1997e/y1997e00.htm (accessed on 11/15/2011)

Weller, S., R. Baer, J. De Alba Garcia, M. Glazer, R. Trotter, L. Pachter, and R. Klein. 2002. Regional variation in Latino descriptions of susto. Cult Med Psychiat 26:449–472.

World Health Organization (WHO). 2002. WHO traditional medicine strategy 2002–2005. WHO EDM/TRM/2002.1. Geneva, Switzerland.

World Wildlife Fund (WWF). 2004. Living planet report 2004. World Wildlife Fund. Gland, Switzerland.

# 34

# BIOLOGICAL DIVERSITY AND HUMAN HEALTH

Using Plants and Traditional Ethnomedical Knowledge to Improve
Public Health and Conservation Programs in Micronesia

Michael J. Balick, Katherine Herrera, Francisca Sohl Obispo,
Wayne Law, Roberta A. Lee, and William C. Raynor

In this chapter we seek to establish a linkage between ecosystem and human health and to show how the latter can be improved through greater emphasis on sustainable use of natural resources, ecosystem conservation, and maintenance of traditional cultural practices, using an example from the island of Pohnpei, Federated States of Micronesia (FSM). There is great change occurring in habitats around the world, and the majority of these are negative, influencing the health of many living organisms (Aguirre et al. 2002). Micronesia, in the western Pacific Ocean, is a region of vast biodiversity, with many unique plant species—76% of the plants native to these islands are found nowhere else on earth (Balick et al. 2009), making this region the eighth most important biodiversity hotspot based on rates of plant endemism. This is the result of the isolation of the individual islands, their diverse topography, and the fact that they are found very far from continents that would otherwise dominate their floras. Micronesia contains over 2,000 individual islands spread throughout 9.1 million km² of ocean, an area approximately the size of the continental United States, and is located about 4,000 km southwest of Honolulu, Hawaii (Fig. 34.1).

Beginning in 1997, a group of institutions and individuals began studying three of these island groups, investigating the botany and traditional use of plants across an east-to-west transect about 2,400 km long beginning in the islands of Kosrae and Pohnpei, FSM, to the Republic of Palau. The initial goals of the research included the documentation of useful plants, including as local medicines, an understanding of their distribution and diversity, and possible threats of extinction due to habitat destruction, overharvest, global change, or neglect. As these are extremely isolated sites, traditional lifestyle and values are still present, although at different levels on each island. The program has grown and now operates under the name "Plants and People of Micronesia Program." Locally based institutions collaborating in this project include the Conservation Society of Pohnpei, the Kosrae Conservation and Safety Organization, the Belau National Museum/Natural History Section, the Micronesia Office of the Nature Conservancy, the College of Micronesia, FSM, Pohnpei Council of Traditional Leaders, and agencies of state and national government in each region. International collaborators include The New York Botanical Garden (NYBG), the Continuum Center for Health and Healing at Beth

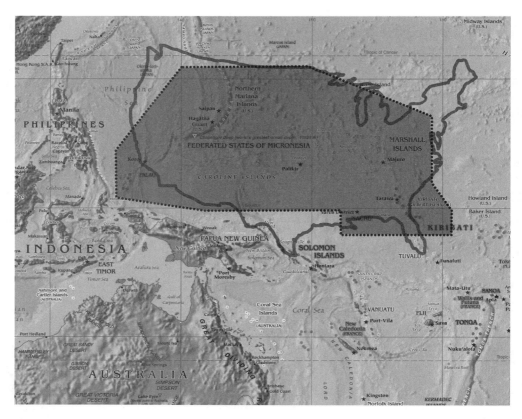

**Figure 34.1:**
Map of Micronesia, showing its position within Oceania. The area shaded dark gray is considered Micronesia, with an overlay of the continental United States for scale. (Source of base map: https://www.cia.gov/library/publications/the-world-factbook/graphics/ref_maps/pdf/oceania.pdf)

Israel Medical Center, the National Tropical Botanical Garden, and the University of Arizona College of Medicine Program in Integrative Medicine. Our earliest work was on Pohnpei, and this brief review focuses on some of the results to date, particularly as they relate to issues involving biodiversity and human health.

Pohnpei is a small volcanic island (344 km²) and one of the four member states of the FSM. As of 2010 a team of three local collectors, based at the Conservation Society of Pohnpei, is carrying out surveys of critical habitats and interviewing local people about traditional uses of plants, and the project has built the first herbarium on the island, at the College of Micronesia, FSM—Palikir Campus. In addition to fieldwork, the project developed the first ethnobotany course in the FSM, taught at the college.

It is such a popular offering that it is taught each semester and attracts students from all of the island states of the FSM (for more details on the course refer to http://www.comfsm.fm/~dleeling/ethnobotany/ethnobotany.html).

The "Plants and People of Micronesia Program" involves floristic studies on each of the three sites, and published its first book, *Ethnobotany of Pohnpei: Plants, People and Island Culture* (Balick et al. 2009), containing a provisional checklist of the plants, information on plant biodiversity and distribution, including agrodiversity, traditional health systems and beliefs, and an ethnobotanical inventory discussing traditional uses for many of the native and introduced plants on Pohnpei and surrounding atolls. Working with three Pohnpeian linguists, the book also established a set of "accepted" local names for plants in an attempt to standardize folk taxonomy on the island. As the project is being carried out under the direction of the traditional leaders of the island's five kingdoms, the book was copyrighted by the *Mwoalen Wahu Illeilehn Pohnpei* (Pohnpei Council of Traditional Leaders), who also stated their responsibility for continuing to gather traditional information and keep it alive through practice.

An updated checklist has been published, *Checklist of the Vascular Plants of Pohnpei, Federated States of Micronesia, With Local Names and Uses* (Herrera et al. 2010), which also includes one photo per genus, as well as habit, status, literature sources, voucher specimen citations, and personal observations. Based on this checklist, 935 vascular plants and infra-species have been recorded from Pohnpei. Of these, 397 taxa are considered native (indigenous and endemic), whereas naturalized and cultivated plants together comprise 538 taxa or 57.5% of the total flora. The rate of endemism is 12.8% for the native flora. Of the 935 species of plants known to occur on the island, 215 have been documented as being used to treat health conditions.

The program has an ethnomedical training component designed to build an understanding of the relationship between public health and biodiversity. One component of this is the preparation of primary health care manuals for each island; with the help of local and international physicians and experts in local traditional knowledge and healing practices, plants are evaluated for their potential use in treating common health conditions. While the goal is providing information on evidence-based use of local remedies, few botanicals—including many of these native and introduced species—have been subjected to Western clinical studies. However, we have listed closely related species that have been studied, particularly those with published bioactivity. Thus, the goal is to list plants that have been used in traditional clinical practice locally and to avoid those that might have toxic compounds or might result in adverse reactions. Another component is the training of U.S. and internationally based physicians, medical students, and students in the medical sciences (e.g. pharmacology) in ethnomedicine and its potential application to clinical practice during a 4- to 6-week on-site experience working with the project. This includes cultural competency training that can transfer to other regions of the world where the physician might practice.

## POHNPEI PRIMARY HEALTH CARE MANUAL

Several years into the program on Pohnpei, we were interviewing a local healer during a cholera epidemic (Balick and Lee 2003). Although the local village dispensary was empty of imported antidiarrheal medications, one of the plants traditionally used to treat this condition—*Psidium guajava* L.—grew all around the facility. However, the young medical officers—Micronesian islanders trained in the United States—were unaware of the traditional therapeutic uses of this plant. We then introduced the healer, and her knowledge of how this plant could be used to help treat diarrhea, to the medical officers, who agreed to employ this therapy in the future. See Figure 34.2.

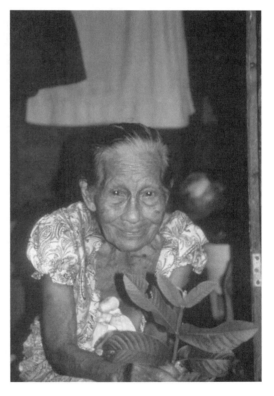

**Figure 34.2:**
The late Maria David, a well-respected traditional healer on Pohnpei, with leaves of *Psidium guajava* L. (Photo by Michael J. Balick)

The concept of developing a reference on the use of local plants in treating primary health care issues was developed shortly thereafter and proposed to the Annual Meeting of the Pacific Basin Medical Association, where it was endorsed as a way to encourage self-sufficiency, biodiversity conservation, and sustainable medical care for the islands. Recently, *The Pohnpei Primary Health Care Manual: Health Care in Pohnpei, Micronesia—Traditional Uses of Plants for Health and Healing* was published (Lee et al. 2010). Written by a team of ethnobotanists, physicians, Pohnpeian traditional healers, public health specialists, and graduate students, it provides information on conditions commonly encountered on Pohnpei, based on the most frequent diagnoses reported by healers as well as the *Annual Health Services Profile* (Pohnpei State Department of Health Services 2006) and documents local knowledge of plants useful for these conditions. The manual is organized into 11 chapters: (1) Bites and Stings; (2) Diarrhea and Gastric Disorders; (3) Skin Disorders; (4) Cuts and Wounds; (5) Colds and Flu; (6) Pain; (7) Stress and Related Symptoms; (8) Infectious Diseases; (9) Culture-Bound Syndromes; (10) Women's Health; and (11) Men's Health.

Within each chapter, there is an introduction to the condition from a physician's standpoint, including information on life-threatening symptoms (e.g., when immediate medical care is necessary). Following the introduction are the plants used to treat the specific conditions, organized by local name, scientific name, plant description and distribution on Pohnpei and in the Pacific, traditional recipes and formulas, pharmacological properties when known, and evidence of toxicity when known (Fig. 34.3).

There is also a glossary of terms, a list of the herbarium specimens and the collectors who provided the information in the manual, and an extensive bibliography. The manual is being widely distributed throughout Pohnpei, particularly at the level of village dispensaries, healthcare practitioners and the hospital and network of clinics.

The formulas and applications for the uses of plants were selected based on frequency of uses reported by local traditional healers who were considered experts in their field. In addition to their expertise, we considered multiple reports of the same plant used to treat the same condition in a similar preparation to have a higher priority for inclusion in the manual than a single report by an individual healer. This publication was also sponsored by the *Mwoalen Wahu Illeilehn Pohnpei*, who gave the team permission to collect information from their subjects. At the beginning of each conversation, a thorough explanation of the purpose of the manual was provided to the local experts, and informed consent was obtained from each (Appendix 34.1). In addition, we avoided recording proprietary family recipes or those that contained sensitive or secret knowledge. An example of the selective collection of generalist data was the case of one expert healer, who practiced at the local state hospital, who approached us and asked one of the team to codify her knowledge in a notebook, for use in teaching her family. The notebook was not duplicated, nor was any of the information used for the project, at the request of the healer. Thus, it is acknowledged that the *Pohnpei Primary Health Care Manual* is simply a collection of basic traditional knowledge and its application to common health conditions.

The *Mwoalen Wahu Illeilehn Pohnpei* and NYBG hold joint copyright to this volume (C. McManus, Washington University St. Louis School of Law, personal communication 2009). This gives both the traditional leaders and NYBG the right to bring lawsuits in the case of copyright infringement and helps to protect inappropriate use of the manual's contents. Under the U.S. and Micronesian legal frameworks, the manual also established the concepts of prior art and prior knowledge involving the use of these plants for these health conditions, and thus provides some level of protection as well.

The manual was based on more than 400 interviews with local Pohnpei community members and 1,300 plant collections made from 1998 to 2008. Approximately 190 individuals shared medicinal plant knowledge of more than 72 genera and 82 species ranging across 46 families.

A literature review of each species and its pharmacological and toxicity properties was conducted using the following online databases: PubMed (www.pubmed.gov), TOXNET (www.toxnet.nlm.nih.gov), ScienceDirect (www.sciencedirect.com), and Biological Abstracts (http://www.columbia.edu/cu/lweb/eresources/databases/2257200.html).

The pharmacology, toxicity, and adverse effects data for each of the plant species, when known, were specifically included to help inform medical personnel of the information available. However, much of

**Figure 34.3:**
*Curcuma australasica* Hook. f. tuber is scraped and wrapped in *inipal* (coconut husk) for use in traditional Pohnpeian medicine. (Photo by Michael J. Balick)

the time these data have significant limitations and may not relate to clinical use. There is a lack of good clinical data and other *in vivo* studies that directly relate to the use of many of the plants discussed in the manual, particularly when considering the dosage and preparation as is undertaken on Pohnpei. Therefore, this book is not designed to be an evidence-based guide to all of the species presented. The following databases were used to cross-check facts, definitions, and botanical nomenclature: International Plant Names Index (http://www.ipni.org/), W³ TROPICOS (http://www.tropicos.org/), and HerbClip (http://abc.herbalgram.org).

## CULTURE-BOUND SYNDROMES ON POHNPEI

Some health conditions are a unique combination of environmental and cultural manifestations occurring in a specific region or within a specific culture. Cultural and spiritual practices often define these unique syndromes according to their belief system. From a medical perspective, these illnesses are referred to as "culture-bound syndromes," and in the field of psychiatry they are defined by the *Diagnostic and Statistical Manual IV, Text Revision* as "generally limited to specific societies or culture areas and are localized, folk, diagnostic categories that frame coherent meanings for certain repetitive, patterned . . . sets of experiences and observations" (American Psychiatric Association 2002, pp. 898–903).

Roger Ward, an anthropologist who studied the Pohnpeian medical practice, wrote *Curing on Ponape: A Medical Ethnography* (1977). He considered three categories of culture-bound syndromes on Pohnpei to be present: bodily illnesses, spirit/sorcery illnesses, and Pohnpeian sicknesses. Bodily illnesses include women's sickness, man's sickness, skin sickness, sicknesses of the eye, jumping sickness, and feeling bad sickness. Spirit/sorcery illnesses are caused by spirits or sorcery and include phenomena such as the mangrove demon (mangrove sickness), family ghosts (ghost sickness), and other spirits (spirit sickness). Pohnpeians distinguish categories of illnesses that can be treated only by local medicines versus illnesses that can be treated with Western medicine.

Shaking sickness—a Pohnpeian illness—presents with symptoms of shaking from high fever. It can be treated with a beverage made from the roots of the palm *Metroxylon amicarum* (Wendl.) Becc. Mangrove sickness is when a person feels bad and has joint pain that moves around the body, as well as headache, stomachache, and backache. If medicine—such as the juice of *Cordyline fruiticosa* (L.) Chev.—is put on one knee, for example, the pain can go to other areas of the body. Mangrove sickness is also intensified by tide— that is, if a person has a headache, it is more intense during high or low tide.

## CONSERVATION RATIONALE AND EFFORTS

Conservationists around the world are in a race against time, and in Micronesia there is great concern about the future of the environment—with significant on-the-ground activity to ensure its protection. The Polynesia-Micronesia hotspot has been recognized as the "epicenter" of the current global extinction crisis (Costion et al. 2009). For four decades, the International Union for Conservation of Nature (IUCN) Species Program, working with the IUCN Species Survival Commission (SSC), has been assessing the conservation status of species on a global scale to identify taxa threatened with extinction and thus promote their conservation. Only two species indigenous to Pohnpei (*Aglaia mariannensis* Merr. and *Intsia bijuga* [Colebr.] Kuntze) and one endemic (*Parkia korom* Kaneh.) species are listed in the 2010.4 online version of the IUCN Red Data List of Threatened Species (http://www.iucnredlist.org), which represents only 0.75% of the island's total native species (Herrera et al. 2010); evaluation and assessment of species in Micronesia is urgently needed in the near future. Also, scientists at NYGB's Geographic Information Systems Laboratory have developed a quick and efficient method for identifying species at high risk of extinction; the threshold to be classified as threatened under IUCN criteria for extent of occurrence is 20,000 km² (http://sweetgum.nybg.org/caribbean/plantsatrisk.php). Thus, endemic species in a small island like Pohnpei (344 km²) fall under species at high risk for extinction. Even though there are no endemic species discussed in the *Pohnpei Primary Health Care Manual*, 47 out of 82 (57%) species are indigenous to Pohnpei, which shows the importance of assessing and conserving the native flora.

In 2006, five Micronesian governments—the Republic of Palau, the Federated States of Micronesia, the Republic of the Marshall Islands, the U.S. Territory of Guam, and the Commonwealth of the Northern Mariana Islands—committed to the Micronesia Challenge, a pledge to "effectively conserve at least 30 percent of the near-shore marine resources and 20 percent of the terrestrial resources across Micronesia by 2020." Our program to identify and document Micronesian plant diversity and the ways it may be used by local people on the three island groups provides support to this conservation effort by identifying centers of plant diversity and threats from invasive species and understanding how the flora can be used to ensure self-sufficiency and reduce dependence on outside inputs. By showing that native biodiversity and cultural traditions can improve public health, a new constituency, the medical community, now recognizes the importance of biodiversity conservation. *The Primary Health Care Manual* has helped to engage this group of influential professionals in Micronesia and is an important example of how the application of biodiversity and ethnobotanical and ethnomedical studies can support local and international conservation programs—in this case, the goal of long-term environmental stability in the Micronesian region. By setting the standards for equitable partnerships, and widely disseminating the results of those collaborations, this program has developed an important and policy-relevant network of supporters in the Pacific region, an area that represents one third of the globe. The *Pohnpei Primary Health Care Manual* and its use as part of an integrative approach to providing and improving healthcare in Micronesia are admittedly first steps in what promises to be a lengthy process. There is no way of predicting whether the outcome will be successful, but at a minimum, the protection of biodiversity and preservation of traditional knowledge through its practice are now recognized as important to re-establishing a more sustainable lifestyle and healthier environment in these remote island nations.

## SAKAU (PIPER METHYSTICUM) CULTIVATION ON POHNPEI: THREATS TO CONSERVATION

*Sakau* (*Piper methysticum* G. Forst.), as it is locally known on Pohnpei, is a long-lived, slow-growing shrub that can grow to more than 6 m in height; it is also commonly known throughout the South Pacific Islands as *kava-kava*. It does not produce seeds (sterile) and is generally propagated by planting of its stalks. The part of greatest interest to humans is the roots because they are used to prepare medicinal and ceremonial beverages and extracts. The roots are stout brown structures that when split open are yellowish in color; they can be harvested after the plant reaches 2 to 3 years in age and grows to a height of approximately 2 to 3 m. The plant grows well in lowland and upland forests (Balick et al. 2010) (Fig. 34.4).

*Sakau* has traditionally been used ceremonially for its relaxing, mood-calming effects and has long been used to promote dispute resolution in group settings. The ritual of using *P. methysticum* is widespread throughout the Pacific, and on Pohnpei its use is considered a cultural keystone practice essential to maintaining the traditional lifestyle and belief system (Balick et al. 2009).

The major *sakau* cultivar on Pohnpei is known locally as *Rahmwahnger* and can be distinguished by a series of purplish or blackish spots on the stem. The second cultivar, known as *Rahmedel*, is characterized by a smooth, unspotted stem. The most common variety of *sakau* in use today is *Rahmwahnger* (95% of all preparations), perhaps because the local people report that it is stronger, with the effect lasting much longer than the second cultivar. The principal use is for a beverage that is used ritually, ceremonially, and, in contemporary times, recreationally. From an ethnomedicinal perspective, the leaves, stems, and roots are important local medicines used in traditional healing to treat both physical and supernatural (culture-bound) ailments for many different conditions. Preparing *sakau* for drinking can be done in many different ways, depending on the use, and it involves methodologies that are particular to the island, such as squeezing the freshly pounded roots with a press made from the inner bark of another plant, *Hibiscus tiliaceus* L. (*keleu* on Pohnpeian) (Fig. 34.5).

As *sakau* is now consumed recreationally by two thirds of the adult population of Pohnpei, cultivation of this plant poses a significant threat to the biodiversity of the main island of Pohnpei. Between 1975 and 1995, the main island lost 47.5% of its native forest cover (2.3% per year), primarily due to an increase in planting *sakau* in the upland forest habitats (Fig. 34.6). From 1995 to 2002, the rate of habitat destruction slowed significantly (1.2% per year) due to

**Figure 34.4:**
*Sakau* (*Piper methysticum*) growing in lowland agroforest. (Photo by Michael J. Balick)

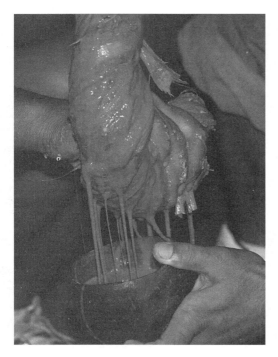

**Figure 34.5:**
Squeezing the press (made from *Hibiscus tiliaceus* L. bark) to release the traditional *sakau* beverage into a coconut cup. (Photo by Michael J. Balick)

community-based efforts that discouraged upland forest conversion for *sakau* production and encouraged traditional cultivation techniques in lowland agroforestry areas. However, upland clearing of native forest is still occurring in areas of the island where upland forests have not been demarcated as a protected area by the locally set watershed boundary line.

Aside from the obvious loss of upland forest, the threats this cultivation practice poses to plant, fungal, and stream community diversity, water quality, and sedimentation are not well understood. In an attempt to study the impacts of upland clearings and document biodiversity, a multi-institutional effort has been undertaken on Pohnpei. Data such as plant, fleshy fungi, freshwater fish, and aquatic invertebrate diversity, water quality, and sediment loads have been gathered to further understand the impact of these human activities on the ecosystem. While analysis is ongoing, preliminary findings indicate that there have been significant changes in biodiversity and increased sediment loads in upland forest areas where clearing has occurred compared to those native forests.

Increasingly, Micronesians are recognizing the importance of ecosystem health as a key to improving public health. By documenting and preserving

**Figure 34.6:**
Area of Pohnpei upland forest with trees cut down to provide extra sunlight for *sakau* being planted. (Photo by Wayne Law)

traditional knowledge, and further understanding the impacts of human activities on these fragile island ecosystems, we aim to help devise new and locality-specific conservation strategies to mitigate the negative impacts on Pohnpei's ecosystem health and the health of the residents. The approach outlined here is currently being implemented in the Republic of Palau in collaboration with the Ministry of Health. The publication on Palauan plants will be based on local health conditions, plant biodiversity, and local priorities. Ultimately, once the research protocols and templates have been worked out on several Micronesian islands, this model could be exported to other Pacific Island nations, and elsewhere as needed. Key to the success of this model is the establishment of full and transparent partnerships with local agencies, traditional leaders, and knowledgeable individuals—and of course the financial support of visionaries who understand the relationship between a viable environment and improved public health. There is clearly value in attracting the support of broader constituencies, such as the medical community and patients, for environmental conservation, and this program is working towards that goal.

Because their natural resources are finite, small islands have always been incubators of novel approaches to conservation. Centuries before Europeans and North American colonists recognized the need for sustainable resource management, Pacific Island residents carefully recorded information on the species biology, diversity, abundance, and reproduction of the animals and plants that sustained them, and adjusted their harvests, mindful of the future needs of their children, grandchildren, and the generations to come.

## ACKNOWLEDGMENTS

Since its inception in 1997, the "Plants and People of Micronesia Program" has received generous support from the V. Kann Rasmussen Foundation, the Overbrook Foundation, the Gildea Foundation, CERC (the Consortium for Environmental Research and Conservation) at Columbia University, the MetLife Foundation, Marisla Foundation, Edward P. Bass and the Philecology Trust, the Germeshausen Foundation, the Prospect Hill Foundation, the National Science Foundation, the Pohnpei Council of Traditional Leaders, the Pohnpei State Government, and the John Simon Guggenheim Memorial Foundation through a Guggenheim Fellowship to Michael Balick during 2006. Balick also served as a MetLife Fellow at The New York Botanical Garden while working on this project. The local names of the plant species and their uses presented in this manual were documented by the following people working on Pohnpei: Elipiana Albert, Michael J. Balick, Alfred Dores, Ben Ekiek, Primo Emos Eperiam, Robert Gallen, Mark Kostka, Molly Hunt, Roberta Lee, Relio Lengsi, William Raynor, Ally Raynor, Pelihter Raynor, Valentine Santiago, Francisca Sohl Obispo, Francisco Sohl, Clay Trauernicht, and Mayorico Victor. The identification of scientific and local plant names was primarily based on the research, collections, observations, and photographs compiled by Roberta Lee, Michael J. Balick, Nieve Shere, Tim Flynn, Wayne Law, and David Lorence. These collections were used to supplement utilization data where appropriate. We would also like to thank the dedicated staff members who served as research and editorial assistants and helped make this work possible: Ryan Huish, Joshua M. Simpson, Sherry Chow, Irina Adam, and Eleanor Stein.

Some of the information presented in this chapter is excerpted or summarized from Balick et al. (2009) and Lee et al. (2010).

## APPENDIX 34.1

## English translation of Pohnpeian version: Plants and People of Micronesia Program—Permission to Participate and Consent Form (Prior Informed Consent)

We are conducting a survey of the plants of Micronesia and their local uses, with the goal of producing a book on this topic. *The Plants and People of Micronesia Program* has as its goal the understanding of the plants found on Micronesia, both native and introduced, and their traditional and modern uses. We represent a consortium of private and governmental groups both local and international, working together to gather and preserve this knowledge, along with the biodiversity of the natural environment it is based upon, as part of a worldwide effort to try and stop the loss of traditional knowledge and the destruction of the natural environment as called for by the Convention on Biological Diversity. The organizations involved in this project on Micronesia include The Conservation Society of Pohnpei, The New York Botanical Garden, Pohnpei State Government, The Pohnpei Council of Traditional Leaders, The College of Micronesia, the Continuum Center for Health and Healing at Beth Israel Hospital in New York City, the Program for Integrative Medicine at the University of Arizona, The Nature Conservancy, The National Tropical Botanical Garden, Belau National Museum, and the Belau Office of the Council of Chiefs.

We will be collecting plants from various places in Micronesia with the permission of the local government and traditional leaders. The plants collected in this survey will be pressed, dried and sent to specialists for identification, and become part of the collections of the herbaria of The College of Micronesia, The New York Botanical Garden, and The National Tropical Botanical Garden. The materials and information collected during this project will be used for scientific studies, not for commercial purposes, and provide a reference that is freely available for consultation, both in Micronesia and internationally, by those interested in this topic. Information collected on the uses of these plants will be noted on the labels, and both the plant and its uses be contained in a book or manual we intend to write. We are seeking general information on uses of plants as food, medicine, for construction, in local storytelling, clothing and similar purposes. We are not collecting secret family information as part of this study, and do not wish you to reveal anything that is not intended for general knowledge of the community. As stated above, the knowledge and plants collected in this survey will be included in a book about this aspect of Micronesian culture, as a record for present and future generations.

If you share our concern about the importance of Micronesian culture and the local uses of plants we would welcome your participation in this project. Your name will be recorded with each plant that we collect with you and along with the information you contribute, to be listed in the final publication.

By signing below, you grant us permission to interview you for this project.

Thank you very much.

Agreed and Accepted,

Name:

Date:

# REFERENCES

Aguirre, A.A., R.S. Ostfelf, G.M. Tabor, C. House, and M.C. Pearl, eds. 2002. Conservation medicine: ecological health in practice. Oxford University Press, New York.

American Psychiatric Association. 2002. DSM-IV-TR. Diagnostic and statistical manual of mental disorders, 4th ed. American Psychiatric Association, Washington, D.C.

Balick, M.J., K. Herrera, and S.M. Musser. 2010. Kava. *In* Coates P.M., J.M. Betz, M.R. Blackman, G.M. Cragg, M. Levine, J. Moss, J.D. White, eds. Encyclopedia of dietary supplements, 2nd ed., pp. 459–468. Informa Healthcare, New York.

Balick, M.J., K. Albert, J. Daniells, L. Englberger, T. Flynn, W. Law, R.A. Lee, D. Lee Ling, A. Levendusky, D.H. Lorence, A. Lorens, J. Phillip, D. Ragone, and B. Raynor. 2009. Ethnobotany of Pohnpei: plants, people and island culture. University of Hawaii Press and The New York Botanical Garden (co-publishers), Honolulu, HI.

Balick, M.J., and R. Lee. 2003. Stealing the soul, *Soumwahu en Naniak*, and *Susto*: understanding culturally-specific illnesses, their origins and treatment. Altern Ther Health Med 9:106–109.

Brosi, B.J., M.J. Balick, R. Wolkow, R. Lee, M. Kostka, W. Raynor, R. Gallen, A. Raynor, P. Raynor, and D. Lee Ling. 2007. Cultural erosion and biodiversity: canoe-making knowledge in Pohnpei, Micronesia. Conserv Biol 21:875–879.

Costion, C.M., A.H. Kitalong, and T. Holm. 2009. Plant endemism, rarity, and threat in Palau, Micronesia: a geographical and preliminary red list assessment. Micronesica 41:131–164.

Herrera, K., D.H. Lorence, T. Flynn, and M.J. Balick. 2010. Checklist of the vascular plants of Pohnpei, Federated States of Micronesia, with local names and uses. Allertonia 10. Allen Press, Lawrence, Kansas.

Lee, R., N. Shere, M.J. Balick, F. Sohl, A.S. Roberts, K. Herrera, S. Dahmer, M. Lieskovsky, A. Dores, W. Raynor, P. Raynor, E. Albert, M. Hunt, C. Trauernicht, L. Offringa, I. Adam, and W. Law. 2010. Pohnpei primary health care manual. Health care in Pohnpei, Micronesia: traditional uses of plants for health and healing. CreateSpace, Charleston, South Carolina.

Pohnpei State Department of Health Services. 2006. Annual Health Services Profile: FY 2006. Pohnpei State Government. Kolonia, Pohnpei, FSM.

Ward, R. 1977. Curing on Ponape: a medical ethnography. Ph.D. thesis. University Microfilm International. Ann Arbor, Michigan.

# PART SIX

APPLIED TECHNIQUES OF CONSERVATION MEDICINE

# 35

## HUMAN HEALTH IN THE BIODIVERSITY HOTSPOTS

Applications of Geographic Information System Technology and Implications for Conservation

## Larry J. Gorenflo

As we enter the second decade of the 21st century, our planet seems out of balance on several fronts. One of the most apparent problems is persisting widespread poverty, and associated human misery, throughout much of the world (World Bank 2010). Despite the definition by the United Nations of "Millennium Development Goals" to improve the human condition, progress toward meeting fundamental human needs and broader achievement of basic human rights have been uneven, and in many countries likely will fall well short of 2015 targets (UN 2010). Another serious problem is the deterioration of natural systems, a consequence of a rapidly expanding human footprint as natural resources are extracted at unprecedented rates to support the increasing demands of Earth's human inhabitants (Wackernagel and Rees 1996; UNDP et al. 2000; Sanderson et al. 2002; Chivian and Bernstein 2008). Conservationists have long pointed to the biological implications of such pressure on nature, noting that species loss at rates 1,000 times or more greater than historical background levels indicates mass extinction of a magnitude witnessed only a few times in our planet's entire history (Pimm et al. 1995). As global human

population continues to grow by more than 200,000 per day (Gorenflo 2006), the challenges of improving human well-being at a large scale and maintaining key natural components of our world will grow accordingly in coming decades (J.E. Cohen 1995, 2003; Cincotta and Engelman 2000).

At first glance, the plights of humans and non-humans appear largely at odds, with meeting human needs seemingly compromising the needs of nature (Ferraro 2002; Sanderson and Redford 2003; Roe and Elliott 2004; Chan et al. 2007). Apparent competition between people and nature emerges on a global scale in terms of the growing human appropriation of the Earth's primary productivity (Vitousek et al. 1997) and in modification of roughly 50% of the planet's surface for human use, the total converted anticipated to increase to 70% in coming decades (FAO 2002; UNDP 2002). However, the seventh Millennium Development Goal, "Ensuring Environmental Sustainability," alludes to a necessary connection between people and nature in the form of mutual benefits, arguing for improvements to the natural environment in the interest of promoting long-term human well-being (UN 2000). In recent years, this relationship between the human condition and nature has been defined more broadly in terms of *ecosystem services*—the benefits to people of functioning ecosystems categorized as provisioning (e.g., food, water), regulating (e.g., climate regulation), cultural (e.g., spiritual, aesthetic), and supporting (e.g., soil formation) services (Millennium Ecosystem Assessment 2005; Melillo and Sala 2008). The benefits to humans from nature, through maintaining natural cycles upon which humans and other species rely (Fisher 2001), and the consequences of interrupting such cycles, are increasingly accepted, providing a link between people and the natural environment that introduces potentially tangible contributions of conservation to human well-being (Rosenzweig 2003). But how these relationships play out can vary, depending on the human systems, associated natural conditions, and links between them.

This chapter examines the relationship between people and the natural environment by focusing on human health in 34 biodiversity hotspots, regions of global importance for conserving the diversity of life on our planet. In addition to their role in conservation, hotspots contain large numbers of people who affect their surroundings and in turn are affected by those surroundings. The approach used here explores human health in hotspots by estimating values for selected health indicators, both to define general health conditions in individual regions and to enable comparisons among regions. The study begins by examining health status in the hotspots, revealing a wide range of variability. It then examines possible connections between human health and the natural environment at a regional scale, considering apparent benefits of maintaining natural habitat amid the broad influence of poverty. Attention then shifts to subregional analyses of infant mortality to explore the health implications of natural habitat in the hotspots and the potential connections between maintaining habitat, diarrheal diseases, and the compromised water sources that often transmit these diseases. The chapter closes by proposing more geographically focused analyses to identify specific settings where conservation

can complement more conventional development interventions that emphasize public health.

## SELECTED HUMAN WELL-BEING INDICATORS IN THE BIODIVERSITY HOTSPOTS

This chapter focuses on biodiversity hotspots (Fig. 35.1). Biodiversity is the diversity of life on Earth, measured in terms of genes, species, populations, and ecosystems (Wilson 2002; Pimm et al. 2008). Hotspots represent one of several templates proposed to define global biodiversity conservation priorities (Myers et al. 2000; Brooks et al. 2006), here focusing on a combination of unique biological contents (species *endemic* to each hotspot) and human threat. Originally conceived by Myers (1988), who identified 10 such regions, conservationists currently define 34 hotspots as regions containing minimally 1,500 endemic vascular plant species and having lost at least 70% of their original habitat (Mittermeier et al. 2004a). Totaling only about 2.3% of the Earth's terrestrial surface, the remaining original habitat in 34 hotspots contains more than 50% of the world's vascular plant species and at least 42% of all terrestrial vertebrate species as endemics. Hotspots are important to biodiversity conservation precisely because of the high levels of endemism they contain. Loss of an endemic species in a hotspot marks its extinction, and in light of high levels of threat in the hotspots widespread loss is imminent without conservation.

Given the large amount of habitat loss in the hotspots, clearly these regions all have a substantial human presence. Studies using geographically referenced global population data have yielded estimates of population in these regions. In 1995, approximately 1.1 billion people inhabited the 25 hotspots defined at the time of that analysis (Cincotta et al. 2000). Using data tied to the most recently available round of decennial censuses (CIESIN and CIAT 2005), a subsequent analysis reported that by 2000 population in the 34

---

1   Several datasets on global population exist, presenting data in gridded map format at resolutions as fine as 1-km grid cells for years as recent as 2010. Although the 2000 global data are more than a decade old, they are the most recent available data tied to a large number of censuses (generally conducted at the beginning of a decade). More recent global population datasets are based on estimates rather than censuses, introducing an additional source of error in many cases that is desirable to avoid. Results of the most recent round of censuses, conducted in or around 2010, were not available when I completed this study.

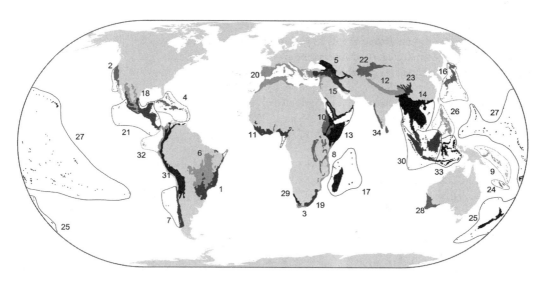

**Figure 35.1:**
Biodiversity hotspots. 1: Atlantic Forest; 2: California Floristic Province; 3: Cape Floristic Region; 4: Caribbean Islands; 5: Caucasus; 6: Cerrado; 7: Chilean Winter Rainfall-Valdivian Forests; 8: Coastal Forests of Eastern Africa; 9: East Melanesian Islands; 10: Eastern Afromontane; 11: Guinean Forests of West Africa; 12: Himalaya; 13: Horn of Africa; 14: Indo-Burma; 15: Irano-Anatolian; 16: Japan; 17: Madagascar and the Indian Ocean Islands; 18: Madrean Pine-Oak Woodlands; 19: Maputaland-Pondoland-Albany; 20: Mediterranean Basin; 21: Mesoamerica; 22: Mountains of Central Asia; 23: Mountains of Southwest China; 24: New Caledonia; 25: New Zealand; 26: Philippines; 27: Polynesia-Micronesia; 28: Southwest Australia; 29: Succulent Karoo; 30: Sundaland; 31: Tropical Andes; 32: Tumbes-Chocó-Magdalena; 33: Wallacea; 34: Western Ghats and Sri Lanka.

hotspots totaled 1.9 billion, or roughly one third of the global population at the time (Mittermeier et al. 2004b; Fig. 35.2).[1] In 22 of the 34 hotspots, population density exceeded the global average in 2000 of 45 persons/km², while in 23 cases population growth exceeded the 1.4% worldwide annual rate of increase. The presence of so many people, their numbers in many cases steadily growing, indicates that biodiversity conservation in the hotspots will have to occur in the context of considerable human occupation. With growing demand for limited resources, understanding key dimensions of human occupation in the hotspots is essential to developing conservation strategies that benefit people as well as nature and contribute to the long-term maintenance of biodiversity. One such dimension is human health, an essential component of the human condition.

Data on two health indicators have been compiled at a sub-national level for most of the areas covered by hotspots—infant mortality and percent of children underweight (CIESIN 2005). These sub-national estimates are particularly valuable for present purposes, providing a direct means of calculating their values for each hotspot through the use of geographic information system (GIS) technology. Infant mortality rate is the number of children who die in their first year for every 1,000 live births, with the global data analyzed generally associated with the year 2000 (base data spanning 1990 to 2002). Figure 35.3a shows infant mortality for the hotspots; the resulting values vary widely, from about 1 for Japan to nearly 110 for the coastal forests of East Africa.[2] A closer examination of these results reveals a pattern generally repeated by the other health indicators considered in this study, namely a broad range of values from more desirable levels in hotspots located in developed countries to levels much less desirable in hotspots located in less-developed countries. In the case of infant mortality,

---

2  For infant mortality rate, no data existed for infant deaths in New Caledonia. Because the hotspot corresponds to the entire nation, I inserted the 2000 infant mortality rate of 7.0 to complete the bar chart summarizing regional data (World Bank 2002).

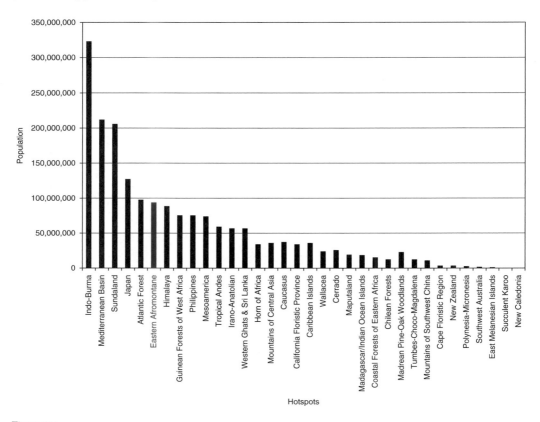

**Figure 35.2:**
Population in the biodiversity hotspots, 2000.

extremely low rates in the developed world contrast markedly with high rates in the less-developed world, the latter including two sub-Saharan Africa hotspots that lose 10% or more of their children in the first year of life.

Children underweight represents the percentage of children aged five years or less whose weight is two standard deviations or more below the median weights established for an international reference population by the U.S. National Center for Health Statistics, U.S. Centers for Disease Control and Prevention, and the World Health Organization (CIESIN 2005). Data again are generally for 2000 (base data covering the years 1990 to 2002). Though focusing on a subset of total population, children underweight is a major risk factor leading to death, particularly in low- and middle-income countries (Skolnik 2008). Figure 35.3b presents the results of this analysis for 25 of the 34 hotspots, with estimates of number of underweight

children, number of children aged five years or less, or both lacking in at least part of the areas covered by the remaining nine hotspots. Once again, we see a broad range of values, from less than 1% of children in the California Floristic Province underweight to more than 40% of the children in the Himalaya hotspot. The contrast between developed and less-developed country continues: hotspots featuring high infant mortality rates also contain higher percentages of children underweight, and vice versa.

Apart from the above two indicators, global data on human health unfortunately tend to be available only at national levels for many of the countries that are partially or totally in the hotspots. However, by using GIS software and global datasets of population in 5-km grid cells (CIESIN and CIAT 2005) to calculate the percentage of each hotspot population contributed by individual countries, one can estimate the value of several indicators of human health in the

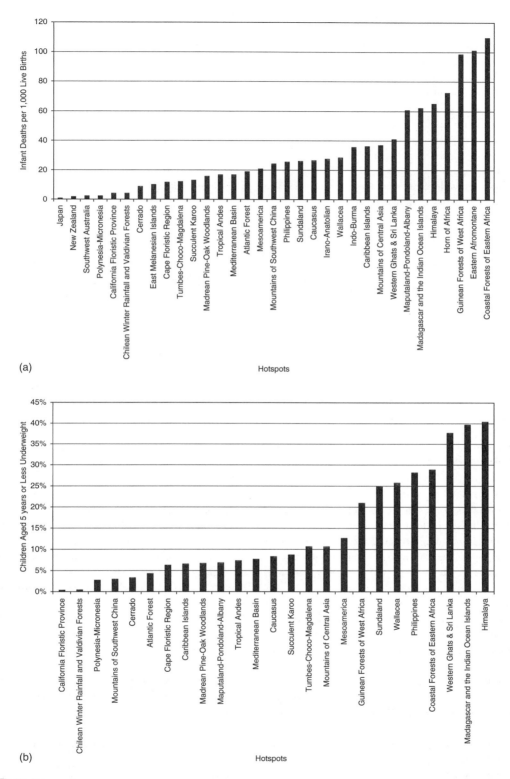

**Figure 35.3:**
Infant mortality rate (a) and children underweight (b) in the biodiversity hotspots, approximately 2000.

34 hotspots—the proportion of an indicator that a given country contributes to a particular hotspot commensurate with the proportion of total hotspot population contributed by that country. This approach is not ideal—it assumes that values of a given health indicator are uniform across an entire nation—but it provides a systematic means of estimation based on the socioeconomic (demographic) presence of individual countries in each hotspot.[3] This method enables use of national-level data compiled by the World Bank (2002) to develop estimates for life expectancy at birth by hotspot in 2000 (Fig 35.4a). Once again, values of this broad indicator of human health vary greatly among the regions, maintaining the contrast between hotspots composed of developed and less-developed countries.

The strong similarities in analytical results for the three health indicators thus far considered—notably the persisting distinction between developed and less-developed regions—suggest that health status in the hotspots is greatly influenced by poverty. Public health has long recognized a central role for poverty in human health (McMichael 2001; MacDonald 2005; Holtz et al. 2008; Skolnik 2008). Inadequate food, poor sanitation, limited access to vaccination and other components of modern medical technology, and insufficient public health programs contribute to health problems in poor nations. Using data compiled at a national level, one can show differences in poverty in the hotpots through estimating per capita gross national income purchasing power parity, a measure expressed in hypothetical international dollars as a common currency available to purchase goods and services at U.S. prices in a given year (in this case, 2000). The resulting graph is consistent with those showing health indicators presented above (Fig. 35.4b). Statistical analyses confirm links between poverty and life expectancy, infant mortality, and percentages of underweight children, the association among these variables borne out by generally strong, significant correlations (Table 35.1).

Using remaining original habitat as a proxy, conservation at a regional scale seems to be much less important to human health in the hotspots. Recall that, by definition, all hotspots have lost minimally 70% of their original habitat (Mittermeier et al. 2004a), thereby likely compromising many of their ecosystem services. But human health in these regions varies widely, and low (and often insignificant) statistical correlations reveal the tenuous connection that health has with the percentage of original habitat remaining in the hotspots (Table 35.1). Characterizing hotspots not solely in terms of original habitat, but also in terms of habitat that had not been converted to human use for settlement, intensive agriculture, or agriculture mixed with other land use (as defined in the Global Land Cover 2000 database, discussed further below; see European Commission, Joint Research Centre 2003), enables further consideration of the contribution of habitat to human health—assuming land cover that may be *natural* though not necessarily *original* provides more ecosystem services than land cover characterized by human use. Again, the statistical association between health and habitat is weak, here lacking statistical significance as well (Table 35.1).

In the case of human health in the biodiversity hotspots, it appears that the penalties of sacrificing natural habitat and functioning ecosystems currently depend to a large extent on poverty. At a regional scale, both percentage of original habitat and percentage of habitat unconverted to human use show little association with any of the three health indicators considered in this study. In contrast, the relationships between poverty and infant mortality rate, children underweight, and life expectancy at birth are strong and statistically significant for three different measures of correlation. These results convey a sense of *decoupling* people from natural systems, suggesting that human health depends primarily on access to modern technology, including improved living conditions, food security, and medical care (McMichael 2001). However, analyses to this point have focused on health indicators measured at a regional scale. Although global studies inevitably encounter constraints of data availability, let us revisit patterns of infant mortality rate in terms of current habitat, using two global datasets whose spatially explicit characteristics enable more precise, sub-regional inquiries.

---

3  The approach used here differs from a recent effort to estimate poverty in the hotspots (Fisher and Christopher 2006). I opted for the present method because it ties estimates of indicators of human well-being to a socioeconomic weighting factor, namely population.

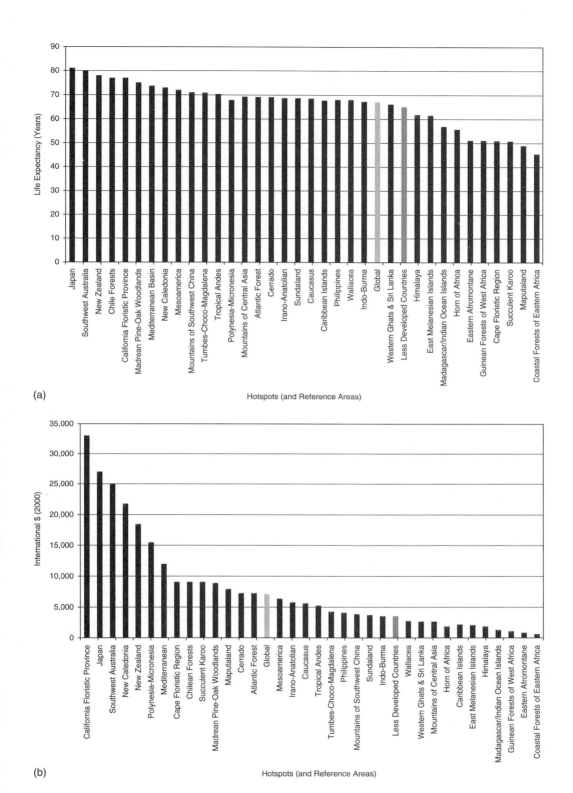

**Figure 35.4:**
Life expectancy at birth (a) and gross national income purchasing power parity (b) in the biodiversity hotspots, approximately 2000 (note the inclusion of overall global values, and values for less-developed countries, of each indicator for reference).

**Table 35.1  Correlations Between Selected Human Health, Income, and Environmental Indicators in the Biodiversity Hotspots**

| Variables | Correlation Measure (if significant, value given in parentheses) | | |
| | Pearson's r | Kendall's tau | Spearman's rho |
| --- | --- | --- | --- |
| IMR × GNI | −0.59 (0.01) | −0.72 (0.01) | −0.86 (0.01) |
| CUW × GNI | −0.54 (0.01) | −0.55 (0.01) | −0.73 (0.01) |
| LEX × GNI | 0.57 (0.01) | 0.55 (0.01) | 0.65 (0.01) |
| IMR × OHAB | −0.35 (0.05) | −0.26 (0.05) | −0.38 (0.05) |
| CUW × OHAB | −0.26 | −0.17 | −0.22 |
| LEX × OHAB | 0.13 | 0.07 | 0.11 |
| IMR × CHAB | 0.06 | 0.00 | 0.01 |
| CUW × CHAB | −0.14 | −0.10 | −0.12 |
| LEX × CHAB | −0.22 | −0.06 | −0.08 |

IMR, infant mortality rate; CUW, percentage children aged five or less underweight; LEX, life expectancy at birth; GNI, gross national income per capita purchasing power parity; OHAB, percentage of original habitat remaining in hotspot; CHAB, percentage of habitat unconverted for human use currently in hotspot.

## INFANT MORTALITY AND HABITAT IN THE HOTSPOTS: POTENTIAL HEALTH BENEFITS OF UNCONVERTED LAND

Global datasets provide important opportunities to conduct large-scale inquiries, and gain new insights, on many key issues across the face of our planet. Unfortunately, as noted, most data on health and related development indicators are available only at national levels. Although one can develop and apply methods to estimate values of these indicators for areas composed of portions of countries, studies based on such estimates are limited when they involve phenomena that require consideration of more precise spatial relationships of co-occurrence or proximity. This is certainly the case with the analyses discussed above. Although at a regional scale health in the hotspots appears unrelated to more natural types of habitat, health status in particular areas with specific habitat conditions, for example, remains unknown. More spatially precise data enable analyses that consider such sub-regional questions. In the present research setting, sub-national datasets on infant mortality and land cover, examined above in regional contexts, support further examination of the relationship

between human health and habitat in the biodiversity hotspots.

Sub-national data on infant mortality were generated by a project at the Center for International Earth Science Information Network (CIESIN) at Columbia University to map selected poverty indicators at sub-national levels (CIESIN 2005). Using data for more than 10,000 national and sub-national geographic units, that project developed global datasets on infant deaths (deaths in the first year of life) and live births generally for the year 2000. Data are in the form of a 0.25-degree global grid, each grid cell measuring roughly 28 km to a side at the equator, becoming slightly smaller towards the poles. These data provide information on the locations of births and infant deaths globally—allowing the calculation of infant mortality for entire hotspots, as discussed above, as well as for specific locations within the hotspots, with the exception of New Caledonia (which lacks sub-national data).

Sub-national data on land use and land cover, essential to understanding global environment at the onset of the new millennium, were generated by the European Commission's Global Land Cover 2000 Project (Fritz et al. 2003; European Commission Joint Research Centre 2003). That project used satellite imagery from two brief periods in the year 2000—primarily SPOT-4 Vegetation Vega2000 data, augmented by Moderate Resolution Imaging Spectroradiometer (MODIS) and Advanced Very High Resolution Radiometer (AVHRR) data to fill in certain gaps—and employed the Food and Agriculture Organization's Land Cover Classification System to categorize the imagery (DiGregorio and Jansen 2000). Project organization consisted of more than 30 teams of regional specialists who examined one or more of 19 regions where they had expertise, yielding a global database that integrates considerable local knowledge. Resulting data consisted of a 30 arc-second global grid, each cell measuring about 1 km to a side at the equator (slightly smaller towards the poles) and classified as one of 22 types of land cover. These data enable identification of land cover and land use for each grid cell within the biodiversity hotspots.

Using GIS software and the Global Land Cover 2000 database enabled definition of areas within the biodiversity hotspots that had been converted to human use, here comprising land cover categorized as artificial surfaces and associated areas (e.g., human settlements), cultivated and managed areas (primarily intensive agriculture), mosaics of cropland and tree cover, and mosaics of cropland and shrub or grass cover. GIS analysis of the sub-national data on births and infant deaths, in turn, enabled calculation of infant mortality in each hotspot for areas converted to human use and for areas that have not been converted (Fig. 35.5). Of the 33 hotspots for which sub-regional infant mortality data exist, 21 have lower infant mortality rates in areas that have not been converted to human use. Among the hotspots where infant mortality is lower in unconverted habitat are the majority of biodiversity regions with high poverty, the main

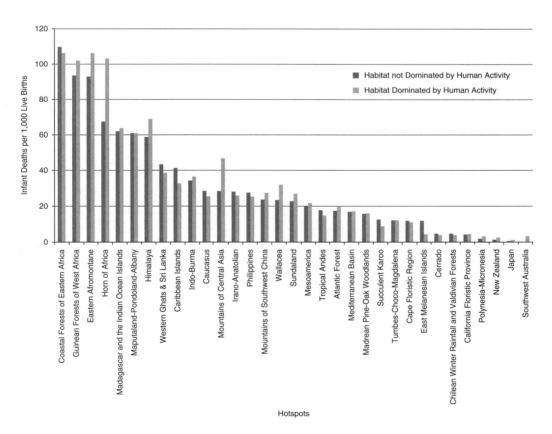

**Figure 35.5:**
Infant mortality rate in areas converted for human use and areas not converted for human use, approximately 2000.

exception being the Coastal Forests of Eastern Africa; ironically, this region also features the highest infant mortality rate of all the hotspots.

The results shown in figure 35.5 reveal a tendency for infant mortality rate in the biodiversity hotspots to be lower in areas characterized by some form of natural habitat. Given the correlation between infant mortality and other health indicators estimated for the hotspots at a regional level (Table 35.2), this pattern might be the case for human health in general, though as those other indicators are unavailable sub-nationally on a global scale, this proposition is impossible to confirm. Localities that have not been converted to human use likely provide more ecosystem services than areas of converted habitat, and lower infant mortality may represent benefits of these services. Of course, one cannot rely on a simple explanation of these results. For instance, as a prime example of areas converted to human use, cities tend to have fewer functioning ecosystem services than areas of unconverted habitat. However, throughout much of human prehistory and history, concentrations of people in urban settings frequently experienced poorer health conditions than their counterparts living elsewhere for a range of reasons, often due to the generally poor conditions that accompany large numbers of humans living in close proximity and the increased ease with which communicable diseases are transmitted in such settings (M.N. Cohen 1989; Paine and Storey 2006). This pattern continues in many cities today, particularly in less-developed countries where conditions are poor due to cramped living conditions, accumulation of human and other forms of waste, and inadequate access to safe water (McMichael 2000; Stephens and Stair 2007). However, cities also often provide improved access to better healthcare and other services than rural localities, thereby at least partially countering the tendency for this type of converted habitat to be less healthy. Based on the data available, it is impossible to explain precisely what processes underlie the results shown in Figure 35.5. Nevertheless, those outcomes indicate a clear tendency for localities with natural habitat to have lower infant mortality than localities where habitat has been converted.

Is it reasonable to expect that natural habitat might lead to lower infant mortality in some hotspots? Part of the answer to this question may lie in the maladies that account for many infant deaths, and in the ecosystem components associated with them. Infant mortality is a complex phenomenon with several causes. Historically, diarrheal diseases have been a major cause of infant death (Hall and Drake 2006).

**Table 35.2  Correlations Between Selected Human Health and Development Indicators in the Biodiversity Hotspots**

| Variables | Correlation Measure (if significant, value given in parentheses) | | |
|---|---|---|---|
| | Pearson's r | Kendall's tau | Spearman's rho |
| IMR × CUW | 0.62 (0.01) | 0.57 (0.01) | 0.75 (0.01) |
| IMR × LEX | −0.75 (0.01) | −0.59 (0.01) | −0.69 (0.01) |
| DDIAR × IMR | 0.90 (0.01) | 0.68 (0.01) | 0.80 (0.01) |
| DDIAR × $H_2O$ | −0.81 (0.01) | −0.55 (0.01) | −0.71 (0.01) |
| DDIAR × GNI | −0.50 (0.01) | −0.61 (0.01) | −0.76 (0.01) |
| IMR × $H_2O$ | −0.77 (0.01) | −0.54 (0.01) | −0.71 (0.01) |
| CUW × $H_2O$ | −0.52 (0.01) | −0.44 (0.01) | −0.55 (0.01) |
| LEX × $H_2O$ | 0.67 (0.01) | 0.51 (0.01) | 0.67 (0.01) |
| GNI × $H_2O$ | 0.68 (0.01) | 0.66 (0.01) | 0.83 (0.01) |

IMR, infant mortality rate; CUW, percentage children aged five or less underweight; LEX, life expectancy at birth; DDIAR, age-adjusted death rate due to diarrheal disease; IMR, infant mortality rate; $H_2O$, percentage population with access to improved water; GNI, gross national income per capita purchasing power parity.

Despite recent progress in reducing mortality from these diseases through improved nutrition, better care-seeking, and rehydration therapy, such maladies continue to plague much of the world (Skolnik 2008). Indeed, diarrheal diseases remain the second leading cause of mortality among children globally (accounting for 18% of child deaths worldwide) and tied for the leading cause of death among children in Africa (WHO 2010b). Diarrhea can be caused by bacteria, viruses, and parasites, including *Shigella* sp., *Salmonella* sp., *Vibrio cholera*, *Escherichia coli*, and rotavirus (Friedman 2008; Skolnik 2008), transmitted by contaminated water or food as well as inadequate sanitation. A plot of age-adjusted death rates (per 100,000 people in the population) due to diarrheal disease estimated for the hotspots in 2004, based on WHO (2009) data, indicates high mortality rates due to these diseases in several high biodiversity regions, the resulting distribution reminiscent of the patterns

shown above for other health indicators (Fig. 35.6; see also Figs. 35.3 and 35.4). The high statistical correlation between diarrheal death rate and infant mortality rate in the hotspots suggests that many infant deaths in these regions are due to diarrheal illnesses (Table 35.2), consistent with observations throughout much of the world (Sampat 2000; Thapar and Sanderson 2004; WHO 2007; Clasen and Haller 2008; WHO and UNICEF 2010).

High correlations between reduced diarrheal death rate and access to improved water sources in the hotspots lend credence to the public health strategy of providing safe water to reduce mortality from diarrhea (Skolnik 2008; Table 35.2). Improving water sources, along with closely related efforts to improve sanitation, are key development interventions used to help reduce the 2.5 billion instances of diarrhea in children each year, and the 1.5 million child deaths that result (Gleick 2002; WHO 2007, 2010a; WHO and UNICEF

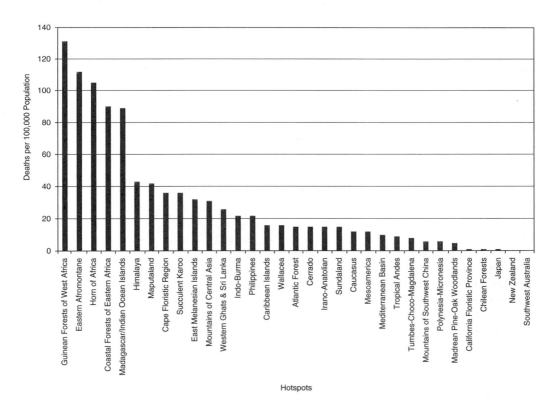

**Figure 35.6:**
Age-standardized mortality rate due to diarrheal disease in the biodiversity hotspots, 2004.

Children's Fund 2010). As seen in Table 35.2, access to improved water sources also tends to be strongly correlated with poverty. Although treatment of many contaminants in water is not difficult, it requires funds that often are unavailable in poor countries (Friedman 2008). In places lacking adequate financial resources to provide improved water sources through technical means, conservation may in particular contribute to human health by helping to maintain natural processes that purify water (IUCN 2000; Pattanayak and Wendland 2007; Vörösmarty et al. 2010).

Natural habitat helps to purify water by supporting natural purification processes (Postel and Richter 2003; Millennium Ecosystem Assessment 2005; Vörösmarty et al. 2005; Postel 2005, 2007). Water quality is particularly susceptible to the benefits of maintaining natural habitat, with certain types, such as forest, found to be especially effective (Bruijnzeel 2000, 2004; Kaimowitz 2004; Scott et al. 2005; Schmoll et al. 2006). Much of water purification as an ecosystem service involves components of the hydrological cycle, slowing runoff (and the erosion and siltation of surface hydrology that often accompanies it) while promoting infiltration that improves the quality of both surface and subsurface water. Although it is impossible to state with certainty based on current data, analysis results presented above augmented by ancillary information on causes of infant death suggest that improved water quality in portions of hotspots where habitat has not been converted to human uses may account for reductions in the infant mortality rates in many of these areas.

## CONCLUSIONS

This study has focused on human health in 34 biodiversity hotspots with the dual aims of describing basic health patterns in these regions and identifying any possible roles that conservation might play in improving health status. In principle, functioning natural ecosystems essential for biodiversity conservation provide important services to people, and some of those services should contribute to better health by purifying air and water, improving agricultural production through nutrient replenishment, protection from certain natural hazards, and so on. But analyses conducted at a regional scale point to poverty—a broad indicator of human well-being—as the key

determiner of health in the hotspots. Neither total percentage of original habitat nor total percentage of unconverted habitat, as proxies for intact nature and ecosystem function, seems to show a strong relationship with the health indicators considered in this study at a regional scale. In the case of original habitat, essential for maintaining the enormous volume of biological diversity in the hotspots, these results challenge the contribution of biodiversity conservation to human health. However, a sub-regional analysis of infant mortality (the one health indicator available at sub-national levels for nearly all of the hotspots), contrasting areas converted for human use against areas not so converted, provides evidence of nature benefiting human health in most hotspots. Given the close connection between infant mortality rate and diarrheal diseases, the tendency for such diseases to involve contaminated water, and the importance of natural processes in improving water quality, maintaining habitat unconverted for human use may contribute to better health in the biodiversity hotspots by providing safer water.

One issue in this chapter that remains unresolved is the role of *biodiversity conservation* in the context of human health. The analysis of infant mortality with respect to habitat types within the hotspots does not specifically target conserved *original* habitat—that is, habitat characteristic of the hotspots prior to broad human impact where much of the resident biodiversity occurs. Instead, it focuses on habitat that has not been converted for human use—habitat that might not be original and, ultimately, might be of limited importance for the conservation of biodiversity. Examples of such habitat include monoculture forests, planted as commercial sources of pulp or timber, represented in the Global Land Cover 2000 dataset as some type of forest but of little value to maintaining biodiversity. Unfortunately, available global data do not allow one to distinguish between original habitat and other forms of habitat that have not been converted for human use, but might have changed considerably from their original state. As shown in research on specific illnesses, the benefits of nature to human health often are tied to functioning ecosystems through complex sets of relationships that help to contain, dilute, or otherwise reduce the transmission of diseases (Patz et al. 2005; Ostfeld et al. 2005; Molyneaux et al. 2008; Keesing et al. 2010). The benefits of conserving original habitat in the hotspots generally will be greater for human health than maintaining unconverted habitat, because of the likelihood that many ecosystem functions in the former will be more intact. In the hotspots, high biological diversity and endemism make original habitat much more important for conservation. Although unable to distinguish between original and other forms of unconverted habitat, analyses in this chapter indicate that natural habitat, whether original or not, can improve human health. Although improvements, in the form of reduced infant mortality rate, often are modest, such results provide additional impetus for maintaining habitat in the hotspots—including key habitat essential for conserving the considerable (and often unique) biodiversity found in each.

One important outcome of the analyses presented above is a sense of how maintaining unconverted habitat can complement more conventional approaches to improving human health in less-developed countries. In the biodiversity hotspots, poor health often is linked to poverty, and both characterize many of the regions examined. In lieu of increasing national income or growing investments in healthcare, and in recognition that development organizations cannot address health problems everywhere, conservation may provide an alternative to more conventional strategies for improving human health, particularly in rural settings featuring natural habitat but unlikely to attract development interventions. Moreover, as a complement to conventional strategies, maintaining unconverted habitat (and conserving any biodiversity that occurs there) provides a sustainable solution susceptible to policy decisions and regional planning, interventions increasingly important as global population (and the challenge of maintaining the health of that population) grows to levels previously unknown (McMichael 2001).

Based on analyses of global data, health improvements by maintaining habitat in the biodiversity hotspots that has not been converted for human use often are marked but not dramatic. In most cases, more conventional interventions presumably will be essential to achieve large reductions in infant mortality. Decisions about where to focus habitat conservation actions to improve human health will require careful analyses of health conditions and natural characteristics at a small geographic scale for specific localities, both to confirm the relationships between health and conservation and to maximize benefits to people

and nature. Such studies should employ detailed data on the human and natural environments that help to identify precise relationships between health and conservation in specific settings. In a world of persisting poverty and rapidly declining biodiversity, actions that potentially benefit both people and nature are certainly actions worth considering.

## REFERENCES

Brooks, T.M., R.A. Mittermeier, G.A.B. da Fonseca, J. Gerlach, M. Hoffmann, J.F. Lamoreux, C.G. Mittermeier, J.D. Pilgrim, and A.S.L. Rodrigues. 2006. Global biodiversity priorities. Science 313:58–61.

Bruijnzeel, L.A. 2000. Forest hydrology. In J.C. Evans, ed. The forests handbook, pp. 301–343. Blackwell Scientific, Oxford, United Kingdom.

Bruijnzeel, L.A. 2004. Hydrological functions of tropical forests: not seeing the soil for the trees? Agr Ecosyst Environ 104:185–228.

Center for International Earth Science Information Network (CIESIN). 2005. Global distribution of poverty datasets (available at http://sedac.ciesin.columbia.edu/povmap/ds_global.jsp#hunger, accessed March 25, 2011).

Center for International Earth Science Information Network, Columbia University (CIESIN) & Centro Internacional de Agricultura Tropical (CIAT). 2005. Gridded population of the world version 3 (GPWv3): Population grids. Socioeconomic Data and Applications Center, Columbia University, Palisades, NY (available at http://sedac.ciesin.columbia.edu/gpw, accessed March 25, 2011).

Chan, K.A., R.M. Pringle, J. Ranganathan, C.L. Boggs, Y.L. Chan, P.R. Ehrlich, P.K. Haff, N.E. Heller, K. Al-Khafaji, and D.P. Macmynowski. 2007. When agendas collide: Human welfare and biological conservation. Conserv Biol 21:59–68.

Chivian, E., and A. Bernstein. 2008. How is biodiversity threatened by human activity? In E. Chivian and A. Bernstein, eds. Sustaining life: how human health depends on biodiversity, pp. 29–73. Oxford University Press, New York.

Cincotta, R.P., and R. Engelman. 2000. Nature's place: Human population and the future of biodiversity. Population Action International, Washington, D.C.

Cincotta, R.P., J. Wisnewski, and R. Engelman. 2000. Human population in the biodiversity hotspots. Nature 404:990–992.

Clasen, T.F., and L. Haller. 2008. Water quality interventions to prevent diarrhea. Cost and cost effectiveness. World Health Organization, Geneva, Switzerland.

Cohen, J.E. 1995. How many people can the earth support? W.W. Norton, New York.

Cohen, J.E. 2003. Human population: the next half century. Science 302:1172–1175.

Cohen, M.N. 1989. Health and the rise of civilization. Yale University Press, New Haven, Connecticut.

DiGregorio, A., and L. Jansen. 2000. Land cover classification system, classification concepts and user manual. Food and Agriculture Organization of the United Nations, Rome, Italy.

European Commission, Joint Research Centre. 2003. Global Land Cover (2000) Database. Data available at http://bioval.jrc.ec.europa.eu/products/glc2000/products.php.

Ferraro, P.J. 2002. The local costs of establishing protected areas in low-income nations: Ranomafana National Park, Madagascar. Ecol Econ 43:261–275.

Fisher, B., and T. Christopher. 2006. Poverty and biodiversity: Measuring the overlap of human poverty and the biodiversity hotspots. Ecol Econ 62:93–101.

Fisher, G.W. 2001. An earth science perspective on global change. In J.L. Aron and J.A. Patz, eds. Ecosystem change and public health, pp. 233–250. Johns Hopkins Press, Baltimore, Maryland.

Food and Agriculture Organization (FAO). 2002. Global forest assessment 2000. Main report. Paper 140. FAO, Rome, Italy.

Friedman, M. 2008. Global perspectives on environmental health. In C. Holtz, ed. Global health care: issues and policies, pp. 419–436. Jones and Bartlett, Boston, Massachusetts.

Fritz, S., E. Bartholomé, A. Belward, A. Hartley, H-J Stibig, H. Eva, P. Mayaux, S. Bartalev, R. Latifovic, S. Kolmert, P.S. Roy, S. Agrawal, W. Bingfang, X. Wenting, M. Ledwith, J-F. Pekel, C. Giri, S. Mücher, E. de Badts, R. Tateishi, J-L. Champeaux, and P. Defourny. 2003. Harmonisation, mosaicing and production of the Global Land Cover 2000 Database (beta version). Directorate-General, Joint Research Centre, European Commission. European Commission, Brussels, Belgium.

Gleick, P. 2002. Dirty water: estimated deaths from water-related disease 2000–2020. Pacific Institute Research Report, Oakland, California.

Gorenflo, L.J. 2006. Population. In E.W. Sanderson, P. Robles Gil, C.G. Mittermeier, V.G. Martin, and C.F. Kormos, eds. The human footprint: challenges for the conservation of biodiversity and wilderness, pp. 63–67. CEMEX, Mexico City, Mexico.

Hall, E., and M. Drake. 2006. Diarrhoea: the central issue? In E. Garrett, C. Galley, N. Shelton, and R. Woods, eds.

Infant mortality: a continuing social problem, pp. 149–168. Ashgate, Burlington, Vermont.

Holtz, C., K. Plitnick, and M. Friedman. 2008. Global perspectives on nutrition. In C. Holtz, ed. Global health care: issues and policies, pp. 369–397. Jones and Bartlett, Boston, Massachusetts.

IUCN (World Conservation Union). 2000. Vision for water and nature. IUCN, Gland, Switzerland.

Kaimowitz, D. 2004. Forests and water: A policy perspective. J Forest Res 9:289–291.

Keesing, F., L.K. Belden, P. Daszak, A. Dobson, C.D. Harvell, R.D. Holt, P. Hudson, A. Jolles, K.E. Jones, C.E. Mitchell, S.S. Myers, T. Bogich, and R.S. Ostfeld. 2010. Impacts of biodiversity on the emergence and transmission of infectious disease. Nature 468:647–652.

MacDonald, T.H. 2005. Third world health: hostage to first world health. Radcliffe, Seattle, Washington.

McMichael, A. 2000. The urban environment and health in a world of increasing globalization: Issues for developing countries. B World Health Org 78:1117–1126.

McMichael, T. 2001. Human frontiers, environments, and disease: past patterns, uncertain futures. Cambridge University Press, New York.

Melillo, J., and O. Sala. 2008. Ecosystem services. In E. Chivian and A. Bernstein, eds. Sustaining life: how human health depends on biodiversity, pp. 75–115. Oxford University Press, New York.

Millennium Ecosystem Assessment. 2005. Ecosystems and human wellbeing. Island Press, Washington, D.C.

Mittermeier, R.A., P. Robles Gil, M. Hoffman, J. Pilgrim, T. Brooks, C.G. Mittermeier, and J. Lamoreux, compilers. 2004a. Hotspots revisited. CEMEX, Mexico City, Mexico.

Mittermeier, R.A., T. Brooks, G.A.B. da Fonseca, M. Hoffman, J. Lamoreux, C.G. Mittermeier, J. Pilgrim, P. Robles Gil, P. Seligmann, K. Alger, F. Boltz, K. Brandon, A. Bruner, J.M. Cardoso da Silva, A. Carter, R. Cavalcanti, D. Church, M. Foster, C. Gascon, L. Gorenflo, B. Gratwicke, M. Guerin-McManus, L. Hannah, D. Knox, W.R. Konstant, T. Lacher, P. Lannghammer, O. Langrand, N. Lapham, D. Martin, N. Myers, P. Naskrecki, M. Parr, D. Pearson, G. Prickett, D. Rice, A. Rylands, W. Sechrest, M.L. Smith, M. Stuart, M. Totten, J. Thomsen, and J. Ward. 2004b. Introduction. In R.A. Mittermeier, P. Robles Gil, M. Hoffman, J. Pilgrim, T. Brooks, C.G. Mittermeier, and J. Lamoreux, compilers. Hotspots revisited, pp. 19–69. CEMEX, Mexico City, Mexico.

Molyneaux, D.H., R.S. Ostfeld, A. Bernstein, and E. Chivian. 2008. Ecosystem disturbance, biodiversity loss, and human infectious disease. In E. Chivian and A. Bernstein, eds. Sustaining life: how human health depends on biodiversity, pp. 287–324. Oxford University Press, New York.

Myers, N. 1988. Threatened biotas: "hotspots" in tropical forests. Environmentalist 8:1–20.

Myers, N., R.A. Mittermeier, C.G. Mittermeier, G.A.B. da Fonseca, and J. Kent. 2000. Biodiversity hotspots for conservation priorities. Nature 403:853–858.

Ostfeld, R.K., G.E. Glass, and F. Keesing. 2005. Spatial epidemiology: An emerging (or re-emerging) discipline. Trends Ecol Evol 20:328–336.

Paine, R.E., and G.R. Storey. 2006. Epidemics, age at death, and mortality in ancient Rome. In G.R. Storey, ed. Urbanism in the preindustrial world: cross-cultural approaches, pp. 69–85. University of Alabama Press, Tuscaloosa.

Pattanayak, S.K., and K.J. Wendland. 2007. Nature's care: diarrhea, watershed protection, and biodiversity conservation in Flores, Indonesia. Biodivers Conserv 16:2801–2819.

Patz, J.A., U.E.C. Confalnieri, F.P. Amerasinghe, K.B. Chua, P. Daszak, A.D. Hyatt, D. Molyneux, M. Thomson, Dr. L. Yameogo, Mwelecele-Malecela- Lazaro, P. Vasconcelos, and Y. Rubio-Palis. 2005. Human health: ecosystem regulation of infectious disease. In Hassan, R., R. Scholes, and N. Ash, eds., Ecosystems and human wellbeing: current state and trends, vol. 1, pp. 391–415. Island Press, Washington, DC.

Pimm, S.L., G.J. Russell, J.L. Gittleman, and T.M. Brooks. 1995. The future of biodiversity. Science 269:347–350.

Pimm, S.L., M.A.S. Alves, E. Chivian, and A. Bernstein. 2008. What is biodiversity? In E. Chivian and A. Bernstein, eds. Sustaining life: how human health depends on biodiversity, pp. 3–27. Oxford University Press, New York.

Postel, S. 2005. Liquid assets: the critical need to safeguard freshwater ecosystems. Paper 170, Worldwatch Institute, Washington, D.C.

Postel, S. 2007. Aquatic ecosystem protection and drinking water utilities. J Am Water Works Assoc 99:52–63.

Postel, S., and B. Richter. 2003. Rivers for life: managing water for people and nature. Island Press, Washington, D.C.

Roe, D., and J. Elliott. 2004. Poverty reduction and biodiversity conservation: Rebuilding the bridges. Oryx 38:137–139.

Rosenzweig, M.L. 2003. Win-win ecology: how the earth's species can survive in the midst of human enterprise. Oxford University Press, New York.

Sampat, P. 2000. Deep trouble: the hidden threat of groundwater pollution. Paper 154, Worldwatch Institute, Washington, D.C.

Sanderson, E.W., M. Jaiteh, M.A. Levy, K.H. Redford, A.V. Wannebo, and G. Woolmer. 2002. The human footprint and the last of the wild. BioScience 52:891–904.

Sanderson, S.E., and K.H. Redford. 2003. Contested relationships between biodiversity conservation and poverty alleviation. Oryx 37:389–390.

Schmoll, O., G. Howard, J. Chilton, and I. Chorus, eds. 2006. Protecting groundwater for health. IWA Publishing, Seattle, Washington.

Scott, D.F., L.A. Bruijzeel, and L.A. Mackensen. 2005. The hydrological and soil impacts of forestation in the tropics. In M. Bonnell and L.A. Bruijnzeel, eds. Forests, water and people in the humid tropics, pp. 622–651. Cambridge University Press, New York.

Skolnik, R. 2008. Essentials of global health. Jones and Bartlett, Boston, Massachusetts.

Stephens, C., and P. Stair. 2007. Charting a new course for urban public health. In L. Starke, ed. State of the world (2007): our urban future, pp. 135–151. W.W. Norton, New York.

Thapar, N. and I. Sanderson. 2004. Diarrhoea in children: An interface between developed and developing countries. Lancet 21:641–653.

United Nations (UN). 2000. United Nations Millennium Declaration. A/RES/55/2, United Nations, New York.

United Nations (UN). 2010. The millennium development goals report 2010. United Nations, New York.

United Nations Development Programme (UNDP). 2002. Global environmental outlook 3. Earthscan, London.

United Nations Development Programme (UNDP), United Nations Environment Programme (UNEP), World Bank, and World Resources Institute (WRI). 2000. A guide to world resources 2000–2001. People and ecosystems: the fraying web of life. World Resources Institute, Washington, DC.

Vitousek, P.M., H.A. Mooney, J. Lubchenco, and J.M. Melillo. 1997. Human domination of earth's ecosystems. Science 277:494–499.

Vörösmarty, C.J. C. Lévêque, C. Revenga, R. Bos, C. Caudill, J. Chilton, E.M. Douglas, M. Meybeck, D. Prager, P. Balvanera, S. Barker, M. Maas, C. Nilsson, T. Oki, and C.A. Reidy. 2005. Fresh water. In R. Hassan, R. Scholes, and N. Ash, N., eds. Ecosystems and human wellbeing: Current state and trends, vol. 1, pp. 165–207. Island Press, Washington, DC.

Vörösmarty, C.J., P.B. McIntyre, M.O. Gessner, D. Dudgeon, A. Prusevich, P. Green, S. Glidden, S.E. Bunn, C.A. Sullivan, C. Reidy Liermann, and P.M. Davies. 2010. Global threats to human water security and river biodiversity. Nature 467:555–561.

Wackernagel, M., and W. Rees. 1996. Our ecological footprint: reducing human impacts on the earth. New Society Publishers, Philadelphia, Pennsylvania.

Wilson, E.O. 2002. The future of life. Knopf, New York.

World Bank. 2002. World development indicators 2002. Compact disk. The World Bank, Washington, D.C.

World Bank. 2010. Global monitoring report 2010: the MDGs after the crisis. The World Bank, Washington, D.C.

World Health Organization (WHO). 2007. Combating waterborne disease at the household level. World Health Organization, Geneva, Switzerland.

World Health Organization (WHO). 2009. Health statistics in support of the global burden of disease, 2004 update. Department of Measurement and Health Information, World Health Organization, Geneva, Switzerland.

World Health Organization (WHO). 2010a. UN-water global annual assessment of sanitation and drinking water (GLAAS) 2010: targeting resources for better results. World Health Organization, Geneva, Switzerland.

World Health Organization (WHO). 2010b. World health statistics 2010. Part II: global health indicators. World Health Organization, Geneva, Switzerland.

World Health Organization (WHO) and United Nations Children's Fund (UNICEF). 2010. Progress on sanitation and drinking water. WHO-UNICEF Joint Monitoring Programs for Water Supply and Sanitation, Geneva.

# 36

# DETERMINING WHEN PARASITES OF AMPHIBIANS ARE CONSERVATION THREATS TO THEIR HOSTS

## Methods and Perspectives

Trenton W. J. Garner, Cheryl J. Briggs, Jon Bielby, and
Matthew C. Fisher

Theoretical evidence and empirical evidence show that costs of parasites to individual hosts scale up: parasites can regulate host populations (Boots et al. 2009). Because parasite extinction is related to the probability of host extinction, host population responses should exhibit thresholds and host extinction due to parasitism should be rare. Parasites contribute directly to biodiversity, and the interactions between hosts and parasites can drive speciation (Little 2002; Vale et al. 2008): as a result, parasites seem unlikely threats to biodiversity. Nevertheless, in recent years, parasitic infections have become widely accepted as primary drivers of host population declines and extinctions (Lafferty 2003; Smith et al. 2009). Whether or not declines due to parasitism are a conservation concern is not always clear, but if the link between infectious disease and decline is through human activities (Ahmed 2000; Cleaveland et al. 2000; Harvell et al. 2002; Forson and Storfer 2006; Krkosek et al. 2007), we must consider infectious disease as we consider habitat alteration, invasive species, pollution, and other recognized threats to biodiversity (Cunningham et al. 2003). Accordingly, infectious disease is increasingly treated as a conservation issue and subject to mitigation for conservation

purposes (Haydon et al. 2006; Jones et al. 2007; Delahay et al. 2009).

Amphibian parasites have become a high-profile case for conservation medicine. At least one, *Batrachochytrium dendrobatidis* (*Bd*), has gained substantial notoriety as a cause of local extirpation and possible extinction of amphibian species (Laurance et al. 1996, 1997; Berger et al. 1998; Bosch et al. 2001; Muths et al. 2003; La Marca et al. 2005; Schloegel et al. 2005; Lips et al. 2006, 2008). Others are notable for causing substantial mortality and sustained declines in host populations (Blaustein et al. 1994; Green et al. 2002; Brunner et al. 2004; Ariel et al. 2009; Teacher et al. 2010). Human activities have influenced the prevalence and virulence of amphibian parasites (Forson and Storfer 2006; Fisher and Garner 2007; Picco and Collins 2008; Rohr et al. 2008; St-Amour et al. 2008; Walker et al. 2008). In response, the amphibian research community has mobilized substantial efforts into describing the distribution and prevalence of amphibian parasites and mitigation in the field (Muths and Dreitz 2008; Lubick 2010).

But does research generally support the hypothesis that amphibian parasites are conservation threats? Conclusions are inconsistent. In a few cases irretrievable population declines can be clearly attributed to an infectious disease, but the greater picture is that amphibian parasites, like so many others, are contextual threats (Brunner et al. 2004; Briggs et al. 2005, 2010; Rachowicz and Briggs 2007; Bielby et al. 2008; Teacher et al. 2010; Vredenburg et al. 2010; Walker et al. 2010). Context remains poorly understood, not least due to the disturbing habit of sampling for parasites or their DNA without any concurrent effort to understand the nature of the host and parasite interaction or to find solid epidemiological evidence for a host conservation threat (Duffus 2010). This practice has led to, and will continue to lead to, misallocation of finite conservation resources, resources generally inadequate for comprehensive species conservation (Isaac et al. 2007).

In this chapter we outline some familiar concepts to elucidate whether infectious diseases are potential conservation threats to amphibian hosts. It is by no means comprehensive: for example, we do not discuss the necessity for proper postmortem examinations (see Pessier and Pinkerton 2003) or appropriate field and statistical methods for assessing amphibian population dynamics or individual survival (see Heyer et al. 1994 and Dodd Jr. 2010). Our goal is to promote collaborative, integrated research that brings together relevant skill sets. Our collective experience encompasses the four broad topics: risk factor analysis, spatial and molecular epidemiology, experimental assessment of costs to hosts, and modeling transmission and disease dynamics. We hope that this chapter reinforces the belief that transdisciplinary teamwork is required to understand host–parasite dynamics and, where necessary, guide conservation interventions (Skerratt et al. 2009).

## RISK FACTOR ANALYSIS

In the context of wildlife disease, the term "risk assessment" generally refers to an assessment of risks posed by animal movement protocols (Miller 2007). In a wider context, a species risk assessment develops guidelines applied to assessing risk of extinction (http://www.iucnredlist.org/). To avoid confusion over terminology, we restrict ourselves to discussing risk factor analyses, aimed at identifying traits or characteristics that render species or populations susceptible to a given threat, such as disease. The identified

characteristics may then be used to further identify species or populations that are highly susceptible to the threatening process.

Identifying risk factors or taxa and locations that may be particularly susceptible to decline due to infectious disease is an informed, alternative approach to the "triage style" focused on recognizably threatened species, or "hotspots" (Ricketts et al. 2005). Ideally the output would be a set of factors that have both high predictive accuracy and broad generality: as always in biology, it is impossible to meet all goals simultaneously (Morin 1998; Werner 1998). Important risk factors will vary among taxonomic groups, threatening processes, responses of interest, and the geographic scale at which the analysis is conducted (Purvis et al. 2000; Fisher and Owens 2004; Isaac and Cowlishaw 2004). The first two sources of variation are determined by definition for amphibian infectious disease (higher taxonomic group and threatening process). Two other sources of variation of consistent importance are host response and the geographic scale of the analysis.

Responses of amphibian hosts when exposed to parasites tend to scale with decreasing probability (Fig. 36.1), and selecting what host response for which to identify risk factors depends on the ultimate objective of the analysis. For example, determining which species are susceptible to a parasite may focus on risk factors associated with infection and transmission without considering more advanced states of disease. Accordingly, as important risk factors will vary depending on the response of interest, analyses should have clear hypotheses as to why traits may be linked to

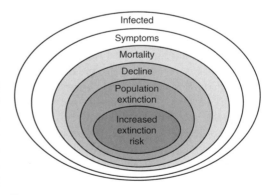

Figure 36.1:
Potential outcomes following exposure of a host to a parasite.

a response. The list of candidate risk factors associated with infection and transmission may include behavioral or evolutionary factors that increase the likelihood of encounter with an infected and permissive host. Alternatively, if an analysis is aimed at identifying species or populations that can or will experience mortality as a result of disease, the environmental drivers of disease or aspects of host biology that affect the parasite's ability to grow after infection has occurred may better predict risk (Brunner et al. 2004; Bosch et al. 2007; Briggs et al. 2010; Walker et al. 2010). Clarity in generating hypotheses is the preferable alternative to "fishing" through a large number of risk factors, as the latter approach inevitably leads to chance positive associations. Although multiple comparison $p$-value adjusters may be employed, these may be somewhat draconian in practice.

The choice of geographic scale requires similar clarity of logic. Bielby et al. (2008) chose a global-scale analysis to maximize the general utility of identified risk factors for determining increased extinction risk due to Bd. They used a broad measure of decline, movement up risk categories of the IUCN Red List, as a response, and all anuran species were selected as the taxonomic unit. The results suggested that components of host life history and geographic distribution on a broad scale associated with increased risk of disease-driven decline. Perhaps more importantly, the identified risk factors were used to model which geographic regions might harbor a large number of highly susceptible species and for which disease data were lacking. Bielby et al. (2008) did make a number of assumptions that limited the predictive accuracy of the study and the ability to account for regional differences in host–parasite dynamics. One of these, homogeneity of species response to Bd, has been invalidated. The outcome of exposure to Bd varies temporally and spatially (Vredenburg et al. 2010), depending upon transmission processes (Briggs et al. 2010) and environmental forcing of disease (Walker et al. 2010).

An alternative approach would be to produce fine-scale analyses on regional species assemblages. Along with better predictive accuracy, it also eliminates the need to account for variation in disease dynamics and epidemiology across broad geographic scales. Since conservation management is conducted at the population level, regional analyses are perhaps more likely to achieve on-the-ground conservation

goals (Fisher and Owens 2004). Several studies (Williams and Hero 1998; Lips et al. 2003; Walker et al. 2010) illustrate how knowledge of local species ecology, life history, and environment inform risk factor analyses for responses associated with infectious diseases. These analyses have produced sets of risk factors that should enable highly accurate predictions within the geographic region of interest but come with an associated loss of generality. Variation in risk factors reported by Williams and Hero (1998) and Lips et al. (2003) illustrates this. More importantly, we usually lack fine-scale knowledge regarding species population trends, population response(s) to different threat processes, or species ecology and life history necessary for conducting detailed, regional analyses.

Comparisons of risk factors over multiple responses and scales are the most appropriate way to identify variables consistently important for disease dynamics. For example, across studies there is a consistent association of Bd disease dynamics with an aquatic life history stage (Williams and Hero 1998; Lips et al. 2003; Bielby et al. 2008), which is consistent with the aquatic zoospore responsible for Bd infections. But should process be inferred from risk factor analyses? Evidence for Bd disease dynamics in strictly terrestrial amphibian hosts suggests that transmission is not strictly dictated by the availability of water (Weinstein 2009). While risk factor analyses may highlight key variables important for risk of disease, they cannot unambiguously identify the process at work.

## SPATIAL AND MOLECULAR EPIDEMIOLOGY

Host–parasite relationships are dynamic, but even Hippocrates recognized two extremes of disease, epidemic versus endemic. While damaging to hosts, endemic relationships are relatively stable, long-term associations often involving significant co-evolutionary histories. Epidemics are characterized by sudden increases in the incidence and intensity of infections. Many well-known epidemic diseases are cyclical, regulated by the immune status of the population, and exhibit classical susceptible–infected–recovered (SIR, Box 36.1) dynamics (Cavanagh et al. 2004; Grassly et al. 2005), but uncommonly virulent epidemics can

---

**Box 36.1  Example of a Simple Susceptible Infected Resistant (SIR) Model**

---

$S$ = number of Susceptibles, $I$ = number of Infecteds, $R$ = number of Recovered and Resistant individuals (if the host develops immunity to the disease), and $N = S+I+R$ is the total population size. $\beta$ is the transmission parameter, $\gamma$ is the per-capita recovery rate (if there is no immunity, recovered individuals return to the susceptible class, rather than the resistant class), $d$ is the background death rate due to causes other than the disease, and $\alpha$ is the increased death rate of infected individuals due to the disease.

The equation for the rate of change of infected individuals in the population is:

$dI/dt = (transmission\ function) - (d + \alpha + \gamma)I$

The basic reproductive ratio, $R_0$, of a disease is the expected number of new infections caused by a single infected individual in a fully susceptible population. The disease has the potential to invade the host population and cause an epidemic if $R_0 > 1$.

With a density-dependent transmission function, $(\beta SI)$, $R_0 = \dfrac{\beta N}{(d+\alpha+\gamma)}$ . This leads to a threshold

population size $N_T = \dfrac{(d+\alpha+\gamma)}{\beta}$ , such that the parasite is not expected to lead to an outbreak in

populations smaller than this threshold. Similarly, when the disease suppresses the number of susceptible individuals in a population to a level below this threshold, then the parasite will start to decline (because $dI/dt < 0$). It is because of this population threshold that microparasitic diseases with density-dependent transmission are not expected to be able to lead to extinction of their host population.

With a frequency-dependent transmission function, $(\beta SI/N)$, $R_0 = \dfrac{\beta}{(d+\alpha+\gamma)}$ . The basic

reproductive ratio of the parasite does not depend on the population size, and no such host population threshold exists. If $R_0 > 1$, the disease can continue to have a positive growth rate even if the host population is suppressed to very low densities. Therefore, simple models predict that microparasitic diseases with frequency-dependent transmission have the potential to lead to extinction of their host.

---

cause significant mortality that can result in population decline, or even extinction. Typically this happens when there is no reservoir of individuals that have preexisting immunity and is often related to parasite introduction (de Castro and Bolker 2005). Some of the greatest disease impacts on wildlife were caused by introduced parasites (Plowright 1982; Berger et al. 1998; Lips et al. 2006, 2008; LaDea et al. 2007). The potential for parasite introduction through anthropogenic activities (Karesh et al. 2005; Fisher and Garner 2007; Picco and Collins 2008; Walker et al. 2008) inevitably leads to emerging infectious disease events, so conservation medicine professionals must reliably discriminate between endemic, cyclical,

and novel emerging patterns of disease dynamics to characterize threat to hosts.

Broadly, two hypotheses account for the emergent nature of a novel, epidemic parasite. The "novel parasite hypothesis" (NPH) is based on a model of invasive parasite spread, usually from an unidentified source, while the "endemic parasite hypothesis" (EPH) describes the emergence of disease as a consequence of an altered relationship between host and parasite due to environmental change (Rachowicz et al. 2005). The ability to distinguish between the NPH and EPH is one key to our understanding of the emergence of disease, but some confusion exists as to what patterns of infectious disease discriminate between the two. Amphibian populations are dynamic and are characterized by a significant mortality (Petranka 1989; Peterson et al. 1991). Even if parasites are involved, mortality and decline may only be a reflection of an EPH process—as populations increase, the prevalence of preexisting endemic parasites can increase to levels where mortalities due to infectious disease will also increase. Seasonal environmental changes also affect the probability of amphibian mortality (Reading 2007; Loman 2009), and amphibian immune responses have a temperature-dependent component (Raffel et al. 2006; Ribas et al. 2009). Consequently, seasonal changes in morbidity and mortality caused by endemic parasites will occur.

If variable levels of disease are not diagnostic, understanding what constitutes "normal," forecastable endemic disease dynamics requires knowledge of what parasites occur and how prevalence fluctuates, but these baseline data for wildlife populations are rare. A few "natural laboratories" exist, such as the long-studied Soay sheep of St. Kilda, where nematode parasite burdens, environmental variation, and mortality have been recorded since the 1980s (Paterson et al. 1998). Long-term datasets are essentially nonexistent for amphibians, so what constitutes "normal" disease dynamics is uncertain. An alternative is to compile epidemiological data to identify the spatial signature of disease spread. As an example, the emergence of *Bd* in amphibians was originally identified in the 1990s by an international consortium of researchers comparing data on simultaneous amphibian declines and mortality in Central America and Australia. They soon identified a common pathology, and an etiological fungal agent (Berger et al. 1998), and the disease, chytridiomycosis, was designated a new phenomenon observed in amphibian communities and regions. New technologies in online mapping and informatics used by the *Bd* Global Mapping Project have simplified the rapid reporting of *Bd* infections, leading to the first global maps of the distribution (Olson and Ronnenberg 2008, www.bd-maps.net). Other amphibian disease mapping projects include the Abnormal Amphibian Monitoring Project (http://www.fws.gov/contaminants/Issues/Amphibians.cfm) and the Frog Mortality Project (http://www.froglife.org/disease/frog_mortality_project.htm). These projects show that when a collective will to contribute surveillance data to a common pool exists, novel insights into the spatial extent of a novel parasite will accrue rapidly. The reporting of any amphibian mortalities is possible using these tools, but without rigorous pathology, mapping databases report only phenomenological, not epidemiological, data. Pathology will always be a bottleneck in tracking parasites and discriminating between endemic and novel parasites, but breakthroughs in parasite identification, detection, and genotyping will rapidly improve our ability to report and map the emergence of infectious diseases.

The initial, non-equilibrium invasion by an emerging parasite is best considered using epidemiological metapopulation transmission models with dispersal kernels inferred from spatial data. However, transmission models have found limited use as a method for inferring rates of spread of emerging infectious disease of amphibians. This is likely due to the limited and incomplete data that accompany the early stages of disease emergence. Reliable estimates require fine-grain knowledge of spread rates in well-studied environments (e.g., Pyrenees, Walker et al. 2010; the Sierra Nevadas, Rachowicz and Briggs 2007 and Vredenburg et al. 2010; southeastern England, Teacher et al. 2010) and an understanding of the vectors that are specific to the transmission of amphibian parasites. This is no small job, as the majority are multihost parasites and may infect non-amphibian species (Mao et al. 1999), so credible efforts to use transmission modeling to forecast disease spread are likely to be difficult for most amphibian parasites. In some cases, the outputs of mapping projects will describe the error that surrounds estimated dispersal kernels, enabling the prediction of spatial disease transmission to be undertaken using the modeling methods used in other fields of disease-outbreak prediction (Riley 2007). On the other hand, equilibrium models using Pavlovsky's framework recognize that (1) disease is geographically proscribed; (2) spatial variation arises due to variation in ecological conditions and host diversity; and (3) mapping of biotic and abiotic variables should enable prediction of contemporaneous and future change in risk (Pavlovsky 1966; Ostfeld et al. 2005). Foremost to the analysis of parasite surveillance data is the demonstration of "spatial heterogeneity" of relative risks. Under the terminology of Wakefield et al. (2000), areas that show elevated levels of residual relative risks are defined as disease clusters.

Surveillance of the Iberian peninsula for *Bd* has shown that the parasite is widespread and occurs in several species (Bosch et al. 2006), at least three of which exhibit mortality. Walker et al. (2010) focused on a single species, the common midwife toad (*Alytes obstetricans*), and used Bayesian statistical approaches to describe relationships among environmental factors determining the prevalence of *Bd* infection and mortality. The Bayesian School of statistical inference provides a means of quantifying uncertainty, a feature particularly valued by epidemiologists (Carabin et al. 2003; Basáñez et al. 2004; Clements et al. 2006). Assuming that *Bd* distribution had reached demographic equilibrium, the authors showed there was no significant relationship between the *presence* of *Bd* and environmental variables. Temperature metrics and ultraviolet radiation predicted the *prevalence* of infection within infected populations, while mortality was highly correlated with temperature and altitude, demonstrating how extrinsic factors have the potential to modulate the host–parasite relationship, including the expression of disease.

Underlying the spatial epidemiology of a parasite is its molecular counterpart. Expectations for the spatial and temporal distribution of genotypes can be generated under the contrasting NPH/EPH hypotheses. If *Bd* spread from South Africa via *Xenopus* trade, as hypothesized by Weldon et al. (2004), the center of genetic diversity for *Bd* should be there. Furthermore, the genetics of the global population of *Bd* derived from an early long-distance dispersal event will be recognizably different from the genetics of *Bd* in demographic equilibrium (EPH). Rapid international dispersal as per the NPH model of emergence should result in no or a weak relationship between multilocus genotypes and geographic distance and *Bd* genotype distributions will be correlated with global amphibian trade (Fisher et al. 2009a). Existing molecular studies have confirmed some of the expectations of the NPH model but were hampered by extraordinarily low nucleotide polymorphism. Nonetheless, *Bd* isolates from temperate North America and Europe exhibit slightly greater genetic diversity than *Bd* from tropical America, Australia, and Africa, refuting somewhat out-of-Africa emergence. Diversity was greater in isolates derived from North American bullfrogs than *Xenopus*, the second of two proposed international vectors of *Bd* (Garner et al. 2006; James et al. 2009). Overall global diversity could be explained by the widespread distribution of a single diploid *Bd* and

subsequent diversification through mitotic recombination, a pattern reflected at finer scales. *Bd* appears to have been widely and repeatedly introduced into Iberia (Walker et al. 2008, 2010) and the Sierra Nevada Mountains (Morgan et al. 2007), and expanded its range in both locations. Local genetic diversification is also reported (Morgan et al. 2007). Notably, none of these studies could clearly identify the source for any of the proposed introductions.

## USING EXPERIMENTS TO MEASURE COSTS TO HOSTS

Published experiments on amphibian hosts and parasites include molecular investigations of host responses, single species or stage exposure trials, and complex, multihost mesocosm experiments examining infection and transmission dynamics (e.g., Warkentin et al. 2001; Schotthoefer et al. 2003; Pearman and Garner 2005; Maniero et al. 2006; Brunner et al. 2007; Cunningham et al. 2007; Rachowicz and Briggs 2007; Woodhams et al. 2007; Greer et al. 2008; Johnson et al. 2008; Morales and Robert 2008; Johnson and Hartson 2009; Murphy et al. 2009). Whether these studies inform conservation varies widely. Experiments are artificial, and conclusions drawn from experiments may not translate to nature (Morin 1998). This is especially true of experiments that are pure tests of theory, where amphibian hosts are selected on the basis of their suitability for hypothesis testing rather than imposed by conservation imperatives. Experiments are more relevant (and often more publishable) when presented with field studies and/or epidemiological models (Johnson et al. 1999; Brunner et al. 2004; Briggs et al. 2005; Rachowicz and Briggs 2007; Storfer et al. 2007; Mitchell et al. 2008; Garner et al. 2009b; Johnson and Hartson 2009).

Following the spirit if not the letter of Koch's postulates (Koch 1884), testing causality should allow a researcher to conclude that exposure to the parasite incurs biologically meaningful costs under biologically relevant circumstances. One possible approach, the use of graded doses to elicit graded host responses, has been used effectively for several amphibian infectious diseases (Schotthoefer et al. 2003; Pearman et al. 2004; Carey et al. 2006). This approach puts host responses into a biologically meaningful framework because increasing density of parasites is expected to relate positively with the probability of infection and disease (Anderson and May 1982; Ebert et al. 2000). The relationship may be continuous (classic dose dependence) or stepwise (threshold responses). Showing cause and effect, though, does not necessarily illustrate the process through which a parasite imposes costs. Voyles et al. (2009) provide robust evidence for the physiological process a host undergoes when dying from infection with *Bd*, but how the fungus elicits this remains unclear. This contrasts sharply with the current experimental evidence for how ranaviruses evade host immunity, inhibit host DNA, RNA, and protein synthesis, and induce host cell apoptosis (Chinchar et al. 2009).

A thorough understanding of the disease process may not be required to identify a conservation threat, but understanding the mechanism is integral to elucidating the disease process and for justifying experimental results, which may suggest conservation approaches (Skerratt et al. 2009). For example, supernumerary limbs detected in several North American amphibian species caused by trematode infection result from metacercariae physically interfering with normal limb development (Johnson et al. 1999; Rohr et al. 2008; Johnson and Hartson 2009). Here a feasible conservation intervention might be to reduce the density of intermediate snail hosts from which transmission to larvae occurs. Experimentally developed information regarding the interaction between ranavirus genotypes and host adaptive immunity (Chinchar and Granoff 1986; Morales and Robert 2007, 2008; Chinchar et al. 2009) are paralleled by field observations suggestive of virus and/or host (co) evolution (Storfer et al. 2007; Ridenhour and Storfer 2008; Teacher et al. 2009). In this case, adaptation should contribute to the probability of host persistence, and it is arguable whether a conservation intervention should be attempted (Greer et al. 2008; Teacher et al. 2010).

Contextualizing host costs requires factorial experimental designs, and because of constraints on the number of independent variables that can be tested in any one experiment, rigorous field observations and knowledge of host ecology and life history and of parasite life history are essential for prioritization. Johnson et al. (1999) and Rachowicz et al. (2006) are examples of how a field study determines factors that can be tested experimentally. Several factors, many discussed above, are likely to be of repeated importance in experimental tests of contextual parasite costs. One example, environmental temperature, has been invoked by many as an important covariate of virulence (Cunningham et al. 1996; Pounds et al. 2006; Bosch et al. 2007; Gray et al. 2007). Some experiments have shown that temperature influences virulence (Rojas et al. 2005; Ribas et al. 2009), which makes biological sense as both amphibian hosts and their parasites are ectothermic. Nevertheless, experimental evidence of an effect of temperature on disease may not provide information relevant for conservation. Rojas et al. (2005) showed that tiger salamanders (*Ambystoma tigrinum*) kept at low temperatures are less able to cope with ranavirus infection, but this does little towards illuminating why the preponderance of data show ranavirus mass mortality events are more common during the warmer summer months (Cunningham et al. 1996; Collins et al. 2004).

Perhaps one of the most valuable uses for experiments is to test for interactions between disease and another threatening process. Correlations have been reported (McKenzie 2007; Rohr et al. 2008; St-Amour et al. 2008) and the default assumption is that multiple stressors are worse (Relyea 2003), but cumulative or synergistic interactions between disease and other threatening processes are not a given. In the study by Rohr et al. (2008), synergisms were supported: pollution increased trematode prevalence and costs to hosts by suppressing host immunity and elevating the number of intermediate hosts. Parris and Baud (2004) also report interactions between infectious disease and pollutants. In contrast, synergisms between altered environmental temperature during overwintering and exposure to *Bd* were weakly supported: warmer overwintering conditions aided infection but impaired post-infection parasite proliferation. Furthermore, infectious disease was not as important a predictor of mortality as were body mass and mass lost during overwintering, both of which were not influenced by either threatening process (Garner et al. 2011).

## MODELS OF TRANSMISSION AND DISEASE DYNAMICS

The application of mathematical models to disease dynamics has a long history (Hamer 1906; Ross 1911; Kermack and McKendrick 1927), and much of the current theory continues to be inspired by work done decades ago (Anderson and May 1978, 1981). Models have been used to capture and explain dynamics of several infectious diseases of humans and livestock as well as designing strategies aimed at controlling infectious diseases of livestock, such as culling, vaccination, or sterilization (Grenfell and Dobson 1995; Barlow 1996; Greenfell et al. 2001; Keeling et al. 2001; Ferguson et al. 2001a, 2001b, 2003, 2005; Koelle et al. 2005; Hall et al. 2007; Nguyen and Rohani 2008). All models are simplifications, but through simplification researchers can distinguish noise from key features that drive disease and host population dynamics. This can be accomplished using very basic models, frequently used to investigate general properties: the dynamics of viral diseases were elucidated this way (Anderson and May 1978, 1981). Making predictions for specific systems more often requires highly detailed models (e.g., Morris et al. 2001), but irrespective of complexity, the language of mathematics provides a framework that forces us to (a) be specific about all of the assumptions that we are making, and (b) be accurate and logical in the conclusions that we come to about the effects of those assumptions.

Transmission and host population dynamics models can be used in a number of ways to determine when and where infectious diseases are conservation threats to amphibians. Simple models illustrate the point that the highly virulent parasites discussed above do not always have persistently large impacts on their host population (Anderson 1979). Many amphibian parasites are broadly distributed and occur at high prevalence, and models show that parasites with little impact on host mortality are more likely to maintain high prevalence in a population (Anderson 1979; McCallum and Dobson 1995). Dynamics of parasites in multihost systems are impossible to infer from prevalence data alone (McCallum and Dobson 1995) and difficult to study experimentally. The great majority of amphibian parasites are multihost, and amphibian hosts commonly support multiple parasites (Prudhoe and Bray 1982; Miller et al. 2008). Modeling may provide the only avenue for understanding the dynamics of multihost parasites and multiple infections in host communities (Begon 2008).

To the best of our knowledge, Bd is the only disease of amphibians where mathematical models were used to understand the impacts of disease on host population dynamics. Briggs et al. (2005) initially used a stage-structured version of an SIR microparasite (Box 36.1) model to understand Rana muscosa population persistence with Bd (endemic dynamics) while most were rapidly driven extinct (epidemic dynamics). Although mass mortality is frequently catastrophic in adult frogs (Vredenburg et al. 2010), survival of some adult frogs was required for endemic disease dynamics, a prediction later confirmed in a capture–mark–recapture (CMR) study (Briggs et al. 2010). Interestingly, survival of infected adults at sites with endemic disease dynamics required incorporating some features common to macroparasite models, specifically the dynamics of the intensity of infection (fungal load), a disease dynamic commonly ignored in microparasite models (Box 36.2). If the intensity of infection of an individual remained below a threshold, then survival of that individual was possible. Threshold host responses to Bd have been confirmed experimentally in Australian (Stockwell et al. 2010) and European (T.W.J. Garner and M.C. Fisher unpublished data) host species. Is it possible that other amphibian microparasites also elicit mortality through host threshold responses? Studies of ranavirus (ATV) dynamics in tiger salamanders are suggestive (Brunner et al. 2004). Larval tiger salamanders suffer high but not comprehensive mortality due to ATV and metamorphosing survivors often maintained infections. Whether the survivors were those individuals that experienced weaker viral loads remains unknown. The application of quantitative methods for estimating burden of ranavirus infection (Picco et al. 2007) to the ATV-tiger salamander system would go far towards assessing this.

An environmental stage has not been described for Bd, but other amphibian parasites have proven ability to persist outside the host (Johnson et al. 2008). Irrespective of how infectious particles are maintained in the environment, the availability of infectious particles does relate to the number of infected individuals in the host population. How the number of infected relates to disease dynamics is described by the transmission function, an important factor determining the conservation threat of diseases. For each susceptible individual, the transmission function (rate at which uninfected and susceptible individuals enter the infected class) depends on (1) the number of contacts with other individuals per unit time; (2) the fraction of contacts that are with infected individuals (if $S$ is the number of susceptibles, $I$ is the

---

**Box 36.2: Microparasites versus Macroparasites**

---

Mathematical models of host–parasite interactions distinguish between two parasites categories (Grenfell and Dobson 1995): microparasites (viruses, bacteria, protozoa, usually fungi) and macroparasites (helminths, arthropods). Microparasites are small, multiply relatively rapidly within their host, and frequently reach large numbers per host. Models of microparasitic diseases typically divide the host population into classes based on infection status (e.g., SIR model) and do not keep track of the size of the parasite population or the number of parasites per host. Susceptible individuals become infected through contact with infected individuals, occasionally through contact with free-living infectious stages, and simple microparasite models assume that all infected individuals are equally infectious. Macroparasites, in contrast, are typically larger and do not replicate inside the host, although they do produce infectious stages that are released to infect additional hosts. Both the infectivity and the impact of a macroparasite on its host are strongly determined by the intensity of infection (infectious burden). Macroparasite models typically keep track of the size of the parasite population and the distribution of parasites across the host population.

A number of diseases do not fit neatly into this dichotomy. Some microparasitic diseases are well characterized by the SIR-type compartment models (Grenfell et al. 2001), but in others the intensity of infection plays an important role in dynamics (Nowak and May 2000). Increasingly, models have been developed for describing within-host dynamics of microparasites, in many cases aimed at understanding the interaction between the population growth of the parasite and host immunity (Anita et al. 1996; Levin and Anita 2001).

---

number of infected, and $N$ is the total population size, then this is simply $I/N$); and (3) the probability that a contact with an infected individual results in infection (frequently assumed to be a constant).

Two idealized transmission functions, density-dependent versus frequency-dependent, make different assumptions about contact rates. Density-dependent transmission assumes that the rate of contact increases linearly with the size of the population, while frequency-dependent transmission assumes that each individual comes in contact with a constant number of individuals per day, regardless of the size of the population. These assumptions are respectively represented mathematically as $bSI$ (density-dependent transmission function) and $bSI/N$ (the frequency-dependent transmission function). Other transmission functions frequently involve forms for which the contact rate initially increases with population size but levels off at high population densities (McCallum et al. 2001). In simple microparasite–host models, the form of the transmission function plays a fundamental role in determining the outcome of host–parasite dynamics (Box 36.2). Density-dependent transmission results in a threshold population size below which

the parasite is predicted to decline and disease outbreaks are not expected in small populations, nor expected to result in host population extinction. Saturating transmission functions that initially increase but then level off with host population size are also expected to have a threshold population size. Frequency-dependent transmission, however, dictates no such threshold, and for this reason parasites transmitted in a frequency-dependent manner are thought to have greater potential for host extinction (de Castro and Bolker 2005).

Estimating transmission rates in the field is exceedingly difficult, and experimentally derived estimates may or may not relate to the natural setting. Model-fitting approaches, however, allow for an effective interpretation of the results of laboratory and field experiments. Fitting more biologically meaningful mechanistic models, rather than standard statistical models (e.g., General Linear Model), to experimental results is the preferred method (Burnham and Anderson 2002) and generates best estimates of parameter values in the process (Johnson and Omland 2004). Woodhams et al. (2008), using models to interpret the results of laboratory experiments,

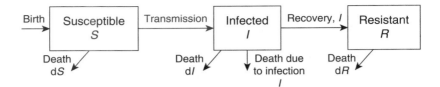

Figure 36.2

demonstrated *Bd* life history tradeoffs at different temperatures. Rachowicz and Briggs (2007) and Greer et al. (2008) both used a maximum-likelihood approach to distinguish among various forms of transmission functions for *Bd* and ranavirus, respectively. Greer et al. (2008) performed transmission experiments on ranavirus and found that the best-fit transmission model was a power function $bSI^q$ (with q = 0.26). Thus, the rate of new infections increased less than linearly with increasing density of infected individuals, leading the authors to conclude that ranavirus is unlikely to lead comprehensively to host extinction, a conclusion supported by field studies (Storfer et al. 2007; Teacher et al. 2009, 2010).

## MITIGATION: IS IT POSSIBLE?

The development of approaches for mitigating amphibian infectious disease is perhaps the most urgent yet most understudied aspect of amphibian disease conservation research. Veterinary research has developed numerous methods for treating a variety of diseases in individual amphibians (Wright and Whitaker 2001), and chemical treatments developed for some high-risk diseases affecting threatened amphibian species (Woodhams et al. 2003; Garner et al. 2009a; Martel et al. 2011) could conceivably be applied to simple amphibian communities occurring in isolated locations (Vredenburg et al. 2010). Nonchemical approaches also show promise (Harris et al. 2006; Banning et al. 2008; Harris et al. 2009a, 2009b), and *in situ* experiments are under way to determine whether lab-based studies can be extended to the field (Lubick 2010; V. Vredenburg, personal communication 2010). Nevertheless, it is undetermined if any of these approaches can be applied to more general field settings. Initial trials at treating individuals in

a simple, isolated amphibian community *ex situ* and returning them to the site were not effective at persistently clearing infection (Lubick 2010), although infectious burdens were lower. This last is reassuring, given the evidence for infectious thresholds associated with host mortality, but subsequent reinfection may render this effect unimportant (Briggs et al. 2010). Applying individual-based treatments to natural situations is likely to prove impractical unless guided by sound epidemiological information.

The methods discussed above can be used to determine what interventions are appropriate and to guide their application. Consider Briggs et al. (2010) and Walker et al. (2010). In both studies prolonged larval stages are implicated as potential reservoirs of disease: reducing the proportion of infected tadpoles could be considered a viable option for reducing transmission (but see Lubick 2010). If the goals include reduction of costs to hosts as well as reducing the transmission rate, then interventions may differ. Reducing the infectious burdens of adult *Rana muscosa* would be the likely option for amphibian welfare in the Sierra Nevadas, but not in the Pyrenees. Here, some would target metamorphic climax due to the increased likelihood of death, but others may argue that prolonged infection during the larval stage also incurs costs, and that these costs contribute to the likelihood of mortality at metamorphosis. A simple experiment testing whether prolonged infections during the larval period contribute to post-metamorphic mortality rates is required.

Strong evidence for disease-driven declines at Iberian high-elevation locations (Bosch et al. 2001, 2007; Walker et al. 2010) suggests several possible interventions. High-elevation populations without evidence of infection are candidates for some form of quarantine, but this is probably impossible. Altering transmission dynamics at infected sites seems a more feasible option, so determining what governs transmission at these sites would be an important research question (Briggs et al. 2010). The occurrence of isolates of hypovirulent *Bd* (Fisher et al. 2009b) suggests that a strategy of proactively infecting natural populations in advance of the spread of *Bd* may stall the advance of a more virulent genotype if exposure generates immunity and prior occupancy lends a competitive edge to *Bd* lineages. The occurrence of cryptic mortality (Tobler and Schmidt 2010), environmental change, and rapid range expansion by *Bd* (Vredenburg et al. 2010) all mean that populations without detectable infections would still require continued surveillance and potential intervention.

Difficulties estimating the transmission functions of amphibian parasites (Rachowicz and Briggs 2007; Greer et al. 2008) and repeated evidence for environmental, host genetic, and parasite genetic effects in amphibian host–parasite dynamics all demonstrate the need for a better understanding of the relative contribution made by these factors to disease-driven amphibian decline. A theoretical framework for this exists in quantitative genetics, where trait variation is partitioned into genetic and environmental components and their interactions ($G \times E$), which illustrates how consistent host responses can be expected to be: a strong interaction suggests that rank orders of genotypes should change as environment changes. Parasite genotype can also be incorporated. Amphibians are amenable to quantitative mating designs (Laurila et al. 2002; Sommer and Pearman 2003) and it may be possible to measure full-sib/half-sib trait variation in the wild. If the call for establishing captive populations of purportedly disease-threatened amphibian species is heeded, pedigree-based quantitative analyses are another option (Storfer 1996).

The current perspective is that "once the genie is out of the bottle, there's no putting it back," an understandable outlook given the tone of much of the recent literature on amphibian infectious diseases. We have recently lost amphibian species to disease, and may lose more. However, rediscoveries of amphibian species presumed extinct due to infectious disease are accumulating (Kolby and McCranie 2009), and eliminating an infectious agent may not be required for managing disease in amphibian populations (Briggs et al. 2010). Instead, disease should be managed like any other threat to biodiversity, albeit using epidemiological research and principles. This includes educating researchers, managers, and the public that infectious diseases of amphibians can potentially be managed, as we have managed the infectious diseases of humans, livestock, and non-amphibian wildlife species.

## ACKNOWLEDGMENTS

During the writing of this chapter, T.W.J.G., M.C.F., and J.B. were supported by the EU BiodivERsA-funded project R.A.C.E. (Risk Assessment of Chytridiomycosis to European amphibian biodiversity). C.J.B. was supported by grant EF-0723563 provided through the NSF Ecology of Infectious Diseases Program. T.W.J.G. was also supported by an RCUK Fellowship. All authors would like to acknowledge the inputs of students, collaborators, supervisors, and colleagues over the years that have influenced their respective research programs. We would like to thank F. Clare for reading an earlier version of the manuscript.

## REFERENCES

Ahmed, S.A. 2000. The immune system as a potential target for environmental estrogens (endocrine disruptors): a new emerging field. Toxicology 150:191–206.

Anderson, R.M. 1979. Parasite pathogenicity and the depression of host population equilibria. Nature 279:150–152

Anderson, R.M., and R.M. May. 1978. Regulation and stability of hostparasite population interactions. I. Regulatory processes. J Anim Ecol 47:219–247.

Anderson, R.M., and R.M. May. 1981. The population dynamics of microparasites and their invertebrate hosts. Phil T Roy Soc B 291:451–524.

Anderson, R.M., and R.M. May. 1982. Coevolution of hosts and parasites. Parasitology 85:411–426.

Anita, R., M.A. Nowak, and R.M. Anderson. 1996. Antigenic variation and the within-host dynamics of parasites. Proc Natl Acad Sci USA 93:985–989.

Ariel, E., J. Kielgast, H.E. Svart, K. Larsen, H. Tapiovaara, B.B. Jensen, and R. Holopainen. 2009. Ranavirus in wild edible frogs Pelophylax kl. esculentus in Denmark. Dis Aquat Org 85:7–14.

Banning, J.L., A.L. Weddle, G.W. Wahl III, M.A. Simon, A. Lauer, R.L. Walters, and R.N. Harris. 2008. Antifungal skin bacteria, embryonic survival, and communal nesting in four-toed salamanders, Hemidactylium scutatum. Oecologia 156:423–429.

Barlow, N.D. 1996. The ecology of wildlife disease control: simple models revisited. J Appl Ecol 33:303–314.

Basáñez, M.G., C. Marshall, N. Carabin, T. Gyorkos, and L. Joseph. 2004. Bayesian statistics for parasitologists. Trends Parasitol 20:85–91.

Begon, M. 2008. Effects of host diversity in disease dynamics. In R. Ostfeld, F. Keesing, and V.T. Eviner, eds. Infectious disease ecology: the effects of ecosystems on disease and disease on ecosystems, pp. 12–29. Princeton University Press, Princeton.

Berger, L., R. Speare, P. Daszak, D.E. Green, A.A. Cunningham, C.L. Goggin, R. Slocombe, M.A. Ragan, A.D. Hyatt, K.R. McDonald, H.B. Hines, K.R. Lips, G. Marantelli, and H. Parkes. 1998. Chytridiomycosis causes amphibian mortality associated with population declines in the rain forests of Australia and Central America. Proc Natl Acad Sci USA 95:9031–9036.

Bielby, J., N. Cooper, A.A. Cunningham, T.W.J. Garner, A. Purvis. 2008. Predictors of rapid decline in frog species. Conserv Lett 1:82–90.

Blaustein, A.R., D.G. Hokit, and R.K. O'Hara. 1994. Pathogenic fungus contributes to amphibian losses in the Pacific Northwest. Biol Conserv 67:251–254.

Boots, M., D. Childs, D.C. Reuman, and M. Mealor. 2009. Local interactions lead to pathogen-driven change to host population dynamics. Curr Biol 19:1660–1664.

Bosch, J., and I. Martinez-Solano. 2006. Chytrid fungus infection related to unusual mortalities of Salamandra salamandra and Bufo bufo in the Penalara Natural Park, Spain. Oryx 40:84–89.

Bosch, J., I. Martinez-Solano, and M. Garcia-Paris. 2001. Evidence of a chytrid fungus infection involved in the decline of the common midwife toad (Alytes obstetricans) in protected areas of central Spain. Biol Conserv 97:331–337.

Bosch, J., L.M. Carrascal, L. Duran, S. Walker, and M.C. Fisher. 2007. Climate change and outbreaks of amphibian chytridiomycosis in a montane area of Central Spain: is there a link? Proc Roy Soc B 274:253–260.

Briggs, C.J., V.T. Vredenburg, R.A. Knapp, and L.J. Rachowicz. 2005. Investigating the population-level effects of chytridiomycosis: an emerging infectious disease of amphibians. Ecology 86:3149–3159.

Briggs, C.J., R.A. Knapp, and V.T. Vredenburg. 2010. Enzootic and epizootic dynamics of the chytrid fungal pathogen of amphibians. Proc Natl Acad Sci USA 107:9695–9700.

Brunner, J.L., D.M. Schock, and J.P. Collins. 2007. Transmission dynamics of the amphibian ranavirus Ambystoma tigrinum virus. Dis Aquat Org 77:87–95.

Brunner, J.L., D.M. Schock, E.W. Davidson, and J.P. Collins. 2004. Intraspecific reservoirs: complex life history and the persistence of a lethal ranavirus. Ecology 85:560–566.

Burnham, K.P., and D.R. Anderson. 2002. Model selection and multimodel inference: a practical information-theoretic approach, 2nd ed. Springer-Verlag, New York.

Carabin, H., M. Escalona, C. Marshall, S. Vivas-Martinez, C. Botto, L. Joseph, and M.G. Basanez. 2003. Prediction of community prevalence of human onchocerciasis in the Amazonian onchocerciasis focus: Bayesian approach. B World Health Organ 81:482–490.

Carey, C., J.E. Bruzgul, L.J. Livo, M.L. Walling, K.A. Kuehl, B.F. Dixon, A.P. Pessier, R.A. Alford, and K.B. Rogers. 2006. Experimental exposures of boreal toads (Bufo boreas) to a pathogenic chytrid fungus (Batrachochytrium dendrobatidis). Ecohealth 3:5–21.

Cavanagh, R.D., X. Lambin, T. Ergon, M. Bennett, I.M. Graham, D. van Soolingen, and M. Begon. 2004. Disease dynamics in cyclic populations of field voles (Microtus agrestis): cowpox virus and vole tuberculosis (Mycobacterium microti). Proc Roy Soc B 271:859–867.

Chinchar, V.G., and A. Granoff. 1986. Temperature-sensitive mutants of frog virus 3: biochemical and genetic characterization. J Virol 58:192–202.

Chinchar, V.G., A.D. Hyatt, T. Miyazaki, and T. Williams. 2009. Family Iridoviridae: poor viral relations no longer. In J. Etten, ed. Lesser known jarge dsDNA viruses: Current topics in microbiology and immunology, Vol. 328, pp. 123–170. Springer-Verlag, New York.

Cleaveland, S., M.G.J. Appel, W.S.K. Chalmers, C. Chillingsworth, M. Kaare, and C. Dye. 2000. Serological and demographic evidence for domestic dogs as a source of canine distemper virus infection for Serengetti wildlife. Vet Microbiol 72:3–4.

Clements, A.C.A., N.J.S. Lwambo, L. Blair, U. Nyandindi, G. Kaatano, S. Kinung'hi, J.P. Webster, A. Fenwick, and S. Brooker. 2006. Bayesian spatial analysis and disease mapping: tools to enhance planning and implementation of a schistosomiasis control programme in Tanzania. Trop Med Int Health 11:490–503.

Collins, J.P., J.L. Brunner, J.K. Jancovich, and D.M. Schock. 2004. A model host–pathogen system for studying infectious disease dynamics in amphibians: tiger salamanders (Ambystoma tigrinum) and Ambystoma tigrinum virus. Herpetol J 14:195–200.

Cunningham, A.A., T.E.S. Langton, P.M. Bennett, J.F. Lewin, S.E.N. Drury, R.E. Gough, and S.K. MacGregor. 1996. Pathological and microbiological findings from incidents of unusual mortality of the common frog (Rana temporaria). Philos T Roy Soc B 351:1539–1557.

Cunningham, A.A., P. Daszak, and J.P. Rodríguez. 2003. Pathogen pollution: defining a parasitological threat to biodiversity conservation. J Parasitol 89:S78–S83.

Cunningham, A.A., A.D. Hyatt, P. Russell, and P.M. Bennett. 2007. Emerging epidemic diseases of frogs in Britain are dependent on the source of ranavirus agent and the route of exposure. Epidemiol Infect 135:1200–1212.

De Castro, F., and B.M. Bolker. 2005. Mechanisms of disease-induced extinction. Ecol Lett 8:117–126.

Delahay, R.J., G.C. Smith, and M.R. Hutchings. 2009. The science of wildlife disease management. In R.J. Delahay, G.C. Smith, and M.R. Hutchings, eds. Management of disease in wild mammals, pp. 1–8. Springer-Verlag, New York.

Dodd Jr., C.K., ed. 2010. Amphibian ecology and conservation. Oxford University Press, Oxford, England.

Duffus, A.L.J. 2010. Chytrid blinders: what other disease risks to amphibians are we missing? EcoHealth 6:335–339.

Ebert, D., C.D. Zschokke-Rohringer, and H.J. Carius. 2000. Dose effects and density-dependent regulation of two microparasites of Daphnia magna. Oecologia 122:200–209.

Ferguson, N.M., C.A. Donnelly, and R.A. Anderson. 2001a. The foot-and-mouth epidemic in Great Britain: pattern of spread and impact of interventions. Science 292:1155–1160.

Ferguson, N.M., C.A. Donnelly, R.M. Anderson. 2001b. Transmission intensity and impact of control policies on the foot and mouth epidemic in Great Britain. Nature 413:542–548.

Ferguson, N.M., M.J. Keeling, W.J. Edmunds, R. Gani, B.T. Grenfell, R.M. Anderson, and S. Leach. 2003. Planning for smallpox outbreaks. Nature 425:681–685.

Ferguson, N.M., D.A.T. Cummings, S. Cauchemez, C. Fraser, S. Riley, A. Meeyai, S. Iamsirithaworn, and D.S. Burke. 2005. Strategies for containing an emerging influenza pandemic in Southeast Asia. Nature 437:209–214.

Fisher, D.O., and I.P. Owens. 2004. The comparative method in conservation biology. Trends Ecol Evol 19: 391–398.

Fisher, M.C., and T.W.J. Garner. 2007. The relationship between the introduction of Batrachochytrium dendrobatidis, the international trade in amphibians and introduced amphibian species. Fungal Biol Rev 21:2–9.

Fisher, M.C., T.W.J. Garner, and S.F. Walker. 2009a. The global emergence of Batrachochytrium dendrobatidis in space, time and host. Annu Rev Microbiol 63:291–310.

Fisher, M.C., J. Bosch, Z. Yin, D.A. Stead, J. Walker, L. Selway, A.J.P. Brown, L.A. Walker, N.A.R. Gow, J.E. Stajich, and T.W.J. Garner. 2009b. Proteomic and phenotypic profiling of the amphibian pathogen Batrachochytrium dendrobatidis shows that genotype is linked to virulence. Mol Ecol 18:415–429.

Forson, D.D., and A. Storfer. 2006. Atrazine increases ranavirus susceptibility in the tiger salamander, Ambystoma tigrinum. Ecol Appl 16:2325–2332.

Fraser, C., C.A. Donnelly, S. Cauchemez, W.P. Hanage, M.D. Van Kerkhove, T.D. Hollingsworth, J. Griffin, R.F. Baggaley, H.E. Jenkins, E.J. Lyons, T. Jombart, W.R. Hinsley, N.C. Grassly, F. Balloux, A.C. Ghani, N.M. Ferguson, A. Rambaut, O.G. Pybus, H. Lopez-Gatell, C.M. Alpuche-Aranda, I.B. Chapela, E.P. Zavala, D.M. Guevara, F. Checchi, E. Garcia, S. Hugonnet, C. Roth, The WHO Rapid Pandemic Assessment Collaboration. 2009. Pandemic potential of a strain of influenza A (H1N1): early findings. Science 324:1557–61

Garner, T.W.J., M. Perkins, P. Govindarajulu, D. Seglie, S.F. Walker, A.A. Cunningham, and M.C. Fisher. 2006. The emerging amphibian pathogen Batrachochytrium dendrobatidis globally infects introduced populations of the North American bullfrog, Rana catesbeiana. Biol Lett 2: 455–459.

Garner, T.W.J., G. Garcia, B. Carroll, and M.C. Fisher. 2009a. Using itraconazole to clear Batrachochytrium dendrobatidis infection, and subsequent depigmentation of Alytes muletensis tadpoles. Dis Aquat Org 83:257–260.

Garner, T.W.J., S. Walker, J. Bosch, S. Leech, J.M. Rowcliffe, A.A. Cunningham, and M.C. Fisher. 2009b. Life history trade-offs influence mortality associated with the amphibian pathogen Batrachochytrium dendrobatidis. Oikos 118:783–791.

Garner, T.W.J., J.M. Rowcliffe, and M.C. Fisher. 2011. Climate, chytridiomycosis or condition: an experimental test of amphibian survival. Global Change Biol 17:667–675.

Grassly, N.C., C. Fraser, and G.P. Garnett. 2005. Host immunity and synchronized epidemics of syphilis across the United States. Nature 433:417–421.

Gray, M.J., D.L. Miller, A.C. Schmutzer, and C.A. Baldwin. 2007. Frog virus 3 prevalence in tadpole populations inhabiting cattle-access and non-access wetlands in Tennessee, USA. Dis Aquat Org 77:97–103.

Green, D.E., K.A. Converse, and A.K. Schrader. 2002. Epizootiology of sixty-four amphibian morbidity and mortality events in the USA, 1996–2001. Ann NY Acad Sci 969:323–339.

Greer, A.L., C.J. Briggs, and J.P. Collins. 2008. Testing a key assumption of host-pathogen theory: density and disease transmission. Oikos 117:1667–1673.

Grenfell, B.T., O.N. Bjornstad, and J. Kappey. 2001. Travelling waves and spatial hierarchies in measles epidemics. Nature 414:716–723.

Grenfell, B.T., and A.P. Dobson, eds. 1995. Ecology of infectious diseases in natural populations. Cambridge University Press, Cambridge.

Hall, I.M., J.R. Egan, I. Barrass, R. Gani, and S. Leach. 2007. Comparison of smallpox outbreak control strategies using a spatial metapopulation model. Epidemiol Infect 135:1133–1144.

Hamer, W.H. 1906. Epidemic disease ii: England. Lancet 1:733–739.

Harris, R.N., T.Y. James, A. Lauer, M.A. Simon, and A. Patel. 2006. Amphibian pathogen Batrachochytrium dendrobatidis is inhibited by the cutaneous bacteria of amphibian species. EcoHealth 3:53–56.

Harris, R.N., R.M. Brucker, J.B. Walke, M.H. Becker, C.R. Schwantes, D.C. Flaherty, B.A. Lam, D.C. Woodhams, C.J. Briggs, V.T. Vredenburg, and K.P.C. Minbiole. 2009a. Skin microbes on frogs prevent morbidity and mortality caused by a lethal skin fungus. ISME J 3:818–824.

Harris, R.N., A. Lauer, M.A. Simon, J.L. Banning, and R.A. Alford. 2009b. Addition of antifungal skin bacteria to salamanders ameliorates the effects of chytridiomycosis. Dis Aquat Org 83:11–16.

Harvell, C.D., C.E. Mitchell, J.R. Ward, S. Altizer, A.P. Dobson, R.S. Ostfeld, and M.D. Samuel. 2002. Climate warming and disease risk for terrestrial and marine biota. Science 296:2158–2162.

Haydon, D.T., D.A. Randall, L. Matthews, D.L. Knobel, L.A. Tallents, M.B. Gravenor, S.D. Williams, J.P. Pollinger, S. Cleaveland, M.E.J. Woolhouse, C. Sillero-Zubiri, J. Marino, D.W. Macdonald, and M.K. Laurenson. 2006. Low-coverage vaccination strategies for the conservation of endangered species. Nature 443:692–695.

Heyer, W.R., M.A. Donnelly, R.W. McDiarmid, L.C. Hayek, and M.S. Foster, eds. 1994. Measuring and monitoring biological diversity. Standard methods for amphibians Smithsonian Institution Press, Washington.

Isaac, N.J., and G. Cowlishaw. 2004. How species respond to multiple extinction threats. Proc Roy Soc B 271:1135–1141

Isaac, N.J.B., S.T. Turvey, B. Collen, C. Waterman, and J.E.M. Baillie. 2007. Mammals on the EDGE: conservation priorities based on threat and phylogeny. PLoS One 3:e296.

James, T.Y., A.P. Litvintseva, R. Vilgalys, J.A.T. Morgan, J.W. Taylor, M.C. Fisher, L. Berger, C. Weldon, L. du Preez, and J.E. Longcore. 2009. Rapid global expansion of the fungal disease chytridiomycosis into declining and healthy amphibian populations. Plos Pathog 5:e1000458.

Johnson, J.B., and K.S. Omland. 2004. Model selection in ecology and evolution. Trends Ecol Evol 19:101–106.

Johnson, P.T.J., K.B. Lunde, E.G. Ritchie, and A.E. Launer. 1999. The effect of trematode infection on amphibian limb development and survivorship. Science 284:802–804.

Johnson, P.T.J., and R.B. Hartson. 2009. All hosts are not equal: explaining the differential patterns of malformations in an amphibian community. J Anim Ecol 78:191–201.

Johnson, P.T.J., R.B. Hartson, D.J. Larson, and D.R. Sutherland. 2008. Diversity and disease: community structure drives parasite transmission and host fitness. Ecol Lett 11:1017–1026.

Jones, M.E., P.J. Jarman, C.M. Lees, H. Hesterman, R.K. Hamede, N.J. Mooney, D. Mann, C.E. Pukk, J. Bergfeld, and H. McCallum. 2007. Conservation management of Tasmanian devils in the context of an emerging, extinction-threatening disease: devil facial tumor disease. EcoHealth 4:326–337.

Karesh, W.B., R.A. Cook, E.L. Bennett, and J. Newcomb. 2005. Wildlife trade and global disease emergence. Emerg Infect Dis 11:1000–1002.

Keeling, M.J., M.E.J. Woolhouse, D.J. Shaw, L. Matthews, M. Chase-Topping, D.T. Haydon, S.J. Cornell, J. Kappey, J. Wilesmith, and B.T. Grenfell. 2001. Dynamics of the 2001 UK foot and mouth epidemic: stochastic dispersal in a heterogeneous landscape. Science 294:813–817.

Kermack, W.O., and A.G. McKendrick. 1927. Contributions to the mathematical theory of epidemics, part I. Proc Roy Soc A 115:700–721.

Koelle, K., X. Rodo, M. Pascual, M.D. Yunus, and G. Mostfa. 2005. Refractory periods and climate forcing in cholera dynamics. Nature 436:696–700.

Kolby, J.E., and J.R. McCranie. 2009. Discovery of a surviving population of the montane streamside frog Craugastor milesi (Schmidt). Herpetol Rev 40:282–283.

Krkosek, M., J.S. Ford, A. Morton, S. Lele, R.A. Myers, and M.A. Lewis. 2007. Declining wild salmon populations in relation to parasites from farmed salmon. Science 318:1772–1775.

Koch, R. 1884. Die aetiologie der tuberkulose. Mitt Kaiser Gesundh 2:1–88

La Deau, S.L., A.M. Kilpatrick, and P.P. Marra. 2007. West Nile virus emergence and large-scale declines of North American bird populations. Nature 447:710–713.

La Marca, E., K.R. Lips, S. Lötters, R. Puschendorf, R. Ibáñez, J.V. Rueda-Almonacid, R. Schulte, C. Marty, F. Castro, J. Manzanilla-Puppo, J.E. García-Pérez, F. Bolaños, G. Chaves, J.A. Pounds, E. Toral, and

B.E. Young. 2005. Catastrophic population declines and extinctions in neotropical harlequin frogs (Bufonidae: Atelopus). Biotropica 37:190–201.

Lafferty, K.D. 2003. Is disease increasing or decreasing, and does it impact or maintain biodiversity? J Parasitol 89:S101–S105.

Laurance, W.F., K.R. McDonald, and R. Speare. 1996. Epidemic disease and the catastrophic decline of Australian rain forest frogs. Conserv Biol 10:406–413

Laurance, W.F., K.R. McDonald, and R. Speare. 1997. In defence of the epidemic disease hypothesis. Conserv Biol 11:1030–1034.

Laurila, A., S. Karttunen, and J. Merilä. 2002. Adaptive phenotypic plasticity and genetics of larval life histories in two Rana temporaria populations. Evolution 56:617–627.

Levin, B.R., and R. Antia. 2001. Why we don't get sick: the within-host population dynamics of bacterial infections. Science 292:1112–1115.

Lips, K.R., J.D. Reeve, and L.R. Witters. 2003. Ecological traits predicting amphibian population declines in Central America. Conserv Biol 17:1078–1088.

Lips, K.R., F. Brem, R. Brenes, J.D. Reeve, R.A. Alford, J. Voyles, C. Carey, L. Livo, Pessier, A.P., and J.P. Collins. 2006. Emerging infectious disease and the loss of biodiversity in a Neotropical amphibian community. Proc Natl Acad Sci USA 102:3165–3170.

Lips, K.R., J. Diffendorfer, J.R. Mendelson, and M.W. Sears. 2008. Riding the wave: reconciling the roles of disease and climate change in amphibian declines. Plos Biol 6: e72.

Little, T.J. 2002. The evolutionary significance of parasitism: do parasite-driven genetic dynamics occur ex silico? J Evolution Biol 15:1–9.

Loman, J. 2009. Primary and secondary phenology. Does it pay a frog to spawn early? J Zool 279: 64–70.

Lubick, N. 2010. Emergency medicine for frogs. Nature 465:680–681.

Maniero, G.D., H. Morales, J. Gantress, J. Robert. 2006. Generation of a long-lasting, protective, and neutralizing antibody response to the ranavirus FV3 by the frog Xenopus. Dev Comp Immunol 30:649–657.

Mao, J., D.E. Green, G. Fellers, and V.G. Chinchar. 1999. Molecular characterization of iridoviruses isolated from sympatric amphibians and fish. Virus Res 63:45–52.

Martel, A., P. Van Rooij, G. Vercauteren, K. Baert, L. Van Waeyenberghe, P. Debacker, T.W.J. Garner, R. Ducatelle, F. Haesebrouck, and F. Pasmans. 2011. Developing a safe antifungal treatment protocol to eliminate Batrachochytrium dendrobatidis from amphibians. Med Mycol 49:143–149.

McCallum, H., N. Barlow, and J. Hone. 2001. How should pathogen transmission be modelled? Trends Ecol Evol 16:295–300.

McCallum, H., and A.P. Dobson. 1995. Detecting disease and parasite threats to endangered species and ecosystems. Trends Ecol Evol 10:190–194

McKenzie, V.J. 2007. Human land use and patterns of parasitism in tropical amphibian hosts. Biol Conserv 137:102–116.

Miller, P.S. 2007. Tools and techniques for disease risk assessment in threatened wildlife conservation programmes. Int Zoo Yearb 41:38–51.

Miller, D.L., R.S. Rajeev, M. Brookins, J. Cook, L. Whittington, and C.A. Baldwin. 2008. Concurrent infection with ranavirus, Batrachochytrium dendrobatidis, and Aeromonas in a captive anuran colony. J Zoo Wildl Med 39:445–449.

Mitchell, K.M., T.S. Churcher, T.W.J. Garner, and M.C. Fisher. 2008. Persistence of the emerging infectious pathogen Batrachochytrium dendrobatidis outside the amphibian host greatly increases the probability of host extinction. Proc Roy Soc B 275:329–334.

Morales, H., and J. Robert. 2007. In vivo characterization of primary and secondary anti- ranavirus CD8 T cell responses in Xenopus laevis. J Virol 81:2240–2248.

Morales, H., and J. Robert. 2008. In vivo and in vitro techniques for comparative study of antiviral T-cell responses in the amphibian Xenopus. Biol Proceed Online 10:1–8.

Morgan, J.A.T., V.T. Vredenburg, L.J. Rachowicz, R.A. Knapp, M.J. Stice, T. Tunstall, R.E. Bingham, J.M. Parker, J.E. Longcore, C. Moritz, C.J. Briggs, and J.W. Taylor. 2007. Population genetics of the frog-killing fungus Batrachochytrium dendrobatidis. Proc Natl Acad Sci USA 104:13845–13850.

Morin, P.J. 1998. Realism, precision, and generality in experimental ecology. In W.J. Resetarits Jr. and J. Bernardo, eds. Experimental ecology: issues and perspective, pp. 50–70. Oxford University Press, England.

Morris, R.S., J.W. Wilesmith, M.W. Stern, R.L. Sanson, and M.A. Stevenson. 2001. Predictive spatial modelling of alternative control strategies for the foot-and-mouth disease epidemic in Great Britain, 2001. Vet Rec 149:137–144.

Murphy, P.J., S. St-Hilaire, S. Bruer, P.S. Corn, and C.R. Peterson. 2009. Distribution and pathogenicity of Batrachochytrium dendrobatidis in boreal toads from the Grand Teton area of Western Wyoming. EcoHealth 6:109–120.

Murray, K., R. Retallick, K.R. McDonald, D. Mendez, K. Aplin, P. Kirkpatrick, L. Berger, D. Hunter,

H.B. Hines, R. Campbell, M. Pauza, M. Driessen, R. Speare, S.J. Richards, M. Mahony, A. Freeman, A.D. Phillott, J.-M. Hero, K. Kriger, D. Driscoll, A. Felton, R. Puschendorf, and L.F. Skerratt. 2010. The distribution and host range of the pandemic disease chytridiomycosis in Australia spanning surveys from 1956–2007. Ecology 91:1557–1558.

Muths, E., and V. Dreitz. 2008. Monitoring programs to assess reintroduction efforts: a critical component of recovery. Anim Biodivers Conserv 31:47–56.

Muths, E., P.S. Corn, A.P. Pessier, and D.E. Green. 2003. Evidence for disease-related amphibian decline in Colorado. Biol Conserv 110:357–365.

Nowak, M.A., and R.M. May. 2000. Virus dynamics. Cambridge University Press, England.

Nguyen, H.T.H., and P. Rohani. 2008. Noise, nonlinearity and seasonality: the epidemics of whooping cough revisited. J Roy Soc Interface 5:403–413.

Olson, D.H., and K.L. Ronnenberg. 2008. Batrachochytrium dendrobatidis mapping project. Partners in Amphibian and Reptile Conservation http://www.parcplace.org/bdmap2008update.html, Accessed March 25, 2011.

Ostfeld, R.S., G.E. Glass, and F. Keesing. 2005. Spatial epidemiology: an emerging (or reemerging) discipline. Trends Ecol Evol 20:328–336.

Parris, M.J., and D.R. Baud. 2004. Interactive effects of a heavy metal and chytridiomycosis on gray treefrog larvae (Hyla chrysoscelis). Copeia 2004:344–350.

Paterson, S., K. Wilson, and J.M. Pemberton. 1998. Major histocompatibility complex variation associated with juvenile survival and parasite resistance in a large unmanaged ungulate population (Ovis aries L.). Proc Natl Acad Sci USA 95:3714–3719.

Pavlovsky, E.N. 1966. Natural nidality of transmissible diseases with special reference to the landscape epidemiology of zooanthroponoses. University of Illinois Press, Urbana.

Pearman, P.B., and T.W.J. Garner. 2005. Susceptibility of Italian agile frog populations to an emerging Ranavirus parallels population genetic diversity. Ecol Lett 8:401–408.

Pearman, P.B., T.W.J. Garner, M. Straub M, and U.F. Greber. 2004. Response of Rana latastei to the ranavirus FV3: a model for viral emergence in a naïve population. J Wildl Dis 40:600–609.

Pessier, A.P., and M. Pinkerton. 2003. Practical gross necropsy of amphibians. Semin Avian Exot Pet Med 12:81–88.

Peterson, C.L., R.F. Wilkinson, D. Moll, and T. Holder. 1991. Premetamorphic survival of Ambystoma annulatum. Herpetologica 47:96–100.

Petranka, J.W. 1989. Density-dependent growth and survival of larval Ambystoma: evidence from whole-pond manipulations. Ecology 70:1752–1767.

Picco, A.M., and J.P. Collins. 2008. Amphibian commerce as a likely source of pathogen pollution. Conserv Biol 22:1582–1589.

Picco, A.M., J.L. Brunner, and J.P. Collins. 2007. Susceptibility of the endangered California tiger salamander, Ambystoma californiense, to ranavirus infection. J Wildl Dis 43:286–290.

Plowright, W. 1982. The effects of rinderpest and rinderpest control on wildlife in Africa. Symp Zool Soc Lond 50:1–28.

Pounds, L.A., M.R. Bustamante, L.A. Coloma, J.A. Consuegra, M.P.L. Fogden, P.N. Foster, E. La Marca, K.L. Masters, A. Merino-Viteri, R. Puschendorf, S.R. Ron, G.A. Sánchez-Azofeifa, C.J. Still, and B.E. Young. 2006. Widespread amphibian extinctions from epidemic disease driven by global warming. Nature 439:161–167.

Prudhoe, S., and R.A. Bray. 1982. Platyhelminth parasites of the Amphibia. British Museum (Natural History) and Oxford University Press, Oxford.

Purvis, A., J.L. Gittleman, G. Cowlishaw, and G.M. Mace. 2000. Predicting extinction risk in declining species. Proc Roy Soc B 267:1947–1952.

Rachowicz, L.J., and C.J. Briggs. 2007. Quantifying the disease transmission function: effects of density on Batrachochytrium dendrobatidis transmission in the mountain yellow-legged frog Rana muscosa. J Anim Ecol 76:711–721.

Rachowicz, L.J., J.-M. Hero, R.A. Alford, J.W. Taylor, J.A.T. Morgan, V.T. Vredenburg, J.P. Collins, and C.J. Briggs. 2005. The novel and endemic pathogen hypotheses: competing explanations for the origin of emerging infectious diseases of wildlife. Conserv Biol 19:1441–1448.

Rachowicz, L.J., R.A. Knapp, J.A. Morgan, M.J. Stice, V.T. Vredenburg, J.M. Parker, anC.J. Briggs. 2006. Emerging infectious disease as a proximate cause of amphibian mass mortality. Ecology 87:1671–1683.

Raffel, T.R., J.R. Rohr, J.M. Kiesecker, and P.J. Hudson. 2006. Negative effects of changing temperature on amphibian immunity under field conditions. Funct Ecol 20:819–828.

Reading, C.J. 2007. Linking global warming to amphibian declines through its effects on female body condition and survivorship. Oecologia 151:125–131.

Relyea, R.A. 2003. Predator cues and pesticides: A double dose of danger for amphibians. Ecol Appl 13:1515–1521.

Ribas, L., M.-S. Li, B. Doddington, J. Robert, J.A. Seidel, J.S. Kroll, L. Zimmerman, N.C. Grassly, T.W.J. Garner, and M.C. Fisher. 2009. Expression profiling the temperature-dependent amphibian response to infection by Batrachochytrium dendrobatidis. PLos One 4:e8408.

Ricketts, T.H., E. Dinerstein, T. Boucher, T.M. Brooks, S.H.M. Butchart, M. Hoffman, J.F. Lamoreux, J. Morrison, M. Parr, J.D. Pilgrim, A.S.L. Rodrigues, W. Sechrest, G.E. Wallace, K. Berlin, J. Bielby, N.D. Burgess, D.R. Church, N. Cox, D. Knox, C. Loucks, G.W. Luck, L.L. Master, R. Moore, R. Naidoo, R. Ridgely, G.E. Schatz, G. Shire, H. Strand, W. Wettengel, and E. Wikramanayake. 2005. Pinpointing and preventing imminent extinctions. Proc Natl Acad Sci USA 102:18497–18501.

Ridenhour, B.J., and A. Storfer. 2008. Geographically variable selection in Ambystoma tigrinum virus (Iridoviridae) throughout the western USA. J Evolution Biol 21:1151–1159.

Riley, S. 2007. Large-scale spatial-transmission models of infectious disease. Science 316:1298–301.

Robert, J., L. Abramowitz, J. Gantress, and H.D. Morales. 2007. Xenopus laevis: a possible vector of ranavirus infection? J Wildl Dis 43:645–652.

Rohr, J.R., A.M. Schotthoefer, T.R. Raffel, H.J. Carrick, N. Halstead, J.T. Hoverman, C.M. Johnson, L.B. Johnson, C. Lieske, M.D. Piwoni, P.K. Schoff, and V.R. Beasley. 2008. Agrochemicals increase trematode infections in a declining amphibian species. Nature 455:1235–1240.

Rojas, S., K. Richards, J.K. Jancovich, and E.W. Davidson. 2005. Influence of temperature on ranavirus infection in larval salamanders Ambystoma tigrinum. Dis Aquat Org 63:95–100.

Ross, R. 1911. The prediction of malaria, 2nd ed. Murray, London.

St-Amour, V., W. Wong, T.W.J. Garner, and D. Lesbarrères. 2008. Anthropogenic influence on prevalence of two amphibian pathogens. Emerg Infect Dis 14:1175–1176.

Schloegel, L.M., J.-M. Hero, L. Berger, R. Speare, K. McDonald, and P. Daszak. 2005. The decline of the sharp-snouted day frog (Taudactylus acutirostris): the first documented case of extinction by infection in a free-ranging wildlife species? EcoHealth 3:35–40.

Schotthoefer, A.M., A.V. Koehler, C.U. Meteyer, R.A. Cole. 2003. Influence of Ribeiroia ondatrae (Trematoda: Digenea) infection on limb development and survival of northern leopard frogs (Rana pipiens): effects of host stage and parasite-exposure level. Can J Zool 81:1144–1153.

Skerratt, L.F., T.W.J. Garner, and A.D. Hyatt. 2009. Determining causality and controlling disease is based on collaborative research involving multidisciplinary approaches. EcoHealth 6:331–334.

Smith, K.F., K. Acevedo-Whitehouse, and A.B. Pedersen. 2009. The role of infectious diseases in biological conservation. Anim Conserv 12:1–12.

Sommer, S., and P.B. Pearman. 2003. Quantitative genetic analysis of larval life history traits in two alpine population of Rana temporaria. Genetica 118:1–10.

Stockwell, M., J. Clulow, and M. Mahoney. 2010. Host species determines whether infection load increases beyond disease-causing thresholds following exposure to the amphibian chytrid fungus. Anim Conserv 13:S62–S71.

Storfer, A. 1996. Quantitative genetics: a promising approach for the assessment of genetic variation in endangered species. Trends Ecol Evol 11:343–348.

Storfer, A., M.E. Alfaro, B.J. Ridenhour, J.K. Jancovich, S.G. Mech, M.J. Parris, and J.P. Collins. 2007. Phylogenetic concordance analysis shows an emerging pathogen is novel and endemic. Ecol Lett 10:1075–1083.

Teacher, A.G.F., T.W.J. Garner, and R.A. Nichols. 2009. Evidence for directional selection at a novel major histocompatability class 1 marker in wild common frogs (Rana temporaria) exposed to a viral pathogen. PLoS One 4:E4616.

Teacher, A.G.F., A.A. Cunningham, and T.W.J. Garner. 2010. Assessing the long-term impact of Ranavirus infection in wild common frog populations. Anim Conserv 13:514–522.

Tobler, U., and B.R. Schmidt. 2010. Within- and among-population variation in chytridiomycosis-induced mortality in the toad Alytes obstetricans. PLoS One 5:e10927.

Vale, P.F., M. Stjernman, and T.J. Little. 2008. Temperature-dependent costs of parasitism and maintenance of polymorphism under genotype-by-environment interactions. J Evolution Biol 21:1418–1427.

Vredenburg, V.T., R.A. Knapp, T.S. Tunstall, and C.J. Briggs. 2010. Dynamics of an emerging disease drive large-scale amphibian population extinctions. Proc Natl Acad Sci USA 107:9689–9694.

Voyles, J., S. Young, L. Berger, C. Campbell, W.F. Voyles, A. Dinudom, D. Cook, R. Webb, R.A. Alford, L.F. Skerratt, and R. Speare. 2009. Pathogenesis of chytridiomycosis, a cause of catastrophic amphibian declines. Nature 326:582–585.

Wakefield, J.C., N.G. Best, L. Waller. 2000. Bayesian approaches to disease mapping. In P. Elliot, J. Wakefield, N. Best, and D. Briggs eds. Spatial epidemiology: methods and applications, pp. 104–127. Oxford University Press, England.

Walker, S.F., J. Bosch, T.Y. James, A.P. Litvintseva, J.A.O. Valls, S. Piña, G. Garcia, G.A. Rosa, A.A. Cunningham, S. Hole, R. Griffiths, and M.C. Fisher. 2008. Invasive pathogens threaten species recovery programs. Curr Biol 18:R853–R854.

Walker, S.F., J. Bosch, V. Gomez, T.W.J. Garner, A.A. Cunningham, D.S. Schmeller, M. Ninyerola, D. Henk, C. Ginestet, A. Christian-Philippe, and M.C. Fisher. 2010. Factors driving pathogenicity versus prevalence of the amphibian pathogen Batrachochytrium dendrobatidis and chytridiomycosis in Iberia. Ecol Lett 13:372–382.

Warkentin, K.M., C.R. Currie, and S.A. Rehner. 2001. Egg-killing fungus induces early hatching of red-eyed treefrog eggs. Ecology 82:2860–2869.

Weinstein, S.B. 2009. An aquatic disease on a terrestrial salamander: individual and population-level effects of the amphibian chytrid fungus, Batrachochytrium dendrobatidis, on Batrachoseps attenuatus (Plethodontidae). Copeia 2009:653–660.

Weldon, C., L.H. du Preez, A.D. Hyatt, R. Muller, and R. Speare. 2004. Origin of the amphibian chytrid fungus. Emerg Infect Dis 10:2100–2105.

Werner, E.E. 1998. Ecological experiments and a research program in community ecology. In W.J. Resetarits Jr., J. Bernardo, eds. Experimental ecology: issues and perspectives, pp. 3–26. Oxford University Press, England.

Williams, S.E., and J.-M. Hero. 1998. Rainforest frogs of the Australian wet tropics: guild classification and the ecological similarity of declining species. Proc Roy Soc B 265:597–602.

Wright, K.M., and B.R. Whitaker. 2001. Amphibian medicine and captive husbandry. Krieger Publishing, Malabar.

Woodhams, D.C., R.A. Alford, and G. Marantelli. 2003. Emerging disease of amphibians cured by elevated body temperature. Dis Aquat Org 55:65–67.

Woodhams, D.C., K. Ardipradja, R.A. Alford, G. Marantelli, L.K. Reinert, and L.A. Rollins-Smith. 2007. Resistance to chytridiomycosis varies among amphibian species and is correlated with skin peptide defenses. Anim Conserv 10:409–417.

Woodhams, D.C., R.A. Alford, C.J. Briggs, M. Johnson, and L.A. Rollins-Smith. 2008. Life history trade-offs influence disease in changing climates: strategies of an amphibian pathogen. Ecology 89:1627–1639.

# 37

# STRATEGIES FOR WILDLIFE DISEASE SURVEILLANCE

## Jonathan M. Sleeman, Christopher J. Brand, and Scott D. Wright

Epidemiologic surveillance is defined by the Centers for Disease Control and Prevention (CDC) as the "ongoing systematic and continuous collection, analysis, and interpretation of health data". The objective of surveillance is to generate data for rapid response to the detection of a disease of concern to apply prevention, control, or eradication measures as well as to evaluate such interventions. This is distinct from disease monitoring, which usually does not involve a particular response to disease detection.

Surveillance for wildlife diseases has increased in importance due to the emergence and re-emergence of wildlife diseases that are threats to human, animal, and ecosystem health, or could potentially have a negative economic impact. It has been estimated that 75% of emerging human diseases are zoonotic in origin, of which the majority originate from wildlife (Taylor et al. 2001). However, there are unique challenges concerning wildlife disease surveillance such that disease and pathogens can be very difficult to detect and measure in wild animals. These challenges have been described previously (Wobeser 2006), but one of the primary issues is that disease in wildlife often goes unrecognized, especially in remote locations. Furthermore, sick and dead animals are very difficult to detect, as animals will disguise the signs of illness or hide when diseased. Carcasses from diseased animals are also rapidly removed by scavengers or will rapidly decompose, rendering them

suboptimal for diagnostic purposes. There is also a lack of validated diagnostic tests for most wildlife disease agents as well as baseline data. The paucity of laboratory capacity with expertise in wildlife disease diagnostic investigation is also an impediment. Finally, surveillance networks for wildlife diseases that perform field investigations and report disease events are under-developed in most regions of the world.

Despite these challenges, a number of very important epidemiological surveillance projects have been ongoing or recently developed, and some examples are described in this chapter. The examples are mostly drawn from the experiences of the U.S. Geological Survey National Wildlife Health Center (NWHC) and are provided to illustrate the different surveillance strategies and sampling techniques that can be used and have proven successful. Some future directions for wildlife disease surveillance are also suggested.

## SURVEILLANCE STRATEGIES

The first goal of any disease surveillance program is to define the objective(s), as the system established may vary depending on the desired outcome—that is, early detection or outbreak response; evaluation of disease management actions; determination of presence or absence of a disease or pathogen; for research or education; or a combination of these objectives.

While it is possible to achieve multiple objectives using the same system, very often the differing objectives may not be compatible. For example, early detection systems should be modified annually to respond to changing exposure risk factors, improved understanding of the epidemiology of the disease, and lessons learned from previous surveillance. However, from a research perspective this would preclude the ability for inter-annual comparability of results. Efforts should target different objectives to be as compatible as possible without compromising the primary goal. The establishment of accurate case definitions for wildlife diseases can also be a challenge, yet this is essential to ensure comparability among data collected by different groups.

Types of surveillance are commonly divided into two major categories, passive versus active and scanning versus targeted surveillance. Active surveillance involves actively searching for particular diseases or information; passive surveillance involves data collected from disease observations on an *ad hoc* basis. Scanning surveillance involves continuously searching for disease within a population, and targeted surveillance involves looking in selected high-risk subsets of the population. These techniques are often combined; for example, scanning passive surveillance involves the continual looking for and investigating wildlife mortality events.

## Passive Surveillance

Passive surveillance takes advantage of previously collected data that are often obtained for different reasons but that are then used for surveillance purposes. Advantages of passive surveillance include cost-effectiveness and the ability to take advantage of convenience sampling and existing databases. Disadvantages include biased sampling and incomplete geographic coverage, precluding the ability to make statistical inferences about the population of interest. Maintenance and ongoing analysis of long-term datasets are necessary to determine baseline data for diseases and susceptible species before any perturbations to the established trends can be detected. Furthermore, wildlife population sizes are often unknown, and this lack of denominator information prevents calculation of disease prevalence and incidence and other basic descriptive epidemiologic parameters. An example of the use of passive surveillance

was the ability to observe an unexpected increase in submissions of raptors to wildlife rehabilitators and diagnostic facilities that was determined to be due to West Nile virus infection (WNV) (Joyner et al. 2006; Saito et al. 2007).

A major use of passive surveillance is to evaluate factors relating to mortality events that can be useful in providing descriptive epidemiologic parameters and generating hypotheses regarding the impact of disease on wildlife populations. For example, a retrospective review of avian mortality events due to salmonellosis in the United States determined that this disease was a significant contributor to mortality in certain passerine species, and identified increased salmonellosis-related mortality in specific geographic regions (Hall and Saito 2008). A 20-year-old manatee (*Trichechus manatus*) database was used to analyze trends in watercraft-related mortality (Ackerman et al. 1995; Wright et al. 1995). Managers used this information to establish manatee protection zones and limit watercraft use in these zones to reduce manatee mortality. Long-term datasets at the NWHC were used to document the effects of lead ingestion by bald eagles (*Haliaeetus leucocephalus*) and waterfowl and provided the scientific information that resulted in the ban on the use of lead shot for waterfowl hunting in the United States (Franson et al. 1986; Friend et al. 1999).

Another use of passive surveillance is to combine two types of data—for example, water quality data and precipitation data with the incidence of red tides to determine whether environmental factors contribute to the emergence or persistence of these events (Landsberg et al. 2007). This analysis determined that red tides thrive in water with high salinity, which occurs in estuaries, especially during droughts. Manatees frequent estuaries because of the abundant grass beds; however, this feeding behavior exposes them to fatal concentrations of brevetoxin (Bossart et al. 1998), and this combination of information provided a better understanding of how red tide events affect manatees.

## Morbidity and Mortality Investigations

Morbidity and mortality investigation of wildlife is a process whereby data are collected and analyzed to determine why an event occurred and if possible how to prevent or control this and similar events in the future. It is the most commonly used type of passive surveillance. These investigations are dependent upon the discovery of sick or dead animals by the public and as a result are biased to events in highly populated or easily accessible areas, pathologic conditions that cause obvious clinical signs or death, or large, highly visible animals. To best determine the cause of wildlife mortality events, carcasses need to be examined by laboratories specializing in wildlife diagnostic investigations. Some species-specific surveillance programs have been developed; for instance, the Amphibian Research and Monitoring Initiative (http://armi.usgs.gov/; accessed March 27, 2011), which is designed to increase surveillance for amphibian mortality events.

As often as possible, disease investigations lead to a management response and are also included as part of larger, more comprehensive surveillance programs. For example, mortality investigations of species known to be susceptible to H5N1 highly pathogenic avian influenza (HPAI) represent an important component of the interagency surveillance strategy for early detection of H5N1 HPAI in migratory birds in the United States (Brand 2009). Enhanced mortality investigations may also be a component of the response to the detection of an important disease by other methods in a surveillance program, such that if HPAI was detected in a hunter-harvested bird, increased testing of dead birds for avian influenza in proximity to this detection would be instituted.

Disease investigations are characterized by the collection of information associated with the event, such as location, species and numbers of animals involved, time progression of the event, habitat type, recent weather, and potentially related human activity. This information is combined with necropsy findings and ancillary diagnostic evaluations (Fig. 37.1) and is used to determine the etiology, describe the circumstances surrounding an event, evaluate the ecological impact and risk to wildlife, human, or domestic animal health, and ultimately provide management recommendations. The investigation also represents a temporal and geospatial record of the particular event and will add to the baseline data, allowing the significance of a similar event in the future to be compared to past events. Furthermore, comparing it to findings from past events can more easily reveal a new disease. In this way, disease investigations provide the opportunity to discover novel pathogens. White nose

syndrome in wild bats (Blehert et al. 2009), WNV in wild birds (Reed et al. 2003), avian vacuolar myelinopathy in American coots (*Fulica americana*) (Thomas et al. 1998), and *Perkinsus*-like organisms in frogs (Davis et al. 2007; NWHC unpublished data 2000) are a few recent examples of new diseases discovered in wildlife that resulted from mortality investigations.

In contrast, targeted surveillance does not require a full examination of the animals collected, thereby using fewer resources. Surveillance programs are often funded for the detection of a single disease agent, and so resources are focused on the work necessary to detect that disease. This was the case during the investigation of WNV in the United States. Thousands of dead wild birds were submitted to the NWHC for WNV testing but no further examination was possible, representing a missed opportunity. However, the selection of the type of diagnostic approach may allow for the identification of additional agents besides the targeted pathogen. If virus isolation rather than PCR is used, then additional agents can be identified through the targeted surveillance program. For example, other viruses such as Eastern equine encephalitis can be detected during WNV surveillance (Beckwith et al. 2002; Dusek et al. 2009).

The value of disease investigations contributing to our knowledge of long-term trends of wildlife diseases cannot be overemphasized. This value is realized when such data are used to predict and perhaps mitigate the affects of environmental factors such as global environmental change on wildlife health. Wildlife diseases such as avian botulism, WNV, avian cholera, and epizootic hemorrhagic disease (EHD) are affected either by seasonal availability of arthropod vectors and/or by host population density. Climate change could dramatically affect vector distribution or change migratory pathways or breeding seasons (Walther et al. 2002). In turn, these changes can affect the presence and distribution of diseases detected through clinical signs or mortality investigations. Using percentage of harvested white-tailed deer (*Odocoileus virginianus*) with hoof-wall growth interruptions as an indicator of the annual incidence of EHD, Sleeman et al. (2009) found that the incidence was greater in years with higher winter and summer average temperatures, and lower summer rainfall. They hypothesized that as temperatures continue to increase there will be more frequent and

**Figure 37.1:**
Pathologist at the U.S. Geological Survey's National Wildlife Health Center performs a necropsy on a gray wolf (*Canis lupus*).

severe outbreaks of EHD as well as spread to new geographic areas.

Success of large-scale disease investigation programs (regional, national, or global) depends upon the participation of many collaborators. Ideally, a surveillance network of trained field partners should exist to maximize the temporal and spatial coverage of the program. In the United States, the professionals most often involved are state and federal government employees who work for wildlife management or public health agencies, and occasionally personnel from universities and wildlife-focused nonprofit organizations. Although some mandates exist for reporting wildlife disease events, and attention to these events is received from highly trained personnel, it is often personal interest from individuals and groups that determines whether information or samples are submitted to a diagnostic laboratory. There is currently no legal requirement to report most wildlife diseases of management or conservation concern. However, professional training can enhance participation and improve quality of samples submitted by providing information on data and sample/carcass collection, shipping protocols, personal protection

equipment (PPE), carcass disposal, and management recommendations. Professional workshops also provide the opportunity to explain why disease investigations are important and how the information collected is used to assist with management of wildlife populations and facilitate communication with stakeholders and the public.

In summary, mortality investigations serve as a "trigger event" to launch a more intense surveillance effort to contain or stop the progression or spillover of a disease. Information gathered is used to describe disease trends over space and time, and these long-term databases are used to generate hypotheses, predict future events, and illustrate the progression and persistence of diseases. As WNV progressed west from the East Coast of the United States, wild bird mortality data were used to indicate the presence of the virus in a new area as well as the change in wild bird species affected over time. By the time WNV arrived in the western half of the United States, the avian sentinel species changed from corvids to small passerines (Marra et al. 2004; NWHC unpublished data 1999–2004). Finally, for rarely encountered species such as cetaceans, much of what is known about these species is gleaned from information collected during necropsies of the rare beach-cast animal.

## Active Surveillance

Active surveillance is a proactive process of surveying for a particular disease, and is usually ongoing. Goals of an active surveillance program are typically (1) early detection of the introduction or occurrence of a disease in a given area or population so that timely and appropriate control measures can be taken; (2) demonstration of the absence of a disease; (3) assessment of the prevalence and spatial distribution of a disease to assist in determining disease management strategies; or (4) monitoring of a disease to determine epidemiological changes in response to disease management actions or other ecological or environmental changes (Thrusfield 1995). Active surveillance involves a more rigorous and complex approach to designing the program so that the results have statistical validity and unbiased inferences about the population of interest can be drawn. This often results in a relatively large sample size, which together with the increased logistics of capturing and handling free-living wildlife makes this form of surveillance expensive relative to passive surveillance. Because of this, large-scale active surveillance in free-living wildlife is usually limited to diseases of high consequence or global concern, such as chronic wasting disease (CWD), HPAI, bovine tuberculosis (*Mycobacterium bovis*), and Ebola virus outbreaks.

Simple probability-based surveillance methods include simple random sampling of the population of interest, stratified random sampling where defined subunits of the population are sampled based on knowledge of risk factors, systematic sampling, and cluster sampling (Ratti and Garton 1994). Random selection of individuals or units to sample within the statistical design framework is a key assumption for most probability-based methods of surveillance. However, randomness is problematic when conducting surveillance in free-living wildlife, as this assumption is often not met, and sampling is more opportunistic or "convenience sampling" (Anderson 2001). Environmental factors, species characteristics, methods of obtaining individuals for sampling, and human influences create a complex set of biases difficult or impossible to control in designing large-scale wildlife surveillance. Additional complexities in

designing a probability-based surveillance program include lack of knowledge or definition of the population at risk, which is especially true of migratory wildlife. In many cases, the prevalence of the disease, or disease agent, is low, requiring relatively large sample sizes to detect an agent or determine significant changes in prevalence or distribution. The sensitivity and specificity of the tests used to determine infection or exposure is a factor that should also be considered in determining sample size requirements (see Aguirre Chapter 39, this volume). Statistical assistance and consultation should be sought in the design stage of an active surveillance program.

Targeted surveillance is a form of active surveillance in which statistical inferences to the population of interest are limited. In targeted surveillance a cohort of the population of interest is targeted for sampling because it has a higher risk for exposure or is more susceptible, or identification of infection or exposure in an individual is easier or more reliable than in the rest of the population. In many regards, targeted surveillance and sentinel surveillance using free-living wildlife are similar in concept, and the terminology is often used interchangeably. For example, several waterfowl species—*Cygnus* spp. (Newman et al. 2009), Eurasian pochard (*Aythya ferina*) and tufted duck (*A. fuligula*) (Keawcharoen et al. 2008)—have been referred to as sentinels for the occurrence of H5N1 HPAI because of their high susceptibility (i.e., mortality) to this virus as well as visibility on the landscape. Surveillance for CWD often targets animals displaying typical clinical signs, such as neurological deficits and emaciation (Samuel et al. 2003). In these examples, the primary goal of surveillance is detection of the disease in an area, rather than a determination of prevalence or distribution. Selection of the targeted populations is to optimize the likelihood of detecting the disease. Inferences about the population of interest from finding one or more positive animals are limited largely to the knowledge that the disease or agent is present, and further studies are needed to elaborate on the prevalence in the population. Under some conditions, selection of target sub-populations can be based on the efficiency of obtaining samples, which also may increase cost-effectiveness.

A nationwide surveillance program for the early detection of the introduction of H5N1 HPAI to the United States by wild birds was initiated in 2006 due to the increased recognition of the potential role of migratory birds in the long-distance expansion of this virus (USDA and USDI 2006). Multiple sampling methods were employed. The first stage applied unequal probability random sampling that was weighted by geographic region. Emphasis was placed on collecting samples from migratory birds in Alaska and the lower Pacific Flyway states because of the number of waterfowl and shorebird species that are known to migrate between North America and Asia, including migratory birds from regions in Asia where H5N1 HPAI was occurring (Brand 2009). Molecular studies of 38 low-pathogenicity avian influenza viruses isolated from Alaska during 2006 and 2007 as part of this surveillance program showed that nearly half of the viruses had at least one gene segment more closely related to Asian than North American strains of viruses (Koehler et al. 2008), indicating a higher degree of intercontinental viral genetic exchange in Alaska than previously reported (Krauss et al. 2007). A total of 72,320 wild birds were tested during three surveillance years between 2006 and 2009 using live-captured and hunter-killed birds; this represents one of the largest wildlife disease surveillance projects undertaken (NWHC, unpublished data 2006–2009). It is important to regularly evaluate large-scale active surveillance programs to ensure that goals are being met as well as to determine cost-effectiveness. For example, results from HPAI surveillance have increased our understanding of the epidemiology of avian influenza viruses that will be useful in the design of new and more effective surveillance programs (Munster et al. 2007).

## WILDLIFE SENTINELS FOR HEALTH AND DISEASE

The concept of using sentinel animals as a surveillance tool has been widely applied for both infectious diseases and environmental toxins (Thrusfield 1995), though is probably underused (Rabinowitz et al. 2005). In its broader sense, a sentinel can be defined as a susceptible animal (or a sentinel unit as a susceptible population) used to detect or quantify the presence or occurrence of a pathogen, disease, or other environmental hazard. The utility of a sentinel is its ability to serve as an indicator of the presence or absence of an agent in a given area in a more timely, sensitive, visible, or cost-effective manner than other

types of surveillance. This is because sentinel animals are either more at risk, sensitive, or susceptible to the specific agent than the species or population of concern; effects of the agent are more easily observed or occur earlier in sentinels than in target populations; sentinels are more easily observed and sampled than other animals; sentinels are the actual source of the agent for the target population; or it is logistically more cost-effective than other forms of surveillance.

Halliday et al. (2007) lay out a framework for evaluating the utility of sentinel animals for infectious diseases based on characteristics of the pathogen, the target population, and the sentinel species or population. Depending on the specific objectives of the surveillance and its ecological context, critical attributes of the sentinel system that must be considered include (1) sentinel response to the pathogen or agent, (2) relationship between sentinel and target populations, and (3) routes of transmission.

Wildlife sentinels in particular have been used to determine the presence of disease agents for zoonotic diseases in which the human population is the "target" of concern (e.g., WNV in crows [Eidson et al. 2001] and sylvatic plague [*Yersinia pestis*] in carnivores [Willeberg et al. 1979]) as well as for diseases of domestic animals and livestock (e.g., *rinderpest* in African buffalo [*Syncerus caffer*] [Rossiter 1994]). However, sentinels have also been used for diseases of concern to wildlife conservation (e.g., the presence of canine distemper virus in domestic dogs in close proximity to wild African carnivores [Roelke-Parker et al. 1996]).

Other examples include use of wing-feather clipped mallards (*Anas platyrhynchos*) as sentinels to determine the onset and course of avian botulism (*Clostridium botulinum* type C) on wetland units (Rocke and Brand 1994). The objective of this work was to determine the site-specific environmental factors related to botulism toxin production and transfer to birds. Using free-flying birds as sentinels for botulism posed uncertainties as to whether ingestion of toxin occurred at the site of morbidity or mortality, or on adjacent wetlands—hence the use of wing-clipped birds. Close monitoring and rapid removal and replacement of moribund and dead sentinels also enabled a quantitative assessment of the magnitude of mortality and relative availability of toxin. Similarly, coyotes (*Canis latrans*) and other carnivores have served as effective sentinels for sylvatic plague in wild rodent populations (Willeberg et al. 1979) and have been used to alert public health agencies to the risk for plague infection in humans. Frölich et al. (1998) demonstrated the utility of red foxes (*Vulpes vulpes*) as sentinels for rabbit hemorrhagic disease virus through their antibody response to the accumulative effect of consumption of infected rabbits.

In certain situations, animal sentinels are deliberately placed in the field to detect infection or exposure to agents. Confinement in cages or restriction of movements allows access to these sentinels for sequential observations and sampling, as well as the ability to account for the sentinel population at risk, quantify morbidity and survival rates, and examine time-series responses. Rocke et al. (2002) used a combination of wild-caught American coots and captive-reared, wing-clipped mallards that were penned as sentinels on a North Carolina reservoir to detect the onset and course, potential source, and etiology of an unknown disease agent causing avian vacuolar myelinopathy (Thomas et al. 1998). In Hawaii, Atkinson et al. (1993) used sentinel chickens and canaries exposed in cages hung in the forest canopy to monitor the transmission of avian pox and avian malaria to determine specific locations and elevations where disease transmission in endemic forest birds was occurring (Fig. 37.2). The deliberate exposure of wild or captive-raised sentinels as described above offers several advantages over the use of "natural" or free-living wildlife sentinels, but also requires precautionary measures. Care should be taken that other diseases potentially affecting wildlife are not introduced into wild populations by the sentinels, and that they do not serve as reservoir or amplification hosts for diseases present in wild populations.

Wildlife sentinels have also been used as indicators of ecosystem or environmental health (NRC 1991). For example, mink (*Neovison vison*) are often used as sentinels for persistent and ubiquitous contaminants such as mercury and polychlorinated biphenyls as they are widely distributed, abundant, and regularly trapped, making them an excellent model to monitor environmental pollution on temporal and spatial scales (Basu et al. 2007). Furthermore, as high-trophic-level, piscivorous mammals, mink bioaccumulate appreciable concentrations of pollutants, increasing the detection of these compounds. For the same reasons, several marine vertebrate species make excellent sentinels for marine ecosystem health (Aguirre and Tabor 2004; Tabor and Aguirre 2004).

## DATA MANAGEMENT, RESPONSE, AND COMMUNICATIONS

Determination of data to collect and systems to use to capture field data is necessary before beginning surveillance. At a minimum, data on sample identification, species, date, age, sex, and location should be collected. Data fields must be standardized to allow comparability, although such standards are rarely used in wildlife disease surveillance. The traditional paper data card is being replaced by PDAs or smart phones, often with GPS capabilities that allow for the electronic capture and transfer of data to a database. This results in fewer transcription errors, among other advantages. Finally, a database system to track, store, retrieve, analyze, and disseminate information is an essential component, and there are a number of database formats, such as SQL server, that allow Internet-based systems with Web access. Response plans should be in place for all diseases for which active surveillance is being conducted. These plans define the actions that will be taken should the disease be detected. It should include communications plans; assessment and monitoring surveillance plans; specific regulatory, disease prevention, control, or eradication actions that may be taken; and how success will be measured.

## TYPES OF SPECIMENS

The type of diagnostic samples collected will be determined by the surveillance technique and sample transport requirements as well as the goal of the surveillance effort. Samples can range from whole carcasses, specific biological samples such as blood, the measurement of biomarkers, use of proxy species, or simple observation of clinical signs, to name a few. Fresh carcasses are advantageous as they provide the maximum amount and diversity of biological materials for diagnostic investigation, which is particularly useful if the etiology is unknown. As discussed previously, moribund and dead wild animals can be very difficult to find, and active searching for carcasses, or "carcass sweeps," in geographic areas at risk for exposure can be useful. These searches are subject to sampling bias and are dependent on species, terrain, and disease of interest (Wobeser 2006). However, the use of volunteer observers, or "citizen scientists," to

**Figure 37.2:**
Federal biologists set up cages containing sentinel chickens and canaries in the forest canopy in Hawaii to monitor the natural transmission of avian pox and avian malaria to determine specific locations and elevations where disease transmission in endemic forest birds is occurring.

collect data on house finches (*Carpodacus mexicanus*) with clinical signs of mycoplasmal conjunctivitis illustrates the usefulness of engaging the public in tracking the spatiotemporal spread of a disease on a large geographic scale (Dhondt et al. 2005).

There has also been increasing interest in the use of syndromic surveillance as part of early detection systems. Syndromic surveillance applies to surveillance using health-related data that precede diagnosis and signal a sufficient probability of an outbreak to warrant further investigation (Buehler et al. 2003). The feasibility of detecting bioterrorism events by investigating wildlife mortality is being explored, especially as several bioterrorism agents of highest concern are also wildlife diseases. Consequently, an unusual die-off of a wildlife species known to be susceptible to a particular bioterrorism agent may be an early warning of risk to human health, especially if clinical signs manifest in animals before humans (Rabinowitz et al. 2006). However, syndromic surveillance systems in wildlife have not been rigorously evaluated.

Collection of animals by lethal methods for diagnostic sampling can also be performed, and with appropriate design this method may eliminate some of the sources of bias and allow for more random sampling. This method is usually employed when random sampling is required to determine the prevalence or geographic distribution of the disease of interest. It also allows for the collection of the widest variety and optimal tissue samples for diagnostic purposes. However, this method can be controversial and cannot be used for threatened and endangered species.

Sources of wildlife convenience samples for carcasses, live animals, and other biological materials include hunter-harvested animals, road-killed animals, animals brought to wildlife rehabilitators, and ongoing research projects. The non-randomness of convenience samples militates against straightforward inference from sample to population, but they have been used for recent surveillance projects such as H5N1 HPAI in migratory birds (Brand 2009).

Radiotelemetry tracking of animals, particularly if fitted with mortality sensors that facilitate the recovery of dead animals, provides unique opportunities to determine the cause of mortality due to the availability of fresh carcasses as well as the population-level effects of disease as the population size at risk is known. For example, an outbreak of EHD in a radio-collared population of white-tailed deer allowed the

detection of an event that would have gone undetected as well as the determination of an accurate
mortality rate (Beringer et al. 2000). In addition, satellite telemetry, though expensive, can provide local
and long-range movement data for migratory animals.
This has been used to provide valuable movement
data for species that are natural reservoirs for important pathogens such as birds with HPAI and bats
with Nipah virus (Epstein et al. 2009). Not only are
these data important for understanding host range,
but they also provide expanded spatial information
about disease distribution that allows for broader risk
assessments.

Biological samples for surveillance purposes can
also be collected from live animals: either samples can
be collected opportunistically during routine operations, or animals can be specifically captured or handled for sampling purposes. The types of specimens
that can be collected from live animals include blood
for serological or molecular analyses, feces for parasitological evaluation, feathers or pelage for heavy metal
analysis, as well as soft tissue or bone biopsies, among
other samples. Fresh urine, feces, and feathers may
also be collected without capturing an animal. The
diagnostic information available from live-captured
samples can be more limited compared to postmortem examination of whole carcasses as well as more
technically challenging and expensive to obtain.
However, this can be a useful technique that allows
targeting of specific populations or when lethal collection is not feasible or desired (Aguirre et al. 2002).

Exposure to noxious substances can be detected
by measuring physiological indicators or biomarkers.
Examples include measurement of enzymes such as
cholinesterase and delta aminolevulinic acid dehy-
dratase to indicate organophosphate or carbamate
pesticide exposure, and lead poisoning, respectively
(Friend and Franson 1999). In addition, activation
of the hepatic enzyme cytochrome P450 occurs after
exposure to various compounds such as polyaromatic
hydrocarbons. Many of these physiological responses
lack specificity and will occur after exposure to a
variety of compounds, limiting their usefulness in
determining etiology. However, they can be useful
in monitoring the long-term health of wildlife populations and ecosystems exposed to contaminants.
Surveying for cytochrome P450 levels in sea otters
(*Enhydra lutris*) after the Exxon Valdez oil spill
has been used to determine population health and

evaluate progress toward near-shore ecosystem recovery (Peterson et al. 2003).

The use of proxy indicators or species takes
advantage of the trophic relationship in which predators or scavengers are examined for evidence of the
disease agent in the prey. This method uses the fact
that predators will be exposed to a large sample of
prey animals as well as that predators are generally
longer-lived than prey. A recent study investigating
the potential of coyotes as sentinels for *M. bovis*,
which is present in white-tailed deer in northeastern
Michigan, found that by focusing on coyotes rather
than deer, 97% fewer animals were sampled and
the likelihood of detecting *M. bovis* increased by 40%
(VerCauteren et al. 2008).

Disease surveillance of hosts that are not the
species of most concern or the direct target of management actions can be a useful technique in assessing
risk to the target wildlife population or in assessing
the impact of management interventions. For example, surveillance for canine distemper virus in domestic dogs has been performed to assess risks to wild
carnivores in contact with their domestic counterparts
as well as to evaluate vaccination campaigns (Bronson
et al. 2008; Cleaveland et al. 2000, 2006).

The questionnaire is a common tool used in public
health and agriculture to obtain surveillance data
(Thrusfield 1995). However, this technique has not
commonly been used for wildlife disease surveillance,
as free-ranging wildlife populations are usually not
closely associated with humans. Surveys of demographic groups who have regular contact with wildlife, such as hunters or wildlife rehabilitators (Kalish
et al. 2005; Schopler et al. 2005), can be a useful technique. Furthermore, this technique can be useful in
evaluating health risks to wildlife populations from
humans or domestic animals. Guerrera et al. (2003)
conducted interview questionnaires of villagers living
in close proximity to mountain gorillas (*Gorilla
beringei beringei*) in Bwindi Impenetrable Forest
National Park, Uganda, to estimate the prevalence
of infectious diseases in this human population and
consequently to evaluate the risk for transmission
from humans to gorillas. Questionnaires are also
useful for meta-analyses (i.e., the collection and analysis of data from a variety of sources for the purpose
of integrating the findings; Gordis 2000) and can
be especially useful for obtaining unpublished data.
The World Organization for Animal Health (OIE)

regularly sends questionnaires to compile wildlife health data from participating countries into a central database.

Surveillance approaches for diseases can also involve detection of disease-causing agents in the environment—the air, water, soil, or other environmental matrices that can serve as sources of exposure to infectious agents or contaminants. Enteric diseases in particular are excreted by infected animals into the water or soil, and can persist for variable but sometimes extended time periods, depending on the pathogen and the physical, chemical, and biological properties of the environmental matrix. Fecal material itself can be used as an environmental surveillance tool, and was used in the H5N1 HPAI early detection surveillance in the United States (USDA and USDI 2006). Advantages of using environmental samples include the relative ease of obtaining samples, the ability to collect relatively large sample sizes, and the site-specific information on the distribution of the disease and exposure risk. However, numerous factors will affect the reliability of this method for detecting pathogens. These include knowledge of factors such as the modes of transmission and excretion of the agent; survivability or persistence of the agent under various environmental conditions; diagnostic methods, quantification methods, detection limits specific for the agent and validated for the environmental conditions under which samples were collected; and the appropriate sampling design. Other disadvantages include lack of assurance of host species when multispecies flock or herd is tested as well as the limited data that can be collected on specific animals, such as age and sex. However, for closely monitored populations in which individuals can be identified and tracked, these detailed demographic data may be available. Sleeman et al. (2000) were able to conduct detailed parasitological surveys of mountain gorillas in which the prevalence of different parasites could be compared among groups, and between age and sex as these animals were closely observed, allowing environmental fecal samples to be linked to specific individuals.

## FUTURE DIRECTIONS

Disease prevention is the desired method to protect the health of wildlife populations, as once a disease has been introduced into a population it can be very difficult, if not impossible, to control or eradicate (Wobeser 2006). There are few effective wildlife disease management tools available (e.g., population reduction, use of vaccines or other biologics, and environmental modification), but they are expensive, often lack any assurance of success, and can be unpalatable to the general public. To increase the probability of successful wildlife disease management, future surveillance efforts should be based on risk analysis, investigation of potential exposure pathways, and improved knowledge of reservoirs of potential emerging pathogens (Haydon et al. 2002). New molecular techniques have opened up avenues for pathogen discovery not previously available (Lipkin 2008), and application of spatially referenced databases such as GIS allows for risk assessments that can assist in targeting surveillance to high-risk populations and geographic locations (Sleeman 2005). Integration and analysis of real-time data from a variety of sources, including human and animal health data with climatic, ecological, hydrological, geological, and socioeconomic data, among other sources, to determine drivers of disease emergence and generate predictive models will help direct resources to geographic areas and populations, so-called hotspots, with the greatest need (Jones et al. 2008). Increased global capacity to detect, diagnose, and provide robust and rapid responses to wildlife disease outbreaks and emerging diseases will also be critical in this effort.

## ACKNOWLEDGMENTS

We thank Joanne Bosch and Kathy Wesenberg for assistance with this chapter as well as Drs. J. Brown and J. Epstein for many helpful comments.

## REFERENCES

Ackerman, B.B., S.D. Wright, R.K. Bonde, D.K. Odell, and D.J. Banowetz. 1995. Trends and patterns in mortality of manatees in Florida, 1974–1992, U.S. National Biological Service Information and Technology Report, pp. 223–258. Florida Fish and Wildlife Conservation Commission, St. Petersburg.

Aguirre, A.A., R.S. Ostfeld, G.M. Tabor, C. House and M.C. Pearl. 2002. Conservation medicine: ecological health in practice. Oxford University Press, New York.

Aguirre, A.A., and G.M. Tabor. 2004. Marine vertebrates as sentinels of marine ecosystem health. EcoHealth 1:236–238.

Anderson, D.R. 2001. The need to get the basics right in wildlife field studies. Wildl Soc Bull 29:1294–1297.

Atkinson, C.T., R.J. Dusek, and W.M. Iko. 1993. Avian malaria fatal to juvenile I'iwi. Hawaii's Forests Wildl 8:1,11.

Basu, N., A.M. Scheuhammer, S.J. Bursian, J. Elliott, K. Rouvinen-Watt, and H.M. Chan. 2007. Mink as a sentinel species in environmental health. Environ Res 103:130–144.

Beckwith, W.H., S. Sirpenski, R.A. French, R. Nelson, and D. Mayo. 2002. Isolation of eastern equine encephalitis virus and West Nile virus from crows during increased arbovirus surveillance in Connecticut, 2000. Am J Trop Med Hyg 66:422–426.

Beringer, J., L.P. Hansen, and D.E. Stallknecht. 2000. An epizootic of hemorrhagic disease in white-tailed deer in Missouri. J Wildl Dis 36:588–591.

Blehert, D.S., A.C. Hicks, M. Behr, C.U. Meteyer, B.M. Berlowski-Zier, E.L. Buckles, J.T.H. Coleman, S.R. Darling, A. Gargas, R. Niver, J.C. Okoniewski, R.J. Rudd, and W.B. Stone. 2009. Bat white-nose syndrome: An emerging fungal pathogen? Science 323:227.

Bossart, G.D., D.G. Baden, R.Y. Ewing, B. Roberts, and S.D. Wright. 1998. Brevetoxicosis in manatees (*Trichechus manatus latirostris*) from the 1996 epizootic: Gross, histologic, and immunohistochemical features. Toxicol Pathol 26:276–282.

Brand, C.J. 2009. Surveillance plan for the early detection of H5N1 highly pathogenic avian influenza virus in migratory birds in the United States: Surveillance Year 2009. General Information Product, U.S. Geological Survey, Reston, Virginia.

Bronson, E., L.H. Emmons, S. Murray, E.J. Dubovi, and S.L. Deem. 2008. Serosurvey of pathogens in domestic dogs on the border of Noel Kempff Mercado National Park, Bolivia. J Zoo Wildl Med 39:28–36.

Buehler, J.W., R.L. Berkelman, D.M. Hartley, and C.J. Peters. 2003. Syndromic surveillance and bioterrorism-related epidemics. Emerg Infect Dis 9:1197–1204.

Cleaveland, S., M.G.J. Appel, W.S.K. Chalmers, C. Chillingworth, M. Kaare, and C. Dye. 2000. Serological and demographic evidence for domestic dogs as a source of canine distemper virus infection for Serengeti wildlife. Vet Microbiol 72:217–227.

Cleaveland, S., M. Kaare, D. Knobel, and M.K. Laurenson. 2006. Canine vaccination—providing broader benefits for disease control. Vet Microbiol 117:43–50.

Davis, A.K., M.J. Yabsley, M.K. Keel, and J.C. Maerz. 2007. Discovery of a novel alveolate pathogen affecting southern leopard frogs in Georgia: description of the disease and host effects. Ecohealth 4:310–317.

Dhondt, A.A., S. Altizer, E.G. Cooch, A.K. Davis, A. Dobson, M.J.L. Driscoll, B.K. Hartup, D.M. Hawley, W.M. Hochachka, P.R. Hosseini, C.S. Jennelle, G.V. Kollias, D.H. Ley, E.C.H. Swarthout, and K.V. Sydenstricker. 2005. Dynamics of a novel pathogen in an avian host: Mycoplasmal conjunctivitis in house finches. Acta Tropica 94:77–93.

Dusek, R.J., R.G. McLean, L.D. Kramer, S.R. Ubico, A.P. Dupuis, II, G.D. Ebel, and S.C. Guptill. 2009. Prevalence of West Nile virus in migratory birds during spring and fall migration. Am J Trop Med Hyg 81:1151–1158.

Eidson, M., N. Komar, F. Sorhage, R. Nelson, T. Talbot, F. Mostashari, R. McLean, and West Nile Virus Avian Mortality Surveillance Group. 2001. Crow deaths as a sentinel surveillance system for West Nile virus in the northeastern United States, 1999. Emerg Infect Dis 7:615–620.

Epstein, J.H., K.J. Olival, J.R.C. Pulliam, C. Smith, J. Westrum, T. Hughes, A.P. Dobson, Z. Akbar, R. Sohayati Abdul, B. Misliah Mohamad, H.E. Field, and P. Daszak. 2009. *Pteropus vampyrus*, a hunted migratory species with a multinational home-range and a need for regional management. J Appl Ecol 46:991–1002.

Franson, J.C., G.M. Haramis, M.C. Perry, and J.F. Moore. 1986. Blood protoporphyrin for detecting lead exposure in canvasbacks. *In* J.S. Feierabend and A.B. Russell, eds. Lead poisoning in wild waterfowl—a workshop, pp. 32–37. National Wildlife Federation, Washington D.C.

Friend, M., J.C. Franson,. 1999. Field manual of wildlife diseases general field procedures and diseases of birds. U.S. Dept. of the Interior, U.S. Geological Survey, Washington D.C.

Frölich, K., F. Klima, and J. Dedek. 1998. Antibodies against rabbit hemorrhagic disease virus in free-ranging red foxes from Germany. J Wildl Dis 34:436–442.

Gordis, L. 2000. Epidemiology. W.B. Saunders, Philadelphia.

Guerrera, W., J.M. Sleeman, S.B. Jasper, L.B. Pace, T.Y. Ichinose, and J.S. Reif. 2003. Medical survey of the local human population to determine possible health risks to the mountain gorillas of Bwindi Impenetrable Forest National Park, Uganda. Int J Primatol 24:197–207.

Hall, A.J., and E.K. Saito. 2008. Avian wildlife mortality events due to salmonellosis in the United States, 1985–2004. J Wildl Dis 44:585–593.

Halliday, J.E.B., A.L. Meredith, D.L. Knobel, D.J. Shaw, B. Bronsvoort, and S. Cleaveland. 2007. A framework for evaluating animals as sentinels for infectious disease surveillance. J Royal Soc Interf 4:973–984.

Haydon, D.T., S. Cleaveland, L.H. Taylor, and M.K. Laurenson. 2002. Identifying reservoirs of infection: A conceptual and practical challenge. Emerg Infect Dis 8:1468–1473.

Jones, K.E., N.G. Patel, M.A. Levy, A. Storeygard, D. Balk, J.L. Gittleman, and P. Daszak. 2008. Global trends in emerging infectious diseases. Nature 451:990–993.

Joyner, P.H., S. Kelly, A.A. Shreve, S.E. Snead, J.M. Sleeman, and D.A. Pettit. 2006. West Nile virus in raptors from Virginia during 2003: Clinical, diagnostic, and epidemiologic findings. J Wildl Dis 42:335–344.

Kalish, M.L., N.D. Wolfe, C.B. Ndongmo, J. McNicholl, K.E. Robbins, M. Aidoo, P.N. Fonjungo, G. Alemnji, C. Zeh, C.F. Djoko, E. Mpoudi-Ngole, D.S. Burke, and T.M. Folks. 2005. Central African hunters exposed to simian immunodeficiency virus. Emerg Infect Dis 11:1928–1930.

Keawcharoen, J., D. van Riel, G. van Amerongen, T. Bestebroer, W.E. Beyer, R. van Lavieren, A. Osterhaus, R.A.M. Fouchier, and T. Kuiken. 2008. Wild ducks as long-distance vectors of highly pathogenic avian influenza virus (H5N1). Emerg Infect Dis 14:600–607.

Koehler, A.V., J.M. Pearce, P.L. Flint, J.C. Franson, and H.S. Ip. 2008. Genetic evidence of intercontinental movement of avian influenza in a migratory bird: The northern pintail (*Anas acuta*). Mol Ecol 17:4754–4762.

Krauss, S., C.A. Obert, J. Franks, D. Walker, K. Jones, P. Seiler, L. Niles, S.P. Pryor, J.C. Obenauer, C.W. Naeve, L. Widjaja, R.J. Webby, and R.G. Webster. 2007. Influenza in migratory birds and evidence of limited intercontinental virus exchange. PLoS Pathog 3:1684–1693.

Landsberg, J.H., G.A. Vargo, L.J. Flewelling, and F.E. Wiley. 2007. Algal biotoxins, infectious diseases of wild birds. In N.J. Thomas, D.B. Hunter, and C.T. Atkinson eds. Infectious diseases of wild birds, pp. 431–455. Blackwell Publishing Professional, Ames, Iowa.

Lipkin, W.I. 2008. Pathogen discovery. PLoS Pathog 4:e1000002.

Marra, P.P., S. Griffing, C. Caffrey, A.M. Kilpatrick, R. McLean, C. Brand, E. Saito, A.P. Dupuis, L. Kramer, and R. Novak. 2004. West Nile virus and wildlife. Bioscience 54:393–402.

Munster, V.J., C. Baas, P. Lexmond, J. Waldenstrom, A. Wallensten, T. Fransson, G.F. Rimmelzwaan, W.E.P. Beyer, M. Schutten, B. Olsen, A.D.M.E. Osterhaus, and R.A.M. Fouchier. 2007. Spatial, temporal, and species variation in prevalence of influenza A viruses in wild migratory birds. PLoS Pathog 3:e61.

National Research Council Committee on Animals as Monitors of Environmental Health (NRC). 1991. Animals as sentinels of environmental health hazards. National Academy Press, Washington D.C.

Newman, S.H., S.A. Iverson, J.Y. Takekawa, M. Gilbert, D.J. Prosser, N. Batbayar, T. Natsagdorj, and D.C. Douglas. 2009. Migration of whooper swans and outbreaks of highly pathogenic avian influenza H5N1 virus in eastern Asia. PLoS One 4:e5729.

Peterson, C.H., S.D. Rice, J.W. Short, D. Esler, J.L. Bodkin, B.E. Ballachey, and D.B. Irons. 2003. Long-term ecosystem response to the Exxon Valdez oil spill. Science 302:2082–2086.

Rabinowitz, P., Z. Gordon, D. Chudnov, M. Wilcox, L. Odofin, A. Liu, and J. Dein. 2006. Animals as sentinels of bioterrorism agents. Emerg Infect Dis 12:647–652.

Rabinowitz, P.M., Z. Gordon, R. Holmes, B. Taylor, M. Wilcox, D. Chudnov, P. Nadkarni, and F.J. Dein. 2005. Animals as sentinels of human environmental health hazards: An evidence-based analysis. EcoHealth 2:26–37.

Ratti, J.T., and E.O. Garton. 1994. Research and experimental design. In T.A. Bookhout. ed. Research and management techniques for wildlife and habitats, pp. 1–23. Allen Press, Lawrence, Kansas.

Reed, K.D., J.K. Meece, J.S. Henkel, and S.K. Shukla. 2003. Birds, migration and emerging zoonoses: West Nile virus, Lyme disease, influenza A and enteropathogens. Clin Med Res 1:5–12.

Rocke, T.E., and C.J. Brand. 1994. Use of sentinel mallards for epizootiologic studies of avian botulism. J Wildl Dis 30:514–522.

Rocke, T.E., N.J. Thomas, T. Augspurger, and K. Miller. 2002. Epizootiologic studies of avian vacuolar myelinopathy in waterbirds. J Wildl Dis 38:678–684.

Roelke-Parker, M.E., L. Munson, C. Packer, R. Kock, S. Cleaveland, M. Carpenter, S.J. Obrien, A. Pospischil, R. HofmannLehmann, H. Lutz, G.L.M. Mwamengele, M.N. Mgasa, G.A. Machange, B.A. Summers, and M.J.G. Appel. (1996. A canine distemper virus epidemic in Serengeti lions (*Panthera leo*). Nature 379:441–445.

Rossiter, P.B. 1994. Rinderpest. In J.A.W. Coetzler, G.R. Thomson, and R.C. Tustin, eds. Infectious diseases of livestock with special reference to southern Africa, pp. 735–757. Oxford University Press, New York.

Saito, E.K., L. Sileo, D.E. Green, C.U. Meteyer, G.S. McLaughlin, K.A. Converse, and D.E. Docherty. 2007. Raptor mortality due to West Nile virus in the United States, 2002. J Wildl Dis 43:206–213.

Samuel, M.D., D.O. Joly, M.A. Wild, S.D. Wright, D.L. Otis, R.W. Erge, and M.W. Miller. 2003. Surveillance strategies for detecting chronic wasting disease in free-ranging deer and elk, CWD Surveillance Workshop U.S. Geological Survey, Madison, Wisconsin.

Schopler, R.L., A.J. Hall, and P. Cowen. 2005. Public veterinary medicine: public health - survey of wildlife rehabilitators regarding rabies vector species. J Am Vet Med Assoc 227:1568–1572.

Sleeman, J.M. 2005. Disease risk assessment in African great apes using geographic information systems. EcoHealth 2:222–227.

Sleeman, J.M., J.E. Howell, W.M. Knox, and P.J. Stenger. 2009. Incidence of hemorrhagic disease in white-tailed deer is associated with winter and summer climatic conditions. EcoHealth 6:11–15.

Sleeman, J.M., L.L. Meader, A.B. Mudakikwa, J.W. Foster, and S. Patton. 2000. Gastrointestinal parasites of mountain gorillas (Gorilla gorilla beringei) in the Parc National des Volcans, Rwanda. J Zoo Wildl Med 31:322–328.

Tabor, G.M., and A.A. Aguirre. 2004. Ecosystem health and sentinel species: adding an ecological element to the proverbial "canary in the mineshaft." EcoHealth 1:226–228.

Taylor, L.H., S.M. Latham, and M.E.J. Woolhouse. 2001. Risk factors for human disease emergence. Royal Soc Phil Tran B 356:983–989.

Thomas, N.J., C.U. Meteyer, and L. Sileo. 1998. Epizootic vacuolar myelinopathy of the central nervous system of bald eagles (Haliaeetus leucocephalus) and American coots (Fulica americana). Vet Pathol 35:479–487.

Thrusfield, M.V. 1995. Veterinary epidemiology. Blackwell, Oxford.

U.S. Department of Agriculture (USDA) and U.S. Department of Interior (USDI). 2006. An early detection system for highly pathogenic H5N1 avian influenza in wild migratory birds—U.S. interagency strategic plan. Accessible at: http://www.nwhc.usgs.gov/publications/other/Final_Wild_Bird_Strategic_Plan_0322.pdf (Accessed March 27, 2011).

VerCauteren, K.C., T.C. Atwood, T.J. DeLiberto, H.J. Smith, J.S. Stevenson, B.V. Thomsen, T. Gidlewski, and J. Payeur. 2008. Sentinel-based surveillance of coyotes to detect bovine tuberculosis, Michigan. Emerg Infect Dis 14:1862–1869.

Walther, G.R., E. Post, P. Convey, A. Menzel, C. Parmesan, T.J.C. Beebee, J.M. Fromentin, O. Hoegh-Guldberg, and F. Bairlein. 2002. Ecological responses to recent climate change. Nature 416:389–395.

Willeberg, P.W., R. Ruppanner, D.E. Behymer, H.H. Higa, C.E. Franti, R.A. Thompson, and B. Bohannan. 1979. Epidemiologic survey of sylvatic plague by sero-testing coyote sentinels with enzyme immunoassay. Am J Epidemiol 110:328–334.

Wobeser, G.A. 2006. Essentials of disease in wild animals. Blackwell Pub., Ames, Iowa.

Wright, S.D., B.B. Ackerman, R.K. Bonde, C.A. Beck, and D.J. Banowetz. 1995. Analysis of watercraft-related mortality of manatees in Florida, 1979–1991. In T.J. O'Shea, B.B. Ackerman, and H.F. Percival eds. Population biology of the Florida manatee (Trichechus manatus latirostris), pp. 259–268. National Biological Service, Washington D.C.

# 38

# WILDLIFE HEALTH MONITORING SYSTEMS IN NORTH AMERICA

## From Sentinel Species to Public Policy

Michelle M. Willette, Julia B. Ponder, Dave L. McRuer, and Edward E. Clark, Jr.

## THE CASE FOR MONITORING WILDLIFE HEALTH

Wildlife has value. Some values are tangible and quantifiable, such as nutritional or economic benefits associated with harvesting and ecotourism. Other values are less tangible, such as contributions to "the web of life" and sociocultural traditions (Chardonnet et al. 2002). Wildlife can also serve as sentinels of problems in the environment, including the presence of contaminants, emerging diseases, altered habitat, and climate change. Emerging diseases such as West Nile virus (WNV), white-nose syndrome in bats, or chytridiomycosis in amphibians can have dramatic effects on wildlife populations and the public benefits associated with them. Other wildlife diseases may have significant consequences for domestic animals and humans, consequences that may rise to the level of affecting national security. The implications of these factors make the ability to monitor wildlife health a critical tool for important public interest endeavors, including the management and conservation of ecosystems; the protection of human health and safety; the stability of regional and national agricultural economies; and the defense of nations from biosecurity and bioterrorism threats.

## EMERGING INFECTIOUS DISEASES

Wildlife is recognized as a key component of ecosystem health, a term reflecting the "ecohealth/one health" concept that integrates human, domestic animal, wild animal, and environmental health (Aguirre et al. 2002). The increased significance of zoonotic disease is most commonly associated with more frequent interfaces between humans and animals within a global community. Livestock and poultry are also vulnerable to the effects of certain infectious diseases transmitted by wildlife. The economic losses from disease outbreaks and trade restrictions can be severe. Unfortunately, although the disease pathogens may be shared between domestic animals and wildlife, the system requirements for monitoring their health are very different. Routine healthcare is nonexistent in wild animals, and reporting of disease occurs primarily in situations where human health is threatened, a risk of economic impact is identified, or a major die-off is reported. Efforts to control disease in livestock can be complicated by the persistence of the pathogen in wildlife reservoirs. Inevitably, there will be inherent conflicts between control methods for enzootic wildlife diseases that affect livestock and wildlife conservation.

## WILDLIFE AS SENTINELS FOR ECOLOGICAL HEALTH

For centuries, wild animals have been recognized as valuable indicators of environmental health (Rabinowitz and Conti 2009). The widely used reference to "the canary in the coalmine" refers to the historical practice of taking caged birds into the shafts of underground mines to detect the presence of toxic gases. Since birds are extremely sensitive to these odorless and colorless gases, their health is affected before human health, thus alerting the miners to leave the area before they too are harmed.

The potential to identify health threats to wildlife populations and forecast a need for action is well demonstrated by dichloro-diphenyl-trichloroethane (DDT). It is generally acknowledged that the documentation of DDT's insidious effects on such species as bald eagles, pelicans, peregrine falcons, and ospreys led to the study of the effects of DDT and its metabolites in humans, which in turn led to the subsequent ban on the manufacture and use of this product in the United States in June 1972. A contemporary example is lead, a heavy metal. Lead toxicity in wildlife has been reported for over a century. Recently, spent lead from ammunition contained in deer gut piles left in the field has been identified as the cause of lead poisoning in avian predators and scavengers. That finding generated interest in possible human lead intoxication in processed venison. Preliminary investigations show that consumers of game meat hunted with lead shot or standard lead rifle bullets are at risk for lead exposure (Hunt 2006).

In 1979, the National Research Council (NRC) held the *Symposium on Pathobiology of Environmental Pollutants: Animal Models and Wildlife as Monitors*, which focused on research approaches, methods, and techniques for using wildlife as environmental indicators. Unfortunately, despite the acknowledged value of monitoring wildlife health as a means of assessing

environmental threats, no systems were established for the comprehensive monitoring of wildlife health (NRC 1979). In 1991, the NRC again examined the issue of *Animals as Sentinels of Environmental Health Hazards*, specifically in the context of using wildlife and fish to evaluate the pervasion and implications of pollutants such as those found in connection with hazardous waste disposal or industrial activities (NRC 1991). Again, while the NRC pointed out many situations in which monitoring wildlife would be extremely valuable, it acknowledged that there were many obstacles to comprehensive and proactive monitoring of wildlife health, beyond the testing of fish and shellfish harvested for human consumption.

## WILDLIFE AS INDICATORS OF BIOSECURITY AND BIOTERRORISM THREATS

In addition to the natural or unintended outbreaks of wildlife disease organisms, there has been increasing concern over the potential for the deliberate release of potent zoonotic pathogens through acts of bioterrorism. According to the Centers for Disease Control and Prevention (CDC 2010), five of the six Category A bioterrorism threats identified are zoonotic pathogens. These Category A threats are Ebola (and other viral hemorrhagic diseases), plague, botulism, tularemia, and anthrax, all of which can be found in wildlife, may be spread by wildlife, or may show up first in wild species with high sensitivity to a particular pathogen. These zoonotic diseases are priority biosecurity/bioterrorism concerns because they (a) can be easily disseminated or transmitted from person to person; (b) result in high mortality rates and have the potential for major public health impact; (c) might cause public panic and social disruption; and (d) require special action for public health preparedness.

## PUBLIC POLICY IMPLICATIONS

Reaser et al. (2002) specifically identified monitoring of wildlife health and the development of a rapid response capability to wildlife disease events as a priority environmental security concern in the United States and other countries. Among the policy recommendations, Reaser and co-authors called for the monitoring of wildlife health to be elevated to the status of air and water quality monitoring, an effort that was acknowledged to be dauntingly dependent on the development of a comprehensive, interdisciplinary approach and the development of rapid response teams to quickly intervene to control and mitigate wildlife disease.

The need for such wildlife disease monitoring has been illustrated many times since the publication of these policy recommendations. Many of these events have been shown to have profound implications for human health and regional or international economies. Among these wildlife disease outbreaks were such zoonotic diseases such as Ebola, severe acute respiratory syndrome (SARS), avian influenza (H1N5), Lyme disease, and monkeypox. In all cases, it was only after the outbreak resulted in human illness and death that any effective response was deployed. The 2002 recommendations are still relevant—and still have not been acted upon.

## SELECT WILDLIFE DISEASE MONITORING/SURVEILLANCE SYSTEMS

Although the concepts of disease monitoring and surveillance often are used interchangeably, distinguishing between them can be helpful in understanding current programs. Monitoring systems are focused more on the health and disease status of populations, by looking for signs of disease in general (Salman 2003). The goal of a monitoring system would be to detect changes in patterns of disease or health, which would then indicate a need for more active surveillance. The term "surveillance" should be used to describe a more active system and implies that some form of directed action will be taken. A disease surveillance system requires three components—a defined disease monitoring system, a predefined threshold for disease level at which action will be taken, and predefined interventions (Salman 2003). Most existing wildlife disease programs are actually surveillance systems—actively looking for a specific disease in a specific species (or group of species) in a defined geographic location. These systems often are research- or policy-based and can be very costly (see also Chapter 37 in this volume).

While there are systems in place for monitoring emerging diseases in humans and some domestic animals, there currently is no comprehensive, integrated, national strategy for the monitoring or surveillance of wildlife health issues in the United States. Multiple agencies, both governmental and non-governmental, have piecemeal programs that provide information in focused areas, but there is no central coordinating keystone. The CDC has regulatory authority for certain reportable zoonotic diseases (e.g., rabies); the U.S. Department of Agriculture (USDA) is responsible for domestic and captive animal disease and health, including captive wildlife; and the U.S. Fish and Wildlife Service and the individual states have management authority over free-ranging wildlife. However, responsibility for managing diseases in free-ranging populations is not always clearly defined.

The Center for Epidemiology and Animal Health maintains the National Center for Animal Health Surveillance for animals regulated by USDA. The National Center for Animal Health Surveillance has two components—the National Animal Health Monitoring System, which monitors livestock and poultry health issues, and the National Animal Health Surveillance System, a network of public and private partners with the goal of establishing a surveillance system for emerging infectious diseases and foreign animal diseases. Examples of the National Animal Health Surveillance System database include surveillance programs for H5 and H7 avian influenza, WNV, chronic wasting disease in wapiti and deer, bovine and bison brucellosis, and the Michigan Bovine Tuberculosis Eradication Project (APHIS 2010).

Also within USDA, Wildlife Services coordinates the National Rabies Management Program as part of the National Wildlife Disease Surveillance and Emergency Response Program. Wildlife Services does not monitor for the disease, instead focusing on using oral rabies vaccine in certain wild animals to attenuate further wildlife spread of rabies virus. Wildlife Services also participates in surveillance programs for plague, tularemia, classical swine fever, pseudorabies, swine brucellosis, avian influenza, chronic wasting disease, WNV, and bovine tuberculosis. A stated goal of the program is to develop a robust and nationally coordinated system to survey for wildlife diseases and to respond to a variety of emergencies, including natural disasters and disease outbreaks (APHIS 2005).

The National Wildlife Health Center (NWHC) was established in 1975 within the U.S. Geological Survey (USGS) and is the largest federal program devoted to addressing wildlife disease problems. The center focuses on issues relating to wildlife health that affect public and domestic animal health; combating wildlife disease emergence and re-emergence are top priorities. As part of its mission, the NWHC monitors disease and assesses the impact of disease on wildlife populations, defines ecological relationships leading to the occurrence of disease, transfers technology for disease prevention and control, and provides guidance, training, and onsite assistance for reducing wildlife losses when outbreaks occur (Chapter 37 in this volume).

An electronic gateway to biological data and information maintained by federal, state, and local government agencies, academic institutions, non-governmental organizations, and private industry, Wildlife Disease Information Node (WDIN) was a component of the National Biological Information Infrastructure. It offered data management tools and services for wildlife disease surveillance through the Wildlife Health Monitoring Network (NBII 2010). The products and tools of the WDIN are now supported by the Wildlife Data Integration Network at the University of Wisconsin Nelson Institute for Environmental Studies (www.wdin.org).

In addition to the wildlife health work being done by federal agencies, there are several collaborative programs working on wildlife disease surveillance. These programs tend to be very regional- or disease-specific. One example of a public–private partnership is the Southeastern Cooperative Wildlife Disease Study at the University of Georgia, the first non-governmental regional diagnostic and research center specifically for wildlife diseases. This university-based program has support from 15 states and Puerto Rico, USDA APHIS, and the Biological Resource Discipline of the USGS. The cooperative acts as a provider of wildlife pathology services for state and federal agencies. Because the services are funded by these agencies, wildlife submissions require approval from the appropriate agency. However, while a tremendous amount of wildlife data is collected and disseminated through this program, legally the information is available only to the submitting agency (SCWDS 2010).

Two recent emerging diseases in birds have stimulated the development of national surveillance programs: WNV and highly pathogenic H5N1 avian influenza (HPAI). During the height of its outbreak WNV was actively tracked in susceptible birds. Health departments and diagnostic laboratories in many states tested dead wild birds for the presence of the virus and the CDC tracked reports of WNV at the county level in wild birds through a collaboration with USGS (ArboNET). As the disease moved into an endemic phase, the number of states and counties reporting data on WNV-positive wild birds decreased substantially, and the CDC concluded that "traditional forms of surveillance had not been predictive of human risk" (McNamara 2007). Beginning in 2001, a novel surveillance program was developed using zoo animals as sentinel species for WNV. Typically located in urban areas, zoos provide a captive population of susceptible animals that are easily observed and can be "captured" for diagnostic sampling. Based on a partnership of zoos, public health departments, and veterinary diagnostic laboratories, a system for standardized testing and reporting of zoo animals was developed. This model of zoo animals as sentinels for disease is one of many that are being integrated into the surveillance systems for HPAI (USDA/AZA 2010). The U.S. Interagency Strategic Plan for Early Detection of Avian H5N1 HPAI in Wild Migratory Birds is a surveillance system developed for avian influenza in wild birds (USDA 2005). The goals of this plan are to describe the essential components of a unified national system for the early detection of HPAI in migratory birds; to provide guidance to federal, state, university, and non-governmental organizations for conducting HPAI monitoring and surveillance of migratory birds in the United States; and to assimilate the data into the Highly Pathogenic Avian Influenza Early Detection Data System national database. The plan incorporates five strategies—investigation of avian morbidity and mortality events, surveillance of live wild birds, surveillance of hunter-killed birds, surveillance of sentinel species, and environmental sampling. This wide-ranging and inclusive plan gives a perspective on the numerous components and level of detail required for surveillance of a single disease; it may provide a template for surveillance of other wildlife diseases.

Another surveillance system for avian influenza in wild birds is run by a public–private cooperative administered primarily through the Wildlife Conservation Society and funded by the U.S. Agency for

International Development (USAID) and the CDC. This program, the Wild Bird Global Avian Influenza Network for Surveillance (GAINS), aims to establish a global surveillance network of wild birds, broaden scientific understanding of avian influenza epidemiology, and disseminate information (see Chapters 1 and 3 in this volume).

While often for different reasons—human health, agricultural impact, recreational impact—the above entities recognize the need for monitoring systems for wildlife disease. Currently, the United States does some active surveillance centering on specific diseases and some passive monitoring of morbidity and mortality events (NWHC 2010). However, the programs are patchwork, overseen by a variety of organizations and regulatory agencies, and lack integration and a comprehensive national strategy. While some programs are publicly funded and others privately funded, available resources are limited for all of them.

## ESSENTIALS OF A WILDLIFE HEALTH MONITORING SYSTEM

In 2004 and 2005 a study was undertaken to evaluate the public health value of wildlife health data. Existing sources of wildlife morbidity and mortality data, including wildlife rehabilitation centers, veterinary clinics, environmental agencies, animal control groups, universities, and trappers, were evaluated, and three pilot active surveillance projects were undertaken to compare and contrast methods for collecting wildlife disease data on Vancouver Island, British Columbia. Of these, the wildlife rehabilitation centers dealt with the greatest variety of wildlife from the largest geographic area. However, few of these organizations had the capacity to collect samples for diagnostic evaluation and fewer still maintained adequate records; only 4% had searchable database records. Other obstacles to participation included financial disincentives, permit restrictions, privacy issues, staff safety, and lack of contact between wildlife and public health agencies (Stitt et al. 2007). Similar findings were identified in a Colorado study investigating the ability and usefulness of various agencies and organizations to detect infectious disease events in wild mammals (Duncan 2008).

## A MANDATE

As the Vancouver Island study noted, there are inherent challenges to creating an effective system for wildlife health monitoring and surveillance, with the most significant being the lack of a program with the mandate to observe, interpret, and report wildlife disease patterns (Stitt et al. 2007). Virtually every group or association affiliated with wildlife has recognized that obstacle and called for such a mandate. The Association of Fish and Wildlife Agencies, an organization that represents all of North America's fish and wildlife agencies from every level of government, proposed a National Fish and Wildlife Health Initiative in September 2005. The goals of the initiative are to develop and enhance state and territorial fish and wildlife management agency capability to effectively address health issues involving free-ranging fish and wildlife, and to develop and implement a national strategy to address health issues involving free-ranging fish and wildlife through management, surveillance, and research. To date, it remains in draft form (National Fish and Wildlife Health Initiative for the United States 2006). The U.S. Animal Health Association issued a resolution urging USDA APHIS Veterinary Services to ensure continued highest priority for integrated and comprehensive surveillance planning and implementation through the National Animal Health Surveillance System, and to have the appropriate agencies initiate and support a legislative effort to create a system for comprehensive and integrated wildlife disease surveillance (USAHA 2007).

Many organizations feel a mandate already has been issued in the form of Homeland Security Presidential Directive Number 9, which directs the Secretaries of the Interior, Agriculture, and Health and Human Services, the Administrator of the Environmental Protection Agency, and the heads of other appropriate federal departments and agencies to build upon and expand current monitoring and surveillance programs, including wildlife disease systems, to provide early detection and awareness of disease, pest, or poisonous agents (HSPD 2004).

Unfortunately, as long as wildlife health monitoring efforts focus primarily at the individual level, the full potential to detect emerging diseases will not be realized. New emphasis on nontraditional innovative approaches should be considered to maximize the ability to monitor wildlife health.

## A STRATEGY

An integral part of a mandate for a wildlife health monitoring or surveillance system is a strategy for use of the resultant information. The ultimate objective of a surveillance program is to collect, analyze, interpret, and disseminate data to be used for policies and procedures for wildlife conservation, domestic animal health, and public health. The governance structure must define roles and responsibilities, determine what triggers a response, and identify what that response will be. Control strategies may include prevention of introduction of disease, control of existing disease, or eradication. Management may be directed at the disease agent, host population, or habitat, or it may be focused on human activities.

The Canadian surveillance system provides a model of a private–public partnership that includes wildlife disease surveillance. The Canadian Animal Health Surveillance Network (CAHSN) is the federal agency with the mandate to fulfill animal disease surveillance goals. By establishing a network of federal and provincial government agencies, university animal health diagnostic laboratories, and practicing veterinarians, the CAHSN has positioned itself to be a centralized resource for data relating to emerging diseases in animals, especially those with zoonotic potential or economic risk. A key partner in the CAHSN is the Canadian Cooperative Wildlife Health Centre, which is focused on wildlife. The Centre is a university-based partnership between the Canadian Colleges of Veterinary Medicine, governmental agencies, and non-governmental organizations. Through the support of the Centre's various partners, the basics of a centralized system for wildlife disease surveillance have been established, encompassing field personnel for specimen collection and submission of biological specimens, diagnostic laboratories for necropsies and testing, a centralized database, and a communications system. Funding is provided by the collaborative partnership of universities, governmental agencies, and non-governmental organizations.

## FUNDING

Funding for wildlife monitoring and surveillance programs is extremely limited. Those programs that have been implemented, such as rabies, bovine tuberculosis, brucellosis, WNV, or HPAI Early Detection System in Wild Migratory Birds, are tied to specific diseases with human health or agricultural interests. The Wildlife Global Animal Information Network for Surveillance (Wildlife GAINS Act H.R. 1405/S. 1246) would establish a comprehensive, worldwide wildlife health surveillance system to enhance awareness of and preparedness for emerging infectious diseases. The program would monitor and track diseases in wildlife that can affect wild animals, livestock, and human populations. At the time the bill was referred to subcommittee for further action, no funding source had been identified. A similar effort on an international basis has been sought involving organizations such as the World Health Organization, United Nations Food and Agriculture Organization, World Organization for Animal Health, and other key stakeholders such as the World Conservation Union. The goal of this effort would be to design and implement a global animal surveillance system for zoonotic pathogens that gives early warning of pathogen emergence, is closely integrated with public health surveillance, and provides opportunities to control such pathogens before they can affect human health, food supply, economies, or biodiversity. Again, no funding source was identified. Of particular concern is that available funding for wildlife monitoring and surveillance is often driven by a crisis or media attention. This reactive system is not sustainable as the funding sources tend to dissipate when the immediate problem is resolved or the public loses interest, making funding for long-term initiatives difficult.

In 2009, the USAID awarded grants totaling about $260 million for an Emerging Pandemic Threats program (USAID 2010). The program consists of four projects—PREDICT, RESPOND, IDENTIFY, and PREVENT—concentrated on geographic "hotspots," including the Amazon region of South America, the Congo Basin of East and Central Africa, and the Mekong region and Gangetic Plain of Asia. With technical support from CDC and USDA, each project consists of a coalition of partners with animal and human health expertise to build local, national, and regional capacities for early disease detection, laboratory-based disease diagnosis, rapid disease response and containment, and risk reduction. It is hoped that this comprehensive, international effort will contribute to the routine management of infectious diseases as well. Another benefit may be the documentation

## OBTAINING SPECIMENS

A surveillance system is dependent on the ability to collect an adequate number of biological samples from a representative wildlife population, analyze those samples for the presence of health markers, interpret the results, identify trends, and respond to emerging issues. Sample collection opportunities are limited in wild animals. Although some systems for surveillance of diseases such as avian influenza actively trap animals for specific sample collection, most sampling is limited to carcasses or animals captured for other reasons. It has long been assumed that opportunistic, or convenience, samples are not representative of a population, generally a requirement for statistically valid estimators of surveillance parameters. While the detection of specific disease agents in individual animals may be important, analysis of epidemiological trends and patterns requires representative sampling of an identified population. In most situations sample availability is typically biased and there is a lack of reliable population data—the missing denominator (Stallknecht 2007). The Vancouver Island study noted that bats, rodents, and marine mammals were rarely found despite the fact that they are common on and around Vancouver Island. "The public's focus on certain wildlife (such as hunted or charismatic species) meant that a number of important zoonotic disease reservoirs, such as rodents and bats, were not included in ongoing surveillance" (Stitt et al. 2007). Baker et al (2004) proposed that road traffic casualties in the United Kingdom could be used to monitor a species' abundance either qualitatively or quantitatively and undertook a study to answer that question and identify confounding factors. The authors concluded that there is a lack of basic population data for most British mammal species (as is true in the United States), that further investigation into the validity and precision of the technique was required for a range of mammal species before it could be used in a national monitoring program, and that alternative methods that might be more suitable may exist for some species. While it is likely that convenience samples will not correlate to a given population for all species, understanding and characterizing the nature of the bias should allow for statistical analysis. Wildlife health monitoring and surveillance present unique challenges that will, no doubt, require unique collaborative approaches. Fortunately, our understanding of natural systems and biostatistics continues to evolve.

To increase the scope of wildlife and biological monitoring, there has been a steady increase in the number of volunteer-based initiatives (Cohn 2008). These "citizen science" projects allow data to be gathered in a prescribed manner and are particularly useful for monitoring temporally and spatially distributed events that would otherwise require significant funds and staff resources if conducted through a traditionally oriented professional approach (Lodge 2006). Many critics of these projects argue the value of data collected in such a manner. Arguments against data collected through citizen science projects include increased sampling and reporting biases, quality assurance issues, long-term inconsistencies in maintaining a volunteer presence, and limited scientific validation of results (Boudreau and Yan 2004; Delaney et al. 2008). However, when study protocols are designed to take citizen science into account and proper education is applied both before and throughout the project, it has been demonstrated that the quantity of the data can be dramatically increased and the quality of the data is not compromised.

Examples of successful citizen science projects are numerous, including the National Audubon Society's annual Christmas bird count, the International Piping Plover Census, several programs at the Cornell Laboratory of Ornithology, and SEANET, a program based at the Tufts Cummings School of Veterinary Medicine. Several studies have been conducted to assess the scientific validity of citizen science-sourced data. One study assessed the ability of volunteers to locate, identify, and determine the gender of invasive and native crabs from 52 sites across a 725-km coastal transect in the northeastern United States (Delaney et al. 2008). This study determined that students in grades three and seven had the ability to differentiate between species of crabs with over 80% and 95% accuracy, respectively. Gender proved to be slightly more difficult, but accuracy exceeded 80% for seventh-grade students, while 95% accuracy was found for students with at least two years of university education.

In another study, a volunteer-based monitoring program designed to detect the non-indigenous spiny water flea, *Bythotrephes longimanus*, found that the program could detect the majority of *Bythotrephes* invasions (Boudreau and Yan 2004).

Beginning in 2001, a novel surveillance program was developed using zoo animals as sentinel species for WNV. Established in 2000, the Zoo Animal Health Network, part of the Davee Center for Epidemiology and Endocrinology at Lincoln Park Zoo, has provided focus and leadership in the development of national surveillance systems for zoonotic diseases using zoological institutions (ZAHN 2010).

Analogous to zoological institutions, animals seen in wildlife hospitals and rehabilitation centers represent another untapped resource for data on free-ranging wildlife. Nationwide, there are estimated to be more than 3,000 permitted wildlife rehabilitators (individuals and organizations). Each year, they collectively receive hundreds of thousands of wild birds, mammals, reptiles, and amphibians, representing a cross-section of wildlife endemic to their respective communities. The U.S. Fish and Wildlife Service annually issues more than 1,500 permits for the rehabilitation of migratory birds (S. Lawrence, personal communication 2010). All but a few states issue separate permits for the rehabilitation of mammals and other wildlife. Nevertheless, little has been done to facilitate the collection of data for purposes beyond the care of the individual patient. Other than annual permit-related activity reports required by federal and state agencies issuing rehabilitation permits (which vary greatly from agency to agency), there is almost no methodical collection of information from these facilities by any group or agency. Many of these care providers are volunteers, who while often well trained may not have academic or scientific credentials. However, there are an increasing number of large, professionally run wildlife hospitals, with veterinarians, technicians, biologists, and other licensed and degreed professionals on staff. In such facilities, questions about the quality and integrity of data collected, which historically have led many wildlife health researchers to dismiss clinical care facilities as a potential source of useful information, are no longer valid. Most of these professionally staffed centers are nonprofit organizations, supported by charitable donations; therefore, many have limited resources for activities beyond their core missions. Given the limited resources of the clinical wildlife care community, the acceptance of additional wildlife health monitoring responsibilities and/or the adoption of new database technologies may be contingent on third-party funding, or direct benefits to the respective centers from an expanded ability to use the data they collect for internal program analysis, evaluation, and planning.

Recognizing the critical role of wildlife in the emergence of disease, as well as the impacts of anthropogenic change on animal health, the Clinical Wildlife Health Initiative was formed in 2009. With leadership provided by The Raptor Center at the University of Minnesota and the Wildlife Center of Virginia, the Clinical Wildlife Health Initiative is a collaboration of public and private academic and scientific institutions focused on using data from animals presented to wildlife rehabilitation centers to monitor the health of free-ranging wildlife. This organization has begun to develop a Web-based surveillance system with data collection tools to integrate the wealth of potential information that can be gleaned from wild animals seen in rehabilitation settings. It is also working to establish a network for addressing emerging issues in wildlife rehabilitation medicine.

Successfully networking the extremely diverse clinical wildlife care community could provide a valuable model for the standardization of data collection and use, if not an actual foundation upon which to build a comprehensive database of anthropogenic wildlife health incidents. If regulatory agencies could adopt a single report format, using standardized terminology, the amount of information collected could be vast. Over time, these data can be compiled and analyzed to detect changes and statistical anomalies in morbidity and mortality that could indicate significant trends in wildlife health, including the emergence of disease or other environmental issues affecting wildlife. For many common species, patient intake records at wildlife care centers may be the only baseline references available to recognize, let alone quantify, changes in the population or distribution of a given species. Ironically, wildlife presented for rehabilitation may be the best sentinels. These animals generally live in close proximity to humans, are often diseased, and, it could be argued, are the most affected by environmental changes, especially anthropogenic causes (Sleeman 2008).

After biological samples are obtained, a surveillance system is dependent on the ability to analyze those samples for the presence of disease agents and interpret the results. Due to a lack of resources, treatment of ill wild animals is often done on a volunteer basis and based on presumptive diagnoses and supportive care protocols. Furthermore, most tests that have been developed for companion animals and livestock have yet to be validated in wildlife, confounding the interpretation of the results. The lack of sensitive and specific diagnostic tests is a serious shortcoming requiring accelerated research in the future (Bengis 2002). These inadequacies led the authors of the Vancouver Island study to conclude that given the current opportunities and obstacles, targeted surveillance for known pathogens in specific host species, rather than general surveys for unspecified pathogens, was judged to be a more effective and efficient way to provide useful public health data (Stitt 2007). However, due to the limited availability of wildlife, others believe we need to maximize information obtained from any "hands-on" opportunities with wildlife, living or dead, including the collection of body fluids, tissues, and excretions (Bengis 2002).

## DATA MANAGEMENT

The Vancouver Island study found that insufficient data collection and management did not allow wildlife health data to be used by public health agencies (Stitt et al. 2007). This is true in the United States as well. Even if a definitive etiological or pathological diagnosis is made in wildlife, there is no centralized reporting structure to capture the information unless it is a reportable disease at the state, federal, or international level. A system based on syndromic surveillance has been proposed as a potential tool for monitoring wild animal health issues. Syndromic surveillance uses pre-diagnostic data to track or signal a change in pattern requiring further exploration (CDC 2008). The development of a strong baseline of admission patterns of wild animals presented to wildlife hospitals, including clinical signs, circumstances of admission, and signalment (species, age, recovery location, date of recovery), has the potential to become a significant tool for monitoring change. Over time, this baseline can be used to predict future expectations, and standard epidemiological methods can be used to compare actual data with predictions. Aberrations or changes in patterns will be detected

and may flag population-level events. This information can then be used to target further studies aimed at etiological identification.

There are many entities that deal with wildlife on a regular basis that could contribute information, both biological data and samples. These entities include wildlife rehabilitators, governmental agencies, private organizations, diagnostic laboratories, zoological institutions, and universities. Many of these entities already collect and record the type of information being discussed. Unfortunately, the information is incomplete and not standardized. Most importantly, it is not accessible. Other obstacles to using this data include privacy and ownership issues and how to communicate the results of any analyses.

## CONCLUSIONS

While there have been repeated calls for increased monitoring of wildlife health since the 1970s, and dozens of high-profile outbreaks of wildlife disease, many with profound human, agricultural, and environmental implications, there has been little in the way of fundamental innovation related to strategies or programs for gathering, collecting, compiling, analyzing, and acting upon wildlife health information. Even within the various agencies of the federal government with a mandate to monitor certain wildlife health issues, there has been a parochial resistance, or at least a lack of effective initiative, to develop standardized wildlife health terminology, definitions and descriptions for wildlife health conditions, or ecumenical systems for collection and sharing data. With few exceptions, such as Zoo Animal Health Network, which uses zoo animals to monitor for the presence of certain zoonotic pathogens, there seems to have been a general lack of openness to nontraditional data sources, or novel data collection techniques.

Nevertheless, an increasing level of attention is being devoted to wildlife health issues with implications for human health, and on the implications of such environmental issues as climate change. In the wake of the massive 2010 Gulf Coast oil spill, there was widespread denunciation of the government's efforts to quantify short-term and long-term damage on wildlife. A chorus of congressional, academic, and private-sector institutions demanded more effective mechanisms to document and evaluate the consequences of such manmade disasters on wildlife health. As the need to gain insights into these matters becomes more urgent, circumstances and sheer necessity may break down the resistance of traditional institutions to new approaches and new sources of data. New pathways to wildlife health monitoring, such as those promoted by the Clinical Wildlife Health Initiative, and those engaging "citizen science" efforts, hold great promise to expand the scope and effectiveness of traditional strategies. The recommendation made by Reaser et al. (2002) to elevate the monitoring of wildlife health to the levels of importance assigned to air and water quality monitoring seems even more appropriate, 10 years later. With adequate financial and programmatic support, increased coordination and standardization of data collection protocols, and an elevated public policy emphasis, the targeted and opportunistic monitoring of wildlife health can be quickly and effectively advanced. This will deepen our understanding of, and ability to respond to, significant wildlife health events and circumstances. Such understanding and response capabilities are critical to the protection of human and environmental health, and the effective management and conservation of valuable and important public trust resources.

## APPENDIX 38.1

## Recommended Reading

Gubernot, D.M., B.L. Boyer, and M.S. Moses. 2008. Animals as early detectors of bioevents: Veterinary tools and a framework for animal–human integrated zoonotic disease surveillance. Public Health Rep 123:300–315.

Halliday, J.E. B., A.L. Meredith, D.L. Knobel, D.J. Shaw, B.M. de C. Bronsvoort, and S. Cleaveland. 2007. A framework for evaluating animals as sentinels for infectious disease surveillance. J R Soc Interface 4:973–984.

McKenzie, J., H. Simpson, and I. Langstaff. 2007. Development of methodology to prioritize wildlife pathogens for surveillance. Prev Vet Med 81:194–210.

Nusser, S.M., W.R. Clark, D.L. Otis, and L. Huang. 2007. Sampling considerations for disease surveillance in wildlife populations. J Wildlife Manage 72:52–60.

Wagner, M.M., A.W. Moore, and R.M. Aryel, eds. 2006. Handbook of Biosurveillance. Elsevier: Burlington, MA.

## Wildlife Health/Emerging Diseases Websites

American Association of Zoo Veterinarians <http://www.aazv.org/>
EcoHealth <http://www.ecohealth.net/>
European Association of Zoo and Wildlife Veterinarians <http://www.eazwv.org/php/>
National Wildlife Health Center <http://www.nwhc.usgs.gov/>
Wildlife Data Integration Network <www.wdin.org>
Emerging Infectious Diseases <http://www.cdc.gov/ncidod/EID/index.htm>
ProMed <http://www.fas.org/promed/>
Society for Conservation Biology <http://www.conbio.org/>
Wildlife Disease Association <http://www.wildlifedisease.org/>

## Listservs

Wildlife Disease News Digest <newsdigest.wdin.org>
ProMed <http://www.fas.org/promed/>
Emerging Infectious Diseases <http://www.cdc.gov/ncidod/EID/index.htm>

## REFERENCES

Aguirre, A.A., R.S. Ostfeld, G.M. Tabor, C. House, and M.C. Pearl, eds. 2002. Conservation medicine: ecological health in practice. Oxford University Press, New York.

Animal Plant Health Inspection Service (APHIS). 2005. Wildlife Disease Surveillance and Emergency Response. Website: http://www.aphis.usda.gov/about_aphis/programs_offices/veterinary_services/ceah.shtml, accessed March 28, 2011.

Animal Plant Health Inspection Service (APHIS). 2010. Center for Epidemiology and Animal Health. Website: http://www.aphis.usda.gov/about_aphis/programs_offices/veterinary_services/ceah.shtml, accessed March 28, 2011.

Baker, P.J., S. Harris, C.P.J. Robertson, G. Saunders, and P.C.L. White. 2004. Is it possible to monitor mammal population changes from counts of road traffic casualties? An analysis using Bristol's red foxes Vulpes vulpes as an example. Mammal Rev 34: 115–130.

Bengis, R.G., ed. 2002. Infectious diseases of wildlife: detection, diagnosis, and management. Off Int Epiz Sci Tech Rev 21 (1 and 2).

Boudreau, S.A., and N.D. Yan. 2004. Auditing the accuracy of a volunteer-based surveillance program for an aquatic invader *Bythotrephes*. Environ Monit Assess 91:17–26.

Centers for Disease Control (CDC). 2008. Syndromic Surveillance: An applied approach to outbreak detection. Website: http://www.cdc.gov/ncphi/disss/nndss/syndromic.htm, accessed March 25, 2011.

Centers for Disease Control (CDC). 2010. Emergency preparedness and response: bioterrorism agents/diseases. Website: http://emergency.cdc.gov/agent/agentlist-category.asp#a., accessed March 25, 2011.

Chardonnet, P.H., B. des Clers, J. Fischer, R. Gerhold, F. Jori, and F. Lamarque. 2002. The value of wildlife. Off Int Epiz Sc Tech Rev 21:15–51.

Cohn, J.P. 2008. Citizen science: can volunteers do real research? BioScience 58:192–197.

Delaney, D.G., C.D. Sperling, C.S. Adams, and B. Leung. 2008. Marine invasive species: validation of citizen science and implications for national monitoring networks. Biol Invasions 10:117–128.

Duncan, C., L. Backus, T. Lynn, B. Powers, and M. Salman. 2008. Passive, opportunistic wildlife disease surveillance in the Rocky Mountain Region, US. Transbound Emerg Dis 55:308–314.

Homeland Security Presidential Directive (HSPD). 2004. Homeland Security Presidential Directive 9. Website: http://www.dhs.gov/xabout/laws/gc_1217449547663.shtm, accessed March 27, 2011.

Hunt, W.G., W. Burnham, C.N. Parish, K.K. Burnham, B. Mutch, and J.L. Oaks. 2006. Bullet fragments in deer remains: implications for lead exposure in avian scavengers. Wildl Soc Bull 34:167–170.

Lodge, D.M., S. Williams, H.J. MacIsaac, K.R. Hayes, B. Leung, S. Reichard, R.N. Mack, P.B. Moyle, M. Smith, D.A. Andow, J.T. Carlton, and A. McMichael. 2006. Biological invasions: Recommendations for U.S. policy and management. Ecol Appl 16:2035–2054.

McNamara, T. 2007. The role of zoos in biosurveillance. Int Zoo Yearbook 41:12–15.

National Biological Information Infrastructure (NBII). 2010. http://www.nbii.gov, accessed March 27, 2011.

National Wildlife Health Center (NWHC). 2010. http://www.nwhc.usgs.gov accessed March 27, 2011.

National Research Council (NRC). 1979. Animals as monitors of environmental health hazards. National Academy Press, Washington, D.C.

National Research Council (NRC). 1991. Animals as sentinels of environmental pollutants. National Academy Press, Washington, D.C.

Rabinowitz, P., and L. Conti. 2009. Human–animal medicine. Elsevier, New York.

Reaser, J.K., E.J. Gentz, and E.E. Clark. Jr. 2002. Wildlife health and environmental security: new challenges and opportunities. *In* A.A. Aguirre, R.S. Ostfeld, G.M. Tabor, C. House and M.C. Pearl, eds. Conservation medicine: ecological health in practice, pp. 383–392. Oxford University Press, New York.

Salman, M.D. 2003. Animal disease surveillance and survey systems: methods and applications. Iowa State Press, Ames, Iowa.

Sleeman, J.M. 2008. Use of wildlife rehabilitation centers as monitors of ecosystem health. *In* M.E. Fowler and R.E. Miller (eds.). Fowler's zoo and wild animal medicine current therapy, pp. 97–104. Saunders Elsevier, St. Louis, Missouri.

Southeastern Cooperative Wildlife Disease Study (SCWDS). 2010. http://www.scwds.org, accessed March 27, 2011.

Stallknecht, D.E. 2007. Impediments to wildlife disease surveillance, research, and diagnostics in wildlife and emerging zoonotic diseases: the biology, circumstances and consequences of cross–species transmission. Curr Top Microbiol 315:445–461.

Stitt, T., J. Mountifield, and C. Stephen. 2007. Opportunities and obstacles to collecting wildlife disease data for public health purposes: Results of a pilot study on Vancouver Island, British Columbia. Can Vet J 48:83–90.

United States Animal Health Association (USAHA). 2007. United States Animal Health Association Committee Resolutions. Website: http://www.usaha.org/committees/resolutions/2007/resolution05-2007.pdf, accessed Feb. 21, 2011.

United States Agency for International Development (USAID). 2010. Emerging program threats overview. website: http://pdf.usaid.gov/pdf_docs/PDACP822.pdf, accessed March 27, 2011.

United States Department of Agriculture (USDA). 2005. An early detection system for highly pathogenic H5N1 avian influenza in wild migratory birds—USA interagency strategic plan. Website: http://www.usda.gov/documents/wildbirdstrategicplanpdf.pdf, accessed March 27, 2011.

United States Department of Agriculture/Association of Zoos and Aquariums (USDA/AZA). 2010. Avian influenza surveillance system for zoological institutions. Website: http://www.zooanimalhealthnetwork.org/ai/Home.aspx, accessed Feb. 21, 2011.

Willette, M., and J. Ponder. 2009. Monitoring wildlife health. *In* K. Shenoy, ed. Topics in wildlife medicine: infectious diseases, pp. 28–40. National Wildlife Rehabilitators Association: St. Cloud, Minnesota.

Zoo Animal Health Network (ZAHN). 2010. Homepage. Website: http://www.zooanimalhealthnetwork.org/Home.aspx, accessed March 25, 2011.

# 39

# EPIDEMIOLOGIC INVESTIGATION OF INFECTIOUS PATHOGENS IN MARINE MAMMALS

## The Importance of Serum Banks and Statistical Analysis

A. Alonso Aguirre, Melinda K. Rostal, B. Zimmerman, and Thomas J. Keefe

Serological assays are frequently used in both human and veterinary medicine and epidemiology to diagnose and determine the prevalence, incidence, and morbidity of disease. In humans, serum banks have been established for the purpose of conducting epidemiological research in oncology and infectious disease. The human serum bank JANUS holds approximately 500,000 serum samples dating from 1973 (Jellum et al. 1993). More recently, in 2007, the United Kingdom established Biobank for long-term storage of 500,000 human samples (Elliott and Peakman 2008). While many zoological and wildlife rehabilitation organizations have banked serum and tissues (Hutchins 1990), fewer serum banks of free-ranging wildlife are available for serological studies. Arctic and U.S. national marine mammal tissue banks have been established to study the toxicological effects of environmental contaminants such as PCBs (Becker et al. 1993; Lillestolen et al. 1993; Becker and Wise 2006; Kucklick et al. 2010). While wildlife serum banks do exist, such as the Alaska Department for Fish and Game wildlife serum library (Becker et al. 1993), a centralized, national, accessible serum bank for wild species does not exist. In this chapter, we will use marine mammals as an example to explore both the need for a national

serum bank and the protocols required to develop, sustain, and use such a serum bank.

Mass mortalities of marine mammals caused by infectious diseases are well documented (Hardwood and Hall 1990; de Bruyn et al. 2008). These mortalities may play an important role in the dynamics of marine mammal populations and have important implications for the genetics and evolution of a species. Also, disease can be a catastrophic event for an endangered species and can lead to extinction (Cunningham and Daszak 1998). The consequences of disease may be important at both the individual and the population levels (Cunningham 1996; Daszak et al. 2000). Diseases may cause severe mortality, increased susceptibility to predation, and lower reproductive potential. However, to date, there is no effective means to determine whether a disease is affecting the population (McCallum and Dobson 1995; Lafferty and Gerber 2002). In this chapter, we define disease as "any impairment that interferes with or modifies the performance of normal [structure] and functions, including responses to environmental factors such as nutrition, toxicants, and climate; infectious agents; inherent or congenital defects; or combinations of these factors" (Wobeser 1994).

Within the past 30 years a number of epizootics in marine mammals have occurred. In 1979–80, hundreds of harbor seals (*Phoca vitulina*) died of influenza A virus at Cape Cod (Geraci et al. 1982). Since 1981, leptospirosis epidemics have occurred in California sea lions (*Zalphus californianus),* and die-offs are common every year (Dierauf et al. 1985; Gulland et al. 1996). The best-publicized infectious disease agent affecting marine mammals is morbillivirus; within the past 20 years there have been at least eight major epizootics (Di Guardo et al. 2005). The first known morbillivirus epizootic of marine mammals was associated with the mortality of thousands of Baikal seals (*Phoca sibirica*) in Lake Baikal, Russia, in 1988 and was consistent with canine distemper virus infection (Mamaev et al. 1996). Another morbillivirus, phocine distemper virus, caused a massive die-off of more than 17,000 harbor seals off the northwestern European coast (Kennedy et al. 1988, 1989; Osterhaus et al. 1990). That same year a morbilliviral infection killed approximately 50% of bottlenose dolphins (*Tursiops truncatus*) along the American Atlantic coast (Geraci 1989; Lipscomb et al. 1994). Dolphin morbillivirus killed thousands of striped dolphins (*Stenella coeruleoalba*) in the Mediterranean Sea in 1990–91 (Domingo et al. 1990) and recently led to a massive die-off in 2007 (Raga et al. 2008). A fourth morbillivirus was isolated from porpoises and designated porpoise morbillivirus (Kennedy et al. 1988). Also, a severe morbillivirus outbreak killed up to 50% of an important endangered Mediterranean monk seal (*Monachus monachus)* colony off the western coast of Africa (Osterhaus et al. 1997; Kennedy 1998). Morbilliviral serologic studies in manatees (*Trichechus manatus*) (Duignan et al. 1997), pilot whales (*Globicephala melas*) (Duignan et al. 1995), and Atlantic walruses (*Odobenus rosmarus rosmarus*) (Duignan et al. 1994) have demonstrated a wide host range. In 2006–07, a morbillivirus outbreak led to the death of several pilot whales off the coast of Spain (Fernández et al. 2008). Though a majority of the epidemics occurred in the European Atlantic, the disease has caused mortality in the American Atlantic, and antibodies to phocine distemper virus have been detected in harp seals (*Phoca groenlandica*), ringed seals (*Phoca hispida*), harbour seals, and grey seals (*Haliichoerus grypus*) along the western Atlantic coasts (Duignan et al. 1997) and in small cetaceans from the Southeast Pacific (Reidarson et al. 1998; van Bressem et al. 1998), illustrating the wide geographic distribution and host range of this group of morbilliviruses.

Serology has been important in diagnosing and determining the scope of these outbreaks. Archived serum samples at the Marine Mammal Center (TMMC) permitted the evaluation of leptospirosis titers in California sea lions over 11 years, demonstrating that the disease not only is endemic but also causes periodic outbreaks (Lloyd-Smith et al. 2007). Without access to banked serum, the true epidemiology and ecology of leptospirosis in California sea lions may not have been elucidated. Also, Zarnke et al. (2006) used sera stored at −40°C to perform serological screens on harbour seals in Alaska for *Brucella* spp. phocid herpes virus 1 and 2 and phocid distemper virus in animals sampled between 1976 and 1999. Finally, serology can be used in endangered species to indicate whether infectious diseases may be playing a role in the decline of the species. A study (Burek et al. 2005) investigating the impact of infectious agents on Steller sea lions (*Eumetopias Jubatus*) provided important data on the prevalence of several pathogens; however, the impact these diseases were having on the population could not be determined due to inconsistencies with the testing, incomplete data sets, and insufficient serum to complete the diagnostics, indicating the importance of having a serum bank with strict protocols for testing, required data, and accessing samples.

The primary objective of banking serum of live, free-ranging marine mammals is to develop a reference collection to provide long-term baseline data on individual and population health. This could be accomplished by using retrospective environmental, molecular, or epidemiologic surveys as new tests are available or new diseases are diagnosed. A blood specimen can be used for a broad spectrum of analyses to evaluate an animal's health or disease status, including plasma or serum biochemistry, hematology, serology, endocrinology, virology, molecular epidemiology, nutritional components, bacteriological culture, hemoparasitology, antibody titers, protein electrophoresis, toxicology, and others. Baseline data are needed to determine what diseases are endemic in a population. These data can be used to establish whether an epidemic is an endemic disease, a pathogen emerging into a new geographic area, or a new disease agent that has been introduced to an immunologically naïve population. These data can be collected by several population-level methods: disease monitoring, the systematic collection of biomedical data to identify diseases; disease surveillance, the continuous assessment of the status of specific diseases through the observation of individuals; and disease investigation, the investigation of disease to eliminate or control its spread (Chapter 37 in this volume). A national serum bank would provide a resource that could be used for real-time biomonitoring or for future retrospective analyses and documentation of infectious diseases in marine mammals.

A pilot serum bank was initiated in 1997: the National Marine Fisheries Service (NMFS) provided two freezers to selected stranding rehabilitation centers located at the New England Aquarium, the University of California, Davis (with TMMC), and the University of Alaska, Fairbanks. Protocols for and designs of these collections have been established but not standardized. Samples are being collected primarily from pinnipeds at this time (K. Beckmen and F. Gulland, personal communication 2001). As an example, the serum bank in Alaska has archived whole blood, buffy coats, packed red blood cells, plasma, and serum, which are kept in 1.5- or 2.0-mL externally threaded Nalgene cryovials and frozen at $-40^\circ$C or $-70^\circ$C. They use MS Access to digitally organize their records (K. Beckmen, personal communication 2011).

## ORGANIZATION OF THE SERUM BANK

The following protocol describes the processing, storage, and analytical characterization of serum specimens collected during live stranding events, live capture-and-release research projects, subsistence hunted animals, animals obtained as part of mortality investigations, and/or live bycatch of free-ranging marine mammals. Its primary objective is to ensure that appropriate and adequate specimens are handled for long-term banking purposes and, if needed, that these specimens reach a diagnostic laboratory in suitable condition for detailed study. Strict adherence to protocol is necessary to ensure that each specimen is preserved in the appropriate and standardized manner; the stored specimens are available for further study; and the resulting analysis is of sufficient validity to be useful in the management of a selected species or disease.

Basic organizational steps that require consideration while creating a national marine mammal serum bank include identification of participants, including governmental and non-governmental archival and storage facilities; creation of a review board; establishment of a policy for accession of previously collected and new samples; development of standard operating procedures (SOPs), including data forms; implementation of a quality assurance program; and creation of a policy for the access of stored samples for epidemiological studies. The serum bank described in this chapter is designed to be compatible with the management of the National Marine Animal Tissue Bank (Becker and Wise 2006) currently in place at NMFS. As such, the following protocol will name specific governmental organizations that we recommend be involved in certain aspects of the serum bank. These recommendations are flexible and do not represent any commitment by the proposed organizations. NMFS is used here as a practical example because of its role in managing marine mammals and its existing programs. This includes maintenance of a serum specimen database to ensure tracking and quality of samples and analyses, coordination of this effort with other national programs, and the dissemination of information about the program. In addition, a serum catalog would be maintained and would be available to scientists and managers via the Internet for assistance in designing retrospective studies in marine species.

Species of interest include sentinel species for specific geographic or trophic areas and species likely to be involved in mortality events (e.g., *Tursiops* for Gulf of Mexico, *Phoca* and *Zalophus* for the Pacific coast). Also, it is recommended that studies targeting specific declining populations, such as Hawaiian monk seals (*Monachus schauinslandi*), Steller sea lions, and beluga (*Delphinaperus leucas*), be included to enhance the understanding of the health and disease of these populations.

## REVIEW BOARD DUTIES

The review board is made up of a team of scientists and should include at minimum an epidemiologist, a veterinarian, an ecologist, and a molecular biologist, as well as representatives of involved organizations. The review board is needed to create and enforce the policies for the serum bank. The review board will determine the requirements for facility involvement in sample storage and accession. Finally, the review board will set up a review committee to process the applications to use the serum for eco-epidemiological studies. The composition of the board will be reviewed annually.

## ACCESSION OF PREVIOUSLY COLLECTED SAMPLES

It is recommended that NMFS collaborates with existing serum banks to incorporate those existing specimens into this proposed National Marine Animal Serum Catalogue (NMASC) and to provide long-term banking of future samples. The accession of samples previously collected into the national serum bank requires the contributing institution to provide basic data collected during animal processing, identify associated serum samples, demonstrate the samples have been stored in accordance with serum bank policies, and agree to follow the bank policies for releasing samples for further studies.

The recommended basic data set required for sample accession of existing sera comprises both collection event data as well as data for each individual sample. Collection event data should include the name, address, and phone number of the collecting institution, including the institution's sera collection custodian, and a brief description/activity code of the collection (e.g., strandings, subsistence hunts, or live capture releases). Sample specific data should include identification (ID) number, date of collection, geographic location or stock ID, species, sex, age, type of sample, as well as any additional information on clinical, epidemiological, and analytical results, disposition of the animal, and the availability of samples.

A computerized inventory database should be established for the serum bank. This includes the development of accession numbers to create a meaningful assigned code and/or barcode. For example, under a coding scheme with ___ ___ (geographic code) ___ ___ (species) ___/__/__ (Date: yyyymmdd)___ ___ (institution code), HI-MS-20091020MM, would identify the sample as: Hawaii, monk seal October 20, 2009, The Marine Mammal Center. For the data to be useful and retrievable, standardized codes must be established and implemented, including codes for: geographic locations, laboratory procedures, species, institution, and stranding, among others. The use of standardized codes, code values, and units of results, as well as the ability to map the data to known and accepted standards, will allow the data to be searchable.

Existing collections or banks shall not be required to have the same level of minimum data elements in the catalog, but the data should be available from the person responsible for the individual collections. Every effort should be made to provide such data for the catalog. Some samples will be rejected for accession into the NMASC. Only qualified personnel, selected by the review board, will be allowed to reject specimens for accession.

A standardized set of rejection criteria (Tilton et al. 1992) will be used to determine sample rejection from the serum bank. The criteria include but are not limited to inadequate or improper specimen ID number, improper collection tube used, insufficient quantity of specimen, specimen is hemolyzed, lipemic, or icteric, anticoagulated blood is clotted, the specimen is improperly transported (a break in the cold chain), purple- or green-top tubes are insufficiently filled. Hemolytic, lipemic, or icteric states are included as they may interfere with testing and may affect certain biochemical factors (including potassium, LDH, and AST). Also, processing errors should be considered when deciding to reject samples from the serum bank (Tilton 1992), including aliquot misidentification, misplaced specimen, specimen repeatedly frozen and thawed, serum/plasma contaminated with red blood cells, or the specimen was stored at the incorrect temperature.

## METHODS FOR COLLECTION AND ACCESSION OF NEW SAMPLES

### Blood Collection and Serum/Plasma Separation

Procedures for venipuncture, collection, processing, and labeling of marine mammal blood have been described by Dierauf and Gulland (2001) and Kucklick et al. (2010). However, it is worthwhile to review certain steps of collection and processing that may affect sample quality. Following collection, the EDTA and lithium heparin tubes should be immediately chilled on blue ice or refrigerated (4°C) until processed. If tubes are processed in the field, one should prepare two blood smears from the EDTA tube, centrifuge all tubes at 10,000 rpm for 10 minutes, aseptically pipette plasma into a cryovial, and ultrafreeze immediately. The red and white blood cells from either tube may be saved in cryovials and ultrafreezed immediately. The red-top or SST serum tube(s) should remain upright at room temperature (18°C to 21°C) or in the shade for 30 minutes, not exceeding one hour. Delaying centrifugation for longer than one hour can lower sodium, potassium, creatinine, glucose, blood urea nitrogen, and globulin values and raise calcium, phosphorus, total protein, and albumin values. Once the clot is retracted, the specimen is centrifuged for 10 minutes at 10,000 rpm. A sterile pipette is used to aliquot the serum into pre-labeled 1-mL cryogenic vials. The cryovials should not be overfilled, to allow for expansion during freezing, and should be securely sealed. Samples should be ultrafreezed immediately.

Cryovials with the capacity of 1.0 to 1.2 mL have been selected in accordance with recommendations by the Centers for Disease Control and Prevention (CDC 1997) and the U.S. Department of Agriculture. These are the most commonly used vials in serum banks. In addition, small aliquots protect against the complete loss of the specimen should a vial break and against the deleterious effects associated with repeated freezing and thawing of a specimen (Lennette and Schmidt 1969). For example, the stability of DNA extracts after repeated freeze–thaw cycles was dependent on stored volume: 100-$\mu$L volumes were not stable after three freeze–thaw cycles, compared to 1-mL total volumes that were stable after 14 freeze–thaw cycles on the accurate quantification of DNA extracts of *Toxoplasma gondii* by hybridization probe techniques (Bellete et al. 2003). Newly obtained aliquots of sera will be divided into three portions: one portion will remain at the local facility, one portion will be sent to National Veterinary Services Laboratory as part of the biomonitoring component, and the last portion will be accessioned into the NMASC for long-term storage.

## Labeling and Data Collection

Cryogenic vials should be labeled with the following information: specimen # (e.g., 1/3, 2/3, and 3/3 if three

aliquots were made), serum or plasma, date collected, animal ID number and/or barcode if available, and species. The minimum data set required for samples to be accessioned into the serum bank includes the contributing institution and custodial contacts; the collection code and date of collection (indicate stranding date vs. collection date if those are different); the species code (based on NMFS PIMMS) and field identification; biological data, including age, sex and reproductive status, geographic location, blubber depth; seasonal factors (i.e., molting, fasting); and health data, including diagnosis, clinical data, treatments, and disposition of animal. In addition to sampling event data, the following information should be available: type of collection (i.e., EDTA) and collection/storage vial; diagnostic tests performed; the results and laboratory that performed the testing; the type, amount, and location of specimens banked; identifying numbers (i.e., NMFS#, USFWS#); and any special collection and processing instructions. As discussed for the accession of previously collected samples, new acquisitions will be required to use standard codes as designed by the review board and should be given a unique accession number. At this time, the accession number, animal ID number, and date of collection should be verified and entered into the computer.

## STANDARD OPERATING PROCEDURES

### Storage and Cold Chain

Specimens entering the bank storage facilities need to be organized into prenumbered cryoboxes. Preferably, the submitting facility will ship specimens pre-organized; however, organization should be ensured on arrival, and the samples may be added into a partially full cryobox. During reorganization, cryoboxes should be kept on blue or wet ice; however, individual cryovials should not directly contact the ice. After filling, the cryobox should be placed immediately in the ultrafreezer. Specimens should be organized by ID number and a locator code should be created. The locator code should include freezer unit number, a code for the freezer section, a box number, and a grid spot within the box. Detailed locator codes will minimize searching for material, which risks warming the ultrafreezer. Records should be maintained both on hard copy and in the computer log and should include the vial locator code as described above, the number of vials per individual animal, the individual specimen ID, and serum/plasma amount. For example: grids 40–50 of cryobox 6 in compartment 2, left side of ultrafreezer 1 at Kewalo Research Facility, NMFS Honolulu Laboratory, has 11 1-mL vials stored for seal YC35. Duplicated inventory records should be maintained in a second location separate from the working records.

Serum samples must be labeled and stored in accordance with a well-designed identification and tracking software system, compatible with computerized and physical retrieval. The barcode system can be easily implemented for tracking marine mammal sera specimens (Holland et al. 2003). EcoHealth Alliance uses a barcode system implemented over three years for Nipah virus research in Malaysia (Epstein personal communication 2011).

If laboratory testing of the specimen will occur upon arrival to the laboratory or if aliquots need to be prepared from older or larger specimens, the samples should be recovered by rapidly thawing vials with constant, gentle motion in a water bath at 35°C to 37°C. The sample should be aseptically removed with a pipette to aliquot or to perform a specific laboratory test. Following aliquoting, storage is as indicated below.

Serum and plasma specimens can be stored by freezing, ultrafreezing, or lyophilization. The difference between short-term storage (freezing) and long-term storage (ultrafreezing and lyophilization) is important, as freezer storage of serum at −3°C, even briefly, can cause a steady and significant decline of some serum enzymatic activities, such as creatine kinase (Lev et al. 1994–1995). Furthermore, antibody titers to some pathogens decrease relatively quickly at −20°C in serum samples (Bothig et al. 1992). Short-term storage is defined as the freezing of serum/plasma samples at −20°C to −3°C without protective agents (Lev et al. 1994-1995). It is considered the simplest and least expensive storage method for serum banks. Long-term storage is the freezing of serum/plasma samples at −70°C (ultrafreezer), −196°C (liquid nitrogen), or −20°C to −30°C (freezer) in the presence of protective agents including thiol compounds.

## Human Safety Considerations

Safety precautions must be observed throughout the handling, preservation, and maintenance process of any biological specimen. Cryogenic temperatures could result in exposure of personnel to extreme cold conditions. Insulated gloves and long-sleeved laboratory coats should be worn to protect skin from exposure. Improperly sealed cryovials may break or explode. A face shield should be mandatory when retrieving vials from liquid nitrogen. It is imperative that laboratory personnel be adequately trained regarding the risks associated with handling animal samples (including serum) and provided the appropriate personal protective equipment to prevent transmission of zoonotic diseases working in at least a Biosafety Level 2 Laboratory.

## Quality Assurance Program

A quality assurance (QA) program is a system of planned actions to ensure consistent quality and integrity of data. A quality assurance program minimizes the potential for variation or errors due to storage and handling affecting the interpretation of results. Components of the quality assurance program include sample quality control, a long-term storage stability program, the development of standardized analytical methods and controls, and document standardization for record-keeping.

## Quality Control

Quality control (QC) is an integral part of quality assurance that uses statistical techniques to continually and actively monitor, analyze, and audit the serum bank to determine if equipment, reagents, and operators are functioning and performing according to predetermined standards. A well-designed and implemented QC program increases confidence that the results generated are accurate (Westgard and Klee 1996; Holland et al. 2003). Quality control includes the maintenance of facilities/equipment, annual or required interval calibration certification, routine performance inspection, training/certification of laboratory personnel, routine monitoring of materials, logbook maintenance of daily temperature (e.g., freezer) records, receipt of materials and discard of expired reagents, and routine internal audits.

It is critical to sample integrity that no interruptions in the cold chain occur, as freeze–thaw cycles may decrease sample quality. Therefore, storage facility temperatures should be continuously monitored. Controlled temperature storage areas that require monitoring include room temperature ($17°C$ to $28°C$), overnight-refrigerated storage ($2°C$ to $8°C$), and ultra-freezing storage ($-70°C$ to $-90°C$).

## Long-Term Stability Program

The long-term stability program is designed to evaluate the suitability/integrity of stored samples by performing analyte and sample testing. A representative number of repository samples should be subjected to periodic testing, according to the prescribed interval (e.g., annually, biennially), to confirm sample integrity. QC tests comparing analyte results from serum of the same donor with a difference of 10 years in storage time from the JANUS serum bank indicate that the primary structure of proteins is retained, although specific enzyme activity declines over time at $-25°C$ storage. In addition, several serum organic acids, amino acids, and carbohydrates are reasonably stable after storage at $-25°C$. Trace metals and many inorganic salts in serum are stable almost indefinitely. Although the JANUS bank stores samples at $-25°C$, storage at $-70°C$ to $-80°C$ is more protective of sample integrity, and thus we recommend ultrafreezer storage (Bothig et al. 1992; Jellum, Anderson et al. 1993; Holland, Smith et al. 2003; Elliott and Peakman 2008). In addition to sample testing, the long-term stability program should include a visual examination of samples for storage integrity and container closure and an annual review of specimen usage to allow approved periodic disposal of excess or unwanted specimens.

## Standardization of Laboratory Tests and Controls

The standardization of laboratory testing and controls is important to ensure that diagnostic results are compatible and comparable (Wells et al. 2004). Components of this standardization include inter-laboratory comparability strategies related to serum analyses, the development, analyses, and distribution

of control sera, reference standards and new techniques for the detection of disease in marine animals, and the maintenance of records demonstrating compliance.

Development of standard analytical methods to be used by all participating laboratories is important to ensure inter-laboratory comparability. Methods, validation protocols, and SOPs should be developed and reviewed by experts in the disease entity/species specialty group and must be adopted by all participating laboratories. In addition, the review board, with the input of expert scientists in laboratory technology and clinical pathology, should create laboratory reference standards for the common tests performed on marine mammal samples and standards for routine stability testing of reference standards.

Control materials should initially be developed for morbillivirus, *Brucella* spp., and *Leptospira* spp. These control materials will be made available to laboratories performing analyses on marine animal sera. As new diseases are detected, appropriate testing or screening methods will be developed and validated through inter-laboratory comparisons and double-blind studies.

## Documentation of Standardization

Documentation of compliance with standardized analytical protocols is important to ensure that all stages of laboratory testing follow the recommended protocols. In addition, validation plans and results need to be documented to support the accuracy of test methods. An annual or biennial report should be compiled and distributed to the review board describing the QC methods that have been implemented and summarizing QC for the serum bank.

## Access Policy

The review board must ensure that the serum bank has a policy, including a request form, for collaborating scientists to access samples for research. The review board will create a review committee consisting of at least three members representing either the team of scientists and/or the contributing institution. The committee may send the request to external experts for scientific review prior to the decision.

The request form should include the following information: the name of the principal investigator and affiliated institution, specific samples (including species and age group) and quantities desired, explanation of proposed research, justification for use of the banked samples, the research facility where analyses will be conducted, analytical QC procedures, and an estimated completion date. In addition, researchers should submit a signed agreement to report the results to the bank for inclusion in the restricted database and that all publications or presentations acknowledge the bank and the original contributing institution.

Similar to the National Marine Mammal Tissue Bank access policy (http://www.nmfs.noaa.gov/pr/health/tissue/), the samples submitted to the long-term storage serum bank will be divided into duplicate samples (A and B). Thus, 50% of each specimen is intended for long-term storage as a more permanent archive and 50% is available to the scientific community for research. Formal requests for B-samples will be submitted to the review committee described above. Partial use of A-samples will be restricted and may be accessed only after all other sources of samples have been depleted (Table 39.1). The experimental design for A-samples will be reviewed by external experts and by managers of the marine animal populations involved in the study.

Access to the collection data will not imply access to the serum specimens or to other individually identifiable data. There will be two levels of data access: public collection information (the minimum database) and restricted information on specific samples, including biological and analytical data. Access to restricted information shall be granted by the custodian of the collection in conjunction with the review board. The NMFS serum specimen tracking and database shall provide the means of collecting and storing relevant and appropriate data at both the collection level and the specimen level.

## STATISTICAL PROCEDURES FOR SAMPLE SIZE REQUIREMENTS IN EPIDEMIOLOGICAL STUDIES

Sample size is an important consideration when determining whether a proposed research project should be allowed access to the banked serum. Sample size is particularly important for studies using serum from endangered marine mammals, or for A-samples. If the sample size is not adequate to detect a disease with a known prevalence, then the study should be restructured to access the banked serum. The following is a discussion of the determination of sample size requirements.

### Estimation of Sample Size

The estimation of the required sample size to detect the presence of antibodies against selected infectious agents via population screening is important for population assessment strategies, including health assessment of wild stocks, unusual mortality events, strandings, or translocations. Sample sizes required to detect the prevalence of infection in a population are presented in Table 39.2. The entries in the table show the probability of detecting at least one diseased animal in the sampled population for several combinations of prevalence and sample size. For example, if the disease has a prevalence of 5% and 10 animals are screened, there is only a 40% probability of detecting a diseased animal. If the prevalence is 1%, the chance

**Table 39.1 Relative Percentages of Serum Classified by Category A (Long-term Archival) and B (Available for Research)**

| (Category) Use | Sample A | Sample B |
|---|---|---|
| (1) Baseline analyses | 10% | 10% |
| (2) Contributor | 25% | 50% |
| (3) Scientific community | 25% | 40% |
| (4) Long-term archival | 40% | 0% |

Modified from the National Marine Mammal Tissue Bank (http://www.nmfs.noaa.gov/pr/health/tissue/).

**Table 39.2  Probability of Detecting Antibody to an Infectious Disease Agent by Prevalence of Infection in the Population and Sample Size**

| Sample Size (n) | Prevalence of Infection (%) | | | | | | |
|---|---|---|---|---|---|---|---|
| | 0.1 | 0.2 | 1 | 2 | 5 | 10 | 20 |
| 10 | 0.010 | 0.020 | 0.096 | 0.183 | 0.401 | 0.651 | 0.893 |
| 20 | 0.020 | 0.039 | 0.182 | 0.332 | 0.642 | 0.878 | 0.988 |
| 30 | 0.030 | 0.058 | 0.260 | 0.455 | 0.785 | 0.958 | 0.999 |
| 40 | 0.039 | 0.077 | 0.331 | 0.554 | 0.871 | 0.985 | >0.999 |
| 50 | 0.049 | 0.095 | 0.395 | 0.636 | 0.923 | 0.995 | >0.999 |
| 75 | 0.072 | 0.139 | 0.529 | 0.780 | 0.979 | >0.999 | >0.999 |
| 100 | 0.095 | 0.181 | 0.634 | 0.867 | 0.994 | >0.999 | >0.999 |
| 125 | 0.118 | 0.221 | 0.715 | 0.920 | 0.998 | >0.999 | >0.999 |
| 150 | 0.139 | 0.259 | 0.779 | 0.952 | >0.999 | >0.999 | >0.999 |
| 175 | 0.161 | 0.296 | 0.828 | 0.971 | >0.999 | >0.999 | >0.999 |
| 200 | 0.181 | 0.330 | 0.866 | 0.982 | >0.999 | >0.999 | >0.999 |

of finding a positive animal when only 10 animals are screened is only 10%. It should be noted that the probabilities in this table were computed under the assumptions that if antibodies are present, they are distributed uniformly across the age classes (i.e., pup, juvenile, adult) and the sampling of animals would be proportional across age classes. If either assumption is incorrect (e.g., if antibody prevalence is higher in adults than in pups), then the sampling strategy would need to be modified to maximize the probability of detecting infected individuals in the population. Such stratified random sampling of the population would require an increase in the total sample size to ensure adequate representation within the various age classes (Aguirre et al. 1999).

To further illustrate the proper requirements for sampling strategies, suppose that a reasonable sample size that could be collected from a prospective donor population is 40. If 40 animals were sampled and their sera examined, one would expect to detect antibody against a specific agent if 20% or more of the animals were infected (Table 39.2). If, however, the prevalence of infection is 2%, then the probability of detecting an infected marine animal is only 55%; and if the prevalence is only 1%, then there is just a 33% chance that an infected animal will be found. Sampling approximately 100 animals is required to achieve the probability of detecting a 2% prevalence of infection with more than 85% certainty.

To minimize the risk of introducing infectious agents into naïve populations when translocating animals, probabilities of detection of less than 85% will not be acceptable. The objective here is not to determine whether the animals being screened are actually shedding the organism at the time of sampling. The utility of the probabilities provided in Table 39.2 are limited to ability to detect prior exposure in a population. Obviously, one would not translocate animals from a population with known infection by an agent to a population that has no prior exposure to that agent. Also, the values presented in Table 39.2 and the discussion above provide further rationale for the recommendation that individual, rather than population, assessment be used to make decisions about the suitability of prospective donor populations for translocation.

## Determining Disease Prevalence

Diagnostic tests occasionally produce erroneous results. The ability of a diagnostic test to accurately identify a diseased or infected individual is known as its sensitivity. The ability of the test to accurately identify a non-diseased or non-infected individual is known as its specificity. Suppose that a test to be used to screen for a disease is not 100% sensitive (i.e., that false negatives occur). It is important to consider these factors when determining the sample size needed to ensure that a population of marine mammals is free of disease.

Let P denote the population prevalence of a disease. For a sample size of n from the population, let X denote the number of animals that test positive. For any one animal, the probability of testing positive is given as:

P+ = P(positive test | disease) × P(disease) + P(positive test | no disease) × P(no disease)

$$= \text{Se} \times P + (1 - \text{Sp}) \times (1 - P) \qquad (1)$$

where Se and Sp denote the sensitivity and specificity of the test, respectively. If the test is 100% sensitive and specific, then P+ = P. If the test results for the n sampled animals can be assumed to be independent, then X has a binomial distribution with parameters n and P+. For large n, X is approximately distributed as a Poisson variable with intensity parameter (mean) μ equal to nP+, so that the probability of X being equal to any non-negative integer is given as:

$$P(X = k) \cong e^{-nP^+}(nP^+)^k / k! \text{ (for k=0,1,....)}$$

Thus, $P(X > 0) = 1 - P(X = 0) = 1 - e^{-nP^+}$

Consequently, the level of significance of the statistical test of the hypothesis that the population is free of disease (i.e., Ho: P = 0), which is rejected if X > 0 is equal to

$$\alpha = P(X > 0 | P = 0) = 1 - e^{-n(1 - \text{Sp})} \qquad (2)$$

and the power of the test is equal to:

$$\text{Power} = P(X > 0 | P > 0) = 1 - e^{-n[\text{Se}P + (1 - \text{Sp}) \times (1 - P)]} \qquad (3)$$

Hence, any α-level test with power greater than or equal to α must satisfy:

$$n(1 - \text{Sp}) \leq -\ln(1 - \alpha) \qquad (4)$$

and

$$n[\text{Se} \times P + (1 - \text{Sp}) \times (1 - P)] \geq -\ln(1 - \alpha) \qquad (5)$$

If the diagnostic test is 100% specific (i.e., Sp = 1). Then, inequality (4) is satisfied for any value of n, and inequality (5) reduces to

$$n \text{ Se } P \geq -\ln(1 - \alpha) \qquad (6)$$

or equivalently

$$n \geq -\ln(1 - \alpha)/(\text{Se} \times P) \qquad (7)$$

Consequently, sample size calculations based on an assumed sensitivity of 100% needs to be increased by a factor equal to (1/Se). Example sample sizes are presented in Table 39.3 for detecting disease when the prevalence of infection in the population ranges from 0.1% to 10% and when using a diagnostic test with sensitivity as low as 90%.

In many other types of studies, the goal is simply to estimate the prevalence of a specific disease in a population of animals with "acceptable" precision for the estimate of prevalence, denoted as P; this objective can be stated more rigorously as that of estimating,

**Table 39.3 Sample Size Required for Sufficient Power to Detect Antibody to an Infectious Agent by Prevalence of Infection in the Population and Sensitivity of the Diagnostic Test**

| Power | Sensitivity (%) | Prevalence of Infection (%) | | |
|---|---|---|---|---|
| | | 10 | 1 | 0.1 |
| 80% | 100 | 16 | 161 | 1609 |
| | 95 | 17 | 170 | 1694 |
| | 90 | 18 | 179 | 1789 |
| 90% | 100 | 23 | 230 | 2303 |
| | 95 | 24 | 244 | 2424 |
| | 90 | 26 | 256 | 2558 |

with probability $\gamma$, the population prevalence of a disease, again denoted as P, within acceptable error E—that is,

$$P(|p-P|\leq E)=\gamma$$

The sample size n needed to achieve this objective is given by:

$$n\geq P(1-P)z^2/E^2 \qquad (8)$$

where z is the tabular value of the standard normal distribution corresponding to central probability (i.e., confidence level) equal to $\gamma$ (e.g., 1.96 for 95% confidence). Often, the acceptable error E is expressed as a percent or fraction of P—that is, E = dP, where d < 1 (e.g., d = 10%), in which case the above expression for the sample size n becomes:

$$n\geq(1-P)z^2/Pd^2 \qquad (9)$$

The sample size requirements for estimating disease prevalence ranging from 0.1% to 20% (as per Table 39.4) are presented in Table 39.4 for 10% to 50% error in the estimate and for both 90% and 95% confidence levels.

## CONCLUSIONS

It is critical that a serum bank be developed in conjunction with current tissue banks already in use with new emerging infectious diseases and toxicant threats, the development of new diagnostic tests, and in consideration of decreasing population levels of marine mammals. Standard SOPs and QC requirements for the serum bank will allow high-quality epidemiological studies to be conducted. In addition, careful analysis by a team of scientists will ensure that the stored samples in the bank are carefully maintained for research that is of sufficient sample size and quality to use these valuable resources.

## REFERENCES

Aguirre, A.A., J. Rief, and G. Antonelis. 1999. Hawaiian monk seal epidemiology plan: health and disease status studies. NOAA-TM-NMFS-SWFSC-280. U. S. Department of Commerce, Honolulu, Hawaii.

Becker, P.R., B.J. Koster, S.A. Wise, and R. Zeisler 1993. Biological specimen banking in Artic research: an Alaska perspective. Sci Total Environ 139/140:69–95.

Becker, P.R., and S.A. Wise. 2006. The U.S. National Biomonitoring Specimen Bank and the Marine Environmental Specimen Bank. J Environ Monitor 8:795–799.

Bellete, B., P. Flori, J. Hafid, H. Raberin, and R. Tran Manh Sung. 2003. Influence of the quantity of nonspecific DNA and repeated freezing and thawing of samples on the quantification of DNA by the Light Cycler®. J Microbiol Meth 55:213–219.

**Table 39.4 Sample Size Required to Estimate, with 90% or 95% Confidence, the Prevalence of a Disease in the Population by Prevalence of Infection in the Population and Allowable Error of the Estimate**

| Confidence Level (%) | Error (%) | Prevalence of Infection (%) | | | | | | |
|---|---|---|---|---|---|---|---|---|
| | | 0.1 | 0.2 | 1 | 2 | 5 | 10 | 20 |
| 90 | 10 | 270332 | 135031 | 26790 | 13260 | 5141 | 2435 | 1082 |
| | 20 | 67583 | 33758 | 6697 | 3315 | 1285 | 609 | 271 |
| | 30 | 30037 | 15003 | 2977 | 1473 | 571 | 271 | 120 |
| | 40 | 16896 | 8439 | 1674 | 829 | 321 | 152 | 68 |
| | 50 | 10813 | 5401 | 1072 | 530 | 206 | 97 | 43 |
| 95 | 10 | 383776 | 191696 | 38032 | 18824 | 7299 | 3457 | 1537 |
| | 20 | 95944 | 47924 | 9508 | 4706 | 1825 | 864 | 384 |
| | 30 | 42642 | 21300 | 4226 | 2092 | 811 | 384 | 171 |
| | 40 | 23986 | 11981 | 2377 | 1176 | 456 | 216 | 96 |
| | 50 | 15351 | 7668 | 1521 | 753 | 292 | 138 | 61 |

Bothig, B., L. Danes, E. Gerike, D. Ditmann, and E. Svandova. 1992. Qualification of long stored samples of serum banks for seroepidemiological studies. J Hyg Epidemiol Microbiol Immunol 36:269.

Burek, K.A., F.M. Gulland, G. Sheffield, K.B. Beckman, E. Keyes, T.R. Spraker, A.W. Smith, D.E. Skilling, J.F. Evermann, J.L. Stott, J.T. Saliki, and A.W. Trites. 2005. Infectious disease and the decline of Steller sea lions (Eumetopias jubatus) in Alaska, USA: insights from serologic data. J Wildl Dis 41:512–524.

CDC. 1997. CDC/ATSDR Specimen and data bank. Centers for Disease Control and Prevention, Atlanta.

Cunningham, A.A. 1996. Disease risks of wildlife translocations. Conserv Biol 10:349–353.

Cunningham, A.A., and P. Daszak. 1998. Extinction of a species of land snail due to infection with a microsporidian parasite. Conserv Biol 12:1139–1141.

Daszak, P., A.A. Cunningham, and A.D. Hyatt. 2000. Emerging infectious diseases of wildlife—threats to biodiversity and human health. Science 287:443–449.

de Bruyn, N.P.J., A.D.S. Bastos, C. Eadie, C.A. Tosh, and M.N. Bester. 2008. Mass mortality of adult male subantarctic fur seals: are alien mice the culprits. Plos One 3:e3757.

Di Guardo, G., G. Marruchella, U. Agrimi, and S. Kennedy. 2005. Morbillivirus infections in aquatic mammals: a brief overview. J Vet Med Assoc 52:88–93.

Dierauf, L.A., and F.M. Gulland. 2001. CRC handbook of marine mammal medicine: health, disease and rehabilitation. CRC Press, Boca Raton, Florida.

Dierauf, L.A., D.J. Vandenbroek, J. Roletto, M. Koski, L. Amaya, and L.J. Gage. 1985. An epizootic of leptopirosis in California sea lions. J Am Vet Med Asocc 187:1145–1148.

Domingo, M., L. Ferrer, M. Pumarola, A. Marco, J. Plana, S. Kennedy, M. McAliskey, and B.K. Rima. 1990. Morbillivirus in dolphins. Nature 348:21.

Duignan, P.J., J.T. Saliki, D J. St Aubin, J.A. House, and J.R. Geracia. 1994. Nuetralizing antibodies to phocine distemper virus in Atlantic walruses (Odobenus rosmarus rosmarus) from Arctic Canada. J Wildl Dis 30:90.

Duignan, P.J., C. House, J.R. Geraci, N. Duffy, B.K. Rima, M.T. Walsh, G. Early, D.J. St Aubin, S. Sadove, H. Koopman, and H. Rhinehart. 1995. Morbillivirus infection in cetaceans of the western Atlantic. Vet Microbiol 44:241–249.

Duignan, P.J., N. Duffy, B.K. Rima, and J.R. Geracia. 1997. Comparative antibody response in harbour and grey seals naturally infected by a morbillivirus. Vet Immunol Immunopathol 55:341–349.

Elliott, P., and T.C. Peakman. 2008. The UK Biobank sample handling and storage protocol for the colelction,

processing and archiving of human blood and urine. Int J Epidemiol 37:234–244.

Fernández, A., F. Esperón, P. Herraéz, A. Espinosa de los Monteros, C. Clavel, A. Bernabé, J.M. Sánchez-Vizcaino, P. Verborgh, R. DeStephanis, F. Toledano, and A. Bayón. 2008. Morbillivirus and pilot whale deaths, Mediterranean Sea. Emerg Infect Dis 14:792–794.

Geraci, J.R. 1989. Clinical investigation of the 1987–88 mass mortality of bottlenose dolphins along the US central and south coast. Final Report to National Marine Fisheries Servies. US Navy Office of Naval Research and Marine Mammal Commission, Washington, D.C.

Geraci, J.R., D.J. St Aubin, I.K. Barker, R.G. Webster, V.S. Hinshaw, W.J. Bean, H.L. Ruhnke, J.H. Prescott, G. Early, A.S. Baker, S. Madoff, and R.T. Schooley. 1982. Mass mortality of harbor seals: pneumonia associated with influenza A virus. Science 215:1129–1131.

Gulland, F.M., M. Koski, L.J. Lowenstine, A. Colagross, L. Morgan, and T. Spraker. 1996. Leptospirosis in California sea lions (Zalophus californianus) stranded along the central California coast, 1981–1994. J Wildl Dis 32:572–580.

Hardwood, J., and A. Hall. 1990. Mass mortality in marine mammals: its implications for population dynamics and genetics. Trends Ecol Evol 5:254–257.

Holland, N.T., M. Smith, B. Eskenazi, and M. Bastaki. 2003. Biological sample collection and processing for molecular epidemiological studies. Mutat Res 543:217–234.

Hutchins, M. 1990. Serving science and conservation: the biological materials request protocol of the New York Zoological Society. Zoo Biol 9:447–460.

Jellum, E., A. Anderson, P. Lund-Larsen, L. Theodorsen, and H. Orjasaeter. 1993. The JANUS serum bank. Sci Total Environ 139/140:527–535.

Kennedy, S. 1998. Morbillivirus infections in marine mammals. J Comp Pathol 119:201–225.

Kennedy, S., J.A. Smyth, P.F. Cush, P.J. Duignan, M. Plattern, S.J. McCullough, and G.M. Allan. 1989. Histopathologic and immunocytochemical studies of distemper in seals. Vet Pathol 26:97–103.

Kennedy, S., J.A. Smyth, P.F. Cush, S.J. McCullough, G.M. Allan, and S. McQuaid. 1988. Viral distemper now found in porpoises. Nature 336:21.

Kennedy, S., J.A. Smyth, S.J. McCullough, G.M. Allan, F. McNeilly, and S. McQuaid. 1988. Confirmation of cause of recent seal deaths. Nature 335:404.

Kucklick, J., R. Pugh, P.R. Becker, J. Keller, J. Day, J. Yordy, A. Moors, S. Christopher, C. Bryan, L. Schwarcke, C. Goetz, R. Wells, B. Balmer, A. Hohn, and T. Rowles. 2010. Specimen banking for marine animal health assessment. Interdisciplinary studies on environmental chemistry, pp. 15–23. Environmental Specimen Bank, Terrapub, Tokyo.

Lafferty, K.D., and L.R. Gerber. 2002. Good medicine for conservation biology: the intersection of epidemiology and conservation theory. Con Biol 16:593–604.

Lennette, E.H., and N.J. Schmidt. 1969. Diagnostic procedures for viral and rickettsial infections. American Public Health Association, New York.

Lev, E.I., I. Hendler, R. Siebner, Z. Tashma, M. Wiener, and I. Tur-Kaspa. 1994-1995. Creatine kinase activity decrease with short-term freezing. Enzyme Protein 48:238–242.

Lillestolen, T., N. Foster, and S.A. Wise. 1993. Development of the National Marine Mammal Tissue Bank. Sci Total Environ 139:97–107.

Lipscomb, T.P., F.Y. Schulman, D. Moffett, and S. Kennedy. 1994. Morbilliviral disease in Atlantic bottlenose dolphins (*Tursiops truncatus*) from the 1987–88 epizootic. J Wildl Dis 30:367–571.

Lloyd-Smith, J.O., D.J. Greig, S. Hietala, G.S. Ghneim, L. Palmer, J. St. Leger, B.T. Grenfell, and F.M. Gulland. 2007. Cyclical changes in seroprevalence of leptospirosis in California sea lions: Endemic and epidemic disease in one host species? BioMed Central Infect Dis 7:1–30.

Mamaev, L.V., S.A. Belikov, N.N. Denikina, B. Edginton, I.K.G. Visser, A.D.M.E. Osterhaus, L. Goatley, T. Barrett, T. Harder, and B.K. Rima. 1996. Canine distemper virus in Lake Baikal seals (*Phoca sibirica*). Vet Rec 138:437–439.

McCallum, H., and A. Dobson. 1995. Detecting disease and parasite threats to endangered species and ecosystems. Trends Ecol Evol 10:190–194.

Osterhaus, A.D.M.E., J. Groen, H. Niesters, M.W. van de Bildt, B. Martina, L. Vedder, J. Vos, H. van Egmond, B. Abou-Sidi, and M.E. Bar-ham. 1997. Morbillivirus in monk seal mass mortality. Nature 388:838–839.

Osterhaus, A.D.M.E., J. Groen, H.E. Spijkers, H.W. Broeders, F.G. UytdeHaag, P. de Vries, J.S. Teppema, I.K.G. Visser, M.W. van de Bildt, and E.J. Vedder. 1990. Mass mortality in seals caused by a newly discovered morbillivirus. Vet Microbiol 23:343–350.

Raga, J.A., A. Banyard, M. Domingo, M. Corteyn, M.F. van Bressem, M. Fernández, F.J. Aznar, and T. Barrett. 2008. Dolphin morbillivirus epizootic resurgence, Mediterranean Sea. Emerg Infect Dis 14:471–473.

Reidarson, T.H., J. McBain, C. House, D.P. King, J.L. Stott, A. Krafft, J.K. Taubenberger, J. Heyning, and T.P. Lipscomb. 1998. Morbillivirus infection in stranded common dolphins from the Pacific Ocean. J Wildl Dis 34:771–776.

Tilton, R.C., A. Balows, D.C. Hohnadel, and R.F. Reiss. 1992. Clinical laboratory medicine. Mosby Year Book, Baltimore.

van Bressem, M.F., K. van Waerebeek, M. Fleming, and T. Barrett. 1998. Serological evidence of morbillivirus infection in small cetaceans from the Southeast Pacific. Vet Microbiol 59:89–98.

Wells, R., H.L. Rhinehart, L.J. Hansen, J.C. Sweeney, F.I. Townsend, R. Stone, D.R. Casper, M.D. Scott, A. Hohn, and T. Rowles. 2004. Bottlenose dolphins as marine ecosystem sentinels: Developing a health monitoring system. EcoHealth 1:246–254.

Westgard, J.O., and G.G. Klee. 1996. Quality management. *In* C. Burtis. Fundamentals of Clinical Chemistry, 4th ed. WB Saunders Company, Philadelphia, pp. 211–223.

Wobeser, G.A. 1994. Investigation and management of disease in wild animals. Plenum Publishing Corporation, New York.

Zarnke, R.L., J.T. Saliki, A P. Macmillan, S.D. Brew, C.E. Dawson, J.M. ver Hoef, K.J. Frost, and R.J. Small. 2006. Serologic survey for *Brucella spp.,* phocid herpesvirus-1, phocid herpesvirus-2, and phocine distemper virus in harbor seals from Alaska, 1976-1999. J Wildl Dis 42:290–300.

# 40

## SORTA SITU

The New Reality of Management Conditions for Wildlife Populations in the Absence of "Wild" Spaces

Barbara A. Wolfe, Roberto F. Aguilar, A. Alonso Aguirre, Glenn H. Olsen, and Evan S Blumer

The rate of species loss today is approaching catastrophic levels. Scientists project that over the next two decades, more than 1 million species of plants and animals will become extinct. E.O. Wilson has estimated that "the rate of loss may exceed 50,000 a year, 137 a day ... this rate, while horrendous, is actually the minimal estimate, based on the species/area relationship alone" (Kellert and Wilson 1993, p. 16; Aguirre 2009). Ever-expanding communities, strained natural resources, changes in land use, and other anthropogenic drivers are compromising ecosystems and rapidly changing the landscape and the availability of "wild" spaces.

One outcome of these changes is the manifestation of a new global reality for wildlife. Where truly "wild" populations are increasingly rare and more animals are managed in protected zones, refuges, and conservation centers, the difference between *in situ* (wild populations in native habitat) and *ex situ* (captive populations in non-native habitat) becomes less distinct (Fig. 40.1). In fact, most wildlife populations of today and tomorrow exist on a continuum between *in situ* and *ex situ*. We define this new reality as *sorta situ*[1]—neither one nor the other—to describe the changing nature of population management in the 21st century (Aguirre and Pearl 2004). This chapter will use current examples to illustrate *sorta situ* populations and circumstances, and discuss considerations that are necessarily becoming part of the strategy of conservation medicine practitioners in managing and caring for wildlife populations in this changing global paradigm.

The continuum of conditions for *sorta situ* populations can be viewed across two key variables: available habitat, including space, habitat quality, and the maintenance of ecosystem processes; and management intensity, including healthcare and protection from outside threats. In the past, wildlife populations lived on large landscapes in their native habitat, without human intervention. Diseases and populations were, for the most part, self-limiting, and terrestrial animals were free to move in response to seasonal and

---

1 Term coined by John Jensen (2001), Environmental Program Director of the George Gund Foundation, describing "sort of" *situ* as an *ex situ* management strategy approximating *in situ* conditions.

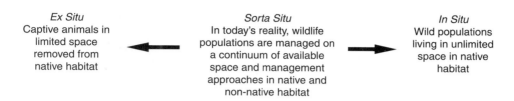

| Ex Situ | Sorta Situ | In Situ |
|---------|------------|---------|
| Captive animals in limited space removed from native habitat | In today's reality, wildlife populations are managed on a continuum of available space and management approaches in native and non-native habitat | Wild populations living in unlimited space in native habitat |

**Figure 40.1:**
The changing nature of wildlife management and conservation from *in situ* to *ex situ*, with *sorta situ* in between.

dietary needs. On the other end of the spectrum, zoos managed animals outside their native habitat in small captive groups, treating disease and injury on an individual basis and controlling nutritional input and reproductive output. Today, fences, borders, and human habitation limit the spaces wild populations can occupy, and small, fragmented populations require careful monitoring and management to avoid devastating population declines and extinction.

This new reality calls for a new approach to conservation medicine and management: a combination of *ex situ* developed skills, including small population management, practical veterinary care, and intensive reproductive management, linked to landscape-scale monitoring and field skills that include habitat restoration, reintroduction techniques, epidemiology, ecological modeling, behavioral ecology, and community-based conservation.

## SPECTRUM OF MANAGEMENT INTENSITY IN *SORTA SITU* POPULATIONS

### Intensive Management of Wild Populations

Increasingly, "wild" populations live in limited spaces of varying sizes in their native habitat and, in some cases, are closely monitored and often visited by wildlife health professionals when necessary. The Mountain Gorilla Veterinary Project of Rwanda, Uganda, and the Democratic Republic of Congo is an example of this type of *sorta situ* model (Cranfield et al. 2002). In this program, the approximately 740 remaining "wild" mountain gorillas (*Gorilla gorilla beringei*) live in fragments of their native habitat, are

visited frequently by humans through ecotourism, and are monitored closely by a team of veterinarians and staff, who respond to each individual gorilla's need for medical care. A majority of "wild" rhinoceroses in Africa similarly live in refuges and parks under armed guard. They are observed frequently, treated for injuries, and translocated in response to poaching and other risks.

Even in geographically unrestricted bird populations, individual veterinary care is being employed for population management. The California condor (*Gymnogyps californianus*) reintroduction program in Vermillion Cliffs, Arizona, California, and Baja California has used veterinary support for lead-poisoned birds since the problem was first detected in the late 1990s. Bullet fragments in carcass remains left by hunters and eaten by condors have markedly increased the number of chronically lead-poisoned birds in recent years. Condors are tested and treated in the field if possible, while severely lead-poisoned birds with crop stasis or other complications are admitted to zoo hospitals for treatment. Once treated, the birds are sent back to the original capture site for release and monitoring.

Whooping cranes (*Grus americana*), both the existing wild flock (migrating between Wood Buffalo National Park, Alberta, Canada, and Aransas National Wildlife Refuge, Texas) and two introduced flocks (a non-migratory flock in central Florida and an eastern migratory flock between Wisconsin and Florida), are monitored and medically managed by a transdisciplinary team belonging to many institutions. This team collaborates to rescue and rehabilitate any sick or injured whooping cranes. Most individual members of the introduced flocks are fitted with radio transmitters and benefit from frequent and intense monitoring by dedicated teams of wildlife biologists.

When a problem is detected or when routine captures are planned, veterinarians participate to examine the individual crane and obtain diagnostic samples, providing information on the health of both the individual and the population.

## Limited Management of Captive Populations

While some "wild" populations receive intensive management and care, some "captive" populations thrive far from native habitat on expansive tracts of land, receiving little management or medical care. Such is the case in many game ranches, wildlife parks, and hunting preserves. In Texas, for instance, over 250,000 exotic ungulates live on game ranches (Teer et al. 1993; Mungall and Sheffield 1994; Demarais et al. 1998). Many of these herds are virtually unmanaged, and free-ranging populations of axis deer (*Axis axis*), fallow deer (*Dama dama*), sika deer (*Cervus nippon*), blackbuck antelope (*Antilope cervicapra*), nilgai antelope (*Boselaphus tragocamelus*), aoudad (*Ammotragus lervia*), and other exotic ungulates have been established through escape and release (Huerta-Patricio et al. 2005).

## POPULATIONS ON THE SPECTRUM OF AVAILABLE SPACE

### From Zoos to Conservation Centers

According to the Conservation Breeding Specialists' Group, there is a need for conservation of threatened species to be shifted from zoos to larger breeding and conservation facilities (CBSG Newsletter Jan. 2010). In such facilities, animals given more space and kept in larger, more natural groupings might exhibit more natural behaviors (Clubb and Mason 2003; Li et al. 2007) and experience reduced stress, improved health, and enhanced breeding success. A recent study found, in fact, that large enclosure size and the opportunity to interact with conspecifics was associated with reproductive success in the southern white rhinoceros, *Ceratotherium simum* (Metrione and Harder 2009).

As zoological institutions become more involved in *in situ* conservation efforts, a new emphasis has been placed on the challenges of current breeding and captive management paradigms and the creation of larger conservation facilities. As established by the recent strategic plan of the Association of Zoos and Aquariums, captive facilities "will ensure the sustainability of diverse wildlife collections in accredited zoos and aquariums; advance high standards of wildlife-focused animal care and welfare; and foster outcome-based conservation by connecting zoos and aquariums to the wild" (http://www.aza.org/StrategicPlan/). A recently formed consortium of such facilities in the United States known as the Conservation Centers for Species Survival comprises five of the largest land-holding institutions: Smithsonian Conservation Biology Institute in Front Royal, Virginia; White Oak Conservation Center in Yulee, Florida; The Wilds in Cumberland, Ohio; Fossil Rim Wildlife Center in Glen Rose, Texas; and the San Diego Safari Park in Escondido, California. This consortium combines research, management, and training efforts to improve conservation of animal species and natural resources. The combined space available for captive animal management in these five institutions is over 20,000 acres, and the consortium is dedicated to cooperatively studying how landscape-scale settings and new techniques in captive management will uniquely benefit certain wild animal populations.

### Wild Game Farms, Ranches, and Preserves

Wildlife and exotic game are farmed and hunted on privately owned lands ranging from less than 100 to over 100,000 acres in many countries. Hudson et al. (1989) described three different management approaches—farming, ranching, and herding—used by managers of native and exotic ungulates. Ranched and farmed populations are both confined by fences, but at different levels of management intensity. While farming involves intensive genetic, medical, and nutritional management of the captive population, ranched animals are managed primarily as confined wild populations. Herding, the least intensive approach and the one used more often on game preserves, describes management of a wild population relying on natural migration patterns to move and control animals. For instance, Rocky Mountain elk (*Cervus elaphus*) and American bison (*Bison bison*) in the Greater Yellowstone Area of the northern United States are managed by herding between seasonal feeding grounds on public and private lands.

## Protected Areas

With species extinction rates threatening to increase to nearly 1,000-fold background rates, establishing protected areas for wildlife management is considered the primary defense against extinction (Joppa et al. 2008). To date, approximately 60,000 protected areas, defined as "[An] area of land and/or sea especially dedicated to the protection and maintenance of biological diversity and of natural and associated cultural resources, and managed through legal or other effective means" (IUCN 1994), cover approximately 13% of the earth's land surface (Phillips 2003). Available land, ecosystem stability, and management strategies within these protected areas, however, may determine species success in protected areas (Chape et al. 2005). Surveys of fragmented forests have shown that forests of less than 100 km² are insufficient to protect small populations of vulnerable birds over long periods of time (Brooks et al. 1999; Ferraz et al. 2003). Strong inverse relationships have been demonstrated between reserve size and extinction rates (Newmark 1996, 2008), and between ecosystem stability and species diversity (Dobson 2009). In some of these cases, certain specialized management techniques, such as the release of captive-bred animals, are being employed to augment fragile populations of single species. However, management of these *sorta situ* environments will likely require more systematic, ecosystem-wide conservation practices for the maintenance of ecological balance.

## Fences, Walls, and Borders

Fences are used for many, and sometimes combined, reasons (Bode and Wintle 2009). Veterinary fencing in southern Africa was historically intended to control the spread of animal pathogens such as foot and mouth disease, which is transmitted from wildlife reservoir hosts to domestic livestock with devastating economic consequences. Many of the wildlife reservoirs of these diseases crossed international borders and intermingled with susceptible stock, creating regulatory difficulties in managing both disease spread and food safety in farmed animals. Veterinary fences provided an anthropogenic disease barrier, easing regional political strain and decreasing disease spread between wild ungulates and livestock. However, fencing and isolation of populations have been found to be detrimental to free-ranging species and ecosystem processes by preventing seasonal migration in large mammals and decreasing species abundance and genetic diversity of wild populations (Bolger et al. 2008; Chase and Griffin 2009), as well as affecting human welfare by excluding populations from their traditional lands and natural resources (Western 2002; Hoole and Berkes 2010). In Scotland, where fences are maintained to protect forest habitat from destruction by deer, capercaillie (*Tetrao urogallis*) and black grouse (*Tetrao tetrix*) have experienced up to 32% annual mortality due to fence collision (Catt et al. 1994). The walls being built between Israel and Palestine and more recently along the U.S.–Mexico border can similarly affect terrestrial populations of wildlife by dividing conservation areas and refuges (Cohn 2007; Sayre and Night 2009; Flesch et al. 2010; Wildlife Society 2010). Disruption of transboundary movement corridors by impermeable fencing has isolated some wildlife populations on both sides of the border.

Clearly, fences and borders, compounding the reduction of wild spaces available to maintain population stability and genetic dispersal, affect connectivity and spatial distribution of populations. Without the application of scrupulous mitigation strategies (Flesch et al. 2010), the result of these barriers is further division and isolation of wild species into subpopulations, increasing the chance that they will require some level of management to survive in the *sorta situ* future.

## Transfrontier Conservation Areas

By far the largest examples in the spectrum of available space are transfrontier conservation areas (TFCAs), recently established in Africa. TFCAs are a cooperative natural resource management strategy spanning numerous parks, reserves, and countries, encompassing over 1.2 million km² of land and providing vast expanses for wildlife. While innovative, TFCAs can be politically and economically challenging from a disease transmission and food safety perspective (Cumming et al. 2007). For protected areas to succeed in conserving wildlife and habitat, they must benefit the neighboring human populations (Phillips 2003; Hoole and Berkes 2010). Their planning and management must involve the local people, including social, economic, and conservation objectives, with long-term goals and political considerations in mind.

## HEALTH CONSIDERATIONS IN THE MANAGEMENT OF *SORTA SITU* POPULATIONS

From intensively managed in unrestricted space to unmanaged in large enclosures in captivity, *sorta situ* populations encounter novel conditions, whether in the form of frequent human exposure, foreign climates, new food sources, new parasites and diseases, new competitors and predators, or even novel soil and substrates. As conservation medicine practitioners, we are now challenged to identify, predict, and assess the health impacts of the host of habitat changes that these populations encounter across the range of management scenarios, and to better manage these changes to conserve both wild populations and their habitats.

### The Human–Wildlife–Domestic Animal Interface

Wildlife populations, regardless of their available space, no longer exist in isolation from humans, domestic animals, and their evolving diseases. Decreasing space and increasing globalization have led to an increase in emerging infectious diseases (Daszak et al. 2000). An ever-increasing human population and decreasing wildlife habitat, in combination with changing animal ecology and climatic conditions, are thought to have led to the emergence of diseases such as Lyme disease, Hendra and Nipah viruses, and hantavirus pulmonary syndrome (Daszak and Cunningham 2002). Animals maintained on game ranches in their native habitat can present a risk to their wild counterparts, due to movement and fence line contact. For instance, the recent spread of chronic wasting disease in wild deer (*Odocoileus* spp.) and elk (*Cervus elaphus*) in the northern United States most likely occurred by transportation of farmed cervids between states and across the U.S.–Canada border by the owners and employees of farming operations (Williams and Miller 2002). With the introduction of deer farming in New Zealand, the presence of malignant catarrhal fever in domestic sheep resulted in outbreaks of the disease in susceptible deer (Wilson 2002).

In southern Africa, ecotourism plays a bigger economic role than agriculture, forestry, and fisheries combined, leading to greater exposure of wildlife populations to human presence (Osofsky et al. 2008). Particularly in developing countries, human populations, and their associated domestic animals, are relatively dense at the borders of protected areas for various reasons (Kalema-Zikusoka 2005). In such areas, emerging and re-emerging zoonotic diseases are of persistent and increasing concern to the health of the wildlife, domestic animals, and humans.

The health problems observed in free-ranging wildlife populations today resemble those seen historically in captive wildlife. Where diseases once existed in ecological "balance" and were self-limiting on a population scale, fragmented populations now are more vulnerable to stress, reproductive suppression, decreasing genetic diversity, malnutrition, and environmental pollutants, all leading to reduced immune protection against disease. The result is that wild animals are more vulnerable than ever to the possibility that a disease could wipe out a local population.

### Animal Density and Translocation: Effects on Disease Risk

In captive populations, common diseases have been managed by exclusion, quarantine, sanitation, and vaccination, while outbreaks tend to be more devastating to larger, less intensively managed populations. The management of native and non-native wildlife species on limited habitat is altering the ecology of pathogens. For instance, the introduction or reintroduction of predators to an area alters the behavior of prey animals, presenting a host of risks to the prey, including decreased foraging and reproduction, increased dispersal, and, potentially, increased spread of disease (Heithaus et al. 2009; Sih et al. 2010). This phenomenon has been studied intensively in the reintroduction of grey wolves (*Canis lupus*) to the Greater Yellowstone Ecosystem (Fortin et al. 2005; Creel and Christianson 2008, 2009). Often, indigenous wildlife harbor diseases to which they are resistant or are unaffected carriers, and transmit them readily to susceptible exotic species (Chomel et al. 2005). Such is the case with the parasite *Parelaphostrongylus tenuis*, which is carried asymptomatically by white-tailed deer in the northeastern United States and has caused lethal central nervous system parasitic migration in a yet incompletely defined number of species of domestic and exotic ruminants and camelids (Nagy 2005).

Animal density can be influenced in both unrestricted spaces and in captivity. Where supplemental feeding of wildlife occurs, diseases like brucellosis in Rocky Mountain elk—which normally calve in seclusion, limiting transmission of *Brucella abortus*—are spread more rapidly due to increased animal density and contact (Cross et al. 2010). Following the spillover of bovine tuberculosis to white-tailed deer in the northern United States, supplemental feeding of deer led to widespread tuberculosis in wild deer populations. White-tailed deer became a reservoir for tuberculosis, leading to spillback and increased incidence of tuberculosis in domestic cattle (Schmitt et al. 1997).

Anthropogenic changes in animal density can have unexpected effects on disease prevalence. Increasingly, ungulate herds are managed as either semi-free-ranging or free-ranging populations with supplemental feeding. In such circumstances, predictive models of transmission and susceptibility to disease are altered. For example, in a study of infection by multiple pathogens in red deer (*Cervus elaphus*) under different management conditions, host body condition was negatively associated with infection with the lungworm *Elaphostrongylus cervi*, as would be expected due to the availability of host resource for use in parasite defense. Conversely, host density was also negatively associated with parasite counts (Vicente et al. 2007). Epidemiological models predict that host population density would correlate positively with abundance of nematode parasites because high population density, and therefore high contact rates, would increase the potential for transmission under natural conditions (Arneberg 2001). In this case, however, population density is artificially manipulated by supplemental feeding, and therefore the animals near feeding stations, although at high density, have improved body condition and therefore improved immune defenses. This is not the case in captive and semi-free-ranging populations in which all animals have access to feed, but are fed and sheltered at a small number of stations, where aggregation will encourage parasite transmission.

## Augmenting Wild Populations Using Captive Management Techniques

Maintaining species genetic diversity is becoming increasingly difficult as fewer and more fragmented "wild" populations exist. Captive populations have been managed in zoological institutions for decades as safety nets or "arks" for supporting wild populations, for repopulating lost populations, and as a management system for sustaining genetic diversity both in the wild and in captivity (Soulé et al. 1986; Lees and Wilcken 2009). Augmentation of isolated wild populations through captive breeding may be fundamental, at least in the short term, to the survival of some wild populations (Conde et al. 2011). This is the case for the Mexican wolf (*Canis lupus baileyi*), the black-footed ferret (*Mustela nigripes*; Box 40.1), the kakapo (*Strigops habroptila*), the sandhill crane (*Grus canadensis*), and whooping cranes at present.

There are an estimated 47 wild Mexican wolves in the Blue Range Reintroduction Area between Arizona and New Mexico. Released wolves have been illegally hunted to the point at which the wild population is not sustainable without intense monitoring and protection. To augment the wild population, seven founder Mexican wolves held in zoos and conservation centers have produced approximately 300 animals for release over the past 20 years. Data gathered through these captive breeding and release efforts are also used to improve both captive management of the species and future reintroduction efforts (http://www.fws.gov/southwest/es/mexicanwolf/).

Perhaps the most dramatic example of *sorta situ* conservation management in New Zealand is the kakapo. This is an example of "bringing the zoo to the animal." This species is one of the most intensively managed in the world: the entire population lives on two predator-free islands off the coast of southern New Zealand and has been managed there since 1983 (Innes et al. 2010). The population reached a nadir of just 48 in 1993, with only 17 females, but has since reached 123. Nest monitoring, translocation, and artificial feeding, along with the removal of all introduced predators and one native predator, have been instrumental in the success of this program (Clout and Craig 1995; Allen and Lee 2006).

For the whooping crane, the optimal captive flock size, allowing for the retention of greater than 90% of the genetic diversity for more than 100 years, has been determined to be 153. Artificial insemination and selective breeding have been applied to disperse genetics and increase fertility in the captive population, resulting in a managed flock that may have more genetic diversity than the last remaining wild flock of whooping cranes (Jones and Lacy 2009).

---

**Box 40.1   A *Sorta Situ* Success: The Black-footed Ferret**

---

The black-footed ferret (*Mustela nigripes*) is considered to be a conservation success story brought about by captive breeding and reintroduction. Thought to be extinct decades ago, the species was repopulated through a small population discovered in 1985. In this case, regular introduction of animals produced by natural and assisted breeding has proven to be a useful strategy for the maintenance of genetic diversity in this small population (Wisely et al. 2008). Through the recovery of the species, disease has played a major roll. Canine distemper in captive ferrets and sylvatic plague in wild prairie dog colonies and recently in black-footed ferrets have severely set back the program's success at re-establishing the species in the past (Matchett et al. 2006). Intense management of wild populations has included health monitoring, vaccination, surveillance of associated indicator species (coyote, fox, and badger), and identification and monitoring of recaptured adults. Since 1991, 19 specific black-footed ferret reintroduction projects have been conducted across eight U.S. states, Canada, and Mexico. All reintroductions from 1991 to 1996 continue to be occupied by ferrets, and half of all introductions to date are considered "successful" (i.e., self-sustaining with 30 or more breeding adults capable of supporting other sites with translocations) or "improving" (i.e., increasing population) (33% and 17%, respectively). As of 2010, an estimated 1,500 ferrets are living and surviving in the wild across prairie dog habitat with no fewer than 30 reproductive adults in each population. This program has meant an investment of over $US30 million since 1981 and the commitment of many federal, tribal, and state biologists, ecologists, veterinarians, non-governmental organizations, zoos, and private landowners (http://www.defenders.org/programs_and_policy/wildlife_conservation/imperiled_species/black-footed_ferret/background_and_recovery.php).

---

## Health Concerns in Small and Augmented Populations

While the augmentation of wild populations using captive management techniques has to date allowed some species to escape extinction, it underscores the delicate balance of *sorta situ* population management, as it carries its own risks to the survival of the population (Swaisgood et al. 2006). In planning reintroduction of a captive-bred species, managers must consider the implications of captive management techniques on survivability in the wild, and the need to monitor health in the species community once captive-bred animals enter the wild population.

It is well established that the smaller the population, the more susceptible it is to devastation due to disease and other factors. Often, metapopulations—regional populations comprising fragmented subpopulations—persist due to the occurrence of limited dispersal between patches, despite the instability of local subpopulations (Levins 1969). Demographic, environmental, and genetic stochasticity and natural

catastrophes have been identified as sources of uncertainty for the determination of a minimum viable population (Shaffer 1981).

When small populations are managed in single-species enclosures in high density, they tend to be susceptible to disease outbreaks much like production animals. Recently, reintroduction efforts for the masked bobwhite quail (*Colinus virginianus ridgwayi*) experienced such a phenomenon (Aguilar et al. 2008; Pacheco et al. 2008). The Buenos Aires Wildlife Refuge is located on the U.S.–Mexico border and was established, in part, to provide habitat for the masked bobwhite quail, populations of which are rapidly declining. In the past 15 years, over 31,000 released masked bobwhite quail were produced at the refuge by a flock of captive quail kept as an assurance colony. In 2007, the flock was reduced by 50% due to a multi-faceted disease outbreak in the holding facility, and devastation of the released flock by avian malaria has further complicated the possibility of continuing reintroduction attempts in the region (Andreína Pacheco et al., in press).

Many captive populations are also proving to be unsustainable—fluctuating based on intense competition for limited resources, trends in species interest, and space restrictions—and subject to unique health concerns. Historically, animals have been genetically managed in captivity in small groups, and when their groups exceed available holding space, breeding is prevented. Males are held in isolation, females develop fertility problems associated with failure to carry and deliver offspring on a regular basis, and the fitness of the population declines. In cooperatively breeding species such as African wild dogs (*Lycaon pictus*), the loss of a single dominant breeder can result in prolonged disruptions in breeding. In many species, such as great apes and wild equids, males managed in bachelor groups suffer injury and social stresses not natural for their species. Often, breeding is prevented on a broad scale across zoological institutions, leading to captive population crises. While the genetic and demographic models for sustainable animal populations are viable, the reality of institutional needs and limitations interfere with these models reaching fruition. Of benefit to global populations would be the consideration of the *sorta situ* scale in designing management programs: the provision of more space, larger and more natural groupings, establishment of source and sink populations, and conditions conducive to the development of more adaptive traits would improve health and reproductive potential, as well as better establish populations that may be destined for reintroduction.

## Understanding Health and Husbandry Needs

Freshwater mussels of the order Unionoidea are the most imperiled group of animals in North America: over 75% of species are threatened, endangered, of special concern, or extinct (Williams 1993). Yet our lack of understanding of their basic physiological needs has limited our ability to improve their survival. Increasingly, mussel populations are relocated to refuges to protect them from construction zones and invasive mussel colonization, or to recolonize following pollution events and extirpation (Cope and Waller 1995). However, while they are normally long-lived animals, a high proportion of these mussels relocated or brought into captive propagation settings die within the first year of translocation (Cope et al. 2003).

Health evaluation of freshwater mussels has historically been limited to behavioral changes and mortality rates. Recently, researchers have begun to develop a systematic approach to the evaluation and monitoring of health and stress in captive freshwater mussels (Wolfe et al. 2008; Burkhard et al. 2009). This diagnostic capability allows better health care, improved understanding of the health concerns of mussels in captivity and in the wild, improved assessment the health of an aquatic habitat through the health of its inhabitants, and improvements in our efforts to conserve these imperiled animals, whose existence is crucial to the health of our freshwater habitats.

## Nutritional Challenges

Populations managed *ex situ* on large landscapes are exposed not only to parasites and infectious diseases to which they are naïve, but also to novel forages. In some cases, a particular species may have an unexpected reaction to a plant or toxin—for example, Eld's deer (*Cervus eldi thamin*) exposed to endophyte-infested tall fescue grass, *Festuca arudinacea* (Wolfe et al. 1998). Tall fescue is a hardy, high-yield, cool-season perennial grass and is the most cultivated grass fed to beef cattle in the United States (Alderson and Sharp 1993). The hardiness of fescue grass is further improved by a symbiotic relationship with the endophyte fungus *Neotyphodium* (*Acremonium*) *coenophialum*, which produces a toxic ergot alkaloid. In cattle, while fescue toxicosis most commonly presents as rough hair coat, heat stress, suppressed appetite, poor growth, or reduced calving rates, some animals experience tail tip or hoof sloughing and fat necrosis due to peripheral vasoconstriction. In Eld's deer, abdominal fat necrosis is the most common manifestation and can be so severe as to cause ureteral blockage, uremic crisis, and death. Interestingly, this phenomenon has been noted only in female Eld's deer, presumably due to seasonal rut-related weight loss in males.

Many decades of feeding wild animals in captivity have provided us with a tremendous database of health problems associated with incorrect assumptions about, and deviations from, native diets. As recently as 2005, a review of health problems in captive giraffes (*Giraffa camelopardalis*) has linked the feeding of common ruminant diets to problems such as rumen acidosis, poor body condition, phytobezoars, urolithiasis, hoof problems, and peracute mortality syndrome (Schmidt 2005). In the wild, the giraffe diet comprises primarily *Acacia* and *Combretum* leaves (Pellew 1984), while in captivity this species has typically been offered a diet of hay supplemented with 16% fiber pellets. This low-fiber, high-carbohydrate diet, when fed to giraffe and other browsing species, has been shown to be an inappropriate replacement for browse, resulting in altered volatile fatty acid production in the rumen and ultimately in a host of metabolic problems. Institutions housing browsing ruminants are therefore challenged to provide browse, or a suitable replacement, as a significant proportion of the diet in temperate areas where browse is unavailable for much of the year.

Even when animals live in their native habitat on native food sources, confinement can lead to health problems and population changes when forage is limited and migration is prevented. Rothschild's giraffe (*G. c. rothschildi*) are limited to five viable populations in Kenya and Uganda. In Lake Nakuru National Park, a 50% population decline of Rothschild's giraffe between 1995 (127) and 2002 (62) has been attributed to a dietary change caused by events following the drought resulting from the 1993–95 El Niño Southern Oscillation. Brenneman et al. (2009) suggested that the drought reduced the availability of browse and limited carrying capacity of the park. Due to the limited availability of forage, giraffes were restricted to smaller areas of acacia woodlands and were forced to overgraze the acacia, eating more bark and therefore consuming higher levels of tannin than would normally be tolerated. High levels of concentrated tannins in their diet led to physiologic compromise, particularly in young giraffe, which underwent increased predation by lions in 2001 and 2002. In this case, isolation of a population and disruption of potential migration routes, in combination with climatic events, led to a rapid population decline for what appear to be dietary reasons.

Many species managed *ex situ* are exposed to forages and other feeds in a quantity or cycle that is unnatural. Persian onagers (*Equus hemionus onager*), in their native semi-desert habitat in Iran, forage on relatively poor-quality grasses during the warm season and browse on trees and shrubs when grass is unavailable. Managed on pastures in captivity, this species often encounters year-round lush grasses and/or hay supplemented with pelleted concentrates. In the

absence of a "lean" season, these animals have a tendency to develop overabundant fat stores. Interestingly, postmortem incidental findings in obese Persian onagers at The Wilds often include excess liver iron stores as well as indications of liver function compromise (B. Wolfe and E. Blumer unpublished data 2010), which may be associated with obesity. In humans, iron overload has been found to be associated with obesity and insulin resistance (Moirand et al. 1997; Ferranini 2000; Fargion et al. 2005), and in domestic horses, obesity and insulin resistance have been linked to systemic inflammation (Vick et al. 2006; Adams et al. 2009), which can lead to changes in iron metabolism (Borges et al. 2007). The black rhinoceros (*Diceros bicornis*) has also been known to develop iron storage disease in captivity (Smith et al. 1995; Paglia et al. 2001; Dierenfeld et al. 2005; Dennis et al. 2007). The association of obesity and iron overload is currently being investigated in captive and wild black rhinoceroses (P.M. Dennis and M.M. Vick personal communication 2010).

## IMPROVING *SORTA SITU* MANAGEMENT FOR SPECIES CONSERVATION

### Creating Buffer Zones Around Protected Habitats

Clearly, habitat fragmentation and destruction are having many serious effects on threatened species. Using wildlife management, veterinary care, training, and education, conservationists are working toward mitigating the impacts of fragmentation on species whose survival will necessarily be within small, often isolated, habitat patches. A key example of this is the Atlantic Forest of Brazil, the most endangered rainforest on the planet, with only 2% of its original extent remaining. Within these forest fragments are some of world's most endangered wildlife and plant species, including the black lion tamarin (*Leontopithecus chrysopygus*). A buffer zone is being developed around the protected area in conjunction with an effort to examine the health of individuals, the risk of disease transmission among fragments, and the viability of black lion tamarin metapopulations inhabiting this rainforest. This buffer zone is made up of extremely small fragments of forest, each one too small to sustain a viable population of tamarins. However, this *sorta situ* population has to date avoided extinction due to meticulous management. Ongoing work in Brazil involves the augmentation of the wild population with captive-bred tamarins and research studies aimed at protecting the remaining forest (Valladares-Padua et al. 2002).

Carnivores and other species that range widely have a unique challenge in protected areas. In these reserves, and particularly those with high perimeter-to-area ratios, the majority of large carnivore mortalities are due to human interaction when the animals roam outside the reserve, creating a population sink for these species. For such populations, reserves of greater size, or buffer zones, are preferable for their long-term conservation (Woodroffe and Ginsberg 1998; Baeza and Estades 2010).

## Restoring Habitat Quality in Protected Areas

In much of the developing world, human population growth, habitat encroachment, and competition with domestic livestock are having an increasing effect on wildlife populations and habitats. The interactions of these natural and "man-made" processes are often unpredictable. However, experiments in habitat improvement for targeted species and populations are showing how agricultural and forestry practices can be used to restore habitat quality. In Indonesia's Ujung Kulon National Park, the International Rhino Foundation and its partners are working to expand the habitat for the critically endangered Javan rhino (*Rhinoceros sondaicus*). Their plan is to expand rhino habitat in Indonesia by creating a 4,000-hectare research and conservation area adjacent to the currently restricted rhino habitat in Ujung Kulon. This effort will manage the new area to increase/improve rhino "necessities"—water, wallows, saltlicks, and appropriate edible vegetation—to ensure that this area can support an expanded rhino population. This will include replanting natural forest vegetation with rhino food plants in some areas, and carefully implementing controlled slash-and-burn patch management in designated and closed forest areas to promote regeneration of edible plants for rhinos. The project also includes aggressive removal of *Arenga* palm, an invasive species that competes with many key rhino food plants (S. Ellis personal communication 2011).

## Using Information Gained from Captive Management for Wild Species Conservation

With the ever-diminishing difference between the ecology of wild and captive populations comes opportunity. As conservation medicine practitioners we have always been charged with using what we learn about wild populations to improve the health and welfare of captive animals. Today, this evolving multi-perspective approach to animal management works in both directions on the spectrum of *sorta situ*. What we learn about captive populations in various management settings can be used to better understand and improve the preservation of wild animals. For instance, a recent study of anesthetic methods in Sichuan takin (*Budorcas taxicolor*) was conducted at The Wilds to prepare for anesthesia of wild takin on a reserve in Sichuan, China, where Chinese and American colleagues are currently studying the behavioral ecology of this little-studied species. In captivity, takin are anesthetized using medetomidine and butorphanol, providing a relatively safe general anesthesia, albeit with slow induction. The large habitats where takin reside at The Wilds allow comparison of anesthetic regimens emphasizing the rate of induction by measuring time and distance traveled from induction to immobilization. This study resulted in an anesthetic regimen that transferred well to the wild takin populations in mountainous Sichuan province, where slow induction and failure to find an animal following the administration of anesthetic could result in mortality.

## CONCLUSIONS

The scientific field of conservation medicine is constantly evolving, and with it the methods and assumptions by which we practice. As the discrepancy between "wild" and "captive" animals becomes indistinct, and the management of *in situ* and *ex situ* populations becomes more commonly a spectrum of "*sorta situ*"—neither one nor the other—we must become better able to predict the response of animal populations to new environments in order to protect and conserve them. Our approach to healthcare in populations on the *sorta situ* spectrum has ranged from individual-oriented intensive management in zoological institutions to simple observation and monitoring of wild populations. Today, we must be more strategic and we must have more of an impact. Our efforts to protect populations through translocation or restriction, regardless of the amount of space available, should therefore include an improved ability to predict the effects of geographic barriers and new habitat inputs on animal populations, both under current circumstances and in the case of changing climatic conditions. Conservation efforts must use transdisciplinary approaches to consider entire ecosystems in their management plans, rather than simply single-species populations. Finally, management of animals on large landscapes in our current global reality must take into account the people, agriculture, domestic animals, and disease vectors incumbent on the landscape. To accomplish this monumental task requires more than a hybrid of wild and captive management approaches: it requires a singularity of purpose encompassing a broad spectrum of scientific specialties and social paradigms, which will challenge our breadth and global cooperation into the next millennium.

# REFERENCES

Adams, A.A., M.P. Katepalli, K. Kohler, S.E. Reedy, J.P. Stilz, M.M. Vick, B.P. Fitzgerald, L.M. Lawrence, and D.W. Horohov. 2009. Effect of body condition, body weight and adiposity on inflammatory cytokine responses in old horses. Vet Immunol Immunopathol 127:286–294.

Aguilar, R.F., D. Cohan, M. Hunnicut, J.P. Carroll, and M. Garner. 2008. Teetering on the brink: a massive mortality episode in the endangered masked bobwhite quail (Colinus virginianus ridgwayii). Proceedings of the American Association of Zoo Veterinarians, pp. 178–179. Los Angeles, California.

Aguirre, A.A. 2009. Biodiversity and human health. EcoHealth 6:153–156.

Aguirre, A.A., and M.C. Pearl. 2004. New technology and sorta situ: Conservation medicine linking captive and wildlife populations. Proceedings of the AAZV/ AAWV/WDA Joint Annual Conference, pp. 453–455. San Diego, California.

Alderson, J., and W.C. Sharp. 1993. Grass varieties in the United States. Agricultural Handbook No. 170. United States Department of Agriculture, Washington, D.C.

Allen, R.B., and W.G. Lee. 2006. Biological invasions in New Zealand. Ecological Studies Series 186, Springer Verlag, Berlin, Germany.

Arneberg, P. 2001. An ecological law and its macroecological consequences as revealed by studies of relationships between host densities and parasite prevalence. Ecography 24:352–358.

Baeza, A., and Estades, C.F. 2010. Effect of the landscape context on the density and persistence of a predator population in a protected area subject to environmental variability. Biol Conserv 143(1):94–101.

Bode, M., and B. Wintle. 2009. How to build an efficient conservation fence. Cons Biol 24:182–188.

Bolger, D.T., W.D. Newmark, T.A. Morrison, and D.F. Doak. 2008. The need for integrative approaches to understand and conserve migratory ungulates. Ecol Lett 11:63–77.

Borges, A.S., T.J. Divers, T. Stokol, and O.H. Mohammed. 2007. Serum iron and plasma fibrinogen concentrations as indicators of systemic inflammatory diseases in horses. J Vet Internal Med 21:489–494.

Brenneman, R.A., R.K. Bagine, D.M. Brown, R. Ndetei, and E.E. Louis, Jr. 2009. Implications of closed ecosystem conservation management: the decline of Rothschild's giraffe (Giraffa camelopardalis rothschildi) in Lake Nakuru National Park, Kenya. Afr J Ecol 47:711–719.

Brooks, T.M., S.L. Pimm, and J.O. Oyugi. 1999. Time lag between deforestation and bird extinction in tropical forest fragments. Conserv Biol 13:1140–1150.

Burkhard, M.J., S. Leavell, R.B. Weiss, K. Kuehnl, H. Valentine, G.T. Watters, and B.A. Wolfe. 2009. Analysis and cytologic characterization of hemocytes of freshwater mussels (Quadrula sp.). Vet Clin Path 38:426–436.

Catt, D.C., D. Dugan, R.E. Green, R. Moncrieff, R. Moss, N. Picozzi, R.W. Summers, and G.A. Tyler. 1994. Collisions against fences by woodland grouse in Scotland. Forestry 67:105–118.

Chape, S., J. Harrison, M. Spaulding, and I. Lysenko. 2005. Measuring the extent and effectiveness of protected areas as an indicator for meeting global diversity targets. Phil Trans Royal Soc B 360:443–455.

Chase, M.J., and C.R. Griffin. 2009. Elephants caught in the middle: impacts of war, fences and people on elephant distribution and abundance in the Caprivi Strip, Namibia. Afr J Ecol 47:223–233.

Chomel, B.B., A. Belotto, and F.-X. Meslin. 2005. Wildlife, exotic pets and emerging zoonosis. Emerg Infect Dis 13:6–11.

Clout, M.N., and J.L. Craig. 1995. The conservation of critically endangered flightless birds in New Zealand. Ibis 137:S181–S190.

Clubb, R., and G. Mason. 2003. Captivity effects on wide-ranging carnivores. Nature 435:473–474.

Cohn, J.P. 2007. The environmental impacts of a border fence. Bioscience 57:doi:10.1641/B570116.

Conde, D.A., N. Flesness, F. Colchero, O.R. Jones, and A. Scheuerlein. 2011. An emerging role of zoos to conserve biodiversity. Science 331:1390–1391.

Cope, W.G., and D.L. Waller. 1995. Evaluation of freshwater mussel relocation as a conservation and management strategy. Reg Rivers-Res Manag 11:147–155.

Cope, W.G., M.C. Hove, D.L. Waller, D.J. Hornback, M.R. Bartsch, L.A. Cunningham, H.L. Dunn, and A.R. Kapuscinski. 2003. Evaluation of relocation of unionid mussels to in situ refugia. J Molluscan Stud 69:27–34.

Cranfield, M., L. Gaffikin, J. Sleeman, and M. Rooney. 2002. The mountain gorilla and conservation medicine. *In* Aguirre, A.A., R.S. Ostfeld, G.M. Tabor, C. House, and M.C. Pearl, eds. Conservation medicine: ecological health in practice, pp. 282–296. Oxford University Press, New York.

Creel, S., and D. Christianson. 2008. Relationships between direct predation and risk effects. Trends Ecol Evol 23:194–201.

Creel, S., and D. Christianson. 2009. Wolf presence and increased willow consumption by Yellowstone elk: implications for trophic cascades. Ecology 90:2454–2466.

Cross, P.C., E.K. Cole, A.P. Dobson, W.H. Edwards, K.L. Hamlin, G. Luikart, A.D. Middleton, B.M. Scurlock, and P.J. White. 2010. Probable causes of increasing brucellosis in free-ranging elk of the Greater Yellowstone Ecosystem. Ecol Appl 20:278–288.

Cumming, D., H. Biggs, M. Kock, N. Shongwe, S. Osofsky and members of the AHEAD-Great Limpopo TFCA Working Group. 2007. The AHEAD (Animal Health for Environment and Development)-Great Limpopo Transfrontier Conservation Area (GLTFCA) Programme: Key questions and conceptual framework revisited. Wildlife Conservation Society, The Bronx, New York. http://wcs-ahead.org/workinggrps_limpopo.html

Daszak, P., and A.A. Cunningham. 2002. Emerging infectious diseases: a key role for conservation medicine. *In* A.A. Aguirre, R. Ostfeld, G.M. Tabor, C. House, and M.C. Pearl, eds. Conservation medicine: ecological health in practice, pp. 40–61. Oxford University Press, New York.

Daszak, P., A.A. Cunningham, and A.D. Hyatt. 2000. Emerging infectious diseases of wildlife –threats to biodiversity and human health. Science 287:443–449.

Demarais, S., J.T. Baccus, and M.S. Traweek, Jr. 1998. Nonindigenous ungulates in Texas: long-term population trends and possible competitive

mechanisms. Trans North Am Wildl Nat Res Conf 63:49–55.

Dennis, P.M., J.A. Funk, P.J. Rajala-Schultz, E.S Blumer, R.E. Miller, T.E. Wittum, and W.J.A. Saville. 2007. A review of some of the health issues of captive black rhinoceroses (*Diceros bicornis*). J Zoo Wildl Med 38:509–517.

Dierenfeld, E.S., S. Atkinson, A.M. Craig, K.C. Walker, W.J. Streich, and M. Clauss. 2005. Mineral concentrations in serum/plasma and liver tissue of captive and free-ranging rhinoceros species. Zoo Biol 24:51–72.

Dobson, A. 2009. Food web structure and ecosystem services: insights from the Serengeti. Phil Trans Royal Soc B 364:1665–1682.

Fargion, S., P. Dongiovanni, A. Guzzo, S. Colombo, L. Valenti, and A.L. Fracazani. 2005. Iron and insulin resistance. Aliment Pharm Therapeut 22:61–63.

Ferranini, E. 2000. Insulin resistance, iron and the liver. Lancet 355:2181–2182.

Ferraz, G., G.J. Russell, P.C. Stouffer, R.O. Bierregaard, Jr., S.L. Pimm, and T.E. Lovejoy. 2003. Rates of species loss from Amazonian forest fragments. Proc Natl Acad Sci 100:14069–14074.

Flesch, A.D., C.W. Epps, J.W. Cain III, M. Clark, P.R. Krausman, and J.R. Morgart. 2010. Potential effects of the United States-Mexico border fence on wildlife. Conserv Biol 24:171–181.

Fortin, D., H.L. Beyer, M.S. Boyce, D.W. Smith, T. Duchesne, and J.S. Mao. 2005. Wolves influence elk movements: behavior shapes a trophic cascade in Yellowstone National Park. Ecology 86:1320–1330.

Heithaus, M.R., A.J. Wirsing, D. Burkholder, J. Thompson, and L.M. Dill. 2009. Towards a predictive framework for predator risk effects: the interaction of landscape features and prey escape tactics. J Anim Ecol 78:556–562.

Hoole, A., and F. Berkes. 2010. Breaking down fences: recoupling social–ecological systems for biodiversity conservation in Namibia. Geoforum 41:304–317.

Hudson, R.J., K.R. Drew, and L.M. Baskin. 1989. Wildlife production systems: economic utilisation of wild ungulates, Cambridge University Press, United Kingdom.

Huerta-Patricio, E., K.D. Cameron, G.N. Cameron, and R.A. Medellin. 2005. Conservation implications of exotic game ranching in the Texas Hill Country. *In* V. Sanchez-Cordero and R.A. Medellín. eds. Contribuciones mastozoologicas en homenaje a Bernardo Villa. Ch. 21, pp. 237–252. Instituto de Biología and Instituto de Ecología, UNAM, CONABIO, México.

Innes, J., D. Kelly, J.M. Overton, and C. Gillies. 2010. Predation and other factors currently limiting New Zealand forest birds. New Zeal J Ecol 34:86–114.

IUCN. 1994. Guidelines for protected area management categories. International Union for the Conservation of Nature. Gland, Switzerland.

Kalema-Zikusoka, G.K., J.M. Rothman, and M.T. Fox. 2005. Intestinal parasites and bacteria of mountain gorillas (*Gorilla beringei beringei*) in Bwindi Impenetrable National Park, Uganda. Primates 46:59–63.

Kellert, S.R., and E.O. Wilson. 1993. The biophilia hypothesis. Cambridge Island Press/Shearwater, Washington, D.C.

Lees, C.M., and J. Wilcken. 2009. Sustaining the ark: the challenges faced by zoos in maintaining viable populations. Int Zoo Yearbook 43:6–18.

Levins, R. 1969. Some demographic and genetic consequences of environmental heterogeneity for biological control. Bull Entomol Soc Am 15:237–240.

Li, C.W., Z.G. Jiang, S.H. Tang, and Y. Zeng. 2007. Influence of enclosure size and animal density on fecal cortisol concentration and aggression in Pere David's deer stags. Gen Comp Endocrinol 151:202–209.

Jones, K., and B. Lacy. 2009. Whooping crane master plan. International Whooping Crane Recovery Team, Baraboo, Wisconsin.

Joppa, L.N., S.R. Loarie, and S.L. Pimm. 2008. On the protection of "protected areas." Proc Natl Acad Sci USA 105:6673–6678.

Matchett, M.R., D.E. Biggins , V. Carlson, B. Powell, and T. Rocke. 2010. Enzootic plague reduces black-footed ferret (*Mustela nigripes*) survival in Montana. Vector-borne Zoonot 10:27–35.

Metrione, L., and J.C. Harder. 2009. The effects of environment on the social development and reproductive success of adolescent and adult female southern white rhinoceros. Sixth Rhino Keeper Workshop, Rhino Keeper Association, Bush Gardens, Tampa, Florida.

Moirand, R., A.M. Mortaji, O. Loréal, F. Paillard, P. Brissot, and Y. Deugnier. 1997. A new syndrome of liver iron overload with normal transferrin saturation. Lancet 349:95–97.

Mungall, E., and W.J. Sheffield. 1994. Exotics on the range. Texas A&M University Press, College Station, Texas.

Nagy, D.W. 2005. Parelaphostrongylosis and other parasitic diseases of the ruminant nervous system. Vet Clin North Am 20:393–412.

Newmark, W.D. 1996. Insularization of Tanzanian parks and the local extinction of large mammals. Conserv Biol 10:1549–1556.

Newmark, W.D. 2008. Isolation of African protected areas. Frontiers Ecol Environ 6:321–328.

Osofsky, S.A., D.H. Cumming, and M.D. Kock. 2008. Transboundary management of natural resources and the importance of a "one health" approach: perspectives on southern Africa. *In* E. Fearn and K.H. Redford, eds. State of the wild 2008–2009: a global portrait of wildlife, wildlands, and oceans. Island Press, Washington, D.C. http://www.wcs-ahead.org/print.html

Pacheco, M.A., A.A. Escalante, M.M. Garner, G.A. Bradley, and R.F. Aguilar. 2011. Haemosporidian infection in captive masked bobwhite quail (*Colinus virginianus ridgwayi*), an endangered subspecies of the northern bobwhite quail. Vet Parasitol 182:113–120.

Pacheco, M.A., A. Escalante, K. Orr, M. Garner, and R.F. Aguilar. 2008. Acute mortality and malaria in captive masked bobwhite quail. Proc Wildl Soc New Zeal Vet Assoc Conf., p. 11. Palmerston North, New Zealand.

Paglia, D.E., D.E. Kenny, E.S. Dierenfeld, and I.H. Tsu. 2001. Role of excessive maternal iron in the pathogenesis of congenital leukoencephalomalacia in captive black rhinoceroses (*Diceros bicornis*). Am J Vet Res 62:343–349.

Pellew, R.A. 1984. The feeding ecology of a selective browser, the giraffe (*Giraffa camelopardalis tippelskirchi*). J Zool 202:57–81.

Phillips, A. 2003. Turning ideas on their head: the new paradigm for protected areas. Paper written for Fifth World Parks Congress. www.uvm.edu/conservationlectures/vermont.pdf. Last accessed April 18, 2011.

Sayre, N.F., and R.L. Knight. 2009. Potential effects of United-States-Mexico border hardening on ecological and human communities in the Malpai borderlands. Conserv Biol 24:345–348.

Schmidt, D. (ed.). 2005. Giraffe Nutrition Workshop Proceedings, PMI Nutrition International, Lincoln Park Zoo, Chicago, Illinois.

Schmitt, S.M., S.D. Fitzgerald, T.M. Cooley, C.S. Bruning-Fann, L. Sullivan, D. Berry, T. Carlson, R.B. Minnis, J.B. Payeur, and J. Sikarskie. 1997. Bovine tuberculosis in free-ranging white-tailed deer from Michigan. J Wildl Dis 33:749–758.

Shaffer, M.L. 1981. Minimum population sizes for species conservation. BioScience 31:131–134.

Sih, A., D.I. Bolnick, B. Luttbeg, J.L. Orrock, S.D. Peacor, L.M. Pintor, E. Preisser, J.S. Rehage, and J.R. Vonesh. 2010. Predator-prey naivete, antipredator behavior, and the ecology of predator invasions. Oikos 119:610–621.

Smith, J.E., P.S. Chavey, and R.E. Miller. 1995. Iron metabolism in captive black (*Diceros bicornis*) and white (*Ceratotherium simum*) rhinoceroses. J Zoo Wildl Med 26:525–531.

Soulé, M., M. Gilpin, W. Conwa, and T. Foose. 1986. The millennium ark: how long a voyage, how many staterooms, how many passengers? Zoo Biol 5:101–113.

Swaisgood, R.R., D.M. Dickman, and A.M. White. 2006. A captive population in crisis: Testing hypotheses for reproductive failure in captive-born southern white rhinoceros females. Biol Conserv 129:468–476.

Teer, J.G., L.A. Renecker, and R.J. Hudson. 1993. Overview of wildlife farming and ranching in North America. Trans North Am Wildl Nat Res Conf 58:448–459.

Valladares-Padua, C., S.M. Padua, and L. Cullen Jr. 2002. Within and surrounding the Morro do Diablo State Park: biological value, conflicts, mitigation and sustainable development alternatives. Environ Sci Policy 5:69–78.

Vicente, J., U. Höfle, I. García Fernández-De-Mera, and C. Gortazar. 2007. The importance of parasite life history and host density in predicting the impact of infections in red deer. Oecologia 152:155–164.

Vick, M.M., B.A. Murphy, D.R. Session, S.E. Reedy, E.L. Kennedy, D.W. Horohov, R.F. Cook, and B.P. Fitzgerald. 2006. Effects of systemic inflammation on insulin sensitivity in horses and inflammatory cytokine expression in adipose tissue. Am J Vet Res 69:130–139.

Western, D. 2002. In the dust of Kilmanjaro. Island Press, Washington.

Wildlife Society. 2010. Final position statement: the impact of border security measures on wildlife, 2 pp. TWS, Washington, D.C. http://joomla.wildlife.org/index.php?option=com_content&task=view&id=117&Itemid=299. Last accessed April 18, 2011.

Williams, E.S., and M.W. Miller. 2002. Chronic wasting disease in deer and elk in North America. Rev Sci Tech Off Int Epiz 21:305–316.

Williams, J.D., M.L. Warren, K.S. Cummings, J.L. Harris, and R.J. Neves. 1993. Conservation status of freshwater mussels of the United States and Canada. Fisheries 18:6–22.

Wilson, P.R. 2002. Advances in health and welfare of farmed deer in New Zealand. New Zeal Vet J 50:S105–S109.

Wilson, K.J. 2004. Flight of the Huia: ecology and conservation of New Zealand's frogs, reptiles, birds and mammals. Canterbury University Press, Christchurch, New Zealand.

Wisely, S.M., R.M. Santymire, T.M. Livieri, S. Mueting, and J.G. Howard. 2008. Genotypic and phenotypic consequences of reintroduction history in the black-footed ferret (*Mustela nigripes*). Conserv Gen 9:389–399.

Wolfe, B.A., M. Bush, S.L. Monfort, S.L. Mumford, A. Pessier, and R.J. Montali. 1998. Abdominal lipomatosis attributed to tall fescue toxicosis in deer. J Am Vet Med Assoc 213:1783–1785.

Wolfe, B.A., M.J. Burkhard, S. Leavell, R.B. Weiss, K. Kuehnl, H. Valentine, G.T. Watters, and B. Bergstrom. 2008. The clam exam: clinical pathology of freshwater mussels, p. 186. Proc Am Assoc Zoo Vet, Los Angeles, California.

Woodroffe, R., and J.R. Ginsberg. 1998. Edge effects and the extinction of populations inside protected areas. Science 280:2126–2128.

# 41

## MODELING POPULATION VIABILITY AND EXTINCTION RISK IN THE PRESENCE OF PARASITISM

Patrick Foley and Janet E. Foley

"In the grand pattern of evolution. nothing is more dramatic than the prevalence of extinction" (Simpson 1953).

"So, naturalists observe, a flea has smaller fleas that on him prey; and these have smaller still to bite 'em; and so proceed ad infinitum" (Jonathan Swift 1733 *On Poetry: A Rhapsody*).

Although extinction is a natural process, since the arrival of modern *Homo sapiens*, extinction rates have risen sharply above geological background rates (Martin and Klein 1984). In the Cenozoic a typical mammal species persisted for about 1 million years (Simpson 1953). Recent historical extinction rates suggest that mammal and bird species lifespans have been reduced 100- to 1,000-fold (May et al. 1995).

Parasites in the broad sense, including pathogenic viruses, bacteria, fungi, protozoa, flatworms, roundworms, and arthropods, are common and by definition diminish host fitness. Human beings enjoy at least 434 species of eukaryote parasites, including 83 protozoa, 130 flukes, 54 tapeworms, 114 nematodes, and 49 arthropods (Ashford and Crewe 2003). In one review, parasites cited as *obviously* threatening IUCN Red-Listed mammals included 13 viruses, eight bacteria, five helminths, three arthropods, one protozoan, and one fungus (Pedersen et al. 2007). Parasites can cause local to global extinction of their

hosts. Plague caused by *Yersinia pestis* can eradicate local populations of prairie dogs (*Cynomys* spp.) in Colorado (Stapp et al. 2004), and fungal pathogens of sudden oak death, chestnut blight, and white nose syndrome in bats cause local and potentially global extinctions. Lowland Hawaiian honeycreepers (*Vestiaria coccinea*) have been essentially extirpated by bird malaria (van Riper et al. 1986). Because human activity (i.e., host habitat loss and deterioration, pathogen transport, vector introduction in the case of the honeycreepers) has contributed to the extreme population consequences of these pathogens, it is our responsibility to attempt to reduce the extinction risks our fellow travelers have been obliged to take.

Effects of parasitism are exacerbated in small host populations. Caughley (1994) describes two paradigms that account for the endangerment of species: the declining species paradigm and the small population paradigm (Fig. 41.1). A declining species has a mean growth rate that is less than zero ($r_d < 0$): a species with an otherwise positive $r_d$ could decline if parasites sufficiently damage host fitness. However, the more common means by which parasites increase the probability of host extinction is to operate within populations that are very small, even if they have $r_d > 0$. This small population paradigm is our main concern in this chapter.

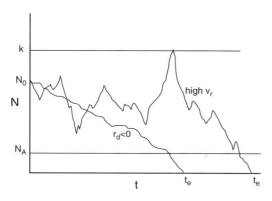

**Figure 41.1:**
Two populations go extinct, illustrating Caughley's two paradigms of conservation biology. The low $r_d$ population dies off almost deterministically. The high $v_r$ population does a random walk between the carrying capacity K and the Allee threshold $N_A$. Even though its $r_d > 0$ for N between $N_A$ and K, environmental and demographic stochasticity lead to the population's demise. The deterministically doomed population has a carrying capacity of 0 and no meaningful $N_A$; it is a sink (Pulliam 1988), with $r_d < 0$. The expected time to extinction of a population $T_e$ is the mean of all the particular realizations $t_e$. For $r_d < 0$, $T_e \sim \ln(N_0)/(-r_d)$. For $r_d = 0$, $T_e \sim \ln^2(K/N_A)/v_{re}$.

If small local populations become extinct, encompassing metapopulations have fewer occupied patches and are more likely to go extinct. Population viability analyses (PVAs) attempt to predict such extinctions. However, PVAs often fail in some way, most acutely for epidemiologists who are trying to predict extinction, because most omit the effects of other organisms that interact with the target species, such as predators, disease reservoirs, or parasites. These ecological interactions can lead to oscillatory dynamics in the target species, increased variance in population growth rates, and increased small population vulnerability, due to environmental stochasticity.

Useful, robust models of parasitized host populations, whose parameters can be estimated with some accuracy, even with scanty data, help inform efforts to understand and manage such populations. In this chapter, we present robust local and metapopulation models that effectively incorporate parasite effects on vulnerable host populations (Foley 1994, 1997, 2000; Hanski 1996; Hanski et al. 1996, Hanski 1999). The local population parameters include the host carrying capacity K, the mean host population growth rate $r_d$, the variance in this growth rate due to environmental stochasticity $v_r$, the inter-annual correlation in growth rate ρ, and the Allee threshold $N_A$ at the lower end of population viability. We show that parasites can realistically contribute to species endangerment even if their hosts decline to the point where the parasites themselves could perish. And we show how a metapopulation model can cope with oscillatory host–parasite population dynamics that defy standard population viability analysis.

## DO ENDANGERED SPECIES SHED PARASITES?

Patchily distributed, increasingly rare hosts are lousy hosts. In such populations, parasite transmission is expected to decline, parasite endemicity to fade out, and the parasite to disappear before its host. For example, parasite richness among anthropoid apes and carnivores increased with geographic range and population density (Nunn et al. 2003; Lindenfors et al. 2007). But the IUCN Red List of Threatened Species lists 36 threatened and 81 non-threatened primate species, and the Global Mammal Parasite Database revealed that threatened primates carry about half as many parasite species as their less endangered cousins, regardless of whether parasites were helminths, protozoa, or viruses (Altizer et al. 2007).

The problem of small host population sizes maintaining parasites can be evaluated with a Kermack-McKendrick microparasite model with density-dependent transmission and demography (Kermack and McKendrick 1927; Anderson and May 1991). A population of hosts of size N has susceptible S, infectious I, or resistant R individuals. A SIR model with some demography is:

$$N = S + I + R \tag{1}$$

$$\frac{dS}{dt} = bN - \beta SI - d_s S$$
$$\frac{dI}{dt} = \beta SI - \gamma I - d_I I \tag{2}$$
$$\frac{dR}{dt} = \gamma I - d_R R$$

where b is the population birth rate, $d_S$ is the death rate among uninfected hosts, $\beta$ is the parasite transmission rate, and the recovery rate $\gamma$ is the reciprocal of the mean time to recovery. Table 41.1 lays out the main variables and parameters of SIR-type models.

An epidemic occurs in a naïve host population when $R_0$, the basic reproductive number, satisfies

$$R_0 = N\frac{\beta}{\gamma+d_I} > 1. \qquad (3)$$

When there is herd immunity, an epidemic is harder to start since the initial number of susceptibles S must satisfy the same equation, replacing N with S. The "effective reproductive number" following (Cintron-Arias et al. 2009), $R_e$, is

$$R_0\frac{S}{N} = S\frac{\beta}{\gamma+d_I}. \qquad (4)$$

The conditions for endemicity can be derived by setting the system (2) to equilibrium and requiring I > 1. We assume that overall birth compensates for overall death and that all the rates are constant. At equilibrium we get

$$S^* = \frac{\gamma+d_I}{\beta}$$

$$R^* = \frac{\gamma}{d}I \qquad (5)$$

$$I^* = \frac{N - \frac{\gamma+d_I}{\beta}}{1+\frac{\gamma}{d}}$$

To keep an endemic alive, $I^*$ must be at least 1. Demographic (or other forms of) stochasticity may require maintenance of a higher value, denoted $I_e$. Thus we get this threshold for endemicity:

$$N_e = \frac{\gamma+d_I}{\beta} + (1+\frac{\gamma}{d})I_e \qquad (6)$$

In many diseases, $\gamma$ is only a few times larger than d, so that the endemic threshold for N is not much higher than the epidemic threshold N. Endemic fadeout is often due to stochastic variation of I(t) around the equilibrium value $I^*$.

**Table 41.1  SIR Model Variables and Parameters**

| | |
|---|---|
| N | host population size N = S + I + R |
| S | number of susceptibles in the host population |
| E | number of exposed, latent individuals in the host population |
| I | number of infectives in the host population |
| R | number of recovered individuals in the host population |
| $\beta$ | the transmission rate |
| $\gamma$ | rate of recovery |
| b | birth rate |
| $b_m$ | maximum birth rate at low population density |
| $K_b$ | birth rate carrying capacity |
| K | equilibrium population size due to birth and death $K_b(1 - ((d + d)/b_m)$ |
| $d_S$ | normal mortality rate for susceptibles |
| $d_I$ | normal mortality rate for infecteds $d_I = d_S + \delta$. |
| $d_R$ | normal mortality rate for recovereds. Usually we assume $d_R = d_I$. |
| $\delta$ | additional mortality rate among infecteds |
| $R_0$ | basic reproductive number when S = N (well, almost N; say N – 1) |
| $R_e$ | effective reproductive number for S of any size. This depends on S. |
| $L_j$ | parasite load $L_j = \sum \delta_j I_j / N$, summing over all parasites |
| T | period between epidemics |

Based on simulations we show that I(t) is approximately normal around I*; that is

$$I(t) \sim Normal(I^*, \sigma_I^2) \qquad (7)$$

where $\sigma_I^2$ rises quickly with environmental stochasticity (Foley and Foley 2010). These I(t) values are not independent, of course, since they change only slightly each day. The extinction of the disease in the host population will occur when the I(t) drops below 1. The most important quantities for predicting the time to extinction of the disease are then $I^*/\sigma_I$ and the effective time between independent I(t) values. High I* and low stochasticity forestall endemic fadeout. And as seen in Equation 5, high I* depends on high N with this classic model. Small populations are expected to shed microparasites.

However, small host populations do maintain parasites, most often because of frequency-dependent transmission and alternative hosts. Parasites transmitted in frequency-dependent fashion, such as some sexually transmitted and vector-borne diseases, do not depend on high N to persist (McCallum et al. 2001; de Castro and Bolker 2005). A recent evaluation of six years of field vole/cowpox data showed that transmission was slightly more frequency- than density-dependent (Smith et al. 2009). Keeling and Rohani's (2008) recent book assumes frequency-dependent transmission for most of its analyses.

The dynamics of frequency-dependent transmitted disease are

$$\frac{dS}{dt} = bN - \beta SI/N - d_S S$$

$$\frac{dI}{dt} = \beta SI/N - \gamma I - d_I I \qquad (8)$$

$$\frac{dR}{dt} = \gamma I - d_R R$$

The threshold for I to increase occurs when dI/dt = 0, and

$$0 = \frac{dI}{dt} = \beta SI/N - \gamma I - d_I I \qquad (9)$$

$$R_0 = \frac{\beta}{\gamma + d_I} \qquad (10)$$

$R_0$ in this case does not depend on N, so even endangered populations can retain their parasites in a deterministic world.

Perhaps the most important way a rare host species retains a parasite community is by spillover from more common reservoir hosts (de Castro and Bolker 2005). Alternate hosts are strongly implicated in disease problems for several captive populations of mammals. A cross-comparison of the IUCN Red List of Threatened Mammals and the Global Mammal Parasite Database showed that threatening parasites were less host-specific than expected (Pedersen et al. 2007). Useful PVAs will include parasites, alternative hosts, and various modes of parasite transmission.

## POPULATION VIABILITY ANALYSIS

PVA is an applied branch of population biology with an immediate goal of predicting a species or population's expected time to extinction, $T_e$, or its probability of extinction within a certain time frame, for example 100 years. Quantitative PVA is usually limited to the one species of immediate concern: of 407 papers reviewed, Sabo (2007) found only nine with time series data for two interacting species. Treatments of parasite effects in PVA are mainly confined to verbal discussion.

Overviews of the mathematics, statistics, and ideas used in PVA can be found in several excellent books (Caughley and Gunn 1996; Morris et al. 1998; Morris and Doak 2002). Commonly PVA relies on Leslie matrix manipulation in software such as RAMAS (Akçakaya et al. 1997) and VORTEX (Miller and Lacy 2005), which can give clues about life history evolution, past booms and busts, and the sensitivity of $\lambda_1$, the dominant eigenvalue of the population's Leslie matrix, to mortality at different ages (Caswell 1989; Menges 1990; Tuljapurkar 1990; Burgman et al. 1993; Ebert 1999). However, most (88%) in the PVA studies surveyed by Groom and Pascual (1998), required laborious estimation of survivorship and fecundity in order to estimate $\lambda_1$. If $\lambda_1 < 1$, the population declines. Missing from these analyses may be important biologically realistic features such as density dependence, influence of environmental stochasticity, and temporal autocorrelation in survivorship and fecundity.

In a world where data are costly and ecosystems are fragile, transient, and stochastic, we need models that capture essential features. Time series PVAs (Gerber et al. 2005), or count-based PVAs (Morris and Doak 2002), attempt to do so, although the earliest ones avoided a consideration of carrying capacity, focusing instead on the linear model estimation of $r_d$ and $\sigma^2$. Later models explicitly incorporated carrying capacity, demographic and environmental stochasticity, Allee thresholds, and so forth (Lande 1993; Foley 1994, 1997; Lande et al. 2003).

## LOCAL HOST POPULATION MODELS: DISCRETE TIME FORMULATION

A simple general model for local population dynamics keeps track of population size $N(t)$ and a population multiplier $\lambda(t)$ that may vary with time and depend on N.

$$\lambda(N,t) = \frac{N(t+1)}{N(t)} \tag{11}$$

$\lambda(N,t)$ measures population fitness and sometimes goes by the name of "fundamental net reproductive rate" (Begon et al. 1996). This can be confusing because PVA analyses often use the symbol $\lambda$ for the *constant* eigenvalues of a Leslie matrix. Ecologists usually describe population growth in terms of another growth rate

$$r(N,t) = \ln \lambda(N,t) = \ln \frac{N(t+1)}{N(t)} \tag{12}$$

If our initial population size is $N_o$,

$$N_t = N_0 \lambda^t = e^{rt} \tag{13}$$

The parameter r is known as the "intrinsic rate of natural increase" or "intrinsic growth rate," although r, which depends explicitly on the relationship of the population to its environment, could be best called the per capita population growth rate; r is a compound interest rate, and when $\lambda$ is close to 1, r is close to 0.

Although $\lambda$ and r depend on a population's interactions in a community and abiotic world, this can be simplified by distinguishing a deterministic component of r, the mean per capita growth rate $r_d$, and a chance component, $V_r$ also influenced by N and t. Even if $r_d > 0$, the population will fluctuate in a band approximately delineated by its carrying capacity K

and its Allee threshold $N_A$. Above and below this band, $r_d < 0$ by definition. If the fluctuations are strong enough, or the band narrow enough, the population eventually falls below $N_A$, and extinction can be expected. The population's drunken walk is over.

Since Shaffer (1981), conservation biologists have recognized four distinct forms of chance that can increase a population's vulnerability to extinction: demographic stochasticity, environmental stochasticity, catastrophes, and random genetic drift. These have been called the Four Horsemen. Environmental stochasticity includes the random effects of weather, community interactions, and other effects external to the species. Catastrophes are rare, extreme forms of environmental stochasticity. Disease can critically influence most if not all of these forms of chance in host populations.

The seminal work *Island Biogeography* (MacArthur and Wilson 1967) modeled local extinction on demographic stochasticity using birth and death processes. May (1974) coined the terms "demographic stochasticity" and "environmental stochasticity" in another classic while investigating the stability of communities. Catastrophes have been most entertainingly treated in (Mangel and Tier 1993, 1994). A unified approach to three forms of stochasticity was proposed by Foley (1997). The probability density of $r(N,t)$ takes the form

$$r(N,t) \sim$$
$$\left\{ \begin{array}{ll} \text{Normal}(r_d(N), v_r + v_1/N) & \text{with probability } (1-q) \\ \ln(1-\Delta) & \text{with probability } q \end{array} \right\} \tag{14}$$

where q gives the probability of a catastrophic year, $\Delta$ gives the proportion of the population destroyed in a catastrophe (Hanson and Tuckwell 1981; Mangel and Tier 1993), and $v_1$ represents the variance in r due to demographic stochasticity in a single female.

A population goes quasi-extinct (doomed to quick extinction) when N goes below $N_A$. If q (or $\Delta$) and $v_1$ are small enough, then the environmental stochasticity model is an adequate approximation of the population dynamics. Analytic results for extinction times and population sizes have been developed for this model (Lande 1993; Foley 1994; Middleton et al. 1995; Hanski et al. 1996; Foley 1997). We make the following simplifying assumptions.

$r_d(N) = r_d \to$ constant on the interval $[N_A, K]$

$V_r(N) = v_r \to$ constant on the interval $[N_A, K]$ (15)

$v_1 = 0 \to$ negligible demographic stochasticity

$q = 0 \to$ no catastrophes.

Local extinction times depend on the deterministic per capita growth rate $r_d(N)$, which we assume to be approximately constant over the viable population range, $N_A$ to $K$. $N_A$, the Allee threshold (Table 41.2), is the hardest extinction parameter to estimate since populations do not stick around $N_A$ for long (Foley 2000).

The simplest ways to include density dependene at high N include (1) the logistic equation or its discrete-time variant the Ricker equation (Royama 1992; Burgman et al. 1993) and (2) a ceiling at K (MacArthur and Wilson 1967; Foley 1994). There is no standard way to model Allee effects, which include difficulties in finding mates, difficulties in forming defensive social units, genetic impoverishment effects, and many others (Allee et al. 1949; Franklin 1980; Ralls et al. 1988; Foley 1992; Lande 1995, 1998). The simplest way to deal with Allee effects is to develop a lower threshold population size $N_A$, below which we expect the population to decline deterministically. Although it is hard to estimate $N_A$ in practice, many studies

simulate times until populations reach some arbitrary threshold, at which point a quasi-extinction is said to occur.

# LOCAL HOST POPULATION MODELS: STOCHASTIC DIFFERENTIAL EQUATION FORMULATION

The appropriate stochastic differential equation to model the change in n = ln(N) is

$$dn = r_d dt + \sqrt{v_r} \, dW(t). \tag{16}$$

The drift term $r_d$ and the diffusion term $v_r$ we have met. W(t) is the Wiener or white noise process, $dW \sim$ Normal(0,dt). If $v_r$ is very small, the solution to Equation (16) approaches the deterministic result

$$N_t = N_0 e^{r_d t} \tag{17}$$

A population ceiling at N = K cuts off this exponential growth at the plateau of the carrying capacity. In the more familiar logistic growth model, population size can exceed the carrying capacity; in the ceiling model it cannot. Since we are working on the natural scale n, the diffusion has a constant diffusion

| Table 41.2 The Critical Extinction Model Parameters | |
|---|---|
| **Parameter** | **Definition and comments** |
| $r_d(N)$ | expected r(t), drift component. This is $r_d(N) = \ln \lambda(N)$. |
| $r_m$ | maximum r at low population density |
| $v_r$ | variance of r(t), the diffusion component |
| K | carrying capacity of the habitat for our population |
| $\rho$ | correlation of r(t) and r(t+1) |
| $N_A$ | Allee threshold. Below $N_A$, $r_d(N)$ is negative. |
| $N_0$ | starting population size |
| $v_{re}$ | $= v_r(1 + \rho)/(1 - \rho)$ effective variance in r, incorporating temporal correlation. |
| s | $= r_d/v_{re}$ relative importance of the deterministic and the chance effects. |
| $T_e$ | expected time until local extinction |
| e | local extinction rate in a metapopulation. e = $1/T_e$ |
| c | colonization rate |
| P | patch occupancy fraction |
| $T_M$ | expected time to extinction of the metapopulation |
| q | probability of a catastrophe |
| $\mu$ | probability of death in case of a catastrophe (including a disease) |

term $v_r$, and there is no distinction between the Ito and Stratonovich interpretations (Kloeden and Platen 1992). The logarithm of K (= k) is a reflecting boundary and o is an absorbing boundary for the diffusion of n.

The fullest knowledge that we can obtain about the future history of our population, which starts off at n = $n_o$, is the probability density p(n,t). The density p(n,t) satisfies the Kolmogorov forward and backwards equations (Karlin and Taylor 1981, Gardiner 1985). The backward equation (KBE)

$$\frac{\partial p(n,t)}{\partial t} = \frac{v_r}{2}\frac{\partial^2 p(n,t)}{\partial n^2} + r_d \frac{\partial p(n,t)}{\partial n} \qquad (18)$$

is especially useful to obtain results about sojourn and extinction times (Foley 1994; Middleton et al. 1995; Hanski et al. 1996). The expected time to extinction Te($n_o$) satisfies the following ordinary differential equation, which follows from KBE (Ewens 1979, Gardiner 1985).

$$-1 = \frac{v_r}{2}\frac{\partial^2 T_e(n_0)}{\partial n_0^2} + r_d \frac{\partial T_e(n_0)}{\partial n_0}$$
$$T_e(0) = 0 \qquad (19)$$
$$\frac{dT_e(k)}{dn_0}$$

## CALCULATING EXPECTED EXTINCTION TIME $T_E$

The expected time to extinction is well approximated by these equations (Foley 1994; Foley 1997):

$$T_e(N_0) = \frac{2\ln N_0}{v_{re}}\left(\ln K - \frac{\ln N_0}{2}\right) \to \text{when } r_d = 0 \quad (20)$$

$$T_e(N_0) = \frac{1}{2sr_d}\left(K^{2s}(1-N_0^{-2s}) - 2s\ln N_0\right) \quad (21)$$

$\to$ when $r_d$ is not 0

where

$$v_{re} = v_r \frac{1+\rho}{1-\rho} \qquad (22)$$

is the effective variance in r, incorporating temporal autocorrelation correlation (r(t) to r(t+1)),

$$s = \frac{r_d}{v_{re}} \qquad (23)$$

measures the relative importance of the deterministic and the chance effects, and both K and $N_0$ should be discounted by $N_A$ in this way:

$$K_e \leftarrow K/N_A$$
$$N_{0e} \leftarrow N_0/N_A \qquad (24)$$

Equations 20 and 21 can be simplified. A population with low growth rate ($r_d$) near its carrying capacity has persistence time

$$T_e = \frac{\ln^2(K/N_A)}{v_{re}} \qquad (25)$$

and this rises quickly when $r_d$ is well above zero. In fact,

$$T_e \propto K^{2s} \qquad (26)$$

a form assumed by Hanski (1992, 1994) in his incidence function papers.

From these last five equations we can glean a few useful points. Population persistence depends critically upon the ratio of K to the Allee threshold $N_A$ and on a version of environmental stochasticity, $v_{re}$, which takes into account temporal autocorrelation r. And of greatest importance, if $r_d$ dominates $v_{re}$ (that is, s >> 1), population persistence times explode.

Numerical analysis shows that a logistic model of density dependence leads to a population persistence time that is close to that of the ceiling model, especially when $r_d$ is small compared to $v_{re}$ (Foley 1997):

$$T_e logistic \cong T_e ceiling(1-0.7s) \qquad (27)$$

## ESTIMATING EXTINCTION

Early efforts to estimate extinction parameters using time series ignored carrying capacity focusing on the mean growth rate $r_d$ and environmental stochasticity $v_r$ (Dennis et al. 1991). More recent approaches have used sophisticated state space models that focus on distinguishing observation error from real population fluctuations (Holmes et al. 2007). Cruder but biologically based approaches are discussed in Foley (2000). We briefly mention a few main points.

Given a time series of host population sizes, a regression analysis of r(N,t) on N(t) can provide estimates of $r_m$, $v_r$, and K using Equation 28, as can be seen in Figure 41.2.

$$r(t) = r_d(t) + \varepsilon_t = r_m\left(1 - \frac{N(t)}{K}\right) + \varepsilon_t \tag{28}$$

$\rightarrow$ where $e_i \sim$ iid Normal $(0, v_r)$

But the value of these estimates depends on (1) how realistic the Ricker model that underlies them is, (2) the accuracy of estimates of N(t) that are observed, and (3) whether time series used in the estimates are long enough to be representative. In particular, density dependence is unlikely to exactly follow a logistic-type model, in part because Allee effects are likely to interfere. We also do not often have the 20 to 30 years needed (Holmes et al. 2007) to get good time

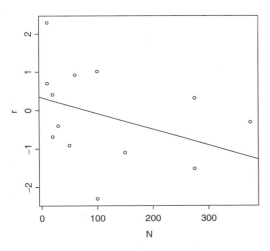

**Figure 41.2:**
Density dependence is not very clear in the Rainbow Bay chorus frog. The regression line r = a + bN shows p = 0.231 with $R^2$ = 0.109. If you can believe this model and if the population size estimates are accurate, $r_m$ = a = 0.327, K = $-r_m/b$ = 79.5, and $v_r$ = 1.97. It seems that a more honest estimate of $r_d$ is very close to zero. Assuming $r_d$ = 0, K = 79.5, $v_r$ = 1.97, $\underset{\tilde{}}{r}$ = 0, and $N_A$ = 1, Equation 25 gives us an expected time to extinction of $T_e$ = 9.72 years. In fact the Rainbow Bay chorus frog population went locally extinct six years after these data were available for analysis, in 2000 (Daszak et al. 2005). Climatic fluctuation, *not* chytridiomycosis, was held responsible.

series estimates of the extinction parameters. Surely the most data-efficient approach is to start our estimation with empirical Bayesian priors based on data in closely related species or even populations. Rather than focus only on sparse data from target species, why not use data from related organisms as Bayesian priors? Then further surveys of the targets give improved posterior estimates using standard Bayesian methods, as can be seen in an example estimating $v_r$ with frogs in Foley (2000).

## METAPOPULATION MODELS

In 1969, Richard Levins was working on the theory of biological control, a theory he hoped to apply to Cuban agriculture. As a broadly trained population biologist, Levins was steeped in the current island biogeography ideas and in Sewall Wright's shifting balance theory of evolution, which assumed that species are often distributed across "demes" (i.e., isolated local populations) (Wright 1931). So it is not surprising that Levins sought a way of describing a collection of demes or local populations as a "metapopulation" (Levins 1969, 1970) using a strategically simplified model. A strict sense Levins metapopulation comprises a very large number of similar populations, each equally likely to send out colonists to each patch, and each equally likely to go extinct. Extinction rates e and colonization rates c are constant over time.

A Levins metapopulation has deterministic dynamics that describe the change in P, the fraction of patches occupied

$$\frac{dP}{dt} = cP(1-P) - eP \tag{29}$$

where, using Hanski's (1999) terminology, c is the colonization rate and e the extinction rate for each patch. At equilibrium, the solution to Equation 29 is

$$P^* = 1 - \frac{e}{c} \tag{30}$$

which tells us a couple of things: (1) In a very large metapopulation with constant rates, populations may wink out and into existence, but the proportion of occupied patches should remain constant, and *2) if extinction rates exceed colonization rates, the metapopulation is doomed. A Levins metapopulation never goes extinct if c exceeds e. A finite Levins metapopulation (with H habitat patches) need not persist indefinitely. There is a random walk between M = 0 and M = H, and a kind of demographic stochasticity occurs that Hanski calls "immigration-extinction stochasticity." Nisbet and Gurney give an approximate formula for metapopulation persistence (Nisbet and Gurney 1982):

$$T_M \cong T_L \exp\left[\frac{HP^*}{2(1-P^*)}\right] \quad (31)$$

Clearly, the larger the number of patches and the higher the deterministic patch occupancy, the better for the metapopulation.

If local extinction is correlated among patches, $T_M$ is much decreased. Hanski calls this effect "regional stochasticity." For example, American pika (*Ochotona princeps*) populations near Bodie in California showed spatially clustered extinction and colonization events over four years (Smith and Gilpin 1997).

Regional climatic stresses and regional epidemics (Keeling et al. 2004) create interpatch extinction rate correlation. In models of structured metapopulations with regional stochasticity, a basic approach is to find an equivalent unstructured metapopulation that has the same persistence properties as the more complicated real metapopulation (Frank and Wissel 2002, Drechsler and Johst 2010).

## PARASITES ENTER THE PVA PICTURE

Metapopulation and local host population persistence depend on the critical extinction parameters K, $N_A$, $r_d$, $v_r$, and r. Most or all of these parameters are affected by parasite presence (Table 41.3). A diversity of metapopulation models has been developed for specific parasites—for example, measles in humans (Bolker and Grenfell 1995; Grenfell and Harwood 1997), plague in rats and great gerbils (*Rhombomys opimus*) (Keeling and Gilligan 2000a; Davis et al. 2007), and

Table 41.3 Chorus Frog (*Pseudacris ornata*) Population Fluctuation at Rainbow Bay, South Carolina
N(t) is the number of breeding females. r(t) = ln(N(t+1)/N(t)) measures population growth. Fluctuations in r(t) can be explained by observation error + environmental stochasticity (largely due to climate in this case) + demographic stochasticity (which is a small effect here).

| Year | N(t) | r(t) |
|------|------|------|
| 1979 | 100 | −2.303 |
| 1980 | 10 | 2.303 |
| 1981 | 100 | 1.012 |
| 1982 | 275 | 0.310 |
| 1983 | 375 | −0.310 |
| 1984 | 275 | −1.522 |
| 1985 | 60 | 0.916 |
| 1986 | 150 | −1.099 |
| 1987 | 50 | −0.916 |
| 1988 | 20 | −0.693 |
| 1989 | 10 | 2.303 |
| 1990 | 100 | −2.303 |
| 1991 | 10 | 0.693 |
| 1992 | 20 | 0.405 |
| 1993 | 30 | −0.405 |
| 1994 | 20 | |

Data estimated from Figure 4e of Semlitsch et al. 1996.

the SIS disease caused by feline enteric coronavirus (Foley et al. 1999). More general models tend to avoid the internal structure of local populations (Hess 1996a; Gog et al. 2002). We try to construct a local model reasonably consistent with standard SIR models and with local extinction driven by environmental stochasticity. Local extinction rates can then provide the e parameter for a Levins metapopulation.

Rodent and other wildlife host populations often manifest high growth rate, annual population fluctuations, density dependence, and seasonality. In some cases they also show population cycles, possibly due to predator–prey cycles (Turchin 1993, 2003). The simplest way to capture most features (except seasonality and cycles) uses a stochastic Ricker equation.

$$N(t+1)=N(t)e^{r(N,t)}=N(t)e^{r_m(1-N/K)+\varepsilon_t} \quad (32)$$

The growth parameters $r_m$ (the maximum r at low population density), K, and $v_r$ can be estimated from time series. But Equation 32 does not handle parasites well because death rates, which are implicitly incorporated in r(t), are often different for infected and uninfected hosts. This is implied in Equation 2, where $d_s$ need not equal $d_I$.

Many SIR models either ignore demography (because a short epidemic is the event of interest) or assume the host population is at equilibrium such that births exactly balance deaths. Neither approach is suitable for endangered hosts, where extinction is the main event of interest. We can write

$$r(t)=b(t)-d(t) \quad (33)$$

which suggests

$$b(t)=b_m(1-N(t)/K_b)+\varepsilon_t \quad (34)$$

Here $b_m$ is maximum birth rate at low population density and $K_b$ is the birth rate carrying capacity. If death rate is constant over the SIR classes, this leads to the model

$$N(t+1)=N(t)e^{r(N,t)}=N(t)e^{b(t)-d(t)}$$
$$=N(t)e^{b_m(1-N/K_b)+\varepsilon_t-d(t)} \quad (35)$$

The death rate of a non-diseased host, d, can be estimated as 1/lifetime of the host, where γ is

1/recovery time. The mortality rate attributable to disease δ is slightly trickier. The usual information we have about a disease is its probability of causing death μ. Disease-caused death is only one of three possibilities an infected host can experience at a given moment (the other two are to die of some other cause d, or to recover $\tilde{\gamma}$ Then μ should satisfy

$$\mu=\frac{\delta}{\gamma+d+\delta} \quad (36)$$

which gives us an estimate for d,

$$\delta=\frac{\mu}{1-\mu}(\gamma+d) \quad (37)$$

If environmental stochasticity is low ($v_r \sim 0$) and seasonality can be neglected, then the population size N reaches the equilibrium

$$N^*=K=K_b(1-\frac{d}{b_m}) \quad (38)$$

If d is much smaller than $b_m$, $K_b$ is the carrying capacity according to our usual intuitions. But if mortality due to the disease is high enough, the species can go locally extinct. This model has the suspicious feature that when N > $K_b$, births become negative. But this is not a problem for populations near the lower equilibrium K, nor for populations near extinction.

In light of these considerations, we now write out a SIR model with density dependence and higher mortality for infected individuals:

$$\frac{dS}{dt}=b_m(1-N/K_b)N-\beta SI-dS$$
$$\frac{dI}{dt}=\beta SI-\gamma I-(d+\delta)I \quad (39)$$
$$\frac{dR}{dt}=\gamma I-dR$$

These equations sum to

$$\frac{dN}{dt}=\left[b_m(1-N/K_b)-d-\delta\frac{1}{N}\right]N \quad (40)$$

so that, in effect, the parasite reduces the host population carrying capacity to

$$N^*=K_b\left[1-\frac{d}{b_m}-\frac{\delta}{b_m}\frac{I^*}{N^*}\right] \quad (41)$$

Equation 40 indicates that the deterministic component of population fluctuations should be written as a function of both I and N.

$$r_d(N,I)=b_m(1-N/K_b)-d-\delta\frac{I}{N} \quad (42)$$

Thus, parasites diminish not only the carrying capacity but also the "intrinsic" growth rate of the host. It would be possible to model a SIR system for any specific parasite to obtain mean values of I/N, but this effort is unrealistic. Many parasites of an endangered host are spillover from other reservoir hosts, and most hosts have more than one parasite. As a consequence it may be more useful to consider the overall parasite load and its contributions to the diminishment of K and $r_d$. Suppose that a host has J parasites, and that $I_j/N$ is the fraction of hosts infected by the j'th parasite. The parasite load $L_J$ is

$$L_J=\sum_{j=1}^{J}\delta_j\frac{I_j}{N} \quad (43)$$

Extinction parameter changes become

$$N^*=K_b(1-\frac{d}{b_m}-\frac{1}{b_m}L_J^*) \quad (44)$$

$$r_d(N,I)=b_m(1-N/K_b)-d-L_J^* \quad (45)$$

where I is the vector of infected covering all J parasites. Even if each parasite has a small effect on the host, the effect of the overall load on *extinction* can be nonlinear. For example, $r_d$ can drop below zero even when N is small, dooming the host population.

Even when $r_d$ remains above zero for small N, Equation 26 shows the significant nonlinear effect of perturbing $r_d$, which appears in the exponent $s = r_d/v_{re}$. K is also shrunk by parasite load (Equation 44). The overall parasite load will usually also vary over time, contributing to the environmental stochasticity parameters $v_r$ and $\rho$ and thus to $v_{re}$.

Parasitism can have an important effect on local extinction times, but metapopulation persistence also is damaged in a nonlinear way. Patch occupancy is linearly diminished by a rise in e, but Equation 31 shows that $T_M$ is strongly nonlinearly affected by this rise. Moreover, local epidemics are likely to spread regionally, creating regional stochasticity with even greater metapopulation effects than in the Levins metapopulations.

Environmental stochasticity among years is correlated. Drought years tend to follow each other, and so can bad parasite years. Autocorrelation $\rho$ helps deal with this problem. Its estimation is the standard one, once the researcher has decided what the residuals are for r(t)—that is, what model is used to get the $\rho_i$. Adjusting $v_{re}$ to incorporate $\rho$ allows a more honest treatment of environmental effects, including the damped epidemic oscillations theory often predicts. Moreover, $\rho$ helps deal with observational error. If there were no environmental stochasticity, and $v_r$ arose entirely from observational error, then $\rho$ would be negative. Negative $\rho$ shrinks $v_{re}$, diminishing the effect of observation error.

## DO OSCILLATORY DYNAMICS SKEW THE SYSTEM?

Mechanisms underlying infectious disease cycles include interactions with host immunity, environmental drivers, or classic epidemiological cycles as recovered individuals accumulate during epidemics but are replaced by susceptible hosts due to vital dynamics. An example of a disease with a daily cycle is relapsing fever, where the spirochete has greatest activity in the blood nightly in order to capitalize on the activity of the tick vector, which typically feeds at night (Southern and Sanford 1969). Pathogens such as those that cause anaplasmosis, syphilis, and trypanosomiasis have approximately biweekly cycles due to antigenic variation, a phenomenon in which pathogens experience recombination in their major antigens, which serves to provide them with several more weeks of ineffectual host immunity. A two- to four-week period cycle comes about because vertebrate hosts commonly produce antigen-specific adaptive immune responses on about that time frame. Seasonally varying diseases arise due to seasonal changes in host behavior such as increased crowding as weather becomes cooler (a very common factor in the annual epidemics of childhood disease), and seasonal peaks in vector activity are often seen in tick-, flea-, and mosquito-borne disease (Altizer et al. 2006).

Finally, epidemiological cycling driven by inherent fluctuations in susceptible and recovered/resistant hosts occurs as well. Some best-studied examples are found in morbilliviruses such as measles, where, in unvaccinated populations, recovered people increase in number during epidemics but as susceptible children are born, a critical threshold for the propagation of a new epidemic is eventually exceeded and another epidemic can ensue—that is, $R_e$ >1 again. There is some chance as to when such an epidemic is "seeded" (i.e., invaded by a new infectious individual), but on the whole, pre-vaccination measles exhibited cycles due to SIRS dynamics superimposed on the expected annual cycles due to behavior change. Plague, which exhibits seasonal cycles due to the pathogen and vector differential survival over the course of the year, also may recur in multi-annual cycles if susceptible hosts are consumed but eventually repopulate (Parmenter et al. 1999; Keeling and Gilligan 2000b), and rabies, which has no particular seasonal peaks, classically represents the phenomenon of cycling due to decreases followed by increases in the pool of susceptible individuals (Childs et al. 2000). Grouse infections with trichostrongyle worms also cycle (Hudson et al. 1998). In small but interconnected host populations, those diseases where cycling arises from temporary exhaustion of the pool of susceptibles would commonly die out unless the disease is a spill-over event from some interacting more abundant host.

Many SIR models including macroparasite models predict that, if a parasite persists, the ratio I/N tends towards oscillations that damp down to an equilibrium value (Bailey 1975, Anderson and May 1982). In our model and notation, the approximate period T is

$$T \cong 2\pi \sqrt{\frac{1}{\gamma d}(R_0 - 1)} \qquad (46)$$

The mean disease duration is $1/\gamma$ and the mean animal lifetime is $1/d$. So Equation 46 predicts that longer interepidemic periods arise from longer lifespans (it takes longer for susceptibles to be recruited), longer recovery times, and larger $R_0$ (it takes longer for a population to recover from a severe epidemic). Simple theoretical models predict that these oscillations will usually dampen, and yet do not for many diseases. Using more complicated SIR models, Dietz and Grossman showed that seasonal parameter variation could act to make the disease

pendulum swing more widely, creating multiple-year oscillations (Dietz 1976; Grossman 1980). Examples are discussed in Altizer et al. (2006). Bartlett (1956) showed that stochasticity could also create multiple-year cycles.

Most PVAs depend on population density time series taken on an annual basis. In some ways this allows us to neglect seasonal cycles. One possible approach to dealing with multiple-year cycles is to construct a more abbreviated time series using data only at the low point of each cycle, say every two or three years. Parameter estimates for $r_d$, K, $v_r$ and $\rho$ from the abbreviated time series might be more useful, and they should not be cyclic. We know of no work using this approach so far.

## METAPOPULATION CONNECTIVITY BLUES

When parasites are specific to their endangered hosts, too much metapopulation connectivity might seem to be a bad thing, since parasites can follow the same corridors as their hosts (Hess 1994, 1996b). Hess' models assume that infected host patches cannot recover until the hosts go locally extinct. A more complicated picture develops when recovered host patches are possible (McCallum and Dobson 2002). Higher connectivity rarely increases host vulnerability in the case of host-specific parasites. However, alternative reservoir hosts do, in many cases, lead to higher disease prevalence and more persistent disease in a well-connected metapopulation.

## THE CONSERVATION OF THREATENED HOSTS

Habitat loss and degradation are main threats to endangered species. Anything we can do to improve carrying capacity is obviously good, and our extinction models show how to calculate the improvement in population persistence times given an improvement in K. The main contribution parasites make to host extinction threat is to diminish $r_d(N)$. Already threatened hosts may have an $r_d$ hovering near zero. New parasites or immunologically compromised host populations can turn a source patch into a sink. The implications for patch occupancy and metapopulation viability are obvious, and this is borne out even in simple metapopulation models. Thus a conservation biologist should be moved to improve habitat quality (raising other components of $r_d$), raising host immunological capacity, perhaps with influx of more genetically diverse individuals as has been done with the Florida panther (*Puma concolor coryi*) with some success (Hedrick 1995; Gross 2005), or even immunizing hosts.

Climate change, host density reduction, and changes in host population connectivity interact with disease risk to influence the vulnerability of hosts to extinction. Certainly some parasites will be shed as hosts become rare. But at the same time the threatened host may lose immunological capacity with genetic impoverishment. It is tricky to determine the optimal amount of connectivity among patches in a metapopulation of hosts. Higher connectivity leads to more patch occupancy and to rescue effects (Brown and Kodric-Brown 1977), but it also leads to greater parasite transmission between patches, keeping host species from shedding parasites. This may not be an issue if most of the parasite load comes from reservoir hosts that are less patchily distributed than our species of concern (McCallum and Dobson 2002).

Parasite faunas on a particular host species (component communities) become less similar as a function of increasing distance (Poulin 2003); this has been shown clearly for fleas (Krasnov et al. 2005). This contributes of course to regional stochasticity and increased metapopulation vulnerability. For example, as the Sierra Nevada of California warms, American pikas (*Ochotona princeps*) are likely to find their host range increasingly overlaps with those of rodents whose ranges are shifting upwards. Pikas in western North America have four host-specific fleas, while they share 18 flea species with rodents (Hubbard 1947). Little is known about the potential for parasites to spillover to pikas. Estimating the future change in the parasite load $L_J$ (as J increases) will help to quantitatively predict the increased vulnerability of the host to extinction. This is true both for parasites as components of environmental stochasticity and as catastrophes.

## CONCLUSIONS

PVA attempts to predict extinction times or rates using information about a species and its environment.

Parasites can have substantial effects on species persistence. Endangered species do not shed all of their parasites, mainly because of alternative host reservoirs and frequency-dependent transmission. Parasites sometimes act as catastrophes, but predicting catastrophes is problematic, since they fall outside the usual scope of normal statistics. Quantitative approaches include using Bayesian prior estimates of catastrophe probability q and mortality fraction $\mu$ from phylogenetically related organisms. Parasite spillover from domesticated and wild animals might also be predicted to some extent.

Even when virulence is low for any given parasite, overall parasite load can have a powerful nonlinear effect on local host time to extinction $T_e$, and because these population effects are propagated across a metapopulation, parasitism has an even more powerful nonlinear effect on metapopulation patch occupancy P and metapopulation extinction time $T_M$.

A PVA that omits disease underestimates the risk that a metapopulation will go extinct. We raise the question how conservation biologists can practically incorporate parasite load into their PVA and consider how different management strategies and different global trends might affect local and metapopulation extinction rates. Details of the biology of disease in specific hosts must be considered. For example, accurate prediction of host population fluctuations in a system with strong disease cycles that drive host population cycles may require more explicit modeling of disease dynamics than for a system lacking such cycles. However, an emphasis on density dependence, Allee effects, and spatial autocorrelation of disease across a metapopulation can allow us to develop highly useful estimates for the trajectories of vulnerable populations even where all of the intricacies of disease have yet to be quantified.

## REFERENCES

Akçakaya, H.R., M.A. Burgman, and L.R. Ginzburg. 1997. Applied population ecology: principles and computer exercises using RAMAS EcoLab 1.0. Applied Biomathematics, Setauket, New York.

Allee, W.C., A.E. Emerson, O. Park, T. Park, and K.P. Schmidt. 1949. Principles of animal ecology. W.B. Saunders, Philadelphia.

Altizer, S., C.L. Nunn, and P. Lindenfors. 2007. Do threatened hosts have fewer parasites? A comparative study in primates. J Anim Ecol 76:304–314.

Altizer, S., A. Dobson, P. Hosseini, P. Hudson, M. Pascual, and P. Rohani. 2006. Seasonality and the dynamics of infectious diseases. Ecol Lett 9:467–484.

Anderson, R. 1979. Parasite pathogenicity and the depression of host population equilibria. Nature 279:150–152.

Anderson, R., and R. May. 1982. Directly transmitted infectious diseases: Control by vaccination. Science 215:1053–1060.

Anderson, R., and R. May. 1991. Infectious diseases of humans. Oxford University Press, Oxford.

Anderson, R.M. 1982. The population dynamics and control of hookworm and roundworm infections. In R.M. Anderson, ed. Population dynamics of infectious diseases theory and applications, pp. 67–108. Chapman and Hall, London.

Anderson, R.M., and R.M. May. 1978. Regulation and stability of host-parasite population interactions I. Regulatory processes. J Anim Ecol 47:219–247.

Ashford, R.W., and W. Crewe. 2003. The parasites of Homo sapiens: an annotated checklist of the protozoa, helminths and arthropods for which we are home. Taylor & Francis, London.

Bailey, N. 1975. The mathematical theory of infectious diseases and its applications. Griffin, London.

Bartlett, M. 1956. Deterministic and stochastic models for recurrent epidemics. In J. Neyman, ed. Proceedings of the Third Berkeley Symposium on Mathematical Statistics and Probability, pp. 81–109. University of California Press, Berkeley.

Begon, M., J. Harper, and C. Townsend. 1996. Ecology: individuals, populations and communities. Blackwell Scientific Publications, Boston.

Bolker, B., and B. Grenfell. 1995. Space, persistence and dynamics of measles epidemics. Phil Trans Royal Soc London B 348:309–320.

Brandle, M., and R. Brandl. 2001. Species richness of insects and mites on trees: expanding Southwood. J Anim Ecol 70:491–504.

Brown, J.H., and A. Kodric-Brown. 1977. Turnover rates in insular biogeography: effect of immigration on extinction. Ecology 58:445–449.

Burgman, M.A., S. Ferson, and H.R. Akçakaya. 1993. Risk assessment in conservation biology. Chapman & Hall, New York.

Caswell, H. 1989. Matrix population models. Sinauer, Sunderland, Massachusetts.

Caughley, G. 1994. Directions in conservation biology. J Anim Ecol 63:215–244.

Caughley, G., and A. Gunn. 1996. Conservation biology in theory and practice. Blackwell, Cambridge, Massachusetts.

Childs, J.E., A.T. Curns, M.E. Dey, L.A. Real, L. Feinstein, O.N. Bjornstad, and J.W. Krebs. 2000. Predicting the local dynamics of epizootic rabies among raccoons in the United States. Proc Nat Acad Sci USA 97:13666–13671.

Cintron-Arias, A., C. Castillo-Chavez, L.M.A. Bettencourt, A.L. Lloyd, and H.T. Banks. 2009. The estimation of the effective reproductive number from disease outbreak data. Math Biosci Engineering 6:261–282.

Daszak, P., D.E. Scott, A.M. Kilpatrick, C. Faggioni, J.W. Gibbons, and D. Porter. 2005. Amphibian population declines at Savannah River site are linked to climate, not chytridiomycosis. Ecology 86: 3232–3237.

Davis, S., N. Klassovskiy, V. Ageyev, B. Suleimenov, B. Atshabar, A. Klassovskaya, M. Bennett, H. Leirs, and M. Begon. 2007. Plague metapopulation dynamics in a natural reservoir: the burrow system as the unit of study. Epidemiol Infect 135:740–748.

de Castro, F., and B. Bolker. 2005. Mechanisms of disease-induced extinction. Ecol Lett 8:117–126.

Dennis, B., P.L. Munholland, and J.M. Scott . 1991. Estimation of growth and extinction parameters for endangered species. Ecol Monogr 61:115–143.

Dietz, K. 1976. The incidence of infectious diseases under the influence of seasonal fluctuations. Lecture Notes Biomath 11:1–15.

Drechsler, M., and K. Johst. 2010. Rapid viability analysis for metapopulations in dynamic habitat networks. J. Royal Soc London B 277:1689, 1889–1897.

Ebert, T.A. 1999. Plant and animal populations methods in demography. Academic Press, San Diego.

Ewens, W. 1979. Mathematical population genetics. Springer-Verlag, Berlin.

Foley, J., P. Foley, and N. Pedersen. 1999. The persistence of an SIS disease in a metapopulation. J Appl Ecol 36:555–563.

Foley, P. 1992. Small population genetic variation at loci under stabilizing selection. Evolution 46:763–774.

Foley, P. 1994. Predicting extinction times from environmental stochasticity and carrying capacity. Conserv Biol 8:124–137.

Foley, P. 1997. Extinction models for local populations. In I. Hanski and M. Gilpin, eds. Metapopulation biology, pp. 215–246. Academic Press, New York.

Foley, P. 2000. Problems in extinction model selection and parameter estimation. Environ Manage 26:S55–S73.

Foley, P., and J. E. Foley. 2010. Modeling susceptible infective recovered dynamics and plague persistence in California rodent–flea communities. Vector-borne Zoon 2010:59–67.

Frank, K., and C. Wissel. 2002. A formula for the mean lifetime of metapopulations in heterogeneous landscapes. Amer Natur 165:374–388.

Franklin, I.R. 1980. Evolutionary change in small populations. *In* M.E. Soule and B.A. Wilcox, eds. Conservation biology, pp. 135–149. Sinauer, Sunderland, Massachusetts.

Gardiner, C. 1985. Handbook of stochastic methods. Springer, New York.

Gerber, L.R., H. McCallum, K.D. Lafferty, J.L. Sabo, and A. Dobson. 2005. Exposing extinction risk analysis to pathogens: is disease just another form of density dependence? Ecol Appl 15:1404–1414.

Gog, J., R. Woodroffe, and J. Swinton. 2002. Disease in endangered metapopualtions: the importance of alternative hosts. P R Soc B 269:661–676.

Grenfell, B., and J. Harwood. 1997. (Meta)population dynamics of infectious diseases. Trends Ecol Evol 12:395–404.

Groom, M.J., and M.A. Pascual. 1998. The analysis of population persistence: An outlook on the practice of viability analysis. *In* P.K. Fiedler and P. M. Kareiva, eds. Conservation biology, 2nd edition, pp. 4–27. Chapman and Hall, Inc., New York.

Gross, L. 2005. Why not the best? How science failed the Florida panther. PLoS Biol 3:e333. doi:10.1371/journal.pbio.0030333

Grossman, Z. 1980. Oscillatory phenomena in a model of infectious diseases. Theor Pop Biol 18:204–243.

Hanski, I. 1992. Inferences from ecological incidence functions. Am Nat 139:657–662.

Hanski, I. 1994. A practical model of metapopulation dynamics. J Anim Ecol 63:151–162.

Hanski, I. 1996. Metapopulation ecology. *In* O.E. Rhodes, R.K. Chesser, and M.H. Smith, eds. Population dynamics in ecological space and time. First Savannah River Symposia on Environmental Science, Aiken, South Carolina, USA, May 3–7, 1993, pp. 13–43. University of Chicago Press, Chicago, Illinois.

Hanski, I. 1999. Metapopulation ecology. Oxford University Press, New York.

Hanski, I., P. Foley, and M. Hassell. 1996. Random walks in a metapopulation: How much density dependence is necessary for long-term persistence? J Anim Ecol 65:274–282.

Hanson, F.B., and H.C. Tuckwell. 1981. Logistic growth with random density-independent disasters. Theor Pop Biol 19:1–18.

Hedrick, P.W. 1995. Gene flow and genetic restoration: the Florida panther as a case study. Conserv Biol 9:996–1007.

Hess, G. R. 1994. Conservation corridors and contagious disease: a cautionary note. Conserv Biol 8:256–262.

Hess, G. 1996a. Disease in metapopulation models: implications for conservation. Ecology 77:1617–1632.

Hess, G. R. 1996b. Linking extinction to connectivity and habitat destruction in metapopulation models. Am Nat 148:226–236.

Holmes, E.E., J.L. Sabo, S.V. Viscido, and W.F. Fagan. 2007. A statistical approach to quasi-extinction forecasting. Ecol Lett 10:1182–1198.

Hubbard, C.A. 1947. Fleas of western North America, their relation to the public health. Iowa State College Press, Ames.

Hudson, P.J., A.P. Dobson, and D. Newborn. 1998. Prevention of population cycles by parasite removal. Science 282:2256–2258.

Karlin, S., and H.M. Taylor. 1981. A second course in stochastic processes. Academic Press, New York.

Keeling, M.J., and C.A. Gilligan. 2000a. Bubonic plague: a metapopulation model of a zoonosis. P R Soc B 267:2219–2230.

Keeling, M.J., and C.A. Gilligan. 2000b. Metapopulation dynamics of bubonic plague. Nature 407:903–906.

Keeling, M.J., and P. Rohani. 2008. Modeling infectious diseases. Princeton University Press, Princeton, New Jersey.

Keeling, M.J., O.N. Bjornstad, and B.T. Grenfell. 2004. Metapopulation dynamics of infectious diseases. *In* I. Hanski and O.E. Gaggiotti, eds. Ecology, genetics and evolution of metapopulations, pp. 415–445. Elsevier, Amsterdam.

Kennedy, C.E.J., and T.R.E. Southwood. 1984. The number of species of insects associated with British trees: a re-analysis. J Anim Ecol 53:455–478.

Kermack, W., and A. McKendrick. 1927. Contributions to the mathematical theory of epidemics. Royal Stat Soc J 115:700–721.

Kloeden, P.E., and E. Platen. 1992. Numerical solution of stochastic differential equations. Springer-Verlag, Berlin.

Krasnov, B.R. 2008. Functional and evolutionary ecology of fleas: a model for ecological parasitology. Cambridge University press, Cambridge.

Krasnov, B.R., G.I. Shenbrot, D. Mouillot, I.S. Khokhlova, and R. Poulin. 2005. Spatial variation in species diversity and composition of flea assemblages in small mammalian hosts: geographical distance or faunal similarity? J Biogeog 32:633–644.

Lande, R. 1993. Risks of population extinction from demographic and environmental stochasticity and random catastrophes. Am Natur 142:911–927.

Lande, R. 1995. Mutation and conservation. Conserv Biol 9:782–791.

Lande, R. 1998. Demographic stochasticity and Allee effect on a scale with isotropic noise. Oikos 83:353–358.

Lande, R., S. Engen, and B.-E. Saether. 2003. Stochastic population dynamics in ecology and conservation. Oxford University Press, Oxford.

Levins, R. 1969. Some demographic and genetic consequences of environmental heterogeneity for biological control. Bull Entomol Soc Amer 15:237–240.

Levins, R. 1970. Extinction. In M. Gerstenhaber, ed. Some mathematical problems in biology, pp. 77–107. American Mathematical Society, Providence, Rhode Island.

Lindenfors, P., C.L. Nunn, K.E. Jones, A.A. Cunningham, W. Sechrest, and J. Gittleman. 2007. Parasite species richness in carnivores: effects of host body mass, latitude, geographic range and population density. Global Ecol Biogeog 16:496–509.

MacArthur, R., and E. Wilson. 1967. The theory of island biogeography. Princeton Univ. Press, Princeton.

Mangel, M., and C. Tier. 1993. A Simple direct method for finding persistence times of populations and application to conservation problems. Proc Nat Acad Sci USA 90:1083–1086.

Mangel, M., and C. Tier. 1994. Four facts every conservation biologist should know about persistence. Ecology 75:607–614.

Martin, P.S., and R.G. Klein, eds. 1984. Quaternary extinctions a prehistoric revolution. University of Arizona Press, Tucson.

May, R. 1974. Stability and complexity in model ecosystems. Princeton University Press, New Jersey.

May, R.M., J.H. Lawton, and N.E. Stork. 1995. Assessing extinction rates. In J.H. Lawton and R.M. May, eds. Extinction rates, pp. 1–24. Oxford University Press, England.

McCallum, H., and A. Dobson. 2002. Disease, habitat fragmentation and conservation. P R Soc B 269:2041–2049.

McCallum, H., N. Barlow, and J. Hone. 2001. How should pathogen transmission be modelled? Trends Ecol Evol 16:295–300.

Menges, E.S. 1990. Population viability analysis for an endangered plant. Conserv Biol 4:52–62.

Middleton, D.A.J., A.R. Veitch, and R.M. Nisbet. 1995. The effect of an upper limit to population size on persistence time. Theoret Pop Biol 48:277–305.

Miller, P.S., and R.C. Lacy. 2005. VORTEX: a stochastic simulation of the extinction process. Version 9.50 user's manual. Conservation Breeding Specialist Group (SSC/IUCN), Apple Valley Minnesotta.

Morris, W., D. Doak, M. Groom, P. Kareiva, J. Fieberg, L. Gerber, P. Murphy, and D. Thompson. 1998. A practical handbook for population viability analysis. The Nature Conservancy, Washington, D.C.

Morris, W.F., and D. Doak. 2002. Quantitative conservation biology: the theory and practice of population viability analysis. Sinauer, Sunderland, Massachussetts.

Nisbet, R., and W. Gurney. 1982. Modelling fluctuating populations. Wiley, New York.

Nunn, C.L., S. Altizer, K.E. Jones, and W. Sechrest. 2003. Comparative tests of parasite richness in primates. Am Natur 162:597–614.

Parmenter, R.R., E.P. Yadav, C.A. Parmenter, P. Ettestad, and K.L. Gage. 1999. Incidence of plague associated with increased winter-spring precipitation in New Mexico. Am J Trop Med Hyg 61:814–821.

Pedersen, A.B., K.E. Jones, C.L. Nunn, and S. Altizer. 2007. Infectious diseases and extinction risk in wild mammals. Conserv Biol 21:1269–1279.

Poulin, R. 2003. The decay of similarity with geographic distance in parasite communities of vertebrate hosts. J Biogeog 30:1609–1615.

Poulin, R., and S. Mourand. 2004. Parasite biodiversity. Smithsonian Press, Washington, D.C.

Pulliam, H.R. 1988. Sources, sinks and population regulation. Am Natur 132:652–661.

Ralls, K., J.D. Ballou, and A. Templeton. 1988. Estimates of lethal equivalents and the cost of inbreeding in mammals. Conserv Biol 2:185–193.

Royama, T. 1992. Analytical population dynamics. Chapman & Hall, London.

Sabo, J.L. 2007. Population viability and species interactions: Life outside the single-species vacuum. Biol Conserv 141:276–286.

Semlitsch, R.D., D.E. Scott, J.H.K. Pechmann, and J.W. Gibbons. 1996. Structure and dynamics of an amphibian community. In M.L. Cody and J.A. Smallwood, eds. Long-term studies of vertebrate communities, pp. 217–250. Academic Press, San Diego.

Shaffer, M.L. 1981. Minimum population sizes for species conservation. Bioscience 31:131–134.

Simpson, G.G. 1953. The major features of evolution. Columbia University Press, New York.

Smith, A.T., and M. Gilpin. 1997. Spatially correlated dynamics in a pika metapopulation. In I.A. Hanski and M.E. Gilpin, eds. Metapopulation biology: ecology, genetics, and evolution, pp. 407–428. Academic Press, San Diego.

Smith, M.J., S. Telfer, E.R. Kallio, S. Burthe, A.R. Cook, X. Lambin, and M. Begon. 2009. Host-pathogen time series data in wildlife support a transmission function between density and frequency dependence. Proc Nat Acad Sci USA 106:7905–7909.

Southern, P., and J. Sanford. 1969. Relapsing fever: a clinical and microbiological review. Medicine (Baltimore) 48:129–150.

Stapp, P., M.F. Antolin, and M. Bull. 2004. Patterns of extinction in prairie dog metapopulations: plague outbreaks follow El Nino events. Frontiers Ecol Environ 2:235–240.

Tuljapurkar, S. 1990. Population dynamics in variable environments. Springer-Verlag, New York.

Turchin, P. 1993. Chaos and stability in rodent population dynamics evidence from non-linear time-series analysis. Oikos 68:167–172.

Turchin, P. 2003. Complex population dynamics : a theoretical empirical synthesis. Princeton University Press, Princeton, N.J.

van Riper, C. III, S.G. van Riper, M.L. Goff, and M. Laird. 1986. The epizootiology and ecological significance of malaria in Hawaiian land birds. Ecol Monogr 56:327–344.

Wright, S. 1931. Evolution in Mendelian populations. Genetics 16:97–159.

# 42

## USING MATHEMATICAL MODELS IN A UNIFIED APPROACH TO PREDICTING THE NEXT EMERGING INFECTIOUS DISEASE

Tiffany L. Bogich, Kevin J. Olival, Parviez R. Hosseini, Carlos Zambrana-Torrelio, Elizabeth Loh, Sebastian Funk, Ilana L. Brito, Jonathan H. Epstein, John S. Brownstein, Damien O. Joly, Marc A. Levy, Kate E. Jones, Stephen S. Morse, A. Alonso Aguirre, William B. Karesh, Jonna A. K. Mazet, and Peter Daszak

Emerging infectious diseases (EIDs) pose a significant threat to human health, global economies, and conservation (Smolinski et al. 2003). They are defined as diseases that have recently increased in incidence (rate of the development of new cases during a given time period), are caused by pathogens that recently moved from one host population to another, have recently evolved, or have recently exhibited a change in pathogenesis (Morse 1993; Krause 1994). Some EIDs threaten global public health through pandemics with large-scale mortality (e.g., HIV/AIDS). Others cause smaller outbreaks but have high case fatality ratios or lack effective therapies or vaccines (e.g. Ebola virus or methicillin-resistant *Staphylococcus aureus*). As a group, EIDs cause hundreds of thousands of deaths each year, and some outbreaks (e.g., SARS, H5N1) have cost the global economy tens of billions of dollars. Emerging diseases also affect plants, livestock, and wildlife and are recognized as a significant threat to the conservation of biodiversity (Daszak et al. 2000). Approximately 60% of emerging human disease events are zoonotic, and over 75% of these diseases originate in wildlife (Jones et al. 2008). The global response to such epidemics is frequently reactive, and the effectiveness of conventional disease control operations is often "too little, too late". With rising globalization, the ease with which diseases spread globally has increased dramatically in recent times. Also, interactions between humans and wildlife have intensified through trade markets, agricultural intensification, logging and mining, and other forms of development that encroach into wild areas. Rapid human population growth, land use change, and change in global trade and travel require a shift toward a proactive, predictive, and preventive approaches for the next zoonotic pandemic.

The key emergence event for most infectious diseases is a change in transmission dynamics within or between host populations. The interconnectedness of humans, domestic animals, and wildlife facilitates the spillover of pathogens between hosts (Daszak et al. 2000). External forces, such as agricultural intensification, global travel, and the accidental trans-location of pathogens, augment this interaction. The role of zoonotic pathogens in causing human disease may be particularly important because when these diseases first emerge, humans have no acquired immunity to novel pathogens, resulting in sometimes highly lethal infections (e.g., AIDS/HIV, Ebola virus disease).

Despite the huge social, demographic, and economic impact of EIDs, there has been little advancement in understanding how anthropogenic changes drive disease emergence and in developing proactive, predictive, and preventive approaches (Hufnagel et al. 2004; Weiss and McMichael 2004; Ferguson et al. 2005; Wolfe et al. 2005). In this chapter, we describe a strategy to create a unifying predictive model for the zoonotic and pandemic potential of a given region by integrating predictive models of each stage of the process of zoonotic disease emergence. The three stages of emergence that we address are (1) a "pre-emergence" phase, where anthropogenic changes cause animal populations to come into contact, leading to cross-species transmission of their pathogens, (2) a spillover stage, where animal pathogens enter human populations, and (3) pandemic emergence, where pathogens are able to exploit human travel and trade networks to emerge across international and regional boundaries. Each stage of the emergence process requires a different approach and analyses at different scales. Each of these modeling exercises is then linked to data collection on the ground. Models are then parameterized through effective active and passive surveillance of wildlife, monitoring of key-words in media, and analysis of published literature.

This modeling approach also helps to increase surveillance efficiency by facilitating spatial and species-specific (e.g., phylogenetic) targeting of wildlife to sample for likely zoonotic pathogens. Our strategy is designed for the early detection of novel pathogens with human pandemic potential, to allow animal and human health professionals the opportunity to predict emergence and prevent spread. It also provides the tools to target important sentinel species at active human interfaces to improve on the efficiencies of previous surveillance for rare pathogens of interest. Our vision is to expand on lessons learned in order to better assess local capacity, increase the value of infectious disease modeling, implement targeted and adaptive wildlife disease surveillance systems, develop and deliver new technologies to improve efforts in hotspots, and use cutting-edge information management and communication tools to bring the world closer to realizing an integrated, globalized approach to controlling emerging zoonotic diseases.

In this chapter, we focus in particular on three key steps in designing this integrated modeling and field surveillance approach: (1) the selection of geographic sites for surveillance, (2) the selection of target species for sampling, and (3) the construction of predictive models of spread and future emergence (Table 42.1).

## DEFINITIONS, DRIVERS, AND BIASES

### History and Debate over the Definition of an EID

In the introduction to this chapter, we defined EIDs as diseases that have recently increased in incidence, have moved from one host population to another, are caused by recently evolved strains, or exhibit a change in pathogenesis. We use this definition because, despite the widely accepted importance of EIDs, there is little agreement on the exact properties that classify a disease as "emerging." While the term has generally been used to emphasize the novelty of a given infectious disease, closer inspection reveals that there is no consensus on what defines this novelty. With an increasing number of studies investigating the phenomenon of emergence and the underlying environmental and anthropogenic drivers (e.g., Taylor et al. 2001; Jones et al. 2008), it is important to agree on a medically and biologically meaningful definition of emergence. Such a definition should, in principle, allow one to decide not only whether a given infectious disease can be called "emerging", but also when and where exactly it emerged, and to do so via rigorous and quantifiable criteria.

The first mention of EIDs that can be found on MEDLINE was provided by Oster (1961), who concentrated solely on animal diseases but supplied a definition that can be generalized to human diseases. He describes the "sudden invasion by epizootic diseases into countries where they have never before struck" and mentions that these "have been described as 'emerging diseases', a new term which would seem to indicate new infectious disease situations."

**Table 42.1 Summary of Questions, Approaches, and Results Related to Three Components of EID Surveillance and Prediction**

| Objective | Questions | Approach | Result |
|---|---|---|---|
| 1. Selecting geographic sites for surveillance | - What is the risk of transmission to humans?<br>- What is the distribution of undiscovered pathogens? What areas have been undersampled?<br>- What are the spatial drivers of disease emergence?<br>- How will the risk of disease emergence change geographically? | Spatial and temporal general linear models | - Geographically refined surveillance strategies<br>- Refined "hotspot" maps<br>- Sub-regional "hotspot" maps |
| 2. Selecting species for sampling | - Which wildlife species are the greatest risk of being the source for zoonotic disease emergence? | Spatial and temporal general linear models | - Refined surveillance strategies according to phylogenetic relatedness and contact opportunities |
| 3. Predicting spread and future emergence events | - Can the potential of a region to produce pandemic pathogens be measured?<br>- Can the vulnerability of a region to the spread of an EID be determined? | Matrix-based population simulation | - A global emerging infectious disease vulnerability map |

There are two ways in which a disease can be considered new (Table 42.2). In the first instance, the definition can be relatively specific. A disease may be "emerging" in that it has crossed the species barrier to infect a novel host, or that its clinical signs or symptoms or pathogenicity has changed. In other words, the disease is genuinely new to a host. In this sense, every disease can emerge only once in each host. Some diseases, such as measles (Babbott and Gordon 1954), sleeping sickness (Steverding 2008), and bubonic plague (Hays 2006), emerged in prehistoric or ancient times, whereas others, such as Ebola virus (World Health Organization 1978), Nipah virus (Chua et al. 2000), and SARS (Guan et al. 2003), emerged in recent years.

Some authors, on the other hand, have proposed defining EIDs in the wider purview of all diseases that are increasing in incidence (Institute of Medicine 1992; Morse 1993; Levins et al. 1994; Morse 1995; Jones et al. 2008). This approach includes not only diseases that are genuinely new in a host and are increasing in incidence by virtue of being recognized in the first place, but also diseases that were previously present at a lower level or are expanding to new areas. In this sense, a disease can emerge and re-emerge multiple times and in different locations.

With increasing interest in emerging infectious diseases, it is important to agree on the meaning of the term, which has been used for a variety of different and sometimes seemingly unrelated phenomena. In previous definitions, it has been interpreted in two ways: as the appearance of a new pathogen in humans or as a disease becomes a growing concern. These two scenarios can be distinguished by differentiating between primary and secondary emergence. For this chapter, we limit our focus to those EIDs that can infect humans. On the basis of the distinction between primary and secondary emergence, the following definitions are proposed for an emerging infectious disease (Table 42.2):

- *Primary emergence:* A novel infectious disease appears in humans by means of transmission from

**Table 42.2 Previous Definitions of EIDs**

| | Primary Emergence | | Secondary Emergence | | |
| --- | --- | --- | --- | --- | --- |
| | New host | New symptoms | Detection | Increased incidence | Expansion |
| Oster (1961) | | | | | • |
| Lederberg et al. (1992) | | | | • | |
| Morse (1993) | | | • | • | • |
| Levins et al. (1994) | • | • | • | • | |
| Morse (1995) | | • | | • | • |
| Garnett and Holmes (1996) | • | • | | | |
| Kilbourne (1996) | • | • | • | | |

Included factors of primary emergence were crossing of the species barrier to adapt to a new host (humans), the appearance of new symptoms or new pathogenicity, and new detection of a disease. Included factors of secondary emergence were an increase in incidence and expansion to a new area. Morse (1995) lists the appearance of an infection "for the first time" as emergence, which fits all categories of primary emergence, without being explicit about the mechanisms.

animals or the environment and adaptation to infecting humans, or through evolution within human hosts to develop new pathogenicity or resistance to treatment. In this case the first recorded cluster in humans is taken as the EID event. If an earlier case than the previously earliest known case is found retrospectively (as has happened for HIV), the timing of the event should be corrected accordingly.

• *Secondary emergence:* An existing infectious disease increases in incidence in a population in a way that constitutes a significant change with respect to a baseline incidence. This is the case when a disease occurs where it has never previously been reported (and the baseline incidence was zero), or when a disease displays a trend of increasing incidence with respect to a non-zero incidence. The timing of the emergence event, in this case, should be the beginning of the increase.

## Characterizing the Drivers of Emergence

Despite the threat posed by EIDs, we still do not fully understand the mechanism of emergence; instead, we rely heavily on a reactive approach of responding to pathogens after they have emerged. We must first take a broad-scale, ecological approach to understanding

the processes driving emergence. The process of disease emergence is complex and generally driven by factors that "provide conditions that allow for a select pathogen to expand and adapt to a new niche" (Smolinski et al. 2003). These factors are largely environmental, ecological, political, economic, and social forces, which function on a range of different scales. During the past two decades, numerous studies have classified emerging diseases according to the factors underlying their emergence, commonly referring to these factors or processes as *drivers* of emergence.

The first attempt to classify drivers of emergence was published by the Institute of Medicine (IOM) in 1992 (Lederberg et al. 1992). This report identified six factors in the emergence of infectious diseases: (1) human demographics and behavior; (2) technology and industry; (3) economic development and land use; (4) international travel and commerce; (5) microbial adaptation and change; and (6) breakdown of public health measures. These factors are not mutually exclusive and are relevant to different stages of emergence (e.g., spillover or an increase in incidence). Seven additional drivers were added in a follow-up IOM report in 2003 (Smolinski et al. 2003): "human susceptibility to infection," "climate and weather," "changing ecosystems," "poverty and social inequity," "war and famine," "lack of political will," and "intent to harm." Other studies have found that disease emergence from animal hosts to humans is driven mainly by anthropogenic forces, such as land use change (Patz et al. 2004) or global trade and travel across ecological and environmental boundaries (Hufnagel et al. 2004). The classification of these "factors in emergence" paved the way for research with respect to the underlying drivers of infectious disease emergence.

At larger spatial scales, datasets are freely available for many of these drivers (e.g., human population density or land use change). Analyzing these datasets allows us to move beyond a correlative approach for testing drivers of disease to a predictive framework (Jones et al. 2008; Dunn et al. 2010). Datasets for each driver are often correlated, so it is important to check for independence among variables when using multiple driver datasets in a single analysis. Determining, quantifying, and ranking drivers of emergence at smaller spatial scales can be more complicated. Often an emergence event arises from multiple drivers interacting simultaneously or sequentially. Further, the time lag between the driver acting directly on a host, pathogen, or environment and the origin of the emergence event can vary. The duration of this time lag may scale with organism generation time; for example, a driver acting directly on pathogens (short generation time) would have a much smaller lag in effect than a driver acting on a mammalian host species (longer generation time).

The spread of genetically based resistance will always lag behind the emergence of a pathogen and may be affected by other drivers. One could estimate a probability curve for this, and estimate lag time based on the slope of the curve. While drivers of emergence are indeed complicated, we can still make inferences on the role of multiple drivers acting simultaneously or sequentially, the time lag between drivers and emergence, and the possibility of unintentional drivers, those that were originally thought to be mitigating forces.

## Quantifying Missing Reports and Biases in Reporting

Existing datasets have identified over 350 infectious diseases that have emerged in the past 70 years (Woolhouse and Gaunt 2007; Jones et al. 2008; Dunn et al. 2010). It is likely, however, that there have been numerous unreported cases of novel diseases. Whether the numbers of emerging infectious diseases are on the rise or health officials have merely grown more aware of these events is debatable and can only be estimated against the backdrop of the highly uneven surveillance capabilities across the globe. EID surveillance has become a high-priority issue for both local and global health authorities, thereby making reporting more equitable. Thorough, accurate disease surveillance reporting relies on comprehensive, unbiased participation of all national and sub-national health agencies. This has been highlighted in recent years by the SARS epidemic, H5N1 highly pathogenic avian influenza, the global H1N1 influenza pandemic, and, most poignantly, the ongoing HIV pandemic.

These diseases, whose spread may have at one time been constrained locally, are increasingly transcending national boundaries (Institute of Medicine 2009). Local outbreaks are of concern to the global community because of their potential for pervasive spread. We rely on human reports of these types of local events to detect epidemics with pandemic potential that require global action. However, this type of participatory reporting is incomplete and biased due to an uneven distribution of health systems, detection mechanisms, and communication infrastructure. Disincentives to reporting, such as negative political and economic consequences of control measures, may also result in reporting bias. When trying to determine the underlying drivers for global disease emergence events, the source of biases in reporting must be accounted for to ensure that true differences are reported rather than artifacts of sampling or reporting.

A number of factors may affect the probability of detecting novel EIDs or influence the lag time between infection and detection of a novel pathogen. Factors intrinsic to both the pathogen and the exposed individual—such as the pathogen's virulence and the individual's socioeconomic status—will determine whether the individual seeks medical attention and whether the medical examiner identifies the infection as novel. Unusually infectious or virulent pathogens may have a greater chance of being reported due to large numbers of infected individuals or more detrimental health effects. Long latency periods, during which individuals are asymptomatic, lead to temporal biases due to the lag time between the initial case and detection, as was the case with variant Creutzfeldt–Jakob disease and HIV/AIDS, which is suspected to have emerged in the United States more than a decade before it was identified in 1981 (Gilbert et al. 2007).

Socioeconomic factors also play a role both as a driver of disease emergence and as a source of reporting bias. Lower-income countries have higher rates of malnutrition and reduced access to potable drinking water, sanitation, immunizations, and health services (Ruger and Kim 2006; World Health Organization 2010). Furthermore, many low-income countries, particularly in sub-Saharan Africa, are faced with double-digit HIV infection rates. These populations are more susceptible to EIDs due to greater exposure to infective agents and depressed immunity. Whether or not infected individuals in low-income countries receive medical attention depends also on the availability, accessibility, and appropriateness of medical services and the individual overall ability to use them (Ensor and Cooper 2004). GDP and population density are the strongest correlates with the supply of qualified medical staff, healthcare facilities, diagnostics, and treatments (World Health Organization 2010).

The first step in correcting for this reporting bias of EIDs is to identify the sources for potential bias in the data. Then, proxies may be determined that help account for this non-random bias (i.e., distance to nearest hospital, use of traditional medicine, or per capita spending on healthcare). Reporting of disease is also non-random throughout the world because of local capacity to conduct and publish research, and the dearth of investigation taking place in underdeveloped and hard-to-reach areas. To control for this when building their model of global EID risk, Jones et al. (2008) constructed an index of sampling bias based on author addresses of publications in the *Journal of Infectious Disease* from 1973 to 2008.

Caution must be taken in choosing potential datasets to act as proxies for bias measures. There must be quantification or evidence supporting a mechanistic link between the proxy and the outcome. Using this approach, we posit that the number of infectious diseases to have emerged over the past half-century is likely much greater than we had previously anticipated. Others have also suggested that recent exposure events are more common, as a result of more suboptimal attempts by pathogens to invade novel populations in the past—sometimes termed "viral chatter" (Antia et al. 2003; Woolhouse et al. 2005; Wolfe et al. 2007).

# SITE SELECTION, SPECIES SELECTION, AND PREDICTIVE MODELING

## Select Geographic Sites for Surveillance

Jones et al. (2008) provided an example of a comprehensive approach to identifying sites as priority areas for sampling for the next EIDs. These sites have been dubbed "hotspots" and represent areas of higher EID risk. This process of identifying EID hotspots began with an exhaustive literature search to collect biological, temporal, and spatial data for EID "events" in human populations between 1940 and present. Jones et al. (2008) based their database of EIDs on previous work (Taylor et al. 2001) and updated it with additional information on microbial pathogens. All types of pathogens found in humans were entered into the database, including sexually transmitted diseases (STDs), zoonoses, drug-resistant microbes, vector-borne diseases, and food- and water-borne infections. Information on time, location, pathogen type, transmission mode, other hosts, and pathogen life history traits was added. Further, the most commonly cited causes of emergence for each pathogen were determined (Daszak et al. 2000; Smolinski et al. 2003; Morens et al. 2004; Patz et al. 2004; Weiss and McMichael 2004). Finally, shape files defining the published boundaries of the initial emergence event were created in ArcGIS (ESRI 2005).

The final published database covered global events between 1940 and 2004 and reported 335 EID events in humans. Using these 335 EID events, a risk model was constructed using logistic regression to determine the probability of an EID event in every 1-degree grid cell of the world. These estimates are based on historical patterns of EID events and their environmental and biological drivers (including human population density and growth, mammal diversity, precipitation, temperature, latitude, and reporting effort). Then, an EID risk value was calculated for every 1-degree grid cell of the world using human population density and growth, mammal density, latitude, and rainfall with the coefficients of the multivariate logistic regression model (Jones et al. 2008).

Previous efforts to understand patterns of EIDs have highlighted viral pathogens (particularly negative-stranded RNA viruses) as a major threat because of their high rates of nucleotide substitution, often poor copy-editing, and higher capacity to adapt to new hosts (higher "evolvability"; Burke 1998). However, Jones et al. (2008) found that a majority of EID pathogens were bacterial, specifically novel drug-resistant strains. Controlling for reporting effort, the number of EIDs still showed a highly significant relationship with time (generalized linear model with Poisson errors, offset by log(JID articles) (GLM$_{PJID}$), $F_{1,57} = 96.4$, $p < 0.001$), supporting the widespread claim that the threat of EIDs to global health is increasing (Fauci 2001; Smolinski et al. 2003; Morens et al. 2004; King et al. 2006). Even after controlling for reporting effort, the number of EID events originating in wildlife reached the highest proportion in the most recent decade, highlighting the importance of understanding the factors that increase the contact between wildlife and humans in developing any predictive model. The strong relationship between high wildlife host biodiversity—primarily found in low-latitude developing countries—and EID events caused by zoonotic pathogens from wildlife (e.g., SARS, Ebola) suggests that these geographic regions will continue to be a key source of novel EIDs in the future. It also reinforces the need for pathogen surveillance in wild animal populations as a forecasting measure for EIDs (Karesh and Cook 2005; Kuiken et al. 2005; King et al. 2006). Jones et al. (2008) found that areas of the planet with the greatest EID risk also had the lowest levels of surveillance effort, therefore highlighting the importance of this approach for public health resource allocation.

We have since updated the driver data and spatial resolution of the risk model in Jones et al. (2008). The original spatial resolution was approximately 100-km$^2$ grid cells of the world; using the native resolution of the driver datasets, we have reduced this resolution to 1 km$^2$, allowing for country-level EID risk maps to be drawn at a resolution useful for regional-level planning. Mammal diversity per 1-km$^2$ grid cell was calculated using range maps based on Mammal Species of the World 2005. Human population density and growth were updated according to the Global Rural-Urban Mapping Project and the Gridded Population of the World (http://sedac.ciesin.columbia.edu/gpw). At the global scale, the 1-km risk map was developed using the same model coefficients as in Jones et al.

(2008), but incorporates new driver datasets as described above at their native resolution, so the distribution of wildlife zoonotic EID risk (Fig. 42.1) is qualitatively comparable to that of the original risk map. At the country level, the improved datasets allow us to examine the influence of the two main drivers, mammal diversity and human population density, on EID risk.

EID risk maps can allow us to select sites for sampling that we believe to be more likely to harbor the next EID-causing pathogen in wildlife. We can also test the hotspots model by sampling in paired "hot" and "cold" sites. This allows for the constant feedback of field data into models to revise and update the prediction of EID risk.

## Select Species to Target for Sampling

### Life-History Traits

Species are not equal in their ability to harbor and transmit infectious diseases. For example, there is some debate as to whether certain characteristics of bats (e.g., their longevity, colonial roosting habits, and ability to fly and hibernate) may make them better viral reservoirs than other groups of mammals (see Chapter 14 in this book). A recent analysis of bat hosts and viruses (Turmelle and Olival 2009) shows that some species in a given area will be more likely to harbor a greater number of viruses than others, and that population genetic structure ($F_{ST}$; related to migratory capacity and mixing of genetic populations) significantly correlates with their known viral diversity. $F_{ST}$ is a measure of the genetic mixture of individuals between populations. Turmelle and Olival (2009) used a combined model that includes $F_{ST}$, the International Union for Conservation of Nature (IUCN) species threat status, and a measure of research sampling bias, and found that these variables account for 33% of known viral diversity in bats ($p = 0.02$). Approaches similar to this, which account for species-specific ecological and evolutionary traits, may be useful for identifying species with the highest projected pathogen viral richness. We can combine this approach with a geographically targeted one to identify the most cost-effective species (bats and other species) and locations to target for active wildlife surveillance.

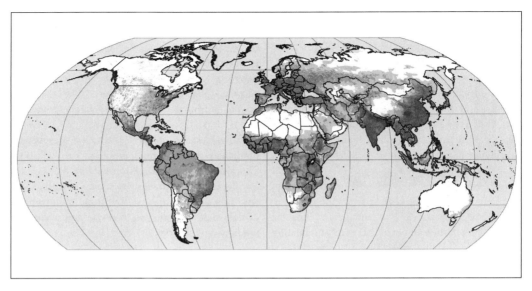

**½ Standard Deviation**

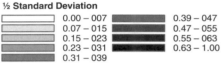

| | |
|---|---|
| 0.00 – 007 | 0.39 – 047 |
| 0.07 – 015 | 0.47 – 055 |
| 0.15 – 023 | 0.55 – 063 |
| 0.23 – 031 | 0.63 – 1.00 |
| 0.31 – 039 | |

**Figure 42.1:**
Global map of zoonotic emerging disease hotspots risk from wildlife based on the Jones et al. (2008) model and updated mammal diversity and human population density and growth driver datasets. Risk is given by a scale from low (0.00, white) to high (1.00, black) risk.

## Phylogenetic Relatedness

Another factor in the process of emergence is host relatedness with humans. Potential similarities that arise from shared ancestry, such as receptors that allow a virus to enter a cell, may play a major role in facilitating spillover of pathogens. To date, this assumption has not been explicitly tested in a phylogenetic framework, especially for viruses. Using host and pathogen data from the Jones et al. (2008) database, we have examined the distribution of wildlife and domestic hosts for pathogens known to cause human disease. Mammals appear to host the greatest proportion of pathogens emerging from wildlife to infect humans (Fig. 42.2). We constructed a database of all known mammal–virus associations to test the importance of phylogeny in estimating the probability of a virus being shared between a non-human mammalian host and humans. The final mammal–virus association database consisted of over 1,200 pairs, including over 300 unique mammal species and over 200 unique virus species. We also tested whether the probability of a virus being shared between mammal hosts and humans increased with increasing human–host contact.

After correcting for biases in reporting effort, we found that the probability of humans and non-human mammal hosts sharing a virus increased with increasing phylogenetic relatedness. Further, the probability of humans and non-human mammal hosts sharing a virus also increased with increasing contact opportunity, either through domestication or shared habitat. These results, combined with life-history trait targeting and hotspot mapping, improve our understanding of host-pathogen transmission and help to provide basic guidance in the identification of wildlife species most likely to be the source of the next EID in humans. This understanding lays the groundwork for us to begin to predict the consequences of anthropogenic activities that increase interaction between humans,

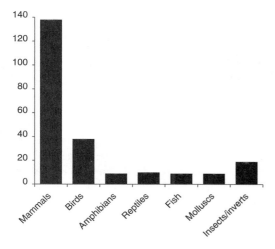

**Figure 42.2:**
The number of human EID events identified by Jones et al. (2008) by host species, as recorded in the original database. Mammals are responsible for by far the greatest number of human EIDs recorded thus far.

domestic animals, and wildlife, such as logging, hunting, or building roads.

Future research is also necessary to understand the relative importance of host phylogeny versus contact opportunity with humans. This will allow for a better surveillance strategy that targets wildlife and domestic host species most likely to be the source of the next EID in humans. Using the model of phylogenetic relatedness and contact opportunities described, these findings could be advanced further by using a Gap Analysis, a tool used to assess decision-making in conservation to identify areas that have been undersampled for pathogens relative to mammalian (and phylogenetic) diversity.

*Contact Opportunities and Risk Interfaces*

Human contact with wildlife species, both direct and indirect, is undoubtedly an important factor in the transmission and emergence of new human pathogens from wildlife. High-risk contact interfaces could be the starting point for investigating pathogen diversity and prevalence (total number of cases of a disease in a population at a given time) in wildlife. Using estimates of the range and distribution of pathogen prevalence and incidence of every known EID family, we can use power calculations to look at how many individuals of

each reservoir species need to be sampled within a given set of species in a specific interface. Calculating an expected prevalence of known EID families allows us to recognize unusual events during routine sampling.

Our vision is that sampling of high-risk interfaces could be conducted over multiple seasons to obtain a baseline species diversity dataset. Then teams can determine the number of individuals per species needed for sampling to increase detection probability using estimated prevalence values for known pathogens (see Chapter 39 in this book). Next, a set of target species in the risk interface could be sampled, using the minimum number of individuals required for improved detection. Then if the prevalence is unusually high, teams could conduct follow-up sampling of species identified and appropriate potential spillover hosts, in intact or native range where possible.

## Construct Predictive Models of Spread and Future Emergence

Finally, once we have a good grasp of historical disease data, current disease risk, and the socioeconomic, environmental, and biodiversity profile of a given region, we can analyze the likelihood that a given pathogen could break out and become truly pandemic (as defined by cross-continental transmission). Our group has developed a vulnerability map of this type for avian influenza (Hosseini et al. 2010) that examined travel routes, airplane travel capacity, and connections between all major airports using ten years of information from Freedom of Information Act requests to the U.S. Fish and Wildlife Service on the global wildlife trade, trade data from the United Nations Food and Agriculture Organization, and data from the International Airline Transport Alliance. Trade routes, export and import statistics, travel, and wildlife trade patterns were examined to determine how these factors increase the risk of H1N1 spreading from a hotspot region into major global population centers. This model can be generalized to the country and airport level to determine which locations are most vulnerable to importation of EIDs through trade and travel (Hosseini et al. 2010). This methodology could be crucial for identifying airports or transportation centers where pathogen monitoring and intervention will be particularly effective in preventing disease spread.

## TECHNOLOGICAL ADVANCES

Recent technological advances have improved our ability to identify high-risk interfaces for disease transmission and to detect novel pathogens before widespread spillover occurs. These advances include improvements in information technology, molecular diagnostics, and risk modeling. Further advances in communications technology will serve to bring countries traditionally isolated from international health networks into the global fold. Developments over the past 15 years allow us to gather reports from disparate sources and use the Internet as a common platform for exchanging information. Examples include the Global Public Health Intelligence Network, the Program for Monitoring Emerging Diseases, and HealthMap (http://www.healthmap.org). The greatest limitations of these networks are the underlying limitations of the national reporting systems and their bias towards English-speaking countries (Keller et al. 2009). Telemedicine, or cell phone-mediated medical diagnoses, will allow technologically underserved areas to leapfrog ahead without enduring massive infrastructure changes, due in great part to the near-ubiquitous use of cell phones in much of the developing world. Systems are now being created to allow medical care providers to text coded reports to be analyzed *en masse* (Yang et al. 2009). Similarly, monitoring the frequency of specific disease-related terms in daily Internet postings, search queries, or SMS text messages is now providing alternative forms of disease surveillance (Ginsberg et al. 2009). The extent to which telemedicine and the Internet decrease the disparity between countries in regard to access to health information and capacity to detect EIDs remains to be seen, but our increased capacity to reach understudied areas suggests that this will be significant.

Platforms for pathogen discovery and our ability to follow footprints of infectious agents require the laboratory and computational infrastructure sufficiently powerful to dissect complex host–microbe interactions. For example, MassTag PCR is a multiplex platform that allows animal and human health specialists and epidemiologists to simultaneously test one sample for the presence of up to 30 different agents. MassTag PCR is a powerful tool for genomics, molecular virology, computational biology, surveillance, pathogen discovery, outbreak detection, and epidemiological investigations (Lipkin 2010).

## CONCLUSIONS

EIDs are a growing and complex threat to global public health. Diseases emerge when socioeconomic or environmental changes provide the optimal conditions for pathogens to exploit new host populations, increase in pathogenicity, or otherwise amplify transmission. We present a broad-scale, strategic approach for selecting geographic sites and species for sampling and then present a framework for making predictions about the future risk of EIDs from wildlife. In our view, the best approach to detecting and preventing the next emerging infectious disease before it becomes a pandemic threat is through building a broad coalition of partners to discover, detect, and monitor diseases at the wildlife–human interface using a localized, risk-based approach. These efforts can integrate predictive modeling, digital sensing, on-the-ground surveillance, and advanced molecular techniques at critical points for disease emergence, which then feed back to models for testing and refinement.

## ACKNOWLEDGMENTS

We wish to acknowledge the Research and Policy for Infectious Disease Dynamics (RAPIDD) program of the Science and Technology Directorate, U.S. Department of Homeland Security, and the Fogarty International Center, NIH and the generous support of the American people through the United States Agency for International Development (USAID) Emerging Pandemic Threats PREDICT. The contents are the responsibility of the authors and do not necessarily reflect the views of USAID or the United States Government.

## REFERENCES

Antia, R., R.R. Regoes, J.C. Koella, and C.T. Bergstrom. 2003. The role of evolution in the emergence of infectious diseases. Nature 426:658–661.

Babbott, F.L., and J.E. Gordon. 1954. Modern measles. Am J Med Sci 228:334–361.

Burke, D.S. 1998. The evolvability of emerging viruses. *In* A.M. Nelson and C.R. Horsburgh, eds. Pathology of emerging infections. American Society for Microbiology, pp. 1–12 Washington D.C.

Chua, K.B., W.J. Bellini, P.A. Rota, B.H. Harcourt, A. Tamin, S.K. Lam, T.G. Ksiazek, P.E. Rollin, S.R. Zaki, W.J. Sheih, C.S. Gouldsmith, D.J. Gubler, J.T. Roehrig, B. Eaton, A.R. Gould, J. Olson, H. Field, P. Daniels, A.E. Ling, C.J. Peters, L.J. Anderson, and B.W.J. Mahy. 2000. Nipah virus: a recently emergent deadly paramyxovirus. Science 288:1432–1435.

Daszak, P., A.A. Cunningham, and A.D. Hyatt. 2000. Emerging infectious diseases of wildlife—threats to biodiversity and human health. Science 287:443–449.

Dunn, R.R., T.J. Davies, N.C. Harris, and M.C. Gavin. 2010. Global drivers of human pathogen richness and prevalence. Proc Royal Soc B 277:2587–2595.

Ensor, T., and S. Cooper. 2004. Overcoming barriers to health service access: influencing the demand side. Health Policy Plann 19:69–79.

ESRI. 2005. Environmental Research Systems Institute, Inc., Redlands, California.

Fauci, A.S. 2001. Infectious diseases: considerations for the 21st century. Clin Infect Dis 32:675–685.

Ferguson, N.M., D.A.T. Cummings, S. Cauchemez, C. Fraser, S. Riley, A. Meeyai, S. Iamsirithaworn, and D.S. Burke. 2005. Strategies for containing an emerging influenza pandemic in Southeast Asia. Nature 437:209–214.

Garnett, G.P. and E. C. Holmes. 1996. The ecology of emergent infectious disease. Bioscience 46:127.

Gilbert, M.T.P., A. Rambaut, G. Wlasiuk, T.J. Spira, A.E. Pitchenik, and M. Worobey. 2007. The emergence of HIV/AIDS in the Americas and beyond. Proc Nat Acad Sci 104:18566–18570.

Ginsberg, J., M.H. Mohebbi, R.S. Patel, L. Brammer, M.S. Smolinski, and L. Brilliant. 2009. Detecting influenza epidemics using search engine query data. Nature 457:1012–1014.

Guan, Y., B.J. Zheng, Y.Q. He, X.L. Liu, Z.X. Zhuang, C.L. Cheung, S.W. Luo, P.H. Li, L.J. Zhang, Y.J. Guan, K.M. Butt, K.L. Wong, K.W. Chan, W. Lim, K.F. Shortridge, K.Y. Yuen, J.S.M. Peiris, and L.L.M. Poon. 2003. Isolation and characterization of viruses related to the SARS coronavirus from animals in Southern China. Science 302:276–278.

Hays, J. 2006. Epidemics and pandemics: their impacts on human history. ABC-CLIO, Santa Barabara, CA.

Hosseini, P., S.H. Sokolow, K.J. Vandegrift, A.M. Kilpatrick, and P. Daszak. 2010. Predictive power of air travel and socio-economic data for early pandemic spread. PLoS One 5:e12763.

Hufnagel, L., D. Brockmann, and T. Geisel. 2004. Forecast and control of epidemics in a globalized world. Proc Nat Acad Sci USA 101:15124–15129.

Institute of Medicine. 1992. Emerging Infectious microbial threats to health in the United States. National Academy Press, Washington, D.C.

Institute of Medicine. 2009. Sustaining global surveillance and response to emerging zoonotic diseases. National Academy Press, Washington, D.C.

Jones, K.E., N. Patel, M. Levy, A. Storeygard, D. Balk, J.L. Gittleman, and P. Daszak. 2008. Global trends in emerging infectious diseases. Nature 451:990–994.

Karesh, W.B., and R.A. Cook. 2005. The human–animal link. Foreign Affairs 84:38–50.

Keller, M., M. Blench, H. Tolentino, C.C. Freifeld, K.D. Mandl, A. Mawudeku, G. Eysenbach, and J.S. Brownstein. 2009. Use of unstructured event-based reports for global infectious disease surveillance. Emerg Infect Dis 15:689–695.

Kilbourne, E.D. 1996. The emergence of "emerging disease": A lesson in holistic epidemiology. Mt Sinai J Med 63:159.

King, D.A., C. Peckham, J.K. Waage, J. Brownlie, and M.E.J. Woolhouse. 2006. Infectious diseases: preparing for the future. Science 313:1392–1393.

Krause, R.M. 1994. Dynamics of emergence. J Infect Dis 170:265–271.

Kuiken, T., F.A. Leighton, R.A.M. Fouchier, J.W. LeDuc, J.S.M. Peiris, A. Schudel, K. Stohr, and A. Osterhaus. 2005. Pathogen surveillance in animals. Science 309:1680–1681.

Lederberg, J., R.E. Shope, and S.C.J. Oakes. 1992. Emerging infections: microbial threats to health in the United States. Institute of Medicine, National Academy Press, Washington D.C.

Levins, R., T. Awerbuch, U. Brinkmann, I. Eckardt, P. Epstein, N. Makhoul, C. Albuquerque de Possas, C. Puccia, A. Spielman, and M.E. WIlson. 1994. The emergence of new diseases. Am Sci 82:52–60.

Lipkin, W.I. 2010. Microbe hunting. Microbiol Mol Biol R 74:363–377.

Morens, D.M., G.K. Folkers, and A.S. Fauci. 2004. The challenge of emerging and re-emerging infectious diseases. Nature 430:242–249.

Morse, S.S. 1993. Emerging viruses. Oxford University Press, New York.

Morse, S.S. 1995. Factors in the emergence of infectious disease. Emerg Infect Dis 1:7–15.

Oster, M.S. 1961. Emerging animal diseases of usareur-usafe interest. Med Bull US Army Eur 18:176–180.

Patz, J.A., P. Daszak, G.M. Tabor, A.A. Aguirre, M. Pearl, J. Epstein, N.D. Wolfe, A.M. Kilpatrick, J. Foufopoulos, D. Molyneux, and D.J. Bradley. 2004. Unhealthy landscapes: policy recommendations on land use change and infectious disease emergence. Environ Health Persp 112:1092–1098.

Ruger, J.P., and H.J. Kim. 2006. Global health inequalities: an international comparison. J Epidemiol Comm Health 60:928–936.

Smolinski, M.S., M.A. Hamburg, and J. Lederberg. 2003. Microbial threats to health: emergence, detection, and response. National Academies Press, Washington D.C.

Steverding, D. 2008. The history of African trypanosomiasis. Parasites & Vectors 1:3.

Taylor, L.H., S.M. Latham, and M.E.J. Woolhouse. 2001. Risk factors for human disease emergence. Phil Trans Roy Soc B 356:983–989.

Turmelle, A.S., and K.J. Olival. 2009. Correlates of viral richness in bats (Order Chiroptera). EcoHealth 6:522–539.

Weiss, R.A., and A.J. McMichael. 2004. Social and environmental risk factors in the emergence of infectious diseases. Nat Med 10:S70–S76.

Wolfe, N.D., P. Daszak, A.M. Kilpatrick, and D.S. Burke. 2005. Bushmeat hunting, deforestation and prediction of zoonotic emergence. Emerg Infect Dis 11:1822–1827.

Wolfe, N.D., C.P. Dunavan, and J. Diamond. 2007. Origins of major human infectious diseases. Nature 447:279–283.

Woolhouse, M.E., and E. Gaunt. 2007. Ecological origins of novel human pathogens. Crit Rev Microbiol 33:231–242.

Woolhouse, M.E.J., D.T. Haydon, and R. Antia. 2005. Emerging pathogens: the epidemiology and evolution of species jumps. Trends Ecol Evol 20:238–244.

World Health Organization. 1978. Ebola haemorrhagic fever in Sudan. Bull World Health Organ 56:247–270.

World Health Organization. 2010. World health statistics. Geneva, Switzerland.

Yang, C., J. Yang, X. Luo, and P. Gong. 2009. Use of mobile phones in an emergency reporting system for infectious disease surveillance after the Sichuan earthquake in China. Bull World Health Organ 87:619–623.

# INDEX

Page numbers followed by f indicate figures; t, tables; b, boxes.